Springer Series in Geomechanics and Geoengineering

Series Editor

Wei Wu, *Universität für Bodenkultur, Vienna, Austria*

Geomechanics deals with the application of the principle of mechanics to geomaterials including experimental, analytical and numerical investigations into the mechanical, physical, hydraulic and thermal properties of geomaterials as multiphase media. Geoengineering covers a wide range of engineering disciplines related to geomaterials from traditional to emerging areas.

The objective of the book series is to publish monographs, handbooks, workshop proceedings and textbooks. The book series is intended to cover both the state-of-the-art and the recent developments in geomechanics and geoengineering. Besides researchers, the series provides valuable references for engineering practitioners and graduate students.

Indexed by SCOPUS, EI Compendex, INSPEC, SCImago.

Alessio Ferrari · Marco Rosone ·
Maurizio Ziccarelli · Guido Gottardi
Editors

Geotechnical Engineering in the Digital and Technological Innovation Era

Springer

Editors
Alessio Ferrari
Engineering Department
University of Palermo
Palermo, Italy

Laboratory of Soil Mechanics
École Polytechnique Fédérale de Lausanne
Lausanne, Switzerland

Maurizio Ziccarelli
Engineering Department
University of Palermo
Palermo, Italy

Marco Rosone
Engineering Department
University of Palermo
Palermo, Italy

Guido Gottardi
Department of Civil, Chemical,
Environmental, and Materials Engineering
University of Bologna
Bologna, Italy

ISSN 1866-8755 ISSN 1866-8763 (electronic)
Springer Series in Geomechanics and Geoengineering
ISBN 978-3-031-34760-3 ISBN 978-3-031-34761-0 (eBook)
https://doi.org/10.1007/978-3-031-34761-0

This Springer imprint is published by the registered company Springer Nature Switzerland AG
The registered company address is: Gewerbestrasse 11, 6330 Cham, Switzerland

Preface

The sustainable deployment of energy and raw materials, the conscious development of buildings and infrastructures and the coexistence with extreme climatic events are nowadays upmost themes that must be combined with ecological and economic goals, as well as with the safety of the territory and of the built environment. Geotechnical engineering necessarily plays a leading role in this context; by using innovative measurements and implementing new analysis techniques, it is possible to interpret and intervene in complex processes, at different scales and in the presence of multiple climatic and environmental forcing.

The 8th Italian Conference of Researchers in Geotechnical Engineering 2023 (CNRIG'23) was organized to highlight how the latest developments in geotechnical research can address global and emerging challenges and to promote the interaction among geotechnical engineering and applied sciences, with special focus on technological and digital innovations. CNRIG'23 was held on July 5–7, 2023, at the University of Palermo (Italy), under the auspices of the National Group of Geotechnical Engineering (GNIG). Researchers were invited to present their most recent works in this context, highlighting the innovative potential of modern technological and digital tools and the growing opportunities offered by the synergic interaction with other disciplines.

This book collects the keynote contributions and the ninety-six papers presented at the conference. The papers cover a wide range of emerging themes in Geotechnics, including innovation in laboratory testing, new trends and applications for in-situ monitoring, thermo-hydro-chemo-mechanical behavior of geomaterials, novelties in computational geomechanics, analyses of instability processes in seismic conditions, probabilistic approaches for data analyses, resilience of critical infrastructures, advances in risk mitigation strategies, eco-friendly solutions for soils and rocks stabilization, environmental protection and the ecological transition.

We would like to express our warmest thanks to the Scientific and the Review Committees of the event, as well as to all the authors for the high quality of their contributions and to all the anonymous reviewers for their valuable work. We are grateful to Marco Starvaggi for his assistance with the preparation of this book.

Alessio Ferrari
Marco Rosone
Maurizio Ziccarelli
Guido Gottardi

Committees

Organizing Committee

Alessio Ferrari	Università degli Studi di Palermo
Maurizio Ziccarelli	Università degli Studi di Palermo
Marco Rosone	Università degli Studi di Palermo
Antonio Casella	Università degli Studi di Palermo
Giovanni Sapienza	Università degli Studi di Palermo
Marco Starvaggi	Università degli Studi di Palermo
Giusy Floriana Valentino	Università degli Studi di Palermo
Francesco Castelli	Università degli Studi di Enna "Kore"
Valentina Lentini	Università degli Studi di Enna "Kore"
Maria Rossella Massimino	Università degli studi di Catania
Salvatore Grasso	Università degli studi di Catania
Ernesto Cascone	Università degli Studi di Messina
Giovanni Biondi	Università degli Studi di Messina

Scientific and Review Committee

Guido Gottardi	Alma Mater Studiorum Università di Bologna
Sara Amoroso	Università degli Studi "G. D'Annunzio"
Monica Barbero	Politecnico di Torino
Giovanni Biondi	Università degli Studi di Messina
Ernesto Cascone	Università degli Studi di Messina
Francesco Castelli	Università degli Studi di Enna "Kore"
Federica Cotecchia	Politecnico di Bari
Sabatino Cuomo	Università degli Studi di Salerno
Alessio Ferrari	Università degli Studi di Palermo
Anna Maria Ferrero	Università di Torino
Salvatore Grasso	Università degli Studi di Catania
Valentina Lentini	Università degli Studi di Enna "Kore"
Claudia Madiai	Università degli Studi di Firenze
Maria Rossella Massimino	Università degli Studi di Catania
Francesco Silvestri	Università degli Studi di Napoli Federico II

Contents

Keynote Lectures

Some Remarks on the Response of Framed Structures to Tunnelling in Coarse-Grained Soils

Daniela Boldini(✉)

Sapienza University of Rome, 00184 Rome, Italy
daniela.boldini@uniroma1.it

Abstract. The paper summarises a series of researches aimed at investigating the response of framed structures to tunnelling in coarse-grained soils. In particular, experimental data obtained by centrifuge tests as well as numerical results of advanced 3D finite element calculations are described. Finally, a case-history related to the excavation of the Milan metro line 5 is also discussed, with a focus on monitoring records and possible optimisation of settlement sensors by inverse analysis.

Keywords: Tunnelling · Soil-structure interaction · Framed structures · Coarse-grained-soils · Centrifuge data · Numerical analysis · Monitoring data

1 Introduction

While in the last decades the response of load-bearing masonry wall structures to tunnelling was thoroughly studied [e.g., 1–5], only limited attention has been dedicated to framed structures so far [e.g., 6–9]. In the following the main results of a cooperative research activity in this field between the University of Nottingham and the University of Bologna, with further contributions from the Cambridge University and Sapienza University of Rome, are described.

2 Results of the Experimental and Numerical Studies

2.1 Centrifuge Tests

41 centrifuge tests, whose results are summarised in [10, 11], were performed at 68 g in the 4-m diameter geotechnical centrifuge at the University of Nottingham (Fig. 1a). The prototype tunnel scenario is characterised by a diameter $D_t = 6.1$ m and a cover-to-diameter ratio $C/D_t = 1.3$. Tunnel excavation was simulated in a homogeneous layer of dry Leighton Buzzard Fraction E fine-grained silica sand, prepared at a relative density of $I_d = 90\%$ or 30%. Tests were carried out in greenfield conditions as well as in presence of a plane-strain aluminium frame founded on rafts or separated footings resting on the ground surface (Fig. 1b). Different structural models having variable number of stories (2 or 5) and bays (3 or 6), bay length b, structural thickness t, eccentricity over width e/B (0, 0.2 or 0.5) and weight (SW, corresponding to the self-weight of the frame, or 2SW, corresponding to doubled weight obtained by dead weights added at the top of it) were employed.

A. Ferrari et al. (Eds.): CNRIG 2023, SSGG, pp. 3–11, 2023.
https://doi.org/10.1007/978-3-031-34761-0_1

Figure 2 shows the results obtained for a tunnel volume loss of $V_{lt} = 1\%$ in an experiment with sand prepared at $I_d = 90\%$ and with the frame F2t3b6L (two stories 2.6 m each, thickness of structural element $t = 0.22$ m, bay length $b = 5.2$ m, frame width $B = 30$ m) centred above the tunnel. Concerning measured settlements (Fig. 2a), it can be observed a gap formation at the soil-foundation interface as well as local changes in the slope of the curves at the frame wall positions. In addition, horizontal displacements of the frame are negligible due to the occurrence of slippage at the soil-foundation interface (Fig. 2b).

Comparison with centrifuge tests using plates to simulate the presence of a structure [12] indicates that framed structures are more flexible than plates having the same equivalent bending stiffness and that the former are always characterised by sagging and hogging regions regardless of their stiffness and width. As a matter of fact, framed structures are characterised by a shear-dominated behaviour, as clearly evident from the deformed configuration shown in Fig. 3.

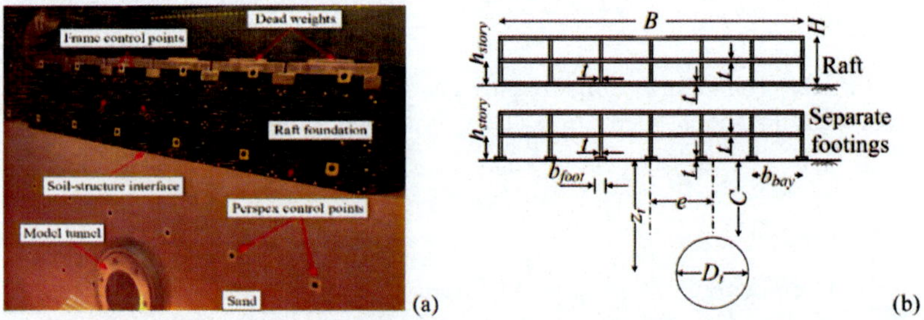

Fig. 1. Experimental results (a) and experimental layout (b) (modified from [10]).

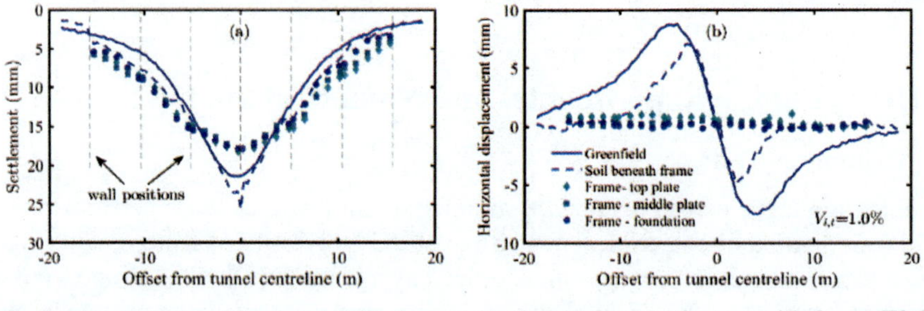

Fig. 2. Settlement (a) and horizontal displacements (b) obtained in the test with frame F2t3b6L at a tunnel volume loss $V_{l,t} = 1\%$.

Fig. 3. Deformed configuration of the frame in the test with frame F2t3b6L.

2.2 Finite Element Analyses

Finite element analyses were carried out with the commercial code Abaqus and are fully illustrate in [13]. The numerical domain was defined to perfectly match the centrifuge box dimensions, as shown in Fig. 4. Soil behaviour was described by the SANISAND constitutive model [14], with material constants summarised in Table 1 [10]). They were selected considering the results of laboratory tests conducted on the Leighton Buzzard Fraction E sand as well as those of greenfield centrifuge tests by Farrell [12]. For the frame a linear elastic law with reduced stiffness to account for the partial welding of the structural components was considered ($E = 53.8$ GPa, $\nu = 0.334$, $\gamma = 27$ kN/m^3), while a no-penetration/sliding-friction interaction law was adopted at the contact with the soil to allow for gap formation and relative sliding. After the activation of geostatic stress in the soil ($K_0 = 0.5$) and self-weight in the structure, tunnel excavation was simulated by imposing incremental displacements at the tunnel boundary nodes considering as fixed the point at the tunnel invert.

Figure 5 proposes a comparison between centrifuge data with the frame F2t3b6L and numerical results in terms of settlements and horizontal displacements at the soil-frame interface for a tunnel volume loss of $V_{lt} = 1\%$. It is apparent the good performance of the model in predicting the greenfield response as well as the changes in the displacement field due to the presence of the structure. A slight underestimation in soil settlement and frame horizontal displacement can however be observed in the interaction analysis. A synthetic picture of numerical results for frames on raft foundations and their comparison with centrifuge data is given in Fig. 6, showing the increase in the maximum angular distortion of the frame with the increase in the tunnel volume loss. Stiffer frames and eccentric configurations are characterised by lower values of β_{max}, while the effect of structural weight and width is clearly detrimental.

The presence of masonry infill panels, not discussed here, has a strong influence on the response of framed structures. Further details can be found in [15].

Fig. 4. Numerical domain and mesh discretization for the simulation of test with frame F2t3b6L.

Table 1. Material constants of the SANISAND constitutive model.

Elasticity
$G_0 = 400$; $\nu = 0.05$
Critical state
$M_{c}=1.287$; $c = 0.780$; $\lambda_c = 0.00178$
$e_0 = 0.8191$; $\xi = 2.4352$
Yield surface
$m = 0.01$
Plastic modulus
$h_0 = 4.05$; $c_h = 1.1$; $n^b = 2.8$
Dilatancy
$A_0 = 0.55$; $n^d = 2.564$
Fabric-dilatancy tensor
$z_{max} = 0$; $c_z = 0$

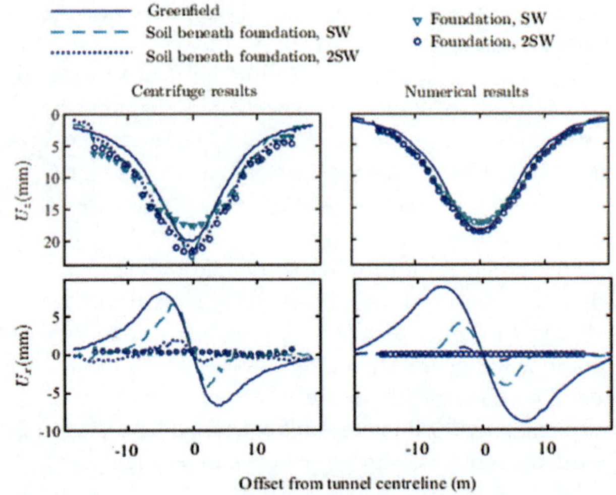

Fig. 5. Comparison between centrifuge and numerical results in greenfield conditions and with frame F2t3b6L for a tunnel volume loss $V_{l,t} = 1\%$ (modified from [13]): settlements U_z and horizontal displacements U_x at the soil-frame interface.

2.3 Modification Factors for Angular Distortion

Figure 7 shows design charts for a preliminary risk assessment of the engineering problem at hand. Modification factors, in contrast to the original proposal of Potts & Addenbrooke [2], are here given in terms of the ratio between the maximum angular distortion of the structure, defined with respect to the building bays, and the maximum average

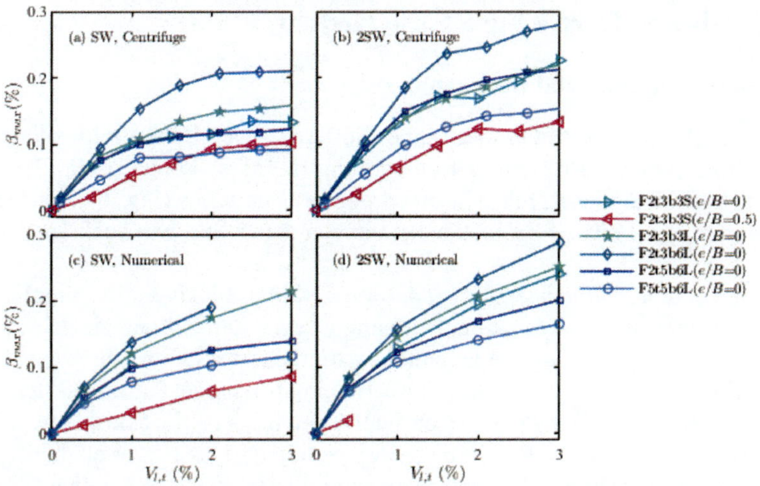

Fig. 6. Comparison between centrifuge and numerical results in terms of maximum angular distortions of the frame (modified from [13]).

greenfield slope. The soil-to-structure relative stiffness κ was carefully assessed by considering the representative soil Young's modulus for the tunnelling-induced level of shear strain and the equivalent building shear stiffness for 1-m run using the Timoshenko beam theory. The figure indicates that numerical results well fit within the empirical envelopes drawn by [10] based on centrifuge test results. More specifically, numerical values tend to concentrate towards the upper envelope for centred frames and towards the lower enveloped for the eccentric tunnels.

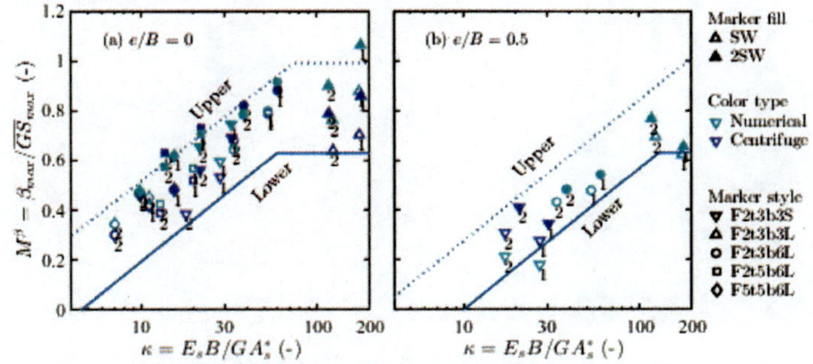

Fig. 7. Modification factor of angular distortion for rafts: (a) central and (b) eccentric tunnels (modified from [13]. Envelope data are from [10]). Numbers "1" and "2" near the symbols indicate the percentage of tunnel volume loss.

3 The Milan Metro Line 5 Case History

3.1 Monitoring Data and Back-Analysis

The case history of the Milan metro line 5 offered several possibilities to investigate the response of framed structures to tunnelling in coarse-grained soils. In particular, the behaviour of a nine-storey reinforced concrete building (Fig. 8a) affected by the excavation of two TBM-EPBs, 6.69 m in diameter at the face, was back-analysed in [16, 17].

A quite complex finite element model was set up in Plaxis for the simulation of the progressive twin tunnel excavation, imposing at each tunnel boundary a displacement field capable of reproducing the greenfield settlement trough measured just before the building (i.e., section S35 in Fig. 8b). The hardening soil model with small strain stiffness was adopted for the soil strata, with a particular emphasis on the calibration of the initial shear stiffness and its decay. For the building a combination of beams elements for beams and columns, plate elements for floor slabs, internal panels and retaining walls, and non-porous volume elements for foundations (five strip footings and three rafts at a depth of 4.0 m below the ground level) were adopted. Figure 9 illustrates a comparison in terms of settlements between numerical results and measurements recorded over three sides of the building at different stages of the twin tunnel excavation. The finite element predictions are capable of well capturing the monitoring data and the progressive transition of the settlement curve from hogging to sagging over the structure longitudinal sides resting on strip foundations. The model response, at least for this particular case, is essentially controlled by the buried portion of the structure, having the above-ground portion only a limited influence on the calculate settlements.

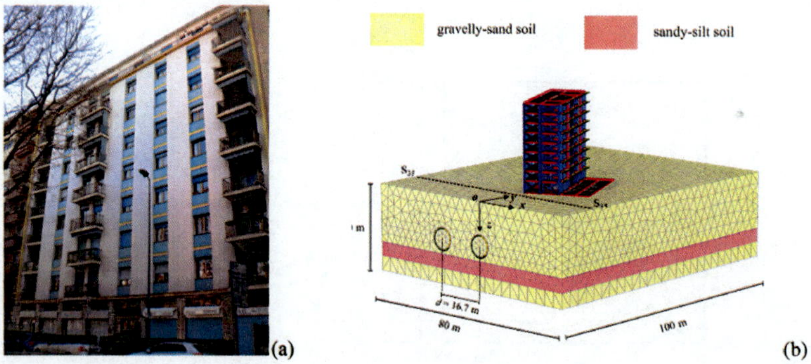

Fig. 8. The selected case history of the Milan metro line 5: (a) picture of the building and (b) view of the numerical model (modified from [17]).

3.2 A Numerical Exercise for Optimisation of Monitoring Sensors

The case history described in the previous paragraph was employed to assess the influence of uncertainties in soil and TBM parameters on the computed results, as discussed in

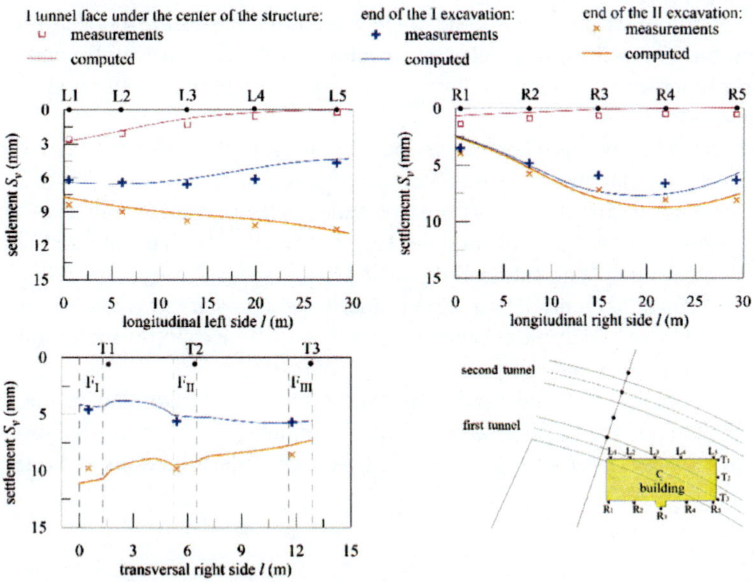

Fig. 9. Comparison between settlements measured along different sides of the building and numerical results (modified from [17]).

[18]. The numerical exercise required the following steps: i) the substitution of the finite element model by a corresponding meta-model to reduce the computational effort, ii) the execution of a global sensitivity analysis for problem optimization to establish the dominant model parameters, iii) the implementation of an inverse analysis technique to improve the values of the selected parameters for model calibration and iv) the design of an optimized monitoring set-up by performing a second global sensitivity analysis. The sensitivity analysis at point ii) revealed that initial shear modulus, the secant stiffness for primary loading and the volume loss are the parameters with the heaviest influence on the foundation settlement. The second sensitivity analysis (point iv) allowed to identify the 6 more relevant sensors for the identification of model parameter values (i.e., L4, L5, T1, T2, R1 and R2 of Fig. 9). This helped in improving the agreement between finite element results and in situ measurements even with a reduced number of sensors.

4 Conclusions

The main conclusions obtained from the research activity described in the previous paragraphs can be summarised as follows:

- framed structures are characterised by a shear-dominated behaviour. Horizontal displacements are negligible for structures on continuous foundations, while they can be significant in the case of separated footings;

- reliable numerical models should correctly predict the soil response to tunnelling in greenfield conditions, estimate the structural stiffness and weight, and account for the presence of an interface at the soil-structure contact to allow sliding and gap formation;
- structural damage should be described by appropriate deformation indicators, i.e. angular distortion and, if applicable, horizontal strain;
- higher values of angular distortions are typically associated to frames of larger width and weight, whereas stiff frames and frames eccentric with respect to the tunnel axis suffer lower tunnelling-induced distortions;
- properly calibrated and detailed 3D numerical models are nowadays capable of simulating even very complex interaction problems by explicitly including the tunnel, the soil, and the framed structure with its foundation;
- numerical models can be used to optimise the sensors adopted in the monitoring system. Values of model parameters can be updated on the basis of monitoring data to provide better predictions of tunnelling-induced deformations and damage.

References

1. Burland, J.B., Broms, B.B., De Mello, V.F.B.: Behaviour of foundations and structures. In: 9th International Conference on Soil Mechanics and Foundation Engineering, pp. 495–546. Japanese Geotechnical Society, Tokyo (1977)
2. Potts, D.M., Addenbrooke, T.I.: A structure's influence on tunnelling-induced ground movements. Proc. Inst. Civ. Eng. Geotech. Eng. **125**(2), 109–125 (1997)
3. Burd, H.J., Houlsby, G.T., Augarde, C.E., Liu, G.: Modelling tunnelling-induced settlement of masonry buildings. Proc. Inst. Civ. Eng. Geotech. Eng. **143**(1), 17–29 (2000)
4. Pickhaver, J., Burd, H., Houlsby, G.T.: An equivalent beam method to model masonry buildings in 3D finite element analysis. Comput. Struct. **88**(19), 1049–1063 (2010)
5. Yiu, W.N., Burd, H.J., Martin, C.M.: Finite-element modelling for the assessment of tunnel-induced damage to a masonry building. Géotechnique **67**(9), 780–794 (2017)
6. Goh, K.H., Mair, R.J.: Response of framed buildings to excavation-induced movements. Soils Found. **54**(3), 250–268 (2014)
7. Son, M., Cording, E.J.: Estimation of building damage due to excavation-induced ground movements. J. Geotech. Geoenviron. Eng. **131**, 162–177 (2005)
8. Boldini, D., Losacco, N., Bertolin, S., Amorosi, A.: Finite element modelling of tunnelling-induced displacements on framed structures. Tunnel. Undergr. Space Technol. **80**, 222–231 (2018)
9. Elkayam, I., Klar, A.: Nonlinear elasto-plastic formulation for tunneling effects on superstructures. Can. Geotech. J. **56**(7), 956–969 (2019)
10. Xu, J., Franza, A., Marshall, A.M.: The response of framed buildings on raft foundations to tunnelling. J. Geotech. Geoenviron. Eng. **146**(11), 04020120 (2020)
11. Xu, J., Franza, A., Marshall, A.M., Losacco, N., Boldini, D.: Tunnel-framed building interaction: comparison between raft and separate footing foundations. Géotechnique **71**(7), 631–644 (2021)
12. Farrell, R.: Tunnelling in sands and the response of buildings. Ph.D. thesis, Engineering Dept., Cambridge University (2010)
13. Boldini, D., Losacco, N., Franza, A., Dejong, M., Xu, J., Marshall, A.: Tunneling-induced deformation of bare frame structures on sand: numerical study of building deformations. J. Geotech. Geoenviron. Eng. **147**(11), 04021116 (2021)

14. Dafalias, Y.F., Manzari, M.T.: Simple plasticity sand model accounting for fabric change effects. J. Eng. Mech. **130**(6), 622–634 (2004)
15. Franza, A., Seyedmohsen, M., Boldini, D., Losacco, N.: An equivalent beam approach for assessing tunnelling-induced distortions on frames with infills. Tunnel. Undergr. Space Technol. **129**, 104686 (2022)
16. Fargnoli, V., Boldini, D., Amorosi, A.: Twin tunnel excavation in coarse grained soils: observations and numerical back-predictions under free field conditions and in presence of a surface structure. Tunnel. Undergr. Space Technol. **49**, 454–469 (2015)
17. Fargnoli, V., Gragnano, C.G., Boldini, D., Amorosi, A.: 3D numerical modelling of soil-structure interaction during EPB tunnelling. Géotechnique **65**(1), 23–37 (2015)
18. Schoen, M., Hölter, R., Boldini, D., Lavasan, A.A.: Application of optimal experiment design method to detect ideal sensor positions: a case study of Milan metro line 5. Tunn. Undergr. Space Technol. **130**, 104723 (2022)

Demand Hazard Curves for the Assessment of Seismic Induced Slope Displacements in Ischia Island

Anna d'Onofrio(✉), Francesco Gargiulo, and Rocco Ceres

Università di Napoli Federico II, Napoli, Italy
donofrio@unina.it

Abstract. This study is part of wider research aimed at the evaluation of ground instability and permanent deformation related to landslides and liquefaction in the volcanic island of Ischia adopting a multi-hazard approach. Ischia was frequently interested by instability caused by both hydro-meteoric and seismic events, the latest of which occurred with an unusual spatial and temporal concentration in 2009, 2017 and 2022 in the same area of Casamicciola. The research is aimed at tentatively evaluate potential cascading effects related to ground instabilities considering the variable susceptibility to seismic instability related to the groundwater fluctuation induced by meteoric events. Here the results of a probabilistic assessment of seismic induced slope displacements carried out with reference to the north-western part of Ischia Island, recently hit by the earthquake, are presented. The study allowed the construction of specific demand hazard curves expressed in terms of annual rate of exceedance of permanent earthquake induced slope displacement. These curves can be a useful tool to be used by both planners and designers. The resulting hazard maps, based on a probabilistic approach, represent the first step toward a multi hazard analysis.

Keywords: Earthquake induced instability · Multi-hazard · Demand hazard curve

1 Introduction

Ischia is an active volcanic island in the gulf of Naples (Italy); it represents the emerged top of a large volcanic complex that is part of the Phlegrean volcanic district. The island extends over an area of about 42 km^2, morphologically dominated by Mt. Epomeo. The island is densely populated, with more than 60,000 inhabitants that, in touristic seasons almost double. It is characterized by an active hydrothermal system mainly located in the North and South-West sectors of the Island. The same sectors of the island have been historically hit by several earthquakes which caused extensive structural damage as well as landslides, especially localized between the municipalities of Casamicciola Terme, Lacco Ameno and Forio. Macroseismic observations show that both historical and recent earthquakes are mostly characterized by shallow hypocentral depths (1–2 km) and high intensities rapidly decreasing with distance. This is the case of the catastrophic

A. Ferrari et al. (Eds.): CNRIG 2023, SSGG, pp. 12–19, 2023.
https://doi.org/10.1007/978-3-031-34761-0_2

event of 1883 (Imax = XI degree MCS), with an inferred magnitude between 4.3 and 5.2 [1] and of the strongest seismic event instrumentally recorded occurred on 21st August 2017 (Mw 3.9 - Md 4.0) [2] as well. This latter caused relevant damages in a narrow area of Casamicciola Terme [3] where fractures on ground, infrastructure and roads have been recognized as well as widespread sliding of drywalls and small landslides in volcanoclastic deposits. The North-western area of the island has been also historically subjected to numerous landslides mainly triggered by intense rainfall events [4, 5]. As a matter of fact, the northern side of Monte Epomeo is characterized by very steep slopes covered by thin layers of unsaturated pyroclastic soils with varying degree of permeability. These peculiar characteristics represent significant predisposing factors to slope instability occurrences as, for instance, the catastrophic weather-induced flow slides and debris flows in pyroclastic soils occurred in 2009 and, more recently in 2022. The current research takes its cue from the latest sequence of catastrophic events occurred in the north-western sectors of Ischia Island that, in the last decade, has been affected by significant instability phenomena due to both seismic and meteoric events. Starting from this evidence a multi-hazard approach will be adopted to account for the possible combination of predisposing factors, preparatory phenomena and triggering events. This approach will be tentatively applied for the evaluation of ground instability and permanent ground deformation induced by earthquake, also considering for the possible interaction with events like intense rainfall.

In this paper the preliminary hazard assessment of the seismic induced slope displacements carried out for the municipalities of Casamicciola, Lacco Ameno and Forio is presented. Nine maps of expected permanent slope displacements referred to as many return periods have been drawn using a GIS platform. These in turn, allowed to assign a displacement hazard curve to each point of the grid map, expressing the annual rate of exceedance of seismic induced permanent displacements in the study area. The displacement hazard curves represent a very attractive tool to obtain a preliminary indication of the seismic performance of a slope. In fact, they allow to evaluate the frequency of occurrence of a given slope displacement in a specific site or, alternatively, the slope displacement associated to a given return period.

2 Outline of the Methodology

The methodology adopted to evaluate the seismic induced slope instability involves a stepwise procedure including: 1) a Probabilistic Seismic Hazard approach adopted to evaluate the seismic demand parameter (PGA at bedrock) associated to different annual rate of exceedance in a given reference period; 2) a careful characterization of the study area in terms of hydrogeological, mechanical and topographic properties, aimed at the evaluation of the predisposing factors to both static and seismic slope instability; 3) the evaluation of the expected slope response expressed in terms of slope displacements associated with different return periods, adopting a classic sliding rigid block analysis for slope instability (Newmark's approach); 4) the construction of hazard demand curves in terms of mean annual rate of exceedance of the slope displacements at any grid point of the study area. The proposed procedure has been applied to evaluate the seismic induced slope instability hazard of the municipalities of Casamicciola, Lacco Ameno and Forio

(Fig. 1a). It has been implemented in a Geographic Information System (GIS) adopting a cell grid of 5x5m, determined by the resolution of the Digital Terrain Model (Fig. 1b), that in turn allowed to draw the slope angle map shown in Fig. 1c.

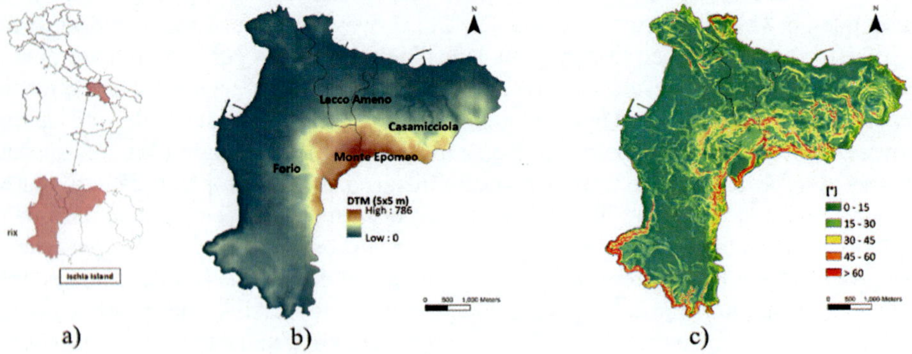

Fig. 1. Study area (a), DTM (b) and slope angle map (c) of the area

3 Geological and Geomorphological Setting

The island of Ischia is mainly constituted by volcanic rocks, by landslide deposits and, subordinately, by sedimentary rocks. The volcanic rocks present on the island come from lava flows and domes and explosive products, which have generated tuff formations and extensive tephra of ash and lapilli. The extreme complexity of the covering formations is well described in the detailed geological maps at scale 1:10.000 by [6, 7].

Figure 2a shows the resulting geo-lithological map where lithotypes have been represented on the map according to the basic units encoding suggested by [8]. The correlation scheme between lithostratigraphic [6, 7] and geo-lithological units is reported in [9]. Soil cover units are represented by younger sub-aerial pyroclastites, widespread epiclastites re-sedimented from the older volcanites, ancient landslides (mostly debris flow and avalanches), and coastal-alluvial plain deposit [9]. It is worth observing that moving from Mount Epomeo flanks toward the coastal areas the soil grading passes from coarse grained soils to fine sands of volcanic origin, sometimes showing a silty fraction. The patterns of groundwater flow are particularly complicated, due to the complex volcano-tectonic structures that affect ground water conditions. Figure 2b shows a map of the seasonal mean groundwater table depth computed from ground level, obtained from [10]. Physical and mechanical properties have been assigned to each of the soil units identified in Fig. 2, based on the results of 24 laboratory tests (20 direct shear and 4 triaxial tests) collected within the activities of Grade III Seismic Microzonation, *MS III* [11, 12] and on the results of further shear tests carried out within this research work.

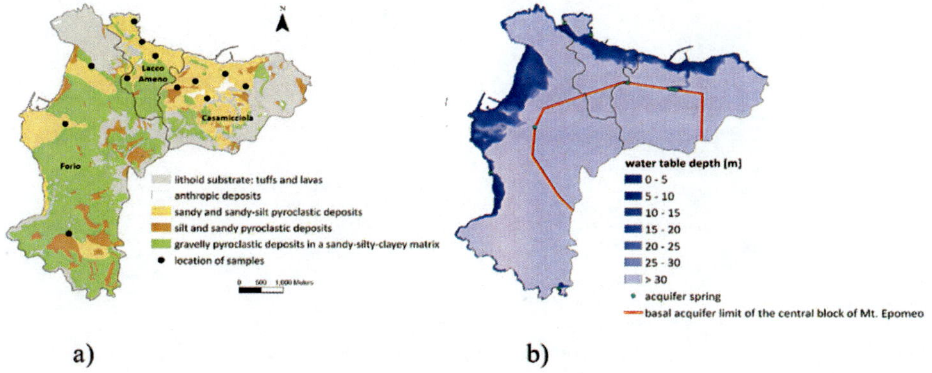

Fig. 2. (a) Geolithological map and (b) groundwater table map of the N-W sector of Ischia Island

Table 1 summarizes the values of unit weight and strength parameters corresponding to the 50th percentile of the relevant distribution and adopted for the evaluation of the susceptibility of the study area to seismic induced instability. More details about soil characterization can be found in [13].

Table 1. Unit weight and strength parameters for the main lithologies identified in Fig. 3

Gravelly pyroclastic deposits in sandy, silty clayey matrix			Sandy and sandy silt pyroclastic deposits			Silt and sandy pyroclastic deposits			Anthropic deposits		
γ (kN/m^3)	c' (kPa)	ϕ' (°)	γ (kN/m^3)	c' (kPa)	ϕ' (°)	γ (kN/m^3)	c' (kPa)	ϕ' (°)	γ (kN/m^3)	c' (kPa)	ϕ' (°)
14.9	10.3	32.4	15.7	13.2	32.3	15.0	10.0	30.0	14.0	0.0	25.0

4 Seismic Demand

The seismic demand has been evaluated adopting the Probabilistic Seismic Hazard model proposed by [14]. The demand parameter, expressed in terms of reference Peak Ground Acceleration at bedrock, a_r, has been calculated for nine return periods, T_R ranging between 30 to 2500 yr. Figure 3 shows the map of a_r relevant to a return period of 475 years, i.e. a probability of exceedance of 10% in fifty years and the corresponding map reporting the acceleration at surface, a_s. This latter has been evaluated considering both stratigraphic and topographic effects, as follows:

$$a_s = S_S \cdot S_T \cdot a_r \tag{1}$$

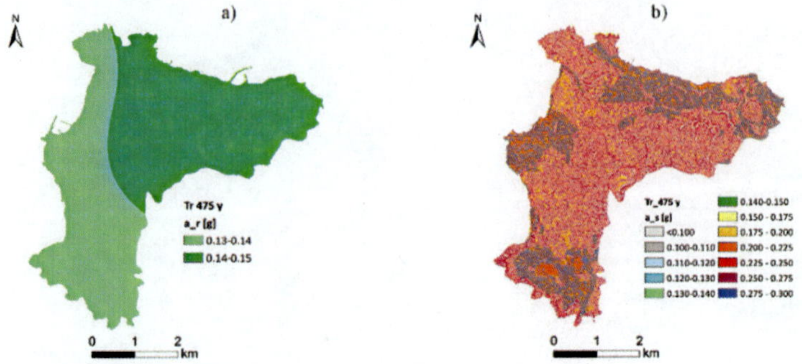

Fig. 3. Maps of peak ground acceleration expected (a) at bedrock, a_r and (b) at surface, a_s for a return period of 475 yrs.

In Eq. 1, S_S is the nonlinear stratigraphic amplification factor, evaluated as variable between 1.3 and 1.9 with the relationships proposed by [15], whose distribution is reported in Fig. 4b, with reference to $T_R = 475$ yr. It was assigned based on the classification map obtained by integrating the soil classification of the area reported within the *MSIII* [11, 12] with the soil type classification map proposed by [16] for the whole Italian territory (Fig. 4a). S_T is the topographic factor, evaluated as variable between 0.5 and 1.4 and assigned on the basis of the slope curvature (Fig. 4c) to account for the amplification or attenuation of the seismic motion in case of convex or concave geometry.

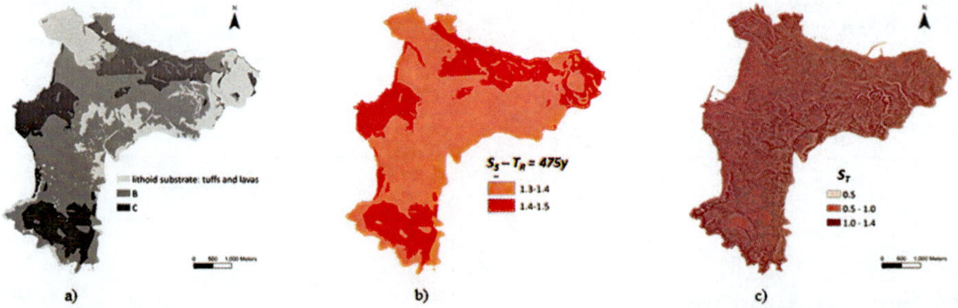

Fig. 4. Maps of (a) soil classification, (b) stratigraphic and (c) topographic amplification factors

5 Results

The seismic performance of slopes has been evaluated using the Newmark displacement-based approach, modelling the slope as a rigid friction block sliding on an inclined plane. Even if this latter hypothesis is appropriate for the study of shallow translational mechanisms, this method is preferred in practice since it represents a good compromise between simplicity and accuracy of the results. Furthermore, it can be simply implemented in a GIS environment for the evaluation of slope stability on a pixel basis. Following this method, the displacements are triggered when the acceleration overpasses the 'critical acceleration' threshold, ac. This latter can be seen as a parameter synthetically expressing the predisposing factors characterizing the slope. A map of the computed critical accelerations has been drawn (Fig. 5a) based on the geometrical, mechanical, and hydraulic parameters described above. No significant information has been available on the thickness of seismic induced landslides in the study area. An average value of 3 m has been assumed, as suggested by [17] based on morphological and stratigraphic evidence.

By combining the susceptibility to seismic induced instability and the seismic demand, expressed in terms of peak ground acceleration expected at surface with different return periods, the permanent displacements have been calculated in a simplified way by adopting the semi-empirical relationship proposed by [18] and here reported:

$$d = B \cdot e^{-A\frac{a_c}{a_s}} \tag{2}$$

where A and B are coefficients depending on the soil classes (Fig. 4a) and on the range of values of the acceleration expected at surface relevant to the upper-bound curve (94th percentile). Nine maps of earthquake induced displacements has been obtained; those relevant to the return periods of 475 and 2475 years are reported in Fig. 5b, c. In each map the displacements are classified referring to the threshold displacement d = 2 cm, related to instability in rock-like subsoil, d = 5 cm, related to a brittle soil behaviour, and d = 15 cm, related to a free-field ductile soil behaviour [19]. It is worth observing that the areas where attention needs to be focused are located between the municipalities of Casamicciola and Lacco Ameno, in the same zones where historical earthquake-induced landslides occurred in 1883.

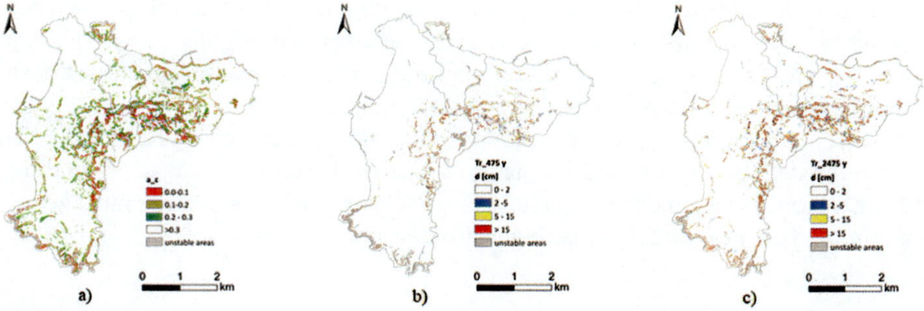

Fig. 5. Maps of (a) critical acceleration, a_c (grey zones are characterised by unstable conditions under static loading) and earthquake induced displacements relevant to (b) $T_R = 475$ yrs and (c) $T_R = 2475$ yrs.

Based on these maps it has been possible to build a demand hazard curve in terms of mean annual rate of exceedance of seismic induced slope displacement for each grid point characterized by a non-zero value of slope displacement. Examples of the resulting curves are reported in Fig. 6 with reference to three points of the grid map (Fig. 6a), characterized by different susceptibility (variable a_c). The susceptibility characterizing each grid point strongly influences the resulting hazard curve as clearly shown in Fig. 6b:

the lower the critical acceleration the higher the resulting displacement associated to a given annual rate of exceedance. Based on the hazard curves it can be possible to produce multiple hazard curves in terms of distribution of return periods relative to a given displacement threshold.

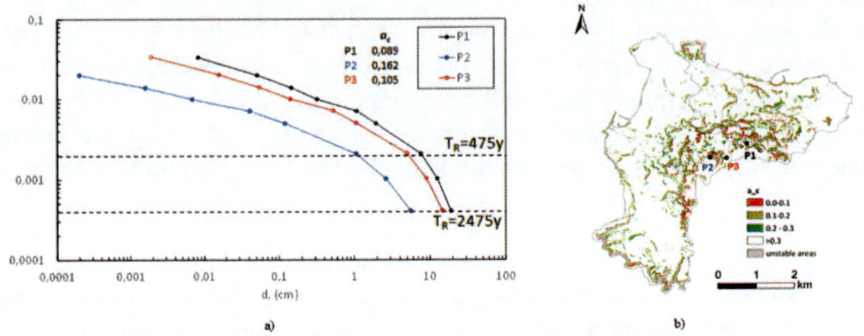

Fig. 6. (a) Displacements hazard curves and (b) related points on the critical acceleration map

6 Conclusions

The proposed displacement hazard curves represent a useful tool for a preliminary evaluation of the seismic performance of slopes, providing site-specific information about the return period associated to different values of seismic induced displacements. These demand curves can be suitable input for the evaluation of the induced risk on exposed elements whose vulnerability must be characterized by proper fragility curves. The hazard maps corresponding to different return periods can be suitable for practitioners and government agencies for a regional planning to identify and monitor zones that are potentially susceptible to earthquake-induced slope instability, thus requiring further detailed, site-specific studies. The probabilistic nature of curves also enables them to be combined with other hazards for a more complete hazard analysis, useful for disaster management to minimize damages caused by earthquake-induced landslides.

References

1. Cubellis, E., Luongo, G.: Il terremoto del 28 luglio 1883 a Casamicciola nell'isola d'Ischia. "Il contesto fisico". Monografia n.1, Servizio Sismico Nazionale, pp. 49–123. Istituto Poligrafico e Zecca dello Stato, Roma (1998)
2. Gruppo di Lavoro INGV sul terremoto dell'isola di Ischia: Rapporto di sintesi preliminare sul Terremoto dell'isola d'Ischia (Casamicciola) M4.0 del 21/08/2017 (2017). https://doi.org/10.5281/zenodo.886045
3. Azzaro, R., et al.: QUEST- Rilievo macrosismico per il terremoto dell'isola di Ischia del 21 agosto 2017 (2017). https://doi.org/10.5281/zenodo.886047
4. Del Prete, S., Mele, R.: Il contributo delle informazioni storiche per la valutazione della propensione al dissesto nell'Isola d'Ischia. Rend Soc Geol It Nuova Serie. **2**, 29–47 (2006)

5. Santo, A., Di Crescenzo, G., Del Prete, S., Di Iorio, L.: The Ischia island flash flood of november 2009 (Italy): phenomenon analysis and flood hazard. Phys. Chem. Earth **49**, 3–17 (2012)
6. Vezzoli L.: Island of Ischia. In: Vezzoli L, editor. CNR Quaderni de "La ricerca scientifica"; p. 114–0, 122 (1988)
7. Sbrana, A., Toccaceli, R.M.: Carta Geologica della Regione Campania – Foglio 464 - Isola di Ischia, Progetto CARG Regione Campania - Litografia Artistica Cartografica, Firenze. 216 pp + 1 carta: 10.000 (2011)
8. Commissione tecnica per la Microzonazione sismica: Standard di rappresentazione e archiviazione informatica. Vers. 4.2. Presidenza Consiglio dei Ministri-Dip. della Protezione Civile, Conferenza delle Regioni e delle Provincie Autonome, Roma, p. 138 (2020)
9. Mancini, M., et al.: Seismic microzonation in a complex volcano-tectonic setting: the case of northern and western Ischia Island (southern Italy). Italian J. Geosci. **140**(3), 382–408 (2021)
10. Piscopo, V., Lotti, F., Formica, F., Lana, L., Pianese, L.: Groundwater flow in the Ischia volcanic island (Italy) and its implications for thermal water abstraction. Hydrogeol. J. **28**(2), 579–601 (2019). https://doi.org/10.1007/s10040-019-02070-4
11. Riello, G., Petricone, A., Di Grazia, A., Iannotta, A., Miragliuolo, F., Lotito, G.: Microzonazione sismica di III livello del Comune di Lacco Ameno. Relazione Illustrativa. Piano degli studi di microzonazione sismica di III livello dei Comuni di Casamicciola, Lacco Ameno e Forio (2019). http://www.commissarioricostruzioneischia.it/Esiti-Microzonazione.html
12. Toscano, A., Cuccurullo, F., D'Anna, A.: Microzonazione sismica di III livello del Comune di Casamicciola Terme. Relazione Illustrativa. Piano degli studi di microzonazione sismica di III livello dei Comuni di Casamicciola, Lacco Ameno e Forio (2019). http://www.commissarioricostruzioneischia.it/Esiti-Microzonazione.html
13. Gargiulo, F.: Multi-level analysis of seismic ground instability in the volcanic island of Ischia (Italy) Ph.D. thesis Università degli studi di Napoli Federico II (2023)
14. Stucchi, M., Meletti, C., Montaldo, V., Crowley, H., Calvi, G.M., Boschi, E.: Seismic hazard assessment (2003–2009) for the Italian building code. Bull. Seismol. Soc. Am. **101**(4), 1885–1911 (2011)
15. Tropeano, G., Soccodato, F., Silvestri, F.: Re-evaluation of code-specified stratigraphic amplification factors based on Italian experimental records and numerical seismic response analyses. Soil Dyn. Earthq. Eng. **110**, 262–272 (2017). https://doi.org/10.1016/j.soildyn.12.030
16. Forte, G., Chioccarelli, E., Falco, M., Cito, P., Santo, A., Iervolino, I.: Seismic soil classification of Italy based on surface geology and shear-wave velocity measurements. Soil Dyn. Earthq. Eng. **122**, 79–93 (2019). https://doi.org/10.1016/j.soildyn.04.002
17. Caccavale, M., Matano, F., Sacchi, M.: An integrated approach to earthquake-induced landslide hazard zoning based on probabilistic seismic scenario for Phlegrean Islands (Ischia, Procida and Vivara) Italy. Geomorphology **295**, 235–259 (2017)
18. Gaudio, D., Rauseo, R., Masini, L., Rampello, S.: Semi-empirical relationships to assess the seismic performance of slopes from an updated version of the Italian seismic database. Bull. Earthq. Eng. **18**(14), 6245–6281 (2020). https://doi.org/10.1007/s10518-020-00937-6
19. Idriss, I.M.: Evaluating seismic risk in engineering practice. In: Proceedings of the 11th International Conference on Soil Mechanics and Foundation Engineering. San Francisco, CA, pp. 255–320 (1985)

Check for updates

On Multiphysical Couplings in Energy Geotechnics: Relevance and Applications

Alessio Ferrari[1,2](✉)

[1] Università degli Studi di Palermo, Palermo, Italy
alessio.ferrari@unipa.it
[2] École Polytechnique Fédérale de Lausanne, Lausanne, Switzerland

Abstract. Energy Geotechnics includes several applications related to the recovery and storage of energy from and into the ground, energy transportation, and the management of waste and carbon dioxide generated from energy use. These applications often involve the need to consider the behaviour of the involved geomaterials under complex - and often extreme – conditions, involving a series of coupled mechanical, hydraulic, chemical and thermal phenomena. The paper reports selected examples with the aim to present the variety of applications that Energy Geotechnics may cover, and to show some advances on the development of experimental, numerical and constitutive modelling tools for analysing the complex series of couplings involved in those applications.

Keywords: Energy geotechnics · Multiphysical couplings · THCM behaviours of geomaterials

1 Introduction

Geomechanics and geotechnical engineering are at the core of the energy scientific and technological challenges of this century. Most of new and conventional technologies for energy production, transportation, storage and disposal of related waste require in fact a deep understanding of the mechanical behaviour of geomaterials, which are often subjected to complex sequences of multiphysical coupled processes; these processes may involve non-isothermal conditions, cycles of wetting and drying, and perturbations of the original pore fluid composition.

The understanding and the prediction capability of the thermo-hydro-chemo-mechanical response of the involved geomaterials are therefore fundamental requirements in order to design, analyse and assess the long-term behaviour of energy-related geotechnical systems.

The paper presents selected examples of applications and shows some of the fundamental roles that multiphysical couplings play in the analysis and prediction of the behaviour of the systems.

A. Ferrari et al. (Eds.): CNRIG 2023, SSGG, pp. 20–27, 2023.
https://doi.org/10.1007/978-3-031-34761-0_3

2 Selected Examples of Multiphysical Processes in Energy-Related Geotechnical Applications

2.1 Thermo-Hydro-Mechanical Couplings in Energy Geostructures

Energy geostructures are an innovative technology which can significantly contribute to reduce the environmental impacts of the energy needs of the built environment. They integrate heat exchangers (usually plastic pipes) within structural elements such as piles, walls, tunnels and shafts. The exchangers are connected to a heat pump in order to form a ground-source heat pump system, in which soils function as extraction and storage media. The use of such structures started in the 1980s mainly for piles and extended progressively to energy walls and tunnels (see for example [1] and [2]).

Activation of energy geostructures will impact to a certain extent the temperature of the ground and of the structural elements. In this sense, the ability to predict the response of the involved geomaterials in non-isothermal conditions is a fundamental requirement to design and analyse the performance of energy geostructures. From a constitutive response perspective, temperature variations in clayey soils can induce significant volumetric deformations (that can be irreversible for normally-consolidated geomaterials), as well as changes in their shear strength and in the concrete-soil interface behaviour [3].

Hydrothermal couplings must be considered also to assess the overall performance of an energy geostructure, which is strictly related to its interactions with the surrounding environment. An example is given in Fig. 1, where an underground structure equipped with heat-exchangers in parts of the walls is shown. The contours of temperature have been obtained numerically with coupled thermo-hydraulic simulations in the case of a groundwater flow perpendicular to the structures (see [4] for details); the contours show how the groundwater flow clearly affects the temperature field, with relevant heat propagation along the main direction of flow.

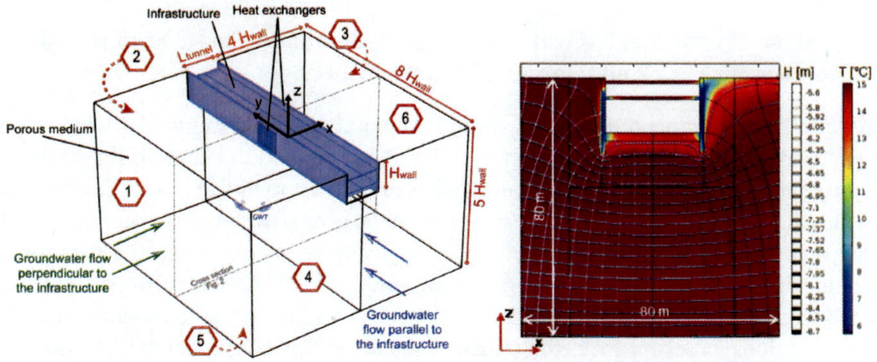

Fig. 1. 3D model of an underground structure equipped with heat exchangers in its walls (H_{wall} = 25.5 m) (*left*), and temperature contour plot for the case of groundwater flow perpendicular to the structure (*right*).

Temperature cycles will result in cyclic deformations of the structural elements and cyclic variations of the stress characteristics; these variations need to be anticipated in order to evaluate their potential impacts on the definition of all possible ultimate and serviceability limit states. An example is given in Fig. 2, which depicts the strains measured along an energy wall of an underground station during operations in heating and cooling modes [5]; it can be clearly seen how the geostructure deforms as a result of the thermal loads in a non-uniform manner, as deformations are partly restrained by the soil and the structural connections.

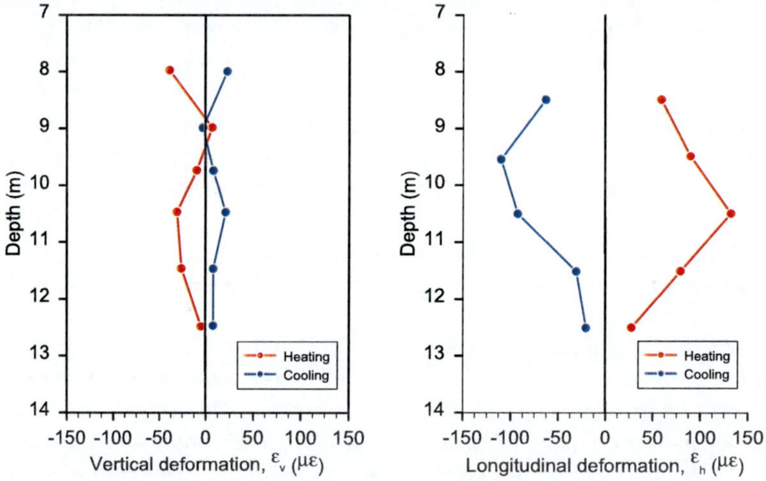

Fig. 2. Deformational behaviour of the wall intrados facing the tunnel in an underground station, as a result of the thermal loads in heating and cooling operations.

2.2 Thermo-Hydro Mechanical Couplings in Bentonite-Based Engineered Barriers for the Confinement of Nuclear Wastes

Bentonite-based engineered barriers are fundamental elements in nuclear waste repository systems. Their low permeability, high retention capabilities and ability to develop swelling pressures in the order of several MPas are key elements to ensure limited transport of radionuclides, isolation of the waste from the groundwater and structural stability of the repository. The ability to predict the response of bentonite upon saturation and upon temperature variations is of outmost importance in this context. In particular, it is needed to anticipate the evolution of the dry density of the bentonite, since most of the safety functions expected from the barrier depend on this characteristic. Ensuring this ability is a challenging task, as the volumetric response of bentonite upon wetting and temperature change is strongly stress-path dependent.

The density of different portions of the barrier will evolve differently depending on the constrains provided by the surrounding elements. As an example, Fig. 3-left shows the development of swelling pressure and swelling heave on a sample of Mx80 granular bentonite subjected to saturation in isochoric condition (similarly to the behaviour in the inner part of the barrier) and under-low-stress condition (similarly to the behaviour in proximity of technological gaps), respectively [6]. Also, a complete analysis of the long-term performance of the repository requires a good understanding of the mechanical evolution of the bentonite upon heating from the radioactive waste. Figure 3-right summarizes the results of finite elements modelling of the void ratio evolution in 100 years for a bentonite barrier, as a result of the hydration from the surrounding host-rock and the thermal wave produced by the nuclear waste. The simulations required to use of a fully coupled constitutive model in which the water retention behaviour of the bentonite plays a major role in the definition of the effective stress and the yielding of the material upon hydration (details in [7]).

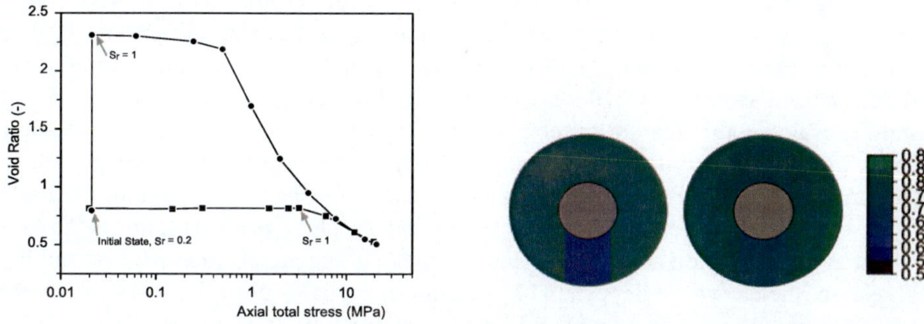

Fig. 3. Volumetric response of bentonite along different stress paths: swelling and compression experiments in high-pressure oedometric testing (*left*). Modelling of barrier homogenization with a coupled thermo-hydro-mechanical constitutive model, from the initial state up to 100 years from the waste emplacement (*right*).

2.3 Hydro-Mechanical Couplings in Shales

Shales are fine-grained sedimentary geomaterials mainly composed of clay minerals with additional larger particles of other minerals, especially quartz and calcite. Their fabric is characterized by fissility along bedding planes which contributes to their marked anisotropic behaviour. Their low porosity and small dominant pore size (often in the order of a few of dozens of μm) result in very low permeabilities and good sealing capacity. These characteristics make shales fundamentals materials in many energy-related applications; they are investigated as sources for entrapped natural gas (shale gas), as potential host rocks for deep nuclear waste disposal systems, or as caprocks in the context of CO_2 geological sequestration. In any such application, a deep understanding of the hydro-mechanical behaviour of these materials is of primary significance.

The clayey nature of shales makes their mechanical response highly dependent on their hydration state. In this regard, the experimental assessment of their properties requires an explicit control of their degree of saturation (or equivalently of their suction), especially when the materials are expected to undergo drying and wetting processes; this is the case of the ventilation of tunnels for waste emplacement, or imbibition associated with hydraulic fracturing of unconventional gas reservoirs. As an example, Fig. 4 shows a clear dependency of the elastic properties of a gas shale on the total suction to which the sample was equilibrated to. Data were obtained with a modified UCS test in which the real-time control of the relative humidity of the sample was guaranteed (see [8] for details). A non-linear decrease of the secant Young's modulus higher than 50% can be detected upon a wetting process. Similar decreases of the uniaxial compressive strength are also observed. Also, the nonlinear elastic behaviour with hysteresis which is typical of these materials is significantly affected by the wetting or drying processes prior to the unloading/reloading paths in terms of total stress, with an important reduction of the non-linearity of the response when the same hydration state is reached in wetting rather than in drying [9].

Swelling of shales upon hydration is responsible for significant water loss during flowback operations in hydraulic fracturing operations of gas shale reservoirs [10] as only about 20% of the injected fracturing fluids are usually recovered after stimulation. At the same time, swelling of shales is a key element for their use as host rocks in nuclear waste disposals, as this feature ensures self-sealing capacity versus possible fractures that may open in the very lifespan of the repository (in the order of 10^5 years).

Several applications require the study of the hydro-mechanical behaviour of shales has to be extended to the case in which a pressurized gas phase is present; this is the case of potential gas migration as a result of the corrosion of the metallic canister containing the waste in nuclear waste deposits [11, 12], or when shales are used as caprocks for structural trapping in the context of carbon geological sequestration [13, 14].

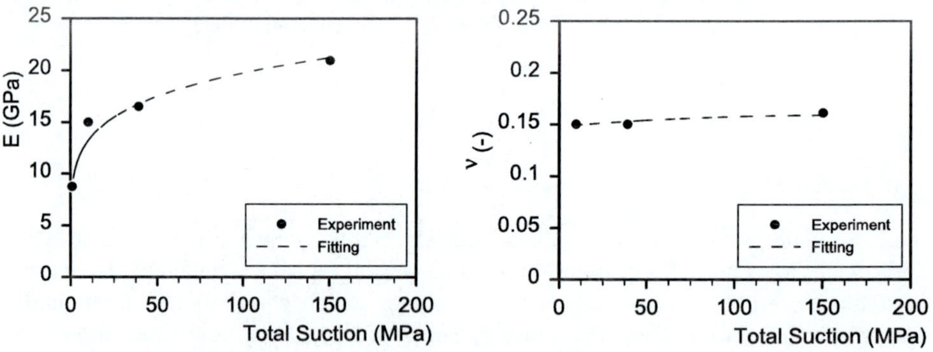

Fig. 4. Evolution of Young's modulus perpendicular to bedding planes and Poisson's ratio for a gas shale sample tested in a controlled total suction device.

2.4 Chemo-Mechanical Couplings: Changes in Pore Water Composition

Several geotechnical engineering applications require more and more the role of the chemical composition of the pore water to be explicitly considered. These situations may concern both environmental geotechnics, such in the case of the salinification of groundwater, or energy-oriented applications such as wellbore stability problems or possible long-term changes of pore water composition in nuclear waste disposal systems.

Experimental evidence has been collected, both for the case of highly active clays (see for example [15], or [16]) and less active clays (see for example [17]) on the impacts that changes in the chemical composition of the pore water have in the hydro-mechanical behaviour. In these experiments, the conventional stress paths which are used to test geomaterials are typically expanded by controlling an additional variable, which is usually the osmotic suction of the geomaterial; this latter is controlled by fixing the ion concentration of salts in the pore water. Chemical composition impacts both stiffness and strength of clays. Figure 5 depicts as an example the variation of the initial oedometric modulus and yield vertical effective stress observed on samples prepared at the same target void ratio, obtained by mixing illite powder with solutions of sodium chloride in distilled water at different concentrations (0.0–6.0 M).

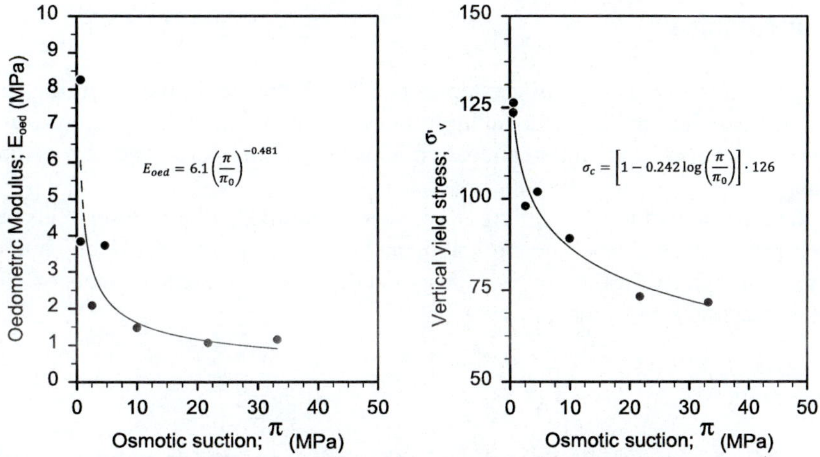

Fig. 5. Evolution of the initial oedometric modulus and vertical yield stress on samples of illite prepared at different osmotic suctions. As the samples were prepared at initial similar void ratio, the difference in the mechanical properties is consequence of the different pore water composition.

In terms of constitutive modelling, the chemo-mechanical couplings can be conveniently considered by introducing an additional state variable (as the osmotic suction) into elasto-plastic frameworks; as an examples, equations in Fig. 5 can be used to make explicit the dependency of the stiffness and the size of the elastic domain on the chemical variable; in particular, the latter dependency would make it possible to reproduce irreversible volumetric compaction for clayey experiencing an increment of the salinity in their pore water, when they are in a normally-consolidated condition.

Alternatively, the role of the pore water composition on the mechanical behaviour can be considered by a suitable definition of the effective stress (see [18]). Implication of using this second approach, among others, is the uniqueness of the failure envelope regardless of the chemical composition of the pore water (Fig. 6).

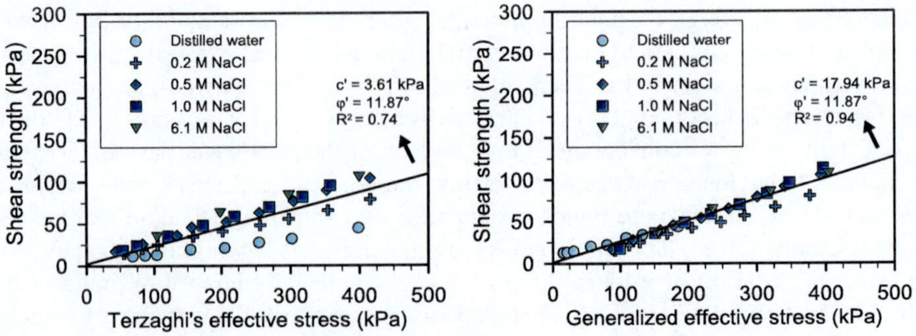

Fig. 6. Representation of shear strength data on Ponza bentonite with varying salt concentrations (experimental data from Di Maio, 1996) with conventional (a) and an extended effective stress definition accounting for the solute suction of the material as suggested by [18] (b).

3 Conclusions

The paper summarized some of the topics and developments in the emerging field of energy geotechnics. In this area multiphysical couplings, including changes in temperature, saturation state, and chemical composition of the pore water are of upmost importance.

Although many of these couplings have been identified, still new research is needed for their proper experimental quantification and for refining constitutive and numerical tools to be integrated in the analysis, design and performance assessments of energy-related geotechnical systems.

References

1. Adam, D., Markiewicz, R.: Energy from earth-coupled structures, foundations, tunnels and sewers. Géotechnique **59**(3), 229–236 (2009)
2. Brandl, H.: Energy foundations and other thermo-active ground structures. Géotechnique **56**(2), 81–122 (2006)
3. Di Donna, A., Ferrari, A., Laloui, L.: Experimental investigations of the soil–concrete interface: physical mechanisms, cyclic mobilization, and behaviour at different temperatures. Can. Geotech. J. **53**(4), 659–672 (2016)
4. Zannin, J., Ferrari, A., Pousse, M., Laloui, L.: Hydrothermal interactions in energy walls. Undergr. Space (China) **6**(2), 173–184 (2021)
5. Zannin, J., Ferrari, A., Kazerani, T., Koliji, A., Laloui, L.: Experimental analysis of a thermoactive underground railway station. Geomech. Energy Environ. (2022)
6. Ferrari, A., Bosch, J.A., Baryla, P., Rosone, M.: Volume change response and fabric evolution of granular MX80 bentonite along different hydro-mechanical stress paths. Acta Geotech. **17**(9), 3719–3730 (2022). https://doi.org/10.1007/s11440-022-01481-0
7. Bosch, J.A., Ferrari, A., Laloui, L.: Coupled hydro-mechanical analysis of compacted bentonite behaviour during hydration. Comput. Geotech. **140**, 104447 (2021)
8. Ferrari, A., Minardi, A., Ewy, R., Laloui, L.: Gas shales testing in controlled partially saturated conditions. Int. J. Rock Mech. Min. Sci. **107**, 110–119 (2018)

9. Minardi, A., Ferrari, A., Ewy, R., Laloui, L.: Nonlinear elastic response of partially saturated gas shales in uniaxial compression. Rock Mech. Rock Eng. **51**(7), 1967–1978 (2018). https://doi.org/10.1007/s00603-018-1453-x
10. Minardi, A., Ferrari, A., Ewy, R., Laloui, L.: The impact of the volumetric swelling behavior on the water uptake of gas shale. J. Nat. Gas Sci. Eng. **49**, 132–144 (2018)
11. Senger, R., Romero, E., Ferrari, A., Marschall, P.: Characterization of gas flow through low-permeability claystone: Laboratory experiments and two-phase flow analyses. Geol. Soc. Special Publ. **400**, 531–543 (2014)
12. Gonzalez-Blanco, L., Romero, E., Marschall, P., Levasseur, S.: Hydro-mechanical response to gas transfer of deep argillaceous host rocks for radioactive waste disposal. Rock Mech. Rock Eng. **55**(3), 1159–1177 (2022)
13. Iglauer, S., Al-Yaseri, A.Z., Rezaee, R., Lebedev, M.: CO2 wettability of caprocks: Implications for structural storage capacity and containment security. Geophys. Res. Lett. **42**(21), 9279–9284 (2015)
14. Minardi, A., Stavropoulou, E., Kim, T., Ferrari, A., Laloui, L.: Experimental assessment of the hydro-mechanical behaviour of a shale caprock during CO2 injection. Int. J. Greenhouse Gas Control **106**, 103225 (2021)
15. Di Maio, C.: Exposure of bentonite to salt solution: osmotic and mechanical effects. Géotechnique **46**(4), 695–707 (1996)
16. Manca, D., Ferrari, A., Laloui, L.: Fabric evolution and the related swelling behaviour of a sand/bentonite mixture upon hydro-chemo-mechanical loadings. Géotechnique **66**(1), 41–57 (2016)
17. Witteveen, P., Ferrari, A., Laloui, L.: An experimental and constitutive investigation on the chemo-mechanical behaviour of a clay. Géotechnique **63**(3), 244–255 (2013)
18. Tuttolomondo, A., Ferrari, A., Laloui, L.: Generalized effective stress concept for saturated active clays. Can. Geotech. J. **58**(11), 1627–1639 (2021)

Rock Masses Characterization with Advanced Measurement Systems for Reliability-Based Design

Maria Rita Migliazza(✉)

Department of Structural, Geotechnical and Building Engineering, Politecnico di Torino, Corso Duca Degli Abruzzi 24, 10129 Turin, Italy
maria.migliazza@polito.it

Abstract. The design processes involving rock engineering problems must take into account the rock mass discontinuous nature, that strongly affects the mechanical response of this medium to any perturbation. Stress-strain behaviour of rock masses is linked to both the mechanical characteristics of rock matrix and to the rock mass degree of fracturing. The characterization of rock masses requires approaches at different scales (microscale, laboratory, site and, sometimes, regional scale), each of which provides different information that can be very relevant from a design point of view. At the microscale, information regarding petrographic and mineralogical features can be acquired and analysed in relation to the mechanical behaviour of rock matrix and discontinuities studied with laboratory tests. At the site scale, information related to discontinuity geometry, such as orientation, persistence and spacing, can be acquired by different measuring methods. Rock masses, by their nature, are discontinuous, inhomogeneous, anisotropic materials and the parameters characterizing them show a natural randomness due to both aleatory variability and epistemic uncertainties. In order to reduce the epistemic uncertainties a representative and sound number of measurements must be collected, and, simultaneously, probabilistic analyses are required to represent the aleatory nature of each variable. This contribution deals with the application of survey advanced techniques useable at different scale with the aim to measure the parameters necessary for a sound characterization of a rock mass useful to apply of reliability-based design approaches in line with the reference standards for the geotechnical design.

Keywords: Rock masses · Reliability-based design · Advanced measurement system

1 Introduction

The rock mass nature is inherently complex due to its Discontinuous, Inhomogeneous, Anisotropic and Non-linear Elastic (DIANE) behaviour, related to the behaviour of rock matrix and the presence of fracture network (Hudson and Harrison 1997) even if sometimes it is considered as a Continuous, Homogeneous, Isotropic, Linear, and Elastic

A. Ferrari et al. (Eds.): CNRIG 2023, SSGG, pp. 28–35, 2023.
https://doi.org/10.1007/978-3-031-34761-0_4

(CHILE) material. Because of its fractured nature, the rock mass can be studied as a continuous, discontinuous or equivalent continuous medium, depending not only on the degree of fracturing but also on the scale of the engineering problem (Barton 1998). In design practice, whatever the rock mass model referred to is, physical, mechanical and geometric parameters are defined by in situ surveys, laboratory tests at small scale or in situ tests. The spatial distribution and in-situ mechanical properties of the discontinuities generally govern the behaviour of the rock mass, and the parameters defining these properties cannot be captured through small scale sampling or testing, nor through in situ tests which, generally, involve a volume of rock mass smaller than REV (representative element volume). An epistemic uncertainty related to measurements, the interpretation of errors or/and the models used to estimate and analyse them must be added to the natural aleatory variability of the parameters characterizing the rock mass (Bedi 2013).

The aleatory variability is related to the natural randomness and can be characterised by stochastic models and propagated by using probabilistic models; on the contrary, the epistemic uncertainty is resulting from a lack of knowledge and it is better represented by non-stochastic approaches. The latter can be reduced or eliminated through additional information or knowledge and improving both the data quality and quantity; the former must be taken into account because it describes the real nature of the analysed data (Fig. 1).

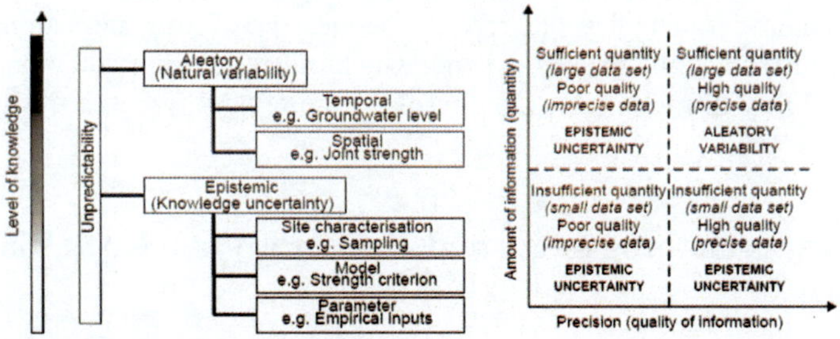

Fig. 1. Uncertainty and variability as a function of quality and quantity of available information (Bedi and Harrison 2013).

As it is currently for other engineering materials, for rock engineering as well the design procedures are evolving towards the application of concepts as geotechnical ultimate or serviceability Limit State Design (LSD), typical of current reference standards (Bozorgzadeh et al. 2028). Next version of the Eurocode 7 (EC7), currently under revision, will be taking into account rock and rock masses too, and reliability-based design (RDB) approaches will be proposed, among others, as design tools for these mediums. RDB uses probabilistic analysis to correlate the loads or actions acting on a system and its resistance, defining a reliability index associated to an uncertainty level and a probability of failure. Both the action and the resistance are characterized by a statistical variability related to the aleatory variability of the design parameters (e.g. strength of intact rock, strength of rock masses, strength of discontinuities, volume of blocks, …) and their uncertainty.

Reliability-based processes are based on the definition of a probability of exceeding and the statistical variability of the quantities involved must necessarily be taken into account. Whether continuous, discontinuous or equivalent continuous approaches are used for the analysis of rock mass, the analysis processes must deal with geometrical and mechanical quantities that have a clear aleatory character, which is combined with heterogeneity and the frequent anisotropy of these media.

More and more it is evident how the understanding of the mechanical response of rock masses can or must be approached at different scales. The microscale provides us with mineralogical and petrographic information useful for understanding the mechanical behavior of the rock matrix at the laboratory scale. The site scale, certainly gives us information about the heterogeneity of the rock masses whose degree of fracturing can be better understood by evaluating regional scales. The techniques and tools available today and in particular the metric analysis of images allow us to have more and more information at different scales of observation; information that tends to reduce the epistemic uncertainties because, on the one hand, they improve the measuring capabilities and accuracy, on the other, they allow to increase the number of data that can be acquired, leading us towards the definition of the random variability of the studied quantities.

The results of laboratory and site scale analyses based on the use of photogrammetric techniques as measuring tool, are following described. They refer to two emblematic examples in the field of rock engineering and show how the use of these advanced technology can help the assessment of the heterogeneity, anisotropy and aleatory variability of peculiar quantities. The first one regards the measure of roughness along natural discontinuities surface, while the second one, which illustrates the evaluation of the blocks volume variability in rock masses, is useful for the reliability-based design of rockfall protection works.

2 Roughness Heterogeneity and Anisotropy of a Rock Discontinuity Surface

The shear strength of discontinuities plays a fundamental role in the stability condition of rock slopes and the joint surface morphology represents one of the most important geometric features strongly influencing the shear resistance. The morphological irregularity of these surfaces represents the amplitude and the shape of the asperities and it is naturally dependent on the considered scale (Barton 1973): roughness is defined at small-laboratory scale and waviness at large-site scale. It is commonly assessed on small portions of the outcropping joint surfaces or at laboratory samples, by using traditional manual and subjective tools (e.g. Barton's profilometer) or highly sophisticated remote and automatic optical methods (laser profilometer, terrestrial or aerial laser scanning and photogrammetric survey, also by using UAV). The former ones require physical contact with the surface and are usually capable of measuring directional profiles at a small scale (usually 10–30 cm); the latter ones obtain morphological information remotely, through processing point clouds and 3D models at different scales (from laboratory to on-site scale).

The roughness measurement is classically expressed through several methods that can be divided in: geometrical descriptors (Myers 1962; Tse-Cruden 1979; Tatone and

Grasselli, 2010), fractal approaches (Mandelbrot 1983; Kulatilake and Um 1997; Odling 1994) and geostatistical methods (Belem et al. 1997; Roko et al. 1997). Many authors proposed correlations for deriving Joint Roughness Coefficient of Barton's criterion from the above mentioned geometrical descriptors (Li and Zhang 2015). Most of them are empirical correlations defined by digitizing the 10 Barton and Choubey's profiles, each one corresponds to a JRC value, and by applying the different methods to obtain the corresponding value of the geometric descriptor.

The roughness and waviness of a discontinuity surface are strictly connected to the microstructure of the rock (type, crystals dimension, level of cementation and connection), the stress state inducing the fractures in a rock mass (shearing or tensile joints) and the geological history that can affect the large scale curvature of the discontinuity surface. But, by a morphological point of view, it is necessary to take into account that the roughness is an uneven, anisotropic and size-dependent property of a discontinuity surface and these factors must be considered and analysed during roughness measurement, in order to estimate discontinuity shear strength and reducing the connected uncertainties.

The use of advanced technologies, the precision of the instruments (performance of digital cameras, both in geometry and radiometry, or laser scanning equipment) and the image processing algorithms used for the reconstruction of 3D models allows to improve the roughness measure giving information regarding unevenness, anisotropy and scale effect and reducing the uncertainty level.

With the aim to measure the rock surface roughness anisotropy and unevenness, several discontinuities at different scales have been surveyed using photogrammetric techniques (Ferrero et al. 2019; Carriero et al. 2022) by taking digital images elaborated with Agisoft Metashape software in order to obtain points clouds and the relative DSMs. In Fig. 2 are reported the results obtained with the analysis carried out on a discontinuity surface of about 1m*1m in a Rhyodacites rock mass. In order to analyse the surface roughness along linear profiles, the scattered points clouds were processed, extrapolating the morphological data according to a regular grid of points having a constant node distance SI (Fig. 2-b). In this way several longitudinal profiles along both x and y directions were obtained and analysed to estimate the roughness parameter Z_2 according to (Myers 1962):

$$Z_2 = \left[\frac{1}{N(SI)^2} \sum_{i=1}^{N} (z_{i+1} - z_i)^2 \right]^{\frac{1}{2}} \tag{1}$$

where N is the number of nodes along a profile, SI is the scanning step and z is the profile height measured in relation to the average plane.

A first analysis was carried out in order to understand the dependency of the results from the SI (Fig. 2-c). It is clear that a SI decreasing induces an increase in the measurement of roughness (Z_2) up to a limit value of SI, below which Z_2 remains almost constant. It is considered as a scanning step for further analysis. More than 200 profiles were analysed in both directions in order to analyse the roughness variability along the surface. Figure 2-d shows that the roughness is not characterized by a unique value but, on contrary, by a variable distribution (well described by a normal distribution) with higher roughness values along the x direction (the orange data). The roughness anisotropy was analysed considering profiles with directions rotated around the centre of the discontinuity according to the scheme of Fig. 2-b. The results (Fig. 2-e) show a clear anisotropy with a minimum roughness along the y direction and a maximum value along a direction inclined of about 45° from the y axes.

This simple example shows how the definition of the JRC parameter by comparison with a measure taken with a manual profilometer along a direction subjectively chosen by the operator, cannot take into account the inherent properties of the roughness. This might induce a high degree of uncertainty in the discontinuity mechanical properties assessment. On the contrary, the use of a no-contact survey techniques shows how the measurement of this quantity can be improved, defining their heterogeneity and anisotropicity.

3 Rock Mass Block Volume Evaluation

Rockfall phenomena are common instabilities occurring in mountainous regions and protective works, as flexible barriers or rockfall embankments, are usually adopted to mitigate the impact of falling boulders on vulnerable structures (structure, infrastructure, buildings, ...). The design of these structure is an emblematic example that highlights the complexity of the application of EC7 principles, especially in relation to the choice of design parameters and the definition of characteristic actions. This latter is strictly connected to the kinematic parameters of the impacting block and it is therefore affected by a number of uncertainties related to slope characterization (geometry, return coefficients), discontinuities survey technique, the definition of the design block volume and the methods used to simulate the block trajectories along the slope.

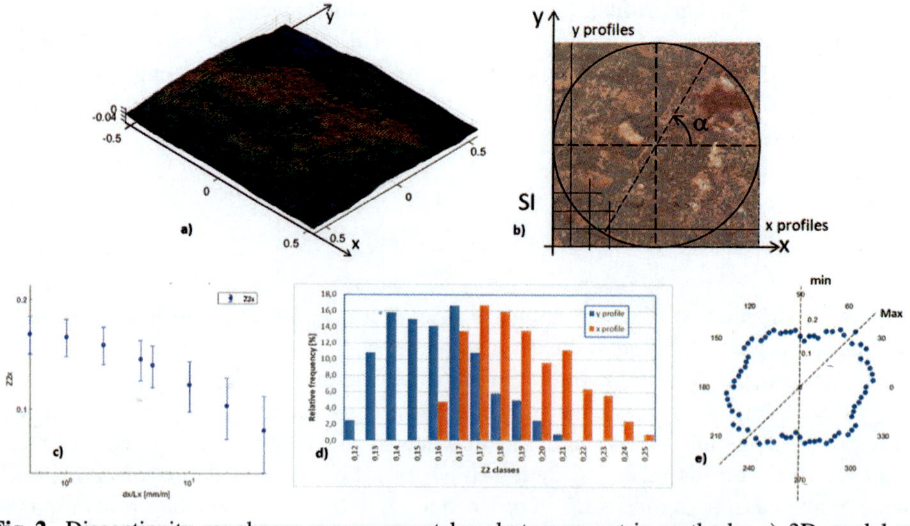

Fig. 2. Discontinuity roughness measurement by photogrammetric methods: a) 3D model of discontinuity surface; b) scheme for longitudinal and anisotropy profile analysis; c) Z_2 vs scan –step SI; d) unevenness of longitudinal profiles along x and y directions; e) roughness anisotropy.

In particular, trajectory, velocity and energy of the falling boulders are closely related to the volume and shape of blocks, which are strictly connected to some discontinuity systems characteristics as orientation and spacing. These geometric properties are characterized by a natural aleatory variability and they play a fundamental role in the analysis of the rockfall phenomena. The surveys carried out to collect the geometrical characteristics of the discontinuity systems must be sufficiently extended in order to get a number of data able to statistically describe the aleatory variability of orientation and spacing and reduce the epistemic uncertainties.

In recent years, the use of non-contact surveying techniques, as laser scanning or photogrammetric methods, allows to collect a large amount of data in a relatively short period of time: these data are, moreover, characterized by a significant increase in quality and accuracy. It follows that assessments of the geometrical characteristics of rock masses are more reliable. The results of rock mass characterization and block trajectory are the input data of the protection structure design phase; for this reason, adequate rock mass data processing and the consequent reduction of uncertainties, have a positive impact on the effectiveness and reliability of protection works.

Number, persistence, spacing and orientation of the discontinuity define the blocks dimension, while the number of sets, spacing ratio and orientation determine their shape. By defining the statistically variation of these geometric parameters and combining them, an In-situ Block Size Distribution curve (IBSD), representing a cumulative distribution of block volume, can be defined (Umili et al. 2020). Starting from these aspects, the definition of an IBSD allows to overcome the traditional approach of the characteristic or of the design block, based on a single value and therefore deterministic. The method is based on the relationship that correlate spacing and orientation of discontinuities with the volume of blocks delimited by them, as, for example, the one proposed by Palmstrong (1996) valid for three sets of discontinuities:

$$V = \frac{S_1 \cdot S_2 \cdot S_3}{sin\gamma_{12} \cdot sin\gamma_{13} \cdot sin\gamma_{23}} \tag{2}$$

where V is the block volume, *S1*, *S2* and *S3* are the discontinuity spacing, while γ_{ij} are the relative angles between each couple of discontinuities.

To define the IBSD curve, it is possible to replace the individual values of spacing and orientation of the above equation with the corresponding statistical distributions. As regards the spacing, its variability can be described by identifying a Probability Density function (PDF) obtainable by defining the best-fitting curve of the collected data, while the orientation variability through the Fisher distribution.

This approach was applied to study the rockfall phenomon (Ferrero et al. 2016) involving a rock face located at Rovenaud, in the Grand Paradiso National Park (Valsavaranche, Aosta, Italy) that hangs over a building housing the "Environmental Information Centre" of the park. The geostructural survey has been performed through the acquisition of terrestrial digital images and the photogrammetric reconstruction of a Digital Surface Model (DSM). About 250 data regarding discontinuity orientation and spacing were collected and analysed in order to define the statistical distribution and the correspondent parameters. In this case, a Monte Carlo simulation, which takes into account the frequency distribution of the traces, allowed to determine the frequency distribution for a considered volume of block obtained from spacing data surveyed on the rock wall. Such distributions can be validated by means of a comparison with the frequency distribution of the volume of the rock blocks observed at the base of the slope (Fig. 3).

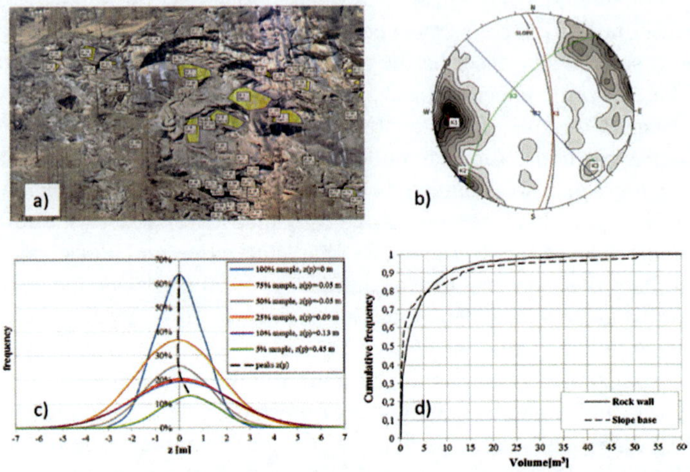

Fig. 3. Rovenaud case histories results: a) Remote survey of discontinuities; b) Stereonet with the discontinuity sets projections; c) Frequency distribution curve of spacing data; d) In-situ Block Size Distribution curve

4 Conclusion

Rock mass are natural medium characterized by a strong heterogeneity and anisotropy. The measurement of the physical, geometrical and mechanical quantities able to describe their mechanical behaviour and the failure conditions have to be carried out at different scales involving a large portion of rock mass and with a high level of accuracy to evaluate their aleatory variability and reduce the epistemic uncertainties.

A great help in this direction is given by the advanced techniques of surveying, which, with an increasingly high precision of the instruments, a great versatility of application at different scales, can provide large amounts of data capable of representing robust and statistically representative samples. This is very important for applying reliability-based design to rock engineering problem, according to the reference standards for geotechnical design.

References

Barton, N.: Quantitative description of rock masses for the design of NMT reinforcement. In: Choubey, V.D. (ed.) International Conference on Hydropower Development in Himalayas, 20–22 April, Shimla, India (1998)

Bedi, A.: A proposed framework for characterising uncertainty and variability in rock mechanics and rock engineering, Ph.D. thesis, Imperial college, London (2013)

Bedi, A., Harrison, J.P.: A comparison of Bayesian techniques and non-probabilistic models in rock engineering design. In: Rock Mechanics for Resources, Engineering and Environment, pp. 105–110 (2013)

Belem, T., Homand-Etienne, F., Souley, M.: – Fractal analysis of shear joint roughness. Int. J. Rock Mech. Min. Sci. **34**(3–4), 130 (1997)

Bozorgzadeh, N., Escobar, M.D., Harrison, J.P.: Comprehensive statistical analysis of intact strength for reliability based design. Int. J. Rock Mech. Min. Sci. **106**(2018), 347–387 (2018)

Carriero, M.T., Ferrero, A.M., Migliazza, M.R., Umili, G.: Evaluation of progressive damage of discontinuity asperities due to shearing by means photogrammetric survey. In: IOP Conference Series Earth Environment Science, vol. 1124 p. 012053 (2022)

Hudson, J.A., Harrison, J.P.: Engineering Rock Mechanics: An Introduction to the Principles. Elsevier, Oxford (1997)

Ferrero, A.M., Migliazza, M.R., Pirulli, M., Umili, G.: Some open issues on rockfall hazard analysis in fractured rock mass: problems and prospects. Rock Mech. Rock Eng. **49**, 3615–3629 (2016)

Ferrero, A.M., Migliazza, M.R., Umili, G.: Comparison of methods for discontinuity roughness evaluation. Rivista Italina di Geotecnica **3**, 5–15 (2019)

Kulatilake, P.H.S.W., Um, J.: Requirements for accurate quantification of self-affine roughness using the roughness-length method. Int. J. Rock Mech. Min. Sci. **34**(3–4), 166.e1–66.e15 (1997)

Li, Y., Zhang, Y.: Quantitative estimation of joint roughness coefficient using statistical parameters. Int. J. Rock Mech. Min. Sci. **77**, 27–35 (2015)

Mandelbrot, B.B.: – The Fractal Geometry of Nature. Freeman, San Francisco (1983)

Myers, N.O.: Characterization of surface roughness. Wear **5**, 182–189 (1962)

Odling, N.E.: Natural fracture profiles, fractal dimension and joint roughness coefficients. Rock Mech. Rock Eng. **27**(3), 135–153 (1994). https://doi.org/10.1007/BF01020307

Palmstrom A.: The weighted joint density method leads to improved characterization of jointing. In: International Conference on Recent Advances in Tunnelling Technology, New Delhi, India, p. 6 (1996)

Roko, R.O., Daemenj, J.K., Myers, D.E.: Variogram characterization of joint surface morphology and asperity deformation during shearing. Int. J. Rock Mech. Min. Sci. **34**(1), 71–84 (1997)

Tatone, B.S.A., Grasselli, G.: A new 2D discontinuity roughness parameter and its correlation with JRC. Int. J. Rock Mech. Min. Sci. **47**(8), 1391–1400 (2010)

Tse, R., Cruden, D.M.: Estimating joint roughness coefficients. Int. J. Rock Mech. Min. Sci. Geomech. Abstracts **16**, 303–307 (1979)

Umili, G., Bonetto, S., Mosca, P., Vagnon, F., Ferrero, A.M.: In situ block size distribution aimed at the choice of the design block for rockfall barriers design: a case study along Gardesana road. Geosciences **10**, 223 (2020). https://doi.org/10.3390/geosciences10060223

Trends, Techniques, Testing, Tribulations, Tasks, Trajectories: The Saga of Data in the Evolution of Geotechnical Design

Marco Uzielli(✉)

Department of Civil and Environmental Engineering, University of Florence, Via di Santa Marta 3, 50139 Firenze, Italy
marco.uzielli@unifi.it

Abstract. Modern engineering design codes have evolved from deterministic to non-deterministic to formats which account explicitly for uncertainties in geotechnical systems. Though non-deterministic approaches have widely proved to be instrumental in cost-performance optimization, they are still not well understood and implemented by geotechnical researchers and practitioners. The main difficulties encountered in the transition to the non-deterministic paradigm may be related to a sub-optimal approach to data acquisition and processing for uncertainty modelling. This is paradoxical given the momentous digital revolution which is ongoing worldwide and unprecedented available computational power. This extended abstract addresses synthetically the role of data in approaching and conducting best-practice non-deterministic design, providing a set of conceptual and operational motivations to promote a more data-centric approach to geotechnical practice. Cultural and operational factors which prevent the full exploitation of non-deterministic design potential are discussed.

Keywords: Geotechnical data · Geotechnical uncertainty · Non-deterministic design · Site characterization · Statistics · Probability · Bayesian probability

1 Trends

Engineering design codes have been evolving from the primordial empirical and deterministic format to formats which rely increasingly on non-deterministic approaches in which uncertainties in the parameters and in the models included in the design process are defined and processed explicitly and quantitatively. The non-deterministic paradigm is particularly suited for geotechnical engineering, which deals largely with natural materials.

This evolution is motivated by the fact that non-deterministic analyses prove to be advantageous at least in terms of achieving consistent levels of conservatism and performance in design; and optimizing costs of design, construction, monitoring, and maintenance with respect to such levels. Nonetheless, there are still difficulties in implementing truly non-deterministic methods in engineering practice. Notwithstanding the

A. Ferrari et al. (Eds.): CNRIG 2023, SSGG, pp. 36–43, 2023.
https://doi.org/10.1007/978-3-031-34761-0_5

inherent complexity of geotechnical systems, such difficulties can arguably be reconducted mainly to the geotechnical community's insufficient acknowledgment of the role and importance of investing resources in data acquisition and interpretation, and to the abuse of the so-called "engineering judgment". This is paradoxical given the momentous digital revolution which is ongoing worldwide and unprecedented available computational power.

Notwithstanding this distorted perception, the transition from determinism (represented by allowable-stress design approaches) is underway and has insofar led, in broad terms, initially to partial factor-based design and, subsequently, to reliability-based design approaches, including important ramifications such as performance-based design. The future, long-term trajectory of geotechnical design cannot be established as new technologies and conceptual approaches extend its scope and perspectives beyond foreseeable boundaries. However, it is undeniable that regulatory design approaches have become increasingly "data-centric", i.e., they are responsive to data and are able to self-improve with data. Future design paradigms can be expected to assign data increasingly ever pivotal role [1].

"Data-centric" and "non-deterministic" are not synonymous concepts, but the adoption of the first facilitates and fosters the latter. In broad terms, both share the willingness to acknowledge and embrace the complexity and indetermination of geotechnical systems, transcending their subjective idealization and awarding centrality to objective information, observation, and data.

This extended abstract addresses synthetically the role of data in approaching and conducting best-practice non-deterministic design, providing an insight into the nature and components of geotechnical uncertainty. Referring specifically to the convenience of the Bayesian approach to geotechnical uncertainty modelling, the paper contributes a set of conceptual and operational motivations to promote a more rigorous though practically feasible data-centric approach to non-deterministic design. Cultural and operational factors which prevent the full exploitation of non-deterministic design potential are discussed, and the most frequent criticisms of non-deterministic approaches discussed and refuted.

2 Techniques

Geotechnical uncertainty is most frequently modelled using statistical and probabilistic approaches. Within this well-established framework, geotechnical uncertainty results as the aggregation of inherent variability, measurement uncertainty, statistical uncertainty, and transformation uncertainty. Inherent variability represents the "real" spatial variability of geotechnical properties which exists as a result of the natural heterogeneity and complexity of soils, which are generated and continuously modified by natural geologic and geomorphologic processes. In-situ effects due to stress state and stress history also lead to spatially variable measurements even for compositionally homogeneous deposits. Measurement uncertainty stems from technological limitations in in-situ and laboratory instrumentation. Statistical uncertainty results from the impossibility of estimating the "true" statistics of geotechnical measurements and derived parameters with full confidence due to the limited size of datasets resulting from site investigation. Transformation uncertainty arises because all models, whether empirical, theoretical, analytical, or data-driven, used to obtain geotechnical design parameters from qualitative information or quantitative data are affected by indetermination, in the form of bias and/or scatter.

Geotechnical practice involves predictions that are usually based on probabilistic characterizations of uncertainties in states of geotechnical systems (as parameterized, e.g., by a set of geometric and geotechnical properties). The state of a geotechnical system at a given time may not be fully known but is unique, not variable. Geotechnical uncertainties are thus mostly epistemic, i.e., related to lack of information. They are best described by degrees-of-belief rather than by frequencies and, thus, best assigned and processed using a Bayesian approach, which updates "prior" probabilities with "evidence" (e.g., data and observations) to obtain "posterior" probabilities.

In terms of mathematical formalism, Bayes' Theorem can be expressed as:

$$P(H|D) = \frac{P(H) \cdot P(D|H)}{P(D)} \tag{1}$$

in which H is a generic hypothesis (e.g., related to a probability distribution, an event, etc.); $P(H)$ is the prior probability of the hypothesis; $P(H|D)$ is the posterior probability of the hypothesis accounting for the observed data; $P(D|H)$ is the likelihood of the hypothesis based on the observed data, i.e., the conditional probability of observing or measuring the data D assuming that the hypothesis H is true; and $P(D)$ is the marginal probability of data irrespective of the hypothesis. The latter term is a normalizing constant which can be obtained through the total probability theorem or, most often, through computational methods. The Bayesian approach is particularly suited to geotechnical engineering as it: (a) fully accommodates the progressive increase in the quantity and quality of available data; (b) is fully harmonized with the observational approach, which is an essential feature of geotechnical practice; (c) allows the integration of "global" and "local" information and databases, thereby circumventing barriers related to site-specificity; and (d) allows the conduction of reliability-based design even when only small data samples are available.

Despite the exquisite formal simplicity of Eq. (1), real geotechnical problems are complex and most often require computational methods for the implementation of Bayes' Theorem. Fortunately, the current digital revolution and the steady increase in computational power allows the processing of huge amounts of data. Irrespective of the data processing approach, aleatory and epistemic uncertainties can be assigned subjectively or objectively. Geotechnical engineers traditionally rely largely on subjective "engineering judgment". However, as highlighted by Prof. Greg Baecher in his 2021 Terzaghi Lecture [2], a vast corpus of research in the geotechnical engineering domain, as well as in numerous other technical and humanistic disciplines, shows that expert judgment is a surrogate for quantitative probabilistic methods to assign uncertainty, and that even very simple algorithms most often outperform expert judgment [3]. In a present and, especially, in a future perspective, artificial intelligence and deep machine learning will broaden the gap in favor of the diagnostic and predictive capabilities of algorithmic approaches with respect to expert-based approaches.

3 Testing

In applying Bayesian approaches to geotechnical systems, prior probabilities are most often necessarily assigned on a subjective basis from qualitative geological models because geotechnical testing data are typically not yet available at the early stages of geotechnical projects, though literature values can sometimes be used [4]. These prior probabilities are then updated using data from geotechnical site characterization.

Geotechnical engineers are trained to acknowledge the role of in-situ and laboratory testing to compose a reliable and useful geotechnical model. However, an "adequate" level of site characterization is seldom attained in practice. This is at least partially due to a well-rooted cultural resistance to investing adequately in data acquisition, despite solid scientific evidence from research and commercial projects attesting to the benefits of site characterization in achieving cost- and performance-effective design. Moreover, such benefits are maximized if non-deterministic design approaches are adopted [5].

The availability of good and sufficiently numerous data is beneficial for the quantification and, in most cases, the reduction of uncertainties. For instance, data plays a central role in characterizing inherent variability as it allows the implementation of quantitative spatial interpolation techniques (e.g., random field modelling and geostatistical kriging) for the estimation of geotechnical properties at unsampled locations (and associated uncertainties) and for the planning and optimization of geotechnical investigation campaigns. The quantitative characterization of inherent spatial variability in geotechnical systems is important for design at least because: (a) many geotechnical designs involve large significant horizontal and/or vertical soil volumes, and spatially variable soil properties may require different designs to optimize cost-performance; and (b) the characterization of spatial variability allows the quantitative estimation of variance reduction effects which can reduce excess conservatism.

The magnitude of statistical uncertainty affects geotechnical design significantly because a higher statistical uncertainty brings an increase in the probability of non-performance of the object of design. A well-structured site characterization is instrumental in reducing statistical uncertainty.

There is a widespread misconception that the characterization of smaller sites requires few data and limited site characterization efforts. The achievement of a sufficiently low level of statistical uncertainty relies on the availability of a sufficiently large dataset, regardless of the geographic extension of the site. Statistical theory is clear in indicating that inferential uncertainty in sampling is not related to the fraction sampled but to sample size, and the possibly lower magnitude of horizontal spatial variability which can be observed at small sites does not do away with the necessity to ensure sufficient data numerosity. Thus, even sites of limited extension require sufficiently conspicuous geotechnical investigation campaigns.

The availability of "local", site-specific information brings a reduction in the transformation uncertainty between a test index and a design parameter and, consequently, to better design in terms of more safety and less excess conservatism if fully probabilistic, reliability-based approaches are adopted. In contrast, partial factor approaches and, especially, non-deterministic approaches are less sensitive to the quantity and quality of site characterization outputs. This has been demonstrated quantitatively in the geotechnical literature ([5] among others).

The level of information resulting from site characterization depends on the quantity (i.e., on the number of test types) and quality (i.e., the degree of precision in measured values for each test type) of data. Improved site characterization can be pursued by: (a) taking field measurements/samples at more locations, thereby increasing the amount of data for a given test type; (b) employing in-situ testing methods with smaller measurement interval; (c) using more test types at proximal spatial locations; and (d) using testing methods with high repeatability and low measurement error. The first two approaches increase information quantitatively, reducing the scatter in testing outputs and improving precision, while the third and fourth increase information qualitatively, reducing bias and improving accuracy.

In-situ testing methods such as CPT and DMT are particularly suited for the modeling of spatial variability because of their small measurement interval, which allows, jointly with their high testing repeatability, more reliable estimation of small-scale variability in the vertical direction, for instance through semivariogram-based modelling. These attributes also result in the reduction of statistical uncertainty. Moreover, transformation uncertainty associated with correlations from high-quality tests is also generally much lower than that pertaining to correlations for lower-quality methods such as SPT.

4 Tribulations

Despite their surging popularity and centrality, the path is not smooth for geotechnical data. Many researchers and practitioners are reluctant to acknowledge its importance and adopt correct practices to implement data in non-deterministic analysis and design. The rugged transition from partial-factor design to fully probabilistic, reliability-based design in geotechnical engineering may be attributed to a fallacious comprehension of the importance of data and to their less-than-optimal management.

A recurring criticism of non-deterministic approaches is that there is no particular motivation to use them because they seem to produce designs comparable to existing practice. This criticism ignores numerous literature contributions demonstrating exactly the opposite. The geotechnical literature abounds with examples attesting to the improvements in cost-benefit ratio brought by non-deterministic approaches ([6–8] among others). This misconception is rooted in the erroneous artificial switching of cause and effect, whereby the quantification of uncertainty is often conducted (whether inadvertently or not) to achieve a non-deterministic design which is reassuringly similar to deterministic design. The current version of Eurocode 7 [9] serves as a good example as discussed in detail in [10]. Clause 2.4.5.2 states that "the characteristic value of a geotechnical parameter shall be selected as a cautious estimate of the value affecting the occurrence of the limit state." This is, conceptually, a subjective definition which is related to: (a) the relative magnitude of the soil volumes whose characterization can be deemed suitable; and (b) the significant volume with respect to mechanisms related to a specific limit state. The same clause states that "if statistical methods are used, the characteristic value should be derived such that the calculated probability of a worse value governing the occurrence of the limit state under consideration is not greater than 5%." This lower-bound fractile requires the definition of an underlying probability density function which, however, is often difficult to characterize reliably in practice using the prevalent frequentist approach. As a result, characteristic values end up being typically assigned subjectively and calibrated to replicate design outputs which would be achieved by non-deterministic approaches [11]. In essence, due to the neglection of the Bayesian approach (especially in cases with limited amounts of data), non-deterministic codes are often implemented as deterministic ones in disguise. The revision of Eurocode 7 may bring advancements with respect to this issue.

A second recurring criticism is that there are not enough data to confidently apply non-deterministic methods. Leaving aside the fact that data scarcity often results from the underestimation of the importance of site characterization and the consequent lack of investment in site characterization, a small sample size is not a valid argument to dismiss estimation of geotechnical property statistics. Even in the case of a small sample size, geotechnical engineers can and should process uncertainties rigorously throughout the design process by implementing Bayesian approaches rather than resorting non-critically to frequentist approaches which are often statistically unreliable or, yet worse, to subjective assignment relying solely on engineering judgment.

Thus, the main criticisms of non-deterministic methods can be refuted on the basis of both theory and evidence. Tribulations related to the collection and utilization of data are, to a considerable extent, self-inflicted by the critics themselves.

5 Tasks

The geotechnical community must assign itself and address several fundamental tasks to evolve both philosophically and technically to overcome current problems and obstacles related to data management. Prof. Greg Baecher [2] stated that "currently, most geotechnical engineers deal with uncertainty using heuristics that ignore many of the most basic rules of probability." Geotechnical engineering education of the future should require structured training in statistical and probabilistic methods, specifically in Bayesian approaches, data analysis, and machine learning methods.

Geotechnical researchers focusing on non-deterministic and data-centric methods should strive to make their promising findings increasingly accessible through sharing of open-source and free software as well as of data. To this purpose, new geotechnical databases providing high-quality, multivariate, validated information are being made publicly available. An example is the ISSMGE TC304 database – Project 304dB at http://140.112.12.21/issmge/tc304.htm. This paper wishes to encourage the geotechnical community to actively contribute to these databases. Sharing could range well beyond the sharing of new data, because a huge wealth of "dark" information from past projects, which is currently dormant in hard copy or digital format, could be resumed and shared to leverage the power of data-centric approaches [1].

The full migration of the geotechnical discipline to the non-deterministic paradigm will also require the upgrading of geotechnical correlations to non-deterministic formats to allow reliable quantitative assignment of transformation uncertainty. This action would also rely significantly on the availability of data because this process would be conducted optimally through data-driven approaches.

6 Trajectories Ahead

Fostered by the ongoing digital revolution, technological achievements, and socio-cultural transformations, engineering design has evolved first from being a celebration of deity to a celebration of human intellect. Now, it is further evolving into an algorthmic process which benefits from the synergy between human and artificial actors. The current migration of geotechnical design codes towards non-deterministic formats and the seemingly irreversible global digital transition suggest that embracing a data-centric perspective in geotechnical practice is no longer a choice.

This should come as no surprise, because the data-centric paradigm arguably provides the natural evolution of the geotechnical discipline [12]. There is no discontinuity with respect to previous steps in the evolution of engineering disciplines, which has consistently relied on using the best possible observations and the most effective available methods and tools offered by knowledge and technology. To further exploit the opportunities offered by the global digital revolution and improved data acquisition measurement technologies, the geotechnical community must persist in the transition towards data-centric approaches as these are instrumental in enabling full compliance to current regulations. Despite difficulties and asperities, this scenario is coming true. Data-driven site characterization and non-deterministic data processing and interpretation are accruing their relevance in the geotechnical discipline and promise to support the further evolution of geotechnical design codes.

This transition need not be troublesome nor conflictual. The surge of data-driven approaches does not undermine or threaten the central role of geotechnical culture. Rather, new engineering paradigms and technological evolutions now require that such culture also include the comprehension of uncertainties and the capability to analyze and proficiently use data. The comprehension of the physical reality of the object of design and the ability on the part of geotechnical professionals to ask the right questions with respect to limit states, performance and, ultimately, the goals of design (performance, conservatism, etc.) transcend data availability and design paradigms. The journey of geotechnical design approaches is in effect a positive one, since, as in the words of Prof. Kok-Kwang Phoon, "for the first time, we are starting to realize we can combine experience and data in even more clever ways" [1].

In his illuminating insight on future global perspectives and on the trajectories of humankind, Yuval Noah Harari stated that "intelligence" is decoupling from "conscience" as a consequence of the development of superior artificial algorithms which utilize unprecedented computing power and giant databases. Moreover, all scientific disciplines "are converging on the all-encompassing dogma of dataism, which says that the universe consists of data flows, and that the value of any phenomenon or entity is determined by its contribution to data processing" [13]. While these statements are factual and irrefutable, humans are not at risk of losing their value. Human consciousness is not optional because engineering design is scoped by ethical principles and strategic choices relying on humanist criteria (for instance, the value of human life, sustainability, and environmental awareness). Modern society believes in humanist dogmas and uses science, including technical sciences, not in order to question these dogmas, but rather in order to implement them. Yet more importantly, as in the inspirational words of Prof.

Peter Lumb [14], “Engineering is precise, it is a science. Yet, in reality, it is vague. It is not a science. It is more an art.”

References

1. Phoon, K.-K.: The story of statistics in geotechnical engineering. Georisk: Assess. Manag. Risk Eng. Syst. Geohazards **14**(1), 3–25 (2020). https://doi.org/10.1080/17499518.2019.1700423
2. Baecher, G.B.: Terzaghi lecture: geotechnical systems, uncertainty, and risk. J. Geotech. Geoenviron. Eng. **149**(3) (2021). https://doi.org/10.1061/JGGEFK.GTENG-10201
3. Kahneman, D., Sibony, O., Sunstein, C.S.: Noise: A Flaw in Human Judgment. Collins, London (2021)
4. Cao, Z., Wang, Y., Li, D.: Quantification of prior knowledge in geotechnical site characterization. Eng. Geol. **203**, 107–116 (2016). https://doi.org/10.1016/j.enggeo.2015.08.018
5. Ching, J., Phoon, K.-K.: Value of geotechnical site investigation in reliability-based design. Adv. Struct. Eng. **15**(11), 1935–1945 (2012). https://doi.org/10.1260/1369-4332.15.11.1935
6. Duncan, J.M.: Factor of safety and reliability in geotechnical engineering. J. Geotech. Geoenviron. Eng. **126**(4), 307–316 (2000). https://doi.org/10.1061/(ASCE)1090-0241(2000)126:4(307)
7. Fenton, G.A., Naghibi, F., Griffiths, D.V.: On a unified theory for reliability-based geotechnical design. Comput. Geotech. **78**, 110–122 (2016). https://doi.org/10.1016/j.compgeo.2016.04.013
8. De Koker, N., Day, P.: Assessment of reliability-based design for a spectrum of geotechnical design problems. Proc. Inst. Civ. Eng. Geotech. Eng. **171**(2), 147–159 (2018). https://doi.org/10.1680/jgeen.17.00047
9. CEN (European Committee for Standardization): EN 1997-1:2004 Eurocode 7: geotechnical design – part 1: general rules. CEN, Brussels, Belgium (2004)
10. Prästings, A., Spross, J., Larsson, S.: Characteristic values of geotechnical parameters in Eurocode 7. Proc. Inst. Civ. Eng. Geotech. Eng. **172**(4), 301–311 (2019). https://doi.org/10.1680/jgeen.18.00057
11. Wang, Y.: Selection of characteristic value for rock and soil properties using Bayesian statistics and prior knowledge. In: ISSMGE Joint TC205/TC304, London (2017)
12. Tang, C., Phoon, K.-K.: Model Uncertainties in Foundation Design. CRC Press, Boca Raton (2021). https://doi.org/10.1201/9780429024993
13. Harari, N.Y.: Homo Deus – A Brief History of Tomorrow. Signal, Toronto (2016)
14. Lam, K., Li, W.K.: Excerpts from Interview with Professor Peter Lumb. Hong Kong Stat. Soc. Newsl. **9**(1), 2–5 (1986)

Laboratory Testing: Innovation in Technologies and Equipment

Cyclic Behavior of Sand Stabilized by Colloidal Silica: Effects of Sample Preparation and Energy–Based Approach

Giovanni Ciardi[1](✉) and Claudia Madiai[2]

[1] Department of Civil and Environmental Engineering, University of Perugia, via G. Duranti 93, 06125 Perugia, Italy
giovanni.ciardi@unipg.it

[2] Department of Civil and Environmental Engineering, University of Florence, via di S. Marta 3, 50139 Florence, Italy

Abstract. Colloidal silica grouting is one of the most promising and innovative techniques for the mitigation of seismic liquefaction. It consists of injecting liquefiable soils with a harmless, low–viscosity, silica–based grout, forming a silica gel among soil particles after a certain time. This study discusses the effects of two techniques used for sample preparation on the cyclic response of stabilized sand by means of undrained cyclic triaxial test. Moreover, part of an existing test database by the same authors is analyzed within the framework of the dissipated energy concept. This energy–based approach provides an alternative way of describing the cyclic response of the stabilized sand. Experimental results show that: i) the cyclic strength of stabilized specimens prepared by moist tamping followed by grout injection is higher than that of stabilized specimens prepared by pluviation of dry sand into liquid grout; ii) the dissipated energy to liquefaction is greater for stabilized sand than for untreated sand; iii) characteristic patterns of dissipated energy are exhibited by stabilized sand samples, resembling the way they achieve the failure condition.

Keywords: Colloidal silica · Soil liquefaction · Laboratory tests · Soil fabric · Dissipated energy

1 Introduction

In recent years, colloidal silica (CS) grouting has been promoted as an effective remedial measure against seismic liquefaction [1]. This soil improvement technique consists of treating the liquefiable soils by the injection of a low–viscosity, time–hardening, nanosilica–based mixture. Once hardened, a colloidal silica gel is formed among the soil grains, resulting in an overall increase of the cyclic resistance of the soil [2–6].

Relevant experimental laboratory tests carried out on different sands stabilized by colloidal silica showed distinctive features of the behavior of the stabilized soil: broadly speaking, while in the untreated liquefiable soils the failure condition is achieved due to the increase and build–up of extra pore water pressure during cyclic loading, the

A. Ferrari et al. (Eds.): CNRIG 2023, SSGG, pp. 47–54, 2023.
https://doi.org/10.1007/978-3-031-34761-0_6

stabilized materials never exhibit "true" liquefaction collapse, and the failure condition is achieved due to the development of sizeable amount of cyclic strain. This indicates that the development and build–up of pore water pressure do not have the same physical effect in treated and in untreated soil [6, 7].

It is known that, in case of laboratory tests on untreated sands, different techniques used for sample preparation can produce different fabrics, resulting in different soil properties under similar testing conditions [8, 9]. In case of sand stabilized by colloidal silica, however, the impact of the sample preparation on the cyclic response of the treated material has not been investigated yet. Among the techniques used for the preparation of samples treated by colloidal silica, the followings have been thoroughly used in previous research: i) moist (or dry) tamping, followed by grout permeation; ii) pluviation of dry sand into liquid grout.

In most of previous experimental studies, the failure condition for stabilized sands in laboratory tests has customary been identified in terms of accumulated strain. In case of cyclic triaxial tests, 5% double amplitude (DA) axial strain has been often used, coherently with the current practice on untreated sand [10]. However, given that stabilized samples show progressive strain development during cyclic loading – and that the possible annulment of effective stress during the cyclic stage do not lead to collapse – it seems useful to analyze the cyclic behavior of samples stabilized by CS in terms of strain energy dissipated during cyclic loading.

The concept of dissipated energy [11] is the base for the so–called "energy–based" liquefaction approach [12, 13], in which the earthquake energy demand and the soil energy capacity are compared. The cumulative dissipated energy per unit volume (energy density), W, can be evaluated in cyclic triaxial tests according to Eq. 1:

$$W = \sum_{i=1}^{k-1} \frac{1}{2}(q^i + q^{i+1}) \cdot (\varepsilon_a^{i+1} - \varepsilon_a^i) \quad (1)$$

where q is the applied deviatoric stress, k is the number of loading increments and ε_a is the axial strain. As shown in Eq. 1, the energy density considers both the stress and the strain developed during loading and could be an important tool in providing an alternative point of view of the cyclic response of stabilized sands.

In this research, the effects of using two different techniques for the sample preparation on the cyclic behavior of stabilized sand are evaluated by means of cyclic undrained triaxial test. Then, test results from treated sand and untreated sand are discussed within the framework of the dissipated energy concept, providing a preliminary assessment of dissipated energy to liquefaction in case of stabilized samples. For these purposes, part of an existing test database [6] is used.

2 Materials, Methods and Test Database

Part of the test database described in Ciardi and Madiai [6] is used in this research. The used tests consist of stress–controlled cyclic undrained triaxial tests carried out at a loading frequency of 0.1 Hz on untreated sand and on sand stabilized by 5% CS (by weight). The material used by Ciardi and Madiai [6] is a clean, mainly siliceous sand.

The sand has grain size distribution within the boundaries suggested for liquefiable materials, mean particles diameter 0.30 mm, uniformity coefficient 1.6, specific gravity 2.65, maximum/minimum void ratios 0.839/0.559. MasterRoc MP 325 (BASF SE) was used as the colloidal silica product: it is a clear solution with 15 ± 1% (by weight) silica content, density of ≈ 1.1 kg/L (20 °C), pH of 10 ± 1 (20 °C) and viscosity of ≈ 10 mPa·s (20 °C). The gel formation was triggered by using a sodium chloride (NaCl) solution (1/10 NaCl/demineralized water, by weight). A gel time of 24 h was chosen, while the curing time (*i.e.*, the time between the formation of the stabilized sample and the beginning of the test) was set in 5 days (4 days if one considers the beginning of one night water–saturation in triaxial cell). After saturation (Skempton's B coefficient > 0.96), the samples were isotropically consolidated at a mean effective initial confining stress, $p_0' =$ 100 kPa. Treated/untreated sand samples were prepared by the moist tamping technique, compacting moist sand (5% water content) in layers to achieve the desired void ratio [6]. In case of stabilized samples, moist tamped specimens were prepared outside the triaxial cell, in a triaxial–like device, allowing the colloidal silica solution to permeate by gravity into samples that were previously flushed with distilled and deaired water and subjected to about 20 kPa confining pressure. Samples were then moved to the triaxial cell once curing time had expired. More details on the moist tamping procedure used for stabilized samples formation are given in [6]. In that work, the specimens' failure occurred after a number of loading cycles at failure, N_f, associated with the first occurrence of double amplitude axial strain ($\varepsilon_{DA} \geq 5\%$), for both untreated sand and treated sand. The relative density of all specimens (Diameter/Height, D/H = 1/2) after consolidation (D_{rc}) was about 30%.

In this study, undrained cyclic triaxial tests on stabilized sand were carried out using the same testing materials, procedures and failure criterion described above, except for the technique used for sample preparation. A pluviation technique was adopted: a fixed oven–dried amount of sand was poured into an external split mold containing the liquid grout. After preparation, the samples were sealed to avoid loss of moisture and moved to the triaxial pedestal at the age of testing. The moist tamping and pluviation techniques are the most used methods used for the preparation of stabilized samples [4, 14, 15] and were thus adopted in this work.

The test database of the present research is summarized in Table 1, where CS_W is the colloidal silica content (by weight), ID is the test identification code and CSR is the cyclic stress ratio ($CSR = q/2/p_0'$).

3 Results

3.1 Effects of Sample Preparation

Figure 1 shows the test results on two stabilized samples, prepared by moist tamping (ID M5_b) and by pluviation (ID P5_b), and subjected to similar CSR. In Fig. 1, p' is the mean effective confining stress, r_u is the pore water pressure ratio (the ratio of the extra pore water pressure, Δu, to p_0') and N is the number of loading cycles.

Broadly speaking, the stress and strain paths are those typical of sand stabilized by CS, characterized by a fast extra pore water pressure increase, by the non–symmetric

Table 1. The test database.

Reference	ID (–)	CS_W (%)	CSR (–)	D_{rc} (%)	N_f (–)	Formation method*
[6]	M0_a	0	0.203	30	79	MT
	M0_b	0	0.232	31	35	MT
	M0_c	0	0.253	29	20	MT
	M0_d	0	0.303	31	4	MT
	M5_a	5	0.228	31	263	MT
	M5_b	5	0.253	35	64	MT
	M5_c	5	0.277	32	36	MT
	M5_d	5	0.302	31	13	MT
This study	P5_a	5	0.233	35	52	PV
	P5_b	5	0.260	34	34	PV
	P5_c	5	0.310	33	8	PV

* MT = moist tamping; PV = pluviation

axial strain development and progressive strain build–up, and by the non–flow–like failure. Not significant differences in the stress and strain curves can be detected depending on the used formation method. However, it worth observing that a greater number of loading cycles was required to induce failure in the specimen prepared by the moist tamping technique (Fig. 1a–d). This trend can be easily detected by plotting the liquefaction resistance curves, as shown in Fig. 2, where data for untreated sand samples – prepared by the moist tamping method – are also shown. The curves for stabilized sand lie above the curve of untreated sand, as expected [1].

By the analysis of test results, it could be stated that also in the case of treated sand the distinctive fabrics obtained by wet pluviation or moist tamping are responsible for the difference in liquefaction resistance (see, *e.g.*, [16] for untreated sand). However, because no information at the microscale is available at this stage of research, it is not still clear whether and how the presence of the gel influences the distinctive fabrics obtained by the two methods used for the preparation of the specimens.

3.2 Dissipated Energy

In this paragraph, test results are discussed introducing the concept of dissipated energy. Figure 3 shows the results of cyclic triaxial tests on one untreated sample and on two stabilized samples prepared by moist tamping and pluviation, under similar *CSR*. For untreated sand, the hysteresis loops in the axial strain–deviatoric stress plane are narrow, becoming much wider as liquefaction is approached. Coherently, the (normalized) dissipated energy builds–up at a slow rate before an abrupt curvature change in the N–W/p_0' plane occurs (Fig. 3a, b). The observed behavior is in agreement with that expected for

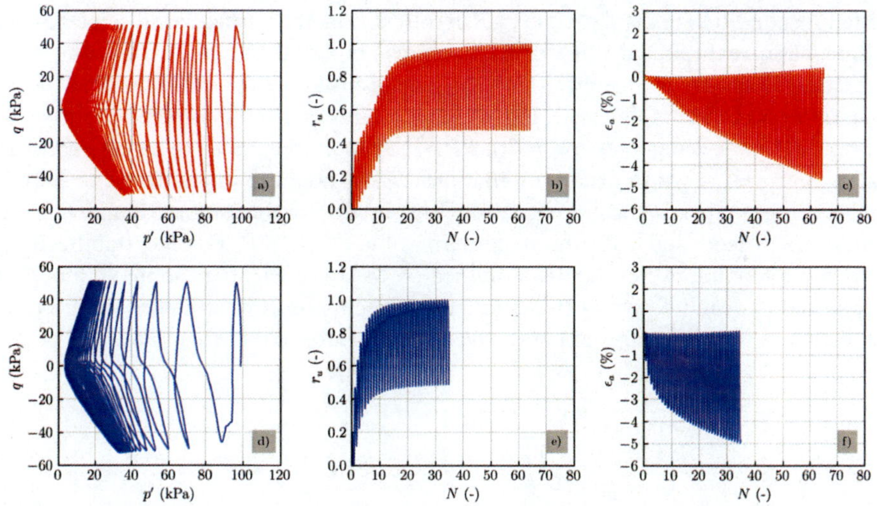

Fig. 1. Test results for stabilized sand under similar *CSR*: moist tamping, ID M5_b (a–c); pluviation, ID P5_b (d–f).

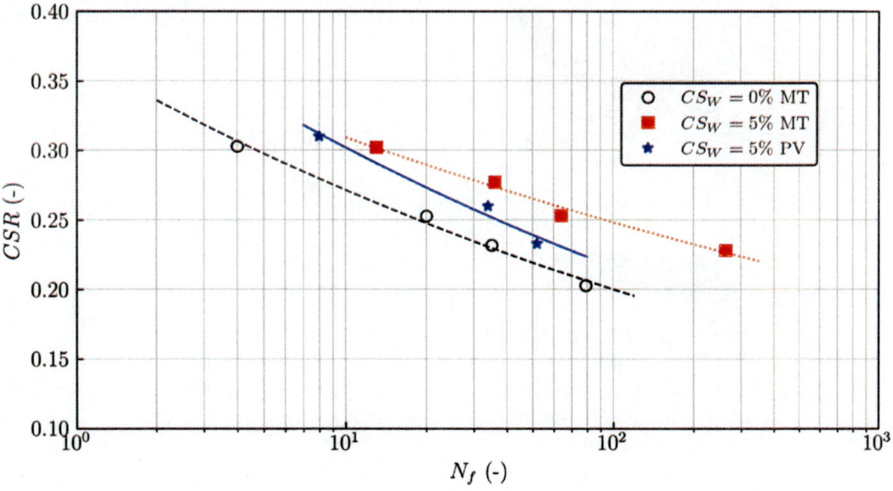

Fig. 2. Cyclic resistance curves.

loose sand subjected to cyclic loading conditions [17]: the axial strain develops gradually, becoming exponentially larger as a state of instability, triggered by the build–up of significant extra pore water pressure, is approached.

In case of stabilized samples (Fig. 3c–f), instead, the axial strain develops steadily since the early stage of cyclic loading (see also Fig. 1): this causes the *W/p₀'* curve to be smoother, since no significant, abrupt strain development occurs before failure is reached. Because many more loading cycles can be sustained by the treated soil before reaching the failure condition under the same *CSR*, it follows that the normalized dissipated energy to liquefaction exhibited by the treated material is higher than the corresponding one for the untreated sand (note the different W/p_0' scales in Fig. 3 for treated/untreated sand).

For the sake of completeness, Fig. 4 shows the normalized dissipated energy against the residual pore water pressure ratio, *RPWP*, namely the r_u value at the end of each loading cycle. It is worth observing that the *RPWP* increase rate is faster in stabilized sand (Fig. 4e–k) than in the untreated one (Fig. 4a–d). For the untreated sand, the failure condition is associated with the maximum value of *RPWP*. For the stabilized sand, many more loading cycles can be sustained by the samples after the maximum *RPWP* is achieved. It is more than ever clear that the mechanism of extra pore water pressure generation in treated sand is not at the basis of the failure of the samples.

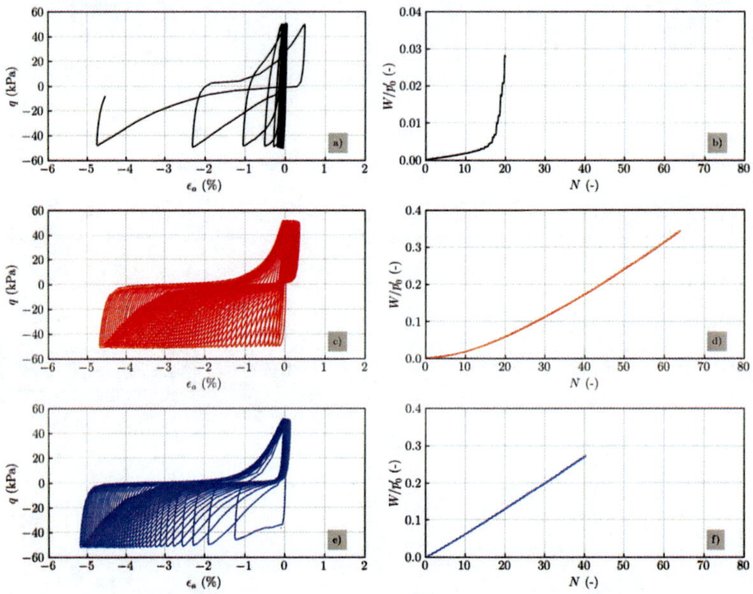

Fig. 3. Results from cyclic triaxial tests under similar *CSR*: untreated sand, ID M0_c (a, b); stabilized sand, moist tamping, ID M5_b (c, d), and pluviation, ID P5_b (e, f).

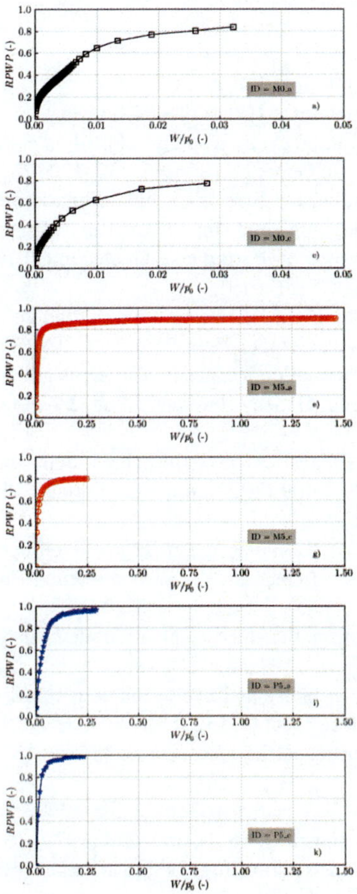

Fig. 4. Residual pore water pressure ratio: untreated sand (a–d); stabilized sand (e–k).

4 Conclusions

In light of the experimental evidence discussed in this study, the following conclusions can be drawn: i) the liquefaction resistance of treated samples prepared by the moist tamping technique followed by grout injection is higher than that exhibited by samples prepared by pluviation of dry sand into liquid grout; ii) the dissipated energy at failure is higher for treated sand than for untreated sand; iii) the curves of dissipated energy for stabilized sand and for untreated sand are significantly different in shape; iv) the build–up of extra pore water pressure during cyclic loading is not at the basis of failure in stabilized samples.

To clarify some unresolved issues of this study, future research should investigate: i) the effects of sample preparation at the microscale when colloidal silica grout treatment is used and ii) the applicability of the energy–based liquefaction approach to predict failure in stabilized sands.

References

1. Ciardi, G., Vannucchi, G., Madiai, C.: Effects of colloidal silica grouting on geotechnical properties of liquefiable soils: a review. Geotechnics **1**(2), 460–491 (2021)
2. Porcino, D., Marcianò, V., Granata, R.: Undrained cyclic response of a silicate–grouted sand for liquefaction mitigation purposes. Geomech. Geoengineering **6**, 155–170 (2011)
3. Ciardi, G., Bardotti, R., Vannucchi, G., Madiai, C.: Effects of high–diluted colloidal silica grout on the mechanical behavior of a liquefiable sand. In: Silvestri, F., Moraci, N. (eds.) Proceedings of the 7th International Conference on Earthquake Geotechnical Engineering, pp. 1820–1827. CRC Press, Boca Raton, FL, USA (2019)
4. Ciardi, G., Bardotti, R., Vannucchi, G., Madiai, C.: Effects of high–diluted colloidal silica grouting on the behaviour of a liquefiable sand. Geotech. Res. **7**(4), 193–208 (2020)
5. Pavlopoulou, E.M.E., Georgiannou, V.N.: Effect of colloidal silica aqueous gel on the monotonic and cyclic response of sands. J. Geotech. Geoenvironmental Eng. **147**(11), 04021122 (2021)
6. Ciardi, G., Madiai, C.: Effects of initial static shear stress on cyclic behaviour of sand stabilised with colloidal silica. Acta Geotech. **18**, 2389–2409 (2022). https://doi.org/10.1007/s11440-022-01737-9
7. Conlee, C.T., Gallagher, P.M., Boulanger, R.W., Kamai, R.: Centrifuge modeling for liquefaction mitigation using colloidal silica stabilizer. J Geotech. Geoenvironmental Eng. **138**(11), 1334–1345 (2012)
8. Kuerbis, R., Vaid, Y.P.: Sand sample preparation–the slurry deposition method. Soils Found. **28**(4), 107–118 (1988)
9. Sze, H.Y., Yang, J.: Failure modes of sand in undrained cyclic loading: impact of sample preparation. J. Geotech. Geoenvironmental Eng. **140**(1), 152–169 (2014)
10. Ishihara, K.: Liquefaction and flow failure during earthquakes. Géotechnique **43**(3), 351–451 (1993)
11. Nemat-Nasser, S., Shokooh, A.: A unified approach to densification and liquefaction of cohesionless sand in cyclic shearing. Can. Geotech. J. **16**(4), 659–678 (1979)
12. Berrill, J.B., Davis, R.O.: Energy dissipation and seismic liquefaction of sands: revised model. Soils Found. **25**(2), 106–118 (1985)
13. Kokusho, T.: Liquefaction potential evaluations: energy-based method versus stress-based method. Can. Geotech. J. **50**(10), 1088–1099 (2013)
14. Gallagher, P.M., Mitchell, J.K.: Influence of colloidal silica grout on liquefaction potential and cyclic undrained behavior of loose sand. Soil Dyn. Earthq. Eng. **22**(9–12), 1017–1026 (2002)
15. Vranna, A., Tika, T., Papadimitriou, A.: Laboratory investigation into the monotonic and cyclic behaviour of a clean sand stabilised with colloidal silica. Géotechnique **72**(5), 377–390 (2022)
16. Mulilis, J.P., Seed, H.B., Chan, C.K., Mitchell, J.K., Arulanandan, K.: Effects of sample preparation on sand liquefaction. J. Geotech. Eng. Div. **103**(2), 91–108 (1977)
17. Yang, Z.X., Pan, K.: Energy–based approach to quantify cyclic resistance and pore pressure generation in anisotropically consolidated sand. J. Mater. Civ. Eng. **30**(9), 04018203 (2018)

Tensio-Inclinometer: A Deployable Wireless Device to Underpin Early Warning Systems for Rainfall-Induced Shallow Landslides

Lucia Coppola[1](✉), Alfredo Reder[2], Alessandro Tarantino[3], and Luca Pagano[1]

[1] University of Naples "Federico II", Naples, Italy
lucia.coppola@unina.it
[2] Regional Model and Geo-Hydrological Impacts-REMHI, Centro Euro-Mediterraneo Sui Cambiamenti Climatici, Caserta, Italy
[3] University of Strathclyde, Glasgow, Scotland, UK

Abstract. Most of the Landslides Early Warning Systems (LEWS) currently in operation are based on monitoring rainfall data only. This feature limits their performance due to false alarms generated by rainfall thresholds inevitably conservative. The accuracy of LEWS may be significantly enhanced by monitoring soil-based variables associated with the stress-strain response of the ground. The paper presents a novel Tensio-inclinometer specifically developed to measure suction changes and suction-induced deformation in shallow covers. The device is made of a MEMS accelerometer mounted on the shaft of a commercial tensiometer. On-board electronics and battery-based power supply make the device fully wireless. The Tensio-inclinometer is therefore easy to deploy and install allowing the design of a flexible and adaptive monitoring network to underpin early-warning systems. The device was tested in a slope physical model where instability of a shallow silt layer was triggered by artificial rainfall. It is shown that pre-failure deformation, as detected by the tilting of the tensiometer shaft, is an adequate landslide precursor and that combined suction and rotation measurements can provide soil-based thresholds for early warning systems.

Keywords: shallow landslides · silty volcanic soils · soil suction · tilting · slope pre-failure deformation

1 Introduction

In coarse-grained and silty sloping covers, repeated rainfalls during the wet season induce progressive suction loss that may eventually lead to shallow landslides. These phenomena involve cover thicknesses not exceeding a few meters and can be classified as debris avalanches, debris flows, or flowslides depending on the geomorphological context where they occur [1]. They cause significant damage and fatalities throughout the world [2]. In Italy alone, a large number of catastrophic shallow landslides have occurred in different regions over the last years including Campania [3], Piedmont (Villar Pellice in 2008, [4]), Liguria (Cinque Terre in 2011, [5]) and Sicily (Messina in 2009,

A. Ferrari et al. (Eds.): CNRIG 2023, SSGG, pp. 55–62, 2023.
https://doi.org/10.1007/978-3-031-34761-0_7

[6]). LEWS represent the mostly adopted strategy [7] to reduce shallow landslide risk. LEWS are based on the monitoring of landslide precursors, in order to warn communities well ahead of landslide initiation, not once landslide has triggered, so alert starts once precursor-informed thresholds are exceeded. Most of the LEWS currently in operation are based on monitoring rainfall data [8], as rainfall is the most convenient variable to monitor. However, rainfall represents only the starting point of the cause-effect chain eventually leading to landslides. So, monitoring additional variables representing 'in series' the different phenomena leading to instability would enhance considerably the performance of the LEWS.

The literature shows a variety of approaches adopted to integrate rainfall data. Measurements of slope kinematics have been incorporated in EWS, including inner displacements via acoustic emission [9] and surface displacements via total station, Global Positioning System (GPS), and photogrammetric techniques [10]. However, there are a number of limitations of using these methods: interpretation of acoustic emissions is not always straightforward, line of sight between observer and target in total stations is affected by rain, and rain can weaken the GPS signal. Volumetric water content measured by dielectric-based sensors has also been included as additional precursor variable as well [11]. Pore-water pressure is however a more significant variable to measure compared to volumetric water content. Slopes are often made of a cohesionless granular materials characterised by inclinations close to the friction angle. As a result, pore-water pressures triggering slope failures are generally in the range of a few kilopascals either in the negative or positive range [12]. In this interval volumetric water content is characterised by poor sensitivity in the negative range (the water retention curve tends to level off when approaching saturated conditions) and no sensitivity in the positive range of pore-water pressure. In contrast, tensiometers generally show accuracy to less than 1 kPa and can measure pore-water pressure in both negative and positive ranges. This makes pore-water pressure measured by tensiometers a relevant precursor of rainfall-induced shallow landsides. So far measurements of slope stress variable (pore-water pressure) and slope kinematics have never been combined to inform early-warning systems. This paper presents a 'Tensio-inclinometer' designed to record simultaneously i) the pore-water pressures in negative and positive ranges using a conventional tensiometer and ii) the pore-water pressure-induced kinematics via the measurement of the inclination of the tensiometer shaft. Its design assumes that tilting of the tensiometer shaft installed vertically into the cover is a suitable proxy measurement of landslide pre-failure kinematics. As a wireless standalone instrument, it can be easily installed at different slope locations to measure precursor landslide variables to underpin early-warning systems for rainfall-induced shallow landslides.

This paper presents Tensio-inclinometer and its validation in a slope physical model, where a silty volcanic layer is placed and wetted until failure occurred. This test is aimed at investigating whether the combined measurements of suction and suction-induced kinematics can be successfully exploited as landslide precursors to inform Landslide Early Warning Systems.

2 The Tensio-Inclinometer

The Tensio-inclinometer consists of a commercial tensiometer (codification T4, UMS production) [13] and a rigid steel beam equipped with an accelerometer installed at its top (Fig. 1). The acceleration measurement system is based on MEMS (Micro-Electro-Mechanical-Systems) accelerometer mounted on a semiconductor chip. In addition, the 'all-in-one' semi-conductor chip includes power supply, power management, data storage and signal digitization data wireless transmission. The tensiometer consists of an acrylic-glass shaft of variable length in-between 0.15–2 m, developed to monitor pore-water pressure from − 85 kPa to 100 kPa (Fig. 1), with an accuracy of about ± 0.5 kPa (depending on cable length). A high air-entry value saturated ceramic cup is positioned at the bottom of the shaft to allow water (under tension) to flow from the soil to the tensiometer water reservoir or vice versa. Water pressure in the tensiometer water reservoir is measured by a piezo-electric pressure sensor positioned at the top of the tensiometer water reservoir. The back of the sensing diaphragm is vented to the atmosphere via the electrical cable and, as a result, the tensiometer measures gauge pore-water pressure. The electrical cable carrying the power supply and the output analog signal is connected to the semi-conductor chip. Data storage and transmission for the tensiometers are achieved via the same components installed on the chip used for tiltmeter data. The chip is housed in a small metal box positioned at the top of the rigid tensiometer shaft. The metal box is clamped to the shaft via a clamping hook and can be easily removed for maintenance or replacement.

3 Experimental Activity

3.1 Slope Physical Model and Tested Soil

The Tensio-inclinometer has been experimentally validated carrying out some tests in a slope physical model simulating wetting-induced landslides in sloping volcanic layers. The slope physical model is made of a tilting tank with a rectangular base 2 m long and 1.5 m wide (Fig. 2). The height of the side walls allows accommodating a soil layer of 0.5 m maximum height (in the direction orthogonal to base). The downslope wall is made of a perforated steel sheet to maintain the soil layer in place once the tank is tilted and allow for water drainage. In addition, a geotextile is interposed between the base of the tank and the soil layer to increase the interface frictional resistance and promote the development of a failure surface well within the soil layer.

The tank base is supported by a steel frame with a hinge located at the tank mid-length. A hydraulic actuator operated manually allows tilting the tank up to 45°. The base structure can be moved on wheels or stand on feet of adjustable height. The tank is surmounted by steel portals carrying three brass nozzles to generate nebulised rain.

The tested soil is a non-plastic, cohesionless silty sand of volcanic origin (Fig. 3). Drained isotropic-consolidated triaxial tests (not yet published) yielded a friction angle of $\phi' = 33°$ and permeameter laboratory tests (not yet published) yielded a saturated hydraulic conductivity of 3×10^{-7} m/s. Surface displacements have been measured by a motorized total station. For sake of brevity, only one of the three tests will be presented in the following. Taking the tank horizontally, a reconstituted layer 35 cm thick has been put in place by pluvial deposition and then vegetated to verify whereas roots could or not nullify pre-failure kinematics. The tank has then been tilted and the layer has then been wetted by an artificial rain of 28 mm/h until inducing global instability.

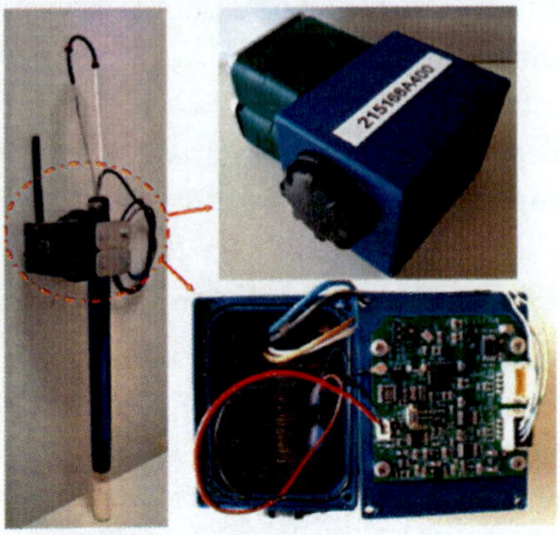

Fig. 1. The Tensio-inclinometer, the metal box and its internal view.

Fig. 2. Slope physical model.

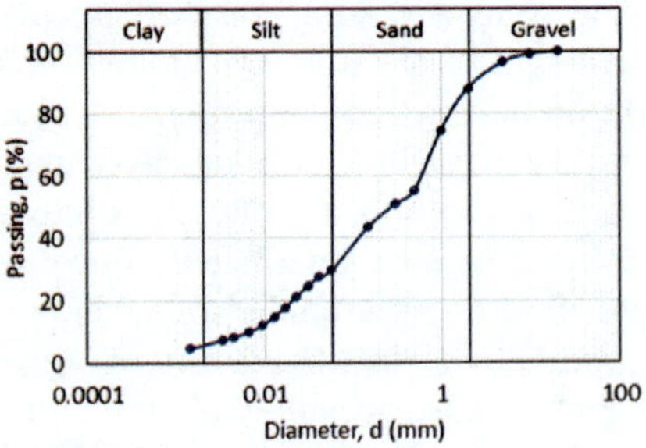

Fig. 3. Grain size distribution of the tested soil.

3.2 Results

For the presented test the tank has been first tilted to 36°. Six Tensio-inclinometers have been installed (Fig. 4a), pushing the ceramic cups to the depths of 35, 25 and

10 cm (Fig. 4a). For each Tensio-inclinometer, the box has been placed at the top of the shaft turned upslope, as close as possible to the ground surface to minimise overturning moments. The devices D_{A1}, D_{B1} and D_{C1} have been placed in the soil quasi-vertically, while the other ones have been positioned orthogonally to the slope surface. All the Tensio-inclinometers have been placed in the upper zone of the slope to minimise the effect of the kinematic restraint acting downslope. Instruments and markers have been concentrated along three different sloping lines labelled as A, B and C. The motorized total station has targeted 4 markers placed at the top of same number of shafts pushed into the layer near the Tensio-inclinometers of the alignment A and C.

In this first stage of the test, rainfall has been applied long enough to generate slightly positive pore-water pressures. Figure 4b shows the evolution of tilting and suction measured by the D_{C1} device. The increase in rotation corresponds to a drop in suction (negative suction indicates positive pore-water pressure).

A failure surface has become visible upslope through the side surface (Fig. 4c). A rotational movement has occurred along the slip surface as indicated by the device counter tilting at the end of the test (Fig. 4b). However, the rotational movement has stopped, and no collapse mechanism has occurred. This has essentially been due to the geometry of the model, implying that the 'effective' inclination of the slope is lower than the inclination of the tank. In fact, the downslope wall imposes a constraint on the kinematics of global instability and forces the failure surface that forms parallel to the base of the tank in the upper portion of the slope to flatten mid-slope to reach the rim of the downslope wall. The 'effective' inclination of the slope ('overtaking angle') is given by the inclination of the segment joining the bottom of the upslope wall with the top of the downslope wall. It results equal to 27.5° and actually governs the global instability. The mobilised 'effective' angle of 27.5° was lower than the critical state angle (33°) but also lower than the liquefaction 'instability' line pulled up by the lower porosity. As a result, neither liquefaction nor global instability took place.

Rainfall has then been interrupted and the slope has been left exposed to the atmosphere for over a week, subject to evapotranspiration. At the end of this period, suction levels have risen to about 4 kPa. Tank tilting has then been increased to 45°, in order to enhance the 'effective' (overtaking) inclination from 27.5° to 35°, i.e. greater than the friction angle of 33° to promote global instability of the slope.

Rainfall has been applied again, adopting the same intensity as the previous test. Global instability has been observed about 70 min after the start of the rainfall. Figure 5 shows the sequence of the evolution of the slope over the test. Figures 6, 7 and 8 show the evolution of rotation and suction recorded by the Tensio-inclinometers combined, if possible, with the surface displacements of the markers recorded by the total station. It worth be noted that the alignment B has been equipped only with tension-inclinometers and one of them malfunctioned, probably due to interference issues during data download; displacement measurements are not available since no marker has been positioned in this alignment. It can be noted that, generally, rotation increases as suction decreases and that rotation increments are recorded well ahead of global instability. However, the stiffer response of the soil 'reinforced' by the root system resulted in rotations in the

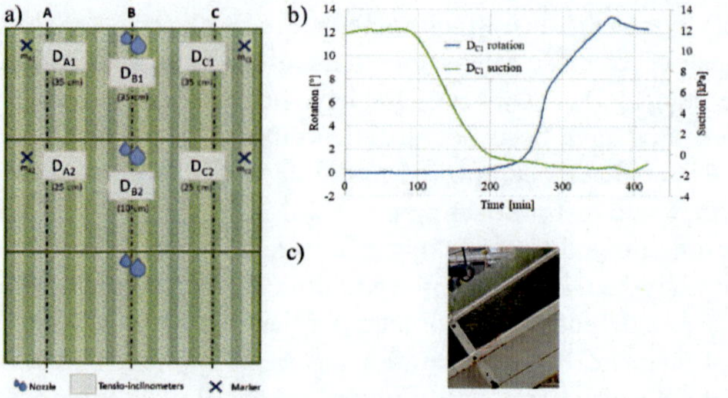

Fig. 4. a) Position of nozzles, Tensio-inclinometers, and markers; b) evolution of rotation and suction during the first stage of the test; c) failure surface [14].

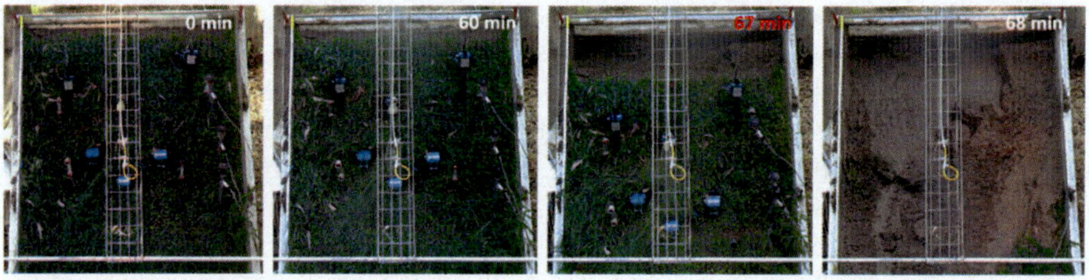

Fig. 5. Pictures taken at different stages of the test [14].

range 1–5°. The evolution of rotations is highly consistent with the evolution parallel-to-slope displacements delivered by total station measurements. Suction paths show that global instability has taken place under slightly positive pore-water pressures.

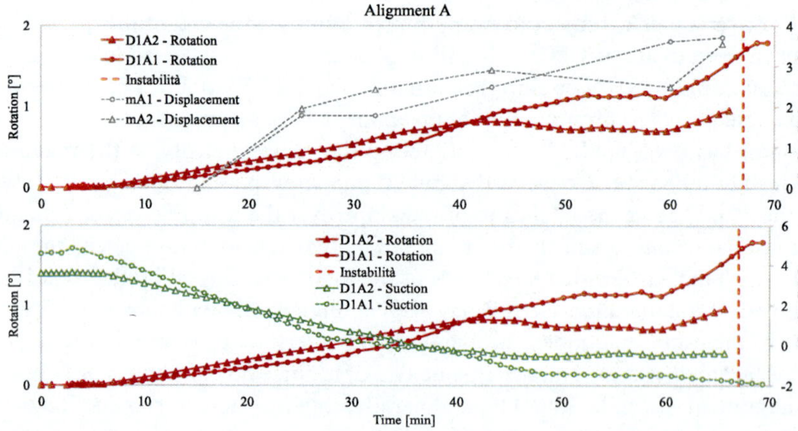

Fig. 6. Rotation and comparison with markers surface displacements and simultaneous measurements of rotation and suction for the alignment A [14].

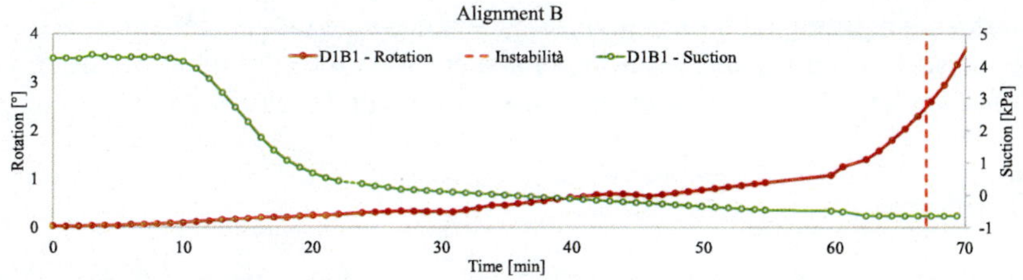

Fig. 7. Rotation and comparison with markers surface displacements and simultaneous measurements of rotation and suction for the alignment B [14].

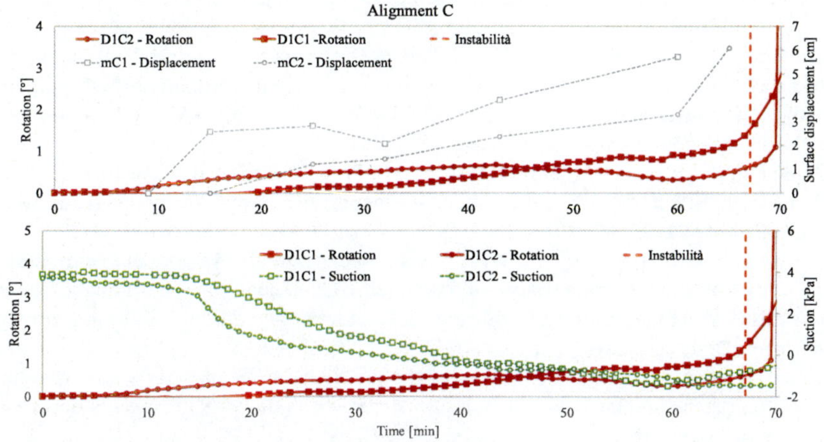

Fig. 8. Rotation and comparison with markers surface displacements and simultaneous measurements of rotation and suction for the alignment C [14].

4 Discussion and Conclusions

The results of the test presented clearly indicate that the silty volcanic covers subject to rainfalls experience pre-failure suction-induced kinematics. This kinematics is characterized by rotational movement in the direction of the slope. It is well detectable even considering a 35 cm thick model and may be considered a landslide precursor in synergy with rainfall and suction records. Moreover, the rotation movement exhibits a significative increase for suction null variations, therefore it appears a more effective landslide precursor.

The same test also indicates that the suction-induced kinematics is adequately captured by tilting evolution. Therefore, its measurement can therefore successfully substitute the measurements of absolute surface displacements. This finding is crucial in designing and implementing light and effective LEW monitoring systems because measuring the rotation of a tensiometer shaft installed in the slope (with the added benefit of suction measurement) is simpler than setting of displacement monitoring system (expensive and difficult to install and manage). In addition, techniques for monitoring displacements tend to become highly inaccurate under conditions of persistent rainfalls.

By comparison, tilting monitoring is easier to incorporate into a wireless standing-alone device rapid-to-install. The Tensio-inclinometer offers the additional benefit of suction measurement and is expected to operate trouble-free under adverse weather conditions.

References

1. Hungr, O., Leroueil, S., Picarelli, L.: The Varnes classification of landslide types, an update. Landslides **11**, 167–194 (2013)
2. Haque, U., et al.: Fatal landslides in Europe. Landslides **13**(6), 1545–1554 (2016)
3. Pagano, L., Picarelli, L., Rianna, G., Urciuoli, G.: A simple numerical procedure for timely prediction of precipitation-induced landslides in unsaturated pyroclastic soils. Landslides **7**, 273–289 (2010)
4. Arattano, M., Conte, R.C., Franzi, L., Giordan, D., Lazzari, A., Luino, F.: Risk management on an alluvial fan: a case study of the 2008 debris-flow event at Villar Pellice (Piedmont, N-W Italy). Nat. Hazards Earth Syst. Sci. - NAT HAZARDS EARTH SYST SCI. **10**, 999–1008 (2010). https://doi.org/10.5194/nhess-10-999-2010
5. Agnoletti, M., Errico, A., Santoro, A., Dani, A., Preti, F.: Terraced landscapes and hydrogeological risk. Effects of land abandonment in Cinque Terre (Italy) during severe rainfall events. Sustainability **11**, 235 (2019)
6. Maugeri, M., Motta, E.: Slope failure. Effects of heavy rainfall on slope behaviour: the October 1, 2009 disaster of Messina (Italy). In: lai, S. (ed.), Geotechnics and Earthquake Geotechnics towards Global Sustainability, Geothecnical, Geological and Earthquake Engineering, pp. 15. Springer, Dordrecht (2011)
7. Greco, R., Pagano, L.: Basic features of the predictive tools of early warning systems for water-related natural hazards: examples for shallow landslides. Nat. Hazards Earth Syst. Sci. **17**, 2213–2227 (2017)
8. Keefer, D.K., et al.: Real-time landslide warning during heavy rainfall. Science **238**, 921–925 (1987)
9. Dixon, N., Spriggs, M.P.: Landslide hazard evaluation by means of several monitoring techniques, including an acoustic emission sensor. Available online: https://repository.lboro.ac.uk/ (2012)
10. Barla, M., Antolini, F.: An integrated methodology for landslides' early warning systems. Landslides **13**(2), 215–228 (2015)
11. Orense, R.P., Towhata, I., Farooq, K.: Investigation of failure of sandy caused by heavy rainfall. In: Proceedings of the International Conference om Fast Slope Movement – Prediction and prevention for risk mitigation (2003) (FSM2003) Sorrento
12. Balzano, B., Tarantino, A., Ridley, A.: Preliminary analysis on the impacts of the rhizosphere on occurrence of rainfall-induced shallow landslides. Landslides **16**, 1–17 (2019)
13. UMS GmbH (last check My, the 2nd, 2021). http://library.metergroup.com/Manuals/UMS/T4_Manual.pdf
14. Coppola, L., Reder, A., Tarantino, A., Mannara, G., Pagano, L.: Pre-failure suction-induced deformation to inform early warning of shallow landslides: proof of concept at slope model scale. Eng. Geol. **309**, 106834 (2022)

First Experiences with Gel-Push Sampler for Testing Coarse Alluvial Soils Under a River Levee

Nicola Fabbian[1](✉), Paolo Simonini[1], Fabio De Polo[2], and Simonetta Cola[1]

[1] DICEA, University of the Study of Padua, Padua, Italy
nicola.fabbian@phd.unipd.it
[2] Mountain Water Authority, Province of Bolzano, Bolzano, Italy

Abstract. In the last ten years, the 'Gel-Push' sampling technique has been establishing itself as an alternative to the expensive freeze sampling method to obtain high quality samples in granular materials. The use of a polymeric gel reduces the friction between the tube wall and soil sample, so minor shear stresses and less disturbance of the in-situ physical soil state are induced during the sampling. This paper describes an experience with one type of this technique applied to the Adige River embankment, near Bolzano (Italy), where a very complex stratigraphy, with soils ranging from clayey silts to medium-large sands with presence of pebbles, exists. The quality of samples collected with the Gel-Push and Shelby samplers were evaluated and compared. Laboratory permeability tests were also conducted to investigate the effect of the gel on the test result. The Gel-Push technique has proven to be reliable for obtaining good samples in almost all the different types of soils, with some limitations in sand mixed with gravel. At the same time, the small amount of gel which permeates the soil during sampling can easily be removed during laboratory tests without effects on the parameter determination.

Keywords: Gel-Push sampling · high-quality undisturbed sample · silty sand · granular material · gel effects

1 Introduction

Obtaining clean or high quality samples in low cohesive soils has always been a difficult task. The literature on undisturbed sampling in sands and silty sands below the water table is mostly limited to ground freezing methods. Groundwater freezing fixed the sand particles and their matrix to the frozen ground. Sand samples were collected by drilling and remained frozen until laboratory shear testing [1]. However, the process of soil freezing is time consuming and prohibitively expensive.

Høeg et al. (2000) [2] attempted to sample 3 m of natural silt at ambient temperature in the capillary zone just above the water table using a 50 mm internal diameter piston sampler at the Swedish Geotechnical Institute, but had limited success. As reported by Høeg et al. [2], vibration during transport and extrusion of the sample from the sample tube can cause disturbance to the silty-sand soil samples. Another experience in

A. Ferrari et al. (Eds.): CNRIG 2023, SSGG, pp. 63–70, 2023.
https://doi.org/10.1007/978-3-031-34761-0_8

which sand sample was taken from below the ground water table without freezing was performed by Konrad et al. (1995) [3], who used a Laval sampler collecting samples with a diameter of 200 mm and a height of 500 mm. To avoid damage to soil structure during transportation of low-cohesion sand, Konrad et al. (1995) [3] developed a method for freezing Laval samples above ground, without causing volume change. Currently, frozen sampling remains the "gold standard" for obtaining sand samples, even though the associated costs make it impractical for most projects [4].

This article provides a brief overview of a new sampling technique called "Gel-Push" (GP) and reports some first experiences of it. The purpose of the study was to test this new technique in Italy, in a very complex stratigraphy, in order to collect even coarse soil samples to be subjected to common laboratory tests.

In this study, soil collected with Gel Push Triple Tube Sampler (GP-Tr) and Shelby samplers were compared through different test results. If the sample appears to be in good visual condition through granulometry analysis, Atterberg limit determination and permeability tests in triaxial and oedometer cells were carried out. Special care was also placed on investigating the effects of gel on the test results.

2 Gel-Push Sampling Methodology

The Gel Push sampling methodology was first developed in 1999 in Japan [5] with the aim to obtain high quality samples in granular materials at reasonable cost. The methods were then gradually developed and modified by Kiso-Jiban Consultants and the Taiwan Construction Research Institute [6]. The technique was used by several researchers around the world, including Japan, New Zealand, Taiwan and Bangladesh [4, 7–11]. This relatively new technique is still under development, so the designers have occasionally made small changes to the samplers to address specific issues reported by end users [4].

Conventional downhole samplers, like the Osterberg and Shelby ones, create disturbance to the sample, above all due to friction mobilized on the side of the soil sample as it enters the core-liner barrel. The use of a polymeric low-friction gel reduces the friction between the tube wall and soil sample, so minor shear stresses are induced in the samples with less disturbance of the in-situ physical soil state. There are currently three different types of Gel push sampler available, specifically named Gel push Static penetration (GP-S), Gel push Triple Tube Sampler (GP-Tr) and Gel push Rotary single tube sampler with core catcher (GP-D). All of them are conceptually similar, but with some modifications that allow the gel to be delivered to the bottom of the sampler. According to Kiso-Jiban Consultants [6], the GP-S sampler should be used for soils composed of silt/soft clay and loose sand with fine sand, the GP-Tr sampler for soils composed of silt/hard clay and dense sand, while GP-D is suitable for hard material (e.g. sandy gravel, fill material, fractured rock). Stringer et al. (2016) [4] recommend using GP-S sampler in soil with an upper limit on cone tip resistance qc of 5MPa and switching to GP-Tr sampler above this limit.

This paper focuses on the GP-Tr sampler, as it was the most suitable sampler for this case study.

2.1 Gel -Push Type Tr Sampler

The GP-Tr sampler is a rotary triple tube device, in which drilling mud is pumped through a rotating shoe to clear the bottom of the borehole and allow the sampler to advance. A schematic of the operation of the sampler is shown in Fig. 1a. A spring-loaded, non-rotating cutting shoe protrudes slightly in front of the clearing shoe (Fig. 1b). Soil that has passed through the cutting shoe is collected in a PVC core liner pipe (outer diameter/inner diameter = 89/83.5mm). A floating piston moves over the top of the trapped soil column (Fig. 1c), forcing the polymer gel down an annulus between the core liner tube and the center tube, out over the cutter shoe, and coating the soil sample. Excess gel is discharged into the main borehole. The total collection with GP-Tr sampler is about 100 cm and ideally takes at least 10 min. The authors' experience with this sampler has shown a strong variability in the sampling length and the time (up to one hour and half for sandy/gravelly soils) taken according to the soil sampled.

The GP-Tr is quite a complicated tool to use; the equipment should be utilized by an expert driller, who must be able to control the feed rate, mud flow rate and rotational speed to reach 'full' potential of the techniques.

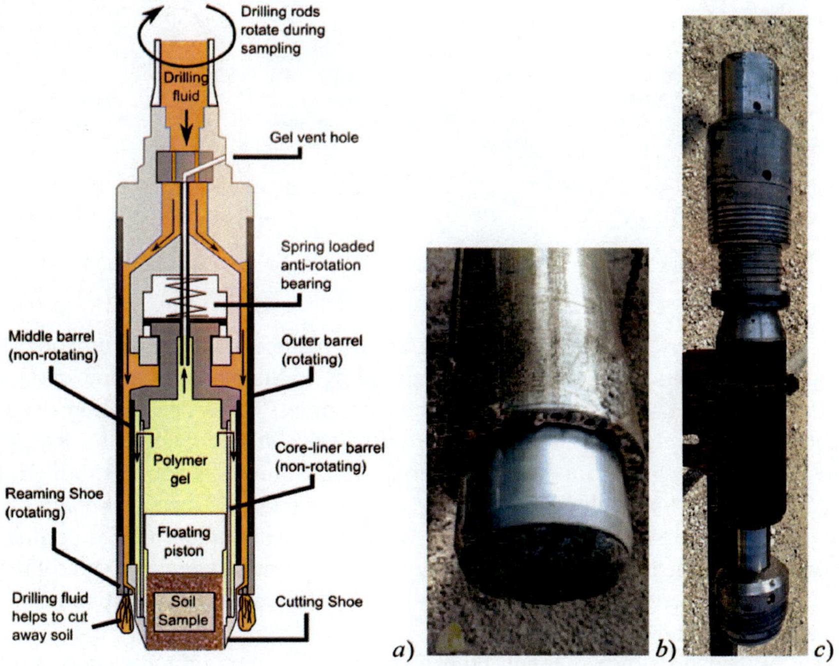

Fig. 1. Schematic of GP-Tr type samplers (*a*) [4], cutting shoe (b) and valves at the top of the sampler for the passage of the lubricant (c);

3 The Field Test Sites

As shown in Fig. 2, the samples were collected in two test-sites along the right embankment of the Adige River, near Bolzano (Italy). The sites were chosen for the proximity to the A22 Highway, which runs parallel, because the final goal of the study is to evaluate the effects of a possible collapse of the embankment on the highway. In the 19th century, several anthropic interventions were performed in the Adige valley and exceptionally straight embankments were realized to confine and rectify the river, previously free to drift in a larger riverbed. In many sections, the new right levee crosses the ancient paleo-river and it has been observed that in a relatively recent past some levee collapses have occurred at those intersection sections [12]. In the selected sites, the embankment structure and its foundation were accurately characterized through several in-situ and laboratory tests. This activity was preliminary to the installation of some traditional and advanced monitoring sensors for analyzing the levee's hydraulic behavior [13].

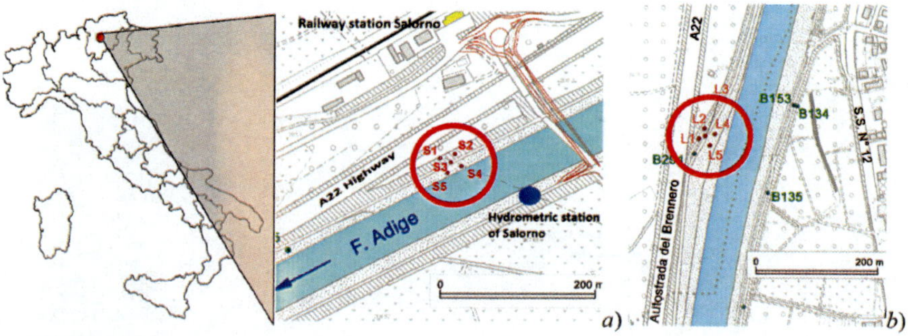

Fig. 2. Test site location from national to local scale: Salorno (*a*) and Laghetti (*b*).

In both the sites, respectively in May and September 2021, five boreholes were drilled at the corners and the center (corresponding to the axis of the embankment) of the 20m square area (Fig. 3). In order to reach more or less the same elevation, the central boreholes (respectively S3 and L3) are approximately 30m deep, while the ones located at the corners are about 25m deep, as the river embankment is about 5 m high. In S3 and L3 20 samples were obtained using the GP-Tr technique, while 19 Shelby samples were extracted from the corner boreholes. During the perforation, in-situ permeability tests were performed using Lefranc tests and an experimental device properly realized to determine the permeability of coarse-grained soils. Standard penetration tests (SPT) were also conducted on site.

Due to their proximity, the boreholes show a similar stratigraphic succession, differentiated only by the composition of the embankment structure and by minor lateral variations in the sediment granulometry.

4 Result of Laboratory Tests

As reported in Table 1, 12 Gel Push samples (GPsam) and 3 Shelby samples (Ssam) are currently opened and examined in laboratory. Given the proximity of the central borehole to the corner ones, the samples extracted with different techniques at the same depth can be compared, for a total of 3 couples.

The extraction of GPsam from the punches was quite easy also manually, due to the presence of gel on the external surface of the soil. In mainly fine-grained samples the gel penetrated into the sample for only 1mm; in granular media, instead, the gel completely saturated the sample. Since in granular samples the gel provided the necessary cohesion for the specimen preparation, it was possible to perform TXCD tests (Fig. 4c). To create as little disturbance as possible, the small specimens (D = 35mm; H = 75mm) for the triaxial texts were obtained by manually inserting some metal punches inside the GPsam during their extrusion. The sample recovery has a particularly variable percentage (20%-100%), with a higher percentage in granular soils than in the more cohesive samples. This agrees with the indications given by the developers of the GP-Tr technique.

The gel is declared soluble in water; the authors therefore tried to eliminate it by "washing" the sample, that is by applying a seepage flow to the specimen with a hydraulic gradient of around 3, after assembling the triaxial cell.

Table 1. Tested samples for the sites of Salorno (S) and Laghetti (L).

Sample	Depth (m)	Type	Composition	Recovery (%)	Gel Penetration
S4-CI1	4.5–5.0	Shelby	Weakly sandy silt	100	-
S3-CI1	8.0–9.0	GP-Tr	Weakly sandy silt	56	Not
S3-CI2	10.0–11.0	GP- Tr	Clayey silt	20	Not
S3-CI3	12.0–13.0	GP- Tr	Weakly sandy silt	100	Not
S3-CI4	14.0–15.0	GP- Tr	Medium coarse sand	80	Yes
S3-CI5	15.0–16.0	GP- Tr	Medium coarse sand	89	Yes
S3-CI6	16.0–17.0	GP- Tr	Gravel and sand	65	Yes
S3-CI8	20.0–21.0	GP- Tr	Clayey silt	29	Not
S3-CI11	27.5–28.5	GP- Tr	Medium sand	52	Yes
L2-CI1	6.0–6.5	Shelby	Medium-fine sand with silt	96	-
L4-CI2	19.5–20.0	Shelby	Fine sand with silt	90	-
L3-CI1	6.0–7.0	GP- Tr	Fine silty sand	80	Yes
L3-CI3	10.5–11.5	GP- Tr	Gravel and sand	90	Yes
L3-CI4	16.5–17.5	GP- Tr	Fine sand	56	Yes
L3-CI8	24.0–25.0	GP- Tr	Medium coarse sand	60	Yes

On the basis of the granulometry (Fig. 3), it is possible to subdivide the analyzed samples into 3 groups. The first group includes five samples mainly composed of silty soil. They were subjected to oedometric and triaxial tests with measure of permeability at different confinement pressure. From the comparison between S4-CI1 and S3-CI1, collected from the same layer, a variability of one order of magnitude around a mean value of permeability k = 5^{-10} m/s was found. Moreover, on S3-CI3 (Fig. 4a), the permeability was obtained with a constant head test in triaxial cell at a confinement of 150 kPa, using two identical specimens, one "washed" and one "unwashed". Since the permeability obtained is practically the same ($1.35*10^{-7}$ m/s and $1.38*10^{-7}$ m/s, respectively), it seems possible to deduce that the gel does not significantly influence permeability, at least for silty soils, since it does not penetrate into the sample.

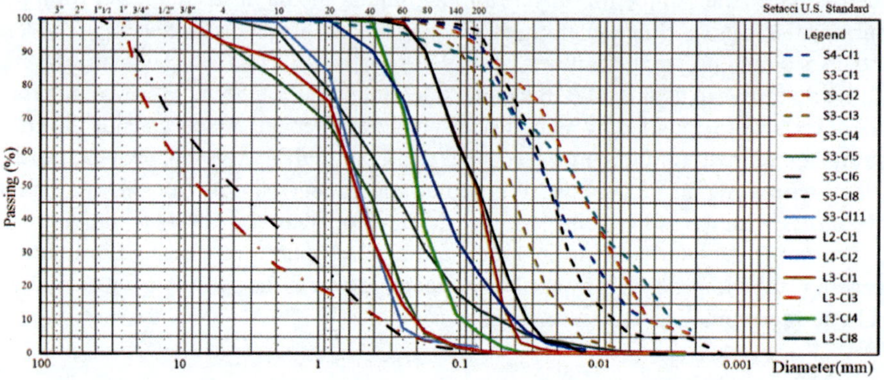

Fig. 3. Granulometric curves of tested sample: 1st group in dotted line, 2nd group in continuous line and 3rd group in dash-dot line.

In the 2nd group there are 8 samples mainly composed of sand. Two pairs of samples seemed to exist for the comparison: L3-CI1 vs. L2-CI1 and L3-CI8 vs. L4-CI2. Unfortunately, the soils in L3-CI1 and L2-CI1 have granulometries that are too different to be compared, while the Ssam L3-CI8 liquefied during extrusion, making it impossible to obtain a sample with the original soil structure.

Fig. 4. S3-CI3 after extrusion and recovery of specimens for triaxial test (*a*); Cross Section of S3-CI8 (*b*); a specimen obtained from S3-CI4 during the setup of a triaxial test (*c*).

Like for S3-CI3, two specimens are obtained from the GPsam S3-CI4 and from S3-CI5 in order to carry out permeability tests in triaxial cell (Figs. 4c and 5a) on "washed" and "unwashed" soils. The permeability was equal to $2.21*10^{-6}$ m/s and $4.61*10^{-7}$ m/s for the medium coarse sand of S3-CI4, and $1.63*10^{-6}$ m/s and $3.22*10^{-6}$ m/s for the gravelly-sandy soil of S3-CI5 (the first data referring to the washed specimens, the second to the unwashed). Even if it can be deduced that for the coarse-grained soil the influence of the gel on permeability is modest, it is advisable to eliminate the gel as much as possible by washing the soil before carrying out the permeability tests. Finally, in the 3rd group, there are two samples mainly composed of gravelly soils. For these samples, two large specimens (D = 85mm) to be tested in triaxial cell were obtained. The GPsam S3-CI6 was extruded by moving it directly inside an openable hollow cylinder with the membrane already mounted on it. Subsequently, after cutting the upper and lower bases, the specimen was set in the tri-axial cell and was saturated. Finally, a small suction was applied. Unfortunately, the sample collapsed when the cylinder was removed (Fig. 5b), because the cohesion provided by the gel was not sufficient to support the specimen's weight. Given the impossibility of obtaining a sample with an intact structure, a permeability test was carried out with a constant load permeameter on the reconstructed material, obtaining $K = 1.04*10^{-3}$ m/s, a higher value than expected. To avoid this, before cutting and installing the specimen in the triaxial cell, the other available sample (GPsam L3-CI3) was previously frozen (Fig. 5c). Once it had thawed, the specimen was washed and subjected to a permeability test at a confinement pressure of 150 kPa. Since the value obtained ($8.38*10^{-8}$ m/s) is very low considering the granular nature of the tested soil, other samples may soon be opened to better investigate this type of soil.

Fig. 5. GPsam S3-CI5 after extrusion (a), Liquefaction of S3-CI6 specimen during installation in triaxial cell (*b*); frozen specimen obtained from L3-CI3 during installation (*c*).

5 Conclusion

Based on the experiences carried out during the on-site collection of samples and their opening in the laboratory, some considerations on GP-Tr sampling techniques can be drawn. They are certainly more challenging and complex than traditional undisturbed sample techniques. Generally, the extracted samples were shorter than the declared length of 1m. In some cases, the Authors recorded a sampling rate of only 20%. The extrusion of the samples can be carried out manually thanks to the lower ground-hollow punch friction due to the presence of the gel. The sampled soil is somewhat compressed and, in some cases, it moves slightly inside the die. In coarse-grained soils, the gel penetrated inside the sample and, if not carefully removed by washing, its viscosity influenced the permeability measurements. The gel is soluble in water and can therefore be eliminated by washing the sample, carried out by applying a flow to the sample with a hydraulic gradient of about 3. In silty-clayey soils the gel does not penetrate inside the samples in most cases, but only permeates the first 10 mm outside layer. However, in some of the samples analyzed it is difficult, if not impossible, to execute the Atterberg limits given the influence of the Gel. To date, other experiences carried out in New Zealand [4] have highlighted the incapacity of GP-Tr to produce high quality specimens. The results obtained in this work, and summarily reported here, provided some initial information on the possibility to characterize the coarse soils under study through laboratory tests from GP-Tr samples, which is impossible with other economic techniques.

References

1. Huang, A.B.: The seventh James K. Mitchell lecture: Characterization of silt/sand soils. Proc. Geotechn. Geophys. Site Characteris. **5**, 3–18 (2016)
2. Høeg, K., Dyvik, R., Sandbækken, G.: Strength of undisturbed versus reconstituted silt and silty sand specimens. J. Geotechn. Geoenvironm. Eng. **126**(7), 606–617 (2000)
3. Konrad, J.M., St-Laurent, S., Gilbert, F., Leroueil, S.: Sand sampling below the water table using the 200 mm diameter Laval sampler. Can. Geotech. J. **32**(6), 1079–1086 (1995)
4. Cubrinovski, M., Stringer, M., Haycock, I.: Experience with gel-push sampling in New Zealand (2016)
5. Yusa, M., Bowman, E.T., Cubrinovski, M.: Observation of microstructure of silty sand obtained from gelpush sampler and reconstituted sample. In: EPJ Web of Conferences, vol. 140, p. 12017. EDP Sciences (2017)
6. Huang, A.B., Mayne, P.W., (eds.): Geotechnical and geophysical site characterization. In: Proceedings of the 3rd International Conference on Site Characterization (2008)
7. Lee, W.F., Ishihara, K., Chen, C.C.: Liquefaction of silty sand—preliminary studies from recent Taiwan, New Zealand and Japan earthquakes. In: Proceedings of the International Symposium on Engineering Lessons Learned from the 2011 Great East Japan Earthquake. Tokyo, Japan (2011)
8. Jamiolkowski, M., Masella, A.: Geotechnical characterization of a tailings deposit in Poland—an update. In: Proc. 3rd Int. Symp. Cone Penetration Testing. Las Vegas (2014)
9. Taylor, M.L., Cubrinovski, M., Haycock, I.: Application of new'Gel push'samplingprocedure to obtain high quality laboratory test data for advanced geotechnical analyses. In: 2012 New Zealand Society for Earthquake Engineering Conference, pp. 1–8. NZSEE, Christchurch, New Zealand (2012, April)
10. Chiaro, G., Kiyota, T., Umehara, Y., Hosono, Y., Yagiura, Y., Chiba, H.: Evaluation of cyclic resistance of high quality undisturbed Chiba silty sand samples retrieved by "Gel-Push" sampling technique. Japanese Geotechnical Society Special Publication **2**(32) (2016)
11. Lunne, T., Long, M.: Review of long seabed samplers and criteria for new sampler design. Mar. Geol. **226**(1–2), 145–165 (2006)
12. Cola, S., Girardi, V., Bersan, S., Simonini, P., Schenato, L., De Polo, F.: An optical fiber-based monitoring system to study the seepage flow below the landside toe of a river levee. J. Civ. Struct. Heal. Monit. **11**(3), 691–705 (2021). https://doi.org/10.1007/s13349-021-00475-y
13. Schenato, L., et al.: Distributed optical fiber sensors for the soil temperature measurement in river embankments. In: 2022 IEEE International Symposium on Measurements & Networking (M&N), pp. 1-6. Padua, Italy (2022). https://doi.org/10.1109/MN55117.2022.9887664

Small Scale Toppling Tests on Simplified Tree Root Prototypes

Andrea Galli[1], Giacomo Marrazzo[1(✉)], Andrea Marsiglia[2,3], Alihossein Ezzati[1], Matteo Oryem Ciantia[2], and Riccardo Castellanza[4]

[1] Politecnico di Milano, Milan, Italy
giacomo.marrazzo@polimi.it
[2] University of Dundee, Scotland, UK
[3] Agroservice s.r.l, Cormano, Italy
[4] Università degli Studi di Milano-Bicocca, Milan, Italy

Abstract. The root plate of trees, from a Geotechnical point of view, plays the role of the "living" foundation of a tall structure, subject to complex loading histories essentially deriving from environmental actions (e.g., wind loads). Its mechanical response to toppling loads (which represent a noticeable source of risk in urban areas), is not yet fully understood by the current interpretative models. In the paper, with the aim of highlighting the main features of the mechanical behavior of such systems, some small-scale 1g tests are presented. A simplified root prototype has been conceived, by combining elementary structural elements aimed at exhibiting both flexural and pullout properties of real-like flat root systems. Toppling loads (at constant vertical load and zero horizontal load) have been applied to the prototype and three different granular materials have been employed to reproduce the soil layer. The results highlight a certain positive correlation between a representative secant stiffness of the tests and their ultimate toppling resistance in case of very deformable soils. More uncertain trends are instead observed in case of stiffer granular materials. Although still not referred to real working conditions of tree roots, the experimental campaign may contribute to a deeper geotechnical understanding of the tree toppling phenomenon.

Keywords: Small scale tests · tree toppling mechanism · lateral load · foundations

1 Introduction

Toppling failure mechanisms of tall trees represent a noticeable source of risk (not only for structures and infrastructures, but also for goods and peoples), which is nowadays also increased by current climate changes, with more and more frequent intense weather events and extreme wind gusts. The protection against such failure mechanisms, frequently responsible of sudden tree collapses, requires a large multidisciplinary approach, combining botanic, agronomic and engineering competences. In this perspective, Geotechnical Engineers may provide a valuable contribution in understanding the behavior of the root plate system when subject to complex loading paths. Roots play in fact

A. Ferrari et al. (Eds.): CNRIG 2023, SSGG, pp. 71–78, 2023.
https://doi.org/10.1007/978-3-031-34761-0_9

the role of a "living foundation" for the tree, whose behavior is governed by the mechanical interaction between the soil and the roots. A large amount of scientific literature is already available on this topic (Schwartz et al., 2010a), in particular when the overall strengthening effects of vegetation and of tree roots on soil layers is considered (Mickovski and Beek, 2009; Schwartz et al., 2010b). When the stability of trees is addressed, however, the attention should rather be focused on the anchoring properties of the root plate system within the soil, and to its overall resistance to toppling actions. To this goal, several experimental works have been proposed (Guitard and Castera, 1995; Zhang et al., 2020 and 2022), investigating the moment-rotation curve of tree prototypes (see also the PhD theses of Zhang, 2021, and Sala, 2021). Numerical modelling techniques, based on finite element analyses, have also been proposed (Dupuy et al., 2005; Yang et al., 2014) proving the good accuracy of the adopted approaches in reproducing the experimental results. From a practical point of view, real scale stability assessment procedures against tree toppling are usually conducted by professional agronomists by interpreting on site non-destructive tests. Among other possible techniques (for a comprehensive discussion, see e.g. the book of Sani, 2017), on site static pulling tests are also performed, whose experimental toppling curves are usually interpreted according to the procedure proposed by Wessolly and Erb (1998). In particular, the limit value of the toppling moment for the tree is extrapolated from the results of on-site experimental tests run until a limited value of tree rotation (usually not exceeding 0.2°, in order to prevent significant damage to the tree root system).

Despite such a large amount of work, both at the scientific and at the professional level, several aspects of the mechanical behavior of root plates are still unclear (and primarily the estimation of the limit toppling moment) and suffer of a lack of information about the influence of the main geometrical and mechanical parameters characterizing the system. The paper aims at giving some contributions in this direction, by providing the results of a small scale testing campaign on a simplified prototype, mimicking a flat root system of a tree subject to toppling loads at constant vertical load.

2 Experimental Apparatus and Loading Program

The root plate of a tree is generally subject to a complex set of loads along the vertical, V, and horizontal, H, directions and to a toppling moment, M, deriving from the combination of the weight of the tree, W, and of possible additional horizontal actions, F, , induced e.g. by wind gusts (Fig. 1a). In the paper, flat root plate systems will only be considered, characterized by a dominant horizontal growing direction, at shallow depth below the ground surface. Both first order and second order roots are in general present, exhibiting essentially a flexural and a diffused pullout resistance, respectively.

With the aim of reproducing in a simplified manner the aforementioned essential characteristics, a small-scale experimental apparatus has been conceived. A uniform layer of 120 mm of a granular material is deposited by means of a pluviation method in a rigid testing box (355 × 165 × 180 mm; Fig. 1b), and a simplified flat root model is placed in the center of the box. An additional covering layer of granular material is then deposited on both sides of the model, in order to completely embed the root system for a thickness of 30 mm. The root model is then rigidly connected to a toppling loading system, constituted by two plastic tanks (each one with a volume of 2 L, and initially mid-filled with water), placed on a wooden plate at a distance $L_0 = 165$ mm from the centerline; the two tanks are connected to each other by means of a peristaltic pump. The vertical component V of the resulting load is then given by the sum of the dead weight of the model ($W_0 = 1.25$ kg) and of the water in the tanks ($W_1 + W_2 = 2$ kg), and it is kept constant during the test. The loading program consisted in applying an increasing value of toppling moment M (at zero horizontal load H), as a function of the difference of the water weights in tanks 1 and 2, controlled by the pump flow rate (a low constant flow rate of 1.1 ml/s has been chosen for all tests). The rotation φ of the system was recorded at a frequency of 10 Hz by means of an inclinometer fixed to the model, and placed just above the final elevation of the ground surface (Fig. 1b).

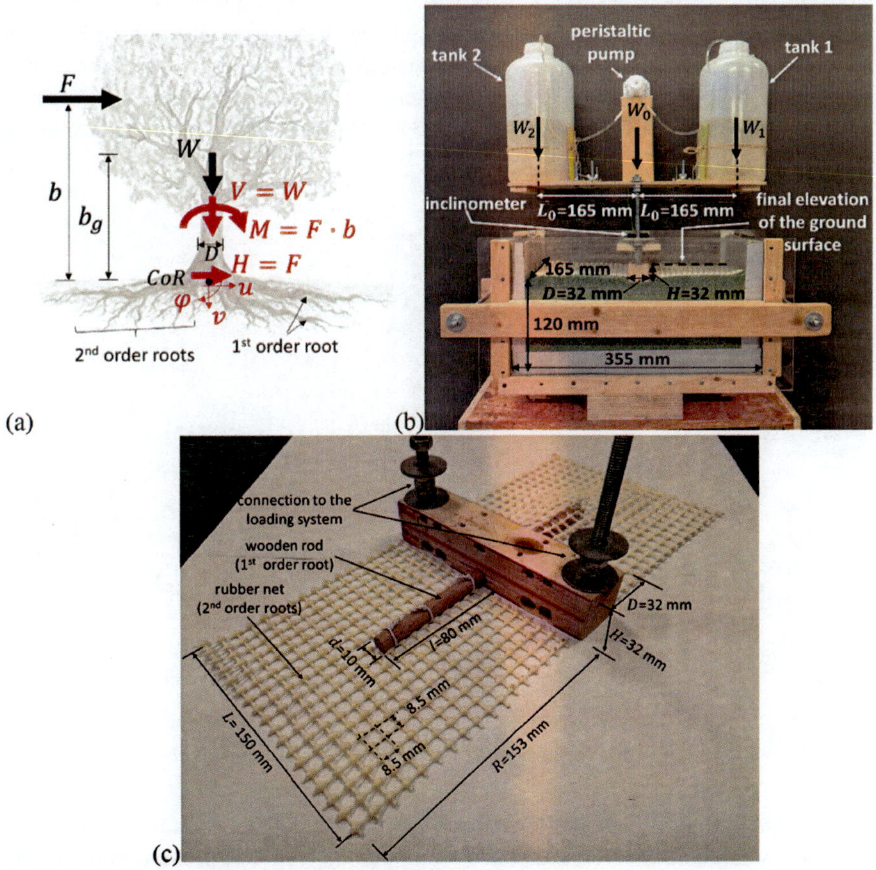

Fig. 1. (a) definition of the loading actions on a tree root plate; (b) view of the experimental setup before the deposition of the 30 mm thick embedment layer; (c) detail of the flat root system.

The flat root system (see also Marsiglia et al., 2022, where the same prototype was tested in vertical compression and pullout tests) is constituted by a wooden beam

(32 × 32 × 150 mm; Fig. 1c) connected to the loading system. First order roots are represented by horizontal wooden rods (diameter $d = 10$ mm, length $l = 80$ mm on each side of the model) introduced at uniform spacing s into the beam. The effect of second order roots have instead been investigated here by fixing to a rod a rubber net (mesh size 8.5 × 8.5 mm), covering a total area of about 230 cm^2 on each side of the model (Fig. 1c). In this work, first order roots have been considered as rigid whilst rubber nets, mimicking second order roots, to be characterized by negligible axial deformability and no bending stiffness. The results have been interpreted in terms of an equivalent plane strain homogenized strip foundation, and the loads normalized with respect to the out of plane dimension (resulting toppling moment is then expressed as Nm/m).

Fifteen tests have been executed, by varying the number of first order roots (1, 2 or 3, i.e. relative spacing $s/d = 15, 7.5$ and 5, respectively), by considering also the condition of no roots ($s/d \rightarrow \infty$, , in order to estimate the contribution of the tree base solely), by introducing second order roots, and by testing three different granular materials. Ticino sand, uniform rubber grains and sawdust were employed (a scaled view of three samples of the materials is shown in Fig. 2a, whilst their grain size distributions are presented in Fig. 2b), in order to consider values of relatively high, medium and low soil stiffness, respectively.

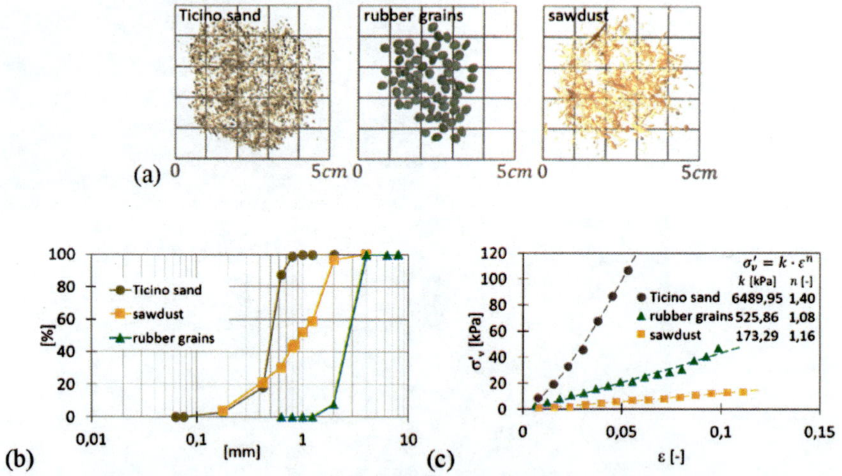

Fig. 2. (a) samples of the three granular materials; (b) grain size distributions; (c) stress-strain relationships in dry oedometric compression tests (dashed lines represent best fitting trends).

Some preliminary dry oedometric tests at low vertical stress on the three materials were also conducted. The vertical stress-strain curves (Fig. 2c) prove that the materials have remarkably different stiffness, as witnessed by the parameters of the equation $\sigma_v' = k \cdot \varepsilon^n$ employed to interpolate the experimental points (fitting values of k and n are also reported in Fig. 2c). The sand is characterized by a stiffness of about one order of magnitude higher than the rubber grains, whilst the stiffness of the sawdust is about 1/3 of that of the rubber grains. The tested dry unit weights of the materials (both in oedometric tests and toppling tests) were about 14 kN/m^3, 6.8 kN/m^3 and 1.3 kN/m^3 for sand, rubber grains and sawdust, respectively. No further mechanical characterization was at the moment possible, since non-standard tests (specifically conceived at low confining stress with respect to standard procedures) would be required.

3 Experimental Results of the Toppling Tests

The results of the tests are summarized in Fig. 3. Subfigures a, c, and e report the time histories of the applied toppling moment M and of the tilting angle φ for tests on sand, rubber grains and sawdust, respectively; subfigures b, d, and f show the corresponding $M - \varphi$ curves.

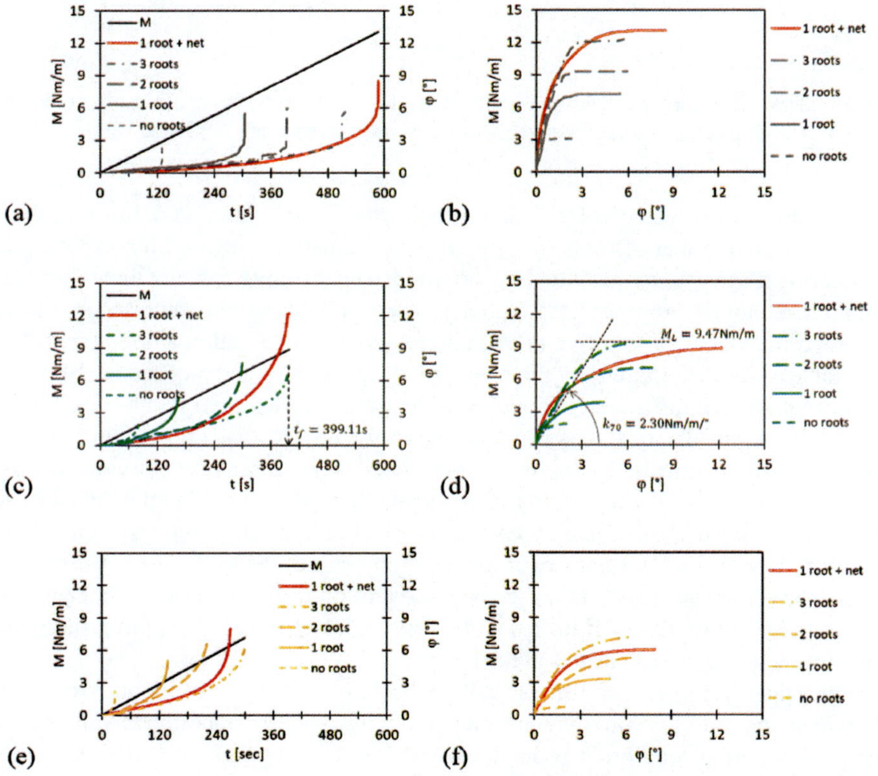

Fig. 3. (a, c, e) loading and rotation histories for the tests on sand, rubber grains and sawdust, respectively; (b, d, f) moment-rotation curves of the tests on the three different materials.

The failure condition was accurately identified for each test as the point where the controllability of the test is lost, i.e. when the curve representing the rotation history becomes vertical (as an example, in Fig. 3c the failure instant t_f for the test on rubber grains with three roots is indicated). All the tests show a marked nonlinear behavior up to failure, and the presence and the number of the first order roots remarkably influence the value of the limit toppling moment. At least for the tests on sand, on the contrary, it does not significantly modify the stiffness of the curves. In order to objectively compare the results, a preliminary interpretation of the $M - \varphi$ curves is hereafter proposed in terms of values of the limiting toppling moment (M_L) and of the 70% secant stiffness (k_{70}; both definitions are also sketched in Fig. 3d with reference to the example tests).

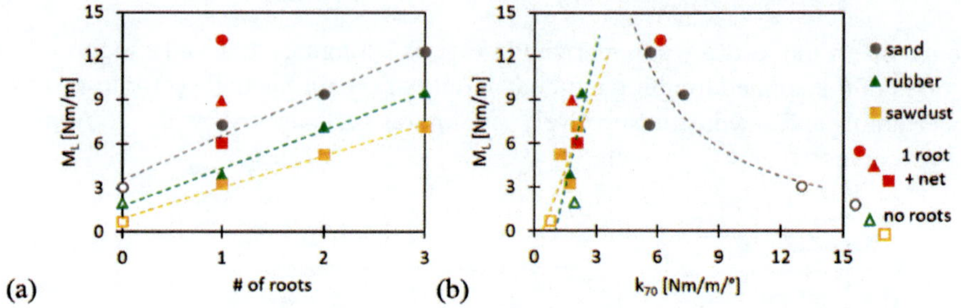

Fig. 4. (a) values of the limit toppling moment M_L for the 15 tests; (b) relationship between M_L and k_{70}. In both figures, dashed lines represent best fitting trends of the plotted values.

The obtained values of M_L for the three materials are summarized in Fig. 4a as a function of the number of roots (empty markers indicate the tests with no roots, i.e. by considering the resistance provided by the wooden beam only). Evident linear increasing trends are recognized for the tests without net (best fitting trends are reported in the figure as dashed lines), suggesting that there is no remarkable interaction among the different roots (at least for the tested s/d values) and between the roots and the wooden beam. This latter provides a significant contribution to the overall resistance (about 1/2 of the case with one root, and 1/4 of the case with three roots) for the tests on sand and on rubber grains, whilst its contribution is much less evident for the sawdust. In view of on-site applications to real scale trees, this could suggest that only in case of relatively stiff soils the size of the base of the tree may give a contribution to the toppling resistance. In case of very deformable soils, on the contrary, the geometry of the root plate is the dominant factor. The presence of second order roots in addition to one first order root (indicated in Fig. 4a by red markers) almost doubles the values of the limit toppling moment, at least for the net extension tested in the experiments.

The values of the secant stiffness k_{70} are instead reported in Fig. 4b and, for the ease of comparison with Fig. 4a, they are plotted against the aforementioned values of M_L. The presence of second order roots apparently does not remarkably affect the considered stiffness. Red markers, in fact, approximately fall within the same range of stiffness values of the other tests. This result indirectly proves that the net behaves as a reinforcing membrane; its contribution is then mobilized only for relatively large rotations φ, consistently with the development of its out of plane (i.e. vertical) displacements. A second observation considers instead the fact that for rubber grains and sawdust (although with some scatter) evident increasing relationships are recognized between k_{70} and M_L (dashed lines represent the proposed linear fitting for the two sets of tests). In case of sand, on the contrary, such relationship is much more uncertain and without a clear trend. If the test with no roots is also considered, moreover, this relationship assumes even a decreasing trend (dashed line in the figure), i.e. opposite to what was observed for the other two materials.

4 Conclusions

Fifteen small-scale 1g experimental tests have been presented, investigating the toppling behavior of simplified tree root prototypes. Monotonic toppling loads have been imposed at constant loading rate, for a fixed vertical load and zero horizontal load. Flat root systems have been modelled by combining different elementary structural components (a continuous wooden beam, wooden rods, rubber net), mimicking the different parts of a real tree, such as the base of the trunk, first and second order roots, respectively. First order roots were here assumed to be flexionally rigid, whilst second order roots to behave like a membrane, with infinite tensile stiffness and no bending stiffness. The influence of different root geometries and of the different granular materials employed to reproduce the foundation soil have been investigated, and the obtained toppling curves have been interpreted in terms of their limit toppling resistance and 70% secant stiffness.

For the considered prototype, independently on the soil characteristics, the geometry of the root system dominates the overall toppling resistance, with linear increasing trends with the number of the first order roots. In case second order roots are also added to the system, the limit toppling resistance significantly increases (up to a factor of two, for the geometry of the net considered in the present campaign). The effect of the wooden beam seems instead to be negligible in case that very deformable soil layers are considered. As far as the secant stiffness is concerned, a clear influence of the number of first order roots can again be observed (and a linear increasing trend with the limit values of toppling moment was suggested), but only in case that low to medium soil stiffness is considered. In case of high soil stiffness, on the contrary, beyond the experimental scatter, remarkably different trends may be observed.

Although a more precise geotechnical characterization of the employed granular materials is still required (specifically conceived at low confining stress), it can however be pointed out that soil properties largely affect the toppling behavior of the tested system (both in terms of limit resistance and overall stiffness of the toppling curve). Depending on soil properties, moreover, different (and even opposite) trends can be observed as far as the influence of the root geometry is concerned. It is Authors' opinion that, in view of applications to real trees, such considerations may contribute in provide more accurate interpretations techniques of the results of on-site pulling tests, being these latter usually limited to very small rotations of the tree, i.e. to a range where soil properties play a dominant role and the contribution of second order roots is not yet fully mobilized.

Acknowledgements. The authors thank the Laboratory of Analysis and Geotechnical Modelling (GeoT-LAM) at Polo Territoriale di Lecco of Politecnico di Milano. The support provided by Agro Services s.r.l. and the Scottish Research Partnership in Engineering (SRPe), through the Industry Doctorate Programme research grant SRPe-IDP/011, is also acknowledged.

References

Dupuy, L., Fourcaud, T., Stokes, A.: A numerical investigation into the influence of soil type and root architecture on tree anchorage. Plant Soil **278**, 119–134 (2005)

Guitard, D.G.E., Castera, P.: Experimental analysis and mechanical modelling of windinduced tree sways. In: Coutts, M.P., Grace, J. (eds.) Wind and trees 182–194 (1995)

Marsiglia, A., Ciantia, M., O., Galli, A., Canepa, D.: Vertical loading tests on a simplified tree root prototype. In: Proceedings of the X International Conference on Physical Modelling in Geotechnics, pp. 832–835. Korean Geotechnical Society, Seoul, Korea (2022)

Mickovski, S.B., van Beek, L.P.H.: Root morphology and effects on soil reinforcement and slope stability of young vetiver (Vetiveria zizanioides) plants grown in semi-arid climate. Plant Soil **324**, 43–56 (2009)

Sala, C.: Tree stability analysis: experimental pulling tests and analytical interpretation. PhD Executive program in Chemical, Geological and Environmental Sciences. Department of Earth and Environmental Sciences, Università degli Studi di Milano-Bicocca, Milano, Italy (2021)

Sani, L.: Statica delle strutture arboree, p. 945. Gifor (2017). ISBN-13: 979-1220016698

Schwarz, M., Lehmann, P., Or, D.: Quantifying lateral root reinforcement in steep slopes - from a bundle of roots to tree stands. Earth Surf Process Landforms **35**, 354–367 (2010)

Schwarz, M., Cohen, D., Or, D.: Root-soil mechanical interactions during pullout and failure of root bundles. J Geophys Res Earth Surf **115**, 1–19 (2010)

Wessolly, L., Erb, M.: Handbuch der Baumstatik und Baumkontrolle. Patzer, Berlin (1998)

Yang, M., Défossez, P., Danjon, F., Fourcaud, T.: Tree stability under wind: simulating uprooting with root breakage using a finite element method. Ann Bot **114**, 695–709 (2014)

Zhang, X., Knappet, J.A., Leung, A.K., Ciantia, M.O., Liang, T., Nicol, B.C.: Centrifuge modelling of root-soil interaction of laterally loaded trees under different loading conditions. Géotechnique (2022). https://doi.org/10.1680/jgeot.21.00088

Zhang, X., Knappett, J.A., Leung, A.K., Ciantia, M.O., Liang, T., Danjon, F.: Small-scale modelling of root-soil interaction of trees under lateral loads. Plant Soil **456**(1–2), 289–305 (2020). https://doi.org/10.1007/s11104-020-04636-8

Zhang, X.: Physical modelling of root-soil interaction of trees under lateral loads. PhD Thesis. Department of Civil Engineering, School of Science and Engineering University of Dundee, UK (2021)

Small Scale Experimental Tests and Simplified Modelling of Horizontal Loading Tests on Embedded Foundations

Andrea Galli[1](✉) and Giuseppe Mortara[2]

[1] DICA, Politecnico di Milano, Milan, Italy
andrea.galli@polimi.it
[2] DICEAM, University Mediterranea of Reggio Calabria, Reggio Calabria, Italy
giuseppe.mortara@unirc.it

Abstract. The paper presents some small-scale 1g experimental tests on a model foundation within a dry sand deposit at different embedment ratios. Horizontal loading paths at constant vertical load have in particular been considered, with specific reference to operational conditions characterized by relatively high values of the static factor of safety with respect to vertical bearing capacity (approximately from 2 to 10), as it often happens for common direct foundations in Civil Engineering applications. An objective definition of the failure points has been proposed, based on the monitoring of displacement increments. The tested conditions could represent a particularly critical condition, since marked coupling effects arise among the different loading components and evident non-linear and geometrical effects dominate the overall mechanical behaviour. The results have been comprehensively interpreted in the light of the macroelement theory by identifying the shape of a possible interaction domain for the system in the $V - H$ load space. A single mechanism plastic-strain hardening model has also been employed and a simple calibration strategy has been adopted. The model has also been validated on some tests, showing a good predicting accuracy.

Keywords: embedded foundations · bearing capacity · constitutive relations · model tests

1 Introduction

The research on the mechanical behaviour of direct foundations is a central topic of interest for geotechnical researchers and a large amount of the scientific literature investigated this field in the last fifty years (see e.g. [1–5]) by proposing comprehensive approaches capable of accounting complex loading conditions. Although there is a good agreement among scientists on the definition of the failure locus of such systems in the space of the generalized loading actions, several aspects of their mechanical response in operating conditions (owing to the highly non-linear behaviour of soils, to the development of irreversible strains well before the achievement of the failure condition, and to the marked path-dependency) are not yet fully understood. In this paper the results of a small

A. Ferrari et al. (Eds.): CNRIG 2023, SSGG, pp. 79–86, 2023.
https://doi.org/10.1007/978-3-031-34761-0_10

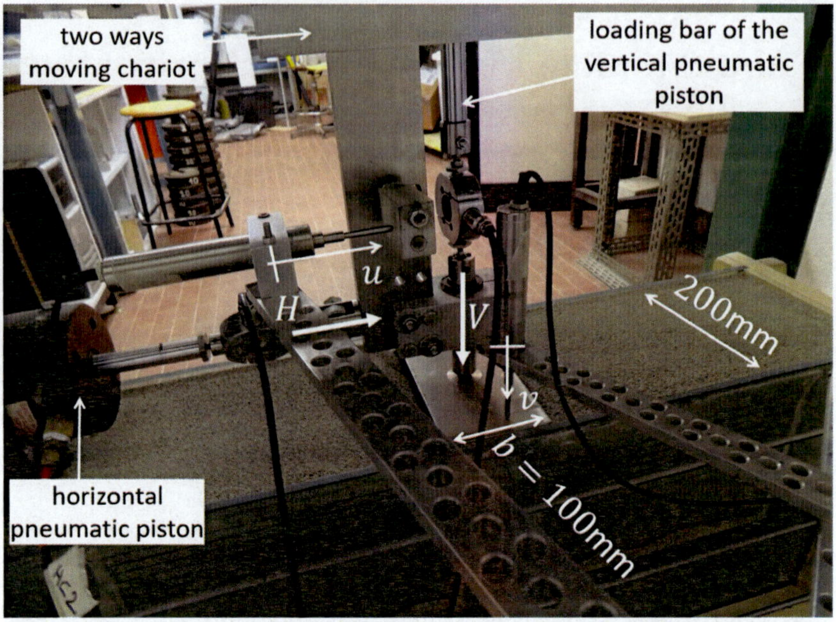

Fig. 1. Experimental setup.

scale 1g experimental campaign on a model foundation at different embedment ratios are presented. In particular, monotonic horizontal loading paths have been considered at constant vertical load. The tests allowed to investigate the progressive failure mechanisms developing at low vertical load (with respect to the vertical bearing capacity, i.e. a common operating condition for civil engineering structures) once a horizontal load is applied and an objective definition of the failure point has been provided.

2 Experimental Apparatus and Testing Programme

A model strip foundation was considered (width $b = 100$ mm, length $l = 198$ mm, and thickness $t = 20$ mm) consisting of a rigid steel plate. Plane strain testing conditions were created in a rigid box ($900 \times 400 \times 200$ mm), by pluviating a thickness $h = 300$ to 400 mm of Ticino sand. Figure 1 shows the experimental setup while a sketch of the system geometry is shown in Fig. 2. The model foundation was posed directly on the sand level. In case that embedded foundations were considered, a wooden box 100 mm thick was connected to the foundation before positioning it on the sand level, and an additional layer of sand of thickness d was deposited. Three final geometric conditions were considered with embedment values $d = 0$, 50 and 100 mm (corresponding to relative embedment ratios $d/b = 0$, 0.5, and 1, respectively). The foundation was connected to a bi-directional loading system, consisting in two pneumatic pistons along the vertical and horizontal directions. Two loading cells were used to measure the total applied components of the vertical, V, and horizontal, H, loads. Similarly, two

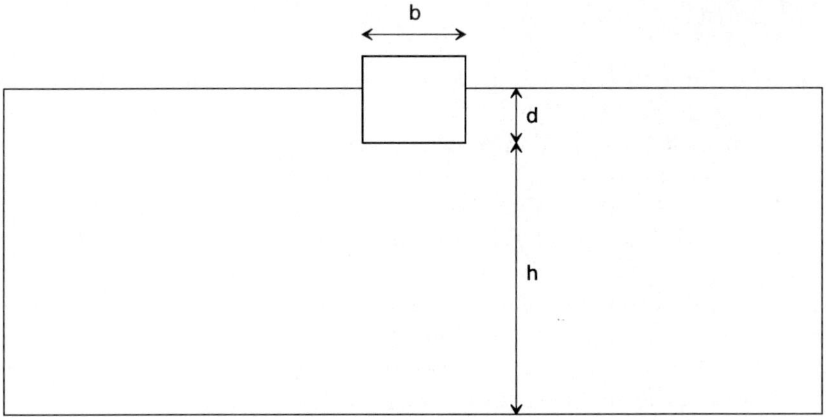

Fig. 2. Geometry of the system.

displacement transducers measured the corresponding horizontal, u, and vertical, v, displacements of the system. The rotation of the model foundations was prevented since the vertical loading system was mounted on a rigid two-ways moving chariot (additional details on the experimental setup and on the employed granular material can be found in [5]). The testing program, summarized in Table 1, consisted in a preliminary phase of vertical centred load at zero horizontal load, up to the prescribed operating conditions denoted by a vertical load V_i; then a horizontal load H was monotonically applied at constant vertical load until the failure of the system ($H = H_f$).

Table 1. Experimental program (lengths in m and loads in kN/m).

test	h	d/b	V_i	H_f	test	h	d/b	V_i	H_f
D1-1	0.4	0.0	1.14	0.49	D2-5	0.3	0.0	7.52	1.12
D1-2	0.4	0.0	1.39	0.59	D3-1	0.3	0.5	1.46	1.16
D1-3	0.4	0.0	2.32	0.67	D3-2	0.3	0.5	2.95	1.53
D1-4	0.4	0.0	2.93	0.77	D3-3	0.3	0.5	4.42	1.97
D1-5	0.4	0.0	3.49	0.92	D3-4	0.3	0.5	5.99	2.37
D1-6	0.4	0.0	4.39	0.99	D3-5	0.3	0.5	7.51	2.73
D1-7	0.4	0.0	4.68	0.99	D3-6	0.3	0.5	9.02	3.16
D1-8	0.4	0.0	5.00	1.13	D4-1	0.3	1.0	1.49	1.54
D1-9	0.4	0.0	5.88	0.99	D4-2	0.3	1.0	2.98	2.26
D2-1	0.3	0.0	1.46	0.57	D4-3	0.3	1.0	4.46	2.72
D2-2	0.3	0.0	2.96	0.81	D4-4	0.3	1.0	6.00	3.66
D2-3	0.3	0.0	4.45	1.08	D4-5	0.3	1.0	7.51	3.92
D2-4	0.3	0.0	6.04	1.24	D4-6	0.3	1.0	9.01	4.29

3 Interaction Diagrams

The determination of the interaction diagrams requires the identification of failure points. As the tests are performed by driving the horizontal force, the failure point is identified by a sudden increase in velocity of the horizontal displacement. This corresponds to the "first failure" of the system. Figure 3 shows, for an example test, the trend of displacements and horizontal force with time that makes evident the failure of the system although the horizontal force is still increasing owing to the geometrical effect caused by the sand accumulating on the front side (with respect to the applied load direction) of the foundation.

The values H_f of the horizontal load can be interpreted, when plotted against the failure values of V, as the locus of failure points by assuming the following generalization of the well known expression proposed in [1]

$$H_f = \mu\,(V - V_t)\left(1 - \frac{V - V_t}{V_m - V_t}\right)^{\beta} \tag{1}$$

where V_m and V_t represent the vertical bearing capacity and the uplift capacity, respectively. Parameters μ and β rule the shape of the interaction diagram. As suggested by [1] the value $\beta = 0.95$ will be assumed and thus the calibration will be performed only on μ, V_t, and V_m. Nevertheless, due to the fact that the maximum vertical force per unit length (for $b = 100$ mm) allowed by the loading system is about 10 kN/m, the vertical bearing capacity cannot be obtained and for this reason it will be computed as $V_m = q_{lim} b$ where q_{lim} is obtained from the classical bearing capacity formula for a cohesionless soil

$$q_{lim} = \frac{1}{2}\gamma_d b N_\gamma d_\gamma + q N_q d_q = \gamma_d b \left(\frac{1}{2} N_\gamma d_\gamma + \frac{d}{b} N_q d_q\right) \tag{2}$$

where $\gamma_d = 15.31\,kN/m^3$ is the dry unit weight of the sand for the selected value of relative density (60%). The bearing capacity factor N_q and N_γ can be found under different geometrical conditions (see e.g. [6,7] and references therein). For the simple conditions analysed here $N_q = (1 + \sin\phi) / (1 - \sin\phi) \exp\{\pi \tan\phi\}$ and $N_\gamma = 2\,(N_q - 1)\tan\phi$, being ϕ the friction angle that becomes a parameter

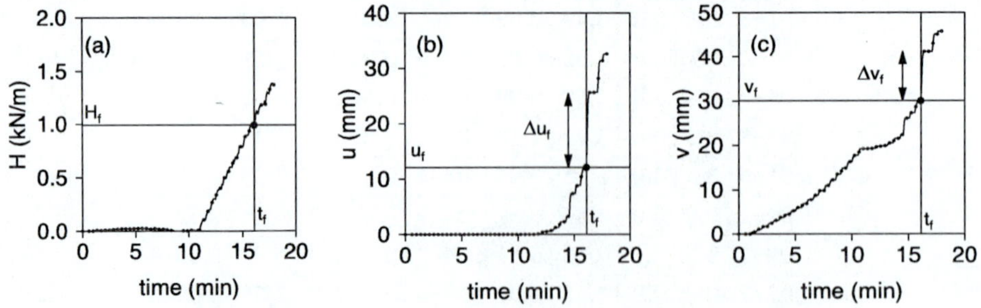

Fig. 3. Identification of the failure point for test D1-7.

Table 2. Parameters of the failure loci for all datasets.

		D1-D2	D3	D4
ϕ	(°)	42.2	42.2	42.2
μ	(-)	0.36	0.60	0.70
V_t	(kN/m)	0.00	−0.20	−0.60
V_m	(kN/m)	12.1	18.8	25.5

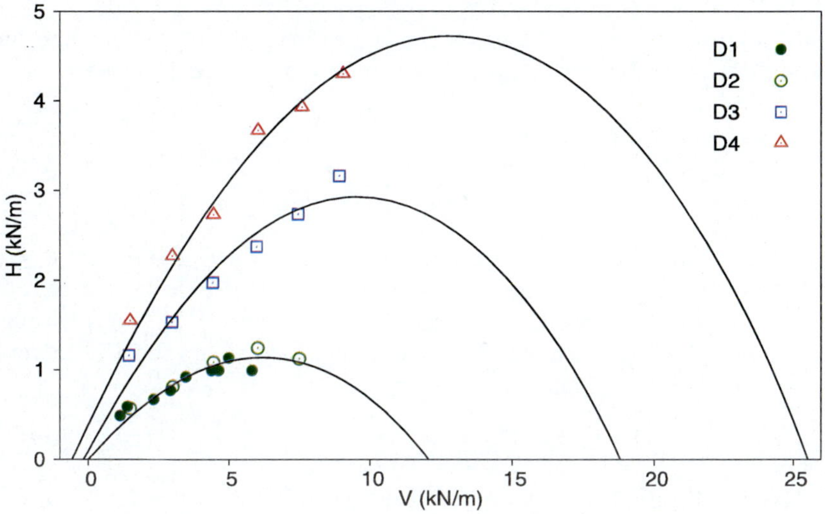

Fig. 4. Failure loci.

to be calibrated. The depth factors d_q and d_γ are set to 1. The determination of interaction diagrams starts from the analysis of failure points from datasets D1 and D2 (related to d/b=0) for which no uplift resistance is expected (V_t=0). A least square method on the values of μ and ϕ can then be applied comparing the experimental values of H_f with those predicted by (1). For the other two datasets (D3 and D4) the least square procedure is applied only on the values of V_t and μ as ϕ has been already obtained from the previous analysis. Table 2 reports the parameters of the three failure loci while Fig. 4 shows the comparison between experimental points and Eq. (1).

4 Simplified Modelling

The experimental results will be interpreted by a bidimensional hardening plastic model. Yield function f and plastic potential g have the same mathematical form of (1)

$$f = H - \mu_f \left(V - V_{tf}\right) \left(1 - R_f\right)^{\beta_f} \tag{3}$$

$$g = H - \mu_g \left(V - V_{tg}\right) \left(1 - R_g\right)^{\beta_g} \tag{4}$$

where

$$R_f = \frac{V - V_{tf}}{V_{0f} - V_{tf}} \qquad R_g = \frac{V - V_{tg}}{V_{0g} - V_{tg}} \tag{5}$$

The hardening law governs the evolution of variable V_{0f} and is represented by a simplification of a more complex evolution law [8]. The hardening law is postulated to be dependent on both the vertical and horizontal displacements

$$V_{0f} = V_{mf}\left[1 - \exp\left\{-\alpha_v\left(\frac{v}{s}\right)^{m_v} - \alpha_u\left(\frac{u}{s}\right)^{m_u}\right\}\right] \tag{6}$$

where s is a unit reference displacement (1 mm in this work). The experimental evolution of the hardening variable V_{0f} can be determined from (3):

$$V_{0f} = V_{tf} + \frac{V - V_{tf}}{1 - \left[\dfrac{H}{\mu_g\left(V - V_{tf}\right)}\right]^{1/\beta_f}} \tag{7}$$

Parameters α_v and m_v are calibrated from the first part of the test ($0 \leq V_{0f} = V \leq V_i$) by comparing the vertical force with Eq. (6). Parameters α_u and m_u are instead calibrated from the horizontal loading phase ($V_{0f} > V_i$) through the comparison of Eq. (6) with Eq. (7). The relationship between the incremental components of displacements is given by

$$\dot{v} = D\dot{u} \tag{8}$$

where the rate of dilation $D = \partial g/\partial V$ (since $\partial g/\partial H = 1$) is

$$D = \mu_g\left(1 - R_g\right)^{\beta_g - 1}\left[R_g\left(1 + \beta_g\right) - 1\right] \tag{9}$$

The previous equation requires the value of V_{0g} that is analougous to (7)

$$V_{0g} = V_{tg} + \frac{V - V_{tg}}{1 - \left[\dfrac{H}{\mu_g\left(V - V_{tg}\right)}\right]^{1/\beta_g}} \tag{10}$$

The plastic potential is completed by the assumption $V_{tg} = V_{tf}$. In the first part of the test the theoretical value of v can be obtained from Eq. (6) as

$$v = \left[\frac{1}{\alpha_v}\ln\left(\frac{V_{mf}}{V_{mf} - V_{0f}}\right)\right]^{1/m_v} \tag{11}$$

The second part must be performed in incremental form: computing the increments $\dot{H}$ using Eqs. (3) and (6), and considering (8), the increment of the vertical displacement is given by

$$\dot{v} = \frac{s\left(1 - R_f\right)^{1-\beta_f}}{\mu_f\beta_f R_f^2\left(V_{mf} - V_{0f}\right)\left[\alpha_v m_v\left(\dfrac{v}{s}\right)^{m_v - 1} + \alpha_u m_u\left(\dfrac{u}{s}\right)^{m_u - 1}\dfrac{1}{D}\right]}\dot{H} \tag{12}$$

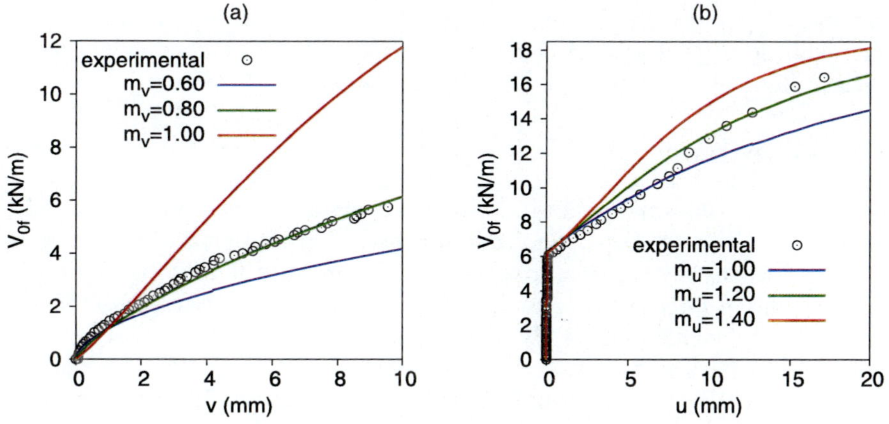

Fig. 5. Calibration of parameters m_v and m_u for test D3-4.

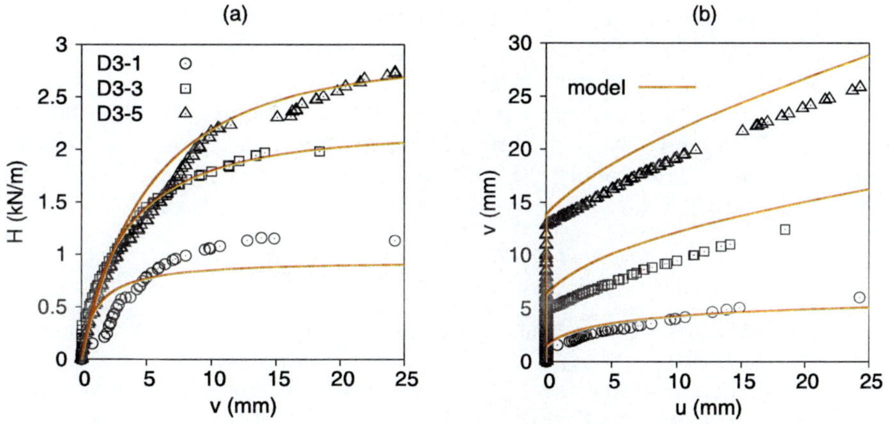

Fig. 6. Comparison between experiments and model for tests D3-1, D3-3, and D3-5.

while the increment of u is found with (8). As can be seen from the previous equations the model requires the calibration of 10 parameters.

The parameters of the model were determined for set D3. Parameters α_v, m_v, α_u and m_u are determined by comparing experimental and model values of variable V_{0f}. In particular, Fig. 5(a) shows the effect of parameter m_v on the quality of simulations for the first part of the test while Fig. 5(b) shows the effect of parameter m_u on the second part of the simulations. As $\beta_f = \beta_g = 0.950$, the last parameter to be calibrated is μ_g that takes the value 1.100. Summarizing, the values of parameters are $V_{tf} = -0.20$ kN/m, $V_{mf} = 18.80$ kN/m, $\mu_f = 0.596$, $\beta_f = 0.950$, $\alpha_v = 0.062$, $m_v = 0.800$, $\alpha_u = 0.040$, $m_u = 1.200$, $\mu_g = 1.100$, and $\beta_g = 0.95$. From Fig. 6, that shows the comparison between experimental data and model predictions for tests D3-1, D3-3, and D3-5, it can be observed that a satisfactory agreement is obtained. Similar results can be obtained, with the same calibration strategy, also for the other two tested embedment ratios, as shown in [9].

5 Conclusions

The paper describes the results of an experimental campaign of 1g laboratory tests on a small-scale strip foundation prototype. In particular, horizontal loading paths at constant vertical load (and with zero rotation of the model) have been considered. The tests were specifically conceived to be representative of typical working conditions for civil engineering application, characterized by relatively high values of the safety factors with respect to applied vertical loads (values 2 to 10 have been tested) and took into account different embedment ratios. Marked non-linear and path dependent mechanical responses have been observed in the tests with the development of progressive failure mechanisms owing to important geometrical effects. In the paper, failure points have been then identified upon a careful interpretation of the model displacement records. The results have been interpreted in the light of the well-known macroelement theory with specific extensions to account for the effect of the embedment ratio, and a single plastic model with non associate strain hardening law has been formulated. The model, once calibrated upon a series of tests, proved to be relatively accurate in reproducing the data and to have good predicting capabilities with respect to other test loading paths and embedment ratios.

References

1. Nova, R., Montrasio, L.: Settlements of shallow foundations on sand. Géotechnique **41**, 243–256 (1991)
2. Butterfield, R., Gottardi, G.: A complete three-dimensional failure envelope for shallow footings on sand. Géotechnique **44**, 181–184 (1994)
3. di Prisco, C., Nova, R., Sibilia, A.: Shallow footing under cyclic loading: experimental behaviour and constitutive modelling. In: Maugeri M., Nova R. (eds.) Geotechnical Analysis of the Seismic Vulnerability of Historical Monuments, Pàtron, Bologna, pp. 99–122 (2003)
4. Gourvenec, S.: Shape effects on the capacity of rectangular footings under general loading. Géotechnique **57**, 637–646 (2007)
5. Galli, A., Farshchi, I., Caruso, M.: Influence of loading path on cyclic mechanical response of small-scale shallow strip footing on loose sand. Can. Geotech. J. **52**, 1228–1240 (2015)
6. Mortara, G.: An exercise on the bearing capacity of strip foundations. Géotech. Lett. **10**, 141–148 (2020)
7. Mortara, G.: Limit analysis solutions for the bearing capacity of strip foundations under seismic conditions. Géotechnique. https://doi.org/10.1680/jgeot.21.00150
8. Mortara, G.: A simple model for sand-structure interface behaviour. Geotech. Eng. **174**, 33–43 (2021)
9. Galli, A., Mortara, G.: Experiments and modelling of horizontal loading tests on small scale foundations in sand at constant vertical load. Géotech. Lett. **11**, 171–178 (2021)

Physical Modelling of Backward Erosion Piping for the Development of Natural-Based Mitigation Strategies

Carmine Gerardo Gragnano(✉), Federico Camiletti, Federica Forbicini, Guido Gottardi, Michela Marchi, and Laura Tonni

Department of Civil, Chemical, Environmental and Materials Engineering, University of Bologna, Bologna, Italy
Carmine.gragnano2@unibo.it

Abstract. River embankments are designed to face hydraulic actions produced by river level fluctuations. Breach occurrence on these earth structures could be driven by various factors, all possibly concurring in the collapse during an extreme event. Among the possible mechanisms that may lead the structure to failure, the onset and progression of erosion phenomena are certainly relevant for the performance of the embankment. When erosion occurs in a sandy aquifer underlying a water-retaining structure during a sufficiently high and persistent hydrometric peak, and an outflow area on the landside is close enough to the river, backward erosion piping (BEP) could produce the removal and transport of sand particles from the aquifer. As pipes start to be created, the safety conditions of the river embankment are threatened. The experimental strategies reproducing this phenomenon at the laboratory scale are already rather established. A newly designed physical modelling approach is currently under development at the Bologna University geotechnical laboratory. The main aim of the model is to investigate the effect of innovative and sustainable countermeasures against BEP. The experimental activities described in this paper rely, in particular, on the use of a small-scale model for the development of a suitable procedure to effectively reproduce BEP. The design schemes and the preliminary results so far obtained are presented to validate the devised physical modelling strategy and the reliability of the adopted monitoring tools.

Keywords: River Embankment · Backward Erosion Piping · Physical Modelling · Hydraulic Gradient · Sand Boil

1 Introduction

The processes that involve the detachment and transport by seepage flow of soil particles are generally classified as internal erosive and can ultimately lead to the instability of water retaining embankments if not timely tackled. According to the ICOLD [1], internal erosion may arise as soon as (1) the hydraulic shear stresses are larger than the resistant contact forces between particles, (2) the water flow is sufficient to carry

A. Ferrari et al. (Eds.): CNRIG 2023, SSGG, pp. 87–94, 2023.
https://doi.org/10.1007/978-3-031-34761-0_11

the eroded particles and (3) the spaces within the porous medium are large enough for the detached particles to pass through. Among other mechanisms (concentrated leak erosion, suffusion and contact erosion), internal erosion may occur in the form of backward erosion piping (BEP), which strongly affects the stability of many earth retaining structures. This phenomenon is widespread in Italy [2–4]. In particular, it has been demonstrated to be a critical failure mechanism for Po River embankments [5]. BEP occurs at the contact between a shallow sandy aquifer and the impermeable blanket above it (the foundation of the embankment, generally made of fine-grained soil); during its progress, shallow pipes are formed in the direction opposite to the flow underneath water-retaining structures by the gradual removal of sandy material by the action of water flowing from the river to the groundwater level landside. The continuous flow due to repeated hydrometric levels may cause the pipe to widen and deepen to such an extent that the water-retaining structure becomes unstable. However, BEP can be easily identified since the process is accompanied by the formation, at the ground surface, of "sand boils", located at the exit points of the formed pipes. Being a significant hazard in the risk assessment of river embankments, BEP has been extensively investigated in the laboratory using different setups [6]. In the last decades, the most used types of testing equipment are basically rectangular boxes filled with sand subjected to a horizontal hydraulic gradient until piping occurs; usually, a horizontal cover (made of cohesive and impermeable material or transparent acrylic) is placed on top of the sand sample to confine the sand layer and act as a stable roof for the pipes, while various exit types may be investigated (area, slope, ditch, hole). In the present research project, an experimental study has been set up to test a natural-based solution to mitigate flood risk due to sand boils reactivations along the Po River (Italy).

To this scope, the physical modelling activities preliminarily focused on the replication of the BEP at the laboratory environment. Several tests have been performed on a small-scale laboratory apparatus, providing clear evidence of the occurrence of the erosion process, insights into the devised experimental procedures and the effectiveness of the proposed mitigation strategy. Part of these tests are presented in this paper and their results have been used to design a larger apparatus, currently under development.

2 Backward Erosion Experiment

2.1 Experimental Set-Up

The small-scale apparatus consists of a rectangular box having plane dimensions $B = 35\text{cm}$ and $L = 50\text{cm}$ and a height $H = 50\text{cm}$. The inlet area is constituted by a flow distribution chamber with a filter to separate water and sand, placed at the upstream seepage section of the model, and connected to an external reservoir, which provides the hydraulic load; the box has a 5 cm thick acrylate cover with a 1.2 cm diameter exit hole located 35 cm downstream with respect to the distribution chamber and producing a flow concentration toward the exit. Above the hole, a cylinder has been designed to collect the eroded sand and to define the hydraulic outflow condition (i.e., constant head). In the center line and along a transversal axis of the top, 9 cm spaced pressure ports are used to monitor the hydraulic head, using three PPTs (pore pressure transducers).

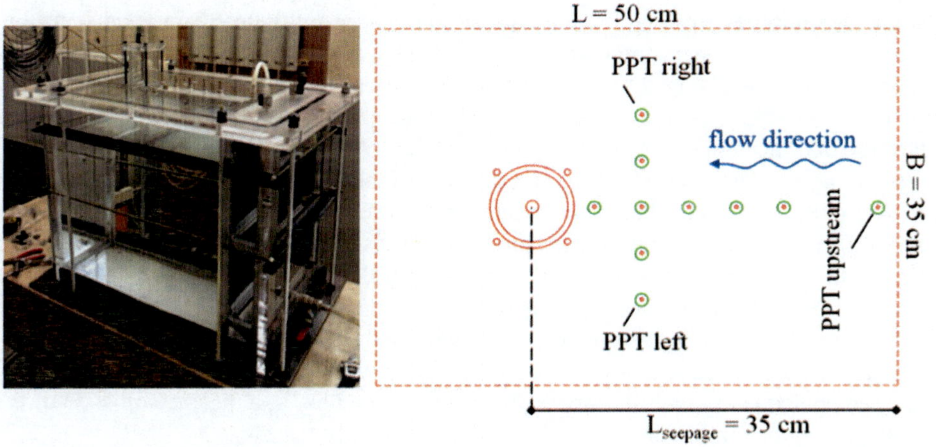

Fig. 1. Picture of the laboratory apparatus for BEP testing (left) and scheme of the top (right).

The influence of the ports on progression is expected to be negligible, given their limited cross-section ($\approx$7 mm^2) and considering that they do not deepen into the sand. The acrylic box has been reinforced by means of an external steel frame. In Fig. 1 are presented a picture of the box (left) and a sketch of the top (right).

2.2 Sample Preparation and Test Procedure

All the experiments have been performed on poorly graded silica sand of the Padano aquifer, characterized by $D_{50} = 0.34$mm, $C_u = 2.1$ and $C_c = 1.2$. The sample permeability varies from 1 to $5 \cdot 10^{-4}$ m/s, as a function of the relative density obtained for the sample.

The minimum and maximum dry unit weights of the soil have been determined according to ASTM [7, 8] that provides $\gamma_{d,min} = 13.7$kN/m^3 and $\gamma_{d,max} = 16.2$kN/m^3, while the specific gravity G_s is 2.680. The soil sample has been prepared with pluviation of dry sand inside the box, gradually filled with water. This technique simulates a natural deposition process and results in optimal saturation of the sample. Being the investigated sand uniform and ensuring that the sand will deposit under 10 cm of water, possible particle sorting effects are limited. The full height of the model is achieved by a series of oven-dried sand throws, about 10kg each, forming a sequence of layers 5 cm thick. The described procedure enables the creation of samples characterized by a relative density of about 20%, which is the degree of packing adopted in the tests described herein. Tests with denser samples, which can be easily prepared by means of external vibration, are currently underway. Once the sample preparation is complete, the acrylic cover is placed on top of the sample. The test procedure consists of applying a horizontal hydraulic gradient to the sand sample. Then the BEP is investigated by means of visual observation of the phenomenon through the transparent cover and the simultaneous acquisition of hydraulic heads and discharge. The head fall (Δh) across the sample is increased in small steps, by raising (with sudden or gradual variation) the upstream water level, while the downstream head is kept constant. The applied hydraulic load is kept constant during a single step, as long as erosion takes place. When erosion stops (after about five minutes from the hydraulic load application), the upstream head is further increased. In every step, the load is increased by 2 cm. The flow rate is measured by progressively weighting the water leaving the model using load cells (HBM RSCC, accuracy 0.048kg). At the same time, the pressure head is measured by PPT (Druck UNIK 5000, accuracy ± 5mm, frequency 0.2Hz). A camera, placed above the setup, records the erosion process, providing a top view of the sand sample. Video frames (1 fps, 3264x2448) and graphical notes collected during the experiments are used to describe the evolution of the erosion process (i.e., pipe geometries). In this paper, the results obtained from two tests are presented: in the first case (a), BEP was free to develop along the contact surface between the aquifer and cover; in the second case (b), a filter barrier (a geocomposite strip, about 2 cm deep, obtained by thermobonding a three-dimensional drainage core of extruded monofilaments between two filtering non-woven geotextiles) has been placed along a full transversal section of the box, at a distance of ≈1/3 of the seepage length from the inlet.

2.3 Backward Erosion Piping Stages

Observations of the backward erosion process in laboratory environments show the existence of three main phases [9]: boiling (initiation of the process), regressive backward erosion (sand volcanoes are formed) and progression (pipe development). These phases have been identified (Fig. 2) in all the experimental tests carried out using the apparatus described in Sect. 2.1 and can be described as follows:

Boiling phase (initiation): the initiation of the backward erosion piping in experiments corresponds to the fluidization of the sand bed in the exit hole, where soil particles continuously rise and fall as in a small boiling, as the effect of local gradients (Fig. 2a). Fluidization occurs when the seepage pressure matches the weight of the soil particles and can only take place at certain flow velocities, that change according to the type of exit point. This phase may start at relatively low average gradients, i_{avg} (0.05 to 0.1 in the field).

Regressive or equilibrium phase (sand volcanoes are formed): increasing the hydraulic fall, the sand particles are carried outside the exit point forming a "sand volcano" of increasing size (Fig. 2b). Shallow pipes a few millimeters high are initially formed (Fig. 2c). Eroded particles expose the others to seepage forces and detachment, resulting in the formation of one or more channels growing upstream. This process is named primary erosion, while secondary erosion is determined by the flow velocity in the pipe and causes particle transport within the channel walls and bed [10]. Keeping constant the imposed head fall, the pipe tips may eventually stop progressing and their lengths remain fairly constant. This is referred to as the equilibrium stage.

Progression. In case the equilibrium is not reached, the pipe progression does not stop and develops in a backward direction at a constant head. A main pipe tends to prevail and eventually reaches the inlet section (Fig. 2d), without the need for a further load increment. The experiment is generally stopped at this stage.

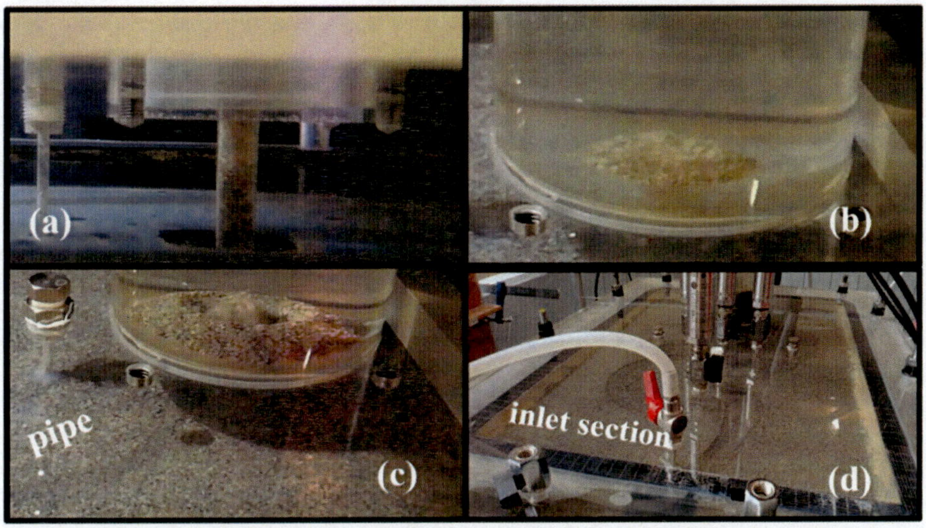

Fig. 2. Picture of the backward piping stages: (a) initiation with boiling in the exit hole, (b-c) formation of the sand volcano during a regressive phase, (c) detail of an erosion pipe and (d) progression phase, with the main pipe reaching the inlet section.

3 Experimental Observations

The results of two different tests are shown in Fig. 3: the first (a) without the barrier and the second (b) with a vertical barrier. Sample preparation in both cases was carried out excluding vibration. In Fig. 3, water discharge and pressure head are plotted as a function of the hydraulic loading steps (Δh). All values are averaged over 1-min intervals. The descriptions of BEP stages are also indicated on the diagram. The first loading step ($\Delta h = 2$ cm) generates piping initiation (boiling phase) in both tests. The corresponding average hydraulic gradient (i_{avg}) is equal to 0.06. No remarkable transport of sand particles outside the vertical exit pipe is observed in this stage. A sand volcano is formed soon after the beginning of the second step. As soon as the erosion process takes place, the pressure head decreases when a pipe grows toward a monitoring point. This trend can be clearly observed in test (a) around the 30th minute when a pipe tip underpasses the right PPT (see also the scheme of pipe development in Fig. 4). During tip progression, the channel tends to meander and sometimes widen, intermittently, with a gradual rise in the water discharge through the exit as the eroded volume increases. For test (a), the progressive stage starts at the beginning of the third load step, with an $i_{avg} = 0.17$. The presence of the barrier (test b) causes i_{avg} to increase to 0.34. In both tests, the higher the erosion, the higher the water discharge.

As a further effect, the barrier results in a significant drop of the pressure head as revealed by PPT located upstream and downstream with respect to the barrier itself. A similar response has been observed by various Authors [11] and could be attributed to a change in the model permeability due to the introduction of the barrier. Consistently, the water discharge measured during test (b) is higher than that in test (a), since the early steps of loading. A further comparison between the two tests can be made with respect to the evolution of the erosion patterns. The pictures, taken from the top view of the cover, are analyzed with respect to a reference system in which the box exit hole represents the origin of the axis and pipe contours are drawn, following their progress (Fig. 4). The geometry of the pipes (i.e., direction, length and pipe meandering) can be drawn with respect to the time intervals with respect to the different stages of the process.

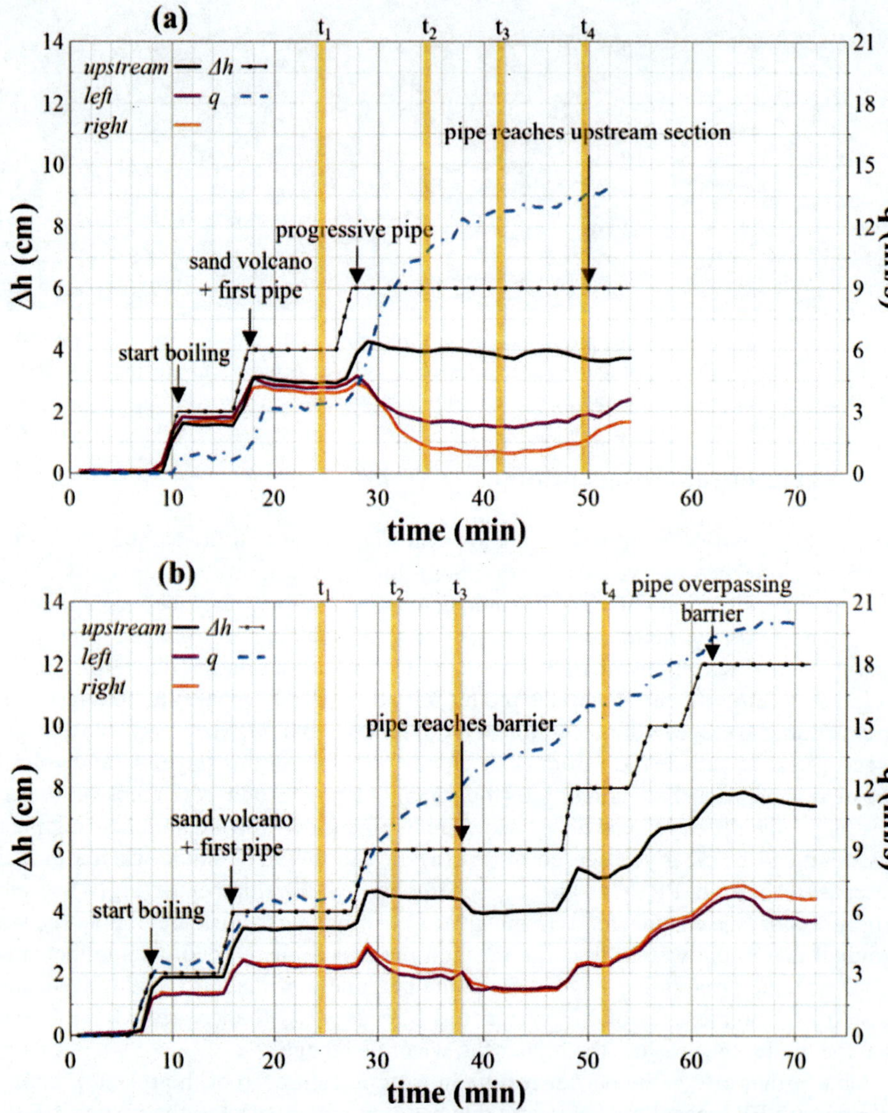

Fig. 3. Monitoring data (water discharge and pressure head at the contact between the cover and the soil) collected during two tests performed (a) without and (b) with barrier.

In Fig. 4, the erosion patterns of the experiment performed (a) without and (b) with the barrier are compared. The geometries of the pipes corresponding to the different time steps (intervals from t_1 to t_4 in Fig. 3) are distinguished using colors. The boundaries between the time intervals are specified in Fig. 3 with vertical yellow lines. The maximum pipe length reached at t_1 (during the equilibrium stage) is slightly higher for test (b) ($\approx$10cm) than that obtained in test (a) ($\approx$8cm). A higher water discharge is associated with the observed higher extension of the pipes. Without a barrier, the pipe meandering appears to be more extensive than that developed with the barrier, with a consequent more uniform distribution of the pipes in case (a). For the two tests, the barrier position ($\approx$22cm from the exit section) is reached roughly at the same time ($\approx$35min) and within the same step of loading (n°3). When the main pipe touches the barrier, it tends to become wider and branches out in a direction parallel to the barrier itself. In the sixth load step, the pipes overpass the barrier at the contact between the barrier and the wall of the box.

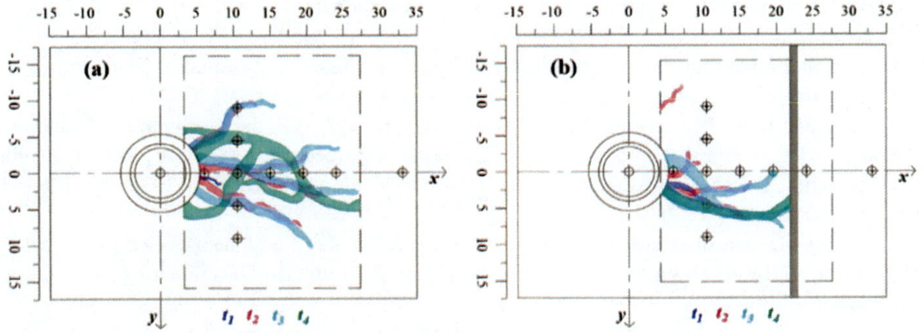

Fig. 4. Pipe progression during the tests performed (a) without and (b) with barrier.

4 Conclusions and Future Studies

The paper describes the preliminary outcomes of an experimental study carried out with the aim of assessing, through physical modeling, the effectiveness of a mitigation strategy on the BEP processes. The investigated mitigation strategy consists of a vertical barrier devised to stop the progression of erosion pipes generated during this phenomenon. During the tests, monitoring data are collected to support experimental evidence and to analyze the effect of the barrier: hydraulic heads, water discharge and pipes geometry are recorded as a function of the different erosion phases. In this paper, two tests are compared: one with and the other without the barrier. Results clearly show that a gradual increase in the hydraulic load generates widespread meanders in the absence of a barrier, evidencing a somewhat uniform distribution of pipes over the entire model width. On the contrary, with the barrier, a predominant pipe is observed from the initial stages of the test. The value of i_{avg} leading to failure is doubled when the mitigation strategy is introduced.

In the next future, tests will be devoted to investigating the process through the application of smaller load increments ($\Delta h \leq 1$cm). In addition, the effects of the barrier on the development of the hydraulic heads, discharge and critical gradients will be further explored as a function of different geometry, manufacturing and position of the barrier. The laboratory activities presented in this paper are part of a larger experimental program, which includes also laboratory medium scale tests and full-scale field tests, in order to overcome potential scale effects that are worthy of being investigated.

Acknowledgments. This research was funded under the scheme for LIFE programme of the European Union, Grant Agreement no. LIFE19 ENV/IT/000071, research project LIFE Sand Boil.

References

1. ICOLD. Internal erosion of existing dams, levees, and dikes, and their foundations. Bulletin 164, International Commission on Large Dams, Paris (2017)
2. García Martínez, M.F., Tonni, L., Marchi, M., Tozzi, S., Gottardi, G.: A numerical tool for the prediction of sand boil reactivations near river embankments. Journal of Geotechnical and Geoenvironmental Engineering **146**, e02380 (2020)
3. García Martínez, M.F., Gottardi, G., Marchi, M., Tonni, L.: On the reactivation of sand boils near the Po River major embankment. In: Calvetti, F., Cotecchia, F., Galli, A., Jommi, C. (eds.) Geotechnical Research for Land Protection and Development CNRIG 2019, pp. 328–337. Springer, Cham (2020)
4. Marchi, M., Martínez, M. F. G., Gottardi, G., Tonni, L.: Field measurements on a large natural sand boil along the river Po, Italy. Quarterly Journal of Engineering Geology and Hydrogeology 54(4), (2021)
5. Gottardi, G., Marchi, M. and Tonni, L.: Static stability of Po river banks on a wide area. In: Geotechnical Engineering for Infrastructure and Development XVI European Conference on Soil Mechanics and Geotechnical Engineering (ECSMGE 2015), pp. 1675–1680 (2015)
6. Robbins, A.B., van Beek, V.M.: Backward Erosion Piping: A Historical Review and Discussion of Influential Factors. In: Association of State Dam Safety Officials - ASDSO - Conferences and Seminars on Dam Safety, New Orleans, Louisiana (2015)
7. ASTM Designation: 4254–00. Test Method for Minimum Index Density and Unit Weight of Soils and Calculation of Relative Density (2017)
8. ASTM Designation: 4253–16. Test Methods for Maximum Index Density and Unit Weight of Soils Using a Vibratory Table (2019)
9. van Beek, V.M., Bezuijen, A., Sellmeijer, H.: Backward Erosion Piping. In: Bonelli, S. (eds) Erosion in Geomechanics Applied to Dams and Levees, pp. 193–269, Wiley, London (2013)
10. Hanses, U., Mueller-Kirchenbauer, H., Savidis, S.: Mechanics of Backward Erosion Under Dikes and Dams. Bautechnik **62**(5), 163–168 (1985)
11. Rosenbrand, E., et al.: Multi-scale experiments for a coarse sand barrier against backward erosion piping. Géotechnique **72**(3), 216–226 (2022)

A Prototype to Measure Water Content in Pyroclastic Soil Covers

Simona Guglielmi[1(✉)], Marianna Pirone[2], Nicola Amatucci[2], Umberto Cesaro[3], Mauro D'Arco[3], and Gianfranco Urciuoli[2]

[1] Formerly Scuola Superiore Meridionale, Largo S. Marcellino, 10, 80138 Napoli, Italy
simona.guglielmi@unina.it

[2] Università degli studi di Napoli Federico II, Dipartimento di Ingegneria Civile, Edile e Ambientale, Via Claudio, 21, 80125 Napoli, Italy

[3] Università degli studi di Napoli Federico II, Dipartimento di Ingegneria Elettrica e delle Tecnologie dell'Informazione, Via Claudio, 21, 80125 Napoli, Italy

Abstract. The research deals with the design of a new device for soil water content measurements in pyroclastic soil covers. Being the latter notoriously location of rainfall-induced flow-slides and debris-flows, the research target is to enhance the monitoring of factors predisposing to failure for their integration in early warning systems. The experimental activity performed on a prototype geometry sensor will be presented.

Electric impedance spectroscopy is used herein to apply an alternate signal over a large frequency range to a soil specimen between two electrodes. As the soil impedance depends on the soil humidity, an on-purpose experimental programme has been performed using the prototype geometry sensor, in order to obtain the calibration function relating the measured impedance to the soil water content. Results on the prototype geometry show a unique monotonic correlation between the impedance modulus and the water content with remarkable repeatability performances, confirming the efficacy and reliability of the technique for the geotechnical application under study.

The novel device appears to be a promising alternative to traditional water content field measuring instruments, with the advantage of satisfying affordability and versatility requirements.

Keywords: Water content measurements · Field monitoring · Pyroclastic slope

1 Introduction

Pyroclastic soil covers are well known to be partially saturated throughout the year and to be commonly affected by disastrous flow-like landslides triggered by heavy rainfall events (e.g. [1–3]).

In this case, field monitoring of factors predisposing landslide activation, such as suction and/or water content into subsoil, is crucial for the development of early warning systems intended as risk mitigation strategies.

A. Ferrari et al. (Eds.): CNRIG 2023, SSGG, pp. 95–102, 2023.
https://doi.org/10.1007/978-3-031-34761-0_12

Within this framework, the paper discusses some preliminary results of an experimental plan devoted to the development and optimization of a prototype system for field water content measurements, based on impedance spectroscopy. With respect to most common techniques as Time Domain Reflectometry, TDR, and capacitance techniques, CT [4], the system proposed here results cheaper because of the electronic components involved and it works at multiple low frequencies from 500 Hz to 50 kHz.

2 Fundamentals of Impedance Spectroscopy

Impedance spectroscopy is commonly used to characterize electrical properties of materials through the application of an electrical stimulus (a known voltage or current). In detail, the technique measures the complex impedance, Z(f), representing the opposition of the medium to the alternate electric current passing through it and depends on frequency, f [5]. Conventional impedance spectroscopy consists of measuring Z as a function of f over a wide frequency range. The resulting structure of the Z(f) vs. f response allows inferring the electrical properties of the electrode–material system.

Z(f) generalizes Ohm's law to circuits in sinusoidal regime and it is expressed in rectangular coordinates as:

$$\frac{V(f)}{I(f)} = Z(f) = Z' + jZ'' \quad (1)$$

$$Re(Z) = Z' = |Z|cos\theta; \; Im(Z) = Z'' = |Z|sin\theta \quad (2)$$

where: Z' is the real part of impedance, associated to the dissipation process and representing the resistive component (resistance); Z" is the imaginary part, corresponding to charge accumulation and representing the capacitive component (reactance).

In polar coordinates, phase angle and modulus equal, respectively:

$$\theta = tan^{-1}\left(\frac{Z''}{Z'}\right) \quad (3)$$

$$|Z| = \sqrt{(Z')^2 + (Z'')^2} \quad (4)$$

3 Laboratory Testing

3.1 Materials and Methods

The material used to carry out the experimental activities is a pyroclastic soil from the site of Nocera Inferiore (SA) in the Lattari Mounts (Fig. 1a), where the typical stratigraphic sequence is made of a maximum 2 m-thick pyroclastic cover resting on a carbonate bedrock (Fig. 1b). The sequence includes the deep layer C, made of finer deposits of the ancient Phlegraen pyroclastic fall, resting directly on the bedrock, underlaying coarse pumices, layer B, and recent top soils, layer A, deposited after the Somma Vesuvius eruption of 79 d.C. [6, 7]. The sampled soil belongs to level A and is mainly sandy (SF = 49%; Fig. 1c), with a significant amount of gravel (GF = 31%) and silt (MF = 18%). The tested soil is non plastic and has: porosity n = 0,6–0,7; particle unit weight γ_s = 25,1 kN/m^3; dry unit weight γ_d = 10,09 kN/m^3.

The soil specimens were prepared layer by layer through dried soil air pluviation [8, 9] to reproduce the field porosity (around n = 0,6; [7]), inside jars of shape and size compatible with the sensor geometry (Fig. 2). After complete oven-drying of the material, a pyroclastic fall was generated by hand with a filling funnel from a distance of 20 cm to the current layer surface. The dried specimens were then wetted using a sprinkler-like system (Fig. 2a) to allow homogeneous distribution of water and preserve the specimen from volumetric collapse.

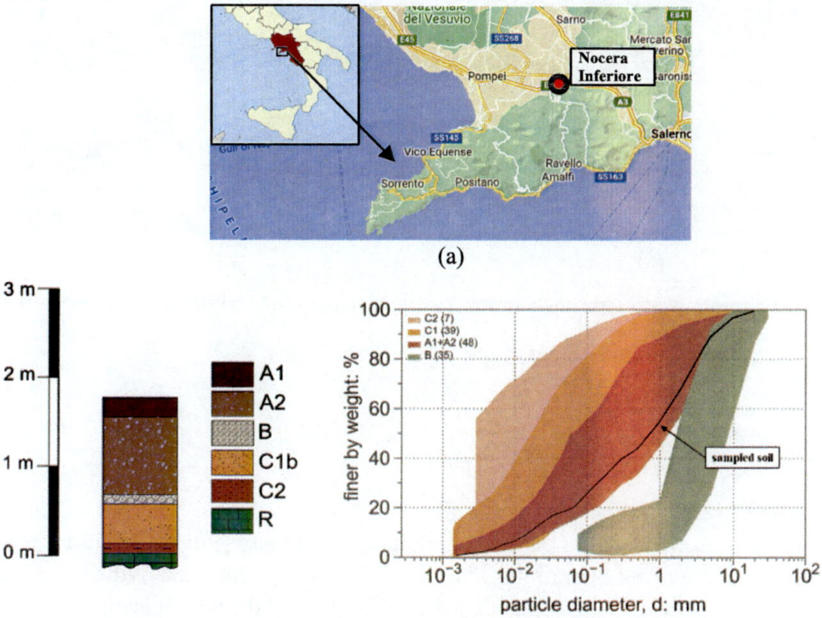

Fig. 1. (a) Location of the sampling site of Nocera Inferiore; (b) stratigraphic sequence and (c) envelopes of grading curves for the different soil layers recognized at the sampling site (in parentheses the number of samples analysed; after [7]); in black the grading curve of the sampled soil.

Following this procedure, specimens were prepared in triplets (Fig. 2b) in order to reproduce each water content on three specimens and investigate the repeatability of the measurements. Four triplets were wetted to obtain the water contents: w = 21%, 31%, 42% and 57%, corresponding to saturation degrees: S_r = 36%, 54%, 71% and 96%, respectively. Another on-purpose triplet was instead prepared for stepwise wetting, with the aim to assess the soil response at the same water contents mentioned above and to add unexplored water content values. Preliminary impedance analyses carried out on this triplet at progressive wetting from w = 16% to w = 26%, including two additional water contents, will be reported in Sect. 4.

Impedance analyses were always performed at different time intervals from wetting, up to the reach of a steady impedance measurement, as will be discussed in more detail in Sect. 4. After wetting and after every metering operation, the specimens were sealed and weighted as to minimize and control evaporation losses.

(a) (b)

Fig. 2. (a) Sample wetting through sprinker-like system and (b) triplets of specimens wetted at different water contents.

3.2 Prototype System

The prototype sensor is made of two parallel stainless steel plates (10 × 6 × 0,3 cm) acting as conducting electrodes, lodged at 4 cm distance on a hard polymeric support. During the impedance metering operations, the two electrodes are inserted in the specimen and the sensor is connected to the impedance measuring circuit, which is equipped with an on-board signal generator and an analog-to-digital converter (Analog device: AD5943, 2017; [10]). Then the impedance spectroscopy analysis is carried out by means of an external computer interfaced via USB.

Through the signal generator, the impedance meter applies to the electrodes within the specimen an alternate voltage, whose frequency is linearly increased over the range [500 Hz - 50 kHz]. This interval allows minimizing environmental interference affecting impedance analyses for higher frequency values. The application of a voltage between the electrodes causes a current intensity across the soil specimen, which is affected by the soil impedance. The current stemming across the specimen is then transduced, digitalized and processed by the impedance meter to obtain the impedance modulus and phase angle.

The performance of the prototype system has been tested by means of an experimental setup made of a decade resistor and a sample capacitor assembled in series. A set of nominal values of impedance has been obtained as a combination of the resistance and the capacitance provided by the two components. Then, the impedance measured through the prototype system has been compared to the nominal assigned one at discrete frequency values. Figure 3 illustrates the results for frequency values set at 1 kHz, 10 kHz and 20 kHz, in terms of both measured against nominal impedance modulus (Fig. 3a) and measuring error (Fig. 3b).

The procedure has shown that the prototype system has a linear response and is satisfactorily accurate, since the deviation between measured and nominal impedance values exceeds ± 2% only at high frequency when the nominal impedance is larger than 70 kΩ.

Further testing of the prototype measuring system was developed by comparing its impedance measurements on specimens at different water contents to those of benchtop impedance meters, conventionally assumed as reference devices in electrical engineering practice. These latter generally operate at discrete frequency, which was set equal to 1 kHz.

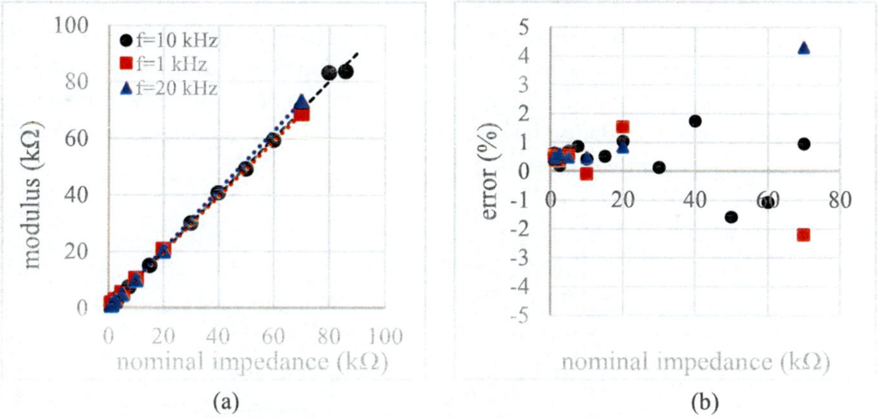

Fig. 3. (a) Linear response and (b) error of the prototype system against an experimental setup (decade resistor and sample capacitor in series) for frequencies f = 1 kHz, f = 10 kHz, f = 20 kHz.

4 Preliminary Results

Some preliminary results on the triplets of specimens shown in Fig. 2 will be presented herein. For each specimen, the impedance measurements were repeated at different time intervals after wetting (2h, 4h, 8h, 24h, 48h, 3, 7, 21 days) in order to allow homogeneous distribution of humidity in the specimen and the reach of steady conditions. However, it is worth specifying that results of impedance analyses are yielded immediately, being the metering technique electrical. Hermetic seal of the holding jars was controlled by recurrent weighting of the specimens.

Figure 4 illustrates the impedance modulus over the whole operational frequency range for the triplets of specimens wetted at w = 21%, 31%, 42% and 57%.

For all specimens, the impedance modulus is seen to increase fast at very low frequencies (below 1 kHz) and then assume a constant frequency-independent value up to very large frequencies. The impedance analyses for specimens of the same triplet (i.e., at the same water content) provide very similar results, except for the triplet at the lowest water content (w = 21%), which is probably more affected by differences in porosity or micro-structure upon limited wetting.

It is also shown that as the water content w increases, the impedance modulus decreases, as expected due to the highest electrical conductivity of the pore water compared to the grains (e.g. [5, 11]).

The impedance phase for the 4 triplets of specimens, which could not be shown here for brevity, varies linearly with frequency up to values of f higher than 45 kHz. As discussed for the impedance modulus, the impedance phase values measured on specimens of the same triplet are highly repetitive.

Figure 5a reports, in linear scale, the values of the impedance modulus, which is frequency-independent, against the corresponding specimen water content (full circles). The empty circles refer to the impedance measured by the benchtop impedance meter (Agilent 4263B LCR Meter), set at fixed operational frequency. The impedance analyses achieved using the prototype measuring system agree with the benchtop impedance metering results, thus confirming the validity of the impedance analyses and the effectiveness of the prototype. Although preliminary, these results suggest the existence of a non-linear monotonic correlation between the impedance modulus and the gravimetric water content of the specimens tested.

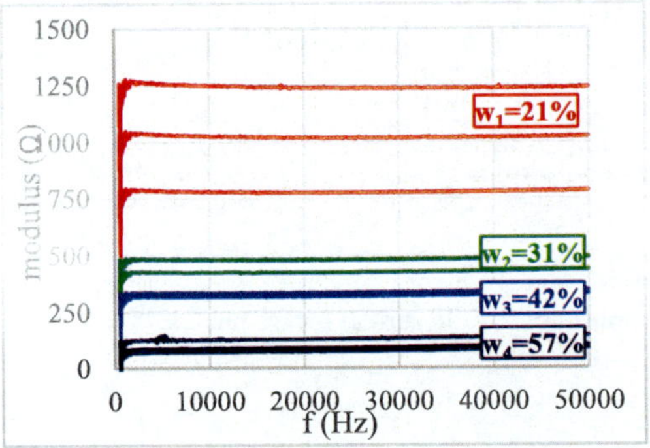

Fig. 4. Impedance modulus over the operational frequency range, for water contents ranging from w = 21% to 57%.

Figure 5b reports the first results obtained on the triplet subjected to stepwise wetting: in this case, the three specimens have been progressively wetted in order to reach w = 16%, 21% and then 26%. For w = 21%, the benchtop impedance measurements are also reported in the figure as empty triangles.

The first results suggest a good match with the trend envisaged for the modulus vs. impedance correlation. At very low water content (w = 16%), the impedance modulus appears again more scattered among specimens at identical water content, confirming that the impedance analysis may be more affected by small structural differences among the specimens at such low water contents. When w = 21% is achieved, the data on the stepwise wetted triplet become more accurate and overlap with those obtained on the single-phase wetting triplet.

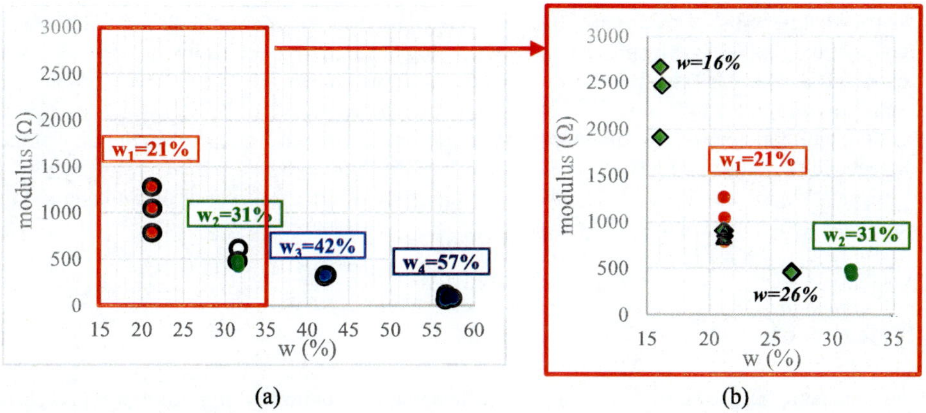

Fig. 5. (a) Preliminary correlation between water content and impedance modulus and (b) enlargement with first results on the stepwise wetted triplet (empty symbols for benchtop impedance measurements).

5 Conclusions

The research presented aims to develop a device for measuring water content in partially saturated soils, based on impedance spectroscopy. The new system is conceived for field applications in pyroclastic soils covering steep slopes susceptible to rapid flowslides [12, 13] and it may be even implemented in real-time monitoring nets within early warning strategies. The experimental activities devoted to test the efficiency of the method and to gain the calibration function of the innovative system in pyroclastic soils have been described.

Preliminary results show that the system is able to characterize the electrical properties of the soil tested. A non-linear monotonic correlation is identified between the soil specimen response as impedance modulus and the soil gravimetric water content, with satisfactory repeatability. Further investigation is required at low saturation degrees, where impedance analyses provide less accurate results, although such low water contents are seldom recorded on site and might be less relevant for slope stability purposes.

The prototype device hence appears as a promising alternative to traditional field water content measuring techniques, with the advantage of reducing costs significantly.

Future work shall include deeper investigation of the correlation between water content and impedance phase angle, other than impedance modulus: indeed, the phase angle is seen to vary linearly over the operational frequency range and may allow improving the predictive capabilities of the device. Alternative sensor geometries are currently being manufactured and, after proper testing, will allow exploiting the versatility of the device as regards customization of the system and adaptation to field measuring operations.

Acknowledgements. This research has been funded by PON Programma R&I 2014–2020 (MITIGO, ARS01_00964) and by project "New technologies for in-time prediction of flowslide occurrence" (TEMPO, P.I. Prof. M. Pirone) conducted in the framework of the STAR programme, financially supported by UniNA and Compagnia di San Paolo.

The authors acknowledge the support of the present study in the framework of the Cooperation Agreement (coordinated by Prof. A. Santo and Prof. M. Pirone) between the Dicea - UniNA, the Campania Region and ITAL SUD, devoted to the study of the triggering of flow-like landslides within the geological context of the Lattari Mts.

References

1. Damiano, E., Olivares, L., Picarelli, L.: Steep-slope monitoring in unsaturated pyroclastic soils. Eng. Geol. **137–138**, 1–12 (2012). https://doi.org/10.1016/j.enggeo.2012.03.002
2. Papa, R., Pirone, M., Nicotera, M.V., Urciuoli, G.: Seasonal groundwater regime in an unsaturated pyroclastic slope. Geotéchnique **63**(5), 420–426 (2013). https://doi.org/10.1680/geot.11.P.049
3. Cascini, L., Sorbino, G., Cuomo, S., Ferlisi, S.: Seasonal effects of rainfall on the shallow pyroclastic deposits of the Campania region (southern Italy). Landslides **11**(5), 779–792 (2014). https://doi.org/10.1007/s10346-013-0395-3
4. Susha Lekshmi, S.U., Singh, D.N., Tarantino, A., Baghini, M.S.: Evaluation of the performance of TDR and capacitance techniques for soil moisture measurement. Geotech. Test. J. **41**(2), 292–306 (2018)
5. Barsoukov, E., Macdonald, J.R.: Impedance Spectroscopy. Theory, Experiment, and Applications. Wiley (2005)
6. Santo, A., Pirone, M., Forte, G., De Falco, M., Urciuoli, G.: Slope stability assessment of the test site in Pagani (Campania, Southern Italy). In: Rainieri, C., Fabbrocino, G., Caterino, N., Ceroni, F., Notarangelo, M.A. (eds.) CSHM 2021. LNCE, vol. 156, pp. 359–365. Springer, Cham (2021). https://doi.org/10.1007/978-3-030-74258-4_23
7. Forte, G., Pirone, M., Santo, A., Nicotera, M.V., Urciuoli, G.: Triggering and predisposing factors for flow-like landslides in pyroclastic soils: the case study of the Lattari Mts. (southern Italy). Eng. Geol. **257**, 105–137 (2019)
8. Battaglio, M., Bellotti, R., Pasqualini, E.: La deposizione pluviale come mezzo per la preparazione dei provini di sabbia. Riv. Ital. Geotec. **2**, 106 (1979)
9. Lo Presti, D.C.F., Pedroni, S., Crippa, V.: Maximum dry density of cohesionless soils by pluviation and by ASTM D 4253–83: a comparative study. Geotech. Test. J. **15**(2), 180–189 (1992)
10. AD5934. 250 kSPS, 12-Bit Impedance Converter, Network Analyzer, Data sheet (2017)
11. Tarantino, A., Ridley, A.M., Toll, D.G.: Field Measurement of suction, water content, and water permeability. Geotech. Geol. Eng. **26**, 751–782 (2008). https://doi.org/10.1007/s10706-008-9205-4
12. Pirone, M., Urciuoli, G.: Cyclical suction characteristics in unsaturated slopes. In: Volcanic Rocks and Soils - Proceedings of the International Workshop on Volcanic Rocks and Soils, Ischia, 24–25 Settembre, 2015 (2016)
13. Pirone, M., Papa, R., Nicotera, M.V., Urciuoli, G.: Analysis of safety factor in unsaturated pyroclastic slope. In: Landslides and Engineered Slopes Experience. Theory and Practice, vol. 3, CRC Press, pp. 1647–1654 (2018)

Investigating the Effects of Fire on Rooted Pyroclastic Soil Properties by Laboratory Burning Treatments

Luca Iervolino(✉), Vito Foresta, and Dario Peduto

Department of Civil Engineering, University of Salerno, Via Giovanni Paolo II, 132, 84084 Fisciano, SA, Italy
liervolino@unisa.it

Abstract. Wildfires increase hillslope instability phenomena by damaging vegetation and altering soil properties. For the present study, a laboratory experimental program was implemented to investigate the physical properties of grass-vegetated pyroclastic soils after their exposure to relatively high temperatures and flames. The soil samples were collected along the Sant'Angelo Creek watershed of the Sarno municipality (Pizzo d'Alvano massif, Campania Region, Italy), which is frequently affected by both wildfires and fast-moving landslides. Owing to the high variability of soil properties (i.e., grain size distribution, porosity, organic matter, root content) in the shallow layers, reconstituted soil samples were tested to different burning treatments to single out the fire-related effects with respect to *i)* the burning duration and *ii)* the time passed after the fire. Results reveal that only the near-surface portion of the soil temporarily undergoes significant temperature variations; soil organic matter decreases as the temperature of exposure increases; soil shear strength and soil infiltration characteristics show relevant changes as both the duration of the treatment and the time after the fire increase. The collected information can provide novel insights toward the understanding of post-fire slope stability and erosion.

Keywords: Wildfires · Slope Instability · Laboratory Testing

1 Introduction

Wildfires can affect the hydro-mechanical characteristics of hillslopes (Rengers et al. 2020). The scientific literature reports increased post-fire slope instabilities (i.e., soil erosion and debris flows) primarily associated with both the altered vegetation condition and the changes in soil properties.

When the soil is burned temperatures rise and a steep gradient is established along the surficial soil layer (DeBano 2000).

Fire may alter several physical soil properties, such as soil structure, texture, porosity, wettability, infiltration rate, and water-holding capacity. The extent of fire effects on these soil properties mainly depends on the intensity, severity, and frequency of fire.

A. Ferrari et al. (Eds.): CNRIG 2023, SSGG, pp. 103–110, 2023.
https://doi.org/10.1007/978-3-031-34761-0_13

Very little information is available on the fire effects on soil mechanical properties and their implication on debris flows occurrence. Besides, the hydro-mechanical characteristics of soil-root systems (SRS) affected by wildfires remain poorly understood (Lei et al. 2022). The SRS are directly related to slope stability; consequently, during the period of low values of root strength, shallow landslides are more likely to occur.

The soil hydraulic behavior is still not well understood in fire-affected areas, since it can be very different mainly depending on fire burn severity (Wieting et al. 2017). Soil-Water Repellency (SWR) refers to the possible formation along the soil profile of a water-resistant layer, which can limit the infiltration rate. When a fire-induced water-repellent layer is formed, the shallower wettable soil layer can be easily saturated and prone to erosion even during moderate intensity precipitations, leading to post-wildfire debris flows (DeBano 2000). Studies demonstrate that SWR increases in low severity burned soils. In contrast, very high temperatures result in SWR decreases (Wieting et al. 2017).

Wildfires also impact the Soil Organic Matter (SOM) content depending on fire type, intensity, and local slope (Gonzaléz Pérez et al. 2004). Loss of SOM generally induces reductions in soil structure and water content. Indeed, SOM holds soil particles in place and helps retain soil moisture. Consequently, the removal of these materials can result in significant soil particle movement in response to rainfall (Gonzaléz Pérez et al. 2004).

Owing to the high site-specificity, the variability of the hydro-mechanical response of the topsoil layer, and the difficulty in measuring soil properties in the field before wildfires, which are generally unexpected, laboratory-controlled tests are needed to provide quantitative information on wildfire effects on soil properties.

Here, we explore the fire-related changes in the properties of the pyroclastic soils covering the Pizzo d'Alvano massif area (Sarno municipality, Campania Region, Italy) by i) utilizing reconstituted grass-vegetated pyroclastic soil samples; ii) using laboratory burning methods that accurately represent changes in soil properties because of heating; iii) evaluating soil property changes with the duration of heating and time after the burning treatment. Referring to the pyroclastic soils under study, research focused on the fire-induced effects on their physical-hydro-mechanical properties are still lacking (Peduto et al. 2022a, b), although they would be fundamental to set up physically-based models of post-wildfire soil behavior at the watershed scale.

2 Methods

Laboratory-controlled burning treatments were performed to investigate possible changes in the physical properties of remolded grass-vegetated pyroclastic soil samples. Unsorted soil was collected from the topsoil layer ($0 \div 15$ cm depth) belonging to the Sant'Angelo Creek watershed of the Sarno municipality (Pizzo d'Alvano massif, Campania Region, Italy). In this area, many slopes covered by pyroclastic soils, originated from the past explosive phases of the Somma-Vesuvius volcano, are exposed to geo-hydrological hazards due to geological, topographic, and climatic conditions and are affected by wildfires mostly in the period between June and September. The vegetation in the area is mainly formed by oaks, chestnuts, and pines. Also, herbaceous species are present (i.e., grasses and shrubs), which represents a strong protection against surface erosion. The unsorted and disturbed soil was sieved at 2 mm to remove coarse fragments. Two 30-cm side concrete cube boxes were filled with the collected soil to resemble the unburnt (U box) and burnt (T box) scenarios, respectively, ensuring a uniform smooth surface and depth of soil profile (Fig. 1). For each box, a 15-cm depth layer of dry gravel was laid out at the bottom with a piece of permeable geosynthetic layer on top; then, 15 kg of soil were progressively added proceeding by 1-cm layers to reach an initial porosity value of 0.67 (Peduto et al. 2022b). After reconstituting the soil samples as above, a total of 10 g of seeds of *graminae* grasses were homogeneously spread on the soil surface of each box. This type of vegetation was selected because of its feasibility for performing laboratory tests (i.e., average root diameter: 0.5 mm), and its representativeness of the herbaceous species of the study area.

Test Boxes - Initial Condition

P_0 [g]	V_0 [cm³]	w_0 [-]	P_s [g]	G_s [-]	V_s [cm³]	$V_w=V_v$ [cm³]
15000	13500	0.3	11538	2.57	4489	9010
n [-]	**P_w [g]**	**S_r [-]**	**LOI [%]**	**G [%]**	**S [%]**	**M&C [%]**
0.67	3461.5	0.4	8.7	15	65	20

Fig. 1. a) Experimental set-up of the T box. b) U box, and c) T box, with the indication of the tested samples. The Table shows the main characteristics of the adopted geomaterial for the two tested boxes (P_0: total weight of soil used for reconstitution; V_0: initial volume occupied by soil samples; w_0: initial water content; P_s: dry weight of soil; G_s: initial value of soil specific gravity; V_s: dry soil volume; V_w: water volume; V_v: void volume; n: porosity; P_w: water weight; LOI: initial value of Loss-on-Ignition; G: gravel, S: sand, M&C: silt and clay fractions).

After 4 months, the burning treatment began for the T box, whereas the U box was left untreated as control reference. The T box sample was heated from above using a propane burner placed 10 cm above the soil surface to account for the direct effects of flames. A concentrated flame was adopted because only the tip of the thermocouple probe measured the temperature, and all tips were aligned on the very center of each burned soil column. K-type thermocouples were installed horizontally and connected to data loggers to monitor soil temperatures at 1-cm intervals from the surface down to 4 cm. Soil temperatures were recorded at 1-min intervals, capturing the burning experiment including a 30-min cooling period. After burning, samples were left to cool overnight.

The different sample classes were labeled as UB-i and Bj-k, with UB and B standing for unburned/burned samples, respectively; i indicates the progressive number of unburned samples; j identifies the duration (in minutes) of the burning treatment (10-,

20-, and 30-min duration of active flame on top of the soil surface), k stands for the actual time (in days) after the execution of the burning treatment (1, 7, and 15 days passed after the burning treatments). Finally, soil sampling was carried out on the shallow fire-affected layer ($0 \div 5$ cm depth) employing a steel sampler.

2.1 Experimental Program and Testing Procedures

Pre- and post-burn soil properties were assessed by using laboratory methods including i) Loss-on-Ignition (LOI) tests (ASTM D7348-08, 2011), ii) Pycnometer Method (PM) tests (ASTM D 854, 2014), iii) Direct Shear (DS) tests (ASTM D3080, 2011) iii) Water-Drop Penetration Time (WDPT) tests (Doerr et al. 2004), and iv) Mini Disk Infiltrometer (MDI) tests (Robichaud et al. 2008).

Loss-on-Ignition (LOI) tests and Pycnometer Method (PM) tests were carried out to study any changes in SOM and G_s because of burning. LOI tests involves the destruction by heat (i.e., exposure time of two hours at 550 °C) of all organic matter in the soil. DS tests were performed on saturated specimens at very low vertical stresses (i.e., 2 kPa), given that the fire can affect only the very surficial part of the topsoil layer. The WDPT method was used to give a qualitative approach to any fire-related changes in SWR. For this purpose, an 80 μL droplet of distilled water was placed onto the soil surface, and the amount of time needed to completely infiltrate the soil was recorded; this time determines the soil repellency rating. Moreover, MDI tests were carried out to investigate how infiltration characteristics change for different post-burn conditions. The Decagon Model S Mini Disk Infiltrometer was used. First, the experimental program consisted of testing intact specimens collected from the U box to assess the infiltration, water-repellency, and shear strength characteristics. Second, specimens from the T box were subjected to burnings of different durations, and their hydro-mechanical characteristics were investigated over time. In addition, the experimental program focused on the influence of high temperatures on SOM through LOI and G_s measurements of soil treated in a muffle furnace for a fixed duration (i.e., 30 min) at increasing temperatures (i.e., 100 to 600 °C).

3 Analysis of Test Results

3.1 Temperature Profile and SOM Content

Studies available in the literature report that the temperatures during a fire generally do not exceed 150 °C at 5 cm depth below the soil surface (e.g., DeBano 2000).

During the burning treatments, the maximum temperatures reached at 1-cm depth were 130, 225, and 300 °C for the B10, B20, and B30 samples, respectively (Fig. 2). They fell to 75, 90, and 100 °C at 2-cm soil depth, respectively, remarking the low thermal conductivity of soils. On the other hand, although characterized by lower maximum temperatures, the heat was retained during long periods at depths greater than 1 cm.

For all the treatments, the temperature started decreasing when the propane flame was turned off. Remarkably, an increase in "cooling gradient" (i.e., the slope of the soil temperature profile computed from the time of propane flame extinction) was observed as the duration of the burning treatment increased due probably to the greater difference in temperature (Δt) between the atmosphere and the soil treated at longer durations. In contrast, the "heating gradient" (i.e., the slope of the soil temperature curve in the active flame section) did not significantly differ among the burning treatments.

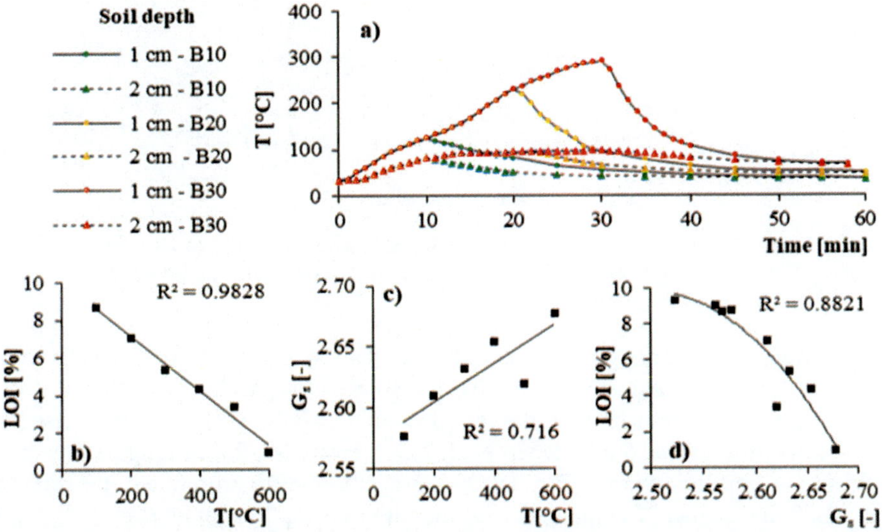

Fig. 2. a) Temperature profiles recorded in soil burned in the laboratory. Profiles are average values calculated at each time step (1 min) from three samples at each burning duration. Temperatures were measured at 1- and 2-cm from the surface. b) LOI, and c) G_S vs. T [°C]. d) LOI vs. G_S.

Figures 2a-c show the results of the LOI and G_s tests carried out to investigate the changes induced by exposure to high temperatures on SOM content. Decreasing LOI and increasing G_s values are found as temperature increases (Fig. 2b, c). Moreover, a decreasing trend in LOI is observed as G_s increases (Fig. 2d). Relatively low LOI values (i.e., relatively high G_s values) are generally associated with a low SOM content (Schulte and Hopkins 2015). Results reveal that the SOM decay increases at increasing temperatures, so burned soil is expected to contain less organic matter than unburned soil. The structure of the surface soil is strictly dependent upon SOM content (Gonzaléz Pérez et al. 2004). By consuming SOM, fire can impact soil structure. This results in increased bulk density, reduced porosity, and decreased water storage capacity of the soil (Peduto et al. 2022a). In this regard, along with the removal of ground cover, surface litter, and/or duff, a loss of SOM can significantly accelerate runoff resulting in increased susceptibility to erosion.

3.2 SWR and Infiltration Characteristics

The results of the WDPT and MDI tests provided preliminary insight into the burning treatments impact on SWR and infiltration rate characteristics.

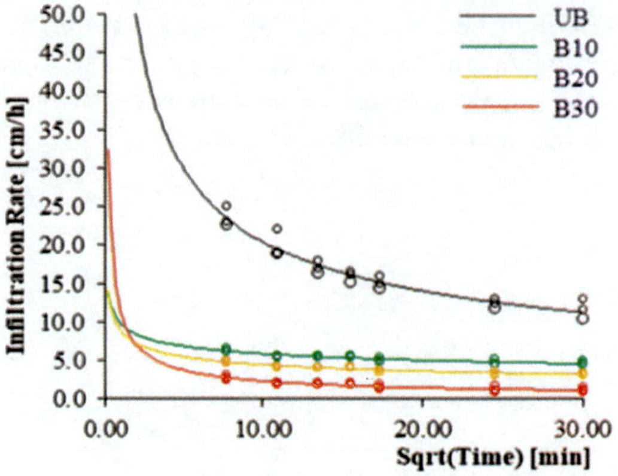

Sample	WDPT [s]	n [-]
UB-1	<5	0.67
UB-2	<5	0.67
UB-3	<5	0.67
B10-1	<5	0.62
B20-1	<5	0.61
B30-1	<5	0.59
B10-7	<5	0.64
B20-7	<5	0.63
B30-7	<5	0.60
B10-15	<5	0.62
B20-15	<5	0.61
B30-15	<5	0.61

Fig. 3. Curves of infiltration rates derived by MDI tests carried out on three samples of each tested soil class (UB, B10, B20, B30). In the Table, measurements of WDPT and porosity (n).

The Table in Fig. 3 shows the measurements of WDPT for unburned and laboratory-treated soil samples. All samples can be classified as wettable (i.e., WDPT < 5 s). Accordingly, we did not find for the tested remolded grass-vegetated pyroclastic soil samples the fire-induced SWR (Doerr et al. 2004). However, the burning treatments did affect the soil infiltration rate (Fig. 3). On average, the infiltration rate of the unburned soil was found higher if compared with the burned soil. This evidence could be due to the higher organics in the unburned soil that enhance water infiltration. Indeed, soils characterized by relatively higher organic matter content are distinguished by a structure that provides the macropore spaces needed for water movement and storage (Gonzaléz Pérez et al. 2004). In contrast, decreasing soil infiltration rates were found as the duration of burning increased. Based on the results of the LOI, G_S, and WDPT tests, we argue that this decrease in infiltration rate was caused by the burning of organic materials rather than the condensation of organic hydrophobic coatings largely reported in the literature (e.g., DeBano 2000). In this regard, since organic matter holds soil particles into aggregates, a loss of organic matter results in a loss of soil structure. The decrease in soil structure is evident in the reduced soil porosity, implying less available space for the water-flow movement along the soil profile, and eventually reducing the preferential water flow paths due to the SRS burning. Finally, the results revealed that the time after the fire did not affect the soil infiltration rate.

3.3 Soil Shear Strength

Figure 4a shows the results of the DS tests carried out at very low vertical stresses (i.e., 2 kPa) on samples thermally treated in the laboratory. The average shear strength found for the UB samples is 40.15 kPa. It decreases to 28.90 kPa and 8.07 kPa for the B10–1 and B30–15 samples, respectively.

Figure 4b shows the ratio (i.e., F ratio) values among the shear strength of each burned sample class and the average shear strength of the unburned control samples (i.e., 40.15 kPa), related to the time of burning and the time after the fire. In particular, the F ratio decreases as either the duration of burning or the post-burn time increase. The minimum reduction in soil shear strength (i.e., 30%) concerned the samples burned for 10 min and tested 1 day after (i.e., B10-1), whereas the maximum reduction (i.e., 80%) pertained to the samples burned for 30 min and tested 15 days post-burn (i.e., B30-15).

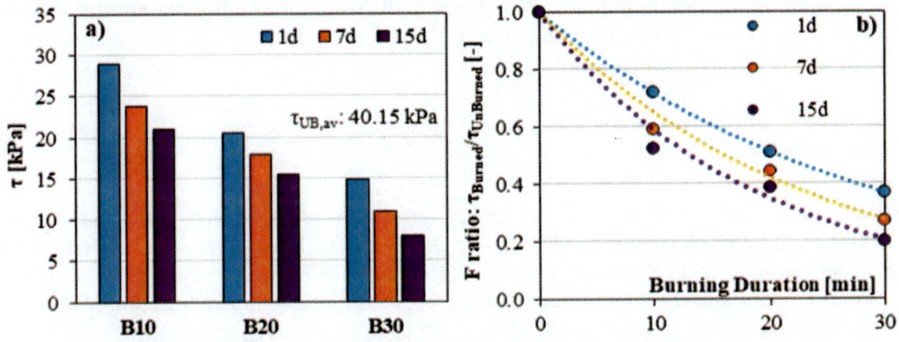

Fig. 4. Effects of the performed burning treatments in terms of a) shear strength values, and b) F ratio, for each tested soil class over time. $\tau_{UB,av}$ is the average shear strength value found for 3 tested UB samples. As there was no peak strength in all the stress-strain curves, the shear stress at 10 mm shear displacement was taken as a failure.

In general, the fire-related changes in soil shear strength are due to the altered soil structure. The progressive decrease in shear strength as the duration of the burning treatment increases may be related to the progressive fire-induced weakening of SRS. Indeed, it is known that roots can reinforce soil and increase its strength primarily by increasing soil cohesion. The high temperatures of the performed laboratory treatments likely led to reductions in soil reinforcement provided by the SRS by reducing their mechanical contribution to shear strength, in agreement with the literature (e.g., Lei et al. 2022). On the other hand, a key role in decreasing soil shear strength is played by the burning of SOM. As a result, a loss of soil structure and a decrease in soil cohesion occur. Additionally, the obtained results suggest that soil shear strength decays with increasing post-fire time due to the death and subsequent decomposition of SRS, leading to reductions in their associated contribution in terms of cohesion. This is in line with the findings of Lei et al (2022), who observed reduced root number and degradation of SRS mechanical properties, leading to the rapid decline of soil cohesion with time after the fire. The decayed roots and the reductions in SOM content may result in the nearly complete decline of the apparent soil cohesion. This normally leads to less resistance against slope instability as shown by Peduto et al. (2022b).

4 Conclusions

The experimental results showed changes in the soil properties because of the performed laboratory-controlled burning treatments. Decreases in SOM content, soil infiltration, and soil shear strength were found as both the duration of the burning treatments and the time after the fire increased. We found that a key role is played by SOM and SRS content. By exposing the soil to increasing temperature, a reduction of soil organic material occurs, along with a weakening/burning of the SRS, with consequences on both the mechanical and hydraulic soil properties. Besides, the process evolves because of the death and decay of roots as well as changes in soil structure. Future studies may involve analyzing the long-term post-fire evolution of soil and vegetation characteristics. The presented study provides preliminary information for understanding the initiation mechanism and assessing the susceptibility to post-fire debris flows and shallow landslides.

References

DeBano, L.F.: The role of fire and soil heating on water repellency. Soil Water Repell. Occur Conseq. Amelior **232**, 193–202 (2003). https://doi.org/10.1016/B978-0-444-51269-7.50020-5

Doerr, S.H., Blake, W.H., Shakesby, R.A., et al.: Heating effects on water repellency in Australian eucalypt forest soils and their value in estimating wildfire soil temperatures. Int. J. Wildl. Fire **13**, 157–163 (2004). https://doi.org/10.1071/WF03051

González-Pérez, J.A., González-Vila, F.J., Almendros, G., Knicker, H.: The effect of fire on soil organic matter - a review. Environ. Int. **30**, 855–870 (2004). https://doi.org/10.1016/j.envint.2004.02.003

Lei, M., Cui, Y., Ni, J., et al.: Temporal evolution of the hydromechanical properties of soil-root systems in a forest fire in China. Sci. Total Environ. **809**, 151165 (2022). https://doi.org/10.1016/j.scitotenv.2021.151165

Peduto, D., Iervolino, L., Esposito, G., et al.: Clues of wildfire-induced geotechnical changes in volcanic soils affected by post-fire slope instabilities. Bull. Eng. Geol. Environ. **81** (2022a). https://doi.org/10.1007/s10064-022-02947-x

Peduto, D., Iervolino, L., Foresta, V.: Experimental analysis of the fire-induced effects on the physical, mechanical, and hydraulic properties of sloping pyroclastic soils. Geoscience **12** (2022b). https://doi.org/10.3390/geosciences12050198

Rengers, F.K., McGuire, L.A., Oakley, N.S., Kean, J.W., Staley, D.M., Tang, H.: Landslides after wildfire: initiation, magnitude, and mobility. Landslides **17**(11), 2631–2641 (2020). https://doi.org/10.1007/s10346-020-01506-3

Robichaud, P.R., Lewis, S.A., Ashmun, L.E.: New procedure for sampling infiltration to assess post-fire soil water repellency. USDA For Serv - Res Note RMRS-RN 1–14 (2008)

Wieting, C., Ebel, B.A., Singha, K.: Quantifying the effects of wildfire on changes in soil properties by surface burning of soils from the Boulder Creek Critical Zone Observatory. J. Hydrol. Reg. Stud. **13**, 43–57 (2017). https://doi.org/10.1016/j.ejrh.2017.07.006

Schulte, E.E., Hopkins, B.G.: Estimation of soil organic matter by weight loss-on-ignition. Soil Org. Matter Anal. Interpret **049**, 21–31 (2015). https://doi.org/10.2136/sssaspecpub46.c3

Calibration Tests of a Shaking Table Apparatus for Testing Large Scale Geotechnical Models

Salvatore Ingegneri, Ernesto Cascone, Giovanni Biondi(✉), Giuseppe Di Filippo, and Orazio Casablanca

University of Messina, Contrada di Dio, S. Agata, Messina, Italy
giovanni.biondi@unime.it

Abstract. A new shaking table apparatus consisting of a shaking table connected to a servo-hydraulic actuator, a large flexible soil container and an automated system for soil deposition, has been set up at the University of Messina and several calibration tests have been carried out. The paper describes the results of those aimed at (i) calibrating the soil deposition system and (ii) checking that the servo-hydraulic control system is capable to accurately reproduce a prescribed acceleration time-history at the shaking table platform, regardless the amplitude, frequency and energy content of the desired motion.

Keywords: Shaking table · Laminar box · Soil pluviation · Plane strain conditions

1 Introduction

The laboratory of Geotechnical Engineering of the University of Messina houses a single degree of freedom shaking table equipped with a large soil container for testing models of geotechnical systems under plane strain seismic conditions (Fig. 1a).

The shaking table (a steel plate mounted on a rigid steel frame 1.5 m wide and 7 m long) is hinged to a servo-hydraulic actuator (operating in a displacement control mode) able to shake with a ± 255 mm stroke and accelerations up to 1 g under the maximum payload of 314 kN. A controller of the servo-hydraulic system allows applying strong motions, minimizing the effect of the soil specimen inside the container.

The large rectangular flexible soil container (inside length 6 m, width 1.5 m and height 2 m) was designed to deform under horizontal shaking according to a shear beam mode, reproducing the free-field condition of a soil layer under vertically propagating shear waves. The side-walls are transversally restrained by rigid steel frames, stiffened by steel counterforts, and by a system of adjustable steel rollers which (i) limits lateral (out of side-walls plane) deformations to reproduce (as close as possible) initial at rest condition when the container is filled of soil and (ii) constrain the motion in the longitudinal direction during shaking, preventing torsional movements.

The system for pluviation of dry sand (Fig. 1b) consists of a hopper (with a slot at the bottom), which can be moved back and forth, above the soil container, and up

A. Ferrari et al. (Eds.): CNRIG 2023, SSGG, pp. 111–118, 2023.
https://doi.org/10.1007/978-3-031-34761-0_14

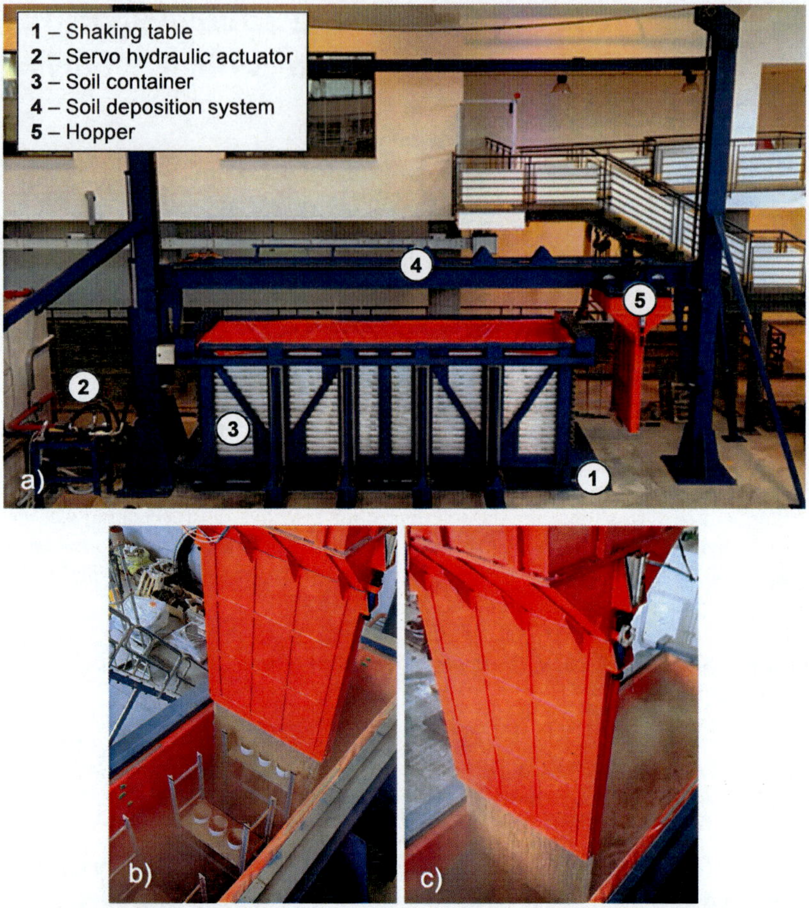

Fig. 1. a) General view of the equipment; b,c) pluviation system: a stage of the calibration/ocedure (b) and a typical deposition stage (c).

and down with respect to its bottom; the capability of the system to control the velocity of the hopper (600–12.000 cm/min), its slot width (1 to 15 mm) and the drop height (30–3000 mm) allows achieving a high degree of spatial uniformity both in terms of grain size distribution and relative density of the soil model.

A detailed description of the novel equipment and of the calibration tests carried out to check its static and dynamic features, can be found in [1–3]. This paper focuses on the tests aimed at (*i*) calibrating the pluviation system and (*ii*) checking that the servo-hydraulic control system accurately reproduce a desired acceleration time-history, regardless the amplitude, frequency and energy content of the desired motion.

2 Calibration of the Pluviation System

To calibrate the pluviation system, a set of identical steel cylindrical molds have been laid inside the container, aligned in three rows along the length of the container and placed at different heights from the bottom of the container (Fig. 1c). A soil consisting of a relatively well-graded sand and fine gravel (Fig. 2a) was used in the calibration procedure: for the selected soil, the minimum and maximum unit weight are 16.4 kN/m^3 and 19.3 kN/m^3, while the maximum diameter ranges from 4.75 mm to 9.5 mm and the effective size are $D_{60} = 2.35$ mm, $D_{30} = 1.12$ mm and $D_{10} = 0.45$ mm, leading to a curvature and a uniformity coefficient equal to 1.17 and 5.2. During the tests the hopper filled with soil was moved at a fixed height (varied in the range 50–250 cm) above the container, dropping a thin curtain of sand and filling the molds (Fig. 1c). The molds were eventually weighted, and the relative density D_R of the soil specimens obtained for each drop height H_d were evaluated. This procedure was repeated for hopper travelling velocity V_h equal to 3.000, 6.600 and 10.000 mm/min. The results of some of these tests are plotted in Fig. 2b and, as expected, show that D_R increases with H_d and V_h, reaching, for the considered soil, a maximum value of about 65%. A linear best fit of the results (continuous lines in Fig. 2b) allows deriving simple predictive equations giving D_R as a function of H_d and V_h [3].

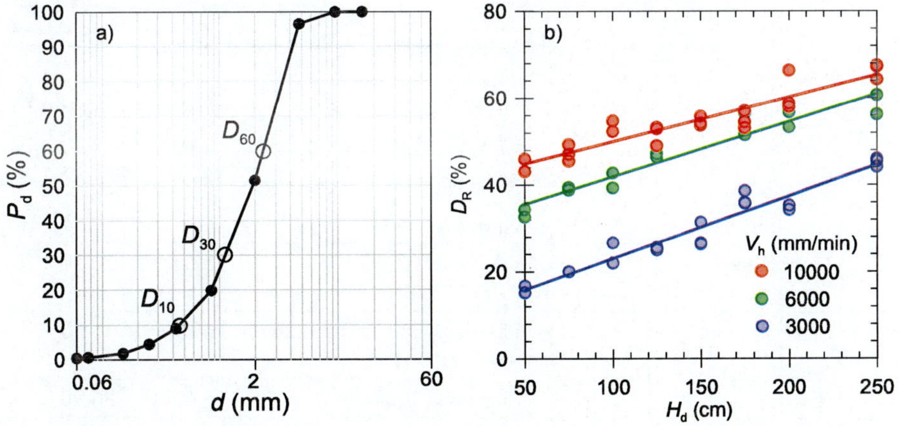

Fig. 2. a) soil grain size distribution; b) results of the calibration procedure.

3 Signal Reproduction Capability

Sets of dynamic tests were carried out on the container full of soil to check the effectiveness of the tuning process implemented in the controller of the servo-hydraulic system in tracking a desired motion. In these tests the desired displacement time-histories (that were applied as command signals to the servo-hydraulic actuator system) were obtained through double integration of a set of eight horizontal component of the ground motion accelerations recorded (at site-to-source distances ranging from 0.92 to about 36.9 km) during earthquakes occurred worldwide with moment magnitude in the range 5.60 – 6.93 (Table 1). For the corresponding acceleration time-histories, Table 2 lists also the peak acceleration a_{max}, the Arias intensity I_a, the number of equivalent loading cycles N_{eq} evaluated according to [4], the mean period T_m evaluated according to [5] and the predominant frequency f_P. These target motions span over wide ranges of amplitude, frequency and energy content (thus allowing to obtain results of general validity). To get an overview of the performance of the servo-hydraulic control system, the target and measured motions have been compared in terms of displacement and acceleration time-histories, 5% damping elastic acceleration response spectra, Fourier amplitude spectra and displacement and acceleration total intensities. Also relative errors were computed with reference to the peak acceleration (ε_A) and displacement (ε_D) and to the cumulated value of the displacement total intensity (ε_{ID}) and relative root mean square errors were also used to compare the target and the measured motion in terms of displacement (ε_d) and acceleration (ε_a) time-histories, corresponding intensity plots (ε_{Ia}, ε_{Id}), acceleration response spectra ($\varepsilon_{S,a}$) and corresponding Houner intensity ($\varepsilon_{H,a}$). A comprehensive discussion of the comparison can be found in [1–3]. Herein, some results are presented and discussed. Specifically, Fig. 3 shows some of the computed errors while Figs. 4 and 5 show the comparison in terms of displacement total intensities (I_d) and the 5% damping elastic response spectra (S_a), respectively.

Table 1. Earthquake records adopted in the dynamic tests.

#	Earthquake	Station (Date)	M_W	R_{ep} (km)
1	Irpinia	Bisaccia (23/11/1980)	6.90	28.30
2	Kobe Japan	Kobe University (16/01/1995)	6.90	0.92
3	Loma Prieta	Gilroy Array #1 (18/10/1989)	6.93	9.64
4	Northridge-01	LA-Wonderland Ave (01/17/1994)	6.69	20.29
5	San Fernando	Pasadena-Old SeismoLab (09/02/1971)	6.61	21.50
6	Sicily	Sortino (13/12/1990)	5.60	36.90
7	Umbria Marche 3rd shock	Cesi Monte (14/10/1997)	5.60	8.70
8	Friuli 1st shock	Tolmezzo (06/05/1976)	6.40	27.70

The data shown in Figs. 3–5, together with the other data presented and discussed in [1–3], point out that that the selected target motions are very well reproduced by the equipment despite the input signal was previously modulated in amplitude and frequency by the feedforward control system.

With reference to the data presented herein it can be observed (Fig. 3) that the errors computed for the displacement peaks (ε_D), time histories (ε_d), intensity plots (ε_{Id}) and total intensity (ε_{ID}) are generally lower than about 3%, 0.1%, 0.02% and 3.7%, respectively; consistently, the intensity plots relative to the measurements closely

Table 2. Seismic parameters of the acceleration time-histories adopted in the dynamic tests.

#	a_{max} (g)	I_a (cm/s)	N_{eq} (–)	T_m (s)	f_p (Hz)
1	0.10	28.53	11.39	0.62	0.44
2	0.31	81.70	5.90	0.38	0.73
3	0.49	169.00	7.59	0.27	2.69
4	0.16	20.40	7.90	0.09	2.05
5	0.21	34.20	8.90	0.09	3.76
6	0.11	5.52	5.33	0.22	1.59
7	0.18	11.41	4.74	0.23	3.42
8	0.32	120.55	9.70	0.39	1.49

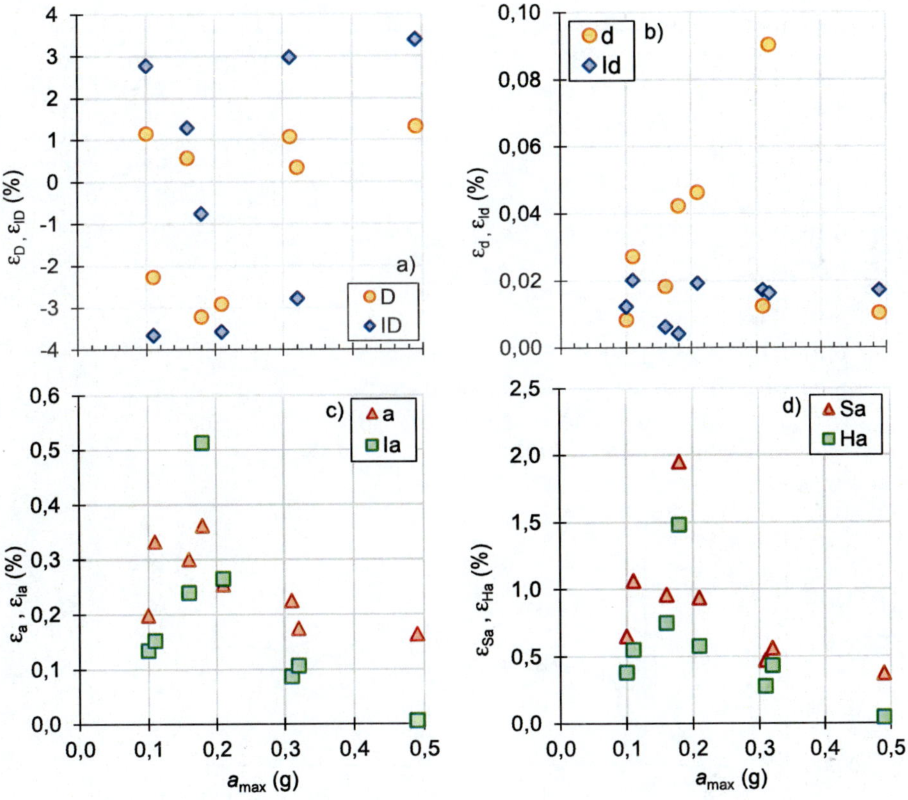

Fig. 3. Signal reproduction capability (container full of sand): relative errors and relative root mean square errors.

approximate those relative to the target motions (Fig. 4). Similarly (Fig. 3), the errors computed on the acceleration time-histories (ε_a) and intensity plots (ε_{Ia}), are always lower than about 0.4 and 0.5%; correspondingly, an overall agreement is observed between the acceleration response spectra (Fig. 5) and the errors ε_{Sa} and ε_{Ha} are always less than about 2% and 1.5%, respectively, ensuring that the motion applied at the base of the soil model is characterized by energy and frequency content very close to the selected targets.

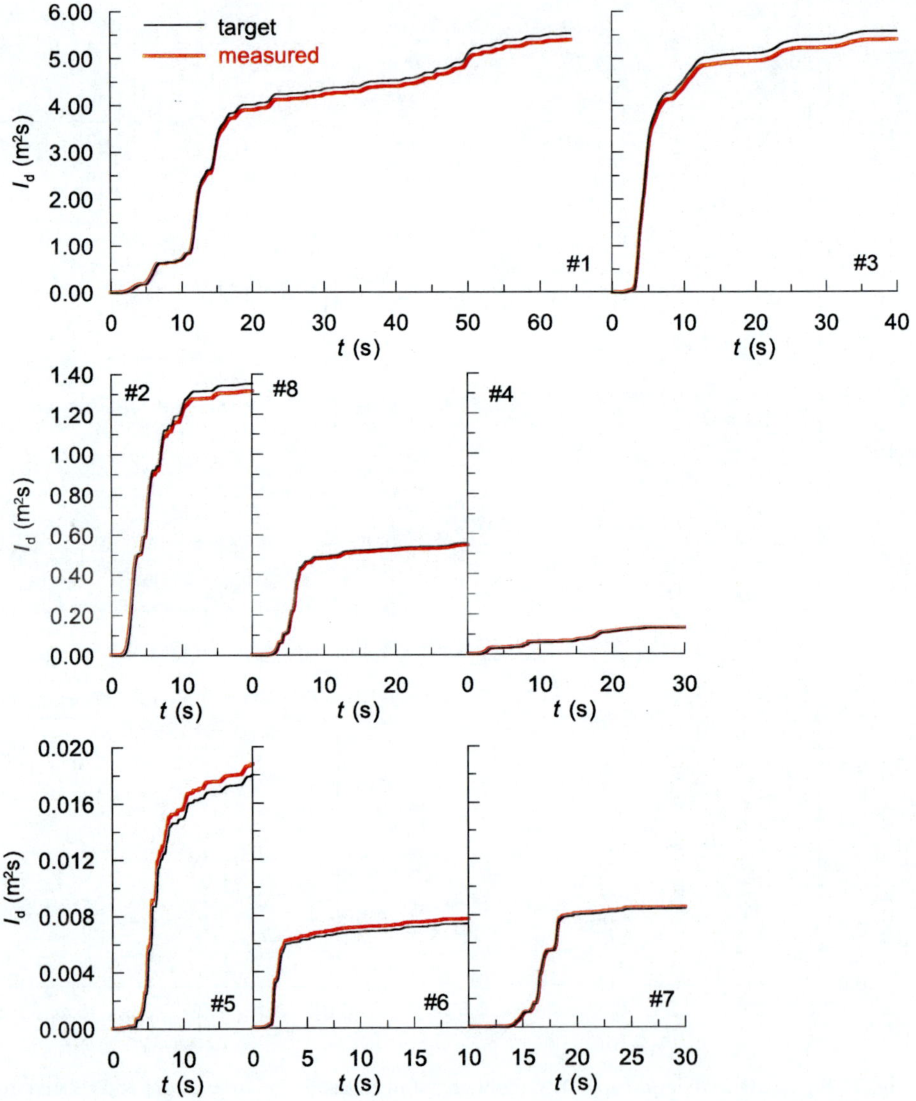

Fig. 4. Comparison between total intensities of target and measured displacements.

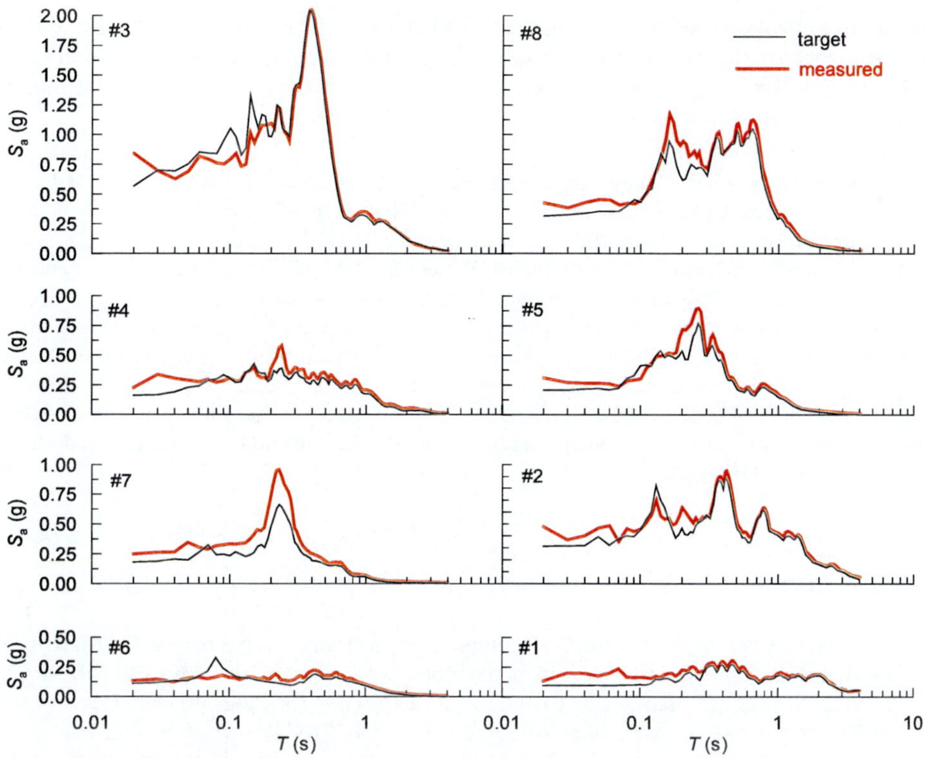

Fig. 5. Comparison between 5% damping elastic response spectra of target and measured accelerograms.

4 Conclusions

In this paper a description of a new equipment for dynamic physical modelling of geotechnical systems housed in the laboratory of Geotechnical Engineering of the University of Messina is provided, giving details on the geometry and mechanical features of the shaking table, of the laminar box and of the soil deposition system.

To calibrate the pluviation system, a well-graded sand and fine gravel was used. The calibration procedure of the pluvial deposition system confirmed the remarkable effect of the sand drop height and of the velocity of the moving hopper spreading sand into the container on the relative density of the sand specimen.

Sets of dynamic tests were carried out on the container full of soil to check the effectiveness of the tuning process implemented in the controller of the servo-hydraulic system in tracking a desired motion. The dynamic tests demonstrated the capability of the shaking table to reproduce the desired acceleration input. Relative errors evaluated between measured and target motions for a set of accelerograms resulted reasonably acceptable.

Dynamic tests carried out on large scale physical models under laboratory-controlled conditions can provide a valuable insight into the system response and also usefully serve as a mean of validation of the numerical analyses aimed at predicting the seismic performance. In this vein the new large shaking table with laminar box described in the paper together with the pluviation system and the control system of the input motion represents a promising equipment.

Acknowledgements. The design and construction of the shaking table equipment was funded by the Italian Ministry of Education University and Research (MIUR), "Programma operativo nazionale ricerca e competitività 2007–2013", Avviso del 18.05.2011, Asse I, Obiettivo Operativo 4.1.1.4, Azione "Rafforzamento Strutturale", Project n. PON A03_00422 "CERISI - Centro di Eccellenza Ricerca e Innovazione Strutture e Infrastrutture di grandi dimensioni".

The experimental activities described in this paper were carried out in the framework of the research project n. 08CT2790090212 entitled "Sistema multisensoriale a basso costo finalizzato alla diagnosi per la tutela e conservazione del patrimonio storico/culturale", funded by Regione Sicilia through PO-FESR 2014/2020, Azione 1.1.5. "Sostegno all'avanzamento tecnologico delle imprese attraverso il finanziamento di linee pilota e azioni di validazione precoce dei prodotti e di dimostrazioni su larga scala".

References

1. Bandini, V., Cascone, E., Biondi, G., Di Filippo, G., Ingegneri, S., Casablanca, O., Aliberti, D., Genovese, F.: The shaking table with laminar box of the University of Messina. Silvestri, F., Moraci, N. (eds.) In: Earthquake Geotechnical Engineering for Protection and Development of Environment and Constructions 2019, pp. 1289–1296. CRC Press/Balkema (2019)
2. Cascone, E., et al.: A new shaking table apparatus for large scale physical modelling of geotechnical systems. In: Sigursteinsson, H., Erlingsson, S., Bessason, B. (eds.) XVII European Conference on Soil Mechanics and Geotechnical Engineering 2019, pp. 786–793. The Icelandic Geotechnical Society, IGS (2019)
3. Ingegneri S.: Calibrazione di un simulatore geotecnico sismico di grandi dimensioni. Ph.D. thesis, Mediterranea University of Reggio Calabria (2020)
4. Biondi, G., Cascone, E., Di Filippo, G.: Reliability of empirical relationships for the evaluation of the number of equivalent loading cycles. Rivista Italiana di Geotecnica **46**(2), 9–39 (2012)
5. Rathje, E.M., Abrahamson, N.A., Bray, J.D.: Simplified frequency content estimates of earthquake ground motions. J. Geotech. Eng. ASCE **124**(2), 150–159 (1998)

Hydro-Mechanical Characterization of a Shale by Unusually High-Pressure Oedometric Tests

Marco Rosone[1], Alessio Ferrari[1,2](✉), Eleonora Crisci[3], and Silvio Giger[4]

[1] Engineering Department, University of Palermo, Palermo, Italy
alessio.ferrari@unipa.it
[2] Laboratory of Soil Mechanics, Ecole Polytechnique Fédérale de Lausanne, Lausanne, Switzerland
[3] Nesol - Numerical Engineering Solutions, Lausanne, Switzerland
[4] National Cooperative for the Disposal of Radioactive Waste (NAGRA), Wettingen, Switzerland

Abstract. The note presents selected results of an experimental campaign conducted with the aim to investigate the hydro-mechanical behaviour of a shale recovered at a depth of about 900 m below the ground. High-pressure oedometric tests were performed to investigate the stress-strain behaviour and the swelling potential. Moreover, constant-head permeability tests were performed using a specially modified oedometric cell oedometric cell. The obtained results highlighted the anisotropy of the behaviour of the geomaterial. Samples recovered parallel to bedding are characterized by yielding stress, compressibility index and swelling index higher than the ones obtained on the sample recovered perpendicular to the bedding. At the same time, the swelling potential increased in the direction parallel to the bedding and with the decrease in the initial density of the geomaterial. However, regardless of the flow direction, the geomaterial showed consistently very low hydraulic conductivity. The collected results provided fundamental elements for the safety case of underground construction of deep radioactive waste repositories in deep Opalinus Clay shale.

Keywords: Opalinus Clay shale · Stiffness · Hydraulic conductivity · Swelling potential

1 Introduction

The underground storage of radioactive waste generated as by-products of nuclear reactors, fuel processing plants, hospitals, research activities and decommissioning of nuclear facilities is a worldwide issue. Therefore, the design of a safe, deep geological repository is a challenge for the international scientific community operating in the field of geotechnical engineering. Nowadays, deep clay shale formations are considered as suitable host rocks for radioactive waste disposal, due to their capability to confine radionuclides on a geological timescale [1, 2]. Opalinus Clay shale is a deep clayey formation which is found in the Northern part of Switzerland and in the Southern part of Germany. Following the

A. Ferrari et al. (Eds.): CNRIG 2023, SSGG, pp. 119–126, 2023.
https://doi.org/10.1007/978-3-031-34761-0_15

investigations conducted over about 30 years, Opalinus Clay shale has been selected as the host rock for the Swiss radioactive waste repository, thanks to its favourable thermo-hydro-mechanical properties [3–6] and self-sealing capacity of fractures [7]. Recently, Opalinus Clay has been deeply investigated with the aim of identifying its geomechanical properties on the sites under consideration for the repository. Opalinus Clay shale has been directly investigated by means of a series of deep boreholes which allowed an in-depth geomechanical characterization campaign on high-quality core samples recovered for laboratory testing.

In this note, selected results from a laboratory testing programme are presented with the aim to highlight how advanced oedometric testing can reveal some of the main hydro-mechanical properties of this shale. Unconventional high-pressure oedometric tests were performed to investigate the stress-strain behaviour and the swelling potential experienced by Opalinus Clay shales upon saturation. Moreover, constant-head permeability tests were performed within the oedometric cell. Considering the intrinsic anisotropy of the material [e.g. 8,9], which is related to the presence of bedding planes, tests were performed on specimens with different orientations with respect to the bedding direction.

2 Material and Methods

The tested material was recovered by an exploratory borehole drilled using a wire-line technique in the Northern Alpine Molasse Basin (Switzerland). The final depth of the borehole was 1310 m and the Opalinus Clay was here detected between 816 and 927 m below the ground level. Due to the depths and the specific nature of the geomaterial, experimental procedures applied to Opalinus Clay samples had to be conducted according to specifically developed advanced protocols, e.g. [10, 11]. After extraction, all cores were immediately inserted into a PVC tube and encapsulated with a fast-curing resin. Then, to evaluate their integrity, the cores were analyzed using X-Ray computer-tomography (XCT). Virtual grayscale cross-sections and CT number profiles were obtained by processing individual core scans. This was useful for the detection of small cracks, for the representation of the variability of the material in terms of density and mineralogy, and for sub-coring selection for laboratory testing. Keller and Giger [12] proved that the CT number is a proxy for material density and mineralogy. Dark parts in the CT images are expected to be low in grain density and clay-rich, while brighter zones are characterized by high grain density minerals such as quartz and calcite minerals. Figure 1 shows the CT image of an Opalinus Clay core sample recovered between 878.11 m and 878.54 m from the ground. The core sample belongs to the mixed clay-silt-carbonate subunit of Opalinus Clay which extended at depths between 860 m and 913 m. As reported in the optimized image of the core (Fig. 1a), to characterize both the higher-density sections (in blue) and the average-lower density sections (in red) recognized in the core scanning, a series of physical (identification tests, Id) and geomechanical testing (oedometric tests, Oed, and swelling tests, Sw) were conducted. Oedometric tests were performed using two unconventional experimental set-ups. The first one consists of a dead-weight loading frame, which allows instant loading up to a vertical force of 60 kN; the second one is composed of a hydraulic jack and a high-rigidity frame capable to apply a maximum vertical load of 80 kN. Unconventional high-pressure oedoemetric cells for specimens having a diameter $D = 35$ mm and height $H = 12.5$ mm were used [13]. These apparatuses allowed the application of a maximum vertical total stress of 61 MPa and 80 MPa, for the first and second layouts respectively.

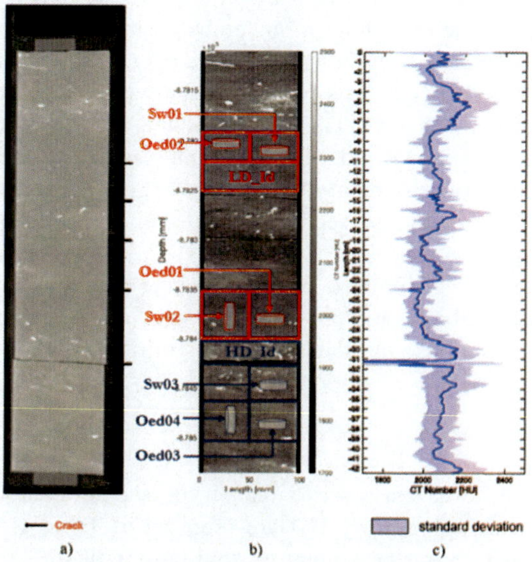

Fig. 1. (a) Core sample in a PVC tube and resin and embedded spacers (cracks are indicated by black arrows); (b) optimized image for the core; (c) vertical profile of CT number (in Hounsfield Units) in the core. LD = average-lower density in red; HD = higher density in blue; Id = identification test; Oed = high-pressure oedometric test; Sw = swelling test.

Pore water pressure control was guaranteed through two independent drainage lines connected to the two ends of the specimen. A Pressure/Volume (PV) controller was used to impose the back pressure and to monitor the pore water volume exchanges with a resolution of 1 mm^3. A second PV controller was connected to perform permeability tests applying different outlet pore water pressures. An artificial porewater was used in the experiments according to the recipes derived from previous investigations on Opalinus Clay [13]; since the artificial water is saturated with respect to calcite and dolomite, pore water interfaces were placed between the PV controllers and the water inlet/outlet of the cells. Axial displacements were measured by means of two LVDTs. An ad-hoc procedure was adopted to ensure adequate preparation of the oedometric specimens. Specimens were prepared by using a mechanical lathe operating in a room with a controlled temperature ($T = 22$ °C). The rotation speed and cutter advancement of the lathe were regulated to reduce heat and stress generation on the specimen. The specimen was then put into the oedometric ring by means of a press. Initial vertical stress of about 0.1 MPa was applied to the specimen to assure proper contact with the loading rod. The resaturation phase was initiated by applying a pore water pressure of 50 kPa at the bottom base of the specimen. The volume was kept approximately constant by adjusting the applied total vertical stress and keeping the change in height within ±

10 μm. Once this process stabilized, a water pressure of 50 kPa was applied at the top of the specimen and further stabilization of displacement was allowed. Mechanical loading was then performed in steps; a first loading path was applied to achieve axial stress comparable with the ones expected in situ (10–20 MPa); at the end of this phase, a permeability test was also performed in some cases. Next, an unloading-reloading cycle was carried out. The reloading, followed by an unloading phase, was continued up to the maximum vertical stress allowed by the apparatuses. Data of the settlement evolution vs time of oedometric tests were analysed with an analytical solution of the theory of consolidation for shales [11]. For each loading step the method allowed for the identification of the primary consolidation settlement ρ', related to the water overpressure dissipation, the secondary settlement due to creep ρ'', and the system compressibility. So, the results reported in the next section will refer only to the first two contributions to the measured displacements.

Free swelling tests in oedometric conditions were performed by applying limited axial stress (55 kPa) and back pressure (50 kPa). Axial displacement was continuously monitored during the process until the complete stabilization.

3 Test Results

The result of identification tests carried out on the section characterised by average lower-density LD and high-density HD are reported in Table 1 in terms of particle density ρ_s, bulk density ρ, water content w, void ratio e, degree of saturation S_r, total suction Ψ, liquid limit w_L; plastic limit w_P, plasticity index PI, sand fraction f_{sa}, silt fraction f_{si} and clay fraction f_{cl}. Mineralogical analysis showed that the main components characterizing the LD material are clay minerals (53%), quartz (24%), calcite (10%), K-feldspar (5%) and siderite (4%). In HD materials, lower amounts of clay minerals (47%) and K-feldspar (4%) were detected to the advantage of the siderite (9%). These data are in good agreement with previous research on Opalinus Clay [6, 10, 15, 16].

Table 1. Main properties of tested samples.

Test	ρ	ρ_s	w	e	S_r	Ψ	w_L	w_P	PI	f_{sa}	f_{si}	f_{cl}
	(g/cm^3)	(g/cm^3)	(-)	(-)	(-)	(MPa)	(-)	(-)	(-)	(%)	(%)	(%)
LD	2.49	2.72	0.041	0.144	0.78	42.6	0.34	0.22	0.12	2	69	29
HD	2.52	2.75	0.040	0.131	0.84	42.0	0.34	0.21	0.13	4	63	33

The oedometric curve obtained on sampled Oed01, Oed02 and Oed03 are reported in Fig. 2. The three specimens were recovered with cylinder axis perpendicular to the bedding (S specimen) but in two different core sections: Oed01 and Oed02 in the low-average density section (LD) and Oed03 in the high-density section (HD). Passing through sample Oed01 and Oedd02 to sample Oed03, the oedometric curves point out a decrease in the yielding vertical effective stress (13.5, 9.4 MPa and 8 MPa respectively), compressibility index at the virgin line C_c (from 0.025 ÷ 0.026 to 0.021) and swelling index C_s

both at the intermediate unloading/reloading cycle (from 0.006 to 0.003) and at the final unloading (from 0.012 to 0.009). The yielding vertical effective stress was estimated by intersecting the initial trend of the oedometric curve with the prolongation of the virgin compression line. When this construction did not provide a suitable result (e.g. test Oed04 in Fig. 3), the vertical yield stress was assumed to be equal to the swelling pressure. For the latter, the difficulties in the experimental determinations for stiff and clay shales can be confirmed [17, 18]. In fact, due to the cumulated strains during the equalization upon saturation, the swelling pressure varied from 3 MPa for Oed03 to 6.6 MPa for Oed01.

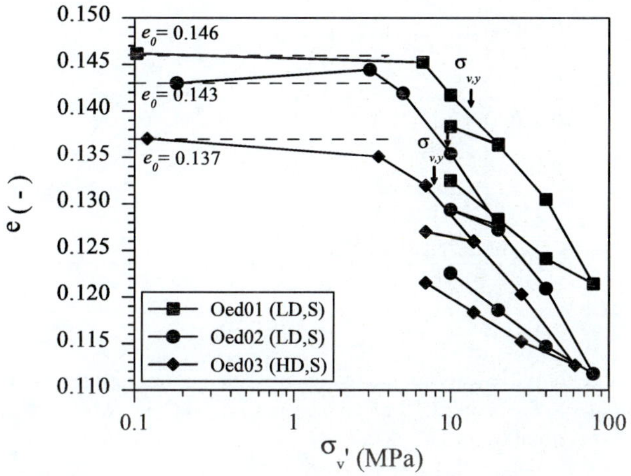

Fig. 2. Results of oedometric tests conducted on specimens Oed01(LD), Oed02 (LD) and Oed03 (HD) recovered with cylinder axis perpendicular to the bedding (S specimen).

Figure 3 shows the comparison between the results of oedometric tests carried out on specimens recovered from the same HD core section, but with different orientations with respect to the bedding plane. The specimen Oed03 recovered with the cylinder axis parallel to the beddings (P sample) clearly shows a higher swelling pressure (6 MPa) compared to the ones measured on the S sample Oed02. However, lower yielding pressure (6 MPa), compressibility index at the virgin line C_c (0.018), as well as swelling index C_s at the final unloading (0.005) were measured. Instead, the swelling index C_s at the intermediate unloading/reloading cycle (0.004) appeared slightly higher than the one measured in the S sample, because of the higher secondary settlement measured in the reloading step, which lasted longer. It is worth mentioning that the difference in the mechanical response between the two samples reduced with the increasing vertical effective stress. In fact, at the maximum applied pressure (61 MPa) the void ratios were almost the same (the difference was lower than 0.001). Then, the anisotropy which characterizes the elastic response of the Opalinus Clay is limited at high confining stress.

The extended solution of consolidation theory for clay shale allowed to compute the hydraulic conductivity at each loading step. These results are plotted in Fig. 4 as a function of the void ratio. In the same diagram, the result of the constant-head permeability test carried out on the specimen Oed03 at the vertical effective stress equal to 14 MPa, is reported. Data of indirect and direct measurements are in good agreement. Moreover, they show a clear dependency on the void ratio and a limited influence on the flow direction. At the equal void ratio, the hydraulic conductivity in the direction perpendicular to the bedding appears to be almost equal to the one in the direction parallel to the bedding.

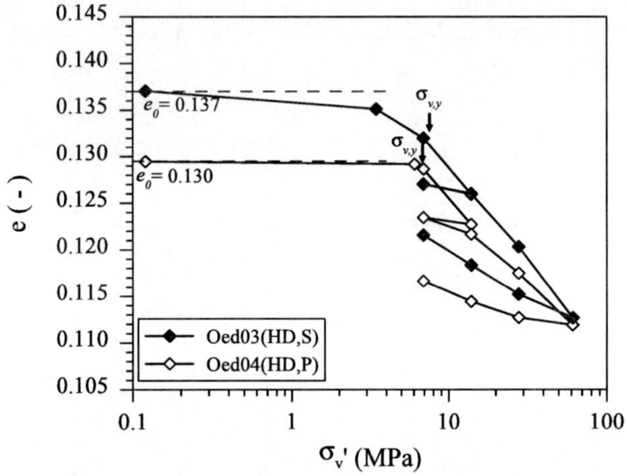

Fig. 3. Results of oedometric tests conducted on specimen EOe03(HD) recovered with the cylinder axis perpendicular to the bedding (S specimen) and Oed04 (HD), whose cylinder axis is parallel to the bedding (P specimen).

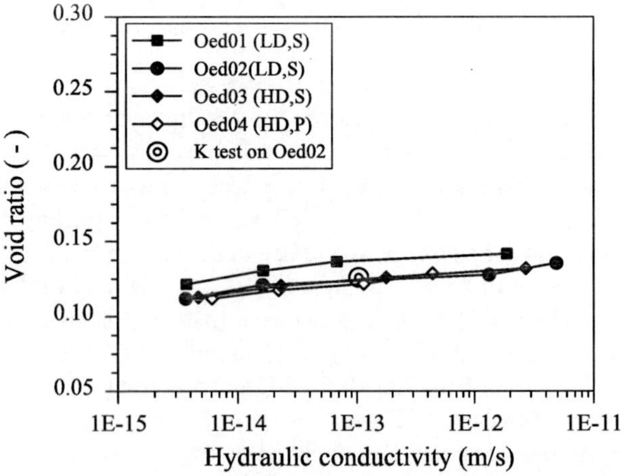

Fig. 4. Hydraulic conductivity values obtained from the analysis of consolidation in oedometric tests and constant-head tests performed during the oedometric test.

The anisotropy in the hydro-mechanical response of Opalinus Clay is also highlighted in Fig. 5, where the results of the swelling tests are reported. As shown, the P sample experienced a significant swelling behaviour compared to the ones observed in the S samples. For the latter, a dependency on the void ratio of the swelling potential can be also addressed.

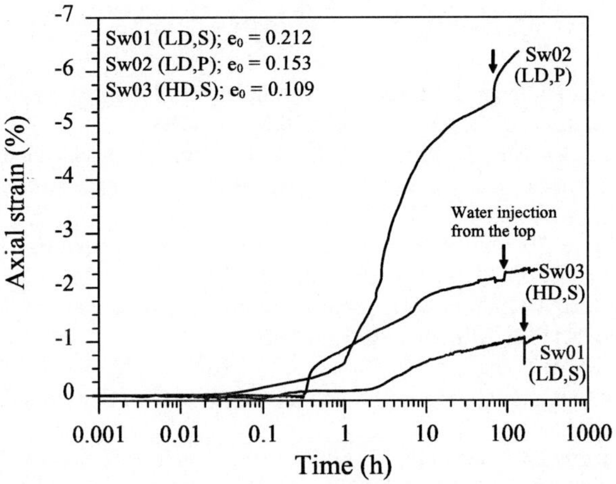

Fig. 5. Results of unconstrained swelling tests in terms of axial strain vs time. Water injection from the base started immediately while from the top it differed with time (black arrows).

4 Conclusions

Preliminary results of an experimental activity carried out on Opalinus Clay core samples were presented with the aim to characterize the geomaterial which is the designated host rock for a deep geological repository in Switzerland.

By performing oedometric tests with equipment specifically modified for high-pressure, the main properties of Opalinus Clay shale were determined and summarised as follows:

- The anisotropy deeply affected the volumetric behaviour under loading. The samples recovered parallel to bedding experienced higher yielding stress, compressibility and swelling index. The latter is further increased when in the geomaterial the porous clay matrix is diffused.
- The geomaterial is characterized by very low hydraulic conductivity and a limited effect of influence on the flow direction.
- High swelling behaviour characterizes the volumetric response of the material upon saturation. The swelling potential increases in the direction parallel to the bedding and with the decrease in the initial density of the geomaterial.

The test results contribute to the site selection and safety case of a deep geological repository.

Acknowledgements. Dr Antonio Casella and Mr. Giovanni Sapienza are acknowledged for their support in the laboratory investigations. This study was funded by the National Cooperative for the Disposal of Radioactive Waste (Nagra), Switzerland.

References

1. Zheng, L., Li, L., Rutqvist, J., Liu, H.-H., Birkholzer, J.: Modeling radionuclide transport in clays. United States (2012). https://doi.org/10.2172/1173163
2. Bossart, P., Bernierm, F., Birkholzerm, J., et al.: Mont Terri rock laboratory, 20 years of research: introduction, site characteristics and overview of experiments. Swiss J. Geosci. **110**(1), 3–22 (2017)
3. Ferrari, A., Favero, V., Marschall, P., Laloui, L.: Experimental analysis of the water retention behaviour of shales. Int. J. Rock Mech. Min. Sci. **72**, 61–70 (2014)
4. Favero, V., Ferrari, A., Laloui, L.: Thermo-mechanical volume change behaviour of Opalinus Clay. Int. J. Rock Mech. Min. Sci. **90**, 15–25 (2016)
5. Giger, S.B., Ewy, R.T., Favero, V., Stankovic, R., Keller, L.M.: Consolidated-undrained triaxial testing of Opalinus Clay: results and method validation. Geomech. Energy Environ. **14**, 16–28 (2018)
6. Ferrari, A., Rosone, M., Ziccarelli, M., Giger, S.B.: The shear strength of Opalinus Clay shale in the remoulded state **21**, 100142 (2020)
7. Gautschi, A.: Safety-relevant hydrogeological properties of the claystone barrier of a Swiss radioactive waste repository: an evaluation using multiple lines of evidence. Grundwasser **22**(3), 221–233 (2017)
8. Corkum, A.G., Martin, C.D.: The mechanical behaviour of weak mudstone (Opalinus Clay) at low stresses. Int. J. Rock Mech. Min. Sci. **44**(2), 196–209 (2007)
9. Minardi, A., Crisci, E., Ferrari, A., Laloui, L.: Anisotropic volumetric behaviour of opalinus clay shale upon suction variation. Géotech. Lett. **6**, 144–148 (2016)
10. Favero, V., Ferrari, A., Laloui, L.: On the hydro-mechanical behaviour of remoulded and natural Opalinus Clay shale. Eng. Geol. **208**, 128–135 (2016)
11. Minardi, A., et al.: Benchmark study of undrained triaxial testing of Opalinus Clay shale: results and implications for robust testing **25**, 100210 (2021)
12. Keller, L.M., Giger, S.B.: Petrophysical properties of Opalinus Clay drill cores determined from Med-XCT images. Geotech. Geol. Eng. **37**(4), 3507–3522 (2019). https://doi.org/10.1007/s10706-019-00815-2
13. Ferrari, A., Favero V., Laloui. L. One-dimensional compression and consolidation of shales. Int. J. Rock Mech. Min. Sci. **88**, 286–300 (2016)
14. Wersin, P., et al.: Rock and porewater characterisation on drillcores from the Schlattingen borehole. In: Nagra Arbeitsbericht NAB 12-054 (2013)
15. Rosone, M., Ferrari, A., Ziccarelli, M., Giger, S.B. The residual shear strength of the shaly and sandy facies of the opalinus clay. In: Ferrari, A., Laloui, L. (eds.) Energy Geotechnics. SEG 2018. Springer Series in Geomechanics and Geoengineering, pp. 426–433. Springer, Cham (2018). https://doi.org/10.1007/978-3-319-99670-7_53
16. Crisci, E., Ferrari, A., Giger, S.B., Laloui, L.: Hydro-mechanical behaviour of shallow Opalinus Clay shale. Eng Geol. **251**, 214–227 (2019)
17. Cui, Y.J., Nguyen, X.P., Tang, A.M., Li, X.L.: An insight into the unloading/reloading loops on the compression curve of natural stiff clays. App. Clay Sci. **83**, 343–348 (2013)
18. Zhang, F., Cui, Y.-J., Conil, N., Talandier, J.: Assessment of swelling pressure determination methods with intact Callovo-Oxfordian claystone. Rock Mech. Rock Eng. **53**(4), 1879–1888 (2019). https://doi.org/10.1007/s00603-019-02016-y

On the Post-peak Behaviour of Remoulded and Jointed Clay Samples During Triaxial Compression Tests

Marco Rosone[1], Esmaeel Rahbari[1(✉)], and Alessio Ferrari[1,2]

[1] Department of Engineering, University of Palermo, 90138 Palermo, Italy
mohammadesmaeel.rahbari@unipa.it

[2] Soil Mechanics Laboratory, Ecole Polytechnique Fédérale de Lausanne, 1015 Lausanne, Switzerland

Abstract. The note presents a selection of the results of a large experimental program conducted with the aim of characterizing the shear strength of Opalinus Clay and Speswhite Kaolin in remoulded and reconstituted conditions, through the execution of unconventional triaxial tests. The collected results clearly show the influence of the degree of pre-consolidation and the joints on the mechanical behaviour of the geomaterials, and in particular on the shear strength. The presence of pre-existing joints inside the specimen completely alters the original behaviour of the material, which becomes ductile and contracting. Moreover, interpreting the results with a sliding blocks model, it has been proved that, under these conditions, the shear strength can be drastically reduced to the residual strength values.

Keywords: clay · shear strength · artificial joints · ultimate state

1 Introduction

The mechanical behaviour of rock and soil masses is deeply affected by the presence of discontinuities and structural joints such as bedding, tectonic joints, shear fractures and exfoliation joints. This becomes particularly relevant in the case of highly persistent joints in over-consolidated and high-plasticity clays due to the pronounced strain-softening behaviour. Opalinus Clay has been selected as a geological barrier for the deep repository for high-activity radioactive wastes. This material was chosen for its ability to prevent the migration of radionuclides [1], to self-seal fractures [2], and its favourable thermo-hydro-mechanical properties [3, 4]. As it is known, the mechanical behaviour of complex geomaterials strongly depends on the microstructure and its evolution with thermal, hydraulic, chemical, and mechanical loads. The microstructure of Opalinus Clay is the result of an articulated geological history, which began with the deposition in the marine environment and continued with intense diagenetic processes. This material is characterized by a significant spatial variability of its properties, which makes it difficult to choose unique geomechanical parameters for the entire formation [1, 5]. Moreover, bedding joints are also very frequent and, due to the complex geological history, it is also possible to identify tectonically disturbed zones and faults. For example,

A. Ferrari et al. (Eds.): CNRIG 2023, SSGG, pp. 127–134, 2023.
https://doi.org/10.1007/978-3-031-34761-0_16

at the Mont Terri Underground Research Laboratory (URL, Canton Jura - Switzerland) three different facies have been identified (shaly, sandy and carbonate-rich) and a particularly disturbed tectonic area, known as the “Main Fault” [1], has been detected. From a geomechanical point of view, the Opalinus Clay behaviour is characterized by a complex elastoplastic constitutive law, whose main characteristics are non-linear elasticity, anisotropy, and brittle failure [6, 7]. In this complicated geotechnical context, it is fundamental to characterize some reference parameters of the shear strength behaviour of Opalinus Clay. Ferrari et al. [8] defined the lower limit of the shear strength parameters of Opalinus Clay by analyzing the mechanical behaviour of the different facies at the remoulded state. The extension of this approach to other reference geomechanical parameters can provide significant indications on the intact material response, useful data for constitutive modelling, and information on the mechanical behaviour of Opalinus Clay in the fault zones.

This note presents a selection of the results of a large experimental program aimed to characterize the shear strength behaviour of Opalinus Clay samples. In particular, the results of unconventional triaxial compression tests performed on completely remoulded samples are discussed with the aim to investigate the brittle response, which characterizes the post-peak behaviour. Moreover, complementary results from a triaxial tests campaign on a reference geomaterial (Speswhite Kaolin) are discussed to highlight the influence of different joint inclinations on the mechanical response.

2 Experimental Materials

The Opalinus Clay tested samples were recovered at the Mont Terri URL (Canton Jura, Switzerland) and belong to the shaly facies. To extend the consideration made on the results obtained on the first series of tests, the Speswhite Kaolin was also tested. The main geotechnical properties of the remoulded Opalinus Clay and Speswhite Kaolin samples, expressed in terms of soil-specific mass ρ_s, liquidity limit w_l, plasticity limit w_p, plasticity index *PI*, activity *A* and granulometric fractions are shown in Table 1; the data reported for Opalinus Clay are similar to those of previous research on Opalinus Clay [4, 8].

Table 1. Index properties of the samples tested.

Soil type	ρ_s	w_l	w_p	*PI*	*A*	Particle size fraction (%)		
	(Mg/m^3)	(%)	(%)	(%)	(%)	Sand	Silt	Clay
Opalinus Clay	2.69 ÷ 2.71	35 ÷ 42	22 ÷ 33	10 ÷ 13	0.22 ÷ 0.33	4 ÷ 5	51 ÷ 56	40 ÷ 45
Speswhite Kaolin	2.6	59	32	27	0.35	0	24	76

The experiments were conducted on the completely remoulded and reconsolidated material. The procedure for preparing the remoulded material proposed by Burland [9] was taken as a reference for identifying the best technique to implement in the laboratory. The samples were initially crushed with a mortar to obtain aggregates of clayey material smaller than 4.75 mm. Then, the material was mixed with distilled water to obtain a water content equal to $1 \div 1.5$ times the liquidity limit and stored in a closed environment for $24 \div 48$ h. After equalization to the desired water content, the material was worked with a metallic spatula on a glass sheet. The completely remoulded material was consolidated under 1D conditions inside a consolidometer (initial height $h = 130$ mm and diameter $d = 100$ mm). For the Opalinus clay samples, the maximum effective vertical stress, applied for successive load increments, was equal to $\sigma'_{v,max} = 320$ kPa. After the loading phase, a stepwise unloading phase was performed up to the vertical stress $\sigma'_v = 40$ kPa and final unloading in undrained conditions at zero total vertical stress was applied. This experimental procedure produced remoulded and saturated clay samples with an overconsolidation ratio OCR ($OCR = \sigma'_{v,max}/\sigma'_{v,0}$) equal to 8. Experiments on Speswhite Kaolin were also conducted on the reconstituted material. The procedure for preparing the reconstituted material was the same as the one mentioned before, but with two main differences: first, this material was already available in powder, and second, due to the softness of the material, after the loading phase conducted up to $\sigma'_{v,max} = 320$ kPa the unloading phase was completely applied in undrained conditions.

3 Experimental Program

The experimental program involved a first series of triaxial compression tests carried out on completely destructured and remoulded Opalinus Clay specimens. Each sample coming from the consolidometer was used to prepare two cylindrical specimens having $d = 38$ mm and $h = 76$ mm. The specimens, when installed in the triaxial cell, were subjected to a constant back-pressure $u_{w,0}$ in the range of $150 \div 250$ kPa and an effective confining stress p'_0 equal to $20 \div 40$ kPa. Once the saturation phase was completed, i.e. Skempton's parameter B was greater than 0.97, isotropic loading/unloading paths were applied to induce $OCR = p'_c/p'_0$ values (p'_c is the maximum mean effective stress applied in the triaxial cell and p'_0 is the mean effective stress applied at the end of unloading stage) between 2 and 20. Basically, after the saturation procedure, the samples were consolidated at the effective confining pressure p'_c equal to 400 kPa or 1200 kPa and then unloaded at the designated value of p'_0. During these phases, drainage was allowed from both specimen ends, and the volumetric strains were calculated through the variations in the water content of the specimen. The same measurements were used to calculate the consolidation coefficient c_v both during the loading paths ($c_v = 4 \div 7 \times 10^{-4}$ cm^2/s) and during the unloading processes, ($c_v = 4 \times 10^{-4}$ cm^2/s $\div 1 \times 10^{-3}$ cm^2/s). Based on these parameters, a properly axial strain rate ($v = 6 \times 10^{-6}$ s^{-1}) was selected, which allowed the equalization and correct measurements of the pore pressures even during the failure phase in undrained conditions [10].

The second set of unconventional triaxial tests was conducted with the aim of evaluating the effect of pre-existing joints on the mechanical behaviour of tested samples. Drained consolidated triaxial compression tests were performed on both Opalinus Clay and Speswhite Kaolin specimens. After being subjected to the experimental procedure defined in the Sect. 2, an Opalinus Clay specimen was cut along a plane inclined at 59°

to create an artificial joint within the clayey mass. Instead, Speswhite Kaolin specimens were cut along different planes inclined of 0°, 15°, 30°, 60° and, 75° with respect to the horizontal. To do this, a specially prepared device and a saw with a very thin blade were used. After cutting, the two specimen parts were brought back into contact and tested in the triaxial cell. The triaxial test on Opalinus Clay specimen was conducted in multi-stage mode. Basically, this test foresaw the halting of the failure stage once the stabilization of the deviatoric stress $q = \sigma_1 - \sigma_3$ was obtained and increasing the effective confinement stress up to the next value. Then, the failure phase continued after verifying the conclusion of the consolidation phase by monitoring the volumetric strain. The selected effective confinement stresses p'_0 were equal to 100, 200, 400 and 800 kPa. This allowed to obtain shear strength envelope in a wide range of stress from a single specimen. Instead, the triaxial tests on Speswhite Kaolin were performed in single-stage mode, applying the same effective confinement stress p'_0 equal to 80 kPa. These tests were aimed to highlight the role of different inclination of pre-existing joints on the shear strength. For the drained failure phase, vertical strain rates v equal to 1×10^{-7} s^{-1} and 4×10^{-7} s^{-1} were selected for Opalinus Clay and Speswhite Kaolin specimens, respectively. To verify the adequacy of this strain rate with respect to the drainage capacity of the specimen, the procedure proposed by Escario and Saez [11] was used. This procedure involved interrupting the deviator phase for 24 h, to allow any dissipation of the induced interstitial overpressures, and verifying that there was no change in shear strength in the restart phase. Given the complexity of the multi-stage test, this verification was only conducted in a few tests.

4 Analysis of the Results

The results of the consolidated undrained triaxial compression tests carried out on the homogeneous specimens (i.e. without artificial joint) are presented in Fig. 1 in terms of deviatoric stress q (Fig. 1a) and pore pressure variations Δu_w (Fig. 1b) as the axial strain ε_a varies. In general, the curves q vs ε_a plotted show a not pronounced peak of strength for an axial strain $\varepsilon_a = 10 \div 12\%$. Only the OPAR4A sample showed a pronounced peak of strength. In any case, the stationary condition of ultimate strength is clearly not reached at the end of the tests. At the beginning of the failure phase, pore pressure has an increasing tendency, but before the peak of strength, the trend completely reversed, becoming downward. The maximum pore pressure value is reached in correspondence of an axial deformation equal to $\varepsilon_a = 2 \div 4\%$. So, the pore overpressure resulted negative or slightly greater than zero in correspondence with the failure, i.e. at the maximum value of the deviatoric stress. The minimum pore pressure value was recorded at the end of the tests ($\varepsilon_a = 18 \div 20\%$), even if the stationary conditions were not clearly reached. As it is known, the reduction of the pore pressure at failure is the experimental evidence of the volumetric expansion behaviour of the material, which in undrained conditions is inhibited. This behaviour is associated with the formation of a clear failure surface inclined to the horizontal at an angle of $57 \div 60°$. The stress paths on the plane (p', q) are represented in Fig. 1c, where the points corresponding to the maximum deviatoric stress have been interpolated collecting them for the different pre-consolidation stress p'_c. In the same figure, the shear strength envelope for the normally consolidated samples is also reported [8]. Data shows that there is clear experimental evidence of the increased cohesion with the pre-consolidation stress p'_c, while the shear strength angle φ' assumes a slight tendency to decrease with p'_c. The mechanical interpretation on the plane q vs p' of the post-peak conditions is not consistent, because of the non-stationary conditions reached at the end of the tests and the formation of the inclined clear failure surface.

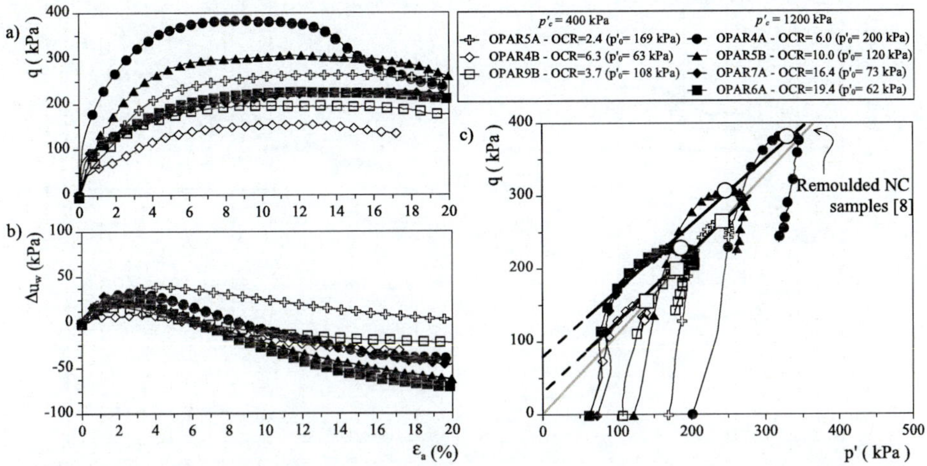

Fig. 1. The undrained failure stage of the triaxial compression tests carried out on homogeneous Opalinus Clay specimens: deviatoric stress q and of the pore overpressure Δu_w (b) as a function of axial strain (a) and stress paths on the plane deviatoric stress q vs mean effective stress p' (c).

To better highlight the role of these inclined surfaces, the results of tests conducted on artificially jointed samples are discussed in the following. The results of the drained triaxial compression test conducted on the Opalinus clay samples artificially jointed with a cut inclined of 59° respect to the horizontal are reported in Fig. 2. In the diagrams the total principal stresses σ_1, σ_3, the pore pressure u_w (Fig. 2a) and of the volumetric strain ε_v (Fig. 2b) are plotted as a function of the applied axial strain ε_a. From these diagrams it is clear that, because of the artificial joint, the mechanical behaviour of the specimen is ductile and contractive during all the shear phases performed. Taking into account the presence of the joint, it was preferred to interpret the test results by calculating the shear stresses τ and the effective normal stress σ'_n acting on the surface of the joint. To do this, a mechanical sliding blocks model, similar to the ones proposed by Chandler [12], Webb [13] and Burland [9] was applied. The evolution of the stresses τ and σ'_n is plotted in Fig. 2c as a function of the relative displacement δ between the two halves of the specimen. Plotting these data to the Mohr-plane (σ'_n, τ), it was possible to directly obtain the shear strength envelope for the Opalinus Clay specimens containing the artificial joints (Fig. 2d). The obtained shear strength parameters ($c' = 0$ and $\varphi' = 9.9°$) substantially coincide with those obtained by Ferrari et al. [8] through ring shear tests performed on completely remoulded specimens of Opalinus Clay in shaly facies. The sliding blocks model was then applied to data collected on Fig. 1 and the stress paths (σ'_n, τ) resulting on the failure plane are plotted in Fig. 3. As shown, the stress paths in the post-peak stages are mainly aligned with a strength envelope having $c' = 0$ and $\varphi' = 29°$, which then represent the ultimate shear strength. So, the sliding blocks interpretation is consistent with data reported by Ferrari et al. [8] for a clay fraction equal to 42%.

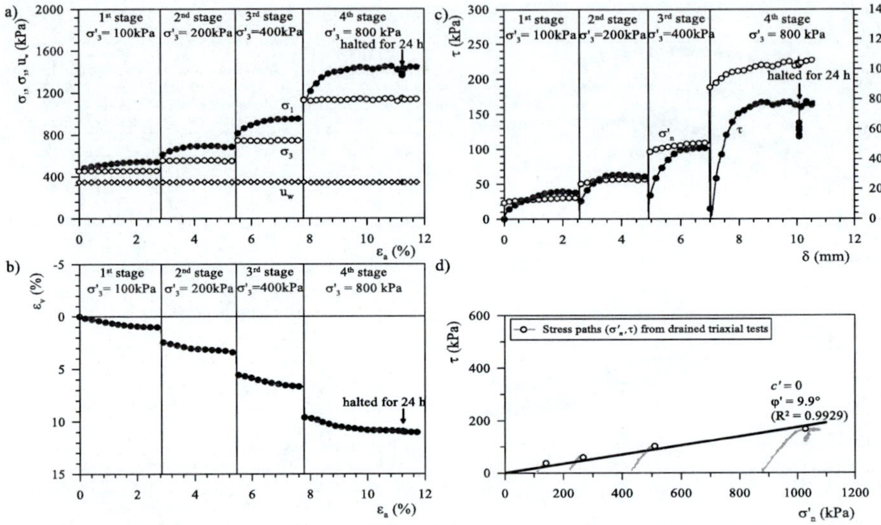

Fig. 2. Total principal stresses σ_1, σ_3, pore pressure u_w (a) and volumetric strain ε_v (b) as a function of axial strain ε_a during the test on specimen with joint. c) Shear stresses τ and effective normal stresses σ_n' acting on the surface of the joint as a function of the relative displacement of the joint δ. d) Strength envelope based on the stress (σ_n', τ) acting along the joint surface.

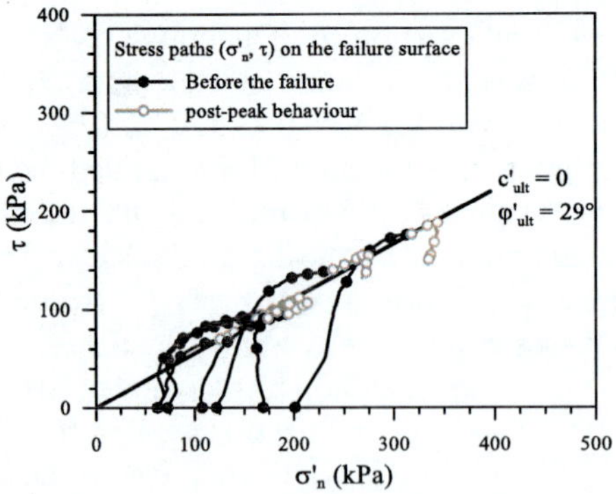

Fig. 3. Stress path (σ_n', τ) along the failure surface of triaxial samples in remoulded Opalinus Clay.

The results of the triaxial test carried out on the artificially jointed samples of Speswhite Kaolin are shown in Fig. 4, where the deviatoric stresses q (Fig. 4a), and the volumetric strain ε_v (Fig. 4b) are plotted as a function of the applied axial strain ε_a.

In Fig. 4c the trends of peak shear strengths (q_{max}) and post-peak shear strengths (q_{ult}) as a function of artificial joints inclination has been plotted. From these diagrams, it is quite evident that, by increasing the inclination of the joint, the mechanical behaviour of the specimen changes from brittle to ductile. By increasing the inclination of the joint, the specimens are characterized by lower peaks of strength, but post-peak strengths are close to each other. The only exception is for the specimen with 75° joint; in this case, the plane of the joints passes through the ends of the specimen, so the deviatoric stress compresses both blocks. Considering that the ratio between height to diameter of triaxial specimens is equal to 2, the reduction in the peak shear strength is not expected to be continued for angles greater than 63°.

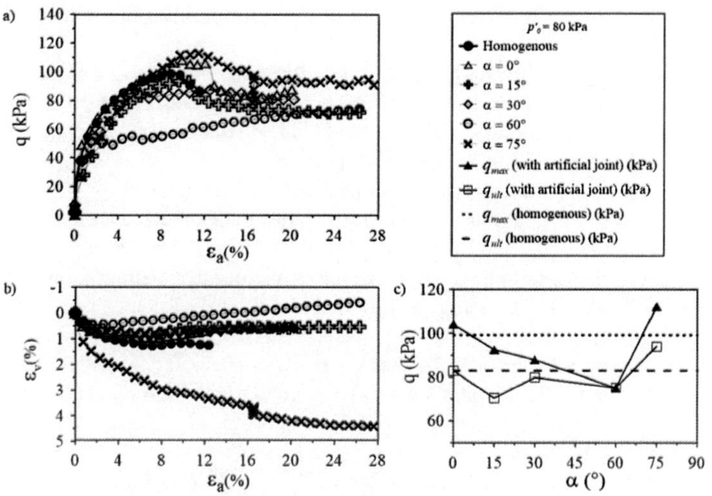

Fig. 4. Deviatoric stress q (a) and volumetric strain ε_v (b) as a function of axial strain ε_a during the triaxial tests on Speswhite Kaolin specimens with joint. c) Peak deviatoric stress q_{max} and post-peak shear stress q_{ult} as a function of the inclination of the artificial joints (α).

5 Concluding Remarks

The note presented a selection of the results of a large experimental program conducted with the aim of characterizing the shear strength of remoulded samples of Opalinus Clay and reconstituted samples of Speswhite Kaolin. Unconventional triaxial tests on homogenous and artificially jointed samples were conducted to investigate how the presence of pre-existing surface with lower strength affect the mechanical behaviour of geomaterials. The results presented for Speswhite Kaolin demonstrate that by increasing the inclination of the joint, the mechanical behaviour of the specimen changes from brittle to ductile. The results presented for Opalinus Clay demonstrate that the over-consolidated specimens show a weak strength peak and a dilatant behaviour which, being inhibited in undrained conditions, results in pore pressures at failure that are negative or slightly greater than zero. A mechanical sliding blocks model was used to interpret in a consistent way the post-peak behaviour of triaxial samples. The presence of joints within the Opalinus Clay homogeneous sample has a significant effect on the mechanical behaviour which becomes ductile and contracting. The shear strength of Opalinus Clay measured under these conditions substantially coincides with the residual strength measured with ring shear tests on remoulded samples.

References

1. Bossart, P., et al.: Mont Terri rock laboratory, 20 years of research: introduction, site characteristics and overview of experiments. Swiss J. Geosci. **110**(1), 3–22 (2017)
2. Gautschi, A.: Safety-relevant hydrogeological properties of the claystone barrier of a Swiss radioactive waste repository: an evaluation using multiple lines of evidence. Grundwasser **22**(3), 221–233 (2017)
3. Ferrari, A., Favero, V., Marschall, P., Laloui, L.: Experimental analysis of the water retention behaviour of shales. Int. J. Rock Mech. Min. Sci. **72**, 61–70 (2014)
4. Favero, V., Ferrari, A., Laloui, L.: Thermo-mechanical volume change behaviour of Opalinus Clay. Int. J. Rock Mech. Min. Sci. **90**, 15–25 (2016)
5. Rosone, M., Ferrari, A., Ziccarelli, M., Giger, S.B.: The residual shear strength of the shaly and sandy facies of the opalinus clay. Springer Series in Geomechanics and Geoengineering, p. 426 – 433 (2019)
6. Favero, V., Ferrari, A., Laloui, L.: Anisotropic behaviour of Opalinus Clay through consolidated and drained triaxial testing in saturated conditions". Rock Mech. Rock Eng. **51**(5), 1305–1319 (2018)
7. Minardi, A., Giger, S.B., Ewy, R.T., Stankovic, R., Stenebråten, J., Soldal, M., Rosone, M. Ferrari, A., Laloui, L.: Benchmark study of undrained triaxial testing of Opalinus Clay shale: Results and implications for robust testing. Geomechanics for Energy and the Environment 25. 100210 (2021)
8. Ferrari, A., Rosone, M., Ziccarelli, M., Giger, S.B.: The shear strength of Opalinus Clay shale in the remoulded state. Geomechanics for Energy and the Environment **21**, 100142 (2020)
9. Burland, J.B.: On the compressibility and shear strength of natural clays. Géotechnique **40**(3), 329–378 (1990)
10. Head K.H.: Manual of soil laboratory testing. vol. 3: effective stress tests, John Wiley & Sons, New York (1986)
11. Escario, V., Saez, J.: Shear strength of partly saturated soils versus suction. Proceedings of the 6th international conference on expansive soils, New Delhi (1987)
12. Chandler, R.J.: The Measurement of Residual Strength in Triaxial Compression. Géotechnique **16**(3), 181–186 (1966)
13. Webb, D.L.: Residual strength in conventional triaxial tests. In: Proceedings of 7th International Conference on Soil Mechanics and Foundation Engineering, Mexico, vol. 1, pp. 433–441 (1969)

On the Fabric of a 3D Printed Soil

Marco Starvaggi[1](✉), Silvia La Rosa[1], Marco Rosone[1], and Alessio Ferrari[1,2]

[1] Department of Engineering, University of Palermo, 90138 Palermo, Italy
{marco.starvaggi,marco.rosone}@unipa.it

[2] Soil Mechanics Laboratory, Ecole Polytechnique Fédérale de Lausanne, 1015 Lausanne, Switzerland

Abstract. 3D printing of earthen materials is on the rise in the field of Additive Manufacturing (AM) due to the opportunity this technique offers to realize structures and manufacts with low environmental impacts, but very little is known about the hydromechanical behavior of soils made with this innovative process. Soils prepared by AM typically present a well-marked double-structured fabric; the characterization of the different structural levels is needed to improve the creation and, consequently, the hydromechanical behavior of 3D printed soils. In this study a simple procedure to estimate the index properties of a 3D printed soil is presented. Then, by means of Scanning Electron Microscopy (SEM) and Mercury Intrusion Porosimetry (MIP), it was possible to take a closer look to the fabric of a 3D-printed soil, highlighting the link between microstructure and the printing process.

Keywords: 3D printed soils · Fabric · Microstructure

1 Introduction

3D printing of soils has recently gained a lot of attention in the building industry due to its low environmental impact and the high availability and recyclability of earthen materials. This trend may lead to an increase in sustainability in the construction process, which is responsible for 40% of the CO_2 emission and 36% of global fine energy use, as reported in a survey made in 2018 [1]. Recent technical developments have increased the potential for large-scale use of additive manufacturing (AM) of soils [2, 3]. Moreover, this technique could be used to construct future carbon-negative buildings, i.e. self-supporting structures that can support the germination and growth of plants and absorb CO_2 via photosynthesis [4]. However, some limitations in this promising manufacturing technique applied to soil structures still remain, such as the high sensitivity to environmental actions and slow construction process [5]. When the AM process involves soils as printing material, geomechanics plays a fundamental role; so it is useful to investigate the hydromechanical behaviour of 3D printed soils to identify the key factors of the printing process that can be optimized.

Fabric is one of the major factors influencing the hydromechanical behaviour of clayey soils [6–9]. This fact seems highly relevant for 3D printed soils since the fabric is strongly controlled by the printing process.

A. Ferrari et al. (Eds.): CNRIG 2023, SSGG, pp. 135–142, 2023.
https://doi.org/10.1007/978-3-031-34761-0_17

This paper initially describes a 3D printing process for soil at the laboratory scale. The index properties of a 3D printed soil are defined, along with the procedure for their determination. Finally, the fabric of a 3D printed soil sample is shown and discussed thanks to the use of Scanning Electron Microscope (SEM) and Mercury Intrusion Porosimetry (MIP).

2 Example of 3D Printing Process of a Geomaterial

The schematic representation of the 3D printer for geomaterials used in this work is shown in Fig. 1a. The process begins with the extrusion of the soil from a tank using a ram pressurized by compressed air. The extruded soil is then convoyed to a second extruder with an endless screw driven by a stepper motor. The soil is steered through a nozzle and finally released as a filament. Robotic arms allow the movement of the extruder. The volume of the printed soil with the defined geometry is obtained as a superposition of different layers [10]. Figure 1b shows a soil sample being 3D printed. At the end of the process, the final structure of the printed volume is the result of the path followed by the screw extruder, as well as of the cross-section of the nozzle of the printer that shapes the filaments [5].

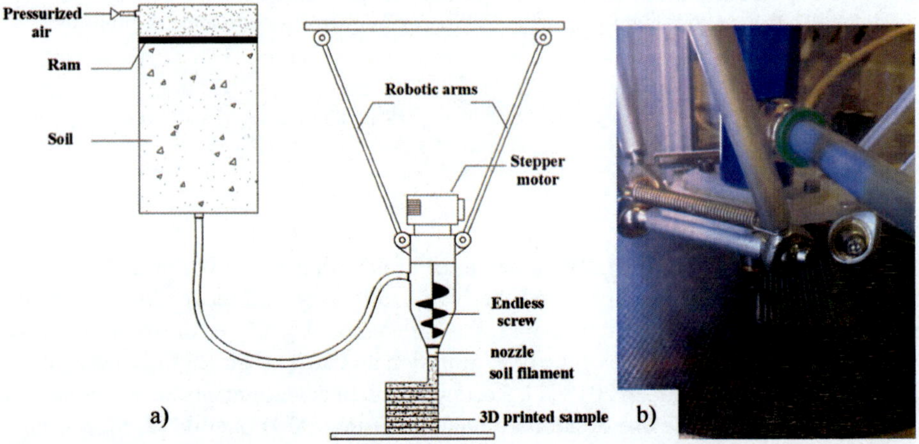

Fig. 1. Schematic representation of the basic component of a 3D printer for geomaterials (Drawing not in scale) (a). Photo taken during printing operation of a soil sample (b).

3 Definition of Index Properties for a 3D Printed Soil

The concentrated phases diagram of Fig. 2 summarizes the terms needed to describe the index properties for a 3D printed soil. The figure distinguishes voids that are found within the filaments (micropores) and voids that are in between them (macropores). In this sense, the microporosity (n^m) is the ratio of the volume of the voids within the filaments $\left(V_v^m\right)$ to the total volume of the 3D printed sample:

$$n^m = \frac{V_v^m}{V_{tot}} \tag{1}$$

The macroporosity of a 3D printed sample represents the ratio of the volume of macropores $\left(V_v^M\right)$ to the total volume of the 3D printed sample (V_{tot}):

$$n^M = \frac{V_v^M}{V_{tot}} \tag{2}$$

Macroporosity is a controlled aspect of the 3D printing process, being related to the volume of the voids resulting from the specific path followed by the extruder during the deposition of the filaments. Thus, it can be determined from geometrical considerations.

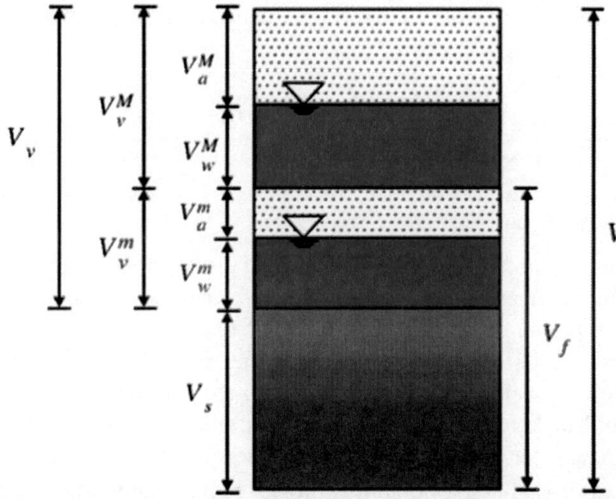

Fig. 2. Schematization of the different phases for a 3D printed soil. V_{tot} is the total volume; V_f is the total volume of the filaments; V_s is the volume of solid phase, V_w^m is the volume of water inside the micropores; V_a^m is the volume of air inside the micropores; V_w^M is the volume of water inside the macropores; V_a^M is the volume of air inside the macropores; V_v^m is the total volume of micropores; V_v^M is the total volume of macropores; V_v is the total volume of pores.

The total porosity of a 3D printed soil (n) is the sum of the microporosity (n^m) and macroporosity $\left(n^M\right)$:

$$n = n^m + n^M \tag{3}$$

The porosity of a single filament $\left(n^f\right)$ is the ratio of the volume of voids within the. filament $\left(V_v^m\right)$ to the total volume of the filament $\left(V_f\right)$ itself:

$$n^f = \frac{V_v^m}{V_f} \tag{4}$$

For what follows in the next section, it is useful to write the relationship among the microporosity (n^m), the porosity of a single filament $\left(n^f\right)$, and macroporosity $\left(n^M\right)$:

$$n^m = n^f\left(1 - n^M\right) \tag{5}$$

4 Material and Method

The 3D printer used to obtain the analysed samples was a *Delta-WASP2040-Clay* equipped with a nozzle having a circular cross-section and a diameter of 1.4 mm. The soil used as printing material was a commercial clayey soil; its geotechnical characteristics are listed in Table 1.

Table 1. Geotechnical properties of 3D printed soil (Specific Gravity G_s, liquid limit w_l, plastic limit w_p, water content of the 3D printed soil w_n, clay fraction f_{clay}, silt fraction f_{silt}, sand fraction f_{sand}).

G_s [-]	w_l [-]	w_p [-]	w_n [-]	f_{clay} [%]	f_{silt} [%]	f_{sand} [%]
2.77	0.45	0.21	0.30	52	46	2

First, the porosity and the degree of saturation of the filament $\left(n^f\right)$ have been assessed by means of a fluid displacement method [11]. Kerdane was used as apolar liquid for the measurements and filaments long at least 1 m were tested. The filament was 3D printed within a specified graduated beaker which was then entirely filled with kerdane. The difference between the weight of the beaker full of kerdane with the 3D printed filament, and the beaker full of kerdane, allows to compute the total volume of the 3D printed filament $\left(V_f\right)$. Then, after desiccation, knowing the specific gravity (G_s) and the water content (w_n), it was possible to obtain the porosity of the filament $\left(n^f\right)$ and its degree of saturation (S_r).

Next, four samples were 3D printed with a prescribed geometry. Figure 3a shows the printing operation where each layer is composed of adjacent parallel filaments obtained as a serpentine. Successive layers have orthogonal orientation one to each other. When the soil is 3D printed following this path, the macroporosity can be geometrically estimated considering the filaments as cylinders with circular cross-section and indefinitely rigid, filling a parallelepiped with a square base. From the analysis of the variation of porosity with the sample dimension, it was demonstrated that when the side of the square basis of the parallelepiped is one order of magnitude bigger than the diameter of the single filament, the macroporosity remains constant and equal to 0.21. Specimens for the measurement of porosity were obtained cutting the 3D printed soil into parallelepipeds with square base (Fig. 3b).

For each specimen, the total volume (V_{tot}) was measured and the volume of the voids (V_v) was determined after desiccation knowing the specific gravity (G_s) and the water content (w_n).

An analysis of the equivalent pore diameter was possible thanks to Image segmentation of Scanning Electronic Microscopy (*SEM*) observation and Mercury Intrusion Porosimetry (*MIP*) on freeze-dried 3D printed specimens.

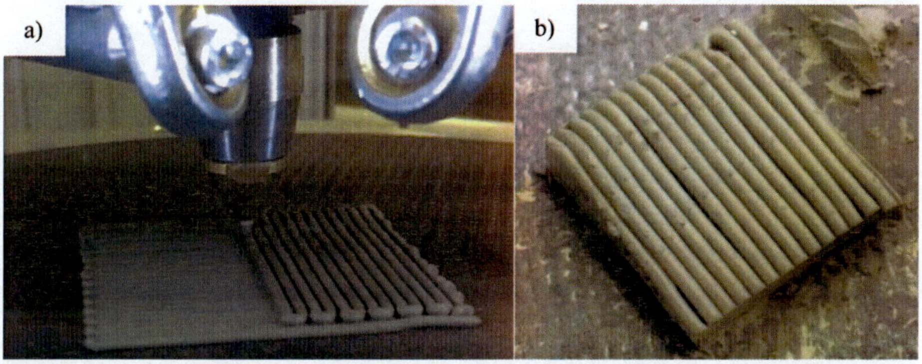

Fig. 3. Photo taken during printing operation (a). A specimen used for the measurement of porosity (b).

5 Results

The tests performed on the filaments resulted in an average filament porosity $\left(n^f\right)$ equal to 0.44 and a degree of saturation S_r equal to 1. Considering macroporosity $\left(n^M\right)$ as a constant and equal to 0.21, according to Eq. (5), microporosity (n^m) results equal to 0.34; hence, from Eq. (3) the estimated total porosity of the 3D printing specimens can be predicted as equal to 0.56. This value is compared with the porosity of four specimens assessed according to the laboratory procedure presented in the previous section. Table 2 shows the obtained results, which seem to be in good agreement with the estimated total porosity of the specimens since these latter have shown an average value of 0.55.

Figure 4a shows the SEM photomicrograph of the cross-section of a specimen obtained with a magnification factor of 60X. It is noticeable that an ovalization of the cross-section has occurred with respect to the circular cross-section of the nozzle used for 3D printing, probably due to its own weight and the weight of the upper layers.

Interestingly, from some measures made with image segmentation, the area of the ovalized cross-section coincides with the one of a circle having the diameter of the nozzle ($d_m = 1.4$ mm), suggesting that the deformation of the filament occurs in undrained conditions. Moreover, the ovalized cross-section does not change the value of the macroporosity $\left(n^M\right)$ geometrically estimated, which remains equal to 0.21. The expected equivalent diameter of macropore derived from geometrical consideration is 715 μm, while the one obtained from image segmentation of SEM photomicrograph results equal to 640 μm. This little discrepancy is due to the saddle effect of the upper filament when laying above two adjacent filaments, clearly visible from Fig. 4a, while the underlying filament is out of focus, and hence considered as straight. If the saddle effect would be

Table 2. Porosity of four 3D printed samples. V_{tot} total volume of the sample, W_s dry weight of sample, V_s volume of solid fraction, V_v volume of voids, n porosity of sample.

Sample	V_{tot} [cm^3]	W_s [g]	V_s [cm^3]	V_v [cm^3]	n [-]
1	3.18	4.31	1.55	1.62	0.51
2	6.15	7.52	2.71	3.43	0.56
3	11.55	12.78	4.61	6.94	0.60
4	8.14	10.49	3.78	4.35	0.53
Average					0.55

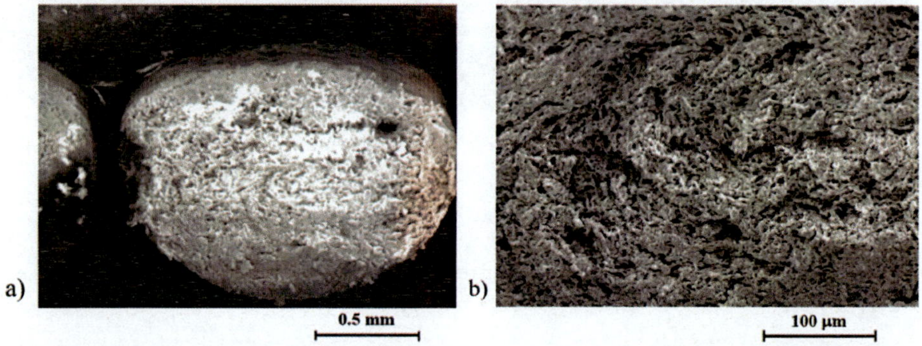

Fig. 4. SEM photomicrographs of the 3D printed soil. Cross-section of the specimen with a magnification of 60X (a). Cross-section of a soil filament with a magnification of 300X (b).

replicated also for the underlying filament, the equivalent diameter of the pore returns to about 700 μm.

Figure 4b shows the cross-section of a soil filament with a magnification of 300X. A spiral trend can be observed in the particle arrangement due to the rotational movement of the screw in the final extruder of the 3D printer. Here, both pores with a maximum diameter of about 30 μm, and pores 1–2 μm wide and 10 ÷ 70 μm long at the contacts between the clay aggregates can be observed.

Figure 5 presents the Pore Side Density PSD (PSD $= -\Delta e_i/\Delta(log\ d)$) as a function of the computed pore diameter d. Pores with an entrance pore diameter of 120 μm or greater are not detected because mercury filled the macropores before the test started properly. Thus, the intruded volume is only related to the pores within the soil filaments. The main modal diameter of pores was 1 μm, which reasonably represents the mean value of the micropores.

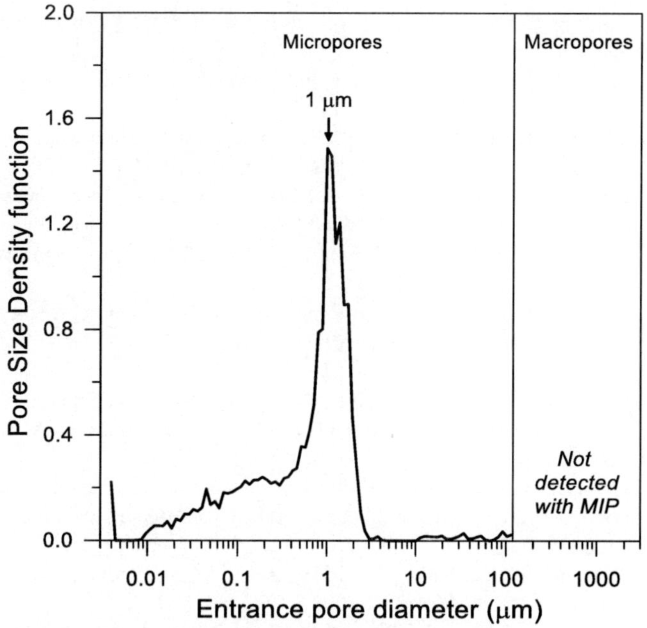

Fig. 5. Pore Size Density as a function of the Entrance Pore Diameter from MIP tests.

6 Conclusion

Results showed that knowing the porosity of the filament $\left(n^f\right)$ and the macroporosity $\left(n^M\right)$ generated by the path followed by the 3D printer, the porosity of the specimens can be estimated with good confidence. Microstructural investigations have highlighted the features of the double porosity network which characterizes 3D printed soil. In particular, the investigation on the equivalent diameter of pores within the 3D printed soil showed a difference of over 2 orders of magnitude between micro and macropores dimensions. Macropores dimension only depends on the geometry imposed during printing operation and it was observed that the deformation of the filaments, due to the superimposed layers, happens in undrained conditions; this does not seem to affect the value of the equivalent diameter of macropores. Particle arrangement within the soil filaments seems to be also highly affected by the printing process, showing a spiral trend in the cross section due to the screw extruder.

Even though further studies are necessary to clearly understand the hydromechanical response of 3D printed soils, this study lays the foundations for better describing and managing the creation of fabric-controlled samples, in the growing field of additive manufacturing of geomaterials.

References

1. Alhumayani, H., Gomaa, M., Soebarto, V., Jabi, W.: Environmental assessment of large-scale 3D printing in construction: a comparative study between cob and concrete. J. Clean. Prod. **270**, 122463 (2020)

2. 3D WASP printing architecture web page. https://www.3dwasp.com/en/3d-printing-architecture. Accessed 10 Jan 2023
3. Izard, J.-B., et al.: Large-scale 3D printing with cable-driven parallel robots. Constr. Robot. **1**(1–4), 69–76 (2017)
4. Barnes, S., Kirssin, L., Needham, E., Baharlou, E., Carr, D.E., Ma, J.: 3D printing of ecologically active soil structures. Addit. Manuf. **52**, 102670 (2022)
5. Perrot, A., Rangeard, D., Courteille, E.: 3D printing of earth-based materials: processing aspects. Constr. Build. Mater. **172**, 670–676 (2018)
6. Collins, K., McGown, A.: The form and function of microfabric features in a variety of natural soils. Géotechnique **24**(2), 223–254 (1974)
7. Romero, E., Simms, P.H.: Microstructure investigation in unsaturated soils: a review with special attention to contribution of mercury intrusion porosimetry and environmental scanning electron microscopy. Geotech. Geol. Eng. **26**, 705–727 (2008)
8. Ferrari, A., Bosch, J.A., Baryla, P., Rosone, M.: Volume change response and fabric evolution of granular MX80 bentonite along different hydro-mechanical stress paths. Acta Geotech. **17**, 3719–3730 (2022)
9. Rosone, M., Megna, B., Celauro, C.: Analysis of the chemical and microstructural modifications effects on the hydro-mechanical behaviour of a lime-treated clay. Int. J. Geotech. Eng. **15**, 447–460 (2021)
10. Ferrari, A., Rosone, M., La Rosa, S., Sapienza, G.: Microstructural characterization of a 3D-printed soil. Soils Rocks **45**(4), 1–6 (2022)
11. Péron, H., Hueckel, T., Laloui, L.: An improved volume measurement for determining soil water retention curves. Geotech. Test. J. 30(1), 100167 (2007)

New Trends and Applications for Measurements and In-Situ Monitoring

Surface Wave Testing with Distributed Acoustic Sensing Measurements to Estimate the Shear-Wave Velocity and the Small-Strain Damping Ratio

Mauro Aimar[1](✉), Brady R. Cox[2], and Sebastiano Foti[1]

[1] Politecnico di Torino, Corso Duca Degli Abruzzi 24, 10129 Turin, Italy
mauro.aimar@polito.it

[2] Utah State University, Old Main Hill, Logan, UT 84322, USA

Abstract. An accurate in-situ estimate of the shear-wave velocity and the small-strain damping ratio profiles is paramount for quantifying the response of soil deposits to dynamic loading. An effective approach relies on the multichannel analysis of surface waves (MASW), which measures the phase velocity and attenuation of Rayleigh waves to derive the stiffness and damping parameters of the soil deposit. This contribution presents results from a MASW survey carried out at Hornsby Bend (Texas), wherein waveforms were recorded simultaneously with a geophone array and a fiber-optic distributed acoustic sensing (DAS) array. DAS is an innovative technology in seismic measurements and monitoring, whose use in geophysics is still limited yet promising. DAS waveforms were processed to extract the Rayleigh wave propagation parameters. Experimental data were then mapped into suitable shear-wave velocity and damping ratio profiles, by means of a Monte Carlo-based inversion algorithm. This study represents the first joint characterization of stiffness and dissipative parameters of a soil deposit based on a fiber-optic array, to our knowledge. The comparison with results from the geophone array demonstrates the reliability of the DAS technology for subsurface characterization. Besides, DAS data exhibit low variability, entailing a high level of accuracy. Therefore, the DAS technology can be successfully used for the joint derivation of the shear-wave velocity and the small-strain damping ratio. It is believed that the diffusion of this technology in geophysical characterization will improve the quality of the in-situ estimates of soil parameters, thus enhancing the reliability of the predicted seismic ground response.

Keywords: Viscoelasticity · MASW · Distributed acoustic sensing

1 Introduction

The small-strain shear-wave velocity and the damping ratio quantify the stiffness and the internal energy dissipation by the soil at low strains, and they have great relevance in the soil response to dynamic loading (e.g., [1]). A promising technique for obtaining these parameters relies on the Multichannel Analysis of Surface Waves (MASW; [2]).

A. Ferrari et al. (Eds.): CNRIG 2023, SSGG, pp. 145–152, 2023.
https://doi.org/10.1007/978-3-031-34761-0_18

This approach relies on the measurement of the spatial phase lag and attenuation of Rayleigh waves along linear arrays with active sources. In this case, the S-wave velocity and damping ratio profiles are jointly estimated through an inversion scheme, where a theoretical soil model is calibrated to match the experimental dispersion and attenuation data.

This study aims at investigating the effectiveness of using fiber-optic cables as measuring devices in MASW testing. Indeed, this technology is gaining relevance in invasive geophysical testing and near-surface characterization (e.g., [3, 4]). However, at the current state of knowledge, there is no application focusing on the joint characterization of stiffness and dissipative characteristics. Therefore, this contribution describes a suite of MASW surveys at the Hornsby Bend (HB) site, in the United States, wherein waveforms were recorded through a geophone array and a fiber-optic cable, implementing the distributed-acoustic sensing (DAS) technology. On the one hand, the research investigates the capability of this new system in retrieving velocity and attenuation data, with a focus on the issues linked with the processing scheme and the reliability of the estimated R-wave parameters. Then, experimental data are jointly inverted to derive S-wave velocity and damping ratio profiles at the investigated site. This contribution represents the first application of fiber-optic data to retrieve the R-wave phase attenuation.

The paper starts with an overview of the DAS technology and it presents the experimental dataset. Then, it addresses the estimation of the Rayleigh phase velocity and phase attenuation and the comparison between the derived propagation parameters obtained from geophone and fiber-optic data at the HB site is presented. The final part of the paper focuses on the derivation of the earth models from experimental data.

2 The Distributed Acoustic Sensing (DAS) Technology

Distributed acoustic sensing (DAS) records a spatially-averaged axial strain induced on the fiber optic by the passing wavefield. Indeed, the passage of mechanical waves induces an axial strain in the fiber-optic cable, that is coincident with the horizontal, in-line strain in the ground when a proper coupling is ensured. An interrogator unit reads the consequent shift in phase lag of a laser pulse traveling in the cable, induced by the variation in the length of the cable itself. However, the device reads the variation in phase difference over a reference length $2g$, called gauge length, around the investigated location, from which the average strain is derived (Fig. 1; [5]). The gauge length affects the quality of spatial sampling, potentially affecting the high-frequency components of the R-wavefield, especially when $2g$ is large. On the other hand, an increase in $2g$ results in an improvement of the signal quality (e.g., [5]). For this reason, it is recommended to carry out multiple measurements, where the gauge length is modified in each step [6].

In MASW testing, the main advantage of the DAS technology with respect to conventional acquisition devices is the enhanced spatial resolution using low-cost instrumentation (e.g., [7]). Applications of this technology to MASW surveys demonstrated that the phase velocity estimates well match those obtained from geophone measurements. However, fiber-optic systems are uniaxial devices, recording only perturbations acting in the longitudinal direction, and the correct location of measurement points may be uncertain in some cases [6]. Also, the signal-to-noise ratio of measured data is lower compared to geophones.

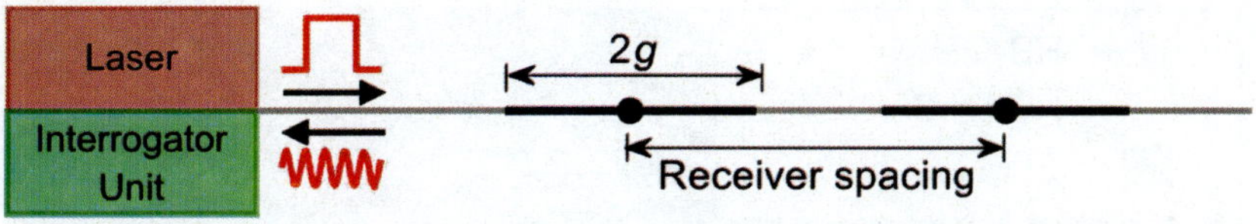

Fig. 1. Schematic model of the DAS system, modified from [6].

3 Site Description and MASW Survey

The site of Hornsby Bend (HB, 30°13.918′N, 97°38.631′W – in the WGS84 Datum) is located on the outskirts of the city of Austin, Texas. The site is a flat area next to the Colorado River, with a stratigraphy mostly characterized by layered alluvial soils. Specifically, a moderately stiff 15 m thick clayey layer overlies a loose-to-medium dense layer with fine sands. At greater depths, stiff clays are found.

The geophysical investigation at the HB site involved two one-dimensional arrays, along the side of a small unpaved road (Fig. 2). The first layout was an array of 48, regularly spaced vertical geophones and 48 horizontal geophones. The inter-receiver spacing was equal to 2 m, hence the total extent of the array was 94 m. This testing setup is hereafter labeled as HB-GP. The receivers were GeoSpace GS-11D 4.5-Hz vertical geophones, whereas four interconnected 24-channel Geometric Geode seismographs recorded waveform data. The second acquisition scheme was a DAS array. This survey utilized a 200-m long NanZee Sensing Technology (NZS-DSS-C02) fiber-optic cable, which was installed adjacent to HB-GP. The cable was buried inside a trench, backfilled with compacted soil to ensure an appropriate coupling of the cable with the ground. Axial strain data were recorded by the ODH4 OptaSense Interrogator Unit, according to a gauge length equal to 2.04 m and a channel separation of 1.02 m, i.e., measurements of the wavefield were provided approximately every 1 m along the cable. This testing setup is hereafter identified as HB-DAS.

HB-GP and HB-DAS simultaneously recorded waveforms generated through the NHERI@UTexas Thumper vibroseis truck [8], which generated a 12-s long chirp signal, with frequency shifting from 5 Hz to 200 Hz. Source locations are on the North-East side, with offsets equal to 5 m, 10 m, 20 m, and 40 m from both HB-GP and HB-DAS. At each shot point, three repetitions were run.

4 Extraction of R-Wave Parameters

The interpretation of experimental data from the HB-GP array was carried out through the Cylindrical Frequency-Domain BeamForming – Attenuation algorithm with Modal Filtering (CFDBFaMF; [9]). This approach is a generalization of the Frequency-Domain BeamForming (FDBF) method [10] for the attenuation estimate, and it is suitable for extracting estimates of modal phase velocity and attenuation, accounting for the cylindrical shape of the Rayleigh wavefront. For simplicity, only the data from vertical geophones were considered for the HB-GP array.

As for the HB-DAS, the extraction of the R-wave parameters adopts a modified version of the CFDBFaMF, which implements an average strain-based beamforming

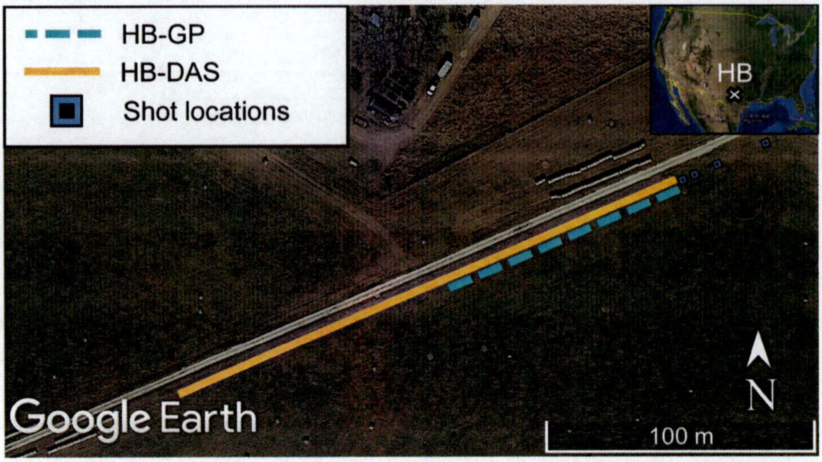

Fig. 2. Hornsby Bend site and acquisition setup of HB-DAS and HB-GP.

(further details are available in [9]). In this way, the procedure properly models the spatial variation of the amplitude and phase of the average radial strain. In this case, R-wave parameters were extracted considering a 94 m-long portion of the fiber-optic cable coincident with the geophone array, so that they sample a comparable volume of the soil deposit.

Figure 3 compares the estimated modal dispersion and attenuation data for the fundamental mode (labeled as "R0") and the first higher mode ("R1"), obtained from the interpretation of the HB-DAS and the HB-GP data. In this case, the data distribution is represented by the interval around the median value, the width of which equals one logarithmic standard deviation [9]. Data statistics are obtained by combining results from different source offsets, in consistency with the multi-offset approach [11].

In general, dispersion and attenuation data well match with each other, although the DAS data do not allow to obtain reliable values at low frequencies. This partially limits the capability of the DAS system in characterizing deeper layers. However, the corresponding degree of data variability is generally less or equal to the one affecting geophone-based estimated parameters. A possible reason behind the low data scatter can be the remarkably larger number of measurement points that the DAS system includes, which provides a more exhaustive dataset of wavefield values to better constrain the velocity and the attenuation estimates.

5 Joint Inversion of Dispersion and Attenuation Data

Experimental R-wave parameters were used to estimate the S-wave velocity and damping ratio profiles, through the joint inversion of phase velocity and attenuation data. This study adopts the experimental data from HB-DAS. Note that the set of experimental data represents a quite challenging condition, as multiple R-wave propagation modes are jointly inverted.

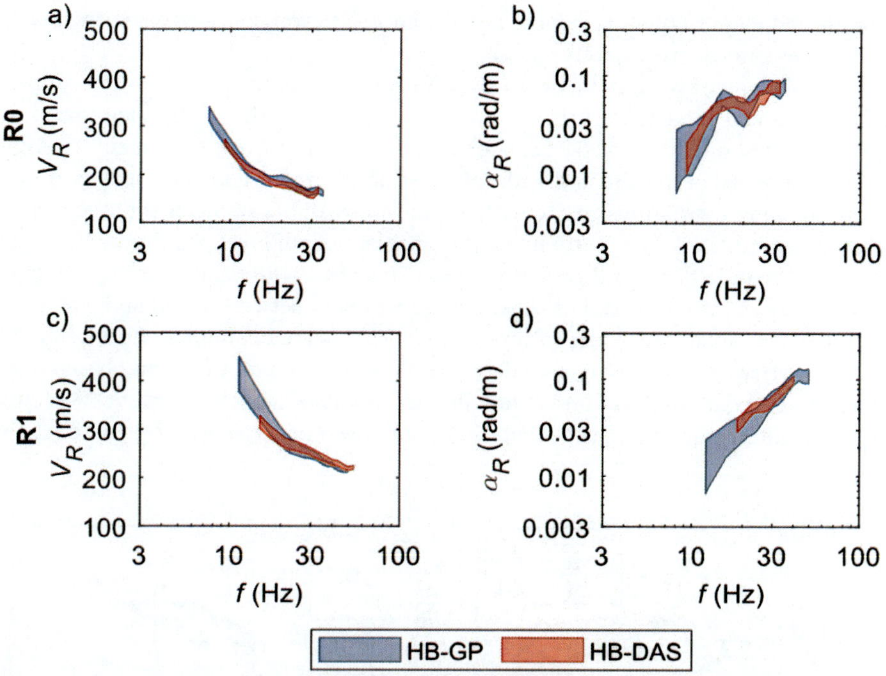

Fig. 3. Comparison between the estimated dispersion data (phase velocity V_R vs. frequency f) and attenuation data (phase attenuation α_R vs. frequency f) from the HB-DAS and the HB-GEO data: a-b) Resulting dispersion (a) and attenuation (b) data for the fundamental mode; c-d) Resulting dispersion (c) and attenuation (d) data for the first higher mode. Estimated data are represented in terms of intervals given by one logarithmic standard deviation around the median value.

The inversion adopted an improved Monte Carlo scheme, which exploits the scaling properties of the Rayleigh eigenvalue problem in linear, viscoelastic media [9]. This strategy allows to optimize the available random samples, ensuring a good quality result with a moderately small number of generated ground models. The randomization of earth models investigated an adequate range of layer thicknesses, S-wave velocities, and damping ratios, based on a three-layer ground model. This choice relies on the stratigraphy inferred by cone penetration soundings carried out close to the investigated site [12]. Instead, the Poisson's ratio and mass densities were fixed at realistic values.

The inversion was run using 10,000 trial earth models. Forward dispersion and attenuation modeling was carried out through the ElastoDynamics Toolbox [13]. The degree of fit to experimental data is quantitatively measured through the following misfit function:

$$S(\mathbf{m}) = \frac{1}{2Mn}\sum_{j=0}^{M-1}\sum_{i=1}^{n}\left[\frac{\left(\ln V_{i,\mathrm{R}j,e} - \ln V_{i,\mathrm{R}j,t}\right)^2}{\sigma^2_{\ln V,i,\mathrm{R}j}} + \frac{\left(\ln \alpha_{i,\mathrm{R}j,e} - \ln \alpha_{i,\mathrm{R}j,t}\right)^2}{\sigma^2_{\ln \alpha,i,\mathrm{R}j}}\right] \quad (1)$$

The definition compares theoretical dispersion data $V_{i,\mathrm{R}j,t}$ and attenuation data $\alpha_{i,\mathrm{R}j,t}$ and observed values $V_{i,\mathrm{R}j,e}$ and attenuation data $\alpha_{i,\mathrm{R}j,e}$, for each considered propagation mode Rj and each frequency sample i. Differences are weighted by the corresponding standard deviations $\sigma_{\ln V,i,Rj}$ and $\sigma_{\ln\alpha,i,Rj}$.

Figure 4 shows results for the best-fitting 30 models. Inverted S-wave velocity profiles exhibit a gradual increase in stiffness with depth. The depths of the identified layers interfaces are about 4 m and 12 m. This result is consistent with the main geological interfaces inferred at the site and with information from past geophysical surveys [12]. Instead, although variations in D_S with depth follow the same layer interfaces as V_S, D_S increases from 5% to 7 ÷ 9% in the intermediate-depth layer. As for the half-space, the variability in both V_S and D_S dramatically increases with respect to shallow layers. However, D_S spans over a rather broader range, mostly between 0.5% and 5% (that is, the variation is about one order of magnitude). This is the combined effect of the high $\sigma_{\ln\alpha}$ in low-frequency experimental data, the relevant influence of V_S on phase velocity and attenuation data, and the moderately low sensitivity of theoretical attenuation curves to D_S at great depths (e.g., [14]), that does not allow a constraint on D_S as effective as

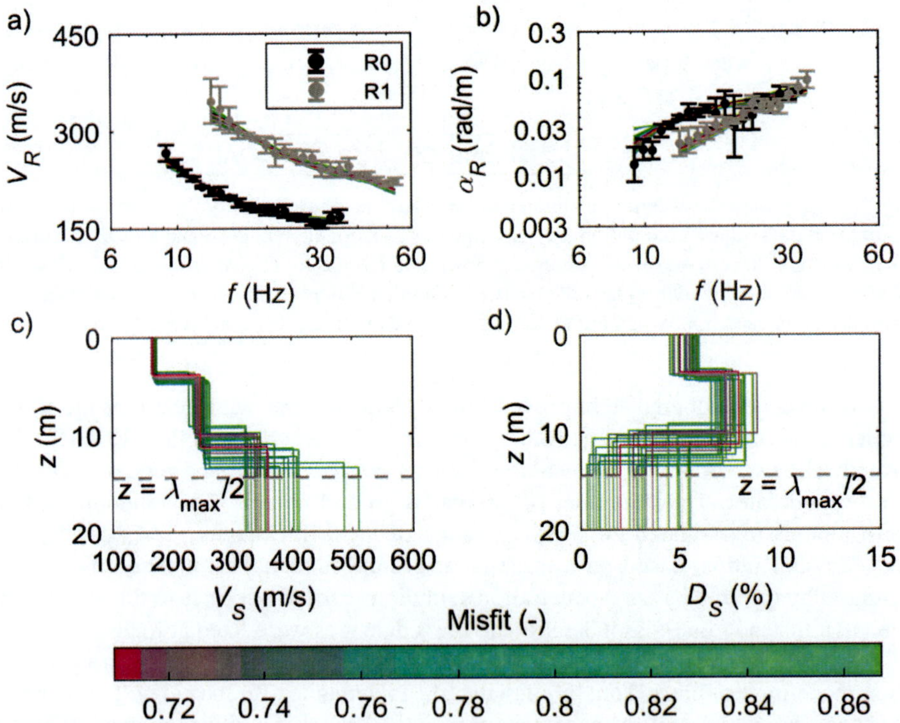

Fig. 4. Best fitting ground models to HB-DAS experimental data: a-b) Theoretical vs. experimental data, for the phase velocity (a) and phase attenuation (b); c-d) Resulting S-wave velocity (c) and damping ratio (d) profiles. The width of error bars in the experimental data is related to the logarithmic standard deviation. The boundary $z = \lambda_{max}/2$ is an approximated value of the maximum investigable depth, that can be achieved from the available experimental data – layer interfaces beneath it are usually less reliable.

in the stiffness modeling. However, it should be remarked that both the velocity and damping ratio profiles exhibit well defined trends, especially in the near-surface layers.

6 Conclusions

This contribution addressed the potential of fiber-optic cables for near-surface characterization of the small-strain S-wave velocity and damping ratio through MASW testing. For this purpose, the study compared estimated phase velocity and phase attenuation data obtained from geophone and fiber-optic Distributed Acoustic Sensing (DAS) data at the Hornsby Bend site. In this way, the capability of this new system to jointly retrieve phase velocity and attenuation data was addressed.

In order to process DAS data, the CFDBFaMF algorithm was adapted to account for the peculiar geometric attenuation that the recorded average strain by the fiber-optic assumes. The comparison demonstrated that, on average, the resulting phase velocity and phase attenuation data are compatible with each other. However, DAS data exhibit lower variability. This improvement is the effect of the remarkably larger number of measurement points that the DAS system includes and the consequently improved spatial resolution, which provides a more exhaustive dataset of wavefield values to better constrain the velocity and the attenuation estimates. Therefore, the DAS technology can be successfully used to jointly estimate the phase dispersion and attenuation data, obtaining the same level of reliability of geophone arrays and improved accuracy.

Finally, experimental data obtained from the HB-DAS survey were jointly inverted to obtain S-wave velocity and damping ratio profiles. The improved sampling scheme implemented in the inversion algorithm and the presence of multi-mode observed data resulted in well-constrained S-wave velocity and damping ratio profiles, especially in the near-surface layers. However, the estimated ground models are affected by greater variability at depth, especially in terms of the damping ratio. The great scatter is the combined effect of the high $\sigma_{\ln\alpha}$ in low-frequency experimental data and the moderately low sensitivity of theoretical attenuation curves to D_S at great depths. This study represents the first joint inversion of dispersion and attenuation data extracted from a fiber-optic array, to our knowledge.

References

1. Foti, S., Aimar, M., Ciancimino, A.: Uncertainties in small-strain damping ratio evaluation and their influence on seismic ground response analyses. In: Latest Developments in Geotechnical Earthquake Engineering and Soil Dynamics, pp. 175–213. Springer, Singapore (2021)
2. Foti, S., Lai, C.G., Rix, G.J., Strobbia, C.: Surface Wave Methods for Near-Surface Site Characterization. CRC Press, Boca Raton (2014)
3. Galan-Comas, G.: Multichannel analysis of surface waves using distributed fiber optic sensors. Mississippi State University (2015)
4. Mateeva, A., et al.: Distributed acoustic sensing for reservoir monitoring with vertical seismic profiling. Geophys. Prospect. **62**, 679–692 (2014)
5. Bakulin, A., Silvestrov, I., Pevzner, R.: Surface seismics with DAS: an emerging alternative to modern point-sensor acquisition. Lead. Edge **39**(11), 808–818 (2020)

6. Bakku, S.K.: Fracture characterization from seismic measurements in a borehole. Massachusetts Institute of Technology (2015)
7. Jousset, P., et al.: Dynamic strain determination using fibre-optic cables allows imaging of seismological and structural features. Nat. Commun. **9**(1), 1–11 (2018)
8. Stokoe, K.H., II., Cox, B.R., Clayton, P.M., Menq, F.-Y.: NHERI@UTexas experimental facility with large-scale mobile shakers for field studies. Front. Built Environ. **6**, 575973 (2020)
9. Aimar, M.: Uncertainties in the estimation of the shear-wave velocity and the small-strain damping ratio from surface wave analysis. Ph.D. thesis, Politecnico di Torino (2022)
10. Lacoss, R.T., Kelly, E.J., Toksoz, M.N.: Estimation of seismic noise structure using arrays. Geophysics **34**(1), 21–38 (1969)
11. Cox, B.R., Wood, C.M., Teague, D.P.: Synthesis of the UTexas1 surface wave dataset blind-analysis study: inter-analyst dispersion and shear wave velocity uncertainty. In: Geo-Congress 2014, Atlanta, pp 850–859 (2014)
12. Kallivokas, L.F., Fathi, A., Kucukcoban, S., Stokoe, K.H., II., Bielak, J., Ghattas, O.: Site characterization using full waveform inversion. Soil Dyn. Earthq. Eng. **47**, 62–82 (2013)
13. Schevenels, M., Degrande, G., François, S.: EDT: an elastodynamics toolbox for MATLAB. Computers & Geosciences (2009)
14. Verachtert, R.: Deterministic and probabilistic determination of dynamic soil characteristics. Ph.D. thesis, KU Leuven (2018)

Distributed Fiber-Optic Sensors for Monitoring Slow Landslides and Anchors for Their Stabilization

Lorenzo Brezzi[1(✉)], Emilia Damiano[2], Luca Schenato[3,4], Martina De Cristofaro[2], Nadia Netti[5], Lucio Olivares[2], and Simonetta Cola[1]

[1] DICEA University of Padua, Padua, Italy
lorenzo.brezzi@unipd.it
[2] Department of Engineering, University of Campania "L. Vanvitelli", Naples, Italy
[3] Research Institute for Geo-Hydrological Protection, CNR, Padua, Italy
[4] Department of Information Engineering, University of Padua, Padua, Italy
[5] DEMI University of Naples "Federico II", Naples, Italy

Abstract. Over the last few decades, both natural and man-made territories in Italy have shown great vulnerability due to the impact of climate change and the lack of maintenance of the territory itself. This makes it necessary to adopt surveillance and alert systems able to detect, in real time, the imminence of critical phenomena. In this context, optical fiber-based sensors are proving to be an innovative and appealing instrumentation in many applications due to their low cost-effectiveness and their possible large-scale use. In the geotechnical field, their employment in the monitoring of landslide deformation fields, or of elements interacting with landslides, has been tested both in the laboratory and at full scale. This paper presents two recent applications of distributed fiber-optic sensors on slow-moving landslides: in one application, they are the basic element of a smart inclinometer; in the other, they are the tool for monitoring passive composite anchors properly designed for slope stabilization. Measurements taken over more than one year show their feasibility in realizing a stand-alone real-time monitoring system for both landslides and interacting structures, but also reveal some difficulties and limitations inherent in their use.

Keywords: DFOS · smart inclinometer · smart passive anchors · landslides · innovative monitoring · risk management · early-warning systems

1 Introduction

After earthquakes, landslides cause the greatest number of fatalities and damages in Italy [1], with significant social and economic impacts. If stabilization of a slope cannot be effective or economically sustainable, the protection of structures and people can be pursued by installing monitoring systems to keep the status of the landslide under control. Even when interventions are possible, monitoring is useful to assess their effectiveness in stabilizing slopes and the condition of the reinforcement over time. In this context,

A. Ferrari et al. (Eds.): CNRIG 2023, SSGG, pp. 153–160, 2023.
https://doi.org/10.1007/978-3-031-34761-0_19

much research has focused on the development of new sensors based on fiber optic sensing technology. Due to their characteristics of durability, cost-effectiveness, immunity to electromagnetic interference and ability to operate remotely [2], these sensors are increasingly used in geotechnical applications and many other fields of civil engineering. The most widely adopted sensors are based on the Brillouin scattering phenomenon (BOTDA or BOTDR), which makes it possible to detect changes in strain or temperature along an optical cable with an almost continuous spatial distribution, hence the name Distributed Fiber Optic Sensors (DFOS).

Two recent applications of DFOS technology for monitoring slow-moving landslides are presented in this paper: an intelligent inclinometer and an intelligent composite anchor. Recently developed DFOS-based inclinometers [3, 4] mainly consist of a fiber cable disposed along the outer surface of a vertical pipe that serves as the inclinometer's casing. At the lower end of the pipe, the fibre cable is laid in a loop, as a closed circuit allows better performances when the DFOS is queried with BOTDA technique. However, previous research has shown that the accuracy of measured strain is strongly influenced by the spatial resolution of the sensor, the connection method and the composite structure of the fiber. Furthermore, difficulties arise in the installation of such inclinometers in the field when pipes tens of meters long have to be assembled.

Among the numerous reinforcements used for landslide stabilization, composite anchors [5] have proven to be the most versatile as they are capable to withstand massive forces as required in this application. Composite anchors are sub-horizontal passive reinforcements composed of 3–6 m long hollow threaded carbon steel rods, installed using the self-drilling technique and connected together with coupling nuts up to reach the desired depth. After installation, some harmonic steel strands are introduced and cemented inside the bars to increase the anchor's strength, while the anchor head is fixed externally to a precast concrete bearing plate. When the unstable mass develops displacements, shear stresses increase along the soil-anchor interface, promoting a synergistic reinforcing effect. Some early attempts to measure axial strain along the anchor profile were performed with strain gauges, but good results were obtained using Distributed Fiber Optic Sensors [6], which can be easily installed in the anchor cavity together with the strands. Among the interrogation methodologies, the Brillouin method is the most widely used in geotechnical applications as it can be used with cables up to 50 km long, providing a spatial resolution of approximately 20–50 cm.

The work aims to present the strengths as well as the criticalities of using FO sensors to monitor landslides or reinforcements for their stabilization. Two case studies are presented for this purpose: the smart inclinometer was tested in the Centola landslide, while the Fantoni landslide was used to study the smart composite anchor.

2 Materials and Methods

2.1 The Smart Inclinometer

The NSHT is a strain transducer designed to overcome the main limitations of using FOS in both geotechnical and structural monitoring [7]. It consists of a common fiber for telecommunication buried in a resin core and packaged by two tapes of composite materials. In the smart inclinometer presented here, two NSHTs are externally installed along an inclinometer casing tube (Fig. 1a) so that traditional inclinometer measurements can be performed in the meantime. To evaluate the new device, the strain profiles derived from the lateral deflection measured by the smart inclinometer and the conventional inclinometer probe are compared. It is worth noting that the evaluation of displacements from the strain measured by the smart inclinometer using the classical conjugate beam method involves many assumptions and can lead to an error that increases with the calculation distance [8]. Two prototypes of NSHTs were initially tested in the laboratory to select the most effective type of fiber cable and packaging tapes for the current application and to calibrate the sensor [9]. Based on the laboratory results, the current setup was chosen, which consists of two NSHTs realized with a low bend loss fiber cable (Thorlabs CCC1310-J9) and packaged by two tapes in glass-fiber composite material. The fiber was interrogated using a commercial BOFDA analyzer (Optosensing srl) with a maximum sampling of 5 cm and a spatial resolution of 50 cm.

2.2 The Smart Passive Anchor

The smart passive anchor is a composite anchor with one DFOS cable for strain measurement cemented inside the central cavity. The cable is a BruSens® V9 (Solifos), and, as in the previous case, a commercial BOFDA Analyzer (FibrisTerre, Germany) with a maximum sampling and spatial resolution of 5 and 20 cm, respectively, is used for its interrogation. Together with the strain cable, at least one anchor of a group of monitored anchors must also be equipped with a temperature DFOS sensor (BRUSens® DTS STL PA, Solifos), which is used to determine the temperature compensation of the strain measurements. The accuracies of the strain and temperature measurements are of 2 $\mu\varepsilon$ and 0.1 °C, respectively. Since the fiber is installed in a loop configuration (a single fiber runs in two ways inside the bar), the measurements allow the detection of occurrence of any bending moments. Once the deformations along a bar are obtained, given its stiffness, the axial force acting in the anchor and the stress transferred by the soil to the reinforcement can be derived. This makes it possible to assess how much the anchor is working and how far it is from the yield or structural failure limits, during its entire life. Moreover, the position of the sliding surface can be determined, thus assessing the adequacy of the reinforcement's length.

3 Field Monitoring by Using DFOS

3.1 Centola Landslide Case Study

The site selected for testing the smart inclinometer is located in the municipality of Centola, near the Cilento coast, where a slow active landslide affects a small urban settlement (Fig. 1a). The landslide was previously investigated through a geological field survey and the analysis of borehole core samples. It is part of a complex landslide system [10] in which several small landslides, classified as debris or clay rotational slides–earthflows, develop above the old complex deep-sited one. In the area, under a 30–40 m thick layer of landslide debris, the conglomeratic and arenaceous formation of the Cilento Group extensively outcrops above the marl-clayey Mesozoic formation.

The landslide was monitored starting in September 2021: two borehole sites, namely S1 and S2, which are 100 m apart (Fig. 1a), were set. In S1, a 40 m long inclinometer tube was equipped with two NSHTs disposed with a rotation of 45° from each other (Fig. 2b). After the drilling operation, when the inclinometer casing tube was dropped down, the two NSHTs were fixed to the tube through screws at a 2 m interspace (Fig. 1a). The hole was finally cemented and after 15 days the zero reading was taken. In S2, another 40 m long conventional inclinometer tube was installed. An open pipe and two Casagrande piezometers completed the monitoring system. During the first part of the monitoring, temperature profiles along the depth were measured by using a thermocouple placed in the piezometer tube due to the unavailability of a temperature-sensitive FOS. In September 2022, another borehole (S3 in Fig. 1a) was instrumented with a new smart inclinometer and an additional cable for temperature measurements (Fibrasens DTS 2/2f, s4u®), which allowed the recovery of soil temperature profiles during the last months of the monitoring period. The results of the latter smart inclinometer are not reported here as it has not recorded any significant slope movement to date.

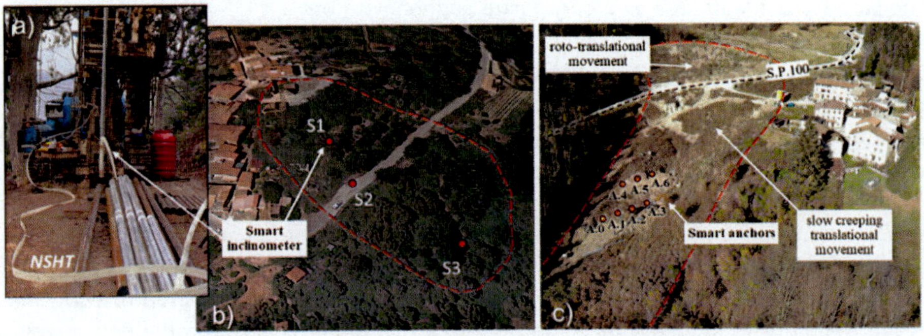

Fig. 1. a. Smart inclinometer during installation; b. Centola landslide; c. Fantoni landslide.

3.2 Fantoni Landslide Case Study

Seven smart composite anchors were installed in a test site within the Fantoni landslide in the municipality of Recoaro Terme (province of Vicenza). The Fantoni landslide (Fig. 1b) is divided into two portions by a provincial road, which over the years has shown clear signs of deterioration due to slope instability. The upstream portion is a predominantly roto-translational movement, mainly stabilized by four structural wells and two sheet pile walls built in 1990 and 2015. The downstream portion, instead, is a slow translational creeping movement that is still active. Two standard inclinometers were used to check the displacement trend of the landslide. The first manual device, which indicated a displacement rate around 80 cm/y, worked for only one month before going out of service due to excessive deformation. A second automatic inclinometer was installed immediately after the realization of the test site to assess the condition of the slope with the anchors. A debris cover, from 4 m to over 40 m thick, forms the moving volume. It consists of a medium density mixture of gravel and silty to slightly silty sand, presenting also rock clasts of calcareous-dolomitic nature. Slightly to heavily weathered Ladinian Vulcanic rocks or a sedimentary rocky substrate mainly consisting of limestone, dolomite and calcareous conglomerate lie beneath the debris.

The smart anchors were installed in the downstream portion of the landslide (Fig. 1b). They have sub-horizontal inclination angles between 15° and 20° and lengths from 18–42 m to ensure that a portion of at least 8–10 m is anchored to the bedrock. They are composed of a hollow bar with a 76 mm external diameter and with 4 tendons cemented inside. Six anchors (A.1 - A.6) were instrumented with a single optical fiber cable for strain measurement (BruSens® V9, Solifos). This cable, approximately 500 m long, has a loop configuration, entering and exiting each anchor, starting at A.1, passing through the others (A.2, A.3, etc.) and returning again to A.1, where it can be connected to the analyzer device. This installation allows easy and close access to both ends of the thread for a double-ended measurement. The A.6 anchor was instrumented with an additional cable for temperature compensation (BRUSens® DTS STL PA, Solifos), a specific cable made with the fiber hosted in a gel-filled stainless-steel loose tube to prevent the optical core of the fiber from being affected by mechanical strain. Finally, the A.0 anchor, selected for a pull-out test, was equipped with a separate 80 m long fiber cable for strain measurement. These latter results are not shown here for brevity.

4 Results and Discussions

The installation of the smart inclinometer was completed in November 2021. Conventional inclinometer data, collected over the course of about 18 months and shown in Fig. 2a and b in terms of cumulative displacements, revealed that a sliding zone can be recognized at a depth of 40 m at inclinometer I2 (S2 site in Fig. 1a), while it is not discernible at inclinometer I1 located at a higher elevation of about 25 m. This would suggest that the lower end of the I1 has not reached stability, remaining all within the landslide body characterized, in this portion, by a kinematic representative of an earthflow with overlapped landslide bodies. In any case, the landslide moved at displacement rates of the order of a few cm/y, with the acceleration related to fluctuations in the water table reflecting the rainfall regime.

The strain profiles were derived from the changes of the Brillouin Frequency Shift (BFS) along the fiber cables of the two NSHTs installed in I1 after temperature compensation. Figure 2c shows the strain data collected during the first month of monitoring along the NW face of the tube (dashed lines) and compared to those retrieved by conventional inclinometer data (continuous lines): a positive or negative strain corresponds to a tensile or compressive state in the NSHT, respectively. The measurements collected in December 2021 revealed a fairly good agreement between the traditional and innovative strain profiles. However, the NSHT showed a shortening along its entire length, as an almost constant strain of a few tens of $\mu\varepsilon$ was recorded; this would suggest that the inclinometer tube assumed a profile entirely characterized by a constant deflection, being suspended within the landslide body.

The strain increased over time during the following months, indicating that the inclinometer was subject to ongoing deformation, especially after the heavy period of rain in November-December 2021 characterized by a cumulative rainfall of 560 mm. Observing the March 2021 measurements, it can be seen that although the two profiles do not match perfectly, the NSHT one resembles the shape of the traditional one, especially in the most superficial part of the soil deposit, up to a depth of 15 m.

Unfortunately, comparison of the two data sets is very difficult at site S1 because the pipe has not reached stable soil formation and the data retrieved by the inclinometer probe are affected by the assumption of fixed end of the tube. Moreover, at such a small level of strain, innovative measurements can be affected by some errors due to temperature compensation, which was initially based on local temperature data. To overcome these difficulties, as previously mentioned, site S3, located about 200 m downslope of S2, was also instrumented in September 2022. The slope movements stopped during summer of 2022. The subsequent rainy period in October-November caused the activation of superficial movements involving about 11 m of soil depth. The temperature profiles collected at site I3 during these months were used to compensate the NSHT measurements. The obtained strain profiles (Fig. 2d), which cannot be compared to the previous ones as another analyzer with a different basal Brillouin frequency was used, showed a trend more compatible with the kinematics of the slope. Interestingly, a progressive increase in strain occurred only in the upper part of the NSHT, with a peak located at a depth of around four meters, where a strong change in curvature was recorded by the conventional inclinometer (N-22 profile in Fig. 2b).

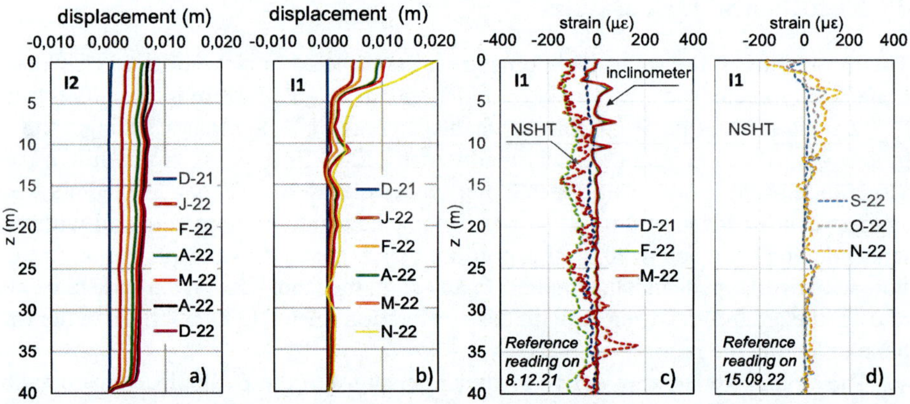

Fig. 2. Trends of cumulative displacements at (a) inclinometer I2 and (b) inclinometer I1; strain profiles along the NW face of I1 during (c) the first period and (d) autumn 2022.

The installation of the smart anchors was completed on 12 December 2020 (used as a reference reading). Again, the strain and temperature were derived from the BFS measured in the cable, using the temperature and strain coefficients provided by the cable manufacturer. The temperature, measured exclusively in anchor A.6, is used to compensate the thermal effects affecting the strain measured in all anchors. In Fig. 3 only the up-down measurements are reported due to lack of space. Anchors A.1, A.2 and A.6 are more evidently deformed, with peak values of more than 4500 µε, while A.3, A.4 and A.5 show lower maximum values of strain. Although the strain measurements have different shapes along the reinforcements, all show a shallow part with almost zero deformation, and an evident increase up to the depth where the slip surface is located.

The different positions of the peaks in the anchors confirm that the sliding surface is located near the bedrock, which is reached at a greater depth close to the landslide center (A.3 and A.6) and at lesser depth near the lateral border (A.1 and A.4). In fact, anchor A.3 shows extremely low strain, possibly because it is positioned laterally and in the upper row. On the contrary, A.6 exhibits the highest strain, corresponding to the highest load, most likely because the sliding layer is thicker in the center.

From the technical data sheets provided by the manufacturers, the yield and ultimate strengths of the bars and strands are obtained, thus deriving the equivalent stiffness of the composite anchor. Once the strain is obtained, the force loading each anchor is immediately derived. In particular, the maximum force values reported in Fig. 3 show that almost all anchors have exceeded the yield limit (1208 kN) even though they are still far from their ultimate load (2134 kN). Finally, it is interesting to note that from March 2021 to September 2022 the curves show variations less than 25%. The continuous inclinometer measured an average displacement rate of 4 mm/y in the same period, much lower than the pre-intervention 80 cm/y. This underlines how these anchors actually contribute significantly to the stability of the landslide and that their structural dimensioning was well executed.

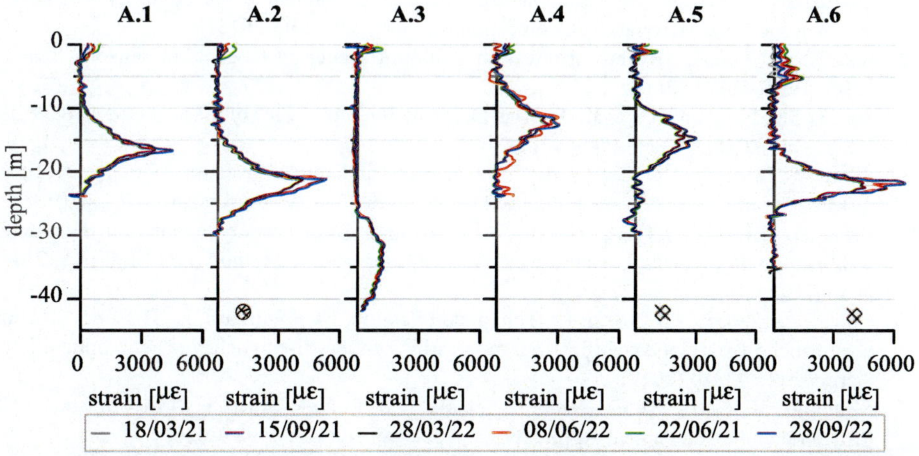

Fig. 3. Strain measured in the up-down fiber portion in anchors A.1 – A.6 with time. Maximum strain and maximum force values measured on 28/09/2022 for each anchor, and their position along the anchor, are also reported. Reference reading on 02/12/2020.

5 Conclusions

The results presented underline the effectiveness of the smart reinforcement system instrumented with optical fiber. The almost two-year long monitoring allows us to recommend that an integrated monitoring system such as this become standard practice to effectively measure the working conditions of reinforcements during their entire service life. Preliminary results from the Centola site are also encouraging. Although the complex kinematics of the landslide does not allow for a complete evaluation of the innovative instrument as of yet, the study suggests that it can capture the main features of the phenomena, while revealing that an accurate analysis is required to compensate for the thermal effect along the cable. In fact, the strain measured by the optical cable derives from both tensile and temperature variation, making it necessary to subtract the temperature-dependent strain from the total strain to identify the mechanical component. This becomes even more important in applications where deformations are especially small, like in the smart inclinometer.

Acknowledgement. The research on smart inclinometer was partially funded by the Program VALERE: "VAnviteLli pEr la RicErca", DDG n. 516–24/05/2018. The research on smart anchors was funded by the project POR-FERS INMOSTRA by the Veneto Region and by a research accord with the Province of Vicenza.

References

1. Trigila, A., Iadanza, C., Esposito, C., Scarascia Mugnozza, G.: Comparison of logistic regression and random forests techniques for shallow landslide susceptibility assessment in Giampilieri NE Sicily. Italy. Geomorphology **249**, 119–136 (2015)
2. Bao, X.Y., Chen, L.: Recent progress in Brillouin scattering based fiber sensors. Sensors **11**(4), 4152–4187 (2011)
3. Sun, Y., Shi, B., Zhang, D., et al.: Internal Deformation Monitoring of Slope Based on BOTDR. J. Sensors. (2016). Article ID 9496285
4. Zhou, Y., Dongjian, Z., Zhuoyan, C., et al.: Research on a novel inclinometer based on distributed optical fiber strain and conjugate beam method. Measurements 153 (2020)
5. Brezzi, L., Bisson, A., Pasa, D., Cola, S.: Innovative passive reinforcements for the gradual stabilization of a landslide according with the Observational Method. Landslides **18**, 2143–2158 (2021)
6. Cola, S., Schenato, L., Brezzi, L., Tchamaleu Pangop, F.C., Palmieri, L., Bisson, A.: Composite anchors for slope stabilization: monitoring of their in-situ behavior with optical fibre. Geosciences **9**, 240 (2019)
7. Minutolo, V., Cerri, E., Coscetta, A., et al.: NSHT: New Smart Hybrid Transducer for Structural and Geotechnical Applications. Appl. Sci. **10**(13), 4498 (2020)
8. Zhang, L., Shi, B., Zhu, H., Yu, X., Wei, G.: A machine learning method for inclinometer lateral deflection calculation based on distributed strain sensing technology. Bull. Eng. Geol. Env. **79**(7), 3383–3401 (2020). https://doi.org/10.1007/s10064-020-01749-3
9. Di Gennaro, L., Damiano, E., De Cristofaro, M., et al.: An innovative geotechnical and structural monitoring system based on the use of NSHT. Smart Mater. Struct. 31, 065022 (2022)
10. Valiante, M., Guida, D., Della Seta, M., Bozzano, F.: A spatiotemporal object-oriented data model for landslides (LOOM). Landslides **18**(4), 1231–1244 (2020). https://doi.org/10.1007/s10346-020-01591-4

Smart Monitoring by Fiber-Optic Sensors of Strain and Temperature of a Concrete Double Arch Dam

Lorenzo Brezzi[1](✉), Luca Schenato[2,3], Simonetta Cola[1], Nicola Fabbian[1], Paolo Chemello[4], and Paolo Simonini[1]

[1] DICEA, University of Padua, Padua, Italy
lorenzo.brezzi@unipd.it
[2] Research Institute for Geo-Hydrological Protection, CNR, Padova, Italy
[3] Department of Information Engineering, University of Padua, Padua, Italy
[4] ENEL Green Power, Rome, Italy

Abstract. Concrete arch dams are large constructions aimed at producing hydroelectricity, providing water for irrigation and controlling flooding. Since concrete dams in Italy are quite old-fashioned structures, it is a challenge to properly evaluate their conditions, manage maintenance interventions and optimize energy production for a more sustainable development. Traditional field monitoring carried out through visual inspections, topographical measurements and point sensors is complex and gives only discontinued spatial information. Recently, numerous innovative techniques have progressed greatly, and, among these, the use of Distributed Optical Fiber Sensors (DFOS) as detector of strain and temperature can be considered an attractive option, as it allows spatially dense measurements over large distances and with high resolution. To estimate the reliability and potentiality of this innovative system in monitoring the extremely low strains sustained by dams, a concrete double arch dam, namely the Ponte Cola dam in North Italy, was recently instrumented with two different types of DFOS (one for strain and one for temperature measurement) both in the foundation and along the crown of the dam. Several measurement campaigns were carried out and the data collected are briefly presented in this paper and are compared with those obtained through traditional monitoring techniques.

Keywords: Distributed Optical Fiber Sensor · Smart monitoring · Low strain · Structural monitoring

1 Introduction

1.1 Dam Monitoring

In Italy, dams play a central role in electrical power production, temporal and spatial management of water resources, as well as in the mitigation of river floods. The exercise conditions of a dam must be constantly adapted to the environmental conditions (rainfall,

A. Ferrari et al. (Eds.): CNRIG 2023, SSGG, pp. 161–168, 2023.
https://doi.org/10.1007/978-3-031-34761-0_20

local temperatures and upstream water inflow) [1] and monitoring is a priority for real-time assessment of the structure's health state and safety conditions, also in relation to exposure to unexpected dangers. Commonly, monitoring is realized with many punctual sensors of different types and the large amount of collected data are then analyzed with methods that allow a reliable interpretation of dam body deformation [2–4]. In this stage the spatial interpolation of information is fundamental not only to obtain a complete vision of the over-all dam strain behavior but also to evidence local anomalies. To overcome limitations related to punctual measurements, the use of new advanced sensors based on optical fiber technology and, in particular, Distributed Fiber Optic Sensors (DFOSs) is now increasing [5]. They provide measurements of strain and/or temperature with high spatial resolution over long distances. The operating principle of DFOSs is based on the injection of a light wave into an optical fiber and on the analysis of the retroreflected light signal generated by the scattering effects in the silica which constitutes the fiber core. When the fiber cable is connected to a structure which deforms over time, the fiber develops axial strain which proportionally changes the backscattering signals. Analyzing the variation of the optical signal, the axial strain and temperature profiles along the fiber are obtained [6, 7].

This paper presents the preliminary results of a monitoring activity carried out on an Italian concrete dam using two types of DFOSs, one for strain and one for temperature measurements. The research is finalized to estimate the reliability and potentiality of such innovative sensors in monitoring the very low strains developed by dams in their exercise. To this aim, the DFOS data are compared with those collected using other, more traditional systems.

2 Case Study: Ponte Cola Dam

Fig. 1. **a.** Aerial view and **b.** upstream facing of Ponte Cola dam

The case study is Ponte Cola dam, a concrete dam located close to Garda Lake in the Valvestino valley (Brescia). The dam has a double curvature arch shape (Fig. 1) and is built on the rock formation of Upper Trias dolomite. The transition from the double arches to the rock is formed by a massive buffer, almost 24 m wide and 26 m high, which deeply fits into the rocky foundation. Laterally and below it, a waterproof screen is obtained through cement injections executed beyond 60 m of depth. The dam body, made up of 21 ashlars, is 122 m high (crowning height 505 m a.s.l.), has a development at the crest of 286.27 m and a total volume of 239′300 m^3. It was built between July 1960 and October 1962, with the purpose of storing water from the Toscolano and S. Michele streams in a reservoir, providing an energy reserve of about $45x10^6$ kWh.

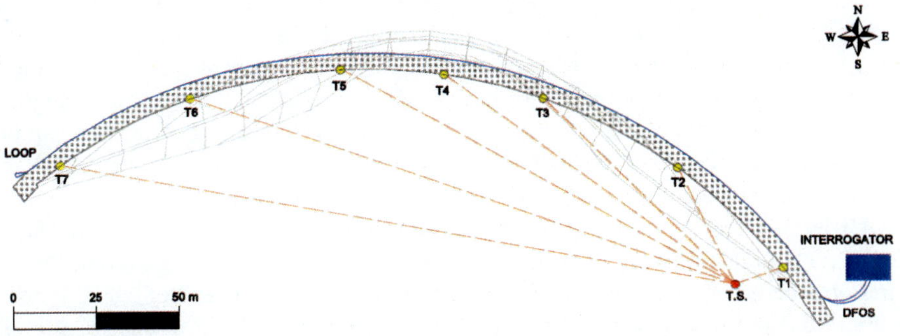

Fig. 2. Positioning scheme of the topographic targets (T1–T7), of the total station (T.S.) and of the DFOS (blue line) on the crown of the dam.

Since its construction, Ponte Cola dam has always been monitored with several instruments. In particular, at the beginning, within the dam body 22 thermocouples (10 thermocouples on the downstream face, 4 on the upstream one and 18 in the concrete casting) were installed, along with 65 removable strain gauges, 29 distance dilatometers and 124 thermo-extensometers. Despite the fact that only some of them are still functional, maintenances and upgrades of the system have been made periodically. The last important upgrade was realized in March 2021, when a topographic survey system capable of automatically detecting the position of about 30 targets on the downstream face was installed. Seven of these targets are distributed symmetrically at the top of the concrete body (Fig. 2). The system provides measurements with a fixed frequency and millimeter precision. In the same spring, an 80 m deep borehole, crossing about 40 m of concrete and 40 m of calcareous rock, was executed starting from a niche, specifically created in the dam body at the base of the central ashlar. Inside the borehole, 5 rod extensometers, aimed at measuring the displacements of the dam foundation and of the underlying soil layers, were installed in April 2021.

2.1 DFOS Installation

Between January and February 2021, two pairs of DFOSs were installed at the dam: one pair was positioned in a small groove created along the dam crown (Figs. 2 and 3a); the other was inserted in a borehole parallel to the borehole hosting the rod extensometers (Fig. 3b). The groove and borehole hosting the fiber cables were filled with mortar to ensure an efficient coupling between the fiber and the structure. To stay within the requirements of length, this paper presents only the measures related to the crown.

The installed DFOS are of two types: one is an armored corrugated optical fiber cable for strain measurement (BruSens® V9, Solifos), while the other is a cable for temperature measurement (BRUSens® DTS STL PA, Solifos) placed parallel to the first. The length of the cables is approximatively 200 m and 600 m for those installed in the dam foundation and at the crown, respectively. Temperature measurement is crucial for understanding the thermo-mechanical behavior, but is also the key for evaluating the thermal-optical effects on the cable strain. In fact, since the optical cable can deform due to traction or temperature variations, the cable's temperature must be known in order to estimate its thermal strain; this value can then be detracted from the total measured strain, therefore providing the mechanical component. This aspect is even more relevant in applications such as the one here described, where the expected strain ranges are modest (below 10^{-4}) but the temperature variations are relevant. In fact, the dam body is a massive structure that does not exhibit large deformations, despite the fact that the outside temperature can vary by more than 30 °C from summer to winter.

All the DFOSs are interrogated with a Brillouin Optical Frequency Domain Analyzer (BOFDA) from FibrisTerre (Germany) adopting a double-end configuration: the cable starts at a certain point, runs along the profile and, after a curve of 180°, it returns to the initial point, where the two cable ends are connected to the interrogator. This configuration allows a spatial resolution of 20 cm. The declared accuracies for strain and temperature measurement are of 2 $\mu\varepsilon$ and 0.1 °C, respectively.

Special precautions were followed to avoid any damage to the fibers during installation. Although the fibers are coated with special protective cores, they are unable to withstand excessive concentrated stresses or very small radii of curvature. Since the cables were installed in a loop configuration, they were protected through a specifically made device from developing excessive deformations at the loop point. In any case, even if the fiber breaks at one point, the double-end configuration still allows interrogation of the cable at both ends adopting a single-end configuration, ensuring measurements up to the point of breakage, albeit increasing the spatial resolution.

Fig. 3. Localization of the fibers installed on crown (**a**) and niche (**b**).

2.2 Measurements Campaigns

Some preliminary manual measurement campaigns were carried out after the installation of sensors. In this phase, two calibration tests (here not show for lack of space) were performed on the cables for temperature detection. Subsequently, three long surveys with automatic interrogation of DFOSs were conducted in August 2021 (26 days), in November-December 2021 (27 days) and in July-September 2022 (40 days). In these long surveys, the DFOSs were connected through standard communication cables to the interrogator located in the dam control house. In this way, it is possible to contemporary interrogate all four fibers and automatically record the data to a digital memory, without interruption and without the need for the presence of a technician.

The data obtained by the DFOSs are variations of strain or temperature with respect to an initial reference measurement. Thanks to the loop configuration, all measures are obtained in both "up-down" and "down-up" directions, which generally show high consistency with one another. Four typical temperature and strain profiles determined with fibers at the dam crown are shown in Fig. 4. To make their interpretation easier, they belong to the second and third periods and all vary with respect to the first reference measurement obtained on 02/08/2021 at noon.

It can be immediately noticed in Fig. 4a that for each date the temperature is approximately constant along the entire extension of the fiber. The two profiles recorded at November 2021 show a decrease of about 15 °C and 20 °C with respect to August 2021. In August and September 2022, instead, the temperature trend is again consistent with the reference measurement. Although the measurement is approximately constant, small peaks with an interspace of about 10–12 m are noticeable. Similar peaks are even more evident in the strain profiles. Consistent with the temperature variation, the strain measurements seem to depend on seasonality, emphasizing how the dam deformations are mainly due to variations in air temperature.

Figure 4 also shows high measurement consistency in the two directions, up-down and down-up, confirming the reliability of the measurements.

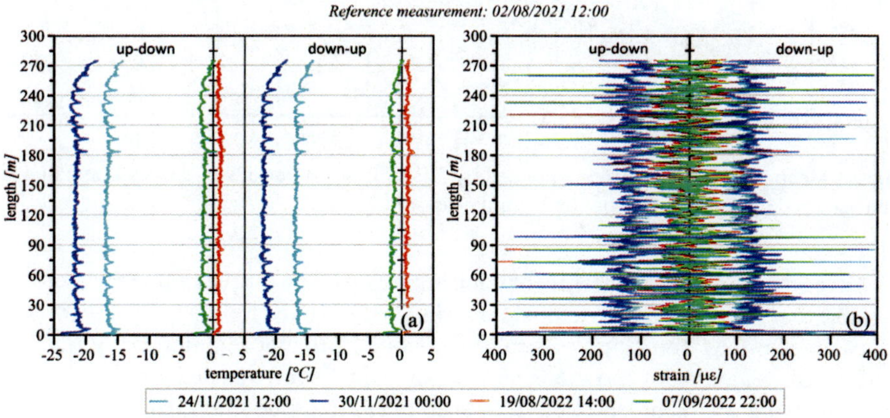

Fig. 4. Temperature (**a**) and strain (**b**) profiles acquired with DFOSs at the dam crown in 4 single measurements. Reference measurement: 02/08/2021 at noon.

3 Data Analysis

A first interpretation of measurements is carried out by comparing the temperatures measured by DFOSs with those obtained from thermocouples installed on the dam (Fig. 5). In this regard, three different sensors can be considered significant for comparison: one installed at an elevation of 490 m asl on the upstream side of the dam, measuring air temperature; one located at 440 m asl on the same side and measuring water temperature in the reservoir; one drowned in the concrete at an elevation of 490 m asl, recording temperatures inside the dam. The variability of air temperature is evident, both in the day-night and summer-winter cycles. On the other hand, the daily temperature variations in water and concrete are much less pronounced, underlining the great thermal inertia of these materials compared to that of air. It is important to note that the comparison here discussed is not reported to evaluate the reliability of the data acquired through thermal DFOS, which was already verified using the calibration tests, but only to show the high variability of temperatures in this structure with respect to the external temperature, and to highlight the importance of detecting the temperature in exactly the same position as the strain cable in order to obtain a proper measurement of strain.

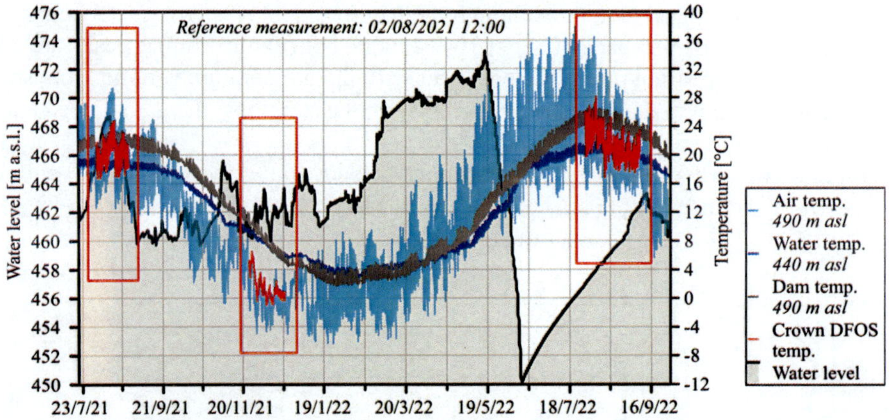

Fig. 5. Mean temperature in time obtained by DFOS in the 3 periods of survey compared with data acquired by some traditional temperature sensors.

Given the evident homogeneity of temperature variation profiles along the crown, it is possible to refer to an average temperature variation obtained for the entire DFOS profile for each recording. To compare its trend over time with that acquired using traditional sensors, the temperature variations obtained by DFOSs are translated by a value equal to the temperature recorded at the reference time (12.00 noon on 02/08/2022) by the concrete sensor. The results, shown in Fig. 5, show high consistency in the temperature trends. At the reference hour, the temperatures recorded in the concrete and air were 21.7 °C and 21.5 °C, respectively, indicating near thermal equilibrium. The data obtained with the DFOSs show greater temperature variability between concrete and water, but this can be explained considering that the fiber is installed on the outside surface of the dam, while the sensor is buried several meters inside the dam body. The measurement obtained from the DFOSs is more consistent with those recorded in air, but with less variability daily, indicating that contact with the dam surface and the thin layer of grout cementing the fiber to the dam allows for some thermal inertia.

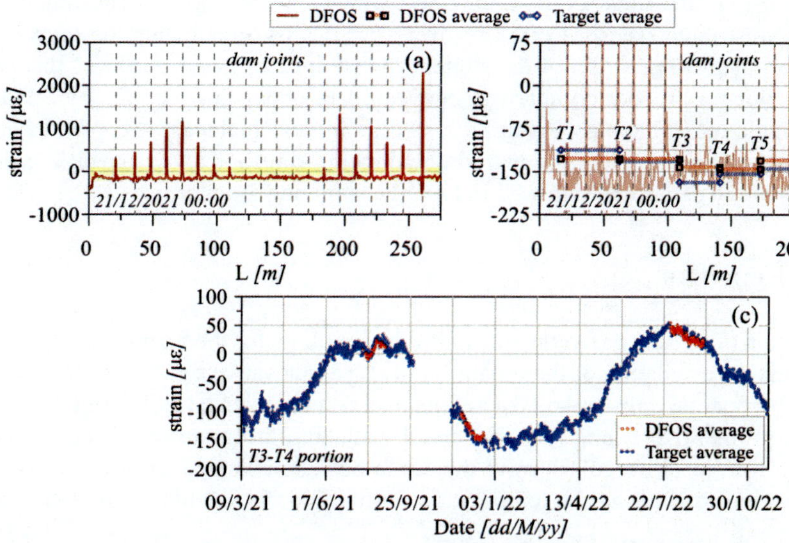

Fig. 6. **a.** Strain measurement obtained by the DFOS along the crown at 21/12/2021 00:00; **b.** zoom of plot a in the range of strain (75 με; + 225 με) with mean strain values in the portion between two nearby targets obtained with DFOS (red segments) and topographic data (blue segments). The position of dilatation joints is indicated by vertical segmented lines. **c.** Average strain of the segment T3–T4 measured with DFOS and topographic survey plotted vs time.

In order to assess the reliability of the strains measured with DFOSs, a comparison with the deformations obtained with topographic survey is attempted. In this regard, the average deformations of the sections between two consecutive targets are calculated based on the displacements measured by the topographic system. Even in this case, the deformations are obtained by assuming as reference configuration the position of targets at noon on 02/08/2022. A translation of the data acquired by topographic system is also applied, since the targets are located on the downstream side, while the DFOSs are glutted on the up-stream side of the dam. The spatially dense measurements provided by DFOSs are then cut into portions between two neighboring targets, and the average strain data in each interval are calculated. To give an example of the strain profiles obtained for each measure, Fig. 6a shows the strain obtained with DFOSs along the crown (ruby red line) on 02/12/2021. For sake of comparison, in Fig. 6b a zoom of the same plot in the range of strain (–75 με; + 225 με) is shown, in which the average strain measured by DFOS in the portion of crown between two nearby targets and the average strain calculated on the base of topographic surveys in the same portion are indicated respectively with red and blue segments. In all the portions the coherence between the data obtained with the two systems is evident, also considering the very small value of the average strains (less than 50 $\mu\varepsilon = 0.5 \times 10^{-4}$).

Another interesting feature clearly visible in the same figure is the presence of several strain peaks measured with the DFOSs, distributed at fairly regular distances. The strains associated with these peaks reach values exceeding 900 $\mu\varepsilon$, higher on the lateral sides and lower in the central part of the crown. Plotting the location of the expansion joints of the dam, it is evident that each strain peak corresponds to a joint, underlining the ability of the distributed sensor to pick up even such details. Finally, Fig. 6c shows the time trend of the average strain of the stretch T3–T4, located in the center of the dam (Fig. 2), allowing a broader comparison between data obtained over time by DFOSs and topographic measurements. Again, the measurements obtained with optical fibers show high consistency with data from traditional monitoring, and allow expansions and contractions of the dam to be followed in both the short and long term.

4 Final Remarks

The application of sensors based on optical fiber technology for the monitoring of Ponte Cola dam permits evaluation of the reliability of these innovative sensors for the in-site monitoring of modest strain (up to 10 $\mu\varepsilon$). The data demonstrate the ability of DFOSs to obtain strain profiles over a long span with small spatial resolution, an ability that permits detection of even local irregularities in the strain behavior. The paper also underlines the importance of measuring the temperature profile in order to obtain a proper strain measurement.

Acknowledgements. The research was performed with funding of Green Power spa and cofounded by the University of Padova within the program UNI-IMPRESA 2019 (project TITANO).

References

1. Dhandre, N.M., Kamalasekaran, P.D., Pandey, P.: Dam parameters monitoring system. In: 7th India International Conference on Power Electronics - IICPE, pp. 1–5. IEEE (2016)
2. Scaioni, M., Marsella, M., Crosetto, M., Tornatore, V., Wang, J.: Geodetic and remote-sensing sensors for dam deformation monitoring. Sensors **18**(11), 3682 (2018)
3. Su, H., Wen, Z., Sun, X., Yan, X.: Multisource information fusion-based approach diagnosing structural behavior of dam engineering. Struct. Control. Health Monit. **25**(2), e2073 (2018)
4. Lin, P., Li, Q., Fan, Q., Gao, X.: Real-time monitoring system for workers' behaviour analysis on a large-dam construction site. Int. J. Distrib. Sens. Netw. **9**(10), 509423 (2013)
5. Schenato, L.: A review of distributed fibre optic sensors for geo-hydrological applications. Appl. Sci. **7**(9), 896 (2017)
6. Soga, K.: Understanding the real performance of geotechnical structures using an innovative fibre optic distributed strain measurement technology. Riv. Ital. Geotech **4**, 7–48 (2014)
7. Cola, S., Schenato, L., Brezzi, L., Tchamaleu Pangop, F.C., Palmieri, L., Bisson, A.: Composite anchors for slope stabilization: Monitoring of their in-situ behaviour with optical fibre. Geosciences **9**(5), 240 (2019)

Numerical Modelling of Sant'Anna Flood Control Reservoir (Panaro River, North Italy): A Tool for Predicting the Behavior of Flood Control Structures During Flood Events

Maria Teresa Carriero[1], Renato Maria Cosentini[1](✉), Daniele Costanzo[1], Maria Rita Migliazza[1], Stefano Parodi[2], and Massimo Valente[3]

[1] Politecnico di Torino, Corso Duca degli Abruzzi, 24, 10129 Turin, Italy
renato.cosentini@polito.it

[2] Agenzia Interregionale del Po, Via Attiraglio, 24, 41122 Modena, Italy

[3] Agenzia Interregionale del Po, Str.da Garibaldi, 75, 43121 Parma, Italy

Abstract. In the last years, several flooding phenomena have occurred, mainly due to climate change. A flood control reservoir is one of the most widely used structural means of managing flood events. The main purpose of these reservoirs is to temporarily store the flood water and release it slowly at a safe rate after flooding so as not to cause damage downstream. These reservoirs consist of a system of levees and a flood wall built longitudinally and transversally to the river, respectively. A monitoring system (piezometers and hydrometers) is generally installed to control the response of both structures to flood events. The definition of a numerical model of the levees and the flood wall can be a powerful tool to simulate the behavior of the structures under flood events, through which any anomalies can be identified by comparing the analysis with the in-situ measurements. This approach makes it possible to identify any criticality of the structures and to define all the mitigation actions necessary to preserve their integrity, preventing tragic collapses. In this context, this note presents the numerical model of levees and flood dam that define the Sant'Anna flood control reservoir (Panaro river, north Italy). The geotechnical models were built according to both laboratory and in situ tests and calibrated using the monitoring results of some flood events that happened between 1997 and 2020. The note, therefore, presents some preliminary analyses.

Keywords: Monitoring · numerical model · flood events · risk mitigation

1 The Case History: The Sant'Anna Flood Control Reservoir (Panaro River, North Italy)

The Sant'Anna flood control reservoir was built to ensure the hydraulic safety of that area of Padana Plain close to the city of Modena and lessen the intensity of flood occurrences in the valley section of the Panaro river.

The Sant'Anna flood reservoir covers a total area of 3 km^2, with a storage area oriented in a SW-NE direction delimited by main and secondary levees and it is divided

A. Ferrari et al. (Eds.): CNRIG 2023, SSGG, pp. 169–177, 2023.
https://doi.org/10.1007/978-3-031-34761-0_21

into two distinct basins: a larger in-stream basin straddling the watercourse (78% of the entire basin) and a subsidiary one on the orographic right.

The flood dam consists of a fully overflowing concrete gravity barrier (Fig. 1). The outflow is provided not only by the spillway sill, but also by nine rectangular spans all guarded by flat gates. Downstream of the main structure is the dissipation basin equipped with four rows of staggered ties. The downstream plate on which the structure is built is headed on three longitudinal diaphragms with a length of 19.50 m, of which the two below the retaining structure are also transversally connected.

Five geotechnical investigation campaigns (1998, 2006, 2008, 2016 and 2020) were carried out to identify and characterize the ground and the material involved in hydraulic processes induced by the temporarily store of the flood water and its release. Many in situ investigations (boreholes, CPT and SPT tests, permeability tests), classification analysis (grain-size distribution, Atterberg limits, natural water contents) and mechanical laboratory tests on undisturbed and reconstructed samples (shear direct, triaxial, oedometric and resonant column tests) were carried out. The analysis of the experimental results on the basis of physical and hydro-mechanical characteristics allowed the identification of 6 soil macro-categories whose characteristic parameters have been used in the numerical modeling as described below (Table 1).

In 2016 a monitoring system consisting of piezometers and hydrometers along the levees and at the flood dam, was installed. Nine sections along the main and secondary banks (Fig. 2a) were equipped with Casagrande piezometers, 2 along the central vertical axes of the banks and two more at the toe of the countryside face (Fig. 2b); 4 piezometers were installed along the river shaft and, at the barrage 10 piezometers and 2 hydrometers, one upstream and one downstream, were positioned (Fig. 2c).

Several studies have been conducted to assess the hydraulic efficiency of this flood control reservoir (e.g. [1, 2]), but few analyses have been performed to establish its geotechnical behavior.

Assessing the geotechnical performance and stability of these structures under the action of flood events allows the early detection of possible damage. The integration of numerical models and field-monitoring data helps to implement this process of geotechnical evaluation of the structures [3]. In this context, this paper presents the numerical model of levees and flood dam that define the Sant'Anna flood control reservoir (Panaro River, Modena, Northern Italy).

2 Geotechnical and Numerical Model Levees and Flood Dam

Numerical models of the levees and flood dam of the Sant'Anna flood control reservoir were built by gathering all available data from historical documents on the flood control reservoir project, survey of land (performed in 2016), and geotechnical investigation campaigns.

Nine 2D geotechnical models of the levee sections with piezometers were reconstructed (Fig. 2a). The dimension of each model is about 100 m in length and 30 m in height. The domain was discretized, adopting a 15-node triangular mesh, into about 8000 elements to ensure excellent reconstruction of flow phenomena. Static and hydraulic boundaries conditions were defined as follows: model base fixed in the vertical direction

Fig. 1. Flood dam of the Sant'Anna flood control reservoir [4].

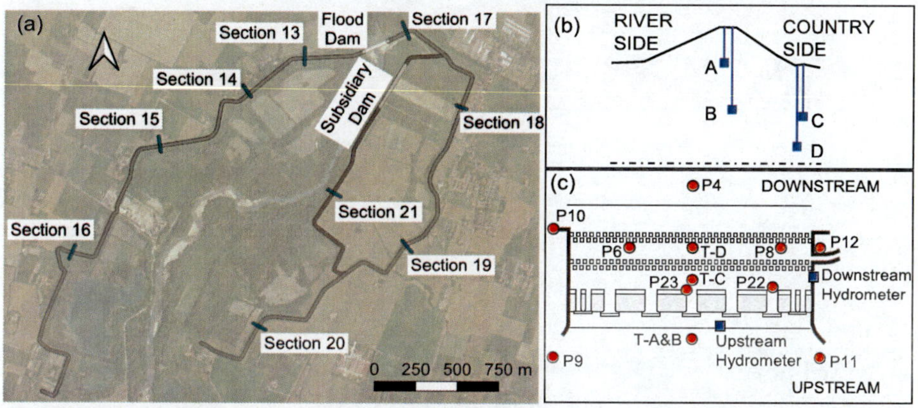

Fig. 2. (a) Levees and flood dam layout; (b) Levee section with piezometers location; (c) Location of piezometers (dot) and hydrometers (square) at the flood dam.

and impermeable; lateral sides fixed in horizontal direction and free flow. A Mohr-Coulomb elastoplastic constitutive model was adopted for all soils and the plastic diaphragm. An example of one levee model (section n. 14 – see Fig. 2a) is shown in Fig. 3.

The 2D numerical model of flood dam is shown in Fig. 4. The whole model measures 135 m in length and 50 m in height. It was discretized in 15323 elements adopting a 15-node triangular mesh. The retaining structure and the downstream plate were modelled as a no-porous elastic material. To adequately account for the presence of the holes equipped with gates to regulate the river flow, the unit weight assigned to the dam was appropriately calculated. The three diaphragms under the plate were modelled as no-porous elastic beam elements, whereas foundation soils were modelled as Mohr-Coulomb elastoplastic model. Interfaces elements were also implemented between the soil and the structures: diaphragms and plate. The properties of the interface elements were assigned by applying a reduction factor (R) to the parameters of the soils in contact with the structures. For the diaphragms-soil interfaces, an adhesive-attritive behavior was adopted with a reduction factor R = 0.7. A simple attritive model characterized by the same friction angle of the soil foundation (R = 1) was assumed for the plate-soil interface. The hydraulic and static boundary conditions of the flood dam model were adopted similarly to the levee models.

Soil geotechnical parameters adopted in the analyses for both models are shown in Table 1. Table 2 reports the structural parameters of the flood dam.

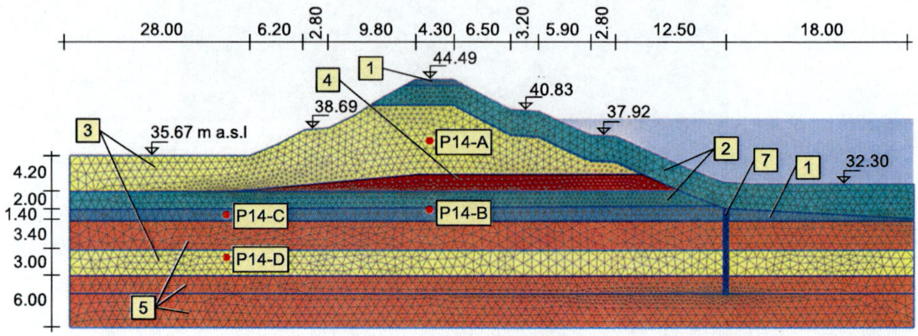

Fig. 3. Numerical model of a levee (section 14)

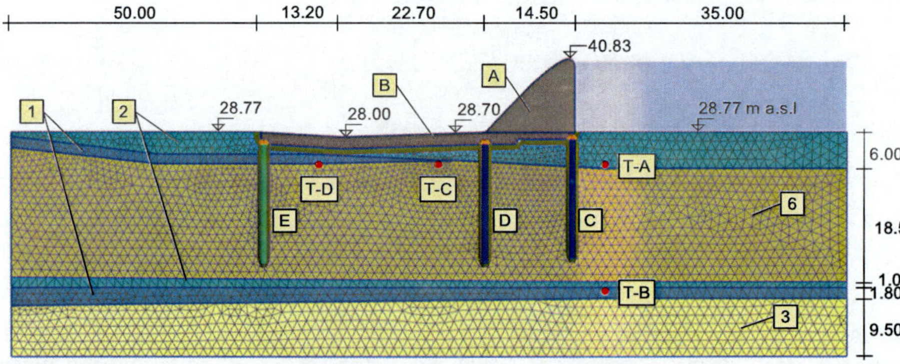

Fig. 4. Numerical model of the flood dam

3 Preliminary Analysis and Results

The previous described models were used to carry out numerical simulations using the commercial finite element, plane strain software PLAXIS 2D® [5]. Coupled hydromechanical analyses were performed on each levee and flood dam models to simulate their response during reservoir and discharge operations adopted to manage flood events. Flood events occurring between 1997 and 2020 were considered for the calibration of the models. This note reports only the preliminary analyses performed on the structures under the action of the historical flood event occurred in December 2020. The reservoir

Table 1. Soil geotechnical parameters

	Gravel and Sand	Sand with clayey silt	Sandy silt	Clay	Clay and silt	Sandy Silt and Clay	Cement bentonite
ID	1	2	3	4	5	6	7
γ_n: kN/m^3	20	18.6	19.6	17.5	19.0	19.0	19.0
φ': °	34	32	28	18	27	24	27
c': kPa	1	3	14	14	29	18	29
K: m/s	$2.8 \cdot 10^{-4}$	$1.2 \cdot 10^{-6}$	$1.0 \cdot 10^{-8}$	$1.9 \cdot 10^{-8}$	$1.5 \cdot 10^{-8}$	$1.4 \cdot 10^{-8}$	$1.0 \cdot 10^{-12}$
υ: -	0.3	0.3	0.3	0.3	0.3	0.3	0.3
E: kPa	42190	25550	11920	5525	11680	2920	11680

Table 2. Structural parameters of flood dam

	Dam	Plate	Diaphragma	Diaphragma	Diaphragma
ID	A	B	C	D	E
γ_n: kN/m^3	22	24	–	–	–
w: kN/m/m	–	–	19.2	19.2	24
d: m	–	–	0.8	0.8	1.0
υ: -	0.25	0.25	0.25	0.25	0.25
E: kPa	$30 \cdot 10^6$	$30 \cdot 10^6$	–	–	–
EA: kN/m	–	–	$24 \cdot 10^6$	$24 \cdot 10^6$	$30 \cdot 10^6$
EI: kN/m^2/m	–	–	$1.3 \cdot 10^6$	$1.3 \cdot 10^6$	$2.5 \cdot 10^6$

storage and outflow discharge curves in terms of upstream water level recorded during the flood event are shown in Fig. 5. The average velocity of reservoir storage was estimated at 0.37 m/h, while the average velocity of flow discharge is 0.18 m/h.

The analyses of the levees were performed in three steps: an initial phase, a reservoir storage phase, and an outflow discharge phase. The initial phase was carried out to assess the initial stress state condition of the model. During this phase, the water level of the model was set based on the piezometer values recorded before the flood event.

The last two steps of the analysis were conducted by means of flow simulations imposing the progressive raising or lowering of the reservoir level based on reservoir storage and outflow discharge curves. Slope stability analyses were also conducted.

The numerical results of the increase in pore-pressure were compared with those measured by the installed piezometers (Table 3). Figure 6 shows the comparison of measures of the piezometers during the reservoir storage and the flow discharge with the corresponding value obtained by the numerical model.

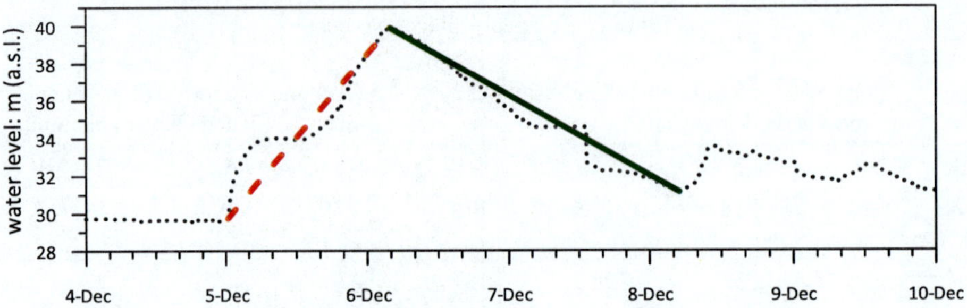

Fig. 5. Recorded reservoir storage and outflow discharge curves (dot line), average velocity of reservoir storage (dashed red line), and average velocity of flow discharge (solid green line)

Table 3. Comparison of the increment of pore pressure model and piezometer measurements

Piezometer (Fig. 3)	End of reservoir storage phase		End of flow discharge phase	
	P14 – B	P14 – C	P14 – B	P14 – C
Numerical Model: m	1.70	0.85	−0.41	−0.96
Real measurement: m	0.55	0.59	0.82	0.74

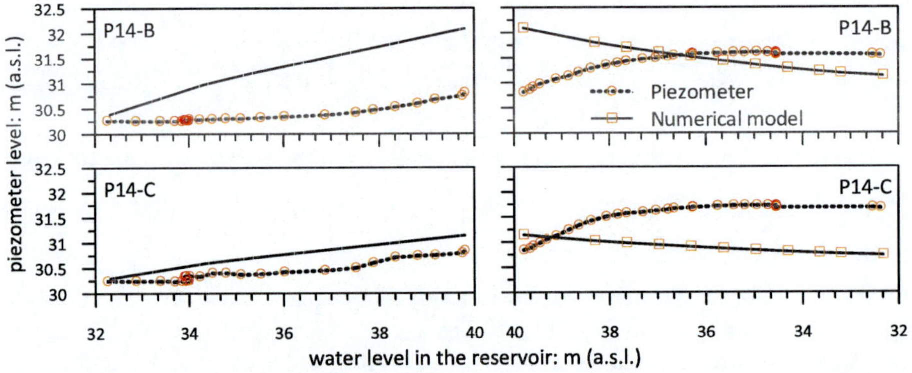

Fig. 6. Comparison of the measured piezometer levels with the corresponding numerical results

The differences observed between the numerical simulation and actual field measurements are due to the way the reservoir phase was simulated. This phase was modeled through steady state analysis, increasing the water level in the reservoir step by step. This choice made it possible to define an upper bound condition that can only be reached if the flooding phase takes place at a very low speed. The slope stability analyses showed Safety Factors higher than 1.5 for all phases analysis.

As shown in Table 3 and Fig. 6, the comparison between pore pressure computed in the model and the measured pressure does not consider either piezometer P14 - A or P14 - D. The first because it is located above the groundwater level, the latter because its measurement is governed by the regime of the deeper aquifer, a hydraulic condition that has not been implemented in the model. A coupled transient flow analysis was conducted on the flood dam applying a progressive increment of the reservoir level till to the maximum value reached during the reservoir storage phase. The numerical values of the increase in pore pressure at the end of reservoir storage were compared with those measured at the point where piezometers are installed (Table 4). Also in this case, the deepest piezometer (T-B – see Fig. 4) was neglected in the comparison, its measurement being governed by the deepest aquifer.

Table 4. Comparison of model pore pressure model increase and piezometer measurements below the flood dam at the end of reservoir storage

Piezometer (Figs. 2(c) and 4)	T-A	T-C	T-D
Numerical Model: m	11.01	3.24	2.98
Real measurement: m	9.08	6.00	5.94

This first analysis showed some differences between numerical results and measurements. The piezometer T-A provides a value that is out of agreement with the actual level in the reservoir, which is better captured by the numerical model. In terms of hydraulic head loss under the plate, the numerical results show higher reductions than real measurements. Better correlations between measurements could be obtained by inspecting the condition of the installed instruments and refining the hydraulic conditions of the model by carrying out a parametric analysis to take into account the local variability of hydraulic parameters.

Furthermore, to assess the stability of the structure against uplift phenomena induced by the pore pressures, transient flow analyses were performed considering different holding durations of the maximum water level in the reservoir. Finally, a steady state flow analysis was carried out to evaluate the maximum possible pore pressure regime under the structure. Figure 7 shows the pore pressure distribution under the structure at different time of maximum reservoir level stationing: at the end of the reservoir storage, 24 h, 75 h, 7 days and 15 days after reaching the maximum storage level and under steady state condition. Based on these results, an uplift stability analysis of the retaining structure was conducted according to NTC18 [6]. By applying partial coefficients and neglecting the stabilization contribution provided by the friction of the diaphragms, the destabilizing force (resultant hydraulic under-pressure) remains lower than the stabilizing force (total weight of the structure) as required by the code.

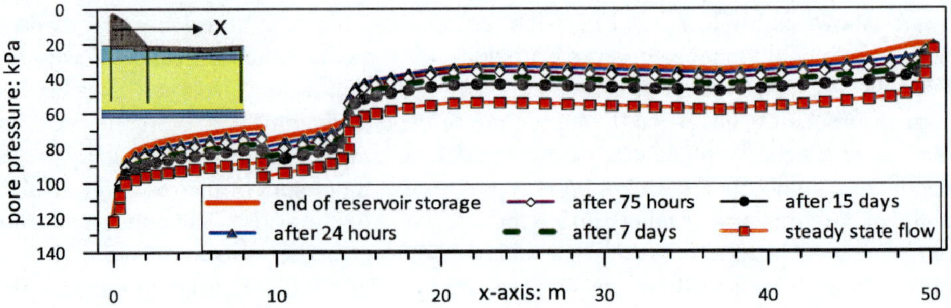

Fig. 7. Distribution of the pore pressure under the flood dam.

4 Final Remarks

Numerical models of levees and flood dam defining the Sant'Anna flood control reservoir (Panaro river, north Italy) have been presented. These models have been used to perform transient and steady state flow analyses, slope stability and uplift assessments of the structures. The preliminary results reported in this note have referred to the response of the structures to the flood event recorded in December 2020. Although the 2D model is not able to account for 3D spatial variability of hydraulic conditions, leading to differences in the comparison between numerical results and data recorded by monitoring devices, a satisfactory response of models in representing the actual response of the structures can be observed. Optimizing the numerical analysis requires both a verification of the installed equipment and an in-depth parametric study to accurately characterize the hydraulic parameters. A 3D model would be more accurate in simulating the spatial variability of hydraulic parameters, but would require a complete reconstruction of the landscape, which cannot be achieved with existing data. Despite the observed differences, these preliminary results highlight that the models can be adopted as potential digital twins, i.e. a combination of the reality and digital models capable of capturing real time behavior of the structures, assessing their current state and predicting their response, detecting any problem in advance and thus preventing dramatic damage. The advantage of these digital twins is that their characteristics can be updated based on additional information that becomes available over time and data from of monitoring system.

References

1. Balistrocchi, M., Ranzi, R., Orlandini, S., Bacchi, B.: Stima dell'Efficienza della Cassa di Laminazione di Sant'Anna sul Panaro. In: XXXVI Convegno Nazionale di Idraulica e Costruzioni Idrauliche-IDRA 2018, pp. 345–349. Ancona (2018)
2. Balistrocchi, M., Orlandini, S., Ranzi, R., Bacchi, B.: Copula-based modeling of flood control reservoirs. Water Resour. Res. **53**(11), 9883–9900 (2017)
3. Rivera-Hernandez, X.A., Ellithy, G., Vahedifard, F.: Integrating field monitoring and numerical modeling to evaluate performance of a levee under climatic and tidal variations. J. Geotech. Geoenviron. Eng. **145**(10), 05019009-1–14 (2019)
4. HyLab-University of Parma. http://hylabnew.unipr.it/wp-content/uploads/2014/07/articolo-panaro-1.png. Assessed 21 Dec 2022

5. PLAXIS. PLAXIS 2D Ultimate Reference Manual. Bentley Systems International Limited, Dublin (2022)
6. Ministero delle infrastrutture e Trasporti. Aggiornamento delle «Norme tecniche per le costruzioni» (2018). Decreto 17/01/2018. Gazzetta Ufficiale Serie Gen. N. 42 del 20/02/2018

Investigating the Effects of Water Levels Measured in Two Nearby Rivers on Groundwater Pore Pressures Regime

Giorgia Dalla Santa[1,2](✉), Nicola Fabbian[1], and Simonetta Cola[1]

[1] Dip. ICEA, Università degli Studi di Padova, Via Ognissanti 39, Padova, Italy
giorgia.dallasanta@unipd.it

[2] Dip. Geoscienze, Università degli Studi di Padova, Via Gradenigo 6, Padova, Italy

Abstract. The paper analyzes the groundwater regime in the embankment foundation of the artificial Gorzone channel located in the Po Plain near Venice, North Italy, under the conditions imposed by the hydraulic regime of the Gorzone itself and the nearby Adige river. Sand boils and blow-outs phenomena occur both at and beyond the levee toe, as the dam located some kilometers downstream is activated to keep the channel water level high for irrigation purposes in spring and summer. In order to design proper interventions, the seepage under the embankment has been investigated and the groundwater regime has been monitored for 2 years. The acquired data revealed that pore pressure regime in the deeper layers is not completely consistent with the Gorzone water level variations; it seems influenced also by the hydraulic regime of the nearby Adige river, that here flows parallel to Gorzone, at the distance of 180 m from the opposite levee.

The paper presents the results of some FEM analysis carried out to better investigate the groundwater regime under the influence of Adige and other local forcings. After a calibration based on the acquired pressure data, the model was used to analyze the levee seepage regime under different boundary conditions. The numerical results highlight that the sand-boils and blow outs are probably caused by a seepage confined in the upper high-permeability deposit induced by Gorzone water level, while the deeper pore pressures are mainly related to the Adige hydraulic conditions and to the activity of some extracting wells in the area.

Keywords: River embankment · seepage · groundwater pore pressure · FEM analysis

1 Introduction

Since many years, at Boscochiaro village (Cavarzere, Venice, Italy) several sand-boils and blow-out phenomena occur at the toe of the left levee of the Gorzone channel, that here flows parallel to the Adige river at a distance of about 180 m. The context is typical of a lowland area, at about 15 km from the sea. The stratigraphy is characterized by a continuous alternation of sandy and silty-clay alluvial deposits hosting a multilayer

A. Ferrari et al. (Eds.): CNRIG 2023, SSGG, pp. 178–185, 2023.
https://doi.org/10.1007/978-3-031-34761-0_22

aquifer, with the water table laying 1 m from the ground level. The Gorzone fluvial regime is dual: in winter, under normal runoff, the free surface is affected also by the tidal fluctuations imposed by the nearby sea, resulting in a mean water level almost at the same elevation of the countryside; in spring and summer, the dam of Buoro, located at about 3 km downstream, is activated in order to keep the channel water level at a constant level that is 1 m higher than the natural one, to ensure the supply of irrigation canals. In this condition, on the left embankment countryside, numerous sand boils and water blow-outs appear at the embankment toe and within the small ditches.

To understand the causes of these phenomena and in order to support the design of proper interventions, on January 2020 a monitoring system was installed at the landside embankment toe to study the pore pressure regime. The monitoring system consists in pressure and temperature sensors inserted in the ground at the depths corresponding to the most permeable levels, previously identified by boreholes, CPTU and geoelectrical surveys. In the same section at the embankment toe, also a Distributed Fiber Optical Sensor (DFOS) cable was inserted in the ground to measure the temperature along three vertical profiles. In addition, a probe for the combined measure of water pressure and temperature was installed inside the Gorzone channel water.

The use of temperature as a tracer of seepage through the embankment is proposed some years ago by the researchers of the University of Padova and Deltareas [1, 2]. The basic principle is to assume that in plain areas under normal flow regime characterized by very low water filtration velocity, the soil temperature within the embankment body and its foundation is determined only by heat exchange with air and residential water mainly by conduction, thus registering a variation only on a daily and seasonal basis. In contrast, in case of a sudden water level rising or to a flood, the consequent abnormal seepage flows induce higher temperature variations due to the rapid arrival of water at different temperature, associated with an increased heat exchange among water and soil particles due to the increased thermal gradient and to the increased advective heat exchange. To this aim, the application of DFOS to detect and track the temperature variations in the soil can be very interesting to integrate traditional field monitoring, as they allow high spatial resolution measurements over large distances and with high sampling frequency [3]. On the base of this principle, some experimental studies were recently carried out [4, 5].

In this case study, the data provided by the monitoring system, associated with those obtained from the previously geotechnical investigations, revealed that water pore pressure regimes were not completely consistent with the Gorzone water level variations. No variations were captured in the pressures data at the artificial alteration of the fluvial regime, as well as no alteration of the natural vertical temperature profile was detached by the DFOS when the blow-outs occurred [6]. Therefore, a finite element model was developed by using the FeFLOW DHI code, to better investigate the seepage conditions and the underground temperature regime.

2 Monitoring System

Over the past 30 years, sensors using the optical fiber technology have experienced considerable development and diffusion in the engineering fields, thanks to the greater opportunities that this technology offers if compared to traditional sensing. A DFOS provides high spatial and temporal resolution measurements, as it is sensitive along its entire length. By means of a proper interrogator, the fiber is stimulated with a laser signal that is refracted along its entire length depending on the characteristics of the medium in which the cable is deployed (the subsoil, the water, etc.). This way, depending on which component of the refracted wave is analyzed (Stokes / anti-Stokes, Raman, Brillouin or Rayleigh components), it is possible to measure the temperature or the deformations of the medium over the entire cable length; the data spatial and temporal resolution depend on the fiber optic cable type and on the used interrogator [3].

In the considered test site, the DFOS cable (BRUsens DTS STL cable) has been arranged in 3 verticals down to a depth of 20 m. In each vertical the cable is installed with a U-shaped configuration, with a descending and an ascending span and a 180° bend at the bottom. The connecting sections have been buried in a small trench [6]. The verticals are 3 m spaced along a line parallel to the toe of the landside embankment (orange point in Fig. 1). Special attention was paid to the installation methodology: the 20-ton static penetrometer used for the installation was equipped with a disposable drill bit, designed and manufactured on purpose, to protect the cable while drugging it deep into the underground. This methodology was developed to achieve a good thermal contact between DFOS and soil with a minimal disturbance, which is crucial in the filtration-affected soil temperature monitoring. In addition, it is a fast installation methodology [6].

Fig. 1. Map of the investigated area. The orange symbol indicates the position where DFOS and pressure/temperature sensors are installed; the blue one indicates the location the diver for channel water level and temperature monitoring. The red ellipse identifies the downstream dam. The black line represents the section represented in the FE model. The green symbol represents the extracting wells in Adige floodplain.

The cable is interrogated with a Sensornet's Oryx SR DTS instrument in Raman mode, which can provide a spatial resolution of 1 m and temperature measurement accuracy of 0.5 °C, up to a maximum cable length of about 2 km and with an acquisition time of about 2 s. The measurement campaign is performed on a monthly basis.

Previous geotechnical investigations indicated that the subsoil stratigraphy is characterized by a continuous succession of sandy and fine alluvial deposits, closely linked to the transport and deposition actions of the Adige river flowing nearby. Previous geotechnical surveys reveal that, in the monitoring system location, after a first superficial layer of silty clay constituting also the levee body, a fine sandy deposit hosting a superficial phreatic aquifer lays among the depth of −2.0 and −3.0 m a.s.l., while a deeper sandy layer lays between −3.05 and −3.8 m a.s.l; the two aquifers are separated by an impermeable level of grey clays. Thus, in addition to the DFOS cables, in the same location, 2 pressure and temperature sensors have been installed at the depth of 10 m and 18 m, corresponding to the more permeable layers, in order to monitor the pressure field and the temperatures in the embankment foundation soils. They provide data with hourly frequency and a precision of ±0.05kPa or ±0.1 °C. Finally, a diver was installed for continuous acquisition of the level and water temperature in the channel (Fig. 1).

3 Experimental Results

Figure 2 compares the total head measured from 25/1/2021 to 30/09/2022 at the depths of 10 m and 18 m with the water levels measured both in the Gorzone channel and in the Adige river. The following observations can be pointed out:

1. It is evident the seasonal behavior of Gorzone water level (black line), that is affected by tidal excursion in autumn-winter and remains quite constant in spring-summer.
2. The total heads measured both at the depths of 10 and 18 m are lower than the channel water level. This indicates that the blow-out observed in the countryside, close to the levee, are not caused by a pressure rising occurring in the deep aquifers associated to seepage in the levee foundation [6].
3. The piezometric heights measured at the depth of 18 m are always lower of 80–120 cm than that measured at 10 m. This indicates the hydraulic separation among the two aquifers, that is provided by the 4 m thick gray clay layer. In addition, it points out that the expected hydrostatic pore pressure distribution is altered, probably due to water extraction in some wells in the area.
4. The total heads variation in the sandy layers are not consistent with the Gorzone fluvial regime, given that no significant changes are recorded when the downstream dam is activated and the Gorzone water level increases of about 1 m. This observation suggests that the aquifers hosted in the sandy layers are not in close hydraulic connection with Gorzone channel.
5. The observed blow-outs are more likely ascribable to seepage in a thin superficial sandy layer confined at the embankment base, strictly caused by the water level rising and disconnected by the pore pressure regime in the deeper layers.

Even the registered temperature data shown in Fig. 3 supports these observations, given the discordance between the temperatures measured in the channel water and the ones registered by the two probes installed in the sandy layers. While the river water varies in temperature from 5 to 25 °C over the whole year, the soil temperature at the depth of 10 m shows only a light and delayed seasonal temperature variation, complying with the installation depth. Consistently, the temperature sensor at higher depth (18 m) displays a constant temperature equal to 12.5 °C. Also, the temperature vertical profiles captured by the DFOS during the monthly campaigns are mainly affected by the penetration in the ground of the seasonal air temperature variation thus indicating a substantial temperature stability starting from the depth of 10 m. Consequently, in this case, temperature measurements are not able to capture sudden changes that can be addressed as seepage fluxes indicators.

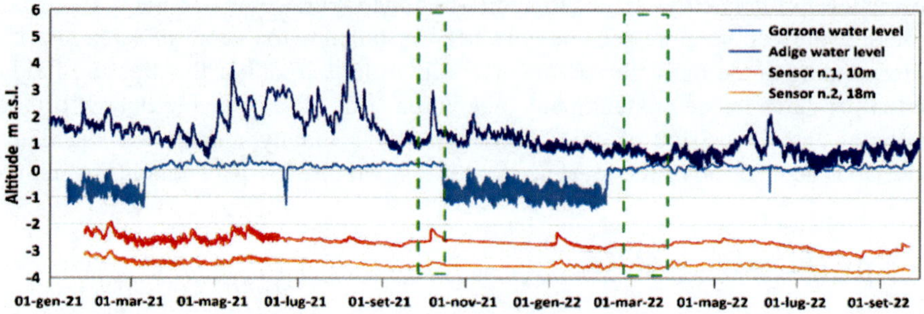

Fig. 2. Water levels and total head measured by the monitoring system; the boxes indicate the periods reproduced with the FE model.

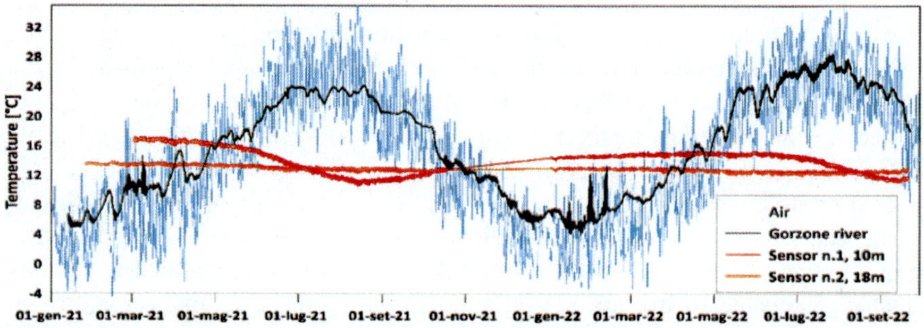

Fig. 3. Temperature dataset: the air temperature (light blue) is compared with the Gorzone water temperature, and with the temperatures measured at the depths of 10 and 18 m, in the sandy layers.

4 Hydro-Thermal Modelling

To verify these hypotheses, supported by the experimental data, a FE model was implemented by using the FeFLOW software to represent the Gorzone embankment in the section corresponding to the monitoring system and its extension to the nearby Adige river, in order to evaluate the relations among the recorded pressures and the Adige fluvial regime (Fig. 1). The stratigraphic sequence provided by the local geotechnical investigations (vertical A in Fig. 4) have been recognized in other available boreholes carried out in the nearby Adige floodplain (vertical B in Fig. 4) to realize a group of water extraction wells. The wells have been inserted in the model as boundary condition. Figure 4 depicts the model section, reporting the location of the significant elements (boreholes, wells, boundary condition, etc.).

Permeability, porosity and thermal properties of each layer were initially estimated from field investigations or assumed from tabled values. In a second time, they have been optimized (Table 1) through the model calibration performed by using the monitoring dataset registered from 23/2/2022 to 24/3/2022 (Fig. 2). In this phase, it was observed that the only way to match the pressure data recorded in vertical A is to assume the presence of some additional (undeclared) extraction wells located next to the monitoring system (supposed in vertical C), probably serving a nearby garden center. In fact, only an additional water extraction of almost 0.8l/s can justify the measured groundwater pressure regimes under the observed forcing water levels.

Table 1. Main characteristics of the layers represented in the model stratigraphic sequence.

Layer	Soil classification	k [m/s]	n
1	Silty clay	10^{-9}	0.2
1′	Fine sand	10^{-5}	0.3
2	Fine sand	10^{-5}	0.3
3	Gray clay	10^{-9}	0.2
4	Medium sand	$5 * 10^{-5}$	0.3
5	Silty clay	10^{-9}	0.2

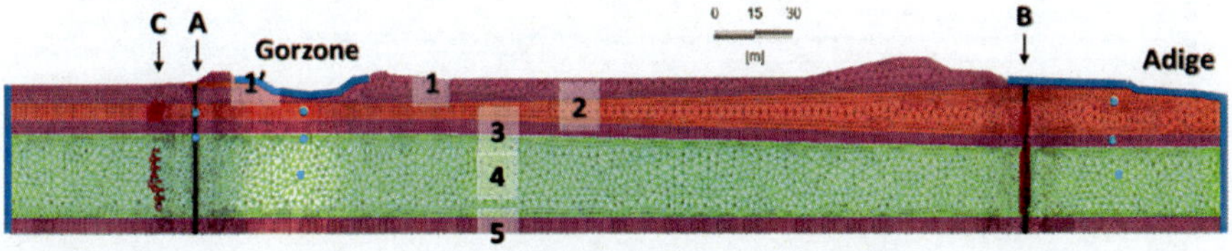

Fig. 4. FeFLOW model. There are indicated the layers (characteristics listed in Table 1), the applied hydraulic boundary conditions (blue lines), the verticals where stratigraphic data are available (A, B), the control points (light-blue points) and the identified extracting wells (B, C).

The calibrated model was used to reproduce the seepage regime in the period 24/9/2021–15/10/2021 and verify the hydraulic interconnection among the pressure distribution in the sandy aquifers, the Adige river and the extracting wells. In that period, the Gorzone water level was kept high by the dam, while the Adige water level registered an increase of about 2 m due to a small flood. In the meanwhile, the piezometric head at the depth of 10 m registered a sudden and steep increase, while the one at 18 m displays only a very slight variation. In the model, the imposed hydraulic boundary conditions are represented by the measured water levels and, in the far field, the water table was congruent with the water level in the ditches (1 m deep from the ground).

Figure 5 shows the obtained modelling outputs in terms of flow lines and piezometric head distribution (referred to the model local zero, equal to 50 m above sea level). Also in this situation, the only way to match the model outputs with the pressure regime recorded by the sensors during the flood event is to assume that the extraction wells suddenly interrupted their activity, restarting the day after. In fact, the water level increase due to the flood registered in the Adige river is not sufficient to justify the steep and sudden groundwater pressure alteration, that, moreover, slight anticipated the flood arrival. The interruption of the pumping activities may be justified by the non-necessity of irrigation water for the garden center, thanks to the occurrence of the rain.

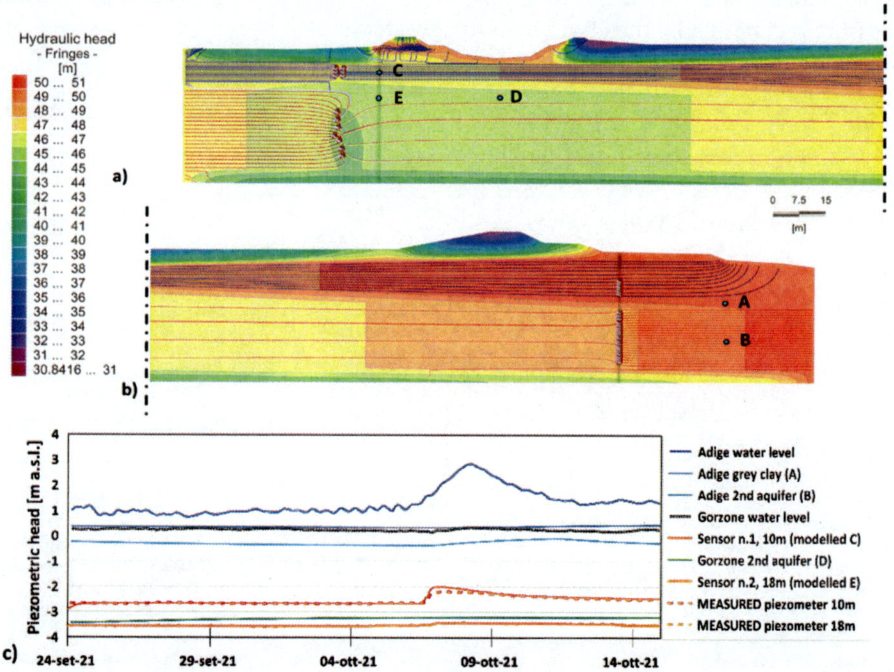

Fig. 5. Model outputs at the end of the period 24/9/2021–15/10/2021, a) in the Gorzone embankment section and b) in the Adige river section. Blue and red lines represent the flow lines in the upper and deeper sandy layers respectively. In c) the piezometric heads measured by the sensors are directly compared with the model outputs.

5 Conclusions

The combination of pressure and temperature datasets acquired by the monitoring system, data from on-site geotechnical investigation and FEM simulations outputs permit to have a better comprehension of the hydraulic phenomena occurring in this singular area and to draw the following conclusions.

1. It can be assumed that there is no direct connection among the observed blow-outs at the toe of the Gorzone embankment and the seepage in the deeper aquifers. It is more likely that the blow-outs occur as a consequence of a more superficial connection

between the channel and the countryside, probably due to the presence of a thin sandy layer at the levee base. As a consequence, possible interventions can be limited to the construction of a diaphragm wall crossing the levee body.

2. The observed alteration of the natural hydrostatic pressure trend expected in low plain areas, as the case study location, can be ascribable to the operations of several extracting wells: in addition to the declared extracting wells located in the Adige floodplain, it is very likely the presence of other extracting wells in the vicinity of the monitoring section, probably for irrigation purposes.
3. Temperature data and vertical profiles provided by punctual sensors and DFOS are in accordance with a natural seasonal penetration of air temperature in the subsurface. Nevertheless, in this case study, the temperature measurements did not provide useful information to localize seepage fluxes, probably because here the seepage fluxes are limited to the localized superficial sandy layer and so constrained that the DFOS spatial resolution (1 m) could be not sufficient to detect them.

Further studies are planned to analyze the effectiveness of DFOS temperature measurements to detect underground seepage and seepage in levee. Experimentally, the DFOS will be tested by applying a new measurement technique that uses hybrid fiber-optic cables and involves heating the cable at the beginning of the measurement campaign, as it is usually done to evaluate thermal properties of soil layers [7] by recording the temperature deployment over time along the cable, thus localizing seepage flows identified by more rapid temperature decreases. In addition, further modelling activities will be developed to analyze how geometric and hydraulic conditions as well as soils permeability and thermal properties may affect the effectiveness of DFOS in detecting subsurface seepage through temperature measurements.

References

1. Bersan, S., Koelewijn, A.R., Simonini, P.: Effectiveness of distributed temperature measurements for early detection of piping in river embankments. Hydrol. Earth Syst. Sci. **22**, 1491–1508 (2018). https://doi.org/10.5194/hess-22-1491-2018
2. Bossi, G., et al.: Multidisciplinary analysis and modelling of a river embankment affected by piping. In: Bonelli, S., Jommi, C., Sterpi, D. (eds.) Internal Erosion in Earthdams, Dikes and Levees: Proceedings of EWG-IE 26th Annual Meeting 2018, pp. 234–244. Springer, Cham (2019). https://doi.org/10.1007/978-3-319-99423-9_22
3. Schenato, L.: A review of distributed fibre optic sensors for geo-hydrological applications. Appl. Sci. **7**, 1–42 (2017)
4. Cola, S., Girardi, V., Bersan, S., Simonini, P., Schenato, L., De Polo, F.: An optical fiber-based monitoring system to study the seepage flow below the landside toe of a river levee. J. Civ. Struct. Heal. Monit. **11**(3), 691–705 (2021). https://doi.org/10.1007/s13349-021-00475-y
5. Fabbian, N., et al.: Innovative and traditional monitoring system for characterizing the seepage inside river embankments along the Adige River in Salorno (Italy). ISFMG, London (2022)
6. Dalla Santa, G., Fabbian, N., Schenato, L., Tedesco, G., Cola, S.: Misure di temperatura con fibra ottica e sensori tradizionali per modellare il regime di filtrazione arginale: il caso studio del fiume Gorzone. Incontro Annuale dei Ricercatori di Geotecnica (IARG) (2021)
7. Dalla Santa, G., Pasquier, P., Schenato, L., Galgaro, A.: Repeated ETRTs in a complex stratified geological setting: high-resolution thermal conductivity identification by multiple linear regression. J. Geotech. Geoenviron. Eng. **148**(4), 04022007 (2022)

Fiber Optic Sensing for Sinkhole Detection in Cohesionless Soil

G. Della Ragione[1](✉), T. Möller[2], C. N. Abadie[2], X. Xu[2], T. S. da Silva Burke[3], and E. Bilotta[1]

[1] Department of Civil, Architectural and Environmental Engineering, University of Naples, Naples, Italy
gianluigi.dellaragione@unina.it
[2] Department of Engineering, University of Cambridge, Cambridge, UK
[3] Department of Civil Engineering, University of Pretoria, Pretoria, South Africa

Abstract. Sinkholes are geo-hazards which can form suddenly without humanly visible pre-indication from soil surface settlement. An early warning system located below ground would enable to detect sinkhole-induced settlement before it reaches the surface, preventing possible damage to infrastructure and protecting lives. This paper focuses on the use of Distributed Fiber Optic Sensing (DFOS) as a solution for early detection of sinkhole formation. A series of small-scale experiments in controlled conditions are used to simulate the formation of a sinkhole. Fiber optics cables are laid in the soil specimen, and strains are collected using the LUNA ODiSi 6100 analyzer. The soil movements are observed through a Perspex window, collected with a camera and analyzed using Particle Image Velocimetry (PIV). Results indicate the ability of DFOS in detecting soil movements and underline the typical signature strain profile expected during sinkhole formation, indicating that at an early stage in the sinkhole formation, horizontal movements govern the strain profile within the cable. In conclusion, our study suggested that the DFOS technology can be used to detect, locate and estimate the size of a sinkhole, even though it has not possible to monitor the real strain level in the soil, due to a lack of shear transfer at the interface between the soil and the cable.

Keywords: Sinkhole · Trapdoor Test · Fiber Optic Sensing

1 Introduction

A sinkhole is created by subsidence of a depression or hole in the ground, eventually creating catastrophic failure at the soil surface. This can be the result of (i) progressive sagging of the overlying deposit above the void, (ii) sudden collapse of brittle formations along failure planes [1]. The genesis of a sinkhole depends on the complex geological structure and the hydrogeological setting of the soil. Nevertheless, the formation of a sinkhole implies that a void must develop at a certain depth and that a disturbing agent must arise to cause the soil to collapse within the void. The areas primarily exposed to these phenomena are karst terrains, which comprises limestone, dolomite, evaporites

A. Ferrari et al. (Eds.): CNRIG 2023, SSGG, pp. 186–193, 2023.
https://doi.org/10.1007/978-3-031-34761-0_23

and gypsum. These are prone to being dissolved with underground water flow, due to their chemical composition, thus allowing the formation of a cavity which may collapse and cause a sinkhole.

Sinkholes can cause severe damage to the existing infrastructures and buildings [2–4], and in rare occasions can lead to fatalities [5, 6]. Monitoring systems are needed to prevent or reduce the risk of any damage. Existing monitoring techniques, such as satellite-monitoring (DInSAR), laser scanner, photogrammetry, total stations and high precision levelling, are focused on the measurement of movements at surface level. These techniques must allow some settlements and deformations in the soil body, before being able to detect the formation of a sinkhole. Therefore subsurface methods such as embedded Distributed Fiber Optic Sensors (DFOS) could detect sinkhole-induced strains and movements earlier, since they can be installed at depth in the ground.

This paper shows an experimental campaign carried out as part of the SINEW project [7] to investigate the efficacy of this latter monitoring technique in detecting strain changes within the soil, induced by the formation of a sinkhole.

2 Methodology

2.1 Experimental Setup

Sinkholes can be simulated in the laboratory by means of trapdoor tests, both at $1g$ [8] or in the geotechnical centrifuge [9, 10]. In this paper, the experimental study was performed at $1g$ using a 790 × 200 mm rectangular rig, equipped with a trapdoor of width, $B = 100$ mm that can mimic the formation of a sinkhole, as shown in Fig. 1.

Samples were prepared using Hostun sand, with an average particle size, $d_{50} = 0.356$ mm [10]. Tests were prepared at increasing relative densities ($D_R = 20\%$, 52% and 88%) to allow comparison of sinkhole formation at different soil conditions. The sand was poured manually for $D_R = 20\%$, and using an automatic sand pourer for the higher densities. The sample height was $H = 200$ mm ($H/B = 2$).

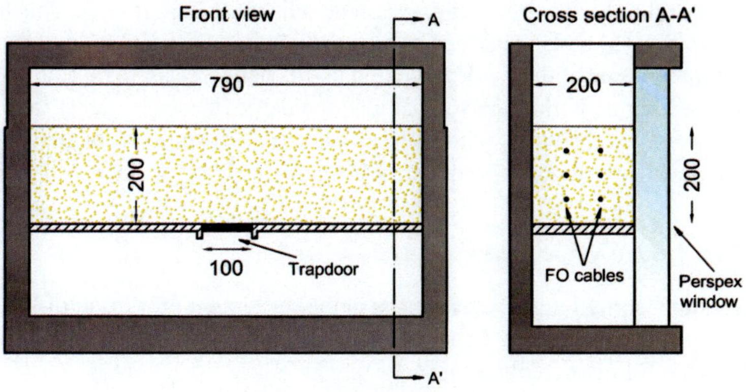

Fig. 1. Experimental rig with plane strain trapdoor

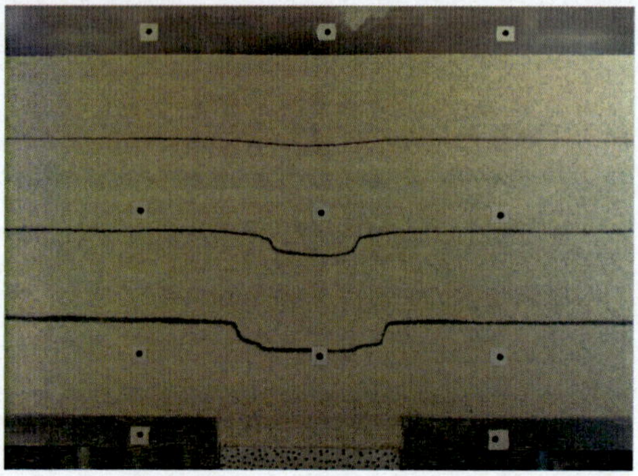

Fig. 2. Image acquired from the Canon Powershot G10, used for PIV analysis

The DFOS cables were installed at heights of $z = 50$, 100, and 150 mm above the trapdoor and these layers were highlighted using dyed black sand to highlight the shear band during failure (Fig. 2). The cables were pinned at the edge of the box, and coated in sand to increase soil-cable interface coupling in selected experiments [11].

The DFOS analyzer used to collect strain within the cable is the Luna ODiSI 6100 [12]. It is well adapted for small scale experiments in the laboratory due to the small spatial resolution and precise gauge length (2.6 mm used in this case). With a sample width of L = 790 mm, this provides ~300 sampling points in the FO cable at each level, enabling the establishment of a precise continuous distributed strain profile.

Images for the PIV analysis were acquired using a Canon Powershot G10 camera and the analysis was carried out using GeoPIV_RG [13].

3 Results

For a trapdoor displacement δ lower or equal to 1–2% of the trapdoor width, the soil is able to adjust and redistribute the stresses on an adjacent support [14]. This behavior is known as arching [15] and is regarded as the *early* phase of the sinkhole formation. This phase is the main focus of the work presented in this paper. Despite measuring the strain profile at three heights within the soil sample, the cable of most interest in this study is the one located at z = 150 mm, which represents a realistic embedded height that could be achieved during earthwork (~1 m depth from original soil surface at full scale).

3.1 Particle Image Velocimetry Analysis

The main results from PIV analyses consist in the settlement profiles at different heights above the trapdoor, which have been fitted using a modified Gaussian distribution [16] according to the following equations:

$$S_v(x) = S_{max} \cdot n / \left\{ (n+1) + \exp\left[\alpha \cdot (x/i)^2 \right] \right\} \tag{1}$$

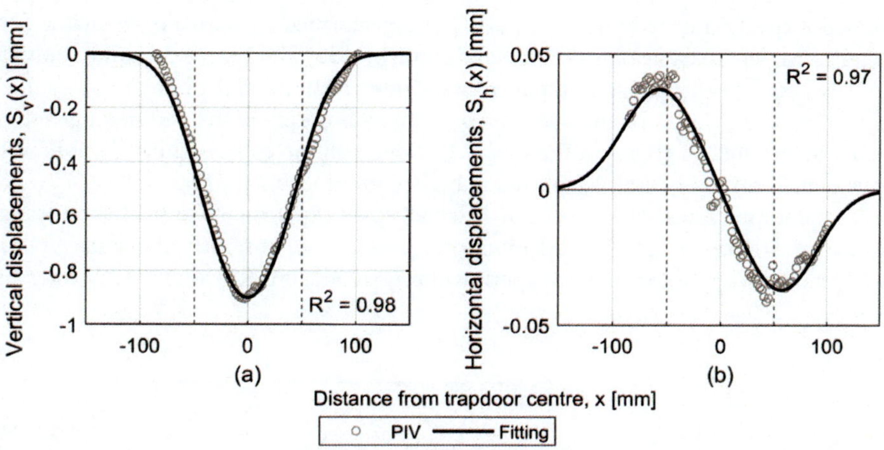

Fig. 3. (a) Vertical displacements, (b) Horizontal displacements from PIV for $D_R = 52\%$ $\delta = 2$ mm and $z = 150$ mm

$$n = 1 + \exp(\alpha) \cdot (2\alpha - 1)/(2\alpha + 1) \tag{2}$$

where S_{max} is the maximum settlement at the trapdoor centerline, i is the distance from the trapdoor centerline of the inflection point, α is a shape factor to alter the location of the inflection point. Horizontal soil displacements from PIV are interpolated using a double Gaussian distribution, according to the following equation:

$$S_h(x) = S_{hmax} \cdot \{\exp[-(x - b)/c] - \exp[-(x + b)/c]\} \tag{3}$$

where S_{hmax} is the maximum horizontal displacement while b and c are fitting parameters. Displacement profiles from PIV are shown in Fig. 3, demonstrating the adequateness of Eq. (1) and (3).

From the displacement profiles, it is then possible to evaluate the strain which are representative of those experienced by the fibre optic cable using:

$$\varepsilon = \left\{\left[\left(x_{i+1} - x_i + S_{h,i+1} - S_{h,i}\right)^2 + \left(S_{v,i+1} - S_{v,i}\right)^2\right]^{0.5}\right\}/(x_{i+1} - x_i) \tag{4}$$

This is needed for comparison with DFOS data, where horizontal and vertical displacement cannot be accessed.

3.2 Strain Profiles from DFOS

The experimental soil strain evaluated from fitted displacement profiles (Eq. (4)) are then compared with the experimental strain experienced by the fiber optic cables, obtained from the Luna ODiSI 6100 interrogator. Figure 4 shows a difference in strain magnitude between the fiber optic cable and the corresponding strain in the soil. This discrepancy might be related to a lack of shear transfer between the soil and the cable, induced by a lack of coupling and the low stress level at 1g. To further understand this behavior, detailed inspection of the fiber optic cable was performed numerically to enable better visualization of the cable deformation and strain profile [17]. Figure 5 shows a comparison between the experimental output from the interrogator and the numerical results for validation. The results concur with the experimental observations and demonstrate the impact of testing at 1g, and of the low stress level, on the cable's ability to follow soil movement, a difficulty that disappears at higher stress levels [7, 17].

This does not invalidate the use of 1g small scale experiments to further understand the use of DFOS for geo-hazard monitoring, as this is currently the most practical and cheapest way of generating experimental results using Rayleigh backscatter light analyzer.

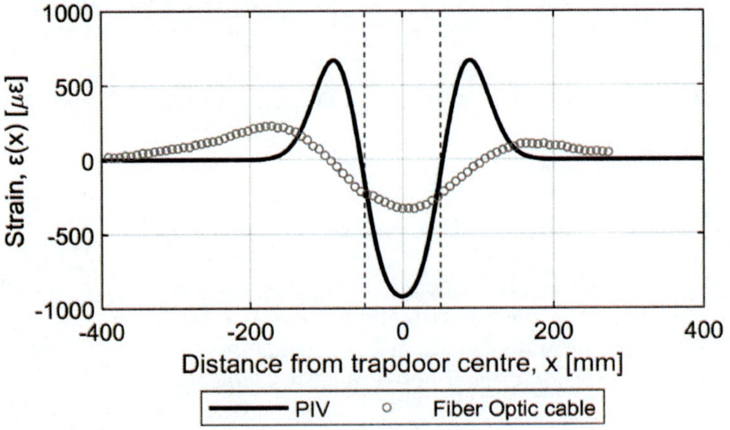

Fig. 4. Soil strain from PIV and strain within the fiber optic cable for $D_R = 20\%$ $\delta = 1$ mm and $z = 150$ mm

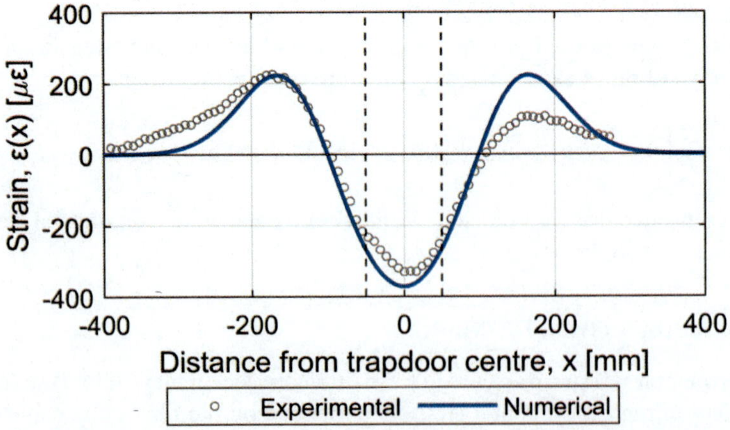

Fig. 5. Strain within the cable for DR = 52% $\delta = 1$ mm and $z = 150$ mm. Experimental strain obtained from the interrogator vs Numerical strain from FE analysis (modified after [17])

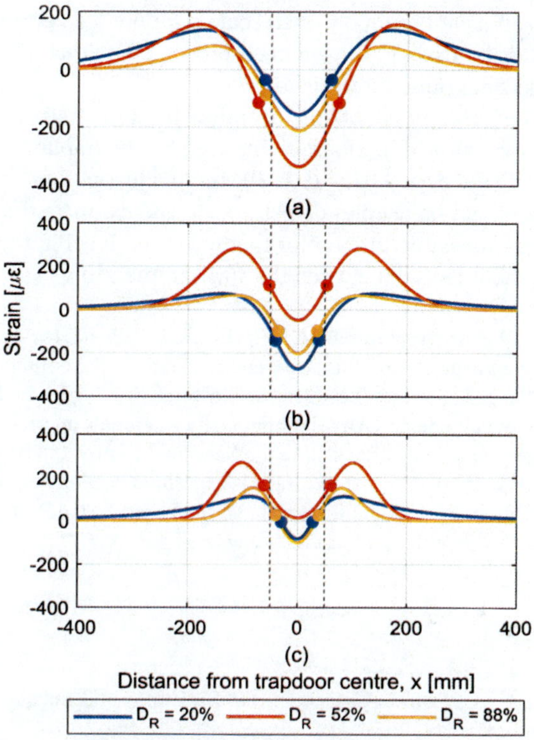

Fig. 6. Inflection point position on the experimental cable's strain profiles for $\delta = 1$ mm: (a) $z =$ 150 mm, (b) $z = 100$ mm, (c) $z = 50$ mm above the trapdoor

Moreover, the singular shape of the strain profile indicates the expected signature that one would expect from a sinkhole detection. The strain profiles obtained can reveal the position and the size prediction of the sinkhole, by looking at the position of the inflection point, as shown in Fig. 6, which pertains to experimental results.

4 Conclusions

Sinkholes are highly dangerous events for the health of infrastructures. Hence, to minimize or avoid the possible damages induced by a sinkhole, it is important to monitor soil deformations. A way to address this, is the use of an inexpensive fiber optic cable, buried at shallow depths, in the soil mass, that 'senses' the change in strain in the soil before its effects reaches the surface.

Different from civil structural applications (e.g. concrete structures, steel structures, etc.), where the fiber can be simply glued onto the surface, the effectiveness of DFOS in following soil movements is strongly dependent from the soil-cable interface and this limit has been considered as the main barrier to the use of this technology for geotechnical applications.

This paper illustrated the effectiveness of a relatively new technology (i.e. DFOS with Raileigh backscattering) in detecting soil movements.

Our study focused its attention on small scale trapdoor test, performed in controlled conditions at 1*g*. Results from PIV have been processed so that can be compared with data obtained by the fiber optic interrogator.

Despite DFOS technology is not able to follow the soil strain as expected, it is able to pick up an unequivocal profile of strain, induced by the trapdoor lowering, since the beginning of the test (i.e. $\delta \leq 0.01 - 0.02 \cdot B$). In conclusion, our study suggested that the DFOS technology can be used to detect, locate and estimate the size of a sinkhole, even though it has not possible to monitor the real strain level in the soil, due to a lack of shear transfer at the interface between the soil and the cable.

Acknowledgements. We are grateful to CSIC (EPSRC (EP/N021614/1) and Innovate UK (920035)) for funding and supporting this work and Dr Jennifer Schooling and Prof. Giulia Viggiani for facilitating the collaboration between the University of Naples Federico II and the University of Cambridge, which enabled Mr Gianluigi Della Ragione's internship at CSIC within the Erasmus+ Programme. We are also thankful to the Schofield Centre and its Director, Prof. Gopal Madabhushi, for their support with the laboratory experiments. Fruitful advice and discussion with Prof. Malcolm Bolton, Prof. Sadik Oztoprak, Dr Sam Stanier and Prof. Giulia Viggiani were also very valuable for the development of this project.

References

1. Jennings, J., Brink, A., Louw, A., Gowan, G.: Sinkholes and Subsidence in the Transvaal Dolomite of South Africa (1965)
2. Gutiérrez, F.: Sinkhole hazards. Oxford Research Encyclopedia of Natural Hazard Science (2016)
3. Johnson, K.S.: Salt dissolution and subsidence or collapse caused by human activities. Rev. Eng. **16**, 101–110 (2005)
4. Buchignani, V., D'Amato Avanzi, G., Giannecchini, R., Puccinelli, A.: Evaportie karst and sinkholes: a synthesis on the case of Camaiore (Italy). Environ. Geol. **53**, 1037–1044 (2008)
5. Bezuidenhout, C.A., Enslin, J.S.: Surface subsidence and sinkholes in the dolomitic areas of the Far West Rand, Transvaal, Republic of South Africa. In: Land Subsidence: Proceedings of the Tokyo Symposium (1970)
6. Cable News Network (CNN). http://edition.cnn.com/2013/03/01/us/florida-sinkhole/index.html. Accessed 12 Jan 2023
7. Abadie, C.N., da Silva Burke, T.S., Xu, X., Della Ragione, G., Bilotta, E.: SINEW: SINkhole early warning. In: 2nd International Conference on Construction Resources for Environmentally Sustainable Technologies (CREST), Fukuoka, Japan (2023)
8. Möller, T., da Silva Burke, T.S., Xu, X., Della Ragione, G., Bilotta, E., Abadie, C.N.: Distributed fibre optic sensing for sinkhole early warning: experimental study. Géotechnique (2022)
9. Jacobsz, S.: Trapdoor experiments studying cavity propagation. In: Proceedings of the First Southern African Geotechnical Conference (2015)
10. da Silva, T.S.: Centrifuge modelling of the behaviour of geosynthetic-reinforced soils above voids. Ph.D. thesis, University of Cambridge, Department of Engineering (2017)
11. Xu, X., Abadie, C.N., Moller, T., Della Ragione, G., da Silva Burke, T.S.: On the use of high-resolution distributed fibre optic sensing for small-scale geotechnical tests at 1g. In: 10th International Conference of Physical Modelling in Geotechnics (ICPMG), Daejon, South Korea (2022)

12. Luna Innovation Incorporated: Optical Distributed Sensor Interrogator Model ODiSI 6100: Data Sheet. Blacksburg, VA, USA (2020)
13. Stanier, S., Blaber, J., Take, W., White, D.: Improved image-based deformation measurement for geotechnical applications. Can. Geotech. J. 727–739 (2017)
14. Dewoolkar, M., Santichaianant, K., Ko, H.-Y.: Centrifuge modeling of granular soil response over active circular trapdoors. Soils Found. (47), 931–945 (2007)
15. Terzaghi, K.: Stress distribution in dry and saturated sand above yelding trap-door. In: 1st International Conference on Soil Mechanics and Foundation Engineering (1936)
16. Vorster, T.E., Klar, A., Soga, K., Mair, R.: Estimating the effects of tunneling on existing pipelines. J. Geotech. Geoenviron. Eng. 1399–1410 (2005)
17. Della Ragione, G., Bilotta, E., Xu, X., da Silva Burke, T.S., Möller, T., Abadie, C.N.: Numerical investigation of fibre optic sensing for sinkhole. Géotechnique (2023)

GIS-Based Analysis of the Potential Effectiveness and Efficiency of Mobile Terrestrial LiDAR to Survey and Monitor Rockfall Areas Along 15 km of Highway E45

Edgar Ferro[1], Francesco Cemin[1], Leonardo De Rosa[2], Alessandro Corsini[2], Francesco Ronchetti[2], Francesco Lelli[2], Alfonso Vitti[1], and Lucia Simeoni[1](✉)

[1] Department of Civil, Environmental and Mechanical Engineering, University of Trento, Trento, Italy
lucia.simeoni@unitn.it

[2] Department of Chemical and Geological Sciences, University of Modena and Reggio Emilia, Modena, Italy

Abstract. A method was developed in a GIS environment to assess the potential effectiveness and efficiency of a mobile terrestrial LiDAR survey at detecting the geometrical changes due to rockfalls from the cliffs facing a 15 km stretch of a highway in an alpine valley. The elements exposed to rockfalls were automatically classified as viaduct or open-sky ground supported road by comparing the DTM with the DSM. The most critical cliffs were then identified through the analysis of the rockfall trajectories. For these cliffs two mobile terrestrial LiDAR surveys from a vehicle travelling on the highway were simulated: a stop&go survey and a kinematic survey. The effectiveness of the surveys was assessed in terms of sensed area and density of the measured points. Their efficiency was specified in terms of feed rate. For both surveys at least 70% of the cliffs was visible with a point density higher than 400 points/m^2. The proportion of sensed area and density of the points provided by the stop&go survey was slightly higher compared with the kinematic survey, but the feed rate for the kinematic survey was higher.

Keywords: Rockfall · highway · mobile terrestrial LiDAR · effectiveness · efficiency

1 Introduction

The European Route E45 runs through the Brenner Pass in the Alps and enters Italy through the Isarco Valley, taking the national designation A22. Here, the A22 highway runs often on viaducts and through tunnels. The Isarco Valley has the typical U-shaped profile of glacial valleys and cliffs that, locally, are steeper than 70°. This usually occurs where the valley crosses the Permian ignimbrite and tuff.

Rockfalls are major natural hazards in this valley [1] and may interact with the highway, causing disruption and posing harm to its users. Periodical surveys of the rock

A. Ferrari et al. (Eds.): CNRIG 2023, SSGG, pp. 194–201, 2023.
https://doi.org/10.1007/978-3-031-34761-0_24

cliffs can potentially identify the areas susceptible to rock detachment and allow to estimate the volume of the unstable blocks. This information could be used as an input for risk assessment and protection work design [2]. This study investigates the potential of mobile terrestrial Light Detection And Ranging (LiDAR) from a vehicle moving along a 15-km long stretch of the A22 highway to survey periodically the rock cliffs, identify those most susceptible to rockfalls that may interact with the highway and hence provide the infrastructure manager with a tool that informs risk mitigation strategies. A rockfall-highway interference analysis was carried out to identify the elements of the highway infrastructure exposed to rockfalls. Then numerical simulations in a Geographic Information System (GIS) environment were performed to assess the potential effectiveness and efficiency of stop&go and kinematic LiDAR surveys. Other techniques, such as LiDAR surveys from an Unmanned Aerial Vehicle (UAV) or helicopter, were not considered due to the flight restrictions above the highway and the unfavourable morphology of the Isarco Valley with very steep cliffs.

2 Rockfall-Highway Interferences

2.1 Classification of Elements Exposed to Rockfalls

The open-source GIS software QGIS was used to identify the three main types of highway infrastructure: (1) tunnel, (2) viaduct and (3) open-sky, ground-supported road.

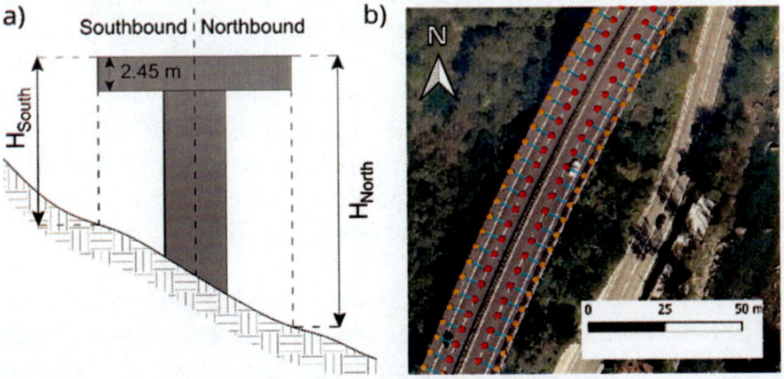

Fig. 1. Height H to recognize viaducts and ground supported roadways. a) Sketch showing the heights for southbound and northbound as the difference between roadway and ground elevations, b) couples of points used to calculate the heights: orange point = ground point from DTM, red point = roadway point from DSM.

The tunnels were indicated in a shapefile freely available from [1] and their position was verified by comparison with orthophotos. The viaducts and stretches of open-sky ground-supported road were distinguished based on the value of H, the minimum between H_{South} and H_{North} shown in Fig. 1a. These were calculated as the difference between the elevations of a roadway point and a ground point, i.e. the coupled red and orange points, respectively, shown in Fig. 1b. The road point elevation was taken from the Digital Surface Model (DSM) and the ground point elevation from the Digital Terrain Model (DTM), both with spatial resolution of 0.5 m and freely downloaded from [1]. Values of H greater than 2.5 m (slightly higher than the height of the decks of the viaducts) were associated with viaducts, lower values with open-sky, ground-supported road. For the viaducts, the positions of the piers were also represented in QGIS.

It was found that, of the 15 km length of the A22 highway between the 58-km and 72-km markers, 55% runs on viaducts, 13% in tunnels and 32% consists of open-sky, ground-supported road.

2.2 Potential Impact of Rockfalls Against Exposed Elements

The expected trajectories of single-block rockfalls that can potentially detach from subvertical rock cliffs were simulated using RocPro3D software [3]. Cliffs sloping more than 70° were identified as the most susceptible to rockfalls, then simulations were carried out for a total of 55 different potential source zones using as input data: a DTM at 2.5 m resolution [1]; a map of the potential source zones; a map of soil types with associated reference normal and tangential restitution coefficients estimated on the basis of values proposed by [4]; a map of rockfall barriers and protections (with location, height and impact resistance from the VISO inventory [5]). Rockfall simulations were carried out for blocks with reference diameters of 0.5 m and 2.0 m, and allowed to map and assess trajectories, bounce heights and impact energies with respect to these two block dimension scenarios. The simulations adopted a uniform probabilistic approach, i.e. variation of normal and tangential restitution coefficients and other soil physical parameters such as dynamic friction, lateral deviation and rebounds flattening on the basis of the velocity of the rockfall, which considers that each parameter value has the same occurrence probability within its variation range, hence: $\varepsilon_p = \varepsilon + \Delta\varepsilon \cdot U(-1, +1)$, where $\Delta\varepsilon$ is the parameter variation and $U(-1, +1)$ is the uniform distribution between -1 and $+1$ sampled from a random number generator (with a seed of about $1.84e^{19}$). The parameter variation $\Delta\varepsilon$ is in turn dependent on the incident velocity following a model, which allows taking into account a larger uncertainty at low velocities compared to high velocities.

Depending on the extension of potential rockfall source areas, a different number of possible detachment points and associated trajectories were simulated (<4000 m^2: 1000 trajectories; 4000–13000 m^2: 1500 trajectories; 13000–20000 m^2: 2000 trajectories; >20000 m^2: 2500 trajectories). The simulation results for blocks with 0.5 m and 2.0 m reference diameters were exported to GIS, as 2.5-m resolution raster maps of rockfall trajectories, energies and bounce heights.

To assess the possible interaction between the rockfalls and the highway, the positions of the viaduct piers were rasterized at 2.5 m spatial resolution. Also, the height of the roadway with respect to the ground was calculated by subtracting the DSM elevation from the DTM elevation.

By GIS-based spatial intersection between the maps of rockfall energy and the map of the pier positions, maps of the points (i.e. cells) where rock blocks can collide with the piers with a given energy were obtained for the 0.5 m and 2 m diameter scenarios. Moreover, the intersection between the maps of rockfall bounce height and the map of the roadway elevation, allowed to identify the points where rockfalls blocks can bounce over and onto the road pavement.

In the 15-km stretch of the highway, 10 cliffs were recognized as sources of the most dangerous rockfalls. Examples of simulation results are shown in Fig. 2. Overall, the simulations showed that the interaction between rockfalls and viaduct piers is possible at several locations along the 15-km stretch of the A22 highway considered in this study. Less likely are the rock blocks bouncing onto the road. These results are obtained by assuming, in the simulations, that the existing rockfall barriers are capable of stopping the intercepted rock blocks whose kinematic energy is lower than the nominal energy that can be absorbed by the barriers.

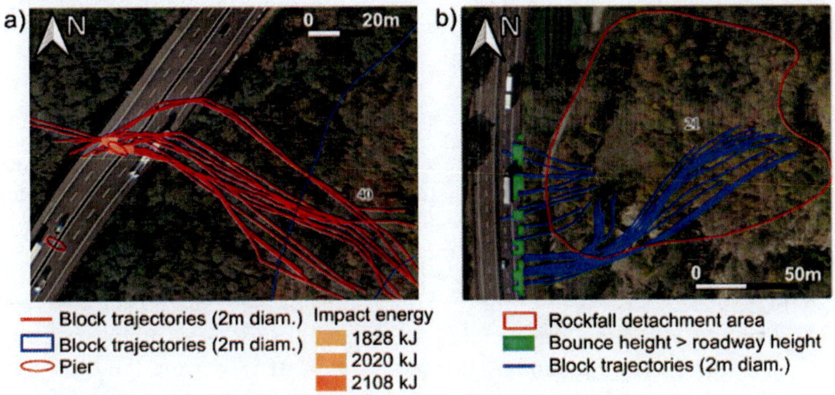

Fig. 2. Rockfall-highway interference for the 2-m block diameter scenario: a) impact between the rock blocks and the piers; b) rock blocks bouncing onto the roadway.

3 Simulations of Mobile Terrestrial LiDAR Surveys

Using GIS software tools, two different types of LiDAR survey from a vehicle moving on the breakdown lane (the outer lane) of the A22 highway were simulated: (1) a stop&go survey [6, 7] and (2) a kinematic survey [8] at 15 km/h. The surveys were simulated for the 10 cliffs recognised as the source of the most critical rockfall-infrastructure interferences. Below, the algorithm used in the simulations is briefly described, while greater detail can be found in [9].

The simulations were performed in GRASS GIS environment [10] with the support of Python scripts. The DEM used in the analyses was created by combining the DTM and the DSM provided by [1]: the elevations from the DSM were assigned to the cells located on the roadway, while the DTM elevations were assigned to the remaining cells. The DSM and DTM used as input and the output DEM had spatial resolution (cell size) of 0.5 m. For each cliff, points corresponding to the positions of the vehicle equipped with the laser scanner were created along the breakdown lanes with a pitch L = 1 m for the kinematic survey and L = 20 m for the stop&go survey; the height h of the laser scanner (the *observer point*) from the pavement was assumed equal to 1.75 m (Fig. 3a).

The GRASS GIS command *r.viewshed* was used to calculate the angle φ between the direction of the horizontal at the *observer point* and the direction of the segment joining the *observer point* to a cell on the cliff (the *observed point*) for each cell of the DEM visible from the *observer point* (Fig. 3b). A lower bound value of $-5°$ was assumed for φ to account for the guardrail, road pavement and other obstructions.

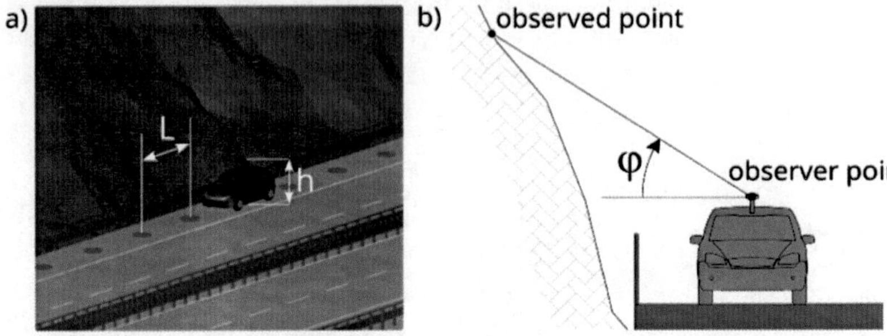

Fig. 3. Mobile terrestrial LiDAR from the breakdown lane: a) vehicle positions (blue points on the breakdown lane) with pitch L equal to 1 m for the kinematic survey and 20 m for the stop&go survey; b) angle φ used to detect the visible cells of the DEM.

A RIEGL VZ-2000i was used for the simulations of the stop&go survey, a RIEGL VUX-1HA for the kinematic survey. Their instrumental variables - minimum and maximum LiDAR distances ($d_{\min}$ and $d_{\max}$), vertical field of view from the the horizontal (FOV), sample frequency (nr_{pps}) and scanning speed (nr_{rps}) - are listed in Table 1.

Table 1. Technical specifications of the laser scanners RIEGL used for the simulations.

Instrument variable	RIEGL VZ-2000i (stop&go survey)	RIEGL VUX-1HA (kinematic survey)
Min-Max distances $d_{\min} - d_{\max}$ [m]	1–600	1–235
Vertical field of view FOV [°]	100 (+60, −40)	360°
Sample frequency nr_{pps} [pts/s]	Up to 500.000	Up to 1.800.000
Scanning speed nr_{rps} [rps]	Up to 240	Up to 250

The distance *dp* along a scan line between points located at a distance *d* from the laser scanner can be calculated as:

$$dp = \frac{nr_{rps}}{nr_{pps}} \cdot \mathrm{FOV} \cdot d \tag{1}$$

where the FOV is expressed in radians.

For the stop&go survey with the RIEGL VZ-2000i, all the cells within the field of view (φ between $+60°$ and $-40°$) and with d between $d_{\min}$ and $d_{\max}$ (i.e. in the range 1–600 m) were assumed visible. The angle β and the distance *ds* between the scan lines were calculated assuming the instrument to perform a 360° (2π) horizontal rotation in 60 s:

$$\beta = \frac{2\pi/60\,\mathrm{s}}{nr_{rps}} \tag{2}$$

$$ds = d \cdot sen\beta \tag{3}$$

For the kinematic survey with the RIEGL VUX-1HA, all the cells with a distance d from the scanner between $d_{\min}$ and $d_{\max}$ (i.e. in the range 1–235 m) and an angle $\varphi > -5°$ were considered visible. The distance between the scan lines ds was calculated as:

$$ds = v/nr_{rps} \tag{4}$$

with $v = 15$ km/h (4.17 m/s).

Finally, for both the kinematic and stop&go surveys the density of the points that can be potentially sensed per m^2 ($nr_{points\_m^2}$) was calculated as:

$$nr_{points\_m^2} = \frac{1}{ds \cdot dp} \cdot cos\psi \tag{5}$$

where ds and dp are expressed in meters and ψ is the angle between the vector V_1, parallel to the segment joining the *observer point* and the *observed point*, and the vector V_2, perpendicular to the *observed point*.

4 Results

4.1 Effectiveness

Considering all the 10 investigated cliffs, the stop&go survey was able to sense more than 90% of the surface area sloping more than 70°. For more than 90% of this area, the point density resulted higher than 400 points/m^2. For the cliff close to the 70-km marker, the area sensed by the stop&go survey is shown in Fig. 4a and Fig. 4b, where the sensed cells of the DEM are shown in blue (a darker blue indicates a higher point density). Figure 4a shows the point density for the whole visible area of the cliff, Fig. 4b only for the parts of the cliff sloping more than 70°.

The kinematic surveys detected more than 85% of the area sloping more than 70°, with more than 70% of this area characterised by a point density larger than 400 points/m^2 and almost all of it with a point density larger than 100 points/m^2. For the cliff near the 70-km marker, the area sensed by a kinematic survey is shown in red in Fig. 4c and Fig. 4d, where darker red corresponds to higher point densities. Figure 4c shows the overall sensed area, Fig. 4d only the parts of the cliff sloping more than 70°.

4.2 Efficiency

The two LiDAR instruments considered in the simulations can achieve similar performances in terms of 3D position accuracy and precision. For the kinematic survey positioning is mainly constrained by GPS/IMU performance which, however, should be adequate for the specific purpose of this study. Key differences between the two survey methods are:

- the scan speed, higher for the kinematic survey;
- the feed rate, higher for the kinematic survey;
- the maximum distance and accuracy/precision, higher for the stop&go survey;
- the point density, higher for the stop&go survey.

Owing to the higher feed rate, the kinematic survey is expected to reduce the time required to execute the survey and hence, the man hours and the occupation time of the breakdown lane. Moreover, the safety measures are simpler for the kinematic survey, in which the vehicle equipped with the laser scanner does not need to stop and the surveyors are not required to leave the vehicle, exposing themselves to a greater danger. Although the kinematic survey is characterized by a smaller distance of the *observed points* and a lower accuracy/precision, it remains suitable for monitoring rock cliffs susceptible to rockfalls that might endanger the highway. Overall, the kinematic survey seems more efficient.

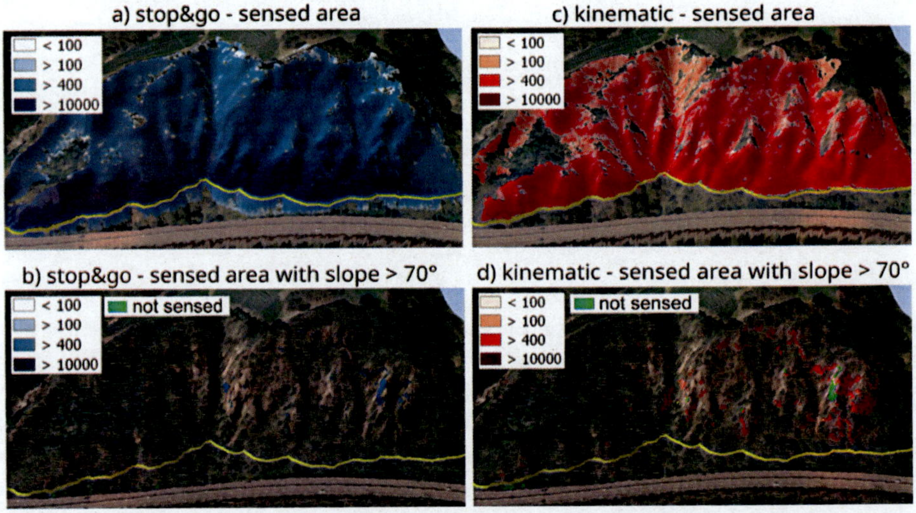

Fig. 4. Simulated LiDAR surveys of the cliff at the 70-km marker; a) cliff area sensed by stop&go survey; b) cliff area sloping more than 70° sensed by stop&go survey; c) cliff area sensed by kinematic survey; d) cliff area sloping more than 70° sensed by kinematic survey; different point density (points/cell) are indicated with different color intensity, as indicated in the legend.

5 Conclusion

Numerical simulations in a GIS environment have shown that mobile terrestrial LiDAR from a vehicle moving slowly along a breakdown lane can be potentially used to survey periodically the rock cliffs susceptible to rockfalls that may be present close to a highway.

A stop&go survey was found able to sense over 90% of the cliff areas sloping more than 70°, 90% of which with a point density greater than 400 points/m^2. A similar performance was achieved with a kinematic survey, which could sense over 85% of the cliff area sloping more than 70%, 70% of which with a point density larger than 400 points/m^2 and almost all of it with a point density larger than 100 points/m^2. Despite the smaller point density, the kinematic survey seems the most convenient, as it can be carried out more quickly and with simpler safety measures. Moreover, for the kinematic survey the point density could be increased by performing multiple scans of the most critical cliffs.

Funding. This research was funded by Autostrada del Brennero S.p.A (Brenner Autobahn AG) and partly by MUR PON R&I 2014–2020 Program (project MITIGO, ARS01_00964).

References

1. Rete Civica Alto Adige. http://geoportale.retecivica.bz.it. Accessed 13 Dec 2022
2. Macciotta, R., Martin, C.D., Morgenstern, N.R., Cruden, D.M.: Quantitative risk assessment of slope hazards along a section of railway in the Canadian Cordillera—a methodology considering the uncertainty in the results. Landslides **13**(1), 115–127 (2015). https://doi.org/10.1007/s10346-014-0551-4
3. RocPro3D. http://www.rocpro3d.com. Accessed 29 May 2023
4. Crosta, G.B., Agliardi, F.: A methodology for physically based rockfall hazard assessment. Nat. Hazard. **3**, 407–422 (2003). https://doi.org/10.5194/nhess-3-407-2003
5. VISO. https://maps.civis.bz.it/?context=PROV-BZ-HAZARD. Accessed 13 Dec 2022
6. Abellán, A., Calvet, J., Vilaplana, J.M., Blanchard, J.: Detection and spatial prediction of rockfalls by means of terrestrial laser scanner monitoring. Geomorphology **119**, 162–171 (2010). https://doi.org/10.1016/j.geomorph.2010.03.016
7. Kromer, R.A., Hutchinson, D.J., Lato, M.J., Gauthier, D., Edwards, T.: Identifying rock slope failure precursors using LiDAR for transportation corridor hazard management. Eng. Geol. **195**, 93–103 (2015). https://doi.org/10.1016/j.enggeo.2015.05.012
8. Lim, S., Thatcher, C.A., Brock, J.C., Kimbrow, D.R., Danielson, J.J., Reynolds, B.J.: Accuracy assessment of a mobile terrestrial lidar survey at Padre Island National Seashore. Int. J. Remote Sens. **34**(18), 6355–6366 (2013). https://doi.org/10.1080/01431161.2013.800658
9. Cemin, F.: Interazione pendio-autostrada a scala regionale: sviluppo di un sistema per la pianificazione di rilievi LiDAR e definizione delle condizioni al contorno idrauliche per lo studio di frane estremamente lente. Master's thesis, Department of Civil Environmental and Mechanical Engineering, University of Trento, Italy (2021)
10. GRASS Development Team. Geographic Resources Analysis Support System (GRASS) Software, Version 7.8. Open Source Geospatial Foundation (2020). https://grass.osgeo.org. Accessed 21 Feb 2023

Analysis of Temporary Deep Landslide Reactivation with Interferometric Monitoring Technique

Enzo Fontanella(✉) and Augusto Desideri

Department of Structural and Geotechnical Engineering, Sapienza University of Rome, via Eudossiana n.18, 00187 Rome, Italy
{Enzo.fontanella,Augusto.Desideri}@uniroma1.it

Abstract. The paper presents the case of a slow landslide temporarily reactivated by the excavation of twin tunnels in the Tuscan-Emilian Apennines. After a description of the case study, the paper illustrates the most significant outcomes of the monitoring system characterized by a large quantity of data which yielded an important opportunity to study the interaction of tunnels along the slope.

The monitoring system consisting of traditional monitoring data, and satellite measurements (SAR interferometric technique) allowed analyzing the characteristics of the landslide movements before, during, and after tunnel reactivation.

Keywords: Landslide monitoring · Twin tunnels · SAR interferometry technique

1 Introduction

This paper illustrates the locally well-known Ripoli - Santa Maria Maddalena (hereinafter, indicated as SMM, in brief) landslide, reactivated in 2011 during the excavation of twin tunnels for the new motorway crossing the Apennines between Bologna and Florence (Italy).

Over 70,000 landslides bodies have been identified in Emilia– Romagna region, most of which are complex landslides and involving structurally complex formations [1–3].

In areas prone to instability, such as in the case here reported, the excavation of a tunnel can trigger or accelerate landslide movements, even on large scale [4–6].

Due to particular significance of the project a constant surveillance of the evolution of movements has been performed. After describing the study site, the paper is basically aimed to show the importance of satellite measurements in the analysis of deep gravitational movements. The interferometric analysis is used in particular to evaluate the actual stability condition of landslide showing its usefulness in long-term monitoring of extremely slow landslides especially when combined with topographic measurements.

A. Ferrari et al. (Eds.): CNRIG 2023, SSGG, pp. 202–209, 2023.
https://doi.org/10.1007/978-3-031-34761-0_25

2 The Study Site

The Val di Sambro twin tunnels project concern the motorway crossing the Apennines between Bologna and Florence. Each tunnel, identified, in Fig. 1, as "North tunnel" (for the traffic towards Bologna) and "South tunnel" (for the traffic towards Florence) has a section of approximately 180 m^2 with a maximum span of 16 m and a distance between the tunnel axes of 14 m. The tunnels were driven with the full-face method using traditional techniques and with fiberglass dowel reinforcement of the face. The excavation in the area close to SMM toke place between May 2011 and July 2014.

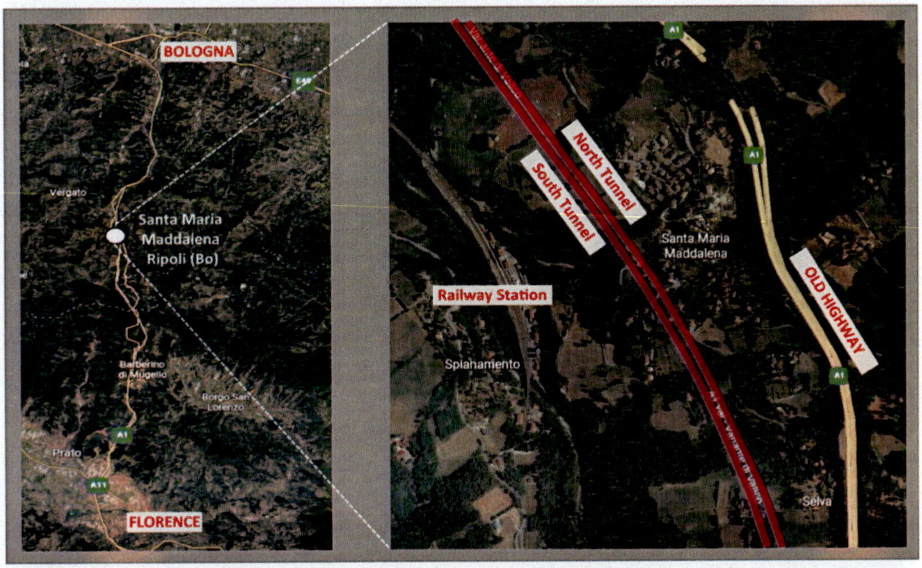

Fig. 1. Location of the "Valico Variant" project and of the Santa Maria Maddalena tunnels

The subsoil in the SMM area is mainly characterized by shallow debris, 20 m thick, overlying the Monghidoro Formation (defined as MOH hereafter), which is represented by a turbidite succession dating from Upper Cretaceous to Paleocene.

In particular the MOH consists of arenaceous-pelitic turbidities with a ratio between sandstone and marl larger than 1. This formation includes a clayey-arenaceous lithozone (MOHa hereafter) defined by clayey/arenaceous facies with a ratio between sandstone and marl ranging from 1/3 to 1/2 (Fig. 2).

3 Slope Monitoring

Figure 3 shows the map of the area of SMM with location of twin tunnels and a part of the installed inclinometers just used to represent the slope movement.

As it can be observed in Fig. 4 the elaboration and interpretation of the integral inclinometer movements highlights the presence of a fully-developed sliding surface having a maximum depth of about 70 m b.g.l..

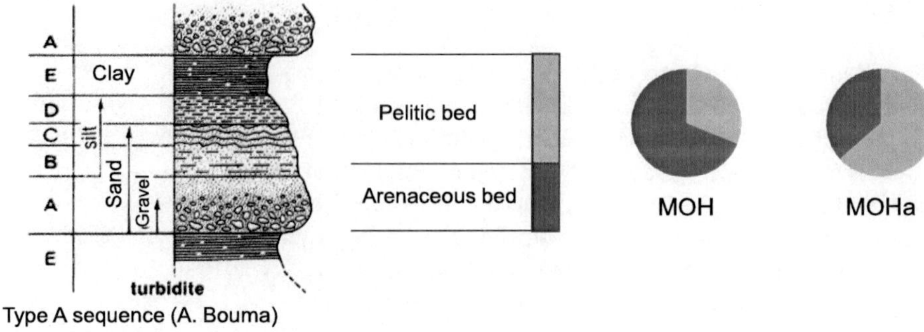

Fig. 2. Example of turbidity layer for Monghidoro geological formation [7]

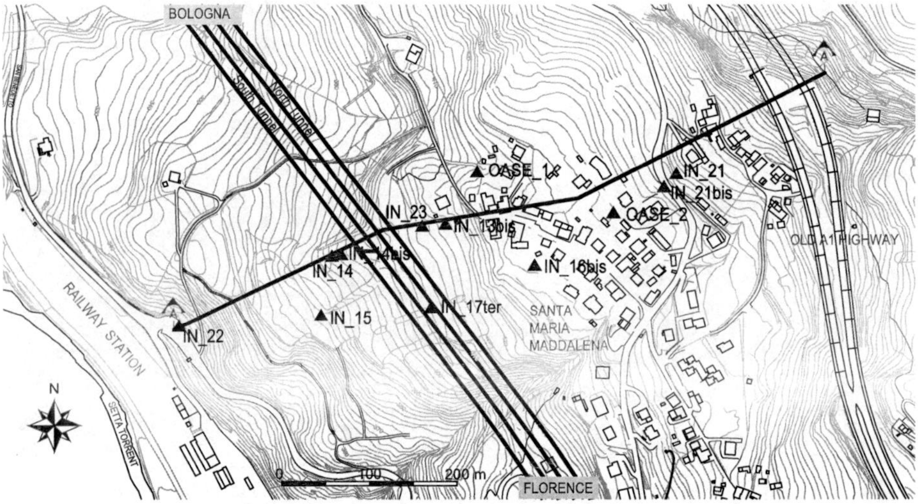

Fig. 3. Map of the area in proximity of Santa Maria Maddalena with the location of the inclinometers installed [4].

As regards to identify the evolution of slope movements at different phase of tunnels excavation (during and after excavation), in Fig. 5 are reported the differential displacements on sliding surface with time.

The inclinometric measurements show that the displacements along the sliding surfaces have gradually ceased after the excavation was completed on December 2014.

The monitoring data presented mainly began after the start of tunnel excavation. In order to investigate the slope activity even before tunnelling, SAR interferometry analyses performed in the area are also considered.

The satellite images datasets were processed to compute the displacements magnitude and rate of natural targets available on the ground surface. The algorithm provides the component of the displacement vector along the line of sight (LOS), i.e. the line joining the radar and the target. Positive displacement values indicate movements towards the satellite (shortening along the LOS), while negative displacements denote movements away from the sensor (lengthening along the LOS). In order to SMM case study, taking into account the maximum slope line (assumed as the preferential direction for sliding phenomena), and the direction of orbital geometry by the two satellite (RADARSAT and TERRASAR X), negative displacements denote downstream movements.

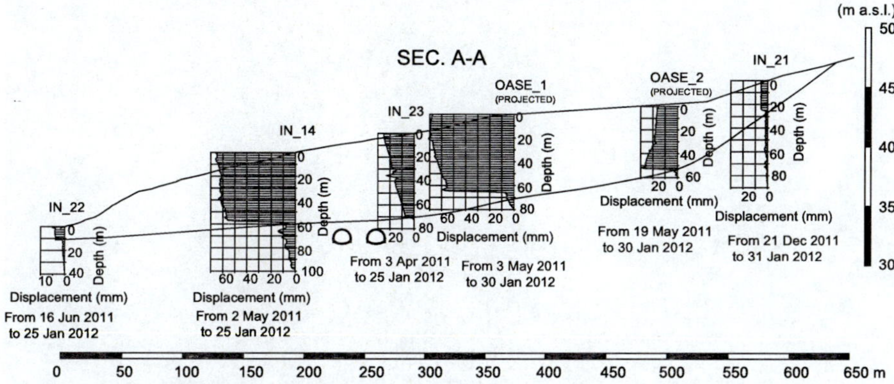

Fig. 4. Interpretation of the inclinometers along Section A-A of Fig. 2, [4].

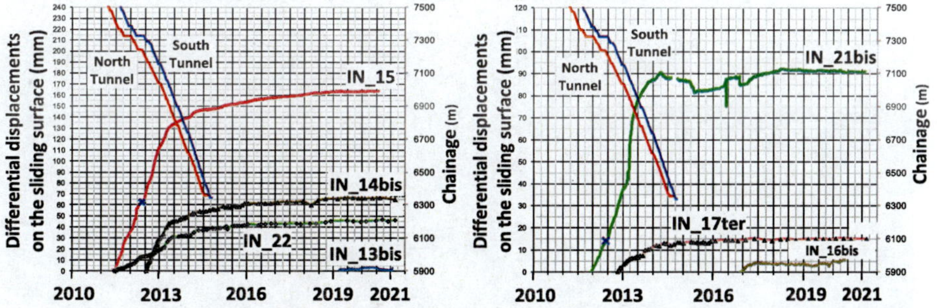

Fig. 5. Differential displacements on the sliding surface

In Fig. 6a,b,c are presented the measurements performed by satellite on the investigated area of SMM for three different period. The first one 2003–2009 is referred to the slope movements before tunnelling excavation. The second one 2012–2013 is referred to the period of sliding reactivation of slope due to tunnel excavation. Lastly the third one 2019–2020 shows the current condition of slope.

The colored dots in Fig. 6 represent the natural targets that the satellite can detect. They are concentrated mainly on the side of the old A1 highway, on SMM village, and SMM railway station.

The repeated passages of the satellite over the same area allow to detect the movements undergone by the different points. The coloring of the points provides an indication of the displacements undergone in a year (the color scale of legend expressed in terms

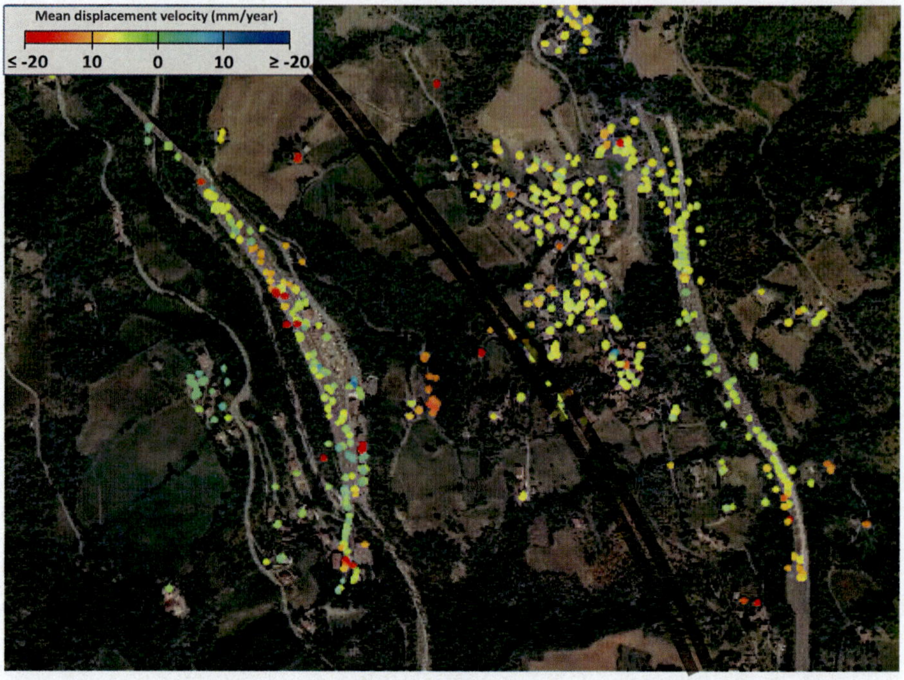

(a) RADARSAT satellite 2003 – 2009 (before tunnelling excavation)

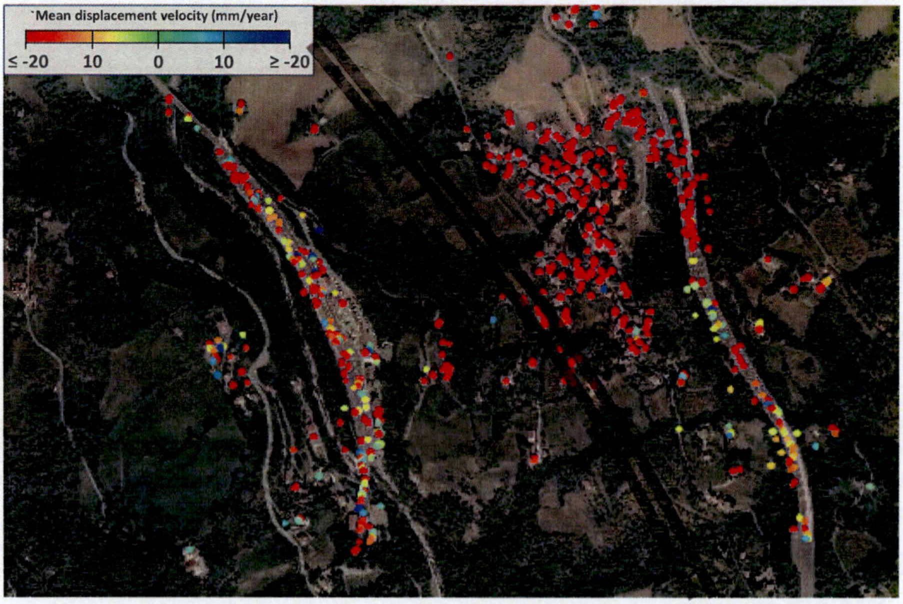

(b) RADARSAT satellite 2012-2013 (during tunnelling excavation)

Fig. 6. a,b,c Spatial distribution and mean LOS displacement rate of the coherent targets in the area surrounding Santa Maria Maddalena

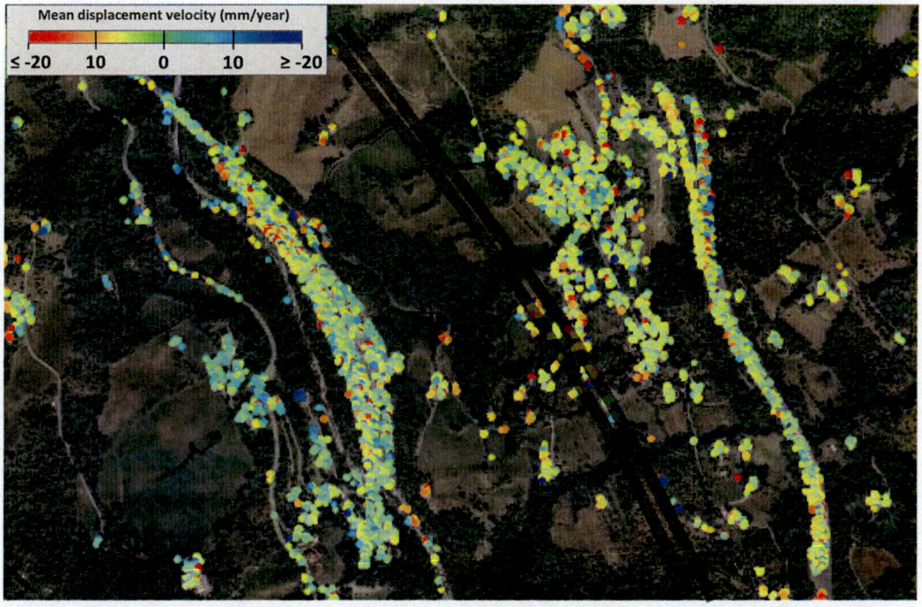

(c) TERRASAR X satellite 2019 – 2021 (after tunnelling excavation)

Fig. 6. (*continued*)

of mm/year is provided in the figure): the colors from green to red refer to downstream displacements, the colors from green to blue refer to upstream displacements.

The mean displacements rate during the period of 2003–2009 is reported in Fig. 6a. It can be noted that the area near the SMM village is interested by displacements rate of few mm per year before the tunnel excavation. A condition that is found similar in many areas of Apennines.

The image in Fig. 6b clearly highlights the period (2012–2013) of reactivation of the movement with surface displacements rate equal or greater than 20 mm/year.

The analysis of Fig. 6c reveals how the current situation (2019–2021) is completely analogous to that in the period 2003–2009 (before tunnelling excavation) with downstream displacements rate less than 5 mm/year, moreover, the satellite measurements are congruent with the topographic measurements at the ground surface [4, 6].

Evolution in time of the displacements of 15 targets near the village of SMM is presented in Fig. 7. The dashed lines in the graph, during 2013, are due to a variation of the satellite from which the measurements were taken. It can also be noted a reduced data scatter for measurements of the new satellite (TERRASAR X) than the previous one (RADARSAT).

In the period 2003–2009 the displacements rate of the target is about 5 mm/year, and significantly increase during tunnelling excavation in 2012–2013. After 2015, is registered a progressive reduction in the ground movement rate with mean values in 2019–2021 of few mm/year.

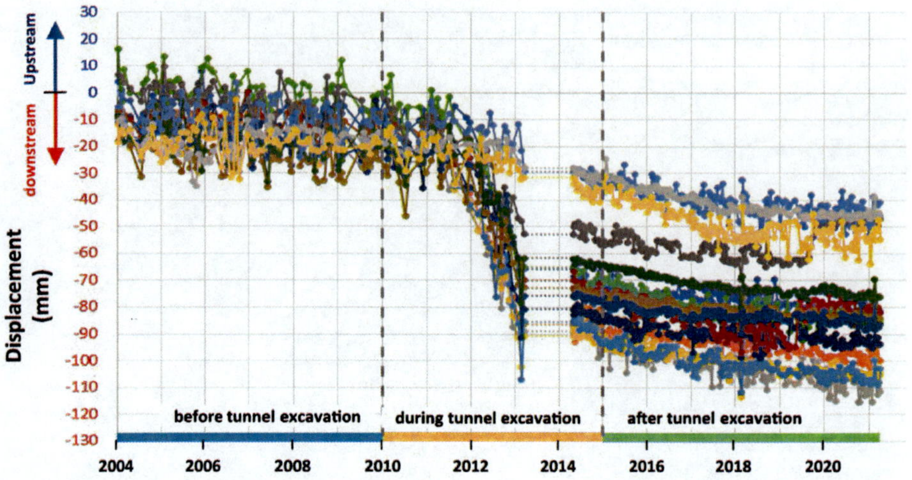

Fig. 7. Evolution in time of the displacements of 15 targets near the village of Santa Maria Maddalena

4 Discussion and Final Remarks

The paper presented the case study of the Santa Maria Maddalena characterized by excavation of twin tunnels in an area affected by deep landslide.

Temporary reactivation of deep landslide is investigated with the large amount of traditional monitoring data and the interferometric technique INSar. The INSar technique is used in particular to analyze the behavior of the slope before excavation and after excavation (long term).

The results of interferometric analysis indicate that very slow landslide movements, with a mean displacement rate of 5 mm/year, were recorded before the tunnel excavation. The tunnelling caused a displacement acceleration along the sliding surfaces, reaching a maximum movement rate of about 20 mm/month.

The slope surface movements continued in time after tunnel faces excavation with a mean displacement rate similar to that registered before excavation.

The inclinometric measurements show the absence of movements along the sliding surface. Confirming that the slope is no longer affected by instability phenomena.

Finally, the slope of Santa Maria Maddalena has returned to its *ante operam* behavior, and the very slow rate of surface movement (a few mm/year) measured in the period 2019–2021 are typical of the slopes of the Tuscan-Emilian Apennines. Other stable areas of this Apennines are characterized by similar rate of movement [4]. It is therefore reasonable to believe that these displacements are not related to instability phenomena.

References

1. Esu, F.: Behavior of slopes in structurally complex formations. In: International Symposium the Geotechnics of Structurally Complex Formations, Capri, Italy, vol. II, pp. 157–169 (1977)

2. D'Elia, B., Picarelli, L., Leroueil, S., Vaunat, J.: Geotechnical characterization of slope movements in structurally complex clay soils and stiff jointed clays. Rivista Italiana di Geotecnica **32**(3), 5–47 (1998)
3. Bandini, A., Berry, P., Boldini, D.: Tunnelling-induced landslides: the Val di Sambro tunnel case study. Eng. Geol. **196**, 71–87 (2015)
4. Desideri, A.: Gallerie e movimenti di versante. Rivista Italiana di Geotecnica **1**, 5–41 (2021)
5. D'Effremo, M.E., Fontanella, E.: The use of advanced remote sensing techniques for monitoring of slopes affected by slow movements. In: Life-Cycle and Sustainability of Civil Infrastructure Systems - Proceedings of the 3rd International Symposium on Life-Cycle Civil Engineering, IALCCE 2012, pp. 2008–2014 (2012)
6. D'Effremo, M.E., et al.: Analysis and monitoring of a tunnelling-induced deep landslide reactivation. In: Landslides and Engineered Slopes. Experience, Theory and Practice, vol. 2, pp. 735–742 (2016)
7. Bosellini, A., Mutti, E., Ricci Lucchi, F.: Rocce e successioni sedimentarie. UTET Torino (1989)

A Macro-element for Pile Groups Subjected to Vertical Eccentric Load

Chiara Iodice[1], Maria Iovino[2](✉), Raffaele Di Laora[1], Luca de Sanctis[2], and Alessandro Mandolini[1]

[1] University of Campania Vanvitelli, Via Roma 29, Aversa, CE, Italy
[2] University of Napoli Parthenope, Centro Direzionale isola C4, Napoli, Italy
maria.iovino@assegnista.uniparthenope.it

Abstract. This work proposes a mathematical approach to derive the displacement pattern of pile groups subjected to vertical eccentric load. To this end, a new macro-element is formulated in the framework of strain-hardening elasto-plasticity. The equations are developed by assuming an associated flow rule for the evaluation of the plastic deformations and an exponential relationship for the load-displacement curve of the group. The collapse domain is approximated by a parabolic equation while the yield surface has the same shape of the collapse surface and is scaled homothetically through a hardening function accounting for the past load history of the foundation. The evolution of this state variable is expressed by a pertinent hardening law. The proposed approach is validated against data from centrifuge experiments on pile groups embedded in clay.

Keywords: Pile groups · Macro-element · Strain-hardening

1 Introduction

A large class of foundation engineering problems can be examined by idealizing the multi component resultant action transmitted by the structure as a purely vertical, eccentric load (e.g., water tanks, chimneys, silos or tall wind turbines). For example, Iovino *et al.* [1] reported a case study of a wind farm in South Italy where the horizontal load under extreme wind action on 95 m high turbines was only 5% of the dead load of the structure and could thus be neglected. In such applications, piles are usually designed to avoid bearing capacity failure or excessive rotation at foundation level. To analyse the 'soil-foundation-structure' system, 3D numerical analyses have no practical appeal as they involve large volumes of supporting and surrounding soil to be modelled. There is, instead, a clear trend toward the use of 'lumped' plasticity models or macro-elements, in which the soil-foundation behavior is described by an incremental relationship between generalized forces and work-conjugated displacements. The advantage of this approach over the so-called direct approach is manifold [2–7]. The research on this subject has primarily focused on shallow foundations, including the case of inertial interaction problems [8, 9], while piled foundations have received much less attention. The paper is aimed at filling this gap for monotonic, vertical eccentric load. A new macro-element (ME)

A. Ferrari et al. (Eds.): CNRIG 2023, SSGG, pp. 210–217, 2023.
https://doi.org/10.1007/978-3-031-34761-0_26

model for pile groups is presented and discussed. Model parameters can be derived from closed form equations and few additional data that could be determined numerically or experimentally. The proposed model is finally validated against centrifuge experiments carried out on annular shaped groups of piles embedded in soft clay [10].

2 Mathematical Framework

The macro-element formulation for pile groups under monotonic loading presented herein relies on the classical theory of isotropic strain-hardening plasticity. The relationship between generalized force vector, $\boldsymbol{V}$, and generalized displacement vector, $\boldsymbol{v}$, is formulated in rate form as:

$$\begin{gathered} d\boldsymbol{V} = \boldsymbol{K} d\boldsymbol{v} \\ \boldsymbol{V} = \begin{Bmatrix} Q \\ M/R \end{Bmatrix} \\ \boldsymbol{v} = \begin{Bmatrix} w \\ \theta R \end{Bmatrix} \end{gathered} \tag{1}$$

where (Q, M) are axial load and moment, (w, θ) the work-conjugate displacement, R a characteristic length of the foundation and $\boldsymbol{K}$ the tangent stiffness matrix of the pile group. The basic assumption of the model is that the total incremental displacement of the foundation can be decomposed into elastic and plastic components as follows:

$$d\boldsymbol{v} = d\boldsymbol{v}^e + d\boldsymbol{v}^p \tag{2}$$

The elastic displacement increment, $d\boldsymbol{v}^{\mathrm{e}}$, occurs whenever there is a change in load on the foundation. Hooke's law establishes the relationship between the recoverable displacement and the load increments:

$$d\boldsymbol{V} = \boldsymbol{K}^e d\boldsymbol{v}^e \tag{3}$$

where $\boldsymbol{K}^{\mathrm{e}}$ is the elastic stiffness matrix.

At any combination of vertical load and moment, a yield surface in the (Q, M) plane is established. The plastic deformation occurs only when the actual load state, $\boldsymbol{V}$, lies on the yield surface and d$\boldsymbol{V}$ is directed outwards.

The failure locus is derived by approximating the exact solution by Di Laora *et al.* [11] with a parabola imposing that the two functions intersect at $(Q_{\mathrm{c}}, 0)$, $(Q_{\mathrm{t}}, 0)$ and $(0.5 \cdot [Q_{\mathrm{c}} + Q_{\mathrm{t}}], M_{\max})$, where Q_{c} and Q_{t} are the axial capacities of the pile group in compression and in tension and $M_{\max}$ is the maximum value of the moment capacity:

$$f(\boldsymbol{V}) = 4\frac{(Q - Q_c)(Q - Q_t)}{(Q_c - Q_t)^2} + \frac{M}{M_{\max}} = 0 \tag{4}$$

The shape of the yield surface is the same as that of the failure locus, while its size may vary depending on where the load state lies and the incremental load vector points toward. In particular, it is scaled homothetically with reference to the axes origin through a hardening function accounting for the past load history of the foundation, ρ_c, as follows:

$$f(V, \rho_c) = 4\frac{(Q - \rho_c Q_c)(Q - \rho_c Q_t)}{\rho_c (Q_c - Q_t)^2} + \frac{M}{M_{\max}} = 0 \tag{5}$$

The evolution of ρ_c is governed by the following hardening law:

$$\begin{aligned} \frac{\partial \rho_c}{\partial w_p} &= -\frac{1-\rho_c}{\rho_c[\ln(1-\rho_c)+\rho_c]} \cdot \left(\frac{\alpha_Q K_v}{Q_c}\right)^2 |w^p| \\ \frac{\partial \rho_c}{\partial \theta_p} &= -\frac{1-\rho_c}{\rho_c[\ln(1-\rho_c)+\rho_c]} \cdot \left(\frac{\alpha_M K_{mm}}{M_{\max} R}\right)^2 |\theta^p R| \end{aligned} \tag{6}$$

where K_v, K_θ are the initial vertical and rotational stiffnesses and α_Q, α_M are hardening parameters.

The increment of the plastic displacement vector is:

$$dv^p = \Lambda \frac{\partial g}{\partial V} = \frac{\left(\frac{\partial f}{\partial V}\right)^T \boldsymbol{K}^e dv}{\left(\frac{\partial f}{\partial V}\right)^T \boldsymbol{K}^e \frac{\partial g}{\partial V} - \frac{\partial f}{\partial \rho}\left(\frac{\partial \rho}{\partial v^p}\right)^T \frac{\partial g}{\partial V}} \frac{\partial g}{\partial V} \tag{7}$$

where Λ is the plastic multiplier, a non-negative scalar quantity whose expression is derived from the combination of the consistency condition and the derivatives of the plastic potential, g. Notably, the macro-element is formulated under the assumption of associative flow rule ($g = f$).

3 Validation Against Experimental Benchmarks

The macro-element was validated using data from centrifuge tests on pile groups under combined axial-moment loads carried out at the Schofield Centre of the University of Cambridge [10]. The experiments were performed at an increased gravity of 50g on annular shaped pile groups consisting of 8 aluminum piles and on isolated piles embedded in Speswhite kaolin clay. For sake of brevity, reference is only made to the experiments referred to as set A including a pile group under centered load (A1), a pile group under highly eccentric load (A2) and two isolated piles, one in compression and one in uplift. The arrangement of the model foundations is schematically shown in Fig. 1. Model piles were 1 mm thick closed-ended hollow cylinders, with an outer diameter of 10 mm and a length of 240 mm, coated with a film of Hostun sand to increase their unit shaft resistance. They were connected by spherical hinges to a circular rigid raft so that they could only withstand axial loads. Pile group A2 was equipped with a cantilever beam for the application of the eccentric load 145 mm away from the center of the foundation.

The layout of the recording devices installed to monitor settlements, rotations, axial loads on piles and pore water pressures within the soil mass is shown in Fig. 1. As each raft behaves like a rigid body, the settlement of any point belonging to it was evaluated by combining the vertical displacement recorded by the Linear Variable Differential

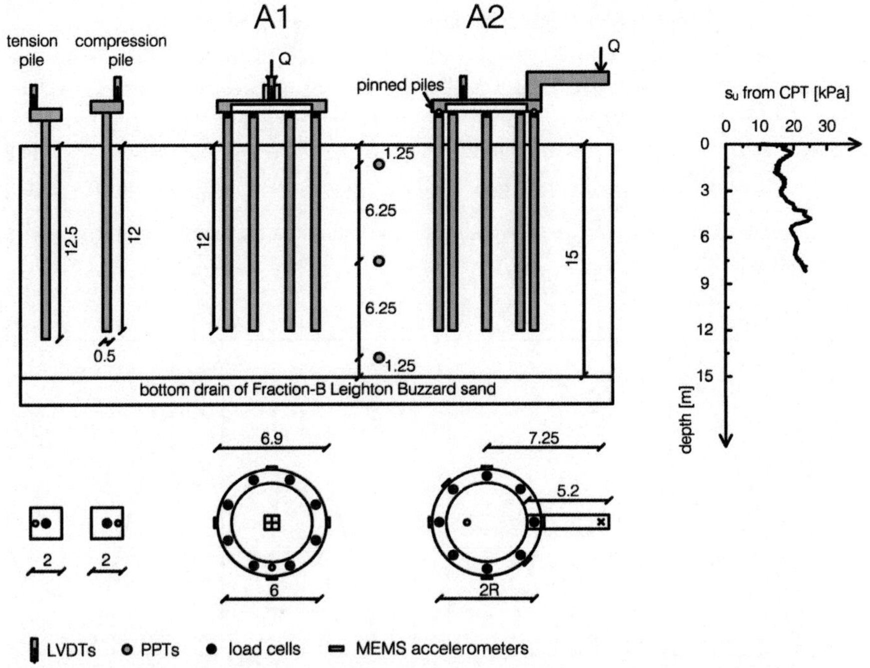

Fig. 1. Plan view and cross section of the model foundations tested in the centrifuge (set A) and soil properties; dimensions (m) are given at prototype scale.

Transformer (LVDT) and the rotations derived from the recordings of the Micro-Electro-Mechanical System (MEMS) accelerometers.

The axial capacity in compression of the pile group, Q_c, was derived from test A1, while, since no experiment was carried out to obtain the one in tension, Q_t, it was evaluated from the single pile test in uplift ($S_u = 267$ kN) considering a unitary efficiency factor. The maximum moment was evaluated as the vertex of the parabola passing through the points (Q_c, 0), (Q_t, 0), (Q_{A2}, M_{A2}), where Q_{A2} and M_{A2} are the maximum axial load and moment capacities derived from test A2. Using the data from test on single pile, the initial stiffness of the isolated pile, K_c, was estimated to be 45 MN/m. The vertical and rotational stiffnesses of the pile group, K_v and K_θ, were calculated from K_c using superposition as in Iovino *et al.* [1] and modelling the pile-to-pile interaction effects with the approximate solution by Dobry & Gazetas [12]. All the parameters used in the analysis with the ME are summarized in Table 1.

The failure loci of the model foundations A1 and A2 are plotted in Fig. 2a and 2b, respectively. The load paths followed in the centrifuge are also shown for comparison; note that they only refer to the external load applied on the piles and, thus, do not include their weight, W_{piles}, which therefore was not considered in the analyses.

The first point of each path is the preload on the foundations due to the weight of the cap, W_{cap}. For pile groups A2, this quantity is slightly eccentric because of the cantilever beam used to apply the eccentric vertical load. Notably, experimental data do not include the displacements due to W_{cap}.

Table 1. Constitutive parameters of the elasto-plastic ME.

Centrifuge test	W_{piles} [kN]	W_{cap} [kN]	M_0 [kNm]	$Q_c + W_{piles}$ [kN]	$Q_t + W_{piles}$ [kN]	M_{max} [kNm]	K_v [MN/m]	K_θ [MNm]
A1	834	379	0	3264	−2023	5648	129	1668
A2	815	481	440	3264	−2023	5648	129	1668

W_{piles} = weight of the piles, W_{cap} = weight of the cap including the cantilever arm, M_0 = initial moment owing to W_{cap}.

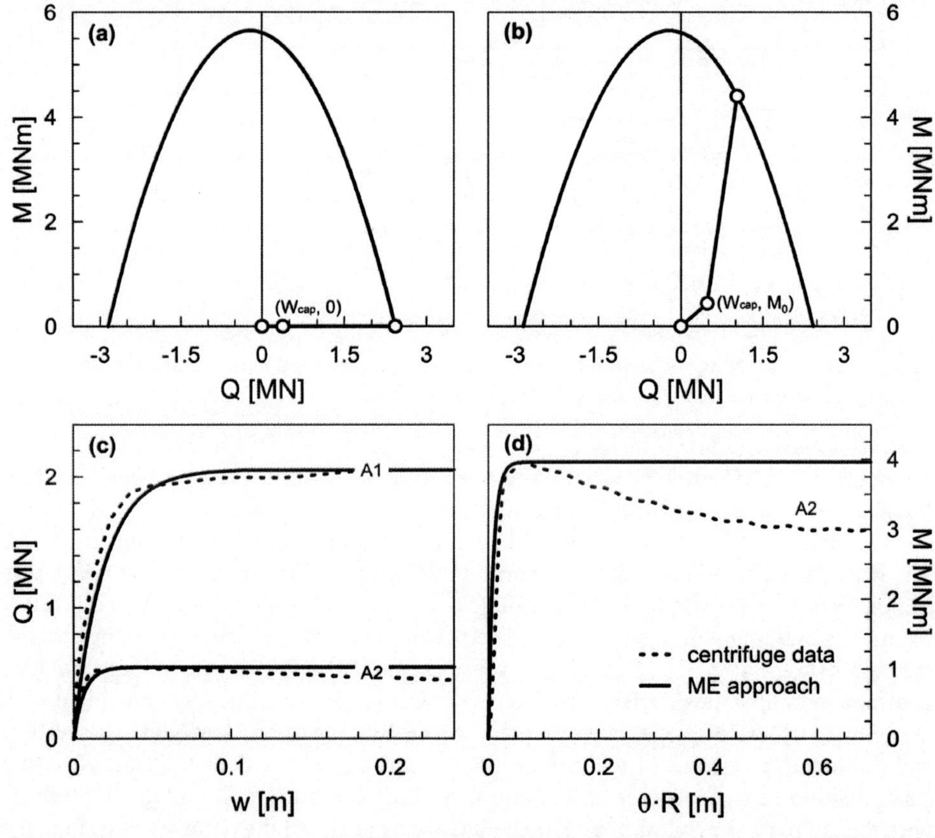

Fig. 2. Failure loci and load paths followed in the centrifuge in tests A1 (a) and A2 (b). Axial load-settlement curves (c) and moment-rotation response (d).

The axial response determined experimentally from test on pile group A1 is compared to that evaluated through the macro-element model in Fig. 2c under the assumption of $\alpha_Q = 1$. The prediction of the proposed model satisfactorily matches the observed behavior, so that $\alpha_Q = 1$ is also taken for the simulation of test A2. The axial load-settlement, *Q-w,* and moment-rotation, *M*-θ, curves predicted by using (α_Q, α_M) = (1, 1) are compared to those gathered from test A2 in Figs. 2c and 2d, respectively, in which *R* is the radius of a circle passing through the piles. The response predicted with the macro-element approach matches in a satisfactory way the rotational response as well as the *Q*-*w* curve observed in the experiment.

4 Response to Complex Load Paths

The performance of the ME to complex load paths, which can be defined as those where the yield surface doesn't expand continuously in consequence of a load reversal, is explored hereafter. To this aim, reference is made to the two load paths shown in Fig. 3, referred to as A3 and A4. In both situations, the pile group is first loaded vertically until $\rho_c = 0.5$ (line AB) and then by a moment component under constant axial load - a usual scenario in many foundation problems - until $\rho_c = 0.75$ (line BC).

The macro-element parameters in Table 1 and $(\alpha_Q, \alpha_M) = (1, 1)$ are taken in the simulations. The vertical displacement thereby increases from zero to a certain value in the first part of the load path involving only the vertical load, developing from the very beginning both elastic and plastic components given that the yield surface expand with the load path and the initial $\rho_c = 0$. When the moment is applied at constant *Q*, both

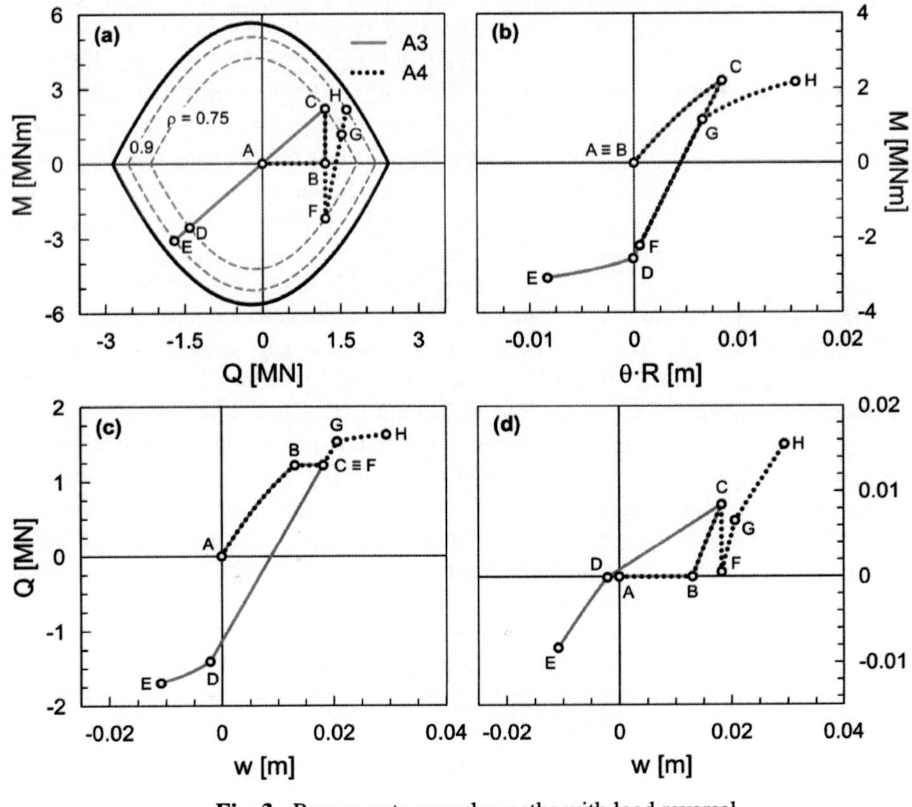

Fig. 3. Response to complex paths with load reversal.

displacement and rotation occur due to the shape of the yield surface, with no elastic component of the vertical displacement.

In load path A3 the group is then subjected to unloading under constant eccentricity, passing through the origin of the (Q, M) plane, until $\rho_c = 0.9$ (line CE). The unloading process is entirely elastic until the load path reaches again the last yield surface (point D, $\rho_c = 0.75$), and thereby elastic-plastic displacement and rotation are developed in consequence of the additional yielding in DE.

The load path A4 is identical to A3 until point C, after which it involves the application of a moment of opposite sign. This phase is entirely elastic and under constant settlement until reaching the yield surface (point F, $\rho_c = 0.75$). Then, the load is applied with a high eccentricity and both elastic displacement and rotation develop until reaching again the yield surface (point G, $\rho_c = 0.75$). Finally, the last part of the path is characterized by elastic and plastic settlements and rotations in consequence of the further yielding until $\rho_c = 0.9$ (line GH). For the sake of completeness, the displacement curves w-θ for the two simulations are also reported in Fig. 3.

5 Conclusions

An elasto-plastic macro-element for pile groups under combined axial load-moment framed in the classical theory of strain-hardening plasticity has been presented and discussed. Only few parameters are needed, namely the bearing capacities of the pile group under compression, tension and moment loads and the vertical and rotational stiffnesses, which anyhow are required in routine design. Two additional constants referred to as hardening parameters, α_Q and α_M, and governing the evolution of the plastic displacements are also necessary, which and can be calibrated experimentally or numerically. The yielding surface and the failure locus have the same shape, i.e. a simple parabola derived by approximating the closed form solution by Di Laora *et al.* [11], while the hardening law allows the yield surface to homothetically expand with reference to the axes origin. The hypothesis of associated flow rule is considered.

The ability of the ME to predict the response of pile groups in the (Q, M) plane is shown with regard to a series of centrifuge tests carried out on model pile foundations embedded in kaolin clay. The comparison with the experimental data proves very satisfactory in terms of load-displacement and moment-rotation curves in monotonic loading paths. Moreover, it is shown that the normality conditions holds, as well as, the set of hardening parameters (α_Q, α_M) = (1, 1) seems a suitable choice, at least in soft clay.

Our purpose for future research is to expand this conceptual framework to the generalized space of force, i.e. by considering the interaction with the horizontal component of the resultant action, and to cyclic loads.

Acknowledgments. This work has been carried out under research project PRIN 2017 'A new macro-element model for pile groups under monotonic, cyclic and transient loadings' funded by Italian Ministry of University and Research.

References

1. Iovino, M., Di Laora, R., de Sanctis, L.: Serviceability limit state analysis of piled foundations under combined axial-moment loading. Acta Geotech. **16**(12), 3963–3973 (2021)
2. Nova, R., Montrasio, L.: Settlements of shallow foundations on sand. Géotechnique **41**(2), 243–256 (1991)
3. Gottardi, G., Houlsby, G.T., Butterfield, R.: Plastic response of circular footings on sand under general planar loading. Géotechnique **49**(4), 453–469 (1999)
4. Houlsby, G.T., Cassidy, M.J.: A plasticity model for the behaviour of footings on sand under combined loading. Géotechnique **52**(2), 117–129 (2002)
5. Salciarini, D., Tamagnini, C.: A hypoplastic macroelement model for shallow foundations under monotonic and cyclic loads. Acta Geotech. **4**(3), 163–176 (2009)
6. Marchi, M., Butterfield, R., Gottardi, G., Lancellotta, R.: Stability and strength analysis of leaning towers. Géotechnique **61**(12), 1069–1079 (2011)
7. Pisanò, F., Flessati, L., Di Prisco, C.: A macroelement framework for shallow foundations including changes in configuration. Géotechnique **66**(11), 910–926 (2016)
8. Chatzigogos, C.T., Figini, R., Pecker, A., Salençon, J.: A macroelement formulation for shallow foundations on cohesive and frictional soils. Int. J. Numer. Anal. Meth. Geomech. **35**(8), 902–931 (2011)
9. Figini, R., Paolucci, R., Chatzigogos, C.T.: A macro-element model for non-linear soil–shallow foundation–structure interaction under seismic loads: theoretical development and experimental validation on large scale tests. Earthquake Eng. Struct. Dynam. **41**(3), 475–493 (2012)
10. de Sanctis, L., Di Laora, R., Garala, T.K., Madabhushi, S.P.G., Viggiani, G.M.B., Fargnoli, P.: Centrifuge modelling of the behaviour of pile groups under vertical eccentric load. Soils Found. **61**(2), 465–479 (2021)
11. Di Laora, R., de Sanctis, L., Aversa, S.: Bearing capacity of pile groups under vertical eccentric load. Acta Geotech. **14**(1), 193–205 (2018). https://doi.org/10.1007/s11440-018-0646-5
12. Dobry, R., Gazetas, G.: Simple method for dynamic stiffness and damping of floating pile groups. Geotechnique **38**(4), 557–574 (1988)

Uprooting Safety Factor of Trees from Static Pulling Tests and Dynamic Monitoring

A. Marsiglia[1], A. Galli[2], G. Marrazzo[2], R. Castellanza[3], and Matteo Oryem Ciantia[1](✉)

[1] School of Science and Engineering, University of Dundee, Dundee, UK
m.o.ciantia@dundee.ac.uk

[2] Department of Civil and Environmental Engineering, Politecnico di Milano, Milan, Italy

[3] Department of Earth and Environmental Sciences, University of Milano-Bicocca, Milan, Italy

Abstract. Tree fall events are acknowledged as important causes of damage in urbanized areas and tree risk assessment is now part of general risk management analyses. Preserving arboreal heritage also implies biomass and biodiversity conservation, reduction of superficial soil erosion, improving the stability of potentially unstable slopes. In this context, tree risk assessment techniques used by professional agronomists may benefit of a deeper mechanical understanding of the soil-root interaction mechanisms controlling tree stability. From a Geotechnical point of view, the tree root plate is conceptually assimilated to the foundation of a slender structure subject to eccentric loads. In the paper results of static pulling tests and wind dynamic monitoring on two liquidambar trees are described; estimations of their uprooting safety factor are then derived according to the currently available techniques used in the agronomic field. The pull tests were performed using an advanced loading-unloading scheme, along two orthogonal directions. The preliminary interpretation shows that the system is characterized by a marked non-linear and irreversible behaviour, even at low loading levels as usually tested in agronomic practice. The results suggest that dynamic tree stability testing could be a valuable tool for stability assessment, albeit the interpretation of the results requires expertise in the dynamic behaviour of trees.

Keywords: Pull tests · tree risk assessment · lateral load · dynamic monitoring

1 Introduction

Climate change is likely to significantly impact the basic transport infrastructures and the built environment worldwide due to a more arid, erratic, and storm-prone weather pattern [1]. Within a bio-engineering context [2], stability assessment of trees is a key issue for the management of urban areas, where accurate analysis techniques can substantially help in protecting natural heritage, human activities (and also human lives) against sudden tree collapses. In this perspective, Geotechnical Engineers may contribute to a deeper mechanical understanding. At the micro scale the uprooting resistance is governed by the interaction between the roots and the soil. At the macroscale the anchoring

A. Ferrari et al. (Eds.): CNRIG 2023, SSGG, pp. 218–225, 2023.
https://doi.org/10.1007/978-3-031-34761-0_27

capacity of the system subject to complex and combined loading actions, as those provided by wind excitations, may be considered as governed by a generalised soil root plate interface strength. From a Geotechnical point of view, the root plate can be hence considered as a "living shallow foundation" of a slender structure, subject to highly eccentric toppling loads. The behaviour will be very complex as affected by soil partial saturation, the high compressibility (typical of vegetated soils) and cohesive strength provided by root anchorage. The interpretative frameworks commonly adopted in agronomic practice in general disregard all these aspects, and it is Authors' willing with this paper to promote the debate on a topic that – potentially – can be of large interest for the Geotechnical scientific community. The results of non-standard static and dynamic tests on two *liquidambar* trees are hereafter presented. Bi-directional pulling tests with load-unloading phase are first described. The same trees have also been instrumented with root plate inclinometers and wind data in proximity of the trees is collected for the same timeframe. The two experimental approaches are then used to calculate and compare the static and dynamic uprooting safety factors.

1.1 State of the Art

In the last twenty years, the methods used in standard agronomic practice to assess the tree safety have evolved from the traditional Visual Tree Assessment [3], towards the use of quantitative geometry-based indexes and on-site non-destructive tests. Among other techniques, static pulling tests are often performed and predictions on the limit toppling resistance of the tree are obtained via a phenomenological interpretation of the test based on the procedure proposed by Wessolly and Erb [4]. Since then, to improve the geotechnical engineering understanding of tree-soil interaction during uprooting, field experimental research [5–7], laboratory physical modelling [8–11] and numerical studies [12–14] have been performed. More recently, the possibility of monitoring trees under actual wind load conditions has been matter of investigation [15], and (i) tree canopy geometrical size, (ii) wind load characteristics and (iii) soil-root system anchorage have been recognized as the main factors influencing the uprooting of trees under wind loads. In such dynamic conditions, the kinetic energy transferred from the wind is partly dissipated through damping of the tree and its branches, and partly transferred to the soil-root plate. Jackson *et al.* [15], by analysing the power spectrum of tree base oscillations, which is related to the wind–tree energy transfer, recently found that the relationship between wind loading and tree deflection appears to mainly be related to wind speed in the high wind speed range. This provides a theoretical basis for the dynamic assessing methodology used in agronomic practice (hereafter presented in Sect. 3.2), which, in principle, bypass also the need of calculating the nature of wind loading, natural frequency of the tree and its damping ratio [16].

2 Uprooting Safety Assessment

2.1 Static Pull Tests

Static pull tests are executed by applying a force F at a height H and with an inclination α with respect to the horizontal ground surface, and by recording the corresponding rotation φ of the root plate (Fig. 1a). The test is usually limited to small rotations values, not exceeding 0.2°. The experimental data are fitted by means of the tipping equation

$$\varphi[°] = \frac{1}{3}\tan\left(\frac{100}{73.85}\frac{F}{F_L}\right) + \frac{1}{3}\left(\frac{F}{F_L}\right)^2 - \frac{1}{10}\frac{F}{F_L} \tag{1}$$

proposed by Wessolly and Erb [4], where the parameter F_L is interpreted as the limit value of the force F inducing uprooting. The safety factor is then determined as

$$F_s^s = M_L / M_{wind}^{max} \tag{2}$$

with M_L corresponding to limit overturning moment and M_{wind}^{max} calculated from:

$$M_{wind}^{max} = A_{cr} p_w^{max} C_w h_{cr} = A_{cr}\left(\frac{1}{2}\mu v_{max}^2\right) C_w h_{cr} \tag{3}$$

where A_{cr} is the crown area, p_w^{max} is the maximum wind pressure (function of air density μ and wind velocity v_{max}); C_w is the aerodynamic drag factor and h_{cr} is the crown centre point height.

2.2 The Dynamic Tree Testing Principle

The technique is based on statistically correlating wind speed and tree inclination measurements over time windows of several minutes [16]. The process is repeated for several more periods, and the mean value pairs are used to create pressure-inclination curves for the tree. The curves are then interpreted by means of the same tipping Eq. (1), where wind pressure p_w takes the place of the force F. A limit wind pressure value p_w^L is in then derived, to be compared to the maximum wind pressure value p_w^{max} the tree will experience in high wind conditions. Bejo *et al.* [17] propose a simplified linear fitting, based on a 1st order Taylor expansion of Eq. (1), for the evaluation of p_w^L. Equation (2) can then be reduced to the ratio of the limit to maximum wind pressure values, avoiding the need for geometric parameters and empirical aerodynamic drag factors:

$$F_s^d = p_w^L / p_w^{max}. \tag{4}$$

3 Experimental Field Tests

3.1 Test Set-Up and Methodology

Pull tests on two *liquidambar* trees were performed at the Lecco campus of the Politecnico di Milano (Fig. 1b). The load was applied by means of a manual winch, and a very limited rotation at the base (usually not exceeding about 0.2°) was imposed. Differently from the standard practice, the same trees were subsequently tested along two perpendicular directions, and for all tests the unloading phase was also recorded. These two latter features are rarely performed in common practice, and very limited data is available in the literature. Table 1 summarizes the tree typologies and the main geometrical characteristics for the tests. The pulling force was measured by a commercial load cell

(Fig. 1d; maximum capacity 5 tons), and root plate rotations were measured by means of small-scale inclinometers fixed at the base of the trunk (Fig. 1e; measurement range ±2°; resolution 0.001°) with a 10 Hz data acquisition frequency. Wind data was collected using a FA4 Ultrasonic anemometer positioned 8 m above the ground as indicated in Fig. 1a and g.

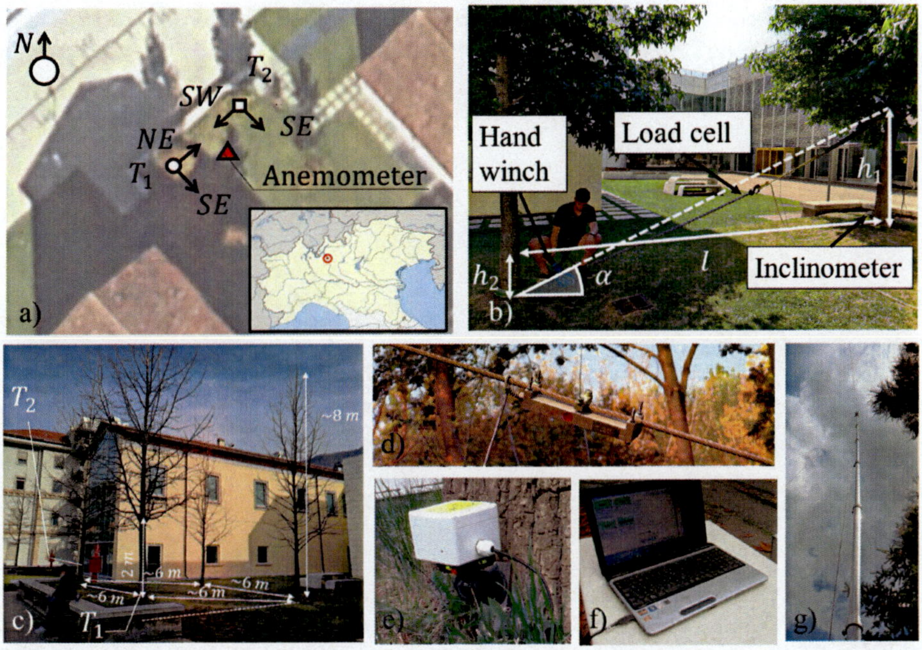

Fig. 1. a) Aerial view of the two trees tested, including pull test directions and anemometer positioning, b–c) geometrical quantities characterizing the static pull test d–g) instrumentation used.

Table 1. Pull test ID and geometrical characteristics of the experimental tests.

Test ID	D at 130 cm from the ground (cm)	h_1 (m)	h_2 (m)	l (m)	α (°)	Pull direction (-)	F_L (kN)
T1-SE	20.7	2	0.45	5.84	22.8	SE	17.5
T1-NE	20.7	2	0	5.8	19.0	NE	18.5
T2-SE	17.2	2	0.45	5.95	22.4	SE	13.3
T2-SW	17.2	2	0	5.8	19.0	SW	14.2

3.2 Test Results

Moment-rotation curves for the root plate of the static pull tests are reported in Fig. 2, where an influence of the loading direction is clear. This is sign of a non-axisymmetric strength of both trees, which may be related the existence of a wind dominant direction between the campus building or to the interference of the root systems with some walls on two sides of the trees (Fig. 1). The unload curves show in all cases a permanent rotation of the root plate, clearly indicating that, despite a rotation of 0.25° is never exceeded, plastic deformations develop in the ground.

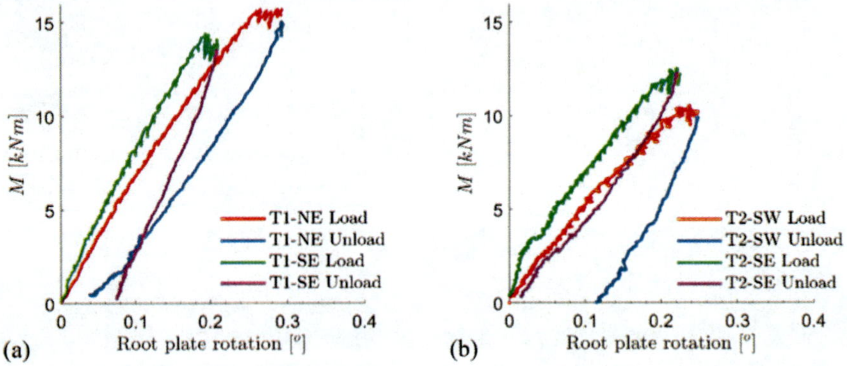

Fig. 2. T1 (a) and T2 (b) pull tests results in terms of moment-rotation curves.

Figure 3 reports wind speed and root plate rotation data recorded during a three-day dynamic tree monitoring. Both wind and tree movement data show large variability and scatter, and data can hence be investigated with various statistical methods. In this work, moving average techniques (period of 40 and 20 min have been chosen) were used. Inclinometer installed on tree T2 unfortunately did not record inclinations in the windier portions of the tests and therefore it has been excluded from the dynamic safety factor evaluation.

4 F_s Calculations

Uprooting safety assessment of trees T1 and T2 were performed using the procedure described in Sect. 2. Figure 4 reports the fitting curves of Eq. (1) to evaluate F_L and p_w^L required to calculate F_S. Table 2 summarizes the obtained factor of safety values (following [4] values $C_w = 0.25$ and $v_{max} = 33$ m/s were assumed).

Figure 4a and b show for the tree T1 the fitting of Eq. (1) over the loading data of the pull tests, and the fitting on the rotation-pressure pairs values from dynamic monitoring. In Fig. 4c and d the same equations have been extrapolated until their limit values (continuous lines). In order to investigate also their prediction capabilities, two shaded regions have been added, representing the range of predicted root plate rotations for F_L (or p_w^L) variations of $\pm 10\%$ and $\pm 25\%$, respectively. For the dynamic tests our

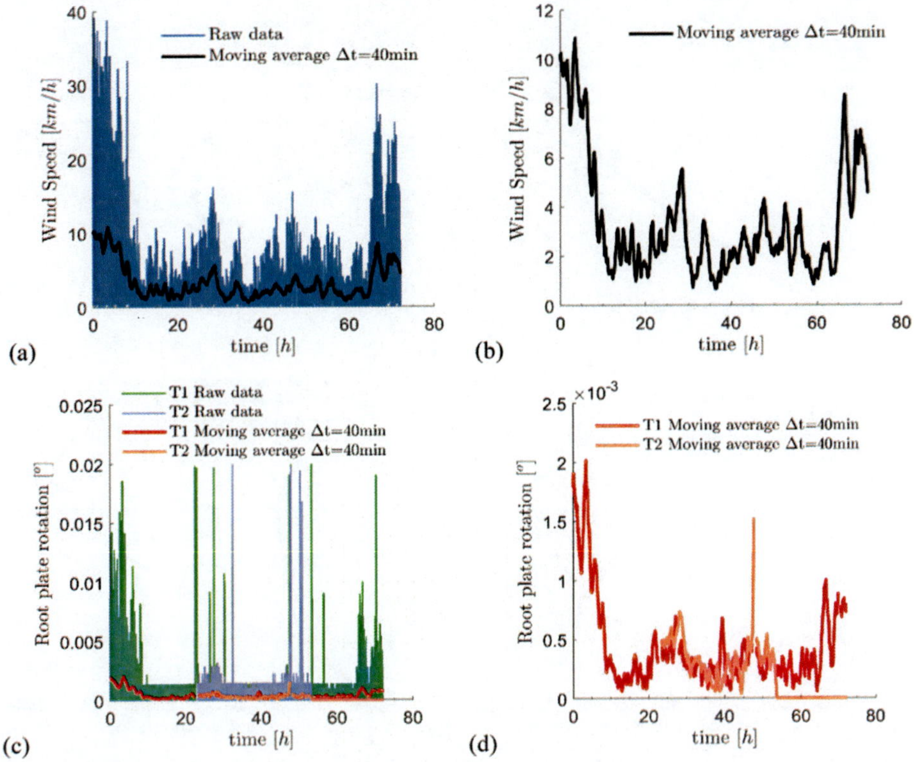

Fig. 3. Raw (a and c) and moving average data (b and d) of wind speed and root plate deflections.

Table 2. Summary of static and dynamic safety factors.

Tree	A_{cr} (m^2)	h_{cr} (m)	M_{wind}^{max} (kNm)	M_L^{SE} (kNm)	M_L^{NE} (kNm)	M_L^{SW} (kNm)	p_w^L (Pa)	F_S^{SE} (-)	F_S^{NE} (-)	F_S^{SW} (-)	F_S^d (-)	$F_{S,lin}^d$ (-)
T1	8.0	2.7	3.6	32.3	35.0	-	1239	9.0	9.7	-	1.85	7.2
T2	9.8	3.0	4.9	24.6	-	26.9	-	5.0	-	5.5	-	-

results are also compared with the linear fitting approach suggested by Bejo *et al.* [17]. Obtained F_S values indicate in all cases that the trees are substantially stable, but they are clearly observed to depend on the pull direction, which is not usually considered in the standard agronomic practice. Interestingly, F_S value from the dynamic monitoring is more conservative and therefore may provide a safe alternative to standard pull tests. On the other hand, the linear fit proposed by [17] largely overestimates the F_S^d obtained with a nonlinear fit of the data.

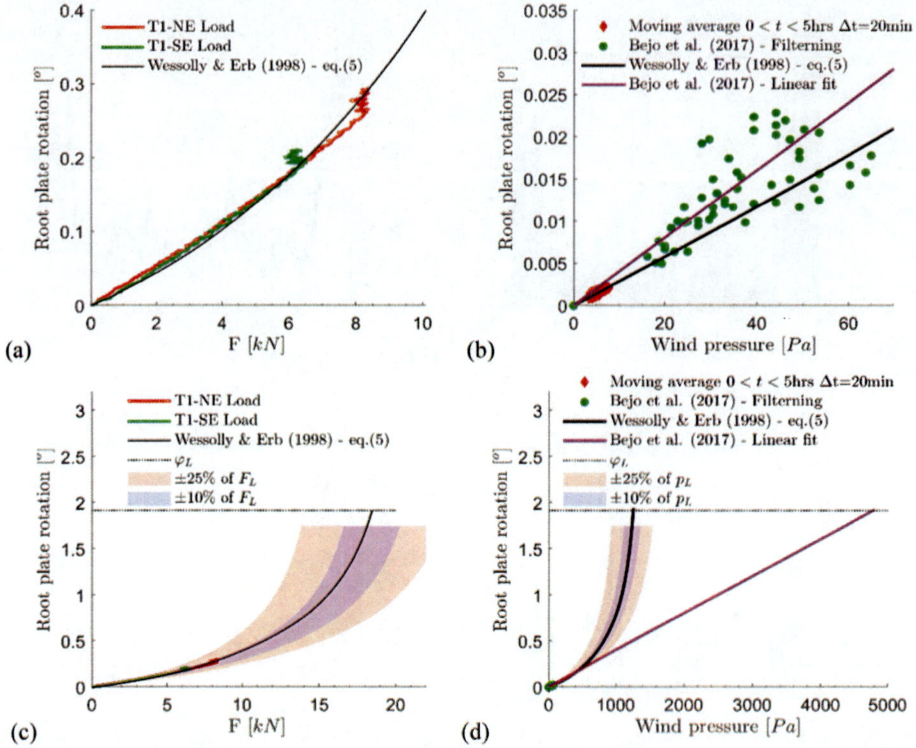

Fig. 4. Fitting curves for (a) static pull test and (b) dynamic monitoring data. (c–d) Extrapolation of experimental data until the limit condition (F_L and p_w^L, respectively; if a linear interpolation is used $p_w^L = 4800Pa$; see Table 2).

5 Conclusions

The results of original bi-directional static pulling tests including the unloading cycles and of dynamic monitoring of two trees are described. The results show that the moment-rotation curves of the system are characterized by marked non-linear and irreversible behaviours, and that a strong correlation exists between wind pressure and root plate deflection in dynamic monitoring tests. In the tested cases it was found that F_S, if computed with the standard available techniques, may significantly depends on the pulling direction and that F_S from dynamic monitoring is more conservative. It should however be noted that these results are based on the extrapolation of the tree response up to failure, and comparison of dynamic monitoring interpretation with complete uprooting tests would be necessary to give more robust conclusions.

Acknowledgments. The support by Agro Services s.r.l. and the Scottish Research Partnership in Engineering (SRPe), through the SRPe-IDP/011 research grant, is acknowledged.

References

1. Satterthwaite, D.: Adapting to climate change in urban areas: the possibilities and constraints in low-and middle-income nations, vol. 1. IIED (2007)
2. Martinez, A., Dejong, J., Akin, I., et al.: Bio-inspired geotechnical engineering: principles, current work, opportunities and challenges. Géotechnique **72**(8), 687–705 (2022)
3. Mattheck, C., Breloer, H.: Field guide for Visual Tree Assessment (VTA). Arboric. J. **18**, 1–23 (1994)
4. Wessolly, L., Erb, M.: Handbuch der Baumstatik und Baumkontrolle. Patzer, Berlin (1998)
5. Nicoll, B.C., Gardiner, B.A., Rayner, B., Peace, A.J.: Anchorage of coniferous trees in relation to species, soil type and rooting depth. Can. J. For. Res. **36**(7), 1871–1883 (2006)
6. Tsen-Tieng, D.L., Rahardjo, H., Choon, L.E., King, F.Y.: Anchorage and stability of tree root–soil plates. Environ. Geotech. **7**(5), 330–337 (2018)
7. Galli, A., Sala, C., Castellanza, R., Marsiglia, A., Ciantia: Lesson learnt from static pulling tests on trees: an experimental study on toppling behaviour of complex foundations. Acta Geotecnica (2023)
8. Mickovski, S.B., Bransby, M.F., Bengough, A.G., Davies, M.C.R., Hallett, P.D.: Resistance of simple plant root systems to uplift loads. Can. Geotech. J. **47**(1), 78–95 (2010). https://doi.org/10.1139/T09-076
9. Kamchoom, V., Leung, A.K., Ng, C.W.W.: Effects of root geometry and transpiration on pull-out resistance. Geotech. Lett. **4**, 330–336 (2014)
10. Zhang, X., Knappett, J.A., Leung, A.K., Ciantia, M.O., Liang, T., Nicoll, B.C.: Centrifuge modelling of root–soil interaction of laterally loaded trees under different loading conditions. Géotechnique 1–15 (2022)
11. Marsiglia, A., Ciantia, M.O., Galli, A., Canepa, D.: Vertical loading tests on a simplified tree root prototype. In: Physical Modelling in Geotechnics, pp. 832–835 (2022)
12. Dupuy, L., Fourcaud, T., Stokes, A.: A numerical investigation into the influence of soil type and root architecture on tree anchorage. Plant Soil **278**, 119–134 (2005)
13. Yang, M., Défossez, P., Danjon, F., Fourcaud, T.: Tree stability under wind: simulating uprooting with root breakage using a finite element method. Ann. Bot. **114**, 695–709 (2014)
14. Dattola, G., et al.: A macroelement approach for the stability assessment of trees. In: Calvetti, F., Cotecchia, F., Galli, A., Jommi, C. (eds.) CNRIG 2019. LNCE, vol. 40, pp. 417–426. Springer, Cham (2020). https://doi.org/10.1007/978-3-030-21359-6_44
15. Jackson, T.D., Sethi, S., Dellwik, E., et al.: The motion of trees in the wind: a data synthesis. Biogeosciences **18**, 4059–4072 (2021)
16. Bejo, L., Sumegi, I., Divos, F.: Dynamic tree stability: improved testing methodology and indications of reliability. In: Proceedings of the 22nd NDTE Symposium. Quebec City, Canada, pp. 175–183 (2022)
17. Bejo, L., Divos, F., Fathi, S.: Dynamic root stability assessment – basics and practical examples. In: Proceedings of the 20th NDTE Symposium. Madison, Wisconsin, USA, pp. 262–269 (2017)

Multidisciplinary Study on a Landslide Area Individuated by Using Statistical Methodologies Before and After the Last Reactivation

Chiara Martinello[1], Marco Rosone[2], Chiara Cappadonia[1](✉), and Giampiero Mineo[1]

[1] Department of Earth and Sea Sciences (DiSTeM), University of Palermo, Palermo, Italy
chiara.cappadonia@unipa.it

[2] Department of Engineering (DI), University of Palermo, Palermo, Italy

Abstract. In this work, a detailed study was conducted to characterize an area selected by applying statistical methodologies for landslide susceptibility modelling of the Imera River basin (Northern Sicily). The area, classified with a very high susceptibility level, was affected by a recent landslide reactivation and it was thoroughly analyzed from the geological and geotechnical points of view. Detailed investigations were performed in the landslide, before and after its last reactivation, in order to evaluate the mechanisms (roto-translational sliding in the upper part and earth-flow in the lower part) which affected a wide and thick body of stiff and highly fissured clays belonging to the Varicolori clay Formation. Remotely sensed datasets, such as orthophotos and digital terrain models, were used for mapping the landslide processes and landforms, as well as to obtain the topographic elements useful to integrate the models into GIS software. The latter was also used for the definition of the moved soil mass after reactivation. The geological and geotechnical models were defined for 2D slope stability analyses. Back-analyses of the two sliding surfaces proved that the mobilized shear strength angle is slightly higher than the upper bound of the residual shear strength angle measured by means of direct shear tests. The obtained results prove that a multidisciplinary study like this represents a promising method for local verification of the reliability of landslide susceptibility analyses conducted at a basin scale with a purely statistical approach.

Keywords: Varicolori clays · Limit equilibrium method · GIS analysis · Landslide susceptibility

1 Introduction

This study focuses on landslide susceptibility and hazard zoning at different scales, highlighting the importance of a correct characterization of the processes and related instability landforms for reliable susceptibility and hazard-zoning maps.

The assessment of landslide susceptibility of a territory can be evaluated by exploiting different approaches, from statistical to physically based analysis (e.g., [1–4]) and

A. Ferrari et al. (Eds.): CNRIG 2023, SSGG, pp. 226–233, 2023.
https://doi.org/10.1007/978-3-031-34761-0_28

from small to detailed scale. Stochastic methods are largely employed to define the probability of occurrence of a landslide in a specific area based on its characteristics [5] and by determining the relationships between the spatial distribution of a landslide scenario (e.g. frequently linked to a specific trigger) and specific geo-environmental variables. This research has been carried out in the contest of the SUFRA (SUscetibilità da FRAna) project which develops a multi-scale methodology for the assessment and management of landslide susceptibility for the whole regional territory of Sicily. For the Imera Settentrionale River Basin, a slide susceptibility map involving all the stages of landslide (activation, propagation, and arrest) was defined by Martinello et al. [6]; following the susceptibility levels identified in this map, remote and on-field detections of new and reactivated landslides that occurred in the most susceptible areas were carried out. In particular, in-depth analyses of the geomorphological and geotechnical conditions were conducted on a slope where on December 2021 a landslide was reactivated by rainfall. This area is known as *Gole del Drago* and it is close (about 1500 m in distance) to a well-documented large landslide, known in the literature as the *Cerda* landslide. The latter was triggered by an earthquake in 2002 [7–10] and is frequently reactivated by rainfall occurring in the wet period of the year [11, 12]. Given the similar environmental conditions between these two nearby areas and considering the several common features they have, the highly accurate investigations reported by Rosone et al. [11] were here used as a reference to extend the ongoing characterization works of the soils involved in landslides. Geological and Geotechnical modelling was also used to characterize the reactivation process of the *Gole del Drago* landslide.

2 Landslide Susceptibility Map in the Study Area

With an area of 343 km^2, the Imera Settentrionale river basin is one of the most important catchments in Sicily (Fig. 1a): it hosts the A19 motorway, which connects the capital (Palermo) to the second main town (Catania) and other strategic infrastructure, such as wells and aqueducts yet damaged in the past and now potentially threatened by the instability phenomena [13]. At the basin scale, carbonate, siliceous-carbonate, and siliciclastic successions are largely outcrop as imbricate geometric structures. This peculiar geological setting, in conjunction with environmental factors such as climate, topography, and seismicity of the area, favours instability phenomena and water erosion landforms.For the whole Imera river basin, a susceptibility map was provided by Martinello et a [6] as the result of a trial test for detecting the optimal slope unit partitioning. Adopting a MARS (Multivariate Adaptive Regression Splines, [14]) statistical method, landslide susceptibility was evaluated on the basis of twelve predictor factors (elevation, landform classification, steepness, northerness, easterness, plane and profile curvatures, topographic wetness index, terrain ruggedness index, stream power index, lithology map and land use) and a database of 1608 slides. The susceptibility model was obtained by using the pixel as mapping units, while the final map was elaborated by zoning the pixel-based output in the optimal slope units selected. As reported in Fig. 1b, the map is characterized by four susceptibility levels: null, low, high and very high. According to this slide classification, the *Gole del Drago* area is classified as "very high" susceptible (Fig. 1c).

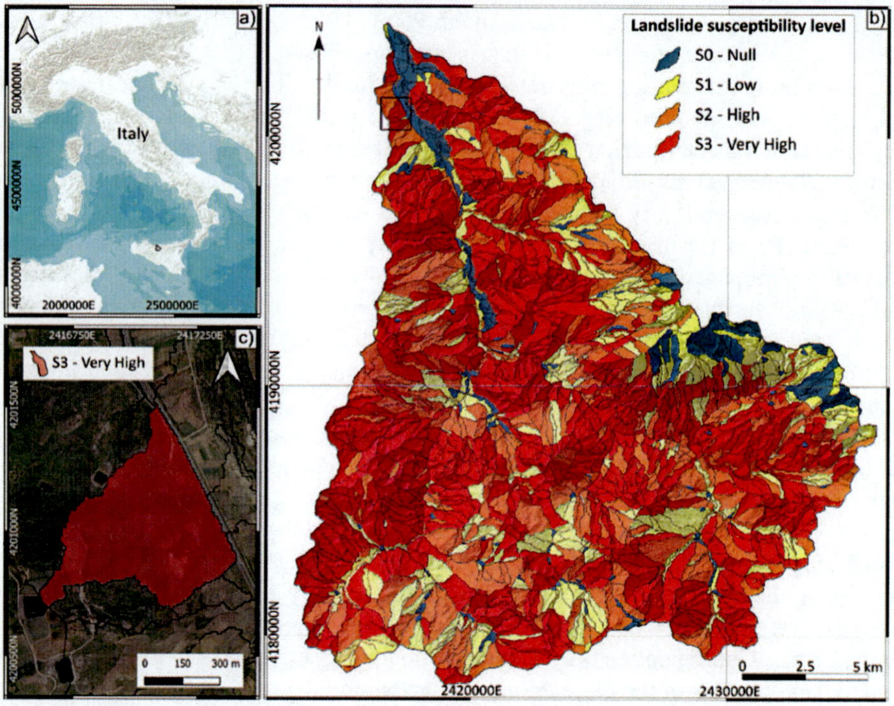

Fig. 1. a) Location of the study area; b) susceptibility map of the whole Imera river basin; c) a particular of the very high susceptibility level of the *Gole del Drago* area (in red).

2.1 The *Gole Del Drago* Area: A Reference Geological Model

As reported in Fig. 2, the *Gole del Drago* area is characterized mainly by the remolded Varicolori clays (AVF) arising from old landslide deposits and the clayey marls and the clayey marls of Polizzi Formation (POZ), outcropping in the highest part of the slope; the whole morphological unit has an average inclination of 13°, calculated exploiting available DTM. The bedrock consists of undisturbed Varicolori clays: stiff and highly fissured clays and marls having a colour variable from grey to red, with interposed levels of calcilutes, quartzous sandstones, calcarenites, and calcirudites. The clays are intensely fissured and frequently included rock fragments, prevalently rounded, and rock layers, with a thickness of about 10 cm or less and low persistence. Fissures subdivide the clays into small stiff fragments (scales) with a length lower than one centimeter and a thickness of some millimeters.

Using remotely sensed datasets available from institutions and specifically derived from surveys using Unmanned Aerial Vehicle (UAV) following the recent reactivation that occurred in December 2021, it was possible to map the landslide area. So, the geometry and the main topographic elements characterizing the landslide were detected and integrated to model the phenomena into Geographic Information System (GIS) software. The main landslides scarps and the directions of the movement were reported in Fig. 3a. Shiny and slicken-sided sliding surfaces have been clearly recognized where the failure surface outcropped on the slope. Generally, these surfaces develop within bands several decimeters thick of completely remoulded clay, which has lost its original structure [15]. However, in the area where the landslide was frequently reactivated in the past, the landslide body completely consists of remoulded clay. Comparing the two DTM related to 2008 (available from institutions) and 2022 (derived from UAV surveys in January 2022) the geomorphological modifications were highlighted and the quantification of the moved soil masses after reactivation was accomplished. More specifically, as shown in Fig. 3b, swelling areas can be observed in the sector affected by the reactivation; the most depressed areas testify to the presence of a new scarp and an emptying linked to the movement of the material that spreads at the bottom of the slope. Counter-sloping areas in the landslide head point out to typical characteristics of a roto-translational sliding mechanism (landslide A), while the depletions in the landslide channel testify to an earth-flow mechanism at the toe (landslide A′).

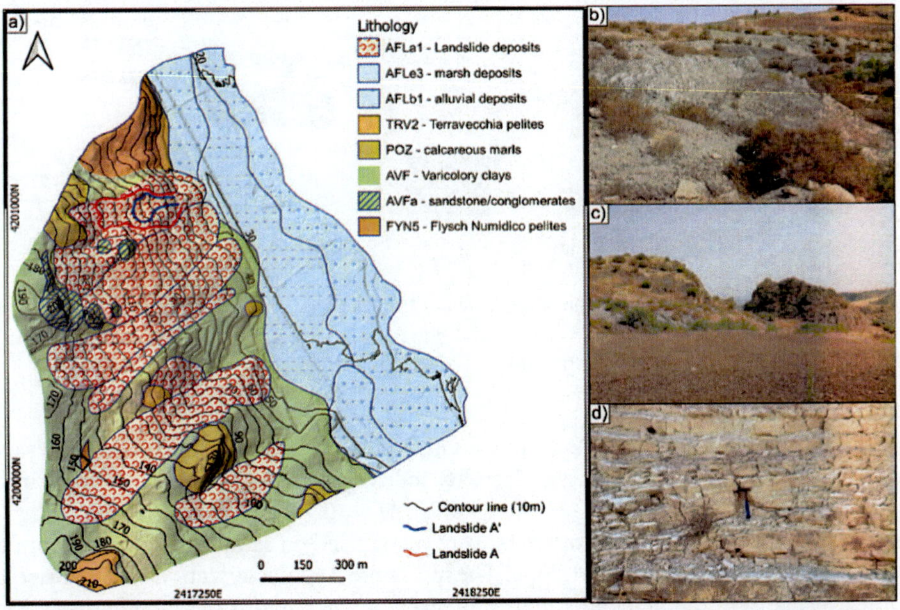

Fig. 2. Geological map of the study area and boundaries (red and blue line) of the landslide (A) involved in a reactivation (A′) on December 2021 (a); Varicolori clays AVF (b), Sandstone and conglomerates AVF (c) and Calcareous marls POZ (d) outcropping on the slope.

3 Geotechnical Characterization and Landslide Modelling

At the laboratory scale, Varicolori clays samples recovered from the landslide body include stony fragments with a diameter equal to or lower than one centimetre. To this regard, Rosone et al. [11] already reported several grain size distributions of samples recovered from the landslide body of the *Cerda* Landslide. They prove that the soil

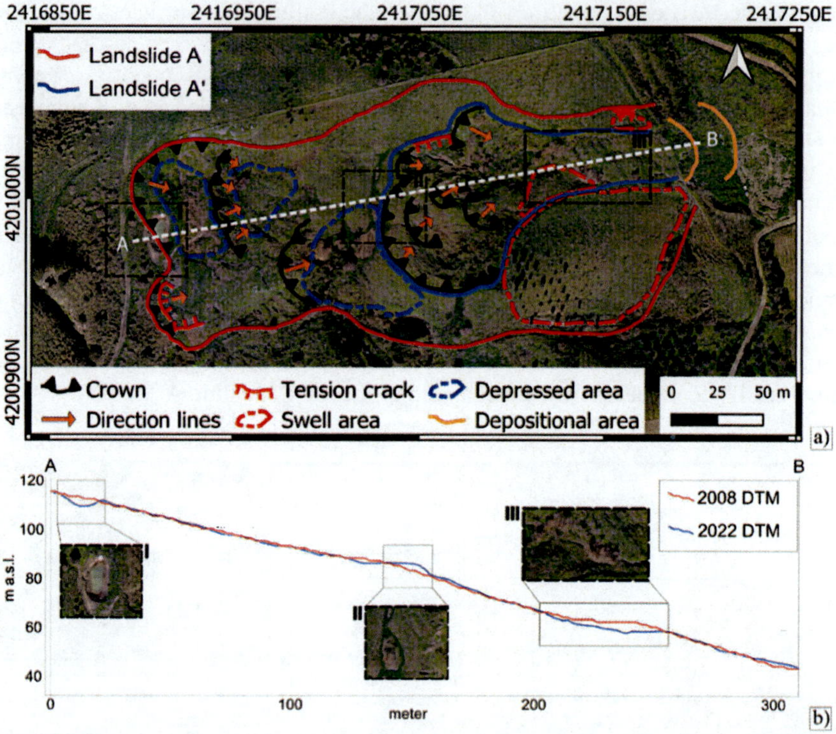

Fig. 3. Geomorphological map and boundaries of the landslide reactivations on the UAV survey acquired in January 2022 (a). The white dotted line refers to the A-B cross-section plotted in (b), where the profiles extracted from the 2008 DTM of 2022 DTM are compared.

presents gravel fractions up to 20% in weight, while the clay fraction varies in the range of 20 ÷ 70%. The liquid limit w_l, plasticity index I_p and activity index AI were in the ranges of $w_l = 50 \div 93\%$; $I_p = 30 \div 68\%$; $AI = 0.75 \div 1.5$. As shown in Fig. 4a and b, very similar physical properties were recognised on samples recovered from the landslide on the *Gole del Drago* slope. The mechanical characterization of the Varicolori clay, especially in the landslide area, was problematic due to stony nodule inclusions. The results of the direct shear tests conducted on remoulded samples recovered from the subject area were reported in Fig. 4c in terms of shear stress τ vs normal effective stress σ'_v. In the same figure, the results of direct shear tests carried out on samples of the scaly clay matrix from *Cerda* landslide are also depicted. Data show a good agreement, and the following parameters could be taken: peak shear strength parameters in the range of cohesion $c'_p = 5 \div 50$ kPa and shear strength angle $\varphi'_p = 22 \div 23°$; residual shear strength parameters: $c'_r = 0$; $\varphi'_r = 12 \div 14°$. It is worth mentioning that ring shear tests performed on reconstituted clayey samples gave lower residual strength parameters ($c'_r = 0$ and $\varphi'_r = 7 \div 9°$) because the incomplete iso-orientation of clay particles on the shear surface in the specimens subjected to the reversal direct shear tests [16–18]. However, it is widely confirmed by the literature (e.g., [11, 19, 20]) that the

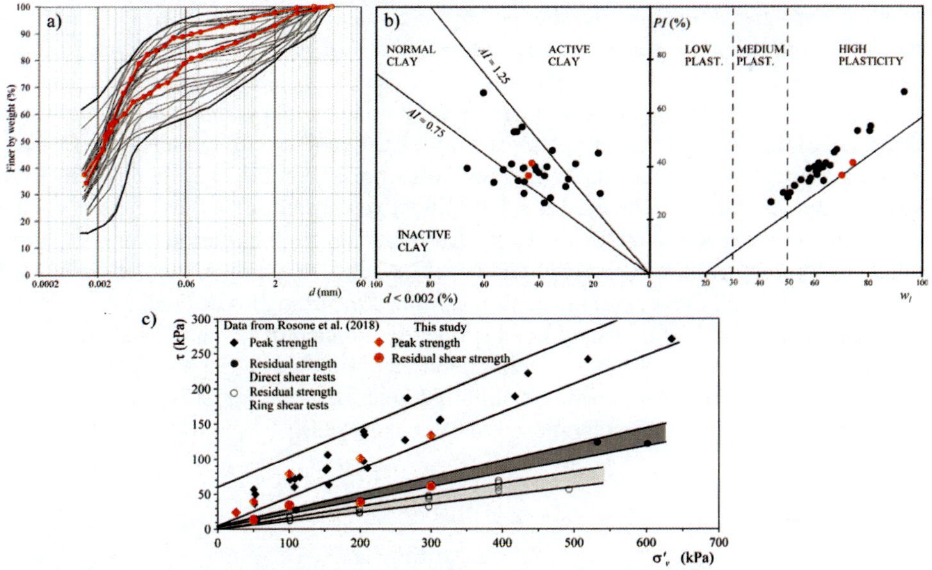

Fig. 4. Grain size distribution (a), activity diagram with plasticity chart (b) and shear strength envelopes of the Varicolori scaly clay samples recovered from the *Gole del drago* landslide (in red) compared with the ones obtained from *Cerda* landslide [11].

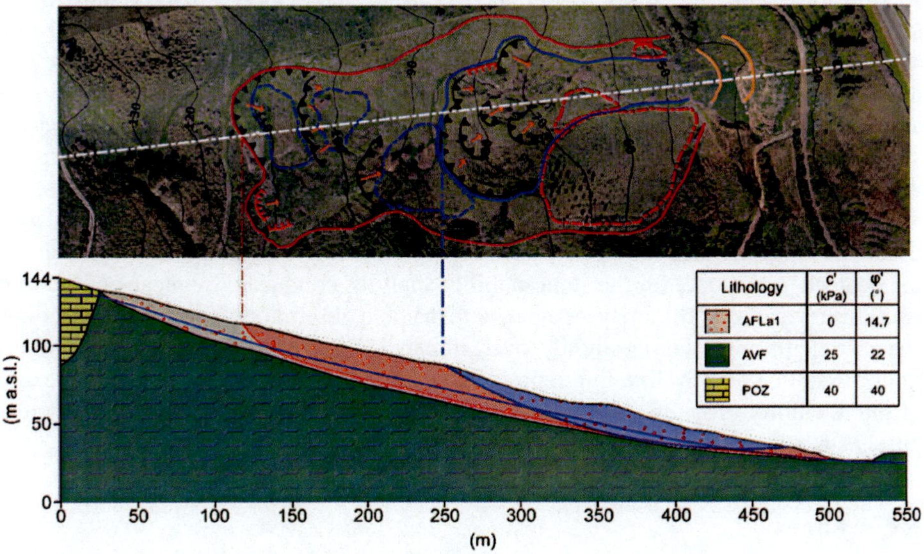

Fig. 5. Comparison between the landslide map derived from UAV surveys in January 2022 and the back-analyzed sliding surfaces plotted on the 2D geotechnical model.

residual shear strength parameter obtained by reversal direct shear tests well agrees with the shear strength mobilised along the failure surface of reactivated landslide in scaly clays.

In order to provide a geomechanical interpretation of the reactivation mechanism, a 2D model of the landslide was implemented and limit equilibrium model analyses were conducted (Fig. 5). Soil stratigraphy and pore water pressure regimes were defined on the base of direct evidence on the field and assuming a geo-structural and hydraulic scheme similar to the one recognized for the Cerda Landslide through numerous surveys, boreholes, exploratory trenches and piezometers [11]. As shown in Fig. 5, the back-analyzed sliding surfaces well agree with the contours of the two sliding bodies recognized in the landslide map derived by means of UAV surveys conducted in January 2022. However, the mobilized shear strength angle of remoulded Varicolori clay along the sliding surface ($\varphi'_{mob} = 14.7°$) in both cases resulted slightly higher than the upper bound of residual shear strength angle measured by means of direct shear tests ($\varphi'_r = 12 \div 14°$).

4 Conclusions

The *Gole del Drago* sector is defined as a "very high" susceptible area according to the tiering of the map of slide landslide susceptibility made through statistical methods for the whole Imera River basin (Northern Sicily). A recent reactivation that affected the slope unit identified confirmed the reliability of the basin scale's hazard zoning. At the slope scale, the study of the geological reference model made it possible to identify the processes and associated diagnostic landforms of a roto-translational slide at the head and an earth-flow mechanism at the toe. The reactivation features were measured and mapped with the support of the aerial data from UAV surveys, useful to integrate GIS modelling to better quantify the moved soil mass after reactivation. The multi-temporal analysis, carried out by comparing available DTMs, highlighted some topographic changes related to the reactivation along the middle sector of the slope unit. Detailed investigations from geological and geotechnical points of view were performed highlighting the main properties of the complex mechanism and defining the geotechnical model for the slope stability analyses conducted by means of the limit equilibrium method. The analyses agree with the field observation, despite the mobilized shear strength angle being slightly overestimated. The results obtained prove that a multidisciplinary study like this represents a promising method for local verification of the reliability of landslide susceptibility analyses conducted at a basin scale with a purely statistical approach.

References

1. Martinello, C., Cappadonia, C., Rotigliano, E.: Investigating the effects of cell size in statistical landslide susceptibility modelling for different landslide typologies: a test in central–northern sicily. Appl. Sci. **13**, 1145 (2023). https://doi.org/10.3390/app13021145
2. Martinello, C., et al.: Investigating limits in exploiting assembled landslide inventories for calibrating regional susceptibility models: a test in volcanic areas of el salvador. Appl. Sci. **12**, 6151 (2022)

3. Cafiso, F., Cappadonia, C., Ferraro, R., Martinello, C.: Rockfall hazard assessment of the monte gallo oriented nature reserve area (Southern Italy). In: Proceedings of the IOP Conference Series: Earth and Environmental Science, Beijing, China, 6 September 2021, vol. 833. IOP Publishing Ltd., Bristol (2021)
4. Cappadonia, C., Cafiso, F., Ferraro, R., Martinello, C., Rotigliano, E.: Rockfall hazards of mount pellegrino area (Sicily, Southern Italy). J. Maps **17**, 29–39 (2021)
5. Brabb, E.E.: The world landslide problem. Episodes **14**, 52–61 (1991). https://doi.org/10.18814/epiiugs/1991/v14i1/008
6. Martinello, C., Cappadonia, C., Conoscenti, C., Agnesi, V., Rotigliano, E.: Optimal slope units partitioning in landslide susceptibility mapping. J. Maps **17**, 152–162 (2021)
7. Azzaro, R., et al.: The earthquake of 6 September 2002 and the seismic history of Palermo (Northern Sicily – Italy): implications for the seismic hazard assessment of the city. J. Seismol. **8**(4), 525–543 (2004)
8. Agnesi, V., et al.: A multidisciplinary approach to the evaluation of the mechanism that triggered the Cerda landslide (Sicily, Italy). Geomorphology **65**(1–2), 101–116 (2005)
9. Bonci, C., et al.: Geological control on large seismically induced landslides: the case of Cerda (Southern Italy). In: Lacerda, W., Ehrlich, M., Fontoura, S.A.B., Sayao, A.S.F. (eds.) Landslides: Evaluation and Stabilization, pp. 985–991. Taylor and Francis Group, London (2004)
10. Bozzano, F., Cet al.: Engineering-Geology model of the seismically-induced cerda landslide. Bollettino Geofisica Teorica Appl. **49**, 205–226 (2008)
11. Rosone, M., Ziccarelli, M., Ferrari, A., Farulla, C.A.: On the reactivation of a large landslide induced by rainfall in highly fissured clays. Eng. Geol. **235**, 20–38 (2018)
12. Rosone, M., Ziccarelli, M., Ferrari, A.: Displacement evolution of a large landslide in a highly fissured clay. Lect. Notes Civ. Eng. **40**, 195–204 (2020)
13. Martinello, C., Cappadonia, C., Conoscenti, C., Rotigliano, E.: Landform classification: a high-performing mapping unit partitioning tool for landslide susceptibility assessment—a test in the Imera River Basin (Northern Sicily, Italy). Landslides **19**, 539–553 (2022)
14. Friedman, J.H.: Multivariate adaptive regression splines. Ann. Stat. **19**, 1–67 (1991)
15. Mandaglio, M.C., Moraci, N., Rosone, M., Airò Farulla, C.: Experimental study of a naturally weathered stiff clay. Can. Geotech. J. **53**(12), 2047–2057 (2016)
16. Rosone, M., Ferrari, A., Ziccarelli, M., Giger, S.B.: The residual shear strength of the shaly and sandy facies of the opalinus clay. In: Ferrari, A., Laloui, L. (eds.) SEG 2018. SSGG, pp. 426–433. Springer, Cham (2019). https://doi.org/10.1007/978-3-319-99670-7_53
17. Rosone, M., Megna, B., Celauro, C.: Analysis of the chemical and microstructural modifications effects on the hydro-mechanical behaviour of a lime-treated clay. Int. J. Geotech. Eng. **15**, 447–460 (2021)
18. Ferrari, A., Rosone, M., Ziccarelli, M., Giger, S.B.: The shear strength of Opalinus Clay shale in the remoulded state. Geomech. Energy Environ. **21**, 100142 (2020)
19. Airò Farulla, C., Cafiso, F., Calvi, F., Rosone, M.: Safeguarding historic towns on hilltops threatened by land sliding: the case of San Fratello in Sicily. Ital. Geotech. **49**(1), 7–28 (2015)
20. Ferrari, A., Ledesma, A., Gonzàlez, D.A., Corominas, J.: Effects of the foot evolution on the behaviour of slow-moving landslides. Eng. Geol. **117**(3–4), 217–2228 (2011)

Recent Developments in Soil Investigation by Medusa SDMT

Paola Monaco[1], Anna Chiaradonna[1(✉)], Diego Marchetti[2], Sara Amoroso[3,4], Jean-Sebastien L'Heureux[5], and Thi Minh Hue Le[5]

[1] University of L'Aquila, L'Aquila, Italy
anna.chiaradonna1@univaq.it
[2] Studio Prof. Marchetti, Rome, Italy
[3] University of Chieti-Pescara, Pescara, Italy
[4] Istituto Nazionale di Geofisica e Vulcanologia, L'Aquila, Italy
[5] Norwegian Geotechnical Institute, Trondheim, Norway

Abstract. The Medusa SDMT is the fully automated version of the seismic dilatometer (SDMT). An extensive in situ testing campaign with the Medusa SDMT was carried out in June 2022 in different soil types at four benchmark test sites in Norway, part of the Geo-Test Sites research infrastructure managed by the Norwegian Geotechnical Institute. The experimental campaign was conducted within the Transnational Access project – JELLYFISh funded by H2020-GEOLAB. This paper presents a preliminary assessment of significant results obtained by Medusa SDMT at the Onsøy test site (soft marine clay), compared to data obtained by traditional (pneumatic) SDMT in past investigations. Furthermore, the comparison is extended to a different benchmark test site in Italy, Fucino-Telespazio (soft lacustrine clay), where Medusa DMT data are available from past studies. Despite significant differences in origin, deposition environment and geological history of the two clay deposits, the profiles of parameters measured and interpreted from Medusa (S)DMT exhibit remarkable similarities. Simultaneous assessment of the soil parameters inferred by Medusa (S)DMT and by the traditional SDMT technology highlights a significant improvement in soil characterization in terms of both the quality and repeatability of the data.

Keywords: Medusa SDMT · Geo-Test Sites · soft clay

1 Introduction

The assessment and mitigation of challenging geo-hazards, frequently exacerbated by extreme events, stimulates an increasing demand for the development of innovative 'smart' technologies to facilitate in-situ soil characterization in a variety of field conditions. One significant recent development in *in situ* testing technology is the Medusa DMT, which represents the last-generation, fully automated version of the flat dilatometer (DMT). The seismic version of the probe (Medusa SDMT) incorporates additional sensors for the measurement of the shear wave velocity V_S.

A. Ferrari et al. (Eds.): CNRIG 2023, SSGG, pp. 234–241, 2023.
https://doi.org/10.1007/978-3-031-34761-0_29

An extensive in-situ testing campaign with the Medusa SDMT was carried out in June 2022 in different soil types at four well-known benchmark test sites in Norway, part of the Geo-Test Sites (NGTS) research infrastructure managed by the Norwegian Geotechnical Institute (NGI) [1]: Halden (silt), Onsøy (soft clay), Tiller-Flotten (quick clay), and Øysand (sand). The experimental campaign was conducted within the Transnational Access project – JELLYFISh funded by H2020-GEOLAB [2].

This paper presents a preliminary assessment of results obtained by Medusa SDMT at the Onsøy soft clay test site, compared to data from previous in situ investigations obtained by traditional pneumatic seismic dilatometer (SDMT). Furthermore, the comparison is extended to a different soft clay benchmark test site (Fucino-Telespazio, Italy), where Medusa DMT and traditional SDMT data are available from past investigations [3, 4].

2 Medusa (S)DMT Equipment and Test Procedure

The Medusa DMT [5, 6] is a self-contained probe able to perform dilatometer tests using a standard blade without the pneumatic cable, the control unit and the gas tank required in the traditional pneumatic DMT (Fig. 1). A motorized syringe, driven by an electronic board powered with rechargeable batteries, hydraulically expands the membrane to obtain the DMT *A*, *B*, *C* pressure readings, which are acquired and stored automatically at each test depth (typically every 0.20 m). The *A*-pressure is recorded when the membrane centre has expanded 0.05 mm against the soil from its initial position. After the *A*-reading the motorized syringe continues to increase the oil pressure, and when the membrane has expanded 1.10 mm at the centre the *B*-pressure is recorded. Soon after the *B*-reading the motorized syringe gradually applies a controlled depressurization, and the *C*-pressure is recorded when the membrane has returned to its initial position. The automatic (volume controlled) hydraulic pressurization of the membrane is highly repeatable and permits to impose a programmable timing to obtain the pressure readings. The probe can operate in cableless mode, which is a significant practical advantage in the offshore industry and for deep investigations. An optional electric cable may be used to obtain real-time data during test execution. The standard Medusa DMT test procedure is the same procedure of the traditional pneumatic DMT test, in accordance with existing standards of the pneumatic DMT [7, 8]. Also the field of application (soil types) is the same.

The Medusa SDMT incorporates additional sensors and components for the measurement of the shear wave velocity V_S in addition to the DMT measurements. The test procedure and interpretation for obtaining V_S measurements using the Medusa SDMT are the same established for the traditional SDMT [9].

3 Medusa SDMT Investigation Campaign at Onsøy

The Onsøy site [10] is located in southeastern Norway, about 100 km from Oslo. The soils are marine clays, deposited during deglaciation and the early postglacial period (Holocene) at times of higher relative sea level. The clay deposit, 25–35 m thick, is normally consolidated, but it exhibits overconsolidation due to aging. The overconsolidation ratio *OCR* decreases from about 4 near the surface to 1.2 at 30 m depth.

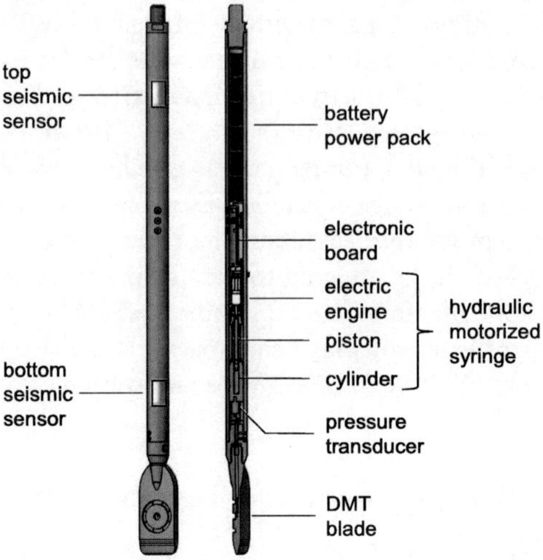

Fig. 1. Main components of the Medusa SDMT.

The natural water content *w* varies between 40% and 70%, and the average plasticity index *PI* varies from about 45% in the upper 8 m to about 25–30% below 8 m. The sensitivity S_t (ratio of peak to remoulded shear strength estimated from fall cone and field vane tests) is constant at around 6 down to about 13 m. Beyond this depth it increases to a value of 45 at approximately 19 m, becoming a quick clay just above bedrock. The percolation of freshwater from the surface has caused an almost linear salinity increase from zero at the surface to 30 g/l at about 7.5 m. Beyond this depth, the salinity remains constant. Organic content values show values around 0.8% in the top 9 m and around 0.6% below this depth.

The field testing program at Onsøy (June 2022) comprised one Medusa SDMT sounding carried out by the standard procedure, and two Medusa DMT soundings carried out using innovative non-standard test procedures (repeated *A*-readings, *A*-reading while penetrating), supplemented by one Medusa DMT dissipation test. All Medusa (S)DMT soundings reached a depth of about 20 m and were located close to one traditional (pneumatic) SDMT sounding performed in 2018 by NGI. Details on the Medusa SDMT testing program and results at Onsøy can be found in [2, 11].

4 Previous Medusa DMT Tests at Fucino-Telespazio

The Fucino-Telespazio site is located in central Italy, about 80 km east of Rome. At the end of the 1980s the site was selected as a national benchmark test site due to its marked spatial homogeneity and simple geological history, and extensively investigated through several in situ and laboratory tests carried out by various research groups [12].

The site is constituted by a thick deposit of soft, homogeneous lacustrine clay, that is highly structured and cemented despite its relatively recent deposition. The calcium carbonate $CaCO_3$ content ranges between 10% and 30% in the upper 25 m, increasing to an average value of about 60% below this depth. The Fucino clay is characterized by high plasticity (plasticity index *PI* mostly between 40% and 70%, natural water content w between 60% and 120%) and is geologically normally consolidated. However, both laboratory and in situ tests indicated a light overconsolidation, which was attributed mostly to diagenetic interparticle bonding due to $CaCO_3$ cementation. Structure and cementation have a dominant influence on stress history, compressibility, consolidation and shear strength properties of the Fucino clay [12].

An experimental program at the Fucino-Telespazio test site, including three Medusa DMT soundings and one traditional pneumatic SDMT sounding to a depth of 30 m, supplemented by one Medusa DMT dissipation test, was carried out in September 2020. One Medusa DMT sounding was carried out by the standard procedure and the other two using non-standard test procedures (repeated *A*-readings, *A*-reading while penetrating). Details on the different Medusa DMT test procedures and results at Fucino-Telespazio can be found in [3] and [4].

5 Medusa (S)DMT Results in Onsøy vs. Fucino Clays

This section shows the comparison of the results obtained by Medusa (S)DMT and by traditional SDMT at the test sites of Onsøy and Fucino-Telespazio. Only the results provided by the same standard test procedure have been retained for comparison and will be discussed further. The test results obtained from Medusa (S)DMT were processed using the same data reduction and interpretation formulae used for the traditional DMT test [13].

Figure 2 shows the depth profiles of the corrected pressures p_0, p_1, p_2 (A, B, C readings corrected for membrane stiffness by calibration) and of the measured shear wave velocity V_S, while Fig. 3 shows the derived intermediate parameters, i.e., the material index I_D, the pore pressure index U_D, the horizontal stress index K_D, and the dilatometer modulus E_D. In the data processing, the in situ u_0 pore pressure distribution at the two sites was assumed as hydrostatic, with a groundwater table at a depth of 1 m at Onsøy (from available piezometer measurements) and 0.6 m at Fucino-Telespazio (as indicated by the p_2 values observed in thin sand layers, see Fig. 2). For a preliminary assessment, the in situ vertical effective stress σ'_{v0} was calculated based on an approximate estimate of the soil unit weight γ obtained from available DMT correlations [13].

One clear piece of evidence emerging from Fig. 2 is the remarkable similarity of the profiles of p_0, p_1, p_2 obtained by Medusa (S)DMT in Onsøy clay and in Fucino clay. A similar trend is observed in Fig. 3 in terms of intermediate parameters except for I_D. In fact, the I_D values at Onsøy are lower than in Fucino clay, although such difference is amplified by the logarithmic scale of the graph. The U_D and K_D profiles are quite similar for both clays, while the E_D obtained in Onsøy clay is slightly lower than in Fucino clay. The V_S profiles at the two sites (Fig. 2) show similar trends.

As to the comparison between results obtained by Medusa (S)DMT and traditional SDMT at each site, Fig. 2 shows that the profiles of p_0, p_1, p_2 obtained by the two probes are very close to each other both at Onsøy and Fucino-Telespazio. Some difference is more evident when these pressures are combined in terms of intermediate parameters (Fig. 3). Indeed, the values of I_D acquired by traditional SDMT appear lower than the I_D provided by Medusa (S)DMT, particularly at Onsøy (again, amplified by the log scale). The same trend is also observed, to a lesser extent, in the profiles of E_D, which depends on the difference $(p_1 - p_0)$ as I_D, and of K_D, which depends only on p_0. The difference is more pronounced at Onsøy than at Fucino-Telespazio. Such discrepancy may be attributed to inherently different technical features of the Medusa (S)DMT and the traditional SDMT equipment: (1) with the Medusa (S)DMT the pressure is generated and measured in the probe at depth, eliminating any pressure equalization problem at the opposite ends of the pneumatic cable that may occur with the traditional equipment; (2) the automated membrane inflation and the incompressibility of the pressurizing fluid (oil) enable the Medusa (S)DMT to enforce the standard pressurization rate with high precision and repeatability. These capabilities of the Medusa (S)DMT permit significantly improving accuracy of the measurements especially in very soft soils.

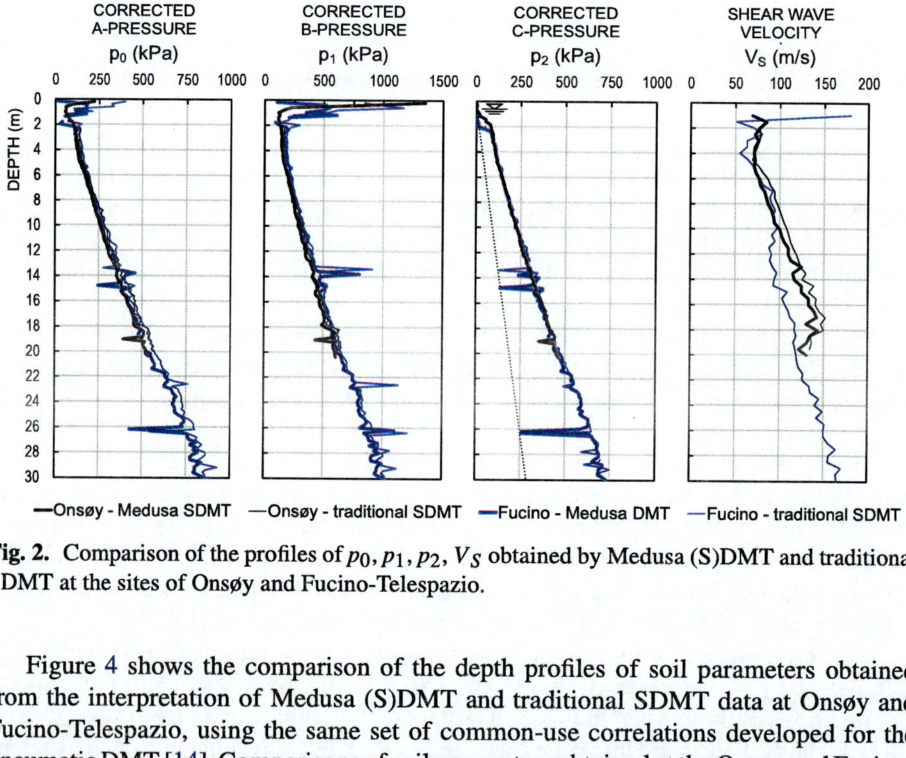

Fig. 2. Comparison of the profiles of p_0, p_1, p_2, V_S obtained by Medusa (S)DMT and traditional SDMT at the sites of Onsøy and Fucino-Telespazio.

Figure 4 shows the comparison of the depth profiles of soil parameters obtained from the interpretation of Medusa (S)DMT and traditional SDMT data at Onsøy and Fucino-Telespazio, using the same set of common-use correlations developed for the pneumatic DMT [14]. Comparisons of soil parameters obtained at the Onsøy and Fucino-Telespazio test sites from a variety of in situ and laboratory tests in past investigations and in the Medusa (S)DMT campaigns are illustrated in [11] and [3]. The profiles of the overconsolidation ratio *OCR* and of the coefficient of earth pressure at rest K_0 estimated from Medusa (S)DMT at Onsøy and Fucino-Telespazio (bold lines in Fig. 4) are nearly coincident and reflect light overconsolidation due to non-mechanical factors (aging at Onsøy, cementation at Fucino-Telespazio), despite significant differences in origin, deposition environment and geological history of the two clay deposits. As shown by [11], the *OCR* profile obtained from Medusa SDMT at Onsøy plot very close to the

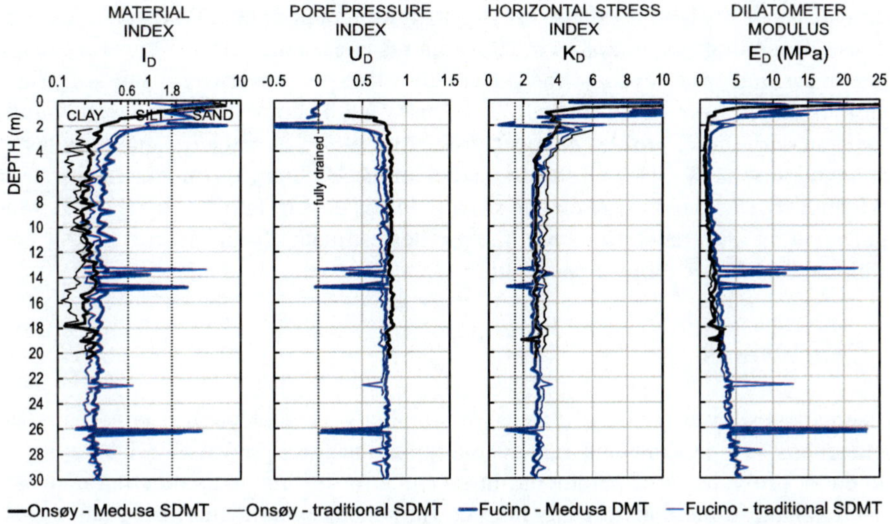

Fig. 3. Comparison of the profiles of the intermediate parameters I_D, U_D, K_D, E_D obtained by Medusa (S)DMT and traditional SDMT at the sites of Onsøy and Fucino-Telespazio.

"best estimate" profile established in previous research (also shown in Fig. 4), while the traditional SDMT appears to overestimate *OCR*.

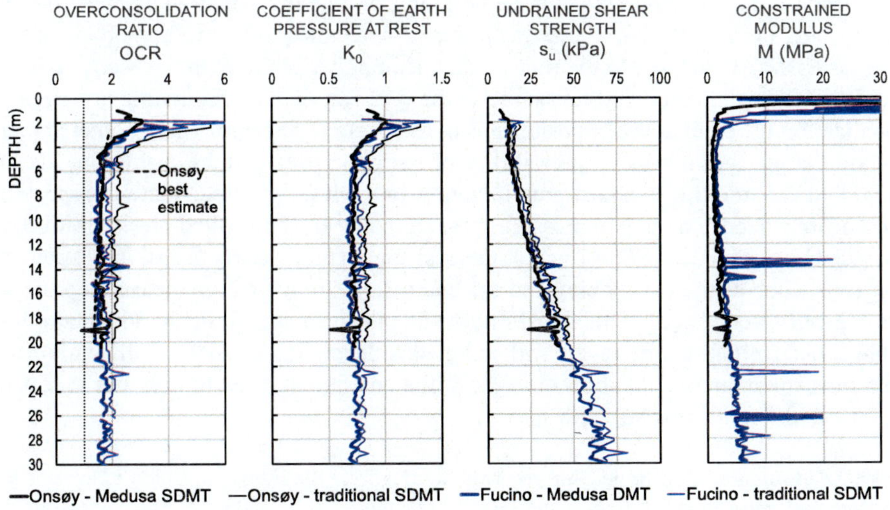

Fig. 4. Comparison of the profiles of parameters obtained from interpretation of Medusa (S)DMT and traditional SDMT at the sites of Onsøy and Fucino-Telespazio.

The profiles of the undrained shear strength s_u obtained from the interpretation of Medusa (S)DMT in Onsøy and in Fucino clays are nearly coincident. Also, such s_u

values are generally lower than the s_u estimated from traditional SDMT. As for OCR and K_0, also depending only on p_0 as s_u, this trend is consistent with the lower A-pressure values measured in absence of the pneumatic cable, as previously discussed. The s_u profiles obtained from Medusa (S)DMT at Onsøy and Fucino-Telespazio were found in agreement with the s_u profiles obtained from a variety of in situ and laboratory tests in past investigations [3, 11]. The constrained modulus M obtained from the interpretation of Medusa (S)DMT in Onsøy clay is slightly lower than in Fucino clay. For M, which depends on the difference $(p_1 - p_0)$, the profiles estimated from Medusa (S)DMT and traditional SDMT are closer to each other.

6 Conclusions

Benchmark test sites, selected to ensure simplicity of geotechnical conditions (i.e., homogenous soil deposit) and ease of interpretation (i.e., pre-existing laboratory and field data), prove to be of paramount importance for testing and validating innovative soil investigation methods. In this respect, the recent experimental program with the Medusa SDMT (the last generation, fully automated version of the SDMT) at the NGTS Onsøy soft clay test site could uniquely benefit from the availability of an existing large and consistent data set obtained in past investigations from a variety of high-quality in situ and laboratory tests. A similar outcome was achieved from a previous investigation campaign at the Fucino-Telespazio soft clay test site. Despite significant differences in origin, deposition environment, and geological history of the two clay deposits, the profiles of parameters measured and interpreted from Medusa (S)DMT exhibit remarkable similarities. One notable feature emerging from the investigation campaigns at the Onsøy and Fucino-Telespazio sites is the superior accuracy of the results obtained from Medusa SDMT compared to the traditional pneumatic SDMT. With the Medusa SDMT the pressure is generated and measured in the probe at depth, eliminating any pressure equalization problem at the opposite ends of the pneumatic cable. The automated membrane inflation and the incompressibility of the pressurizing fluid enable the Medusa SDMT to enforce the standard pressurization rate with high precision and repeatability. The improved accuracy in pressure measurements and controlled pressurization rate provided by the Medusa SDMT makes this instrument particularly useful for testing very soft soils, such as the Onsøy clay and the Fucino clay, in which the measured pressures are typically very small. This capability turns out to be a significant advantage for a reliable determination of relevant soil parameters depending on such small quantities, in particular the overconsolidation ratio OCR, the coefficient of earth pressure at rest K_0, and the undrained shear strength s_u.

Acknowledgments. This project has received funding from the European Union's Horizon 2020 research and innovation program under Grant Agreement No. 101006512. The authors are also grateful to the University of L'Aquila – DICEAA for the financial support, to Luca Agrini (NGI), Nathan Townsend (NGI) and Lorenzo Ferrante (Studio Prof. Marchetti) for their technical support, and to Simone Alber (ETH Zurich) for participating in the field activity within the GEOLAB Early Stage Researcher Internship Program.

References

1. L'Heureux, J.-S., Lunne, T.: Characterization and engineering properties of natural soils used for geotesting. AIMS Geosci. **6**(1), 35–53 (2020). https://doi.org/10.3934/geosci.2020004
2. Monaco, P., Chiaradonna, A., Marchetti, D., Amoroso, S., Le, T.M.H.: GEOLAB – Transnational Access project JELLYFISh – Field testing of Medusa DMT (version 0). Zenodo (2023). https://doi.org/10.5281/zenodo.7695740
3. Marchetti, D., Monaco, P., Totani, G., Totani, F., Amoroso, S.: Comparisons CPT-DMT in soft clay at Fucino-Telespazio GeoTest site. In: 5th International Symposium on Cone Penetration Testing CPT 2022, pp. 190–196. https://doi.org/10.1201/9781003308829-21
4. Monaco, P., Marchetti, D., Totani, G., Totani, F., Amoroso, S.: Validation of Medusa DMT test procedures in Fucino clay. In: 20th International Conference on Soil Mechanics and Geotechnical Engineering ICSMGE 2022, vol. 2, pp. 471–476 (2022)
5. Marchetti, D.: Dilatometer and seismic dilatometer testing offshore: available experience and new developments. Geotech. Test. J. **41**(5), 967–977 (2018). https://doi.org/10.1520/GTJ20170378
6. Marchetti, D., Monaco, P., Amoroso, S., Minarelli, L. In situ tests by Medusa DMT. In: 17th European Conference on Soil Mechanics and Geotechnical Engineering, ECSMGE-2019 (2019). https://doi.org/10.32075/17ECSMGE-2019-0657
7. ASTM D6635-15. Standard test method for performing the flat plate dilatometer. ASTM International, West Conshohocken, PA, USA (2015)
8. ISO 22476-11:2017(E). Geotechnical investigation and testing – Field testing – Part 11: Flat dilatometer test. International Organization for Standardization, Geneva (2017)
9. Marchetti, S., Monaco, P., Totani, G., Marchetti, D.: In situ tests by seismic dilatometer (SDMT). Geotech. Spec. Publ. GSP **180**, 292–311 (2008). https://doi.org/10.1061/40962(325)7
10. Gundersen, A.S., Hansen, R.C., Lunne, T., L'Heureux, J.-S., Strandvik, S.O.: Characterization and engineering properties of the NGTS Onsøy soft clay site. AIMS Geosci. **5**(3), 665–703 (2019). https://doi.org/10.3934/geosci.2019.3.665
11. Monaco, P., Chiaradonna, A., Marchetti, D., Amoroso, S., L'Heureux, J.-S., Le, T.M.: Medusa SDMT testing at the Onsøy Geo-Test Site, Norway. In: 8th International Symposium on Deformation Characteristics of Geomaterials IS-Porto 2023 (2023)
12. Burghignoli, A., et al.: Geotechnical characterization of Fucino clay. In: 10th European Conference on Soil Mechanics and Foundation Engineering, vol. 1, pp. 27–40. Balkema, Rotterdam (2019)
13. Marchetti, S., Monaco, P., Totani, G., Calabrese, M.: The flat dilatometer test (DMT) in soil investigations – A report by the ISSMGE Committee TC16. In: International Conference Insitu Measurement of Soil Properties and Case Histories, pp. 95–131 (2001). Official version approved by ISSMGE TC16 reprinted in 2nd International Conference on Flat Dilatometer, pp. 7–48 (2006)
14. Marchetti, S.: In situ tests by flat dilatometer. J. Geotech. Eng. Divis. **106**(GT3), 299–321 (1980)

Remote Sensing Meteorological and DInSAR Historical Data to Analyse the Kinematic Behaviour of Slow-Moving Landslides at Municipal Scale

Gianfranco Nicodemo[1](✉), Gaetano Pecoraro[1], Guido Rianna[2], Alfredo Reder[2], Davide Luongo[1], Dario Peduto[1], and Michele Calvello[1]

[1] Department of Civil Engineering, University of Salerno, via Giovanni Paolo II, 132, 84084 Fisciano, SA, Italy
gnicodemo@unisa.it

[2] Regional Models and Geo-Hydrological Impacts (REMHI) Division, Fondazione Centro Euro-Mediterraneo sui Cambiamenti Climatici, via T. A. Edison, 81100 Caserta, Italy

Abstract. The present study aims to investigate the kinematic behaviour of slow-moving landslides affecting urban areas at municipal scale, based on remote sensing meteorological and displacement (DInSAR) historical data. The long-term influence of weather patterns on landslide dynamics and their potential variation under the effect of ongoing climate change is taken into account through the reanalysis of the main weather and soil water variables derived by the freely available datasets provided by CERRA within the Copernicus Climate Change Service (C3S). The above-mentioned data are jointly analysed with widespread ground surface displacements data gathered from the processing of very-high resolution Synthetic Aperture Radar (SAR) images acquired by COSMO-SkyMed constellation via Differential Interferometry (DInSAR) techniques. The link between the main weather and soil water variables with the ground deformations sensed from the space was investigated in the territory of Vaglio Basilicata, a municipality in the Basilicata region (southern Italy) widely affected by slow-moving landslides interacting with the built-up environment and, specifically, infrastructure networks. The preliminary results achieved could be valuably used as input to outline a proper procedure aimed at the dynamic evaluation of the infrastructure risk associated with weather-induced reactivations and/or accelerations of slow-moving landslides.

Keywords: Slow-moving landslide · Climate reanalysis · DInSAR · Infrastructure · Risk assessment

1 Introduction

Slow-moving landslides are natural hazards affecting several urban areas all over the world that, even if rarely associated with the loss of human life, can influence both social and economic activities of territory due to the damages induced to structures and

A. Ferrari et al. (Eds.): CNRIG 2023, SSGG, pp. 242–250, 2023.
https://doi.org/10.1007/978-3-031-34761-0_30

infrastructure networks [1–3]. Although the annual cumulative displacements of this type of slope instability are low and typically fluctuate around a practically constant value, the velocity rates measured over shorter time lengths are quite variable. Indeed, this landslide category exhibits relatively slow motions following visco-plastic behaviour over a long period of time [4], with velocity values that do not exceed 1.8 m/h [5]. Nevertheless, weather-induced reactivations and/or accelerations are to be expected. These kinematic features, associated with a permanent or episodic activity, cause damages, whose severity progressively increases over time, to the exposed built-up environment [6]. Therefore, the integrated analysis of the weather patterns influencing the landslide dynamics, the temporal variation of main weather and soil water variables, and the landslide displacement rates, is essential to characterise their kinematic behaviour and to quantify the risk associated with the exposed facilities. In this regard, the assessment of the long-term influence of weather patterns on landslide dynamics and their potential variation under the effect of ongoing climate change is often hindered by the unavailability of spatially and temporally homogeneous and long-lasting weather datasets. A good option to address such limitation is represented by atmospheric reanalysis providing the main weather and soil water variables (e.g., CERRA, freely available within the Copernicus C3S) jointly with ground surface displacements data gathered from the processing of Synthetic Aperture Radar (SAR) images via Differential Interferometry techniques (DInSAR).

The Copernicus European Regional ReAnalysis (CERRA) [7, 8] aims to return a spatially and temporally consistent historical reconstruction of meteorological variables in the atmosphere and at the surface with a horizontal resolution of 5.5 km over Europe and hourly time-step. It is performed by continuously assimilating observations into a physically-based weather model. The same approach, usually known as data assimilation, is widely adopted for weather forecast when a previous forecast is combined with newly available observations to have a new best estimate of the state of the atmosphere (analysis). Specifically, CERRA uses the HARMONIE-ALADIN limited-area numerical weather prediction. In addition, the main inputs are observational data and the lateral boundary conditions provided by ERA5 global reanalysis (with a horizontal resolution of ~32 km) [9].

With reference to the DInSAR data, in the last decades they have widely proved their usefulness in the monitoring of dynamic processes, such as those associated with natural and/or human-induced hazards, and for the estimation of values to be associated with the slow-moving landslide intensity measure (e.g., velocity rate or cumulative displacement). Indeed, the DInSAR-derived landslide deformation patterns are useful for investigating past-landslide evidence and updating landslide boundaries and/or their state of activity [10], detecting unmapped unstable sloping areas [11], performing vulnerability analysis of structures and infrastructure at risk [2, 3, 12]. More recently, DInSAR data were used in detailed analyses comparing the accumulated rainfall and DInSAR displacement time-series with reference to single [13] or several [14] data gathered from a rain gauge. Still, only a few examples concern the correlation between DInSAR, precipitation and soil moisture observations [15] using remote sensing data on large areas.

In this regard, considering the available huge dataset today provided by remote sensing meteorological and DInSAR historical observations, this study analyses – within the territory of the Vaglio Basilicata municipality in Basilicata region (southern Italy) – the relationships between the results of the atmospheric reanalysis of the main weather and soil water variables, and the DInSAR-retrieved ground surface displacements in order to investigate the kinematic behaviour of slow-moving landslides affecting the urban areas and, specifically, infrastructure networks. The preliminary results obtained can provide useful suggestions for the identification of the main variables governing the slope instability and the associated slow-moving landslide dynamics processes, as well as in defining a proper procedure for the dynamic evaluation – at the municipal scale – of risk associated with weather-induced reactivations and/or accelerations of slow-moving landslides.

2 Case Study and Available Dataset

The test area (Fig. 1) is the territory of the Vaglio Basilicata municipality (Basilicata region, southern Italy). From the geological and geomorphological point of view, the area falls within the eastern portion of the southern Apennines, mainly consisting of complex tectonic units characterised by outcropping of the Flysch Rosso formation (Upper Cretaceous-Eocene) and structurally complex clayey marly succession, strongly fractured and deformed, referred to the Lagonegro Unit [16].

This geological unit, strongly fractured and severely deformed, is particularly prone to slope instability phenomena because of its severe tectonic history [16]. Indeed, the whole territory is widely affected by slow-moving landslides (Fig. 1), as reported within the official inventory map drawn up by the Interregional River Basin Authority of Basilicata region (last update 2018), interacting with the built-up environment and infrastructure networks.

For the whole study area, remote sensing displacements data are available, retrieved from the processing via differential interferometric (DInSAR) technique of 124 COSMO-SkyMed Synthetic Aperture Radar (SAR) images acquired on descending orbit from May 2012 to August 2021 (see Acknowledgements). These DInSAR data consist of velocity values – along the Line of Sight (LOS) sensor-target direction – associated with coherent pixels (PS) whose spatial distribution is shown in Fig. 1.

Regarding weather data (Fig. 2), two main variables are considered: cumulative rainfall and volumetric water content (VWC) at the shallower soil depth (2 m). The first provides indications on the maximum water quantities entering the soil, albeit the well-known limitations associated to data retrieved by global datasets, in this case due to limited knowledge of the local hydraulic behaviour. The latter is assumed to be a proxy for the entire soil-atmosphere water balance (VWC also accounts for infiltration and evapotranspiration dynamics).

Figure 2 reports the position of the nine grid points covering the Municipal territory and the monthly values of cumulative precipitation averaged over the period 1991–2020 assumed as the baseline for the current period, as suggested by World Meteorological Organization guidelines. The mean yearly cumulative value is about 830 mm, with significant variations among the values (standard deviation equal to about 65 mm) and

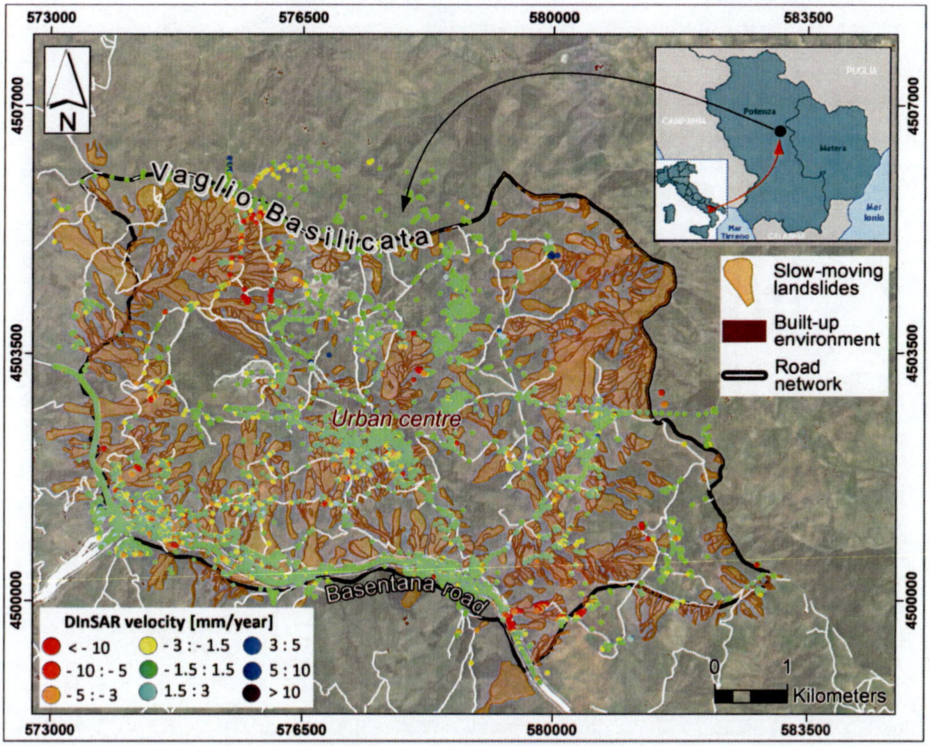

Fig. 1. Vaglio Basilicata study area: the officially slow-moving landslide inventory provided by the Interregional River Basin Authority of Basilicata region, the roads network and LOS average velocities of coherent pixels (PS) derived by DInSAR processing of COSMO-SkyMed images acquired on descending orbit (period: May 2012 – August 2021).

the maximum ones located in the Northern part. Over the year, a two-peaks evolution is observed, with the first maximum during the Spring and the second in late Autumn-early Winter.

3 Results

The remote sensing meteorological and DInSAR data, jointly analysed with reference to a fixed time interval corresponding to the hydrological year (1st September – 31th August) for the eight years of overlap between the two available datasets (2012–2020), has provided the results shown in Fig. 3 and Fig. 4.

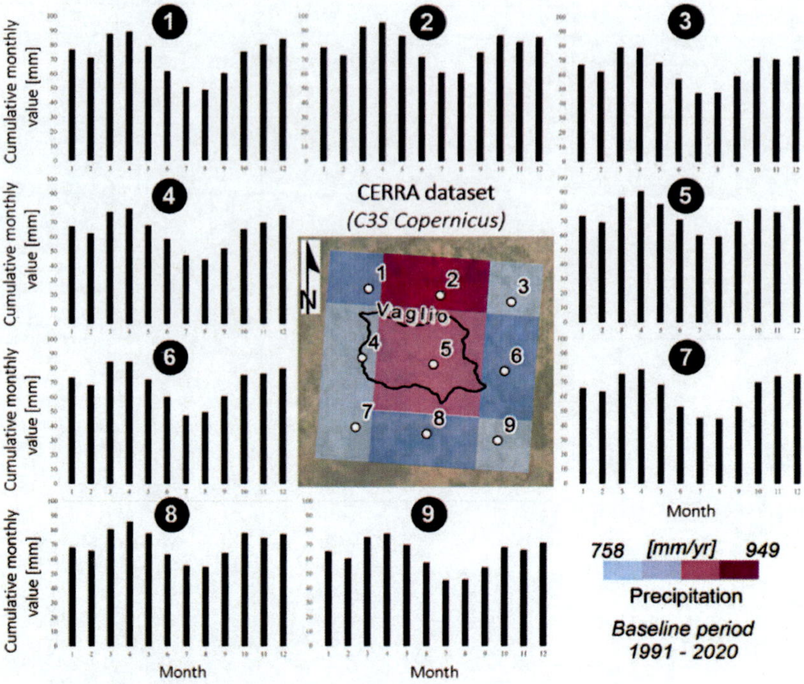

Fig. 2. CERRA dataset with the nine grid points covering the municipal territory of Vaglio Basilicata study area with the monthly values of cumulative precipitation averaged over the period 1991–2020, assumed as the baseline period.

In particular, the DInSAR-retrieved average annual velocities of moving PS (LOS velocity threshold higher than 1.5 mm/yr) covering the slow-moving landslides (Fig. 1), were computed with reference to each hydrological year (from 1 to 8, Fig. 3a) and plotted adopting box-whiskers reporting mean, median, first and third quartile, minimum and maximum values, and relative outliers. This DInSAR-retrieved information was compared with the anomalies in the running cumulative rainfall values over 12 months (Fig. 3b) and anomalies in the running number of months for which the mean volume water content (VWC) is higher that a specified threshold (fixed at 0.27) (Fig. 3c), evaluated by the CERRA dataset in the same observation period. Both anomalies are computed at the monthly scale assuming as baseline the thirty years from 1991 to 2020. The assumed threshold for VWC is quite close to the mean value computed over the entire period 1991–2020.

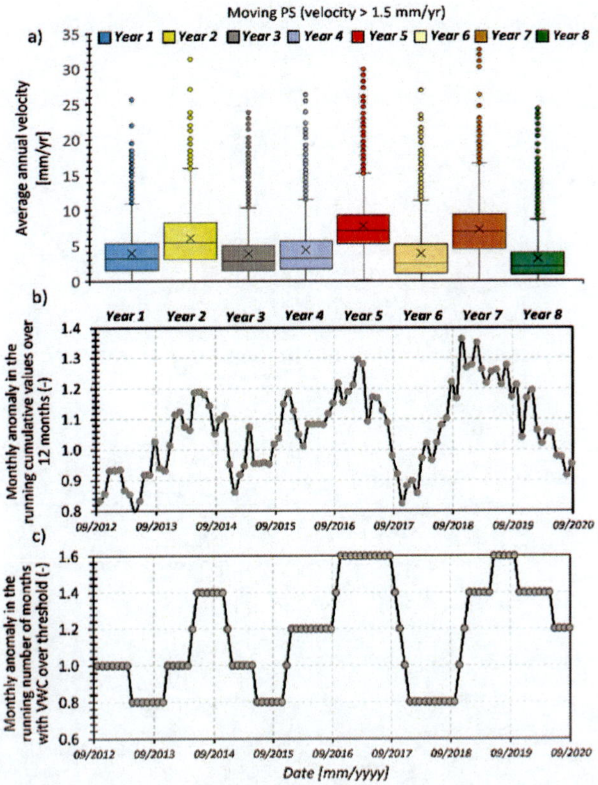

Fig. 3. a) DInSAR-retrieved average annual velocities of moving PS (LOS velocity threshold higher than 1.5 mm/yr); b) anomalies in the running cumulative values over 12 months compared to the baseline 1991–2020; c) anomalies in the running number of months with volume water content (VWC) over threshold (fixed equal to 0.27) compared to the baseline 1991–2020.

The relationship between remote sensing meteorological and DInSAR measurements (Fig. 3) indicates a good match between the two sets of data, with a greater number of monthly anomalies for both cumulative 12-month precipitation values and VWC over the fixed threshold, in correspondence of the three years (2, 5 and 7) for which highest DInSAR average annual velocity are recorded. This result is confirmed by the spatial distribution of slow-moving landslide covered by moving coherence pixel (PS) shown in Fig. 4. Indeed, over the eight investigated hydrological years, 2, 5 and 7 are characterised by a greater number of moving PS (with an average velocity value that exceed the velocity threshold equal to 1.5 mm/yr) as well as higher intensity values.

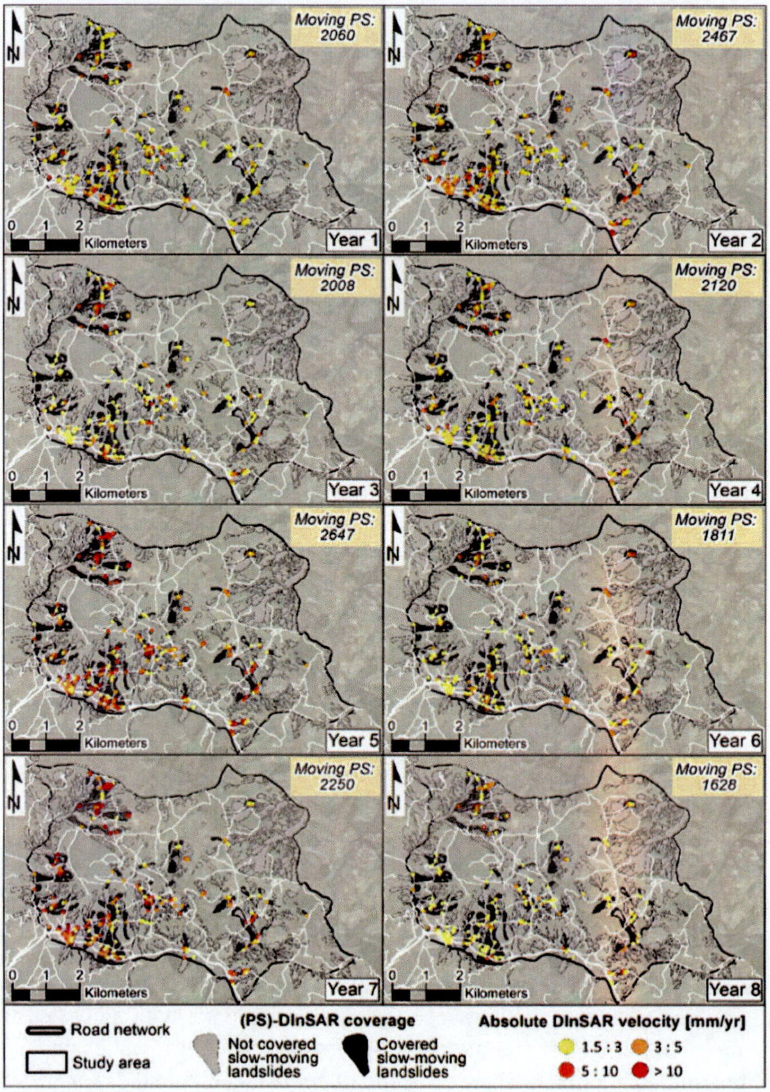

Fig. 4. Spatial distribution of slow-moving landslides covered by high coherence pixel (PS) with an average annual velocity value that exceeds a velocity threshold equal to 1.5 mm/yr over the eighth observation years.

4 Conclusions

The paper presented some preliminary results of a wider research focused on the integrated analysis – over a large area – of remote sensing meteorological and DInSAR historical data, to define a proper procedure for the dynamic evaluation of the infrastructure risk associated with weather-induced reactivations and/or accelerations of slow-moving landslides.

The preliminary outcomes achieved in terms of relationships between the atmospheric reanalysis of the precipitation and soil water variables with the DInSAR-retrieved ground surface displacements over the eight investigated years, highlighted the possibility and feasibility of linking information sensed from space in analyses aimed at studying the kinematic behaviour of slow-moving landslides at municipal scale.

The approach is quite promising since it permits the back-analysis, using freely available data (homogeneous in space and time), of areas or periods not properly covered by weather observations networks in recent decades. In this regard, the high spatial resolution of CERRA reanalysis (5.5 km) also enables to adequately investigate areas that are not too large.

Acknowledgements. This research has been supported by MIUR PON R&I 2014–2020 Program (project MITIGO, ARS01_00964). The authors wish to acknowledge project partner e-GEOS for processing the COSMO-SkyMed data. CERRA data was downloaded from the Copernicus Climate Change Service (C3S) Climate Data Store. Neither the European Commission nor ECMWF is responsible for any use that may be made of the Copernicus information or data it contains.

References

1. Corominas, J., et al.: Recommendations for the quantitative analysis of landslide risk. Bull. Eng. Geol. Environ. **73**(2), 209–263 (2014)
2. Peduto, D., Nicodemo, G., Caraffa, M., Gullà, G.: Quantitative analysis of consequences to masonry buildings interacting with slow-moving landslide mechanisms: a case study. Landslides **15**(10), 2017–2030 (2018). https://doi.org/10.1007/s10346-018-1014-0
3. Ferlisi, S., Marchese, A., Peduto, D.: Quantitative analysis of the risk to road networks exposed to slow-moving landslides: a case study in the Campania region (southern Italy). Landslides **18**(1), 303–319 (2020). https://doi.org/10.1007/s10346-020-01482-8
4. Leroueil, S.: Natural slopes and cuts: movement and failure mechanisms. Géotechnique **51**(3), 197–243 (2001)
5. Hungr, O., Corominas, J., Eberhardt, E.: Estimating landslide motion mechanism, travel distance and velocity. In: Hungr, O., Fell, R., Couture, R., Eberhardt, E. (eds.) Landslide Risk Management, pp. 99–128. Taylor and Francis, London (2005)
6. Peduto, D., Ferlisi, S., Nicodemo, G., Reale, D., Pisciotta, G., Gullà, G.: Empirical fragility and vulnerability curves for buildings exposed to slow-moving landslides at medium and large scales. Landslides **14**(6), 1993–2007 (2017). https://doi.org/10.1007/s10346-017-0826-7
7. Schimanke, S., et al.: CERRA sub-daily regional reanalysis data for Europe on single levels from 1984 to present. Copernicus Climate Change Service (C3S) Climate Data Store (CDS) (2021). https://doi.org/10.24381/cds.622a565a. Accessed 18 Jan 2023
8. Buontempo, C., et al.: The copernicus climate change service: climate science in action. Bull. Am. Meteorol. Soc. **103**(12), E2669–E2687 (2022)
9. Hersbach, H., et al.: The ERA5 global reanalysis. Quart. J. Roy. Meteor. Soc. **146**, 1999–2049 (2020)
10. Peduto, D., Santoro, M., Aceto, L., Borrelli, L., Gullà, G.: Full integration of geomorphological, geotechnical, A-DInSAR and damage data for detailed geometric-kinematic features of a slow-moving landslide in urban area. Landslides **18**(3), 807–825 (2020). https://doi.org/10.1007/s10346-020-01541-0
11. Jovanovski, M., et al.: Landslide characterisation in the Polog region by innovative and conventional methods. Italian Geotech. J. **4**, 7–31 (2021)

12. Nicodemo, G., Ferlisi, S., Peduto, D., Aceto, L., Gullà, G.: Damage to masonry buildings interacting with slow-moving landslides: a numerical analysis. In: Calvetti, F., et al. (eds.) CNRIG 2019. LNCE, vol. 40, pp. 52–61. Springer, Cham (2020). https://doi.org/10.1007/978-3-030-21359-6_6
13. Vecchiotti, F., Amabile, A.S., Clemente, S., Ostermann, M., Nicodemo, G., Peduto, D.: Kinematic and geometric characterization of the Vögelsberg Rockslide (Tyrol, Austria) by means of MT-InSAR data. Geosciences **12**, 256 (2022)
14. Ardizzone, F., et al.: A Procedure for the quantitative comparison of rainfall and DInSAR-based surface displacement time series in slow-moving landslides: a case study in southern Italy. Remote Sens. **15**(2), 320 (2023)
15. Zhong, L., Jinwoo, K.: A framework for studying hydrology-driven landslide hazards in northwestern US using satellite InSAR, precipitation and soil moisture observations: early results and future directions. GeoHazards **2**(2), 17–40 (2021)
16. Sdao, F., Simeone, V.: Mass movements affecting Goddess Mefitis sanctuary in Rossano di Vaglio (Basilicata, southern Italy). J. Cult. Herit. **8**, 77–80 (2007)

An Innovative Holistic GIS-BIM and Artificial Intelligence Based Approach to Manage Mechanized Tunnelling: The Back-Analysis of the Budapest Metro Line4

Luca Paolella[1(✉)], Maciej Ochmanski[2], and Giuseppe Modoni[1]

[1] University of Cassino and Southern Latium, Cassino, Italy
luca.paolella@unicas.it
[2] Silesian University of Technology, Gliwice, Poland

Abstract. The current and continuously increasing demand for urban mobility implies introducing new sustainable and alternative systems to road transport. Where economic viability is established, metro lines are one of the most effective and least impactful solutions if the characteristics of the subsoil are appropriately considered and the construction phases are planned in such a way as to limit the induced ground deformations and not compromise the existing building stock. The excavation of tunnels in loose soils inevitably causes movements in the topsoil resulting in a combination of sagging and hogging, which in an urban environment must be controlled and minimized to avoid damage to the existing structures and infrastructure. Through the back-analysis of the Budapest (Hungary) Metro Line4, in this work, we propose an innovative tool where the design process is based on a GIS-BIM interaction, and the executive phase takes advantage of artificial neural networks capable of adjusting the design choices to the monitoring evidence. The environmental and geotechnical aspects are managed through the GIS Platform; then, 3D subsoil and structural models are developed following the BIM approach. After, the artificial neural network's architecture is first constructed via a trial-and-error process which leads to selecting the best combination of input variables that better correlate to the measured volume loss. Then, real-time analysis is performed, and the transient effect is considered to simulate the excavation advance. The obtained results denote significant effectiveness in predicting the ground deformation and, thus, damage induced at the surface by mechanized excavation.

Keywords: Tunneling · Artificial Neural Networks · GIS-BIM interaction

1 Introduction

The continuous development of built-up areas leads to an increasing demand for efficient and sustainable mobility, which can only be satisfied by exploiting underground space for public transportation systems. However, due to the interactions with the surrounding

A. Ferrari et al. (Eds.): CNRIG 2023, SSGG, pp. 251–258, 2023.
https://doi.org/10.1007/978-3-031-34761-0_31

built environment, excavating tunnels in urban areas implies an articulated, multidisciplinary, and iterative process that unavoidably requires adopting shared operational strategies to execute the works minimizing impacts on existing buildings and infrastructure. For a general design case, firstly, it is necessary to forecast the extent of the zone of deformation; secondly, the magnitude of the induced movement in that zone; and, finally, to assess any beneficial or detrimental modifying effects resulting from the interaction with existing structures and services and minimize the risk of catastrophic loss of ground. All the above requires that the construction method and the whole execution phases must be compatible with the nature of the subsoil and the groundwater conditions and the admissible damage to nearby structures. Hence, a monitoring campaign should always be planned and carried out during and after excavation. Regarding the construction method and implications, the limited effect on the surrounding environment makes mechanized tunneling one of the most efficient methods in urban areas. Because of economic criteria, for tunnels with a length of the order of some kilometers, mechanized excavation with Tunnel Boring Machine (TBM) is often employed, being configured as a self-propelled construction site, and offering a significant speed of advancement, a high ability to control the deformations induced on the surface and, consequently, better safety standards. A TBM consists of a circular milling head, a shell, and a backup of suitable dimensions containing the systems serving the machine (driver's cabin, system wiring, conveyor belt). In the case of Earth Pressure Balance (EPB) and Bentonite Slurry (BS) models, the tiller digs the soil, which is mixed with foams and kept under pressure to ensure the stability of the excavation face; it is then channeled towards the back of the mole and removed via a conveyor belt. Once the tunnel is advanced, the final lining of the tunnel is carried out through a mechanical arm that applies prefabricated reinforced concrete blocks on the excavation walls. Each segment, shaped like an arch, relates to the others by means of special plugs to form a ring representing the tunnel's lining. After the passage of the TBM, the tunnel is ready to accommodate the railway site and the technological systems. Through the back-analysis of the Budapest Metro Line 4, this work offers considerable food for thought regarding an automated and objective procedure implementing an "expert" observational method, i.e., supported by artificial intelligence and addressing the execution of a mechanized tunnel. By pooling geotechnical parameters, design choices, and monitoring evidence (managed in a GIS environment), a detailed parametric study was conducted to identify the most concise combination of input variables that better correlate with the observations without any loss of information. In a subsequent step, the network architecture showing the best performance was tested on a near real-time condition to simulate the development of induced ground deformations during the excavation process.

1.1 GIS-BIM Data Management in Mechanized Tunneling

Due to the amount of data to manage and the different scales of the project, the current operational strategy merges the components of a significant number of approaches, platforms, and tools that describe the whole process, dealing with an increased detail level. The preliminary phase requires studying the surficial and geological/geomorphological contexts at the global scale. The subsequent step consists in defining, with more detail, the geotechnical subsoil model and, consequently, the main features of structural elements.

At the preliminary stage, assumptions about the machine's operational parameters are also introduced, which will be controlled and adjusted step by step during the excavation. Considering the sequential process illustrated in Fig. 1, Geographic Information Systems (GIS) are advantageously employed to manage topographic and territorial aspects. While a GIS platform allows the individuation of simultaneous surficial interferences, showing interactions with existing networks and stations and restricted areas that govern the choice of the track, Building Information Modeling (BIM) enables practitioners to manage the project in its elementary constitutive aspects. From the geotechnical viewpoint, BIM systems provide graphical 3D support to define the subsoil model, coupled with all the attributes of physical and mechanical properties derived from specific interpretations. Borrowing the concept of digitally representing physical and functional characteristics of a facility, which defines a BIM, the breaking down into single and elementary constitutive components, can be carried out concerning the structural elements of tunnels to define the geometry, features, and functional properties of all the parts. All of the above make the BIM-based methodology currently employed to design, visualize, and report the functional characteristics of any transportation system, including the management of the different lifecycle aspects like the safety standards. From the executive viewpoint, the excavation activities are always combined with articulated topographic monitoring, which allows for validating the previously designed assumptions or adjusting them based on the evidence. The iterative process summarized above is worldwide known as the observational method (Peck, 1969), representing a milestone in the tunneling and geotechnical engineering world, when explicitly dealing with complex geotechnical contexts and high uncertainty (Ribacchi, 1993).

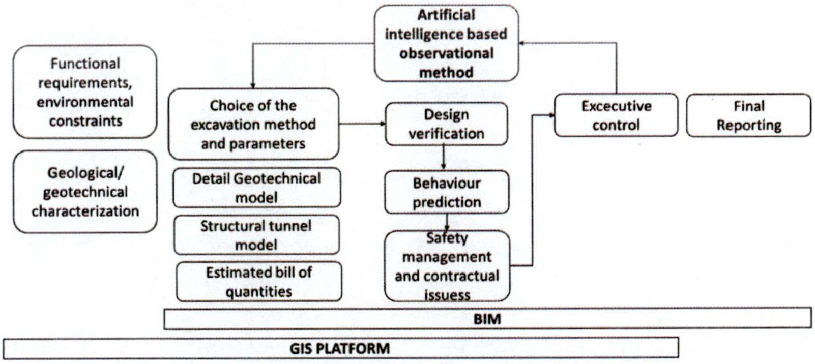

Fig. 1. Schematic representation of the observational method (modified from Ribacchi, 1993).

1.2 Prediction of Ground Deformation with Artificial Neural Networks "ANNs"

Independently of the selected method, the challenge in urban tunneling is to control ground deformations during and after the excavation, preventing damage to the surface manufacts. A first approach to estimate the development of settlements was proposed by Peck (1969), which decoupled the 3D ground deformation into a transversal and longitudinal component, respectively evaluated with the distance from the tunnel axis and the excavation front and modeled via gaussian density and cumulative distributions. After reproducing the most typical scheme of tunnel collapse in centrifuge tests, Kimura and Mair (1981) characterized the observed failure mechanisms with the plasticity theory and proposed an expeditious way to estimate the volume loss, i.e., the ratio between the integral of the subsidence basin and the tunnel cross-section. In recent years novel semi-empirical methods supported by artificial intelligence have become very popular among practitioners. Among these, multilayer perceptron (MLP), Support Vector Machines SVM, and Artificial Neural Networks ANNs have been used to analyze settlements caused by tunnels excavated by shielded machines and sequential excavation methods (Kim et al., 2001). Other works (Ribacchi, 1993; Paolella et al., 2022) pointed out that geotechnical design is significantly affected by varying rock, soil, and underground water conditions and accounted for many variables that were often complex, vague, and uncertain. In fact, due to its capacity to model nonlinear relationships between the involved input and output variables, ANN-based approaches are advantageously and expeditiously applied to predict tunneling-induced ground deformations. An artificial neural network (ANN) can extract a generalized correlation (or regularity) from many individual samples (or experiences) through a learning phase and subsequent validation. The model can also produce appropriate output for new and previously unknown datasets if trained on a significant number of input-output pairs (or pairs of input and target vectors). More specifically, each neuron receives signals from input variables or neurons at earlier stages, combining them linearly in a weighted manner and transmitting them to subsequent layers. Generally, an increase in the number of neurons in the hidden layer means an increase in overall performance, to the detriment of the network's ability to predict outputs for new input datasets. This storage phenomenon, known as overfitting, can be avoided by previously characterizing the root mean square error (RMSE) (Eq. 1) and then choosing as optimal structure the one indicated by the lowest RMSE consistent with the convergence of training and validation trends.

$$RMSE = \sqrt{\frac{1}{N}\sum\nolimits_{n}(Predicted - Observed)^{2}} \tag{1}$$

where N is the total number of observations. In conjunction with the development of high-performance and fast computing tools, several engineering applications have confirmed the efficiency and practicality of neural networks in automatically acquiring correlations between input and output variables. Combining stratigraphic characteristics and seismic action, Paolella et al. (2019) present an ANN-based method capable of estimating the probability of observing ground liquefaction, while Cao et al. (2020) applied ANNs in mechanized tunneling to validate the results of numerical analyses. One of the most popular frameworks exploits the backpropagation learning algorithm based on the generalized delta rule of Rumelhart et al. (1986). The ability to approach markedly non-linear problems, the relatively easy implementation, the robustness in performance, and the flexibility with which connection weights between one neuronal level and the next are corrected, make the algorithm very popular in tunneling (Tang et al., 2018). Other error metrics used to rank the performance of a given network are briefly recalled in the following equations, respectively showing the normalized "ERR_{NORM}" and mean averaged "MAE" errors.

$$ERR_{NORM} = \frac{1}{N}\sum_{i=1}^{N} (Predicted_i - Observed_i)^2/Observed_i^2 \quad (2)$$

$$MAE = \frac{1}{N}\sum_{i=1}^{N} (|Predicted_i - Observed_i|) \quad (3)$$

$$ERR_{MAX}(\%) = (\max(Predicted) - \max(Observed))/(\max(Observed)) * 100 \quad (4)$$

2 The Case Study of Budapest Metro Line 4

Built between 2007–2010, Line 4 of the Budapest Metro (Hungary) extends in a south-west-north-east direction between the Kelenfoldi and Baross stations, covering a length of 7.4 km, connecting the southern part to the city center and including a total of 10 stations. The two tubes forming Line 4 were excavated using an Earth Pressure Balanced TBM having a diameter of 6 m and an overall length of 115 m. During the construction phase, vertical displacements were monitored on more than 500 buildings (Fig. 2a), whose edges were instrumented with around 10,000 rods. Figure 2b shows an example of GIS interpolation, obtained via geostatistical kriging of measured vertical displacements (yellow points of Fig. 2a) at the end of the excavation.

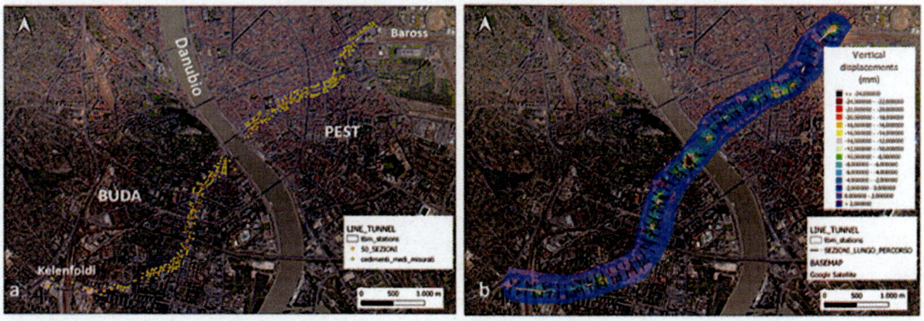

Fig. 2. a) Location of the Budapest Metro4 line and monitored buildings centroids; b) map of the vertical displacements at the end of the excavation: black lines correspond to the 54 cross-sections.

In a preliminary analysis, the track was discretized into 54 middle sections of approximately 125 m, excluding the portion below the Danube River, where no monitoring data is available (Fig. 2b). From the geological viewpoint, three different contexts are crossed. Starting from the Buda side, we observe the predominance of Oligocene clays, alternating with clayey marls and marls covered by young, loose sediments from the Pleistocene and Holocene. Below the Danube, the tunnel cuts through the strongly tectonic Triassic dolomite and clayey marl-marls. Finally, in the Pest area, the Oligocene clay formations are covered by alternating sands and loose sediments thicker than in the Buda area. The scrutiny of around 200 available boreholes, in-situ tests, laboratory tests and geological reports allowed to identify homogenous subsoil layers. For each section, numbered with its "ID", the following properties were defined: the groundwater table (GWT), the thickness of the tunnel cover (C), and the depth of the tunnel axis (Z_0). The total (σ_v) and effective (σ'_v) tensional state at the tunnel axis depth and the averaged (i.e., weighted over the thicknesses) of the undrained cohesion (c_u) and elastic modulus (E) of the soils layers above the excavation were also evaluated. Additionally, the shield diameter (D), which is usually related to the overburden thickness, and the design face pressure (FP) at each section were considered. Finally, following the procedure of Kimura & Mair (1981), the load factor (LF) and the Volume Loss (VL) were estimated. This information was cross-referenced with the monitoring data processing (Fig. 2b). Using applications available in the GIS environment, the cross-sectional profiles for the 54 previously identified sections were first exported. Then, employing algorithms developed ad-hoc in the Python programming code, an automated adaptation of the measured profiles was performed with the theoretical counterparts defined by Peck (1969) as a function of the maximum settlement and the deflection distance ix. The subsequent parametric study consists of 2 phases. The optimal Neural Network architecture is first constructed using a mean square error minimization criterion "RMSE"; the number and size of the hidden layers of the network that guarantee the highest performance is identified, avoiding overfitting phenomena while optimizing the computational burden. Given the amount of data available, which is smaller than to justify the implementation of more sophisticated deep learning models, all architectures are characterized by a single hidden layer. However, parametric analyses agree on an optimal size of 10 neurons for this intermediate layer. In the next step, the prediction efficiency of 21 network structures with gradually decreasing numbers of input variables was evaluated by calculating the normalized error (Eq. 2) obtained over 200 random training and validation steps, respectively, based on 70 and 30% of the data. The basic idea is to consider all stored variables initially and then progressively reduce them by assessing the loss (or redundancy) of information each time. Finally, the correlation between the analyzed networks and the surficial induced deformation, quantified by the volume loss, is schematically shown in Table 1, which reports the top 5 networks and ranks the 11 as the optimal structure. It includes four input variables incorporating geometric information (C/D), geo-mechanical information (averaged c_u and E), and execution aspects (FP).

Table 1. Top 5 performance of nets after parametric analysis.

ID	INPUT VARIABLES	ERR_{NORM}	RANKING
11	c_u, E, C/D, FP	0.0157	1
17	E, c_u, C/D	0.0159	2
8	σ'_v, c_u, E, C/D, FP	0.0159	3
12	σ'_v, c_u, E, FP	0.0160	4
2	GWT, Z_0, γ, σ'_v, c_u, E, N, C/D, FP, LF, VL	0.0167	5

Then, the best-performer network structure was tested on the sector (of length ≈0.9 km) between the Kalvin and Racockzi stations. This track was divided into 90 sections equidistant 10 m from each other. During the excavation, which began in November 2009 and continued until March 2010, the positions of the left and right fronts (T_{dx}, T_{sx}) measured at 12-time steps were known (Fig. 4a). Therefore, in addition to the four network input variables, the relative distances between the left and right tunnel faces and the generic section were introduced (Fig. 3.). The first 2-time steps (t_1, t_2) were used to initialize the network and train it to predict the corresponding volume loss. From step 3, the volume losses measured until step t_{i-1} were considered an additional input reference. The implemented network is schematized in Fig. 3. Figure 4b compares the observed and predicted volume loss at stage 7, while the overall performance is reported in Fig. 4c, where the maximum error defined in Eq. 4 is plotted. It reveals a good agreement between the prediction and observation, especially considering that the maximum errors are affected by other works and excavation activities carried out near the stations and measured by the monitoring system, even if unrelated to the TBM.

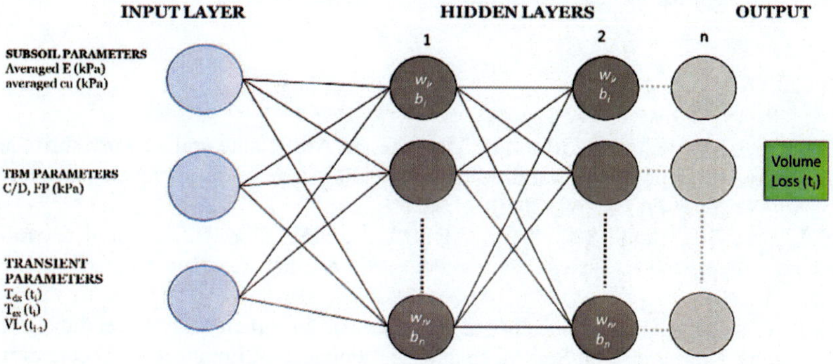

Fig. 3. Scheme of the artificial neural network for transient application.

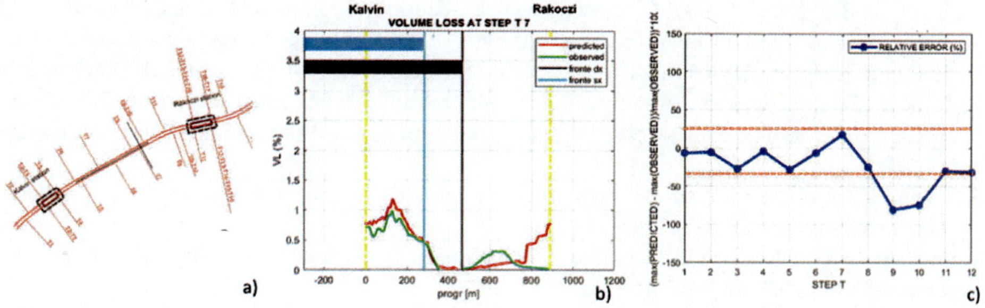

Fig. 4. a) Advance in the excavation between Kalvin and Racockzi stations (the black line refers to shield position at time step 7); b) ANN performance at stage 7; c) overall performance.

3 Results and Future Remarks

Without claiming to be complete on the GIS-BIM interaction that allows for the integrated management of the design, execution, and control phases of tunnels and that binds those involved to use a common language, this work focuses more explicitly on specific execution and monitoring aspects. The contribution of artificial neural networks in predicting the surface deformation induced by mechanized tunnel excavation in an urban environment is tested on the Budapest Metro4 Line. The parametric analyses revealed that the number of input variables of the network could be suitably reduced, containing the computational effort, without affecting the prediction performance. This outcome is due to information redundancy among variables, representing a generalized obstacle to learning neural networks, especially where the starting dataset is limited. After identifying the most relevant properties, a time-related variable is introduced to simulate a real-time application of the observational method. The maximum error confirms the method's robustness in predicting the volume loss, which ranges from ±30%. Once validated in future applications, this framework can provide expert yet objective and technological support to the traditional observational method.

References

Cao, B.T., Obel, M., Freitag, S., Mark, P., Meschke, S.: Artificial neural network surrogate modelling for real-time predictions and control of building damage during mechanised tunnelling. Adv. Eng. Softw. **149**, 102869 (2020)

Kim, C.J., Bae, G.J., Hong, S.W., Park, C.H., Moon, H.K., Shin, H.S.: Neural network based prediction of ground surface settlements due to tunneling. Comput. Geotech. **28**, 517–547 (2001)

Kimura T., Mair R.J.: Centrifugal testing of model tunnels in soft clay. In: Proceedings of the 10th International Conference on Soil Mechanics and Foundation Engineering, Stockholm, vol. 1, pp. 319–322 (1981)

Paolella, L., Salvatore, E., Spacagna, R.L., Modoni, G., Ochmanski, M.: Prediction of liquefaction damage with artificial neural networks. In: Proceedings of 7ICEGE (2019)

Paolella, L., Baris, A., Modoni, G., Spacagna, R.L., Fabozzi, S.: Liquefaction damage assessment using Bayesian belief networks. In: Gottardi, G., Tonni, L. (eds.) Cone Penetration Testing 2022 (2022). www.taylorfrancis.com. ISBN 978-1-032-31259-0, CC BY-NC-ND 4.0 license

Peck, R.B.: Deep excavation and tunneling in soft ground. State of the art report. In: 7th International Conference on Soil Mechanics and Foundation Engineering, Mexico City, pp. 225–290 (1969)

Ribacchi, R.: Recenti orientamenti nella progettazione statica delle gallerie. In: XVIII Convegno Nazionale di Geotecnica, Rimini (1993)

Rumelhart, D.E., Hinton, G.E., Williams, R.J.: Learning representations by back-propagating errors. Nature **323**, 533–536 (1986)

Tang, Y., Xiao, S., Zhan, Y.: Predicting settlement along railway due to excavation using empirical method and neural networks. Soils Found. **59**, 1037–1051 (2018)

Liquefaction-Induced Downdrag on Tapered Piles from Full-Scale Blast Liquefaction Tests

Kyle Rollins[1], Sara Amoroso[2,3](✉), Vincenzo Colella[4], Luca Minarelli[3], and Decker Ure[1]

[1] Brigham Young University, Provo, UT 84602, USA
[2] Univ. of Chieti-Pescara, 65129 Pescara, Italy
sara.amoroso@unich.it
[3] Istituto Nazionale di Geofisica e Vulcanologia, 67100 L'Aquila, Italy
[4] Geofondazioni Ingegneria e Lavori Srl, 30030 Martellago, Italy

Abstract. Frequently, deep foundations extend through potentially liquefiable sand layers near the ground surface and bear on more competent layers at depth. When liquefaction occurs, the skin friction in the liquefied layer would be expected to decrease to some negligible value, but as the liquefiable layer settles, negative skin friction could potentially develop around the pile in this layer as effective stress increases. To investigate the loss of skin friction and the development of negative skin friction, axial load tests were performed on an instrumented full-scale tapered pile before and after blast-induced liquefaction at a site in Mirabello (Ferrara, Italy) that was affected by liquefaction following the 2012 Emilia earthquakes. The test pile was a 16.5 m long concrete pile with a diameter of 0.52 m at the head tapering to 0.26 m at the toe. Following blasting, liquefaction developed within a 6-m thick sand layer below a clay surface layer resulting in significant settlement. Skin friction in the liquefied layer initially dropped to essentially zero. However, as the liquefied sand reconsolidated, negative skin friction became equal to about 50% of the pre-blast ultimate positive skin friction. Negative skin friction in the overlying non-liquefied clay layer was only 80% of the ultimate positive skin friction. This is likely due to the surrounding soil moving slightly away from the tapered pile as the soil settled vertically downward. Despite significant ground settlement, pile settlement was relatively small because of the resistance provided by the toe of the pile.

Keywords: Blast test · tapered piles · liquefaction

1 Introduction

Frequently, deep foundations extend through potentially liquefiable loose to medium dense sand layers and bear on more competent layers at depth as shown in Fig. 1(a). Prior to liquefaction, the applied pile head load, P, is transferred to the soil through upward (positive) skin friction, Q_s and the load in the pile decreases as shown in Fig. 1(b). The load at the base of the pile is carried by end-bearing resistance, Q_b, which requires some settlement to develop as illustrated by the toe resistance vs. settlement (Q_b-z)

A. Ferrari et al. (Eds.): CNRIG 2023, SSGG, pp. 259–266, 2023.
https://doi.org/10.1007/978-3-031-34761-0_32

curve in Fig. 1(d). When liquefaction occurs, skin friction in the liquefied layers is expected to decrease to near zero and many design procedures use this value to evaluate the consequences of skin friction loss and pile settlement as shown in Fig. 1(b). The reduction in positive skin friction in the liquefied layers leads to an increase in load at the toe of the pile and mobilization of the increased end-bearing resistance leads to additional pile settlement as shown in Fig. 1(c and d). As the earthquake-induced pore pressures dissipate in the liquefiable layer and settlement occurs, downward (negative) skin friction develops at the pile-soil interface in the clay layer above the liquefied layer, increasing the load in the pile (see Fig. 1(b)). In addition, the skin friction at the pile-soil interface in the liquefied layer is likely to increase as the excess pore pressure decreases. Therefore, the negative skin friction that ultimately develops in the liquefied layers will likely be higher than zero and will induce even greater load in the pile. The increased negative skin friction in the liquefied and non-liquefied layers further increases the required end-bearing resistance (see Fig. 1(b)) and leads to additional pile settlement as shown in Fig. 1(d).

The neutral plane, shown in Fig. 1(b, c), represents the point where pile settlement and soil settlement are equal. Negative friction develops above the neutral plane and positive friction develops below the neutral plane. As a result, the maximum load in the pile occurs at the neutral plane. The location of the neutral plane is normally obtained by trial and error so that the pile head load, P, plus the negative skin friction above the neutral plane is equal to the positive skin friction and end-bearing resistance, Q_b below the neutral plane.

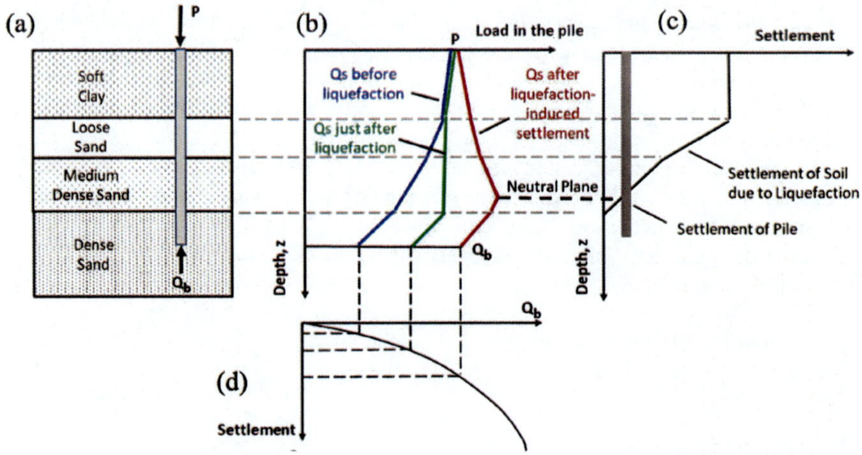

Fig. 1. (a) Soil profile around pile, (b) load in the pile vs. depth, (c) settlement vs. depth, and (d) end-bearing resistance (Q_b) vs. settlement before and after liquefaction-induced settlement.

In the absence of test results, some investigators have used theoretical concepts to predict the behavior of piles when subjected to liquefaction induced drag loads. Boulanger and Brandenberg [1] defined negative skin friction in the liquefied zone in terms of the effective stress during reconsolidation, but concluded that the negative skin friction could be assumed to be zero with little error in the computed pile force or settlement. Fellenius and Siegel [2] applied the Unified Design of Piles approach that was developed for downdrag in clays, to the problem of downdrag in liquefied sand, once again assuming that negative skin friction in the liquefied zone would be zero. They also conclude that liquefaction above the neutral plane would not increase the load in the pile based on the concept that negative friction would already be present prior to liquefaction.

To understand better the development of negative skin friction on piles in liquefied sand and the resulting pile response, several full-scale pile downdrag tests have been performed using blast-induced liquefaction. Blast-induced liquefaction was first used to investigate the lateral resistance of piles in liquefied sands ([3] and [4]) and has become widely used to investigate a number of ground improvement strategies [5–7]. A summary of the blast-induced pile downdrag tests is provided in Table 1.

Table 1. Summary of blast-induced liquefaction pile downdrag tests.

Site location	Pile Type	Soil Profile	Ref
Vancouver, Canada	Driven Steel Pipe: 32.4 cm diameter, 21 m long	6 m of cohesive soil over loose clean sand ($D_r = 40\%$)	[8, 9]
Christchurch, New Zealand	Three Augercast Piles: 61 cm diameter; 8.5 m, 12 m, and 14 m long	1.5 m of cohesive soil over medium dense silty sand ($D_r = 60\%$)	[9, 10]
Mirabello, Italy	Bored Micropiles: 25 cm diameter, 15 m long	6 m of cohesive soil over 3 m of sandy silt and 18 m of silty sand	[11]
Turrell, Arkansas, USA	Three Driven Piles H pile: (H14 × 117), 28 m long Pipe pile: 46 cm diam., 24 m long PSC pile: 46 cm square, 22.5 m long Three Bored Piles 1.22 m diam.27.6 m long 1.82 m dia., 21.3 m long 1.22 m dia. 28 m long	9 m of cohesive soil over silty sand and sandy silt	[12, 13]

Typically, these blast liquefaction tests have shown that the negative skin friction developed following reconsolidation of the liquefied sand layers is about 50% of the ultimate positive skin friction in these layers. However, the downward skin friction in non-liquefied layers is approximately the same as the ultimate positive skin friction. Very little relative displacement (2 to 5 mm) is required to fully mobilize negative friction. As indicated in Table 1, all previous tests have been performed on cylindrical piles with a constant pile diameter versus depth. However, for tapered piles, where the pile diameter decreases with depth, the soil might settle away from the pile as the soil reconsolidates following liquefaction as illustrated in Fig. 2. This could lead to a reduction in the negative skin friction induced on the pile and might be a reasonable approach for mitigating downdrag effects.

To investigate the effect of a tapered pile on the downdrag behavior following liquefaction, a full-scale test was conducted on a 17 m long tapered pile. Liquefaction was induced using controlled blasting while a pile head load was being applied by a hydraulic jack. This paper summarizes the test procedure and the basic results that were obtained.

(a)Before liquefaction (b) After liquefaction-induced settlement

Fig. 2. Illustration of surrounding soil settling away from tapered pile during post-liquefaction settlement and reducing negative skin friction: (a) before and (b) after liquefaction.

2 Soil Profile and Pile Properties

A generalized soil profile for the test site is provided in Fig. 3(a). The profile generally consists of a clayey silt to a depth of about 6 m that is underlain by loose silty sand. A single tapered pile was driven to a depth of 16.5 m with a hydraulic hammer along with two cylindrical reaction piles located about 3 m on either side of the test pile. The test pile had a diameter of 0.51 m at the pile head and 0.26 m at the toe, while the cylindrical reaction piles had a diameter of 0.5 m. The piles were made of reinforced concrete with a hollow section that allowed the insertion of a steel reinforcing bar instrumented with vibrating wire strain gauges at approximately 1.5 m intervals along the length of the pile. Following insertion, the hollow interior was grouted. After grouting, the test pile was monitored for about a month. No significant change of load in the pile was measured during this period, although strain did vary owing to temperature changes near the top of the pile.

3 Static Load Testing

About a month after grouting the pile, a steel reaction frame was attached to the two reaction piles and a hydraulic jack pushed against the frame to measure the applied pile head load vs. deflection curve for the test pile. Load was applied in approximately 240 kN intervals to a maximum load of about 2700 kN where the capacity of the reaction frame was reached. The load in the pile vs. depth is plotted for the maximum load increment and is plotted in Fig. 3 (green line). Side resistance accounted for about 2275 kN or 84% of the total pile capacity while extrapolation of the load vs. depth curve to the pile toe indicates an end-bearing resistance of about 450 kN at the maximum load.

Figure 4 provides a plot of the pile head load-deflection curve obtained from the static load test along with the ultimate load interpreted using the Davisson method. Because the pile capacity exceeded the capacity of the frame, some extrapolation was required to obtain an ultimate static pile resistance of about 2800 kN.

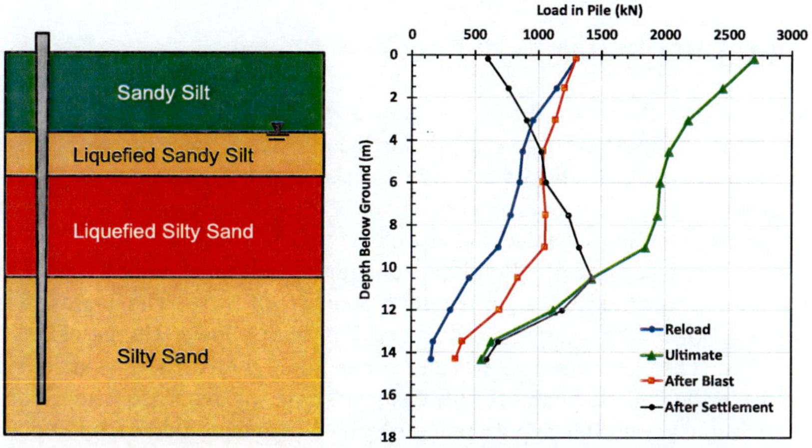

Fig. 3. Idealized soil profile at test site (left side) along with load versus depth curves for the test pile for different conditions before and after blast-induced liquefaction (right side).

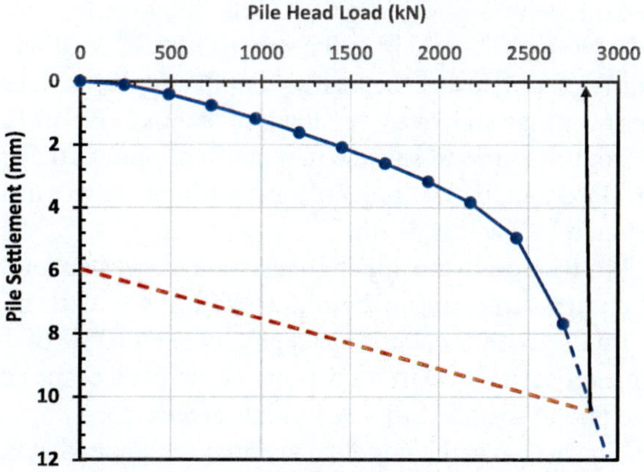

Fig. 4. Measured pile head load-deflection curve from static load test on the tapered test pile.

Static pile resistance was also calculated using a nearby CPT sounding with a recently developed unified CPT approach proposed by Lehane et al. [14]. This method is an attempt to produce a single consensus design equation based on best-practices from four CPT approaches specified by the API code [15]. This unified approach, which is based entirely on load tests involving cylindrical piles, predicts an ultimate pile resistance of only about 1620 kN, which is far below the measured pile resistance. However, research has indicated that tapered piles can produce significantly more skin friction than cylindrical piles [16]. Italian practice typically increases skin friction by a "conicity" correction factor of 1.5 for tapered piles to account for this increased resistance. In this case, a correction factor 2.2 would be needed to match the higher measured capacity for this tapered pile with a 1.5 cm/m taper.

4 Controlled Blasting Downdrag Test

Following the static load test, the pile head load was released and just prior to the blast test a static load of 1300 kN was re-applied to the pile head. This represents a load with a factor of safety of 2.15. The load vs. depth curve after re-application of the 1300 kN load is shown in Fig. 3 (blue curve) and curve is in good agreement with the load vs. depth curve from the initial static load test.

The blast sequence consisted of the detonation of two charges within each of eight blast holes distributed around a circle with a diameter of 14 m. The explosive in each hole consisted of 0.5 kg charge at a depth of 7 m with a lower charge of 2.0 kg at a depth of 11 m. The charges were detonated sequentially with a 1000 ms delay between detonations. Following detonation of the charges, strain was recorded with a high-speed data acquisition system. Pile and ground settlement were measured with a conventional auto-level and with drone structure for motion technology.

Shortly after the completion of the blasting sequence, the load vs. depth curve from about 4 m to 9.5 m became nearly vertical indicating liquefaction with skin friction close to zero (Fig. 3 red line). This reduction in skin friction led to increased mobilization of end-bearing resistance and pile settlement. Pile head settlement gradually reduced the load applied by the hydraulic jack at the pile head load to about 600 kN when soil settlement was nearly complete and the load vs. depth curve reached equilibrium (Fig. 3, black line). At this point, as shown by the load vs. depth curve in Fig. 3, negative skin friction extended from the ground surface to a depth of about 10.5 m, including within the liquefied zone. Based on load in the pile, the neutral plane was at 10.5 m and positive skin friction developed below this depth.

The negative skin friction in the upper cohesive layer was about 80% of the ultimate positive skin friction measured during the static load test while the negative skin friction in cohesive layers for the cylindrical pile has typically been 100% of the ultimate positive friction [9–11]. This reduction in skin friction may be a result of the cohesive soil settling away from the pile due to the pile taper as hypothesized.

The negative skin friction in the liquefied layer varies somewhat versus depth. For the zones within the liquefied layer from 4 to 6 m and 7.5 to 10.5 m the average negative skin friction is 53% of the ultimate positive skin friction. This value (53%) is consistent with several previous blast liquefaction tests with cylindrical piles where the average negative skin friction was 40 to 60% of the ultimate positive skin friction [9–11]. However, from 6 to 7.5 m, the negative skin friction was higher than expected so that the average negative skin friction for the entire liquefied layer was about 67% of the ultimate positive skin friction. These results indicate that the tapered pile was not successful in reducing the negative skin friction in the liquefied layer at the pile-soil interface relative to cylindrical piles. One explanation for this behavior is that the liquefied soil could simply flow towards the pile while in a liquefied state and then exert a similar negative friction to that observed with a cylindrical pile during reconsolidation. In contrast, the undrained shear strength of the surface layer could maintain a small gap at the soil-pile interface and reduce the negative skin friction.

The positive skin friction below 10.5 m was nearly the same as that for the maximum load in the static load test which is consistent with observations from previous pile downdrag tests [9–11]. The negative skin friction that developed following liquefaction and subsequent reconsolidation significantly increased the load in the pile above the applied load. In fact, the load at the toe of the pile was about the same as that which developed during the static load test. After reconsolidation, the soil around the single pile had settled approximately 2.7 cm while the pile head settlement was only 1.07 cm. Therefore, the pile was successful in reducing the pile head settlement relative to that of the surrounding ground despite liquefaction.

5 Conclusions

Based on the results from the static and blast-induced liquefaction tests, the following conclusions have been made:

1. The measured ultimate pile resistance from the static load test on the tapered concrete pile was significantly higher than would be expected using recent CPT-based design methods based on load tests on cylindrical piles. In this case, the computed skin friction would need to be multiplied by a factor of 2.2 to produce agreement with the measured ultimate pile resistance, which is higher than the 1.5 factor typically employed in Italian practice. Increased skin friction on tapered piles has been reported by a number of previous researchers.
2. Following blasting, the skin friction in the liquefied zone was essentially zero; however, as excess pore pressures dissipated and the sand settled round the test pile, negative skin friction developed in both the overlying cohesive soil and the liquefied zone.
3. In the cohesive surface layer, the negative skin friction was about 80% of the ultimate positive friction during the static load test. This result suggests that the cohesive soil may be settling away from the pile thereby reducing the negative skin friction, while the negative skin friction of a cylindrical pile is 100% of the ultimate positive friction.
4. In the liquefied zone following pore pressure dissipation are reconsolidation, the average negative skin friction was 53% of the ultimate positive skin friction during the static load test for most of the zone. This result is consistent with previous tests, performed on cylindrical piles, where the negative skin friction following liquefaction and reconsolidation was typically 40 to 60% of the ultimate positive skin friction before liquefaction at the end of reconsolidation. This result suggests that a tapered pile does not reduce negative friction in liquefied sand, presumably because the liquefied sand tends to flow towards the pile during liquefaction and reconsolidation.
5. In the non-liquefied sand below the liquefied zone, the positive skin friction was approximately the same as the ultimate skin friction measured in the static load test.
6. Although liquefaction induced settlement in the sand around the pile of 2.7 cm, the pile head settlement was only 1.07 cm.

References

1. Boulanger, R.W., Brandenberg, S.J.: Neutral plane solution for liquefaction-induced downdrag on vertical piles. In: Geotech Engineering for Transportation Projects, pp. 470–478 (2004)
2. Fellenius, B.H., Siegel, T.C.: Pile drag load and downdrag in a liquefaction event. J. Geotech. Geoenviron. **134**(9), 1412–1416 (2008)
3. Weaver, T.J., Ashford, S.A., Rollins, K.M.: Response of 0.6 m cast-in-steel-shell pile in liquefied soil under lateral loading. J. Geotech. Geoenviron. **131**(1), 94–102 (2005)
4. Rollins, K.M., Gerber, T.M., Lane, J.D., Ashford, S.A.: Lateral resistance of a full-scale pile group in liquefied sand. J. Geotech. Geoenviron. **131**(1), 115–125 (2005)
5. Wentz, F., van Ballegooy, S., Rollins, K.M., Ashford, S.A., Olsen, M.: Large scale testing of shallow ground improvements using blast-induced liquefaction. In: Proceedings of the 6th International Conference on Earthquake Geotechnical Engineering. New Zealand Geotechnical Society (2015)
6. Ashford, S.A., Rollins, K.M., Lane, J.D.: Blast-induced liquefaction for full-scale foundation testing. J. Geotech. Geoenviron. **130**(8), 798–806 (2004)
7. Gallagher, P.M., Conlee, C.T., Rollins, K.M.: Full-scale field testing of colloidal silica grouting for mitigation of liquefaction risk. J. Geotech. Geoenviron. **133**(2), 186–196 (2007)
8. Rollins, K.M., Strand, S.: Downdrag forces due to liquefaction surrounding a pile. In: Proceedings of the 8th US National Conference on Earthquake Engineering (2006)
9. Rollins, K.M., Strand, S.R., Hollenbaugh, J.E.: Liquefaction induced downdrag and dragload from full-scale tests. In: Iai, S. (ed.) Developments in earthquake geotechnics. Geotechnical, Geological and Earthquake Engineering, vol. 43, pp. 89–109. Springer, Cham (2018). https://doi.org/10.1007/978-3-319-62069-5_5
10. Rollins, K.M., Hollenbaugh, J.: Liquefaction induced negative skin friction from blast-induced liquefaction tests with auger-cast piles. In: Proceedings of the 6th International Conference on Earthquake Geotechnical Engineering. New Zealand Geotechnical Society (2015)
11. Amoroso, S., Rollins, K.M., Lusvardi, C., Monaco, P., Milana, G.: Blast-induced liquefaction results at the silty-sand site of Mirabello, Emilia Romagna region, Italy. In: Geotechnical Earthquake Engineering and Soil Dynamics V, ASCE, p. 10 (2018)
12. Kevan, L., Rollins, K.M., Coffmann, R., Ishimwe, E.: Full-scale blast liquefaction testing in Arkansas USA to evaluate pile downdrag and neutral plane concepts. In: Silvestri, Moraci (eds.) Earthquake Geotechnical Engineering for Protection and Development of Environment and Constructions, pp. 648–655. Associazione Geotecnica Italiana, Rome, Italy (2019). ISBN 978-0-367-14328-2
13. Ishimwe, E., Coffman, R.A., Rollins, K.M.: Analysis of post-liquefaction axial capacities of driven pile and drilled shaft foundations. In: Proceedings of the IFCEE, pp. 272–283 (2018)
14. Lehane, B.M., Bittar, E., Lacasse, S., Liu, Z., Nadim, F.: New CPT methods for evaluation of the axial capacity of driven piles. In: Proceedings of the 5th International Conference on Cone Penetration Testing: (CPT 2022), pp. 3–15. CRC Press (2022)
15. API. ANSI/API RP 2GEO: Geotechnical and Foundation Design Considerations. ISO 19901-4:2003 (Modified), Petroleum and natural gas industries-Specific requirements for offshore structures, Part 4-Geotechnical and foundation design considerations. 1st edn. API Publishing Services, Washington, DC (2011)
16. Nordlund, R.L.: Point bearing and shaft friction of piles in sand. Presented at the 5th Annual Short Course on Fundamentals of Deep Foundations Design, University of Missouri-Rolla (2011)

The Chiaia Station of the Napoli Underground: Observations

Gianpiero Russo[1](✉), Luigi D'Esposito[2], Marco Valerio Nicotera[1], and Ilaria Esposito[1]

[1] DICEA, University of Napoli Federico II, Naples, Italy
pierusso@unina.it
[2] Tecnè spa, Genoa, Italy

Abstract. The focus of this paper is on the construction process of *Chiaia* Station belonging to the final stretch of Line 6 of Napoli's underground network. The station is located in the old town district of *Chiaia*, surrounded by historical buildings and not far from the sea. The main box of the station is constructed within a 32.2 m × 39.5 m rectangular in plan excavation that reaches the maximum depth of 48 m below ground surface. The main shaft is supported in the upper portion by contiguous bored pile walls crossing the top pyroclastic sand layer and embedded in the Neapolitan Yellow Tuff (NYT) formation. The overall geometry of the excavation and many details of the site are such that a complex 3D modelling option is indeed required. A wide range of instruments has been used to supervise the safety of surrounding buildings and to measure ground and groundwater movements: optical survey points, consisting in marks on sidewalks, standard open pipe and Casagrande piezometers, inclinometers and extenso-inclinometers, installed in both NYT layer and bored piles. During excavation several anthropic cavities have been found in the NYT layer, making the excavation process slightly more complicated and risky. In the paper the construction process and the monitoring data are first presented. At the end collected data are compared with some empirical methods.

Keywords: Deep excavation · diaphragm wall · monitoring

1 Introduction

With an urban population of 1.2 million people and more than 5 million in metropolitan area, Napoli is Italy's third more populated city. In 1997 the Municipality of Napoli approved a new City Transportation Plan, which provided for the construction of new underground train lines, stations and car parks [1]. Line 6 is aimed to connect the western area of *Bagnoli* to the city center at *Municipio* station, with a total length of 8 km and twelve stations. The line is already operating between *Mostra* and *Mergellina* stations, and *Chiaia* Station is intermediate along the stretch running from *Mergellina* to *Municipio* [2]. Technical issues made the works slow down in the years and the full line should be operational only at the end of next year.

A. Ferrari et al. (Eds.): CNRIG 2023, SSGG, pp. 267–274, 2023.
https://doi.org/10.1007/978-3-031-34761-0_33

Most of the construction of the station deep shaft was carried out several years ago and only works related to minor finishings are still ongoing. For various reasons the monitoring data could be published only recently.

This paper first briefly describes subsoil conditions and main construction steps. Subsequently some of the most important monitored and recorded data are plotted and discussed trying to outline both a qualitative and quantitative relationship with the construction stages.

2 Site and Subsoil Investigation

Chiaia station has been designed by the Neapolitan architect Uberto Siola and is located in the heart of *Chiaia* borough, a district in the city historical center not far from the multi-awarded *Toledo* Station [3].

The station shaft represents a vertical link between the square of *S. Maria degli Angeli* (at + 40 m *above sea level*) and the street level of the homonymous street (+25 m a.s.l.) from which is separated by a few buildings. The two main levels of the station are connected by a helical ramp while the distinctive feature of the station is represented by a cone of light 50 m deep (*Chiaia* is the deepest station of the Line 6) that will project the sunlight down to the tracks.

In Fig. 1 a plan view of the station site is reported where the main structural supports and the surrounding buildings are also represented. The main box of the station is constructed within a 32.2 m × 39.5 m rectangular excavation at a maximum depth of 48 m below ground level. On the side of the station main box is located the accession shaft - 14.75 m × 18.45 m size in plan 21 m depth - that allows the access to the main pit from the street level.

The subsoil investigations consisted of deep boreholes with soil sampling at different depths and SPT in the uncemented shallow materials. The soil layering at the site consists of three main layers: a) man made soil, b) pyroclastic sand [4], c) yellow tuff [5].

The top boundary of the yellow tuff is represented in Fig. 1 by means of contour lines whose labels correspond to the elevation above the sea level. SPTs showed an approximately linear increase, the Nspt ranging from 5 at 2 m to 25 at 16 m below ground surface. Uniaxial compressive tests were carried out on many tuff samples carefully obtained via a double rotary sampler. The measured uniaxial compressive strength ranged between 1 and 4 MPa showing a clear correlation with the unit weight of the material (15–17 MPa).

In the same plan view the layout of the internal and external monitoring tools is sketched.

3 Excavation Support and Main Construction Step

The top part of the main shaft excavation – embedded in the sandy material - was designed with the provisional support of peripheral diaphragm walls made by contiguous large diameter bored piles.The piles have diameters of 800 and 1200 mm and length ranging between 16 and 23 m. In order to limit settlements during the first stages of excavation the perimetral diaphragms are supported by 3 to 4 levels of pre-stressed anchors.

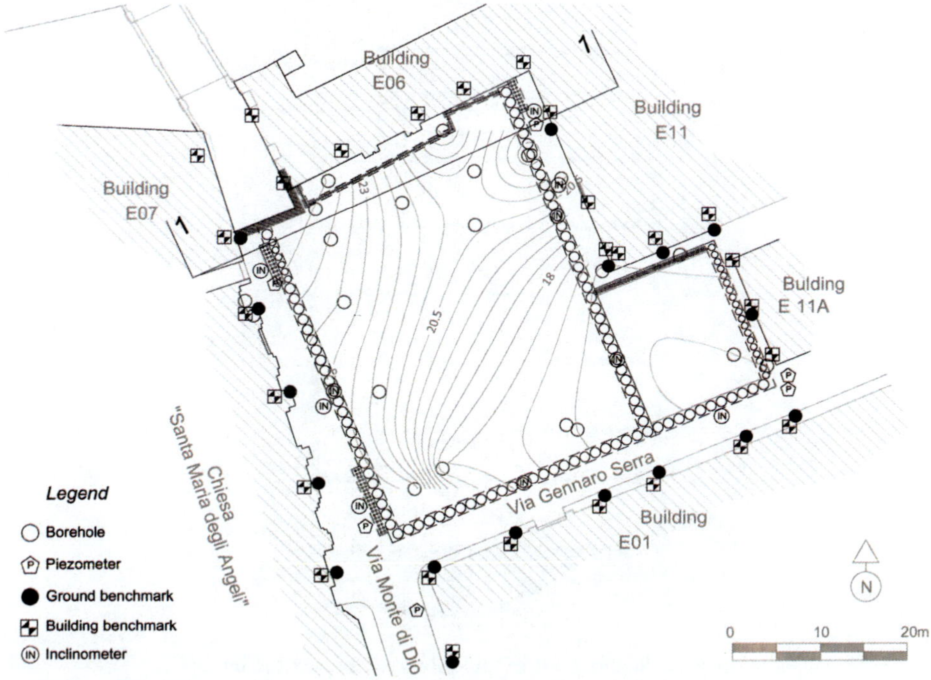

Fig. 1. Chiaia Station site with subsoil investigations, monitoring layout and top boundaries contours of the tuff formation

A graphic representation is given in Fig. 2, where the layered profile of the subsoil based on the average elevations of the boundaries among the various formations, as obtained by the boreholes, is also reported. The excavation of the bottom part did not require any temporary support being located in the yellow tuff layer.

The main line tunnel crossed the station area at relatively large depth before the construction of the deep station shaft. The platform tunnel required a section enlargement which was realized by accessing the tunnel from the main shaft once the prefabricated concrete lining was demolished similarly to previous stations of Line 6 [6]. The crown of the platform tunnels is well below the top boundary of the Neapolitan Yellow Tuff formation and this allowed a relatively straightforward excavation by conventional tunnelling method. Thus the structural box of the station was build using a bottom up procedure.

Only in the area near the school building (at the northern side of the station box), it was not possible to create the first level of pre-stressed anchors due to the presence of underground floors of the surrounding buildings. Two steel props 30 m long, with an outer diameter of 1000 mm and a thickness of 1.2 cm, were installed in order to support the retaining walls.

For the excavation inside the tuff bank, in consideration of the significant excavation heights and the impossibility of punctually defining the presence of fractures in the NYT, which cannot be excluded a priori, it was necessary to install 5 m long and 26 mm diameter dywidag bars in a regular mesh pattern of 2 m × 2 m of, and to protect the excavation face with a 15 cm thick shotcrete lining reinforced with electro-welded mesh.

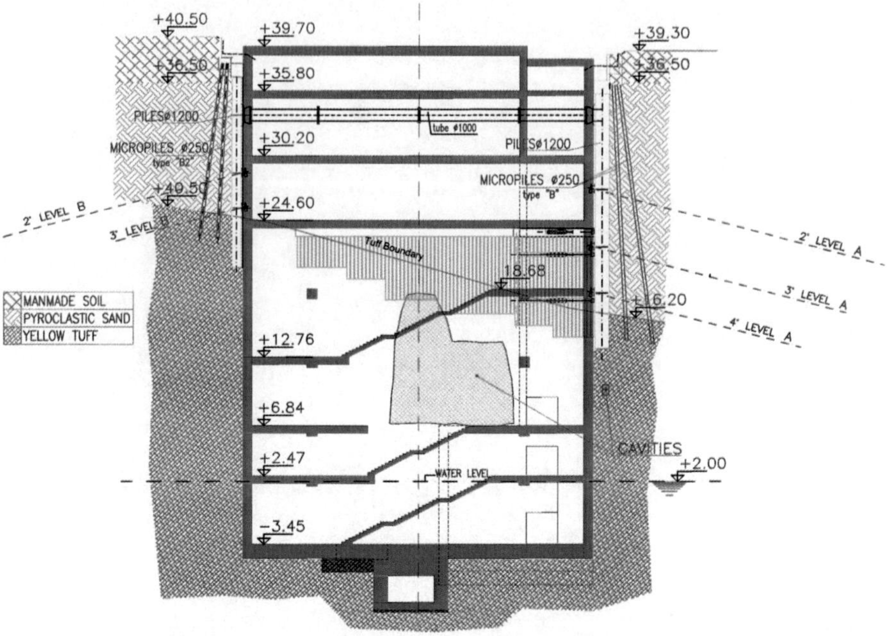

Fig. 2. Sections 1-1 of the Chiaia station shaft (Line 6)

The following list reports just the main construction stages (at least by a geotechnical point of view):

Stage 1. Construction of all the diaphragm walls. The excavation inside the station shaft first reached an average depth of 36.4 m a.s.l. (about – 4 m from p.c.) allowing the so-called archaeological check. A few months later once the archeological step was completed the excavation depth reached first 32.5 m a.s.l. The 1st level of prestressed anchors and the assembly of metal struts were completed. Final depth of the stage 1 + 30 m a.s.l;

Stage 2. Excavation operations continued reaching the tuff in the entire station area and intercepting 3 anthropic cavities. The 2nd and the 3rd level of prestressed anchors were installed. Final depth of the stage 2 + 19 m a.s.l.;

Stage 3. The excavation reached the level of the tunnel (+4.50 m a.s.l.) and the tunnel lining was demolished. The dewatering started and the excavation reached the maximum project depth (−7.95 m a.s.l.). The bottom slab was therefore built.

Stage 4. This stage contains all the operations concerning the ascent phase: the pumps have been closed (March 27, 2015); the anchors were gradually detensioned and removed and the main internal structures were eventually completed (December, 2016).

4 Monitoring Results and Comparison with Empirical Methods

A plan location of the monitoring instruments, and of the boreholes, is reported in Fig. 1. Horizontal and vertical displacements have been measured using inclinometers and benchmarks on sidewalks. The inclinometers investigated by a torpedo, allowed to measure horizontal displacement profiles. Inclinometers were installed both outside the pile wall, reaching a depth of about 50 m, and inside the piles down to the pile tip at a depth slightly variable ranging around 20 m from the ground surface.

During the execution of the works not all the available inclinometers were monitored continuously and only a few of them could be analyzed. For space reasons only the horizontal displacement of the IN3 instrument is presented here (Fig. 3) while a more detailed report of all the available measurement will be published elsewhere.

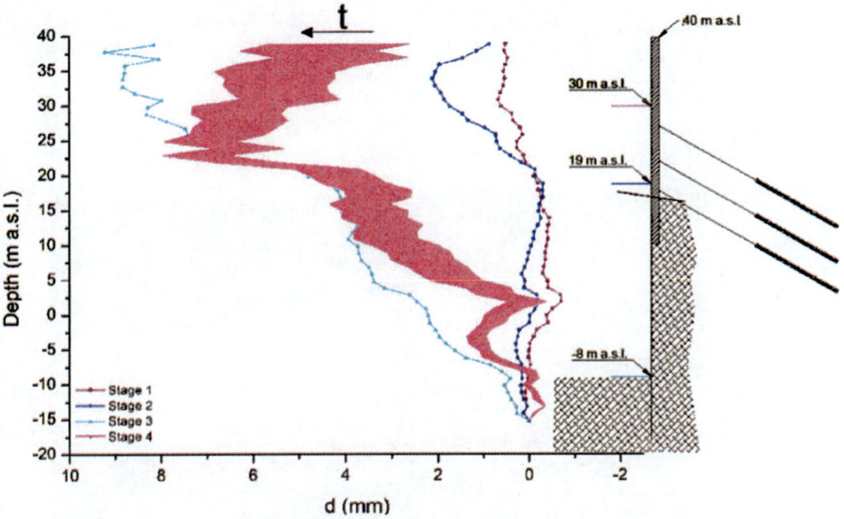

Fig. 3. IN3 horizontal displacement profiles

At the end of stage 1 the displacement profile is almost vertical. Starting from stage 2, corresponding to an excavation depth of about 20 m below the ground level, the horizontal displacements are still very small the maximum displacement reaching + 2 mm (+ towards the excavation) at the depth of 6 m below the ground level. At the end of stage 3 (end of the excavation) the maximum value of horizontal displacement, 9.2 mm, has been measured at the top part of the pile wall. Stage 4 describes the long-term behavior. The evolution of the displacement profile over eight months is represented by a pink shaded area. The displacements slightly increase over time reaching a maximum value equal to 8 mm at the elevation of 22 m a.s.l. On the other hand, the head value is significantly reduced compared to the stage 3. In Fig. 4 the excavation progressive depth is plotted versus time for both the main and the accession shaft. The latter took about one year to be completed while the former took a long time (about 5 years) for a number of reasons. Unforeseen man made cavities in the tuff formation and some minor archeological findings were the main causes of the mentioned delay.

The piezometers outside the excavations recorded the drawdown of the groundwater table when the excavation depth reached the elevation of +5 m a.s.l. The observed lowering was strictly related to the distance of the piezometer from the edge of the excavation. Soon after the pump closure the groundwater table recovered to its undisturbed level.

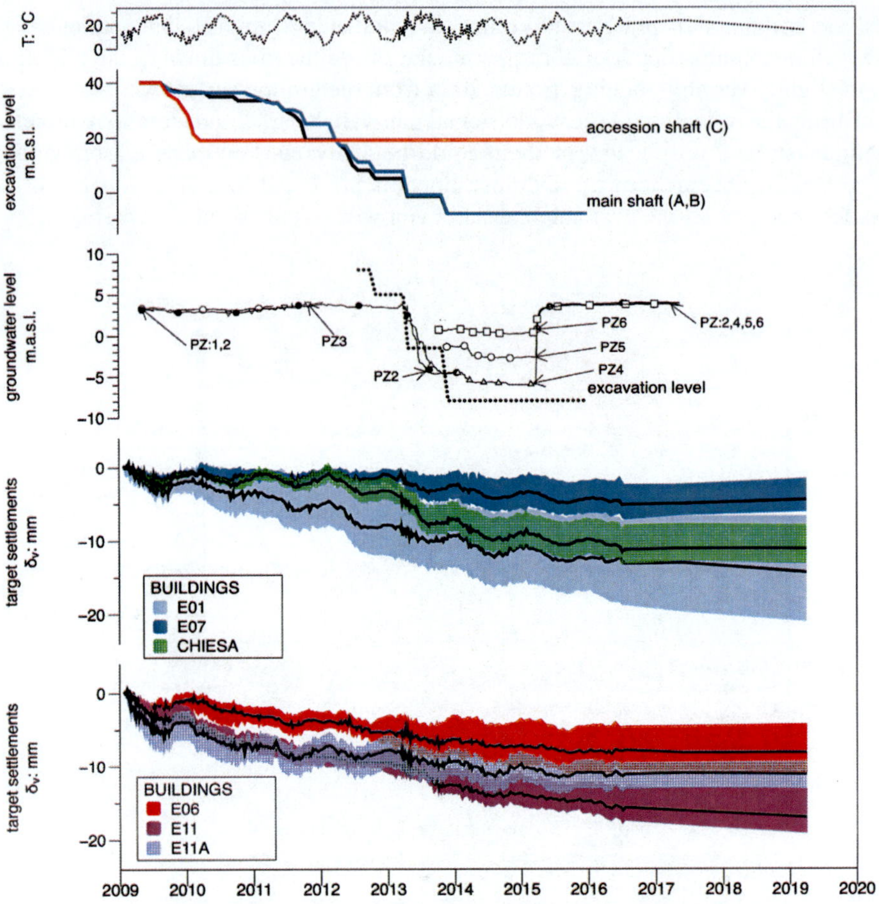

Fig. 4. Building settlement and piezometric head over time

The lower part of the figure reports the settlements of all the surrounding buildings for the whole observation period from 2009 to 2019. The settlements were measured via optical survey using as references both ground and building benchmarks installed in the same plan position. The settlement trend is related to temperature yearly fluctuations shown in the top part of the figure. Of course the excavation deepening is the main factor determining the observed trend. For each color the shaded areas represent the range described by all the benchmarks installed on the same building. It is worthy mentioning that ground and building benchmarks in the same plan position agree. The black lines in the middle of the areas represent the average trend for each building. The longest building E01 shows the wider shaded area and also the largest average settlement the final value in the 2019 approaching 15 mm. At the end of the excavation the E01 average settlement was nearly 10 mm. The minimum settlement is observed at the building E07 which is located close to the corner of the excavation. The building E06 show a small settlement too even being located parallel and close to the northern edge of the excavation shaft. A possible justification of such a behavior may be found in the fact that such building is founded on tuff. All the buildings show a long term increase of the settlement ratio (i.e. the ratio between the settlement at the end of the excavation and the settlement at the end of the observation) being its final value in the range 1.2–1.5. However, displacements maybe considered rather small and this result is in agreement with the findings of previous and relatively recent studies of deep excavations carried out in the past three decades in the Neapolitan area [7–9]. Results in terms of displacement, when compared with the ones provided by empirical predictive methods [10] (continuous line in Fig. 5), are much lower certifying the need for an update of the empirical thresholds on expected settlement and pointing out the need for predictive methods that better describe the behavior of modern supported excavation in this type of subsoil.

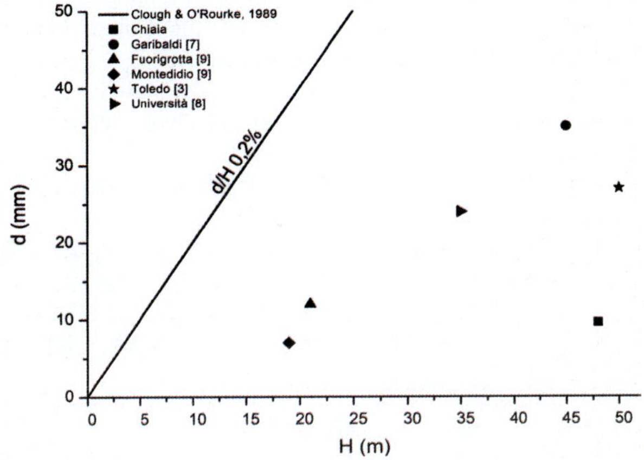

Fig. 5. Maximum horizontal displacement d (case studies in Napoli)

5 Conclusion

Chiaia station is one of the deepest stations along Line 6 and is located in the historical city center close to masonry buildings and facing the main façade of the *Santa Maria degli Angeli* Church, with a very valuable dome of the XV century. More than half of the station was excavated in tuff. No damages at all were observed. It is also important to underline that this is a nearly unique case study in terms of depth of excavation of the tuff close to valuable historical buildings under the simple temporary protection of a sprayed concrete thin lining. The monitoring instruments installed along the excavation sides allowed interesting observations to be made: four distinct realization stages have been identified by observing the horizontal displacement profiles over time, while the temperature and the drawdown of the groundwater table produced affected time-settlement observations. Significant creep effects were recorded after the end of construction with a late increase of the settlement in the range 20%–50%.

References

1. Urban transportation plan of the City of Napoli, unpublished report (1997)
2. Russo, G., Viggiani, C., Viggiani, G.M.B.: Geotechnical design and construction issues for lines 1 and 6 of the Naples underground | Geotechnische Planung und Ausführung der U-Bahnlinien 1 und 6 in Neapel. Geomech. Tunnell. **5**(3), 300–311 (2012)
3. Russo, G., Corbo, A., Cavuoto, F., Manassero, V., De Risi, A., Pigorini, A.: Underground culture: toledo station in Naples, Italy. Proc. Inst. Civ. Eng. **170**(4), 161–168 (2017)
4. Nicotera, M.V., Papa, R., Urciuoli, G.: Hydro-mechanical behaviour of unsaturated pyroclastic soils: an experimental investigation. Eng. Geol. **195**, 70–84 (2015)
5. Amorosi, A., Aversa, S., Boldini, D., Laera, A., Nicotera, M.V.: Application of a new constitutive model to the analysis of plate load tests in a pyroclastic rock. Int. J. Rock Mech. Min. Sci. **78**, 271–282 (2015)
6. Autuori, S., Nicotera, M.V., Russo, G., Di Luccio, A., Molisso, G.: Effects of construction and demolition of a TBM excavated tunnel inside existing diaphragm walls. In: Proceedings of the WTC 2019 ITA-AITES World Tunnel Congress (2019)
7. Viggiani, G.M.B., De Sanctis, L.: Geotechnical aspects of underground railway construction in the urban environment: the examples of Rome and Naples. Geol. Soc. London Eng. Geol. Spec. Publ. **22**(1), 215–240 (2009)
8. Nicotera, M.V., Russo, G.: Monitoring a deep excavation in pyroclastic soil and soft rock. Tunn. Undergr. Space Technol. **117**, 104130 (2021)
9. Russo, G., Nicotera, M.V.: A closed form shape function describing 3D settlement field around a deep excavation in sand. Sci Rep. **12**(1), 18528 (2022)
10. Clough, G.W., O'Rourke, T.D.: Construction induced movements of in situ walls. In: Proceedings of Design and Performance of Earth Retaining Structures. ASCE Special Publication, vol. 15, pp. 439–470 (1989)

Numerical Back-Analysis of In-Situ Constant Head Tests in Partially Saturated Soil Cover to Determine the Permeability Function

Vito Tagarelli(✉), Nico Stasi, and Federica Cotecchia

Polytechnic University of Bari, Bari, Italy
vito.tagarelli@poliba.it

Abstract. The determination of the coefficient of saturated permeability, k_{sat}, of a porous medium is a complex task to carry out, since k_{sat} highly depends on several factors, e.g. nature of the soil, size and the shape of soil particles, void ratio, soil structure, etc., and it also depends on the testing procedure adopted.

Since accurate determinations of the k_{sat} are necessary for numerous applications in the engineering practice, methods for both direct and indirect measurements of the permeability are commonly adopted, mainly by inducing and monitoring either steady-state or transient seepage processes in the soil deposit to investigate.

Typically, in-situ determinations are preferred to those in the laboratory since it is generally believed that in-situ measurements allow to account for any hydraulic heterogeneities in the soil, resulting in more representative testing results; however, during in-situ tests, the initial soil state and the boundary conditions of the induced seepage in the soil are not imposed and controlled, and this poses uncertainties which lead to potentially inaccurate k_{sat} determinations.

In this paper, coupled hydro-mechanical numerical back-analyses of constant head permeability tests, i.e. Guelph's tests, are reported aimed at better estimating the k_{sat}, by computing also the wetting path towards full saturation occurring in the soil.

The comparison between the k_{sat} determinations, obtained from numerical back-analyses, and the results from the semi-empirical relationships commonly adopted for processing the in-situ tests, allow to confirm that the latter always tend to overestimate the saturated permeability values of a non-negligible quantity.

Keywords: Saturated permeability · in-situ permeability test · Guelph permeameter · seepage numerical back-analysis · permeability function

1 Introduction

The academic research activity concerning the study of the soil-vegetation-atmosphere (SVA) interaction has recently gained much attention (e.g. [1]), since this interaction has been identified as a condition at the ground surface affecting the stress-strain states of geotechnical systems; referring to slopes, the SVA interaction has been shown to impact

A. Ferrari et al. (Eds.): CNRIG 2023, SSGG, pp. 275–282, 2023.
https://doi.org/10.1007/978-3-031-34761-0_34

the variations with time of the pore water pressures, either at shallow or large depths even within clayey slopes, and as such, to control weather-induced landslide activity [2, 3]. However, the SVA interaction comprises several processes of different nature and determine a thermo-hydro-mechanical (THM) transient boundary condition at the ground surface of any geotechnical system, with maximum gradients within the first 3–4 m depth, controlling the liquid, gas, and energy exchanges between the soils and both the atmosphere and the vegetation. Hence, the behaviour of a geotechnical system depends on the THM constitutive properties of the soil cover (by up to 3–4 m, [1]), among which the permeability function is one of the most crucial to evaluate.

This paper is intended to contribute to improve the reliability of in-situ permeability tests, through an innovative elaboration of the logged data, which provides more accurate permeability estimations if compared to those resulting from the commonly adopted empirical formulations to interpret in-situ tests. This study may also prompt an advancement in modelling the SVA interaction, by providing more appropriate permeability functions, e.g., better informing the prediction of weather-induced landsliding.

In this view, Guelph's constant head tests [4] were performed in-situ to characterize the permeability function of the heterogeneous clayey soil cover, which outcrops in the toe area of an active slow landslide in the southern Apennines, the Pisciolo landslide [2]. The tests were performed with reference to two portions of the soil cover with different vegetation conditions [5], i.e., Location A and Location B hereafter.

The Guelph's tests discussed in this paper were elaborated through an innovative analysis, accounting for the partially saturated condition and hence, implementing both the water retention properties and the initial state of the soil, both measured in the laboratory on undisturbed soil samples. In particular, the whole permeability function of the soil cover was derived through hydro-mechanical (HM) numerical modelling, by performing the back-analysis of the Guelph's-induced wetting seepage within the soil cover for each in-situ test. The back-analyses also provided evidence of the extents to which the determination of the k_{sat} coefficient using semi-empirical relationships usually adopted for the elaboration of in-situ tests, may results in misevaluations. Indeed, it was already recognized that such empirically based formulations are providing slightly overestimated values of the field coefficient of k_{sat} (e.g. [6]), which also testifies that the elaboration of in-situ permeability tests is not straightforward.

The Guelph's tests here of reference were performed and elaborated within a wider research activity aimed at investigating how different vegetation covers may impact on the hydraulic properties of the soil cover at Pisciolo. It is worth mentioning that the data described hereafter were also included in a work by Tagarelli et al., [7], which was addressed mainly to the hydraulic properties of the rooted soil compared to the bare one, instead, in this manuscript the focus is on how the reliability of the determination of k_{sat} values through in-situ testing may be improved by using numerical modelling.

2 In-Situ Constant Head Permeability Testing

The in-situ k_{sat} determinations of the soil cover were carried out using the Guelph permeameter [4] in two different tests, i.e., May 2021 (Test 1) and in March 2022 (Test 2), each of which performed in two different test locations (i.e., Location A and B). The

Guelph permeameter testing procedure [4] prescribes the application of a water flow to recharge a 30 cm depth cylindrical well drilled in the soil cover, keeping a constant head during the whole test. The flow rate corresponding to the reach of a steady state seepage through the well surface is used to compute k_{sat} with to the Guelph-Richards procedure [8]. Applying the Richard's equation to the water seepage from the well towards the soil in the vadose zone, k_{sat} values can be computed using the single head approach (Elrick et al. [9]), through a closed-form solution of the seepage problem:

$$k_{sat} = \frac{CQ}{2\pi H^2 + C\pi r^2 + 2\pi \frac{H}{\alpha^*}} \tag{1}$$

where, H is the imposed constant water head, r is the radius of the well hole, C is the shape parameter, function of both H and r, and Q is the flow of water necessary to keep the constant water head in the well, α^* is macroscopic capillary length parameter, which is set according to the soil texture-structure category. Equation 1 was used to derive k_{sat} for each of the Guelph tests carried out at Pisciolo, where a water head $h_1 = 15$cm was imposed for about three hours for the tests 1 and 2 within test Location A and for the test 1 within the test Location B; whereas, as for the test 2 within test Location B, a constant water head $h_2 = 25$ cm was imposed for about two hours. For all tests α^* was set to be 0.12, corresponding to "most structured soils, from clay through loam" [4].

Furthermore, for each Guelph test, an undisturbed soil sample representative of the initial state was collected during the excavation of the well hole, to be subjected to soil state determination, in terms of void ratio, e, degree of saturation, S_r, and matrix suction, s. The same determinations were also conducted on samples collected at the end of each Guelph test, representing the final soil state. Such soil state data were of use not only to characterize the Guelph-induced wetting path within the soil cover, but also to inform better the numerical back-analyses of the Guelph test, used to simulate realistically the seepage induced during the Guelph test and to estimate more accurately the k_{sat} of the soil. Indeed, the monitored initial soil states were used to initialize the numerical model consistently; moreover, the specimens were also subjected in the laboratory to retention state determinations along both drying and wetting paths, so that to construct the soil water retention curves to be implemented in the modelling for Location A and B.

The methodology entailed in the numerical back analysis of the Guelph tests is discussed in the following and the resulting k_{sat} values are compared with the those deduced according to the closed form solution reported as Eq. 1, subsequently.

3 Numerical Back-Analysis of the Seepage Process

The transient seepage occurring in the soil cover during each Guelph test was back-analysed through fully coupled hydro-mechanical numerical analyses using the FE code PLAXIS 2D in order to assess the k_{sat} value at the field scale. This FE code [10] implements the Biot's theory [11] to model the transient seepage through the soil, accounting for partial saturation and hydro-mechanical coupling [12]. As for the mechanical constitutive law, the elasto-plastic Mohr-Coulomb constitutive model was adopted, and the partially saturated condition was modelled according to the framework of the single stress variable that implements the generalized formulation of the effective stress, which is expressed through the Bishop's equation [13]. This formulation was already adopted obtaining successful numerical predictions for SVA interaction modelling [3] as well as for the prediction of deep drainage-induced water seepage in the slope [14].

In the present modelling the partially saturated soil behaviour was modelled through the input of the SWRC parameters by adopting the Mualem-van Genuchten model [15, 16] whose parameters were calibrated based upon the data measured in the laboratory during drying-wetting tests on the undisturbed specimens taken in either the Locations A and B following Pedone et al., [17]. All the parameters for the saturated-partially saturated soil behaviour adopted in the test simulations are reported in Table 1.

Table 1. Hydro-mechanical parameters implemented in the numerical back-analyses.

Parameter	Test site A	Test site B
Saturated degree of saturation, $\mathbf{S_{sat}}$ **[-]**	1	1
Residual degree of saturation, $\mathbf{S_{res}}$ **[-]**	0.13	0.13
Van Genuchten parameter, $\mathbf{g_a}$ **[1/m]**	0.0541	0.067
Van Genuchten parameter, $\mathbf{g_n}$ **[-]**	1.346	1.201
Young modulus, **E' [kN/m^2]**	10710	10710
Possion's ratio, **ν' [-]**	0.3	0.3
Effective friction angle, **φ'** [°]	18	18
Effective cohesion, **c' [kN/m^2]**	5	5

The finite element mesh for axial-symmetrical analysis is reported in Fig. 1, which implements an axial symmetrical well of a radius of 3 cm and a depth of 30 cm. The right and bottom boundaries were set both far away so that not to influence the simulated transient water flow in the soil. The initial total stress state in the model was defined by applying the k_0-procedure, adopting a $k_{0,initial} = 1$, according to [3]; instead, the pore water pressure regime was computed by using the head boundary condition type. A steady state seepage state was computed with, both the vertical lateral boundaries set as impervious, and the top and bottom boundaries were location of an applied constant total head value so that to determine across the model a constant suction, consistent with the value monitored at the beginning of the in-situ tests. Indeed, the model initialization allowed to set an initial soil state, e–s–S_r across the model, compatible with that measured at the beginning of each in-situ tests and reported in Table 2.

Once the initialization of the effective stresses was carried out as reported, the excavation of the well hole was modelled by deleting the corresponding soil cluster in the FE model (Fig. 1); furthermore, within the same calculation phase the transient water flow was also applied, by activating a constant pressure head boundary condition; in particular, a total head condition was applied at the boundaries inside the well hole in the numerical model consistent with the constant water level acting in the well during the Guelph test. Impervious hydraulic boundary conditions were set to the right, left and bottom sides for the whole transient phase of the numerical analysis, whereas the upper boundary condition was set as free drainage.

Since the analyses were performed to evaluate the k_{sat} value at the field scale for each Guelph test, for each of these, a parametric analysis for varying k_{sat} values was run, simulating the infiltration performed during the test, for the exact duration of the test. It was then selected as correct the k_{sat} value corresponding to the simulation in which, the water volume infiltrated through the Guelph permeameter during the whole test duration was equal to that measured in situ at the end of the test.

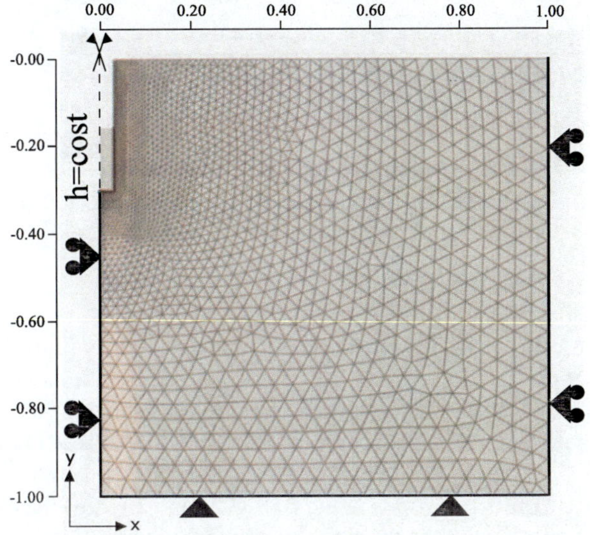

Fig. 1. Finite element mesh adopted for the Guelph test numerical back-analyses.

The numerical back-analyses were performed for the Guelph tests carried out both in May 2021 (Test 1 in Table 2) and March 2022 (Test 2 in Table 2), either in the Location A and B. As anticipated, the initial soil state monitored data (i.e., void ratio, suction and degree of saturation in Table 2) were of use to inform the initialization phase of the HM numerical modelling, which was computed by performing a steady state seepage. Accordingly, given the similarity of the in-situ state values s-S_r (Table 2) before both Test 1 and Test 2 in either Location A or B, in the initialization phase of either Test 1 or Test 2 s = 550 kPa and S_r = 70% were imposed to be uniform in Location A (in accordance with the data in Table 2), whereas s = 100 kPa and S_r = 97% were imposed to initialize the model in the Location B.

Subsequently, the Guelph-induced transient seepage was set to be consistent with a water table of 15 cm above the bottom of the hole for both Test 1 and Test 2 in the V area and for Test 1 in the B area, whereas the water table was set equal to 25 cm above the bottom of the hole for Test 2 in the B area. During the transient seepage phase, the volume of water flowing with time through the submerged boundary was computed till the reach of the total testing time of the in-situ Guelph. The analysis was repeated for different values of the saturated permeability, and the k_{sat} value corresponding to a predicted volume of water infiltrated with time equal to that measured in situ was selected as the back-analysed in-situ k_{sat}.

Table 2. Physical properties and initial state of the soil sampled inside and outside the vegetated area at the beginning of the Guelph's permeability tests.

Soil Property	Test Location A		Test Location B	
	Test 1	Test 2	Test 1	Test 2
Clay fraction, **CF [%]**	14.9	17.4	12.2	11.1
Silt fraction, **MF [%]**	31.22	18.7	18.4	44.6
Sand fraction, **SF [%]**	53.8	43.9	39.4	44.3
Unit weight, **γ [kN/m^3]**	15.6	14.9	14.5	15.4
Void ratio, **e [-]**	0.71	0.78	0.83	0.82
Degree of saturation, **S_r [%]**	68.5	78.7	95.8	97.6
Suction, **s [kPa]**	589	532	106	78

4 Results and Discussion

The in-situ value of k_{sat} coefficient derived using Eq. 1 [8] within the Location A is found to be about constant, i.e., 7.4 * 10^{-08} m/s - 5.9 * 10^{-08} m/s between Test 1 and Test 2 (Table 3). By means of Eq. 1, also the in-situ k_{sat} value for the Location B (Table 3) is found be about constant, ranging from 7.7 * 10^{-09} m/s (Test 1) to 3.9 * 10^{-09} m/s (Test 2).

With reference to both Locations A and B, Fig. 2a and 2b show the measured cumulative water (cm^3) flowed with time through the submerged permeameter hole (i.e., full and empty dots for Test 1 and 2, respectively), compared with the corresponding computed water volumes (i.e., continuous and dashed black lines for Test 1 and 2, respectively) from the numerical simulations. Indeed, the results in Fig. 2 correspond to the simulation in which the water volume infiltrated through the permeameter during the whole test was coherent with that measured in situ (continuous and dashed black lines Fig. 2a and 2b for Test 1 and 2, respectively), which allowed to identify the most proper k_{sat} value for each Guelph test, reported in Table 3.

Table 3. k_{sat} values determined through both Eq. 1 and numerical back-analyses for all tests.

	Test Location A		Test Location B	
	Test 1	Test 2	Test 1	Test 2
$\mathbf{k_{sat}}$ (Eq. 1) [m/s]	7.4 * 10^{-08}	5.9 * 10^{-08}	7.7 * 10^{-09}	6.1 * 10^{-09}
$\mathbf{k_{sat}}$ (Numerical back-analysis) [m/s]	3.0 * 10^{-09}	1.3 * 10^{-09}	6.0 * 10^{-10}	3.0 * 10^{-10}

For all tests, the k_{sat} values resulting from the back-analyses are lower than those computed from to Eq. 1 (Table 3). In particular, within the Location A, k_{sat} from the back-analysis of Test 1 is about 3.0 * 10^{-09} m/s, lower than that computed according to Eq. 1, 7.4 * 10^{-08} m/s, whereas for Test 2 the numerical k_{sat} is 1.3 * 10^{-09} m/s, lower than 5.9 * 10^{-08} from Eq. 1. As for the Location B, the back-analysed k_{sat} value is 6.0 * 10^{-10} m/s for Test 1, instead of 7.7 * 10^{-9} m/s as estimated using Eq. 1, and for Test 2, the back-analysed k_{sat} was found to be 3.0 * 10^{-10} m/s against 6.1 * 10^{-9} m/s as computed from Eq. 1.

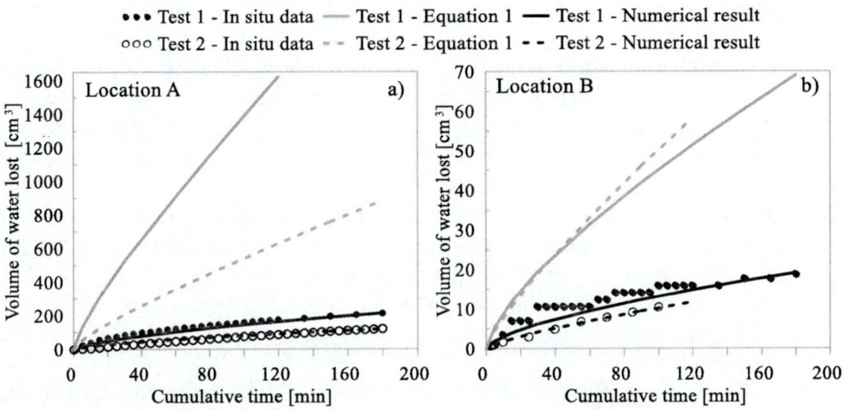

Fig. 2. Predicted water discharge (black continuous and dashed lines) and the monitored data (empty and full dots), for both Guelph's Tests 1 and 2, with reference to Location A (a) and Location B (b), together with the predictions implementing k_{sat} values from Eq. (1) (grey lines).

Since time, it is well-known that when adopting semi-empirical formulations for the elaboration of in-situ permeability testing (e.g. Eq. 1), the shape factor has to be assumed and may play a major role in the k_{sat} estimate [18, 19]; the results here commented give evidence to the size of error that such assumption may determine, since it can be assessed that the numerical back-analyses discussed above reproduce the seepage process taking place during the Guelph's test rather accurately; the only possible source of uncertainty in this modelling would lie in the Mualem model used to simulate the hydraulic function when the soil is partially saturated. However, the closeness of the curves of the measured water volume with time in Fig. 2a and 2b to the corresponding computed curves suggests that such uncertainty does not affect significantly the numerical predictions and may represent only a source of minor error for the numerical procedure to determine k_{sat} with Guelph tests which is proposed here.

It is worthy to mention that the overestimation of k_{sat} from Eq. 1, when compared to the corresponding back-analysed value, is proportional to a factor of about 30 in the Location A, and 16 in the Location B. Such overestimation factors may decrease only up to about 5, if α^*, which concur to determine C [4, 8], is set to be lower than that adopted in this case (i.e., 0,12), even though such new lower α^* would no longer match a soil type coherent with the clayey fractured and rooted soil of reference at Pisciolo.

Figure 2a and 2b also report the water discharge resulted by the numerical modelling when adopting the k_{sat} values derived using Eq. 1 (Table 3). The predicted curves (i.e., grey continuous and dashed lines for Test 1 and 2, respectively in Fig. 2a and 2b) are far higher than those recorded in-situ, giving further evidence to the overestimation of water infiltration that the use of Eq. 1 to estimates the k_{sat} of a soil cover may determine.

Acknowledgements. The authors are grateful for the support provided by PON MITIGO (ARS01_00964) and project PNRR, MISURA M4_C2_1.4, National Centre for HPC, Big Data and Quantum Computing (CN_00000013) - Spoke 5 "Environment and Natural Disasters".

References

1. Elia, G., et al.: Numerical modelling of slope–vegetation–atmosphere interaction: an overview. Q. J. Eng. Geol. Hydrogeol. **50**(3), 249–270 (2017)
2. Cotecchia, F., Tagarelli, V., Pedone, G., Ruggieri, G., Guglielmi, S., Santaloia, F.: Analysis of climate-driven processes in clayey slopes for early warning system design. Proc. Inst. Civ. Eng.-Geotech. Eng. **172**(6), 465–480 (2019)
3. Tagarelli, V., Cotecchia, F.: The effects of slope initialization on the numerical model predictions of the slope-vegetation-atmosphere interaction. Geosciences **10**(2), 85 (2020)
4. Soilmoisture Equipment Corp. Guelph permeameter kit (2008)
5. Tagarelli, V., Cotecchia, F.: Preliminary field data of selected deep-rooted vegetation effects on the slope-vegetation-atmosphere interaction: results from an in-situ test. Ital. Geotech. J. **1**(22), 62–83 (2022)
6. Nam, S., Gutierrez, M., Diplas, P., Petrie, J.: Laboratory and in situ determination of hydraulic conductivity and their validity in transient seepage analysis. Water **13**(8), 1131 (2021)
7. Tagarelli, V., Stasi, N., Cotecchia, F., Cafaro, F.: Root-induced changes in the hydraulic properties of a fine slope cover. In: 8th International Symposium on Deformation Characteristics of Geomaterials IS-Porto2023 (2023, submitted)
8. Reynolds, W.D., Elrick, D.E.: In situ measurement of field-saturated hydraulic conductivity, sorptivity, and the α-parameter using the Guelph permeameter. Soil Sci. **140**(4), 292–302 (1985)
9. Elrick, D.E., Reynolds, W.D., Tan, K.A.: Hydraulic conductivity measurements in the unsaturated zone using improved well analyses. Groundwater Monit. Remediat. **9**(3), 184–193 (1989)
10. Bentley. Plaxis Manuals (2022). https://communities.bentley.com/products/geotech-analysis/w/wiki/46137/manuals---plaxis. Accessed 20 Feb 2023
11. Biot, M.A.: General theory of three-dimensional consolidation. J. Appl. Phys. **12**(2), 155–164 (1941)
12. Galavi, V.: Groundwater flow, fully coupled flow deformation and undrained analyses in PLAXIS 2D and 3D. Plaxis Report (2010)
13. Bishop, A.W.: The principle of effective stress. Tek. Ukebl. **39**, 859–863 (1959)
14. Tagarelli, V., Cotecchia, F.: Coupled hydro-mechanical analysis of the effects of medium depth drainage trenches mitigating deep landslide activity. Eng. Geol. **297**, 106510 (2022)
15. Mualem, Y.: A new model for predicting the hydraulic conductivity of unsaturated porous media. Water Resour. Res. **12**(3), 513–522 (1976)
16. Van Genuchten, M.T.: A closed-form equation for predicting the hydraulic conductivity of unsaturated soils. Soil Sci. Soc. Am. J. **44**(5), 892–898 (1980)
17. Pedone, G., Cotecchia, F., Tagarelli, V., Bottiglieri, O., Murthy, M.B.: An investigation into the water retention behaviour of an unsaturated natural fissured clay. Appl. Sci. **12**(19), 9533 (2022)
18. Chapuis, R.P.: Shape factors for permeability tests in boreholes and piezometers. Groundwater **27**(5), 647–654 (1989)
19. Avci, C.B.: Analysis of in situ permeability tests in nonpenetrating wells. Groundwater **32**(2), 312–322 (1994)

DInSAR Data for Landslides in Basilicata Region: Geotechnical Calibration and Interpretation

Roberto Vassallo[1(✉)], Jacopo De Rosa[1], Caterina Di Maio[1], Gianfranco V. Pandiscia[2], Francesco Trillo[2], Gianluca Cutrera[3], Roberto Murtas[3], and Biagio Lacovara[4]

[1] Università della Basilicata, Potenza, Italy
roberto.vassallo@unibas.it
[2] e-GEOS, Matera, Rome, Italy
[3] Rete Ferroviaria Italiana, Bari, Italy
[4] Geocart, Potenza, Italy

Abstract. The urbanized slopes facing the Basento river valley are affected by many landslides of different types. For several places, DInSAR displacement data as well as digital terrain and surface models (DTM and DSM) were elaborated. SAR COSMO-SkyMed and Sentinel-1 images were processed to evaluate the displacements along the LOS of available scatterers over about the last ten years, LiDAR point clouds were processed for DTM and DSM with 1 m spatial resolution. For two case studies, sufficient inclinometer data were also available, thus it was possible to extend in space and in time reliable information on the landslides and their effects on the built environment. It was therefore possible to locate zones at high risk and to evaluate the effects of remedial measures.

Keywords: landslide · remedial measures · monitoring

1 Introduction

In the last decades, the DInSAR technique has received increasing attention, thanks to its cost effectiveness, accuracy, use of "natural targets" frequently characterized by a high spatial density, and the availability of long data series. If framed within a geological and geotechnical investigation, DInSAR data thus allow to improve the detection and the historical analysis of urbanized areas subjected to landslide movements [e.g.: 1] and analyze the effects of remedial measures.

The MITIGO project, supported by MIUR PON R&I 2014–2020 Program, aims to find innovative solutions to seismic and hydrological risks for a mountainous area located between the cities of Potenza and Matera (Italy) and including the slopes which face the Basento and Bradano river valleys. As for the risk deriving from slow-moving landslides, the study takes advantage of previous geotechnical investigations and monitoring as well as current *ad-hoc* geological and geotechnical studies. Geophysical, hydrological, seismic studies are also being carried out. Furthermore, SAR COSMO-SkyMed

A. Ferrari et al. (Eds.): CNRIG 2023, SSGG, pp. 283–291, 2023.
https://doi.org/10.1007/978-3-031-34761-0_35

(CSK) and Sentinel-1 (S-1) images were processed by the industrial partner e-GEOS to estimate the time evolution, from 2012 to 2021, of displacement components along the satellite line of sight (LOS) of available permanent scatterers (Fig. 1). Two CSK frames, each one approximately 1,700 km^2 wide, were processed (01 ASC and 05 DESC, with ascending and descending data respectively). The frame with descending data includes the territory of Potenza, while that with ascending data includes the territory of Calciano, both considered in this paper. Furthermore, S-1 data were processed on both ascending and descending passes for the whole study area (approximately 3,000 km^2 wide). For some landslides, the remote sensing results, supplemented by Envisat data of the Italian PST-A (High Precision Not-Ordinary Plan of Remote Sensing), provide 20-year data series.

Furthermore, for large parts of the study area, an airborne survey with laser scanner equipped platform was carried out. The LiDAR point clouds were processed by the industrial partner Geocart to obtain digital terrain and surface models (DTM and DSM) with 1 m spatial resolution.

This paper describes two case studies with examples of integrated use of ground-based data (above all inclinometers) and satellite data as means of reconstructing landslide kinematics, understanding instability processes and evaluating the effectiveness of existing remedial measures.

2 Case Studies: Two Large Landslide Systems in Structurally Complex Formations

2.1 Potenza

The slopes east of Potenza city, in Basilicata region, are being extensively monitored and studied [e.g.: 2–6] since they are affected by a number of landslide phenomena, the largest of which being *Costa della Gaveta* and *Varco d'Izzo* earthflows. Both earthflows mainly develop in the structurally complex formation of Varicoloured clays. In the last decades, they have shown continuous and extremely/very slow displacements mainly localized along defined slip zones reaching depths of about 40 m. Displacement rates range from a few mm/year to a few cm/year. Such movements are responsible of frequent damage to buildings and transport infrastructure, including the two main local arteries: the Basentana national road and the national railway. Recent attention has been devoted to the alimentation areas of both earthflows. Such areas, although already noticeably depleted, show important retrogressive instability phenomena, either episodic or continuous, threatening the built environment.

Displacement monitoring of the area by inclinometers, supplemented by a GPS network [4], has been intensely performed since 2005 (and, in limited zones, since the 1990s). It has contributed to reconstruct the geometry of the landslide bodies and to describe and interpret the main kinematic features of the landslides. More recently, DInSAR data have been used to extend the knowledge of displacement distribution in space and time [4, 6], thus reconstructing the displacement history of the slope over at least 20 years. The main issue of satellite data in the area is that, because of the mostly North-South orientation of the slopes, just a small component of displacements

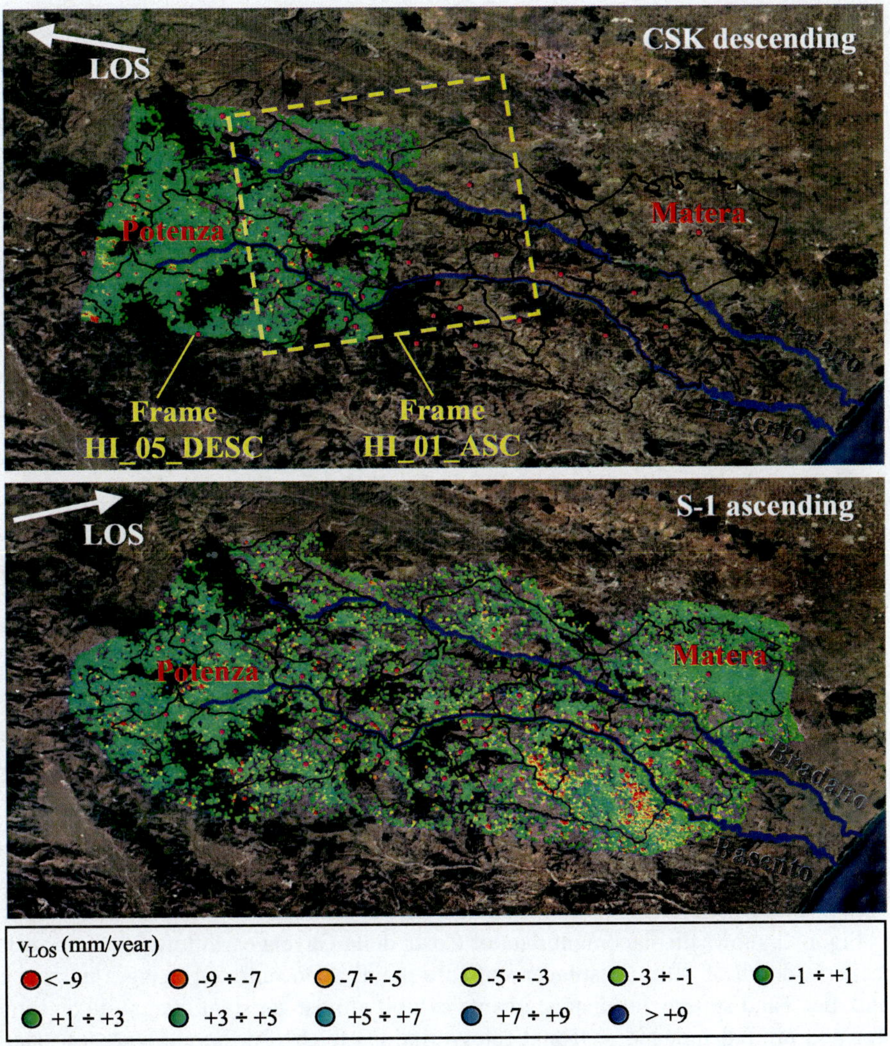

Fig. 1. Average yearly rates of displacement components along the satellite Line Of Sight (LOS) deriving from COSMO-SkyMed descending data (2012–2021) and Sentinel-1 ascending data (2015–2021) within the area of interest of the MITIGO project.

occurs along the satellite line of sight. Such issue has been overcome by exploiting the information on displacement direction derived from ground-based measurements and also using statistical data analysis [6]. So far, satellite data relative to a single constellation (COSMO-SkyMed) and a single pass (ascending) have been used. This work considers data of ascending and descending passes for two constellations (COSMO-SkyMed and Sentinel-1). As an example, descending CSK data are depicted by Fig. 2 in a 3D view of the available high resolution DTM. The analysis is limited to a number of significant buildings and several stretches of the national road.

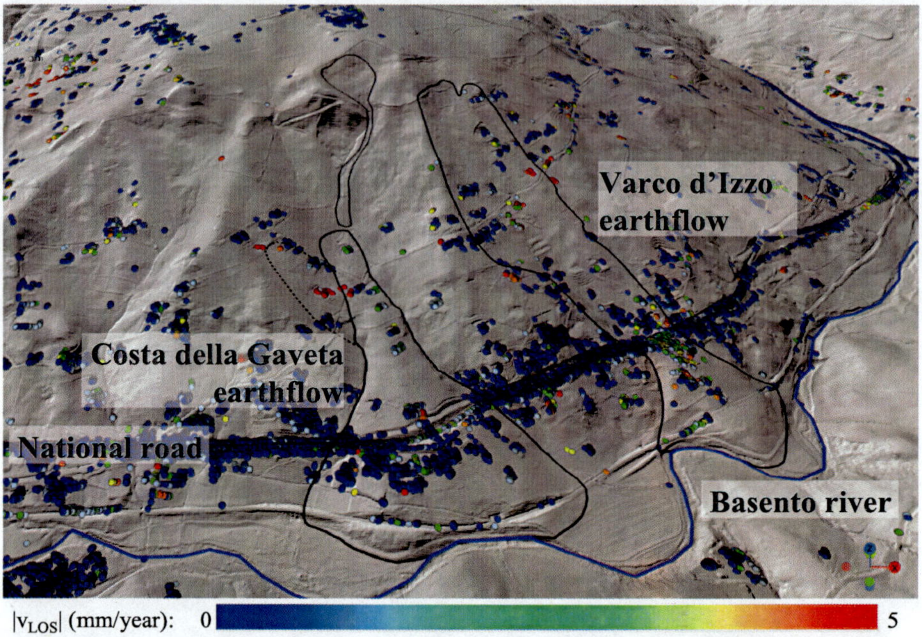

Fig. 2. Average yearly rates of LOS displacements deriving from descending CSK data, represented on a 3D view of the high resolution DTM elaborated by Geocart.

Satellite data are jointly used to obtain information about displacement modulus and inclination in the zones where the azimuth is provided by an inclinometer available in a close range. Thanks to the validation of satellite data by a large number of ground-based measurements carried out in the slope, it is also possible to estimate, in the surrounding areas, displacement intensity, direction and time evolution.

Figure 3 shows displacement data of Costa della Gaveta earthflow. The slip surface reaches a depth of 40 m; displacements are mainly concentrated along a narrow slip band; deep and superficial displacements exhibit almost constant annual displacement rates and rainfall induced seasonal rate variations [5, 6]. In the alimentation zone, a lateral internal landslide is an object of recent study. Satellite data have provided first clear evidence of movement and brought out the need for geotechnical investigations, including boreholes, piezometers and inclinometers, which are currently being installed. Displacement evolution deriving from DInSAR data, reported in Fig. 3 (zone A), suggests higher rates than those detected for the main body of Costa della Gaveta earthflow, though, anyway, of the same order of magnitude.

Figure 4 reports displacement information gathered from satellite data relative to the alimentation zone of Varco d'Izzo earthflow. The state of damage of the internal roads and its rapid evolution makes it probable that some movements are faster than the range visible from satellites. Nevertheless, a number of permanent scatterers provide indication of yearly centimetric movements, approximately in the maximum slope direction. The sign of LOS displacement rates (negative for ascending passes, corresponding to an increase in the distance between the satellite and the scatterer, and positive for descending

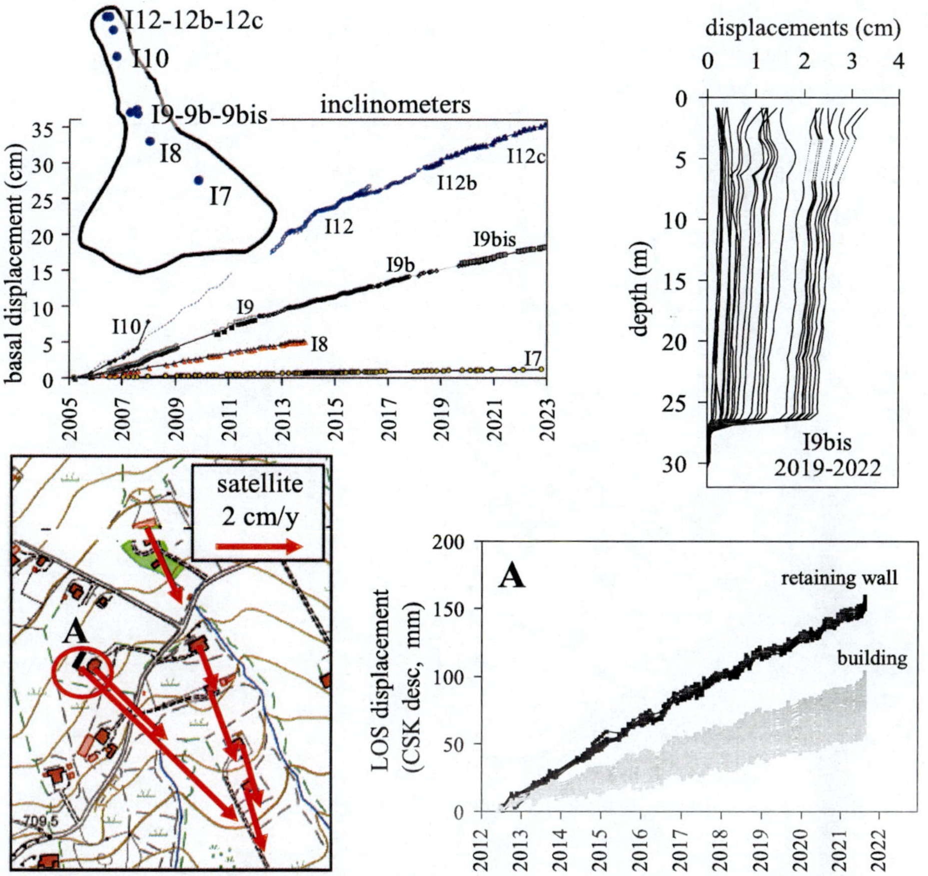

Fig. 3. Costa della Gaveta earthflow: examples of inclinometer and satellite data.

passes) gives clear indication on displacement direction. In the accumulation, movements do not spare the National Road. For the stretch of road crossing Varco d'Izzo earthflow, the landslide footprint is quite evident in descending satellite data reported in the figure, similarly to what was observed for ascending passes [6].

2.2 Calciano

A large landslide system in *Bosco San Domenico,* in the territory of the municipality of Calciano, interacts with the national railway and with the national road Basentana, in the medium Basento river valley (Fig. 5). Two structurally complex Miocene formations outcrop in this area: the Numidian Flysch, which consists of compact quartzarenites with layers of clays and clayey marls, and the Serra Palazzo formation, consisting of an irregular succession of marls, clays and limestones. The landslide is constituted by different bodies made up of soils belonging to both formations, distinct or mixed together. In some zones they consist of sands and silts, in other zones of silts and clays, in both cases

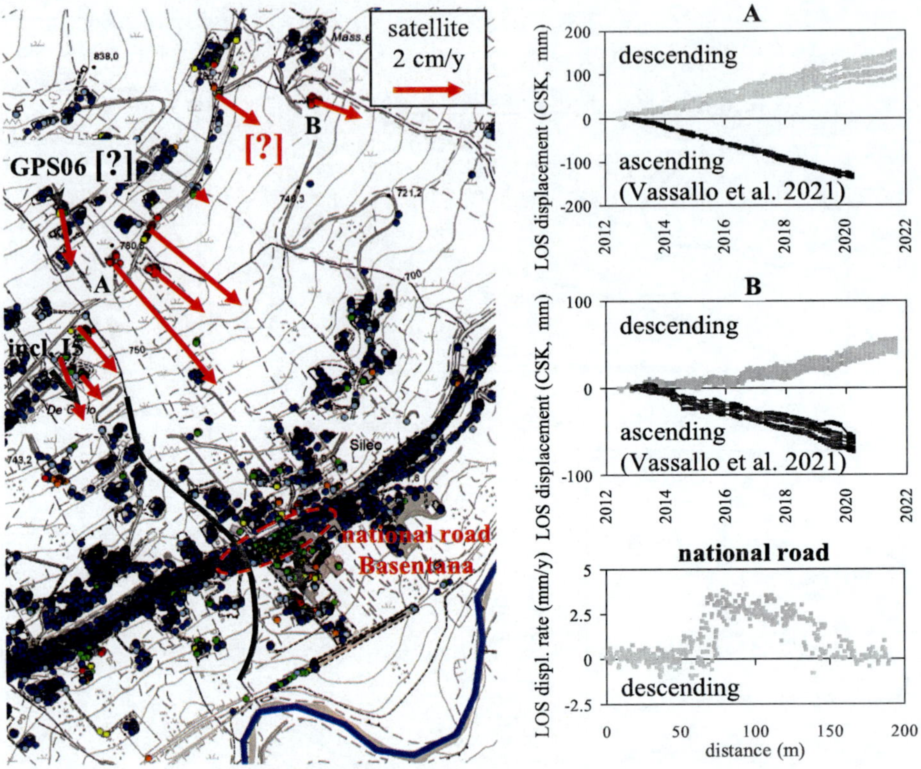

Fig. 4. Varco d'Izzo: displacement rate vectors and LOS components against time.

incorporating rock elements of extremely variable dimensions. Deep drainage systems have been realized in different periods with the aim of reducing the displacement rates. Similarly to Varco d'Izzo slope, satellite data, used in conjunction with ground-based displacement data, allow to obtain a wider picture of the displacement distribution and time evolution. Figure 5 depicts LOS displacement rates deriving from descending S-1 data and ascending CSK data. Due to the direction of displacements, the former data represent a large part of the total rates. Conversely, a small percentage is visible on ascending passes.

The inclinometer and DInSAR data, summarized in Fig. 6, show that in the area located downslope of the railway line, the central well W01 moves with a rate slightly higher than the lateral wells W02 and W03. Upslope of the railway line, wells W2 and W4, placed in the most active landslide body, exhibit rates of 1 ÷ 2 cm/year.

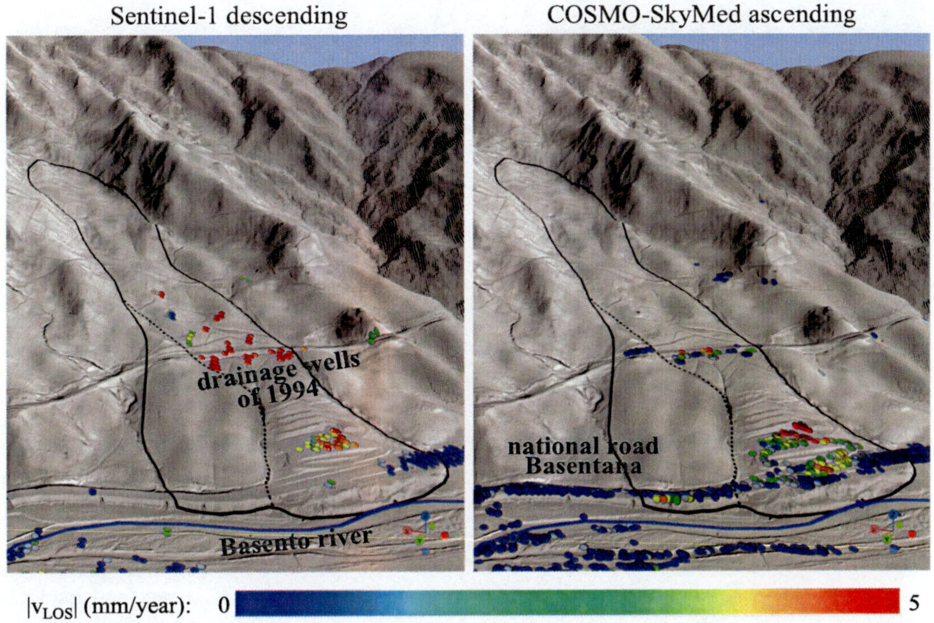

Fig. 5. Calciano: average yearly rates of LOS displacements deriving from Sentinel-1 data and COSMO-SkyMed data, represented on a 3D view of the high resolution DTM.

It can also be observed that the basal velocities detected by inclinometers and the superficial velocities deriving from S-1 and CSK satellite data, detected on deep structures like the wells, range from a few mm/year to 6 ÷ 7 cm/year. The displacement history of well W01 shows, from 2017, a rate decrease to almost half of the previous value. This corresponds to the period of realization of new drainage wells upslope from the railway. A similar decrease is observed for other nearby structures. It is worth noting that no significant variation in the cumulated rainfall time trend occurred after 2017. Slow and continuous displacements are also observed from CSK ascending data where the national road crosses the landslide accumulation.

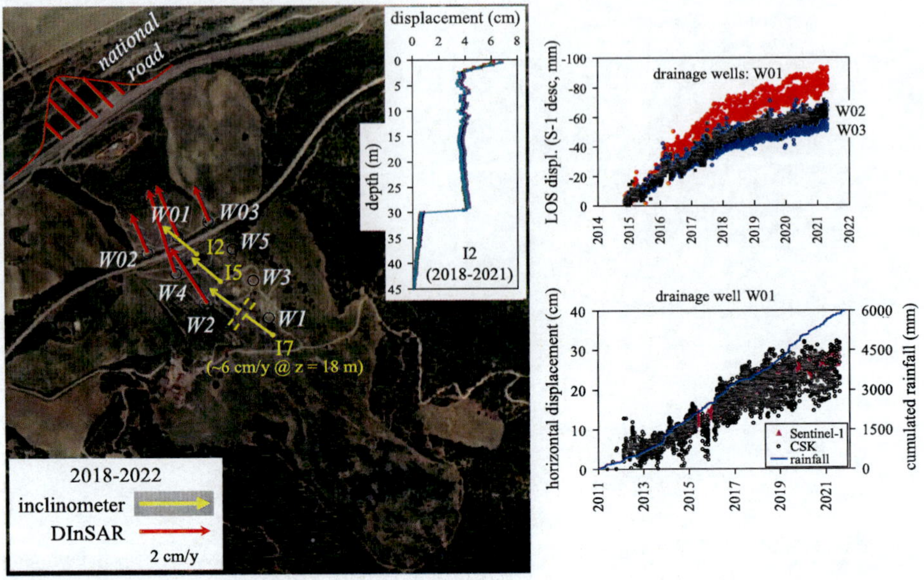

Fig. 6. Calciano: average displacement rates of inclinometers and other scatterers, and displacement time evolution of the drainage wells.

3 Conclusions

DInSAR interferometry estimates the Line-Of-Sight projection of displacements (either towards or away from the satellite) of natural or man-made reflectors. When DInSAR measurements derived from both ascending and descending acquisition geometries are available on an area of interest, it is possible to combine both measurements to extract the vertical and east-west components of the ground displacement. To estimate all three components (e.g. north, east and up) of the displacement vector, it is necessary to use additional ground-based data. In the considered case histories, the integration of DInSAR results with inclinometer data has proved useful to describe more comprehensively the displacements of structures which generally undergo complex interaction with the landslides. The first case shown by this paper refers to a hill east of Potenza city for which satellite data helped in evaluating the zones where the need for geotechnical investigations was maximum. As a consequence, boreholes are currently being drilled, piezometers and inclinometers installed. The second case shows how DInSAR data, given the availability of historical archives covering the last decade, can help evaluating the effectiveness of important remedial measures constructed in the past.

Acknowledgments. This research is supported by MIUR PON R&I 2014-2020 Program (project MITIGO, ARS01_00964) and by R.F.I. Rete Ferroviaria Italiana.

References

1. Peduto, D., Santoro, M., Aceto, L., Borrelli, L., Gullà, G.: Full integration of geomorphological, geotechnical, A-DInSAR and damage data for detailed geometric-kinematic features of a slow-moving landslide in urban area. Landslides **18**(3), 807–825 (2020). https://doi.org/10.1007/s10346-020-01541-0
2. Di Maio, C., Vassallo, R., Vallario, M., Pascale, S., Sdao, F.: Structure and kinematics of a landslide in a complex clayey formation of the Italian Southern Apennines. Eng. Geol. **116**, 311–322 (2010)
3. Di Maio, C., Vassallo, R., Vallario, M.: Plastic and viscous displacements of a deep and very slow landslide in stiff clay formation. Eng. Geol. **162**, 53–66 (2013)
4. Vassallo, R., Calcaterra, S., D'Agostino, N., De Rosa, J., Di Maio, C., Gambino, P.: Long-term displacement monitoring of slow earthflows by inclinometers and GPS, and wide area surveillance by COSMO-SkyMed. Geosciences **10**(5), 171 (2020)
5. Di Maio, C., De Rosa, J., Vassallo, R.: Pore water pressures and hydraulic conductivity in the slip zone of a clayey earthflow: experimentation and modelling. Eng. Geol. **292**, 106263 (2021)
6. Vassallo, R., De Rosa, J., Di Maio, C., Reale, D., Verde, S., Fornaro, G.: In situ and satellite long-term monitoring of two earthflows of the Italian southern Apennines and of the structures built on them. Ital. Geotech. J. **55**(4), 77–95 (2021)

Constitutive Modelling and the Thermo-Hydro-Chemo-Mechanical Behaviour of Geomaterials

One Phase vs Two-Phase Modelling of Infiltration Processes

Mauro Aimar[1] (✉), Gabriele Della Vecchia[2], and Guido Musso[1]

[1] Politecnico di Torino, Corso Duca degli Abruzzi 24, 10129 Turin, Italy
mauro.aimar@polito.it

[2] Politecnico di Milano, Piazza Leonardo da Vinci 32, 20133 Milan, Italy

Abstract. Because of global warming, the frequency and harmfulness of climate extreme meteorological events have increased and are expected to increase. Events such as droughts, heatwaves, storms, floodings strongly impact the functioning of critical infrastructure, as well as the stability and degradation of man-made earthworks and natural slopes. The shallower portions of natural soils and earthworks are thus exposed to increasingly severe dry and hot seasons and to intense rainfalls, which affect the depth of the water table and the hydraulic circulation, triggering material degradation and instabilities. Robust modelling of the soil-atmosphere interaction, correctly accounting for unsaturated flow and boundary conditions, is a requirement of modern geotechnical engineering. The assumptions made when modelling infiltration, in terms of infiltration rate, pore and air pressure distribution and evolution, have in fact a relevant impact on model predictions. As a fraction of the pore space of unsaturated soils is occupied by the gas phase, composed of water vapour and dry air, a complete model of infiltration requires accounting for the inflow of the liquid phase and the outflow of the gas phase. Contrarily, infiltration is very often modelled accounting only for the inflow of liquid water. This work explores the consequences and limitations of such simplification by comparing the predictions obtained by adopting both a two-phase and a one-phase model for the simulation of the infiltration processes.

Keywords: Unsaturated soil mechanics · Infiltration · Multi-phase flow

1 Introduction

Global warming is pushing the weather towards increasingly extreme conditions, with severely dry and hot seasons and intense rainfall events. Such phenomena are expected to impact on the hydro-mechanical state of the shallower portions of soil deposits and geotechnical earthworks. Strong evaporation due to dry seasons might end up altering the integrity of earthworks through the occurrence of drying cracks [1], or reduce the effectiveness of cut-off walls [2]. Rainfall and infiltration lead to an increase in the pore water pressure and in the degree of saturation of the slopes, which favor the onset of instabilities [3]. In particular, in the vadose zone – i.e., the shallower portion of soil deposits resting above the water table – the soil is often unsaturated and the pore water pressure is negative (smaller than atmospheric pressure). Its value is controlled

A. Ferrari et al. (Eds.): CNRIG 2023, SSGG, pp. 295–302, 2023.
https://doi.org/10.1007/978-3-031-34761-0_36

by the interaction between the soil surface and the atmosphere, through evaporation and infiltration, and by the soil hydraulic properties (see, e.g., [4]).

As for infiltration processes, classical modelling approaches only consider the mass transfer of liquid water, whereas no water vapour is accounted for and pore air pressure is assumed as atmospheric (e.g., Richards' equation [5]). However, several studies demonstrated that air entrapment affects the infiltration rate, entailing slower variation in pore water pressure [6–8]. This delay in the pore water pressure increase can be crucial for rainfall-induced slope stability issues, as it may prevent reaching the critical value during the rainfall event. Besides, to the Author's knowledge, the role of water vapour in the infiltration process has not been clearly highlighted yet. This contribution proposes a hydraulic model aimed at reproducing infiltration processes in geomaterials, including the liquid and gas flow of water and air, in order to investigate the influence of water vapour and pore air pressure. Numerical simulations have been performed considering distinct scenarios, including (i) the flow of liquid water only, (ii) both liquid water and water vapour, and (iii) the combined flow of water and air. Model predictions are compared with the transient infiltration problem discussed in Siemens et al. [7].

2 Modelling Water and Air Transport in Unsaturated Soils

Modelling isothermal hydro-mechanical processes in unsaturated soils requires accounting for the linear momentum balance of the solid skeleton and the mass balance of air and water. However, as a first approximation limited to soils whose deformation upon wetting is small (e.g., coarse soils and non-active clays), the linear momentum balance of the solid skeleton can be neglected when modelling infiltration. While the classical description of hydro-mechanical processes is usually performed in terms of fluid phases (*i.e.* liquid and gas), in this study mass balance equations were rather written in terms of chemical species [9].

The mass balance equation for the species water is written by combining the contribution of liquid water and water vapour:

$$\frac{\partial}{\partial t}\left[\phi\left(S_l\rho_l^w + S_g\rho_g^w\right)\right] + \nabla\cdot\left(\rho_l^w\mathbf{q}_l + \rho_g^w\mathbf{q}_g\right) + \nabla\cdot\left[\phi\left(S_l\mathbf{J}_l^w + S_g\mathbf{J}_g^w\right)\right] = 0 \quad (1)$$

where ϕ is porosity, S_l is degree of saturation of the liquid phase, $S_g = 1 - S_l$ is degree of saturation of the gas phase, ρ_l^w is the mass density of the water in the liquid phase, ρ_g^w is the mass density of the water in the gas phase, $\mathbf{q}_l$ is the advective flux of the liquid phase, $\mathbf{q}_g$ is the advective flux of the gas phase, $\mathbf{J}_l^w$ is the diffusive flux of the water in the liquid phase, $\mathbf{J}_g^w$ is the diffusive flux of the water in the gas phase.

The mass balance equation for the species air is written according to a similar formulation:

$$\frac{\partial}{\partial t}\left[\phi\left(S_l\rho_l^a + S_g\rho_g^a\right)\right] + \nabla\cdot\left(\rho_l^a\mathbf{q}_l + \rho_g^a\mathbf{q}_g\right) + \nabla\cdot\left[\phi\left(S_l\mathbf{J}_l^a + S_g\mathbf{J}_g^a\right)\right] = 0 \quad (2)$$

where ${\rho_l}^a$ is the mass density of the air dissolved in the liquid phase, ${\rho_g}^a$ is the mass density of the air in the gas phase, ${\mathbf{J}_l}^a$ is the diffusive flux of the air dissolved in the liquid phase, ${\mathbf{J}_g}^a$ is the diffusive flux of the air in the gas phase.

The water retention curve provides the link between the degree of saturation S_l and matric suction s, defined as the difference between the pressure of the gas phase p_g and the one of the liquid phase p_l:

$$s = p_g - p_l \tag{3}$$

The liquid density ρ_l^w is assumed to be constant and equal to 1,000 kg/m^3, while the vapour density ρ_g^w has been obtained through the psychrometric law, as a function of the total suction ψ and the partial pressure of water vapour at saturation $p_{g,sat}^w$ (e.g., [10]):

$$\rho_g^w = \frac{p_{g,sat}^w}{RT}\exp\left(-\frac{\Psi M_w}{\rho_l^w RT}\right),\ p_{g,sat}^w = 0.6108[kPa] \times 10^{\frac{7.5T[^\circ C]}{T[^\circ C]+273.3}} \tag{4}$$

where $R = 8.3145$ J/(mol × K) is the constant of perfect gases, $M_w = 0.018$ kg/mol is the molar mass of the water, and T is the temperature. For the sake of simplicity, the contribution of the osmotic suction is assumed to be negligible, hence ψ equals s. The gaseous air density ${\rho_g}^a$ is estimated through the law of perfect gases, as a function of the corresponding partial pressure ${p_g}^a$:

$$\rho_g^a = \frac{M_a}{RT} p_g^a \tag{5}$$

The mass density of the air dissolved in the liquid phase ${\rho_l}^a$ is calculated as a function of the liquid water pressure, by means of the Henry's law:

$$\rho_l^a = H\frac{M_a}{RT} p_l^w \tag{6}$$

where $H = 0.031$ is the Henry constant for air dissolved in water.

The advective flux $\mathbf{q}_\alpha$ (with $\alpha = l, g$) can be described through the generalized Darcy law:

$$\mathbf{q}_\alpha = -\mathbf{K}_\alpha \nabla\left(\frac{p_\alpha}{\rho_\alpha g} + h_z\right) \tag{7}$$

where $\mathbf{K}_\alpha$ is the unsaturated hydraulic conductivity tensor, ρ_α is the mass density of the α-phase, and h_z is the elevation head. In isotropic, partially saturated soils, $\mathbf{K}_\alpha$ is an isotropic tensor with magnitude K_α, which can be decomposed into the product between a saturated value $K_{\alpha,sat}$ and a relative permeability coefficient $k_{\alpha,rel}$, dependent on S_α [11].

Instead, the diffusive flux ${\mathbf{J}_\alpha}^\gamma$ (with $\gamma = w, a$) can be predicted through the Fick's law:

$$\mathbf{J}_\alpha^\gamma = -\mathbf{D}_\alpha^\gamma \nabla \rho_\alpha^\gamma \tag{8}$$

where ${\mathbf{D}_\alpha}^\gamma$ is the diffusion tensor of the α-phase of the γ-species in the porous medium. This tensor can be written as follows:

$$\mathbf{D}_\alpha^\gamma = D_\alpha^\gamma \boldsymbol{\tau}_\alpha \tag{9}$$

where ${D_\alpha}^\gamma$ is the diffusion coefficient of the α-phase of the γ-species in "free" conditions, and $\boldsymbol{\tau}_\alpha$ is the tortuosity tensor, which models the tortuosity of the path described by the γ-molecules in the diffusion process. In isotropic, partially saturated soils, $\boldsymbol{\tau}_\alpha$ can be described as an isotropic tensor with magnitude τ_α [12]:

$$\tau_\alpha = (\phi S_\alpha)^{\frac{2}{3}} \tag{10}$$

For the water species, ${\mathbf{J}_l}^w = \mathbf{0}$ due to the assumed constant ${\rho_l}^w$, whereas the diffusion coefficient of the water vapour in free air ${D_g}^w$ is [13]:

$$D_g^w = 0.229 \times 10^{-4}\left[m^2/s\right]\left(1 + \frac{T[K]}{273}\right)^{1.75} \tag{11}$$

As for the air, ${\mathbf{J}_g}^a$ represents the air self-diffusivity and it is assumed to be negligible, whereas the diffusion coefficient of dissolved air in water is ${D_l}^a = 2 \times 10^{-9}$ m^2/s [14]. In this study, the infiltration process is simulated considering three distinct models:

- Model "W$_\mathrm{l}$": only liquid water is included, whereas air remains at atmospheric pressure. In this case, only the mass balance of water is used (Eq. 1), with vapour pressure and density set equal to 0.
- Model "W$_\mathrm{lg}$": the mass balance of both liquid water and water vapour is considered (Eq. 1), whereas the mass balance of air is neglected.
- Model "W$_\mathrm{lg}$ + A$_\mathrm{lg}$": both water and air species are included, and the full set of equations is implemented.

3 Numerical Model

The hydraulic model is used to address the influence of water vapour and dry air on infiltration processes, with reference to the physical and numerical study presented in Siemens et al. [7], who studied the infiltration of an oil along a vertical column filled with a coarse-grained, isotropic transparent soil, with porosity $\phi = 0.5$. The water retention curve of the soil can be described by the Van Genuchten relationship [15]:

$$S_l = S_{l,r} + \frac{1 - S_{l,r}}{\left[1 + (s/P)^n\right]^m} \tag{12}$$

with $S_{l,r} = 0.03$, $P = 0.17$ kPa, $n = 6.17$, and $m = 0.84$ [6]. The saturated hydraulic conductivity for the liquid and gas phase are $k_{l,sat} = 10^{-3}$ m/s and $k_{g,sat} = 8 \times 10^{-3}$ m/s, while the corresponding relative permeabilities are shown in Fig. 1 as a function of S_l and S_g. The soil column is 1,120 mm long and it can be reproduced as a 1-D domain with endpoints A and B, characterized by the spatial coordinate z (Fig. 2). In the considered infiltration process, the column is initially at atmospheric pressure with a uniform suction $s = 1.1$ kPa. This condition is simulated by setting an initial value of $p_g = 0$ kPa and $p_l = -1.1$ kPa throughout the whole domain. At time $t = 0$ s, infiltration is triggered through the application of a constant-head reservoir at the endpoint B, which corresponds to $p_l = 1.2$ kPa. No flow is allowed at the endpoint A. As for the air species, a Dirichlet boundary condition is applied on B, with a pressure equal to the corresponding p_l increased by the soil air-entry value AEV = 0.1 kPa (hence, $p_g = 1.3$ kPa). This constraint forces the suction in B to be at least equal to the AEV, thus allowing air to escape during the process. On A, a Neumann boundary condition forces the air flux to zero.

The model was integrated with the finite-element software Comsol Multiphysics®.

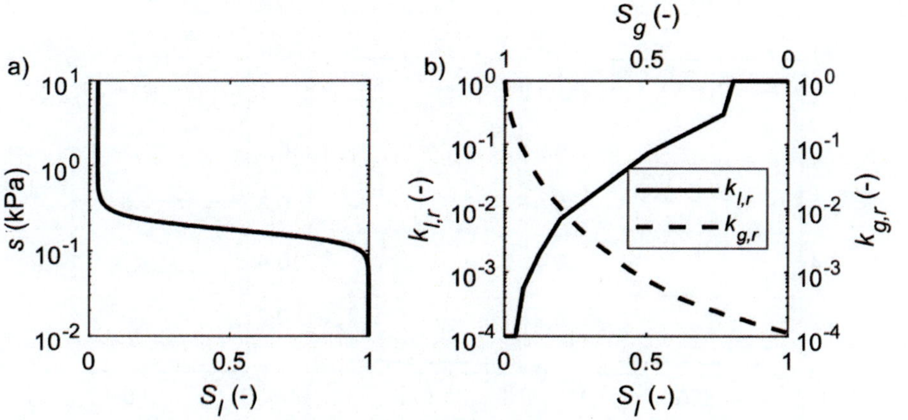

Fig. 1. a) Water retention curve of the soil; b) Relative permeability for the liquid and the gas phase, as a function of the liquid and the gas saturation degree (modified from [7]).

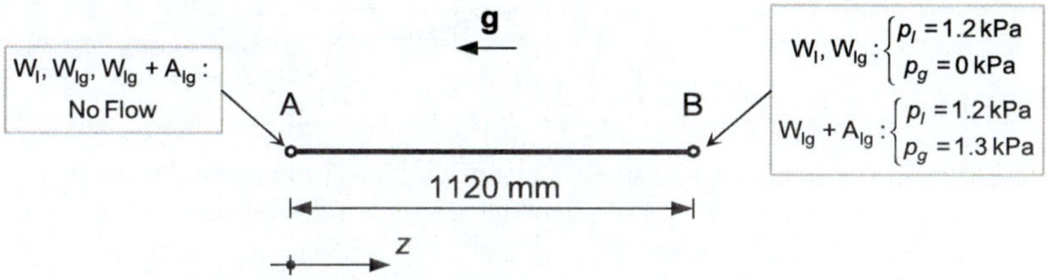

Fig. 2. Geometry of the modeled soil column, including boundary conditions. The vector **g** denotes the gravity acceleration.

4 Results and Discussion

Figure 3a–c show the simulated S_l profiles along column length at different times, for the three models. W_l and W_{lg} models provide rather similar profiles, with a sharp wetting front for S_l increasing from the residual to the full saturated state. For comparison purposes, the wetting front was identified at any time by the point z_f, corresponding to the position where the derivative $\frac{\partial S_l}{\partial z}$ is maximum. Figure 3d reports the time evolution of z_f, which exhibits a linear trend over time and reaches endpoint A at about 400 s. At this time instant, the soil column achieves complete saturation. The strong agreement between W_l and W_{lg} models is a consequence of the rather small variation of $p_g{}^w$ throughout the process (almost equal to $p^w_{g,sat}$ for all the time steps). As for the $W_{lg} + A_{lg}$ model, the shape and the evolution over time of the S_l profiles significantly changes compared with the models involving the water species only. On the one hand, the wetting front advances more slowly, and the time needed to reach endpoint A (900 s) is double with respect to

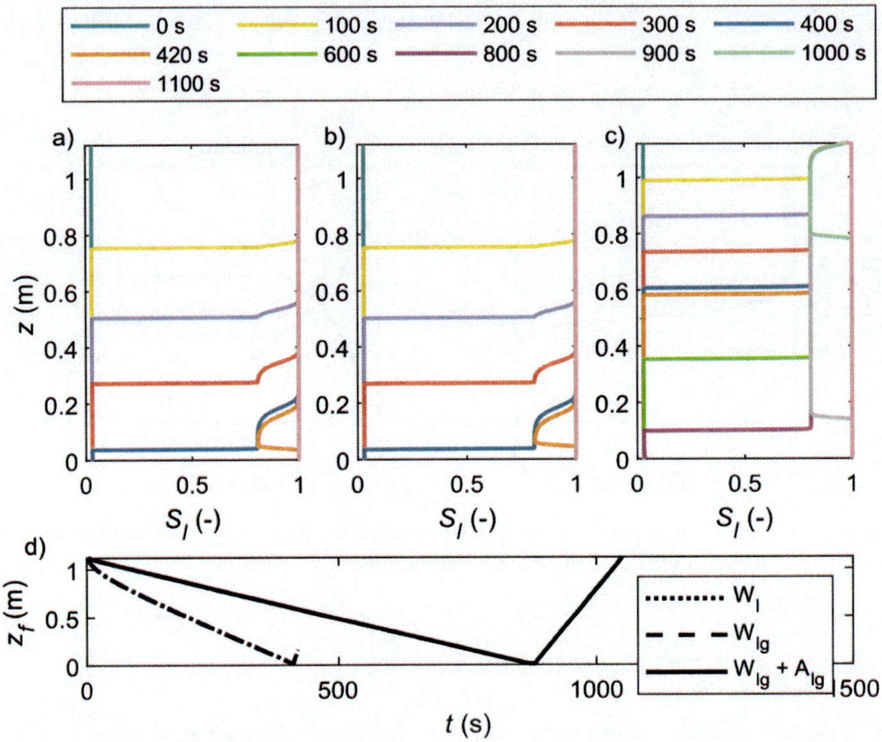

Fig. 3. a–c) Profiles of liquid-phase saturation degree S_l for different time instants, for a) "W_l" model, b) "W_{lg}" model and c) "$W_{lg} + A_{lg}$" model; d) Temporal evolution of the location of the wetting/saturation front, labeled as z_f.

the other cases. Besides, the region above the wetting front is still partially saturated, with $S_l \approx 0.8$. After $t = 900$ s, a saturation front gradually rises along the column, and the portion below it achieves $S_l = 1$. The saturation front proceeding from bottom to top is faster than the wetting front from top to bottom, and saturation completes at 1100 s.

The introduction in the model of the air species thus dramatically modifies the shape and the location of saturation isochrones. Indeed, the liquid water advancement is partially counterbalanced by an increase in the air pressure. Specifically, at each time instant, ${p_g}^a$ profiles exhibit a linear increase in the region above the wetting front, and it remains uniform below (Fig. 4). Indeed, the progressive saturation from one end of the model creates entrapped air, which reacts with an overpressure to the gradual confinement exerted by the advancing wetting front. The air pressure is maximum when the wetting front reaches the bottom of the column. When the wetting front reaches endpoint A, the air pressure gradually reduces from its maximum due a counterflow of air, thus allowing complete saturation of the column.

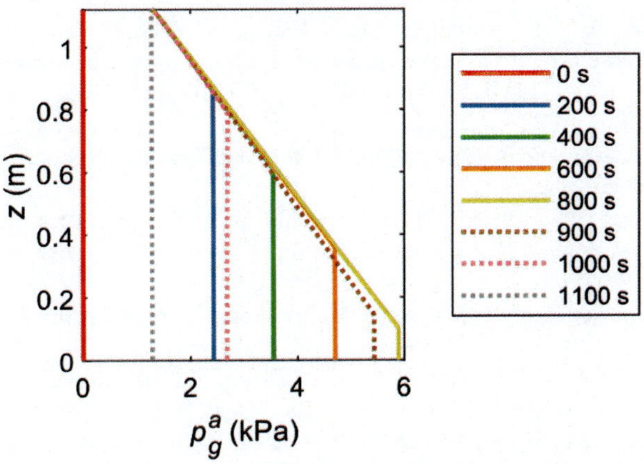

Fig. 4. Profiles of air gas pressure $p_g{}^a$ for different time instants. The dashed lines correspond to the final saturation stage, characterized by the ascent of the saturation front.

5 Conclusions

This contribution addressed the influence of accounting for the gas phase when modelling infiltration. A hydraulic model combining the mass transport of pore water and air was formulated, including both the liquid phase and the gas phase. The model was used to simulate infiltration along an initially dry soil column, considering the flow of liquid water only, the flux of liquid and water vapour, and the flow of both water and air.

Accounting for the presence of air significantly impacts the simulation results. Compared with the cases including only water mass balance, a relevant reduction in the infiltration rate is observed. Pore air partially hinders the water flow, resulting in a longer infiltration process and a slower build-up of pore water pressure. Instead, simulations accounting for water vapour do not substantially differ from those of the liquid-based model. This specific aspect may depend on the small suction range encompassed in the simulations (around 1 kPa), which entails an almost negligible variation of the vapour pressure.

These conclusions apply to the specific material and boundary conditions explored in the present work. Although the main features of the process might be expected to be the same also for fine soils, their relevance might be different depending on the water retention properties of the material. It shall be observed that, at *in-situ* conditions, infiltration is more realistically triggered by a precipitation rate rather than by an imposed water pressure at the ground surface. This different boundary condition might be expected to have a different impact of the air pressure build up, and therefore on infiltration and pressure isochrones.

References

1. Sánchez, M., Manzoli, O.L., Guimarães, L.J.: Modeling 3-D desiccation soil crack networks using a mesh fragmentation technique. Comput. Geotech. **62**, 27–39 (2014)

2. Musso, G., Vespo, V.S., Guida, G., Della Vecchia, G.: Hydro-mechanical behaviour of a cement–bentonite mixture along evaporation and water-uptake controlled paths. Geomech. Energy Environ. 100413 (2022). https://doi.org/10.1016/j.gete.2022.100413
3. Sitarenios, P., Casini, F., Askarinejad, A., Springman, S.: Hydro-mechanical analysis of a surficial landslide triggered by artificial rainfall: the Ruedlingen field experiment. Géotechnique **71**(2), 96–109 (2021)
4. Guida, G., Vespo, V.S., Musso, G., Della Vecchia, G.: The role of hydraulic and thermal properties of soil on evaporation: a numerical insight. Submitted to Environmental Geotechnics, under review (2023)
5. Richards, L.A.: Capillary conduction of liquids through porous mediums. Physics **1**, 318–333 (1931)
6. Siemens, G.A., Peters, S.B., Take, W.A.: Comparison of confined and unconfined infiltration in transparent porous media. Water Resour. Res. **49**(2), 851–863 (2013)
7. Siemens, G.A., Take, W.A., Peters, S.B.: Physical and numerical modeling of infiltration including consideration of the pore-air phase. Can. Geotech. J. **51**(12), 1475–1487 (2014)
8. McWhorter, D.B.: Infiltration affected by flow of air. PhD thesis, Colorado State University (1971)
9. Bear, J., Cheng, A.H.-D.: Modeling Groundwater Flow and Contaminant Transport, vol. 23. Springer, Dordrecht (2010). https://doi.org/10.1007/978-1-4020-6682-5
10. Kendall, C., Caldwell, E.A.: Fundamentals of isotope geochemistry. In: Isotope tracers in catchment hydrology, pp 51–86. Elsevier (1998)
11. Bear, J., Braester, C., Menier, P.C.: Effective and relative permeabilities of anisotropic porous-media. Transp. Porous Media **2**, 301–316 (1987)
12. Lai, S.H., Tiedje, J.M., Erickson, A.E.: In situ measurement of gas diffusion coefficient in soils. Soil Sci. Soc. Am. J. **40**(1), 3–6 (1976)
13. Kimball, B.A., Jackson, R.D., Reginatoe, R.J., Nakayama, F.S., Idso, S.B.: Comparison of field-measured and calculated soil-heat fluxes. Soil Sci. Soc. Am. J. **40**(1), 18–25 (1976)
14. Cussler, E.L.: Diffusion: Mass Transfer in Fluid Systems. Cambridge University Press, New York (1997)
15. Van Genuchten, M.T.: A closed-form equation for predicting the hydraulic conductivity of unsaturated soils. Soil Sci. Soc. Am. J. **44**(5), 892–898 (1980)

The Shear Strength of Two Tectonized Clay Shales

Anna d'Onofrio[1] (✉), Luciano Picarelli[2], and Gianfranco Urciuoli[1]

[1] Università degli Studi di Napoli Federico II, via Claudio 21, 80125 Napoli, Italy
donofrio@unina.it

[2] Centro Euro-Mediterraneo sui Cambiamenti Climatici, CMCC, Lecce, Italy

Abstract. This paper presents the results on an investigation on the shear strength of two clayey facies of the Red Flysch, a tectonized formation that is widely present in the Southern Italian Apennines. The results of the investigation highlight the prominent role played by material fabric and, in spite of the difficulties in soil characterization and modelling due to the structural complexity of the material, appear quite consistent.

Keywords: Tectonized clay shales · laboratory tests · shear strength

1 Foreword

Highly fissured and sheared clay shales are widespread all along the Apennines chain posing severe problems to the stability conditions of pre-existing settlements and to construction of new infrastructures, such as roads and lifelines, of vital importance for local communities. Due to the structural complexity of the material, the knowledge about mechanical behaviour is still modest, nor data provided by the international literature significantly help the geotechnical characterization of the deposits.

The first studies on this subject started in the Seventies thanks to the initiative of outstanding scientists working in the Universities of Naples, Rome and Bari, and continued in the following two decades supported by funding from national research agencies. These studies led to some collective papers rich of data and of research insights [1, 2], which stressed the strong relationship between fabric and mechanical properties of the material highlighting at the same time the long way to run for a deep and complete knowledge of their behaviour. In the same years, parallel studies on hard fissured clays outcropping in other parts of the world, which represented sort of benchmarks, were being carried out also abroad [3, 4]. However, in the following years, despite the need to further improve the knowledge, the research has been declining. Today, only the isolated efforts of some academic groups testify the activity that is still carried out in this field [5–7].

This paper provides some data about the shear strength of two clayey facies of the Red Flysch, a tectonized "structurally complex" formation that outcrops in the Southern Apennines.

A. Ferrari et al. (Eds.): CNRIG 2023, SSGG, pp. 303–310, 2023.
https://doi.org/10.1007/978-3-031-34761-0_37

2 The Micro- and Meso-fabric of Tectonized Clay Shales

The backbone of the Italian peninsula is the result of a complex geological history that eventually led to formation of the Apennines chain [8]. During the Pliocene age, the sedimentary formations that were being deposited in the Thyrrenian Sea experienced a strong tectonic compression and consequent folding; as a result of the induced deviatoric stress field, the fine-grained soils took a special fabric characterized by aggregates of millimeter to centimeter shear lenses usually called "scales". At the end of Pliocene and during Quaternary, further tectonic events caused the uplift of the chain leading to further fracturing and disturbance of soil fabric and formation of large faults and persistent and widespread shear surfaces.

Figures 1b) and c) show some samples of "scaly" clay shale consisting of millimeter to centimeter platy fragments (*shear lenses* according to Skempton & Petley [3]) with polished surfaces (*minor shears*). In some way, the soil might be seen as a gravel characterized by weak and smooth platy particles. Considering that the single shear lenses are in turn made of clay and silt particles, a double fabric can be identified: a microfabric and a mesofabric. The micro-fabric (which is referred to the scale of millimeters to centimeters) is that of the "porous continuum" constituted by the clayey matrix, with its own porosity, which forms the shear lenses: notice however that, accounting for the interparticle bonds between the clayey particles, here the term microstructure will be preferred to microfabric [9]. The mesofabric (scale of centimeters to decimeters) is that of the "discontinuum" constituted by the shear lenses and by the network of fissures. In some cases (as for highly tectonized flysch deposits), the macrofabric may include also small lapideous fragments. The scaly clay shales then present a double porosity consisting of pores, which characterize the microfabric, and minor shears, which characterize the mesofabric. The aggregate of shear lenses, in turn, may be crossed by major (or principal) persistent shear surfaces (Fig. 1d). Therefore, at the scale of the engineering works (meters and more), a third fabric level (the macrofabric), which depends on spacing and distribution of the major discontinuities (faults, shear and slip surfaces) and of rock layers or blocks that are present in many deposits, might be recognized.

Fig. 1. a) Schematic representation of "scaly" clay shale; b) the Red Flysch in the Basento valley; c) the Red Flysch in the investigated area (MC facies); d) a principal shear in the MC facies.

3 General Features of the Investigated Soils

The soil samples analyzed in this research come from two distinct facies of the Red Flysch formation. This is part of the "Lagonegro-Molise basin units", present in a vast area in the Southern Apennines (Fig. 2a), which consists of calcareous and pelitic escarpment successions deposited over older deposits of the Mesozoic age. Typically, these successions display a disordered fabric at the scale of tens and hundreds of meters and more (megafabric), which is the result of mass transport phenomena due to submarine landslides occurred during sedimentary processes, and of tectonic events.

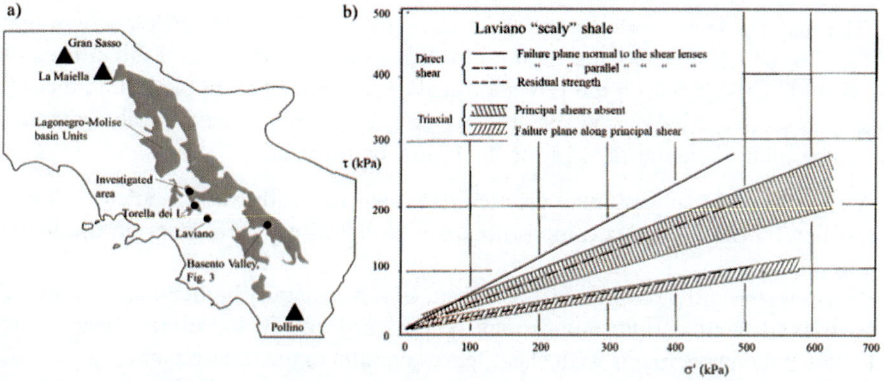

Fig. 2. a) Lagonegro-Molise basin units (after Vitale et al. [11]) and investigated site; b) results of drained direct shear and triaxial tests on the Laviano "scaly" clay shale [2].

The investigations consisted of 110 boreholes dug down to a maximum depth of 200 m within a 5 km long section (Fig. 2a), which allowed to retrieve numerous undisturbed samples. In the same area, several in situ tests were also carried out.

Seven lithological facies were recognized in the Red Flysch [10]; this allowed to address the mechanical characterization of the soil. The Grey Marly Clays (MC) and the Varicoloured Clays (CS) facies are the object of this paper. Both can be recognized as intensely fissured clay shales (class I6 according to the classification proposed by Vitone e Cotecchia [6]), with millimeter to centimeter "scales". The material is crossed by principal shears having a spacing of tens of centimetres. The index properties (see Fig. 3 a, b below) show a high variability; the differences between the two facies, which essentially consist of highly plastic clayey and sandy silts, appear however meaningless. The void ratio (which characterizes the mesofabric) ranges between 0.45 and 0.6 and between 0.5 and 0.7 respectively. The average values of the main index properties of the two facies at hand are presented in Table 1, where they are compared with the corresponding values measured on Red Flysch outcrops present in the Basento Valley and at Torella dei Lombardi (Fig. 2a) [2].

Table 1. I. Average properties of the Red Flysch. Legend: MC and CS: investigated facies; BM: Brindisi di Montagna; TL: Torella dei Lombardi

	N°	C_F(%)	W_L(%)
MC	83	27	66
CS	96	25	62
BM		26	59
TL		24	57

The major role of fabric on the shear strength of "scaly" clay shales is highlighted by Fig. 2b, which presents the failure envelopes obtained in drained direct shear and triaxial tests carried out on the Laviano "scaly" clay shale [2]. In particular, the direct shear tests were carried out on specimens with shear lenses either normal or parallel to the failure plane. The figure suggests the following remarks:

- A significant anisotropy may characterize materials with well-oriented "scales" as testified by the failure envelopes obtained by the direct shear tests on the Laviano clay shale;
- the shear strength measured in direct shear tests is generally higher than that provided by triaxial tests; similar values may be obtained only if the direct shear tests are conducted on specimens with shear lenses parallel to the failure plane;
- the shear strength measured in triaxial tests on specimens including principal shear surfaces is always well below the strength measured in direct shear tests, and very close to the residual value.

Such results indicate a high variability of the shear strength, which strongly depends on fabric. In fact, it ranges between a maximum value, which is measured when failure is imposed in the direction normal to the shear lenses, to a minimum one corresponding to the residual strength available along principal shears, highlighting in this way the major role of macrofabric. Moreover, when the imposed stress path allows a significant modification of the mesofabric leading minor shears to align to the failure planes (as in triaxial tests), the mobilized strength may correspond to the minimum measured in direct shear tests that is however higher than the residual strength.

Figures 3c, d, e show the results of direct shear tests (DS) on the two facies that have been investigated in the present research. These results are compared with those obtained on specimens taken in the Basento valley (Figs. 1b and 2a), where the Red Flysch outcrops as well. These soils present an anisotropic mesofabric (Fig. 1b), which again allowed to perform a series of tests in the direction of clay fragments and in the direction normal to it [2]. Unfortunately, the fabric of the two facies investigated in the present study is more complex and chaotic and cannot be ascribed to one or the other type. As shown in the figure, the peak strength is quite high, and is comparable with the one that has been measured on the Basento valley specimens in the direction normal to the failure plane; the variability of the soil response is also significant and compares quite well with the variability of the index properties (Figs. 3a, b), leading to a not negligible cohesive intercept. It is worth noting that almost all specimens tested in the present investigation showed a post-peak contractive behavior, whatever the normal stress, despite of the drop in strength recorded soon after peak. In this case, the post-peak strength decreases towards the residual value, which is generally due to a dilative soil behavior (void ratio increase), might be due to breaking of clay fragments, thus highlighting the influence of interparticle bonds, and hence of soil microstructure, on the shear strength.

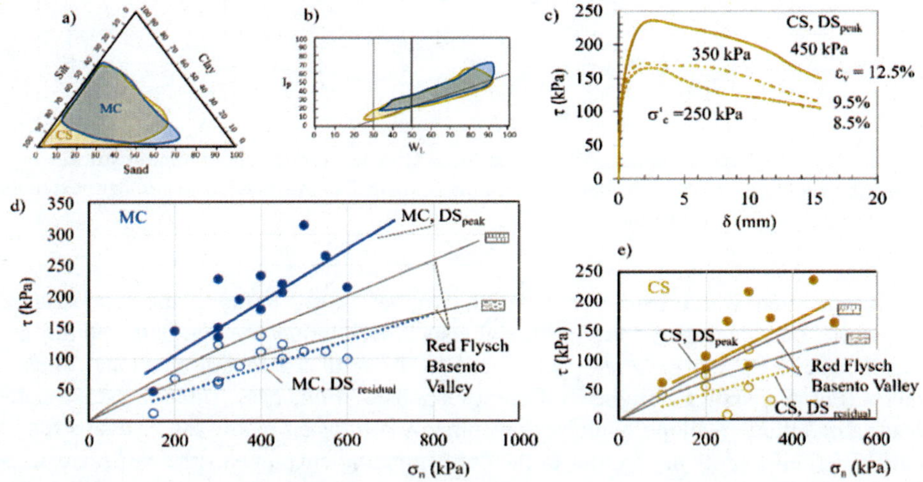

Fig. 3. a) Grain size and b) plasticity of the MC and CS clay shales; c) results of direct shear tests on the CS facies; failure envelopes from direct shear tests obtained on the MC (d) and CS (e) facies, compared with the results of tests on the Basento valley Red Flysch, in the directions normal and parallel to shear lenses.

Figure 4 shows the results of consolidated undrained triaxial tests (TX-CIU) on specimens of the MC facies sampled at depths less than 12 m. These data too show a post-peak shear strength decrease that is as higher as higher the confining stress, thus confirming the results of the direct shear tests.

The stress paths suggest an initially moderate pore water pressure increase followed by a continuous and gradual decrease, which however cannot prevent the observed decrease in strength. Figure 5 plots the results of the TX CIU tests in the q, p' plane. The specimens of the MC facies were consolidated under isotropic effective stresses, ranging between 60 and 600 kPa, consistent with the depth of samples from the ground surface; the varicolored clay (CS), sampled at higher depths, were tested under higher (between 300 and 600 kPa) effective stresses. In order to measure the critical friction angle of the MC facies, some tests were carried out on reconstituted normally consolidated specimens prepared at a water content higher than the liquid limit (1.5 w_L) and then compressed in a consolidometer. It is worth noting that the average index properties of the CS facies are very similar to those of the MC facies (see Figs. 2a, b), thus the critical friction angle is expected to be similar as well. To provide a general framework of all results, the shear strength envelopes obtained in the direct shear tests have been re-plotted in the q, p′ plane.

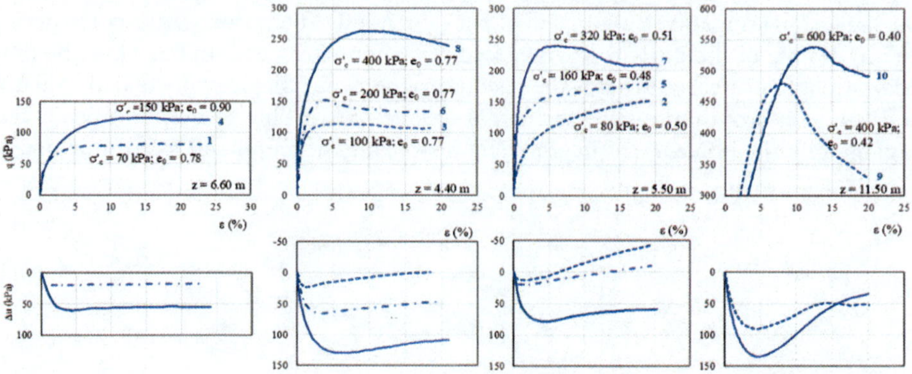

Fig. 4. Results of triaxial undrained tests on the MC facies at different confining stresses (σ'_c); the initial void ratio (e_0) and the depth of sampling (z) are also indicated. The number next to the curves identifies tests in Fig. 5a.

In agreement with previous considerations about the role of fabric, the shear strength of the MC facies measured in the triaxial tests is well below the one obtained through direct shear tests and presents peaks close to the critical strength envelope. The CS facies (Fig. 5b) shows a slightly different response. In this case, since the stress paths reach the failure envelope, which is presumably a little bit below the critical strength envelope (being probably similar to the corresponding envelope of the MC facies), the shear stress remains almost constant while the pore water pressure significantly decreases as sort of dilative shear band formed in the soil; in this way, as the test proceeds, the mobilized friction angle keeps lower and lower values approaching the residual value. This interpretation, however, would deserve to be confirmed by local pore pressure readings since the generation of not uniform excess pore water pressures is not an unlikely event in these terrains [12]. Similar data are reported by Olivares & Picarelli [13] through an in-depth investigation on the "scaly" Bisaccia clay shale. In particular, they compare the material behaviour to that of dense coarse-grained soils, whose friction angle φ' is generally thought to depend on the combination of two different factors, a basic friction angle, which is identified with the critical friction angle, φ'_{cs}, and a dilation angle, i_{eff}, that depends on soil porosity, thus $\varphi' = \varphi'_{cs} + i_{eff}$. As the mean effective stress increases, i_{eff} decreases up to vanish; therefore, at high stresses, $\varphi' = \varphi'_{cs}$. Considering the nature of the interparticle friction of these materials, which is very low and probably not far from the residual value, the Authors suggest assuming $\varphi'_{cs} = \varphi'_r$, φ'_r being the residual friction angle. In the case of "scaly" clay shales, the basic friction angle at high stresses should then tend towards the residual value due to the progressive alignment of shear lenses in the direction of the failure plane. On this point, is worth to note that, working on similar materials, Nardelli et al. [14] measured an inter-scales friction angle close to the critical friction angle.

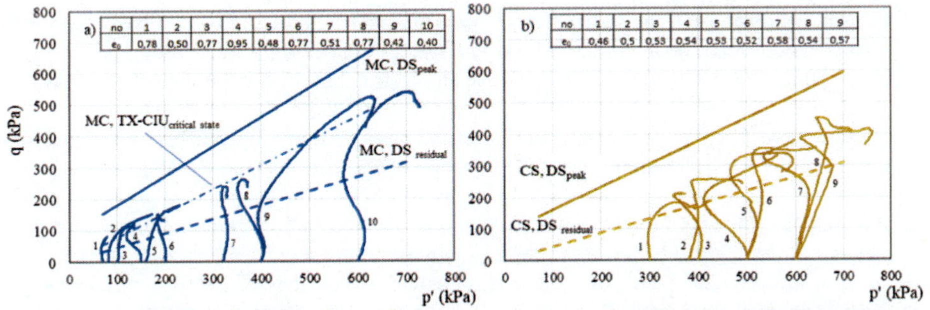

Clay		Shear strength	Test	Acronym	c' (kPa)	φ'(°)
MC	Undisturbed	Peak	DS	MC-DS$_{peak}$	43	21°6'
			TX CIU	MC-TX-CIU$_{peak}$	20	19°8'
		Ultimate	TX CIU	MC-TX-CIU$_{ult}$	0	17°1'
		Residual	DS	MC-DS$_{residual}$	0	12°0'
	Reconstituted	Critical state	TX CIU	MC-TX-CIU$_{critical\ state}$	0	20°0'

Clay		Shear strength	Test	Acronym	c' (kPa)	φ'(°)
CS	Undisturbed	Peak	DS	CS-DS$_{peak}$	41	18°9'
			TX CIU	CS-TX-CIU$_{peak}$	0	16°7'
		Ultimate	TX CIU	CS-TX-CIU$_{ult}$	0	13°1'
		Residual	DS	CS-DS$_{residual}$	0	11°9'

Fig. 5. Stress paths of CIU triaxial tests on the MC (a) and CS (b) facies and critical shear strength envelope of the MC facies (a), compared with the strength envelopes obtained in direct shear tests (Fig. 4).

4 Conclusions

The "scaly" clay shales, which largely outcrop all along the Apennines chain display a complex fabric, which strongly affects the hydraulic and mechanical properties. This paper has presented the results of some laboratory tests aimed at the characterization of the shear strength of two facies of the Red Flysch formation.

The test results suggest sort of interaction between the mechanism of failure induced by testing equipment and soil fabric. In fact, while direct shear tests highlight the role of both microstructure and mesofabric, the results of triaxial tests put into evidence the role of the mesofabric and possibly of the macrofabric, leading in general to lower shear strength parameters. Unfortunately, the problem does not end here since the soil response in boundary value problems is certainly and mostly affected by macrofabric, i.e. by major discontinuities, such as shear surfaces, and by lapideous elements (rock layers, blocks or fragments), when present. This of course rises major uncertainties in the selection of the soil parameters to use in the design. Further research is then needed to give an answer to the pressing requests coming from the labour market about the parameters to use in geotechnical analyses.

Acknowledgments. This research has been supported by MIUR PON R&I 2014–2020 Program (project MITIGO, ARS01_00964).

References

1. A.G.I.: Some Italian experiences on the mechanical characterization of structurally complex formations. In: Proceedings of the 4th International Congress Rock Mechanics, Montreux, vol. 1, pp. 1–20 (1979)

2. A.G.I.: Geotechnical properties and slope stability in structurally complex clay soils. Geotechnical Engineering in Italy. An overview. A.G.I., pp. 189–225 (1985)
3. Skempton, A.W., Petley, D.J.: The strength along structural discontinuities of stiff clays. In: Proceedings of the Geotechnical Conference, Oslo, vol. 2, pp. 29–46 (1967)
4. Morgenstern, N.R., Cruden, D.: Description and classification of geotechnical complexities, pp. 195–203. In: Proceedings of the Geotechnics of Structurally Complex Formations, Capri (1977)
5. Picarelli, L., Olivares, L., Di Maio, C., Silvestri, F., Di Nocera, S., Urciuoli, G. Structure, properties and mechanical behaviour of the highly plastic intensely fissured Bisaccia Clay Shale. In: Tan, T.S., Phoon, K.K., Hight, D.W., Leroueil, S. (eds.) Characterisation and Engineering Properties of Natural Soils, Singapore, vol. 2, pp. 947–982. Springer (2002)
6. Vitone, C., Cotecchia, F.: The influence of intense fissuring on the mechanical behaviour of clays. Géotechnique **61**(12), 1003–1018 (2011)
7. Napoli, L., Festa, A., Barbero, M.: Practical classification of geotechnically complex formations with block-in-matrix fabrics. Eng. Geol. **31**, 106595 (2022)
8. Patacca, E., Scandone, P.: Geology of the Southern Apennines. Bollettino della Società Geologica Italiana **7**, 75–199 (2007)
9. Mitchell, J.K.: Fundamentals of Soil Behaviour. J. Wiley & Sons, New York (1976)
10. d'Onofrio, A., Picarelli, L., Santo, A., Urciuoli, G.: The Red Flysch formation in Southern Apennines: Lithological and structural features and challenges in geotechnical characterization and modelling. Rock Mechanics and Rock Engineering, submitted for publication (2023)
11. Vitale, S., et al.: Structural, stratigraphic and petrological clues for a Cretaceous Paleocene abortive rift in the southern Adria domain (southern Apennines, Italy). Geol. J. **53**(2), 660–681 (2018)
12. Picarelli, L., Olivares, L., Di Maio, C., Urciuoli, G.: Properties and behaviour of tectonized clay shales in Italy. In: Proceedings of "The Geotechnics of Hard Soils - Soft Rocks", vol. 3, pp. 1211–1241 (2000)
13. Olivares, L., Picarelli, L.: Discussion on "A laboratory study of the strength of four stiff clays" by J.B. Burland, S. Rampello, V.N. Georgiannou, G. Calabresi. Géotechnique **46**(3), 491–514 (1999)
14. Nardelli, V., Coop, M.R., Vitone, C., Chen, S.: The inter-scale behaviour of two natural clays. Géotech. Lett. **6**, 205–210 (2016)

An Elastoplastic Framework Accounting for Changes in Matric and Osmotic Suction in Unsaturated Non-expansive Clays

Liliana Gramegna[1(✉)], Ayman A. Abed[2], Wojciech T. Sołowski[3], Guido Musso[4], and Gabriele Della Vecchia[1]

[1] Department of Civil and Environmental Engineering, Politecnico di Milano, P.zza Leonardo da Vinci 32, 20133 Milan, Italy
liliana.gramegna@polimi.it
[2] Department of Architecture and Civil Engineering, Chalmers University of Technology, Gothenburg, Sweden
[3] Department of Civil Engineering, Aalto University, Rakentajanaukio 4A, P.O. Box 12100, 02150, 00076 Espoo, Aalto, Finland
[4] Dipartimento di Ingegneria Strutturale, Edile e Geotecnica (DISEG), Politecnico di Torino, Corso Duca degli Abruzzi 24, 10129 Turin, Italy

Abstract. The mechanical behaviour of clays is significantly influenced by their salinity and degree of saturation, which are expected to change in both natural and anthropogenic environments. This influence is triggered by fabric changes. Capillary forces, related to changes in the degree of saturation, and electrochemical interactions, related to changes in salinity, affect differently the interaction between particles. However, at the continuum scale, it is possible to pinpoint some common characteristics. This paper focuses on the modelling capabilities of an elasto-plastic framework formulated to reproduce the behaviour of unsaturated non-expansive clays exposed to changes in matric and osmotic suction, by introducing osmotic suction on BBM-like models in terms of its effects on the normal compression line. The model, calibrated on experimental data on Boom clay (Mokni et al., 2014) and remoulded loess (Zhang et al., 2022), has been implemented in the Thebes code and used to predict material response under selected chemo-hydro-mechanical paths.

Keywords: Constitutive law · Osmotic suction · Matric suction · Clay fabric · Non-expansive clay

1 Introduction

Changes in the degree of saturation and the chemical composition of the pore fluid have significant effects on clay behaviour, as they affect volume, permeability, retention properties, compressibility and shear strength. In presence of severe safety requirements, the combined effect of the degree of saturation and pore fluid chemistry should thus be considered for design. In particular, unsaturated clays with salts dissolved in the pore

A. Ferrari et al. (Eds.): CNRIG 2023, SSGG, pp. 311–318, 2023.
https://doi.org/10.1007/978-3-031-34761-0_38

fluid are subjected not only to matric suction, but also to osmotic suction, which is related to the molar concentration of the dissolved salt in the pore fluid. Both matric and osmotic suction changes affect the evolution of clay microstructure, and thus the unsaturated clay behaviour. As evidenced by microstructural observations, clay microstructure consists of clay clusters or aggregates, whose size evolves not only with water content but also with pore water salinity, with relevant effects on clay macro-porosity (see, e.g., Musso et al. (2013)). Constitutive models at the continuum scale accounting for microstructure changes have been developed for expansive clays, while less attention has been paid to the hydro-chemo-mechanical behaviour of non-expansive clays. Starting from the experimental evidence on saturated non-expansive clays presented in Torrance (1974) and Musso et al. (2022), this contribution proposes a constitutive framework capable of reproducing the mechanical effects of simultaneous matric and osmotic suction changes. The approach relies on the similarity between the phenomenological effects of matric and osmotic suction on the mechanical behaviour in the compression plane, which in turn depends on fabric evolution (Scelsi et al., (2021)). In this work, a simple constitutive model has been obtained by joining the Barcelona Basic Model (BBM, Alonso et al. (1990)) and the model proposed by Musso et al. (2022) (MSD model). The model, implemented in the Thebes code (Abed and Sołowski, (2017), (2022)), is used to simulate experiments on compacted Boom clay (Mokni et al., (2014)) and a remoulded loess (Zhang et al., (2022)), subjected to different stress paths at varying matric and osmotic suctions.

2 Constitutive Framework

Matric suction is related to capillary forces, acting on solid particles in granular materials and on clusters/aggregates of clay particles. Soil deposition and compaction may induce an 'open fabric' (Alonso et al., (1990)), related to the existence of macro-voids between the clusters/aggregates. Phenomenological evidence of the presence of an open fabric is the fact that, under Normally Consolidated conditions, a higher matric suction allows the material to sustain larger net stresses. Then, the Normal Compression Line (NCL) in the net stress – void ratio plane (σ-u_a, e) moves at higher void ratios with matric suction. In such conditions, a higher matric suction implies then also a larger elastic domain. Consistently, if an unsaturated material with an open fabric is subjected to wetting at constant stress, a reduction in matric suction makes the current void ratio non-compatible with the applied stress and compressive plastic strains develop (the so-called wetting induced collapse). This aspect has been modelled by the so-called Loading-Collapse (LC) curve, as proposed by Alonso et al. (1990) in the BBM model and then recognized as a fundamental feature in any constitutive model for unsaturated soils. However, it is acknowledged that also pore water salinity influences the fabric of non-expansive clays, depending on the soil formation process and the electrochemical environment. When deposition occurs in salty water, non-expansive clays tend to have a "flocculated" fabric, characterized by the presence of larger macro-porosity between clay clusters, with respect to the face-to-face aggregated fabric formed in freshwater. Although less pronounced, similar effects were detected also for compacted clays, at least at sufficiently low-stress levels. Consistently with the evidences for unsaturated soils, Musso et al. (2022) proposed a link between the fabric of non-expansive clayey materials and their phenomenological behaviour, highlighting that the position of the NCL moves towards higher void ratios (open fabric) and its slope changes with increasing pore fluid salinity. Thus, as suggested by experimental evidence (see e.g. Torrance (1974), Musso et al. (2022)), when salinity decreases, compressive irreversible strains are anticipated, in the same way as wetting induces the collapse of unsaturated soils. However, if the material is strongly over-consolidated (i.e. closed fabric), desalinization induces swelling, consistently with the increase of repulsive forces between particles caused by the increase in thickness of the double layer. The analogy with matric suction also works for mechanical loading at constant osmotic suction, as its increase induces an increase in the size of the elastic domain.

2.1 Matric Suction Effect: Barcelona Basic Model

The classical BBM, developed in the framework of hardening elastoplasticity, uses net stress $\sigma'' = \sigma - u_a$ and matric suction $s = u_a - u_w$ as independent variables, where σ, u_w and u_a are the total stress, pore water pressure and pore air pressure, respectively. As the water content influences the soil fabric, BBM defines the Normal Compression Lines (NCL) for different values of suction in the $v - \ln(p'')$ plane as:

$$v = N_s - \lambda_s \ln \frac{p''}{p_r} \tag{1}$$

where N_s is the specific volume for a reference mean stress p_r and λ_s is the slope of the Normal Compression Line. Both N_s and λ_s are functions of suction (see Fig. 5). BBM assumes the loading-collapse (LC) yield curve in the $s - p''$ plane as:

$$p_0 = p_r \cdot \left(\frac{p_0^*}{p_r}\right)^{\frac{\lambda_0 - k}{\lambda_s - \kappa}} \tag{2}$$

where p_0 is the pre-consolidation pressure at matric suction s, p_0^* is the isotropic pre-consolidation pressure at full saturation ($s = 0$) and λ_0 is the slope of the normal compression line in saturated conditions. Equation (2) introduces the dependence of the size of the yield surface on suction. In particular, the larger the matric suction, the larger the size of the yield surface.

2.2 Osmotic Suction Effect: The Musso-Scelsi Della Vecchia Model

Musso et al. (2022) proposed a constitutive model (MSD model in the following) based on the Terzaghi effective stress σ' and on osmotic suction π. Osmotic suction has been set to depend on the chemical activity of the components in solution and, at low concentrations, can be described by the van't Hoff equation, as $\pi = icRT$, being i the number of dissolved species, c the molar concentration of the electrolyte, R the universal gas constant and T the absolute temperature. As the pore fluid chemistry influences the soil fabric, the normal compression line in the semilogarithmic plane v-$\ln(p')$ depends on the saline concentration:

$$v = N_c - \lambda_c \ln \frac{p'}{p'_{rc}} \tag{3}$$

where the intercept N_c and slope of the Normal Compression Line λ_c are both functions of the osmotic suction, while p'_{rc} is a reference pressure (here taken as 1 kPa). The original expression of the yield curve in the $\pi - p'$ plane can be found in Musso et al (2022), but assuming that the dependence of the volumetric virgin compressibility λ on the pore fluid salinity for non-expansive clays is quite limited (i.e. $\lambda_c = \lambda_0$) and that $N_0 = N_c$, the evolution of the size of the yield surface with osmotic suction for an isotropic stress state can be simplified as

$$p'_c = p_0^* \cdot \left(\frac{\pi + \pi_{ref}}{\pi_{ref}}\right)^{\frac{\kappa_\pi}{\lambda_0 - \kappa}} \tag{4}$$

where p'_c is the preconsolidation pressure corresponding to osmotic suction π and κ_π is the elastic logarithmic volumetric stiffness related to changes in osmotic suction. Like in the BBM, plastic strains may develop due to either a reduction of the osmotic suction or a stress change.

2.3 Combining Matric and Osmotic Suction Effects

For soils where the pore liquid is pure water, the BBM links the saturated preconsolidation pressure p_0^* to the unsaturated yield mean net stress, p_0, through Eq. (2). Following the same rationale, the chemo-mechanical model introduces a relation between the preconsolidation pressure for the material saturated with distilled water, p_0^*, and the yield mean effective stress for material saturated with saline solutions, $p'_c(\pi)$, through Eq. (4). Both yield curves in the $p' - s$ plane and in the p' - π plane play the role of the Loading Collapse (LC) curve and have similar trends (see Fig. 1). The approach proposed to combine the effect due to partial saturation and salinity is to add them in terms of LC curve, thus introducing osmotic suction in BBM as an 'equivalent' matric suction s_π. Introducing in the BBM changes in the matric suction s_π equivalent to the osmotic suction would cause similar effects to those caused by the osmotic suction changes in the MSD model. According to the MSD model, when a saline pore fluid is present, the preconsolidation pressure increases from p_0^* to p'_c. The equivalent matric suction s_π is the matric suction that would be obtained from the LC curve of the BBM passing from the saturated pre-consolidation pressure, p_0^*, to p'_c, as depicted in Fig. 3. Therefore, imposing that $p_0 = p'_c$, the expression of $\lambda_s(s_\pi)$ can be obtained from Eq. (2):

$$\lambda_s = (\lambda_0 - \kappa) \cdot \frac{\ln \frac{p_0^*}{p_r}}{\ln \frac{p'_c}{p_r}} + \kappa \tag{5}$$

Exploiting the link between the compressibility dependence on matric suction in BBM, modified to reproduce the increase in compressibility for increasing matric suction (see, e.g. Della Vecchia et al., (2013)), the following expression for the equivalent matric suction is obtained (see Scelsi et al, (2021), for details):

$$s_\pi = -\frac{1}{\beta} \ln \left[\frac{\lambda_s + \lambda_0(r-2)}{\lambda_0(r-1)}\right] \tag{6}$$

where r is a parameter already present in the original BBM formulation. The equivalence of the microstructural effects of both matric and osmotic suction leads to a common LC curve for both effects. BBM is finally enhanced by introducing an equivalent suction s_{ψ}, used in the same way as suction in the original BBM:

$$s_{\psi} = s + s_{\pi} \quad (7)$$

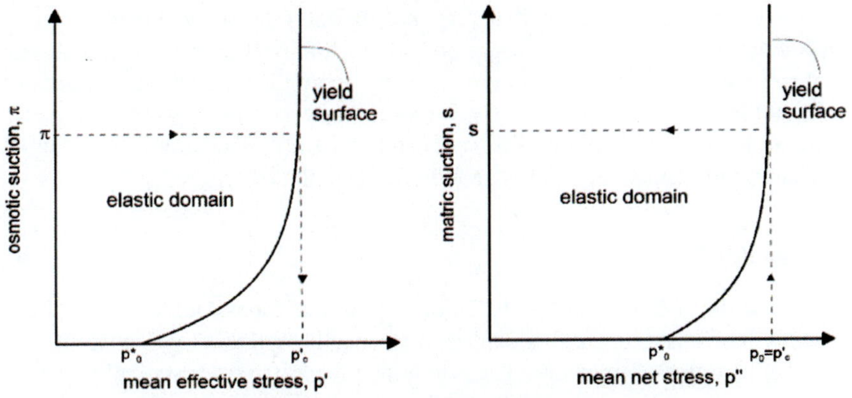

Fig. 1. Procedure to obtain the equivalent matric suction

A yield surface for the first-ever chemical loading has been also introduced, to reproduce the plastic volumetric strain related to the change in soil structure upon the first salinization, as

$$\pi = \pi_0 \quad (8)$$

The maximum osmotic suction ever experienced by the soil π_0 should also be converted to an equivalent matric suction $s_{\pi 0}$ in order to be introduced to model. Once this yield locus is reached, the plastic volumetric strain increment is evaluated as:

$$d\varepsilon_{vs}^{p} = -\frac{\lambda_{\pi}}{v}\frac{ds_{\pi,0}}{\left(s_{\pi,0} + p_{atm}\right)} \quad (9)$$

where λ_{π} is the stiffness parameter for changes in (osmotic) suction for virgin states of the soil.

3 Model Predictions

3.1 Compacted Boom Clay

Mokni et al (2014) performed oedometer tests to investigate the hydro-mechanical response of compacted Boom clay (a kaolinitic-illitic clay of low-medium activity) subjected to different matric and osmotic suction. Figures 2 and 3 show the experimental results and model prediction with respect to the distinct effects of matric and osmotic suction on the compression behaviour of compacted Boom clay. Specimens had been obtained by mixing dry Boom clay powder with saline solutions (4M, 5.4M of $NaNO_3$, corresponding to $\pi = 11000$ kPa and 20000 kPa, respectively). After compaction, specimens have been subjected to either $s = 500$ kPa or $s = 0$ kPa (saturated conditions) and finally loaded and unloaded at constant suction. It is evident that the similar fabric effects are also inducing a similar mechanical response in the compression plane. Model predictions are satisfactory not only along constant suction paths, but also in terms of salinization path ($\pi = 31$ MPa) at constant matric suction ($s = 500$ kPa), as shown in Fig. 4. The chemo-mechanical path applied consists of a mechanical loading at constant matric suction up to the vertical stress of 50 kPa, followed by salinization at constant vertical stress and subsequent loading/unloading. In all the simulations, p_0^* has been set equal to 90 kPa, $p_r = 3000$ kPa and $\pi_0 = 1$ kPa. Model parameters used for the simulations are summarized in Table 1. When not defined in the paper, parameter nomenclature is consistent with Alonso et al. (1990) and Musso et al. (2022).

3.2 Remoulded Loess

Experimental data (Zhang et al., (2022)) on a remoulded loess coming from the Gansu Province (China) have been also exploited to check the predictive capabilities of the proposed constitutive framework. The remoulded loess had been statically compressed in oedometer to a void ratio of 0.53. Oedometer tests were then performed, by loading the specimens up to 25 kPa and then exposing them to a given value of total suction ψ_0 (via relative humidity control). Due to the lack of information related to the as-prepared suction of the specimens, the drying stage has not been simulated and, according to the volume change data presented by Zhang et al. (2022), the change in the void ratio of this stage has been considered negligible. Later, samples had been soaked with different solutions (i.e. distilled water, NaCl solution, Na_2SO_4 solution), corresponding to several values of input osmotic suction π_{in} and finally subjected to a standard oedometer loading/unloading, up to a maximum vertical stress of 1600 kPa. The model has been used to reproduce material response in the compression plane. Material parameters related to unsaturated soil behaviour have been calibrated on the test involving soaking with distilled water. According to the very small deformation upon drying-wetting paths, the parameter κ_s has been set equal to 0.001. Figure 5 shows the experimental data and model predictions for soaking tests with two different saline solutions, i.e. a 0.29 mol/l Na_2SO_4 solution and a 0.17 mol/l NaCl solution (corresponding to input osmotic suctions of $\pi_{in} = 1.39$ MPa and $\pi_{in} = 0.74$ MPa, respectively). Initial total suctions ψ_0 were equal to 14.0 MPa and 20.6 MPa, respectively. In this case, the presence of compressive volumetric strain upon soaking is evident. From the modelling point of view, this volumetric strain is related to the first salinization path, rather than a collapse upon wetting. This evidence is consistent with the elastic reloading path evident in the compression curve after soaking, which would not be present after a wetting induce collapse path. Model parameters used for the simulations are again summarized in Table 1. As for the initial values of the internal variables, p_0^* has been set equal to 50 kPa, $p_r = 3000$ kPa and $\pi_0 = 5$ kPa.

Table 1. Model parameters

	κ	λ_s	ν	M	κ_π	r	β	κ_s	λ_π
Boom C	0.03	0.28	0.33	0.86	0.01	0.65	0.008 kPa^{-1}	0.05	0.26
Loess	0.02	0.062	0.4	1	0.0001	0.65	0.00016 kPa^{-1}	0.001	0.15

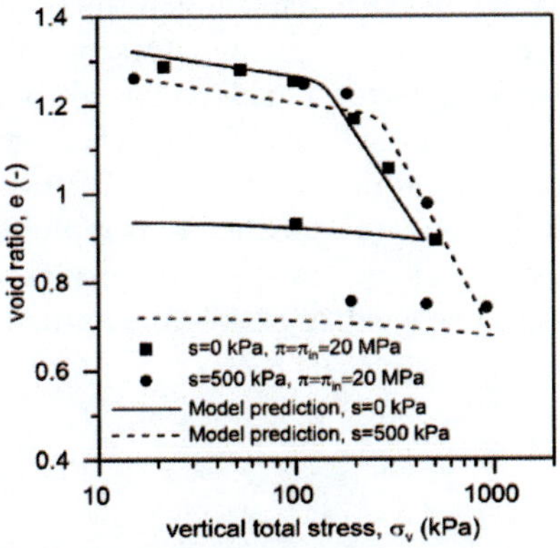

Fig. 2. Effect of matric suction on the compression behaviour of Boom clay (data from Mokni et al, 2014).

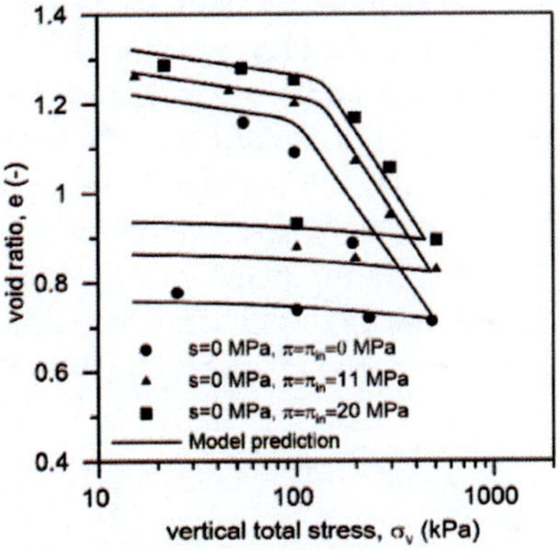

Fig. 3. Effect of osmotic suction on the compression behaviour of Boom clay (data from Mokni et al, 2014).

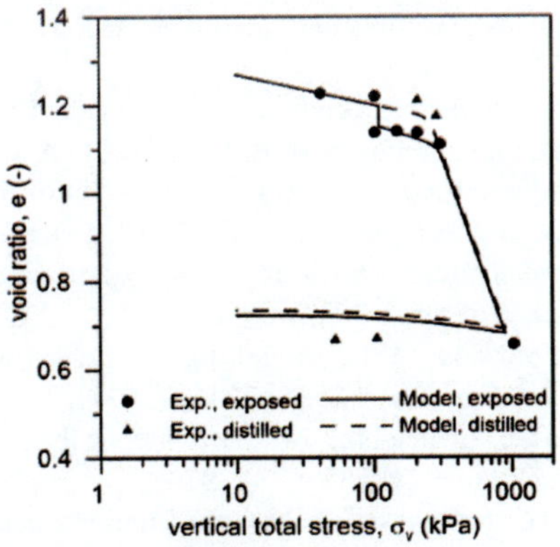

Fig. 4. Chemo-mechanical loading path on compacted Boom clay (data from Mokni et al, 2014).

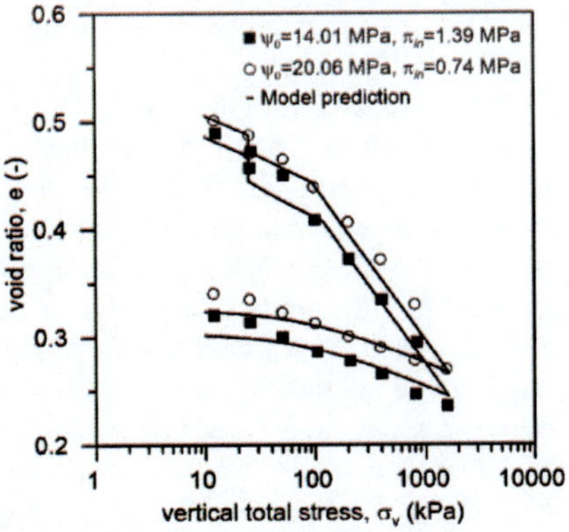

Fig. 5. Chemo-mechanical loading path on remoulded loess (data from Zhang et al, 2022).

4 Conclusions

Microstructural observations on low-medium activity clay materials show that both the increase in matric suction and pore fluid salt concentration may cause a transition from an open to a close fabric. This evidence allows the introduction of a constitutive framework for soils partially saturated with saline solutions, where both salinity and degree of saturation change simultaneously. In this paper, we proposed a simple elastoplastic model, joining together the well-known BBM with a recently published model, MSD, capable of accounting for the salinity of the pore fluid. The model introduces the effect of the osmotic suction in the BBM as an equivalent matric suction, i.e. a matric suction in the BBM which causes the same effects on void ratio as those caused by the osmotic suction in the MSD model. The model is able to well reproduce the stiffness changes of the soil induced by the suction variations and the plastic chemical deformations that develop when the clayey material is exposed for the first time to an increase in osmotic suction. Remarkably, it is formulated within a simple and widespread framework, without requiring the introduction of new parameters. The similar effects of osmotic and matric suctions in fact allow introducing a unique framework for soils partially saturated with saline solutions, where salinity and saturation changes are expected, thus reproducing and predicting the behaviour of moderate plasticity clays. However, the extension of model framework to unsaturated highly expansive clays would require the introduction of a double-porosity approach and further research.

References

Abed, A.A., Sołowski, W.T.: A study on how to couple thermo-hydro-mechanical behaviour of unsaturated soils: physical equations, numerical implementation and examples. Comput. Geotech. **92**, 132–155 (2017)

Abed, A.A., Sołowski, W.T.: Finite element method algorithm for geotechnical applications based on Runge-Kutta scheme with automatic error control. Comput. Geotech. **128**, 103841 (2020)

Alonso, E.E., Gens, A., Josa, A.: A constitutive model for partially saturated soils. Géotechnique **40**(3), 405–430 (1990)

Mokni, N., Romero, E., Olivella, S.: Chemo-hydro-mechanical behaviour of compacted Boom Clay: joint effects of osmotic and matric suctions. Géotechnique **64**, 681–693 (2014)

Musso, G., Romero, E., Della Vecchia, G.: Double-structure effects on the chemo-hydro-mechanical behaviour of a compacted active clay. Géotechnique **63**, 206–220 (2013)

Musso, G., Scelsi, G., Della Vecchia, G.: Chemo-mechanical behaviour of non-expansive clays accounting for salinity effects. Géotechnique (ahead of print, 2022)

Scelsi, G., Abed, A.A., Della, V.G., Musso, G., Sołowski, W.T.: Modelling the behaviour of unsaturated non-active clays in saline environment. Eng. Geol. **295**, 106441 (2021)

Torrance, J.K.: A laboratory investigation of the effect of leaching on the compressibility and shear strength of Norwegian marine clays. Géotechnique **24**, 155–173 (1974)

Zhang, T., Hu, Z., Lan, H., Deng, Y., Zhang, H.: The compression behavior of undisturbed and compacted loess under the controlling of total suction and injected solutions. Front. Earth Sci. **10** (2022)

Numerical Study on Bentonite Permeability Evolution upon Water Hydration

Liliana Gramegna[1(✉)], Robert Charlier[2], and Gabriele Della Vecchia[1]

[1] Politecnico di Milano, Department of Civil and Environmental Engineering, P.zza Leonardo da Vinci 32, 20133 Milan, Italy
liliana.gramegna@polimi.it

[2] Département ArGEnCo - Géotechnique, Géomécanique et Géologie de l'Ingénieur, Université de Liège, 4000 Liège 1, Belgium

Abstract. Water permeability plays a key role in many geomechanical problems, but its experimental determination in unsaturated conditions still represents a challenging task. Such complexity is even more evident when dealing with active clays, like bentonite. Bentonite primarily consists of montmorillonite and, when compacted, is characterized by a multi- porosity structure. Its microstructure evolves upon water saturation, influencing the hydro-mechanical response at the laboratory scale. Experimental evidences highlight how bentonite saturated permeability spans over several orders of magnitude in the range of the most common dry densities and significant permeability variation occurs upon desaturation and pore structure evolution. The present work aims to analyze and reproduce by mean of numerical simulations a set of experimental swelling pressure tests in isochoric conditions performed on bentonites characterized by different initial dry densities and initial microstructures (i.e. granular bentonite and compacted blocks). Taking advantage of numerical simulations, performed via the finite element code LAGAMINE, the hydro-mechanical response of these bentonite assemblies is examined and a better understanding on the influence of the multi-porosity evolution on permeability is provided.

Keywords: Bentonite · numerical modelling · numerical simulations · permeability · double-porosity

1 Introduction

Water transport characteristics are critical in many geotechnical applications, but their experimental determination in unsaturated conditions remains a complex task. In the context of bentonite barriers for underground nuclear waste disposal, an appropriate prediction of permeability evolution upon hydration is considered a priority to safely determine the seal full-saturation time. When barriers are composed of high-density compacted blocks and low-density pellets materials, as in the EB experiment (Alonso et al. 2010), such heterogeneity can result in complex permeability distribution, which can promote preferential flow-paths for radionuclides. This work aims to propose the

A. Ferrari et al. (Eds.): CNRIG 2023, SSGG, pp. 319–326, 2023.
https://doi.org/10.1007/978-3-031-34761-0_39

use of numerical analyses to highlight the role of permeability evolution, allowing the interpretation of laboratory experimental tests as boundary value problems. Isochoric tests hydration on Febex bentonite specimens composed of combined compacted blocks and loose pellets mixture are simulated via the finite element code LAGAMINE (Collin et al. 2002). Numerical simulations of permeability distribution over the specimens are presented and analyzed. Dry density and water content distribution after the tests are compared with post mortem experimental data, showing the satisfactory predictive capabilities of the numerical approach proposed.

2 Hydro-Mechanical Model

The experimental hydration tests, described in Sect. 3, have been reproduced via the integration of two balance equations, namely the linear momentum balance and the water mass balance. The linear momentum balance equation is written as:

$$\nabla \cdot \boldsymbol{\sigma_t} + \boldsymbol{b} = 0 \tag{1}$$

where $\boldsymbol{\sigma_t}$ is the total (Cauchy) stress tensor and $\boldsymbol{b}$ is the body force vector. Water mass balance equation reads:

$$\frac{\partial}{\partial t}(\rho_w \phi S_r) + \nabla \cdot \left(\rho_w \boldsymbol{q_w}\right) = Q_w \tag{2}$$

where ρ_w is the unit weight of liquid water, ϕ is the porosity, S_r is the degree of saturation, Q_w is the external water supply, and $\boldsymbol{q_w}$ is the Darcy velocity vector, which in turn depends on the water pressure u_w gradient via:

$$\boldsymbol{q_w} = -\frac{k_{rw}(S_r) K_w}{\mu_w}(\nabla u_w + \rho_w \boldsymbol{g}), \tag{3}$$

being μ_w the water dynamic viscosity, K_w the water permeability in saturated conditions ($S_r = 1$), and k_{rw} the relative permeability function. The dependence of the relative permeability and S_r is expressed by a power law:

$$k_{rw} = S_r^{n_k} \tag{4}$$

where n_k is a model parameter. The double-porosity structure of the material is accounted for just in terms of hydraulic properties. In particular, the dependence of the permeability on just the evolution of the macro-porosity is reproduced via an enhanced version of the Kozeny-Carman law, modified to introduce explicitly the macrostructural void ratio $e_M = e - e_m$ (being e the total void ratio and e_m the microstructural void ratio), according to Dieudonnè and Charlier (2017):

$$K_w = K_{w0} \frac{(1 - e_{M0})^M}{e_{M0}^N} \frac{e_M^N}{(1 - e_M)^M}, \tag{5}$$

where N and M are model parameters and K_{w0} is a reference permeability (corresponding to a macro-structural void ratio e_{M0}). The evolution of permeability considered in the numerical simulations shown in Sect. 3 (parameters in Table 2) are shown in Fig. 1. Following Romero et al (2011) and according to Dieudonné et al. (2017) and Della Vecchia et al. (2015), the evolution of e_m has been set to depend on the water ratio $e_w = S_r e$, as:

$$e_m = \beta_0 e_w^2 + \beta_1 e_w + e_{m0}. \tag{6}$$

e_{m0} is microstructural void ratio of the dry material, while β_0 and β_1 are parameters dependent on the swelling tendency of the aggregates. As for the water retention curve, the double-porosity model proposed by Dieudonné et al. (2017) has been considered. The model, which includes both capillary water in the macro-porosity and adsorbed water in the microstructure, reads:

$$S_r = \frac{e_m}{e}\exp\left[-(C_{ads}s)^{n_{ads}}\right] + \frac{e-e_m}{e}\left\{1+\left[(e-e_m)\frac{s}{A}\right]^n\right\}^{-m}, \tag{7}$$

where s is matric suction, A, m, and n are the material parameters controlling the macro-structural water retention domain and C_{ads} and n_{ads} are the material parameters controlling the microstructural water retention behavior. Figure 2 presents the water retention curves predicted by the model for Febex bentonite at different dry densities ($\rho_d = 1.60$ Mg/m^3 and $\rho_d = 1.28$ Mg/m^3).

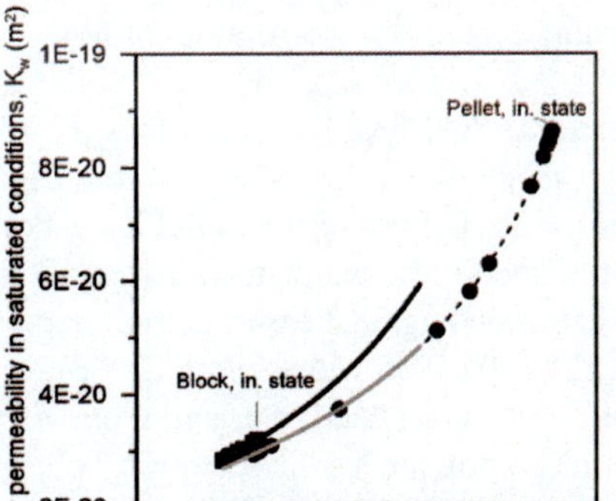

Fig. 1. Permeability evolution as function of macro void ratio.

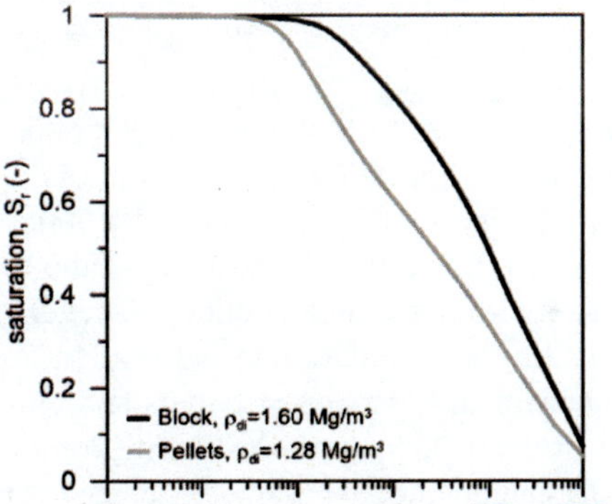

Fig. 2. Water retention curves for Febex bentonite at different dry density.

As for the mechanical behavior of the material, the Barcelona Basic Model (BBM) (Alonso et al., 1990) has been adopted. Despite it has not been explicitly developed for active clays, its predictive capabilities in reproducing bentonite behaviour upon water saturation in various testing conditions have already been proved (Gramegna et al., 2020, 2022).

3 Numerical Modelling

3.1 Experimental Test Description

Villar et al. (2021) performed several hydration tests in isochoric conditions on Febex bentonite specimens composed of contiguous layers of compacted block and pellets mixture, with different initial dry densities and structures. Among those, three tests have been considered in this paper: tests MGR21, 23 and 24 The tests started from almost-identical initial conditions and were stopped for post-mortem analyses after 34, 210 and 14 days, respectively, after the beginning of hydration. The specimens (~100 mm high, ~50 mm for each of the two layers, and 100 mm in diameter, see Fig. 3) were composed of bentonite pellets with an average dry density ρ_d of around 1.30 Mg/m^3 (initial water content $w \approx 6.2\% \pm 3.3\%$), placed in the lower half of the cell, and a bentonite block with a nominal dry density $\rho_d = 1.60$ Mg/m^3 in the upper part (initial water content $w \approx 13.7\% \pm 0.4\%$). Deionized water was injected with a very small hydraulic head of about 140 cm. Water volume was recorded with an automatic volume change apparatus. Water was supplied through porous plates on the bottom side of the oedometer (i.e. in contact with the pellet layer), while the top cover allows air expulsion.

3.2 Numerical Model

The numerical specimens have been subdivided in 800 eight-noded isoparametric elements representing the bentonite materials (400 for each layer of compacted block and pellet). The problem was considered as axisymmetric and oedometer conditions were imposed (Fig. 3). The samples were initially subjected to an isotropic confining stress of 0.1 MPa, and the effect of gravity is neglected. Initial suction in the pellet layer was set equal to 500 MPa for the pellet layers and 80 MPa for the compacted block ones. The hydration of the sample is provided from the lower face (red line Fig. 3), assuming a suction decrease from 500 MPa (u_w = -500 MPa in the pellet layer bottom side) to 0.014 MPa in 1000 s. At the top of the specimen, water flow is not allowed. The hydromechanical model parameters were calibrated against experimental data from swelling-consolidation tests from Hoffman et al. (2007) and Lloret et al. (2003) for the mechanical behavior of pellets and compacted blocks, respectively, and from Alonso et al. (2010), Villar (2000) and Lloret et al. (2005) for the water retention behavior). Mechanical model parameters are given in Gramegna et al. (2023), while water retention constitutive behaviour parameters are reported in Table 1. Constitutive model symbology is consistent with Gramegna et al. (2022). The flow model parameters (Table 2) are calibrated by best fitting the response of the water intake time evolution of test MGR23. Consequentially the model is validated on the comparison between the experimental and numerical results of test MGR21 and MGR24, as well as on the water content and dry density distributions at dismantling of the three tests. Further details on this numerical study are reported in Gramegna et al. (2023).

Table 1. Water retention model parameters.

Symbol	e_{m0}	β_0	β_1	C_{ads}	n_{ads}	n	m	A
Unit	[-]	[-]	[-]	[MPa^{-1}]	[-]	[-]	[-]	[MPa]
Block	0.35	0.15	0.25	0.0028	0.78	3	0.15	0.24
Pellet								

Table 2. Permeability model parameters.

Symbol	$K_{w0}\frac{(1-e_{M0})^M}{e_{M0}^N}$	M	N	n_k
Unit	[m^2]	[-]	[-]	[-]
Block	2.8×10^{-20}	1.2	0.1	3.4
Pellet	2.8×10^{-20}	0.9	0.1	3

4 Results and Discussions

Figure 4 shows the comparison between experimental data and results in terms of water intake time evolution. More than half of the water volume required for complete saturation is consumed in only ten days, due to the rapid initial water intake. The high permeability that characterizes the layers of pellets is primarily responsible for this phenomenon, as confirmed by the several researchers that have analyzed the relationship between pore size distribution (large pore diameters), dry density (very low) and their evolution during hydration (e.g. Cui (2017), Romero (2013) and Villar (2000)).

Permeability and macro void ratio numerical simulations' distribution through the heigth of the sample are presented in Fig. 5 and Fig. 6. The adopted numerical model allows reproducing this response by assuming an initial permeability value equal to $K_w \approx 8 \times 10^{-20}$ m^2 for the pellet layer and $K_w \approx 3 \times 10^{-20}$ m^2 for the compacted block (Fig. 5). These different initial permeabilities are mostly due to the different initial macro void ratio, and they evolve accordingly upon hydration (Fig. 6). A permeability gradient emerges in the pellet layers as the saturation front (shown in terms of numerical and experimental results of water content distribution through the heigth of the sample in Fig. 7) progresses from the bottom. The invasion of macro-porosity due to aggregate swelling related to water saturation is the cause of the following permeability decrease, regardless the fact that the material nearest to the wetting surface tend to expand (as numerical and experimental dry density results through the heigth of the sample testify in Fig. 8). As saturation progresses, the top compacted bentonite block swells, inducing the pellet layer compaction and macro-void ratio reduction. When compared to the pellet layer, the compacted block swelling induces a smaller permeability decrease, due to the lower influence of the micro-structure development process. The numerical permeability prediction in fully saturated conditions is constant throughout the sample, with a value equal to $K_w \approx 3 \times 10^{-20}$ m^2 in both layers.

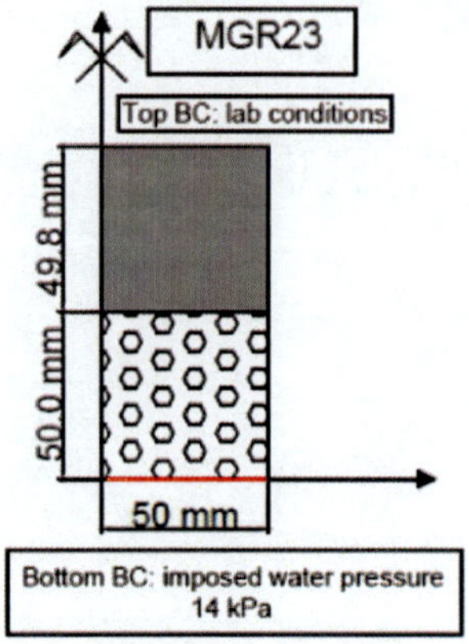

Fig. 3. Characteristics of the considered specimen and boundary conditions of the model.

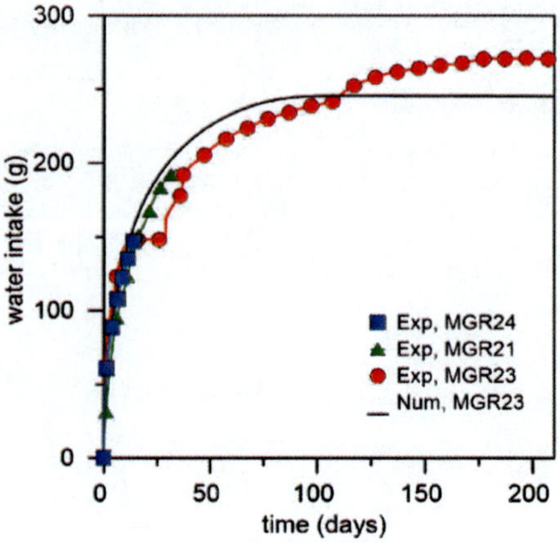

Fig. 4. Water intake evolution over time: experimental and numerical results.

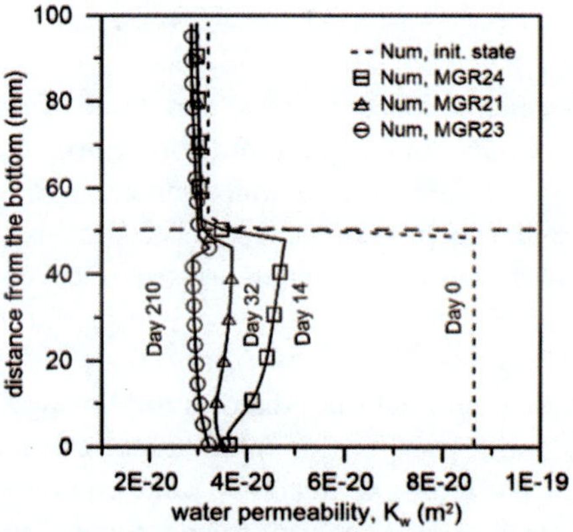

Fig. 5. Permeability evolution through the height of the sample: numerical simulations.

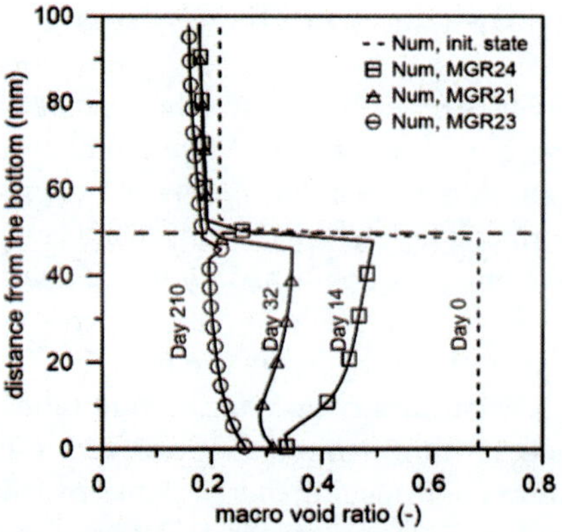

Fig. 6. Macrovoid ratio evolution through the height of the sample: numerical simulations.

Water content (Fig. 7) and dry density (Fig. 8) at different locations from the hydration surface were experimentally determined thanks to post-mortem analyses by cutting the overall assemblies in subsamples. This allowed the comparison with numerical simulations results and the consequential validation of permeability law evolution. In general, as water is supplied, the larger swelling is detected for the pellets in the material in direct contact with the wetting surface and for the compacted block in the one immediately hydrated from the pellet side. Compaction is observed in the upper pellet layer contiguous to the swelling compacted bentonite block. The upper part of the compacted block layer is the last to swell, due to its delayed saturation. As a result, a dry density gradient in the hydration direction develops until the top block swelling and pellet layer compression induce it to vanish. These phenomena are certainly related to water availability in time at the different location of the sample, and thus on permeability evolution. Numerical results compare remarkably well with the experimental water content and dry density distributions, validating the flow model.

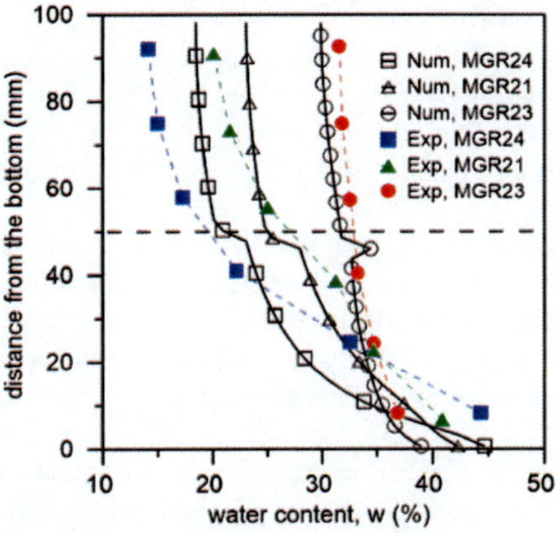

Fig. 7. Water content evolution along the height of the sample: comparison between experimental results and numerical simulations.

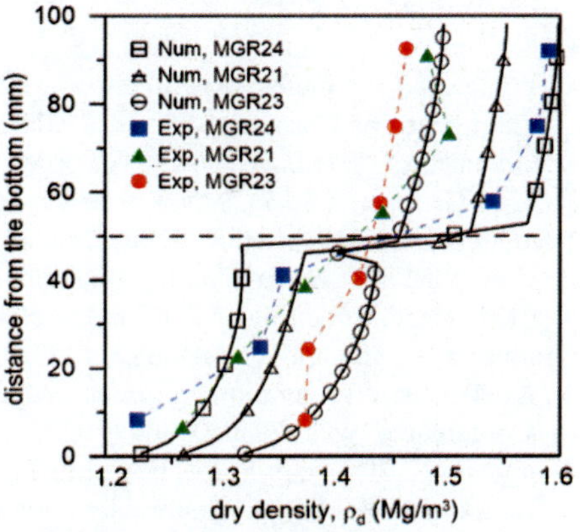

Fig. 8. Dry density evolution along the height of the sample: comparison between experimental results and numerical simulations.

5 Conclusions

Water permeability is critical in many geomechanical applications, and its experimental determination in unsaturated conditions remains difficult. Such complication is magnified when working with expansive and multi-porosity materials such bentonite. In this paper, numerical simulations of soaking tests in isochoric conditions allow evaluating bentonite permeability evolution. The tested samples consist of contiguous layers of Febex compacted block and pellets mixture, whose initial dry densities and structures, and consequentially permeabilities, are significantly different. It is demonstrated that the numerical permeability law evolution calibrated on water intake experimental data is capable to guarantee a proper reproduction and prediction of the strongly-coupled hydro-mechanical phenomena occurring upon hydration. In particular, the numerical results show the predominant impact of material pore-structure on water transfer mechanisms.

References

Alonso, E.E., Gens, A., Josa, A.: A constitutive model for partially saturated soils. Géotechnique **40**(3), 405–430 (1990)

Alonso, E.E., Hoffmann, C., Romero, E.: Pellet mixtures in isolation barriers. J. Rock Mech. Geotech. Eng. **2**(1), 12–31 (2010)

Collin, F., Li, X.L., Radu, J.P., Charlier, R.: Thermo-hydro-mechanical coupling in clay barriers. Eng. Geol. **64**(2–3), 179–193 (2002)

Cui, Y.J.: On the hydro-mechanical behaviour of MX80 bentonite-based materials. J. Rock Mech. Geotech. Eng. **9**(3), 565–574 (2017)

Della Vecchia, G., Dieudonné, A.C., Jommi, C., Charlier, R.: Accounting for evolving pore size distribution in water retention models for compacted clays. Int. J. Numer. Anal. Meth. Geomech. **39**(7), 702–723 (2015)

Dieudonné, A.C., Gatabin, C., Talandier, J., Collin, F., Charlier, R.: Water retention behaviour of compacted bentonites: experimental observations and constitutive model. In: Third European Conference on Unsaturated Soils, E-UNSAT 2016, Paris, France (2016)

Dieudonné, A.C., Della Vecchia, G., Charlier, R.: Water retention model for compacted bentonites. Can. Geotech. J. **54**(7), 915–925 (2017)

Dieudonné, A.C., Charlier, R.: Evaluation of the instantaneous profile method for the determination of the relative permeability function. International Workshop on Advances in Laboratory Testing and Modelling of Soils and Shales, Villars, Switzerland (2017)

Gramegna, L., Imbert, C., Talandier, J., Colln, F., Charlier, R.: Hydro-mechanical behaviour of a pellets based bentonite seal: numerical modelling of lab scale experiments. In: E3S Web of Conferences, vol. 195, p. 04009 (2020)

Gramegna, L., Bernachy-Barbe, F., Collin, F., Talandier, J., Charlier, R.: Pore size distribution evolution in pellets based bentonite hydration: comparison between experimental and numerical results. Eng. Geol. **304** (2022)

Gramegna, L., Villar, M.V., Collin, F., Talandier, J., Charlier, R.: Friction influence on constant volume saturation of bentonite mixed pellet-block samples, a numerical analysis. Appl. Clay Sci. **234**, 106846 (2023)

Hoffmann, C., Alonso, E.E., Romero, E.: Hydro-mechanical behaviour of bentonite pellet mixtures. Phys. Chem. Earth **32**, 832–849 (2007)

Lloret, A., Villar, M.V., Sánchez, M., Gens, A., Pintado, X., Alonso E.E.: Mechanical behaviour of heavily compacted bentonite under high suction changes. Géotechnique **53**(1), 27–40 (2003)

Lloret, A., Romero, E., Villar, M.V.: FEBEX II Project. Final report on thermohydro-mechanical laboratory tests. Technical report, ENRESA (2005)

Romero, E., Della Vecchia, G., Jommi, C.: An insight into the water retention properties of compacted clayey soils. Géotechnique **61**(4), 313–328 (2011)

Romero, E.: A microstructural insight into compacted clayey soils and their hydraulic properties. Eng. Geol. **165**, 3–19 (2013)

Villar, M.V., Iglesias, R.J., Gutiérrez-Álvarez, C., Carbonell, B.: Pellets/block bentonite barriers: laboratory study of their evolution upon hydration. Eng. Geol. **292** (2021)

Villar, M.V.: Caracterizacion Thermo-Hidro-Mecanica de Una Bentonita de Cabo Gata. Universidad Complutense de Madrid (2000)

Hygro-Thermal Modelling of Earthen Materials for Building Applications

Leonardo Maria Lalicata(✉), Agostino Walter Bruno, and Domenico Gallipoli

Department of Civil, Chemical and Environmental Engineering, University of Genoa, 16145 Genoa, Italy
leonardo.lalicata@unige.it

Abstract. Earth is a building material with low carbon emissions compared to conventional concrete or fired bricks. Earthen materials have an excellent capacity to regulate indoor hygrothermal conditions allowing for a better comfort with reduced heating and cooling needs during the life cycle of the building. The present paper presents a theoretical framework to investigate the hygro-thermal response of earthen materials by coupling the principles of unsaturated soil mechanics combined with the thermodynamics of porous media. The degree of coupling between the two variables (temperature and relative humidity or water content) depends on the values of the water and vapour permeability functions which, in turns, depend on the water retention curve of the material. Results shows that, in the hygroscopic domain, the hydro-thermal coupling is more influenced by the saturated permeability than by the vapour diffusion coefficient.

Keywords: rammed earth · sustainability · hygro-thermal coupling

1 Introduction

Earth is an ancient building material that, in recent years, is attracting the interest of both the scientific community and construction industry stakeholders because of its low environmental footprint compared to conventional alternatives, such as concrete or fired bricks. To date, the major issues associated to the earthen building materials are the durability under environmental conditions (rainfall infiltration and capillary) and the absence of well-established methods to estimate the hygro-thermal regulating capacity, [1].

To address this problem, this paper presents a theoretical framework that combines the principles of unsaturated soil mechanics in porous materials and thermodynamics of porous media. The resolving equations have been implemented in a finite element code and the preliminary results of the study are presented in the following sections.

The coupled equations have been derived based on the conservation of both heat and water mass by also accounting for phase changes during evaporation and condensation processes. The model is defined by a water-retention law, the liquid water and water vapour permeability functions, the heat capacity and a moisture-dependent thermal conductivity law.

A. Ferrari et al. (Eds.): CNRIG 2023, SSGG, pp. 327–334, 2023.
https://doi.org/10.1007/978-3-031-34761-0_40

The coupled model is used to investigate the coupled hygro-thermal behaviour of earthen building materials exposed to cyclic variations of temperature and relative humidity. The modelling of capillary rise phenomena from the ground and other water sinks or sources within the material is deemed outside the scope of the present work.

2 Hygro-Thermal Model

Earthen materials are usually made of highly compacted and well-graded soils that remain in a quasi-dry condition (i.e. degree of saturation lower than 10%) during their service life. Despite this rather stable exercise conditions, earthen materials are highly sensitive to environmental agents that may induce a flow of the water mass (in both liquid and vapour form) and a variation of the internal temperature in the wall, which may in turn affect occupants' comfort.

In this study, the soil is treated as a three-phase material consisting of soil grains (solid phase), water (in both liquid and vapour phase) and dry air (in vapor phase). It is assumed that the resistance to gas flow is negligible so that the air pressure is always equal to the atmospheric value and the dry air mass balance can be neglected. Also, the solid skeleton is assumed incompressible so that no variation of the solid phase occurs.

Hence, the problem can be tackled by simultaneously solving the water mass balance and the thermal energy balance in the domain. In earthen materials, these two balance equations are strongly coupled because water fluxes depend on both the changes in relative humidity and temperature. Conversely, heat flow depends on the changes in water content as well. The complete derivation of the coupling terms is outside the scope of this work and will be omitted here. Further details can be found in Bear and Cheng [2].

2.1 Definitions and Constitutive Equations

The relative humidity *hr* (-) is selected as the state variable to describe the hydric state of the material as it is routinely measured in both laboratory and full-scale building applications. The energetic state of the earth depends on the temperature T (K), which is used as the thermal variable.

At equilibrium, relative humidity *hr* and temperature T are related to the total suction ψ by means of the Kelvin's law expressed in Eq. (1). Note that the total suction ψ is given by the sum of the matric suction s (i.e. the positive difference between the air pressure p_a and the water pressure p_l) and the osmotic suction π, which is taken as zero as there are no salts dissolved in the water, thus giving:

$$\psi = \pi + s = s + 0 = p_a - p_l = -\rho_l \frac{R}{M_w} \cdot T \cdot \ln hr \tag{1}$$

where ρ_l is the density of liquid water (1000 kg/m^3), R is perfect gas constant (8.314 J/(mol K)) and M_w is the molar mass of water (0.018 kg/mol). Figure 1 shows the relative dependency of suction s with temperature and relative humidity, as obtained from Eq. (1). Inspection of Fig. 1 indicates that temperature only has a marginal effect on the variation of suction in comparison with the relative humidity. Hence, the dependency of suction on temperature can be neglected especially considering that earth materials typically remain in the hygroscopic domain (i.e. $hr < 0.95$) and at positive temperatures, [3, 4].

The capacity of a material to store/release water due to changes in pore pressure is described by a water retention law that links the amount of water (degree of saturation S_l or water content w_l) to its energetic state (suction $s = -p_l$, or relative humidity hr).

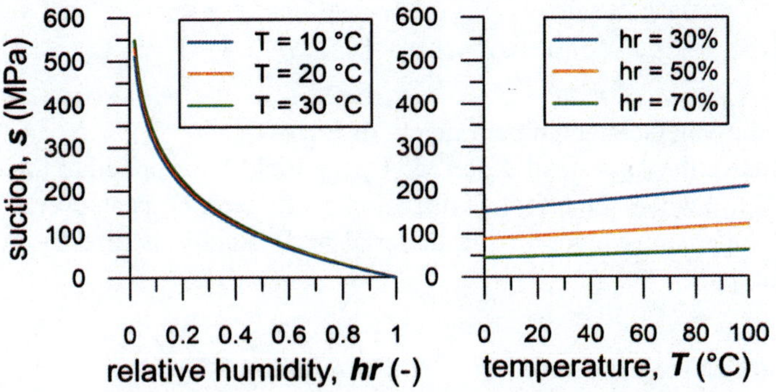

Fig. 1. Influence of suction on relative humidity and temperature.

Here, the well-known Van Genuchten law [5], formulated in terms of gravimetric water content, is adopted:

$$w_l = w_{res} + (w_{sat} - w_{res})\left(1 + \left(\frac{s}{P}\right)^N\right)^{-M} \quad (2)$$

where w_{sat} and w_{res} are the saturated and residual water content respectively, P (MPa), N (-) and M (-) are Van Genuchten fitting parameters.

Relative humidity is defined as:

$$hr = \frac{p_v}{p_{sat}(T)} = \frac{\rho_v}{\rho_{sat}} \quad (3)$$

where p_v is the partial vapour pressure and $p_{sat}(T)$ is the saturated vapor pressure at the current temperature. ρ_v, and ρ_{sat} are the current and saturated vapour densities, respectively. Vapour density and its pressure are linked by the perfect gas relationship in Eq. (4) while the $p_{sat}(T)$, kPa, is defined as in Eq. (5).

$$\rho_v = \frac{M_w}{RT} \cdot p_v \quad (4)$$

$$p_{sat}(T) = 0.6108 \cdot 10^{\frac{7.5T}{T+273.15}} \quad (5)$$

2.2 Water Mass Balance

The water mass balance, Eq. (6), accounts for both the liquid water and the water vapour. The advective flux, on the right hand side of Eq. (6), is described by the generalised Darcy law applied to the liquid water phase only, where $K_w(S_l) = K_{sat}k_{rl}(S_l)$ is the hydraulic conductivity that function of the saturated permeability, K_{sat} (m/s), and the relative permeability, $k_{rl}(S_l)$. The latter depends on degree of saturation, S_l and can be estimated by Mualem's law [6] as in Eq. (7):

$$\frac{\partial}{\partial t}(n\rho_l S_l) = -\nabla\left(\rho_l\left(-\frac{K_w(S_l)}{\rho_l g}\nabla(p_l + \rho_l g z)\right) - D_e\nabla\rho_v\right) \tag{6}$$

$$k_{rl} = \sqrt{S_l}\cdot\left[1 - \left(1 - S_l^{M^{-1}}\right)^M\right]^2 \tag{7}$$

where M is the Van Genuchten parameters of Eq. (4).

The diffusive flux is instead described by the Fick' law applied to the water vapour only, $(-D_e\nabla\rho_v)$, where the effective diffusion coefficient, D_e is an overall quantity that express the vapour resistance of the material and, usually, is directly available from experimental data.

2.3 Thermal Energy Balance

Under the assumption of thermal equilibrium between different phases, the energy balance is imposed via a single differential equation. In the hygroscopic domain, the thermal energy variation due to wetting/drying mechanisms is negligible in comparison with thermal energy variation due to phase changes (evaporation/condensation processes).

According to the above assumptions, the thermal energy balance is written as in Eq. (8):

$$\left(\rho C_p\right)_{eq}\frac{\partial T}{\partial t} = -\nabla(-\lambda\nabla T) - L_v m_{\rightarrow v} \tag{8}$$

where λ is the thermal conductivity, function of the water content, and $\left(\rho C_p\right)_{eq}$ is the equivalent heat capacity of the composite medium defined as:

$$\left(\rho C_p\right)_{eq} = (1-n)\rho_s C_{p,s} + nS_l\rho_l C_{p,l} + n(1-S_l)\rho_v C_{p,v} \tag{9}$$

where the grain density ρ_s and the specific heat of the dry soil $C_{p,s}$ are material properties, the specific heat of liquid water and vapour, $C_{p,l}$ and $C_{p,v}$, are equal to 4.183 and 1.89 kJ/(kg·K) respectively.

Finally, L_v in Eq. (8) is the latent heat of evaporation and is taken equal to $2.5\cdot10^6$ J/kg. $m_{\rightarrow v}$ is the evaporation or condensation mass, which is evaluated from the liquid phase balance as:

$$-m_{\rightarrow v} = \frac{\partial}{\partial t}(n\rho_l S_l) + \nabla\rho_l\left(-\frac{K_w(S_l)}{\rho_l g}\nabla(p_l + \rho_l g z)\right) \tag{10}$$

2.4 The Complete Set of Equations

The water mass balance in Eq. (6) and the thermal balance in Eq. (8) can be recast in terms of temperature and relative humidity by using Eqs. (1, 3, 4, 5 and 10), thus giving the expressions of Eq. (11) and (12). Under exercise conditions (i.e. when the water fluxes are driven by changes in relative humidity or temperature from the outside to the inside of the wall or *vice versa*), the water flux can be assumed horizontal by neglecting the gradient of liquid pressure along the vertical direction z.

$$\rho_d \frac{\partial w_l}{\partial hr}\frac{\partial hr}{\partial t} = -\nabla(-\frac{K_w}{g}\rho_l \frac{R}{M_w}(\nabla hr) + \left(-D_e \frac{M_w}{RT}\left(p_{sat}\nabla hr + hr\left(\frac{dp_{sat}}{dT} - \frac{p_{sat}}{T}\right)\nabla T\right)\right) \quad (11)$$

$$\left(\rho C_p\right)_{eq}\frac{\partial T}{\partial t} = -\nabla(-\lambda \nabla T) + L_v\left(\rho_d \frac{\partial w_l}{\partial hr}\frac{\partial hr}{\partial t} + \nabla(-\frac{K_w}{g}\rho_l \frac{R}{M_w}\left(\frac{T}{hr}\nabla hr\right)\right) \quad (12)$$

3 Numerical Model

The system of partial differential equations has been implemented in the Comsol Multiphysics finite element software. The model simulates the hygro-thermal response of an earth wall with unlimited in-plane extension and subjected to identical cyclic variations of temperature and relative humidity on opposite faces (Fig. 2a). Given the geometry and boundary conditions of the problem, only half wall thickness has been modelled because the middle depth constitutes a symmetry plane. The variations of temperature and relative humidity have therefore been applied only at one end of the one-dimensional finite element model while zero water and heat flow are imposed at the other end representing the symmetry plane, Fig. 2b. The same boundary value problem has already simulated by Soudani *et al.* [7] via an alternative finite element formulation, whose results are the object of comparison in the following part of the paper.

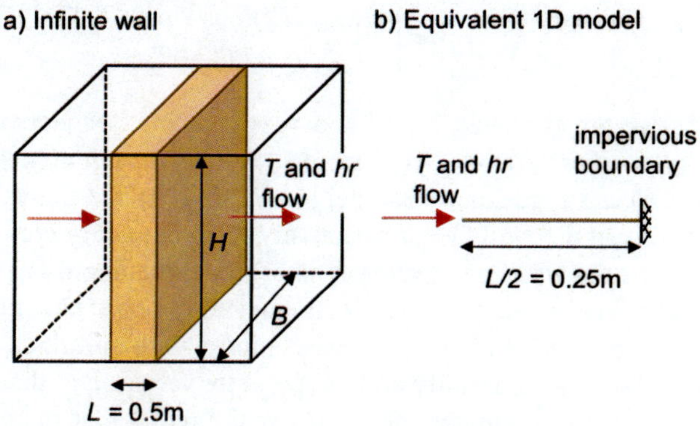

Fig. 2. Numerical model by Soudani *et al.* [5]; b) current simulation.

3.1 Comparison with Soudani *et al.* [7]

A daily sinusoidal variation of ambient relative humidity hr_e, between 0.5 and 0.7, was imposed on both sides of the wall while the ambient temperature T_e was maintained constant and equal to 30 °C. The initial conditions were uniform across the entire wall with a temperature $T_0 = 30$ °C and a relative humidity $hr_0 = 0.7$.

This corresponds to an initial water content of 0.011 and degree of saturation of 0.056, making the earth wall very dry. The overall duration of the simulated experiment was set to 100 h. The parameter values used in the present model are listed in Table 1 and were determined based on the data in Soudani *et al.* [7].

Table 1. Model parameters.

Parameter	symbol	value
Porosity (-)	n	0.35
Dry density (kg/m^3)	ρ_d	1722
Grain density (kg/m^3)	ρ_s	2650
Van Genuchten parameter (MPa)	P	0.55
Van Genuchten parameter (-)	N	1.64
Van Genuchten parameter (-)	$M = 1 - 1/N$	0.39
Saturated water content (-)	w_{sat}	0.20
Residual water content (-)	w_{res}	0
Saturated permeability (m/s)	K_{sat}	1.3 · 10–9
Effective diffusion coefficient (m^2/s)	D_e	2.7 · 10–6
Heat capacity of the grains (kJ/(kg·K))	$C_{p,s}$	0.648
Thermal conductivity (W/(m·K))	λ	$0.6 + 9.22\, w_l$

The moisture and thermal flow at the boundary of the model that can be written as:

$$\begin{bmatrix} (\boldsymbol{g}_l + \boldsymbol{g}_v) \cdot \boldsymbol{n} \\ (\boldsymbol{g}_T - L_v \boldsymbol{g}_v) \cdot \boldsymbol{n} \end{bmatrix} = \begin{bmatrix} \beta\left(p_{sat}(T_e)hr_e - p_{sat}(T_f)hr_f\right) \\ \alpha\left(T_e - T_f\right) \end{bmatrix} \tag{13}$$

where $\boldsymbol{g}_T$, $\boldsymbol{g}_l$ and $\boldsymbol{g}_v$ are the heat, liquid and vapour flows, respectively, α and β are the heat and vapour mass transfer coefficients, which are equal to 8 W/(m^{2}K) and $2.5 \cdot 10^{-8}$ kg/(m^{2}sPa) as suggested by Künzel [6]. T_f and hr_f are the values temperature and relative humidity at the wall face, respectively. This boundary conditions allows for vapour flow only, i.e. the water can leave (or enter) the domain only in the vapour form. The term $L_v \boldsymbol{g}_v$ accounts for the energy flow associated to phase change.

Figure 3 compares results from the present model with simulations from Soudani *et al.* [7] in terms of relative humidity and temperature variation at the symmetry plane. The two models are in good agreement as they both predict that the centre of the wall starts to dry out after 24 h and that the relative humidity reduces from an initial value of 0.7 to a final one of 0.68. The drying process induces heat and moisture flow from the inner wall towards the surrounding environment, thus causing a small temperature reduction on the symmetry plane. The maximum predicted temperature change is 1.5 °C for Soudani *et al.* [7] and 1 °C for the present model. The slight discrepancy between these two values may be due to differences in the finite element formulation and adopted retention law.

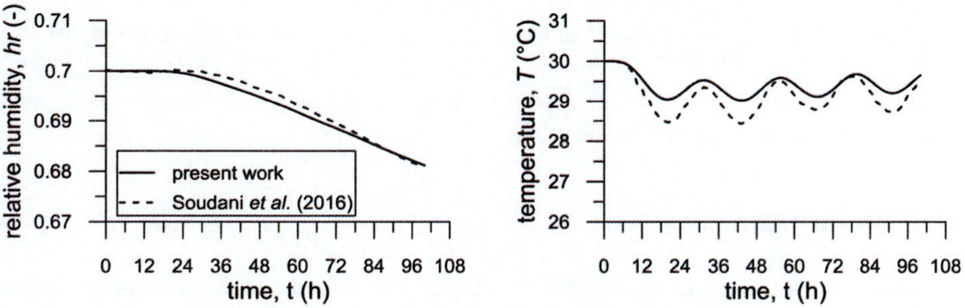

Fig. 3. Validation of the present model.

3.2 Parametric Study

The hygro-thermal response of the earth wall model to cyclic changes of ambient relative humidity has here been calculated for different values of the saturated permeability K_{sat} and the vapour diffusion coefficient D_e. According to Gallipoli *et al.* [1] and Fabbri *et al.* [8], the fine content of earthen building materials can range from 20% up to 80% of total mass, which means that the saturated permeability K_{sat} can vary several orders of magnitude. Conversely, the vapour diffusivity is less affected by the grading curve of the earth material and typically ranges between $1 \cdot 10^{-6}$ m²/s and $5 \cdot 10^{-6}$ m²/s.

To get an insight into the influence of both saturated permeability and vapour diffusivity, the analysis presented in the previous section has been repeated for all possible combinations of the following parameter values:

- $K_{sat} = 1 \cdot 10^{-7}, 1 \cdot 10^{-8}, 1 \cdot 10^{-9}, 1 \cdot 10^{-10}$ (m/s)
- $D_e = 1 \cdot 10^{-6}, 2 \cdot 10^{-6}, 3 \cdot 10^{-6}, 4 \cdot 10^{-6}, 5 \cdot 10^{-6}$ (m²/s)

Figure 4 reports the variation of relative humidity and temperature computed at the wall centre indicating that results are very sensitive to the value of the saturated permeability while the effect of the vapour diffusivity appears rather limited. The initial relative humidity of 0.7 remains virtually unchanged for $K_{sat} = 10^{-10}$ m/s while it reduces to 0.685 for $K_{sat} = 10^{-9}$ m/s and finally drops to about 0.640 for $K_{sat} = 10^{-8}$ 1 $\cdot 10^{-7}$ m/s. Figure 4 also shows that the temperature reduces in all cases with maximum drops ranging from 0.5 °C for $K_{sat} = 10^{-10}$ m/s to 1.75 °C for $K_{sat} = 10^{-7}$ m/s.

Albeit not shown here, the results from the above simulation also confirm that the simplifying assumptions of a negligible variation of water vapour mass and independence of suction on temperature remain valid over the above parameters range.

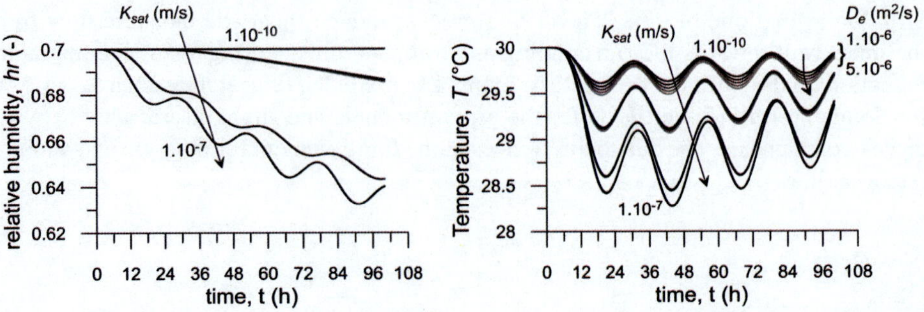

Fig. 4. Relative humidity and temperature variations at the wall center. Influence of saturated permeability K_{sat} and vapour diffusivity D_e.

4 Concluding Remarks

A coupled model has been developed to simulate the heat and mass transfer on earthen building materials. The model accounts for the phase change of water inside the domain and consequently the latent heat released or absorbed by the material during the evaporation or condensation process. Results shows that the hydro-thermal coupling is more influenced by the saturated permeability K_{sat} than by the vapour diffusion coefficient D_e, at least in the explored range of these parameters.

Future work will aim at validating the proposed theoretical framework against experimental data from additional laboratory tests in double climatic chamber of from on-site monitoring of earth buildings.

References

1. Gallipoli, D., Bruno, A.W., Perlot, C., Mendes, J.: A geotechnical perspective of raw earth building. Acta Geotech. **12**(3), 463–478 (2017). https://doi.org/10.1007/s11440-016-0521-1
2. Bear, J., Cheng, A.H.D.: Modeling Groundwater Flow and Contaminant Transport. Springer, Dordrecht (2010). https://doi.org/10.1007/978-1-4020-6682-5
3. Künzel, H.M.: Simultaneous heat and moisture transport in building components. One-and two-dimensional calculation using simple parameters. IRB-Verlag Stuttgart, p. 65 (1995)
4. Lalicata, L.M., Bruno, A.W., Gallipoli, D.: Hygro-thermal coupling in earth building materials. In: Proceedings of the 8 International Conference on Unsaturated Soils, UNSAT 2023, E3S Web of Conferences, vol. 382, p. 23003 (2023). https://doi.org/10.1051/e3sconf/202338223003
5. Van Genuchten, M.T.: A closed-form equation for predicting the hydraulic conductivity of unsaturated soils. Soil Sci. Soc. Am. J. **44**(5), 892–898 (1980)
6. Mualem, Y.: A new model for predicting the hydraulic conductivity of unsaturated porous media. Water Resour. Res. **12**(3), 513–522 (1976)
7. Soudani, L., et al.: Assessment of the validity of some common assumptions in hygrothermal modeling of earth based materials. Energy Build. **116**, 498–511 (2016)
8. Fabbri, A., Morel, J.C., Aubert, J.E., Bui, Q.B., Gallipoli, D., Reddy, B.V: Testing and Characterisation of Earth-based Building Materials and Elements. Rilem State Art Report, vol. 35, p. 296 (2022). https://doi.org/10.1007/978-3-030-83297-1

Experimental and Numerical Investigation on Water Exchange of Opalinus Clay Samples

Qazim Llabjani[1], Vincenzo Sergio Vespo[2(✉)], Eleni Stavropoulou[1], Alessio Ferrari[1,3], and Guido Musso[2]

[1] Laboratory of Soil Mechanics, Swiss Federal Institute of Technology Lausanne (EPFL), Lausanne, Switzerland

[2] Politecnico di Torino, Department of Structural, Building and Geotechnical Engineering, Turin, Italy

vincenzosergio.vespo@polito.it

[3] Dipartimento di Ingegneria, Università degli Studi di Palermo, Palermo, Italy

Abstract. Opalinus Clay is a fine-grained sedimentary geomaterial with a mineral composition consisting mainly of silicates, carbonates and quartz. Due to its favourable barrier properties, this geomaterial has been selected as the host rock for the underground disposal of radioactive waste in Switzerland. In this context, the hydro-mechanical state of Opalinus Clay will be affected by subsequent drying and wetting phenomena. Design procedures require then an appropriate understanding of the processes related to water exchange.

To analyse the behaviour of Opalinus Clay upon suction changes and to define its water retention properties, the vapour equilibrium technique is combined with an accurate assessment of the deformations in the two orthogonal directions using strain gauges. Based on those results, a 3D Thermo-Hydro-Mechanical Finite Element model is implemented and validated. Preliminary analysis shows overall a good agreement between experimental data and modelling results in terms of change in water content, degree of saturation and equilibrium time. The numerical model will be used to define an accurate law for the tortuosity and permeability evolution of Opalinus Clay when subjected to suction variations, that is required to simulate the behaviour of a nuclear waste disposal system.

Keywords: Opalinus Clay · hydro-mechanical behaviour · vapour equilibrium technique · FE-model

1 Introduction

The current or past use of nuclear energy still poses serious environmental concerns related to the disposal of the generated nuclear waste. From a geomechanical perspective, the design of a geological disposal system requires a deep understanding of the thermo-hydro-mechanical behaviour of the involved materials (i.e. the host rock and the engineered barriers). Opalinus Clay has been selected to serve as the host formation for the Swiss underground disposal of radioactive waste thanks to its favourable properties (e.g., low permeability and porosity, high retention and self-sealing properties). Over the

A. Ferrari et al. (Eds.): CNRIG 2023, SSGG, pp. 335–342, 2023.
https://doi.org/10.1007/978-3-031-34761-0_41

lifespan of the repository, the hydro-mechanical state of Opalinus Clay will be affected by the tunnel excavation, ventilation and re-saturation phases, leading to subsequent drying and wetting phenomena [1]. These processes are related to the heat and water fluxes that take place between the geomaterial and the surrounding environment. Such evaluation requires taking into account the couplings between the balance of water mass and energy, together with the relevant boundary conditions. By numerical modelling, the role, the magnitude and the effect of each transport mechanism can be identified. Interestingly, the quantitative evaluation of these fluxes is required in many other relevant applications for civil and environmental engineering, like climate change effects induced by global warming, natural or artificial slope stability and transport of contaminants in the groundwater.

In this context, experimental results on Opalinus Clay have shown high water retention properties and the dependency of the air entry value on the void ratio [2], and have demonstrated an irreversible volumetric response to suction variations [3, 4]. Consistent observation from numerical investigation of the water exchange on Opalinus Clay due to suction changes, suggests that the development of micro-cracks may enhance the flow rate of water vapour [5].

In this paper, a 3D numerical model capable of reproducing the vapour equilibrium technique (VET) is implemented in the commercial Finite Element code Comsol Multiphysics® and validated on tests carried out on four specimens of Opalinus Clay [4]. This numerical model will be used to study in depth the role of advective and diffusive fluxes during the drying and wetting processes to which the Opalinus Clay will be subjected during its lifespan, as well as the effect that the environmental conditions (temperature T, relative humidity h_r, total suction ψ) and the material properties (porosity ϕ, mineralogical composition and fabric) have on the hydro-mechanical behaviour of this material.

2 Material and Methods

Opalinus Clay is a fine-grained sedimentary geomaterial with a mineral composition consisting mainly of silicates, carbonates and quartz. In 2015, a borehole was drilled close to the village of Lausen (north-western Switzerland), where Opalinus Clay was encountered at a shallow depth from 6 to 71 m [6]. The samples considered in this study (namely L3, L8, L11 and L13) were retrieved from this borehole and the initial geotechnical characteristics of the tested specimens are discussed in [4].

To analyse the behaviour of Opalinus Clay upon suction variations and define its water retention behaviour, the VET is combined with an accurate assessment of deformations in two orthogonal directions using strain gauges [4] as shown in Fig. 1. The VET allows the application of a known relative humidity inside a sealed container with saturated saline solutions. Through the so-called 'psychrometric law', relative humidity can be then converted to total suction [7]. Based on the type of salt, different total suction values can be imposed and allow to perform wetting/drying cycles. All tests were conducted at a controlled temperature of 25 °C (T_{env}) and using cylindrical specimens with an initial diameter and height of 25 mm and 20 mm, respectively, with bedding perpendicular to the axis of the cylinder [4]. For each core, two specimens were used for the water retention measurements: one specimen equipped with strain gauges to evaluate volume strains; the second specimen used to monitor the mass variation to assess the water content changes. At each suction step, the achievement of the equilibrium condition is assessed by the stabilization of both strain and mass of the specimen.

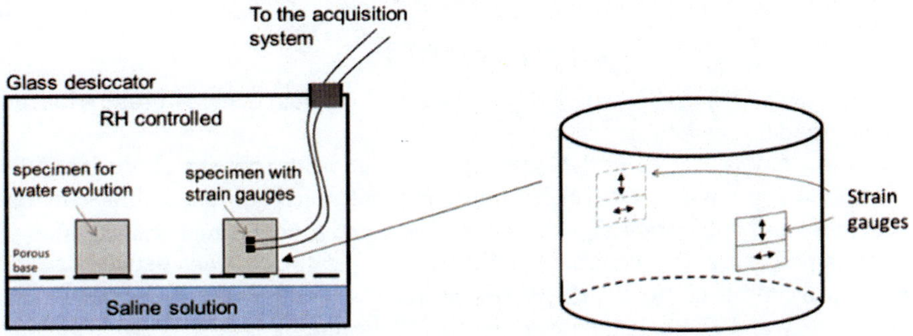

Fig. 1. Experimental setup: sealed container for total suction application (left); strain gauge configuration on the tested specimen (right) (modified after [7])

3 THM Model for Vapour Equilibrium Technique

In order to study the water exchange related to the various water transport mechanisms in the geomaterial, a Thermo-Hydro-Mechanical (THM) model is implemented. It integrates the coupled water mass and energy balance equations. Volume changes upon drying and wetting are evaluated by introducing a suitable dependency of porosity on suction [8]. The mass balance of air is neglected under the common assumption of hydrostatic conditions for the air phase [9]. As a consequence, the air pressure is equal to the atmospheric one and is constant in time and space. The diffusive flux of air dissolved in liquid water is also considered as negligible. The evaporative flow from/to the porous medium is modelled as water and heat fluxes at the soil-atmosphere interface. In the following sections, the nomenclatures 'water' and 'air' indicate the chemical species, while 'gas' or 'liquid' indicate the phases. The water in the gas phase is indicated as 'vapour' and the flow of dry air in the soil is neglected.

Water Mass Balance Equation. The mass balance equation can be expressed in terms of chemical species and, under the above hypotheses, it assumes the form:

$$\frac{\partial}{\partial t}\left[\phi\left(S_l\rho_l + S_g c_g^w\right)\right] + \nabla\cdot\left[-\frac{\mathbf{K}_{sat}k_r}{g}\nabla(u_l + \rho_l g z)\right] + \nabla\cdot\left(-\phi S_g D_v \boldsymbol{\tau}\nabla c_g^w\right) = 0 \tag{1}$$

where ϕ is the porosity; S_l and S_g are the liquid and gas degree of saturation ($S_l = 1 - S_g$); ρ_l is the mass density of the liquid phase; c_g^w is the vapour mass concentration; $\mathbf{K}_{\mathbf{sat}}$ is the saturated hydraulic conductivity tensor; k_r is the relative permeability coefficient; u_l

is the water pressure in liquid phase; g is the gravitational acceleration; z is the elevation head; D_v is the diffusion coefficient of vapour in free air; and τ is the tortuosity tensor.

The hydraulic storage properties of the soil are introduced by setting the dependence of porosity and liquid degree of saturation on suction. In this work, van Genuchten's expression is used, providing the link between effective degree of saturation S_{eff} and total suction:

$$S_{eff} = \left[1 + (\alpha\psi)^n\right]^{-(1-1/n)} \tag{2}$$

where $S_{eff} = (S_l - S_{res})/(1 - S_{res})$, S_{res} is the liquid residual degree of saturation, and α and n are model parameters.

A power law is used to fit the experimental relationship between porosity and total suction ($\phi = a \cdot \psi^b$, where a and b are fitting parameters) and between saturated hydraulic conductivity and void ratio ($K_{sat} = c \cdot e^d$ where c and d are fitting parameters).

The vapour mass concentration c_g^w is defined as the product between the saturated concentration of vapour $c_{g,sat}^w$ and the relative humidity h_r that depends on the temperature T and on total suction ψ via the psychrometric law:

$$c_g^w = c_{g,sat}^w \cdot h_r = c_{g,sat}^w \cdot \exp^{-\psi M^w/\rho_l RT} \tag{3}$$

where M^w is the molar mass of the water and R is the constant of perfect gases.

The vapour diffusivity coefficient in free air D_v and the saturated concentration of vapour $c_{g,sat}^w$ depend on temperature T [10], while the tortuosity is assumed equal to: $\tau = (\phi S_g)^{(2/3)}$ [11].

Thermal Energy Balance Equation. The thermal energy balance equation can be expressed as shown in Eq. (4), accounting for heat flux due to conductive and convective heat transfer [12]:

$$\frac{\partial}{\partial t}\left[E_s\rho_s(1-\phi) + E_l\rho_l S_l\phi + E_g\rho_g S_g\phi\right] + \nabla \cdot \left(\mathbf{i}_c + \mathbf{j}_{E_l} + \mathbf{j}_{E_v}\right) = 0 \tag{4}$$

where $E_s = c_{p,s}T, E_l = c_{p,l}T, E_g = c_{p,g}T + L_w$ represent the energy stored in the solid, in the liquid and in the gas phase, respectively. Such terms depend on the corresponding specific heat capacities ($c_{p,s}$, $c_{p,l}$ and $c_{p,v}$) and on the latent heat of vaporization $L_w(T)$. The conduction of sensible heat is accounted for by the Fourier law, *i.e.* $i_c = -\lambda\nabla T$, where λ is the isotropic thermal conductivity of the medium and is a function of the thermal conductivities of solid particles, liquid water, and vapour respectively weighted by the corresponding volume fractions. Convective energy fluxes related to the movement of the liquid water and vapour are evaluated as follows:

$$\mathbf{j}_{E_l} = E_l\left[-\frac{\mathbf{K}_{sat}k_r}{g}\nabla(u_l + \rho_l g z)\right] \tag{5}$$

$$\mathbf{j}_{E_v} = E_g\left(-D_v\tau\nabla c_g^w\right) \tag{6}$$

Boundary Conditions at the Soil Surface. The boundary conditions for the water mass balance are given by the product between the liquid water mass density and the actual evaporation rate at the soil surface, AE, as in [13]:

$$q_{\mathbf{evap}} = \rho_l AE, \quad AE = PE\frac{p_{g,\mathrm{soil}} - p_{g,\mathrm{env}}}{p_{g,\mathrm{sat}} - p_{g,\mathrm{env}}} \tag{7}$$

where $\boldsymbol{PE}$ is the potential rate of evaporation, normal to the soil surface, $p_{g,soil} = p_{g,sat}(T)h_r(T)$ is the current vapour pressure of the soil at the soil-atmosphere interface, $p_{g,env} = p_{g,sat}(T_{env})h_r(T_{env})$ is the vapour pressure in the surrounding environment. $\boldsymbol{PE}$ depends on the mixing characteristics of the air above the evaporating surface. The actual rate of evaporation $\boldsymbol{AE}$ is equal to $\boldsymbol{PE}$ as long as the soil relative humidity is of 100% *i.e.* $p_{g,soil} = p_{g,sat}$. As $p_{g,soil}$ reduces, the rate of actual evaporation progressively decreases, limiting the flux of water vapour across the soil-atmosphere boundary.

In the thermal energy balance equation, the evaporation process acts as a heat sink, causing a reduction in temperature. Furthermore, the temperature difference between the soil and the environment generates a heat flux, which aims at re-establishing equilibrium. The thermal flux, that is imposed in the portion of the boundary subjected to evaporation, is thus (see [14]):

$$\boldsymbol{q}_{\mathbf{heat}} = L_{\mathrm{w}}\boldsymbol{q}_{\mathbf{evap}} - \varepsilon_{\mathrm{s}}\sigma\left(T_{\mathrm{env}}^4 - T^4\right)\boldsymbol{n} \tag{8}$$

where $\boldsymbol{\varepsilon}_s$ is the soil emissivity, σ is Stefan-Boltzmann constant and $\boldsymbol{n}$ is the unit vector normal to the soil surface.

4 Model Validation

To investigate the water exchange in Opalinus Clay subjected to suction variations, the THM model described in Sect. 3 was implemented in the FE code Comsol Multiphysics® and validated against the results of the VET tests on the specimens reported above. [4] provides the durations and suctions of the VET tests, as well as the initial conditions, the water content and the volumetric strains following the drying and wetting processes. The relationship between the liquid degree of saturation and total suction is established under the assumption of homogeneous distribution of water content and strain on the entire tested specimens. The hydraulic conductivities related to the void ratio are estimated from the one-dimensional consolidations presented in [15] and are assumed isotropic. Table 1 summarizes the initial conditions and the samples' parameters used in the numerical simulations.

Table 1. Initial conditions and parameters used in the numerical simulations (D: main drying path and W: main wetting path)

Core sample	L3		L8		L11		L13	
Hydraulic path	D	W	D	W	D	W	D	W
α *(MPa^{-1})*	0.064	0.260	0.086	0.208	0.262	0.573	0.043	0.126
n	1.51	1.42	1.52	1.46	1.45	1.41	1.69	1.56
a	0.1808	0.1839	0.1513	0.1532	0.1712	0.1725	0.1911	0.1928
b	−0.01	−0.013	−0.007	−0.008	−0.006	−0.007	−0.015	−0.015
c (m/s)	0.0754		1188.1		1.13e7		0.0041	
d	17.3		19.5		27.3		16	
ψ_0 *(MPa)*	4		4		4		10	
T_0 *(°C)*	25		25		25		25	

Figure 2 shows the imposed total suction for all tested specimens as a function of time, as well as the experimental data compared to the results of the numerical simulations, in terms of water content and degree of saturation. These results show that the FE model is able to successfully reproduce the VET tests along the hydraulic drying and wetting paths to which the specimens are subjected. Additionally, the equilibrium time is overall well captured by the model.

Some minor discrepancies can be observed at very high or very low suction levels, mainly due to the non-perfect fitting of van Genuchten's law for the water retention behaviour of the tested specimens. The model can be improved through the back analysis of the transient behaviour considering anisotropic hydraulic conductivity of Opalinus Clay.

This numerical model will be used to study in depth the role of advective and diffusive fluxes during the drying and wetting processes to which the Opalinus Clay will be subjected during its lifespan, as well as the effect that the environmental conditions (T, h_r, ψ) and the material properties (porosity, mineralogical composition and fabric) have on the hydro-mechanical behaviour of this material.

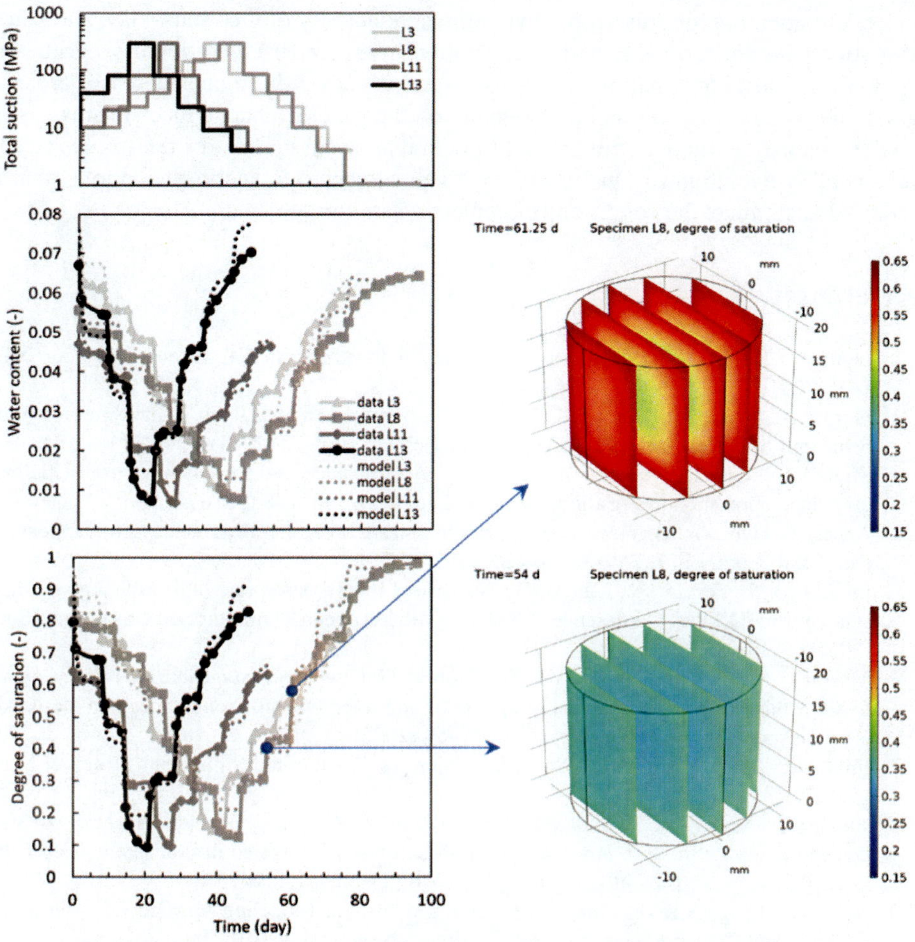

Fig. 2. Comparison between experimental and numerical results of VET tests.

5 Conclusion

In this study, the water exchange that takes place in Opalinus Clay from suction variation is investigated. In this regard, results of VET tests of four samples from Lausen site are used in order to determine the fitting parameters for the water retention properties and the evolution of void ratio and hydraulic conductivity.

The implemented numerical model is composed by the coupled system of water mass and energy balance equations. Appropriate boundary conditions are considered in order to simulate the water and heat flux from/to the geomaterial and its surrounding environment.

Preliminary analysis shows an overall good agreement between experimental data and numerical results in terms of change in water content, degree of saturation and equilibrium time. Perspectives for the improvement of the model are identified, such as the consideration of anisotropic hydraulic conductivity of Opalinus Clay, and better evolution laws for tortuosity and permeability. In this regard, additional experimental data of VET tests on Opalinus Clay from various sites will be considered in order to assess the role of the material heterogeneity on the water exchange mechanisms.

The numerical model will be used to define an accurate law for the tortuosity and permeability evolution of Opalinus Clay when subjected to suction variations, that is required to simulate the conditions of nuclear waste storage.

References

1. Leupin, O.X., et al.: High-level waste repository-induced effects. Nagra Technical report NTB 14-13 (2016)
2. Ferrari, A., Favero, V., Marschall, P., Laloui, L.: Experimental analysis of the water retention behaviour of shales. Int. J. Rock Mech. Min. Sci. (72), 61–70 (2014)
3. Minardi, A., Crisci, E., Ferrari, A., Laloui, L.: Anisotropic volumetric behaviour of Opalinus clay shale upon suction variation. Géotech. Lett. **6**(2), 144–148 (2016)
4. Crisci, E.: Hydro-mechanical response of Opalinus Clay shale: dependency on composition and burial depth, EPFL PhD Thesis (2019)
5. Incollingo, S., Ferrari, A., Musso, G.: Numerical investigation on water exchange of shale samples. In: E3S Web Conference 195, 02025, 4th European Conference on Unsaturated Soils (2020)
6. Mazurek, M., Hurford, A.J., Leu, W.: Unravelling the multi-stage burial history of the Swiss Molasse Basin: integration of apatite fission track, vitrinite reflectance and biomarker isomerisation analysis. Basin Res. (18), 27–50 (2006)
7. Fredlund, D.G., Rahardjo, H.: Soil Mechanics for Unsaturated Soils. John Wiley & Sons, Inc., Hoboken (1993)
8. Musso, G., Vespo, V.S., Guida, G., Della Vecchia, G.: Hydro-mechanical behaviour of a cement-bentonite mixture along evaporation and water-uptake controlled paths. Geomech. Energy Environ. (2022). https://doi.org/10.1016/j.gete.2022.100413
9. Bear, J., Cheng, A.H.D.: Unsaturated flow models. In: Modeling Groundwater Flow and Contaminant Transport, pp. 251–340. Springer, New York (2010). https://doi.org/10.1007/978-1-4020-6682-5
10. Vespo, V.S., Della Vecchia, G., Musso, G.: Modelling evaporation processes of cement-bentonite mixtures. In: E3S Web Conference, vol. 195, p. 02029, 4th European Conference on Unsaturated Soils (2020). https://doi.org/10.1051/e3sconf/202019502029
11. Lai, S.-H., Tiedje, J.M., Erickson, A.E.: In situ measurement of gas diffusion coefficient in soils. Soil Sci. Soc. Am. J. (40), 3–6 (1976)
12. Gens, A.: Soil–environment interactions in geotechnical engineering. Géotechnique **60**(1), 3–74 (2010)
13. Penman, H.L.: Estimating evaporation. EOS Trans. Am. Geophys. Union **37**(1), 43–50 (1956)
14. Ralaizafisoloarivony, N.A., et al.: Experimental and numerical investigation of the drying of an agricultural soil. In: E3S Web Conference, vol. 195, p. 01034, 4th European Conference on Unsaturated Soils. (2020)
15. Crisci, E., Ferrari, A., Giger, S.B., Laloui, L.: Hydro-mechanical behaviour of shallow Opalinus Clay shale. Eng. Geol. (251), 214–227 (2019)

Experimentation of the Thermo-Mechanical Behavior of the Soil-Concrete Interface

Arianna Lupattelli[1](✉), Erica Cernuto[1], Benedetta Brunelli[1], Elisabetta Cattoni[2], and Diana Salciarini[1]

[1] Department of Civil and Environmental Engineering, University of Perugia, 06125 Perugia, Italy
arianna.lupattelli@studenti.unipg.it
[2] eCampus University, 22060 Novedrate, Italy

Abstract. The thermo-mechanical behavior of soils has been the topic of many studies over the last few decades and nowadays is closely related to the promotion of Energy Geostructures (EGs), aimed at reducing energy consumption from fossil fuels and the consequent emissions. During their operation, EGs are subjected to thermal variations due to the exploitation of the low-enthalpy geothermal resource, and this can have an impact on the structure response. To contribute in having a more precise framework of the complex mechanisms that affect the thermo-mechanical behavior of the EGs–soil interface, a modified testing device has been developed at the Laboratory of Geotechnical Engineering of the University of Perugia. A conventional direct shear apparatus has been equipped with a heating plate at the base of the soil samples, where a temperature probe for continuous temperature control has been integrated. In this work, the first results of an experimental campaign conducted to determine the influence of temperature on the shear behavior of soil and soil-concrete interface are shown. Temperature-controlled, direct shear interface tests were carried out at normal stress values ranging from 25 to 100 kPa. By comparing the results obtained for heated and not-heated interfaces (constant thermal load of 30 °C *vs.* room temperature of 20 °C), a limited effect of the temperature variation on the interface shear strength was observed, with a slight increase of the interface friction angle with heating.

Keywords: Soil-Structure Interface · Direct Shear Box · Thermo-Mechanical Behaviour · Shear Strength · Energy Geostructures

1 Introduction

Interaction problems between soil and structures involving contact friction and adhesion take on relevance in geotechnical problems. By definition, the soil-structure interface is a thin soil layer where contact resistance develops, and which can undergo large shear strain ([1, 2], among others). For EGs, the soil-structure interface represents the element through which mechanical and thermal loads are transferred to or from the surrounding soil at the same time. The multi-physical nature of the loads makes the Thermo-Mechanical (TM) behavior of EGs very complex ([3–6] among others) and the

A. Ferrari et al. (Eds.): CNRIG 2023, SSGG, pp. 343–350, 2023.
https://doi.org/10.1007/978-3-031-34761-0_42

role of the soil-structure interface should be addressed for the analysis and design of EGs. Despite currently there exists a broad body of knowledge regarding the influence of temperature variations on the mechanical behavior of soil [7], especially on the soil shear resistance ([8–10], among others), there remains a significant knowledge gap regarding the thermally induced effects on the soil-structure interface. What emerges from previous experimental research is that the differences found in the results can be likely linked to the multitude of experimental configurations, development protocols, and composition of the samples used in the tests [11]. The works of [12] and [13] investigated, respectively, the soil-concrete interface behavior of sandy and clayey soil samples, using a modified direct shear box to reproduce the soil-structure interface conditions. [14] conducted several direct shear tests, using a temperature-controlled direct shear device to evaluate the effects of heating and cooling cycles on soil–concrete interface shear strength, investigating the effects of stress state and history on clayey soil samples. The experimental program of [15] aimed to investigate the TM effects on fine-grained soil-concrete interface subjected to cycling heating in saturated conditions, including some modifications to the direct shear device described by [13].

To enrich the experimental data already available, this paper presents the first tests carried out as part of a wider experimentation program. After presenting the new shear box device with temperature control, the results of the first tests carried out both on dry sandy soils and at the sand-concrete interface, subjected to constant heating conditions are shown.

2 Experimental Setup

To evaluate the TM response of heated soils and at the interface with a structural material, such as concrete, an experimental device has been developed at the Department of Civil and Environmental Engineering of the University of Perugia. A classic shear box has been implemented with: *i)* a thermal resistance that heats the soil sample and a thermal probe for the full control of the temperature during the experimentation (Fig. 1*a*); *ii)* a concrete plate, equipped with thermal resistance, with a height equal to the upper edge

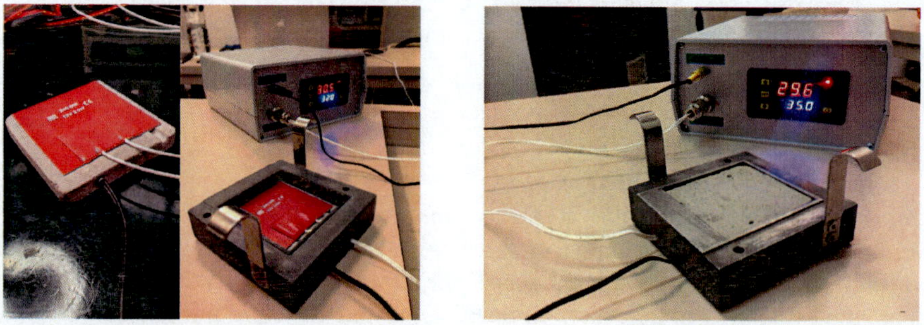

a) *b)*

Fig. 1. Pictures of the new device used to carry out direct shear tests with temperature control for *a)* soil samples, and *b)* at the interface with a concrete plate.

of the lower half-box, to allow the execution of the test at the interface between the two materials (Fig. 1*b*).

Commonly, the configuration of the shear box assembled for the traditional test includes one or more ribbed plates positioned at the base, above which the soil sample is placed for a thickness equal to $t_s = 24.8$ mm, and a further plate on the upper base of the sample.

The new shear box device differs in the absence of ribbed plates at the base of the first half-box and the thickness of the soil sample, which is reduced to $t_s^* = 20.6$ mm due to the positioning of a heating plate with thickness $t_c^* = 9$ mm. Here, a thermal resistance for the application of the thermal load and a temperature probe for continuous temperature control have been integrated (Fig. 2*a*). The device is connected to a control box, with a temperature detector and regulation. Also, a second configuration of this device was realized, to perform shear tests at the soil-concrete interface. In this case, a R_{ck350} concrete plate with a thickness $t_i = 19$ mm and limited roughness has been inserted and fixed in the lower half part of the shear box. A silicon heating mat is integrated at the bottom of the plate (Fig. 2*b*), along with a sensor for temperature control.

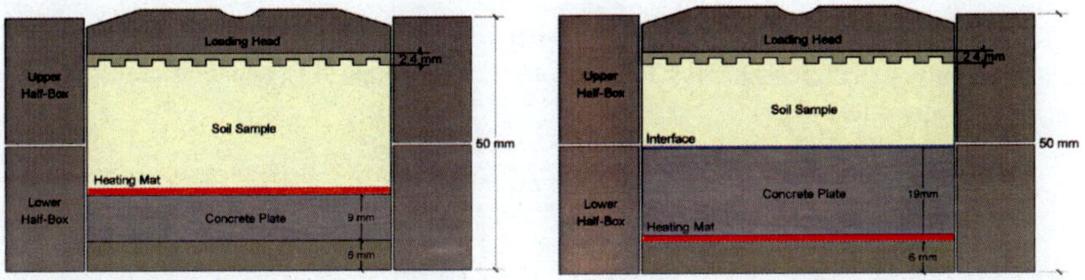

a) *b)*

Fig. 2. Illustration of the device *a)* with the heating plate for thermally-controlled direct shear tests on soil samples, and *b)* with a thicker heating plate for thermally-controlled direct shear tests at the interface between soil and concrete.

2.1 Soil Properties

Two types of soil have been tested. A granulometric analysis (Fig. 3), through the use of 11 sieves (diameter from 0.063 mm to 2 mm), made it possible to classify them, as s_I, silty sand (85.3% sand, and 14.7% silt), and s_{II} sand (100% sand), respectively.

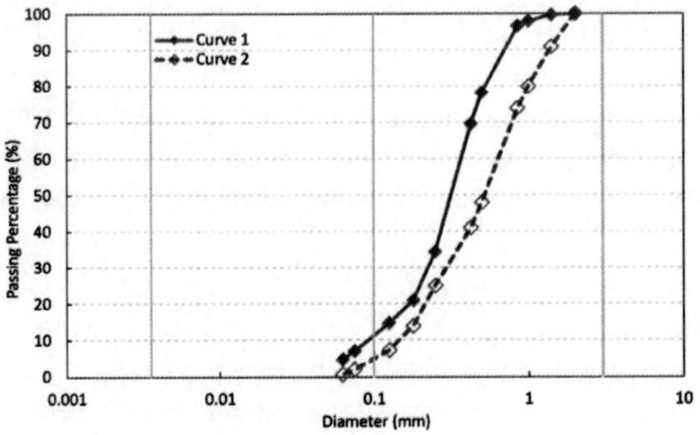

Fig. 3. Grain-size distribution of the two soils tested.

2.2 Testing Program

A total of n.24 direct shear tests (Table 1) has been carried out on dry soil samples and at the interface with a concrete plate.

Table 1. Summary of the direct shear tests performed: traditional direct shear tests (TD), temperature-controlled direct shear test (TD*), interface shear test (TI), and temperature-controlled interface direct shear test (TI*).

N.	Type	Soil	σ_n (kPa)	T_{Fase1} (°C)	T_{Fase2} (°C)	N.	Type	Soil	σ_n (kPa)	T_{Fase1} (°C)	T_{Fase2} (°C)
1	TD	s_I	25	T_{amb}-	T_{amb}-	13	TD	s_{II}	25	T_{amb}-	T_{amb}-
2	TD	s_I	50	T_{amb}-	T_{amb}-	14	TD	s_{II}	50	T_{amb}-	T_{amb}-
3	TD	s_I	100	T_{amb}-	T_{amb}-	15	TD	s_{II}	100	T_{amb}-	T_{amb}-
4	TD*	s_I	25	32.0	31.5	16	TD*	s_{II}	25	28.2	28.6
5	TD*	s_I	50	31.1	32.4	17	TD*	s_{II}	50	29.0	29.7
6	TD*	s_I	100	32.4	32.5	18	TD*	s_{II}	100	29.5	30.3
7	TI	s_I	25	T_{amb}-	T_{amb}-	19	TI	s_{II}	25	T_{amb}-	T_{amb}-
8	TI	s_I	50	T_{amb}-	T_{amb}-	20	TI	s_{II}	50	T_{amb}-	T_{amb}-
9	TI	s_I	100	T_{amb}-	T_{amb}-	21	TI	s_{II}	100	T_{amb}-	T_{amb}-
10	TI*	s_I	25	33.0	33.0	22	TI*	s_{II}	25	29.1	29.3
11	TI*	s_I	50	32.7	33.3	23	TI*	s_{II}	50	28.6	29.0
12	TI*	s_I	100	32.8	33.0	24	TI*	s_{II}	100	28.6	29.1

Among these, tests 1–3 and tests 13–15 were performed using the traditional direct shear (TD) standard procedure, while tests 4–12 and tests16–24 were performed with the new equipment (TD* - temperature-controlled direct shear test, TI - interface shear test, and TI* - temperature-controlled interface direct shear test). The soil samples were prepared to obtain a void index of $e_{sI} = 0.6$ and $e_{sII} = 0.7$. Given the mode of compaction, and the typical range of e_{max}, e_{min} for coarse-grained soils, the samples can be considered as dense, estimating a relative density (D_r) between 50–70%. The temperature and loading conditions of each test are summarized in Table 1, while the TM loading path is shown in Fig. 4.

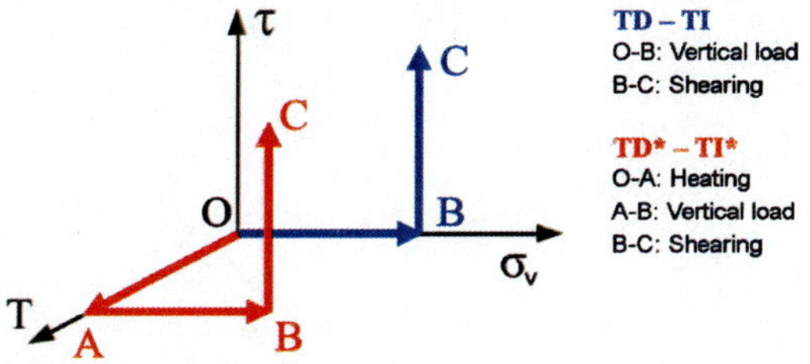

Fig. 4. Stress–temperature paths followed during the tests at environment temperature (TD-TI) and with heating (TD*-TI*).

For the TD and TI tests, the first vertical loading phase (Phase 1) was performed with the application of three different vertical stresses, ranging from 25 to 100 kPa. The subsequent shearing phase (Phase 2) was performed with a rate equal to $v_I = 0.066$ mm/min. In the case of the TD* and TI* (heated tests), a temperature in the heating plate equal to 32.5 °C was set, before the vertical loading phase, and kept approximately constant during the test (Fig. 4).

3 Results and Discussion

The response of soil samples subjected to the testing program summarized in Table 1 has been analyzed to understand and compare their behavior, assessing the resistance parameters, *i.e.*, the peak of shear strength and friction angle. The results of the experimental campaign on soil samples [test n.1–6, 13–18] and on soil-concrete interfaces [test n.7–12, and 19–24] are shown in Figs. 5 and 6, respectively.

The results of the set n.1 [tests n.1–6], carried out on s_I soil (Fig. 5 column *a*) show that the failure behavior is of the brittle type, and the samples show dilation (confirming a typical behavior of dense coarse-grained soils).

At the application of low loads (25 and 50 kPa), the curves $\delta_x - \tau$ coincide for heated and not heated samples, while for higher loads (100 kPa), the peak resistance is slightly greater for the heated samples, with an increase from 99 to 108 kPa. The failure envelopes (adopting the linear Mohr-Coulomb failure criteria) obtained from this set, show friction angle increases from 37.6° to 42.3° (Fig. 7*a*).

The results of the set n.2 [tests n.13–18], carried out on s_{II} soil (Fig. 5 column *b*), show the same failure behavior, and dilatant trend. For a low load of 25 kPa, the heated sample shows a slight lower peak strength than the environmental-temperature sample (from 39 to 27 kPa), while for a higher applied load (50 and 100 kPa) the curves almost coincide in both cases. The failure envelopes, obtained from this set, show that the friction angle increases from to 43.6° to 44.6° (Fig. 7*a*). Limited differences observed in the results of the two sets of tests can be likely re-conducted to different dilatancy angles, being the peak strength likely correlated to soil dilatancy for the coarse grain-size of the examined material.

The results of the set n.3 [tests n.7–12], carried out at the interface between concrete and soil s_I (Fig. 6*a*), show that the failure behavior is of ductile type. When a low vertical load of 25 or 50 kPa is applied, the curves $\delta_x - \tau$ coincide for the heated and not

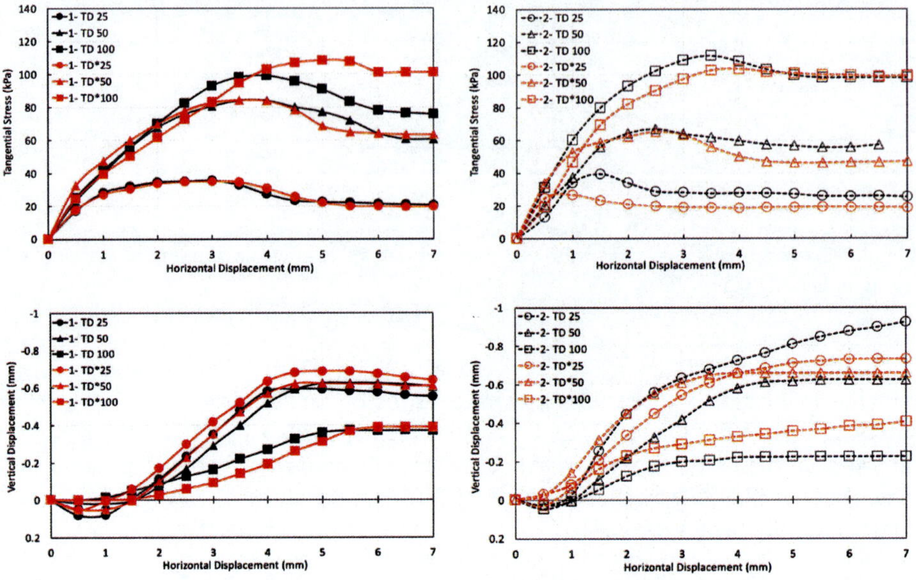

a) *b)*

Fig. 5. Results of TD and TD* tests in terms of horizontal displacements vs. tangential stresses ($\delta_x - \tau$), and horizontal displacements vs. vertical displacements ($\delta_x - \delta_y$) for *a)* s_I soil, and *b)* s_{II} soil.

heated interface test, while for a vertical load of 100 kPa, the peak of shear stress of the heated interface slightly increases (from 70 to 85 kPa). The failure envelopes (adopting the linear Mohr-Coulomb failure criteria) obtained from this set, show friction angle increases from 32.1° to 41.9° (Fig. 7*b*).

The results of the set n.4 [tests n.19–24], carried out at the interface between concrete and soil s_{II} (Fig. 6*b*), show the same failure behavior. It is noted that, with the application of all three vertical loads, the same trend is registered for the curves, with the shear resistance of the heated interfaces slightly increasing with respect to the not-heated ones.

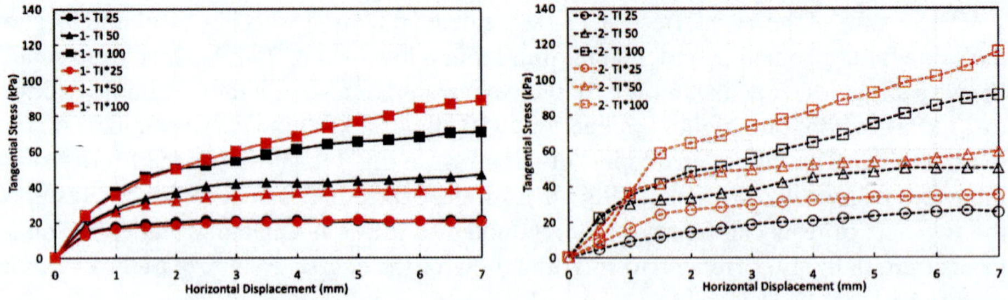

a) *b)*

Fig. 6. Results of TI and TI* tests in terms of horizontal displacements vs. tangential stresses ($\delta_x - \tau$), and failure envelopes for *a)* s_I-concrete, and *b)* s_{II}-concrete interface.

The failure envelopes, obtained from this set, show that the friction angle increases from 41.5° to 47.2° (Fig. 7*b*).

In Fig. 7*c*, the trend of the friction angle for all the considered cases is shown. It is noted that a very limited increase of the friction angle is appreciable for the tests under constant heating.

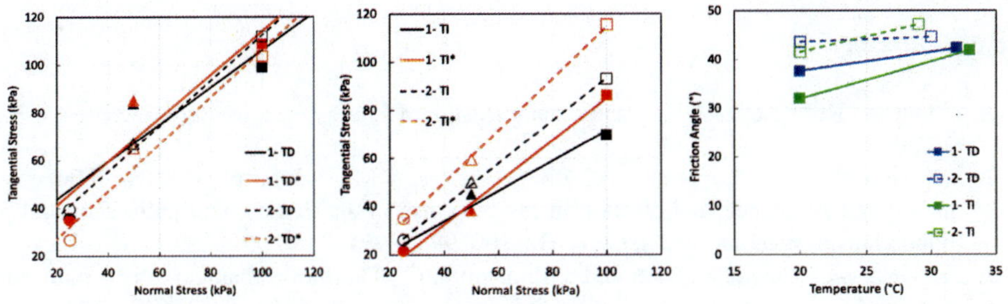

a) *b)* *c)*

Fig. 7. Failure envelopes in the Mohr-Coulomb plane for *a)* soil (TD-TD*), and *b)* interface (TI-TI*) tests, and *c)* variation of the friction angle as a function of temperature for all sets.

4 Conclusions

In the present work, the results of temperature-controlled, direct shear tests on sandy soil samples and at soil-concrete interface are presented, using a modified shear box device developed at the Laboratory of Geotechnical Engineering of the University of Perugia. The aim was to contribute to enriching the understanding of the complex TM behavior of the EGs.

From the TD and TD* experimental results, the following conclusions can be drawn:

- For samples of s_I and s_{II} soil, the peak resistance does not change significantly with temperature. Exceptions are the peak resistance of s_I, that is slightly higher with heating at the maximum applied load, and the peak resistance of s_{II}, that is slightly lower with heating at the minimum applied load.
- The comparison of the failure envelopes shows that the friction angle increases with the heating for both soils, determining a slight increase in shear resistance. In general, such increases and decreases are so small as to be considered negligible, and these trends are confirmed by other studies [11–13].

From the TI and TI* experimental results, the following conclusions can be drawn:

- At the interface, for samples of s_I and s_{II} soil, the shear resistance does not change significantly with heating, even it results slightly higher in the case of the tests with heating.

- The failure envelopes of TI and TI* tests show the same trend as TD and TD* tests, with a slight increase of the friction angle from not-heated to heated interfaces.

Future research works will be dedicated to investigating the behavior of soil samples subjected to cyclic thermal loads and in saturated conditions. Also, efforts will be devoted to the definition of an accurate law for failure envelope.

References

1. Laloui, L., Rotta Loria, A.F.: Analysis and Design of Energy Geostructures. Academic Press (2020)
2. Bourne-Webb, P.J., Lupattelli, A., Bodas Freitas, T.M., Salciarini, D.: The influence of initial shaft resistance mobilisation in the response of seasonally, thermally-activated pile foundations in granular media. GETE **32**, 100299 (2022)
3. Salciarini, D., Ronchi, F., Cattoni, E., Tamagnini, C.: Thermomechanical effects induced by energy piles operation in a small piled raft. Int. J. Geomech. **15**(2), 04014042 (2015)
4. Ronchi, F., Salciarini, D., Cavalagli, N., Tamagnini, C.: Thermal response prediction of a prototype Energy Micro-Pile. GETE **16**, 64–82 (2018)
5. Cecinato, F., Salciarini, D.: Energy performance assessment of thermo-active micro-piles via numerical modeling and statistical analysis. GETE **29**, 100268 (2022)
6. Rotta Loria, A.F., Bocco, M., Garbellini, C., Muttoni, A., Laloui, L.: The role of thermal loads in the performance-based design of energy piles. GETE **21**, 100153 (2020)
7. Vieira, A., et al.: Site characterization for the design of thermoactive geostructures. Soils and Rocks 45(1) (2022)
8. Burghignoli, A., Desideri, A., Miliziano, S.: A laboratory study on the thermomechanical behaviour of clayey soils. Can. Geotech. J. **37**(4), 764–780 (2000)
9. Cekerevac, C., Laloui, L.: Experimental study of thermal effects on the mechanical behaviour of a clay. Int. J. Numer. Anal. Methods Geomech. **28**(3), 209–228 (2004)
10. Laloui, L., et al.: Issues involved with thermoactive geotechnical systems: characterization of thermomechanical soil behavior and soil-structure interface behavior. DFI J. **8**(2), 108–120 (2014)
11. Maghsoodi, S., Masrouri, F., Cusinier, O.: Thermal effects on the mechanical behaviour of the soil-structure interface. Can. Geotech. J. **57**(1), 32–47 (2019)
12. Yavari, N., Tang, A.M., Pereira, J.M., Hassen, G.: Effect of temperature on the shear strength of soils and soil/structure interface. Can. Geotech. J. **53**(7), 1186–1194 (2016)
13. Di Donna, A., Ferrari, A., Laloui, L.: Experimental investigations of the soil-concrete interface: physical mechanisms, cyclic mobilization, and behaviour at different temperatures. Can. Geotech. J. **53**(4), 659–672 (2016)
14. Yazdani, S., Helwany, S., Olgun, G.: Influence of temperature on soil-pile interface shear strength. GETE **18**, 69–78 (2019)
15. Ravera, E., Sutman, M., Laloui, L.: Cyclic thermomechanical response of fine-grained soil-concrete interface for energy piles applications. Can. Geotech. J. **58**(8), 1216–1230 (2020)

Generation of Yield Surfaces and Plastic Potentials in Elastoplastic Modelling of Soils

Giuseppe Mortara(✉)

DICEAM, University Mediterranea of Reggio Calabria, Reggio Calabria, Italy
giuseppe.mortara@unirc.it

Abstract. This paper proposes new strategies for the determination of yield surfaces and plastic potentials for soil elastoplastic modelling. Generation of yield surfaces is based on a yield function with a large number of parameters to account for the many meridian shapes typical of soil modelling. The formulation automatically provides a tension cut-off if a suitable Lode dependence is provided. Generation of plastic potentials is instead based on the adoption of a stress-dilatancy relationship, which may include well known expressions in literature, and of an auxiliary simple meridian function coupled with a desired Lode dependence. By doing so, no specific expressions of plastic potentials need to be implemented and the derivatives are very simple and independent of any stress-dilatancy relationship.

Keywords: elastoplasticity · yield surface · plastic potential · tension cut-off · finite element method

1 Introduction

The solution of boundary value problems related to geomechanics needs the implementation of ad-hoc constitutive models in finite element method codes. The different constitutive models based on elastoplasticity differ for the choice of yield surface, plastic potential, hardening and elastic laws. Extensive research by the author [1–5] has led to yield criteria formulated with different approaches: the first is based on the well known criteria by [6,7]; the second is based on the definition of meridian and deviatoric functions as in [8]; the third is intermediate between the previous and can be called *hybrid* [5]. As far as the plastic potential is concerned, the author [9] has defined an approach that allows to avoid the explicit definition of a specific plastic potential if a stress-dilatancy relationship (SDR) is adopted. Yield surfaces and plastic potentials will be formulated in terms of mean stress p, deviatoric stress q, and Lode angle θ defined as follows

$$p = \frac{1}{3}\sigma_{kk} \qquad q = \sqrt{\frac{3}{2}s_{ij}s_{ij}} \qquad \theta = \frac{1}{3}\arccos\left(\frac{9}{2}\frac{s_{ij}s_{jk}s_{ki}}{q^3}\right) \tag{1}$$

where σ_{ij} is the effective Cauchy stress tensor and s_{ij} is its deviatoric part.

Dedicated to Professor Roberto Nova

A. Ferrari et al. (Eds.): CNRIG 2023, SSGG, pp. 351–358, 2023.
https://doi.org/10.1007/978-3-031-34761-0_43

2 Generation of Yield Surfaces

The main feature of the yield surfaces generated in the following is to have a finite tangent η_t at the origin of the $p - q$ plane. In particular, to avoid tensile stress states the following constraint on η_t must be introduced

$$\eta_t \leq \frac{3}{\cos\theta + \sqrt{3}\sin\theta} \tag{2}$$

Inequality (2) prescribes that $\eta_t \leq 3$ for $\theta = 0$ (triaxial compression) and $\eta_t \leq 3/2$ for $\theta = \pi/3$ (triaxial extension). When the equality holds, the previous condition represents the tension cut-off and is represented by the triangles originated by the trace of coordinate planes $\sigma_x - \sigma_y$, $\sigma_y - \sigma_z$, and $\sigma_z - \sigma_x$ with any deviatoric plane in the space of principal stresses. The fulfillment of Eq. (2) implies that many meridian surfaces in literature, as for example the Cam Clay ones, could not be used because a portion is outside the limit domain $q = 3p/\left(\cos\theta + \sqrt{3}\sin\theta\right)$. Furthermore, even when a simple cone is used as yield surface, condition (2) limits the maximum friction angle ϕ to

$$\phi_{lim} = \arcsin\left(\frac{3}{4\beta + 1}\right) \tag{3}$$

where β is the ratio between the values of q in triaxial extension and compression. The fulfillment of Eq. (2) implies also that the Lode dependence of the surface must depend on the stress-ratio in order to obtain a smooth transition of the surface in the deviatoric representation. The expression of the yield function is put in the form

$$\Phi = q^2 - \Phi_f \Phi_{\rho f}^2 \tag{4}$$

where

$$\Phi_f = (1 - s_t R)\,\Phi_1 + R\Phi_2 + R\Phi_3 \tag{5}$$

$$\Phi_{\rho f} = \frac{\sqrt{3}}{2} \frac{\beta}{\sqrt{\beta^2 - \beta + 1}} \frac{s_\theta}{\cos\left(\dfrac{4 + s_\theta}{3}\pi + \dfrac{\omega}{6}\right)} \tag{6}$$

The functions needed to define the previous equations are:

$$\Phi_1 = s_t\left(p_t + p_c\right)^2 m_1^2\, R^2 \left|1 - s_t R^{n_1}\right|^{2c_1} \tag{7}$$

$$\Phi_2 = s_t\left(p_t + p_c\right)^2 m_2^2\, R^2 \left(1 - s_t R^{n_2}\right)^{2c_2} \tag{8}$$

$$\Phi_3 = s_t\left(p_t + p_c\right)^2 m_3^2\, R^2 \left(1 - s_t R^{n_3}\right)^{2c_3} \tag{9}$$

$$R = \frac{|p_t + p|}{p_t + p_c} \tag{10}$$

$$s_t = \mathrm{sgn}\left(p_t + p\right) \tag{11}$$

$$\beta = \frac{3 + P_m}{3 + m_c + 2P_m} \tag{12}$$

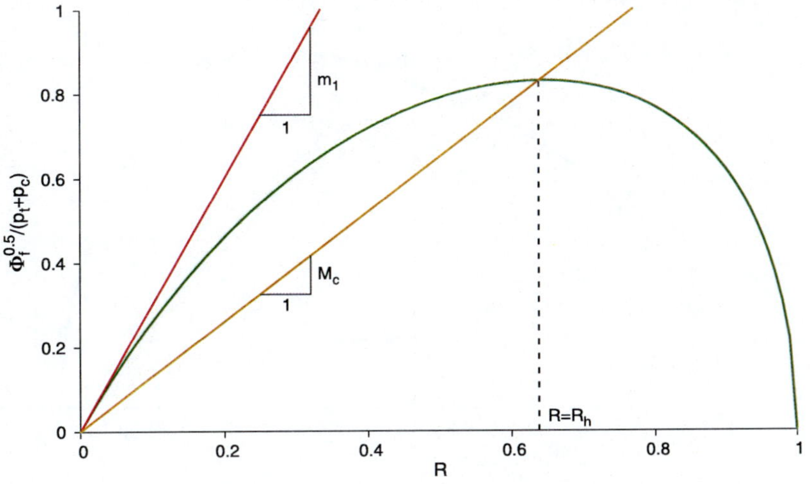

Fig. 1. Geometrical meaning of m_1 and of *ghost parameters* R_h and M_c.

$$s_\theta = 1 - |\mathrm{sgn}\,(\cos 3\theta)| + \mathrm{sgn}\,(\cos 3\theta) \tag{13}$$

$$\omega = \arccos\left[-1 + \frac{27}{2}\frac{\beta^2\,(1-\beta)^2}{(\beta^2-\beta+1)^3}\cos^2 3\theta\right] \tag{14}$$

$$P_m = k_1 m_c + k_2 m_c^3 + k_3 m_c^5 + k_4 m_c^7 \tag{15}$$

$$m_c = \begin{cases} m_1 & \text{for} \quad -p_t \geq p \\ \Phi_f^{0.5}/\,(p_t+p) & \text{for} \quad -p_t < p < p_c \\ 0 & \text{for} \quad p \geq p_c \end{cases} \tag{16}$$

The elastic domain is defined in the mean stress range $-p_t \leq p \leq p_c$ where p_t and p_c are hardening parameters of the criterion. From the analysis of the previous equations it follows that the yield function requires the determination of 9 constants for the meridian trace (m_1, n_1, c_1, m_2, n_2, c_2, m_3, n_3, c_3) and 4 for the deviatoric one (k_1, k_2, k_3, k_4). Criterion (4) will be called 3FCO. As shown in Fig. 1, parameter m_1 is the tangent of the surface at $R = 0$, R_h is the value of R for $\partial\Phi/\partial R = 0$, and $M_c = \Phi_f^{0.5}/[R_h(p_t + p_c)]$. Constants R_h and M_c are not parameters of the surface and are therefore called *ghost parameters* being needed to its definition. Some parameters are either subjected to constraints or imposed: $m_1 \leq 3$, $c_1 > 0.5$, $c_2 = 0.5$ or 1.5, $n_3 = 1$, $c_3 = 0.5$. Parameters m_2 and c_1 are found from the two conditions $\partial\Phi/\partial R = 0$ and $\Phi_f^{0.5}/[R(p_t + p_c)] = M_c$ both computed at $R = R_h$. After fixing a value for c_1 and determining m_2 as

$$m_2 = \sqrt{-\frac{m_1^2 Q_1 + m_3^2 Q_3}{Q_2}} \tag{17}$$

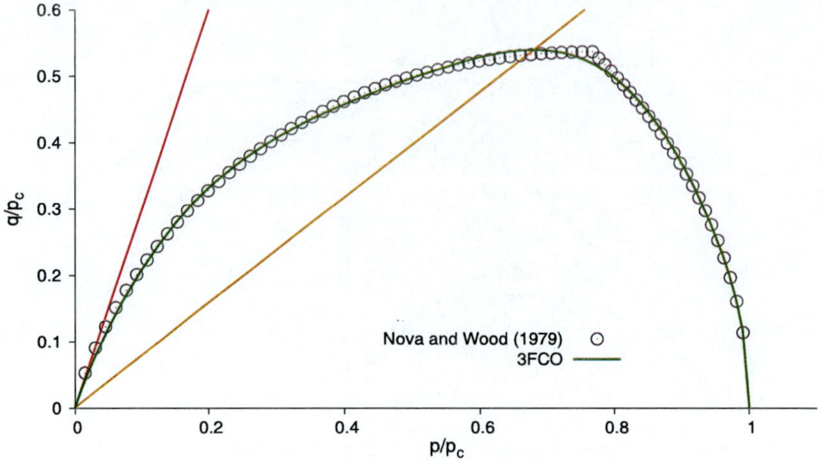

Fig. 2. Comparison between the yield surface by Nova and Wood [10] and 3FCO criterion (*ghost parameters*: $R_h = 0.68$ and $M_c = 0.79$).

where

$$Q_1 = \left(1 - R_h^{n_1}\right)^{2c_1-1} R_h \left\{2\left[1 - R_h^{n_1}\left(1 + c_1 n_1\right)\right]\left(1 - R_h\right) - R_h + R_h^{n_1+1}\right\} \quad (18)$$

$$Q_2 = \left(1 - R_h^{n_2}\right)^{2c_2-1} R_h^2 \left\{2\left[1 - R_h^{n_2}\left(1 + c_2 n_2\right)\right] + 1 - R_h^{n_2}\right\} \quad (19)$$

$$Q_3 = \left(1 - R_h^{n_3}\right)^{2c_3-1} R_h^2 \left\{2\left[1 - R_h^{n_3}\left(1 + c_3 n_3\right)\right] + 1 - R_h^{n_3}\right\} \quad (20)$$

the numerical procedure simply introduces ranges of values for n_1, n_2, c_2, and m_3 while c_1 is found with a bisection algorithm until condition $\Phi_f^{0.5}/[R_h(p_t + p_c)] = M_c$ is not fulfilled. Among the thousands of functions processed, the optimum one is chosen according to the minimum scatter between the input data points and the computed ones. Figure 2 shows the comparison between the composite yield surface by Nova and Wood [10][1] and (4). The surface points are related to $M = 1.40$, $m = 0.70$, and $\mu = 0.80$. The values of *ghost paramaters* are $R_h = 0.68$ and $M_c = 0.79$ while the values of parameters for Eq. (4) resulting from the generation procedure are: $m_1 = 3.000$, $n_1 = 0.604$, $c_1 = 1.148$, $m_2 = 0.738$, $n_2 = 5.284$, $c_2 = 1.500$, $m_3 = 1.192$, $n_3 = 1.000$, $c_3 = 0.500$.

[1] $$q = \begin{cases} p_c R\left[\dfrac{M}{2} - m\ln\left(R\sqrt{1+\mu}\right)\right] & \text{for} \quad \dfrac{q}{p} \geq \dfrac{M}{2} \\ p_c M\sqrt{\dfrac{1-R^2}{4\mu}} & \text{for} \quad \dfrac{q}{p} \leq \dfrac{M}{2} \end{cases}.$$

The values of k_1, k_2, k_3, and k_4 for defining the Lode dependence can be determined from the 3In criterion [5]

$$\xi_0 \left[2\left(\frac{M_c\beta}{9}\right)^3 + \left(\frac{M_c\beta}{9}\right)^2 - \frac{1}{27}\right] + \left[\frac{1}{3} - 3\left(\frac{M_c\beta}{9}\right)^2\right]^{\frac{3-n}{2}} = 0 \tag{21}$$

where

$$\xi_0 = \frac{(3-\sin\phi)^n\,(3+\sin\phi)^{\frac{3-n}{2}}}{(1+\sin\phi)\,(1-\sin\phi)^{\frac{1+n}{2}}} \tag{22}$$

$$M_c = \frac{6\sin\phi}{3-\sin\phi} \tag{23}$$

Fixing a value of n and four values of ϕ (e.g. $\phi = 15, 25, 45, 65°$), four values of β are determined from (21) and, with the aid of Eqs. (12) and (15) for $m_c = M_c$, a linear system is obtained that allows to determine the value of constants k_i for the selected value of n. For $k_1 = k_2 = k_3 = k_4 = 0$ the variation of β as in the Matsuoka and Nakai criterion [6] is obtained.

3 Generation of Plastic Potentials

Many expressions of plastic potentials have been defined in literature that are principally based on stress-dilatancy relationships (SDR) defined on the basis of energy principles or experimental data. The flow rule allows to obtain a differential equation that provides the plastic potential expression. Here, an alternative approach is pursued: a hierarchical SDR is given and the related plastic potential is defined as an auxiliary surface giving the rate of dilation predicted by the SDR. The expression of the stress-dilatancy relationship used in this work is the following

$$d_0 = \frac{M_g^b - \eta_0^b + cM_g\,(\eta_0 - M_g)}{\mu_1 + \mu_2\eta_0} \tag{24}$$

where d_0 and $\eta_0 = q_0/p$, being q_0 the value of q for $\theta = 0$, are the rate of dilation and the stress ratio under triaxial compression conditions while M_g, b, c, μ_1, and μ_2 are parameters of the SDR. The auxiliary plastic potential used in this work has the following expression

$$g = q^2 + \Phi_g \Phi_{\rho g}^2 \tag{25}$$

where Φ_g and $\Phi_{\rho g}$ are the meridian and deviatoric traces of g, respectively. Being convenient to choose an ellipse as meridian shape, Φ_g can be expressed as

$$\Phi_g = -r_e^2\left[k_e^2 - (p - p_g)^2\right] \tag{26}$$

where r_e is the ratio between the ellipse radii along q and p while k_e and p_g are constants obtained by conditions $g = 0$ and $\mathrm{d}q_0/\mathrm{d}p = -d_0$ at (p, q_0)

$$k_e = \frac{q_0}{r_e^2}\sqrt{r_e^2 + d_0^2} \qquad p_g = p - \frac{q_0 d_0}{r_e^2} \tag{27}$$

Table 1. Variables for the auxiliary plastic potentials in Fig. 3.

p	q_0	η_0	d_0	k_e	p_g
130.4	91.3	0.700	0.583	77.0	98.9
187.5	262.5	1.400	0.000	201.9	187.5
333.3	700.0	2.100	−0.583	590.2	575.0

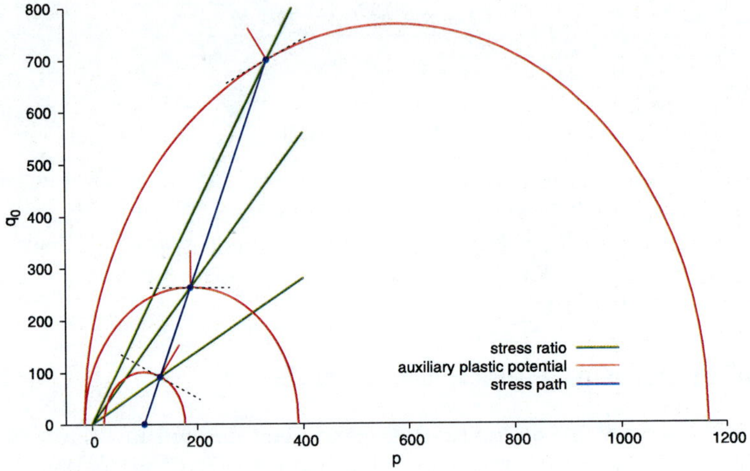

Fig. 3. Auxiliary plastic potentials for $r_e = M_g$ drawn for $\eta_0/M_g = 0.5, 1.0, 1.5$. SDR parameters: $M_g = 1.4$, $\mu_1 = 1.2$, $b = c = \mu_2 = 0$.

Figure 3 shows the evolution of plastic potentials for a linear SDR of parameters $M_g = 1.4$, $\mu_1 = 1.2$, and $b = c = \mu_2 = 0$. The evolution is shown for a conventional triaxial compression starting from $p = 100$ and for $\eta_0/M_g = 0.5, 1.0, 1.5$ while the ellipse shape parameter is $r_e = M_g$. The variables used for defining the auxiliary plastic potentials shown in Fig. 3 are reported in Table 1.

4 Example Modelling

The concepts defined previously are used here for a reinterpretation of the *Sinfonietta Classica* (SC) by Nova [11]. The function used for yield surface and plastic potential of SC has some drawbacks related to its polynomial form

$$f = 3\alpha\,(\gamma - 3)\ln\left(\frac{p}{p_c}\right) - \gamma J_{3\eta} + \frac{9}{4}(\gamma - 1)\,J_{2\eta} = 0 \tag{28}$$

where $J_{2\eta} = \eta_{ij}\eta_{ij}$ and $J_{3\eta} = \eta_{ij}\eta_{jk}\eta_{ki}$, being $\eta_{ij} = s_{ij}/p$ and $\gamma = (9 - M_g^2)/(3 - M_g^2 + 2M_g^3/9)$. The plastic potential is obtained for $\alpha = 3$. Rewritten in terms of p, q and θ Eq. (28) becomes

$$f = 4\gamma q^3 \cos(3\theta) - 27\,(\gamma - 1)\,pq^2 - 54\alpha\,(\gamma - 3)\,p^3 \ln\left(\frac{p}{p_c}\right) = 0 \tag{29}$$

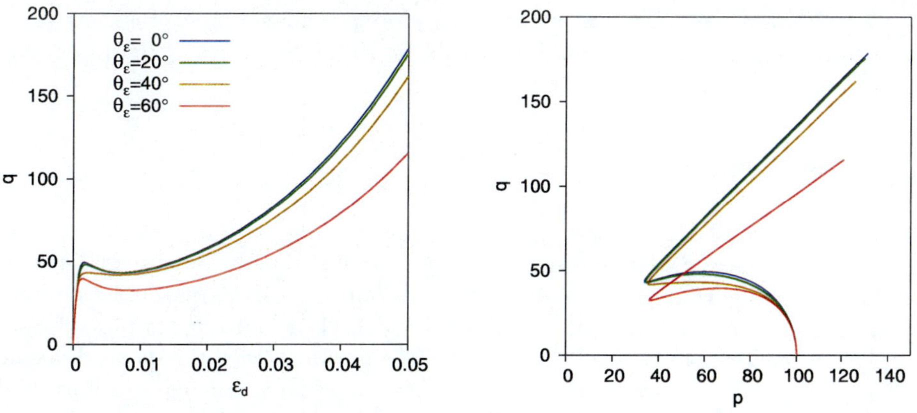

Fig. 4. Simulation of constant volume triaxial tests with a modified version of the *Sinfonietta Classica* by Nova [11].

For given values of p and θ, it is evident that Eq. (29) gives three roots and this in general provides unwanted "false elastic domains". Furthermore, for low values of the mean stress no real solutions are obtained. To overcome this difficulty the stress points of the SC yield surface for $\theta = 0$ obtained for $M_g = 1.28$ and $\alpha = 1.40$ (calibrated in SC for Hostun sand) are used as target for the comparison with surface (4). The generation process provides the following parameters: m_1=1.800, n_1=0.416, c_1=0.606, m_2=1.257, n_2=0.370, c_2=0.500, m_3=0.619, n_3=1.000, c_3=0.500.

The plastic potential obtained through (29) for $\alpha = 3$ is used to determine the target SDR to be compared with (24). An optimization procedure similar to that described for the yield function has been implemented and the obtained parameters are $b = 1.396$, $c = 0.634$, $\mu_1 = 0.000$, and $\mu_2 = 0.421$.

The hardening law used in this work is essentially the same as that used in SC even though it does not consider the hardening term involving the third invariant of plastic strains

$$\dot{p}_c = p_c \frac{1 + e_0}{\lambda^p} \left(\dot{\varepsilon}_v^p + \xi_d \dot{\varepsilon}_d^p \right) \tag{30}$$

where ξ_d is the dilatancy parameter. It is inferred that the plastic logarithmic volumetric compliance of SC is $B_p = \lambda^p/(1 + e_0)$ where λ^p is the slope of the isotropic compression line in the $\ln(p/p_a) - e^p$ plane, being e^p the plastic void ratio and e_0 the void ratio for a reference low value of p (in this work $0.01\,p_a$). The value $B_p = 0.0080 - 0.0026 = 0.0054$ in [11] is obtained through $\lambda^p = 0.0122$ and $e_0 = 1.25$. The dilatancy parameter value is $\xi_d = 0.336$.

The elastic law used in SC is given by the two relationships $\dot{\varepsilon}_v^e = B_e \dot{p}/p$ and $\dot{e}_{ij}^e = L\dot{\eta}_{ij}$ where e_{ij}^e is the deviatoric part of the elastic strain tensor. Being $B_e = 0.00260$ and $L = 0.00218$ the hypoelastic law used here, and equivalent to SC, is consistent with bulk and shear moduli linearly dependent on the mean stress given by $K = k_0 p$ and $G = g_0 p$ with $k_0 = 385$ and $g_0 = 229$.

The *Sinfonietta Classica* is now completely reformulated. Figure 4 shows some simulations of constant volume tests for different values of the Lode angle of strains θ_ε.

5 Conclusions

This paper has illustrated two approaches for the calibration of the parameters of yield surfaces and plastic potentials in elastoplastic models for soils. Different shapes of yield surfaces can be obtained by adopting a single mathematical function of many parameters that are determined with an automatic procedure. A similar procedure can be used to determine the parameters of the stress-dilatancy relationships of the model. Both procedures have been applied to reformulate the *Sinfonietta Classica* model [11] by Professor Roberto Nova.

References

1. Mortara, G.: A new yield and failure criterion for geomaterials. Géotechnique **58**, 125–132 (2008)
2. Mortara, G.: A hierarchical single yield surface for frictional materials. Comput. Geotech. **36**, 960–967 (2009)
3. Mortara, G.: A yield criterion for isotropic and cross-anisotropic cohesive-frictional materials. Int. J. Numer. Anal. Meth. Geomech. **34**, 953–977 (2010)
4. Mortara, G.: A new yield criterion for soils with embedded tension cut-off. Meccanica **54**, 683–696 (2019)
5. Mortara, Giuseppe: Macroscale yield criteria for geomaterials. In: Giovine, Pasquale, Mariano, Paolo Maria, Mortara, Giuseppe (eds.) Views on Microstructures in Granular Materials. AMM, vol. 44, pp. 137–155. Springer, Cham (2020). https://doi.org/10.1007/978-3-030-49267-0_7
6. Matsuoka, H., Nakai, T.: Stress-deformation and strength characteristics of soil under three different principal stresses. Proc. Jpn. Soc. Civ. Eng. **232**, 59–70 (1974)
7. Lade, P.V., Duncan, M.: Elastoplastic stress-strain theory for cohesionless soil. J. Geotech. Eng. Div. **101**, 1037–1053 (1975)
8. Bardet, J.P.: Lode dependences for isotropic pressure-sensitive materials. J. Appl. Mech. **57**, 498–506 (1990)
9. Mortara, G.: Auxiliary plastic potential approach in elastoplasticity for soils. Geotech. Lett. **11**, 221–229 (2021)
10. Nova, R., Wood, D.M.: A constitutive model for sand in triaxial compression. Int. J. Numer. Anal. Meth. Geomech. **3**, 255–278 (1979)
11. Nova, R.: Sinfonietta classica: an exercise on classical soil modelling. In: Saada, A., Bianchini, G. (eds.) Proceedings of the International Symposium on Constitutive Equations for Granular Non-Cohesive Soils, pp. 501–519. Balkema, Rotterdam (1988)

Mean Field Approaches for the Homogenization of Elastic Parameters of Lightweight Cemented Soils

Laura Perrotta[1(✉)], Enza Vitale[2], and Giacomo Russo[2]

[1] Scuola Superiore Meridionale, Largo S. Marcellino 10, 80138 Naples, NA, Italy
laura.perrotta@unina.it

[2] Department of Earth Science, Environment and Resources, University of Naples Federico II, Via Cinthia 21, 80126 Naples, NA, Italy

Abstract. In the field of cement treated soils, the use of Lightweight Cemented Soils (LWCS) i.e., materials that are obtained by mixing air foam with natural soil, water and cement, is of increasing interest. In the paper, the derivation of stiffness modulus of LWCS through a micromechanical approach has been performed. Analytical formulae based on the Mean-Field Eshelby-based Homogenization schemes have been implemented to determine the elastic properties of the homogenized material by taking into account its heterogeneity. The application of Mean Field Approaches requires several input parameters, in particular the foam-induced artificial porosity for LWCS samples, the bulk and shear moduli and Poisson's ratio for cement-treated samples. The artificial porosity has been evaluated by means of X-Ray micro-CT scans, whereas the bulk modulus of the cemented matrix has been derived by experimental results. The homogenized stiffness moduli, computed for different curing times, are in good agreement with those obtained from experimental results. The proposed model predicts with good accuracy the elastic behaviour of the material.

Keywords: Lightweight cemented soils · multiscale analysis · homogenization · Mean Field Approaches

1 Introduction

In the construction process of major civil infrastructures, soils characterized by low mechanical properties generally need to be removed, leading to several economic and environmental drawbacks. Nowadays, this procedure is often avoided thanks to soil improvement techniques, by promoting the sustainability of civil works [1–4].

LWCS are made of soil-cement-water slurry and foam, therefore they are an extremely heterogeneous material. Moreover, LWCS are characterized by different classes of porosity: the artificial porosity induced by the foam (mean diameter about 300 μm) and the porosity of the matrix (mean diameter about 0.3 μm). Due to the different characteristic dimensions, the former can be investigated with X-Ray micro-CT and the latter with Mercury Intrusion Porosimetry tests. Therefore, a micromechanical

A. Ferrari et al. (Eds.): CNRIG 2023, SSGG, pp. 359–364, 2023.
https://doi.org/10.1007/978-3-031-34761-0_44

approach is requested to develop a constitutive model, taking into account the heterogeneities of the material, thus overcoming the limits of conventional approaches based on continuum mechanics.

Taking account of the heterogeneity of the material is strictly related to the length scale chosen for the observation process, since some phases in a heterogenous material can be distinguished only at or below a given length scale [5]. Therefore, the definition of the scale of interest is the first step of a homogenization procedure, which allows to obtain the homogenized properties of a composite material (i.e., a material consisting of two or more different phases that form regions large enough to be regarded as continua and which are usually firmly bonded together at the interface [6]).

The reliable determination of the macroscopic behavior of a heterogeneous material, based on the appropriate and available microstructural information [7], is a big issue in the framework of soils treated with binders and in particular for Lightweight Cemented Soil.

In this study, Mean Field Approaches have been used in order to derive the homogenized stiffness properties of LWCS; the input parameters of the models have been extrapolated from available experimental results, which have also been employed to validate the predicted values.

2 Mean Field Approaches: Mori-Tanaka Scheme

Mean Field Approaches (MFAs) are generally implemented for obtaining the homogenized properties of composite materials [5]. These methods are well adapted for heterogeneous materials with a random microstructure distribution and are used to model composites with one matrix phase and one or multiple inclusion phases with uniform properties for each phase [8]. MFAs are based on Eshelby's solution of a single-inclusion problem; only partial information of the microstructure including the volume fraction (i.e., the ratio between the volume of the considered phase and the total volume of the material), aspect ratio (i.e., the ratio between the major axis and the minor axis of the phase) and the orientation of the inclusions (i.e., isotropy, anisotropy, etc.) are required [9]. MFAs give prediction of the mean fields (averaged response) per each constituent of the material [9].

In this framework, the following assumptions have been made:

- LWCS are regarded as a two-phase material, i.e., matrix and voids induced by foam;
- due to the random microstructure of cement-based materials, all phases have been considered as isotropic;
- the inclusions have been assumed spherical, based on SEM observations of LWCS samples [10];
- the matrix is supposed to be elastic, isotropic and homogeneous;
- bulk and shear modulus k_{foam} and μ_{foam} of the inclusions (i.e., the voids) have been considered equal to 0.

Given a number of phases (p) whose volume fraction is c_p, the homogenized bulk modulus k^{hom} and shear modulus μ^{hom} can be obtained as follows [11]:

$$k^{hom} = \Sigma_p c_p k_p (1 + \alpha_0^{est}(k_p/k_0 - 1))^{-1} x\, [\Sigma_p c_p (1 + \alpha_0^{est}(k_p/k_0 - 1))^{-1}]^{-1} \quad (1)$$

$$\mu^{hom} = \Sigma_p c_p \mu_p (1 + \beta_0^{est}(\mu_p/\mu_0 - 1))^{-1} x\, [\Sigma_p c_p (1 + \beta_0^{est}(\mu_p/\mu_0 - 1))^{-1}]^{-1} \quad (2)$$

where k_p and k_0, μ_p and μ_0 are respectively the bulk and shear moduli of the phase p and of the reference medium denoted by 0. By choosing the matrix as reference, the Mori-Tanaka scheme is applied. Moreover, by considering the LWCS made of only two phases, i.e., the matrix (*m*) and the voids induced by the foam, whose volume fraction is n_{foam}, (1) and (2) can be simplified into [11]:

$$k^{hom}/k_m = 1 + n_{foam}\big((k_{foam}/k_m - 1)/\big(1 + \alpha_m^{est}(1 - n_{foam})(k_{foam}/k_m - 1)\big)\big) \quad (3)$$

$$\mu^{hom}/\mu_m = 1 + n_{foam}\big((\mu_{foam}/\mu_m - 1)/\big(1 + \beta_m^{est}(1 - n_{foam})(\mu_{foam}/\mu_m - 1)\big)\big) \quad (4)$$

where k_m and μ_m, k_{foam} and μ_{foam} are respectively the bulk and shear moduli of the matrix and of the voids. Moreover, the homogenization coefficients $\alpha_0{}^{est}$ and $\alpha_m{}^{est}$ are defined as follows:

$$\alpha_0^{est} \equiv \alpha_m^{est} = 3k_m/(3k_m + 4\mu_m) \text{ and } \beta_0^{est} \equiv \beta_m^{est} = 6(k_m + 2\mu_m)/5(3k_m + 4\mu_m) \quad (5)$$

By setting k_{foam} and μ_{foam} equal to zero, (3) and (4) are transformed into:

$$k^{hom}/k_m = 1 - n_{foam}/\big(1 + \alpha_m^{est}(1 - n_{foam})(k_{foam}/k_m - 1)\big) \quad (6)$$

$$\mu^{hom}/\mu_m = 1 - n_{foam}/\big(1 + \beta_m^{est}(1 - n_{foam})(\mu_{foam}/\mu_m - 1)\big) \quad (7)$$

At first instance $\nu_m = 0.2$ has been considered. In this way, the only influence of n_{foam} on the homogenized results has been investigated [12]:

$$k^{hom}(n_{foam}) = ((1 - n_{foam})/(1 + n_{foam}))k_m \quad (8)$$

$$\mu^{hom}(n_{foam}) = ((1 - n_{foam})/(1 + n_{foam}))\mu_m \quad (9)$$

Then, the complete expression of Mori-Tanaka (6) and (7) has been used.

2.1 Mori-Tanaka Scheme: Input Parameters

The artificial porosity induced by foam (n_{foam}) has been obtained through an image analysis performed on X-Ray micro-CT scans of LWCS samples [13] (Fig. 1).

With regards to the mechanical parameters, the elastic stiffness of the samples prepared at 40% of cement and 40% of foam ($E_{KCF40\%}$) and with only 40% of cement ($E_{KC40\%}$) has been derived from available unconfined compression tests at 1, 7 and 28 days of curing. Poisson's ratio ν_m, has been determined as function of curing time by using the expression (10):

$$E_{oed} = E_{KC40\%}(1 - \nu_m)/((1 + \nu_m)(1 - 2\nu_m)) \quad (10)$$

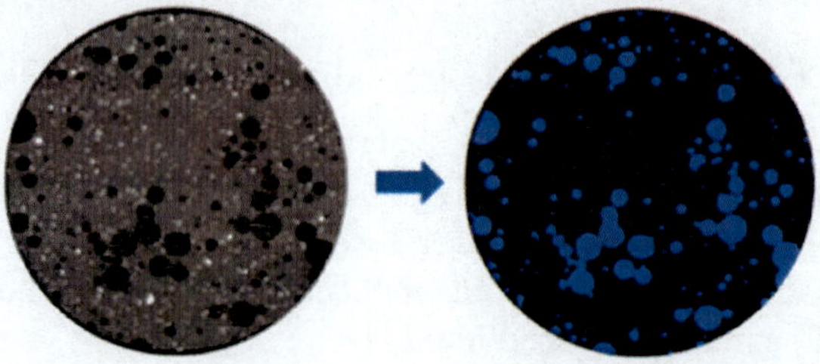

Fig. 1. Artificial porosity segmentation process (Ferriero 2021)

where E_{oed} has been derived from the results of available one-dimensional compression tests performed on LWCS samples at 7 and 28 days of curing [1]. For 1 day of curing ν_m has been assumed equal to 0.2, as usual for cement. The bulk and the shear moduli k_m and μ_m have been computed has follow (Table 1):

$$k_m = E_{KC40\%}/3(1 - 2\nu_m) \tag{11}$$

$$\mu_m = k_m/2(1 + \nu_m) \tag{12}$$

Table 1. Input parameters Mori-Tanaka scheme.

Curing Time [days]	k_m [MPa]	μ_m [MPa]	ν_m [-]	n_{foam} [%]
1	1.820	1.365	0.2	18.42
7	15.991	5.815	0.375	19.23
28	17.311	7.260	0.316	19.45

3 Results and Discussion

A comparison between the elastic stiffness modulus computed with simplified and complete MT schemes and the experimental results is shown in Fig. 2 as function of curing time. The calculation of elastic stiffness adopting the complete formulation of MT scheme leads to values overlapping with the experimental results for all the considered curing times. By using simplified formulae (8) and (9) with constant Poisson's ratio (i.e., $\nu_m = 0.2$) the prediction of the homogenized moduli is less accurate especially for high curing times. The simplified method is not able to catch the ongoing chemo-physical evolution of the system. Conversely, the elastic stiffness determined using the complete formula of MT scheme, taking into account the variation of Poisson's ratio over curing time, fits very well the experimental results.

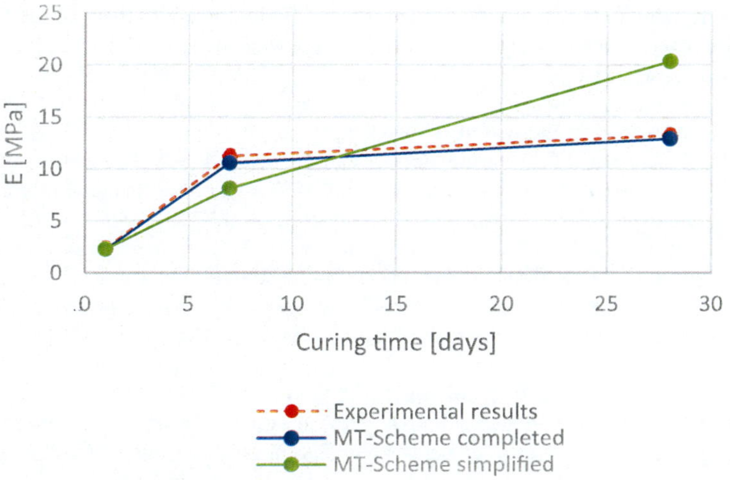

Fig. 2. Evolution of Elastic stiffness modulus over time: MT-scheme completed vs MT -simplified scheme vs. experimental results.

4 Conclusions

Mori-Tanaka scheme belonging to Mean Field Homogenization methods has been used to determine the homogenized elastic stiffness of Lightweight Cemented Soils. The input mechanical parameters of the model have been derived from available experimental tests, whereas the physical parameter of the method (i.e., the artificial porosity n_{foam}) has been computed by performing an image analysis on X-Ray micro-CT scans of LWCS samples.

The obtained results highlight that, by assuming the Poisson's ratio constant and equal to 0.2 and thus considering only the change of the artificial porosity over time, the Mori-Tanaka scheme is not able to predict with good accuracy the evolution of the elastic stiffness of LWCS. Conversely the use of complete Mori-Tanaka scheme, which takes into account the evolution of Poisson's ratio over time, provides homogenized values of elastic stiffness modulus in agreement with those obtained by experimental test results.

In the framework of the homogenization schemes based on Mean Field Approaches, the study highlights the valuable capability of Mori-Tanaka scheme on predicting the elastic properties of an extremely heterogeneous material like the LWCS over time, despite its simple analytical formulation.

References

1. De Sarno, D., et al.: Effects of cement and foam addition on chemo-mechanical behaviour of lightweight cemented soil (LWCS). In: E3S Web of Conferences, vol. 92, p. 11006, EDP Sciences (2019)
2. Guidobaldi, G., et al.: Chemo-mineralogical evolution and microstructural modifications of a lime treated pyroclastic soil. Eng. Geol. **245**, 333–343 (2018)
3. Guidobaldi, G., et al.: Multi-scale analysis of the mechanical improvement induced by lime addition on a pyroclastic soil. Eng. Geol. **221**, 193–201 (2017)

4. Vitale, E., Russo, G., Dell'Agli, G., Ferone, C., Bartolomeo, C.: Mechanical behaviour of soil improved by alkali activated binders. Environments **4**(4), 80 (2017)
5. Ortolano González, J.M., Hernández Ortega, J.A., Oliver Olivella, X.: A comparative study on homogenization strategies for multi-scale analysis of materials. Centre Internacional de Mètodes Numèrics en Enginyeria (CIMNE) (2013)
6. Hashin, Z.: Analysis of composite materials—a survey. 481–505 (1983)
7. Markov, K., Preziosi, L. (eds.): Heterogeneous Media: Micromechanics Modeling Methods and Simulations. Springer, Boston (2000). https://doi.org/10.1007/978-1-4612-1332-1
8. Kanouté, P., Boso, D.P., Chaboche, J.L., Schrefler, B.: Multiscale methods for composites: a review. Arch. Comput. Methods Eng. **16**(1), 31–75 (2009)
9. Ji, H., Mclendon, R., Hurtado, J.A., Oancea, V., Bi, J.: Multi-scale material modeling with the mean-field homogenization method. In: NEFEMS World Congress (2017)
10. Vitale, E., et al.: Chemo-mechanical behaviour of lightweight cemented soils. Acta Geotech. **15**(4), 933–945 (2019). https://doi.org/10.1007/s11440-019-00797-8
11. Bernard, O., Ulm, F.J., Lemarchand, E.: A multiscale micromechanics-hydration model for the early-age elastic properties of cement-based materials. Cem. Concr. Res. **33**(9), 1293–1309 (2003)
12. Miled, K., Sab, K., Le Roy, R.: Effective elastic properties of porous materials: homogenization schemes vs experimental data. Mech. Res. Commun. **38**(2), 131–135 (2011)
13. Ferriero, F.: The sustainable use of lightweight cemented soil: microstructural analysis. Master's Degree Thesis (2021)

Multi-scale Modelling of Natural Composites Using Thermodynamics-Based Artificial Neural Networks and Dimensionality Reduction Techniques

Giovanni Piunno[1(✉)], Ioannis Stefanou[2], and Cristina Jommi[1]

[1] Politecnico di Milano, Piazza Leonardo da Vinci 23, 20129 Milan, Italy
{giovanni.piunno,cristina.jommi}@polimi.it

[2] Nantes Université, Ecole Centrale Nantes, CNRS, Institut de Recherche en Génie Civil et Mécanique (GeM), UMR 6183, 44000 Nantes, France
ioannis.stefanou@ec-nantes.fr

Abstract. Modelling natural composites, as the majority of real geomaterials, requires facing their intrinsic multiscale nature. This allows to consider multi-physics coupling occurring at the microscale, then reflected onto the macroscopic behavior. Geotechnics is constantly requiring reliable constitutive models of natural composites to solve large-scale engineering problems accurately and efficiently. This need motivates the contribution. To capture in detail the macroscopic effects of microscopic processes, many authors have developed multi-scale numerical schemes. A common drawback of such methods is the prohibitive computational cost. Recently, Machine Learning based approaches have raised as promising alternatives to traditional methods. Artificial Neural Networks – ANNs – have been used to predict the constitutive behaviour of complex, heterogeneous materials, with reduced calculation costs. However, a major weakness of ANN is the lack of a rigorous framework based on principles of physics. This often implies a limited capability to extrapolate values ranging outside the training set and the need of large, high-quality datasets, on which performing the training. This work focuses on the use of Thermodynamics-based Artificial Neural Networks – TANN – to predict the constitutive behaviour of natural composites. Dimensionality reduction techniques – DRTs – are used to embed information of microscopic processes into a lower dimensional manifold. The obtained set of variables is used to characterize the state of the material at the macroscopic scale. Entanglement of DRTs with TANN allows to reproduce the complex nonlinear material response with reduced computational costs and guarantying thermodynamic admissibility. To demonstrate the method capabilities an application to a heterogeneous material model is presented.

Keywords: TANN · Thermodynamics · Dimensionality reduction · Multiscale modelling · Composites

A. Ferrari et al. (Eds.): CNRIG 2023, SSGG, pp. 365–372, 2023.
https://doi.org/10.1007/978-3-031-34761-0_45

1 Introduction

Multiscale simulation and homogenization have become crucial tools for modeling complex materials, but require substantial computational resources, rendering their use in industry and engineering practice difficult.

The growth of Artificial intelligence and more advanced computational resources has led to an increasing use of machine learning-based methodologies to improve the efficiency and accuracy of multiscale simulations. Machine learning (ML) algorithms often use Dimensionality Reduction Techniques (DRTs) as an essential step in data pre-processing. There are many works in literature that make use of ML tools for the multiscale modeling of composite materials. [1] focuses on the use of ML techniques to develop multiscale models for multi-permeability porous materials, while [2] focuses on the use of Graph-Informed Neural Networks for solving general multiscale physics problems, among many others (e.g., see [3]).

A limited number of works rely on physics-aware ANNs to speed up multiscale simulations, resulting in black boxes whose results are difficult to justify from a physical standpoint. In this paper, we use Thermodynamics-based Artificial Neural Network (TANN), see [4–6], for the multiscale modeling of micro-structured materials. TANN is coupled with DRTs, the latter applied to microscopic information gathered from numerical simulations to identify a set of Internal State Variables to use at the macroscale. We provide a comparison of several DRTs, including POD, ICA, kernel PCA and autoencoders, and discuss the advantages and disadvantages of each approach for the goal at hand. We propose an application using Drucker-Prager hardening elastic-plastic model with cap, for a heterogeneous model of voxels with spatially correlated constitutive parameters, that mimic a natural composite.

The work is articulated starting from a theoretical section summarizing the main aspects of the employed methods (Sect. 2). In Sect. 3, the realization of the material model and of the numerical database is described. Section 4 details the results obtained. The article ends with conclusions on the obtained results.

2 Theoretical Framework

2.1 Dimensionality Reduction Techniques

The Proper Orthogonal Decomposition (POD) [7] and Principal Component Analysis (PCA) [8] are linear techniques that are used to extract important features or patterns in a dataset. POD, also known as the Karhunen-Loève decomposition, decomposes a dataset into a set of orthogonal modes that represent the most significant features. PCA, on the other hand, finds the directions in a dataset that account for the most variance. The POD modes can be derived from the principal components, and the hierarchy of the modes is determined by the singular values. The method is simply expressed by the following modal decomposition formula:

$$X = \tilde{U}\tilde{S}\tilde{V}^* \rightarrow Z = \tilde{U}^* X \quad (1)$$

Z is the reduced dataset, containing the POD coefficients, projections of X onto the POD modes.

Independent Component Analysis (ICA), see [9], is a technique that aims to identify the underlying independent sources within a multivariate signal by assuming that the data is a linear combination of non-Gaussian sources. Mathematically, ICA can be represented as an optimization problem in which the objective is to uncover a linear transformation (matrix W) that maximizes the non-Gaussianity of the transformed data. Namely, $Z = WX$.

In KPCA [10], a kernel function is used to map the data into a higher dimensional feature space in which they become linearly separable. The first step in KPCA is to compute the kernel matrix, which is a symmetric matrix whose entries are given by the kernel function applied to all pairs of data points. The kernel matrix is then decomposed into its eigenvectors and eigenvalues. The eigenvectors with the largest eigenvalues are selected as the principal components. Finally, the original data is projected onto the principal components, resulting in a lower-dimensional representation of the data.

Autoencoders (AE) are nonlinear dimensionality reduction neural networks, see [8]. AE are unsupervised learning algorithms that map inputs to intrinsic representations and then back to themselves. Given an input $\mathcal{I} \in \mathbb{R}^n$, AE learn an intrinsic representation $\mathcal{R} \in \mathbb{R}^l$, $l \ll n$, which is mapped back into $\mathcal{I}^* \in \mathbb{R}^n$, imposing $\mathcal{I}^* = \mathcal{I}$. The parametrization is implemented by two functions: an encoder, $NN_E : \mathbb{R}^n \to \mathbb{R}^l$, and a decoder, $NN_D : \mathbb{R}^l \to \mathbb{R}^n$.

2.2 Random Field Generation Algorithms

There are numerous published methods for generating realizations of stationary homogeneous spatially correlated random fields. The matrix decomposition method, as described in [11], requires the definition of a discrete set of N_p points at which the random field will be sampled and then to create a covariance matrix, C, quantifying the correlation between all sampling points. With the Cholesky decomposition, the lower triangular matrix of C, L, is obtained. Then, a vector of correlated random variables, $\mathcal{Y}$, is computed by generating a vector of uncorrelated random numbers, $\mathcal{X}$, from a unit normal distribution, calculating $\mathcal{Y} = L\mathcal{X}$. Cholesky decomposition is an exact method, so the simulated Gaussian field follows an exact multivariate Gaussian distribution.

2.3 Thermodynamics-Based Artificial Neural Networks for Multiscale Modelling

In [5, 6], the TANN framework has been used to homogenize the constitutive behavior of a micro-structured heterogeneous inelastic cell.

TANN are based on the thermodynamics theory of Internal State Variables, see [12]. The theory seeks to describe the state of a history-dependent material using a set of variables able to truck the microscopic irreversibilities occurring in the material, so as to permit a description of the state local in time. The material model is obtained by the definition of the Helmholtz free energy density function and the ISV evolution law. Micro-structured heterogeneous materials lack a straightforward definition of macroscopic ISVs. The Authors proposed discovering an a priori unknown set of ISVs, $\mathcal{Z}$, from dimensionality reduction of microscopic state information, encoded in what they refer to as a set of Internal Coordinates (IC), ξ. After defining the macroscopic state space,

$\mathcal{S}$, the TANN framework can be used to learn from data the homogenized behavior of the heterogeneous material in a thermodynamically consistent manner.

The training of the Helmholtz energy network, NN_ψ, is central to the thermodynamic compatibility ensured by TANN. The NN_ψ takes as input the state of the material to output the Helmholtz energy. The latter is automatically differentiated to return the stresses, $\boldsymbol{\Sigma} = \frac{\partial \mathrm{NN}_\Psi}{\partial \boldsymbol{E}}$, (fulfilling the First Principle) and to compute, together with the rates of ISV, $\dot{\mathcal{Z}}$, the rate of energy dissipation, $D = -\frac{\partial \mathrm{NN}_\Psi}{\partial \mathbf{Z}} \cdot \dot{\mathcal{Z}}$. This latter is constrained to be non-negative (fulfilling the Second Principle) using a regularization term included in the definition of the loss function, $\mathcal{L}$, to be minimized during the optimization procedure:

$$\mathcal{L} = \lambda_\Sigma \ell^\Sigma + \lambda_\mathrm{R}^\mathrm{D} \ell^\mathrm{D} \tag{2}$$

with λ_Σ and λ_R^D being weights for regulating the relative magnitudes of outputs, $\ell^\Sigma = \frac{1}{N}\sum_i \left\| \Sigma_\mathrm{i} - \frac{\partial NN_\Psi}{\partial E}_i \right\|_1$ and $\ell_\mathrm{R}^\mathrm{D} = \frac{1}{N}\sum_i \left\| \mathrm{Relu}\left\{ -\left(-\frac{\partial NN_\Psi}{\partial \mathcal{Z}} \dot{\mathcal{Z}} \right)_i \right\} \right\|_1$. The Re-ctified Linear Unit – Relu is defined as $Relu\{x\} = \{x, x > 0; 0, else\}$, $\|\cdot\|_1$ is used for the L_1 norm, N is the number of considered samples.

3 Material Model and Numerical Database

Numerical homogenization is frequently based on the statistical concept of a Representative Volume Element (RVE). The RVE is a region large enough to contain a sufficient number of statistically independent realizations of the field, yet small enough to be computationally tractable. The correlation length is used to define the spatial correlation of the field, but it doesn't dictate the size of the RVE. The RVE size may need to be larger than the correlation length if the latter is small, in order to capture enough independent realizations of the field. In the application at hand, we considered a simplified case, in which the selected RVE is a cube of unitary dimensions. The correlation length of the field has been assumed to be 0.3 times the RVE size. To generate the correlated random field, we used Markov covariance function, with x_{ij} being the lag distance matrix between two points of the domain and θ_c the correlation length.

$$\rho\left(x_{ij}, \theta_c\right) = \exp\left(-\frac{x_{ij}}{\theta_c}\right) \tag{3}$$

The random generation algorithm outputs a standard normal correlated field. The latter can be transformed to match any normal or log-normal distribution. We utilized the procedure to assign constitutive parameters to the geometric field derived from the coordinates of Gauss points of the computational model's elements. Table 1 reports the mean and the standard deviation of the parameters used.

Table 1. Mean ($\tilde{\mu}$) and standard deviation ($\tilde{\sigma}$) of the used constitutive parameters for the elastic-strain hardening plastic Drucker-Prager constitutive model with cap. Parameters without std. Dev. Are considered as homogeneous.

	E(Mpa)	ν(-)	d(kPa)	β(°)	α(-)	R(-)	K(-)	p'_0(kPa)
$\tilde{\mu}$	180	0.3	15	38	0.05	1.2	0.8	100
$\tilde{\sigma}$	1	0.01	0.5	0.1	-	-	-	0.5

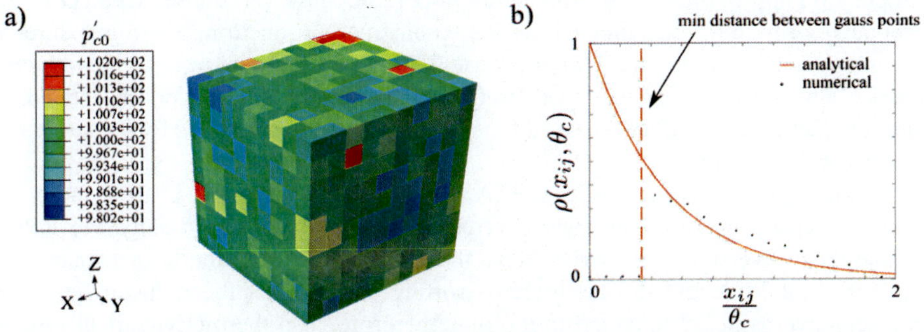

Fig. 1. a) Computational model of the RVE. The initial field of preconsolidation pressure p'_0 is depicted. b) Comparison of the the reconstructed numerical correlation function VS the analytical one.

The description of the constitutive parameters may be found at the link in reference [13], in the ABAQUS user's manual. Figure 1 depicts the computational model and the field of initial pre-consolidation pressure, the model has been initialized with.

For the training of TANNs a dataset of 25000 sample has been generated from 5 dataset of 1000 samples, to which a change of observer has been applied four times. This procedure has been implemented so that the network could learn objectivity from data. The RVE has been subjected to macroscopic random strain paths, after an initial monotonic volumetric compression up to 1e-3.

4 Results

After constructing the numerical database, we trained TANNs. At first, it was necessary to collect the microscopic data in the set of IC, ξ. Elastic and plastic deformations and maximum volumetric plastic deformations were utilized, the latter able to track volumetric hardening. The resulting number of IC's DoFs was 13000. POD was applied for the purpose of obtain ISVs, $\mathcal{Z}$, from the IC set. Figure 2 displays the normalized singular values and their cumulative sum for the first 200 POD modes. The sum approaches one with the considered modes, indicating a high degree of representativeness and a consequent small reconstruction error. The use of POD gives 0.0153 compression ratio. Starting from the reduced field, we applied further DRTs to obtain an extremely reduced set of ISVs. The ability to reconstruct the high-dimensional IC field determined the choice of DRT to use.

According to the Coleman and Gurtin's theory, the set of ISVs must trace and represent the material's microscopic irreversible processes. Estimating the reconstruction error is necessary to assess the ISV's representativity. The latter is, however, difficult to compute. In general, it is not possible to define an analytic inverse function for nonlinear DRTs. In these cases, the reconstruction of the original field is accomplished by fitting an additional function onto data. If the reconstruction error, obtained by the composition of the reduction and reconstruction functions is small, it is possible to conclude that the composition approximates well the identity operator. However, the reduced set is representative only in a sense dictated by the reconstruction function; nothing ensures its representativeness in a thermodynamic sense. To achieve the latter property, the reduction function's fitting should be incorporated into the TANN training procedure. Indeed, the thermodynamic representativeness of the reduced set is so guaranteed by the successful training of the networks.

Following this reasoning, ICA, KPCA, and AE were considered. In ICA, there is an inverse function, so the reconstruction error can be calculated analytically. For KPCA, an analytical inverse doesn't exist. Nevertheless, after defining the kernel matrix, the application of PCA and the hierarchical sorting ensures that the reduced field is as representative as possible of the initial one, therefore, also thermodynamically representativeness is ensured. Finally, encoders do not have analytical inverse, but are simple enough to be incorporated into the TANN training procedure. In this case, it is not strictly necessary to define a decoder in order to achieve field reconstruction. Indeed, if training succeeds, it is concluded that the obtained reduced set is representative, no matter its reconstruction. However, if there is the need of reconstructing the microscopic field from the macroscopic one, it is possible to train an additional decoder, outside (or inside) the TANN training procedure.

A reduction dimension equal to 15 was chosen for all the DRTs to compare. This was chosen after ensuring a corresponding ICA's mean absolute reconstruction error of 1e-5. The training was done considering 10000 epochs, mini-batches of 1000 samples and Nesterov accelerated Adam's optimizer with learning rate 5e-5. The training was successful for all DRTs. In the case of ICA and KPCA, where the dimensionality reduction was performed outside of the TANN training, each training epoch required 3 ms, whereas the coupled training of the encoder required 15 ms per epoch, on a machine with 32 cores. The encoder had 3 hidden layers of decreasing size (150, 100, 50 neurons) and one output layer of 15 neurons, with tanh activation function. ICA fitting required 0.7 s. KPCA took 13 min and fifty seconds, including the fitting of an additional reconstruction function.

Trained TANNs were evaluated in inference mode on an unseen dataset. Figure 3 depicts the results of the prediction on a 3D strain-controlled path to which a volumetric strain was initially applied, followed by random increments. A very good agreement is achieved. The encoder's output contains some outliers. This is attributed to the coupling of the encoder's training with TANN, which causes a delay in achieving learning convergence in 10000 epochs, resulting in a decreased accuracy. With ICA and KPCA, the asymptotic learning value is found very early, between 300 and 200 epochs.

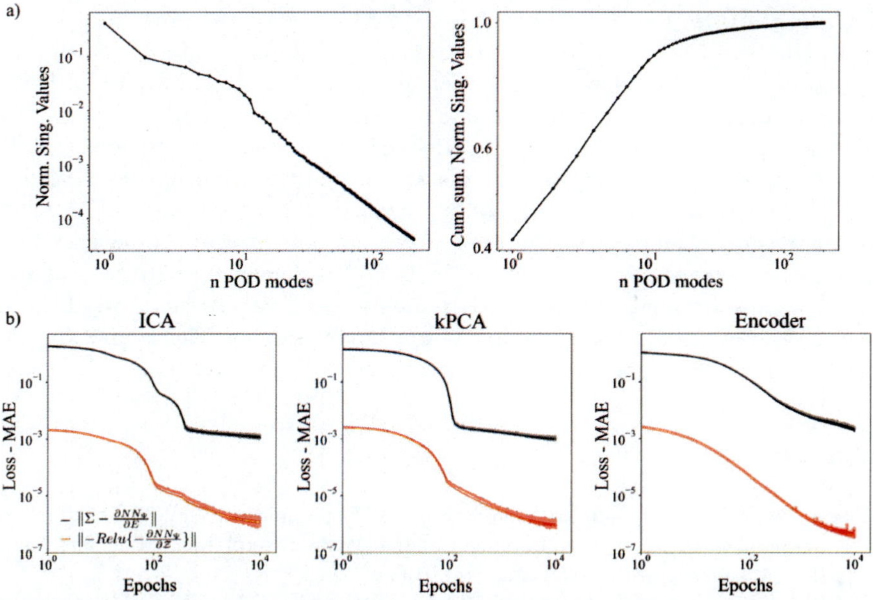

Fig. 2. a) Normalized singular values and their cumulative sum considering the first 200 POD modes obtained from the set of IC, ξ. b) Learning curves expressed in terms of the Mean Absolute Error – MAE, considering ICA, kPCA and the Encoder dimensionality technique. 10000 epochs have been used. Solid lines represent results on the training set, shaded lines below the solid ones on the validation set.

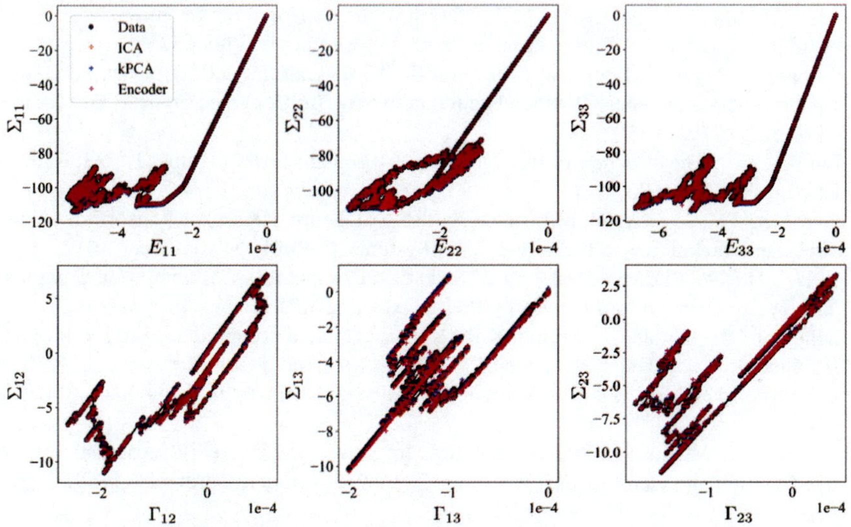

Fig. 3. TANN predictions in inference mode on a set of unseen data. The network's predictions obtained considering the set of ISV obtained with ICA, kPCA and the encoder are compared.

5 Conclusions

In this study, three DRTs have been applied on top of POD to define a set of ISVs for the macroscopic state space definition of a heterogeneous medium. TANNs were subsequently trained on a dataset of data derived from numerical simulations. The RVE was obtained by assigning spatially correlated constitutive parameters to the elements of the model's structured mesh. All DRTs produced satisfactory results. ICA returned excellent results while requiring the least amount of computational time, ensuring the possibility of reconstructing microscopic fields. KPCA demonstrated to be a valid alternative, attractive if a linear method, like ICA, fails to generate satisfactory results. Unlike encoders, KPCA fitting can be decoupled from TANN training. This is time saving and makes it easier to fine-tune neural networks in the TANN architecture.

References

1. Wang, K., Sun, W.: A multiscale multi-permeability poroplasticity model linked by recursive homogenizations and deep learning. Comput. Methods Appl. Mech. Eng. **334**, 337–380 (2018). https://doi.org/10.1016/j.cma.2018.01.036
2. Hall, E.J., Taverniers, S., Katsoulakis, M.A., Tartakovsky, D.M.: GINNs: graph-informed neural networks for multiscale physics. J. Comput. Phys. **433**, 110192 (2021). https://doi.org/10.1016/j.jcp.2021.110192
3. Sorini, A., Pineda, E.J., Stuckner, J., Gustafson, P.A.: A convolutional neural network for multiscale modeling of composite materials. In: AIAA Scitech 2021 Forum, p. 0310 (2021)
4. Masi, F., Stefanou, I.: Multiscale modeling of inelastic materials with Thermodynamics-based Artificial Neural Networks (TANN). Comput. Methods Appl. Mech. Eng. **398**, 115190 (2022)
5. Masi, F., Stefanou, I.: Evolution TANN and the discovery of the internal variables and evolution equations in solid mechanics. arXiv preprint arXiv:2209.13269 (2022)
6. Piunno, G., Masi, F., Stefanou, I., Jommi, C.: Multi-scale modelling of natural composites via thermodynamics-based artificial neural networks. In: Congrées Français de Mècanique – CFM, AFM (2022)
7. Lumley, J.L.: The structure of inhomogeneous turbulent flows. Atmos. Turbul. Radio Wave Propag. 166–178 (1967)
8. Géron, A.: Hands-on Machine Learning with Scikit-Learn, Keras, and TensorFlow: Concepts, Tools, and Techniques to Build Intelligent Systems. O'Reilly Media, Inc. (2019)
9. Kutz, J.N.: Data-Driven Modeling & Scientific Computation: Methods for Complex Systems & Big Data. Oxford University Press, Oxford (2013)
10. Schölkopf, B., Smola, A., Müller, K.R.: Kernel principal component analysis. In: Gerstner, W., Germond, A., Hasler, M., Nicoud, JD. (eds.) Artificial Neural Networks (ICANN 1997). LNCS, vol. 1327, pp. 583–588. Springer, Heidelberg (1997). https://doi.org/10.1007/BFb0020217
11. Myers, D.E.: Vector conditional simulation. In: Armstrong, M. (ed.) Geostatistics. Quantitative Geology and Geostatistics, vol. 4, pp. 283–293. Springer, Dordrecht (1989). https://doi.org/10.1007/978-94-015-6844-9_21
12. Coleman, B.D., Gurtin, M.E.: Thermodynamics with internal state variables. J. Chem. Phys. **47**, 597 (1967)
13. ABAQUS: https://classes.engineering.wustl.edu/2009/spring/mase5513/abaqus/docs/v6.6/books/usb/default.htm?startat=pt05ch18s03abm29.html

Reinterpreting the Bishop's Parameter χ in the Light of the Drying Collapse of Clays: From Phenomenology to Numerical Implementation

Vito Tagarelli(✉), Francesco Cafaro, and Federica Cotecchia

Polytechnic University of Bari, Bari, Italy
vito.tagarelli@poliba.it

Abstract. In the last two decades several researchers investigated all the factors playing a role in determining the water retention properties of deformable clayey soils, either compacted or consolidated, with the aim to depict a general behavioural framework clarifying the interplay between retention properties of the deformable clays and their mechanical properties in saturated conditions. According to the behavioural framework reported by Cafaro and Cotecchia [1], the retention behaviour of clays, and in particular, the clay response to drying was found to depend on its current void ratio, volumetric stiffness, and over consolidation ratio. If the onset of major desaturation is defined as Gross Air Entry (GAE), it was also found that the GAE most often corresponds to the Gross Yield (GY) state in isotropic compression.

Correspondingly, in this contribution, the results of a laboratory free drying test were interpreted in the light of this framework, and back-analysed by adopting the Bishop's effective stress concept where the χ function was calibrated, by varying the residual degree of saturation ($S_{r,res}$). The corresponding hydro-mechanical (HM) numerical modelling was successful to provide good prediction of the volumetric straining of the clay upon drying if compared to the measured one.

It is believed that good performances in predicting the partially saturated HM soil behaviour may be obtained only when the χ-parameter is reinterpreted or adjusted to contemplate the desaturation behaviour of the over-consolidated soils.

Keywords: Partially saturated soil behaviour · Bishop's effective stress · χ-parameter calibration · Numerical back-analysis · Free drying test

1 Introduction

Recently, a strong research effort has been put on deepening the scientific knowledge on thermo-hydro-mechanical (THM) interaction occurring between the soil and the atmosphere, aimed at better tackling several issues became relevant nowadays, such as weather-induced landsliding, climate change adaptation, geo-thermal structures design, nuclear waste disposal [e.g., 2, 3]. Irrespectively to the addressed issue, the interaction

A. Ferrari et al. (Eds.): CNRIG 2023, SSGG, pp. 373–381, 2023.
https://doi.org/10.1007/978-3-031-34761-0_46

processes are most intense within the shallow soil layers (i.e. the soil cover), where the largest temperature and hydraulic gradients occur, and encompass the exchanges of liquid, gas and energy between the vegetated soil and the atmosphere (Soil-Vegetation-Atmosphere interaction, SVA, [4]); at that depths the soil is usually above the water table, being in partially saturated condition, so that also the soil THM properties for the modelling [5] differ than those recorded when in full saturation condition. As result, the THM modelling of the soil cover, when predicting the response of geotechnical structures interacting with the environment may be successful only if partial saturation is accounted for. This step towards a better phenomenological diagnosis and numerical modelling is necessary to design more sustainable and resilient structures and measures.

An historical and unsolved debate in the academic community has been discussing in finding the best solution for describing the effective stress in partially saturated condition, since the constitutive models cannot be formulated using the effective stress as proposed by Terzaghi [6] as single stress variable, but it requires the suction to be included in the effective stress tensor of reference, given by the Bishop's equation [7]:

$$\sigma' = \sigma - P_g \boldsymbol{I} + \chi s \boldsymbol{I} \tag{1}$$

where χ is the Bishop's effective stress parameter, P_g is the gas pressure and s is the suction. The experimental results have shown that the parameter χ depends on factors such as the history of wetting/drying of the soil, the void ratio and the soil structure, but it is usually taken as a function of the degree of saturation, S_r, even if the uniqueness of the relationship between χ and S_r has been always debated.

As such in this manuscript, a framework for the phenomenological behavior of soil along drying path [1] is recalled by reporting laboratory data of different clays, and correspondingly, the numerical analysis of the partially saturated HM soil behaviour is presented by describing the numerical simulation of a free drying test in the laboratory.

2 How Over-Consolidated Clays Behave During Drying

In this section, the macro-response of different fine-grained materials when subjected to drying is briefly discussed, with references to both retention properties and mechanical behaviour during compression, by reporting data already available in the literature.

In particular, clays of different nature and stress-strain history were tested showing that, despite the different structure, the saturated/partially saturated HM behaviour of the clays follow the framework by Cafaro and Cotecchia [1, 8]. As such, the latter is briefly recalled hereafter along with the experimental results of both drying laboratory tests and compression data in saturated condition with reference to reconstituted, diagenized and disturbed natural clays as reported in Table 1. This framework is here considered since it is one of the few referring to the response to drying of clays deposited and consolidated in saturated conditions, rather than compacted soils [e.g., 9].

Through the analysis of several test data, Cafaro and Cotecchia [1] demonstrated that the clay response to drying depends on the void ratio, volumetric stiffness, and overconsolidation ratio. In this framework the state of an initially saturated over-consolidated clay moves along a recompression line (i.e., coincident with the swelling line, of gradient κ in e - ln p' plane) when it is either isotropically compressed by external loading and keeps being saturated, or when it is compressed due to a suction increase upon unconfined drying, but still keeping $S_r = 1$. In both cases, if the effective stress state is described according to the single stress variable framework (i.e., Bishop effective stress, [7]), the Bishop χ-parameter is equal to 1; however, in the first case u_w is positive while p increases, whereas p equals zero and u_w is negative and decreases in the second case. Cafaro and Cotecchia [1] discuss how this is also the case when S_r reduces from 1 to about 0.9 with drying, i.e., the soil state is quasi-saturated. The maximum value of suction, s, by which $S_r \geq 0.9$ is named s_{des}. When s increases beyond s_{des}, the clay starts experiencing major desaturation named as Gross Air Entry (GAE), by Cafaro Cotecchia [1], and a limited and sudden drop in void ratio generally occurs. Beyond GAE, instead, a major S_r drop occurs with minor or no reduction in the clay void ratio, since the clay moves towards its shrinkage limit at about constant void ratio. Moreover, as shown by Cafaro and Cotecchia [1, 8] for over-consolidated clays the GAE corresponds to the Gross Yield (GY) state in isotropic compression, p'_{yis}, so that p'_{GAE} is about p'_{yis}.

To summarize, following the framework, for $s < s_{des}$, i.e., for saturated or quasi-saturated soil, χ must provide p' values consistent with the clay straining along the recompression path, controlled by the pre-yield stiffness. For s about s_{des} there is a sudden and limited volumetric collapse and for $s > s_{des}$, i.e., $S_r < 0.9$, χ must provide p' values consistent with the prediction of no volumetric strain increments under unconfined drying.

Table 1. Clays and tests herein considered

Clay type	Clay name	Oedometric compression	Isotropic compression	Free drying
Diagenized (natural) clay	Montemesola grey diagenized clay	✓ [1]	-	✓ [1]
Disturbed (natural) clay	Montemesola yellow weathered clay	✓ [1]	✓ [1]	✓ [1]
	Pisciolo fissured clay	-	✓ [4, 10]	✓ [10, 11]
Reconstituted clay	Montemesola clay	-	✓ [1]	✓ [1]

The drying response of a diagenized clay is here reported by referencing to the Montemesola grey clay [1], being a grey over-consolidated stiff clay (Subapennine Clays) of mainly illite composition (i.e., about 50 to 70%) [1], whose deposition occurred in the Montemesola Basin (TA, Italy). Its average plasticity index, PI, is about 27%, with medium activity ($A_{av} = 0.56$), so that this clay is classified as medium to high plasticity clays (CL-CH). After the deposition in a marine environment, the clay was subjected to diagenesis (i.e., formation of bonding), which increased the clay strength [1]. The data in Fig. 1 result from a drying test (full squares) and an oedometric compression (empty squares) carried out on a specimen trimmed from an undisturbed sample of this clay [1], showing the correspondence between the GAE and the oedometric GY, which testifies that the desaturation process is controlled by the GY and hence, depends on both stress history and structure for a natural clay since both control GY of clays. Indeed, before GY the slope of the drying curve in Fig. 1 is about the slope of the swelling lines, κ, of the same clay when isotropically recompressed. Beyond the GAE≈GY, further increases of s cause no decrease of the void ratio, whereas, along the oedometric path the volume change is controlled by the virgin compression index, λ.

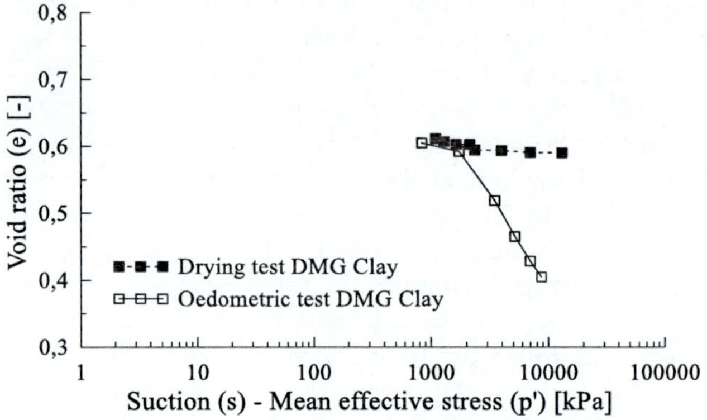

Fig. 1. Oedometric test results (e-log p') and free drying test results (e-log s) with reference to the Diagenized Montemesola Grey Clay (Table 1); data from [1, 8].

The drying response of natural clays with disturbed structure is here recalled by referencing to two clays, whose disturbance derived from different processes, indeed the Montemesola yellow clay [1] is reported as exemplification of a weathered stiff clay, whereas the Pisciolo fissured clay [10] is referred to as an example of tectonized clay.

The Montemesola yellow clay is a fine-grained material (CF = 42–49%) resulting from the weathering of Pleistocene Montemesola clay, described above; both its average PI and A_{av} are about those of the grey original unweathered clay. After the diagenesis, the clays underwent unloading due to erosion, and it was exposed to a semiarid climate as it is nowadays, that caused the material to undergo weathering processes. Figure 2 reports data from two drying tests (grey and black full dots), together with data from an oedometric test (black empty dots) carried out on specimens trimmed from an undisturbed Montemesola yellow clay samples [1].

The Pisciolo clay is a fine-grained material (CF = 37–62%; SF = 30–40%), belonging to a sedimentary succession deposited in a marine basin (Cretaceous-Miocene), thereafter involved in the Apennine orogenesis [10–12] so that its meso-fabric resulted to be highly fissured (I6) with either random (F3) or single (F1) fissure orientation (according to [13]); as result, the matrix of the Pisciolo clay is characterized by a disturbed and highly fissured fabric (i.e., scaly clay [13]). The average PI is 40%, with a high activity (A_{av} = 0.85) so that this clay can be classified as high plasticity clay (CH), whose description has been thoroughly reported by Cotecchia et al. [4, 12] and Pedone [10]. Figure 2 reports data from a drying test (black full squares) and an isotropic compression (black empty squares) carried out on a specimen trimmed from an undisturbed Pisciolo clay sample [4, 10]. On overall, also disturbed clays exhibit GY≈GAE,

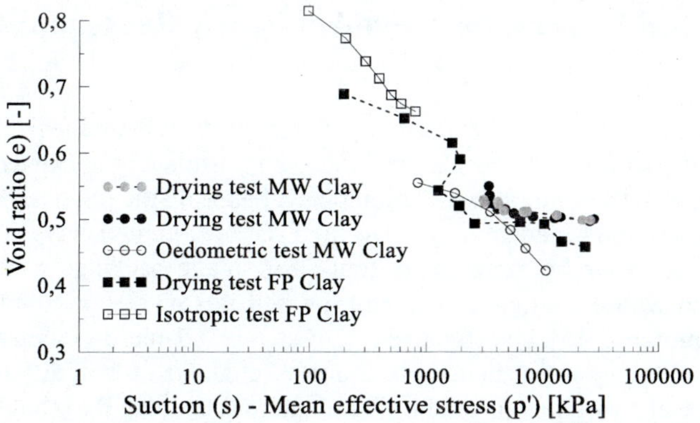

Fig. 2. Oedometric and isotropic test data (e-log p') and drying test data (e-log s) for the weathered Montemesola Clay (data from [1, 8]) and for the Fissured Pisciolo Clay (data from [4, 10]).

further testifying that the desaturation HM process of clays is controlled by both the clay structure, even when disturbed, and the stress-strain history.

Hereafter, the typical response to drying of reconstituted clays is described by referring to the Montemesola clay [1, 14], which was reconstituted in the laboratory and consolidated from slurry up to a $\sigma'_v = 1100$ kPa, and then unloaded to $\sigma'_v = 50$ kPa [1]. Figure 3 reports the isotropic normal compression line of the reconstituted clay, as well as data from a free drying test in the laboratory, showing that even a reconstituted clay exhibits a sudden and limited drop in volume (i.e. drying collapse), which also in this case is about the INCL of the material, further confirming the framework adopted.

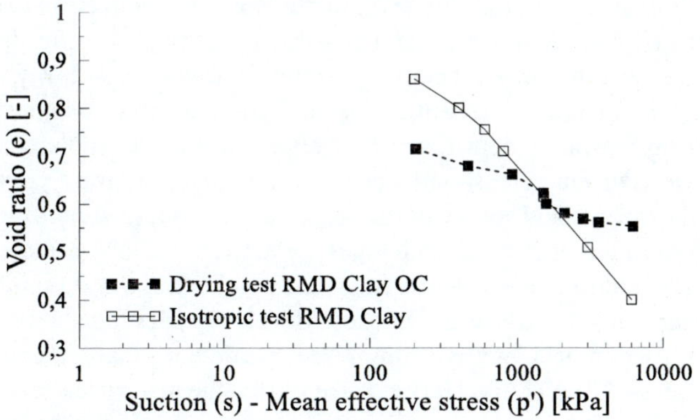

Fig. 3. Isotropic test results (e - log p') and free drying test results (e - log s) with reference to the Reconstituted Montemesola Clay (data from [1, 8]).

3 Mechanical Reasons for a Sudden Fabric Reorganization Under Drying

Evidence of metastable states for normally consolidated fine-grained soils during drying was shown and discussed by Cafaro [14], who postulated the existence of a Fabric Permitted Space in the void ratio – suction plane, based on the observed offset between Intrinsic Compression Line and Virgin Drying Line of saturated clays. Theoretical reasons for this space can be more easily found for sphere packings, which are usually hyperstatic and whose configurational entropy depends on the friction coefficient and the confining pressure [15]. Investigating micro-statics and micro-kinematics of capillary phenomena in dense granular materials, Chareyre et al. [16] found different volumetric strain as effect of two types of loading, i.e., external and capillary loading. For equal contact stress increment, the volumetric strain due to external loading was higher than that induced by the capillary one. This is consistent with the different packing stability expected for the two mechanisms of stress transmission according to a grain column analogy [17].

It is therefore expected that a particles assembly subjected to drying may suddenly experience a volumetric collapse: for fine-grained soils this is due to immediate slippage of clay aggregates, triggered at a suction remaining almost constant [1]. The data in Fig. 2 concerning the drying data on Pisciolo Clay suggests that the drying collapse may occur even with a temporary drop in suction, probably due to the sudden increase of S_r for unchanged water content. Statistical mechanics of porous media should confirm this behaviour. Xu and Louge [18] showed that a saturated mesoscopic sample with a disordered void space progressively drains as the capillary pressure rises, up to a collective first-order phase transition that empties its wetting fluid abruptly.

Interpreting the fabric rearrangement of clays with granular physics may be not rigorous but helpful. Indeed, although for fine-grained soils the interaction between particles is characterized by electrical bonds (and possibly by cementation for diagenized soils) much more than by body forces, it could be argued that in a silty-clayey soil undergoing gross-yielding the kinematic freedom of grains and particles is somehow increasing, allowing packing rearrangement towards energetically more stable configurations. In this respect, the recurrent correspondence of drying collapse, gross yield, and gross air entry reflects the mechanical source of the desaturation process. According to Cafaro and Cotecchia [1], once major structural changes are activated at GY, that over-consolidated clays reach during shrinkage in quasi-saturated conditions, discontinuities in the contractile air-water contact surfaces are likely to develop, with sudden onset of GAE. With further drying, suction acts locally as high inter-particle normal load but is statistically unable to compress the soil bulk. Hence, it seems that the history and/or structure dependence of the GAE may represent an opportunity to inform calculation approaches based on the Bishop's χ-parameter, as discussed in the following.

4 Improving the Performance of the Bishop's Effective Stress

Based on the evidences at the macro-scale, as well as the cited theoretical framework in the literature [1, 14–18], a reinterpretation of the Bishop's effective stress is proposed aimed at improving the performance of the single-variable effective stress in boundary value problems mainly characterised by negligible total stress increments; two different strategies may be followed, resulting in two different options, as reported hereafter.

The option A is providing a new formulation of the Bishop's χ parameter, based on the recalled framework [1]. A tentative formulation may be represented by a new fitting function, associated with a low-cost and not time-consuming calibration procedure (isotropic compression test to measure elastic saturated compressibility and pre-consolidation stress), concerning the Bishop's χ parameter to be such that: (i) when s $= (u_a - u_w) < GAE$; $\chi\ (u_a - u_w) = (u_a - u_w) \Longrightarrow \chi = 1$; (ii) when $s = (u_a - u_w) \geq GAE$; $\chi\ (u_a - u_w) = \text{constant}$.

The option B may be addressed by working on the formulations already available and implemented in the FEM codes; e.g., in Plaxis 2D the χ parameter is computed as a function of the SWRC and the current S_r through a simple equation as follows:

$$\chi = S_{r,eff} = (S_r - S_{r,res})/(S_{r,sat} - S_{r,res}) \quad (2)$$

As such, it is in principle possible to propose a reinterpretation of the parameter $S_{r,res}$ of the equation to compute the Bishop's χ parameter, so that to obtain as result the respect of the conditions imposed within the behavioural framework abovementioned.

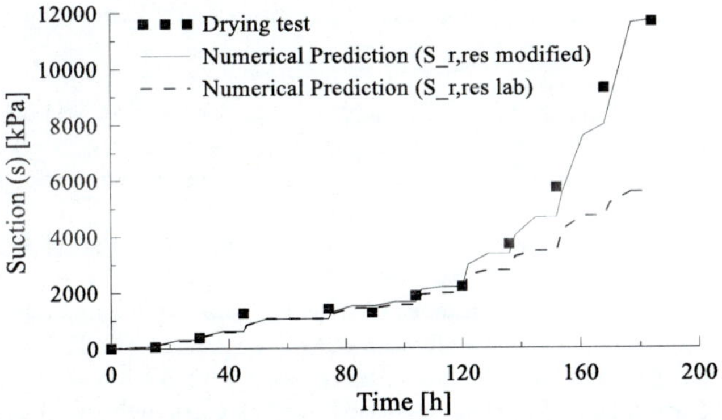

Fig. 4. Laboratory free drying test for a specimen of the Pisciolo Fissured Clay (data from [4]), plotted in the suction against time together with the corresponding numerical prediction.

However, an improvement of the drying numerical prediction attained by "tuning" the value of $S_{r,res}$ would imply a different physical meaning for this parameter, being a fitting parameter to be calibrated rather than a value to be measured at residual moisture; hence, $S_{r,res}$ should be renamed differently. In this perspective, the existing formulation of χ available in Plaxis (Eq. 2) was manipulated to make the Bishop's stress working properly. In this view, Tagarelli and Cotecchia [11] thoroughly described the HM modelling of a free drying test of a Pisciolo clay specimen (Pedone et al. [10]), by adopting the Soft Soil constitutive model, by re-interpreting the $S_{r,res}$ (Fig. 4), which is indeed was calibrated to comply with the framework by Cafaro and Cotecchia, implementing then the following conditions: (i) for $s \leq s_{des}$, and corresponding $p' \leq p'_{GAE}$, the prediction of void ratio variations is controlled by the elastic compression stiffness; (ii) for $s > s_{des}$, the prediction of zero increase in p', keeping $p' = p'_{GAE}$; (iii) the assumption of $p'_{GAE} = p'_{yis} = 980$ kPa, as measured for the Pisciolo clays; the value of $S_{r,res}$ achieved through the calibration was 0.45, which clearly do not anymore represent the real residual degree of saturation, but rather it is calibrated parameter so that to make the Bishop's effective stress properly working in modelling a boundary value problem.

The success of the modelling implementing the calibrated value of $S_{r,res}$ in simulating the drying path is testified by showing the agreement of the predicted suction (black line) with the measurements (black squares) with time in Fig. 4, which also reports the prediction obtained if the residual degree of saturation is set as the physical value (i.e., $S_{r,res} = 0.06$, dashed line), rather than what being calibrated ($S_{r,res} = 0.45$).

Acknowledgements. The authors are grateful for the support provided by PON MITIGO (ARS01_00964).

References

1. Cafaro, F., Cotecchia, F.: Influence of the mechanical properties of consolidated clays on their water retention curve. Ital. Geotech. J. **2**, 13–29 (2015)
2. Cui, Y.: Soil–atmosphere interaction in earth structures. J. Rock Mech. Geotech. Eng. **14**(1), 35–49 (2022)
3. Li, T., Shiozawa, S., McClure, M.W.: Thermal breakthrough calculations to optimize design of a multiple-stage Enhanced Geothermal System. Geothermics **64**, 455–465 (2016)
4. Cotecchia, F., Tagarelli, V., Pedone, G., Ruggieri, G., Guglielmi, S., Santaloia, F.: Analysis of climate-driven processes in clayey slopes for early warning system design. Proc. Inst. Civ. Eng. Geotech. Eng. **172**(6), 465–480 (2019)
5. Elia, G., et al.: Numerical modelling of slope–vegetation–atmosphere interaction: an overview. Q. J. Eng. Geol. Hydrogeol. **50**(3), 249–270 (2017)
6. Terzaghi, K.V.: The shearing resistance of saturated soils and the angle between the planes of shear. In: First International Conference on Soil Mechanics, vol. 1, pp. 54–59 (1936)
7. Bishop, A.W.: The principle of effective stress. Tek. Ukebl. **39**, 859–863 (1959)
8. Cafaro, F., Cotecchia, F.: Structure degradation and changes in the mechanical behaviour of a stiff clay due to weathering. Géotechnique **51**(5), 441–453 (2001)
9. Alonso, E.E., Gens, A., Josa, A.: A constitutive model for partially saturated soils. Géotechnique **40**(3), 405–430 (1990)
10. Pedone, G., Tsiampousi, A., Cotecchia, F., Zdravkovic, L.: Coupled hydro-mechanical modelling of soil–vegetation–atmosphere interaction in natural clay slopes. Can. Geotech. J. **59**(2), 272–290 (2022)
11. Tagarelli, V., Cotecchia, F.: The effects of slope initialization on the numerical model predictions of the slope-vegetation-atmosphere interaction. Geosciences **10**(2), 85 (2020)
12. Cotecchia, F., Pedone, G., Bottiglieri, O., Santaloia, F., Vitone, C.: Slope-atmosphere interaction in a tectonized clayey slope: a case study. Ital. Geotech. J. **1**(14), 34–61 (2014)
13. Vitone, C., Cotecchia, F.: The influence of intense fissuring on the mechanical behaviour of clays. Géotechnique **61**(12), 1003–1018 (2011)
14. Cafaro, F.: Metastable states of silty clays during drying. Can. Geotech. J. **39**(4), 992–999 (2002)
15. Schröter, M.: A local view on the role of friction and shape. In: EPJ Web of Conferences EDP Sciences, vol. 140, p. 01008 (2017)

16. Chareyre, B., Scholtès, L., Darve, F.: Micro-statics and micro-kinematics of capillary phenomena in dense granular materials. In: AIP Conference Proceedings American Institute of Physics, vol. 1145, no. 1, pp. 927–930 (2009)
17. Ridley, A.M.: The measurement of soil moisture suction. PhD thesis, University of London (1993)
18. Xu, J., Louge, M.Y.: Statistical mechanics of unsaturated porous media. Phys. Rev. E **92**(6), 062405 (2015)

Some Improvements of a Visco-Plastic Constitutive Model for Snow

Gianmarco Vallero(✉), Monica Barbero, Fabrizio Barpi, Mauro Borri-Brunetto, and Valerio De Biagi

Politecnico di Torino, 10129 Turin, TO, Italy
gianmarco.vallero@polito.it

Abstract. Snow is a peculiar example of a granular and low density geomaterial that exists at environmental conditions very close to its melting point. Once snowflakes deposit onto the ground, they start to evolve under the effect of both temperature and stress conditions (i.e., snow metamorphism): the result is therefore a complex three-phase material where an ice skeleton (i.e., snow microstructure) is encompassed by voids filled with air and liquid water. From a mechanical point of view, seasonal snow is therefore characterized by bonding/degradation processes between grains, large inelastic deformations and rate-sensitivity. Moreover, in nature, snow can be found in different shapes and structures having significant differences in terms of mechanical strength and physical properties. Therefore, the need for a constitutive model that can be representative of different types and conditions of snow is of paramount importance. Snow mechanics is indeed a topic of wide interest for many application fields, such as: design and management of structures and infrastructure in cold environments; study of new materials for winter sports and leisure activities; avalanche forecast, release and propagation, etc.

In this work, we report on some improvements to an existing constitutive model for snow that was developed in the framework of the nonlinear theory of elasto-visco-plasticity. The numerical implementation was achieved via a fully implicit integration algorithm and a local nonlinear resolving scheme. Finally, some preliminary results are described referring to literature experimental data on snow.

Keywords: Snow mechanics · Finite Element Analysis · Constitutive modelling

1 Introduction

Snow is a natural material composed of an ice skeleton encompassed by voids filled with liquid water and water vapour [1]. The mechanical behaviour of this heterogeneous material depends on a number of factors, such as: snow microstructure, thermal metamorphisms, water changes of phase, environmental conditions, loading rate, etc. [2, 3]. All these aspects make the modelling of the mechanical behaviour of snow a complex and difficult task. However, snow mechanics is crucial to address many interesting problems such as the stability of mountain snowpacks, the design of structures and infrastructures in cold environments, the safety of humans and goods in snow covered areas, the assessment of social and physical risk due to snow avalanches, etc. [4–6].

A. Ferrari et al. (Eds.): CNRIG 2023, SSGG, pp. 382–389, 2023.
https://doi.org/10.1007/978-3-031-34761-0_47

In the framework of Continuum Mechanics, the available snow models are generally developed with reference to the elasto-plastic (EP) theory to model both laboratory experiments and, on some occasions, on-site tests [7]. Nevertheless, available models are generally tailored to some specific types of snow (i.e., rounded grains, faceted crystals, etc.) and cannot be adopted for general purposes [8, 9]. Moreover, many models do not consider the bonding and sintering effects, the viscosity of the ice skeleton, the rate-sensitivity of the material, the volumetric collapse of the weak layer, etc.

In this work, we report on some significant improvements to the EP constitutive model originally proposed by [10] and [11]. Therefore, the improved model is based on: i) the framework and a new integration of the existing model [11], and ii) a new flexible yield locus and visco-plastic strain potential. At the moment, the model is able to quantitatively reproduce many experimental data available in the international scientific literature for dry snow with rounded grains [12–14]. The main outcome of this work is to build a solid basis for a new model that will be able to reproduce the behaviour of different types of snow: rounded and cohesive snow (typical of the snow slab) as well as faceted and low resistance snow (weak layer snow).

2 The Model

The improved model proposed in this work lies on the hypotheses of continuity, homogeneity and isotropy, and is based on the following three key points:

- the general framework of the constitutive model for snow developed by [11] which includes a valuable analytical law for sintering and degradation;
- the overstress theory of viscosity proposed by [15] and then modified by [10] to account also for the presence of irrecoverable strains inside the elastic region;
- a new formulation for the yield locus and the visco-plastic strain potential.

The temperature is assumed to be constant during the simulations, therefore, the model is purely mechanical. Here, we follow the small strain theory and the strain rate tensor is additively decomposed to produce the following usual stress-strain vector relationship:

$$\dot{\boldsymbol{\sigma}} = \mathbf{D}_{\mathrm{el}}\left(\dot{\epsilon} - \dot{\epsilon}^{\mathrm{vp}}\right) \tag{1}$$

where $\boldsymbol{\sigma}$ is the stress tensor (written in Voigt's notation), $\dot{\epsilon}$ is the total strain rate vector, $\dot{\epsilon}^{\mathrm{vp}}$ is the visco-plastic strain rate vector, and $\mathbf{D}_{\mathrm{el}}$ is the elastic stiffness matrix. The elastic matrix can be written with reference to the two Lamé coefficients A and B:

$$\mathbf{D}_{\mathrm{el}} = \begin{pmatrix} A & B & B & 0 & 0 & 0 \\ B & A & B & 0 & 0 & 0 \\ B & B & A & 0 & 0 & 0 \\ 0 & 0 & 0 & G & 0 & 0 \\ 0 & 0 & 0 & 0 & G & 0 \\ 0 & 0 & 0 & 0 & 0 & G \end{pmatrix} \tag{2}$$

where: $A = -\frac{vp}{k} + \frac{4}{3}G$, $B = -\frac{vp}{k} - \frac{2}{3}G$, v is the specific volume, p is the mean volumetric stress, k is the elastic compressibility, and G is the shear modulus. In the following subsections a brief description of the main theoretical aspects of the model is presented.

2.1 Yield Locus and Irreversible Strain Potential

In the present work, a new and unsymmetrical yield surface capable of changing its shape in the meridian plane is proposed and tested. The need for such a surface is linked to the generality that an arbitrary shape can provide to the model and the capability of the surface to deform and change its shape following specific numerical requirements and/or experimental findings. The starting point for the new surface is the snow version of the Modified Cam Clay proposed by [10]:

$$f_C(p,q) = \frac{1}{p_{atm}^2}\left\{q^2 - M^2\left[p(p_0 + p_m - p_t) + p_t(p_0 + p_m) - p^2\right]\right\} \tag{3}$$

where q is the equivalent Mises stress, p_0 is the mean consolidation stress for unbonded snow, p_{atm} is the mean atmospheric pressure, M is the slope of the critical state line, p_m and p_t are the two additional strength parameters (always positive) measuring the bonding of the snow in compression and tension, respectively. Following [8] and [11] we assumed their ratio defined by a constant: $p_t = \chi p_m$. Following the theoretical procedure proposed by [16], we introduced a function $\Gamma(p)$ that modulates the invariant q of the original surface allowing the locus to be flexible and obtaining shapes ranging from bullet-like to drop-like ones:

$$f(p,q) = \frac{1}{p_{atm}^2}\left\{q^2\Gamma(p) - M^2\left[p(p_0 + p_m - p_t) + p_t(p_0 + p_m) - p^2\right]\right\} \tag{4}$$

Imposing that the apex of the curve has the coordinates $(\alpha(p_0 + p_m), \alpha M(p_0 + p_m))$, the analytical expression of the new yield function $f(p,q)$ is the following:

$$f(p,q) = \frac{1}{p_{atm}^2}\left\{q^2 - 4\alpha^2 M^2 (p_0 + p_m)^3 \frac{(\alpha - 1)(p - p_t)(p + p_0 + p_m)[p_t + \alpha(p_0 + p_m)]}{\{-p(p_0 + p_m - p_t) + 2p(p_0 + p_m)\alpha + (p_0 + p_m)[-(\alpha - 2)p_t + (p_0 + p_m)\alpha]\}^2}\right\} \tag{5}$$

where α and M are the only two shape parameters (Fig. 1). The former parameter governs the symmetry of the curve around a vertical axis while the latter describes the slope of the critical state line and therefore measures the size of the yield curve. Suggested values for α are in the range [0.15, 0.75] whereas M will be between 0.50 and 3.00 [8, 13]. In Fig. 2 a short parametric analysis of the effect of both α and M is reported.

From a mathematical point of view, the function $f = 0$ describes a curve which is simply convex and smooth in any point of the $p-q$ plane. The convexity could therefore be lost for some value of f greater than 0. Thus, a similar expression for the visco-plastic strain potential is used to obtain the direction of the visco-plastic strains:

$$g(p,q) = q^2 - 4\alpha^2 M^2 p_{g0}^3 \frac{(\alpha - 1)(p - p_{gt})(p + p_{g0})[p_{gt} + \alpha p_{g0}]}{\{-p(p_{g0} - p_{gt}) + 2\alpha p p_{g0} + p_{g0}[-(\alpha - 2)p_{gt} + \alpha p_{g0}]\}^2} \tag{6}$$

where $p_{g0} = p_0 + p_m$ and $p_{gt} = p_t$. This definition ensures that g is null for any stress state (p, q).

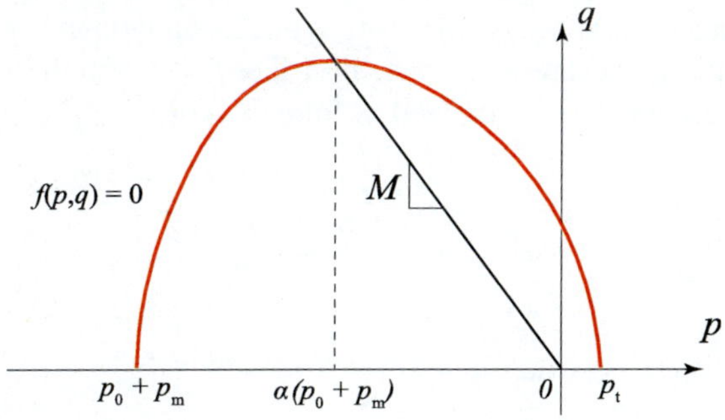

Fig. 1. Sketch of the meridian section of the yield surface proposed in this work.

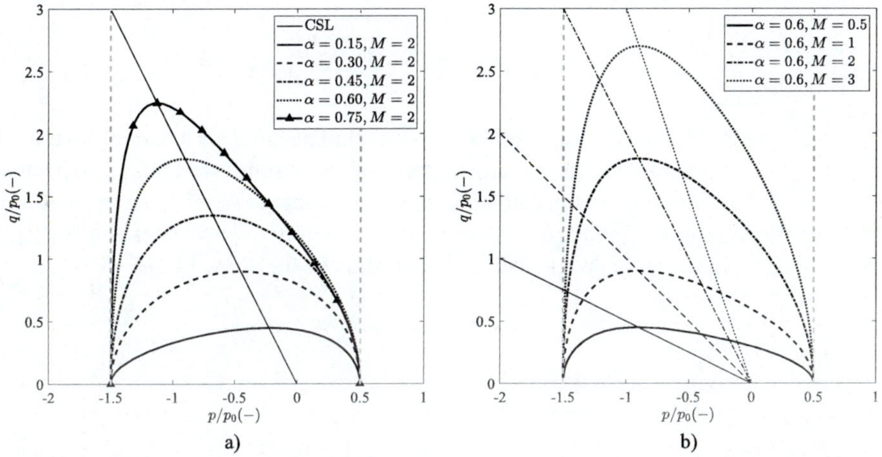

Fig. 2. Parametric analysis on the effect of the shape parameters on the meridian section of the new yield function. a) Effect of α for constant M. The elliptical shape is obtained for $\alpha = 0.475$ where, for different values, the surface progressively unsymmetric. b) Effect of M for constant α.

2.2 Viscosity and Flow Rule

A non-associative flow rule of the Perzyna type was chosen [15]. The model is modified from the original viscosity model to allow viscous-plastic irreversible strains to occur both inside and outside the yield locus. This allows to better reproduce some experimental findings where the viscous behaviour starts from the very beginning of the strain-stress history. In detail, the flow rule can be written as:

$$\dot{\epsilon}^{\mathrm{vp}} = \overline{\psi}\phi(f) = \psi \frac{\sqrt{p^2 + q^2}}{\sqrt{3}p_0} e^{af} \left.\frac{\partial g}{\partial \sigma}\right|_{\mathrm{norm}} \tag{7}$$

where ψ is a constitutive parameter having the dimensions of a strain rate; f is the current value of the yield function; $\left.\frac{\partial g}{\partial \sigma}\right|_{\mathrm{norm}}$ is the normalized first derivative (unit vector) of the visco-plastic potential g; $\overline{\psi}$ is the fluidity parameter defining the rate at which the irrecoverable strains occurs; $\phi(f)$ is the viscous nucleus relating the strain amount to the distance between the stress state and the yield surface $f = 0$. For the sake of conciseness, the dimensionless parameter β is defined as follows [10]:

$$\beta = \psi \frac{\sqrt{p^2 + q^2}}{\sqrt{3}p_0} e^{af} \tag{8}$$

2.3 Sintering and Hardening Laws

The sintering law adopted in this work is the same proposed by Cresseri (2005) based on literature data, for which we refer to their work for further information [10]. The current degree of sintering is given as:

$$S = S_0 \left[1 - \tanh \left(C \int_0^t \sqrt{\left(\dot{\epsilon}_v^{vp}\right)^2 + \left(\dot{\epsilon}_d^{vp}\right)^2 dt} \right) \right] \tag{9}$$

where S_0 is the degree of sintering for the unbonded material, C is a material parameter, $\dot{\epsilon}_v^{vp}$ are $\dot{\epsilon}_d^{vp}$ the volumetric and deviatoric parts of the strain rate tensor, respectively. The amount of sintering is finally related to the additional pressure in compression (p_m) with the following rate relation, in which π_m is a constitutive parameter and is b_{max} the maximum ratio between the neck size and the radius of the snow grain:

$$\dot{p}_m = \pi_m b_{max} \dot{S} \tag{10}$$

The usual volumetric hardening law of the Modified Cam Clay is used to describe the evolution of the mean consolidation stress:

$$\dot{p}_0 = -\frac{v}{\lambda - k} p_0 \dot{\epsilon}_v^{vp} \tag{11}$$

3 Numerical Implementation and Results

The proposed model is composed of a system of 10 nonlinear differential equations to be solved together at any time increment. The constitutive equations are the following:

$$\frac{1}{Z_1} \left(\dot{\sigma} - \mathbf{D}_{el} \dot{\epsilon} + \mathbf{D}_{el} \beta \left. \frac{\partial g}{\partial \sigma} \right|_{norm} \right) = 0 \tag{12}$$

$$\frac{1}{Z_2} g(\sigma) = 0 \tag{13}$$

$$\frac{1}{Z_1} \left(\dot{p}_0 + \frac{v}{\lambda - k} p_0 \dot{\epsilon}_v^{vp} \right) = 0 \tag{14}$$

$$\dot{p}_m - \pi_m b_{max} \dot{S} = 0 \tag{15}$$

$$S - S_0\left[1 - \tanh\left(C \int_0^t \sqrt{\left(\dot{\epsilon}_{\mathrm{v}}^{\mathrm{vp}}\right)^2 + \left(\dot{\epsilon}_{\mathrm{d}}^{\mathrm{vp}}\right)^2 dt}\right)\right] = 0 \tag{16}$$

where the Eq. (12) represents a vector relation of 6 component. Z_0 and Z_1 are two normalizing parameters [11].

The equations were integrated over time by using a fully implicit backward Euler method and the local problem of solving the nonlinear 10-dimension system was solved with an iterative scheme based on the Powell's hybrid method (i.e., a generalized Newton-Raphson method) [17]. The solution of both the general and local problems has been implemented into the UMAT format (written in Fortran 77) for the Abaqus/Standard Finite Element code [18].

To test the capability of the model some numerical analyses were performed on a single finite element with reference to creep tests (Fig. 3a), isotropic compression tests (Fig. 3b) and triaxial compression tests (Fig. 4a and 4b). In general, the model seems able to satisfactorily reproduce the experimental tests. Some issues (related to numerical problems and lack of convergence) can be observed during the isotropic unloading (Fig. 3b) and the triaxial relaxation, especially in quick-time tests (Fig. 4b), where the numerical drop of tension is quicker than the one reported in the experimental findings. In Tables 1 and 2 the model parameters and the initial conditions used in the numerical simulations are reported, respectively.

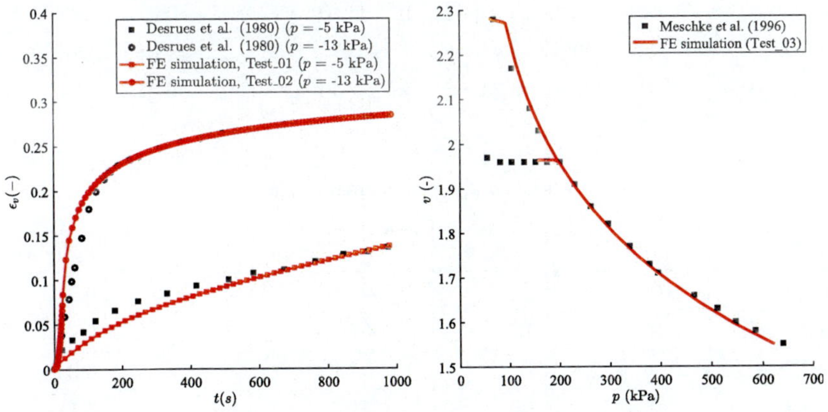

a) b)

Fig. 3. Comparison between the experimental results and the numerical prediction: a) results for volumetric creep tests [12]; b) results for isotropic compression test [13].

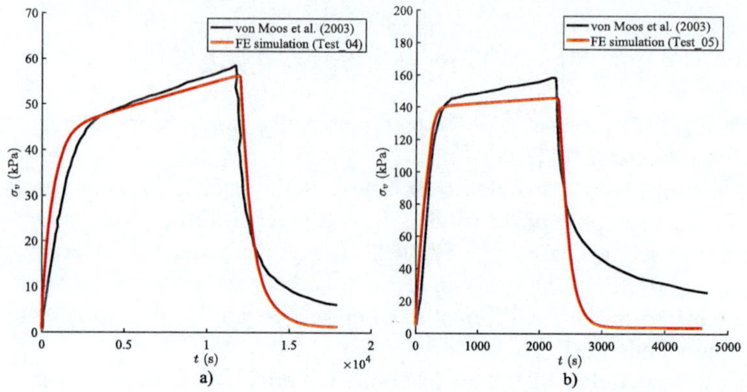

Fig. 4. Comparison between the experimental results and the numerical prediction: a) results for triaxial test in case of long-time test [14]; b) results for triaxial test in case of quick-time test [14].

Table 1. Model parameters used in the simulations.

Test ID	λ (-)	k (-)	G (kPa)	ψ (-)	a (-)	π_m (-)	χ (-)	C (-)	M (-)	α (-)
Test_01	0.35	0.02	2114	1.2e-4	16	40	0.05	0.01	2.88	0.475
Test_02	0.35	0.02	2114	1.2e-4	16	40	0.05	0.01	2.88	0.475
Test_03	0.35	0.02	12000	2.0e-7	16	40	0.05	0.01	2.88	0.475
Test_04	0.35	0.02	8000	4.2e-6	0.35	40	0.05	0.01	2.88	0.475
Test_05	0.35	0.02	20000	2.0e-5	0.35	40	0.05	0.01	2.88	0.475

Table 2. Initial conditions assumed in the simulations.

Test ID	p^0 (kPa)	p_0 (kPa)	v_0 (-)	T (°C)	r_0 (mm)
Test_01	0.0	2	4.58	-5	0.2
Test_02	0.0	2	4.58	-5	0.2
Test_03	-60.0	77	2.28	-5	0.2
Test_04	0.0	25	2.90	-12	0.118
Test_05	-5.0	100	2.44	-12	0.118

4 Discussion and Conclusions

In this work we describe an improved nonlinear visco-plastic model for snow, based on an improved version of the existing framework of the model by [11] and a new unsymmetric yield surface and irreversible strain potential. The new yield locus will be tested in the future to check its reliability and flexibility with respect to a larger collection of data. The model was implemented into the UMAT format for the FE code ABAQUS/Standard and both a fully implicit method and a local iterative algorithm were introduced in the code. The model can reproduce satisfactorily many different laboratory findings as can be deduced in Figs. 3 and 4. Some convergence issues can be found during the unloading isotropic phase (strain control). We are planning in the next future to update the model in order to avoid numerical issues and improve its stability even in these cases. In general, the model is able to replicate both qualitatively and quantitatively the actual behavior of the snow under different stress and strain conditions. This is a significant step towards the complete development of a viscous rate-dependent model with sintering that could be used in the future to model real scale phenomena linked to snow (avalanche triggering, weak layer collapse, etc.).

References

1. Mellor, M.: Engineering properties of snow. J. Glaciol. **19**(81), 15–66 (1977)
2. Petrovic, J.J.: Review mechanical properties of ice and snow. J. Mater. Sci. **38**, 1–6 (2003)
3. Blackford, J.R.: Sintering and microstructure of ice: a review. J. Phys. D Appl. Phys. **40**(21), 355–385 (2007)
4. Mellor, M.: A review of basic snow mechanics. In: IAHS Publ. (eds.) Snow Mechanics. Proceedings of the Grindelwald Symposium April 1974, vol. 114, pp. 251–291 (1975)
5. McCallum, A., White, G.W.: Engineered pavement of snow and ice. In: 8th International Conference on Snow Engineering (2016)
6. Vallero, G., et al.: Experimental study of the shear strength of a snow-mortar interface. Cold Reg. Sci. Technol. **193**, 103430 (2022)
7. Podolskiy, E.A., Chambon, G., Naaim, M., Gaume, J.: A review of finite-element modelling in snow mechanics. J. Glaciol. **59**(218), 1189–1201 (2013)
8. Gaume, J., Gast, T., Teran, J., van Herwijnen, A., Jiang, C.: Dynamic anticrack propagation in snow. Nat. Commun. **9**(1), 3047 (2018)
9. Vallero, G., Barbero, M., Barpi, F., Borri-Brunetto, M., Biagi, V.: Some computational issues in the elasto-plastic modelling of snow. In: 16th Edition of the International Conference on Computational Plasticity (2021)
10. Cresseri, S.: Constitutive modelling of dry granular snow at low strain rates. PhD thesis. Politecnico di Milano, Milano (2005)
11. Cresseri, S., Genna, F., Jommi, C.: Numerical integration of an elastic–viscoplastic constitutive model for dry metamorphosed snow. Int. J. Num. Anal. Methods Geomech. **34**(12), 1271–1296 (2010)
12. Desrues, J., Darve, F., Flavigny, E., Navarre, J., Taillefer, A.: An incremental formulation of constitutive equations for deposited snow. J. Glaciol. **25**(92), 289–307 (1980)
13. Meschke, G., Liu, C., Mang, H.A.: Large strain finite-element analysis of snow. J. Eng. Mech. **122**(7), 591–602 (1996)
14. von Moos, M., Bartelt, P., Zweidler, A., Bleiker, E.: Triaxial tests on snow at low strain rate. Part I. Experimental device. J. Glaciol. **49**(164), 81–90 (2003)
15. Perzyna, P.: The constitutive equations for rate sensitive plastic materials. Q. Appl. Math. **20**, 321–332 (1963)
16. Panteghini, A., Lagioia, R.: An extended modified Cam-Clay yield surface for arbitrary meridional and deviatoric shapes retaining full convexity and double homothety. Geotechnique **68**(7), 590–601 (2018)
17. Powell, M.J.D.: A Fortran Subroutine for Solving Systems of Nonlinear Algebraic Equations. United Kingdom (1968)
18. Abaqus, G.: AA.VV. Abaqus 6.11. Dassault Systemes Simulia Corporation, Providence, RI, USA (2011)

Micromechanical Numerical Modelling of Foundation Punching in Highly Porous Cemented Geomaterials in a Virtual Centrifuge Environment

Jinhui Zheng(✉), Marco Previtali, Matteo Oryem Ciantia, and Jonathan Knappett

University of Dundee, Dundee DD1 4HN, UK
2431370@dundee.ac.uk

Abstract. Foundation design on collapsible soils, such as loess, volcanic soils, tuff or cemented soft calcareous rocks, is a challenging geotechnical problem. When subject to moderate loads, the settlement of the foundation is limited and more or less reversible. However, beyond a load threshold value, soil compressibility increases due to the progressive rupture of intergranular bonds. In some cases, when this threshold is achieved, the bond fracture is so brittle that causes the collapse of the foundation, i.e. a sudden settlement occurring at constant load. To address this problem numerically, the coupled DEM (discrete element method) - FDM (Finite Difference Method) modelling approach is useful because of its excellent ability to deal with non-linear problems with large deformations while also reducing the computational burden in boundary value problems (BVPs). In this work, a coupled model is used to create a virtual centrifuge environment by combining the fast-generation method and particle upscaling. The penetration of a shallow foundation into soft cemented granular materials under different gravity levels is simulated using the two most widely used contact models for cemented materials: the parallel-bonded model (PBM) and the soft-bonded model (SBM). The numerical results show that these contact laws are unsuitable to properly reproduce the collapse-like failure mechanism for highly porous structures efficiently. It is shown that such a feature can be reproduced if a contact law capable of capturing the softening behaviour at the microscale is used.

Keywords: Coupled DEM-FDM · Soft rocks · Virtual centrifuge environment · bond softening behaviour

1 Introduction

Foundations on cemented soils such as loess, volcanic soils or on soft rocks such as tuff, chalk or calcarenites, are characterized by a low bearing capacity, and potential mechanical instability. This is due to their highly porous collapsible microstructure that may be schematized as grains bonded to each other. They show an elastic response under small external loads, while, at higher stresses, the change of voids' volume and the reorganization of grains is caused by the abrupt collapse of the internal structure

A. Ferrari et al. (Eds.): CNRIG 2023, SSGG, pp. 390–397, 2023.
https://doi.org/10.1007/978-3-031-34761-0_48

and the breakage of intergranular bonds leading to a denser granular material (Ciantia et al. 2015). This process is called structure collapse, and drastically affects the loading response of soft rocks.

The bearing capacity of foundations in soft rocks, especially the bearing capacity of shallow foundations, is of great concern to engineering and has been studied by many scholars (Alavi and Sadrossadat 2016; Tajeri et al. 2015). Depending on the failure mechanism assumed, several approaches to predict the bearing capacity of shallow foundations, such as the limit equilibrium method and upper and lower bound limit analysis methods, have been proposed in recent decades. For example, to predict the capacity of a footing on sandstone using the upper bound method, Chang et al. (2008) proposed a revised local shear slip surface where the slip region is divided into several blocks. On the other hand to evaluate the ultimate load of shallow foundations in soft rocks, Serrano et al. (2021) used the mathematical characteristic lines method.

Independently of the analytical method used, the aforementioned studies require a realistic failure mechanism to predict the capacity of shallow foundations. Despite the recent progress using X-ray tomography, which is limited to small samples (Alvarez-Borges et al. 2018), observing the initialization and the development of cracks below footings non-intrusively is challenging.

The DEM is a promising numerical tool able to capture the initiation and development of cracks of bonded geomaterials. It can deal with the large non-linear deformation features better than other traditional continuum method like FEM (Finite element method). However, its limit is the computational burden. For this reason coupled DEM-FDM numerical modelling is starting to be used to investigate the failure mechanism of soft rocks (Jiang et al. 2015; Tu et al. 2017; Zheng et al. 2022) in boundary value problems.

As mentioned above, the microstructure plays an important role in the mechanical response of soft rocks. In particular the softening behaviour of the bonds was shown to be key in predicting foundation bearing capacity on calcarenites (Nova and Parma 2011). To highlight this aspect, in this paper, a coupled DEM-FDM model is used to simulate the penetration of a shallow foundation on a porous calcarenite first by using two traditional DEM contact models that don't include bond softening in the post peak regime. It is shown that the abrupt bond breakage of these contact models fails to reproduce the physical macroscopic behaviour of shallow footings in highly porous soft rocks. It is subsequently shown that by employing a newly developed contact model inclusive of post peak softening behaviour the macroscopic response of the shallow footing is much more realistic.

2 Penetration Simulation of a Shallow Foundation

2.1 Preparation of the Coupled DEM-FDM Model

In view of the computational efficiency, particle upscaling and fast-generation methods are used here to generate the sample (Zheng et al. 2022). For the former, it is essential that the macroscopic mechanical properties of the sample, i.e. Young's modulus and UCS (unconfined compressive strength), remain unchanged when the particle size is enlarged. The details of the above two methods are given in Zheng et al. (2022).

The penetration of a shallow foundation under different gravities using PBM and SBM is simulated. All contact parameters are listed in Zheng et al. (2022). The final coupled model size is shown in Fig. 1. The success of the coupling can be seen in the continuous displacement field shown in Fig. 2.

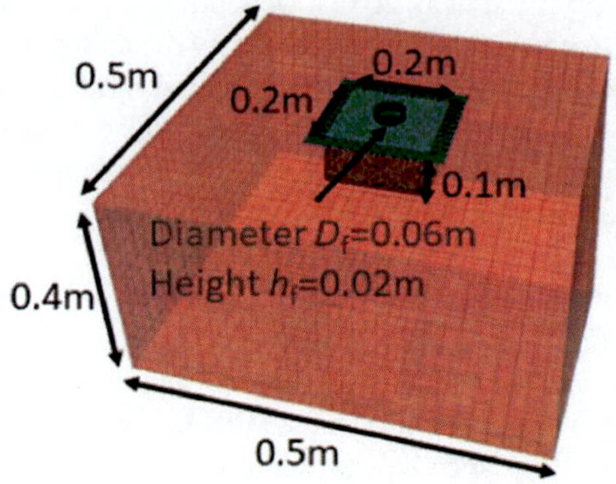

Fig. 1. Geometry size of the coupled model.

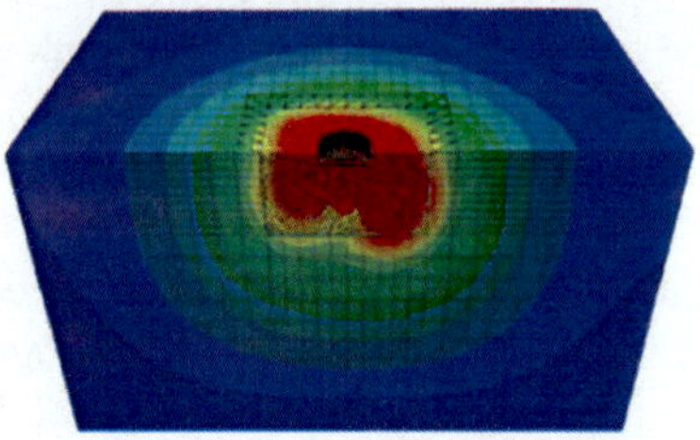

Fig. 2. Displacement continuity in discrete and continuum regions.

The load-displacement curves using PBM and SBM are shown in Fig. 3, where h_1 is the penetration depth. It is shown that the sample generated using SBM results in a higher capacity. This is due to softening occurring under shearing in the SBM sample. Bonds in SBM samples can still carry load even if the shear stress exceeds their peak strengths, instead of brittle failure (as incorporated in PBM). In order to reproduce a more realistic behaviour, the SBM is then used to simulate the soft rocks in the following.

Four different scenarios are used to simulate the shallow foundation failure, under different gravity levels, in order to simulate larger foundations using the same geometry, and these are shown in Fig. 4. This has shown that the failure mechanism depends on the gravity level. Note that the contact forces in the case of 100 g are much higher, producing a significant increment on the reaction force acting on the shallow foundation. The complete load loss during the initial penetration stage is observed in all four cases. This load loss is not only attributed to the structure collapse but also to the non-physical, brittle response of the bond.

The limitation of SBM (the above-mentioned load loss) suggests real softening behaviour of the bond should exist under different loading conditions. Therefore, a damage bond model (DBM) is more suited to reproduce the real mechanical response of the material.

3 Novel Damage Bond Model

3.1 Model Introduction

The basic principles of the bond behaviour are based on traditional beam theory, as done historically (Chen et al. 2022; Potyondy and Cundall 2004). According to multiple beam loading tests, rotational failure is achieved with lower stress levels than compression

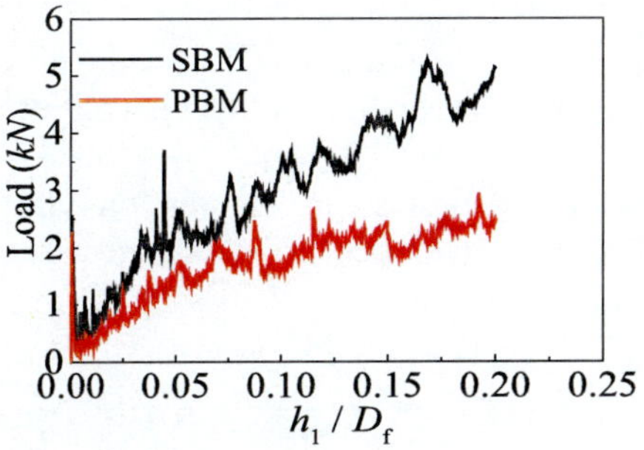

Fig. 3. Load-displacement comparison between SBM and PBM (100 g).

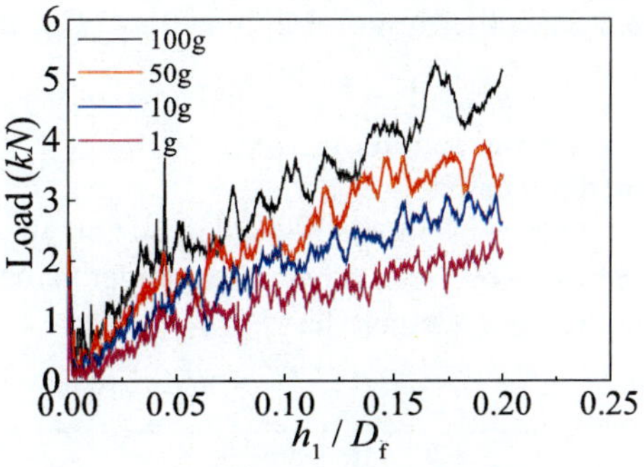

Fig. 4. Load-displacement comparison under different gravity levels based on the SBM.

(Neal 1960). Therefore, it is of importance to also consider the bond damage under rotation.

The generalized loads (Normal and shear forces, and bending moment) acting on the bond are considered linear within a domain controlled by the yield function obtained from laboratory tests (Neal 1960)

$$f_{yield} = \left|\frac{\breve{M}}{\tilde{M}}\right|^{1.001} + \left(\frac{N}{\overline{N}}\right)^2 + \left[\left(\frac{V}{\overline{V}}\right)^4 \Big/ 1 - \left(\frac{N}{\overline{N}}\right)^2\right] - 1 \tag{1}$$

with

$$\begin{bmatrix} \overline{N} \\ \overline{V} \\ \tilde{M} \end{bmatrix} = \begin{bmatrix} \overline{N} \\ \overline{V} \\ \overline{M}/\overline{R} \end{bmatrix} = \begin{bmatrix} N_0 \\ V_0 \\ M_0/\overline{R} \end{bmatrix}(1 - D_d) = \begin{bmatrix} \sigma_0 \overline{A} \\ \tau_0 \overline{A} \\ \sigma_0 \overline{I}/\overline{R}^2 \end{bmatrix}(1 - D_d) \tag{2}$$

where, N and V are the normal and shear force on the bond, $\breve{M}$ is the normalized bending moment on the bond. $\overline{N}$, $\overline{V}$ and $\tilde{M}$ are model parameters to control the surface size. $\breve{M}$ and $\tilde{M}$ are normalized by the bond radius $\overline{R}$. D_d is the bond damage variable. σ_0 and τ_0 are the tensile and shear strength of the bond. $\overline{A}$ and $\overline{I}$ are the area and moment of inertia of the bond.

In Eq. (1), $f_{yield} > 0$ means the bond is in the plastic state resulting in damage. The bond damage variable D_d is calculated in the following form. This form is based on the model of Nguyen et al. (2017), considering the damage caused by rotation.

$$D_d = 1 - e^{-\left(\frac{|u_n^p|}{u_n^c} + \frac{u_s^p}{u_s^c} + \frac{\theta_b^p}{\theta_b^c}\right)} \tag{3}$$

where, u_n^c, u_s^c and θ_b^c are the model parameters to control the softening rate of the bond. u_n^p, u_s^p are the normal and shear plastic displacements and θ_b^p is the plastic bending rotation.

The calculation of the loads on the bond is related to the plastic deformation, the form being similar to that of Nguyen et al. (2017). The details are not shown here due to the limited space.

3.2 Verification of the One-Bond Contact

The response of an individual bond obtained using the new proposed DBM under different loading conditions is shown in Fig. 5. It suggests that rotation has the greatest effect on bond damage.

One thing should be explained here why the degradation rate in compression is faster than tension. This is because a larger value of compressive force produces a greater loss at the same damage level.

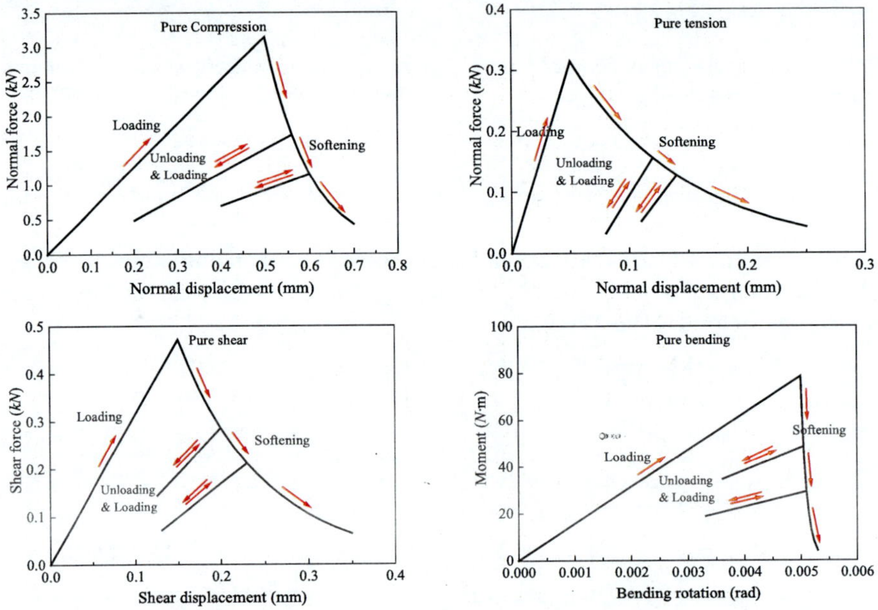

Fig. 5. One-bond contact verification under unidirectional loading conditions.

4 Simulation of the Penetration of a Shallow Foundation Using the Proposed Damage Bond Model

4.1 Simulation Results

In the following, some numerical results are presented to qualitatively show the ability of the proposed damage model to reproduce qualitatively the non-fully-brittle response of soft rocks. Some specific parameters for DBM are shown in Table 1, while the other parameters can be found in Zheng et al. (2022). The load-displacement curve obtained using the proposed model (DBM) are shown in Fig. 6. Due to the softening behaviour, the load loss of the footing is relatively small, which is consistent with the trend of shallow foundation penetration tests in soft rocks. From the micromechanical point of view, the tests that consider softening also present fewer broken bonds (see Fig. 7): after failure, as the particles under the footing still produce a force chain able to provide sufficient reaction force. As shown in Fig. 8, when the footing penetrates to the depth of $0.003D_f$, many rock grains under the footing in the SBM sample are in a non-contact state. On the contrary, the rock grains in the DBM sample are still connected by cement bonds to convey contact forces.

Table 1. Parameters used in DBM.

Parameters	Value
Bond compressive strength	4.5 MPa
Model softening parameters u_n^c	0.000197 *m*
Model softening parameters u_s^c	0.000197 *m*
Model softening parameters θ_b^c	0.00197

One thing to note is that the bearing capacity of the DBM model is smaller than that of the SBM at the end of the penetration. Due to the rapid collapse and densification of the SBM sample, compressive contact forces are generated between the crushed rock particles, which increases the internal contact forces of the sample.

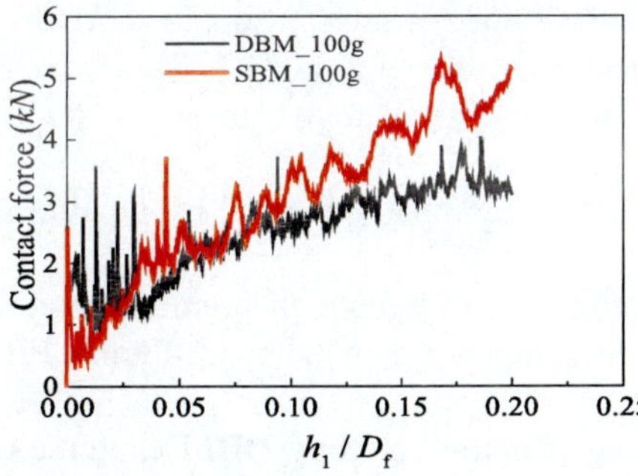

Fig. 6. The load-displacement curve under different contact models.

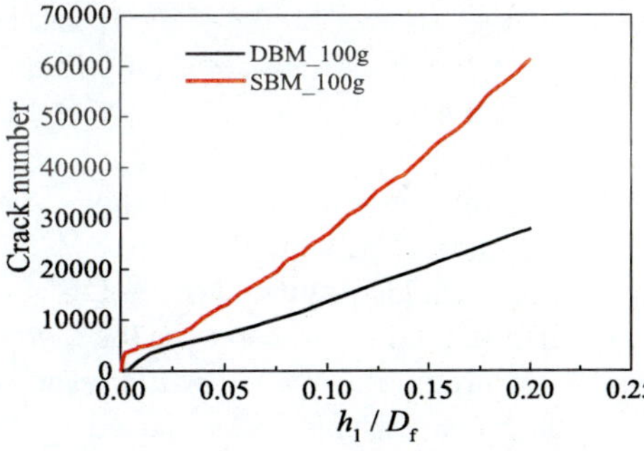

Fig. 7. The crack number during penetration under different contact models.

4.2 Failure Mechanism

The key to evaluating the ultimate bearing capacity of the footing is to determine the failure mechanism of the rock. The displacement vectors of rock particles under footing are displayed in Fig. 9 (the contour represent the particle displacement). The flow mode of rock particles follows the local shear failure mode. This can also indicate that most of the bonds that are broken during penetration are caused by shearing. While the traditional SBM and PBM cannot reflect the real bond mechanical response, only a similar punching failure of footing reported in soft rocks simulated by traditional bonded model (J. Chen et al. 2022).

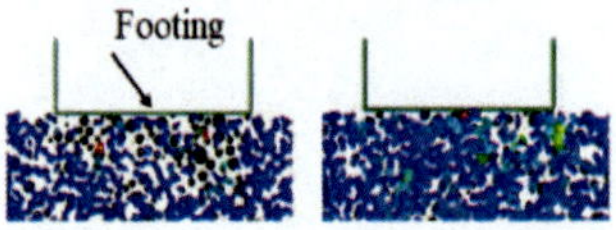

Fig. 8. Particle contacts beneath the footing.

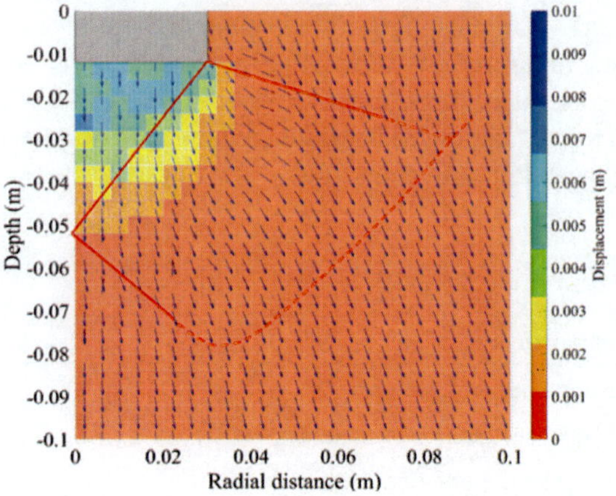

Fig. 9. Local shear failure in the DBM sample.

5 Conclusion

Reported research demonstrates traditional PBM and SBM contact models are unable to capture the mechanical behaviour of bonded materials. To solve the above problem, a novel contact model (DBM) has been developed, which introduces a damage variable to produce strain-softening. Finally, this new contact model has been successfully implemented into a coupled DEM-FDM model to simulate penetration of shallow foundations.

Compared with tension and shear, the bending moment produced by flexure can cause the most damage to the bond.

Rock particles simulated using DBM can form a stable force chain structure at the bottom of the footing to carry the load, while the samples simulated by traditional PBM and SBM produce an overly brittle response.

The failure of soft rocks caused by footing penetration using DBM captures the experimental local shear failure mode, while the failure mode using PBM and SBM is more similar to punching failure.

References

Alavi, A.H., Sadrossadat, E.: New design equations for estimation of ultimate bearing capacity of shallow foundations resting on rock masses. Geosci. Front. **7**(1), 91–99 (2016)

Alvarez-Borges, F.J., Richards, D.J., Clayton, C.R.I., Ahmed, S.I.: Application of X-ray computed tomography to investigate pile penetration mechanisms in chalk. In: Proceedings of the Chalk 2018 Conference Engineering in Chalk, pp. 565–570. ICE (2018)

Chang, J.C., Liao, J.J., Pan, Y.W.: Failure mechanism and bearing capacity of shallow foundation on poorly cemented sandstone. J. Mech. **24**(3), 285–296 (2008)

Chen, J., Bao, N., Sun, R.: Three-dimensional discrete-element-method analysis of behavior of geogrid-reinforced sand foundations under strip footing. Int. J. Geomech. **22**(9), 1–16 (2022)

Chen, X., Wang, L.G., Morrissey, J.P., Ooi, J.Y.: DEM simulations of agglomerates impact breakage using Timoshenko beam bond model. Granul. Matter **24**(3), 1–21 (2022)

Ciantia, M.O., Castellanza, R., di Prisco, C.: Experimental study on the water-induced weakening of calcarenites. Rock Mech. Rock Eng. **48**(2), 441–461 (2015)

Jiang, M., Chen, H., Crosta, G.B.: Numerical modeling of rock mechanical behavior and fracture propagation by a new bond contact model. Int. J. Rock Mech. Min. Sci. **78**, 175–189 (2015)

Neal, B.G.: The effect of shear and normal forces on the fully plastic moment of a beam of rectangular cross section. J. Appl. Mech. Trans. ASME **28**(2), 269–274 (1960)

Nguyen, N.H.T., Bui, H.H., Nguyen, G.D., Kodikara, J.: A cohesive damage-plasticity model for DEM and its application for numerical investigation of soft rock fracture properties. Int. J. Plast. **98**, 175–196 (2017)

Nova, R., Parma, M.: Effects of bond crushing on the settlements of shallow foundations on soft rocks. Geotechnique **61**(3), 247–261 (2011)

Potyondy, D.O., Cundall, P.A.: A bonded-particle model for rock. Int. J. Rock Mech. Min. Sci. **41**(8), 1329–1364 (2004)

Serrano, A., Galindo, R., Perucho, Á.: Ultimate bearing capacity of low-density volcanic pyroclasts: application to shallow foundations. Rock Mech. Rock Eng. **54**(4), 1647–1670 (2021)

Tajeri, S., Sadrossadat, E., Bazaz, J.B.: Indirect estimation of the ultimate bearing capacity of shallow foundations resting on rock masses. Int. J. Rock Mech. Min. Sci. **80**, 107–117 (2015)

Tu, F., Ling, D., Hu, C., Zhang, R.: DEM–FEM analysis of soil failure process via the separate edge coupling method. Int. J. Numer. Anal. Methods Geomech. **41**(9), 1157–1181 (2017)

Zheng, J., Previtali, M., Knappett, J., Ciantia, M.: Coupled DEM-FDM investigation of centrifuge acceleration on the response of shallow foundations in soft rocks. In: 10th International Conference on Physical Modelling in Geotechnics 2022, pp. 264–267. Springer, Daejeon (2022)

Novelties in Computational Geomechanics

On the Evaluation of Indirect Simulations Performance of Multi-parametrical Transient Seepage Models in River Embankments

Ilaria Bertolini(✉) and Guido Gottardi

Alma Mater Studiorum, DICAM Department, Università Di Bologna, Viale del Risorgimento 2, 40136 Bologna, Italy

ilaria.bertolini3@unibo.it

Abstract. The design and vulnerability assessment of earth-filled water retaining structures require the elaboration of reliable seepage analyses. Model predictions, however, can result highly unreliable because of poor parameter estimations. To address this issue, inverse modelling techniques can be implemented, incorporating observations of on-going processes into direct simulations. With reference to river embankments, time series of the monitored soil water content and pore water pressure can be helpful for a proper calibration of hydraulic and retention properties of the materials involved in the seepage problem. For the present study, the Levenberg-Marquardt algorithm implemented in the commercial code Hydrus 2D by Pc-Progress has been applied for the optimization procedure of an instrumented river embankment section along the river Secchia (Modena, Italy), making use of observational data. The seepage analysis is carried out considering various assumptions on the series of monitoring data, weighing distributions and optimized parameters combinations. The study aims to analyze an effective application of combined qualitative and quantitative methodologies for the performance evaluation of indirect simulations in the transient multi-parametrical model under investigation, in order to estimate the optimal set of parameters as a function of the pursued aim of the modelling.

Keywords: Parameter Calibration · Inverse Analysis · River Embankments

1 Introduction

One of the great challenges of seepage modelling in earthen structures is the identification of the correct water retention characteristics of the soils involved, in saturated and unsaturated conditions. An initial estimation of these fundamental input parameters is performed by direct and indirect methods. Direct estimations are based on the simultaneous measurement of water content and matric potential by laboratory tests and/or field monitoring. Indirect estimations rely on pedo-transfer functions that use regression analysis or a neural network approach to obtain, in output, the unknown hydraulic parameters starting from easy-to-measure soil physical properties. In accordance with the chosen method of parameter initial estimation, in the subsequent phase, a calibration

A. Ferrari et al. (Eds.): CNRIG 2023, SSGG, pp. 401–409, 2023.
https://doi.org/10.1007/978-3-031-34761-0_49

with a different degree of accuracy is required, in order to tune the input parameters on field and/or laboratory observed data. More specifically, the calibration of the model parameters is performed using a direct or an inverse procedure. The direct procedure is based on the well-known "trial and error" method, that consists in the iterative change of one or multiple input parameters of a direct seepage simulation until the desired precision (set subjectively by the modeler) is reached in the model output. The inverse procedure (or inverse analysis, hereafter referred to as I.A.) uses observed data to infer model parameters applying again an iterative process by varying the parameters and comparing the real response to the predicted one. At each iteration, an optimization algorithm (as the steepest descent method, the Levenberg-Marquardt's method, etc.) minimizes a suitable objective function which expresses the discrepancy between observed and predicted values. If indirect parameter optimization provides several advantages compared to direct procedures, as a detailed error analysis of the estimated parameters and rapidity, on the other hand, problems of solution uniqueness, computational efficiency and convergence remain to be solved [1]. The performance of indirect simulations can be then evaluated by means of qualitative (i.e., visual comparison) or quantitative methodologies. The latter are based on the application of a set of metrics/indices which expresses numerically the fitting performance. The present work investigates critically the results of an I.A. application to the hydraulic and retention parameters of a river embankment section in a transient seepage simulation, by means of the commercial code Hydrus 2D by Pc-Progress. Both qualitative and quantitative methodologies have been applied to indirect simulations in order to evaluate their performance. Limits and potentialities of each methodology, together with the effectiveness of their combined use in modeler's decision making, have been highlighted in relation to the presented case.

2 Presentation of the Case Study and of the Numerical Model

The chosen river embankment section is located along the river Secchia, the second main right-hand tributary of the Po River in northern Italy, 10 km downstream the location where in January 2014 a disastrous bank collapse was registered. An extensive in situ monitoring system, composed of sensors monitoring volumetric water content and pore water pressure (hereafter indicated with the abbreviations "wc" and "pwp" respectively) has been installed at different depths and distances from the river axis [2]. The combined use of continuous coring boreholes and cone penetration tests enabled a precise definition of the soil stratigraphy (see Fig. 1): Unit A, the embankment, consists of an alternation of silts and sandy silts, Unit A', the fluvial layer, has a coarser texture due to its depositional origin, while unit B, the foundation, is made of finer material compared to the above units. Evaporation tests performed on soil samples belonging to the different soil units enabled the identification of the initial set of the hydraulic and retention parameters to be used in the numerical modelling and their range of variation. Table 1 reports the adopted van Genutchen-Mualem model (hereinafter indicated with the acronym VGM) parameters. The transient flow through the investigated section has been modelled in Hydrus 2D in the period between 12th November 2017 and 23rd June 2018. The model has a width of 100 m and a height of 43 m, while the origin of the reference system is located in the river center (x-axis) and at 10 m a.s.l (z-axis). The applied boundary conditions (B.C.) are summarized in Fig. 1.

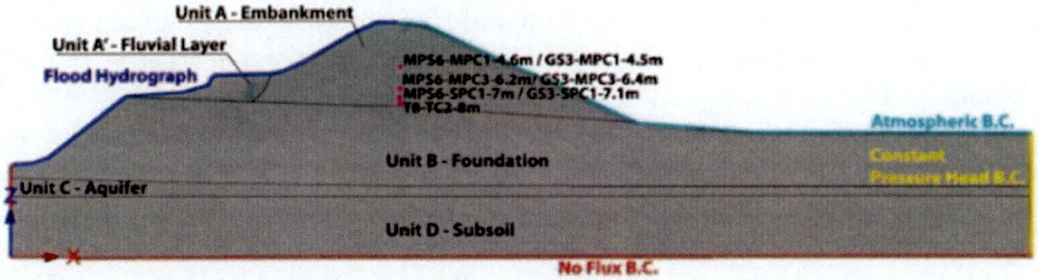

Fig. 1. Cross-section of the river embankment with indication of the boundary conditions (B.C.) applied, of the soil stratigraphy and of the considered 7 sensors installed in Unit A.

Table 1. The initial set of van Genuchten-Mualem parameters adopted in the numerical modelling. θr and θs are the residual and saturated wc, respectively; α is the inverse of the air-entry pressure, n and l are fitting parameters. Ks is the saturated permeability.

VGM	θ_r (m^3/m^3)	θ_s (m^3/m^3)	α (1/m)	n (-)	Ks (m/s)	l (-)
Unit A	0.0040	0.3970	0.8500	1.3040	1.50e−6	0.5
Unit A'	0.0005	0.3190	0.2505	1.2840	4.60e−7	0.5
Unit B	0.0010	0.4240	0.1178	1.1420	3.30e−9	0.5
Unit E	0.0001	0.3662	0.1609	1.3065	4.50e−7	0.5
Unit C	0.0100	0.4300	0.2000	1.2000	1.50e−6	0.5
Unit D	0.0000	0.4300	0.2000	1.2000	1.30e−9	0.5

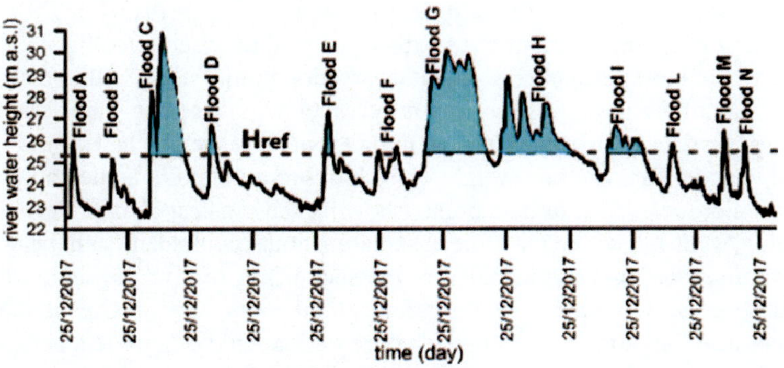

Fig. 2. Hydrograph of the river Secchia in correspondence of the investigated section.

A no flux B.C. is set to the lower boundary of the model located in Unit D (uniform clayey layer) and to the boundary passing through the center line of the riverbed. An atmospheric B.C. is set to the outer slope and to the far field: meteorological conditions are monitored by a 7 km distant station and imposed to the boundaries as outfluxes (evapo-transpiration contribution) and influxes (rainfall). The contours of the inner slope and of the berm are governed by the flood hydrograph (Fig. 2) which is monitored by a stream gauge in the nearby Ponte Motta. In order to subdivide the simulation period in high and low water events, an arbitrary reference river water height ($H_{ref} = 25.3m$) has been set equal to the height a.s.l of the deepest sensor located in Unit A, with the aim of marking the water height lower limit that can possibly trigger relevant changes in the sensor response. Initial conditions have been set according to pwp and wc information available at day 0 from the monitoring devices. A linear interpolation of these data has been assumed above and below the phreatic water table.

3 The Calibration Program of the Model Parameters

Unit A is subjected to changes in pwp and wc distributions as a result of the interaction with atmosphere and the river. As shown in numerous previous studies ([3] among others), these spatial and temporal variations can strongly affect slope stability, thus efforts should be made to predict them and to calibrate at best flow model parameters. The calibration procedure by means of I.A. focuses on the VGM parameters of Unit A. In particular, a Sensitivity Analysis by OFAT (One-Factor-A-Time) Technique has selected only 4 out of 6 parameters to be calibrated (θs, α, n, Ks). OFAT analysis identifies the effects of the variation of one specific independent variable on model output, keeping the remaining ones fixed. 7 sensors located in Unit A have been used as database of observations for the I.A. (see Fig. 1): 4 sensors monitoring pwp and 3 monitoring volumetric wc. As previously mentioned, the inverse parameter estimation has been performed by means of Hydrus 2D which has implemented a Marquardt-Levenberg type parameter estimation technique [4]. 6 different groups of inverse analyses have been carried out, which use:

- 3 different typologies of observations: (1) Pressure Head (PH); (2) water content (wc) and the combined one, (3) Pressure Head + water content (PH + wc).
- 2 different weighing distributions (w.d.) associated to observation data: (1) the same weight (equal to 1) is applied to each observation point falling within the simulation period, (2) different weights based on flood water height and persistence are applied to observation data while points outside flood events (i.e., below the H_{ref} in Fig. 2) have a null weight. The resulting weight of each observation point is then adjusted either to give more relevance to data points showing the greatest change during a certain flooding event and within the same observation node (coincident with the point where a certain sensor has been installed in the section) or to give the same total weight to all observation nodes during a certain flood event, irrespective of the different number of observation points for each observation node. Major details regarding the computation of the final weight for each observation point can be found in [5].

Each of the 6 groups of inverse analyses is composed of 14 indirect simulations, calibrating from 1 to 4 parameters in different combinations.

4 Analysis of the Performance of Indirect Simulations

To evaluate the performance of the successful indirect simulations, the modeler can apply a qualitative and/or a quantitative approach. Both are based on the comparison between simulated and observed data in the considered observation nodes. In the first case, the comparison is graphical, resulting in a straightforward communication, especially suitable in presence of a reduced number of simulations to analyze and of simulations behaving very differently. The major downside is clearly linked to the influence of the modeler's subjective opinion on the final choice of the best performing indirect simulation. The quantitative approach makes use of statistical and mathematical criteria (indices/metrics) to quantify simulation behaviour. Unlike the previous approach, it is objective and independent of modeler's opinion. Moreover, it allows the quantification of the simulation performance within confined time periods (flood peaks, drawdowns, off-flood periods, etc.) or relating to an individual observation node or groups of them (for example all PH sensors or all wc sensors). In the present case study, both approaches have been applied to the performed simulations, but at different stages of the analysis. Initially, a chosen set of metrics/indices has been used to select the indirect simulations that mostly enhance the model performance; subsequently, a graphical qualitative representation has been used to visualize the behaviour of the selected group of simulations in the different observation nodes. Their combined use has guaranteed the identification of the most suitable sets of optimized parameters for the possible different applications of the flow model. Table 2 reports a summary of the metrics/indices adopted in the analysis. The Mean Absolute Error (MAE) and the Root Mean Square Error (RMSE) are well-known accuracy metrics, which consider the average magnitude of the errors in a set of predictions without considering its direction (positive or negative). NSEj and IAj are efficiency criteria derived from modified forms of the Nash-Sutcliffe (NSE) criterion and of the Index of Agreement (IA), respectively. These forms overcome the over-sensibility of NSE and IA to extreme values given by their squared terms.

Table 2. The table reports the set of metrics/indices adopted together with their mathematical formulation and indications of their best value and of the values range.

Mathematical formulation	Best value /range	
$\mathbf{MAE} = \frac{1}{n} \cdot \sum_{1}^{n} \lvert M_i - S_i \rvert$	0 [0 ∞]	M_i observed value S_i simulated value
$IA_j = 1 - \frac{\sum_1^n \lvert M_i - S_i \rvert^j}{\sum_1^n \left(\lvert S_i - \overline{M} \rvert + \lvert M_i - \overline{M} \rvert \right)^j}$	1 [−∞,1]	n° of observed data $\overline{M}$ mean observed values $\overline{S}$ mean simulated values
$NSE_j = 1 - \frac{\sum_1^n \lvert M_i - S_i \rvert^j}{\sum_1^n \lvert M_i - \overline{M} \rvert^j} j = 1$	1 [−∞,1]	$\sigma_M = \sqrt{\frac{1}{n} \sum_1^n \left(M_i - \overline{M} \right)^2}$
$R^2 = \left\{ \frac{\sum_1^n (M_i - \overline{M})(S_i - \overline{S})}{\sqrt{\sum (M_i - \overline{M})^2} \cdot \sqrt{\sum (S_i - \overline{S})^2}} \right\}^2$	1 [0, 1]	$\sigma_S = \sqrt{\frac{1}{n} \cdot \sum_1^n \left(S_i - \overline{S} \right)^2}$
$RMSE = \sqrt{\frac{1}{n} \cdot \sum_1^n (M_i - S_i)^2}$	0 [0, + ∞]	
$KGE = 1 - \sqrt{(R-1)^2 + \left(\frac{\sigma_S}{\sigma_M} - 1 \right)^2 + \left(\frac{\overline{S}}{\overline{M}} - 1 \right)^2}$	1 [−∞,1]	

Another widely used efficiency criterion is the Coefficient of Determination (R^2), that ranges between 0 and 1. A value of 1 means that the dispersion of the predictions is equal to the observations, while a value of 0 means no correlation between them. It is strongly recommended to not use R^2 as the only indicator for simulation evaluation because it only quantifies the dispersion between the values, but it has been demonstrated that if the model systematically over or under predict observation data, this metric could result very close to 1 even if all predictions are incorrect [6]. The Kling-Gupta Efficiency Metric (KGE) is born as an improvement of NSE and it has been elaborated from the decomposition of NSE into 3 different components (correlation, bias and variability) in order to facilitate the analysis of their relative importance. The Mass Balance Error (MBE) results an important indicator of the accuracy with which the numerical solution has been calculated from the finite difference matrix equations. An error equal to 0% means that the mass is perfectly conserved and generally a MBE lower than 1% is required in an accurate numerical modelling.

For each indirect simulation, the value of each adopted metric/index is compared to the value obtained for the base simulation in order to quantify the enhancement given by the parameter calibration. Table 3 reports the indirect simulations that showed the greatest performance. The set of metrics/indices has been applied separately over the whole simulation period and over the major flood events (C, G and H in Fig. 2) to quantify the behaviour of each I.A. in the most relevant periods of the simulation, when major changes in suction distribution and wc are expected. In Table 3 the score is calculated over 3 different databases composed separately of PH, wc and both typologies (PH + wc). This enables to evaluate when a set of parameters calibrated on a certain database (e.g., of typology PH), behaves properly also in the remaining databases (e.g., of typology wc). The score attributed to the combined database (e.g., PH + wc) provides a first indication of the average performance of the indirect simulation. Often the evaluation of the score attributed to the performance of the indirect simulation in each single observation node, as shown in Table 4, provides a final important indication to the modeler. A negative score indicates a worsening of the indirect simulation performance with respect to the base simulation (hereafter called Set 0). When the score is close to zero, it could be read as a minor change with respect to Set 0 and in most cases, no changes are detected from a graphical representation (see for example sensor #2 in Fig. 3). In Table 5 the calibrated VGM parameters are reported for each of the selected indirect simulations. Based on the aim of the numerical modelling, a specific set of calibrated parameters can be chosen. Set 3 should be chosen if the modeler's purpose is an accurate estimation of wc distribution in Unit A, especially during flood events, while the pwp representation remains almost unchanged compared to that of Set 0, as observable in Fig. 3 and Table 3 for all the PH observation nodes. Set 5 should be chosen if the modeler is interested in a good representation of both wc and pwp distributions in Unit A during major flood events. In fact, during these restricted periods, its score is greater than that of Set 1 for both wc and pwp distributions, but poorer than that of Set 3 in wc distribution (see Table 3). Set 1 should be chosen for an accurate estimation of pwp and wc distributions in Unit A during the whole simulation period while its representation during flood events results poorer (in Table 3, compare Set 1 to Set 5 for pwp distribution and Set 1 to Set 3 for wc distribution over the floods restricted period).

Table 3. The table reports the score over the whole simulation period and over the restricted period (floods C, G and H) of the selected 5 Sets of optimized parameters.

Dataset		PH sensors (#1,2,3,4)		wc sensors (#5,6,7)		PH + wc sensors	
		Whole period	Floods C + G + H	Whole period	Floods C + G + H	Whole period	Floods C + G + H
Set 1	PH	0.64	0.56	1.10	0.44	0.38	0.47
Set 2	PH	0.50	0.56	−2.39	−4.23	0.35	0.43
Set 3	wc	−0.09	−0.12	1.33	1.41	−0.06	−0.08
Set 4	PH + wc	0.52	0.58	1.43	-0.26	0.50	0.61
Set 5	PH + wc	0.49	0.58	1.00	0.64	0.48	0.59

Table 4. The table reports the score attributed to each sensor of typology PH and wc for each set of calibrated parameters over the whole simulation period.

	#1 (PH) MPS6-MPC1-4.6m	#2 (PH) MPS6-MPC3-6.2m	#3 (PH) MPS6-SPC1-7m	#4 (PH) T8-TC2-8m	#5 (wc) GS3-MPC1-4.5m	#6 (wc) GS3-MPC3-6.4m	#7 (wc) GS3-SPC2-7.1m
Set1	137.6	−0.52	1.41	0.56	6.23	11.15	0.84
Set2	96.3	−0.39	1.50	0.59	1.73	−14.80	−2.37
Set3	2.8	−0.30	−0.20	0.01	0.62	15.29	0.85
Set4	139.5	−0.66	1.63	0.78	6.48	5.06	2.43
Set5	108.0	−0.55	1.51	0.62	10.53	17.45	−0.25

Table 5. The table reports the calibrated VGM parameters of the best 5 indirect simulations.

Dataset		$\theta_r(m^3/m^3)$	$\theta_s(m^3/m^3)$	$\alpha(1/m)$	n(-)	Ks(m/s)	l(-)
Set 0		0.0040	0.3970	0.8500	1.3040	1.50e−6	0.5
Set 1	PH	0.0040	0.5637	0.8500	1.3040	1.50e−6	0.5
Set 2	PH	0.0040	0.5542	1.104	1.269	1.66e−6	0.5
Set 3	wc	0.0040	0.3739	0.8500	1.3040	1.50e−6	0.5
Set 4	PH + wc	0.0040	0.3501	1.161	1.243	9.95e−7	0.5
Set 5	PH + wc	0.0040	0.3871	1.333	1.340	1.50e−6	0.5

The procedure of I.A. applied has proven valuable in the estimation of all the considered datasets, both during floods and off-flood periods. Nevertheless, the enhancement is not equal between different datasets and even between sensors belonging to the same dataset. An enhancement of all the considered sensors (i.e., observation points) together failed (e.g., any set of parameters enhances Sensor #2, while almost all of them enhance

Sensor #1). The final choice remains with the modeler according to the aim of the calibration; thus a subjective component cannot be eliminated completely from the parameters optimization procedure.

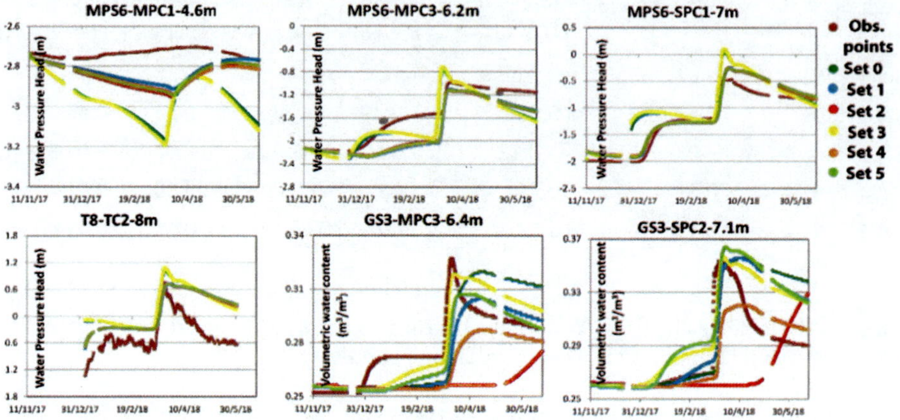

Fig. 3. Graphs reporting the trend with time of the considered Sets of calibrated parameters.

5 Conclusions

I.A. has proven to be a valuable help in parameter calibration of transient seepage models in presence of a multi-layered domain and a long simulation period. The use of different typologies of weighing distribution attributed to observation points enables the modeler to concentrate the calibration efforts on specific simulation periods or to give a different relevance to specific observation data or nodes. The use of a quantitative methodology results of primary importance in a first phase of evaluation of indirect simulation performance. A chosen set of metrics/indices can be applied either to the whole simulation period or to a restricted one, to the whole set of observation nodes or only to a few, providing the modeler all the necessary information to select the best sets of parameters. The qualitative methodology can be subsequently applied to the restricted pool of indirect simulations. For the specific case study herein presented, their combined use resulted successful for the selection of the best performing indirect simulation, whose choice should be always guided by the main purposes of the modelling.

References

1. Bertolini, I.: A methodological approach for the performance optimization of transient seepage models through inverse analysis. Ph.D Thesis (2021)
2. Gragnano, C.G., Rocchi, I., Gottardi, G.: Field monitoring and laboratory testing for an integrated modelling of river embankment under transient conditions. J. G. Geoenviron. Eng. **147**(9), 05021006 (2021)

3. Gragnano, C.G., Bertolini, I., Rocchi, I., Gottardi, G.: On the stability of a fully instrumented river embankment under transient conditions. In: Proceedings of the CNRIG2019, pp 369–378 (2019)
4. Simunek, J., Hopmans, J.W.: Parameter optimization and nonlinear fitting. In: Dane J.H., Topp G.C. (eds.) Methods of Soil Analysis, part 4 Physical Methods (2002)
5. Bertolini I., Gragnano, C.G., Gottardi, G.: On the use of inverse analysis for the estimation of soil hydraulic and retention parameters from monitoring data of a river embankment. In: Proceedings of the ISGSR 2022, Newcastle, Australia, 14–16 December (2022)
6. Krause, P., Boyle, D., Base, F.: Comparison of different efficiency criteria for hydrologic models. Adv. Geosci. **5**, 89–97 (2005)

A 1D Simplified Approach for Liquefaction Potential Evaluation of Soil Deposits

Gabriele Boccieri[1(✉)], Domenico Gaudio[2], and Riccardo Conti[1]

[1] Università Niccolò Cusano, Rome, Italy
gabriele.boccieri@unicusano.it
[2] Sapienza Università di Roma, Rome, Italy

Abstract. Build-up of seismic-induced pore water pressures in saturated sandy soils and the resulting reduction of effective stresses may lead to dramatic consequences. Indeed, as observed during several seismic events occurred over the last decade (Tohoku, Japan and Christchurch, New Zealand 2011; Emilia, Italy 2012; Palu, Indonesia 2018), severe damage due to liquefaction has caused both economic and environment-wise adverse impacts. Therefore, the development of a reliable although simplified tool for the assessment of liquefaction risk may be favorably perceived both in the Academia and in the current practice.

In this framework, the paper presents an improvement of the uncoupled method originally proposed by Seed *et al.* [1], where the excess pore water pressures induced by seismic loading under partially-drained conditions were evaluated. In their work, the Authors modified the well-known Terzaghi one-dimensional consolidation equation by adding a source term, which represents the rate of excess pore pressures generated under fully-undrained conditions. The governing equation is hereby solved using the Finite Difference Method implemented in a homemade *Matlab* script, taking into account the drainage conditions related to soil layering and possible filtering of the input motion caused by soil stiffness degradation, which in turn is induced by the excess pore pressure build-up. The proposed implementation is validated against the results of fully-coupled 1D FE analyses carried out with the Finite Element code *Plaxis* 2D, where the response of liquefiable sandy layers is reproduced through the advanced constitutive model SANISAND [2].

Keywords: Liquefaction · Simplified Approach · Numerical Analyses

1 Introduction

As recognized from post-earthquake surveys (Niigata, Japan 1964; Christchurch, New Zealand 2011; Palu, Indonesia 2018), liquefaction phenomena, which are triggered by seismic-induced excess pore water pressures in saturated sandy soils, may result in catastrophic consequences. This topic was addressed by several researchers over the last decades, who aimed at providing reliable tools for the design practice. Among these, rigorous numerical methods, which make use of advanced constitutive models to reproduce the nonlinear soil behaviour, are typically too onerous, difficult to calibrate and

A. Ferrari et al. (Eds.): CNRIG 2023, SSGG, pp. 410–418, 2023.
https://doi.org/10.1007/978-3-031-34761-0_50

time-consuming. As a result, the liquefaction potential of a given soil deposit is usually assessed using a *decoupled* approach, where the seismic loading is computed through a total-stress Site Response Analysis (SRA), while the earthquake-induced excess pore water pressures are estimated based on semi-empirical relationships.

Starting from the work by Seed *et al.* [1], Boccieri *et al.* [3] developed a *Matlab* [4] routine, which implements the *decoupled* approach using the Finite Difference Method (FDM). This paper introduces two relevant modifications with respect to [3]: (i) the influence of excess pore water pressures on the signal propagation through the soil column is taken into account in a simplified fashion; (ii) a more realistic hypothesis is introduced for the time distribution of equivalent cycles, overcoming the assumption of uniform distribution adopted in the original work.

The new approach is validated against fully-coupled Finite Element (FE) analyses, carried out with *Plaxis* 2D [5], where the mechanical behaviour of the liquefiable sand is simulated through the advanced constitutive model SANISAND [2].

2 Decoupled Approach

The proposed method is based on the *decoupled* approach outlined by Seed *et al.* [1], which basically can be split in two stages: (1) a total-stress 1D SRA to compute accelerations and shear stresses along the soil column; and (2) a consolidation analysis to evaluate the seismic-induced excess pore water pressures, *u,* within the soil deposit. Regarding the second phase, they added a source term to the classical 1D consolidation equation [6], which then reads:

$$\frac{\partial u}{\partial t} = c_v \frac{\partial^2 u}{\partial z^2} + \frac{\partial u_g}{\partial t} \tag{1}$$

where c_v is the consolidation coefficient, and $\partial u_g/\partial t$ represents the rate of the excess pore pressure build-up under fully-undrained conditions, which can be computed based on the results of cyclic undrained tests.

The *decoupled* approach underlying the method allows to compute the pore water pressure build-up in partially-drained conditions via Eq. (1), after assessing the seismic-induced shear stresses through a preliminary total-stress 1D SRA. The source term is linked to the irregular shear stress time history, which in turn is converted to an equivalent cyclic loading with constant amplitude, $\tau_{eq} = 0.65\ \tau_{max}$ (where τ_{max} is the maximum shear stress), equivalent number of cycles N_{eq}, and duration T_d. Hence, the generative term can be rewritten as:

$$\frac{\partial u_g}{\partial t} = \frac{\sigma'_{v0}}{N_L} \cdot \frac{dr_u}{dr_N} \cdot \frac{dN}{dt} \tag{2}$$

where $r_u = u_g/\sigma'_{v0}$ is the pore pressure ratio; $r_N = N/N_L$ is the cyclic ratio; N is the n-th cycle of the loading; and N_L is the number of cycles needed to trigger liquefaction.

Regarding the dissipative term in Eq. (1), $c_v \cdot \partial^2 u/\partial z^2$, the degradation of c_v due to the excess pore water pressures induced by the seismic event was included in the model using a standard empirical equation for sands, in the form [7]:

$$G_0 = F(e) \cdot (p')^{0.5} \tag{3}$$

where G_0 is the small-strain shear stiffness, $F(e)$ is a function of the void ratio, e, and p' is the current mean effective stress, which depends on the actual value of the pore water pressure.

The assumptions introduced for the definition of the curve r_u-r_N, and for the evaluation of N_L and N_{eq}, are discussed in the following, together with an indication for the calibration of the model parameters.

2.1 Excess Pore Water Pressures Relationship and Cyclic Resistance Curve

The r_u-r_N function provides information on the development of excess pore water pressures during an undrained cyclic loading. In this study, a power function was considered:

$$r_u = \chi \cdot (r_N)^{\theta} \tag{4}$$

where χ and θ are two curve-fitting parameters to be determined from cyclic laboratory tests. The functional form of Eq. (4) was chosen because not only it provides a good fit to the experimental data, but it is also easy to differentiate.

At a given depth within the liquefiable sand layer, the number of cycles needed to trigger liquefaction, N_L, can be obtained from a standard laboratory cyclic resistance curve CSR-N_L, where $CSR = \tau_{eq}/\sigma'_{v0}$ is the cyclic stress ratio, σ'_{v0} is the geostatic vertical effective stress and τ_{eq} is related to the maximum shear stress, τ_{max}, induced by the seismic event. The following equation was considered in this work to fit a given CSR-N_L set of experimental data:

$$CSR = CSR_t + \beta \cdot N_L^{-\eta} \tag{5}$$

where CSR_t is the threshold below which no liquefaction occurs when the soil sample is loaded cyclically in undrained conditions.

2.2 Equivalent Cyclic Loading

As mentioned, the irregular shear stress time history induced by the earthquake at a given depth is converted to an equivalent cyclic loading, defined by a constant amplitude τ_{eq}, a number of equivalent cycles N_{eq}, and duration T_d. Following [8], N_{eq} can be evaluated considering the CSR-N_L curve as the locus of same damage level (*i.e.* initial liquefaction), as

$$N_{eq} = \sum_{i=0}^{T_{end}} N_i \cdot X_i \; X_i = \begin{cases} \left(\frac{|CSR_{0.65}|-CSR_t}{|CSR_i|-CSR_t}\right)^{-1/\eta} & \text{if} \, |CSR_i| > CSR_t \\ 0 & \text{if} \, |CSR_i| \le CSR_t \end{cases} \tag{6}$$

where $CSR_{0.65} = \tau_{eq}/\sigma'_{v0}$, $CSR_i = \tau_i/\sigma'_{v0}$, while N_i and T_{end} are the number of cycles with amplitude τ_i and the duration of the input signal, respectively. The equivalent number of cycles was computed through the peak-counting method [8].

Seed *et al.* [1] considered a uniform distribution of the number of cycles over the loading duration T_d, where T_d is the strong motion duration of the input signal. Although simple, this assumption is not representative of the effective energy content distribution over the signal duration. Therefore, the cumulative number of cycles $N(t_k)$ up to the time $t = t_k$ was considered in this work, defined as:

$$N(t_k) = \sum_{i=0}^{k} N_i \cdot X_i \tag{7}$$

Consistently, the derivative *dN/dt* is no longer constant in Eq. (2).

With the aim of computing Eq. (6) and Eq. (7), the irregular shear stress loading at a given depth was obtained by equilibrium from the SRA acceleration time histories, assuming an in-phase acceleration of the 1D soil column.

2.3 Influence of Excess Pore Water Pressures on the Frequency Content of the Earthquake-Induced Soil Accelerations

Soil stiffness degradation due to pore water pressures build-up can modify the frequency content of seismic waves propagating through a saturated sand layer. This was taken into account with an iterative procedure which was implemented in this work to low-pass filter the SRA accelerations consistently with the computed excess pore water pressures. An equivalent natural frequency f_{0eq} of the soil column was introduced, defined as:

$$f_{0\,eq} = \frac{V_{s\,eq}}{4 \cdot H} \tag{8}$$

where H is the total height of the 1D soil column and V_{seq} is the equivalent shear wave velocity. The latter was computed at the time instant t_{95}, corresponding to which $N(t) = 0.95 \cdot N_{eq}$ in the middle of the soil column, as:

$$V_{s\,eq} = \frac{H}{\sum_{i=1}^{n} \frac{h_i}{V_{si}}} \tag{9}$$

where h_i and $V_{si} = (G_i(p')/\rho_i)^{0.5}$ are the thickness and shear wave velocity of the ith sublayer, respectively, and $n = H/h_i$. Based on a preliminary study, a filter frequency $f_{cut} = 3 \cdot f_{0eq}$ was defined. Within the iterative procedure, the value of f_{cut} computed at the jth step is used to low-pass filter the SRA accelerations used to calculate the excess pore water pressures at the $(j + 1)^{th}$ step.

3 Comparison with Coupled FEM Analyses

The proposed FDM *decoupled* approach was validated against the results of *coupled* dynamic FE analyses, carried out with the software *Plaxis* 2D CE v20 [5]. Figure 1a-b shows the 1D soil column considered in the analyses with the profiles of initial shear waves velocity V_{s0} and consolidation coefficient c_{v0}. The soil column is comprised of two layers: the bottom one is a liquefiable sand layer, whereas the top one was modelled as a non-liquefiable layer, consisting of either a gravelly, or a sandy or a clayey crust. The mechanical properties of the soil layers are listed in Table 1, while Fig. 1c-d shows the acceleration time history and Fourier Amplitude spectra of the input signal, corresponding to the Northridge (1994) earthquake, low-pass filtered at 10 Hz.

The mechanical behaviour of the deepest liquefiable layer was reproduced with the SANISAND constitutive model, assuming the parameters reported for Toyoura sand [2], while the *Hardening Soil model with Small-Strain stiffness* (*HSSmall* [9]) was adopted for the top layer. Table 2 reports the relevant parameters for the *HSSmall* model, calibrated based on available literature data.

Table 1. Soil layer mechanical properties. γ_{sat} = saturated soil weight; I_p = plasticity index; e_{min} and e_{max} = minimum and maximum void ratios, respectively; e = void ratio; K_0 = at rest earth pressure coefficient; c' = cohesion; ϕ' = friction angle; k = hydraulic conductivity

Soil	γ_{sat} (kN/m^3)	I_p (%)	e_{min} (-)	e_{max} (-)	e (-)	OCR (-)	K_0 (-)	c' (kPa)	ϕ' (°)	k (m/s)
clay	20	50	-	-	-	1	0.609	0	23.00	$1 \cdot 10^{-6}$
sand	20	-	0.597	0.977	0.650	-	0.500	0	30.00	$5 \cdot 10^{-4}$
gravel	20	-	0.435	0.923	0.740	-	0.500	0	30.00	$1 \cdot 10^{-2}$
liquef. Sand	19	-	0.597	0.977	0.825	-	0.483	0	31.15	$5 \cdot 10^{-4}$

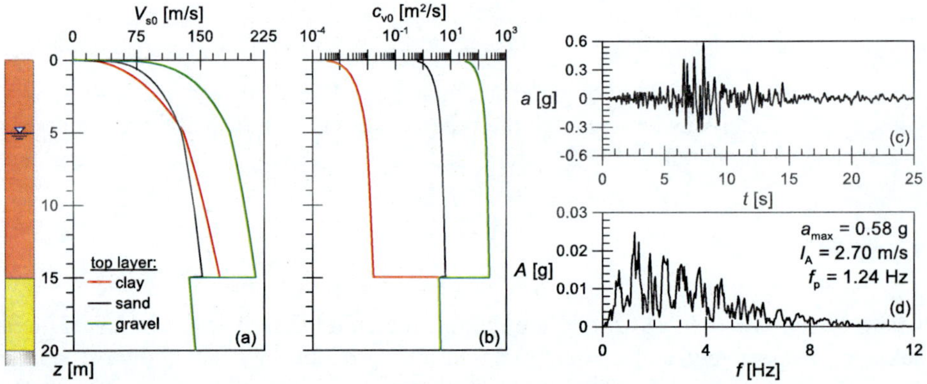

Fig. 1. Profiles of initial shear waves velocities (a) and initial consolidation coefficients (b) in the soil column, (c) input signal, and (d) Fourier Amplitude spectra of the Northridge (1994) earthquake (a_{max} = maximum acceleration; I_A = Arias Intensity; f_p = dominant frequency).

3.1 Calibration of Model Parameters for the Decoupled Approach

To have a consistent comparison between the results of the fully *coupled* FEM analyses and those provided by the *decoupled* FDM method, model parameters required for the simplified approach were calibrated using a series of undrained and drained numerical cyclic shear tests carried out with the *Plaxis Soil Test* tool.

As for the r_u-r_N (Eq. (4)) and the CSR-N_L (Eq. (5)) curves, undrained cyclic shear tests were simulated with a static vertical effective stress σ'_{v0} = 220 kPa, representative of the effective stress state at the centre of the liquefiable layer. The excess pore pressure equation r_u-r_N was calibrated applying *CSR* values in the range 0.25 - 0.05, and the cyclic resistance curve *CSR*-N_L was obtained considering the triggering of liquefaction at $r_u = 0.9$. Figure 2 shows the two curves obtained with the best-fitting values of the coefficients χ and θ (Fig. 2a), and CSR_t, β and η (Fig. 2b).

Table 2. *HSSmall* model parameters assumed for the non-liquefiable layer

Soil	G_0^{ref} (MPa)	m (-)	$\gamma_{0.7}$ (-)	E_{ur}^{ref} (MPa)	ν_{ur} (-)	$E_{ur}^{ref}/E_{50}^{ref}$ (-)	$E_{50}^{ref}/E_{oed}^{ref}$ (-)	R_f (-)
clay	51.8	0.84	$1.0 \cdot 10^{-3}$	46.1	0.2	3.0	1.0	0.9
sand	47.1	0.50	$2.4 \cdot 10^{-4}$	56.5	0.2	3.0	1.0	0.9
gravel	94.7	0.44	$2.4 \cdot 10^{-4}$	113.6	0.2	3.0	1.0	0.9

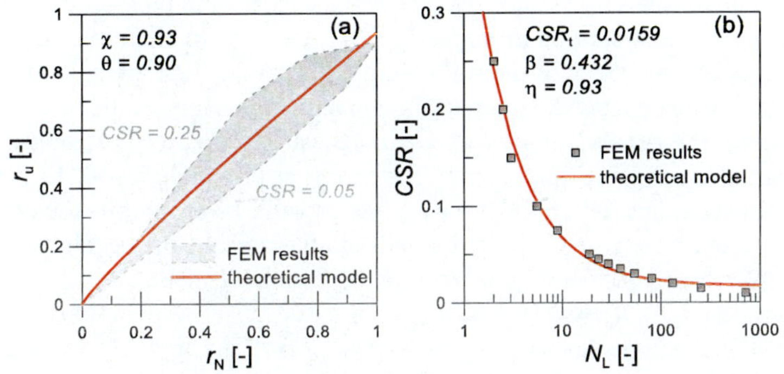

Fig. 2. (a) Excess pore pressure ratio and (b) cyclic resistance curves adopted in the *decoupled* approach.

The 1D SRAs were carried out using the nonlinear soil model proposed by Conti *et al.* [10], which is defined by six parameters: the shear strength, τ_{lim}, and the small-strain shear modulus, G_0; *a* and *b*, defining the shear modulus decay curve; and *c* and *d*, defining the hysteretic damping curve. Parameters *a*, *b*, *c*, and *d* were calibrated based on the results of numerical drained cyclic shear tests, carried out at a static vertical effective stress $\sigma'_{v0} = 150$ kPa, applying a 10 cycles/s sinusoidal shear strain time history. The resulting shear modulus decay and damping curves are plotted in Fig. 3, together with those adopted in the nonlinear total stress SRAs.

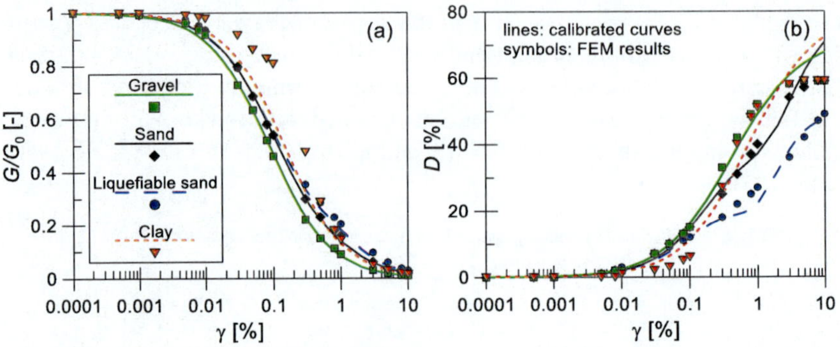

Fig. 3. (a) Shear modulus decay and (b) damping curves adopted in the total stress SRAs (line), calibrated against FEM simulations (symbols).

3.2 Results

Figure 4 compares the space-time contours of the excess pore pressure ratio r_u obtained with the *coupled* (FE) and the *decoupled* (FD) approach. In all cases, complete liquefaction occurs at the bottom of the soil column, due to the high intensity of the seismic event and the distance from the drainage boundary. Conversely, at the interface between the two layers, the development of excess pore water pressures is strongly influenced by the hydraulic condition imposed by the top, non-liquefiable, soil. In the first configuration, the presence of a highly-permeable gravelly layer inhibits the occurrence of complete liquefaction at the top of the underlying sand. In the second configuration, the excess pore water pressures generated in the liquefiable sand spread out through the upper sandy layer. As a result, the diffusion process causes a reduction in strength and stiffness also in the upper layer during the early stages of the applied earthquake. Finally,

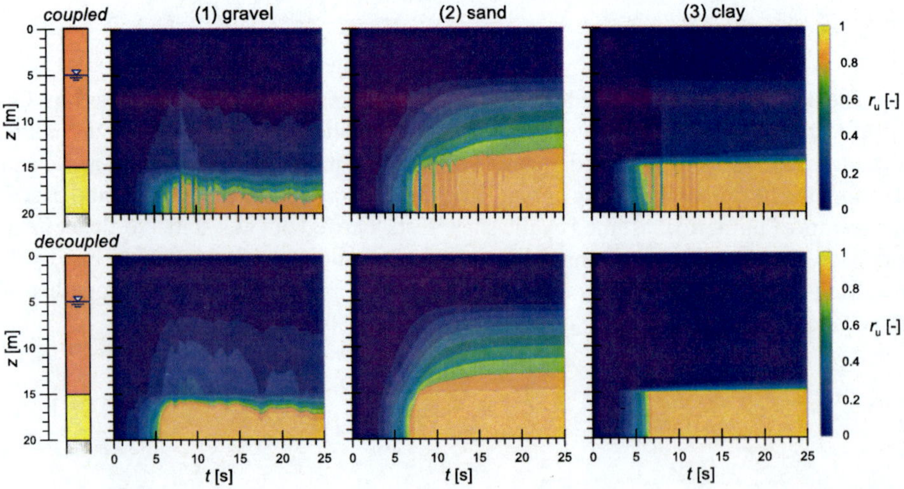

Fig. 4. Comparison of contours of r_u obtained with the *coupled* and *decoupled* analyses.

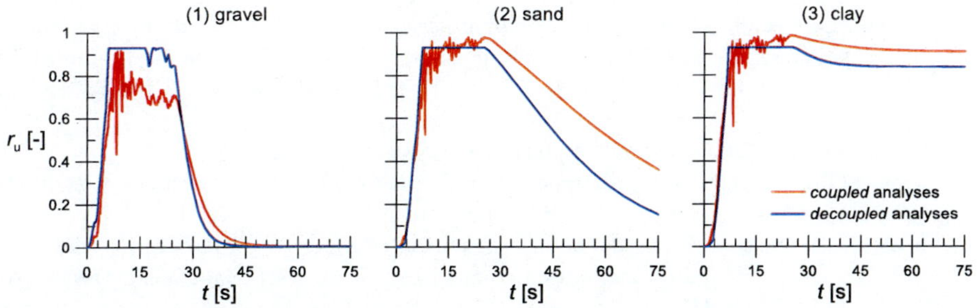

Fig. 5. Comparison of time histories of r_u in the middle of the liquefiable layer ($z = 17.50$ m) obtained with the *coupled* and *decoupled* analyses. The time interval was considered three times that of the seismic event duration.

when the upper layer consists of a low-permeability clayey soil (third configuration), the deepest layer completely liquefies and, as expected, a negligible redistribution of excess pore water pressure takes place in the top soil. Furthermore, Fig. 5 shows the time histories of the excess pore pressure ratio r_u in the middle of the liquefiable sand layer, obtained with FD and FE analyses. The time interval was extended to 75 s, to compare the purely dissipative phase following the seismic event. The importance of the hydraulic conductivity of the top layer is also highlighted in this figure, which shows a decreasing dissipation rate going from the case with a gravelly layer to the one with clayey crust. The *decoupled* approach provides slightly faster dissipation processes than the FE analyses but, despite its simplicity, the results are in a good agreement with the FE ones.

4 Conclusions

In this paper, a *decoupled* approach was presented to provide a tool for assessing the liquefaction hazard of sandy soils, based on the work by Seed *et al.* [1]. An equivalent natural frequency of the soil column was introduced, depending on the excess pore water pressures, so as to consider the shift in frequency content of the propagated signal due to the soil stiffness degradation. Moreover, the assumption of uniform time distribution of the number of equivalent cycles was replaced with a more realistic energy hypothesis. The proposed approach, implemented in a *Matlab* routine through the FDM, was validated against the results of *fully-coupled* dynamic FE analyses. The results of FDM and FE analyses turned out to be in a very good agreement, providing confidence in the predictive capability of the proposed simplified approach.

References

1. Seed, H.B., Martin, P.P., Lysmer, J.: The Generation and Dissipation of Pore Water Pressures During Soil Liquefaction. University of California, College of Engineering (1975)
2. Dafalias, Y.F., Manzari, M.T.: Simple plasticity sand model accounting for fabric change effects. J. Eng. Mech. **130**(6), 622–634 (2004)

3. Boccieri, G., Gaudio, D., Conti, R.: A simplified method for the estimation of earthquake-induced pore pressure. In: Geotechnical Engineering for the Preservation of Monuments and Historic Sites III. CRC Press, pp. 812–822 (2022). https://doi.org/10.1201/9781003308867-62
4. Mathworks Inc. Matlab version 9.10.0 (R2021a). Natick, Massachusetts (2021)
5. Bentley. Plaxis 2D CE v20 – Reference Manual. Delft University of Technology. Delft, The Netherlands. (2020)
6. Terzaghi, K.: Die berechnung der durchassigkeitsziffer des tones aus dem verlauf der hydrodynamischen spannungs. erscheinungen. Sitzungsber. Akad. Wiss. Math. Naturwiss. Kl. Abt. 2A **132**, 105–124 (1923)
7. Wichtmann, T., Triantafyllidis, T.: Effect of uniformity coefficient on G/Gmax and damping ratio of uniform to well-graded quartz sands. J. Geotech. Geoenv. Eng. **139**(1), 59–72 (2013)
8. Hancock, J., Bommer, J.J.: The effective number of cycles of earthquake ground motion. Earthq. Eng. Struct. Dyn. **34**(6), 637–664 (2005)
9. Benz, T.: Small-strain stiffness of soils and its numerical consequences. Ph.D. thesis, University of Stuttgart (2006)
10. Conti, R., Angelini, M., Licata, V.: Nonlinearity and strength in 1D site response analyses: a simple constitutive approach. Bull. Earthq. Eng. **18**, 4629–4657 (2020)

Simulation of Rainfall-Induced Landslides from Small to Large Displacements with an Efficient Sequential Use of FEM and MPM

Francesca Ceccato[1](✉), Meng Lu[2], Matteo Camporese[1], Davide Vallisari[1], and Lorenzo Brezzi[1]

[1] DICEA, University of Padua, Padua, Italy
francesca.ceccato.1@unipd.it

[2] Department of Geotechnical Engineering, Tongji University, Shanghai, China

Abstract. The Finite Element Method (FEM) is widely used to simulate geotechnical engineering problems at small deformations, but it typically fails to capture large displacements and landslide runout. In contrast, the Material Point Method (MPM) is well suited for large displacement problems, but when using explicit time integration schemes, it may be computationally inefficient for long-time processes. In this paper, the two methods are used sequentially to simulate a laboratory experiment of rainfall-induced landslide from small to large displacements and a novel methodology to map results from FEM to MPM is proposed. The numerical results are compared with experimental evidence in which displacements are determined using digital image correlation techniques.

Keywords: MPM · FEM · landslide · rainfall

1 Introduction

Rainfall-induced landslides are common geohazards and have caused extensive casualties and property losses worldwide. The phenomenon can be subdivided in an initiation stage, in which the decrease of suction reduces the soil shear strength and a slip surface develops through the soil mass with little overall movement, and a post-failure stage characterized by a sudden acceleration of the failed soil mass. The material can also fluidize and propagate for long distances reaching high velocities before impact against obstacles or deceleration and deposition in a flat zone (arrest stage).

The initiation phase can be investigated with Finite Element Methods (FEM), but the post-failure behavior cannot be captured because of issues with element distortions. The propagation stage can be successfully simulated with depth-averaged models (DA), but their reliability is questionable in the arrest stage, where vertical decelerations are not negligible and 3D models are better-suited [1]. To simulate the 3D post-failure stage, including the arrest phase, the use of Eulerian methods, such as Lattice-Boltzmann Methods (LBMs) [2], and particle-based methods, such as Particle Finite Element Methods

A. Ferrari et al. (Eds.): CNRIG 2023, SSGG, pp. 419–426, 2023.
https://doi.org/10.1007/978-3-031-34761-0_51

(PFEM) [3], Smoothed Particles Hydrodynamics (SPH) methods [4], and Material Point Methods (MPMs) [5], is becoming, at least in the scientific community, quite common. Among these approaches, MPM is increasingly popular, thanks to the development of multiphase formulations for saturated and unsaturated soils, see e.g. [6–9], its ease of use, due to similarities with FEM, and the availability of open source software, e.g. *Anura3D* (www.anura3d.com), *AMPLE* (https://wmcoombs.github.io/), *Krathos* (https://github.com/KratosMultiphysics/Kratos).

Most of the available multiphase MPM formulations are dynamic, i.e. account for inertial terms, and use explicit time discretization algorithms, which need small time increments to ensure the numerical stability [8]. For these reasons, these approaches are well suited to simulate fast dynamic problems such as slope collapse and impact problems, but they are computationally inefficient to simulate long time processes such as consolidation and infiltration processes or very long propagation of debris flows.

In order to exploit the potentialities of these numerical tools, a new strategy for the initialization of the MPM simulations is proposed. The algorithm maps field variables from a generic distribution of points, e.g. the results of a previous analysis, to the initial position of material points (MPs) for the subsequent MPM simulation.

The methodology is illustrated in Sect. 2 and is applied to the simulation of a laboratory test of rainfall induced landslide described in [10]. The main features of the experiment are summarized in Sect. 3. FEM is used to analyze the rainfall infiltration up to the initiation stage and then the obtained water pressures at failure are mapped to the MPM model to analyze the large deformation post-failure stage. The features of the numerical model are described in Sect. 4. The simulation results are presented in Sect. 5.

2 Numerical Strategy

The standard FEM used in geotechnical practice applies a Lagrangian framework. Since all state variables are computed on the element nodes or Gauss point, which deform with the soil body, FEM cannot tackle large displacement problems because of mesh distortion. In contrast, MPM can simulate large displacement by material points moving through the mesh. On the other hand, FEM is largely available in commercial software and it has a higher computational efficiency compared with the MPM. It is evident that FEM is optimal to simulate small deformation problems and long-time processes while MPM is good for short-time large displacement problems.

In order to exploit the potentialities of both methods to simulate a rainfall-induced landslide, seepage analyses with the FEM software SEEP/W available in Geostudio 2012 are performed to simulate the long-time infiltration process, and fully coupled multiphase MPM simulations with the open source software Anura3D are performed to simulate the post-failure behavior. The two-phase one-point MPM formulation for unsaturated soils presented in [9] is used. This formulation has been successfully used to simulate collapse of water retaining structures [11] and wetting-induced landslide of layered slopes [12]. It considers as primary variables the acceleration of the liquid and solid phases. The governing equations are the momentum balance equations of the liquid and the mixture, the mass balance of the solid skeleton and the liquid. The soil is discretized with one set of material points (MPs) that represent a portion of the mixture and move according to the displacement of the solid phase, but carry the information of both solid and liquid.

The failure time ($t_{failure}$) at which information is mapped from FEM to MPM is determined based on the value of the factor of safety (FS) calculated with a slope stability analysis performed with SLOPE/W at each time increment of the seepage analysis.

The slope stability analysis applies the Mongenstern-Price method [13]. The onset of failure can be identified when FS becomes slightly lower than 1. Total stresses at the failure time are computed with SIGMA/W.

A novel methodology to map quantities from a FEM to MPM is proposed as explained in Sect. 2.1. The proposed mapping procedure is very general and can have a wide applicability in all the cases where field variables, known only at a discrete set of points, e.g. from a previous simulation or even monitoring data, have to be mapped to the MPs to carry out an MPM simulation.

2.1 Mapping Procedure Between FEM and MPM

The results of the FEM analysis at the failure time establish the initial conditions of the MPM model for the large displacement analysis.

Since spatial variables are available at the nodes or gauss points and their coordinates do not coincide with the position of the MPs in the MPM model, a spatial interpolation procedure is implemented based on the kernel interpolation method. The principle of the kernel interpolation is to predict an unknown point based on the observed points close to the unknown point. The kernel interpolation can be written using the following equation:

$$f(x) = \frac{\sum_{i=1}^{m}\left[w\left(\frac{d_i}{h}\right)f(x_i)\right]}{\sum_{i=1}^{m}\left[w\left(\frac{d_i}{h}\right)\right]} \tag{1}$$

where *f(x)* is the unknown point at *x*; $f(x_i)$ is the observed point at x_i; *w* is the kernel function; d_i is the distance between *x* and x_i; *h* is the smoothing parameter; and *m* is the number of observed points. Let *R* denote d_i/h. The kernel function *w* is expressed based on the cubic spline (Eq. 2)

$$w(R) = \alpha \cdot \begin{cases} \frac{2}{3} - R^2 + \frac{R^3}{2} & 0 \le R < 1 \\ \frac{(2-R)^3}{6} & 1 \le R < 2 \\ 0 & R \ge 2 \end{cases} \tag{2}$$

where the values of α for 1, 2, 3-dimensional space adopt, respectively, $1/h$, $15/7\pi h^2$, and $3/2\pi h^3$. Preliminary analyses showed that the optimum smoothing parameter *h* is 0.9*l*, where *l* is the reference mesh size.

3 Experiment Description

This experiment, conducted at the University of Padova, considered a slope failure triggered by rainfall infiltration. The device used to simulate shallow landslides consists of a reinforced concrete box 6 m long and 2 m wide; the height varies linearly from

3.5 m to 0.5 m, such that a maximum sloping angle of 32° can be assigned to the soil (Fig. 1). The slope toe consists of a porous wall made of hollow bricks, allowing water to drain from the subsurface. The artificial rainfall with a steady intensity of 150 mm/h is applied on the slope surface until the slope collapse. The slope model consists of two layers. The top layer is a loose silty sand, while the bottom layer is a compacted sandy clay.

One camera recorded the experiment, and the video is used to determine the displacement field of the soil surface for comparison with the numerical results. Initially, the frames are extracted from the video, and a mesh containing the coordinates of the 1774 patches which will be used for tracking is determined on the initial image. A PIV algorithm [14] allows us to identify the position of these patches in the second image, in order to determine the field of Eulerian shifts between the first frame and the next. The procedure is then repeated on all frames until the collapse is completed. An interpolation of the Eulerian field allows one to derive the Lagrangian displacements on the image plane, in pixels (black arrows in Fig. 2). Finally, the displacements are projected onto the inclined slope, allowing a metric measurement to be obtained (contour plot in Fig. 2). The displacements are assumed to lie on the initial flat slope surface, which is an acceptable assumption for the central portion of the moving mass.

The infiltration process evolved for more than 1.5 h before the failure of the sand layer. At roughly 1.25 h, subsurface flow initiated across the slope toe, consistent with the formation of a perched water table. The failure occurred suddenly without any warning signs and the collapse lasted about 3 s.

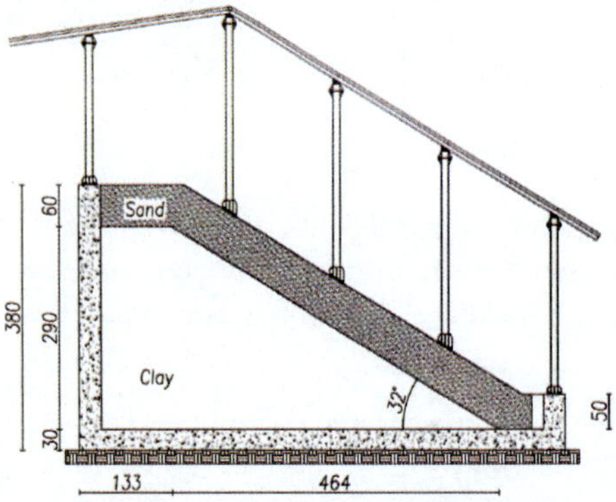

Fig. 1. Geometry, soil layers, and monitoring points of the experimental model (dimensions in cm)

Fig. 2. Final displacements calculated by PIV analysis: the black arrows show the directions, the contour plot the total quantities.

4 FEM and MPM Model

Since failure occurred in the superficial layer and permeability of clay is very low, only the sand layer is modelled and the interface with clay is assumed to be an impermeable boundary.

Figure 3a shows the FEM model of the slope. In this model, the bottom and left side boundary are impermeable, while the right side is a potential seepage face. A flux equal to 150 mm/h is applied on the slope surface to simulate the rainfall infiltration and ponding is not allowed. A triangular mesh is used, with a mesh resolution of 0.05 m.

The transient seepage analysis uses time increments of 0.01 h (36 s) and for each step a slope stability analysis is performed to calculate the FS.

Figure 3b shows the MPM model of the slope. Its hydraulic boundary conditions are consistent with the FEM model. Zero solid displacements are prescribed at the bottom boundary and roller boundary conditions are applied at the lateral surfaces and at the toe of the slope. The computational mesh is triangular. There are 3 MPs in each element of the slope domain.

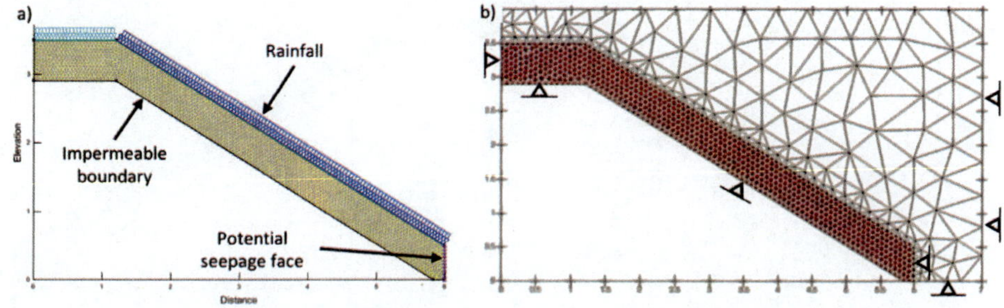

Fig. 3. (a) FEM model of the slope and (b) MPM model of the slope

Table 1 presents the values of soil parameters applied in the FEM and MPM models. In unsaturated conditions, the relation between volumetric water content and suction is given by the soil-water retention curve (SWRC) and the ratio between the hydraulic conductivity and the saturated one, i.e. k/k_{sat}, is given by the Hydraulic Conductivity Curve (HCC). The Van Genuchten model is applied and the SWRC and the HCC are described by Eqs. 3 and 4 respectively

$$\theta = \theta_r + \frac{\theta_s - \theta_r}{\left(1 + \left(\frac{s}{a}\right)^{\lambda}\right)^{m}} \tag{3}$$

$$\frac{k}{k_{sat}} = \frac{\left(1 - \left(\frac{s}{a}\right)^{\lambda-1}\left(1 + \left(\frac{s}{a}\right)^{\lambda}\right)^{-m}\right)^2}{\left(1 + \left(\frac{s}{a}\right)^{\lambda}\right)^{\frac{m}{2}}} \tag{4}$$

where s denotes the suction, θ_r is the residual water content, θ_s is the saturated water content, a and λ are curve fitting parameters, $m = (1 - \lambda)/\lambda$.

The hydraulic constitutive parameters (a, λ, θ_{sat}, θ_{res}, k_{sat} in Eqs. 3 and 4), porosity (n), density of solid grains (ρ_s), liquid density (ρ_L), and friction angle (ϕ) are derived from the original publication [10]; they were obtained by laboratory tests and numerical back analysis. Reasonable values for sandy soils are assumed for the elastic soil parameters (E = Young modulus, ν = Poisson ratio, K_w = water bulk modulus).

Table 1. Material parameters used in the numerical models

a	1.97kPa	K_W	45000 kPa
λ	2.73	E	2500 kPa
θ_{sat}	0.58	ν	0.3
θ_{res}	0.05	n	0.58
k_{sat}	$2.047 \cdot 10^{-4}$ m/s	ρ_S	2718 kg/m^3
φ	33°	ρ_L	1000 kg/m^3

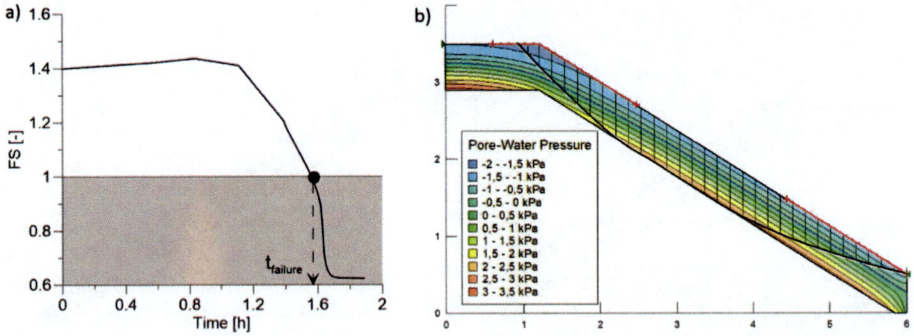

Fig. 4. (a) Factor of safety along time. (b) Critical failure surface at t = 1.56 h (FS = 1.009).

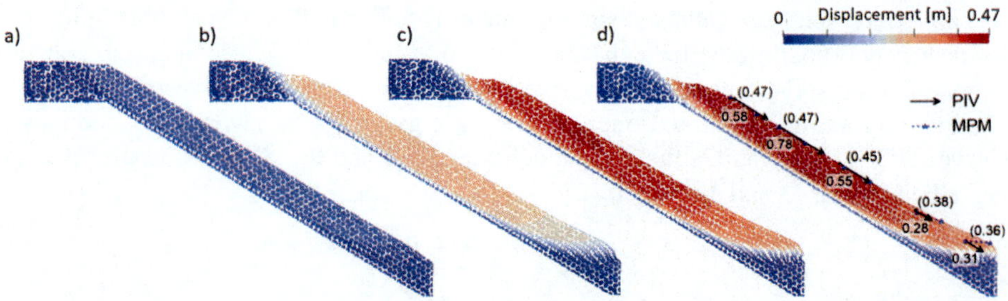

Fig. 5. Displacement at different times after failure a) t = 0.3 s, b) t = 0.8 s, c) t = 1.2 s, d) 3.0 s

5 Results

During wetting, the FS of the slope decreases reaching approximately 1 at 1.56 h (Fig. 4a). This can be assumed as the failure time ($t_{failure}$) and it is in good agreement with the experimental observation. The critical failure surface at t = 1.56 h is shown in Fig. 4b along with the pore pressure distribution.

At the failure time, the pore pressure distribution calculated with SEEP/W and the total stress calculated with SIGMA/W are mapped to MPM. In spite of small deformations being already present, these are neglected and zero displacements and strains are assumed at the beginning of the MPM simulation. After the onset of failure, the slope develops large deformations reaching a new stable configuration in about 3 s (Fig. 5). In Fig. 5d the black arrows indicate the experimental displacement obtained by PIV and the numbers below indicate the displacement magnitude, the blue dotted lines indicate the displacement of reference MPs and the numbers in brackets above indicate the magnitude. It can be observed that there is a relatively good agreement between experiment and simulation.

5.1 Effect of Failure Time

The time $t_{failure}$ at which information is mapped to MPM for the simulation of large displacement is somehow arbitrary, and it could affect the final slope configuration. Figure 6a shows the final slope surface shape at rest obtained with different failure times. Minor differences can be observed for failure times close to 1.56 s ($t_{failure} = 1.58$ s and 1.60 s), corresponding to FS between 1.001 and 0.939. Slightly larger displacements are obtained for $t_{failure} = 1.63$ s (FS = 0.817), in particular larger settlements on the crest and material accumulation at the toe.

The failure time has a much larger effect on the peak kinetic energy as shown in Fig. 6b. Increasing the failure time the kinetic energy increases significantly because the mapped pore pressures are higher, thus decreasing the available shear strength at the failure surface, resulting in higher sliding velocities.

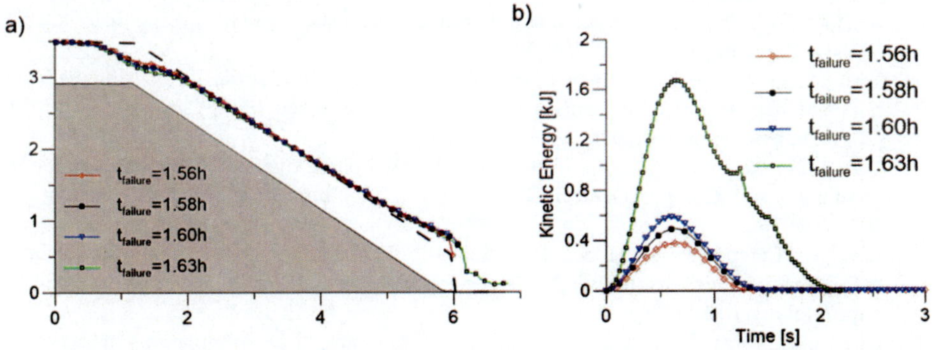

Fig. 6. (a) final slope surface and (b) kinetic energy for different failure times.

6 Conclusions

The paper proposes a new methodology to initialize MP information for large displacement MPM simulation. The methodology is based on the kernel interpolation technique and has the advantage of being easily adapted to initialize any field variable at the MPs. In this paper it is used to map the results of FEM seepage analyses of rainfall infiltration on a slope to MPM for simulating post-failure behavior. This allows us to exploit the potentialities of both methods, e.g. the efficiency of FEM for small deformations and the ability of MPM to simulate large displacements and highly dynamic problems such as landslide propagation.

The laboratory test presented by [10] is simulated, obtaining good agreement with the experimental results. Moreover, the effect of the selection of failure time is investigated showing that minor differences lead to negligible effects on the final shape of the slope, but significant effects are observed on the peak kinetic energy.

References

1. Hutter, K., Wang, Y., Pudasaini, S.P.: The savage-hutter avalanche model how far can it be pushed? Philos. Trans. A. Math. Phys. Eng. Sci. **363**, 1–17 (2005)
2. Pasqua, A., Leonardi, A., Pirulli, M.: Coupling Depth-Averaged and 3D numerical models for the simulation of granular flows. Comput. Geotech. **149**, 104879 (2022)
3. Cremonesi, M., Franci, A., Idelsohn, S., Oñate, E.: A State of the art review of the particle finite element method (PFEM). Arch. Comput. Meth. Eng. **27**(5), 1709–1735 (2020). https://doi.org/10.1007/s11831-020-09468-4
4. Pastor, M., Haddad, B., Sorbino, G., Cuomo, S., Drempetic, V.: A depth-integrated, coupled SPH model for flow-like landslides and related phenomena. Int. J. Numer. Anal. Meth. Geomech. **33**, 143–172 (2009). https://doi.org/10.1002/nag.705
5. Soga, K., Alonso, E., Yerro, A., Kumar, K., Bandara, S.: Trends in large-deformation analysis of landslide mass movements with particular emphasis on the material point method. Géotechnique. **66**, 248–273 (2016). https://doi.org/10.1680/jgeot.15.LM.005
6. Jassim, I., Stolle, D., Vermeer, P.: Two-phase dynamic analysis by material point method. Int. J. Numer. Anal. Meth. Geomech. Anal. Meth. Geomech. **37**, 2502–2522 (2013)
7. Bandara, S., Soga, K.: Coupling of soil deformation and pore fluid flow using material point method. Comput. Geotech. **63**, 199–214 (2015)
8. Yerro, A., Girardi, V., Ceccato, F., Martinelli, M.: Modelling unsaturated soils with the Material Point method. A discussion of the state-of-the-art. Geomech. Energy Environ. **32**, 100343 (2021). https://doi.org/10.1016/j.gete.2022.100343
9. Ceccato, F., Yerro, A., Girardi, V., Simonini, P.: Two-phase dynamic MPM formulation for unsaturated soil. Comput. Geotech. **129**, 103876 (2021). https://doi.org/10.1016/j.compgeo.2020.103876
10. Lora, M., Camporese, M., Troch, P.A., Salandin, P.: Rainfall-triggered shallow landslides: infiltration dynamics in a physical hillslope model. Hydrol. Process. **30**, 3239–3251 (2016). https://doi.org/10.1002/hyp.10829
11. Girardi, V., Yerro, A., Cecato, F., Simonini, P.: Modeling large deformations in levees and water retention structures with a Single-Point 2-Phase MPM approach. In: Proceedings of the Institution of Civil Engineers - Geotechnical Engineering, vol. 174, pp. 577–592 (2021). https://doi.org/10.1680/jgeen.21.00059
12. Girardi, V., Yerro, A., Simonini, P., Gabrieli, F., Ceccato, F.: Wetting induced instabilities in layered slopes: a material point method analysis. Eng. Geol. **313**, 106978 (2023). https://doi.org/10.1016/j.enggeo.2022.106978
13. Morgenstern, N.R., Price, V.E.: The analysis of the stability of general slip surfaces. Geotechnique **15**, 79–93 (1965)
14. Stanier, S.A., Blaber, J., Take, W.A., White, D.J.: Improved image-based deformation measurement for geotechnical applications. Can. Geotech. J. **53**, 727–739 (2016). https://doi.org/10.1139/cgj-2015-0253

G-PFEM Numerical Assessment of Rock Anchor Interface Properties on Pull-Out Capacity for Renewable Offshore Applications

Alessio Genco[1](✉), Matteo Oryem Ciantia[1], Michael Brown[1], Marco Previtali[1], Ana Ivanovic[2], and Nick Cresswell[3]

[1] School of Science and Engineering, University of Dundee, Dundee Scotland, UK
180026675@dundee.ac.uk

[2] School of Engineering, University of Aberdeen, Aberdeen Scotland, UK

[3] Swift Anchors, La Belle Esperance, The Shore, Edinburgh Scotland EH6 6QW, UK

Abstract. With the challenges faced by tidal stream generator application, it is necessary to develop a new generation of Rock Anchors (RAs) that are easy to install and efficient, with accurate, robust, and realistic design procedures. The harsh marine environment requires efficient anchoring solutions capable of sustaining the complex loading conditions experienced. Current RA technologies require simple and robust design methods to overcome the conservatism adopted in classical procedures. Therefore, advanced numerical methods have a crucial role to play in the enhancement of the RA design process. The Geotechnical Particle Finite Element Method (G-PFEM) is an open-source numerical tool suitable for large strain problems and is used herein to investigate the influence of steel-rock interface properties on RA pull-out. A large deformation constitutive model for structured materials, calibrated to replicate the behavior of calcarenite and sandstone, is used to simulate the RA response under axial loading. The Contact Domain (CD) technique is described and was adopted for this work. The parametric analysis considering various combinations of steel-rock interface friction angle (δ), the inclination of the lateral face of the anchor (α) and contact parameters was performed. The results highlight how interface friction affect performance in soft and hard rocks. Due to the large deformation capabilities and the non-local formulation of the numerical approach the RA failure mechanism is realistically investigated.

Keywords: G-PFEM · Offshore foundations · Rock Anchors

1 Introduction

The increasing deployment of offshore renewable energy is leading to the use of novel anchoring solutions. Nevertheless, the tidal environment can exert varying load conditions on floating devices making the deployment of reliable and cost-efficient solutions challenging. Considering that anchoring solutions may reduce the overall cost of the tidal energy systems [1] a better understanding of anchor optimization is crucial. In deep

A. Ferrari et al. (Eds.): CNRIG 2023, SSGG, pp. 427–434, 2023.
https://doi.org/10.1007/978-3-031-34761-0_52

water and hard rocky seabeds, Rock Anchors (RAs) represent an effective alternative to classic piled and gravity-based configurations. However, classic design procedures for RAs are highly conservative and require the assumption/identification of the expected failure mechanism in the rock [2, 3]. Most of the analytical methods available in literature involve simplification of the failure surface geometry and geotechnical parameters adopted to define the load capacity. For this reason, advanced numerical methods are an attractive alternative for RA design optimization. The numerical assessment performed indicate the substantial role of the RA geometry on the load capacity and failure mechanism. A comparison with small strain LA (Limit Analysis) and FEM codes reveals the overestimation of the numerical solution obtained by LA because of the associated flow rule assumption and dilatancy effect. In this study the G-PFEM (Geotechnical Particle Finite Element Method) [9] is adopted to investigate the influence of the rock-anchor interface properties and contact parameters on large deformation axial pull-out of groutless RAs. The numerical method presented is shown to be suitable to handle such large strain problem due to both the contact model and the remeshing technique implemented [4, 5]. Moreover, the non-local regularization adopted allows for mesh independent results despite the softening of the rock constitutive model used [5]. In particular, the large strain structured Modified Cam Clay (MCC) model is herein used to replicate the mechanical behaviour of a sandstone [6]. The rock - anchor interaction requires an accurate definition of the boundary domain. Here the efficiency and effectiveness of the contact model is investigated numerically. The results reveal the substantial influence of the interface friction angle, inclination of the lateral face of the anchor and the penalty parameter, which is related to the penalty contact algorithm adopted, on the load capacity. In particular, although the RA is designed to perform efficiently due to geometrical keying of the anchor head, it is shown that the pull-out capacity increases with the interface friction angle. The numerical method suggested is able to display accurately the failure mechanism.

2 The Structured MCC Model

The structured MCC model adopted here is a non-associative, isotropic hardening finite-deformation plasticity model for structured soils and weak rocks. As detailed in Monforte et al. (2019), the model is equipped with two stress-like internal variables, P_s (preconsolidation pressure, which accounts for the hardening/softening effects due to volumetric and deviatoric plastic strains) and P_t (bond strength, which quantifies the effects of cementation). A representation of the yield surface in the Kirchhoff stress invariants space *Q:P*, highlighting the role of the internal variables, is given in Fig. 1. The analytical expression of the yield and plastic potential are:

$$\begin{cases} f(\tau', P_s, P_t P_m) = (Q/M_f)^2 + P^*(P^* - \mathrm{P}_c^*) \\ g(\tau', P_s, P_t P_m) = (Q/M_g)^2 + P^*(P^* - \mathrm{P}_{\mathrm{c,g}}^*) \end{cases} \tag{1}$$

where $P' = tr(\tau')/3$ and $\mathrm{Q} = \sqrt{3\mathbf{J}_2}$ with τ' and $\mathbf{J}_2$ first and second effective Kirchoff stress tensor respectively, M_f controls the size of the yield surface and M_g the slope of the Critical state line in the triaxial plane. The intersections of the yield locus with positive and negative axis defined by:

$$P_c^* = P_t + P_s + P_m = P_s + (1 + c)P_t \quad (2)$$

where P_s acts as preconsolidation pressure and depends on both volumetric and deviatoric strain rates. The variables P_t and P_m (linear proportional through the model parameter c) account for the effects of interparticle bonding. Whereas the variable P^* is simply expressed by:

$$P^* = P^* + P_t \quad (3)$$

The evolution of the hardening variables is governed by the following expressions:

$$P_s = \rho_s P_s \left[tr(l^p +)\chi_s \sqrt{2/3} \| dev(l^p) \| \right] \quad (4)$$

$$\dot{P}_t = -\rho_t P_t \left[tr(l^p) + \chi_t \sqrt{2/3} \| dev(l^p) \| \right] \quad (5)$$

where $\mathbf{l}^p = \partial g / \partial \tau$; ρ_s, ρ_t, χ_s and χ_t are constitutive parameters which control the rate of degradation during plastic deformations. It must be noted that plastic volumetric compaction enables the increase of P_s whilst plastic dilatation produces a reduction of P_s. The stress ratio dilatancy expression for the structured MCC model follows the expression:

$$d = M_p \left(M_p / \eta^* - 1\right) \quad (6)$$

where $\eta^* = Q/P^*$. It should be noted that for very strong rocks P_t is small with respect to Q of the yield locus making $\eta^* \cong \eta$ for most part of the yield surface. The constitutive model has been calibrated following [7] using experimental data related to drained triaxial test on Berea sandstone from Wong et al. (1997). The experimental yield data points plotted in the triaxial plane are used to calibrate the yield locus dimension Fig. 1a while the experimental dilatancy results are used to calibrate M_g of the plastic potential (Fig. 1b). To calibrate P_s the approach described in [8] is used. The elastic parameters were calibrated using the elastic response of the drained triaxial test (Fig. 1c). Table 1 summarises the final model parameters for Barea sandstone.

Table 1. Elastic, yield surface, plastic potential, and hardening parameters for the non-associated structured MCC calibrated on sandstone experimental data

E (GPa)	ν (-)	k (m/s)	M_f (-)	M_g (-)	P_{s0} (MPa)	P_{t0} (MPa)	P_m (MPa)	ρ_s (-)	ρ_t (-)	χ_s (-)	χ_t (-)
16.8	0.20	1×10^{-15}	1.0	1.25	200	5	180	14	14	0.1	0.0

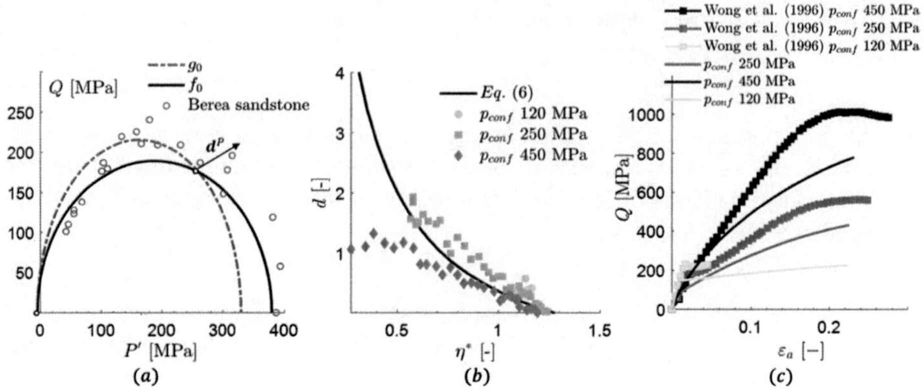

Fig. 1. Calibration of the yield locus and plastic potential for the structured MCC model using drained triaxial tests from Wong et al. (1997).

3 The RA Numerical Model

The geometry of the model is depicted in Fig. 2. The RA developed for tidal renewable applications described by [10] and [3] was modelled in GPFEM using 2D axisymmetric conditions. The RA is defined by a shaft and bottom part with a drill head (Fig. 2) to guarantee the installation process with no grouting. The reference embedment depth H/D of 6 was considered for the simulations since the independence of the load capacity was observed overcome this value. Figure 2 shows displacement boundary conditions imposed in the model. A constant water pressure was imposed at the top surface, while all other boundaries were considered undrained. To model the coupled hydro-mechanical behavior the mixed displacement Jacobian water pressure $(u - \Theta - p_w)$ formulation developed by [4] was used. The pull-out test was simulated by applying a vertical prescribed displacement at the top of the anchor with a constant rate equal 1×10^{-14} m/s. For this rate of displacement and for a rock permeability of 1×10^{-15} m/s drained behaviour is expected according to further numerical investigation conducted. The RA is simulated as a linear elastic body $(E = 210GPa$ and $\nu = 0.3)$ and the sandstone using the structured MCC model as calibrated above. The contact interaction model adopted for the simulations uses the Contact Domain (CD) technique [10] which creates an ancillary contact mesh upon the remeshing through a constrained Delaunay tessellation (Fig. 2a). This large strain-based discretization technique is capable of capturing the contact interaction between deformable bodies. In this work the penalty contact algorithm was used with a Coulomb friction law characterised by an interface friction angle δ. The penalty approach enforces the contact constrains ensuring the partial interpenetration of solid bodies [12]. The normal and tangential gap and the contact forces are linearly proportional through the penalty parameters. The penalty method here was used for its simplicity. The Coulomb contact law defines the stick and slip state according to the stress state in relation to the envelope criterion as shown in Fig. 2b. The contact force for the two tangential states is defined by:

$$F_{TI} = \begin{cases} Cx_1 k_T g_{TI} \hat{t}_I & (stick) \\ Cx_1 \mu k_N g_{NI} \hat{t}_I & (slip) \end{cases} \tag{7}$$

where $\hat{\boldsymbol{t}}_I$ indicates the tangential direction, k_N and k_T the normal and tangential penalty parameter, $C = 2\pi$, x_1 the radial coordinate in axisymmetric conditions, $\mu = \tan\delta$ the interface friction coefficient, whereas g_{NI} and g_{TI} the normal and tangential gap. The constitutive law was validated simulating interface shear tests (Fig. 2b).

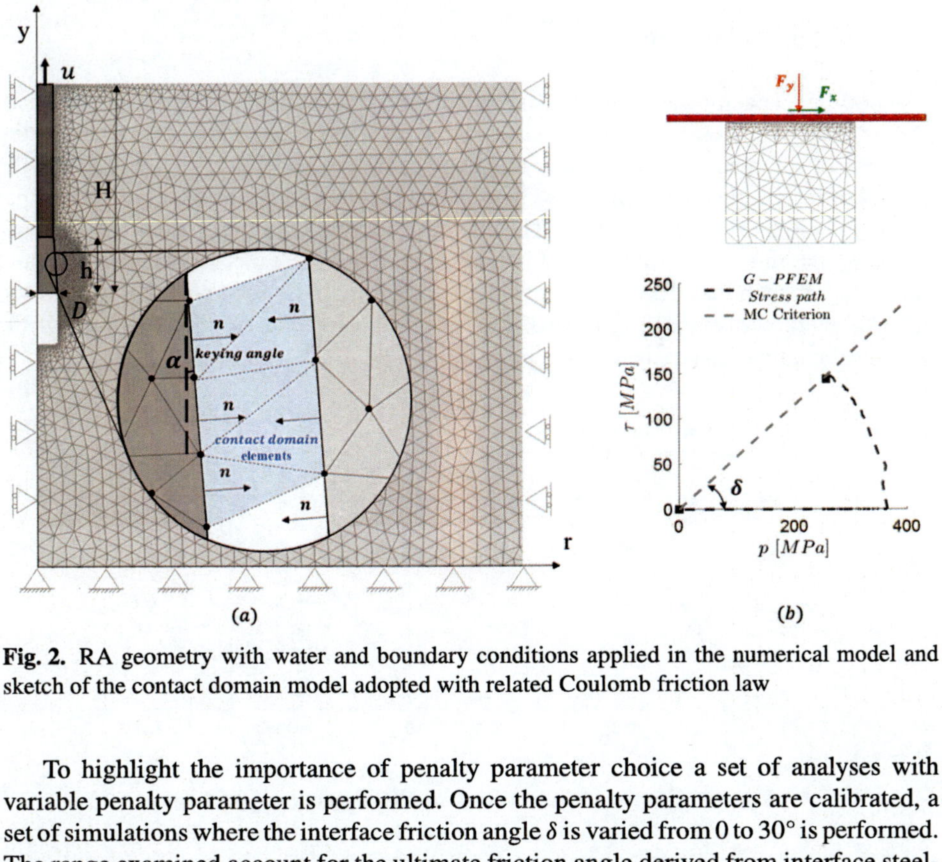

Fig. 2. RA geometry with water and boundary conditions applied in the numerical model and sketch of the contact domain model adopted with related Coulomb friction law

To highlight the importance of penalty parameter choice a set of analyses with variable penalty parameter is performed. Once the penalty parameters are calibrated, a set of simulations where the interface friction angle δ is varied from 0 to 30° is performed. The range examined account for the ultimate friction angle derived from interface steel-rock testing on sandstone [11]. To investigate the effect of the keying angle two other configurations ($\alpha = 6^o$ and $\alpha = 12^o$) are also modelled. Table 2 summarizes the numerical runs.

Table 2. Geometrical and contact parameters used in the numerical parametric study

Test ID	α [°]		δ [°]			K_N [GPa]			K_T [GPa]			
Reference	2.86		30			*E*			*E*			
$T_{\alpha i}$	12	6	30			*E*			*E*			
$T_{\delta i}$	2.86		12	6	0	*E*			*E*			
$T_{K_{N,i}}$	2.86		30			*10E*	*0.1E*	*0.01E*	*E*			
$T_{K_{T,i}}$	2.86		30			*E*			*0.8E*	*0.6E*	0.4*E*	0.2E

4 Numerical Results

The pull-out capacity was computed as the sum of the contact forces of the anchor head. Because of commercial sensitivity issues, all the results presented are normalized by the capacity of reference case defined at the displacement u = $10\%D$. Figure 3 shows the results of tests series $T_{K_{N,i}}$ and $T_{K_{T,i}}$ where the influence of the tangential and normal penalty parameters are highlighted. The results show that if K_N is smaller than the rock stiffness, the numerical solution tends to diverge from the exact solution. K_N should hence be set such that the contact stiffness is larger than the rock stiffness but not too high to avoid convergence problems. K_T does not affect the solution because of the frictional nature (slip) of the contact problem.

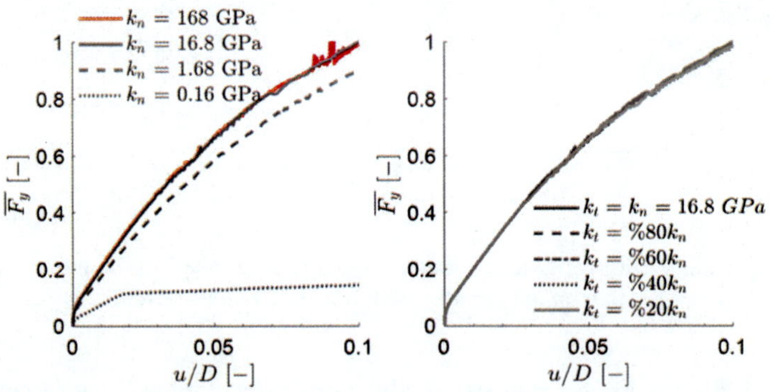

Fig. 3. Normalized pullout force results of tests $T_{K_{N,i}}$ and $T_{K_{T,i}}$

Figure 4a and b present the influence of the interface friction angle with the pull-out capacity. Increasing δ the uplift capacity increases almost linearly for values of u < $10\%D$. For higher values of δ the damage to the rock is more visible as the load capacity curve starts to exhibit an intermittently brittle response and the curve starts to show a non-linear trend. Figure 4c shows that by increasing α, both stiffness and capacity of the RA increase but the response is more brittle and nonlinear. In design practice the limiting load is usually chosen at u = $10\%D$. Figure 5 a, b compares the inclination of the contact force vectors for two simulations with different δ. As expected, the contact force distribution is not homogeneous hence favoring localized damage of the rock at the extremity of the RA head. Rock damage, defined as $D = 1 - P_t/P_{t0}$, function of plastic strains through Eq. (5) is represented in Fig. 5c, d.

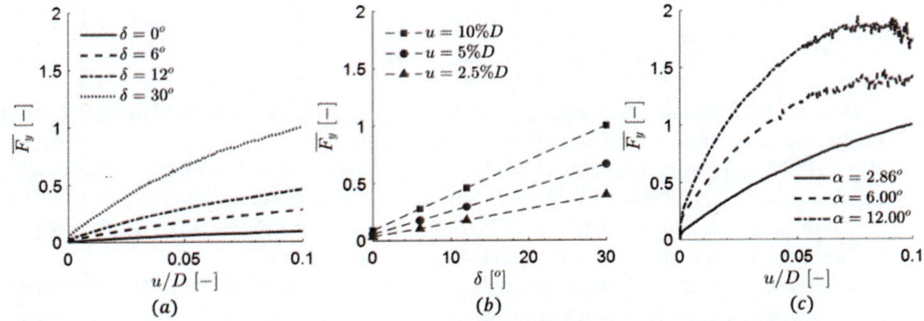

Fig. 4. Results of tests ID: $T_{\delta i}$ and $T_{\alpha i}$

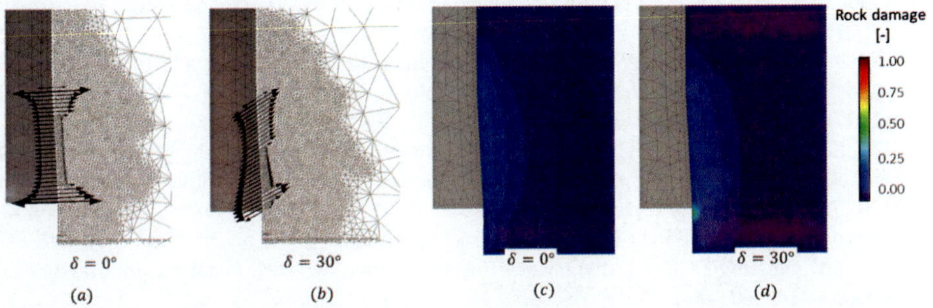

Fig. 5. (a, b) Contact force vector distribution along the bottom part of the anchor and (c, d) rock damage contours at the end of the simulations (u = 10% D) for δ = 0° and 30° respectively.

5 Conclusions

This paper presents a large deformation numerical investigation of the influence of the interface friction properties and contact model parameters on RA pull-out. A non-local large strain based structured MCC model was calibrated to replicate the mechanical behaviour of Berea sandstone. The anchor was modelled as an elastic material and by means of the contact domain approach soil structure interaction was considered. The mechanical contact constraint was enforced by the penalty approach with a frictional Coulomb law capable of efficiently reproducing the large strain displacement behaviour between rock-anchor. The results obtained reveal that interface parameters play a fundamental role in the RA load capacity. In particular (i) the penalty parameter must be selected with care to avoid unrealistic results, (ii) anchor capacity increases linearly with interface friction for displacements < 10%D and that (iii) slight changes to the anchor geometry (increasing the keying angle from 2.86 to 6 and 12°) can increase the capacity by 30% and 80% respectively.

Acknowledgments. I would like to acknowledge the technical and financial support from SWIFT ANCHORS and financial support from the Energy Technology Partnership (ETP), through ETP EIDP #182 research grant.

References

1. Cresswell, N., Jeffcoate, P.: Anchor installation for the taut moored tidal platform PLAT-O. In: 3rd Asian Wave and Tidal Energy Conference (2016)
2. Yap, L.P., Rodger, A.A.: A study of the behaviour of vertical rock anchors using the finite element method. Int. J. Rock Mech. Min. Sci. **21**(2), 47–61 (1984)
3. Cerfontaine, B., Brown, M., Caton, A., Hunt, A., Cresswell, N.: Numerical modelling of rock anchor uplift capacity for offshore applications (2021). 10.2/JQUERY.MIN.JS
4. Monforte, L., Ciantia, M.O., Carbonell, J.M., Arroyo, M., Gens, A.: A stable mesh-independent approach for numerical modelling of structured soils at large strains. Comput. Geotech. **116**(August), 103215 (2019). Elsevier
5. Oliynyk, K., Ciantia, M.O., Tamagnini, C.: A finite deformation multiplicative plasticity model with non–local hardening for bonded geomaterials. Comput. Geotech. **137**, 104209 (2021). Elsevier
6. Wong, T., David, C., Zhu, W.: The transition from brittle faulting to cataclastic flow in porous sandstones: mechanical deformation. J. Geophys. Res. Solid Earth **102**(B2), 3009–3025 (1997). https://doi.org/10.1029/96jb03281
7. Ciantia, M.O., Di Prisco, C.: Extension of plasticity theory to debonding, grain dissolution, and chemical damage of calcarenites. Int. J. Numer. Anal. Meth. Geomech. **40**(3), 315–343 (2016). https://doi.org/10.1002/nag.2397
8. Ciantia, M.O., Castellanza, R., di Prisco, C.: Experimental study on the water-induced weakening of calcarenites. Rock Mech. Rock Eng. **48**(2), 441–461 (2014). https://doi.org/10.1007/s00603-014-0603-z
9. Carbonell, J.M., Monforte, L., Ciantia, M.O., Arroyo, M., Gens, A.: Geotechnical particle finite element method for modeling of soil-structure interaction under large deformation conditions. J. Rock Mech. Geotech. Eng. Chin. Acad. Sci. (2022). https://doi.org/10.1016/J.JRMGE.2021.12.006
10. Oliver, J., Hartmann, S., Cante, J.C., Weyler, R., Hernández, J.A.: A contact domain method for large deformation frictional contact problems. Part 1: theoretical basis. Comput. Meth. Appl. Mech. Eng. **198**(33–36), 2591–2606 (2009). Elsevier B.V. https://doi.org/10.1016/j.cma.2009.03.006
11. Ziogos, A., Brown, M.J., Ivanovic, A., Morgan, N.: Understanding rock–steel interface properties for use in offshore applications. Proc. Ins. Civil Eng. Geotech. Eng. **176**, 1–15 (2021). https://doi.org/10.1680/jgeen.20.00183
12. Wriggers, P.: Computational Contact Mechanics, 2nd edn. Springer, Berlin (2006)

Application of the Material Point Method to the Study of Tailing Dams Failure due to Static Liquefaction

Veronica Girardi[1](✉) and Gianluca Bella[2]

[1] Pini Group Srl, 30122 Trento, Italy
veronica.girardi@pini.group
[2] Pini Group SA, 6900 Lugano, Switzerland

Abstract. This study explores the effects of some relevant physical and numerical factors on the stability of tailing dams. The stability analysis is carried out using the Material Point Method (MPM) which is based on displacements quantification, rather than the Factor of Safety approach implemented by the Finite Element Methods (FEM), or the Limit Equilibrium Method (LEM). A series of numerical simulations is carried out based on a geometry from the upstream construction method. The impact of the undrained strength of the tailings stored inside the basin, and the shear strength properties of the starter dam are preliminary evaluated in terms of static liquefaction susceptibility by quantifying displacements and failure surface. Moreover, the effects of the mesh size is presented to provide some useful guidelines for the design and stability analysis of tailing dams based on the use of MPM.

Keywords: Tailing Dams · Material Point Method · Static Liquefaction · Unsaturated soil

1 Introduction

Mine tailings are byproducts of mineral processing, and they are a mixture of silt and sand with high content of process water, unrecoverable heavy metals and chemical reagents employed during raw materials extraction. These wastes are generally discharged as a slurry into a storage area known as Tailings Storage Facility (TSF) retained by embankments or dams. Tailings dams (TD) are commonly built incrementally over the time and their three main configurations - upstream, downstream and centerline - are mainly driven by geotechnical, topographical and economical considerations. Moreover, tailing dams are heterogeneous systems subjected to several external factors, like the interaction with the atmosphere and as a consequence, the position of the water table within the basin can move upward or downward [1, 2].

The great advances in mining technology over the last century have made it economically feasible to increase the excavated volumes and to mine lower grades of ore. On the other side, the lack of knowledge of tailings physical-chemical properties, together

A. Ferrari et al. (Eds.): CNRIG 2023, SSGG, pp. 435–443, 2023.
https://doi.org/10.1007/978-3-031-34761-0_53

with poor dam management and monitoring practices, are of growing concern and can be considered the main predisposing factors of dam failures. These events are often characterized by tailings flowing downstream with high speed, and able to run across hundreds of meters, with severe consequences in terms of loss of lives, property damages and environmental pollution (Fig. 1a). Based on 358 worldwide failure cases of TD from 1915 to January 2020, the static triggering mechanisms are the predominant cause of the main tailing dams collapses, as shown in Fig. 1b [3, 4]. Moreover, about 45% of worldwide tailing dams have been constructed with the upstream method, which is considered the least stable among the construction techniques.

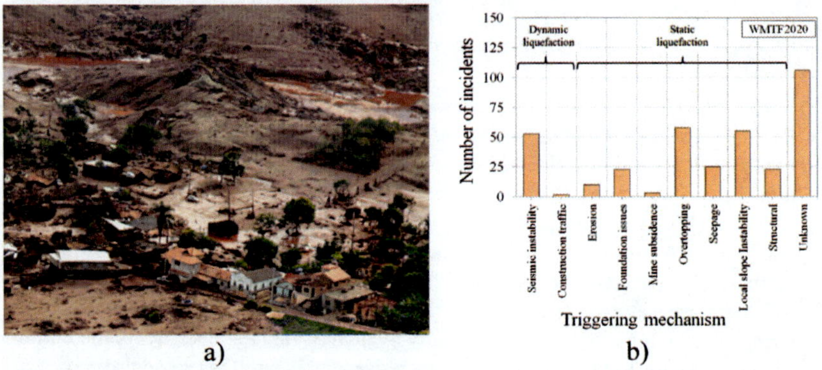

Fig. 1. a) Example of the consequences of TD collapse. District of Bento Rodrigues, Mariana (Brazil): buildings affected by the collapse of Fundão tailings dam [5]. b) Number of failures with the main triggering mechanisms of tailing dams collapse (modified from [4]).

Moving from these reasons, the current study is aimed to preliminary investigate the influence of some factors on the stability of TSF, and to reproduce the tailing dams failure due to static liquefaction occurrence. Numerical analysis are performed with the Material Point Method, a large displacement numerical technique popular in the geotechnical community, but still little-used in stability analysis of tailing dams. The liquefaction occurrence is investigated by considering a simplified approach in terms of undrained shear strength [6]. Moreover, the undrained strength is correlated to the saturation degree of the tailings, based on [7], allowing to take in account in a simplified manner the potential of liquefaction in unsaturated tailings. Variations of the starter dam strength parameters are additionally considered to evaluate the impact on the failure surface and tailings run out. Lastly, the impact of the mesh size on the failure behaviour is also analyzed with the aim of giving practical suggestions to MPM users in charge of design or safety assessment of tailing storage facilities.

2 Numerical Model

The TD numerical model is based on the Material Point Method, a large strain continuum approach developed in 1994 for solid mechanics applications [8] and later on extended to multiphase media, especially to reproduce the behavior of geomaterials (among others [9, 10]). In MPM a background mesh discretizes the computational domains where the material points (MPs) move, simulating the full displacement of a phenomenon under analysis. The MPs are entities representing the matter of bodies analyzed, carrying kinematic and stress/strain variables. In every computational cycle the governing equations which are typical of the continuum description are solved at the nodes of the computational mesh, however no permanent information is stored at the nodes to avoid excessive mesh distortion or entanglement, as in FEM. Hence, mapping procedures are used to project variables from the nodes to MPs. MPM has proven to be an efficient tool in the reproduction of slope instability mechanism, due to the ability of modelling trigger mechanisms, failure onset and post failure [11, 12].

In this work, to mimic the behavior of tailing under liquefaction, we use a formulation able to capture the behavior of soil in undrained conditions implemented in the research code *Anura3D* [13]. The formulation is based on a simplified approach, not accounting for pore pressure dissipation, given the rapidity of load application. The saturated medium is treated as a monophase material, and the momentum balance equation solved (Eq. 1) is the one of the mixture:

$$\rho \boldsymbol{a} = \nabla \boldsymbol{\sigma} + \rho \boldsymbol{g} \tag{1}$$

where ρ is the mixture density, $\boldsymbol{a}$ is the acceleration of the mixture, $\boldsymbol{\sigma}$ the stress tensor and $\boldsymbol{g}$ is the gravity vector. Bold symbols stand for vectors/tensors. The physical description is strongly simplified in this formulation, seepage and the unsaturated soil behavior are not accounted for, with the benefit of significant computational efficiency and simplicity in use. Previous experimental studies proposed empirical correlations between undrained shear strength (c_u) and degree of saturation (S_r). In this work the chosen values of undrained strength are based on laboratory tests on Stava tailings [7] as reported in Fig. 2d, where the experimental points (triangles) are derived from triaxial tests. Three scenarios are considered to investigate the impact of different undrained strength on the failure behavior, associated to different levels of saturation degrees of tailings. In the first scenario (Fig. 2a) the tailings are almost fully saturated ($S_r \approx 98\%$) and the corresponding c_u is equal to 4kPa. In the second case, a scenario of non-uniformity of saturation degree is mimicked, with the top portion of the tailings with $c_u = 12$ kPa ($S_r \approx 89\%$) and the bottom part still saturated, with the same c_u of the first scenario. The third scenario accounts for an even stronger desaturation of the shallow portion of tailings, with $c_u = 16$ kPa ($S_r \approx 83\%$), with lower tailings progressively more saturated, reflected in reducing values of c_u as in the previous scenarios (12kPa and 4 kPa). The second and third scenarios simulate an initial condition with a lower phreatic surface compared to the first scenario, as visible in Fig. 2b and c. In the following, the first scenario will be referred as HP (high phreatic surface), the second and third scenarios as LP1 and LP2 (low phreatic surfaces).

The geometry of the tailing storage reproduces the section of an upstream configuration (Fig. 2), with a mild inclination (4°) of the preexistent topography. The starter dam has slope 2:3 on the external side, whereas the subsequent upstream dikes built on top of the progressively deposited tailings have an average slope of 1:3. For the secondary dikes, the starter dam and the foundation soil drained strength parameters are considered, and Mohr-Coulomb failure criterion is considered. The materials parameters of all the materials included in the MPM model are listed in Table 1. The mesh is made of triangular elements with average edge size equal to 2m for the TD materials and the foundation soil, while a coarser mesh has been adopted for the surrounding domain.

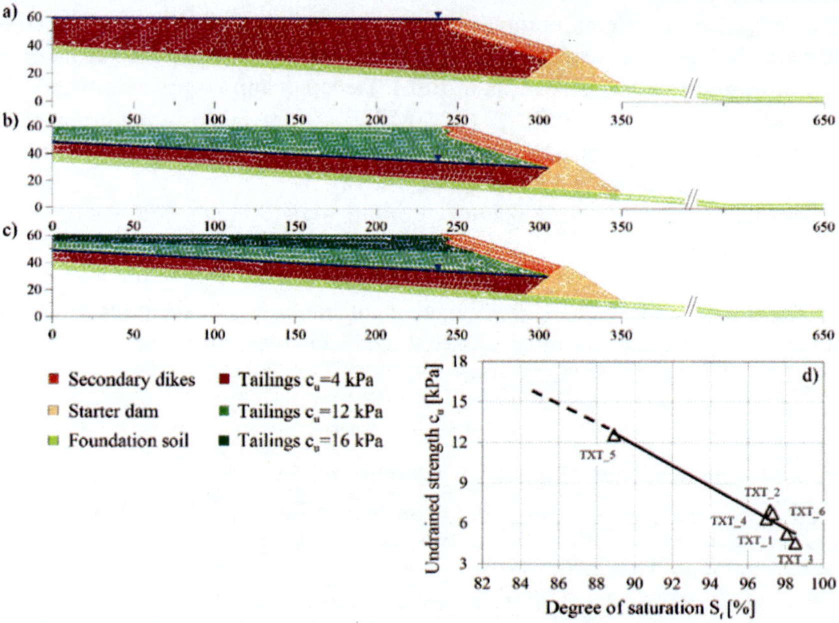

Fig. 2. MPM models of the three investigated scenario, HP a), LP1 b) and LP2 c), with different initial saturation conditions reflected in the undrained strength values; d) undrained strength of tailings at different degree of saturation interpolated from experimental data ([7]).

Fixities along the x-direction are applied on the vertical edges of the model; displacement of the bottom of the model are fully fixed (both in x- and y-direction) while the top horizontal edge has fixities only along y-direction. Each simulation is articulated in an initial step of gravity loading in quasi-static conditions to initialize the soil stresses, followed by the actual failure analysis in fully dynamic conditions. Strain smoothing, a small value of mass scaling (20) and homogeneous local damping equal to 0.05 are used. The simulations last approximately 4min.

Table 1. Material parameters.

Material	Porosity [-]	Solid density [kg/m^3]	Young modulus [kPa]	Poisson ratio [-]	Poisson ratio undrained [-]	Cohesion [kPa]	Friction angle [°]
Tailing S_r = 98%	0.40	2780	20•10^3	0.30	0.495	4	0
Tailing S_r = 89%	0.40	2780	20•10^3	0.30	0.495	14	0
Tailing S_r = 83%	0.40	2780	20•10^3	0.30	0.495	16	0
Secondary dikes	0.40	2700	30•10^3	0.25	0.495	1	32
Starter dam	0.40	2700	40•10^3	0.25	0.495	1	35
Foundation soil	0.40	2700	60•10^3	0.30	0.495	2	40

3 Results

3.1 Impact of Strength Parameters

For HP, i.e. the scenario with tailings fully saturated under the effect of a high phreatic surface, the failure onset is highlighted in Fig. 3, using plastic strain contours. At 5s the accumulation of plastic strain is localized at the interface between the lowermost part of the tailings and the foundation/starter dam, showing immediately the action of the liquefied tailings on the starter dam. At 10s the accumulation of strain at depth continues and at the same time a sliding surface quite rounded starts developing from the inner side of the starter dam toward the tailings surface. At 20s the tailings are fully mobilized, both pushing and flowing above the starter dam. It appears that failure has a retrogressive character, with progressive involvement of tailings on the inner areas of the storage, being poured downstream. This tendency is confirmed if we look at the situation at 55s, with increments in plastic strain magnitude and extension of the tailings zone undergoing failure.

The other two scenarios, LP1 and LP2, show a similar failure mechanism at the beginning of the simulation, with similar shape and slightly lower plastic strains for LPs compared to the HP, as visible in Fig. 4 where the plastic strain contour for the three scenarios at 10s are reported (respectively LP1 and LP2 correspond to Fig. 4b and c). On the contrary, if we consider the evolution of failure, it is possible to appreciate significantly different run out distances for the three scenarios under analysis, as reported in Fig. 5. The contours of final displacements for HP (Fig. 5a), IP (Fig. 5b) and LP (Fig. 5c) show the effect of the progressive lowering of the initial water table: with increasing values of undrained strengths, the propagation of the tailings downstream tends to reduce, with maximum final displacements equal to 317m for HP, 215m for LP1

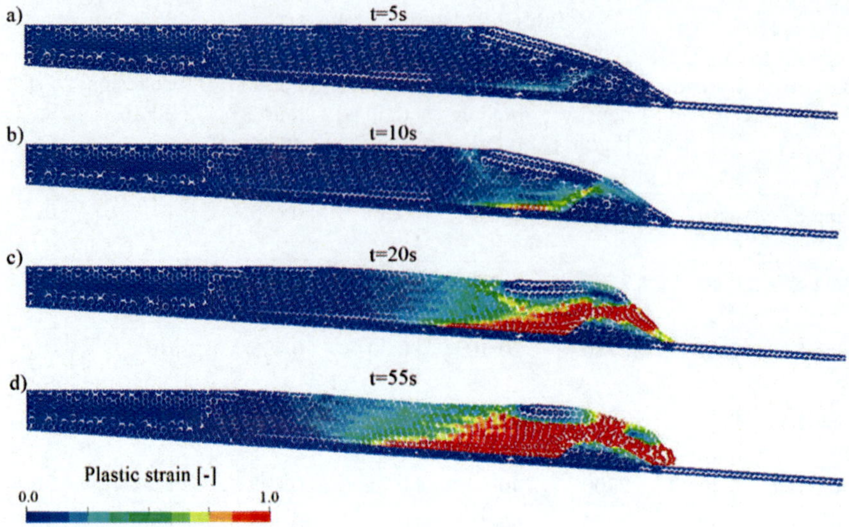

Fig. 3. Plastic strain for HP scenario (high phreatic surface) at four relevant instants of time.

and 191m for LP2. This behavior finds explanation also if we look at the kinetic energy of the systems with time, as presented in Fig. 5d. The HP case shows a significantly higher peak in kinetic energy compared to the other two cases, meaning that larger velocities characterized this first scenario. These results support the importance of maintaining low water tables in tailing storages with efficient draining systems, and therefore the importance of continuously monitor the pore pressure distribution in the tailings, during construction, operation and after depletion of the tailing facility. Concerning potential numerical artifacts, it has to be noted that in all the simulations some MPs are slightly detached from the moving tailing mass.

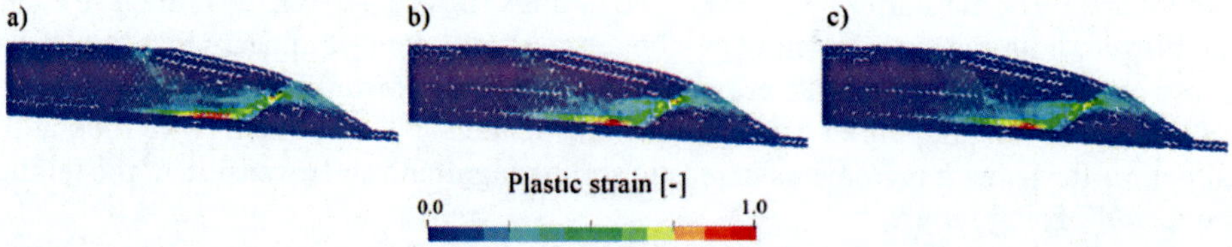

Fig. 4. Plastic strain at t = 10s for the three scenarios of tailing storage analyzed: HP a), LP1 b) and LP2 c).

These MPs can be referred as outliers and represent a consequence of numerical instabilities during the dynamic phenomenon of soil mass propagation. These MPs should be discarded while quantifying in a precise manner the maximum run out.

Failure and post failure behaviors are not only the result of the mechanical properties of the tailings and the saturation state of them, instead, the starter dam and the foundation soil play a crucial role in the failure process. In the present study we will focus only

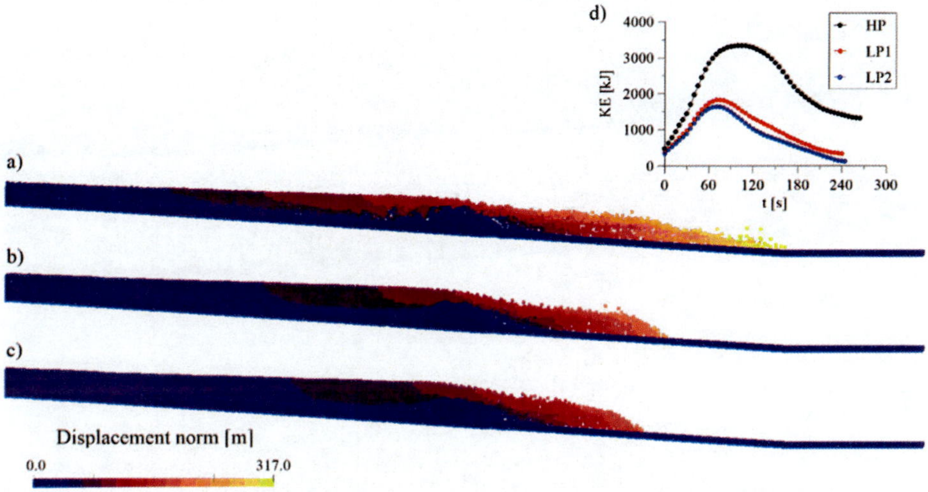

Fig. 5. Final displacement contours (HP: a, LP1: b, and LP2: c) and system kinetic energy vs time for the three scenarios under analysis.

on the starter dam strength parameters impact on failure; the interested readers to the problems induced by strain softening of foundation soil can refer to [6] and references in it. Indeed, two additional combinations of strength parameters of the starter dam are explored: a combination C1 with $\varphi = 38°$ and c' = 1kPa, representing a dam with higher percentage of the coarse fraction, and C2 with $\varphi = 35°$ and c' = 15kPa, simulating a higher fine content.

The differences are very small in terms of final run out, with maximum final displacement of case C1 equal to 291m, while C2 shows the highest values of displacement, equal to 326.m. In terms of failure onset, no significant differences are observed in C1 and C2 if compared to the reference case, since the initial localization of plastic strains is at the interface between the tailings and the inner slope of the starter dam, or between tailings and the foundation soil, leading to state that so the reduced values used in this study doesn't favor initial strain localization in the dam body.

3.2 Impact of Numerical Parameters

The impact of the mesh size on the run out distance is preliminary investigated by considering a finer mesh (5049 elements) if compared to the reference case, as shown in Fig. 6. No appreciable differences in terms of displacements can be observed, and this is probably due to the good quality of the mesh adopted for the reference case (2696 elements). Indeed, upon certain sizes, a finer mesh does not provide better results and, on the contrary just require more computational times, in agreement with several previous studies from the literature.

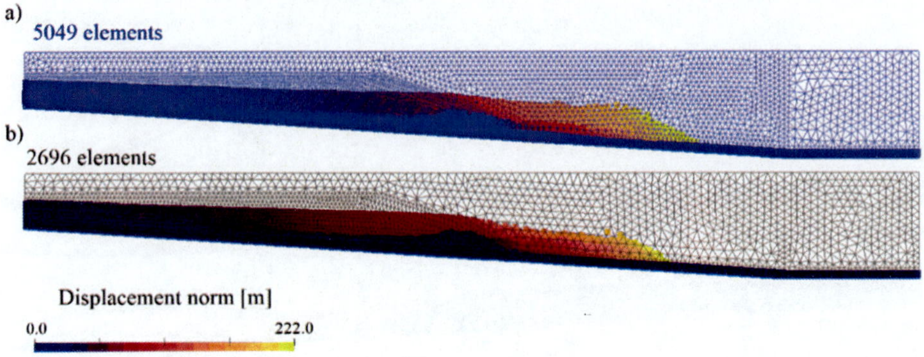

Fig. 6. Final displacements for the LP1 scenario, considering a finer mesh (a) compared to the reference case (b).

4 Conclusions

The current research has preliminary investigated the effects of some relevant physical and numerical factors on the stability of tailing storage facilities by carrying out simplified numerical analysis based on the MPM. Indeed, the model assumptions relate to material behavior (monophase approximation) and the constant undrained strength values once they are assigned to distinct portions of tailing. Nevertheless, this approach can be seen as a fast preliminary tool to predict the behavior of a tailing dam or to back analyze a failure event already occurred, especially by going beyond the solely concept of factor of safety. The changes in undrained strength of the stored tailings impacts significantly on the run out distances, whereas variations of the shear strength of the starter dam as well the mesh size did not significantly affect the failure onset or the final run out. Future developments of this research may account for the contact properties at the interface between tailings and the basement to improve the run-out modelling, combined with the use of more sophisticated constitutive models.

References

1. Bella, G.: Water retention behaviour of tailings in unsaturated conditions. Geomech. Eng. Int'l J. **26**(2), 117–132 (2021)
2. Zanardin, M.T., et al.: The role of capillary water in the stability of tailing dams. Eng. Geol. **105**(1–2), 108–118 (2009)
3. Bella, G.: Evaluation of the static liquefaction behaviour of silty tailings in unsaturated conditions. In: 7th World Congress on Civil Structural and Environmental Engineering (CSEE 2022), pp. 1–5. Avestia Publishing, Lisbona (2022)
4. Rodríguez, R., et al.: How to prevent flow failures in tailings dams. Mine Water Environ. **40**(1), 83–112 (2021). https://doi.org/10.1007/s10230-021-00752-8
5. Do Carmo, F.F., et al.: Fundão tailings dam failures: the environment tragedy of the largest technological disaster of Brazilian mining in global context. Perspect. Ecol. Conserv. **15**(3), 145–151 (2017)
6. Pierce, W. I.: Applying the Material Point Method to Identify Key Factors Controlling Runout of the Cadia Tailings Dam Failure of 2018. Master thesis, Virginia Tech (2021)

7. Bella, G.: Mechanical response of tailings under monotonic triaxial tests in unsaturated and nearly saturated conditions. In: 8th World Congress on Civil, Structural, and Environmental Engineering (CSEE 2023), pp. 1–8. Avestia Publishing, Lisbona (2023)
8. Sulsky, D., et al.: A particle method for history-dependent materials. Comput. Methods Appl. Mech. Eng. **118**, 179–196 (1994)
9. Fern, J., et al.: The Material Point Method for Geotechnical Engineering: A Practical Guide. CRC Press, Boca Raton (2019)
10. Yerro, A., et al.: Modelling unsaturated soils with the material point method. A discussion of the state-of-the-art. Geomech. Energy Environ. **32**, 100343 (2022)
11. Soga, K., et al.: Trends in large deformation analysis of landslide mass movements with particular emphasis on the material point method. Geotechnique **66**, 248–273 (2016)
12. Girardi, V., et al.: Failure of levees induced by toe uplift: Investigation of post-failure behavior using material point method. J. Rock Mech. Geotech. Eng **15**, 970–983 (2023)
13. Anura3D, MPM, Research Community, ANURA3D version 2022 source code (2022.) www.anura3d.com. OpenSource

Zero-Dimensional Seismic Design of Bridge Abutments: A Double Macroelement Approach

Davide Noè Gorini(✉) and Luigi Callisto

Sapienza University of Rome, Rome, Italy
davideno.gorini@uniroma1.it

Abstract. Simplicity is a key element in design. It requires synthesis, in order to provide a straightforward interpretation of a phenomenon. Recent studies have shown that abutments are critical components in the seismic design of bridges, and that many technical provisions oversimplify some dominant factors of their dynamic response, that could be considered only through onerous modelling techniques. Hence, this study proposes a tool that requires a minimal calculation effort for the seismic design of abutments. It consists of the combination of two complementary macroelements: one that describes the response of the soil-abutment system and one simulating the dynamic response of the bridge superstructure. The macroelements are lumped at the abutment-deck contact and reproduce salient features of the dynamic response: frequency dependence, nonlinearity and directional coupling. This approach is implemented in OpenSees and is validated against the results of a fully coupled soil-bridge finite-element model subjected to seismic loading. Finally, a seismic design procedure for bridge abutments based on the use of the double macroelement is outlined.

Keywords: Bridge Abutments · Seismic Conditions · Macroelements · Design · OpenSees

1 Motivation

In semi-integral and integral bridges, the abutments are heavily stressed in the case of strong ground motion because of the participation of soil mass in the dynamic response of the bridge and of the inertial forces transmitted by the deck. Recent studies have shown that these inertial effects can significantly affect the seismic performance of abutments, with important repercussions on the design of the entire infrastructure [1–6].

The seismic design of abutments is conventionally carried out through pseudo-static approaches, in which the inertial effects deriving from the soil and the structure are modelled through equivalent static forces, neglecting the dynamic character of the response and the dependence on the nonlinear soil-structure behaviour. In order to overcome these limitations, Gorini and Callisto [4] proposed an analysis approach based on a structural macroelement for the nonlinear dynamic analysis of the soil-abutment model. With the aim of reaching a minimum in the computational effort, the present study extends

A. Ferrari et al. (Eds.): CNRIG 2023, SSGG, pp. 444–451, 2023.
https://doi.org/10.1007/978-3-031-34761-0_54

the approach above replacing the soil-abutment continuum model with a macroelement simulating nonlinear and inertial features.

In the following, the proposed double macroelement approach is described and its seismic performance compared with the one of a continuum coupled representation of a multi-span girder bridge. Finally, the use of the double macroelement in a seismic design methodology is delineated.

2 Zero-Dimensional Performance-Based Design

The soil-abutment-bridge system is regarded as the combination of two complementary macroelements: one for the soil-abutment system, SAME, and one for the bridge superstructure, TT. The combined SAME-TT macroelement is lumped at the deck-abutment contact providing a zero-length numerical representation of the problem. The SAME-TT is conceived to minimize the computation times of the nonlinear dynamic analysis of soil-bridge systems, taking into account, though in a simplified manner, the key effects influencing the bridge performance: frequency dependence, nonlinear behaviour and multiaxial loading.

The use of the SAME-TT follows the procedure described in [7, 8]: the seismic motion is propagated through the foundation soils by means of a free-field site response analysis, and the motion computed at an effective depth from the abutment foundation is applied to the SAME [6, 7] to carry out a time-domain nonlinear analysis on the SAME-TT at a minimal computational effort.

The proposed approach was implemented in the analysis framework OpenSees [9–11]. The SAME is a multiaxial material that can be assigned to a multiaxial zero-length finite element, ZeroLengthND. The TT can be instead modelled as an assembly of rheological elements available in the library (see Sect. 2.2).

2.1 Macroelement for the Soil-Abutment System

The SAME is schematically shown in Fig. 1a: it simulates the multiaxial response of the soil-abutment system and consists of a relationship between the interaction forces exchanged at the deck-abutment contact, Q_i (i = 1,2,3 being the longitudinal, transversal and vertical direction of the bridge), and the corresponding displacements q_i, such as $\dot{Q}_i = H_{ij} \times \dot{q}_j$ (H_{ij} being the tangent stiffness matrix). It is a multi-surface plasticity law with kinematic hardening, derived within a thermodynamic approach, using hyperplasticity [12].

The plastic response activates from the first yield surface and is delimited by the ultimate limit state surface. The yield surfaces are homothetic and described by a roto-translated ellipsoidal shape [7]. A typical suite of yield surfaces is shown in Fig. 1b, relative to the soil-abutment system analysed in [3, 4, 6, 7] that will be taken as a reference for the subsequent validation of the proposed approach. Under dynamic conditions, the inertial effects developing in the volume of soil interacting with the abutment are simulated by combining the SAME with the modal masses of the soil-abutment system along the three degrees of freedom of the macroelement. This confers to the SAME a frequency-dependent response able to reproduce the fundamental dynamic features of the system. As an example, Fig. 1c shows the variability of the maximum longitudinal displacement of the top of the reference abutment varying the period of the longitudinal load Q_1 obtained with a continuum coupled soil-abutment numerical modelling and with the SAME. The calibration of SAME, requiring the definition of the ultimate yield surface, the initial stiffness matrix and the participating masses of the abutment-soil system, was exhaustively discussed in [7].

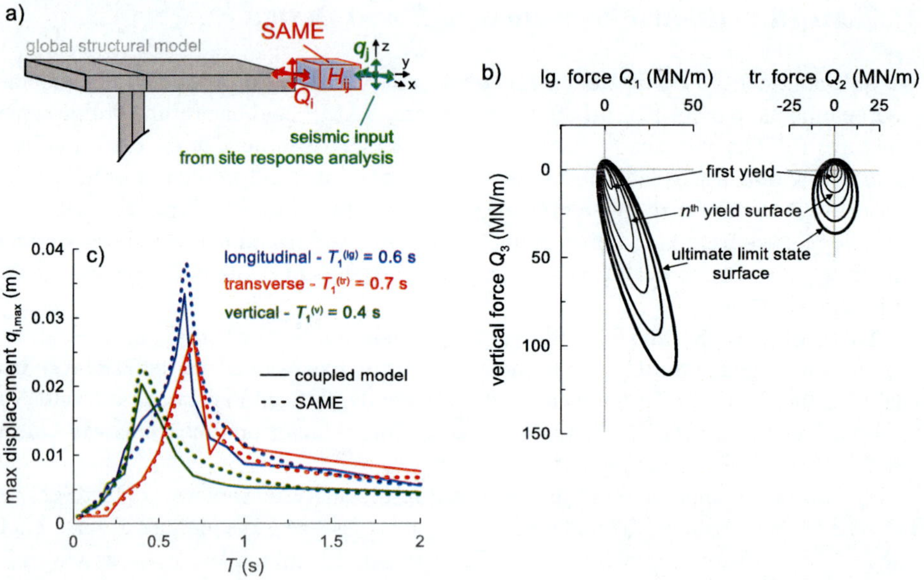

Fig. 1. a) application of the macroelement of the soil-abutment system (SAME) in a nonlinear time-history analysis of a bridge; representation of b) the plastic domain and c) the small-strain frequency-response of the SAME under multiaxial loading.

2.2 Macroelement of the Bridge Superstructure

The macroelement of the bridge superstructure, TT, is conceived to reproduce the transmission, depending on the vibration period T, of the interaction forces Q_i transferred by the deck to the abutment, such that $Q_i(T) = K_{ij}(T) \times u_j(T)$ where $K_{ij}(T)$ is a second order *transfer tensor* and u_j represents the displacement vector of the pier and abutment foundations in direction j. The transfer tensor is conceived to reproduce the dynamic forces transferred by the superstructure to the abutment top: its components, called transfer functions, filter the seismic signal applied to the foundations of the bridge as a function of dynamic characteristics of the entire structure. For a linear structural behaviour, the transfer functions depend on the frequency of the external perturbation, while when the behaviour is nonlinear the stress tensor K_{ij} is also a function of the amplitude of the seismic motion in the foundation. A comprehensive description of the TT macroelement is included in [4], while for brevity in the following we focus on the linear case.

The identification procedure of a linear TT is schematically depicted in Fig. 2. The generic transfer function t_{hh} is obtained by performing a dynamic analysis of the bridge structure: the foundations of the piers and abutments are perturbed by a displacement time-history u_h with progressively increasing frequency (sweep); for each loading frequency the transfer function is computed as the ratio of the maximum interaction force Q_i and the input amplitude u_j. The transfer functions reach their maximum at the resonance periods of the bridge, and can be easily modelled as an assembly of Kelvin-Voight rheological models combined with the participating masses of the superstructure for the modes associated with the peaks of t_{hh}, as detailed in [4].

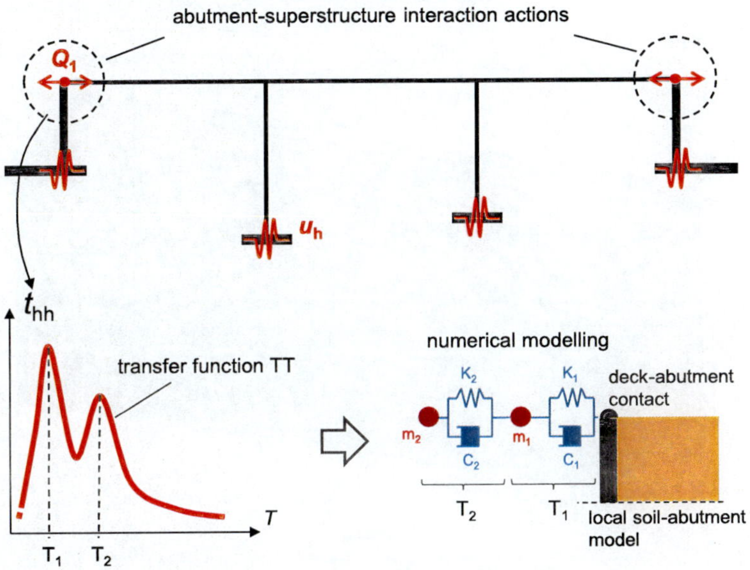

Fig. 2. Numerical modelling and identification of the macroelement of the bridge superstructure (TT) in the longitudinal direction.

3 Validation

The seismic performance of the SAME-TT macroelement is tested on the case study of a multi-span girder bridge [3]. The full soil-bridge model in Fig. 3 was implemented therein in OpenSees and one can refer to [3] for a detailed description of the numerical model. The model includes the bridge structure and a dry, homogeneous soil deposit having the properties of the well-documented Messina Gravels. The present validation considers only the longitudinal component of the seismic motion.

The bridge superstructure has a continuous deck supported by two central piers and by the lateral abutments. An alternating strong and weak contact is assumed between the deck and the vertical elements, intended as a three-directional bearing device (rigid constraint) and a bi-directional device (longitudinal displacement allowed), respectively.

The strong abutment is the one taken as a reference in Sect. 2.1. For simplicity, the structure exhibits a linear visco-elastic behaviour. The mechanical behaviour of the foundation soil was instead described by the PDMY model [13], calibrated according to numerous experimental data [2, 3]. A staged analysis procedure was employed to assemble the model and the use of parallel computing, obtained with the OpenSeesSP interpreter, was needed to get reasonable computation times. The longitudinal component of the no-Collapse Earthquake Tabas record [3] was applied to the lower boundary of the model as a displacement time history.

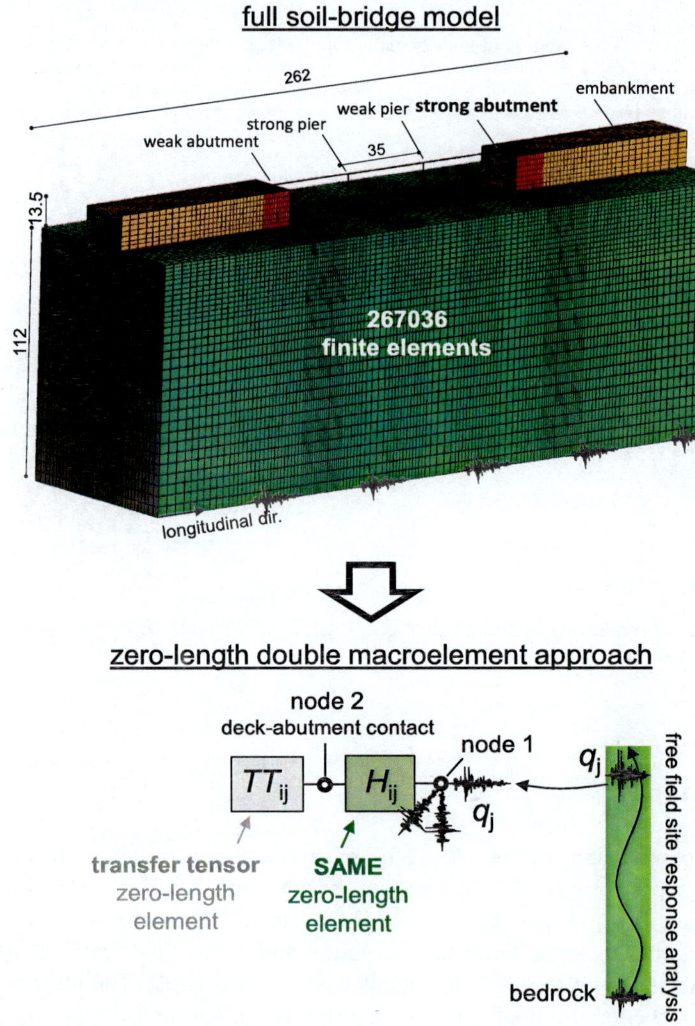

Fig. 3. Reference case study: from the full 3D numerical model to the zero-length SAME-TT macroelement under seismic loading.

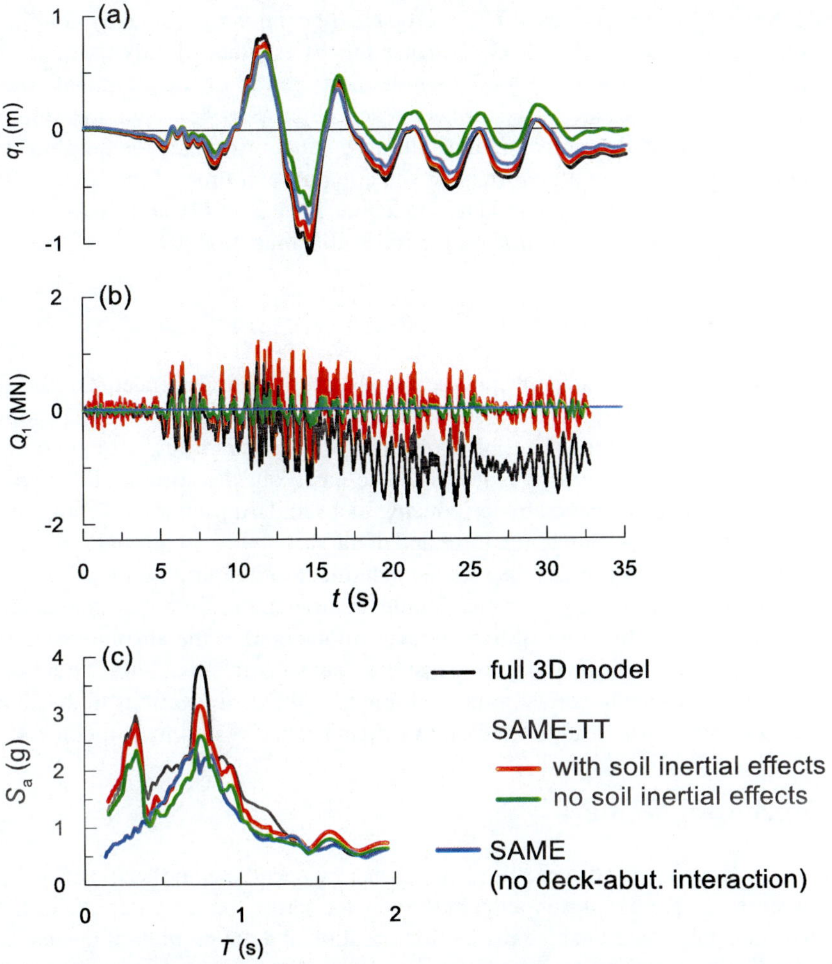

Fig. 4. Reference case study: time evolutions of the longitudinal a) displacements of the abutment top and b) forces exchanged at the deck-abutment contact; c) 5%-damped elastic response spectrum of the longitudinal motion of the abutment top computed with the 3D model and the SAME-TT.

A nonlinear dynamic analysis was also carried out using the SAME-TT approach (Fig. 3), considering the seismic input evaluated through a free-field site response analysis. The SAME and TT macroelements for the case at hand were calibrated in previous works [4, 7]. Figures 4a and 4b compare the time histories of the abutment displacement q_1 and of the interaction force Q_1 at the deck-abutment contact in the longitudinal direction, obtained for different modelling options. Taking the results of the full 3D model as a reference, these figures show that in order to reproduce reasonably well both the displacements and the interaction forces it is necessary to include the participating soil mass in the abutment macroelement (SAME), and at the same time to couple this macroelement with the one that simulated the structural behaviour (TT). It should be noted, however, that the proposed TT macroelement cannot reproduce by definition the final permanent residual value of Q_1, because the TT is attached only to the top of the abutment. The response spectrum of the motion at the top of the abutment computed with the two approaches are very similar to each other (Fig. 4c) when the soil inertial source is included, while neglecting the soil inertia underestimates the amplification of the abutment response. As a final remark, the computation time of the SAME-TT was equal to about 2 min on a common laptop machine, instead of 60 days that were needed to carry out the dynamic simulation on the full soil-bridge model.

4 Use in Abutment Design

The present double macroelement approach can be used to implement an iterative design procedure based on expeditive time-domain analyses, to obtain a system that, subjected to the design seismic motion, develop acceptable displacements and interaction forces.

At the end of this process it is necessary to carry out the structural design of the abutment. This can be achieved by introducing in a standard pseudo-static approach the appropriate values of the interaction force and of the soil-abutment inertial forces. In fact, the load pattern that maximizes the seismic increment of bending moments in the front wall can be obtained directly from the double macroelement time-domain calculation, that provides the time histories of the interaction force and of the absolute acceleration of the soil-abutment system. These time histories can be converted, with the aid of a soil pressure model, in time histories of internal forces at selected locations in the abutment front wall or at the piles head, for a direct verification of the structural members.

5 Concluding Remarks

Compared to a design elastic spectrum, the use of a properly selected set of accelerograms offers a more complete representation of the seismic action, because it includes not only the maximum amplitude but also the motion duration and the number of cycles that can have a significant effect on the seismic performance of non-linear systems. For instance, Fig. 4a shows that the system analysed in this paper (Fig. 3) undergoes a significant permanent displacement at the end of the earthquake that could not be predicted in the absence of information on the entire motion history.

On the other hand, traditional displacement calculations based on the rigid-perfectly plastic assumption [14], neglecting any amplifying effect associated with the deformability of the system, fail to provide a reasonable estimate of the actual response of an abutment which is not connected to the bridge deck [15]. Cases where this connection exists are out of the range of applicability of this method.

On the contrary, the proposed double macroelement approach includes in an essential form the main ingredients that influence the seismic response of a bridge abutment, namely the pre-failure nonlinear behaviour and the inertial response of the soil-abutment system, and the dynamic response of the bridge structure. It may be employed for a direct evaluation of the transient and permanent displacements of the abutment and, with the aid of standard local models, to design its structural members. In this paper, the method has been validated with reference to the sole longitudinal component of the seismic motion. Extension and validation of this approach considering all of the three motion components are currently in progress.

References

1. Elgamal, A., Yan, L., Yang, Z., Conte, J.P.: Three-dimensional seismic response of Humboldt bay bridge-foundation ground system. J. Struct. Eng. **134**(7), 1165–1176 (2008)
2. Gorini, D.N.: Soil-structure interaction for bridge abutments: two complementary macro-elements. PhD thesis, Sapienza University of Rome, Italy. (https://iris.uniroma1.it/handle/11573/1260972)
3. Gorini, D.N., Callisto, L.: A coupled study of soil-abutment-superstructure interaction. In: Calvetti, F., Cotecchia, F., Galli, A., Jommi, C. (eds.) The Geotechnical Research for the Protection and Development of the Territory (CNRIG 2019). Lecture Notes in Civil Engineering, vol. 40, pp. 565–574. Springer, Cham (2020a).https://doi.org/10.1007/978-3-030-21359-6_60
4. Gorini, D.N., Callisto, L.: A macro-element approach to analyse bridge abutments accounting for the dynamic behaviour of the superstructure. Geotechnique **70**(8), 711–719 (2020)
5. Gorini, D.N., Callisto, L., Whittle, A.J.: Dominant responses of bridge abutments. Soil Dyn. Earthq. Eng. (2021). https://doi.org/10.1016/j.soildyn.2021.106723
6. Gorini, D.N., Callisto, L.: La risposta dinamica del ponte nella prestazione sismica delle spalle. In: Proceedings of the XXVII Convegno Nazionale di Geotecnica, Reggio Calabria, Italy, 13–15 July 2022
7. Gorini, D.N., Callisto, L., Whittle, A.J.: An inertial macroelement for bridge abutments. Geotechnique **72**(3), 247–259 (2022)
8. Gorini, D.N., Callisto, L., Whittle, A.J., Sessa, S.: A multiaxial inertial macroelement for bridge abutments. Int. J. Numer. Anal. Meth. Geomech. **47**, 793–816 (2023)
9. Gorini, D.N., Callisto, L.: A multiaxial inertial macroelement for deep foundations. Comput. Geotech. **155**, 105222 (2023). https://doi.org/10.1016/j.compgeo.2022.105222
10. McKenna, F.: Object-oriented finite element analysis: frameworks for analysis, algorithms and parallel computing. Ph.D. dissertation, Univ. of California, Berkeley, CA (1997)
11. McKenna, F., Fenves, G.L., Scott, M.H., Jeremic, B.: Open system for earthquake engineering simulation (2000). http://opensees.berkeley.edu
12. Houlsby, G.T., Puzrin, A.M.: Principles of Hyperplasticity. Springer, Berlin (2006)
13. Yang, Z., Elgamal, A., Parra, E.: A computational model for liquefaction and associated shear deformation. J. Geotech. Geoenviron. Engng. ASCE **129**(12), 1119–1127 (2003)
14. Newmark, N.M.: Effects of earthquakes on dams and embankments. Geotechnique **15**(2), 139–160 (1965)
15. Callisto, L.: On the seismic design of displacing earth retaining systems, in 'Earthquake geotechnical engineering for protection and development of environment and constructions. In: Proceedings of the 7th International Conference on Earthquake Geotechnical Engineering, Associazione Geotecnica Italiana, Rome, Italy, pp. 239–255 (2019)

Modelling Phase Transition in Saturated Granular Materials in MPM

Pietro Marveggio(✉), Matteo Zerbi, and Claudio di Prisco

Dipartimento di Ingegneria Civile e Ambientale, Politecnico di Milan, Piazza Leonardo da Vinci 32, Milan 2013, Italy
pietro.marveggio@polimi.it

Abstract. Describing solid to fluid and viceversa granular material phase transition is crucial when simulating many natural phenomena involving soils and characterized by large displacements, such as landslides or soil liquefaction. Despite the large number of available numerical codes developed to deal with large displacements, such as those based on MPM, PFEM or SPH formulations, evident in the literature is the lack of constitutive approaches capable of simulating phase transition. This work focuses on the implementation of a constitutive model, recently conceived by the authors, in the open source MPM code ANURA3D, modified to account the hydro-mechanical coupling for. Phase transition occurrence is ruled by both void ratio and granular temperature, being this latter a state variable measuring granular agitation. The evolution of granular temperature is governed by the energy balance equation. Under saturated conditions coupling terms are added to account for water damping effects on granular agitation. The implementation is tested by simulating two undrained triaxial tests, performed on an ideal monodisperse granular specimen by changing the imposed strain rate.

Keywords: Phase transitions · Material Point Method · large displacements · constitutive modelling

1 Introduction

In nature, granular materials may behave like solids, when grains are packed in a network of force chains, or fluids, when grains interaction is mainly governed by collisions. A solid-like behavior is generally exhibited when the granular network is confined and when relatively dense assemblies are considered. Under this state, the material is commonly characterized by small strain rates. On the other hand, a fluid like regime is mainly attained under loose conditions and large strain rates. Under saturated conditions, the liquid phase plays a role interacting with both solid skeleton, as interstitial pressure, and colliding particles.

Reproducing this twofold behavior represents a challenge for geotechnical, hydraulic and industrial applications. In the last decade, the use of numerical codes for simulations has become very common, due to consistent developments of numerical tools. Numerical modelling are based on either discontinuum (e.g. [5, 17]) or continuum (e.g. [10])

A. Ferrari et al. (Eds.): CNRIG 2023, SSGG, pp. 452–459, 2023.
https://doi.org/10.1007/978-3-031-34761-0_55

mechanics. The first class of approaches, among which the Discrete Element Method formulation based tools are the most popular, are typically employed when dry conditions are analyzed, as the presence of a liquid phase can be considered, at a relatively small computing cost, only by introducing equivalent interacting mechanisms among particles, without directly solving the full liquid-solid coupled interactions. In contrast, continuum-based numerical codes, capable of accounting for separately solid and liquid phase, like those adopting (i) Eulerian (such as Lattice-Boltzmann Methods (LBMs) [15, 16, 21]), (ii) Lagrangian (such as Particle Finite Element Methods (PFEM) [9] or Smoothed Particles Hydrodynamics (SPH) methods [28]) and (iii) Eulerian-Lagrangian formulations (such as Material Point Methods (MPM) [1, 2, 5, 7, 8, 32]) are becoming quite common.

In order to tackle real problems, the uncertainties still remain numerous. These are related, for instance, to the choice of the constitutive relationship and the calibration of constitutive parameters. In case of dry granular materials, incompressible frictional rigid-plastic constitutive relationships are used and very rarely the so-called μ-I rheologies [14]. In case of saturated either cohesive or granular materials, Bingham rheological laws (or similar) are employed. All these constitutive relationships treat the material as a one phase continuum and they disregard its microstructural evolution due to the increase in strain rate and void ratio [22, 27], fundamental for simulating the passage form solid-to fluid-like material behavior and vice versa ("phase transitions").

Only very recently, a class of constitutive relationships, based on kinetic theories of granular gasses [12], considering the material as a two-phase continuum and capable of describing material phase transition, from quasi static (at inception) to collisional (during propagation) conditions has been conceived [36, 41, 42]. The main purpose of these constitutive relationships is to account for the dependance of both volumetric deformability and increase in dissipations on strain rate. Under saturated conditions, as was previously stated, the most important limitation of standard rheologies derives from the assumption according to which the suspension (made by solid grains and water) behaves like a one phase material. In contrast, especially during force chains arising and collapse processes, dealing with a fully mixture approach is necessary [41].

In this paper, the constitutive model developed by Marveggio et al. [28] and implemented under dry conditions in the open source MPM code ANURA3D (Marveggio et al. [29]), is extended to saturated conditions, by following the same approach suggested by Vescovi et al. [41]. In Sect. 2 the model formulation is briefly described, in Sect. 3 its implementation is discussed, while in Sect. 4 the model is tested by simulating two undrained triaxial tests, performed on an ideal monodisperse granular specimen by changing the imposed strain rate.

2 Constitutive Model

As is discussed in Vescovi et al. [41], the model is formulated according to the mixture theory, thus a parallel scheme is adopted for stress $\boldsymbol{\sigma}$:

$$\boldsymbol{\sigma} = \boldsymbol{\sigma}_s + \boldsymbol{\sigma}_l, \tag{1}$$

where $\boldsymbol{\sigma}_s$ and $\boldsymbol{\sigma}_l$ are the stress second order tensors associated with solid and liquid phases, respectively.

According to Marveggio et al. [22], the solid stress tensor is obtained by adding a collisional ($\boldsymbol{\sigma}_{col}$) to a quasi-static ($\boldsymbol{\sigma}_{qs}$) contribution:

$$\boldsymbol{\sigma}_s = \boldsymbol{\sigma}_{col} + \boldsymbol{\sigma}_{qs}. \tag{2}$$

The collisional contribution is modelled according to the kinetic theories of granular gases [3, 6, 11, 13, 19], developed for ideal, monodisperse spherical deformable granular assemblies. Kinetic theories introduce in the model the dependence on state variables void ratio e and granular temperature T [3], which is a macroscopic measure of the grain agitation at the microscale, ruled by the non-local fluctuating energy E_{fk} balance, suitably modified to account for the presence of the liquid phase, as is described in the following.

The quasi-static stress is evaluated according to a rate-independent elastic-plastic constitutive relationship with a mixed isotropic-kinematic strain hardening, where a closed shaped yield locus is employed, based on the critical state theory, interpreted as the steady condition when granular agitation tends to zero [28].

The liquid phase is assumed to be incompressible and its constitutive relationship to only depend on e, in agreement with what experimentally obtained by Boyer et al. [4], Guazzelli and Pouliquen [12] and Pan et al. [24]:

$$\boldsymbol{\sigma}_l = u_w \boldsymbol{I} + 2\eta(e)\dot{\boldsymbol{\varepsilon}}_d, \tag{3}$$

where u_w is the pore water pressure, $\boldsymbol{I}$ is a second order identity tensor, $\dot{\boldsymbol{\varepsilon}}_d$ is the deviatoric strain rate second order tensor and η a viscosity coefficient, accounting for the dissipation of energy (depending this latter on the presence of the suspended granular phase), function of the solid concentration (or void ratio) as follows:

$$\eta(e) = \eta_0 F_\eta(e), \tag{4}$$

where η_0 is the molecular viscosity of the liquid and:

$$F_\eta(e) = \begin{cases} C_\eta \frac{f_\eta(e)}{f_\eta(e)+C_\eta} & e \geq e_c \\ C_\eta & e < e_c \end{cases}, \tag{5}$$

where C_η is a non-dimensional constitutive parameter and f_η is a non-dimensional function describing the dependency of η on the void ratio for large values of e [18, 30, 33]:

$$f_\eta(e) = \left(1 - \frac{1+e_c}{1+e}\right)^{-2.5/(1+e_c)}. \tag{6}$$

The critical void ratio e_c, introduced in Eqs. 5 and 6 is the void ratio value associated with force chains arise under steady conditions. From a practical point of view, only when force chains are not formed (and the strain rate is sufficiently law), the deviatoric components of the liquid stress tensor are not negligible with respect to the deviatoric stresses associated with both collisional and quasi static contributions [30].

As is sketched in Fig. 1, the liquid phase is assumed to dissipate energy due to (i) geometrical effects (Γ_E), (ii) lubrications (Γ_L) and grain agitation (Γ_{GL}). As is clarified in the following, Γ_{GL} is assumed to affect even the solid phase energy balance.

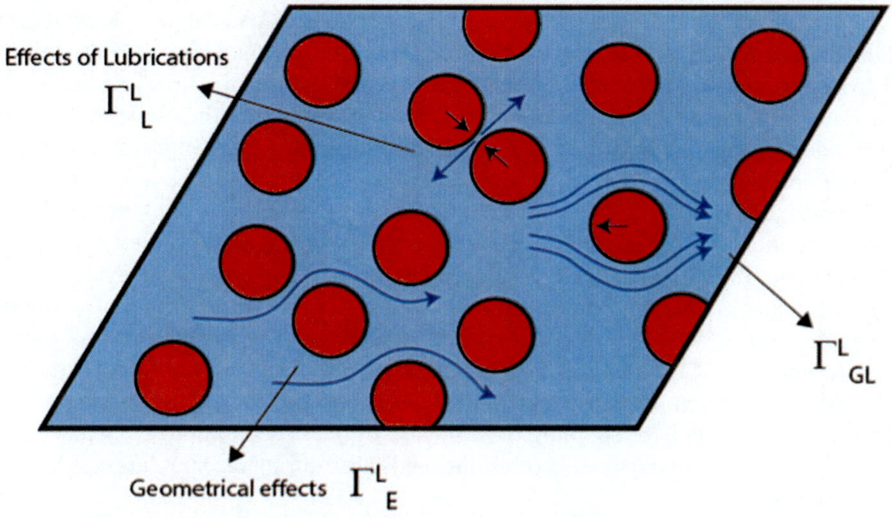

Fig. 1. Sketch enlightening liquid phase energy dissipation mechanisms.

As was previously mentioned, the fluctuating kinetic contributions related to collisional stress definition are ruled by the system agitation, which is measured by the internal variable T. The granular temperature evolution is assumed to be tuned by the balance of internal collisional energy, which can be either stored in terms of fluctuating kinetic ($\dot{E}_{kf}^{col}$) or elastically during particle collisions ($\dot{E}_{el}^{col}$) [22]. The formulation of this equation, originally proposed by Marveggio et al. [22] for dry evolving conditions, has been modified to account the liquid presence for:

$$\dot{E}_{el}^{col} + \dot{E}_{kf}^{col} = \dot{\boldsymbol{\varepsilon}}^T \boldsymbol{\sigma}^{col} - \Gamma^{col} - \nabla \cdot \boldsymbol{q} - \Gamma_{GL}, \tag{7}$$

where $\dot{\boldsymbol{\varepsilon}}^T \boldsymbol{\sigma}^{col}$ is the production term, $\nabla \cdot \boldsymbol{q}$ the advection term and Γ_{col} the energy dissipated during particle collisions [3, 29].

For Γ_{GL} the same expression suggested by Vescovi et al. [41] is adopted:

$$\Gamma_{GL} = \frac{54}{d^2} \eta_0 \frac{1}{1+e}. \tag{8}$$

In Eq. 8, theoretically derived from the Stokes laws, d is assumed to be the mean value of the particle size distribution.

3 Numerical Implementation in MPM

To testify the model applicability to real saturated boundary value problems, the authors implemented the constitutive model in the open-source dynamic explicit MPM code ANURA3D. ANURA3D is a software, using the material point method, developed since 2008 aimed at numerically simulating large deformations and soil–water–structure interaction problems.

The Material Point Method consists in a modified version of the well-known Finite Element formulation, accounting for the possibility for material points (MPs), acting as integration points and in which the material mass is lumped, to "flow" within a fixed mesh, discretizing the spatial domain. Thus, the convergence to the numerical solution is provided by a suitable set in terms of mesh refinement and quantity of discretizing material points [29].

Marveggio et al. [29] discussed the implementation of the dry formulation of the previously mentioned constitutive model, considering a single point formulation [32]. This paper assumes the soil to be a fully saturated mixture of a two constituent system (liquid and solid grains). This two-phase continuum is numerically discretized by considering two distinct sets of MPs, one for each phase, respectively. Despite the two phase single point formulation [44] being more computationally efficient, this choice is required in case of problems in which phase transition takes place, according to the authors, to fully account for the two phase coupling even in case of phase separation, i.e. when part of the domain, initially occupied by both solid and liquid, during the simulation is left with one constituent only.

The formulation of the two-phase double-point approach is based on the mixture theory and considers separately the mass and the linear momentum balance of the liquid and solid constituents.

Since solid and liquid phases are separately represented by two different sets of Lagrangian MPs, each moves along with their own layer such that the mass of all MPs remains constant. This implies that the conservation of both solid and liquid mass is a priori satisfied [10].

On the other hand, the linear momentum balance is enforced separately on the two phases, by considering buoyancy and drag interactions [10].

Once the energy balance in the framework of the mixture theory is formulated and assuming the “quasi static” and “liquid” balances are a priori satisfied as the constitutive model is considered [23], only the one associated with collision terms and their coupling due to the liquid presence must be solved (Eq. 7). To this aim, as was in Marveggio et al. [23], T is added as a material variable on the solid MPs only. T is linearly mapped to the node for the MPM solution scheme, while, as is done for linear momentum balance, a forward Euler approach is used for time integration.

4 Validation of the Model Numerical Implementation

Two undrained standard triaxial compression tests are numerically simulated by imposing two different strain rates (deviatoric strain rate $\dot{\varepsilon}_d = 1\ \text{s}^{-1}$ and $\dot{\varepsilon}_d = 10\ \text{s}^{-1}$) on two identical cubic loose ($e = 0.69$, with $e_c = 0.685$) specimens initially isotropically consolidated under an effective pressure of 50 kPa. The constitutive parameters employed for the granular material are taken from [22, 30]. The ANURA3D numerical results are validated against those obtained by integrating the constitutive law in MATLAB.

In Fig. 2, the numerical results, obtained by imposing $\dot{\varepsilon}_d = 1\ \text{s}^{-1}$ (Fig. 2a) and $\dot{\varepsilon}_d = 10\ \text{s}^{-1}$ (Fig. 2b), are plotted in the deviatoric stress versus deviatoric strain plane. For $0.04 < \varepsilon_d < 0.06$ the data are also magnified by putting in evidence the (i) quasi-static, (ii) collisional and (iii) liquid contributions.

Initially, the quasi static contribution is the one prevailing and, thus, the two curves plotted in Fig. 2a and 2b coincide. For ε_d>0.02, the quasi static contribution is nil and, as was observed in Berzi et al. [3], Redaelli and di Prisco [27] and Vescovi et al. [31] for steady conditions, since force chains disappear, the deviatoric stress is orders of magnitude less then that calculated when force chains develop in the granular material. In this case, the deviatoric stress is equal to the sum of liquid and collisional contributions. In the first case (Fig. 2a), the two contributions are almost equivalent, while in the second one (Fig. 2b) the collisional one is prevailing.

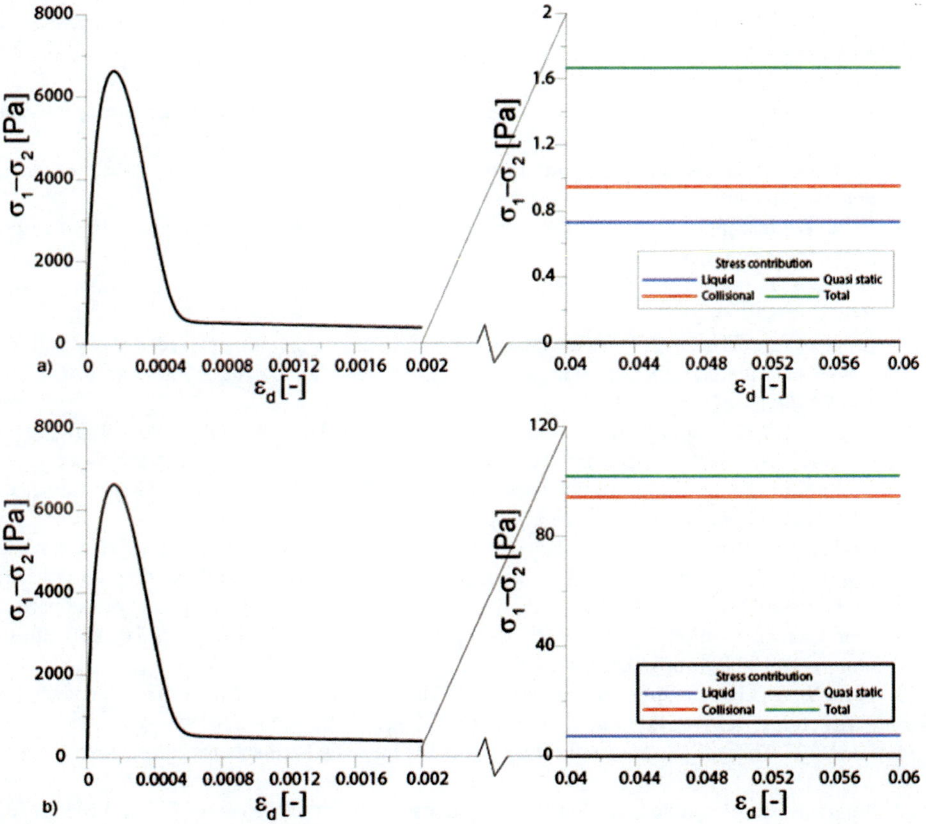

Fig. 2. Undrained standard triaxial compression test results: a) $\dot{\varepsilon}_d = 1s^{-1}$, b) $\dot{\varepsilon}_d = 10s^{-1}$.

5 Concluding Remarks

In this paper, a recent constitutive model, inspired to the work of Marveggio et al. [22] for dry conditions, is extended to saturated conditions and implemented in the open source MPM code ANURA3D. This new version of the model is aimed at numerically simulating phase transition in granular materials under saturated conditions. In the collisional regime, under saturated conditions, the energy balance equation modifies, since liquid-solid phases interact reciprocally: the liquid viscosity damps the granular agitation, affecting the mechanical behavior of the suspension. The model, conceived in the framework of the mixture theory, accounting for the two phases separately, seems to correctly reproduce the physical phenomena associated with phase transition. The implementation of the model, requiring the solution of an additional field equation, has been validated with reference to standard undrained triaxial test performed at different strain rates.

References

1. Abe, K., Soga, K., Bandara, S.: Material point method for coupled hydromechanical problems. J. Geotech. Geoenviron. Eng. **140**, 04013033 (2013)
2. Bandara S.: Material point method to simulate large deformation problems in fluid-saturated granular medium. PhD thesis, University of Cambridge (2013)
3. Berzi, D., Jenkins, J.T.: Steady shearing flows of deformable, inelastic spheres. Phys. Fluids **23**(1), 22–25 (2015)
4. Boyer, F., Guazzelli, É., Pouliquen, O.: Unifying suspension and granular rheology. Phys. Rev. Lett. **107**(18), 1–5 (2011)
5. Calvetti, F., di Prisco, C.G., Vairaktaris, E.: DEM assessment of impact forces of dry granular masses on rigid barriers. Acta Geotech. **12**(1), 129–144 (2016). https://doi.org/10.1007/s11440-016-0434-z
6. Campbell, C.S.: Rapid granular flows. Annu. Rev. Fluid Mech. **22**(1), 57–90 (1990)
7. Ceccato, F., Redaelli, I., di Prisco, C., Simonini, P.: Impact forces of granular flows on rigid structures: comparison between discontinuous (DEM) and continuous (MPM) numerical approaches. Comput. Geotech. **103**, 201–217 (2018)
8. Ceccato, F., Leonardi, A., Girardi, V., Simonini, P., Pirulli, M.: Numerical and experimental investigation of saturated granular column collapse in air. Soils Found. **60**, 683–696 (2020)
9. Cremonesi, M., Franci, A., Idelsohn, S., Oñate, E.: A state of the art review of the particle finite element method (PFEM). Arch. Comput. Meth. Eng. **27**(5), 1709–1735 (2020). https://doi.org/10.1007/s11831-020-09468-4
10. Fern, J., Rohe, A., Soga, K., Alonso, E.: The Material Point Method for Geotechnical Engineering: A Practical Guide, 1st edn. CRC Press, Boca Raton (2019)
11. Garzó, V., Dufty, J.W.: Dense fluid transport for inelastic hard spheres. Phys. Rev. E **59**(5), 5895 (1999)
12. Guazzelli, É., Pouliquen, O.: Rheology of dense granular suspensions. J. Fluid Mech. **852**, P1 (2018)
13. Jenkins, J., Savage, S.: A theory for the rapid flow of identical, smooth, nearly elastic, spherical particles. J. Fluid Mech. **130**, 187–202 (1983)
14. Jop, P., Forterre, Y., Pouliquen, O.: A constitutive law for dense granular flows. Nature **441**(7094), 727–730 (2006)
15. Leonardi, A., et al.: Physical review e - statistical, nonlinear, and soft matter. Physics **92**(5), 052204 (2015)
16. Leonardi, A., Pirulli, M.: Procs. EGU General Assembly 2018, 19494-1 (2018).
17. Lu, C., Tang, C., Chan, Y., Hu, J., Chi, C.: Eng. Geol. **183**, 14–30 (2014)
18. Krieger, I., Dougherty, T.J.: A Mechanism for non-Newtonian flow in suspensions of rigid spheres. Trans. Soc. Rheol. **3**(1), 137– 152 (1959)

19. Lun, C.K.K.: Kinetic theory for granular flow of dense, slightly inelastic, slightly rough spheres. J. Fluid Mech. **233**, 539–559 (1991)
20. Mangeney, A., Heinrich, P., Roche, R.: Analytical solution for testing debris avalanche numerical models. Pure Appl. Geophys. **157**(6–8), 1081–1096 (2000)
21. Marchelli, M., Leonardi, A., Pirulli, M., Scavia, C.: On the efficiency of slit dams in retaining granular flows. Geotechnique **70**(3), 226–237 (2020)
22. Marveggio, P., Redaelli, I., di Prisco, C.: Phase transition in monodisperse granular materials: How to model it by using a strain hardening visco-elastic-plastic constitutive relationship. Int. J. Numer. Anal. Meth. Geomech. **46**, 2415–2445 (2022)
23. Marveggio P., Zerbi M., di Prisco C.: How simulating phase transition during impacts of dry granular flows against rigid barriers by using a MPM approach. In: preparation (2023)
24. Pan, Z., de Cagny, H., Weber, B., Bonn, D.: S-shaped flow curves of shear thickening suspensions: direct observation of frictional rheology. Phys. Rev. E**92**(3), 032202 (2015)
25. Pudasaini, S.P., Hutter, K.: Avalanche Dynamics: Dynamics of Rapid Flows of Dense Granular Avalanches. Springer, Berlin, New York (2007)
26. Redaelli, I., Di Prisco, C., Vescovi, D.: A visco-elasto-plastic model for granular materials under simple shear conditions. Int. J. Numer. Analyt. Meth. Geomech. **40**(1), 80–104 (2016)
27. Redaelli, I., di Prisco, C.: DEM numerical tests on dry granular specimens: the role of strain rate under evolving/unsteady conditions. Granular Matter **23**(2), 1–34 (2021). https://doi.org/10.1007/s10035-021-01091-9
28. Shen, W., Zhao, T., Zhao, J., Dai, F., Zhou, G.G.: Quantifying the impact of dry debris flow against a rigid barrier by DEM analyses. Eng. Geol. **241**, 86–96 (2018)
29. Sulsky, D., Zhou, S., Schreyer, H.: Application of a particle-in-cell method to solid mechanics. Comput. Phys. Commun. **87**(1–2), 236–252 (1995)
30. Vescovi, D., Marveggio, P., Di Prisco, C.G.: Saturated granular flows: constitutive modelling under steady simple shear conditions. Géotechnique **70**(7), 608–620 (2020)
31. Vescovi, D., Redaelli, I., di Prisco, C.: Modelling phase transition in granular materials: from dis-continuum to continuum. Int. J. Solids Struct. **202**, 495–510 (2020)
32. Yerro, A., Alonso, E.E., Pinyol, N.M.: The material point method for unsaturated soils. Geotechnique **65**, 201–217 (2015)
33. Einstein, A.: On the motion of small particles suspended in a stationary liquid, as required by the molecular kinetic theory of heat. Ann. Phys. **322**, 549–560 (1905)

Effect of Soil Permeability on CPTu Test Results in Structured Clay Soils

Kateryna Oliynyk[1,2], Matteo O. Ciantia[1], and Claudio Tamagnini[2](✉)

[1] School of Science and Engineering, University of Dundee, Dundee, Scotland, UK
m.o.ciantia@dundee.ac.uk

[2] Department of Civil and Environmental Engineering, University of Perugia, Perugia, Italy
claudio.tamagnini@unipg.it

Abstract. The PFEM implementation of fully coupled equations governing the solid skeleton deformations and pore fluid flow in porous saturated media is used to investigate the effects of soil permeability on cone tip resistance and excess pore pressure measurements in CPTu tests performed in a structured soil deposit. In the simulations, soil behavior is modeled with the *FD_Milan* model, a finite deformation version of the isotropic hardening model for structured soil originally developed at the Politecnico of Milano by Nova and coworkers. The results indicate that as soil permeability increases, the volumetric deformations around the cone tip become non negligible and the excess pore pressure in correspondence of the cone probes is also affected. The destructuration process caused by the intense deformations around the cone tip and along the shaft is responsible for a significant reduction of the expected net cone resistance based on the apparent value of the soil undrained strength.

Keywords: CPTu Testing; Structured Clays · PFEM Modeling · Finite Deformations

1 Introduction

Recent development in particle-based numerical methods such as MPM and PFEM have allowed to address in a rational, accurate and efficient way the interpretation of the complex coupled deformation and flow processes which occur during a CPTu test in the soil around the advancing cone tip. One important advantage is the possibility of describing soil behavior adopting advanced constitutive models with realistic stress-dilatancy laws. Examples of recent calibration chamber (CC) CPTu test simulations are provided in refs. [3] and [6].

In this work, the geomechanics-oriented PFEM code GPFEM [2] is used to investigate the effects of soil permeability on predicted excess pore pressures in the soil around the penetrating cone and the effect of soil bonding on predicted values of cone tip resistance on a soft natural clay – Osaka clay [1]. The soil is modelled with the *FD_Milan* model, an isotropic hardening elastoplastic model for natural, structured soils developed

A. Ferrari et al. (Eds.): CNRIG 2023, SSGG, pp. 460–467, 2023.
https://doi.org/10.1007/978-3-031-34761-0_56

for finite deformations by Oliynyk et al. [7], adopting a multiplicative decomposition of the deformation gradient into elastic and plastic parts.

The focus of this paper is twofold: a) defining the permeability threshold which marks the practical limit of the undrained deformation of the soil; b) assessing the capability of this particular in-situ test of characterizing the strength properties of natural structured clays with variable levels of initial bonding.

2 The FD_Milan Model

The *FD_Milan* model [7] is a non-associative, isotropic hardening finite-deformation plasticity model for structured soils and weak rocks based on the multiplicative decomposition of the deformation gradient and on the adoption of a suitable free energy function to describe the elastic response of the material.

The model is equipped with two stress-like internal variables. The first, P_s (preconsolidation pressure of the soil if it was unstructured), accounts for the hardening/softening effects due to volumetric and deviatoric plastic strains; the second, P_t (bond strength), quantifies the effects of material fabric and bonding. A representation of the yield surface in the Kirchhoff stress invariants space $Q : P$, highlighting the role of the internal variables, is given in Fig. 1.

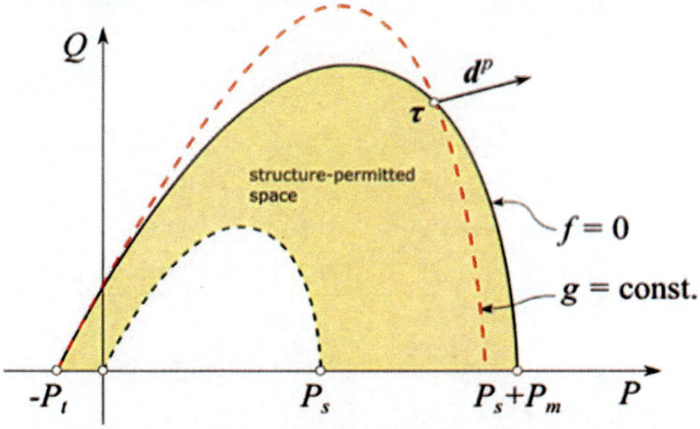

Fig. 1. Yield surface and plastic potential of the FD_Milan model.

The quantities P_s and P_t, in turn depend on the strain-like internal variables E_v^p, E_s^p and N_v^p, representing the accumulated plastic volumetric deformation, the accumulated plastic deviatoric deformation and the accumulated norm of the plastic volumetric deformation, as detailed in [7].

A positive value of P_t results in an expansion of the elastic domain in stress space. On the positive part of the P axis, the isotropic yield stress in compression is increased by a quantity $P_m = kP_t$, with k a material constant. On the negative axis, the vertex of the yield surface is displaced from the origin to $P = -P_t$. In this way the model allows to capture true cohesion and non-negligible tensile strength characteristic of structured soils. The region in the stress space contained between the actual yield surface (full black line in

Fig. 1) and the intrinsic yield surface of the unstructured material (with $P_t = P_m = 0$, , black dashed line in the figure), represents the so-called structure-permitted space, which is accessible only by the combined effects of fabric and intergranular bonding. Irreversible deformations and material progressive destructuration are a direct consequence of soil fabric rearrangements and the concurrent breakage of intergranular bonds. From a macroscopic point of view, this process induces a progressive reduction of P_t with increasing accumulated plastic deformations. This process is described by a suitable monotonic hardening law, yielding $P_t = 0$ as the asymptotic value for very large accumulated plastic deformations. To provide a characteristic length scale to the constitutive equation, with the objective of regularizing the numerical solution in presence of strain localization, the integral non-local approach of refs. [5, 7] has been adopted, in which the internal variables E_v^p, E_s^p and N_v^p are treated as non-local, spatially averaged quantities over a neighborhood Ω of the material point. The size of this neighborhood is controlled by a material constant l_c, called characteristic length.

In this work, the *FD_Milan* model has been used to describe the mechanical behavior of Osaka clay, a soft natural clay. The calibration of the material constants has been made based on the experimental data provided in ref. [1], see Table 1. As the flow rule of the material has been assumed associative, equal values of the constants α, μ and M_c have been adopted for the yield function f and the plastic potential g.

Table 1. Material constants of the *FD_Milan* model for the Osaka clay.

$\hat{\kappa}$ (-)	G_0 (MPa)	α (-)	P_{ref} (MPa)	$M_{f,c}$ (-)	α_f (-)	μ_f (-)	$M_{g,c}$ (-)
1.82	3.0	0.0	5.0	1.10	0.75	1.50	1.10
α_g (-)	μ_g (-)	ρ_s (-)	ρ_t (-)	ξ_s (-)	ξ_t (-)	k (-)	l_c (mm)
0.75	1.50	8.33	15.0	0.0	3.0	5.0	5.0

3 PFEM Model and Simulations Program

The numerical simulations are aimed to model a series of standard CPTu tests in a calibration chamber filled with fully saturated Osaka clay. The piezocone has a radius *R*.

of 17.8 mm and a cone tip angle of 60°. The calibration chamber has a radius R_C = 0.45 m (~25 *R*) and a height *H* of 1.05 m.

Thanks to the geometries of the piezocone and the calibration chamber, the problem has been analyzed assuming axisymmetric conditions. A sketch of the problem geometry along with some images of the PFEM discretization adopted in one of the simulations at 3 different tip advancement depths is given in Fig. 2. It can be noted that the adaptive refinement algorithm is able to track the damage zone around the moving piezocone tip and reduce the element size based on the imposed accuracy requirements. All simulations have been performed adopting the mixed $\boldsymbol{u}^s - \Theta - u$ formulation of Monforte et al. [4], considering the fully coupled hydromechanical problem of deformation and transient pore water flow induced in the soil mass by the piezocone advancement.

The bottom and lateral boundaries of the calibration chamber are assumed as rigid, impervious and perfectly rough. At the top surface of the soil body a uniform normal pressure $q_0 = 100$ kPa and a constant pore water pressure $u = 0$ is imposed. Given the relatively small dimensions of the calibration chamber, gravity is set to zero ignoring the self-weight. Consistently with these assumptions, the initial Cauchy effective stress in the soil mass is uniform, with components $\sigma_z = 100$ kPa and $\sigma_r = K_0\sigma_z$, with $K_0 = 0.5$, and the initial pore water pressure is uniform and equal to zero.

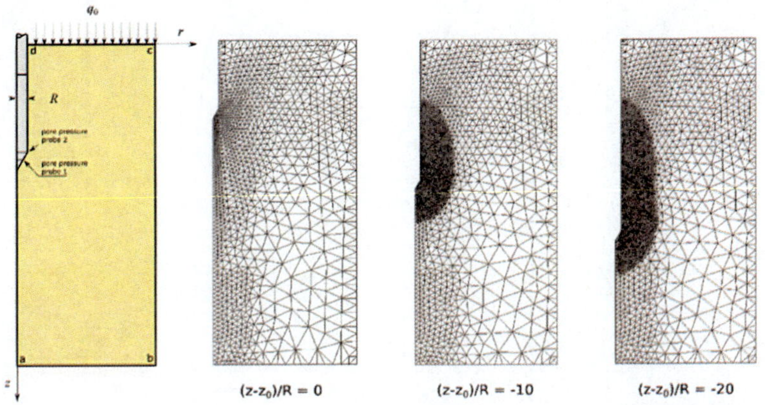

Fig. 2. Problem geometry and spatial discretization at 3 different tip advancement depths.

The initial value of P_s is set to 120 kPa. Six simulations, considering 4 different values for the hydraulic permeability, k_h, in the range 1.0e−9 to 1.0e6 m/s, and 3 different initial values for the bond strength, P_{t0}, ranging from 0 to 60 kPa were performed. The piezocone has been modelled as a rigid smooth impervious body. At the beginning of the simulations, it is wished-in-place at an initial depth $z_0 = 0.25$ m and then displaced vertically at a constant penetration speed of 2.0 cm/s, up to a depth $z = z_0 + 20R$. Table 2 lists the complete set of simulations run.

Table 2. Program of PFEM simulations of CPTu tests.

Run # →	r100	r101	r102	r103	r203	r303
P_{t0}(kPa)	30.0	30.0	30.0	30.0	60.0	0.0
$k_h(m/s)$	1.0e−6	1.0e−7	1.0e−8	1.0e−9	1.0e−9	1.0e−9
h_{crit}/l_c	0.1	0.1	0.1	0.1	0.1	0.1

4 Effect of Soil Permeability

The first group of four simulations (r100-r103) allows to explore the effects of soil permeability on the predicted values of cone tip resistance and excess pore pressures at the two pore pressure transducers, as well as on the distribution of the accumulated plastic deformations, of the internal variables and of the excess pore pressure in the soil. The profiles of net cone tip resistance $q_n := q_c - \bar{\sigma}_{z0}$, where $\bar{\sigma}_{z0}$ is the total vertical geostatic stress at the measurement depth, and of the excess pore water pressure Δu_1 with the normalized penetration depth $Z = (z - z_0)/R$, obtained with the four hydraulic permeability values considered, are shown in Figs. 3a and 3b.

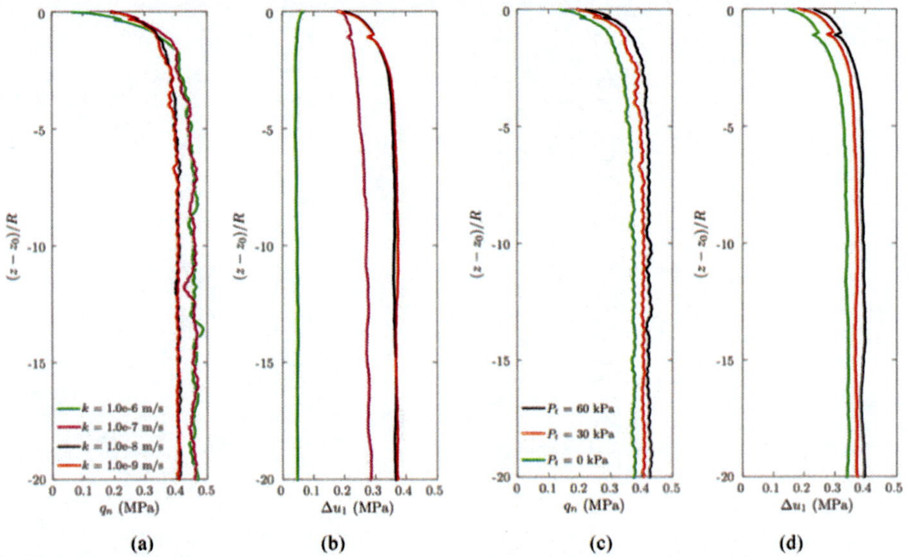

Fig. 3. Simulations r100-r103: a) profiles of net cone tip resistance, q_n, and b) excess pore pressure at transducer 1, Δu_1, with cone tip advancement for different soil permeabilities. Simulations r103, r203, r303: c) profiles of net cone tip resistance, q_n, and d) excess pore pressure at transducer 1, Δu_1, for different initial values of bond strength.

The effect of k on the computed excess pore water pressures is quite significant: as expected, the lower is the soil permeability, the higher is the value of Δu computed at the transducer. The results for $k = 1.0\text{e}{-8}$ m/s and 1.0e−9 m/s are quite similar, indicating that the undrained limit has been reached in these two cases. The values of Δu obtained for the highest permeability are very small, indicating that the soil is approaching the drained limit. On the contrary, the differences observed in the net cone resistance in the four simulations are relatively small. As expected, the largest steady-state value of q_n is obtained for the highest permeability, and the smallest for the lowest permeability. However, the difference in the two steady-state values of q_n are smaller than 15% of the steady-state q_c value computed in the lowest permeability case.

The contour maps of the accumulated plastic volumetric and deviatoric deformations computed in the two simulations r100 ($k = 1.0\text{e}{-6}$ m/s) and r103 ($k = 1.0\text{e}{-9}$ m/s) are shown in Figs. 4a and 4b. The difference in the order of magnitude of the values of E_v^p computed at the maximum cone penetration is clearly apparent. Very large positive volumetric deformations develop around the penetrating cone in the higher permeability case, while in the lower permeability case the values of E_v^p are smaller by an order of magnitude. In the low permeability case, the plastic region around the advancing piezocone is quite large, extending for about 6R both vertically, below the cone tip, and horizontally from the piezocone axis. The plastic zone in characterized by a complex pattern of accumulated volumetric and deviatoric plastic deformations, with clearly visible bands of localized shear and volumetric deformations. In the high permeability case, the collapse mechanism is quite different and more similar to a "punching" process, with a much smaller and homogeneous plastic zone around the cone tip and shaft.

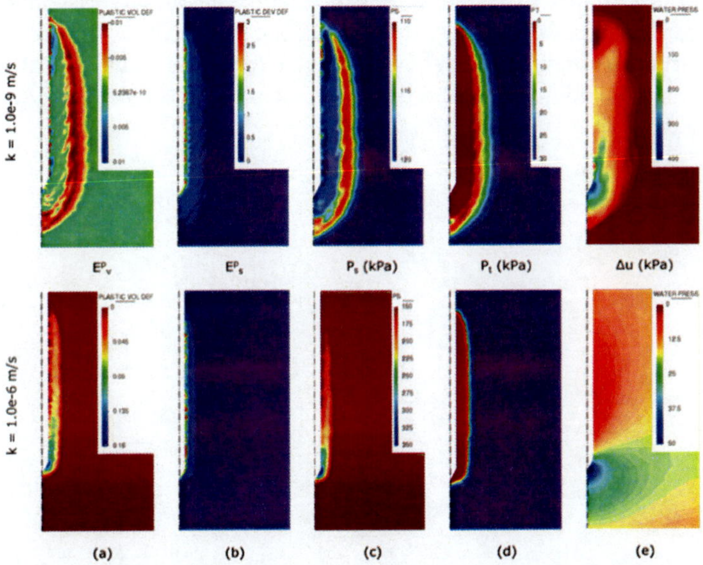

Fig. 4. Simulations r100 and 103: contour maps of the accumulated plastic volumetric (a) and deviatoric (b) deformations, preconsolidation pressure (c), bond strength (d) and excess pore pressure (e) at the maximum cone penetration depth.

Figures 4c to 4e report the contour maps of the preconsolidation pressure, P_s, the bond strength, P_t, and the excess pore water pressure, Δu. As expected, large differences are apparent in the excess pore pressure distributions in the two cases. The spatial distribution of the bond strength confirms the previous observations concerning the fundamentally different nature of the deformation patterns for the low and high permeability cases. In both simulations, the accumulated plastic deformations are sufficient to create a zone of completely destructured soil around the cone tip and shaft.

5 Effect of Soil Destructuration on Net Cone Tip Resistance

The effects of the initial degree of bonding are shown in Figs. 4 and 5, which refer to simulations r103, r203 and r303. In these simulations the soil deforms in almost undrained conditions. Figure 3 shows the profiles of net cone resistance, q_n, and excess pore pressure Δu_1 as a function of Z. After an initial transient stage up to $Z = 5$, all the profiles reach a stationary state up to the final value of the penetration $Z = 20$. It can be observed that the stationary values of q_n, Δu_2 and Δu_1 increase with increasing initial value of the bond strength, P_{t0}. The higher is the initial structure of the soil, the higher are the cone resistance and the excess pore pressures during cone advancement. It is however worth noting that the differences between the stationary values of q_n are relatively limited, in spite of the large differences in the values of P_{t0} considered.

This is to be attributed to the intense destructuration that the large accumulated plastic deformations induce in the soil around the piezocone tip and shaft, see Fig. 4. An insight into this destructuration process is shown in Fig. 5, the stress paths (SP) experienced by the soil in 4 material points located at radial distances of $2R$, $3R$, $4R$ and $5R$, respectively are represented.

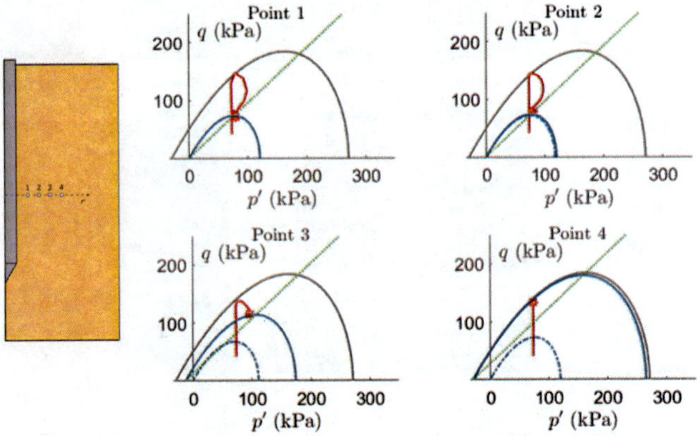

Fig. 5. Stress paths experienced by the soil in 4 material points at different distances from the piezocone axis.

The beginning of the SPs marks the initial state of the soil at the beginning of the penetration process, while their ends – marked with a star – represent the current state of the soil, after a significant advancement of the cone tip from the depth of the four material points. In the figure, the full curve in black is the initial yield surface of the material; the dashed-dotted line in blue is the yield surface of the unstructured soil, and the full curve in blue is the yield surface at the current state.

It is immediately apparent that the soil up to $2R$ form the shaft experiences a complete destructuration, as the current yield surface coincides with that of the unstructured soil. During the initial portions of the SPs, the soil is still in the elastic range and the deformation occurs at constant mean effective stress, as in perfectly undrained conditions. After yielding, the destructuration causes the yield surfaces at points 1 and 2 to shrink while the stress paths bend downwards. As the distance to the piezocone axis increases, the destructuration process becomes less marked. At $4R$ from the shaft (point 4), the final state of the soil has just reached the initial yield surface and no destructuration has occurred.

The main implication of these observations is that the net cone resistance appears to be mainly controlled by the response of the soil in a relatively narrow zone, extending to about 3 radii from the piezocone axis, where the destructuration is intense and the soil strength is not so different from that of the corresponding unstructured soil.

6 Concluding Remarks

The soil response to the penetration of the piezocone in CPTu tests can be modelled quantitatively in a rational and computationally efficient way with the PFEM. The simulations performed with variable permeability indicate that, for soft clay soils, the undrained limit is reached when the soil permeability is lower than 1.0e-9 m/s. For significantly higher soil permeabilities, the excess pore pressure dissipation around the cone tip and shaft is non negligible and the conventional interpretation methods based on the assumption of undrained conditions become questionable.

Comparing CPTu tests performed considering broad values of initial bond strength shows that the net cone resistance tends to be only marginally affected by the in-situ soil structure. Although q_n does increase with increasing P_{t0}, the influence of this initial condition remains relatively modest, due to the intense destructuration processes taking place in the soil region around the piezocone tip and shaft A practical consequence of this phenomenon is that a soil characterization based mostly on CPT tests could lead to an underestimation of the structure of a natural deposit in-situ.

References

1. Adachi, T., et al.: Stress–strain behavior and yielding characteristics of Eastern Osaka clay. Soils Found. **35**(3), 1–13 (1995)
2. Carbonell, J.M., Monforte, L., Ciantia, M.O., Arroyo, M., Gens, A.: Geotechnical particle finite element method for modeling of soil-structure interaction under large deformation conditions. J. Rock Mech. Geotech. Eng. **14**(3), 967–983 (2022)
3. Martinelli, M., Galavi, V.: Investigation of the material point method in the simulation of cone penetration tests in dry sand. Comp. Geotech. **130**, 103923 (2021)
4. Monforte, L., Carbonell, J.M., Arroyo, M., Gens, A.: Performance of mixed formulations for the particle finite element method in soil mechanics problems. Comput. Part. Mech. **4**(3), 269–284 (2016). https://doi.org/10.1007/s40571-016-0145-0
5. Monforte, L., Ciantia, M.O., Carbonell, J.M., Arroyo, M., Gens, A.: A stable mesh-independent approach for numerical modelling of structured soils at large strains. Comp. Geotech **116**, 1–13 (2019)
6. Monforte, L., Gens, A., Arroyo, M., Mánica, M., Carbonell, J.M.: Analysis of cone penetration in brittle liquefiable soils. Comp. Geotech. **134**, 104123 (2021)
7. Oliynyk, K., Ciantia, M.O., Tamagnini, C.: A finite deformation multiplicative plasticity model with non – local hardening for bonded geomaterials. Comp. Geotech. **137**, 104209 (2021)

A Comparison Between Theoretical and Numerical Ultimate Failure Domains for Pile Groups

Francesco Potini(✉) and Riccardo Conti

Università degli Studi Niccolò Cusano, Roma, Italia
francesco.potini@unicusano.it

Abstract. Multiaxial strength envelopes have been recently proposed as a rational approach for the design of pile foundations subjected to inclined and eccentric loads, leading to more economical solutions with respect to standard methods applied in the design practice. Theoretical solutions based on the application of the Upper-Bound (UB) and Lower-Bound (LB) theorem of Limit Analysis have been proposed to derive the ultimate failure domain of pile groups. In this paper, theoretical predictions for a 4 × 1 pile group embedded in a homogeneous clayey soil deposit are validated against the results of numerical analyses carried out with the Finite Element code Abaqus. To take into account the nonlinear behaviour of reinforced concrete piles, a damaged plasticity model was used for the concrete, while the reinforcing bars were modelled as equivalent elastoplastic shell elements. The soil was modelled as a standard elastic-perfectly plastic material with a Tresca failure criterion to represent the undrained soil behaviour in a total stress analysis.

Keywords: Pile Foundations · Ultimate Failure Domains · Finite Element Analyses

1 Introduction

The scientific literature on numerical and experimental modelling of pile foundations in failure conditions is relatively recent. Based on the results of 3D Finite Element (FE) analyses, Gerolymos and Papakyriakopoulos [1] proposed a macro-element model for the analysis of single piles and pile groups under horizontal loads. Psychari and Anastasopoulos [2] analysed the nonlinear response of pile groups in clay, subjected to multiaxial loads. The Authors implemented a 3D FE model in Abaqus [6], using an elasto-plastic constitutive model with kinematic hardening for the soil and the Concrete Damage Plasticity model to describe the mechanical response of reinforced concrete (RC) piles. The same FE model was used by Sakellariadis and Anastasopoulos [3] to reproduce the experimental behaviour of a 2 × 1 pile group in a homogeneous deposit of saturated sand, as observed in small-scale centrifuge models. Centrifuge tests on reduced-scale models were carried out also by de Sanctis *et al.* [4], who analysed the behaviour of a circular pile group embedded in kaolin clay, subjected to a vertical eccentric load.

A. Ferrari et al. (Eds.): CNRIG 2023, SSGG, pp. 468–476, 2023.
https://doi.org/10.1007/978-3-031-34761-0_57

As shown theoretically by Di Laora *et al.* [5] and numerically by Sakellariadis and Anastasopoulos [3], the moment capacity of a pile group subjected to a vertical eccentric load is a result of two concurrent resistance mechanisms. The first one, M_{ax}, is due to the axial loads acting into the piles of the group, while the second one, M_b, stems from the possible contribution given by the formation of plastic hinges at the piles' head. The latter mechanism can be relevant in the case of large-diameter piles.

Di Laora *et al.* [5] proposed an Upper-Bound (UB) solution for the ultimate failure domain of pile groups under vertical eccentric loads, based on two main assumptions, *i.e.*: (i) the piles exhibit a rigid-perfectly plastic behaviour and (ii) the connecting raft is a rigid body not in contact with the soil. The theoretical solution takes into consideration two possible connections between the piles and the cap: either a hinge-connection, thus including only the M_{ax} contribution to the moment capacity of the group, or a rigid-plastic internal fixity, thus taking into account both the M_{ax} and M_b contributions.

This work presents some of the results of 3D FE analyses carried out with the software Abaqus [6]. The main objective of the numerical study is to validate the theoretical solution proposed by Di Laora *et al.* [5].

2 Problem Definition and Numerical FE Model

The case of a 4 × 1 pile group was analysed. The RC piles have a diameter $d = 1.5$ m, length $L = 20$ m, and spacing $s = 3d$. A class C25/30 concrete is assumed for the RC piles, with a typical reinforcement ratio $\rho = 1.8\%$ of B450C steel bars. The piles are embedded in a homogenous clay deposit with undrained shear strength $s_u = 50$ kPa and unit weight $\gamma = 20$ kN/m^3. The initial geostatic stress state is characterised by a ratio $\sigma_{h0}/\sigma_{v0} = 0.8$ between the total horizontal and vertical soil stress.

Total stress analyses were carried out, assuming undrained conditions. 8 node linear brick elements with reduced integration (C3D8R) were used both for the soil and the concrete. Reinforcement bars were modelled as an equivalent steel tube embedded in the concrete continuum elements, using 4 node linear surface elements with reduced integration (S4R).

Figure 1 shows the FE mesh adopted in the analyses, where the size of the elements (piles and soil close to the foundation) was chosen based on a preliminary parametric study in order not to affect the numerical results. To maximize the computational efficiency, symmetrical conditions along the (y-z) plane passing through the piles' axis were considered. Standard constraints were applied to the model boundaries, *i.e.*: normal displacements were restrained along the vertical boundaries, while all displacements were fixed at the base. In order to reduce boundary effects, vertical boundaries were placed at a distance of 10*d* from the piles' edge, while the base of the model was placed at a distance of 5*d* from the pile's end.

2.1 Constitutive Models and Soil-Pile Interface

A standard elastic-perfectly plastic constitutive model with a Tresca failure criterion and an associated flow rule was adopted for the soil. Table 1 summarizes the physical and constitutive parameters used in the analyses: the unit weight, γ; the Young modulus,

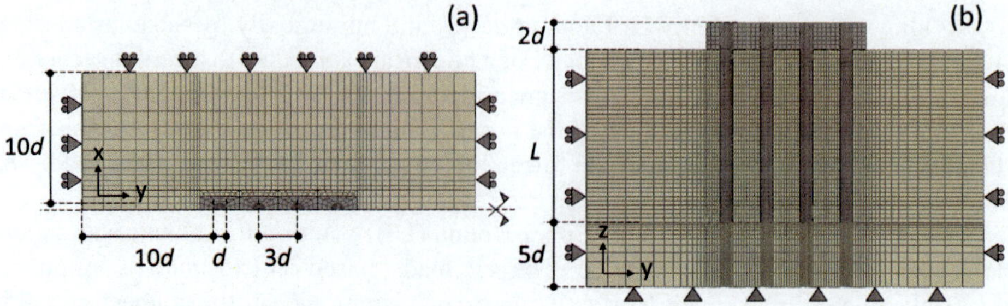

Fig. 1. Plane views of the mesh adopted for the 3D FE model of the 4x1 pile group: (a) horizontal x-y plane; (b) vertical y-z plane.

$E = E_u$; the Poisson's ratio, $\nu = \nu_u$; the friction angle, $\phi = \phi_u$; the cohesion, $c = s_u$ and the angle of dilatancy, ψ.

Table 1. Physical properties and constitutive parameters adopted for the soil

γ [kN/m^3]	E [MPa]	ν	ϕ [°]	c [kPa]	ψ [°]
20	21.6	0.475	0	50	0

The concrete was modelled using the Concrete Damaged Plasticity (CDP) model implemented in Abaqus [6], which allows reproducing the highly nonlinear behaviour exhibited by the concrete under both monotonic compression and tension. Damage evolution laws were not taken into consideration in this work, as the latter define the typical response of concrete under cyclic loading.

Figure 2 shows the stress-strain response assumed for the concrete under uniaxial conditions. The material behaviour in tension is linear until f_{t0}, beyond which a softening branch is detected. The stress-strain response in compression, instead, is linear-elastic until the yielding stress f_{c0}, beyond which a hardening and perfectly plastic behaviour is assumed until the failure stress f_{cu}. After f_{cu}, the material response follows a softening branch until complete failure due to crushing. Table 2 summarizes the physical and constitutive parameters adopted for the concrete: the dilation angle, ψ, and the eccentricity parameter, ε, define the flow rule; K controls the shape of the yield surface; f_{b0} and f_{c0} are the concrete biaxial and uniaxial compressive strength, respectively; μ^{visc} is a viscosity parameter, which allows regularizing the evolution laws of the material in the highly-nonlinear regime.

Finally, an isotropic elastic-perfectly plastic constitutive model was used for the reinforcement steel tube with a von Mises failure criterion and an associated flow rule. The physical and constitutive parameters adopted in the analyses are: $\gamma = 78$ kN/m^3, $E = 200$ GPa, $\nu = 0.2$ and yield strength $f_y = 450$ MPa.

The pile-soil interface was modelled with tensionless contact elements, which allow both sliding and separation. An elastoplastic frictional behaviour was assumed for the

Table 2. Physical and constitutive parameters adopted for the CDP model.

γ [kN/m^3]	E [GPa]	ν	ψ [°]	ε	f_{b0}/f_{c0}	K	μ^{visc}
20	21.5	0.15	30	0.1	1.16	0.667	10^{-4}

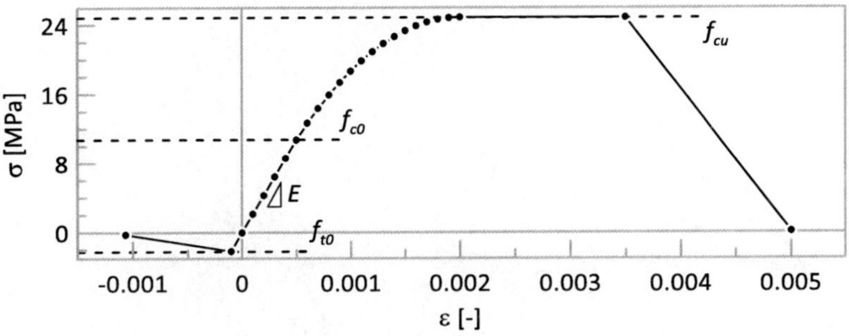

Fig. 2. Stress-strain response assumed for the concrete under uniaxial conditions.

interface, with the shear strength defined as:

$$\tau_{lim} = \mu\, \sigma_n (\leq \tau_{max}) \tag{1}$$

where σ_n is the normal stress acting on the interface, μ is the friction coefficient and τ_{max} is a threshold value for the maximum allowable shear strength. In order to obtain an equivalent cohesive behaviour of the interface along the whole pile length, a relatively high value of the friction coefficient was adopted ($\mu = 2$), together with $\tau_{max} = s_u$.

3 FE Model Validation

3.1 Structural Behaviour of the Single Pile

The simple case of a cantilever beam loaded by a concentrated load at the free edge was analysed to validate the structural behaviour of the RC piles. The beam has a circular transversal section ($d = 1.5$ m, $\rho = 1.8\%$) and a length $L = 15$ m. An axial load, N, is initially applied at the free edge, followed by a transversal displacement, v. The displacement is gradually increased until a plastic hinge forms at the fixed end of the beam.

Figure 3(a) compares the numerical and theoretical moment-displacement curves for three values of N. The dependency of the moment capacity of the RC section on the applied axial load is well reproduced by the CDP model, except for the ultimate failure condition, where, due to the confinement imposed by the steel tube, the numerical model exhibits a higher ductility compared with the theoretical one. As a result, the values of the bending moment at failure are slightly higher than theoretical predictions, as shown in Fig. 3(b).

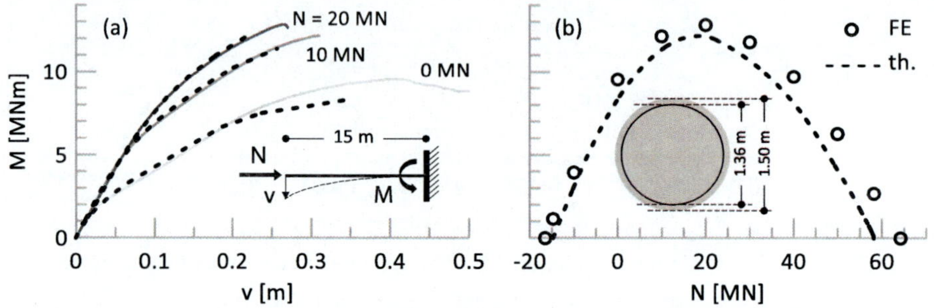

Fig. 3. Cantilever beam loaded by a concentrated load at the free edge. Comparison between numerical and theoretical results in terms of: (a) M-v curves and (b) N-M interaction diagram.

3.2 Vertical Bearing Capacity of the Single Pile

With the aim of validating the soil-pile numerical model, the case of a single pile subjected to a pure axial compression and uplift was modelled. The load is applied by imposing a vertical displacement at the pile head, and the results are compared with the theoretical predictions for the undrained bearing capacity under compression ($N_u = 5.51$ MN) and uplift ($S_u = 5.42$ MN).

Figure 4 represents the load-displacement response of the single pile. The shaft resistance is fully mobilized for a relatively small displacement (about 20 mm), consistently with [7]. In compression, the pile-soil interface exhibits the expected cohesive behaviour. During uplift, instead, a gap between the pile and the surrounding soil forms close to the pile head, and the theoretical value of the shaft resistance is not reached. As expected, the base resistance is mobilized only at large pile displacements, and the numerical plateau of the corresponding load-displacement curve is in good agreement with the theoretical prediction.

4 Analysis of the 4 × 1 Pile Group

In order to focus on the response of the sole pile group, the connecting raft is assumed not in contact with the soil in the 3D FE model.

The response of the pile group under pure axial compression and uplift is firstly investigated and compared with the single pile behaviour. As illustrated in Fig. 5, internal

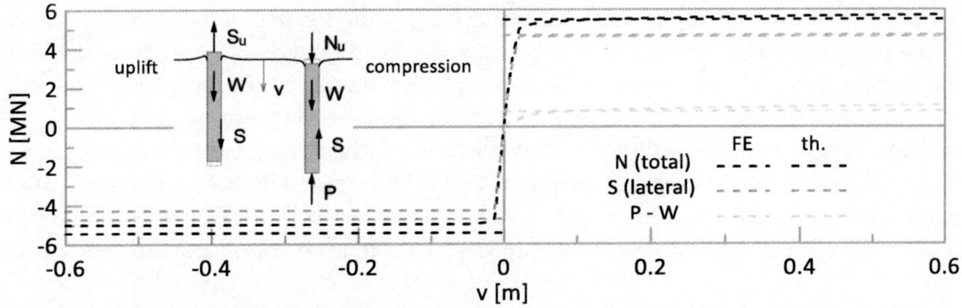

Fig. 4. Single pile subjected to a pure axial load in compression and uplift: comparison between numerical results and theoretical predictions.

(2–3) and external (1–4) piles exhibit a slightly different response during the initial phase of the load-displacement path. When the shaft resistance is fully mobilized, all piles behave in the same fashion, showing a lower bearing capacity compared with the response of the single pile, due to interaction effects.

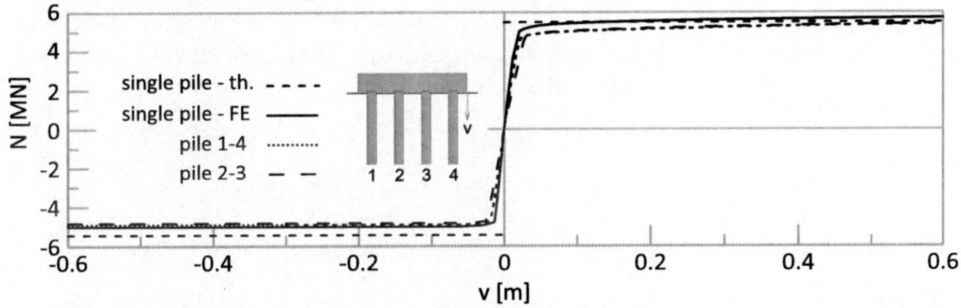

Fig. 5. 4 × 1 pile group: comparison between the single pile and the pile group load-displacement response under pure axial compression and uplift.

The effect of a vertical eccentric load is simulated by imposing a rigid roto-translation to the connecting raft. In order to study the effects of the two resistance mechanisms, M_{ax} and M_b, both hinged and fixed connections are modelled between the pile cap and the raft.

Figure 6 compares numerical and theoretical results for the ultimate (V, M) domain of the pile group. For the sake of consistency, the theoretical domain is computed using the numerical values of the bearing capacity of the single pile (Fig. 4) and the numerical N-M interaction domain for the pile's cross section (Fig. 3b). For each applied V, the distance between the two failure domains provides the theoretical contribution M_b due to the formation of plastic hinges at the pile head. Numerical and theoretical domains are in good agreement, in the case of both hinged and fixed connections. However, due to piles interaction effects, the numerical domain is smaller than the corresponding theoretical one.

For the sole case of fixed pile-cap connection, Fig. 7 shows the key results referring to the five vertexes of the numerical domain shown in Fig. 6 (points (a)–(e)). Specifically, Fig. 7 shows: the axial force computed at the top of each pile (N_i), together with the resultant vertical force on the group, V; the contribution of the two bending resistance mechanisms developed with the cap rotation, θ. For the sake of comparison, Fig. 7 shows also the corresponding theoretical values for the single pile. In cases (a) and (e) all piles are in uplift and compression, respectively. As a result, once all piles reach the same axial load, the moment capacity of the group is due only to M_b. In cases (b), (c) and (d) the M_{ax} contribution is fully mobilised for relatively low cap rotations ($\theta \approx 0.005$ rad), and it remains constant with increasing the displacement. As observed by Psychari and Anastasopoulos [2], the M_b contribution offers on average 30% of the bending resistance of the pile group, and, in all the analysed cases, it requires significant cap rotation to be fully mobilised ($\theta = 0.03 \div 0.04$ rad).

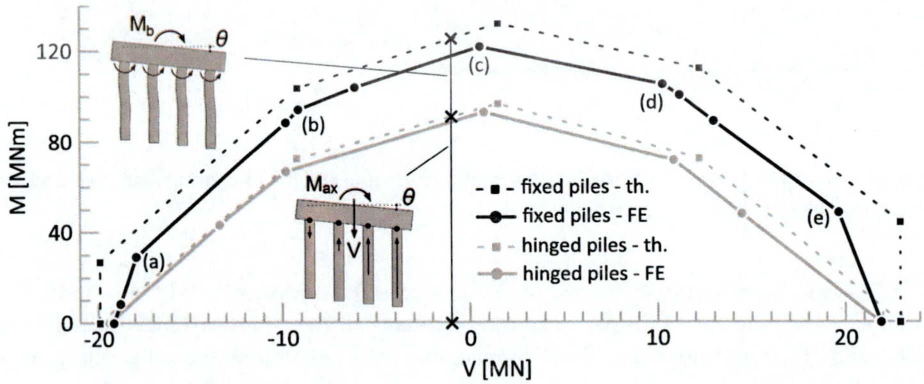

Fig. 6. 4 × 1 RC pile group ultimate failure domain: numerical-theoretical comparison.

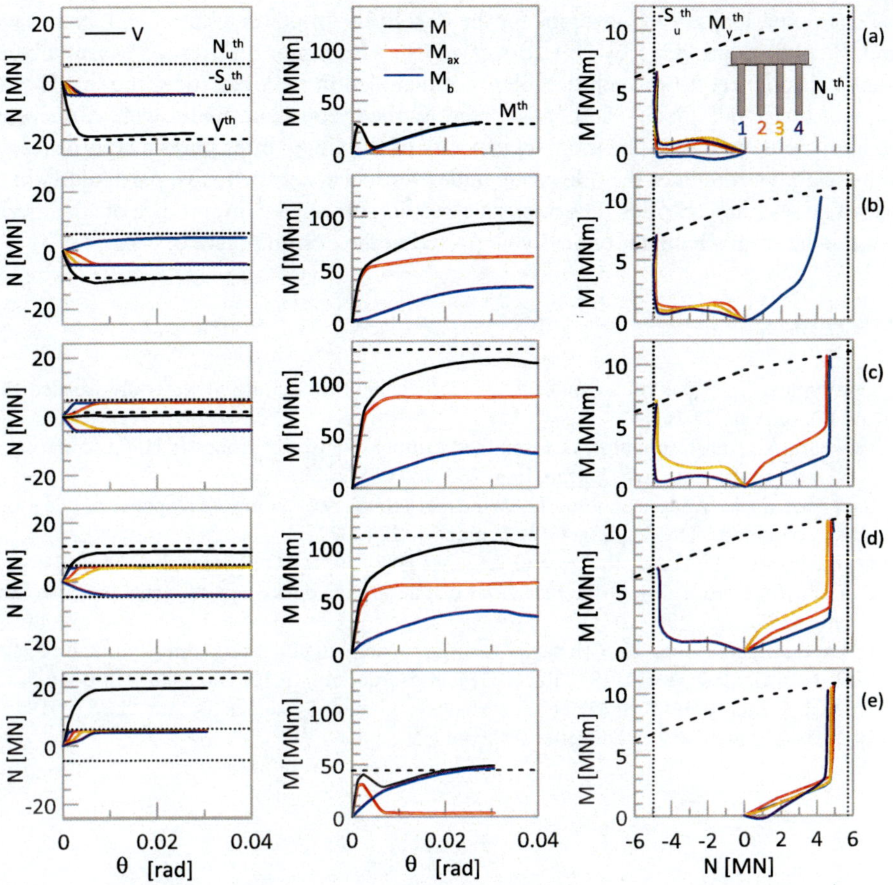

Fig. 7. 4 × 1 RC pile group FE results: axial load and bending moment capacity developed with cap rotation; piles head cross section *N-M* interaction.

5 Conclusions

In order to validate the vertical load-moment failure domain proposed by Di Laora *et al.* [5], a 3D FE model was implemented for a 4 × 1 RC pile group embedded in a clay deposit. The proposed numerical approach takes advantage of appropriate constitutive models for the soil, concrete and reinforcement, allowing to describe the nonlinear response of the foundation subjected to vertical eccentric loads. A preliminary study was conducted to assess the capability of the numerical model to reproduce the structural and geotechnical behaviour of the single pile. The Concrete Damaged Plasticity (CDP) model allows to take into account the *N-M* interaction for RC piles under bending loads: plastic hinges develop and a realistic structural failure occurs. The pile-soil interaction is modelled using an interface that allows both sliding and gap under tension, calibrated to behave as an equivalent cohesive interface.

The numerical failure envelops for the 4 × 1 pile group are in good agreement with the UB solution provided by Di Laora *et al.* [5], with minor differences essentially due to interaction effects between the piles, not included in the theoretical model. The two moment resistance mechanisms postulated by the theoretical model were observed in the numerical analyses. The moment capacity of the single piles plays a significant role in the overall strength of the pile group under vertical eccentric loads, particularly in the case of large-diameter piles. The numerical results showed the importance of considering also this latter mechanism, typically neglected in the design practice.

References

1. Gerolymos, N., Papakyriakopoulos, O.: Macroelement Modelling of Laterally Loaded Piles and Pile-groups (2016)
2. Psychari, A., Anastasopoulos, I.: Combined loading of RC pile groups in clay accounting for N-M interaction. Soil Dyn. Earthq. Eng. **163**, 107490 (2022)
3. Sakellariadis, L., Anastasopoulos, I.: On the mechanisms governing the response of pile groups under combined VHM loading. Géotechnique 1–22 (2022)
4. De Sanctis, L., Di Laora, R., Garala, T.K., Madabhushi, S.P.G., Viggiani, G.M.B., Fargnoli, P.: Centrifuge modelling of the behaviour of pile groups under vertical eccentric load. Soils Found. **61**(2), 465–479 (2021)
5. Di Laora, R., de Sanctis, L., Aversa, S.: Bearing capacity of pile groups under vertical eccentric load. Acta Geotech. **14**(1), 193–205 (2018). https://doi.org/10.1007/s11440-018-0646-5
6. ABAQUS, Inc.. ABAQUS 2019 User's Manual. Dassault Systèmes Simulia Corp (2019)
7. Viggiani, C.: Fondazioni. Helvelius Edizioni (1999)

The Soft-Oedometer: A Simple Test to Calibrate Advanced Constitutive Models for CPT Simulations in Soft Rocks

T. Riccio(✉), M. Previtali, Matteo Oryem Ciantia, and M. Brown

School of Science and Engineering, University of Dundee, Dundee, UK
140007677@dundee.ac.uk

Abstract. Axial pile design approaches based on CPT data have recently been developed for low-medium density chalk ground conditions, while numerical modelling of similar boundary value problems has proved useful in understanding better pile behavior and aiding the interpretation of field and physical tests. In this paper a simple method to calibrate an advanced constitutive model for soft rocks is presented. To this end a newly developed soft oedometer apparatus able to measure both axial and radial stresses is used to perform 1D compression tests of low-medium density chalk samples. The calibrated large strain constitutive model is then used to perform a numerical simulation of a CPT installation in chalk. The results are compared to field data, showcasing the models' capabilities, and highlighting the potential of the soft oedometer for the calibration of constitutive models without the need of triaxial test data.

Keywords: Soft rock · chalk · element testing · constitutive modelling · CPT

1 Introduction

Driven or jacked pile foundation design in low-medium chalk can be challenging due the materials' complex mechanical behavior. The uncertainty associated with pile driving in chalk has led to ambiguity in appropriate design approaches and a reliance on conservative recommendations to account for risk during installation and the potential for low axial capacity [1]. Advancements in continuum mechanics and the reduction in computational burden has allowed numerical models to inform new pile design approaches for a variety of ground condition (e.g. the PISA project [2]). However, this is yet to be achieved for chalk due to its complex mechanical behavior, not typically captured in constitutive models. In addition, models employed in numerical modelling can be challenging to calibrate due to their high number of parameters.

In this work, an innovative calibration technique for the large strain structured modified Cam Clay (s-MCC) model by [3] is presented. The reference material is a soft chalk from the St. Nicholas at Wade (SNW) site, UK. Calibration only requires uniaxial compression tests, indirect tensile tests and 1D compression tests using a bespoke oedometer. After presenting the novel soft oedometer designed at the University of Dundee, model

A. Ferrari et al. (Eds.): CNRIG 2023, SSGG, pp. 477–484, 2023.
https://doi.org/10.1007/978-3-031-34761-0_58

performance is demonstrated through numerical triaxial tests, which are compared to tests performed on intact SNW chalk samples from the literature. Finally, a CPTu test is simulated using the calibrated s-MCC model implemented in the Geotechnical Particle Finite Element Method (GPFEM), an extension of the standard Finite Element Method that uses mixed formulation low-order elements and continuous remeshing to overcome the mesh deformation issues that affect FEM in large deformation problems [3]. The numerical CPTu results are then compared to field CPTs performed at SNW [4] demonstrating how the advanced modelling technique can be used to address the complex hydro-mechanical mechanisms associated with chalk.

2 Standard Element Testing on SNW Chalk

To preliminary characterise the material, index, unconfined compression (UC), and Brazilian tensile (BT) test were performed on both intact and destructured samples. Chalk samples from SNW were cored to a diameter of 38 mm. UC samples were sized with a ratio of 1:2 and prepared using a special mold to ensure flat ends. BT samples were approximately sized 2:1, UC and BT broadly followed [5, 6], respectively. Samples were oven dried at 105 °C for 24h allowing for index properties to be obtained [7], after they were saturated in a vacuum chamber (95 kPa). All tests were displacement-controlled (0.05 mm/min). Destructured material was prepared by crushing air-dried intact chalk material with a mallet, then sieved through a 425 μm sieve, after, reconstituted to 1.2 $\times$ Liquid limit (LL = 28–29%, [8]), in line with [7]. Index properties obtained for the saturated intact material and given in Table 1 were indicative of a highly porous, low density chalk material [1]. Porosity/initial void (n/e_0 [-]) based on weight and dimension, water-immersion (using $G_s = 2.7$) and mercury intrusion porosimetry (MIP) all achieved similar values. Permeability (k_v) of the intact material appeared in the same order of magnitude (10^{-8} m/s) as others [10] Intact dry density (IDD), bulk density (BD) and porosity (n) are useful properties and can be tentatively correlated with strength properties, such as unconfined compressive strength (UC), as has been shown by [9].

Table 1. Intact chalk index properties

Material	BD* [Mg/m^3]	IDD* [Mg/m^3]	n/e_0 [-]	k_v [m/s]
SNW Chalk	1.82–1.92	1.42–1.50	0.44–0.47 / 0.83–0.85	2.55E-8

*Bulk density and Intact dry density obtained via BS1377–2, Clause 9

UC tests indicated an average $\sigma_c = 2.2\,\text{MPa}$, similar to [10] and an average $\sigma_t = 600\,\text{kPa}$ was obtained through BT. UC tests indicated a significantly lower stiffness ($E = 1\,\text{GPa}$) than seen by [10] ($E \approx 4\,\text{GPa}$), highlighting the challenge of non-local sensors for strain measurement, which can lead to underestimation of stiffness [11].

3 The University of Dundee Soft Oedometer Tool

The soft oedometer, originally developed by [12] and then modified [13] to study soft rock behaviour has been re-developed at the University of Dundee (Fig. 1a). The soft oedometer can allow for soft rock samples to be installed into the specimen ring via a cylindrical cutter leading to good confinement, which is otherwise challenging in conventional oedometers [14]. It has also been designed to assess permeability of the material using a simplified constant head style apparatus, as is shown in Fig. 1b. The circumferential strains generated during sample consolidation are recorded through strain gauges placed circumferentially around the flexible confinement ring. Thin tube theory is then used to calculate the radial strains (ε_r). Calibration of the strain gauges is performed using an incompressible Newtonian fluid as a specimen. To assume quasi $1D - k_0$ conditions ε_r should not exceed 0.01% [15]. The soft-oedometer developed therefore has a radial stress (σ_r) working range of up to 3 MPa (Fig. 1c), radial stresses which exceed this level in the experimental data are denoted with a grey-scale line in place of colour. It is notable from Fig. 1c that even at larger radial stress levels, radial strain is still low (i.e., at 6 MPa) it is less than 0.02%.

1. Local LVDT loading piston.
2. Floating brass ring instrumented with 3No. strain gauges.
3. Base plate.
4. Intake pipe to constant head reservoir.
5. Cell sealed with latex + 4No. O-rings. Sample floating between 2No. saturated porous stones.
6. Outtake pipe to collection reservoir.

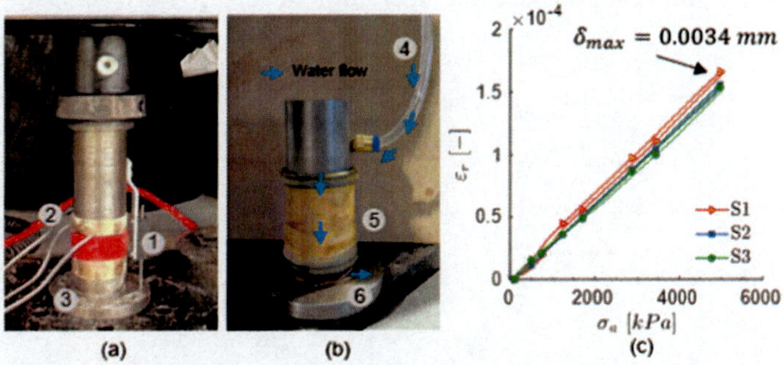

Fig. 1. a) Soft oedometer tool in consolidation mode b) In permeability mode c) radial strains recorded during calibration of the oedometer.

3.1 Oedometric Tests (OED) on Intact and Remoulded SNW Chalk

The soft odometer tool described above was used to test SNW chalk in both intact and remoulded states. As a result of inferred measurements and the use of a floating ring, which limits shear stress development on the ring interface, the full stress tensor was measured in all tests. Test results show that Oedometric intact yield (at knee-point) was reached at an average of $\sigma_a = 5.4$ MPa ($e_{yield} = 0.80$) comparable to results achieved when testing SNW chalk with traditional oedometers [8]. After yielding, a trend towards the normal consolidation line (NCL) obtained for the remolded chalk was noted (see Fig. 2a). Results of the preliminary remoulded test suggested that the putty preparation method did not overly influence the $1D - k_0$ behavior, when comparing to [8] (see Fig. 2b), however it was noted that precision of the initial sample height is important.

For the intact test, the initial k_0 is low indicating the elastic response of the still bonded material and after yield, the radial stress increases such that the k_0 of the remoulded material begins to recover (see Fig. 2c). At high stresses, the intact test begins to trend to a similar k_0 shown by the remoulded material ($k_0 \approx 0.5$) suggesting that the intact oedometric tests fully destructure and may be used to identify both pre- and post-yield properties. It is still of benefit to perform the initially destructured test as the initially intact tests would require significantly higher stress levels to reach the same NCL (and so k_0) defined for the destructured material. This is likely to be a result of intact 1D boundaries and re-arrangement of the chalk fabric during stressing as seen by both [8, 10] and shown in Fig. 3a. From a practical point of view, the soft-OED apparatus can hence characterize both intact and destructured states of the chalk. In the elastic regime knowing the Young's modulus from a UC along with Oedometric modulus, Poisson's ratio can be determined. At high stresses the critical friction angle can be estimated using Jaky formula ($k_0 = 1 - \sin(\phi_{cs})$), which compares well with that obtained in a triaxial [10]. Considering that during a CPT test the intact material is destructured ahead and around the cone, both intact and remoulded characterization would be necessary to better interpret and eventually numerically model CPT in chalk. Large deformation shearing during installation may also lead to the development of a low strength silty material around the shaft, as seen in field scale CPT tests [16].

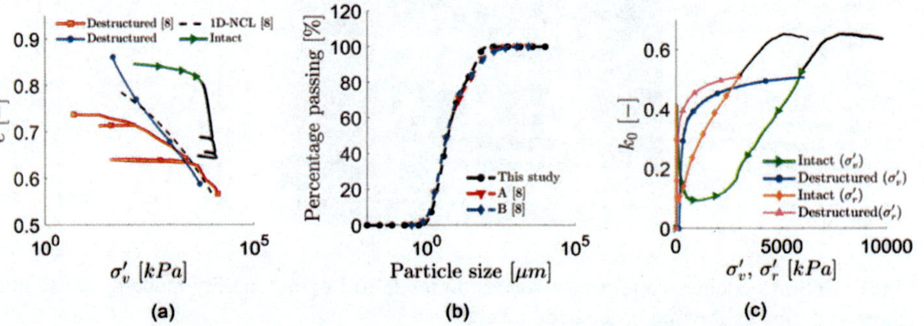

Fig. 2. a) Comparison of 1D-NCL and traditional oedometric to soft oedometric test in $e - \sigma'_v$ b) PSD test results and sensitivity of material preparation, c) k_0 evolution of intact and destructured sample under oedometric loading (intact oedometer and PSD data [17]).

4 Numerical Modelling of CPT Installation in Chalk

4.1 Model Calibration

The s-MCC model used to describe chalk behavior is a non-associative, isotropic hardening finite-deformation plasticity model for structured soils and weak rocks. As detailed in [3], the model is equipped with two stress-like internal variables, P_s (preconsolidation pressure, which accounts for the hardening or softening effects as a result of plastic volumetric and deviatoric strains) and P_t (bond strength, which accounts for the effects of cementitious bonds). The analytical expression of the yield and plastic potential are:

$$\begin{cases} f(\tau', P_s, P_t) = (Q/M_f)^2 + P^*(P^* - P_c^*) \\ g(\tau', P_s, P_t) = (Q/M_g)^2 + P^*\left(P^* - P_{c,g}^*\right) \end{cases} \tag{1}$$

where $P^* = tr\left(\boldsymbol{\tau}'\right)/3 - P_t$ and $Q = \sqrt{3\mathbf{J}_2}$ with $\boldsymbol{\tau}'$ and $\mathbf{J}_2$ first and second effective Kirchoff stress tensor respectively, M_f controls the size of the yield surface and M_g controls the slope of the CSL in the triaxial plane. Intersections of the yield locus on the positive and negative axis is defined by:

$$P_c^* = P_t + P_s + P_m = P_s + (1 + c)P_t \tag{2}$$

The evolution of P_s and P_t follows the expressions:

$$\begin{cases} \dot{P}_s = \rho_s P_s\left[(tr(l^p) + \chi_s\sqrt{2/3}||dev(l^p)||\right] \\ \dot{P}_t = -\rho_t P_t\left[(tr(l^p) + \chi_t\sqrt{2/3}||dev(l^p)||\right] \end{cases} \tag{3}$$

where $\mathbf{l}^p = \partial g/\partial\boldsymbol{\tau}$; with $\boldsymbol{\tau}$ the Kirchoff stress, ρ_s, χ_s, ρ_t and χ_t are constitutive parameters which control the rate of degradation during plastic deformations.

The model was calibrated against the presented experimental results. For the sake of brevity, Table 2 summarizes the parameters of the model which are derived from the simple element tests discussed above and additional literature data. The model, its yield locus, plastic stress like variables and hardening parameters are discussed in greater depth by [3] and [18].

Table 2. Constitutive model parameters

Parameter	Value		Source
e_0	0.83	Density/specific gravity	
$E\ [GPa]$	1.0	UC (this study)	Stiffness of intact material
$\nu\ [-]$	0.2–0.3	OED (this study)	Poisson's ratio from elastic phase soft OED
$M_f\ [-]$	0.9	OED (this study)	Yield surface parameter
M_g [-]	1.2 / 1.25	OED (this study) / TXD [10]	Critical state line (CSL) for plastic potential from soft oedometer in remoulded state
c [-]	20	Back assessed	Proportionality factor to assess p_m
$P_{s0}\,[kPa]$	3000	Isotropic compression (ISO) [19]	Preconsolidation pressure of unbonded chalk
$P_{t0}\ [kPa]$	200	BT (this study)	Approximated from Brazilian tensile tests
$k_v\ [m/s]$	2.55x10^{-8}	OED (this study)	Hydraulic conductivity

Figure 3a shows the yield locus and CSL identified by the destructured soft OED and TXD, along with the literature data, whilst Fig. 3b compares a numerical and experimental triaxial stress path on SNW chalk to validate the soft OED calibration procedure. This highlights that; the model can accurately replicate drained triaxial (TXD) stress paths when calibrated using the simple tests discussed above. One can note in $q - \varepsilon_a$, the models' ability to capture the brittle to ductile shear response for samples with higher confinement.

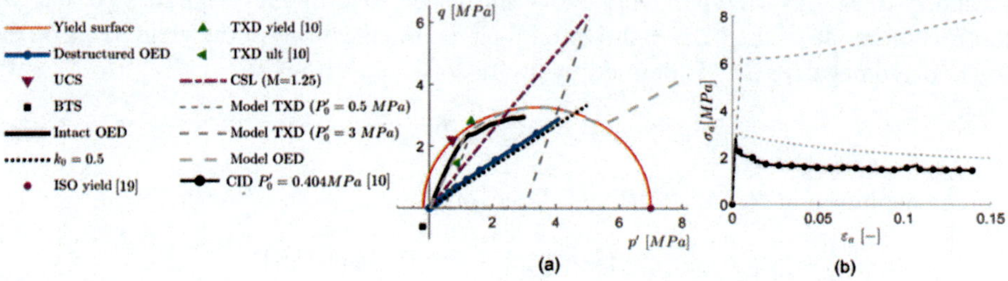

Fig. 3. a) Yield locus (of intact chalk) and b) experimental and numerical CID test.

4.2 Model Application to CPTu in SNW Chalk

To demonstrate the performance of both the constitutive model and numerical framework in replicating large deformation soil-structure interaction problems, a simulation of a standard CPTu (see Fig. 4a) was performed and compared to field-scale CPTs at the SNW site [4]. The 4.3cm diameter cone was pushed at a rate of 0.2 m/s, as in the field. The hydraulic conductivity of the material was set to 1e−7 m/s. While this is one order of magnitude higher than obtained from the falling head test, the latter is based on an intact sample, while the former allows for one to qualitatively account for the presence of discontinuities, i.e., "rock mass" permeability, and it is closer to the reported field permeability values from the literature [20]. Figure 4b compares q_c and u between the tests. These preliminary results appear to capture the experimental data under partially drained conditions and large deformations which may be used to study new design approaches. The authors hypothesize that the slight difference between numerical results and field data in the water pressure field is due to the constant permeability assumption in the model, opposite to the decrease in permeability following material densification encountered in the field.

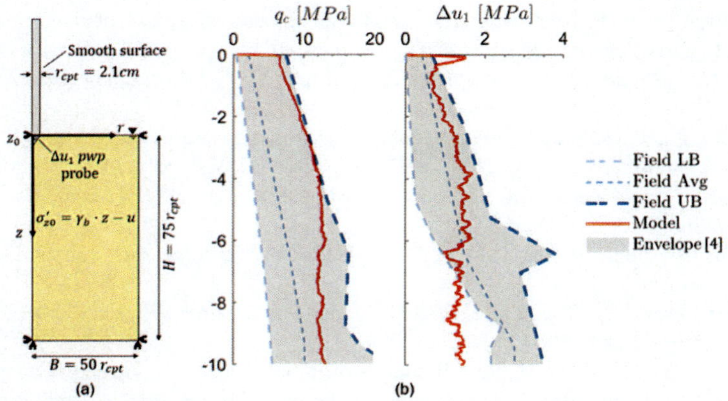

Fig. 4. a) Model geometry and boundary conditions b) Numerical vs field scale CPTs (where depth is normalized by CPT radius ($r_{cpt} = 2.1$cm).

5 Conclusion

In this paper simple element test tools were used to calibrate a constitutive model for bonded materials which was shown to accurately replicate stress paths in triaxial space. The soft oedometer element tests also showed that deformation in a quasi $1d - k_0$ condition leads to structural degradation of the intact materials' bonded structure once stressed beyond the yield, transforming into a soil-like material, but in a dense and compacted state. The calibrated model was used to study coupled hydro-mechanical problems in low-medium density chalk at a well-known site. Numerical simulations showcased the capability of the constitutive model well by matching q_c and u increments obtained from the field CPT tests. Permeability adopted in the model was based upon an intact sample and does not directly account for field characteristics, like discontinuities which influence the in-situ permeability. Future simulations should aim to include this characteristic of the geological unit in the field, perhaps through discrete fracture network analysis or similar approaches. The paper used a constitutive model which has the potential to study coupled large deformation continuum mechanics, which may be used to improve the current practices for the installation in soft rocks.

Acknowledgements. This work is an output from ICE-PICK, funded by NIA EPSRC grant EP/W00013X/1.

References

1. Lord, J.A., Clayton, C.R.I., Mortimore, R.N.: Engineering in chalk CIRIA C574. (2002). ISBN: 978-0-86017-574-2.
2. Byrne, B.W., et al.: PISA design model for monopiles for offshore wind turbines: application to a marine sand. Geotechnique **70**, 1030–1047 (2020). https://doi.org/10.1680/jgeot.18.P.255
3. Monforte, L., Ciantia, M.O., Maria, J., Arroyo, M., Gens, A.: Computers and geotechnics a stable mesh-independent approach for numerical modelling of structured soils at large strains. Comput. Geotech. **116**, 103215 (2019). https://doi.org/10.1016/j.compgeo.2019.103215

4. Buckley, R., et al.: Field investigations into the axial loading response of displacement piles in chalk. In: Offshore Site Investigation Geotechnics 8th International Conference Proceedings, pp. 1178–1185. Society of Underwater Technology (2017). https://doi.org/10.3723/OSIG17.1178
5. BSI: BSI Standards Publication Geotechnical investigation and testing — Field testing. (2018)
6. ASTM Committee D18: ASTM D3967–08: Standard Test Method for Splitting Tensile Strength of Intact Rock Core Specimens 1. Annual Book of ASTM Standards. 1–3 (2016). https://doi.org/10.1520/D3967-08
7. BSI: British standard: BS 1377–2:1990. Methods of test for soils for civil engineering purposes, (1990)
8. Alvarez-Borges, F.J.: The Shaft Capacity of Small Displacement Piles in Chalk (2019) https://eprints.soton.ac.uk/438900/
9. Matthews, M.C., Clayton, C.R.I.: Influence of intact porosity on the engineering properties of a weak rock. In: Anagnostopoulos, A., Schlosser, F., Kalteziotis, N., and Frank, R. (eds.) Geotechnical Engineering of Hard Soils - Soft Rocks. International Symposium, pp. 693–702. Balkema (1993). ISBN: 9054103450
10. Vinck, K., et al.: Advanced in-situ and laboratory characterisation of the ALPACA chalk research site. Géotechnique. 1–41 (2022). https://doi.org/10.1680/jgeot.21.00197
11. Perbawa, A., Gramajo, E., Finkbeiner, T., Santamarina, J.C.: Global vs local strain measurements in triaxial tests – Implications. In: 53rd U.S. Rock Mechanics/Geomechanics Symposium, pp. 3–10 (2019). http://hdl.handle.net/10754/662801
12. Kolymbas, D., Bauer, E.: Soft oedometer. a new testing device and its application for the calibration of hypoplastic constitutive laws. Geotech. Test. J. **16**, 263–270 (1993). https://doi.org/10.1520/gtj10044j
13. Castellanza, R., Nova, R.: Oedometric tests on artificially weathered carbonatic soft rocks. J. Geotech. Geoenviron. Eng. **130**, 728–739 (2004). https://doi.org/10.1061/(ASCE)1090-0241(2004)130:7(728)
14. Nielsen, S.D., Peri, E., Nielsen, B.N.: Evaluation of preparation techniques of chalk samples for oedometer testing. In: 17th European Conference on Soil Mechanics and Geotechnical Engineering, ECSMGE 2019 - Proceedings (2019). https://doi.org/10.32075/17ECSMGE-2019-0242
15. Okochi, Y., Tatsuoka, F.: Some factors affecting k0-values of sand measured in triaxial cell. Japense Soc. Soil Mech. Found. Eng. **24**, 52–68 (1985). https://doi.org/10.3208/sandf1972.24.3
16. Ciavaglia, F., Carey, J., Diambra, A.: Monotonic and cyclic lateral load tests on driven piles in chalk. (2017).https://doi.org/10.1680/jgeen.16.00113
17. Riccio, T., Ciantia, M.: Stress path measruements of chalk under 1D compression. In: 16th BGA Young Geotechnical Engineers Symposium (2022). https://doi.org/10.57711/kz8q-q554
18. Ciantia, M.O.: Installation effects on axial performance of monopiles in chalk for offshore renewables. In: Huynh, D.V.K., Tang, A.M., Doan, D.H., Watson, P. (eds.) VSOE2021 2021. LNCE, vol. 208, pp. 424–432. Springer, Singapore (2022). https://doi.org/10.1007/978-981-16-7735-9_47
19. Priol, G., Gennaro, V. de, Delage, P., Collin, F., Charlier, R.: Suction and time effects on the behaviour of a reservoir chalk, pp. 519–526 (2006). ISBN: 9780415410014
20. Buckley, R.: The Axial Behaviour of Displacement Piles in Chalk. Imperial college, London (2018). https://doi.org/10.25560/78617

Processes Involved in Deformations and Instability in Static and Seismic Conditions

Geotechnical Characterization of the Subsoil at the "*Regina Montis Regalis*" Basilica in Vicoforte, Italy

Sara Bandera(✉), Domenico Gioffrè, and Carlo Giovanni Lai

Department of Civil Engineering and Architecture (DICAr), University of Pavia, 27100 Pavia, Italy
sara.bandera@unipv.it

Abstract. The Italian historical built environment is vulnerable to several geohazards which can compromise its integrity. Therefore, its sustainable preservation and the improvement of its resilience are of utmost importance and need to be carefully investigated. In the framework of the GIANO research project, an in-depth investigation is underway on the "R*egina Montis Regalis*" Basilica of Vicoforte (CN). The construction of the Basilica, which hosts the world's largest oval masonry dome, began in 1596 and lasted for more than 130 years. Structural problems developed since the initial stages of its construction (around 1599).
The Basilica of Vicoforte has been extensively studied over the last 40 years. However, previous studies have been focusing on the structural behavior of the Basilica alone, with little attention to its interaction with the soil.
This contribution aims at presenting the main subsoil model properties determined by reviewing the results of laboratory and field tests performed during three geotechnical and geophysical campaigns carried out at the site since 1976. In fact, an accurate knowledge of the subsoil properties and a proper understanding of its complex behavior are essential to conduct advanced numerical analyses which consider both the soil behavior and the structural response of the Basilica.

Keywords: Cultural Heritage Conservation · Soil Modelling · Geotechnical Characterization

1 Introduction

The "*Regina Montis Regalis*" Basilica of Vicoforte (CN) (Fig. 1(a)) is a monument of great historical, architectural, and, most importantly, structural significance as it hosts the world's largest oval masonry dome ($\sim$ 37.23 m long and $\sim$ 24.89 m wide) [1]. The construction of the Basilica started in 1596, but structural problems emerged already around 1599 (Fig. 2). They were primarily attributed to foundation differential settlements observed on the west side of the building with respect to the east side. Settlements were caused by the heterogeneity of the deposit, as the site is characterized by a compressible clayey-silt layer of increasing depth in the east-west direction, which lies on top of a marlstone layer outcropping on the east side of the Basilica. Investigations revealed that five of the eight pillars of the Basilica are founded on the compressible soft layer, while

A. Ferrari et al. (Eds.): CNRIG 2023, SSGG, pp. 487–494, 2023.
https://doi.org/10.1007/978-3-031-34761-0_59

the remaining three lay directly on the marlstone [2, 3]. The construction of the structure was paused and resumed several times until the Basilica was completed in 1735. However, new concerns over the structural safety of the drum-dome system emerged shortly after as an extended system of asymmetric wide cracks opened (Fig. 1(b)). Initial studies aimed at consolidating the Sanctuary started in 1934 with the extensive work carried out by Eng. Garro [3], who improved the understanding of the evolution of the differential settlements between adjacent columns with time (Fig. 2). Table 1 summarizes the settlements for three pillars (NO, O_d and O_s in Fig. 2) measured with reference to those laying directly on the marlstone. As can be seen, the settlement pillar O_d underwent before the construction of the drum-dome system was ~ 550 mm, thus explaining the structural issues observed on the masonry. Since 1975–76 the Basilica and its subsoil were subjected to structural and geotechnical investigation campaigns targeted to investigate the geometry of the structure and of the foundations and the mechanical properties of both subsoil and masonry. The structural integrity of the Basilica is currently monitored through permanent static (since 2004) and dynamic (since 2015) multi-sensor monitoring systems.

The current study on the Basilica is part of the research project GIANO, which aims at evaluating potential geo-risks and suggesting viable mitigation strategies to improve the resilience of the historical structural heritage. In this case study, the goal is to perform soil-structure interaction analyses on digital models to reproduce the construction sequence of the Basilica, as suggested in Fig. 2. The main objective is to reproduce the settlements as determined by Garro and to investigate the evolution of the stress-strain properties and behavior of the subsoil to ultimately link it to the behavior of the structure. A fundamental role in geotechnical numerical analyses is played by the geotechnical characterization of the site. This contribution then presents the most relevant geotechnical subsoil properties of the site by considering the data gathered during three geotechnical and geophysical investigation campaigns carried out in Vicoforte since 1976.

Fig. 1. (a) View of the "*Regina Montis Regalis*" Basilica in Vicoforte and (b) Pattern of the cracks observed by Eng. Garro on the Basilica of Vicoforte [3].

2 Geotechnical Investigation Campaigns

Three geotechnical and geophysical investigation campaigns were carried out at Vicoforte in 1976, 2004 and 2008. Figure 3 summarizes the in-situ tests cited in this contribution to characterize the subsoil. An overview of test performed is briefly presented in the following. A more comprehensive discussion is presented in [2–7].

The first geotechnical campaign was carried out in 1976 [4] following a survey resulting in concerns over the stability of the Basilica. Both in situ and laboratory tests were conducted and their aim was twofold: defining the stratigraphy and mechanical properties of the subsoil and determining geometrical dimensions and laying ground level of the foundations. Shallow boreholes (S1 to S14 in Fig. 3) were performed up to 20m depth; piezometers were installed at the bottom of 5 boreholes to manually measure the water table level in the period May-June. Undisturbed samples were taken at site and classified; then they were tested in laboratory. The geometry of the foundations was defined by performing additional 29 boreholes on the floor of the Basilica.

A second campaign was carried out in 2004 to evaluate the geotechnical properties of the soil deposit before consolidation [5]. To this aim, four boreholes (SV1 - SV4 in Fig. 3) were drilled up to 20 m depth. Piezometric measurements were also performed within the boreholes in the period January – February and undisturbed samples were taken from the site to perform laboratory tests to measure the mechanical and dynamic properties of soil deposit. Geophysical tests, such as cross-hole (CH), were also carried out on the west side of the deposit (Fig. 3) to define the compression (V_p) and shear (V_s) wave velocity profiles of the unconsolidated portion of the area.

A final investigation campaign was performed in 2008. This campaign was mainly intended to determine the electrical and seismo-stratigraphical properties of the subsoil. Therefore, different types of advanced geophysical tests were performed including 2D seismic and 3D electric tomography [2, 6, 7]. The campaign was then completed by carrying out two cross-hole (CH) tests, which were executed in the same boreholes employed in 2004 (Fig. 3).

3 Geotechnical Characterization

3.1 Soil Characterization and Classification

Boreholes performed in 1976 and 2004 were distributed throughout the footprint of the Basilica, both inside and outside its perimeter (Fig. 3). Logs results highlighted the heterogeneity of the deposit and revealed the presence of two main geological formation: a shallow soft clayey-silt layer with minor sand content under lens form on top of a marlstone formation. A thin layer of alteration material of variable thickness from centimeters to 1–1.5 m was also observed between the two main formations. The thickness of the clayey-silt formation reaches its maximum (~5–6 m) towards the west side of the site; it then reduces moving to the east, becoming zero at boreholes S4 and S5. Therefore, the east side of the Basilica is mostly founded on the bedrock, which has a very high strength and very low compressibility. Conversely, the clayey-silt formation is highly compressible. Furthermore, it is slightly over consolidated as a result of the groundwater seasonal oscillations and possibly of locally applied and removed loads within historical times [4]. This qualitative and quantitative difference in foundation soil type between east and west sides is believed to be the main cause for the differential settlements observed below the plinths of the Sanctuary (Fig. 2) which triggered the asymmetric pattern of cracks observed on the dome-drum system of the Basilica (Fig. 1(b)) [2–4].

Table 1. Summary of the settlements of O_s, O_d and NO measured by Eng. Garro [3]; a, b, c, d and e refer to settlements represented in Fig. 2.

Settlement/Pillar	Os [mm]	Od [mm]	NO [mm]
a	28	42	33
b	72	80	64
c	190	314	218
d	212	330	230
e	346	350	370

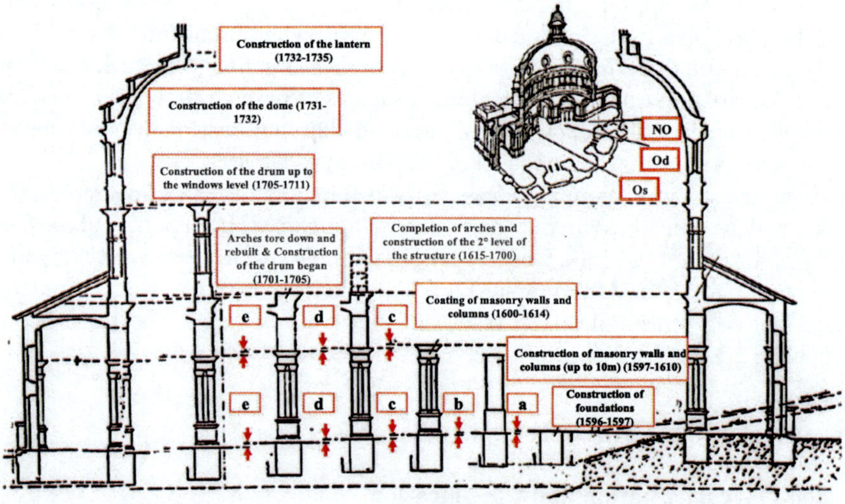

Fig. 2. Schematic representation of the construction sequence and summary of the differential settlements estimated by Eng. Garro in 1962 (Modified after [3]).

One of the properties measured in laboratory was the unit weight γ [kN/m^3] of both clayey silt and marlstone. This quantity varied non-uniformly with depth: it ranged between 18.2 and 19.5 kN/m^3 for clayey silt (with mean value of $\gamma_{\text{clayey-silt}} = 19$ kN/m^3), while $\gamma_{\text{marlstone}}$ was between 19.1 and 21.7 kN/m^3 (with mean value of 21 kN/m^3). This non-uniform variation of γ with depth poses a significant challenge when the small-strain shear modulus – depth profiles need to be determined from the shear wave velocity profiles obtained from geophysics tests.

The hydrogeological condition of the area was also considered to characterize the deposit and it was assessed by means of short-term piezometric measurements carried out in 1976 and 2004 and by examining data collected by three open tube piezometers, part of the permanent static monitoring system installed in Vicoforte. Figure 4(a) plots data measured in May-June 1976 (top) [4], and in January-February 2004 (bottom) [5]. As can be seen, in 1976 the water table level was characterized by negligible oscillations over the measurement period. However, data show that the water table level appeared to be lower on the west side of the Basilica, seemingly following the marlstone profile. Similar results were also obtained in 2004, with the water table having negligible oscillations during the sampling period, except for SV-1, in which the water table level increased by 1m. However, this increase could be attributed to the meteorological conditions during the measurements (i.e., presence of snow close to the borehole). The three open tube piezometers were installed on the west side of the Basilica in the clayey-silt layer. Measurements were collected from 2004 to 2014. Missing data in the 2004–2014 and

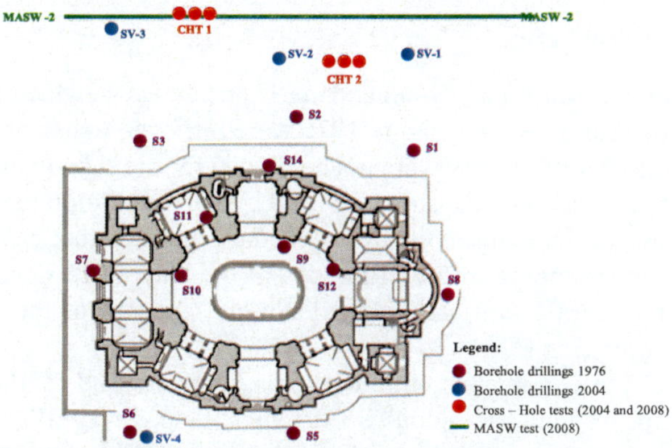

Fig. 3. Summary of the tests carried out in 1976, 2004 and 2008 and which are of interest in this contribution (modified from [2]).

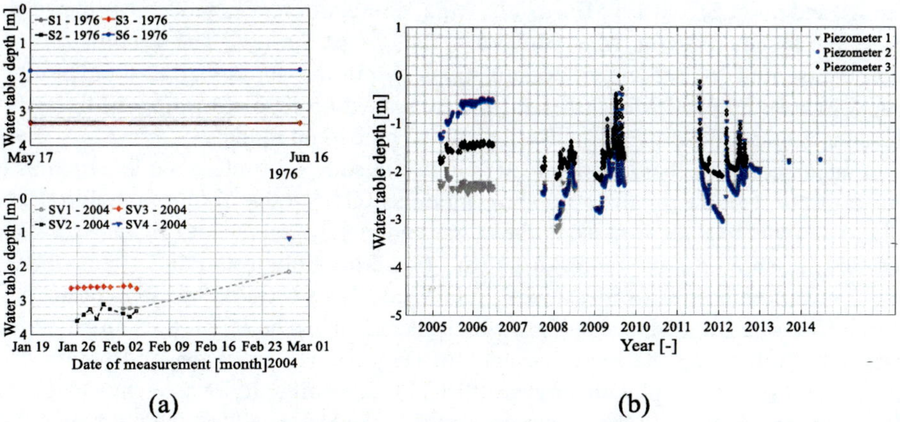

Fig. 4. (a) Water table levels measured in 1976 and 2004 (data from [4] and [5]); (b) water table measurements collected by the permanent piezometers between 2004 and 2014.

2014-to date period are attributed to communication issues between the piezometers and the electronic system installed; an updating of this system is underway. A post-processing of the raw data collected by the permanent piezometers network carried out by *Politecnico di Torino* is shown in Fig. 4(b), where a significant noise is observed. Despite this, the water levels measured are comparable with those of 1976 and 2004. Figure 4(b) shows that piezometer #2 is affected by the most significant change in water table depth, while piezometers #1 and #3 are characterized by relatively constant water level over the years. However, given the very low permeability of the clayey-silt layer ($k = 3.4 \times 10^{-8}$ m/s) [2], data collected by open tube piezometers are not reliable as these instruments fail to respond to changes in a timely fashion. Furthermore, it is also important to point out that the surface of the shallow groundwater is influenced by a drainage system consisting of a network of old trenches and tunnels around and under the footprint of the Basilica [4].

3.2 Mechanical Properties

Mechanical properties of the clayey-silt and marlstone layers were determined by means of laboratory tests. Samples collected in 1976 were retrieved under or very close to the Basilica footprint. Therefore, they were not employed for the subsoil characterization as their response is influenced by the stress induced by the foundation loads: a reliable and detailed assessment of the actual stress history of the deposit cannot be reconstructed [5]. Hence, strength and compressibility properties of the clayey-silt deposit were defined only considering the 2004 samples, deemed as representative of the conditions of the soil prior to consolidation [5].

The results of the oedometric tests (on both clayey silt and marlstone) allowed to determine how the oedometric modulus (M [MPa]) and the coefficient of consolidation c_V [m^2/s] vary with vertical mean effective stress σ_V' [kPa] as shown in Fig. 5(a) and (b). As can be seen in Fig. 5(a), M is approximately equal to 250 MPa at $\sigma_V' = 25$ kPa for marlstone (dot-dashed black line), while for the clayey-silt layer M varies quite significantly at low confining stresses, even for specimens retrieved at similar depths, and ranged from 1.5 MPa to 12.5 MPa. These differences reduced as the vertical stress increased, reaching M ~ 90 MPa at the end of the test. The oedometric modulus is essential for settlement computations within the numerical model. Results in Fig. 5(b) are employed when assessing the consolidation state of the deposit throughout the construction sequence of the Basilica in the numerical model.

Fundamental mechanical parameters to use in soil constitutive models such as the Poisson ratio (ν), the small-strain shear modulus (G) and the Young modulus (E) can be indirectly determined by using the compression (V_P) and shear (V_s) wave velocity profiles measured during geophysics tests. Two cross-hole tests (CHT1 and CHT2 in Fig. 3) were employed to measure V_s and V_p on the west side of the deposit [5, 7]. Results for V_s are shown in Fig. 6(a), which also includes the V_s profile computed from data collected during the MASW-2 test [6] (Fig. 3). The variation of V_s with depth confirmed the stratigraphy already established using borehole logs. V_s has low values up to 5m (approximately the maximum depth of the clayey silt on the west side of the deposit); they then increase indicating an improvement of the mechanical properties of the materials encountered [5]. Figure 6(a) also shows that, while V_s profiles have a

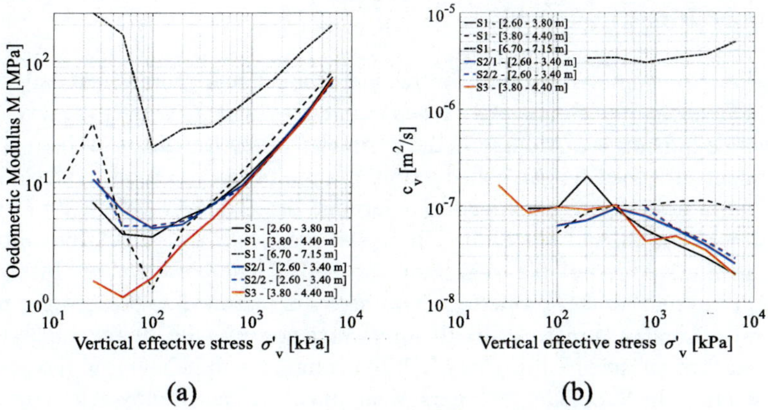

Fig. 5. Variation of (a) M and (b) c_V with σ_V' for clayey-silt and marlstone (data from [5]).

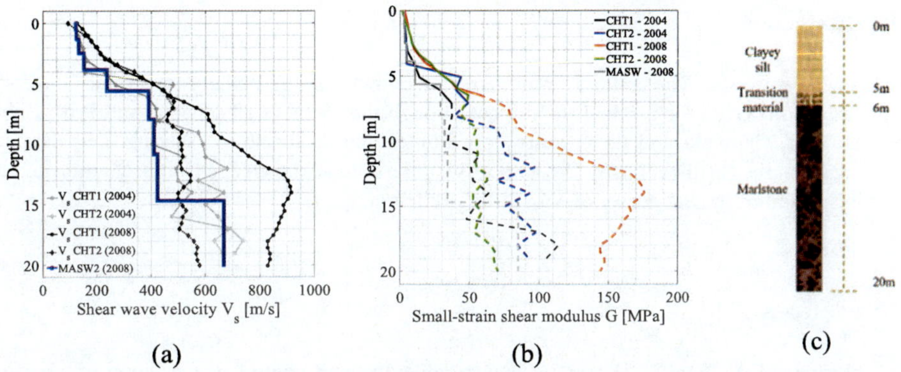

Fig. 6. (a) V_s profiles obtained in 2004 and 2008 ([5, 6]); (b) G computed using V_s in (a). Solid lines indicate the clayey silt; dashed lines represent the marlstone. (c) Representative stratigraphic log for the West side of the deposit (after [2]).

similar trend, there are considerable differences both between results of CHT1 and CHT2 (despite their close location) and between velocity profiles determined in 2004 and 2008. Therefore, the variation of V_s with depth and location added to the non-uniform variation of γ with depth, results in mechanical properties changing with both depth and location. Figure 6(b) presents a preliminary computation of the small-strain shear modulus G [MPa] as a function of depth, derived considering all the V_s profiles in Fig. 6(a), but assuming mean values, constant with depth, of the unit weight for clayey-silt ($\gamma_{\text{clayey-silt}} = 19$ kN/m^3) and marlstone ($\gamma_{\text{marlstone}} = 21$ kN/m^3). A representative stratigraphic log for the West side of the deposit is presented in Fig. 6(c). Figure 6 offers a clear indication on how challenging the characterization of the subsoil at Vicoforte is.

4 Conclusions

This contribution summarizes some of the geotechnical and geophysical data collected during the investigation campaigns carried out at Vicoforte (CN) over the last 40 years. The borehole logs drilled throughout the perimeter of the Basilica and the piezometric measurements are used for the construction of a preliminary subsoil model. The shear wave velocity profiles are employed to determine the variation with depth of both the shear stiffness and the bulk modulus. These soil properties are crucial for calibrating a proper constitutive model able to account for the influence of the stress history of the deposit on the mechanical properties. This study is intended as the starting point for performing advanced numerical simulations which consider the interaction between the Basilica and the subsoil to ultimately link the complex soil behavior to the network of cracks affecting the Sanctuary. The numerical model will be employed to reproduce the history of the cracks that opened in the drum-dome system, and which were possibly caused by the foundation differential settlements. These results will then be employed to design an effective long-term geotechnical monitoring system for the future preservation of the Cathedral.

Acknowledgements. The authors would like to acknowledge the Italian Ministry of University and Research (MUR) for funding the project through the PRIN 2020 scheme, and Prof. Chiorino and Prof. Ceravolo and their teams from Politecnico di Torino for sharing their data and knowledge on the Basilica. Lastly, the authors would like to acknowledge Geom. Martino and the Administration of the Sanctuary of Vicoforte for their support.

References

1. Chiorino, M.A., Fea, G., Losana, G.: Strengthening and control of the dome of Vicoforte sanctuary. In: Proceedings of IASBE International Symposium on Structural Preservation of Architectural Heritage, pp. 723–724, Rome (1993)
2. Lai, C.G., Corigliano, M., Sánchez Lizárraga, H., Scandella, L.: Definition of Seismic Input at the "Regina Montis Regalis" Basilica of Vicoforte. Northern Italy. IUSS Press, Pavia (2009)
3. Garro, M.: Opere di consolidamento e restauro. Relazione riassuntiva 29–6–62, Technical Report (in Italian) (1962)
4. Rodio, S.P.A.: Indagini geognostiche e prove geotecniche di laboratorio, Technical Report (in Italian) (1976)
5. Dipartimento di Ingegneria Strutturale e Geotecnica: Esecuzione ed interpretazione di prove geotecniche per la caratterizzazione del terreno di fondazione sottostante il Santuario di Vicoforte. Technical Report (in Italian), Torino (2004)
6. PROGEO: Indagini geosismiche per la caratterizzazione del sottosuolo della Basilica di Vicoforte. Technical Report (in Italian) (2008)
7. SOING: Caratterizzazione geotecnica del sottosuolo con metodologie geofisiche. Technical Report (in Italian) (2008)

Earthflows in the Basento Valley: Hydraulic Characteristics Influencing Their Kinematics

Caterina Di Maio[1](✉), Jacopo De Rosa[1], Roberto Vassallo[1], Gianluca Cutrera[2], and Roberto Murtas[2]

[1] Università Della Basilicata, Potenza, Italy
caterina.dimaio@unibas.it
[2] Rete Ferroviaria Italiana, Bari, Italy

Abstract. Earthflows, in different evolution phases, are the most diffused landslides in the Lucan Apennines. For very representative cases in the structurally complex Miocene formation of Varicoloured Clays, the paper presents the results of permeability tests carried out in the landslide bodies, in the stable formation and in the slip zones, as well as updated long data series of porewater pressures and basal displacements. The results show that the highest permeability occurs in the slip zone, where also the most significant porewater pressure variations attributable to rain have been recorded. To seepage pressures in the slip zone, deep basal displacement rates seem to be strongly correlated.

Keywords: Landslide · Permeability · Water pressure · Displacement

1 Introduction

This paper reports on some results of the research project MITIGO (PON R&I 2014–2020) aimed to landslide risk mitigation in the Italian Apennine environment. The study area is a very large portion of the Basilicata region, in southern Italy, in which the Apennines thrust on the Bradanic foredeep units. Many different geological formations outcrop in the area which is characterized by diffused and variegated instability phenomena. As well as in the whole Apennine chain, earthflows of different thickness and rates are widespread in clay formations; their evolution is controlled by alternating phases of rapid surges and long dormancy. A typical phase sequence consists of retrogression of the alimentation zone, filling of the landslide channel, emptying of the channel, and accretion of the accumulation (Urciuoli et al. 2016). The earthflows can move very slowly, or remain dormant for many decades, thus allowing urbanization. However, slowly over the years, the movements accumulate, thus causing damage to the built environment with unsustainable cost for the community (Di Maio et al. 2010; 2013; 2020; 2021). The project MITIGO has been funded, on one hand, to understand such landslides' behavior and the causes of their reactivation and, on the other hand, to design effective stabilization measures. One of the studied sites is a hill east of Potenza city, where many landslides of several different types occur. For some of them, geotechnical investigations, monitoring of displacements and porewater pressures are being carried

A. Ferrari et al. (Eds.): CNRIG 2023, SSGG, pp. 495–502, 2023.
https://doi.org/10.1007/978-3-031-34761-0_60

out since 2005. In particular, the case history of the Costa della Gaveta earthflow reveals kinematic, hydraulic and chemo-mechanical features which indicate how movements occur and thus also suggest the most reasonable interventions. The studies begun about 20 years ago are now intensified and extended to the entire hill thanks to the mentioned project (Vassallo et al. 2023).

This paper dwells on the hydraulic features of two earthflows in very slow kinematic phase, and presents updated porewater pressure series relative to landslide body, stable parent formation and slip zones. It also presents the results of in-place permeability tests carried out by different procedures, among which one developed to evaluate seepage rate with continuity along borehole walls. The results, together with a numerical 3D model, allow to understand how the landslide movements can respond rather rapidly to rain, notwithstanding the high clay fraction and the depth (40 m) of the slip surface.

2 Geological Formations, Landslide Material, Kinematic Features

The lithological succession of the *Costa della Gaveta* hill, east of Potenza city, consists of two main systems stratigraphically superposed: the Varicoloured Clays (AV, upper Cretaceous - lower Miocene) and the *Corleto Perticara* (CP, Eocene- lower Miocene) formation. The AV is constituted by a succession of fine-grained soils sedimented in pre-orogenic marine basins and subsequently involved in orogenic processes. Clays, marly clays and marls are interbedded with layers of calcareous marls, calcilutites and bioclastic calcarenite. The overlying CP formation consists of alternating layers and benches of marly limestones and massive calcilutites.

The hill is affected by many landslides and, above all, by large earthflows. The most studied one is the homonymous *Costa della Gaveta* earthflow; it is about 900 m long, 100 to 600 m wide, about 40 m deep (Fig. 1). The landslide is constituted by a chaotic and intensely fissured mixture of the materials of the two formations, the clay fraction c.f. ranges between 25% and 50%; the liquid limit w_L between 50% and 100%; the degree of saturation Sr is about 100%, only shallow soil layers undergo seasonal Sr variations. The average water content w of the saturated zone decreases from 25% in the landslide to about 15% in the stable formation. The shear strength available on the slip surface is close to residual values (Di Maio et al. 2015). The residual friction angle φ'_r determined in laboratory varies in the range 6°–13°, depending on *c.f.*, mineral and pore fluid composition. A decrease in pore solution concentration has been recognized as one of the main causes of the observed soil behavior.

Several inclinometer tubes have been installed in the landslide since 2005 (Fig. 1). Over the whole monitoring period, the movement has consisted in sliding along a thin zone. Internal deformations are noticeable only in the shallowest soil layers. Di Maio et al. (2010; 2013) and Vassallo et al. (2020; 2021) showed that: *a)* yearly displacement rates have been rather constant over the last 20 years with seasonal variations; *b)* displacement rates decrease in the downslope direction from a few cm/year in the head to a few mm/year in the accumulation zone; *c)* constant soil discharge in the main track can explain the observed variations in displacement rates along the longitudinal axis. Analogous behavior is exhibited by the nearby Varco d'Izzo earthflow (Vassallo et al. 2021).

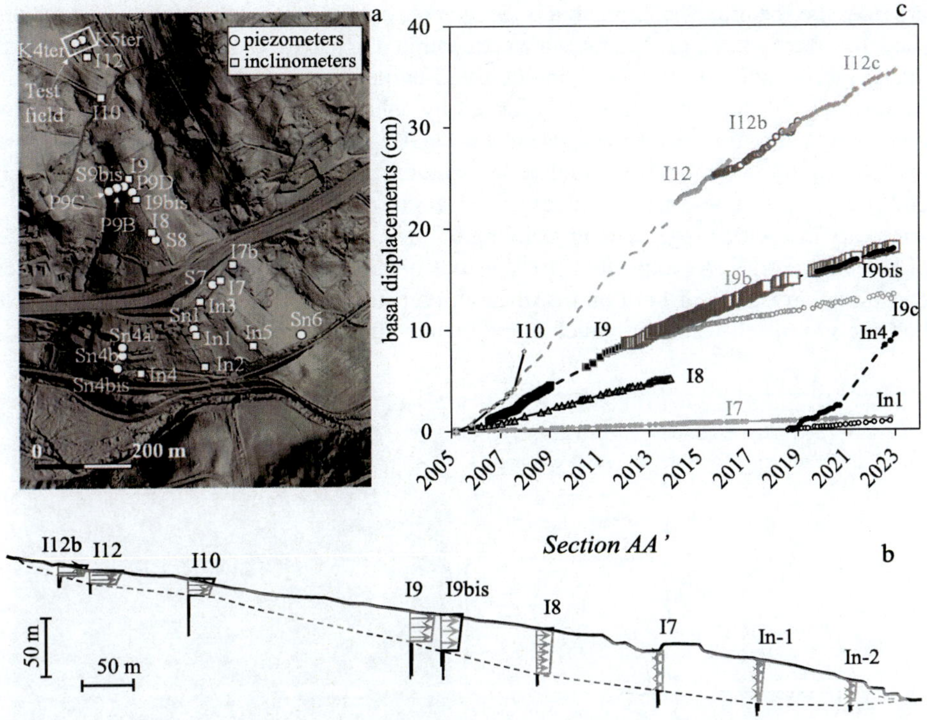

Fig. 1. Costa della Gaveta earthflow map and borehole location (a); longitudinal section AA', inclinometer profiles and slip surface (b); time trend of basal displacement in some verticals (c).

3 Hydraulic Conductivity and Porewater Pressures

The considered slope is subjected to rainfall of long duration and low intensity (Vassallo et al. 2015; 2016). The maximum precipitations occur at the end of winter - beginning of spring, the minimum occur in summer; the deep displacement rates too increase in winter and are almost null in summer. In fact, the displacement rates of very slow landslides in clay slopes are known to undergo seasonal variations associated to hydrological conditions (e.g., Cotecchia et al. 2020). In order to understand how this occurs, permeability was evaluated and porewater pressures were monitored in the landslide body, in the stable parent formation and in the slip zone.

Several different in-place permeability tests were carried out in the two earthflows. Besides falling head tests in the Casagrande piezometers, conventional slug tests were carried out in a number of boreholes deeper than the slip surface, supported by permeable screens. Furthermore, in the same boreholes, seepage tests were carried out by experimental procedures and a permeameter designed for the purpose of estimating seepage rate with continuity along the borehole. This latter is emptied and the water level is kept below the slip surface, the permeameter is installed at the desired depth to collect water inflowing from the surrounding soil through the screen. The inflowing water volume per unit time, divided by the area of the borehole wall intercepted by the permeameter, provides the specific discharge, that is the average seepage rate at that depth. The results show that the highest seepage rates are concentrated in narrow bands about 0.2 m–0.5 m thick corresponding to the slip zone detected by nearby inclinometers. The hydraulic conductivity of the slip zone was evaluated by calibrating numerical FEM water flow models by using the permeability values $k_l = 10^{-9}$ m/s and $k_f = 10^{-10}$ m/s evaluated experimentally in the landslide and in the parent formation respectively (Di Maio et al. 2021). Figure 2 shows that the values of hydraulic conductivity thus obtained in the slip zone vary in a wide range, reaching about 10^{-5} m/s, thus values much higher than those of the other zones. As expected, the k values obtained by interpreting the oedometer consolidation curves of both undisturbed and reconstituted specimens of the landslide body and of the slip zone are much lower than the field values.

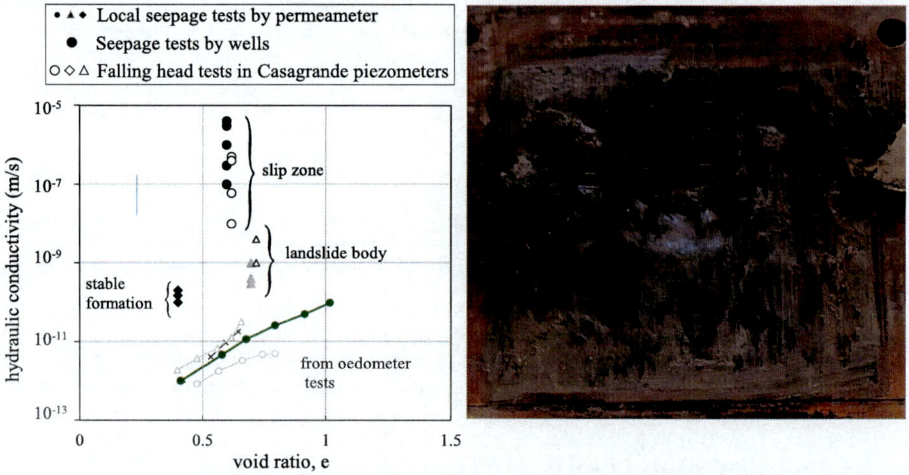

Fig. 2. Hydraulic conductivity in the different zones of the subsoil determined by different types of field tests and by oedometer tests; shear surface of a sample tested in the Casagrande direct shear after many shearing cycles.

The high permeability of the active slip zone is probably due to both its particular fabric and to the persistent sliding in residual conditions. As a matter of fact, the local roughness due to the presence of hard elements of lapideous consistency probably determines, during sliding, formation of fissures and opening of irregular voids. At a different scale, this phenomenon can be observed on the shear surface of laboratory specimens brought to residual conditions in the course of direct shear tests. Figure 2 shows the slip surface of an undisturbed specimen of the considered soil which was furrowed by sand and gravel elements during shear. It seems possible to hypothesize that similar and more important effects can occur *in situ* and that the higher the displacement rate, the higher the shear zone permeability.

Porewater pressures are being monitored in the considered two earthflows by Casagrande piezometers, open tubes and vibrating wire cells since 2005. Figure 3 distinguishes the piezometric level in the landslide body and in the stable formation from those in the slip zones (data relative to *Varco d'Izzo* are indicated as V.I.). The figure shows that the piezometric level in the landslide and in the formation have been practically constant in the whole monitoring period or, at least, varied not seasonally. On the contrary, large seasonal variations have been evaluated in the slip zones.

Figure 4 shows in detail some more recent data relative to the middle *Costa della Gaveta* landslide body (P9D, K9D) and to its head zone, where eleven boreholes supported by permeable screens have been realized to monitor water flow, and to carry out permeability tests. The figure reports the water level from the ground in the boreholes Ki and the water level in three Casagrande piezometers P9D, K4ter and K5ter installed at depths between 8 and 25 m, in correspondence of the shear zone detected by the nearby inclinometers. Since the permeability of the slip zone is much larger than that of the surrounding soil, the water level in the open screened tubes can be considered indicative of the hydraulic head in the slip zone itself. As a matter of fact, the variations in the holes are very similar to those exhibited by the piezometers. The figure also compares basal

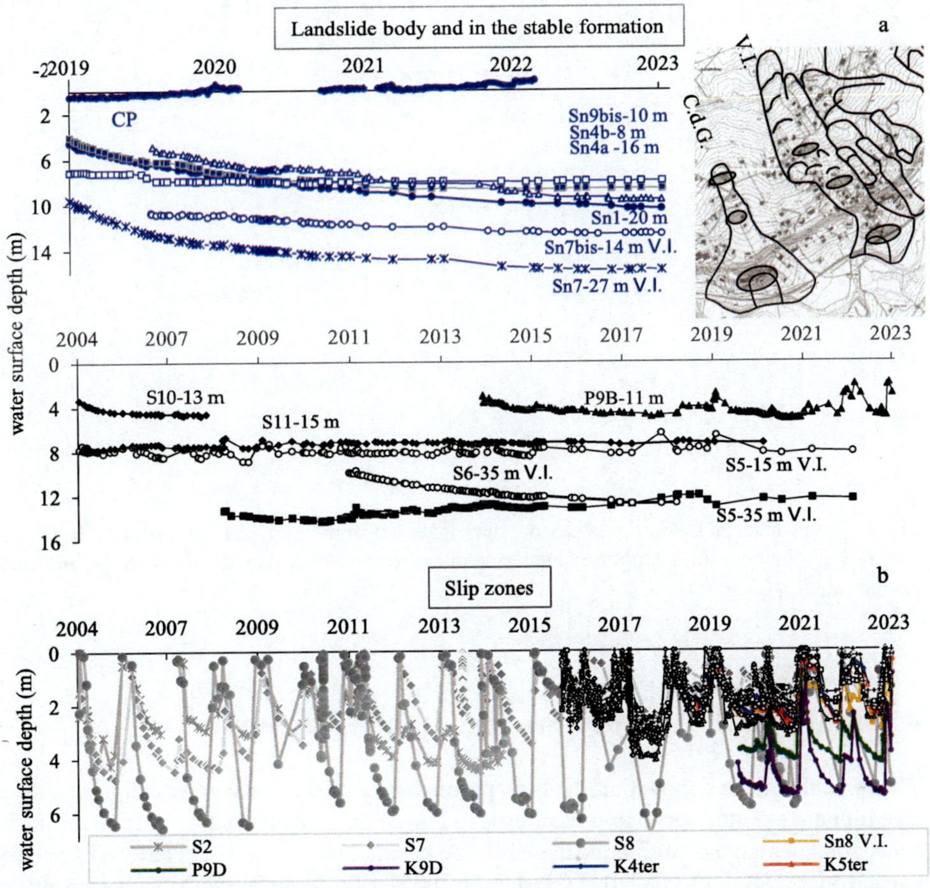

Fig. 3. Piezometric level in the two earthflows evaluated by electric wire cells (a) and Casagrande piezometers (b) in the landslide body and in the formation, and piezometric level in the slip zone evaluated by Casagrande piezometers and open tubes.

displacement rates (i.e., along the slip surface) to porewater pressures in the slip zone. The rate values are divided by the number between brackets to make them comparable to each other. The envelope of water level (grey area) is reported on the displacement rate data to compare the time trends. The figure shows that the seasonal variations of the two physical parameters are similar and in phase. Vassallo et al. (2016), by a data-driven evolutionary modelling technique, found strong correlation between contemporary rainfall and the rates of basal displacements in many inclinometers installed in the landslide. Similar results were found by the code Modflow 3D.

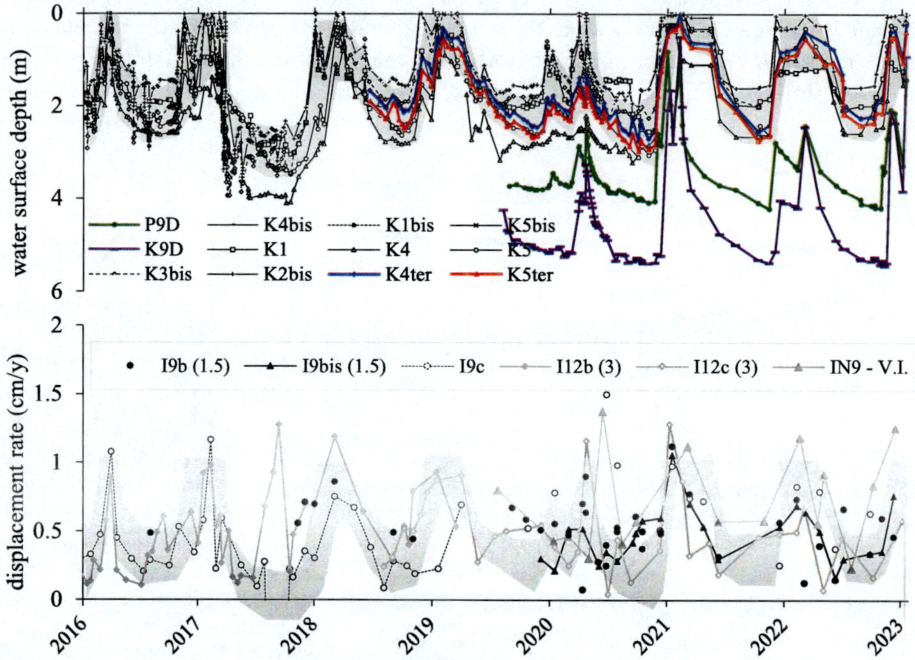

Fig. 4. Water level in Casagrande piezometers installed in the slip zone of *Costa della Gaveta* earthflow and in boreholes supported by permeable screens deeper than the slip surface, and basal displacement rates.

4 3D Model of Transient Seepage Induced by Rain

Theoretical analyses show that the high permeability of the shear zone can have a decisive influence on the seepage pressures throughout the subsoil and, in particular, in the shear zone, where they influence the landslide basal movements. Calculations have been carried out by MODFLOW-2005 (Harbaugh 2005), a finite-difference code which allows to consider the slip zone, with its own properties different from those of the surrounding soil, intersecting the ground surface. Several different transient analyses were carried out applying the historical rainfall series on the ground of *the Costa della Gaveta* earthflow.

The transient analyses were carried out for: i) homogeneous soil with $k = 10^{-9}$ m/s, ii) slip band more permeable than the other two regions, iii) slip band less permeable than the other two regions. The model is described in detail in Di Maio et al. (2021). The results of the analyses indicate that the high permeability along the slip surface provokes a two-fold effect of: *a)* important overall drainage - that makes the average yearly porewater pressure decrease - and *b)* important seasonal porewater pressure variations due to rainfall. The numerical results in the landslide head are compared to the experimental data in Fig. 5. It can be observed that, in the case of more permeable slip band, the numerical data are very close to the experimental ones. On the contrary, the model results relative to both homogeneous soil and less permeable slip zone exhibit trends much different from the experimental ones.

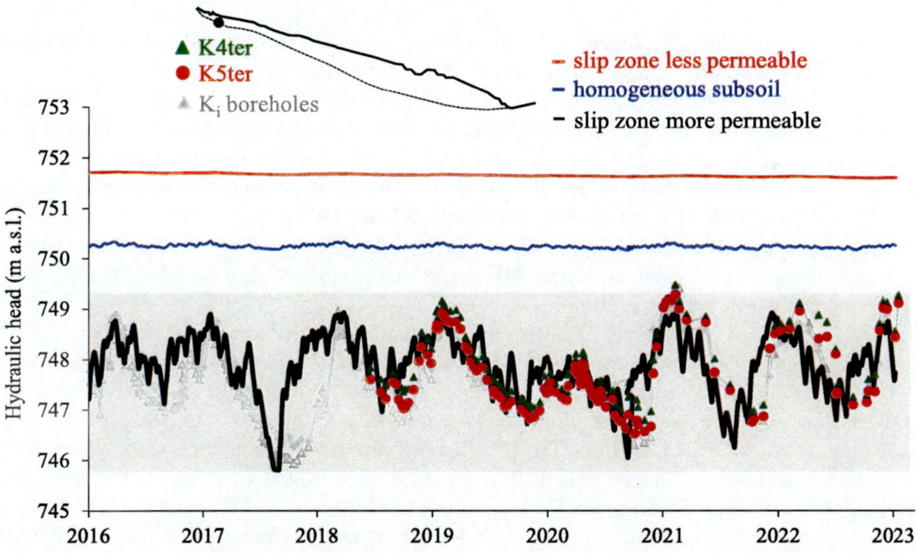

Fig. 5. Hydraulic head against time in the head of the *Costa della Gaveta* landslide: experimental (Ki wells, K4ter, K5ter) and numerical results for the cases of homogeneous subsoil with $k = 10^{-9}$ m/s, permeable slip zone ($k = 10^{-7}$ m/s) and less permeable ($k = 10^{-10}$ m/s).

5 Conclusions

The results of the study of Costa della Gaveta earthflow, a typical landslide of the Lucan Apennine environment, show that the permeability of the slip zone is higher than that of the landslide body and of the stable formation. The high permeability causes a general drainage effect in the whole slope. However, it also provokes important seasonal variations of porewater pressure in the slip band, whereas negligible variations occur far from it and from other discontinuities. Basal displacement rates appear strongly correlated to seepage pressures in the slip zone. The data available at the moment for the Varco d'Izzo earthflow - in the same slope, geological formations and evolution phase -

show similar behavior. So, the experimental data and the theoretical model suggest for deep drainage, which is the most diffused remedial measure for this type of landslides, the design criterium of inducing rapid pore water pressure dissipation directly on the slip surface, because it is hydraulically connected to the atmosphere. The study of the influence of hydraulic parameters on the landslides of the Basento valley is currently also carried out in other large landslides with different permeability distribution.

Acknowledgments. This research is supported by MIUR PON R&I 2014–2020 Program (project MITIGO, ARS01_00964) and by R.F.I. Rete Ferroviaria Italiana.

References

Cotecchia, F., Santaloia, F., Tagarelli, V.: Towards a geo-hydro-mechanical characterization of landslide classes: preliminary results. Appl. Sci. **10**, 7960 (2020)

Di Maio, C., Vassallo, R., Vallario, M., Pascale, S., Sdao, F.: Structure and kinematics of a landslide in a complex clayey formation of the Italia Southern Apennines. Eng. Geol. **116**, 311–322 (2010)

Di Maio, C., Vassallo, R., Vallario, M.: Plastic and viscous displacements of a deep and very slow landslide in stiff clay formation. Eng. Geol. **162**, 53–66 (2013)

Di Maio, C., Scaringi, G., Vassallo, R.: Residual strength and creep behaviour on the slip surface of specimens of a landslide in marine origin clay shales: influence of pore fluid composition. Landslides **12**(4), 657–667 (2015)

Di Maio, C., De Rosa, J., Vassallo, R.: Pore water pressures and hydraulic conductivity in the slip zone of a clayey earthflow: experimentation and modelling. Eng. Geol. **292**, 1–14 (2021)

Di Maio, C., De Rosa, J., Vassallo, R., Coviello, R., Macchia, G.: Hydraulic conductivity and pore water pressures in a clayey earthflow: Experimental data. Geosciences **10**, 102 (2020)

Harbaugh, A.W.: MODFLOW-2005, The U.S. Geological Survey Modular Ground-Water Model -- the Ground-Water Flow Process: U.S. Geological Survey Techniques and Methods (2005)

Urciuoli, G., Comegna, L., Di Maio, C., Picarelli, L.: The Basento valley: a natural laboratory to understanding the mechanics of earthflows. Rivista Italiana di Geotecnica **1**, 14–33 (2016)

Vassallo, R., Doglioni, A., Grimaldi, G.M., Di Maio, C., Simeone, V.: Relationships between rain and displacements of an active earthflow: a data-driven approach by EPRMOGA. Nat. Hazards **81**(3), 1467–1482 (2016). https://doi.org/10.1007/s11069-015-2140-9

Vassallo, R., Grimaldi, G.M., Di Maio, C.: Porewater pressures induced by historical rain series in a clayey landslide: 3D modeling. Landslides **12**, 731–744 (2015)

Vassallo, R., De Rosa, J., Di Maio, C., Reale, D., Verde, S., Fornaro, G.: In situ and satellite long-term monitoring of two earthflows of the Italian southern Apennines and of the structures built on them. Italian Geotech. J. **55**(4), 77–95 (2021)

Vassallo, R., Calcaterra, S., De Rosa, J., Di Maio, C., Gambino, P.: Long-term displacement monitoring of two very slow earthflows by inclinometers, GPS and Cosmo-SkyMed data. Geosciences, **10**(171) (2020)

Vassallo, R., e.: DInSAR data for landslides in Basilicata region: geotechnical calibration and interpretation. Submitted to CNRIG2023 (2023)

Observation and Analysis of a Moving Slope

Alessandra Di Mariano[1,2](✉) and Antonio Gens[1,2]

[1] Centre Internacional de Mètodes Numèrics a L'Enginyeria (CIMNE), 08034 Barcelona, Spain
alessandra.dimariano@upc.edu

[2] Universitat Politècnica de Catalunya (UPC), Campus Nord, 08034 Barcelona, Spain

Abstract. A northbound motorway close to the city of Barcelona has shown various deformation signs in its pavement and, as a consequence, it has been the subject of numerous studies and projects, aimed at reducing the road maintenance costs. In the area object of the present study, the road follows a slope, which is located below an old quarry converted in a waste disposal facility before the deformation signs appeared in the motorway. An extensive instrumentation system was installed along the slope adjacent to the motorway. The available monitoring data has allowed to identify the instability of the slope as the cause of the deformation signs in the motorway, as well as to locate the failure surface in various sections of the slope.

The paper describes the analysis of the evolution of the instability phenomenon and presents the stabilizing measure adopted. The reduction of the observed slope movement rate shows the success of the stabilizing measure.

Keywords: Slope Stability · Monitoring · Slow Movement · Stabilizing Measures · Slope Analysis

1 Introduction

The road network of a country plays a fundamental role in its economic development and therefore, one of the most important and complex tasks that Administrations have to face is ensuring the road network functionality and safety over time [1]. Due to their configuration, roads often run in different geological contexts surrounding slopes that can be prone to instability phenomena with potential risks for road users and socioeconomic impacts that can be significant at times [1]. Properly managing these risks implies carrying out activities that can predict and prevent, to a certain extent, potential instability phenomena, considering operational and economic limitations [1]. In general, the processes involved in landslides or other instability phenomena are highly complex and need to be adequately identified in order to select appropriate control measures [2–7]. An erroneous estimate of these phenomena can lead to the adoption of ineffective or even counterproductive solutions for their stabilization ([4, 5, 7], among others).

This work refers to a case study in the North-East of Spain, related to the section of a road, in the vicinity of a river, that was built by cutting the toe of its surrounding slope. In the upper part of the slope there is an old quarry which was converted years ago into

A. Ferrari et al. (Eds.): CNRIG 2023, SSGG, pp. 503–510, 2023.
https://doi.org/10.1007/978-3-031-34761-0_61

a controlled waste disposal facility. The slope has shown signs of instability that have affected both the functionality and safety of the road. In order to reduce the rate of the movement and increase the stability of the slope, the decision was taken to remove part of the waste material from the disposal facility, thus unloading the crest of the slope. The extensive monitoring system installed at the site allowed to follow the evolution of the phenomenon and verify the effectiveness of the measure adopted.

2 Description of the Case Study

A Catalan regional motorway, built in 1989, has shown various deformation signs in its pavement between the years 2006 and 2018 [8]. The motorway, which was built by cutting the toe of an adjacent slope, connects the city of Barcelona to the north of the Catalan region, crossing two river valleys. It is a dual carriageway road, which is heavily travelled (more than 30,000 vehicles/day) and whose deformation signs exclusively concern a 500 m long section that runs through one of the two river valleys. Studies to determine the origin of these deformation signs began in 2008, initially focusing only on the road pavement. In 2010, a gabion wall was built at the toe of the adjacent slope, as a protection element for the motorway (Fig. 1). Later on, in 2015, a Construction Project for major repair works was drawn up to recover the motorway safety and functionality. On the one hand, the project involved the design of the corrective measures for the signs of instability observed and, on the other, the execution of important repair works on the motorway pavement, including the installation of mechanical anchors and drain pipes in one part of the slope adjacent to the road (Fig. 1). In 2017, during the execution of the project, the Contractor was able to verify that the section of the motorway under study was actually undergoing movements in its transverse direction, produced by the displacement of the slope adjacent to it (Fig. 1).

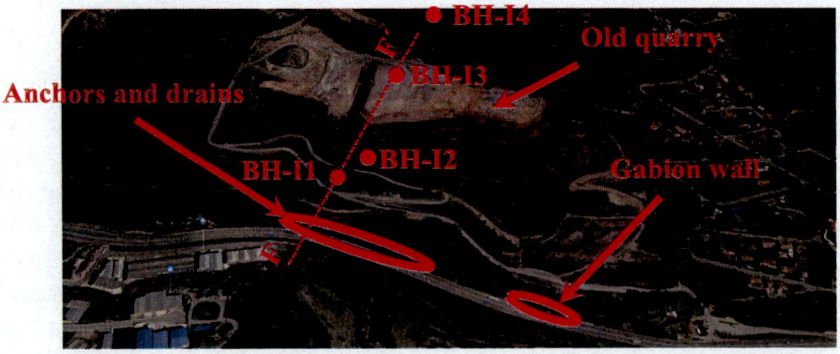

Fig. 1. Section of the motorway referring to the case study and location of the old quarry. The image shows the approximate location of the gabion wall, mechanical anchors and drain pipes present at the toe of the slope surrounding the motorway.

At the crest of the slope, there is an old quarry whose extractive activity was carried out between the years 1960 and 2004 (Fig. 1). The environmental restoration of the slope began in 2004 with the conversion of the quarry into a controlled waste disposal facility. During the execution of the motorway repair works, the construction manager found a cause-effect relationship between the completion of the quarry exploitation and the appearance of the deformation signs on the motorway. Between 2017 and 2018, the second phase of the Construction Project was carried out, which included the execution of new geotechnical investigations and the evaluation of the stability of the slope by numerical analyses. As a result, the conclusion was drawn that the deformation signs observed were the consequence of a phenomenon of slope instability that mobilizes a total volume of about 4,000,000 m^3 of material, according to the most recent monitoring data. With the aim of limiting the risks associated with this instability phenomenon, in 2018 local authorities made the decision to interrupt the placement of waste in the old quarry and, at the same time, monitor both the slope and the entire area of the waste disposal facility, in order to clearly define the limits of the unstable mass.

3 Location and Ground Conditions

The area under study is located on the northern limit of the Catalan pre-coastal mountain range where it meets the Ebro river depression. From top to bottom, the geological sequence of the area, includes the following materials (Fig. 2): anthropic fills (R) and quarry restoration fills (RC), sandy clays with gravels of colluvial origin from the Quaternary Era (Q), Eocene siltstones and sands (E), alternations of decametric to centimetric banks of dolomitic and calcareous rocks from the Middle Triassic (M1) (materials exploited in the old quarry) and finally shale, claystone and sandstone deposits, belonging to the Lower Triassic (B).

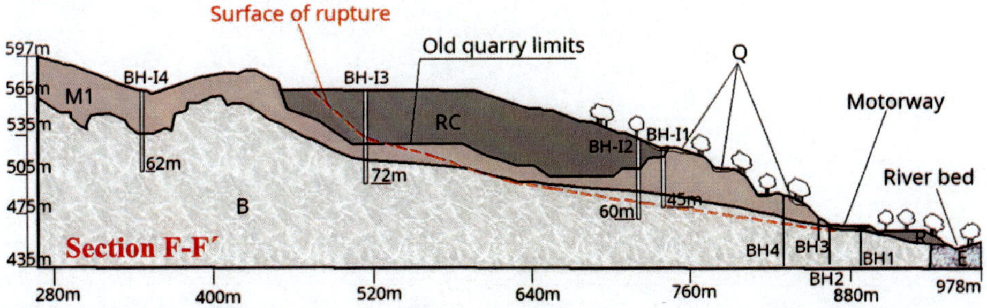

Fig. 2. Geological profile corresponding to one of the cross sections of the slope (Section F-F´ in Fig. 1). The figure shows the location of the failure surface indicated by the inclinometers.

Figure 2 shows the geological profile of section F-F´ in Fig. 1, displaying as well the locations of the old quarry, the motorway, the nearby river and the position of the surface of rupture according to the available monitoring readings. It also shows the location of four biaxial inclinometers (BH-I1-BHI4) and four boreholes (BH1-BH4) used, among other things, for electrical resistivity tomography (Fig. 2). During the execution of the Construction Project, the Contractor carried out some laboratory tests to characterize the materials of the slope and performed numerical analyses to reproduce the history of events and analyze the relative contribution of each of them to the evolution of the instability phenomenon. A simplified plane strain model was used, in which the geometry of the problem was approximated to a uniform cross section. Only two materials were considered, the M1 unit from the Middle Triassic and the deeper B unit from the Lower Triassic (Fig. 2). The behaviour of both units was modelled using the generalized Hoek-Brown criterion (Eq. 1), to consider the different degrees of alteration of the rock masses (Table 1). The m_b parameter in Eq. (1) represents a reduced value of the material constant m_i (Eqs. 2). The parameters s and a are material constants (Eqs. 2), σ_1' and σ_3' represent the maximum and minimum effective principal stresses at failure, while σ_{ci} is the uniaxial compressive strength of the intact material. Finally, GSI represents the Geological Strength Index and D is a factor that depends on the degree of alteration of the material, ranging from 0 to 1. All the analyses were carried out under drained conditions.

$$\sigma_1' = \sigma_3' + \sigma_{ci}\left(m_b \frac{\sigma_3'}{\sigma_{ci}} + s\right)^a \tag{1}$$

$$m_b = m_i exp\left(\frac{GSI - 100}{28 - 14D}\right); s = exp\left(\frac{GSI - 100}{9 - 3D}\right); a = \frac{1}{2} + \frac{1}{6}\left(e^{-GSI/15} - e^{-20/3}\right) \tag{2}$$

Table 1. Geomechanical parameters used in the analysis of the instability phenomenon. The parameter E_i represents the initial value of Young's modulus and ν the Poisson's ratio.

Parameter	M1 unit		B unit		Weathered B unit	
	Peak	Residual	Peak	Residual	Peak	Residual
σ_{ci} (MPa)	75	75	17	17	17	17
GSI	60	30	50	20	35	20
m_i	10	10	8	8	8	8
D	0	0	0	0.9	0.8	0.9
m_b	2.397	0.821	1.341	0.044	0.167	0.044
s	0.0117	0.0004	0.0039	3.06×10^{-6}	0.0001	3.06×10^{-6}
a	0.503	0.522	0.506	0.544	0.516	0.544
E_i (MPa)	12740	12740	1920	1920	1920	1920
ν	0.22	0.22	0.22	0.22	0.22	0.22

Numerical predictions about the location of the slip surface were later confirmed by in situ observations. The value of the friction angle at the depth of the failure surface obtained from the analyses is in the range 15–16°. Actually, in the area of study, the lower Triassic shales (B unit) have a strongly weathered appearance, especially at the depth of the failure surface. To gather more information on the drained shear resistance of unit B at the depth of the sliding surface, ring shear tests were performed under residual conditions at the Geotechnical laboratory of UPC, using the Bromhead equipment

Table 2. Parameters of three soil samples belonging to unit B, taken at the depth of the slope failure surface, tested under residual conditions with annular shear apparatus.

Sample depth	Ring Shear		Atterberg Limits		
(m)	c'_{res} (kPa)	ϕ_{res}°)	w_L (%)	w_P (%)	PI (%)
34.2–34.5	0	17.9	-	-	-
34.8–35.0	0	23.0	28.0	16.2	11.8
45.0–45.5	0	18.1	26.4	16.3	10.1

(Table 2). Laboratory tests led to greater values of unit B residual friction angle with respect to the ones estimated from numerical analyses (Table 2).

According to the latest available monitoring data, the failure surface of the unstable slope has a depth of about 40 m and it is located mainly at the upper part of the lower lithological layer (unit B in Fig. 2). The unstable mass directly affects the regional motorway, as well as nearly the entire waste disposal facility. Indirectly, it also acts on the bed of the nearby river located downhill from the motorway (Fig. 2), an industrial estate located between the road and the river, as well as some telecommunications and electricity supply lines.

4 Instrumentation and Monitoring

In September 2017, an extensive monitoring system was installed in the area. It includes biaxial inclinometers, topographical surveys (GPS levelling), open standpipes and vibrating wire piezometers, automatic prisms, and load cells in the mechanical anchors located at the base of the slope (Fig. 1). While inclinometers are used to follow the evolution of the slope movement with depth and clearly identify the failure surface, topographical surveys allow detecting the 3D movement of the slope surface. The evolution of the water table and the pore water pressures at the depth of the failure surface is observed by means of open and vibrating wire piezometers, respectively. Automatic prisms, on the other hand, register the movement of various structural elements located in the unstable area. Finally, load cells are used to observe the evolution of the load on the mechanical anchors at the toe of the slope (Fig. 1). In order to establish an early warning system, warning, alert and alarm thresholds have also been defined in the monitoring plan for each instrument, based on the information gathered during the construction project.

5 Control Measures

Once established that the damage to the motorway was entirely derived from the movement of its surrounding slope, the engineers in charge of the project analysed different corrective measures, with the aim of applying at least one of the following actions: removal of the load from the upper part of the slope, surcharge application at the toe of the slope and construction of structural elements capable of resisting the observed movement. The evaluation of the different alternatives was carried out in terms of both the increase in the slope Safety Factor (SF) and the economic cost of execution. In addition, for each alternative, other features were considered such as aspects related to possible environmental impacts, consequences on the road users or third parties in general and degrees of reliability as well as life span of the stabilization measures. Once the proposed alternatives were examined, local authorities made the decision to stabilize the slope through the removal of load from its crest by excavating the waste material and pumping the water accumulated in the disposal facility. Compared to the other stabilization measures, this one has the advantage of being a flexible measure, that can be easily carried out in different stages without having to modify the road layout or the configuration of its surroundings. Also, it can be adapted to the response of the slope, based on the evolution of the monitoring data.

The excavation and unloading activities of the upper part of the slope began at the end of 2018, being interrupted due to logistical problems, between the months of January and August of the year 2020. They were later resumed in September of the same year and have continued until the end of February 2022.

6 Evolution of the Slope Movement

The first estimate of the slope movement rate was made in 2016 based on tachymetric data available since 2006. Due to some uncertainties about the accuracy of these data, the estimate is approximate and it only gives information on the movement rate mean values, without offering any data on the evolution of the phenomenon. An average rate of 8 mm/month over a period of 10 years (from 2006 to 2016) is estimated, without considering any possible temporary rate variations associated, for example, with changes in the hydraulic conditions of the slope. From the end of 2017, with the numerous monitoring data available, there are reliable records of the evolution of the instability phenomenon. The data indicate that the slope movement rate is not constant and it changes with any variation in the ground materials stress state related, for instance, with fluctuations in pore water pressures due to rain events (Fig. 3). At the beginning of the monitoring period, the acceleration of the movement occurred with a certain delay with respect to each rainy event (Fig. 3). Over time, this delay started diminishing and the slope movement became increasingly sensitive to any variation of stress (Fig. 3). In the last days of 2018, the unloading of the upper part of the slope began (Fig. 4). As the disposal facility was excavated, the movement rate progressively reduced and the phenomenon of instability became more independent of the stress changes in the slope (Fig. 3–Fig. 4). At the end of January 2020, due to a violent storm (153 l/m^2 in about 3 days), the movement accelerated again, reaching an overall maximum rate of about 20 mm in 5 days. After only four days from the start of the storm, at the depth of the failure surface, an increase in pore water pressure of about 4 m of water column in just over 24 h was observed (Fig. 3–Fig. 4). The rapid increase in pore pressure indicates an increase in the permeability of the lithological formations above the failure surface, possibly due to the evolution of their weathering process. In the same time period, survey prisms registered peak rates of up to 90 mm/month. The storm coincides with the beginning of a period in which the unloading works of the head of the slope were temporarily suspended for logistical reasons and, when the storm ended, the movement progressed with average rates of about 14 mm/month (Fig. 4). When the excavation activities resumed, the movement and its rates gradually reduced, until reaching the current values of approximately 0.3 mm/month.

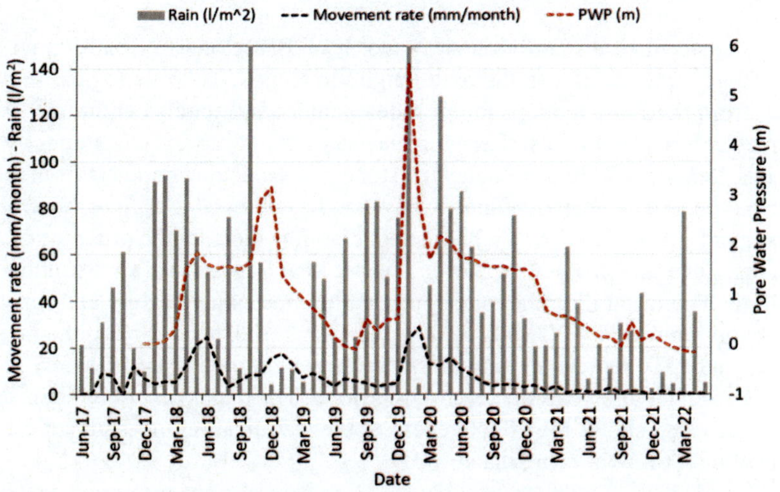

Fig. 3. Evolution of the slope movement rate in millimeters per month and of the pore water pressures in meters of column of water. The graph shows also the histogram of the rain events in liters per square meter. Observation period June 2017-May 2022.

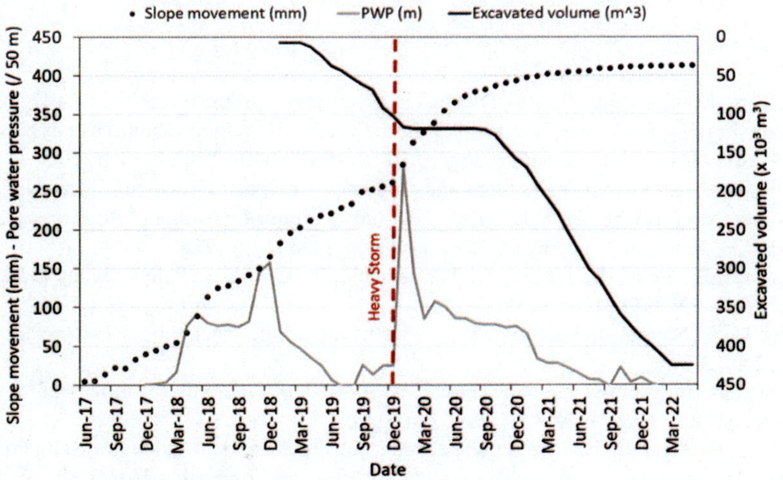

Fig. 4. Evolution of the slope movement, the volume of the excavated waste material and the pore water pressure with time. Observation period June 2017-May 2022.

The graphs of Fig. 3–Fig. 4 clearly show the adequacy and effectiveness of the selected slope stabilization measure. Despite the progressive decrease in movement, the monitoring data indicate that the slope is not fully stable yet, being the slope unloading process still incomplete.

7 Summary and Conclusions

The slope adjacent to a main motorway, north of Barcelona in Spain, moves slowly in a direction perpendicular to the road longitudinal axis. The lithological sequence of the area, from the most modern to the oldest materials, includes anthropic fill layers, Quaternary colluvial materials, alternating layers of dolomitic and calcareous rocks from the Middle Triassic, and finally shale, claystone and sandstone deposits from the Lower Triassic period. An extensive monitoring system has allowed to follow the evolution of the movement since 2017. The inclinometers readings indicate the presence of a failure surface that generally crosses the lower Triassic shales, reaching a maximum depth of about 40 m. The instability phenomenon puts the motorway's safety and functionality at risk, involving nearly 4,000,000 m^3 of ground material, according to the most recent monitoring data. To reduce the rate of the movement and increase its safety factor, the decision was taken to unload the crest of the slope. The paper describes some details of the case study and shows the effectiveness of the stabilization measure adopted, based on the evolution of monitoring data.

Acknowledgements. The work presented in this article is supported by Infraestructures.cat. The support provided by the Severo Ochoa program through the Grant CEX2018–000797-S funded by MCIN/AEI /https://doi.org/10.13039/501100011033 is also acknowledged.

References

1. Ferlisi, S., Marchese, A., Peduto, D.: Quantitative analysis of the risk to road networks exposed to slow-moving landslides: a case study in the Campania region (southern Italy). Landslides **18**(1), 303–319 (2020). https://doi.org/10.1007/s10346-020-01482-8
2. Varnes, D. J.: Slope movement types and processes. In: Special Report 176: Landslides: Analysis and Control Schuster, R.L, Krizek, R.J (eds.), Transportation and Road research Board, pp. 11–33. National Academy of Science, Washington D.C.(1978)
3. Alonso, E.E., Gens, A., Lloret, A.: The landslide of Cortes de Pallas. Spain. Géotechnique **43**(4), 507–521 (1993)
4. Leroueil, S.: Natural slopes and cuts: movement and failure mechanisms. Géotechnique **51**(3), 197–243 (2001)
5. Corominas, J.: El clima y sus consecuencias sobre la actividad de los movimientos de ladera en España. Rev. C & G. **20**(3–4), 89–113 (2006)
6. Urciuoli, G., Picarelli, L.: Interaction between landslides and man-made works. In: Proceedings of the 10th International Symposium on Landslides and Engineered Slopes, vol. 2, pp. 1301–1307 Xi'an (China), CRC Press (2008)
7. Alonso, E.E., Pinyol, N.M. Puzrin, A.M.: Geomechanics of Failures. Advanced Topics. Springer, Dordrecht (2010). https://doi.org/10.1007/978-90-481-3538-7
8. Di Mariano, A., Gost, X., Gens, A. Álvarez, E.: Control and monitoring of an unstable slope. In: International Symposium on Field Monitoring in Geomechanics. Conference proceedings ISFMG 2022, pp. 1–7. London: British Geotechnical Association (BGA) (2022)

OpenSees Analysis of a Tailings Storage Facility in Southern Tuscany (Italy)

Andrea Geppetti[1](✉), Johann Facciorusso[1], Giovanni Ciardi[2], Luis Felipe Prada-Sarmiento[3], and Claudia Madiai[1]

[1] Department of Civil and Environmental Engineering, University of Florence, Firenze, Italy
andrea.geppetti@unifi.it
[2] Department of Civil and Environmental Engineering, University of Perugia, Perugia, Italy
[3] Department of Civil and Architectural Engineering, Aarhus University, Aarhus, Denmark

Abstract. Besides producing raw materials essential for contemporary societies, the mining industry also generates millions of tons of waste every year. Most of this waste is represented by tailings consisting of saturated fine non-plastic soils, generally arranged in structures called tailings dams located at a short distance from the mine. These structures contain the tailings through an embankment made with the coarser fraction of residual materials. The rate of failure for these structures is high and often results in the release of the materials in the downstream territories with catastrophic consequences for the population, the environment, and the local economy. In recent decades, the use of advanced numerical models to predict the mechanical behavior of these structures, subjected to different types of loading, initial and boundary conditions, has become increasingly widespread. Contrary to the more classical methods, such as the limit equilibrium, which only determines the safety factor for stability, these advanced tools can quantify stress distributions, deformations, accelerations and pore water pressures over time. In this work, the liquefaction potential under seismic conditions of a tailings dam located in southern Tuscany was analyzed with a dynamic finite element model implemented in the OpenSees software. Three numerical analyzes were carried out with three different advanced constitutive models describing the dynamic behavior of tailings. The constitutive models were calibrated based on the results of field tests and cyclic laboratory tests carried out on reconstituted specimens.

Keywords: tailings dams · seismic liquefaction · numerical analysis

1 Introduction

Tailings storage facilities (TSF) are structures that arise from the need to store waste material from mining activities. These dam structures are highly susceptible to collapse, with a two orders of magnitude failure rate higher than conventional water retaining dams [1]. One of the most common construction techniques for these facilities is the so-called upstream method, as it is speedy in terms of construction and economical compared to other methods, but it is also the most susceptible to failure in seismic regions. The upstream process involves the gradual construction of the dam in a regressive manner by

A. Ferrari et al. (Eds.): CNRIG 2023, SSGG, pp. 511–518, 2023.
https://doi.org/10.1007/978-3-031-34761-0_62

founding part of the dam directly on the loose and weak tailings contained in the settling basin. The analysis of multiple collapses that have occurred in the past has shown that liquefaction is one of the most frequent causes of failures for tailings dams built with the upstream method [2–4]. In this work, a numerical analysis of a tailings dam located in southern Tuscany (Italy) will be performed to identify the seismic liquefaction potential.

2 Overview of the Tailings Storage Facility

The area where the tailings dam is located has always been subject to mining activities that can be traced back to the Etruscan populations. The main minerals extracted from the mines were sulfides, calcium pyrites, lead and silver. The activities in this area intensified in the second half of the 1900s and ended in 1985. The construction of the examined tailings dam began in the decade of 1960s and is now closed. This structure is the most important of a more extensive system made of numerous embankments and basins in terms of size and the damage that its collapse could cause. The dam was built using the upstream method and today reaches a height of about 36.5 m above ground level and contains approximately 600,000 m^3 slurries. The section is composed of three main geotechnical units (Fig. 1): settling tailings (U1), embankment material (U2) and clay foundations (U3).

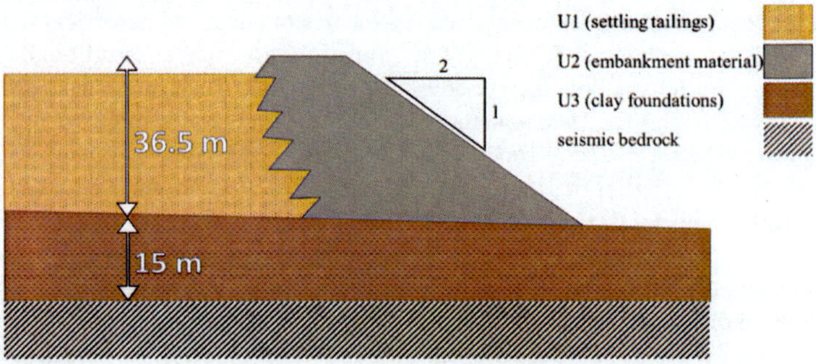

Fig. 1. Cross section of the tailings dam

2.1 Geotechnical Characterization

The geotechnical characteristics shown below are the synthesis of numerous investigation surveys and field and laboratory tests carried out over the years and still in progress. The settling tailings unit (U1) is made of the finest fraction of the waste material deriving from mining activities and consists mainly of sandy silts (Fig. 2a and Table 1). The embankment material (U2) is composed of the coarser fraction of the waste derived from mining activity and consists of gravels and heterometric pebbles in abundant blackish gray sandy matrix (Fig. 2b and Table 1). Finally, the clay foundation (U3) material is the foundation soil on which the structure was built and is essentially made of altered clays (Table 1).

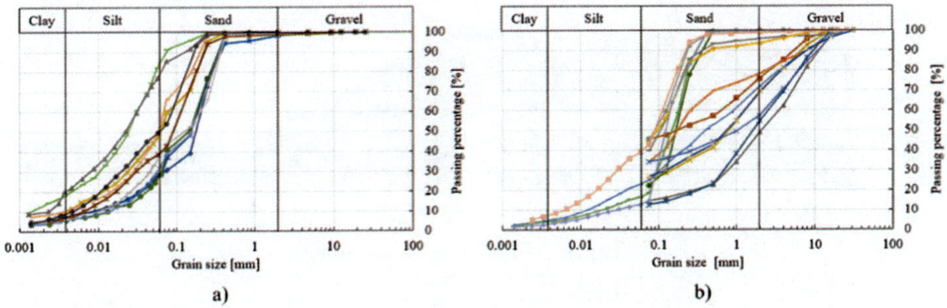

Fig. 2. Grain size curves: a) settling tailings U1 b) embankment material U2

Table 1. Soil characteristics

Material	I_p [%]	e_{min} [-]	e_{max} [-]	γ_{sat} [kN/m^3]	φ' [°]
settling tailings U1	5	0.6	1.375	20.00	33.0
embankment material U2	5	0.6	1.2	20.00	33.0
clay foundations U3	24	0.5	0.86	18.90	30.0

3 Finite Element Model

Three 2D finite element analyses with three different constitutive models used to describe the U1 material were conducted with the open-source software OpenSees [5] to evaluate the seismic liquefaction potential of the basin. Firstly, the geometry of the model was defined, then the most suitable constitutive models to represent the behavior of the materials were identified and calibrated. Finally, the boundary and initial conditions and the seismic actions were specified. Each of the three analyses consisted of two successive steps. In the first step, a static analysis was executed to determine the initial stress state within the tailings storage facility (TSF). In the second step, the seismic analysis was carried out to identify the TSF susceptibility to liquefaction. It is worth noting that it be more realistic to simulate the dam construction phases over time to estimate the initial stress field better. In the present model, a wished-in-place condition with the final configuration of the upstream retaining dam was prescribed. Further work is therefore needed to include this aspect.

3.1 Geometry Definition

The geometry of the structure was based on topographic surveys and information gathered from geotechnical and geophysical investigation. Based on the geometry, the domain was discretized with a finite element mesh using the AUTOMESH-2D/3D software [6] (Fig. 3). The dimensions of the domain have been determined to analyze the part of the basin closest to the embankment without this being affected too much by edge effects.

The OpenSees analyses were conducted with quadrangular elements called Nine Four Node Quad u-p Element [5], which has eight nodes on the perimeter and a central

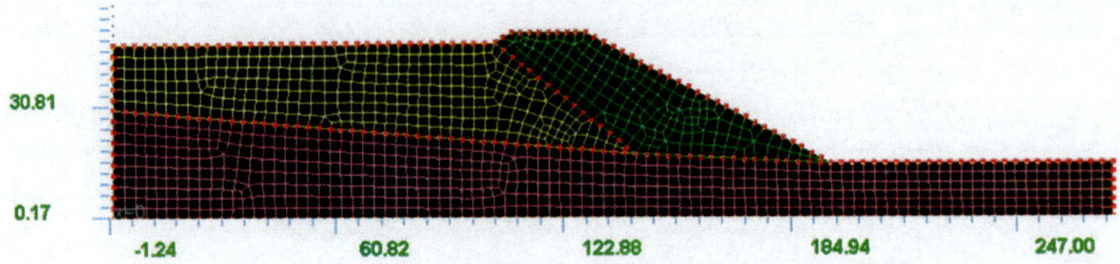

Fig. 3. Model mesh definition

one. AUTOMESH-2D/3D only produces quadrangular elements, so the five missing nodes were later added. The domain consists of a total of 1147 elements 4801 nodes and 5735 Gauss points. The height of the elements was determined using the relation:

$$\Delta h = \frac{V_S}{k \cdot f}$$

where k is a reduction coefficient V_S is the minimum value of the shear waves found in the layer and f is the maximum frequency of the signal that will propagate in the layer. The height of the element obtained with this relationship is 3 m, and the same value has been attributed (where possible) to the width of the element to maintain a uniform aspect ratio [7]. The groundwater level was positioned just below the ground level to simulate a condition of total saturation.

3.2 Constitutive Models Definition and Calibration

Three numerical analyses were carried out, assigning three different constitutive models to the unit called settling tailings U1, namely SANISAND [8], PM4Sand [9], and Pressure Dependent Multi Yield 02 (PDMY02) [10]. For the other two units (U2, U3), the Pressure Independent Multi Yield (PIMY) [10] constitutive model was used for all three analyses. SANISAND is a fully tensorial constitutive model formulated in the frame of the critical state soil mechanics and bounding surface theory. SANISAND incorporates a Drucker-Prager narrow yield surface that evolves kinematically within the bounding, critical and dilatancy surfaces and prescribes a non-associate flow rule. The model accounts for memory effects to remember the most recent reversal points and incorporates a fabric tensor to improve the description of cyclic behavior. The PM4Sand constitutive model is based on the same concepts as SANISAND, but it is formulated in terms of plane stresses and introduces substantial changes to fabric equations. PDMY02 and PIMY are based on Drucker-Prager J2 multi-surface-plasticity modified with a Lade-Duncan failure criterion. The calibration processes for the SANISAND, PM4Sand and PDMY02 models is practically identical to the one described in [11] except for the shear modules for small strains, which have been obtained with the methods proposed by Taiebat [12, 13] and to which the following values have been attributed: $G_{SANISAND} = 243$, $G_{PM4Sand} = 236$ and $G_{PDMY02} = 294$. The results obtained from undrained cyclic triaxial tests on reconstituted specimens have been compared with single-element numerical simulations (element test) to calibrate the constitutive models. The simulation of stress-controlled undrained cyclic triaxial tests was performed both in two (to calibrate the PM4Sand model) and in three dimensions for SANISAND and PDMY02. To represent the sample in two dimensions a quadrangular element called Nine Four Node Quad u-p Element was used, whereas in three dimensions a cubic element called SSPbrickUP Element was used. Figure 4 shows the comparison between the experimental results and the numerical simulations in terms of effective stress paths (p'-q), variation of pore water pressure ratio ($r_u = \frac{\Delta u}{p'}$) and stress-strain cycles ($\varepsilon_a - q$). From the examination of Fig. 4 it is possible to deduce that there is a good agreement between the experimental results and the numerical simulations.

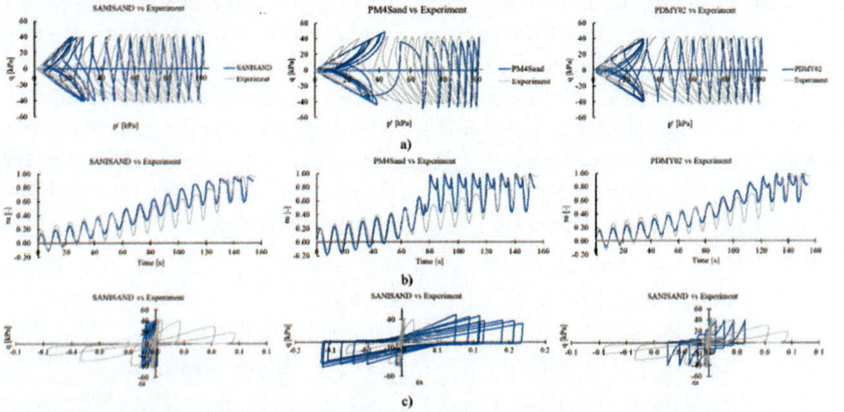

Fig. 4. Comparison between simulations and experimental results of: a) p'-q stress paths, b) pore water pressure ratio r_u and c) axial strain ε_a

As mentioned in Sect. 2.1, the material constituting the unit embankment material U2 has high gravel content with diameters even greater than 1 cm which made it impossible to carry out laboratory tests on reconstituted samples and to proceed with the same calibration procedure as that carried out on the settling tailings U1 material. Therefore, parameters proposed by the authors [5, 10, 14] for different soil types were attributed to the PIMY constitutive model that was used to describe both U2 and U3 units. On the other hand, the liquefaction susceptibility of these units will not be analyzed, although the wave propagation to the U1 unit is relevant. Further work to better characterize these units is needed.

3.3 Boundary Conditions

The boundary conditions described below follow those proposed by McGann-Arduino (2011) [15]. The nodes at the base of the model are free in the horizontal direction while they are fixed in the vertical direction. Furthermore, all the nodes of the base are linked to each other through the OpenSees equalDOF [5] command which forces them to have the same displacement in the horizontal direction. The properties of a Lysmer-Kuhlemeyer (1969) dashpot [16] are assigned to one of the base nodes (and consequently also to the others, as stated above). Dimensions of some orders of magnitude greater than the internal elements were assigned to the elements at the borders of the model to guarantee free-field conditions during the wave propagation. This configuration allowed in the finite element model to simulate the presence of two massive columns of soil at the ends of the model which absorb the seismic waves upon reflections.

3.4 Ground Motion Definition

The input seismic motion used in the analysis was obtained by SCALCONA 3.0 software [17]. The software returns a set of 7 natural seismic inputs recorded on the surface of a class A subsoil (Fig. 5a) and satisfying the spectrum-compatibility with the response spectrum provided at the site by the Italian standards (NTC2018) [18] for a return period of 2475 according to the requirements of the Italian rules for embankment dams 2014 [19]. The dynamic finite element analyses presented in this article were conducted only with reference to the signal number 6 (Fig. 5). Since the accelerograms provided by SCALCONA 3.0 are recorded on outcropping rock, the motion to be applied to the base of the 2D model was computed deconvolving the selected signal. The deconvolution was performed on a 1D linear viscoelastic model of the seismic bedrock using the STRATA software [20].

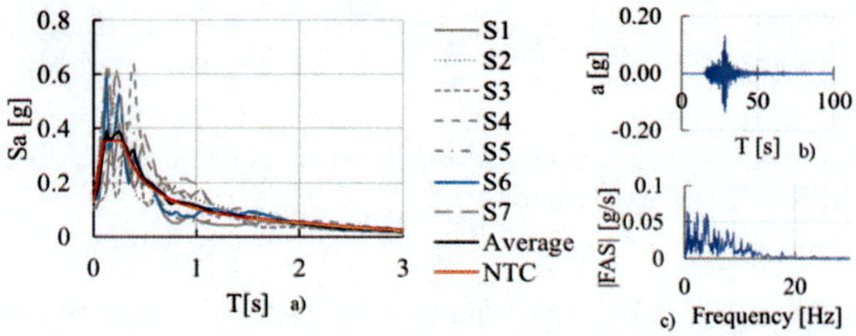

Fig. 5. Characteristics of the seismic input: a) response spectra of: the seven signals from SCALCONA3.0, the average spectrum and the reference spectrum of the Italian regulations; b) acceleration time history of signal 6 assumed in FEM analyses, c) Fourier spectrum of signal 6

4 Results

Three numerical analyses of the structure were conducted, each time assigning a different constitutive model to the material describing the U1 geotechnical unit. The constitutive models used for this unit are SANISAND, PM4Sand and PDMY02 while the PIMY constitutive model was used in the U2 and U3 geotechnical units. Each analysis consisted of two steps: first a gravitational analysis to calculate the initial stress distribution of the TSF and second the dynamic analysis with the application of the seismic loading. In this work the attention is concentrated on the liquefaction potential of the tailings settling basin. The liquefaction potential was evaluated through the pore water pressure ratio:

$$r_u = \frac{\Delta u}{\sigma_0'}$$

where Δu is the excess pore water pressure and $\sigma_0^{'}$ is the initial effective stress. Postprocessing of results was made using Paraview software [21]. From an examination of Fig. 6 which shows the maximum pore water pressure ratio recorded in the settling basin after the strong motion, it can be seen that: for the cases in which the material is modeled with SANISAND and PDMY there are very small areas in which r_u reaches values of 0.96 and 0.66 respectively, while in the case of PM4Sand there are no critical values of r_u. In Fig. 6, a green dot highlights the node with the maximum horizontal displacement. For all models, the maximum horizontal displacement amounted to a value of about 5 cm.

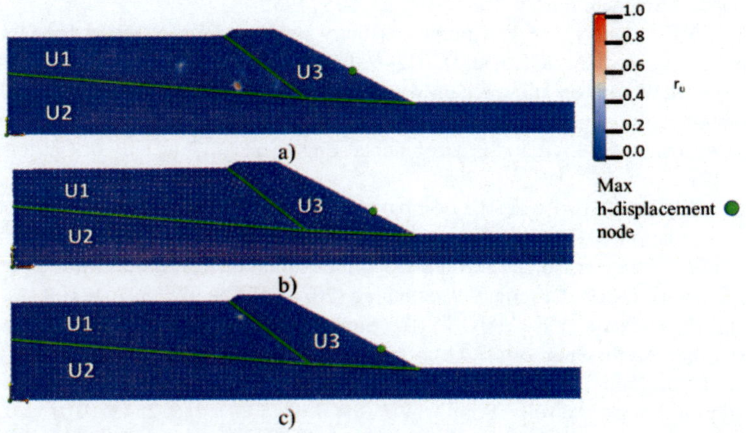

Fig. 6. Analysis results in terms of pore water pressure ratio and max horizontal displacement a) SANISAND b) PM4Sand c) PDMY

5 Conclusions

Three dynamic finite element models were developed to evaluate the seismic liquefaction susceptibility of the basin of a tailings storage facility. Each model employed a different sophisticated constitutive model to describe the mechanical behavior of the material contained in the basin. Results showed that there are no critical zones in the basin in terms of pore water pressure ratio and displacements that could trigger liquefaction. However, it will be necessary to deepen many aspects of the modeling and characterization of materials to obtain results more representative of the behavior of the structure during an earthquake.

References

1. ICOLD.: Tailings Dams - Risk of Dangerous Occurrences, Lessons Learnt from Practical Experiences, Bulletin 121 (2001)
2. Azam, S., Li, Q.: Tailings dam failures: a review of the last one hundred years. Geotech. News **28**(4), 50–54 (2010)

3. Rico, M., Benito, G., Salgueiro, A.R., Díez-Herrero, A., Pereira, H.G.: Reported tailings dam failures: a review of the European incidents in the worldwide context. J. Hazard. Mater. **152**(2), 846–852 (2008)
4. Lyu, Z., Chai, J., Xu, Z., Qin, Y., Cao, J.: A comprehensive review on reasons for tailings dam failures based on case history. Adv. Civil Eng. **2019**, 1–18 (2019)
5. Mazzoni, S., McKenna, F., Scott, M.H., Fenves, G.L.: OpenSees Command Language Manual. University of California, Berkeley (2006)
6. Ma, X.W., Zhao, G.Q., Sun, L.: AUTOMESH-2D/3D: robust automatic mesh generator for metal forming simulation. Mater. Res. Innovations **15**(s1), s482–s486 (2011)
7. Rice, J.: Is the aspect ratio significant for finite element problems? No. 85–535) Department of Computer Science, Purdue University 85, 535 (1985)
8. Dafalias, Y.F., Manzari, M.T.: Simple plasticity sand model accounting for fabric change effects. J. Eng. Mech. ASCE **130**(6), 622–634 (2004)
9. Boulanger, R.W., Ziotopoulou, K.: PM4Sand (version 3.1): A sand plasticity model for earthquake engineering applications. Report No. UCD/CGM-17/01. Center for Geotechnical Modeling, Department of Civil and Environmental Engineering, University of California, Davis, CA (2017)
10. Yang, Z., Elgamal, A., Parra, E.: Computational model for cyclic mobility and associated shear deformation. J. Geotech. Geoenviron. Eng. ASCE **129**(12), 1119–1127 (2003)
11. Geppetti, A., Facciorusso, J., Madiai, C.: Calibrazione di modelli costitutivi per materiali sterili minerari. IARG, Caserta 7–9 settembre (2022). ISBN: 9788897517108
12. Taiebat, M., Dafalias, Y.F.: SANISAND: Simple anisotropic sand plasticity model. Int. J. Numer. Anal. Meth. Geomech. **32**(8), 915–948 (2008)
13. Taiebat, M., Jeremić, B., Dafalias, Y.F., Kaynia, A.M., Cheng, Z.: Propagation of seismic waves through liquefied soils. Soil Dyn. Earthq. Eng. **30**(4), 236–257 (2010)
14. Khosravifar, A.: A note on calibrating PDMY. J. Chem. Inf. Model. **53**, 1689–1699 (2013)
15. McGann, C., Arduino, P.: Dynamic 2D Effective Stress Analysis of Slope. OpenSees user manual (2011)
16. Lysmer, J., Kuhlemeyer, A.M.: Finite dynamic model for infinite media. J. Eng. Mech. Div. ASCE **95**, 859–877 (1969)
17. Lai, C. G., Zuccolo E.: Prosecuzione delle attività di definizione di input sismici sismocompatibili e spettrocompatibili per i comuni della regione Toscana. Scientific Report (2017)
18. DM 17 gennaio 2018. Norme tecniche per le costruzioni. G.U. n.42 del 20 febbraio 2018
19. DM 26 giugno 2014. Norme tecniche per la progettazione e la costruzione degli sbarramenti di ritenuta (dighe e traverse). Gazzetta ufficiale n. 156 dell'8 luglio 2014
20. Kottke, A.R., Wang, X., Rathje, E.M.: Technical Manual for Strata. Architectural, and Environmental Engineering University of Texas, Geotechnical Engineering Center Department of Civil (2013)
21. Henderson, A.: ParaView Guide, A Parallel Visualization Application. Kitware Inc. (2007)

An Integrated Monitoring Network for the Mitigation of the Coastal Risk

Valentina Lentini[1(✉)], Francesco Castelli[1], and Sebastiano D'Amico[2]

[1] Faculty of Engineering and Architecture, Universy "Kore" of Enna, Cittadella Universitaria, 94100 Enna, Italy
{valentina.lentini,francesco.castelli}@unikore.it
[2] Department of Geosciences, Faculty of Science, University of Malta, Msida, Malta
sebastiano.damico@um.edu.mt

Abstract. The southern coast of Sicily (Italy) and the islands of Malta's archipelagos are highly exposed to risks coming from the sea. These coasts are subjected to fast erosion phenomena due to natural causes, which involve failure of cliffs, triggering of localized erosions and the possibility of flooding.

In this context the research project NEWS - Nearshore hazard monitoring and Early Warning System - (INTERREG V-A Program "Italia-Malta"), aimed to implement a strategy, common to the two Countries, which includes: i) monitoring systems on land and offshore, ii) an early warning system to prevent the risks coming from the sea, iii) the proposal of specific actions aimed to increase the resilience of the population against these natural disasters.

The test sites include low and high coasts: i) Granelli (Pachino) and Santa Maria del Focallo (Ispica) in South Sicily, characterized by thin sandy coast; ii) Selmun and Wied Il-Mielah in the archipelago of Malta, characterized by high cliffs and rocky arch.

The proposed solutions adopt an integrated geophysical sensor network on land, as well as, a networked set of wave buoys installed offshore Sicily and Maltese Islands. The goal of the network system is to estimate in advance the possibility that an area of the coast may be affected by a flood or a strong erosion concomitantly with a storm or with the collapse of a cliff, and to promptly activate safeguard measures to avoid damage to populations living close to the coasts.

Keywords: Coastal Risk · Erosion · Cliff Failure · Monitoring

1 Introduction

In the last decades, there was an increasing concern about the occurrence of natural hazards, especially regarding the risks coming from the sea. This is mainly due to the increased human exposure.

This paper would like to be a first step in the direction of improving the general knowledge of erosion problems and cliff failure within two coastal sector in the South Sicily (Italy) and in the Maltese archipelago. Obtained results are useful information

A. Ferrari et al. (Eds.): CNRIG 2023, SSGG, pp. 519–526, 2023.
https://doi.org/10.1007/978-3-031-34761-0_63

to develop an integrated coastal erosion management program aimed to partially solve erosion problems and review location of human activities and arrangements.

Coastal retreat is the landward displacement of shoreline because of marine erosion or flooding [1]. Usually, it is contrasted by protection works constructed on very specific sites, to solve concrete and urgent problems. Most of them constitute a temporary remedial without a general erosion management plan [2, 3]. According to [4–6], coastal erosion management must be organized at fully scales (i.e. at a large, regional scale) in which the dynamics of the whole coastal system is represented.

On the other hand, cliff failures are predominantly linked with the lithology, the structure of the rocky mass [7] as well as the weather conditions and the wave action [8]. The factors, affecting intrinsically the stability of cliffs, are therefore the lithological, stratigraphic, structural and morphological settings and the hydrogeology and the mechanical properties of the rocky mass. However, the external factors are represented by impact of sea waves, currents and tides, as well as meteorological agents, the biological activity of marine micro and macro organisms and by human activities.

2 The Research Project NEWS

2.1 An Overview

The Project NEWS - Nearshore hazard monitoring and Early Warning System - (INTERREG V-A Program "Italia-Malta") focuses on fast erosion due to natural and anthropic causes which involve the failure of cliffs, the triggering of localized erosions and the possibility of flooding with reference to the southern coast of Sicily and the islands of the archipelagos of Malta.

To achieve this goal, it is necessary the acquisition of local knowledge on issues related to coastal erosion and stability of cliffs. The advances in cloud computing have opened new opportunities also in early warning systems against natural disasters and emergency management issues. In particular, the project aims to develop an integrated system, including monitoring, early warning and mitigation of the coastal risks, advising the population about the risk of flooding, littoral sand erosions and cliff failures.

In this context, the main goal of the research Project NEWS was: (i) to encourage interventions of Public Administrations about risk factors; (ii) to contribute for the analysis of the urban hazard; (iii) to define the "best practices" and the strategies for the management and reduction of the risk related to the natural disasters [9–12].

2.2 The Test Sites

The study focuses on different sites including sandy low and rocky high coasts: i) the coasts of Granelli (Pachino) and Santa Maria del Focallo (Ispica) in South Sicily (Fig. 1), characterized by thin sandy coast; ii) the high stretch of coast of Selmun and Wied Il-Mielah (Fig. 2) in the archipelago of Malta, characterized by high cliffs and rocky arch. These areas are located in a geological and geomorphological context that makes them particularly vulnerable to natural events (landslides, floods, coastal erosion, etc.) and that requires an accurate knowledge of the issues related to natural instabilities.

Fig. 1. Granelli and Santa Maria del Focallo (Sicily).

(a) (b)

Fig. 2. Malta: Selmun Bay (a); Wield il-Mielaħ Window (b).

3 Sea Cliff Stability and Coastal Erosion

Wave-induced sea cliff instability and sandy coasts erosion are significant problems. The gradual increase in mean global warming will lead to a rise in the sea level and consequently a marked rise in the coastal erosion of beaches and rocky coasts. For purposes of coastal planning and engineering mitigation, it is therefore necessary to assess the potential increase in erosion processes due to wave action [13].

In the case of cliff, the basic factors controlling erosion are the force of waves at the base of cliffs and the mechanical strength of rock masses. Additional factors are reduction in rock strength owing to weathering by sea spray, tidal action, temperature variations and material fatigue caused by cyclic loading of the wave action. Although the natural process of sea cliff erosion is complex and is the cumulative result of numerous interacting variables that are significant at various spatial and temporal scales, wave erosion at the base of the sea cliff is usually assumed to be the basic controlling factor on the process of coastal cliff retreat. This process can be considered as a major cause of erosion for many sea cliffs of the test sites.

The stability of the coastal cliffs is normally analyzed similarly to that of the rocky slope, by geo-structural analyses [14] and stress-strain modeling, with a pseudo-static analysis of the action of wave motion incident at the cliff toe. In the coastal area, the analysis about the morphological evolution and instability, that affect the cliffs, requires understanding of the complex framework of interactions between the internal characteristics of the coastal system and the external stresses, including sea wave action and weathering. For this reason, methods to assess cliff erosion must take into consideration the possibility to include these aspects in the estimation of the rocky cliff stability.

With this aim, new methods can be elaborated considering a modification of the classification system based on the Slope Mass Rating Index (SMR), through the integration of additional parameters such as wave action and rock mass conditions.

Furthermore, to analyze the destabilizing effect of the waves on the cliff, it is also necessary to examine the frequency of the wave motion [15], since just the not very energetic but extremely continuous waves cause the erosion of the coastal slopes, also associated with mechanisms of failure, due to cyclic stress of the rocky mass.

For this purpose, measures like the ambient noise and strong-motion records can be used to analyze the dynamic behavior of the cliff. The wavelets constituting ambient noise undergo reflections, refractions and trap-ping due to wave guide effects, as well as attenuations, which depend on the nature of the subsoil. Ambient noise, that involves very small oscillations (10 ÷ 15 m/s^2 in acceleration), can be extracted from random noise through appropriate techniques, such as Horizontal-to-Vertical Spectral Ratio (HVSR), as shown in Fig. 3 for one of the test sites located in Malta [16–19].

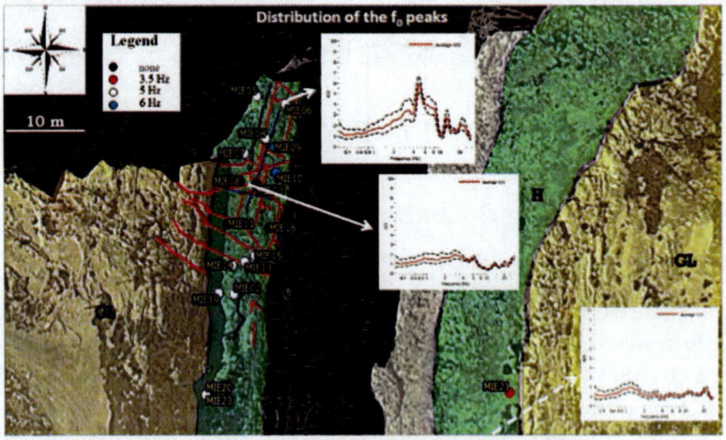

Fig. 3. Seismic noise measurements: distribution of the f_0 peaks.

The interpretation of these measures, related to the effect of the wave pulse and propagation on the cliff, can be supported by the mechanical characterization of the rocky mass (Fig. 4), to individuate a correlation between the dynamic response of the cliff to the wave action and the structural properties of the rocky mass.

Cliffs 6.0 ÷ 7.0 m in height are a quite common morphology also along the Sicilian coast, as for example at "Point Ciriga". These "weak" rocks are mainly affected by rock falls due to wave erosion at the foot of the cliff. Furthermore, at both sites, digital models of the main geological features have been created also using UAVs (unmanned aerial vehicles) equipped with different sensors.

At the same time, the experimental activities involved in the research project, provide the use of GPS directional wave buoys to be installed both in Gozo (Malta) and in Granelli (Sicily) to measures sea waves (height, direction, and period) and sea temperature in real time.

For one of the test site of Maltese islands, in Fig. 5 station position of the wave buoy is indicated by the black spot and empty circles show the annular wave bins measured by the HF radar at SOPU station in Gozo. The background map shows the sea depth contours in this marine area.

In the case of Sicily, mostly characterized by sandy low coastals, the goal of the study is to estimate the flooded areas for the test site (Granelli village). At this aim the wave characteristics (significant wave height, peak period, and direction) and storm tide (i.e. the sum of the storm surge and astronomical tide) must be known. Based on the data

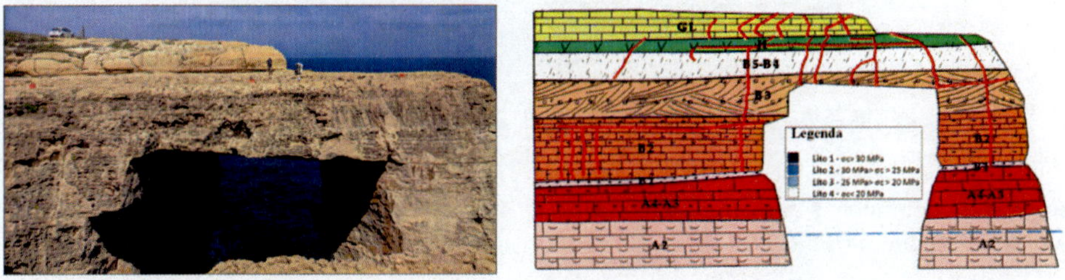

(a) (b)

Fig. 4. (a) Wied il-Mielah̄ (Maltese Islands); (b) engineering geological model.

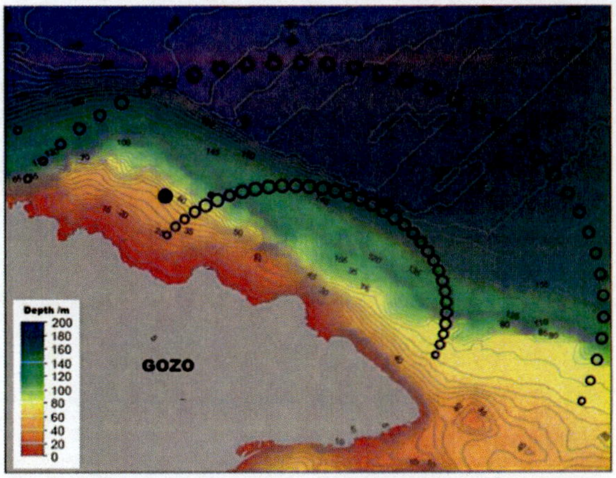

Fig. 5. Station position of the wave buoy in Gozo (Malta).

provided by the buoys, also installed in Granelli, numerical analyses by means of Xbeach have been performed. Figure 6a shows the calculation domain used for the simulations conducted at the Granelli beach. The figure also shows the location of Sect. 1 in which the time-series of the moving waterline was recorded.

Finally, Fig. 7 shows six sectors (A1–A6) in which the village of Granelli was divided according to the population density. The results on flooded surface areas within such sectors will be used to train the Artificial Neural Networks [20, 21].

The presented methodology that plan to use both noise measurements and GPS directional wave buoys can be a useful contribute to assess defense strategies for the Maltese and Sicily coastal, in the frame of managing the erosion risk.

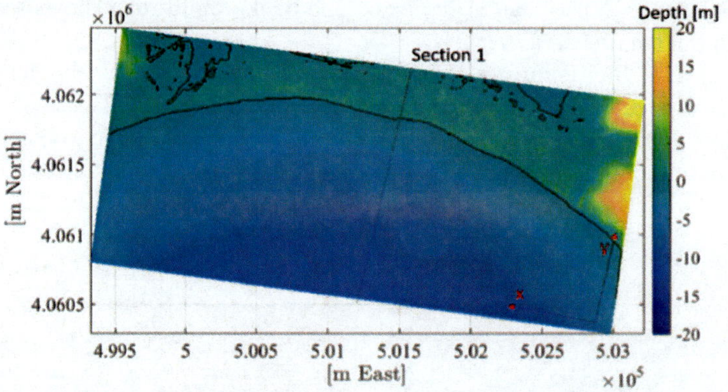

Fig. 6. Computational domains and bathymetry adopted for the numerical simulations using Xbeach at Granelli (Sicily).

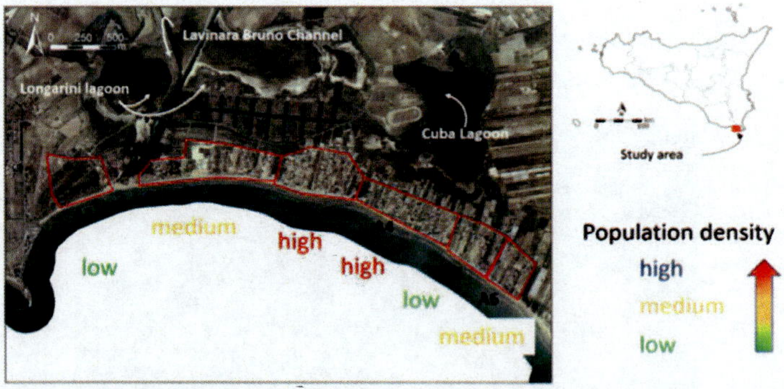

Fig. 7. Population density and location of the six sectors (A1–A6) in which the village of Granelli was divided.

4 Concluding Remarks

This paper aims to improve the general knowledge of erosion problems and cliff failure within two coastal sector in the South Sicily (Italy) and in the Maltese archipelago.

In the Mediterranean area, rocky cliffs represent widespread high-risk landforms, as they are highly frequented touristic places often interested by landslide processes. In particular, Malta represents a significant case study as several cliffs located all around the island are involved in instability processes, as evidenced by wide block-size talus distributed all along the coast line. These diffused instabilities are related to the geological setting of Malta Island, i.e. the over-position of grained limestone on plastic clay deposits that induces lateral spreading phenomena associated to falls and topples of different-size rock blocks and is responsible for a typical landscape with stable plateau of stiff rocks bordered by unstable cliff slopes.

At the same time, in the southern of Sicily (Italy), the coast of "Point Ciriga" (Ispica) characterized by weak rocks is affected by landslides and rock falls due to wave erosion at the foot of the cliff but also Granelli (Pachino) and Santa Maria del Focallo (Ispica) characterized by thin sandy coast are affected by coastal erosion issues.

In the framework of the research project NEWS, for high coasts literature available data were collected and new geological, geophysical and geotechnical investigation have been carried out to characterize the rocky cliff. Based on the surveyed joint spatial distribution, single-station noise measurements has been used to cover both the unstable zone and the stable area. It is possible to associate vibrational evidences to different instability levels, i.e. deriving the presence of already isolated blocks by the local seismic response. For sicilian sandy low coastals, GPS buoys have been also used to mitigate coastal flooding risks.

Acknowledgement. The research work has been supported by the **Project NEWS** "*Nearshore hazard monitoring and Early Warning System*", financed by the INTERREG V-A Program "Italia-Malta", Project Code C1–3.2–60, Priority Axis 3 "*Protecting the environment and fostering an efficient use of resources*", Specific Objective 3.2 "*Mitigation of the effects of the climate changes*". This is a "free copy" and the content of this publication does not reflect the official opinion of the European Union. Responsibility for the information and views expressed lies entirely with the authors.

References

1. Bird, E.C.: Submerging Coasts: The Effects of a Rising Sea Level on Coastal Environments. Wiley J. & Sons, Chichester (1993)
2. Hooke, J.: Coastal Defense and Earth Science Conservation. The Geological Society (1998)
3. Pethick, J.: Coastal management and sea level rise. Catena **42**(2–4), 307–322 (2001)
4. Fischer, D.: Responses to coastal threats: toward an integrated strategy. J. Coastal. Res. **1**(4), 383–388 (1985)
5. Carter, R.W.G.: Coastal Environments. An Introduction to the Physical, Ecological, and Cultural Systems of Coastlines. Academic Press, Cambridge (1988)
6. Bray, M., Carter, D., Hooke, J.: Littoral cell definition and budgets for central southern England. J. Coastal Res. **11**(2), 381–400 (1988)
7. Alejano, L.R., Garcìa-Cortes, S., Garcìa-Bastante, F., Martìnez-Alegrìa, R.: Study of a rockfall in a limy conglomerate canyon (Covarrubias, Burgos, N. Spain). Environ. Earth. Sci. 70(6), 2703–2717 (2013)
8. Sunamura, T.: Geomorphology of Rocky Coasts. John Wiley & Sons, Chichester (1992)

9. Lentini, V., Castelli, F., Distefano, G.: Management of transport infrastructures: a procedure to assess the landslide risk. In: Calvetti, F., Cotecchia, F., Galli, A., Jommi, C. (eds.) Proceedings of Geotechnical Research for Land Protection and Development, pp.121–128. Springer, Cham (2019). https://doi.org/10.1007/978-3-030-21359-6_13
10. Lentini, V., Distefano, G., Castelli, F.: Consequence analyses induced by landslides along transport infrastructures in the Enna area (South, Italy). Bull. Eng. Geol. Env. **78**, 4123–4138 (2019)
11. Castelli, F., Castellano, E., Contino, F., Lentini, V.: A web-based GIS system for landslide risk zonation: the case of Enna area (Italy). In: Proceedings of 12th International Symposium on Landslides. taylor & francis (2016)
12. Castelli, F., Lentini, V., Grasso, S.: Recent developments for the seismic risk assessment. Bull. Earthq. Eng. **15**(12), 5093–5117 (2017). https://doi.org/10.1007/s10518-017-0163-1
13. Budetta, P., Galietta, G., Santo, A.: A methodology for the study of the relation between coastal cliff erosion and the mechanical strength of soils and rock masses. Eng. Geol. **56**, 243–256 (2000)
14. Pappalardo, G., Punturo, R., Mineo, S., Ortolano, G., Castelli, F.: Engineering geological and petrographic characterization of migmatites belonging to the calabria-peloritani orogen (Southern Italy). Rock Mech. Rock Eng. **49**(4), 1143–1160 (2016)
15. Pellicani, R., Miccoli, D., Spilotro, G., Gallipoli, M.R., Mucciarelli, M., Bianca, M.: Dynamic response of a rocky cliff under the sea wave pulse: a study along the Adriatic coast of Polignano (Apulia, Italy). Environ. Earth Sci. **73**(10), 6243–6257 (2014). https://doi.org/10.1007/s12665-014-3848-7
16. Iannuccia, R., Martino, S., Paciello, A., D'Amico, S.: Rock mass characterization coupled with seismic noise measurements to analyze the unstable cliff slope of the selmun promontory (Malta). In: Proceedings of Symposium of the International Society for Rock Mechanics. Procedia Engineering 191, pp. 263–269 (2017)
17. Panzera, F., D'amico, S., Galea, P., Lombardo, G., Gallipoli, M.R., Pace, S.: Geophysical measurements for site response investigation: preliminary results on the island of Malta. Bollettino di Geofisica Teorica ed Applicata **54**(2), 111–128 (2013)
18. Iannucci, R., Martino, S., Paciello, A., D'Amico, S., Galea, P.: Investigation of cliff instability at għajn ħadid tower (selmun promontory, malta) by integrated passive seismic techniques. J. Seismolog. **24**(4), 897–916 (2020). https://doi.org/10.1007/s10950-019-09898-z
19. Iannucci, R., Martino, S., Paciello, A., D'Amico, S., Galea, P.: Engineering geological zonation of a complex landslide system through seismic ambient noise measurements at the Selmun Promontory (Malta). Geophys. J. Int. **213**(2), 1146–1161 (2018)
20. Iuppa, C., Cavallaro, L., Giarrusso, C., Musumeci, R.E., Savasta, G.: Coastal flooding risk assessment through artificial intelligence. In: Ksibi, M., et al. (eds.) EMCEI 2019. ESE, pp. 2005–2009. Springer, Cham (2021). https://doi.org/10.1007/978-3-030-51210-1_314
21. Iuppa, C., Cavallaro, L., Musumeci, R.E., Foti, E.: Coastal flooding risk assessment by a neural network approach. In: Proceedings of the Coastal Engineering Conference (2020)

The Static and Seismic Behaviour of a Slow-Moving Landslide: The Case of Montemartano (Umbria, Central Italy)

Valeria Licata(✉), Mauro Bonasera, Danilo D'Angiò, Alessandro Fraccica, Michele Perrotti, and Saverio Romeo

Istituto Superiore per la Protezione e la Ricerca Ambientale (ISPRA), Rome, Italy
valeria.licata@isprambiente.it

Abstract. The note presents some preliminary results of a study aimed at evaluating the stability of a complex landslide under static and seismic conditions. The examined area is located at Montemartano village (Spoleto, Central Italy), where the combination of structurally complex formations, tectonic faults and the effects of environmental conditions strongly influences the stability of the hillsides, even with low-gradient slopes. The subsoil settings were defined starting from a deepened geological and geomorphological field survey and by geotechnical investigations. In addition, piezometric and inclinometer monitoring campaigns, as well as, satellite radar interferometry (InSAR) technique were used to study the kinematics of the landslide. The latter confirmed the ongoing movement of the landslide which was interpreted and back-analyzed. Finally, the behaviour under seismic loads was evaluated in terms of displacements of the unstable sliding mass.

Keywords: Complex slow-moving landslide · deep movement · complex geological units · slope seismic behaviour · Montemartano village

1 Introduction

Active slow-moving landslides, typically characterized by rotational/translational kinematics, are a widespread issue for the Italian territory, which compromise the safety and the fruition of strategic infrastructures, as well as, of civil buildings.

According to the Italian inventory of landslides, IFFI (https://idrogeo.isprambiente.it), out of a total of 620,558 recorded instabilities (active or quiescent), 36% is represented by rotational/translational slides and 13% by slow flows. The magnitude of the phenomenon, highlighted by the above data, combined with the seismic hazard of Italian territory and with the potential effects of the climate change, puts in light the importance of a correct evaluation of both the static and the seismic behaviour of active slow-moving landslides to properly manage the risk.

In this sense, whereas literature contribution to the static behaviour of slow-moving landslides is very rich and exhaustive, few literature works were dedicated to the evaluation of co-seismic displacements of pre-existing slow-moving landslides [1, 2].

A. Ferrari et al. (Eds.): CNRIG 2023, SSGG, pp. 527–535, 2023.
https://doi.org/10.1007/978-3-031-34761-0_64

The case of Montemartano landslide fits into this context. The landslide is a wide 0.75 km^2 slow-moving mass, characterized by a deep sliding surface (located at 38m from the ground in the deepest point). The unstable mass has been studied over a long period of time [3], since it affects an area with residential buildings, commercial activities and strategic roads. Starting from a geological/geomorphological overview of the area, the data inferred from satellite and site monitoring are analysed and discussed. Following, the static condition is back-analysed by using different constitutive and stratigraphic models. Finally, the propagation of seismic histories allowed to give a preliminary evaluation of the co-seismic displacements and some insights on the applicability of conventional Newmark's method [4], commonly used in the engineering practice.

2 Geological and Geomorphological Frame

The area of interest is located in the Central Italy and is characterized by calcareous and marly terms of the Umbria-Marche succession involved in compressive and extensional tectonic phases. *Scaglia Rossa* formation and older terms of the sequence constitute an anticline structure overthrusting the younger marly formations of the succession, i.e., *Scaglia Cinerea*, *Bisciaro* and *Schlier* formations, shown in Fig. 1a. Field surveys allowed the recognition of geological and geomorphological units both inside and outside the E-W oriented landslide body. The localisation of the thrust plane, in Fig. 1a, remains still uncertain and debated [5].

The Authors [3] reconstructed a detailed cross-section by means of boreholes and geotechnical testing data (Fig. 1b). As shown, the observed movement is characterized by a deep-seated sliding surface intersecting the *Bisciaro* formation, composed by alternating structured clayey and limestone layers (picture in Fig. 1b).

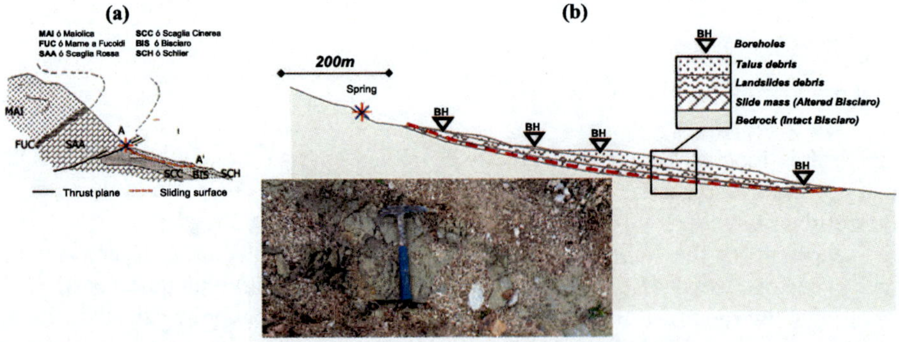

Fig. 1. Geological cross-section [5] (a); Geological cross-section of the landslide according to the stratigraphic model by [3] and image of a geological outcrop in the area of interest (b).

3 Monitoring Data

The satellite radar interferometry (InSAR) technique [6] was used to analyse the historical movement of the landslide. Displacement maps of Persistent Scatterers, PS (Fig. 2), within the so-called area of interest AOI, and time series (Fig. 3) were generated. Table 1 indicates the analysed dataset, in both the ascending and descending orbital geometry, and the reference period.

Table 1. Summary of InSAR data used.

Satellite mission	Reference period	Orbital geometry	Notes
ERS (ESA)	1992 – 2001	Asc/desc	
ENVISAT (ESA)	2002 – 2010	Asc/desc	
COSMO-SkyMed (ASI)	2011 – 2016	Asc/desc	2016 pre-earthquake series
COSMO-SkyMed (ASI)	2016 – 2021	Asc/desc	2016 post- earthquake series

Figure 2a,b shows a good density of PS in the Montemartano village. The scarcity of PSs is noted in the surrounding areas, due to the extensive vegetation cover and the absence of stable reflectors. Figure 2a,b also illustrates the average velocities of PSs to assess the landslide extent and its state of activity. As shown, a stable area is detected in Montemartano old town (green dots in Fig. 2a,b). Conversely, an unstable sector at the new buildings area (in the landslide body) is distinctly noted both in the Ascending (red dots with negative values) and the Descending (blue dots with positive values) orbital geometries. These findings, i.e. velocity with opposite sign in different orbital geometry, clearly indicate a distinct horizontal component of the deformation along the Eastward direction.

(a) (b)

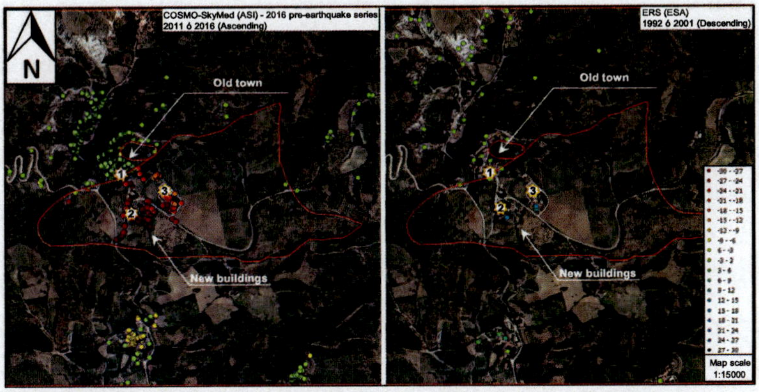

Fig. 2. PSs spatial distribution and their average velocities: (a) ascending and (b) descending geometries.

Figure 3a-c shows the displacement time series (along the Line of Sight, LoS) for the most representative and highly consistent PSs (P1, P2 and P3 shown in Fig. 2a,b). For comparison, the data of the 2016 Central Italy earthquake is also reported in the same charts.

According to Fig. 3, Point 1 (in the old town in Fig. 2) shows negligible displacements. Point 2 and Point 3 (in the new buildings area in Fig. 2) indicate similar evolutions in both the ascending and the descending geometry. Referring for instance to the ascending geometry in Fig. 3b,c, a very clear trend is detected at constant average velocity before (~28 mm/y for P2; ~ 20 mm/y for P3) and after (~23 mm/y for P2; ~ 15 mm/y for P3) the seismic event. Therefore, a general average deceleration of ~ 5 mm/y (for both Point 2 and 3) can be observed after the seismic event.

The analysed satellite dataset did not allow estimating the displacements accumulated during the earthquake sequence. Nevertheless, the surveys of local authorities and field reconnaissance evidenced no damages to constructions due to the seismic events. As a result, it is reasonable to conclude that the landslide experienced irrelevant co-seismic displacements during the 2016 seismic sequence. Similar evidence was also detected in the case of La Sorbella landslide [1].

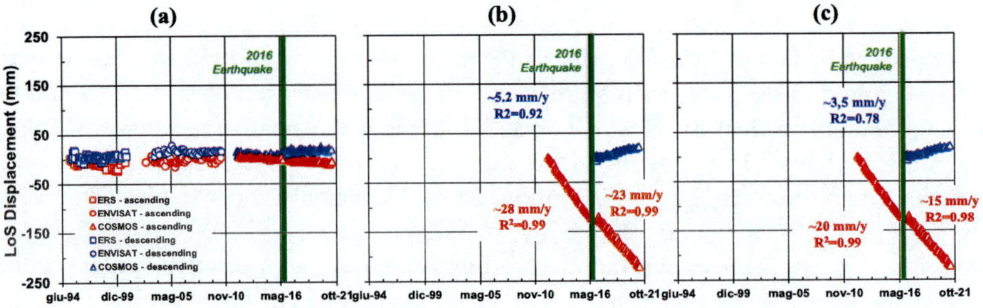

Fig. 3. LoS displacement time series for Point 1 (a), Point 2 (b) and Point 3 (c).

The above satellite monitoring data were used to integrate those inferred from a pre-existent inclinometer campaign (executed between 2002 and 2004), described in detail by the Authors [3]. Briefly, the latter let to define the depth of the sliding surface, located at 38m from the ground level in the deepest point, as well as the thickness of the sliding surface, 2m-thick, as shown in Fig. 1f. Moreover, in the same campaign, piezometric monitoring were used to define the position (reported in Fig. 4) and the average oscillation of the groundwater (±3m).

4 Static Analyses

2D Finite Element (FEM) analyses and the Limit Equilibrium Method (LEM) were used to back-analyze and to assess the interaction between shallow gravitational phenomena and the deep-seated landslide, using RS2 and Slide 2D (Rocscience™) respectively. The modeled section is reported in Fig. 4, where the geotechnical properties (Table 2) were based on the available investigation campaign [3].

Four modelling cases were considered (Table 2), under steady-state hydraulic conditions. In FEM-1 and LEM-1, a 2m-thick layer was included at the bottom of the sliding mass to physically reproduce the failure surface (Unit 3) inferred from the monitoring data. In FEM-2, the unstable mass was simulated using a Mohr-Coulomb failure criterion with peak + residual strength parameters. In LEM and FEM analyses, an initial geostatic stress state was imposed, assuming an earth pressure coefficient, $K_0 = 1$.

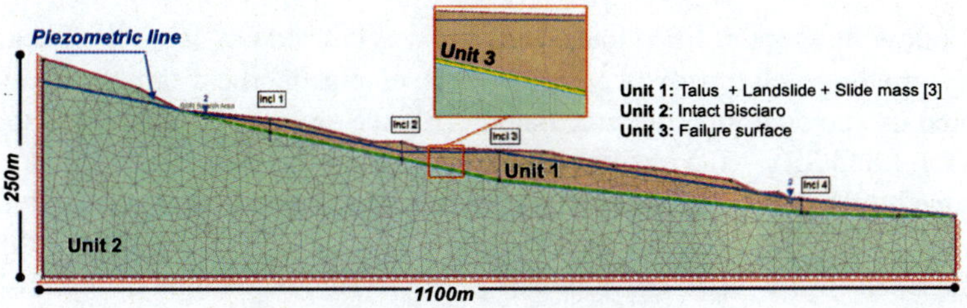

Fig. 4. Numerical FEM model.

Firstly, LEM back-analyses were performed by Morgenstern and Price (1965) theory [7]. For the sake of the brevity, LEM-1 and LEM-2 results are not shown but similar distributions of the failure surfaces are observed. In both cases, a friction angle of approximately 13.4° provided a safety factor, SF ~ 1, at the deep-seated failure surface (Unit 3 in Table 2) and shallower landslide bodies were observed with SF < 1.

Table 2. Simulation cases with different layer configuration and constitutive models.

Case	γ [kN/m^3]	c' [kPa]	φ'$_{peak}$ [°]	φ'$_{res}$ [°]	Model	Notes
LEM-1	19	–	13° ÷ 23°		Unit 1 (RPP)	φ'$_{peak}$ reduced in Unit 3 to get landslide triggering
	19	–	16 → 13.4		Unit 3 (RPP)	
	22	50	28		Unit 2 (RPP)	
LEM-2	19	–	16 → 13.4		Unit 1 (RPP)	φ'$_{peak}$ reduced in Unit 1 to get landslide triggering
	22	50	28		Unit 2 (RPP)	
FEM-1	19	–	23		Unit 1 (EPP)	1° stage: stress initialization 2° stage: SSR analyses
	19	–	13.4		Unit 3 (EPP)	
	22	50	28		Unit 2 (EPP)	
FEM-2	19	0	16	13.4	Unit 1 (PR)	1° stage: stress initialization 2° stage: SSR analyses
	22	50	28		Unit 2 (EPP)	

RPP: ridig perfectly plastic; **EPP**: elastic perfectly plastic; **PR:**peak + residual

FEM results are represented in Fig. 5a. After the first stage of stress initialization, significant shear strains are already observed in both cases (FEM-1/FEM-2). In FEM-1, strains are more localized within the failure surface (Unit 3), being the only one characterized by φ' = 13.4°. In FEM 2, failure mechanisms delimit two soil masses whose sliding surface partially coincides with that observed for the deep-seated sliding surface, in agreement with the failure mechanisms interpreted by the Authors [3]. The observed large strains are also indicative of incipient instability of the deep-seated landslide (i.e., SF ~ 1).

The following stage of FEM back-analysis was carried out using the Shear Strength Reduction method [8] and displayed in terms of plastic shear strains in Fig. 5b. As highlighted by the contours, the calculated critical Strength Reduction Factors (SRF ~ 0.7 in FEM-1 and SRF ~ 0.5 in FEM-2) are associated to a failure mechanism at the toe of the slope but shallower landslide bodies are also activated (as also detected by LEM models). The latter are mobilized by a φ' varying between 22.3° and 23.5°: these values can be ascribed to the different values of plasticity index (as also observed by [3]) as well as to the different degrees of remolding of the layers above the sliding surface.

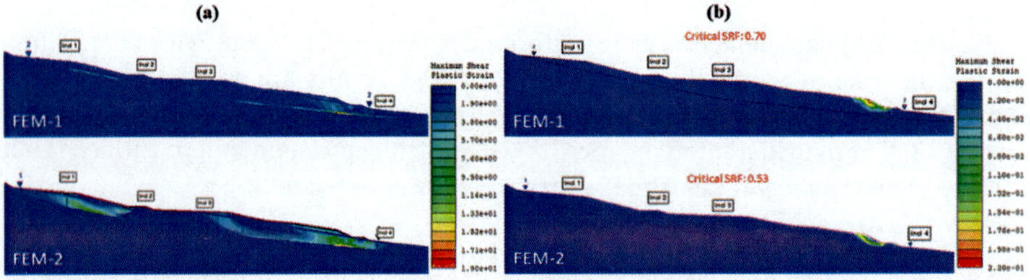

Fig. 5. Plastic shear strain contours observed in FEM-1 and FEM-2 after stress initiation (a) and by SSR method (b).

5 Seismic Response

The seismic behavior was analyzed considering the seismic sequence of the 2016 Central Italy earthquake. As mentioned in §3, this choice was aimed at reproducing the negligible co-seismic displacements recorded during and after the seismic sequence.

Figure 6 shows the EST component (parallel to the sliding direction) of the input signals (blue line), selected from ITACA database [9], recorded on stiff, outcropping and horizonal bedrock at Spoleto Monteluco station (~9 km far from Montemartano village). As shown in Fig. 6, the input signals exhibit a very low value of the acceleration, a_g, which is < 0.1g in all cases.

A Linear Equivalent (LE) method was used to propagate the signals through a 1D soil column corresponding to the central part of the sliding body (synthetized in Table 3). The non-linear and damping properties of Unit1a and Unit1b were inferred from Vucetic & Dobry curves [10]. A visco-elastic behavior was assumed for the outcropping bedrock (Unit 2) with damping ratio $\xi = 3\%$.

Table 3. Soil model used for 1D-LE analyses.

Geotecnical unit	Layer thickness [m]	V_S [m/s]	γ [kN/m^3]
1a: Talus + Debris landslide	10 15	250 400	19
1b: Slide mass (altered Bisciaro)	13	650	19
2: Bedrock (intact Bisciaro)	Half-space	1200	22
V_S: shear wave velocity	*γ: unit weights*		

The results of 1D seismic response analyses (SRA) are plotted in Fig. 6, in terms of acceleration histories computed at the soil surface (green line) and at the centre of gravity (red line) of the landslide (at 19 m from the ground level). Both acceleration time histories were used to calculate the co-seismic displacements by Newmark's method (NM). To this end, the critical acceleration of the system, a_c, was determined by considering the residual strength parameters (LEM-2 model), and by varying the piezometric line (in Fig. 4) up to a 3 m-lower position (accordingly with piezometric monitoring data). The resulting low values of a_c (varying between 0 and 0.02 g) highlight that the slope is balancing at the verge of an incipient instability state, prior to the earthquake.

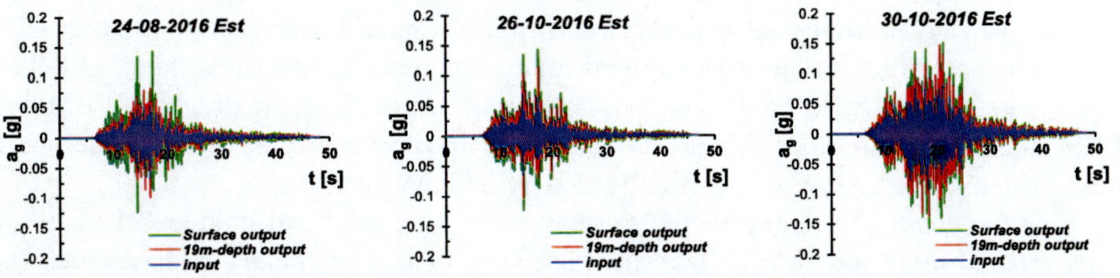

Fig. 6. Accelerometric input motions (recorded at Spoleto Monteluco station) and propagated histories from 2016 Central Italy sequence.

The NM co-seismic displacements, u_{final}, were calculated for the surface (green line) and the centre of gravity (red line) time histories and plotted in Fig. 7. In the same chart, for comparison, damage thresholds by [11] are reported too. By analysing Fig. 7a-b, for the three earthquakes, the co-seismic displacements attain values strongly dependent from the yield acceleration, a_c. Moreover, the calculated displacements, u_{final}, would produce damage from 'repairable' to 'collapse' which are in discordance with the negligible damages observed on site right after the earthquakes. These results are quite in agreement with the conclusions inferred from La Sorbella landslide case [2], where the monitored co-seismic displacements [1] were incompatible with those predicted by Newmark's sliding block analysis.

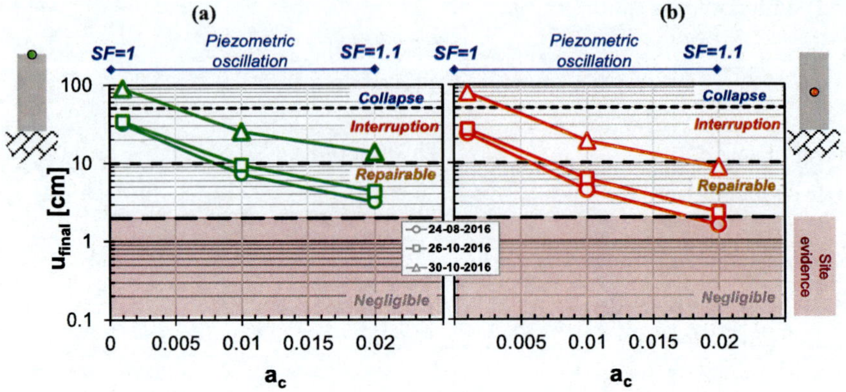

Fig. 7. Accelerometric input motions and propagated histories from the 2016 Central Italy sequence inferred from the surface (a) and the center of gravity (b) of the soil column.

6 Conclusions and Further Developments

The case of Montemartano, a deep and slow-moving landslide, was presented.

InSar monitoring data confirmed the ongoing movement of the unstable mass, even in the new building area of Montemartano (not detected in [3]), and a deceleration after the 2016 Central Italy seismic sequence.

Static FEM and LEM analyses predict the activation of a movement of the entire sliding mass layer, along the slip surface within the shear band when a value of φ' ~ 13° is reached. Other failure mechanisms are detected in the superficial layer of debris cover, as also confirmed from on-site evidence of scarps. Specifically, both FEM and LEM analyses locate a slope shallow instability in the lower part of the slope and a deep-seated movement was also highlighted by both approaches.

One-dimensional seismic response analyses were performed, under 2016 Central Italy seismic sequence, to calculate the co-seismic displacements of the landslide. The results were not in agreement with site evidence, where no significant damages to existing buildings were detected during and after the earthquakes. This conclusion put under light the reliability of conventional Newmark's method, when applied to the prediction of seismic induced displacements of active slow-moving landslides (see also [2]). This latter aspect deserves a better deepening considering the potential role of 2D/3D effects.

References

1. Ruggeri, P., Fruzzetti, V.M.E., Ferretti, A., Scarpelli, G.: Seismic and rainfall induced displacements of an existing landslide: findings from the continuous monitoring. Geosciences **10**(3), 90 (2020)

2. Kohler, M., Puzrin, A.M.: Mechanism of co-seismic deformation of the slow-moving La Sorbella landslide in Italy revealed by MPM analysis. J. Geophys. Res. Earth Surface, **127** (2022)
3. Grana, V., Tommasi, P.: A deep-seated slow movement controlled by structural setting in marly formations of Central Italy. Landslides **11**(2), 195–212 (2013). https://doi.org/10.1007/s10346-013-0384-6
4. Newmark, N.W.: Effects of earthquakes on dams and embankments. Géotechnique **15**(2), 139–160 (1965)
5. Cattuto, C.: Sorgenti di Montemartano. Possibile miglioramento della captazione. Documentazione A.S.E.M. Azienda Servizi Energetici Municipalizzati, Spoleto (1991)
6. Ferretti, A., Novali, F., Burgmann, R., Hilley, G., Prati, C.: InSAR permanent scatterer analysis reveals ups and downs in san Francisco bay area. Eos **85**(34), 1–3 (2004)
7. Morgenstern, N.R., Price, V.E.: The analysis of the stability of general slip surfaces. Géotechnique **15**(1), 79–93 (1965)
8. Matsui, T., San, K.-C.: Finite element slope stability analysis by shear strength reduction technique. Soils Found. **32**, 59–70 (1992)
9. Russo, E., et al.: Italian Accelerometric Archive v 3.2 - Istituto Nazionale di Geofisica e Vulcanologia, Dipartimento della Protezione Civile Nazionale (2022)
10. Vucetic, M., Dobry, R.: Effect of soil plasticity on cyclic response. J. Geotech. Eng. **117**(1), 89–107 (1991)
11. Silvestri, F., et al.: Analisi e zonazione della stabilità dei pendii in condizioni sismiche: applicazioni di metodi tradizionali ed avanzati ad un'area di studio. In: Evelius Edizioni (eds) (2003)

Check for updates

Geotechnical Investigations and Monitoring of the Archaeological Site of Santa Croce in Ravenna (Italy)

Michela Marchi(✉), Ilaria Bertolini, and Guido Gottardi

DICAM Department, Alma Mater Studiorum, Università di Bologna, viale del Risorgimento 2, 40136 Bologna, Italy
michela.marchi@unibo.it

Abstract. The city of Ravenna, in northern Italy, is worldwide famous for the well-preserved late Roman and Byzantine architectures characterized by fabulous mosaics that have earned the UNESCO World Heritage status. The constantly growing attention towards their conservation has promoted research activities that focus on the effect of the considerable cumulated land subsidence on their current conditions. Today, the sum of long-lasting natural and more recent anthropogenic subsidence has brought the level of such ancient archaeological sites and monuments well below the current position of the phreatic water table, still close to the current ground surface of the city centre, as it is the case of the Santa Croce church and its archeologic area. The Church, dating back to the 5th Century AD, is surrounded by ancient Roman floor mosaics, located ~3.2 m from the main street level. In this site, a pumping system must permanently operate to lift to the external sewage network the water collected from the local subsoil by means of a rather old and no longer efficient drainage system. A geotechnical investigation campaign was carried out in 2022, together with the installation of a new piezometric monitoring system, with the aim of investigating the shallow aquifers underneath the monuments. Relevant data interpretation will enable to better understand the current situation and, above all, to devise a proper solution which can combine the long-term conservation of the area with its usability.

Keywords: Land Subsidence · Dewatering · Shallow Aquifers · Preservation · Santa Croce in Ravenna

1 Introduction

The city of Ravenna is located in northern Italy, few kilometers far from the Adriatic Sea (Fig. 1B) and ~60 km from the Po river delta. It is well-known for its precious mosaics preserved inside the eight Early Christian Monuments, which are included in the UNESCO World Heritage List. The city has been always affected by the land subsidence phenomenon of both natural and anthropic origin. A significant deterioration to the urban as well as to the natural environments became evident since the late '70s. In the city centre, the crypt of monumental churches, such S. Francesco's, and the basement

A. Ferrari et al. (Eds.): CNRIG 2023, SSGG, pp. 536–543, 2023.
https://doi.org/10.1007/978-3-031-34761-0_65

of old buildings were permanently flooded. The same problem affects the excavated archaeological area of Santa Croce (see Figs. 1C and 1D), which belongs to the buffer zone of the UNESCO world heritage site of Galla Placidia Mausoleum and San Vitale Basilica (see Fig. 1A). The configuration of this site, an excavated area up to 4 m deep from the surrounding ground level and about 2 m below the current groundwater level, requires continuous dewatering by means of a drainage system, in order to avoid the immediate flooding of the whole area. This condition is the result of the cumulated land subsidence and no doubt represents an emblematic case study of related cultural heritage preservation. The paper describes the investigations and the preliminary studies recently carried out to identify the local stratigraphy and soil properties, the original conception of the existing drainage system and the emerging critical issues produced by the continuous dewatering process.

Fig. 1. A) Aerial view of the Monumental area of S. Vitale-Galla Placidia (on the left) and of the Archaeological area of Santa Croce (on the right). B) Location of Ravenna in northern Italy. C and D) Photos of the Santa Croce Church and of the archaeological excavations around it.

2 Land Subsidence in the Ravenna Area

[1–3] among others, have investigated the evolution of the land subsidence of the Ravenna area in conjunction with the relevant causes, by means of precision levelling of a network of benchmarks scattered throughout the territory, collecting data spanning from 1897 to 2002 (Fig. 2). The land settlement in the Ravenna city centre passed from ~5.5 mm/year in the period 1900–1957 [2] to ~80 mm/year between 1972–73, when the peak was recorded [1]. The causes of the settlement increase registered from the '50 have been ascribed primarily to groundwater pumping for industrial and civil uses (the latter resulting the prevailing in the city centre), and, to a lesser extent, to gas extractions. The greater settlement rates have been observed in the industrial district of Ravenna, where values touched 110 mm/year between 1972–73. The archaeological area of Santa Croce experienced ~1.3 m of land subsidence in the period 1897–2002 [3]. To understand the relevance of the anthropic component with respect to the total settlement, the entity of the natural land subsidence component should be also estimated. To this scope, the data reported in [4] can be used. They provide and estimation of about 1.2 mm/year over the last 2500 years for the Holocene era, which correspond to ~0.10 m of natural subsidence in the period 1897–2002. This value is particularly high, near to the maximum values recorded in the eastern Po river Plain, due to the local higher thickness of Pleistocene sediments (>2500 m).

Corrective actions were undertaken by the Italian government starting in the '70s in order to face the emergency. In the Ravenna area, the construction of a new industrial aqueduct (late '70s) and the activation of the Ridracoli public aqueduct (at the end of the '80s) greatly reduced the water demand [3]. The consequent drastic reduction of groundwater withdrawal generated an extensive pore water pressure (pwp) recovery in the artesian aquifer system underneath the Ravenna city centre, reducing the settlement rate to ~1.3 mm/y in the period 1998–2002 (Fig. 2), a value close to the natural subsidence rate reported by [4]. The positive trend observed at the beginning of the 21st century has been confirmed by more recent data [5]. Despite that, the relative distance between the groundwater table and the ground level has been irreversibly reduced, and, at present, its consequences must be faced.

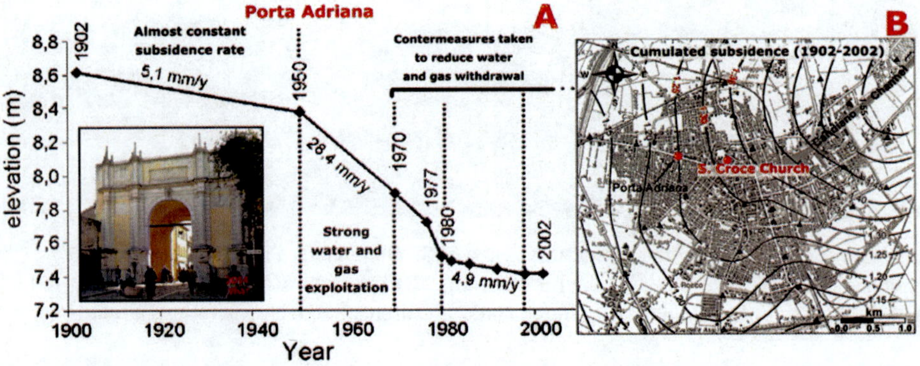

Fig. 2. A) Elevation in time of a benchmark located in Porta Adriana (200 m distant from Santa Croce) (modified from [4]). B) Cumulated land subsidence in the historic centre of Ravenna (in meters) in the period between 1902–2002 (modified from [3]).

3 The Case Study of Santa Croce

Ravenna is located on the ridge of an ancient beach of Etruscan Age formed by the deposits of the ancient Po river delta, and located below another ridge of more recent Roman age [3]. These littoral sandy deposits of Holocene origin are about 20 m deep from the ground surface and they constitute the geologic Unit A. Geologic Unit B is constituted by a 400 m thick alluvial deposit of the Upper Pleistocene, characterized by a succession of sandy and loamy fresh water aquifers separated by clayey and silty-clayey aquitards (thickness ratio between aquifers and aquitards 1:1). Geologic Unit C is constituted by a continuous marine sequence of the Plio-Pleistocene, whose inferior border is found in the

Ravenna area at ~2.6 km from the surface, symptom of a high tectonic subsidence [4]. The settlements of anthropic origin in the Ravenna area are mainly due to water withdrawal and they are confined within the Unit B, settlements caused by gas exploitation have a minor role and are instead confined within the Plio-Pleistocene marine deposits [4].

The archaeological site of Santa Croce is characterized by the remains of an old Roman domus dating 2nd – 3rd century A.D., whose mosaics have been found and preserved onsite, and of a 5th century A.D. Placidian Church, profoundly modified up to the 7th century A.D [6]. Today the Church shows the same configuration inherited from the 17th century, but excavations and restorations carried out between the 19th and the 20th century A.D. have brought to light the stratification of the archaeological remains. The excavations have reached a depth of ~2 m below the mean sea level, corresponding to the Roman remains, while the level of the street is at ~1.5 m above: then, the depth of the Roman ground level with respect to the current one is greater than 3.5 m [6]. The excavations have been left open to appreciate the peculiar archaeological stratification and to admire the beautiful Roman mosaics from Galla Placidia street (Fig. 3A).

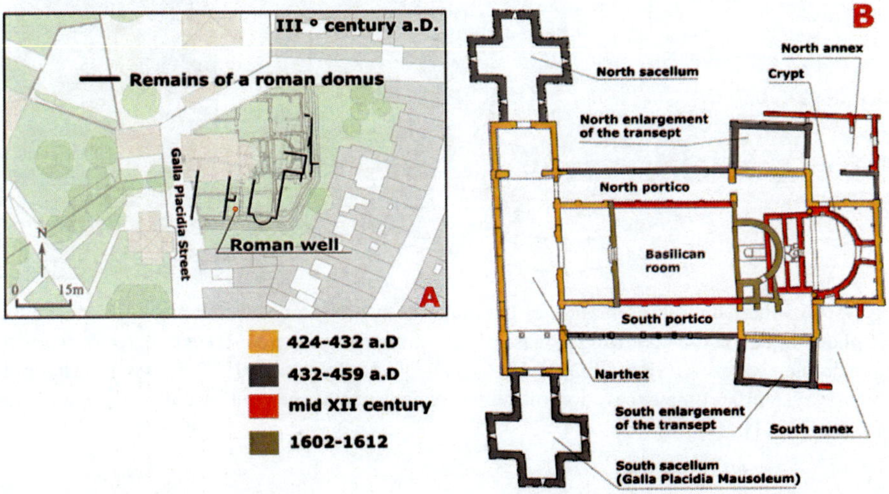

Fig. 3. A) Plan of the Archaeological site of Santa Croce, showing the remains of the Roman domus (II-III century A.D). B) Representation using different colors of the construction phases of the Santa Croce church (from ~400 to ~1600 A.D) (modified from [6]).

4 The Drainage System Operating in the Area

In the early '70s, a temporary drainage system was built to keep the archaeological excavations dry (Fig. 4C). Subsequently, in 1979, a permanent drainage system was built at the foot of the slope, along two sides of the excavated perimeter. The system, schematically represented in Fig. 4A, consists of perforated tubes connecting three wells (P1, P2 and C1). The water flows from P1 to C1; in the latter two submerged pumps lift the water into the public sewer of the higher street (Fig. 4A). In 1984 a concrete tank was added to the system for the collection of the drainage waters and the pumps (see Fig. 4B). At present, the system suffers the lack of maintenance, is locally obstructed and water mainly runs at the ground surface eroding soil particles. Consequent severe differential settlements in the archeological remains of the area around the well C1 are evident, together with clear signs of instability in the close-by slope (see Fig. 5A). In recent years (2002, 2016, 2019, 2021), the archaeological area of Santa Croce was flooded due to sudden breaks down of both pumps. In these occasions, the water table raised to its "undisturbed" level (about 0.03 m below the mean sea level), entirely covering the archaeological remains and causing a dramatic proliferation of aquatic vegetation on the remains (see Fig. 5B).

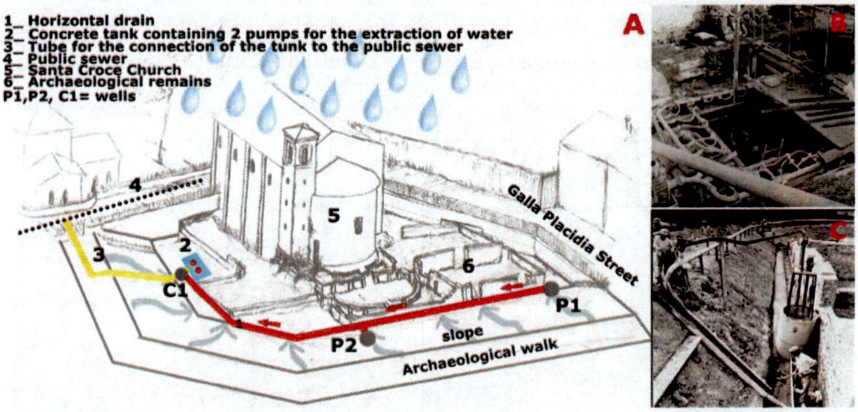

Fig. 4. A) Schematic representation of the drainage system in use in the archaeological site. B) Old photo dated 10/1984 showing a phase of the realization of the concrete tank for the collection of the drained water [AFSRA NEG. N. 63167]. C) Old photo dated 9/1979 showing the temporary drainage system and the excavation of the permanent one at the foot of the soil slope [AFSRA NEG. N. 34901].

5 Geotechnical Investigations

The definition of a detailed geotechnical model of the site is an essential step in the process of devising the preservation strategy of historic buildings [7]. In 2022, the subsoil underneath the archaeological site of Santa Croce was investigated by two continuous coring boreholes (S2 and S3), 2 cone penetration tests (CPTu), one seismic piezocone test (SCPTU1), one dilatometer test (DMT1), several SPT and Lefranc tests. Boreholes S2 and S3 were carried out up to a depth of 36 m and 13 m from the local ground level (~Galla Placidia Street), respectively; CPTU1, CPTU2 and CPTU3 to 30 m, 31 m and 17 m, while DMT1 to 20.6 m. In 1983, a previous extensive geotechnical investigation campaign had been carried out [8] (Fig. 6 Left), even if more focused on the near monumental area of San Vitale and Galla Placidia. The soil profile deduced from the investigations is shown

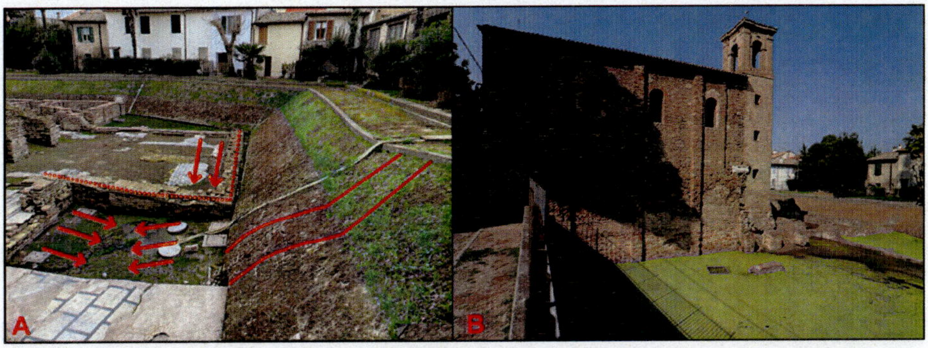

Fig. 5. A) Evidences of the ground settlements and of the slope instability in close proximity of the tank. B) View of the archaeological site entirely flooded during the pump break down in August 2021.

in Fig. 6 (Right). The local stratigraphy is characterized by an anthropic unit, ~5 m thick, rich in archaeological remains (Unit R) belonging to geologic Unit A (alluvial deposit, Holocene era) and constituted mainly by sandy-silt mixtures. The natural water content is close to the Plastic Limit, symptom of a hard soil subjected in the past to superficial desiccation [8]. Traces of oxidation are compatible with the exposure to atmospheric conditions. Unit 1, below Unit R, is characterized by a clayey sand layer up to ~22 m of depth from the ground level, corresponding to geologic Unit A. The average coefficient of saturated permeability of this unit is ~$4.6*10^{-6}$ m/s as estimated from CPT tests [9] of the 2022 campaign and confirmed by subsequent permeability tests carried out in triaxial cell. Unit 2, underneath Unit 1, is constituted by a low to high plasticity clayey, silty-clay material of olive-brown or dark brown color.

This unit is the first aquitard of geologic Unit B with an average coefficient of saturated permeability equal to ~$3.3*10^{-9}$ m/s. Its thickness varies between 3.3 and 6.1 m (average 5.1 m). Below Unit 2, an alternation of grey sands and silty sands, corresponding to the first aquifer of Geologic Unit B, can be observed (Unit 3). The average thickness of this layer is ~5.7 m. This sandy formation has a coefficient of saturated permeability of $2.2*10^{-6}$ m/s as estimated from CPT tests. Below Unit 3, a silty clay layer of grey-green color is observable (Unit 4) which consists of fine-grained mixtures of the same nature of Unit 2 (the average coefficient of permeability of this Unit cannot be estimated from CPT tests that had to stop above). The lower boundary of Unit 4 was reached only in borehole S2 (geotechnical campaign 2022) and it was found at a depth of −35.5 m from the ground level.

During the 2022 campaign, open standpipe piezometers were installed in the Archaeological area of Santa Croce and in the Monumental Area of San Vitale-Galla Placidia. All of them have been equipped with pressure transducers and datalogger, for the real-time monitoring of the pwp regime in the aquifer system. PZ1, PZ3 and PZ5 are placed in the first aquifer, while PZ2 in the second (deeper) aquifer. The collected data show that the shallow aquifer is particularly sensitive to local precipitations (Fig. 7). In proximity to the drainage system there is a remarkable drawdown in the pwp of Unit 1 (1° aquifer), which enables the area to be kept dry. The highest reduction in pore pressure is

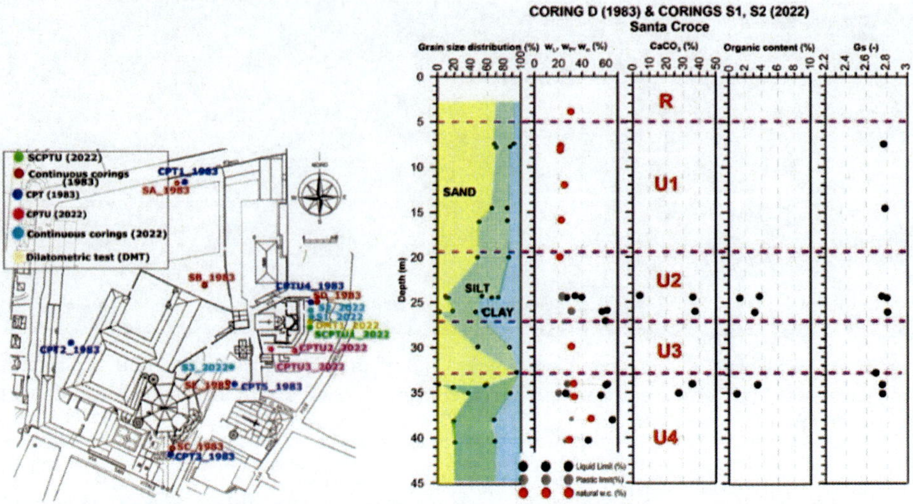

Fig. 6. Left: Plan of the field investigations performed in 1983 and 2022. Right: Soil profile including information on the physical characteristics of the units.

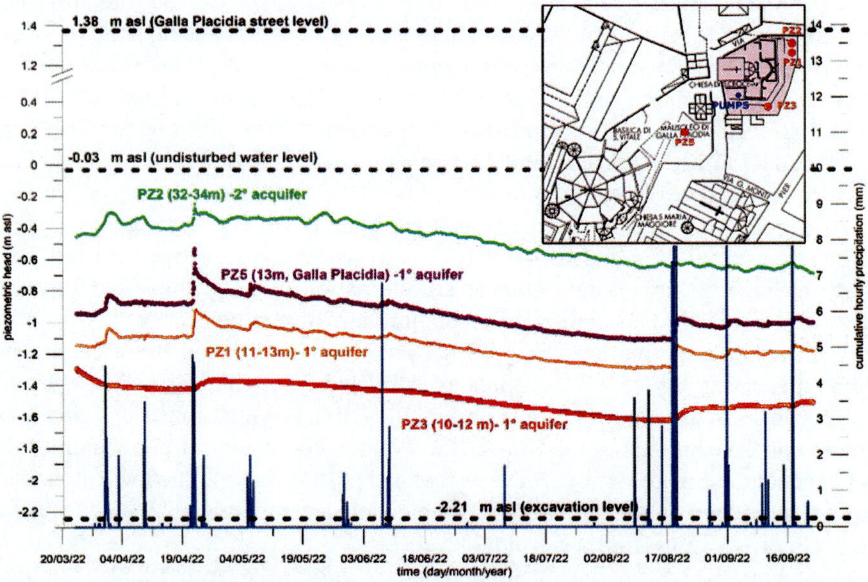

Fig. 7. Trend of the piezometric head (right) and cumulated precipitation data (left) vs. time, monitored in the area of S. Croce-Galla Placidia (time interval: ~6 months). On the top right, plan with location of the instruments.

recorded by PZ3, consistently with its position closer to the pumps. Also, PZ5 appears to be affected by the withdrawal, but to a lesser extent, being outside the area of the drainage system. The hydraulic head in Unit 3 (2nd aquifer monitored from PZ2) tends to the undisturbed water level (black dashed line in Fig. 7), providing evidence of an upward seepage flow from Unit 3 to Unit1.

6 Conclusions

The paper describes some preliminary findings from the case study of the Archaeological Area of Santa Croce, located in the centre of Ravenna, historically subjected to critical issues connected to a significant phenomenon of land subsidence, remarkably worsened in the last decades due to anthropic causes. The new data obtained from a most recent geotechnical investigation campaign have been integrated with those coming from an extensive previous one carried out with the main aim of preserving the nearby monumental area of San Vitale and Galla Placidia (UNESCO World Heritage Site) in the '80s. The definition of a detailed soil profile and of the pore water pressure regime in the aquifer system is the first step for understanding the local hydrogeology affecting the excavation area and for subsequently devising the best strategy for the preservation of the site and for the future fruition of the whole complex.

Acknowledgments. This research has received funding from the EU Horizon2020 under grant agreement n° 821282 (SHELTER).

References

1. Carbognin, L., Gatto, P., Mozzi, G., Gambolati, G.: Land subsidence of Ravenna and its similarities with the Venice case. In: Saxena, S.K. (ed.) Proceedings of the International Conference on Evaluation and Prediction of Subsidence, pp. 254–266, Florida, USA (1978)
2. Salvioni, G.:1 movimenti del suolo nell'Italia centro-settentrionale, Bollettino di Geodesia e Scienze Affini, I.G.M., anno XVI, Firenze (1957)
3. Teatini, P., Ferronato, M., Gambolati, G., Bertoni, W., Gonella, M.: A century of land subsidence in Ravenna. Italy. Environ. Geol. **47**, 831–846 (2005)
4. Bertoni, W., Elmi, C., Marabini, F.: The subsidence of Ravenna. Giornale di Geologia Applicata **1**, 23–32 (2005)
5. Bertolini, I., Marchi, M., Gottardi, G.: Heritage learning from a well-documented geotechnical cold case: the two towers of bologna. Int. J. Architectural Heritage (2022)
6. Bitelli, G., Bonsignore, F., Del Conte, S., et al.: Updating the subsidence map of Emilia Romagna region (Italy) by Integration of SAR interferometry and GNSS time series: the 2011–2016 period. In: Proceedings of IAHS, vol. 32, pp. 39–44. Copernicus Publications (2020)
7. Cassanelli, R., Cirelli, E., Crociati, E., Civiletti, J.P., et al.: La Basilica di Santa Croce: nuovi contributi per Ravenna tardo-antica. Curated by David M. Edizioni del Girasole (2013)
8. Ricceri, G.: Studi e ricerche nell'area di S.Vitale Galla Placidia e Santa Croce in Ravenna. SGEditoriali, Padova (Italia) (1992)
9. Bertolini, I., Marchi, M., Tonni, L., et al.: CPT data interpretation for an improved characterization of the paleosol stratigraphy in the Po River Valley, Italy. In: Cone Penetration Testing 2022. CRC Press, pp. 294–299 (2022)

Finite Element Modelling of Seismic Performance of an Excavation Supported by Propped Diaphragm Walls in a Natural Structured Clay Soil

Kateryna Oliynyk[1,2], Giulia Ferraro[2], and Claudio Tamagnini[2(✉)]

[1] School of Science and Engineering, University of Dundee, Dundee, Scotland, UK
2439697@dundee.ac.uk

[2] Department of Civil and Environmental Engineering, University of Perugia, Perugia, Italy
claudio.tamagnini@unipg.it

Abstract. A series of non-linear FE simulations of the behavior of excavations in natural structured clays supported by diaphragm walls under seismic loading conditions has been performed to investigate the effects of the initial degree of structure and of the destructuration rate on the performance of the structure, quantified in terms of permanent displacements and rotations of the walls. In order to reproduce the main characteristics of the cyclic/dynamic soil response at very small to medium-large strain levels, a recently proposed hypoplastic model for clay has been used to describe the soil behavior. In the model, an additional scalar state variable (sensitivity s) allows to incorporate the effects of soil structure. The results show that, while destructuration effects are almost negligible during the static excavation stage, the progressive degradation induced by the earthquake shaking may induce a significant increase in the accumulated permanent displacements, mostly controlled by the destructuration rate of the material.

Keywords: Diaphragm Walls · Performance-Based Design · Structured Clays · Hypoplasticity

1 Introduction

In recent times, much attention has been given to the evaluation of the seismic response of flexible retaining structures such as sheetpiles and diaphragm walls in terms of their performance, i.e., by comparing the permanent wall displacements/rotations accumulated during the seismic event with some prescribed tolerance limits depending on the limit state under consideration.

Significant progresses have been made in the development of simplified approaches for the evaluation of the permanent wall movements based on extensions of the classical Newmark methods for excavations in dry sands [1–3]. However, the use of advanced numerical analysis tools – e.g., the FE method – for the solution of the dynamic equations of motion and mass conservation, governing the response of the retaining structure and

A. Ferrari et al. (Eds.): CNRIG 2023, SSGG, pp. 544–552, 2023.
https://doi.org/10.1007/978-3-031-34761-0_66

of the saturated soil mass interacting with it, is to date the only approach capable of accounting for the complex features of the cyclic/dynamic response of fine-grained soils as well as for the complex spatial and temporal evolution of the excess pore water pressures generated by the soil deformations under practically undrained conditions.

Recently, Cattoni and Tamagnini [4] have perform a series of FE simulations to investigate the seismic behavior of a deep excavation in a soft clay layer, supported by two propped diaphragm walls, to compare the predictions obtained with the classical Modified Cam Clay model and with an advanced cyclic hypoplastic model for clays, from ref. [5], pointing out that the choice of the constitutive model may have a dramatic effect on predicted permanent wall and ground displacements, due to the impact that the inelastic volumetric deformations accumulated during the undrained seismic shearing of the soil have on the development of the excess pore water pressures around the excavation.

In the work of ref. [4], no account was taken of the fact that natural soils typically display some peculiar features in their fabric and possess interparticle bonding. These features – collectively referred to as the soil "structure" [6] – make their mechanical behavior different from those of very recent deposits or reconstituted soils, for which the original hypoplastic models for clays were developed. The objective of this work is to explore the influence that destructuration phenomena induced by the continuous accumulation of deviatoric strains during the seismic shaking may have on the accumulated permanent displacements of a retaining structure in a structured natural clay. To this end, a new version of the hypoplastic model for clays, recently proposed by Masin [7] is adopted in a series of dynamic simulations of the seismic response of a retaining structure similar to the one considered in ref. [4]. The new hypoplastic model incorporates the small-strain stiffness features of natural clay soils and an additional scalar state variable to describe the effects of soil structure. The results obtained allowed to quantify the influence of the assumed destructuration rate on the seismic performance of the structure.

2 The Hypoplastic Model for Structured Clays

The hypoplastic model for clays proposed in ref. [7] is an evolution of the original hypoplastic model developed by Masin, see ref. [5]. Among its novel features, the following are worth mentioning: a) the incorporation of an anisotropic pseudo-elastic stiffness tensor for very small strain levels which can be characterized with the experimental data obtained from shear wave velocity measurements, and for which the small strain shear modulus is a power function of the mean stress; b) the adoption of a set of explicitly defined asymptotic states; c) the incorporation of an additional scalar state variable – the soil sensitivity s, based on the original proposal of Gens and Nova [8] – to quantify the effects of soil structure and the strain-induced destructuration process by means of a suitable evolution equation for s. Finally, the model incorporates a tensorial internal variable – the intergranular strain $\boldsymbol{\delta}$ – to provide sufficient memory of the recent strain history to capture the soil hysteretic behavior under cyclic loading.

The details of the constitutive equation of the hypoplastic model adopted in this study can be found in ref. [7] and are not reported here due to space limitations.

For the forthcoming discussion, we only mention that, in the model for structured clays, Hvorslev equivalent pressure is controlled by both the current void ratio and the sensitivity according to the following relation:

$$p_e^* = sp_r \exp\{[N^* - \ln(1+e)]/\lambda^*\} \tag{1}$$

where N^* and λ^* are the material constants defining the reference value of $\ln(1+e)$ at $p = p_r$ on the virgin compression line of the unstructured soil, and its slope in the $\ln(1+e) : \ln(p/p_r)$ plane, respectively.

To describe the destructuration process, the sensitivity is provided with the following evolution equation:

$$\dot{s} = -k(s - s_f)\dot{\epsilon}^{d}/\lambda^* \tag{2}$$

where k and s_f are two material constants, while $\dot{\epsilon}^d$ is a function of the volumetric and deviatoric strain rates given by:

$$\dot{\epsilon}^{d} = \sqrt{(\dot{\epsilon}_v)^2[A/(1-A)] + (\dot{\epsilon}_s)^2} \qquad \dot{\epsilon}_v = -\mathrm{tr}(\dot{\epsilon}) \qquad \dot{\epsilon}_s = \sqrt{2/3}\|\dot{\epsilon}\| \tag{3}$$

and A is a third material constant. Equation (2) provides a monotonically decreasing evolution of s with increasing accumulated strains up to a final asymptotic value $s = s_f$. The constants k and A control the destructuration rate and the balance between the destructuration effects associated to the volumetric and deviatoric strain rates, respectively.

Table 1. Material constants of the hypoplastic model adopted in the simulations.

φ_c	λ^*	κ^*	N^*	ν	R	β_r	χ
(°)	(–)	(–)	(–)	(–)	(–)	(–)	(–)
21.9	0.095	0.015	1.344	0.1	5.0e–5	0.1	1.0
A_g	n_g	m_{rat}	ϑ	k	A	s_f	p_t
(–)	(–)	(–)	(–)	(–)	(–)	(–)	(kPa)
270.0	1.0	0.7	10.0	variable	0.2	1.0	5.0
α_G	α_ν	ρ_s	ρ_d	k_{hyd}	C_w		
(–)	(–)	(t/m^3)	(t/m^3)	(m/s)	kPa^{-1}		
1.0	1.0	1.90	1.44	1.08e–8	1.0e–5		

For the problem at hand, the hypoplastic model has been characterized with the material constants reported in Table 1, calibrated in ref. [9] from experimental data obtained for London clay. To keep the formulation of the model simple, the (pseudo)elastic anisotropy has not been considered, setting the constants α_G and α_ν equal to one. The last two constants appearing in the table represent the hydraulic permeability of the soil (k_{hyd}) and the compressibility of the pore water (C_w), respectively. A series of preliminary simulations of cyclic simple shear tests under undrained conditions has been performed to investigate the model ability to reproduce the shear stiffness decay and the damping ratio with increasing cyclic shear strain amplitude. The results, reported in ref. [10], show that the model prediction are in good agreement with the modulus decay and damping curves proposed by Vucetic and Dobry [11] for clays.

3 Finite Element Model and Simulations Program

3.1 Problem Geometry and Discretization

The problem considered in this work is an ideal – although realistic – 8 m deep and 18 m wide excavation in a saturated clay layer, 25 m thick, underlain by a rigid bedrock. The excavation is supported by two reinforced concrete diaphragm walls, 14 m long and 1.0 m thick, propped at the crest by a 1 m thick reinforced concrete slab, assumed hinged at the contact with the two walls. A sketch of the problem geometry, representing only half of the domain, is provided in Fig. 1.

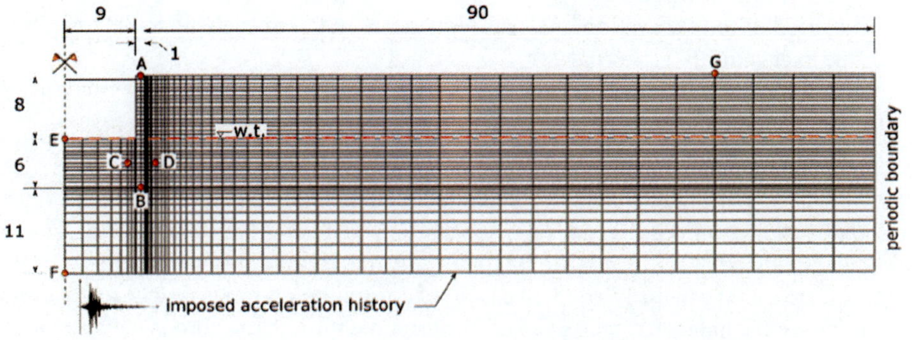

Fig. 1. Geometry of the excavation and of the retaining structure (dimensions are in m).

The structural elements are assumed as linear elastic, with a unit weight $\gamma_{RC} = 25$ kN/m^3, a Young's modulus $E_{RC} = 1.0e7$ kPa and a Poisson's ratio $\nu_{RC} = 0.25$.

The research-oriented FE code Tochnog professional [12] has been used to perform the numerical simulations. The full soil domain, 200 m wide, has been discretized with about 3800 Q4P4 bilinear elements with equal order of interpolation, suitably stabilized to circumvent the stability problems associated with the standard formulation of this element which does not satisfy the Ladyzhenskaya–Babuška–Brezzi (LBB) condition [13]. Particular care has been taken in the selection of the maximum element size, to avoid spurious high-frequency filtering.

In order to allow the two walls to behave as perfectly impervious diaphragms, they have been modeled with 112 one-phase bilinear Q4 elements. Four linear structural beam elements have been used for the prop. The interfaces between the soil and the two walls have been modeled with Q4P4 continuous thin-layer elements. As the walls are considered as cast in place, the interface elements have been given the same behavior and the same properties of the soil layer.

3.2 Initial Conditions and Simulation Program

Geostatic conditions are assumed for the soil layer before the construction of the retaining structure. The coefficient of earth pressure at rest K_0 is set equal to 0.92 and the initial average value of void ratio in the soil layer is set equal to 0.84. The geostatic pore water pressure field is hydrostatic, with a piezometric surface located 8 m below the ground surface (Fig. 1). Under such assumptions, the hypoplastic model reproduces the behavior of a stiff clay, with an overconsolidation ratio in terms of mean effective stress, OCR_p as defined in ref. [12], ranging from about 10 at the bottom of the layer to about 20 at the ground surface, not accounting for soil structure. This last feature is imposed after setting the geostatic conditions (post-depositional structure), by assigning a constant initial value of s, equal to 2.0, for the entire layer.

All the numerical simulations of the seismic behavior of the retaining structure have been performed in the following stages:

1) activation of the diaphragm elements, with their self-weight, as wished-in-place;
2) simulation of a first excavation step, up to a depth of 0.5 m;
3) activation of the strut elements;
4) simulation of the excavation in 15 subsequent steps, 0.5 m each, up to the final depth of the excavation;
5) activation of the periodic boundary conditions on the lateral vertical boundaries of the domain and of the imposed acceleration history at the layer base.

The stages (1)–(4) are performed assuming drained conditions, with fixed horizontal and vertical displacements at the base and fixed horizontal displacements at the lateral boundaries. Their purpose is to set the appropriate values for the stress and the internal variables fields (σ, e and δ) at the beginning of the seismic event (stage 5). The last stage is performed as a fully coupled dynamic consolidation process, whose governing equations are the balance of mass and of linear momentum of the porous medium, under the assumption of "medium speed processes" of Zienkiewicz et al. [14].

To investigate the effect of soil destructuration on the seismic performance of the retaining structure, a group of three different simulations are performed with varying destructuration rate k, as shown in Table 2. In particular, the values adopted for simulations r01 and r03 represent a relatively high and a relatively low value for this material constants.

Table 2. Program of FE simulations of the seismic behavior of the retaining structure.

Run # →	r01stat	r01dyn	r02stat	r02dyn	r03stat	r03dyn
s_0	2.0	from r01stat	2.0	from r02stat	2.0	from r03stat
k	0.8	0.8	0.4	0.4	0.2	0.2

3.3 Seismic Input

The seismic input adopted in the dynamic stages of the simulation is a real accelerogram measured at a station in Assisi (Italy) during the 1997 Umbria-Marche earthquake. Its main characteristics are: peak acceleration $a_{\text{max}} = 0.2$ g; fundamental period $T = 0.32$ s; dominant frequency $f = 3.13$ Hz; Arias intensity $I_A = 0.264$ m/s. The time history of horizontal acceleration and its Fourier spectrum are shown in Fig. 2.

A baseline correction procedure has been employed to eliminate the effects of the numerical sampling of the recorded signal. The resulting acceleration history has been imposed as a boundary condition at the bottom of the soil layer.

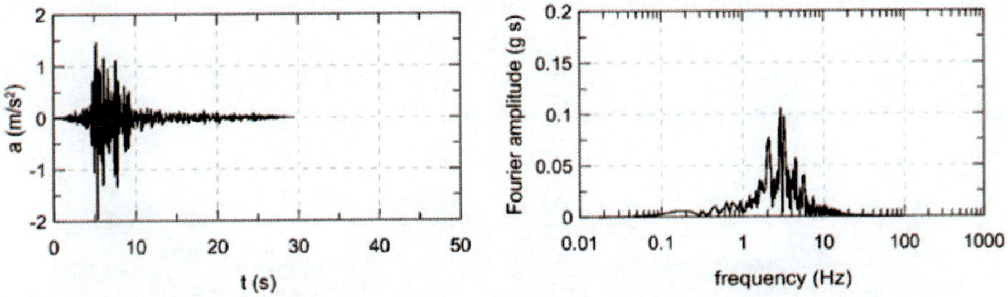

Fig. 2. Time history and Fourier spectrum of the horizontal acceleration of the seismic input adopted.

4 Selected Results

Some selected results of the three dynamic simulations are presented here, focusing on the accumulated permanent wall displacements and rotations of the walls and on the evolution with time of the soil sensitivity.

The evolution with time during the seismic event of the average horizontal displacement and rigid rotation of the right wall, calculated as:

$$u_{x,\text{ave}} = \left(u_x^B + u_x^T\right)/2 \qquad \theta = \tan^{-1}\left\{\left(u_x^B - u_x^T\right)/2\right\} \tag{4}$$

are shown in Figs. 3a and 3b. In Eq. (4), u_x^B and u_x^T are the horizontal displacements at the base and at the top of the wall, respectively.

Since the right wall rotates clockwise, both the average horizontal displacement and rotation provided by Eq. (4) are negative and decrease monotonically with time due to the accumulation of irreversible deformations during the earthquake shaking. Similar patterns of horizontal displacement and rotation are obtained for the left wall, but with positive values due to its counterclockwise rotation.

As expected, both wall displacement and rotation increase as the constant k, controlling the destructuration rate, increases. It is worth noting that the impact of this soil property on the prediction is significant, with an increase of about 100% of the final

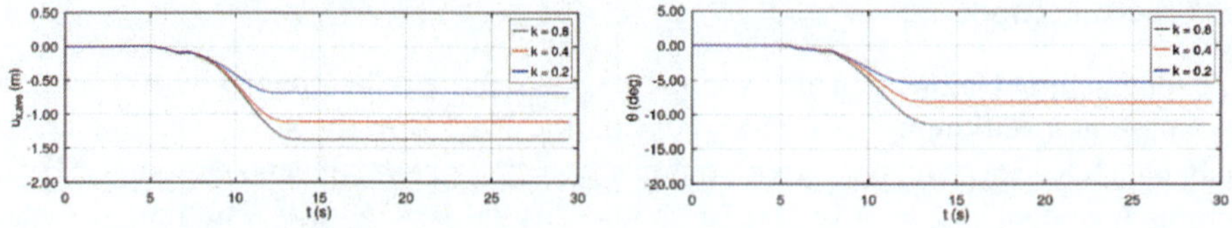

(a) (b)

Fig. 3. Time evolution of: a) average horizontal displacement, and b) rigid body rotation of the right wall, during the earthquake loading stage.

values of $u_{x,\text{ave}}$ and θ for an increase of k from 0.2 to 0.8. In particular, the final average displacement and rotation of the wall computed in simulation r03 are so large that the retaining structure can be considered to have reached collapse.

The reason for this behavior can be found in the time and spatial evolution of the soil sensitivity s for the simulations r01 and r03, shown in Fig. 4.

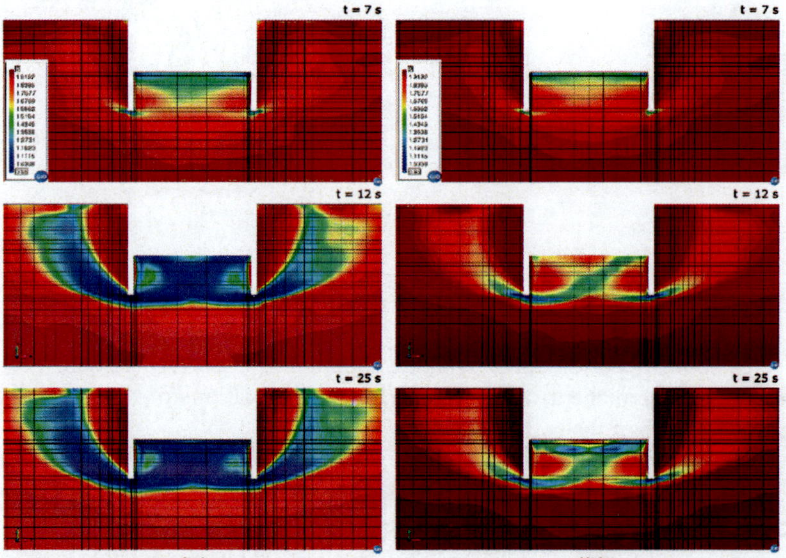

Fig. 4. Contour maps of soil sensitivity at 3 different time stations: a) simulation r01 ($k = 0.8$); b) simulation r03 ($k = 0.2$).

At $t = 7$ s, when the permanent displacement and rotation of the wall are still very small, there is almost no difference in the spatial distributions of s in the two simulations, except perhaps in a shallow region close to the excavation bottom. This implies that the destructuration effects occurred during the excavation stage have been minimal.

On the contrary, the progressive accumulation of strains – mostly deviatoric due to the almost undrained nature of the soil deformation – along the different loading cycles produced by the earthquake motion, determine a progressive reduction of s towards its asymptotic unit value, and with it a progressive reduction in soil stiffness and strength which amplify the soil deformations cycle after cycle. In run01 (Fig. 4a), at $t = 12$ s, the progressive accumulation of deformations is sufficient to achieve full destructuration in the zones where shear deformation concentrate. These zones clearly identify the formation of a complete collapse mechanism.

In the simulation performed with the smaller value of k (run03, Fig. 4b), the destructuration pattern observed is qualitatively similar to the one of run01. However, full destructuration only occurs in a relatively small zone close to the wall tips, and the soil around the retaining structures still possesses some degree of structure at the end of the earthquake event.

5 Concluding Remarks

A series of fully coupled non-linear dynamic FE simulations of the behavior of an excavation in saturated clay supported by propped diaphragm walls, in connection with the adoption of a recent advanced hypoplastic model for natural structured clays, has allowed to identify the role played by the destructuration process associated with strain accumulation on the seismic performance of the retaining structures.

Regardless of the destructuration rate, the loss of initial structure at the end of the excavation process is generally small and limited to a shallow portion of soil below the final excavation level. On the contrary, significant destructuration effects are produced by the cyclic accumulation of (mostly deviatoric) strains during the earthquake loading stage. These, in turn, determine the progressive reduction of the stiffness and strength of the soil and thus have a significant effect on the accumulated permanent displacements of the structure When the soil is characterized by a high destructuration rate, this process may eventually lead to the failure of the retaining structure.

References

1. Conti, R., Madabhushi, S.P.G., Viggiani, G.M.B.: On the behaviour of flexible retaining walls under seismic actions. Géotechnique **62**(12), 1081–1094 (2012)
2. Conti, R., Viggiani, G.M.B., Burali d'Arezzo, F.: Some remarks on the seismic behaviour of embedded cantilevered retaining walls. Géotechnique **64**(1), 40–50 (2014)
3. Oliynyk, K., Conti, R., Viggiani, G., Tamagnini, C.: A generalised Newmark method with displacement hardening for the prediction of seismically induced permanent deformations of diaphragm walls. Géotechnique (2022). https://doi.org/10.1680/jgeot.21.00022
4. Cattoni, E., Tamagnini, C.: On the seismic response of a propped rc diaphragm wall in a saturated clay. Acta Geotech. **15**(4), 847–865 (2020)
5. Mašín, D.: A hypoplastic constitutive model for clays. Int. J. Numer. Anal. Meth. Geomech. **29**(4), 311–336 (2005)
6. Burland, J.B.: On the compressibility and shear strength of natural clays. Géotechnique **40**(3), 329–378 (1990)
7. Mašín, D.: Clay hypoplasticity model including stiffness anisotropy. Géotechnique **64**(3), 232–238 (2014)
8. Gens, A., Nova, R.: Conceptual bases for a constitutive model for bonded soils and weak rocks. In: Geotechnical Engineering of Hard Soils-Soft Rocks, pp. 485–494 (1993)

9. Mašín, D.: 3D modeling of an NATM tunnel in high K_0 clay using two different constitutive models. J. Geotech. Geoenviron. Eng. **135**(9), 1326–1335 (2009)
10. Ferraro, G.: Risposta sismica di strutture di sostegno flessibili in terreni argillosi strutturati. Università degli Studi di Perugia (in italian), Tesi di Laurea Magistrale (2022)
11. Vucetic, M., Dobry, R.: Effect of soil plasticity on cyclic response. J. Geotech. Eng. **117**(1), 89–107 (1991)
12. Tochnog: Tochnog Professional User Manual (2022). https://www.tochnogprofessional.nl/
13. Boffi, D., Brezzi, F., Fortin, M.: Mixed finite element methods and applications, vol. 44, pp. xiv-685. Springer, Heidelberg (2013). https://doi.org/10.1007/978-3-642-36519-5
14. Zienkiewicz, O.C., Chang, C.T., Bettess, P.: Drained, undrained, consolidating and dynamic behaviour assumptions in soils. Geotechnique **30**(4), 385–395 (1980)

A Simplified Method for Relating Rainfall to Movements of Slopes Reinforced by Piles

Antonello Troncone(✉), Luigi Pugliese, Andrea Parise, and Enrico Conte

Department of Civil Engineering, University of Calabria, 87036 Cosenza, Rende, Italy
antonello.troncone@unical.it

Abstract. Using one or more rows of piles embedded in the stable portion of a slope is usually recognized as a suitable measure for slope stabilization. However, the landslide might be still reactivated due to a significant raising of the groundwater level caused by rainfall. In this context, a simplified methodology is presented in this paper to predict the rainfall-induced movements of landslide bodies in the presence of stabilizing piles. Specifically, the proposed method uses a water balance equation to relate rainfall to groundwater level changes, and an equation of motion of a two-block system representing the landslide body, to relate these changes to the landslide movements, accounting for the passive resistance exerted by the piles. This latter force can be evaluated on the basis of the pile arrangement and the possible pile-soil failure mode using some practical solutions available in the literature. The application of the method to an ideal case study is presented to show the effects of the stabilizing piles on the landslide mobility.

Keywords: Landslide mobility · Rainfall · Groundwater level · Stabilizing piles · Analytical method

1 Introduction

Groundwater level fluctuations owing to rainfall often cause landslide movements that are characterized by a sequence of stages of motion and rest. In particular, the landslide is reactivated during rainy periods as a consequence of the increase in the piezometric level, whereas its velocity reduces, eventually until the motion stops, in dry periods when the groundwater level lowers. These landslides are usually characterized by a velocity on the order of few centimeters per year [1]. Therefore, they can be defined as very slow-moving landslides. Another common characteristic of such landslides is that the unstable soil mass moves practically as a rigid body, with an essentially uniform profile of the horizontal displacement and soil deformations concentrated in a thin shear zone located at the base of the sliding mass where the soil shear strength is at residual [1]. The mobility of this type of landslides can be prevented or attenuated by using different stabilization measurements [2, 3]. When the slip surface is located at a relatively shallow depth (generally < 10 m), an effective solution consists in the construction of one or more rows of piles adequately embedded in the stable soil [4]. In this context, the piles should be as largely spaced as possible to reduce the costs, but at the same time they

A. Ferrari et al. (Eds.): CNRIG 2023, SSGG, pp. 553–560, 2023.
https://doi.org/10.1007/978-3-031-34761-0_67

should be close enough to ensure an interaction between them. To this purpose, the pile spacing generally ranges between 2 and 4 times the pile diameter [5]. The efficiency of this stabilization measurement is usually evaluated by calculating the increment in the factor of safety of the slope due to the presence of the piles [4]. However, a prediction of the possible slope movements represents a more rational approach, especially for landslides periodically reactivated by rainfall. The behavior of the slopes reinforced by piles can be undoubtedly analyzed using numerical methods [6–8]. However, such methods could require high computational costs and a large number of parameters that are often of difficult evaluation. Some solutions of practical interest are also available [9, 10]. They however do not provide any information about the landslide displacements. In this context, a method for the prediction of landslide movements in the presence of stabilizing piles was proposed by Galli and Di Prisco [11] who employed the simple scheme of infinite slope and a Winkler-type model to simulate the soil-pile interaction. However, this method does not account for the relationship between rainfall and pile response.

In this paper, a method of practical interest is presented for the prediction of rainfall-induced landslide movements in the presence of stabilizing piles. This method assumes that the landslide body (which has a finite extension) slides along a bi-planar slip surface. On the basis of the rainfall recordings, the method takes advantage of some analytical solutions that allow first the prediction of the groundwater level fluctuations due to rainfall, and then that of the displacement of the considered landslide body. The derived solution is simple to use and requires a limited number of parameters as input. In addition, most of these parameters can be determined from conventional geotechnical tests, whereas the remaining ones could be calibrated using a procedure that is included in the method, on the basis of the available measurements of groundwater level and displacement recorded in a precedent period of observation. After this calibration, a prediction of the landslide movements due to expected rainfall scenarios can be readily performed using the present method. An application of the method to an ideal case study is also carried out to show the effects of the presence of stabilizing piles on the landslide mobility.

2 Calculation Method

The scheme in Fig. 1a shows a slope inclined of an angle α and subjected to a seepage parallel to the ground surface. The groundwater level fluctuations are indicated by $h(t)$ and are measured from a steady state level whose position is defined by z_w. This latter can be assumed equal to the maximum depth in an observation period preceding the prevision. The landslides body is defined by two planar slip surfaces inclined by β_1 and β_2 with respect to the horizontal plane, respectively. One or two rows of piles are installed in the slope, with the piles that are characterized by diameter D and spacing S (Fig. 1b). The groundwater level changes, $h(t)$, can be calculated under the assumption that they occur simultaneously with rainfall (that is often the case when the ground surface is fissured due to the continuous reactivation stages of the landslide) and the resulting water level gradually reduces owing to seepage.

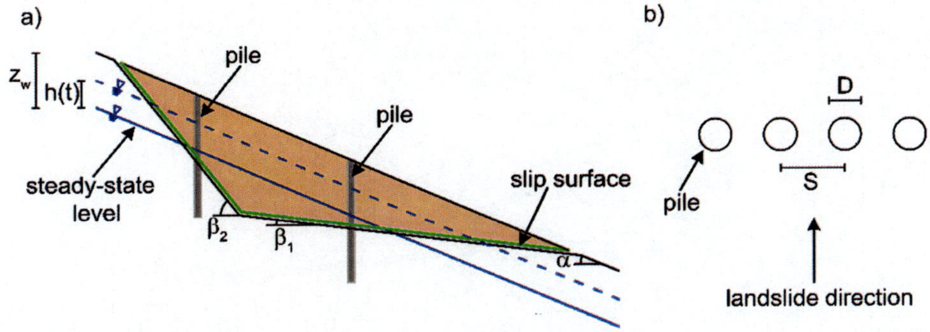

Fig. 1. Scheme showing a slope reinforced by piles: a) cross-section of the slope; b) plan of the pile arrangement.

Under these conditions, the equation proposed by Conte et al. [12] is herein employed:

$$h(t) = h_o e^{-\bar{k}t\sin\alpha\cos\alpha} + \sum_{i=1}^{N} \frac{h_i}{A} e^{-\bar{k}\sin\alpha\cos\alpha(t-t_{oi})} \quad (1)$$

in which h_o defines the initial position of the water level with respect to the steady state level, N is the number of rainfall events considered, t is time, t_{oi} is the time at which the ith event commences, $\bar{k}$ is a model parameter related to the soil permeability, and $A = n(1 - S_r)$ where n is the porosity and S_r is the degree of saturation of the portion of the soil located above the groundwater level. An evaluation of these parameters may be performed by measurements of dilatational and shear wave velocities [13, 14]. Finally, h_i is the water volume (per unit area) infiltrating into the slope, which is related to the rainfall depth R_i of the ith event by the following equations [15]:

$$h_i = R_i \text{ if } R_i < p \quad (2a)$$

$$h_i = p \text{ if } R_i \geq p \quad (2b)$$

in which p defines the maximum water volume (per unit area) than can infiltrate into the slope. For the sake of simplicity, constant values of A and p are assumed in the present study. The unstable soil mass is subdivided into two triangular blocks (Fig. 2) that behave rigidly. One row of stabilizing piles is considered for each block. The forces acting on these blocks are their weights W_1 and W_2, the reactions of the piles F_{P1} and F_{P2}, the effective normal forces N_1' and N_2', the shear forces T_1 and T_2, the water pressure resultants U_1 and U_2, and the viscous forces F_1 and F_2 acting at the slip surface. The viscous forces are activated only when the landslide body moves, otherwise they are nil. Additionally, the reaction force E' and the resultant of pore water pressure U act on the vertical interface between the blocks. Due to the relative displacements occurring between the blocks, force E' is assumed to be inclined of an angle φ' with respect to the horizontal direction, where φ' is the angle of shearing resistance of the soil.

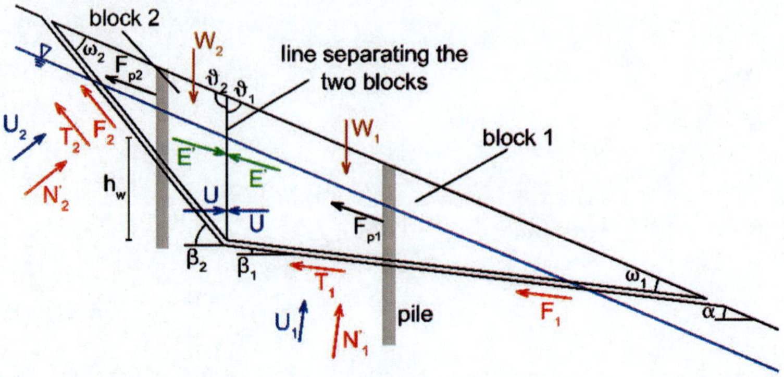

Fig. 2. Two-block model considered in the present study, with indication of the forces acting on each block.

The block weights, W_i, and the forces due to water, U_i, can be calculated using the following expressions, where the index *i* refers to the block number (i.e., 1 or 2):

$$W_i = \gamma A_i \tag{3}$$

$$U = \frac{1}{2}\gamma_w [h_w + h(t)]^2 \cos^2\alpha \tag{4}$$

in which γ is the soil unit weight, γ_w is the unit weight of water, A_i is the area of the ith block and h_w is indicated in Fig. 2. Forces U_i can be calculated as a function of U and the angles ϑ_i and ω_i indicated in Fig. 2, as follows:

$$U_i = U\frac{\sin\vartheta_i}{\sin\omega_i} \tag{5}$$

In addition, the viscous forces, F_i, are expressed by Bingham's law as a function of the coefficient of viscosity μ_s and velocity $v_i(t)$ of the ith block at time *t*:

$$F_i = \mu_s v_i(t) \tag{6}$$

The piles oppose the landslide movements with a resistance force F_{Pi}, as shown in Fig. 3. This latter is calculated in the present method as the resultant of the pile-soil contact stresses acting on the portion of the pile above the slip surface under ultimate conditions. These forces are a function of the spacing between piles and the pile-soil failure mode that in turn depends on the soil properties and the geometrical and mechanical characteristics of the piles. To this purpose, the solutions proposed by Viggiani [9] and Di Laora et al. [10] can be used for piles embedded in cohesive and cohesionless soils, respectively.

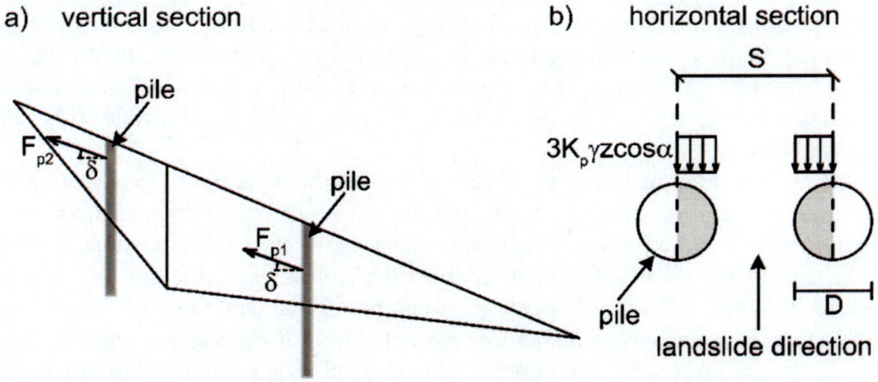

Fig. 3. Forces exerted by the piles on the unstable soil mass: (a) vertical section (b) horizontal section.

As an example, referring to the failure mode C experienced by a pile in cohesionless soil [10], the vertical and horizontal components of F_{Pi} take the following expressions:

$$F_{Phi} = \frac{D}{S}\int_0^{l_1} 3K_P\sigma_v^{'}\cos\alpha dz \tag{7}$$

$$F_{Pvi} = F_{Phi}\tan\delta \tag{8}$$

In these equations, l_1 defines the length of the portion of the pile in the unstable soil mass, δ is the soil-pile interface friction angle, $\sigma_v^{'}$ is the vertical effective stress at the depth z and K_P is the passive earth pressure coefficient. The resultant F_{Pi} forming an angle δ with respect to the horizontal plane, is calculated as:

$$F_{Pi} = \sqrt{F_{Phi}^2 + F_{Pvi}^2} \tag{9}$$

Finally, forces N_1', N_2', T_1, T_2 and *E'* are unknown. However, T_i is related to N_i' by the Mohr-Coulomb failure criterion, i.e.

$$T_i = c_r^{'}L_i + N_i^{'}\tan\varphi_r^{'} \tag{10}$$

in which L_i defines the length of the slip surface of the ith block, whereas $c_r^{'}$ and $\varphi_r^{'}$ are the shear strength parameters of the soil in the shear zone, which is generally at residual for the type of landslide herein considered.

As the force U strongly influences the slope stability, a critical threshold, U_{crit}, can be calculated from the equilibrium of the forces acting on the blocks (excluding the viscous forces) under a condition of incipient failure of the soil mass. The expression of U_{crit} is derived in the paper by Troncone et al. [16]. Motion occurs when $U > U_{crit}$. On the contrary, the landslide body is at rest when $U \leq U_{crit}$. The landslide movements are calculated from the dynamic equilibrium of the two blocks, resulting in the following motion equation [16]:

$$\frac{d}{dt}v(t) + \lambda v(t) = \chi[U(t) - U_{crit}] \tag{11}$$

In this equation, $v(t)$ defines the velocity of the landslide body, and λ and χ are coefficients depending on the slope geometry and geotechnical parameters, the expressions of which can be found in the paper by Troncone et al. [16]. Solution of Eq. 11 provides the function $v(t)$ from which the displacement experienced by the landslide body can be calculated by integration.

The method described in this section requires a limited number of soil parameters. Most of them can be determined from conventional geotechnical tests, such as the unit weight γ, intercept cohesion $c_r^{'}$ and angle of shearing resistance $\varphi_r^{'}$ at residual condition for the soil in the shear zone, angle of shearing resistance of the soil φ', and soil-pile friction angle δ. The other parameters can be instead calibrated by matching the available measurements recorded in a precedent period of observation with the results provided by the present method. Specifically, p, $\overline{k}$ and A are calibrated on the basis of the measurements of groundwater level, whereas μ_s is calibrated on the basis of the measurements of displacement using the procedure suggested by Conte et al. [9]. After performing the calibration of these parameters, the method can be employed to predict future landslide displacements directly from the expected rainfall scenarios.

3 Analysis of a Case Study

In this section, the proposed method is employed to analyze an ideal case study involving a landslide body with $\alpha = 11°$, which is bounded below by two planar slip surfaces forming the angles $\beta_1 = 6°$ and $\beta_2 = 18°$ with the horizontal plane, respectively. The maximum depth of the slip surface is 5.4 m from the ground surface. The landslide body consists of a cohesionless soil characterized by the following geotechnical parameters: $\gamma = 17$ kN/m^3, $\varphi^{'} = 15°$, $c^{'} = 0$ kPa, $\varphi_r^{'} = 12.5°$ and $c_r^{'} = 0$ kPa. It is assumed that a row of piles is installed into the lower block of the landslide body (Fig. 2). Each pile has a diameter $D = 0.80$ m, spacing $S = 1.60$ m and provides a resistant force of 144 kN/m corresponding to the failure mode C [10]. The required model parameters are $A = 0.09$, $\overline{k} = 1.82$ day^{-1}, $p = 125$ mm and $\mu_s = 2.9 \cdot 10^3$ kN·day/m^2. Figure 4 synthetizes the results obtained using the proposed method with and without the presence of the stabilizing piles. Specifically, the groundwater level fluctuations provided by Eqs. (1) and (2) when the rainfall recordings shown in Fig. 4a are used as input, are shown in Fig. 4b. This figure also shows the critical groundwater level $z_{w,crit}$ (measured from the ground surface) corresponding to a condition of limit equilibrium of the landslide body (i.e., when $U = U_{crit}$). In other words, movements of the landslide body occur when the piezometric level rises above this threshold. As can be seen, $z_{w,crit}$ significantly decreases in the presence of piles (Fig. 4b). Consequently, a higher raising of the groundwater table is needed in this case to reactivate the landslide body, than in the case when piles are not installed in the slope. Figures 4c and 4d show the velocity and displacement of the landslide body. The curves have a similar shape, but the displacement magnitude calculated in the presence of the piles is about four times smaller than that obtained when the piles are ignored (Fig. 4d).

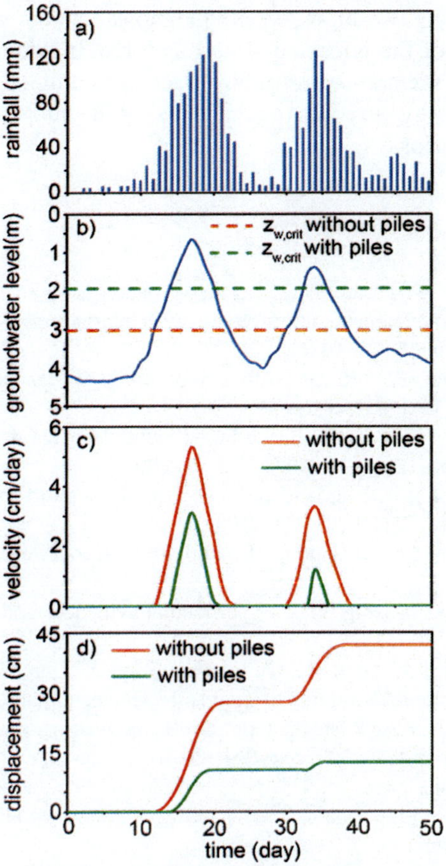

Fig. 4. Relationship between rainfall and landslide mobility: a) rainfall; b) groundwater level changes with indication of the critical groundwater level $z_{w,crit}$ (measured from the ground surface); c) landslide velocity; d) landslide displacement.

Analogous considerations can be made for the landslide velocity (Fig. 4c). Summarizing, installing a row of piles gives rise to an increase in U_{crit} with respect to the case in which the slope is not reinforced with piles. Consequently, the presence of piles is capable of producing a significant reduction in the landslide mobility during the reactivation stages.

4 Concluding Remarks

A simple method of practical interest has been proposed in this paper to predict the movements experienced by landslides periodically reactivated by rainfall, when one or more rows of piles with a stabilizing task are installed in the slope. Specifically, the method relates rainfall to groundwater level changes by a balance equation, and these latter to the landslide displacements solving a motion equation. To this end, the landslide body is schematized by a two-block system the movements of which are contrasted by the passive resistance of the piles and a viscous force, in addition to the soil strength in the shear zone. The method seems to be attractive for engineering applications, as it can be used for forecasting purposes to relate directly the landslide displacements to the expected rainfall scenarios.

References

1. Leroueil, S., Vaunat, J., Picarelli, L., Locat, J., Faure, R., Lee, H.: A geotechnical characterization of slope movements. In: Proceedings 7th International Symposium on Landslides, Trondheim, vol. 1, pp. 53–74 (1996)
2. Cotecchia, F., Lollino, P., Petti, R.: Efficacy of drainage trenches to stabilise deep slow landslides. Géotech. Lett. **6**, 1–6 (2016)
3. Conte, E., Troncone, A.: A performance-based method for the design of drainage trenches used to stabilize slopes. Eng. Geol. **239**, 158–166 (2018)
4. Ito, T., Matsui, T., Hong, W.P.: Extended design method for multi-row stabilizing piles against landslide. Soils Found. **22**(1), 231–238 (1982)
5. Duncan, J.M., Wright, S.G., Brandon, T.L.: Soil Strength and Slope Stability, 2nd edn. Wiley, Hoboken (2014)
6. Lirer, S.: Landslide stabilizing piles: Experimental evidences and numerical interpretation. Eng. Geol. **149–150**, 70–77 (2012)
7. Muraro, S., Madaschi, A., Gajo, A.: On the reliability of 3D numerical analyses on passive piles used for slope stabilisation in frictional soils. Géotechnique **64**(6), 486–492 (2014)
8. Pirone, M., Urciuoli, G.: Analysis of slope-stabilising piles with the shear strength reduction technique. Comput. Geotech. **102**, 238–251 (2018)
9. Viggiani, C.: Ultimate lateral load on piles used to stabilize landslides. In: Proceedings of 10th International Conference on Soil Mechanics and Foundation Engineers, Stockholm, vol. 3, pp. 555–560 (1981)
10. Di Laora, R., Maiorano, R.M.S., Aversa, S.: Ultimate lateral load of slope-stabilising piles. Gèotech. Lett. **7**, 1–8 (2017)
11. Galli, A., di Prisco, C.: Displacement-based design procedure for slope-stabilizing piles. Can. Geotech. J. **50**, 41–53 (2013)
12. Conte, E., Donato, A., Troncone, A.: A simplified method for predicting rainfall induced mobility of active landslides. Landslides **14**, 35–45 (2017)
13. Conte, E., Cosentini, R.M., Troncone, A.: Shear and dilatational wave velocities for unsaturated soils. Soil Dyn. Earthq. Eng. **29**, 946–952 (2009)
14. Conte, E., Cosentini, R.M., Troncone, A.: Geotechnical parameters from Vp and Vs measurements in unsaturated soils. Soils Found. **49**, 689–698 (2009)
15. Conte, E., Pugliese, L., Troncone, A.: A simple method for predicting rainfall-induced shallow landslides. J. Geotech. Geoenviron. Eng. ASCE **148**(10), 04022079 (2022)
16. Troncone, A., Pugliese, L., Lamanna, G., Conte, E.: Prediction of rainfall-induced landslide movements in the presence of stabilizing piles. Eng. Geol. **288**, 106143 (2021)

Back-Analysis of the Post-failure Stage of a Landslide in Sensitive Clays

Antonello Troncone(✉), Luigi Pugliese, Andrea Parise, and Enrico Conte

Department of Civil Engineering, University of Calabria, 87036 Cosenza, Rende, Italy
antonello.troncone@unical.it

Abstract. The landslides involving sensitive clays are generally characterized by very high velocity and long run-out distance. These landslides have captured the interest of the geotechnical community since the pioneering works by Skempton and Northey [1] and Bjerrum and Landva [2] inspired by the impressive events occurred in Norway. However, the unavailability of suitable methods of analysis has been one of the main obstacles to the advancement of knowledge about the kinematics of these natural disasters. In the present study, the Material Point Method (MPM) is used to simulate the run-out process of a landslide in sensitive clays that occurred at Saint-Jude (Québec, Canada) in 2010. To assess the accuracy of the analysis, the final profile and the displacement magnitude detected after the event are compared to those obtained by the numerical simulation. The results provided by MPM are in satisfactory agreement with field observation. The failure mechanism and the development of the failure surface within the slope are also reproduced successfully. These results also show that MPM is an attractive method to analyze the kinematics of landslides in sensitive clays, requiring also a limited number of conventional geotechnical parameters as input data.

Keywords: landslides · sensitive clays · run-out process · material point method

1 Introduction

Slope movements and failure mechanisms are commonly categorized into four distinct stages [3]: pre-failure, failure, post-failure and eventual reactivation. However, slope stability is generally dealt with considering only the failure phase by means of simplified methods [4–7] or the pre-failure and failure stages using traditional numerical techniques (such as FEM or FDM) referring to a Lagrangian approach, under the assumptions of small deformations [8–11]. However, this approach is not capable to deal with the post-failure stage of landslides when large deformations occur. Recently, more advanced numerical techniques have been developed to deal with problems involving large deformations, which take advantage of both Lagrangian and Eulerian approaches. Among them, the Material Point Method (MPM) is one of those that have obtained increasing attention in the recent years [12–19]. Soga et al. [12] highlighted the efficiency of MPM in analyzing the deformation processes of landslides, including the post-failure stage.

In this paper, MPM is used to analyze a landslide that occurred in 2010 at Saint-Jude (Québec, Canada), which mainly involved a formation of sensitive clays. These soils

A. Ferrari et al. (Eds.): CNRIG 2023, SSGG, pp. 561–568, 2023.
https://doi.org/10.1007/978-3-031-34761-0_68

might be subjected to a considerable reduction in shear strength when they are remolded [1]. Consequently, the landslides involving these clays (named also quick clays) are generally characterized by very high velocity and long run-out distance (comparable with those experienced by the lateral spreads due to liquefaction). Such landslides captured the interest of the international geotechnical community after the pioneering works by Skempton and Northey [1] and Bjerrum and Landva [2], which were inspired by the impressive events occurred in Norway. However, the unavailability of suitable methods of analysis has been one of the main obstacles to the advancement of knowledge about the kinematics of these natural disasters.

In the present study, an analysis of the post-failure stage of the Saint-Jude landslide is performed in order to complete the understanding of the deformation processes occurred during this complex landslide event.

2 A Brief Description of the Saint-Jude Landslide

On May 10th 2010, a large landslide occurred at Saint-Jude, in Québec, Canada. This landslide was studied by Locat et al. [20] who provided a detailed description of the event with a reconstruction of the failure mechanism. According to these authors, natural causes should have triggered the Saint-Jude landslide. Specifically, the main triggering factors of the event are considered to be the erosion produced by the Salvail River, located at the base of the slope, in combination with the high pore water pressures existing in the slope. The dimensions of the landslide were about 270 m in width and 210 m in length, with a maximum height of the head scarp of about 7 m. The subsoil in the area of the landslide consists of three geological units, named Unit A, Unit B and Unit C, respectively [20]. Unit A, which can be classified as a dense sand from a geotechnical point of view, is the shallowest one, extending down to a depth of about 4 m from the ground surface. Unit B has a thickness of approximately 22 m and is formed by a uniform sensitive clay with an overconsolidation ratio in the range 1.2–1.9. Finally, Unit C is a stiff silty clay of low sensitivity with a thickness of about 5 m. Below these units, the soil is characterized by high mechanical properties so that it can be reasonably assumed as the bedrock of the overlying soils. The groundwater level is very close to the ground surface. The landslide, which showed the typical characteristics of a spread in sensitive clays, involved only Unit A and Unit B [20]. On the basis of the results of many CPTs, Locat et al. [20] defined the location of the failure surface, the most portion of which developed just above the interface between Unit B and Unit C. Figure 1 shows a representative cross-section of the slope, where the location of the slip surface and the slope profile observed after the landslide are indicated. The displacement vectors of three benchmark points situated on the original ground surface are also shown in this figure, for the sake of completeness. As can be seen, a run-out distance of about 60 m occurred. This latter is the distance between the tip of the displaced soil and the toe of the failure surface (Fig. 1).

According to Locat et al. [20], the slope under consideration was in a condition of precarious equilibrium at the beginning of 2010. Indeed, a slope safety factor of 1.03 was evaluated by these authors using the limit equilibrium method. In the spring of 2010, a portion of the soil at the toe of the slope was eroded by the Salvail River. This

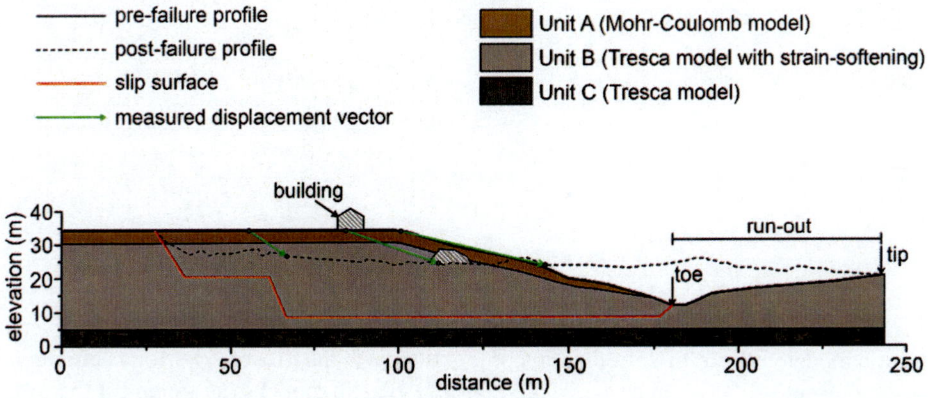

Fig. 1. Representative cross-section of the slope with an indication of the geological units forming the subsoil (after Locat et al. [20]).

phenomenon, in combination with the high pore water pressures already existing in the slope, triggered the failure [20]. Unfortunately, the volume of the eroded material is unknown. However, Locat et al. [20] believed that the size of this volume would not be significant as the slope initially was in precarious stability conditions. The post-failure stage of the landslide was very fast. Therefore, considering that the involved soil is formed prevalently by saturated clay, it is reasonable to assume that this process occurred under undrained conditions [20].

3 MPM Simulation of the Saint-Jude Landslide

In this section, MPM is employed to perform a back-analysis of the post-failure stage of the Saint-Jude landslide. To this purpose, the code Anura3D is employed (www.anura3d.com). The analysis is carried out under the assumption of plane-strain conditions, employing a computational mesh formed by triangular elements with an average size of 1.5 m. Three material points are initially distributed for each active element. Boundary conditions are simulated by preventing both vertical and horizontal displacements at the bottom of the model, whereas only the horizontal displacements are prevented on the vertical sides. No failure surface is preventively imposed. Following Locat et al. [20], the analysis is carried out considering undrained conditions for Unit B and Unit C, and drained conditions for Unit A. The soil parameters used in the analysis are drawn from the paper by Locat et al. [20] and are shown in Table 1.

The presence of the building shown in Fig. 1 is also taken into account in the simulation. This latter is schematized by a cluster of elastic material with unit weight $\gamma_b =$ 12 kN/m^3, Young's modulus $E_b = 50000$ kPa and Poisson's ratio $\nu_b = 0.2$.

The initial stress state of the slope is generated using the well-known procedure of gravity loading. As this procedure represents a quasi-static problem, a coefficient of local damping $\alpha = 0.75$ is used in this stage to accelerate the numerical convergence.

Table 1. Soil parameters used in the analysis (data drawn from Locat et al. [20]).

Soil	γ (kN/m^3)	c_u(kPa)	φ' (°)	ψ (°)	E' (MPa)	E_u(MPa)	ν
Unit A	18.6	–	35	0	25	–	0.30
Unit B	16	45	–	0	–	25	0.45
Unit C	16.8	70	–	0	–	25	0.45

Failure involved mainly Unit B which consists of sensitive clays characterized by a strong shear strength reduction due to remolding. To simulate this behavior, an elasto-plastic model with strain-softening in conjunction with the Tresca criterion is employed in the present study, whereas an elastic perfectly plastic model is assumed for the other soils indicated in Fig. 1. In the above-mentioned strain-softening model, the undrained shear strength of the sensitive clay, c_u, is gradually reduced with increasing deviatoric plastic strain invariant, ε_d^p, according to the following equation [21, 22]:

$$c_u = c_{ur} + (c_{uo} - c_{ur})e^{-\lambda\varepsilon_d^p}, \tag{1}$$

in which

$$\varepsilon_d^p = \sqrt{\frac{2}{3}e_{ij}^p e_{ij}^p}, \tag{2}$$

e_{ij}^p is the deviatoric part of the plastic strain tensor, c_{uo} and c_{ur} are the undisturbed and the remolded values of the undrained shear strength, respectively, and λ is a shape factor that controls the strength decrease rate. The optimal value of λ was preventively calibrated by performing a sensitivity analysis, in order to provide the best agreement between prediction and observation. The resulting value is $\lambda = 50$. In addition, the value of c_{uo} is taken equal to the value of c_u reported in Table 1. Finally, c_{ur} is assumed equal to the completely remolded shear strength, whose value is 1.6 kPa [20].

After applying the gravity loading procedure, the landslide triggering was simulated by switching the behaviour of the sensitive clay from linear elastic to elasto-plastic with strain-softening. The purpose of this stage is to simulate the effects of the river erosion and high pore water pressures in the slope. In this way, as the slope is precariously stable, material points start moving and accumulate deviatoric plastic strains with a consequent reduction in the undrained strength of the sensitive clay, according to Eq. 2. In this condition, the unstable soil mass moves until a new condition of equilibrium is attained. A local damping $\alpha = 0.10$ is employed in this stage.

Figure 2 shows the final configuration of the unstable soil mass provided by the numerical simulation, along with a comparison with the slope profile observed before and after the landslide. This figure shows that the final configuration of the landslide body is satisfactorily predicted by the MPM simulation.

The distance of run-out, the depletion and deposition zones obtained from the numerical simulation are close to the observed ones. In particular, the calculated run-out distance is approximately 51 m, the depletion zone is about 10 m and the maximum displacement on the order of 60 m. Additionally, Fig. 2 also presents a comparison between

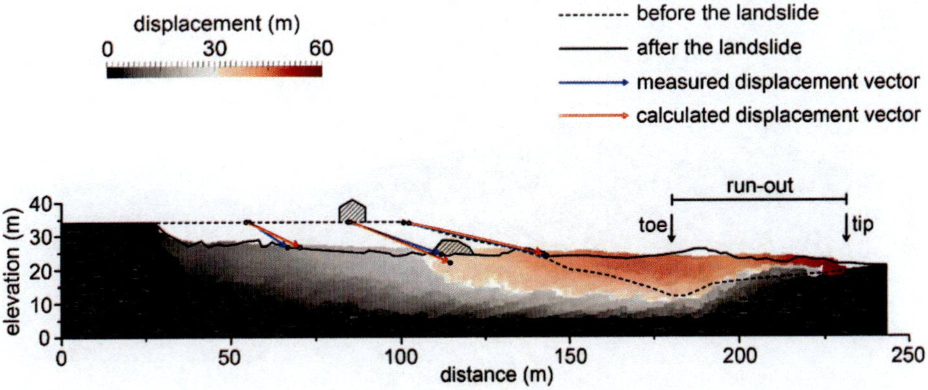

Fig. 2. Comparison of the final profile of the slope provided by MPM to that detected after the landslide.

the calculated displacement at the considered benchmark points with those observed in field [20]. As can be seen, also this comparison is satisfying. Finally, Fig. 3 shows the accumulated deviatoric strains at three different times, to show the evolution of the failure surface during the post-failure stage of the landslide. These results highlight that the failure mechanism is characterized by two failure surfaces. The first failure surface develops from the riverbed to just behind the building (Fig. 3a). In addition, the upper part of the soil mass defined by this failure surface includes some inclined shear zones similar to those found by Locat et al. [20]. Successively (Fig. 3b), the landslide body moves substantially as a translational slide, covering the riverbed along with a portion of the opposite flank. During this stage, the size of the accumulation zone grows, whereas the depletion zone extends upstream. At the same time, a second failure surface forms at a shallower depth than the first one, defining the main scarp of the landslide (Fig. 3c). This means that failure takes place in two consecutive stages, involving two failure surfaces situated at different depths. These results are in agreement with what observed by Locat et al. [20]. Indeed, according to these authors, a failure surface developed almost horizontally for about 100 m upstream from 2.5 m below the river elevation, whereas a second one, located approximately 10 m above the main one, propagated in the upper part of the slope up to the backscarp.

In Fig. 3c, the deviatoric strain field obtained at the last step of the simulation is compared with the failure surface reconstructed by Locat et al. [20] comparing the CPT profiles available before and after the landslide. This latter surface is drawn in green and refers to the slope before the landslide. As can be seen, the agreement between the failure surface observed by Locat et al. [20] and the shear zones provided by the numerical simulation is satisfactory, also considering the manner in which the failure surface was detected in field.

Summarizing, this work demonstrates that MPM is a useful and effective numerical technique for the analysis of landslides involving very sensitive clays which are generally characterized by a very fast post-failure stage and long run-out distances, the prediction of which is a key step in risk assessment and mitigation measure design.

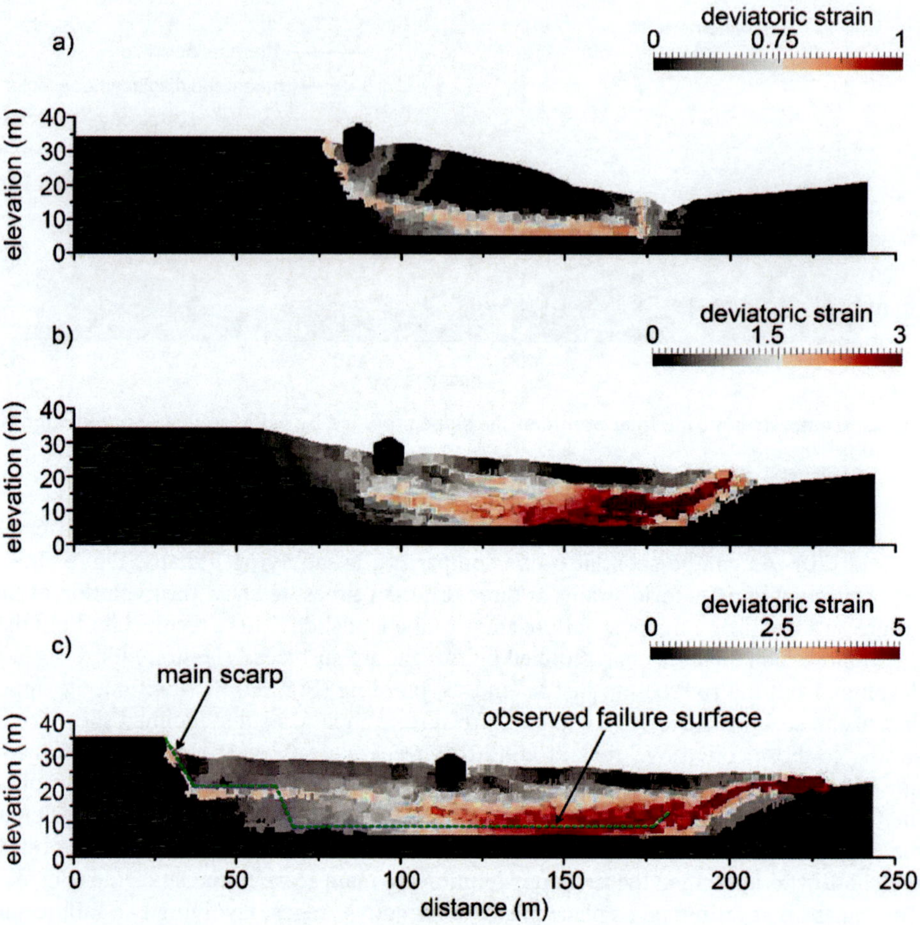

Fig. 3. Development of deviatoric strain field with time t: a) t = 2.5 s; b) t = 15 s; c) t = 90 s (final configuration).

4 Conclusions

In this work, the Material Point Method (MPM) has been used to analyze the post-failure stage of a landslide in sensitive clays that occurred at Saint-Jude (Québec, Canada) in 2010. The Saint-Jude landslide is a well-documented case history. A complete geotechnical characterization of the involved soils is available, along with an accurate description of the occurred failure mechanism. The results provided by the numerical simulation validate the suitability of MPM in reliably simulating the development of the failure surface within the slope, and capturing the run-out process of the considered landslide. Specifically, although the slip surface was not preventively imposed in the numerical model, the failure mechanism obtained from the analysis is in close agreement to that reconstructed by Locat et al. [20]. In addition, the final configuration of the landslide body obtained from the MPM simulation matches fairly well the observed one. To provide a complete understanding of the deformation mechanisms of this complex landslide, some features related to the kinematics of the landslide body after the failure stage are also highlighted. The obtained results demonstrate that MPM is an effective numerical technique to predict reliably both the run-out process and the failure mechanism of landslides in very sensitive clays, requiring also a limited number of conventional geotechnical parameters as input data.

References

1. Skempton, A.W., Northey, R.D.: The sensitivity of clays. Géotechnique **3**(1), 30–53 (1952)
2. Bjerrum, L., Landva, A.: Direct simple shear tests on Norwegian quick clays. Géotechnique **16**(1), 1–20 (1966)
3. Leroueil, S.: Natural slopes and cuts: movement and failure mechanisms. Géotechnique **51**(3), 197–243 (2001)
4. Duncan, J.M.: State of the art: limit equilibrium and finite-element analysis of slopes. J. Geotech. Eng. **122**(7), 577–596 (1996)
5. Conte, E., Donato, A., Troncone, A.: A simplified method for predicting rainfall-induced mobility of active landslides. Landslides **14**(1), 35–45 (2016). https://doi.org/10.1007/s10346-016-0692-8
6. Conte, E., Pugliese, L., Troncone, A.: A Simple method for predicting rainfall-induced shallow landslides. J. Geotech. Geoenviron. Eng. ASCE **148**(10), 04022079 (2022)
7. Troncone, A., Pugliese, L., Lamanna, G., Conte, E.: Prediction of rainfall-induced landslide movements in the presence of stabilizing piles. Eng. Geol. **288**, 106143 (2021)
8. Potts, D.M., Dounias, G.T., Vaughan, P.R.: Finite element analysis of progressive failure of Carsington embankment. Géotechnique **40**(1), 79–101 (1990)
9. Griffiths, D.L., Lane, P.A.: Slope stability analysis by finite element. Géotechnique **49**(3), 387–403 (1999)
10. Troncone, A.: Numerical analysis of a landslide in soils with strain-softening behaviour. Géotechnique **55**(8), 585–596 (2005)
11. Conte, E., Donato, A., Pugliese, L., Troncone, A.: Analysis of the Maierato landslide (Calabria, Southern Italy). Landslides **15**(10), 1935–1950 (2018). https://doi.org/10.1007/s10346-018-0997-x
12. Soga, K., Alonso, E., Yerro, A., Kumar, K., Bandara, S.: Trends in large-deformation analysis of landslide mass movements with particular emphasis on the material point method. Géotechnique **66**(3), 248–273 (2016)
13. Yerro, A., Alonso, E., Pinyol, N.: Run-out of landslides in brittle soils. Comput. Geotech. **80**, 427–439 (2016)
14. Conte, E., Pugliese, L., Troncone, A.: Post-failure stage simulation of a landslide using the material point method. Eng. Geol. **253**, 149–159 (2019)
15. Troncone, A., Conte, E., Pugliese, L.: Analysis of the slope response to an increase in pore water pressure using the material point method. Water **11**(7), 1446 (2019)
16. Ceccato, F., Leonardi, A., Girardi, V., Simonini, P., Pirulli, M.: Numerical and experimental investigation of saturated granular column collapse in air. Soils Found. **60**, 683–696 (2020)
17. Cuomo, S., Di Perna, A., Martinelli, M.: Modelling the spatio-temporal evolution of a rainfall-induced retrogressive landslide in an unsaturated slope. Eng. Geol. **294**, 106371 (2021)
18. Troncone, A., Pugliese, L., Conte, E.: Analysis of an excavation-induced landslide in stiff clay using the material point method. Eng. Geol. **296**, 106479 (2022)
19. Troncone, A., Pugliese, L., Parise, A., Conte, E.: A simple method to reduce mesh dependency in modelling landslides involving brittle soils. Géotech. Lett. **12**(3), 167–173 (2022)

20. Locat, A., Locat, P., Demeres, D., Leroueil, S., Robitaille, D., Lefebvre, G.: The Saint-Jude landslide of 10 May 2010, Quebec, Canada: Investigation and characterization of the landslide and its failure mechanism. Can. Geotech. J. **54**(10), 1357–1374 (2017)
21. Conte, E., Pugliese, L., Troncone, A.: Post-failure analysis of the Maierato landslide using the material point method. Eng. Geol. **277**, 105788 (2020)
22. Troncone, A., Pugliese, L., Parise, A., Conte, E.: Analysis of a landslide in sensitive clays using the material point method. Geotech. Res., 1–11 (2023). https://doi.org/10.1680/jgere.22.00060

Integrated Physical and Numerical Modelling to Study the Hydro-Mechanical Response of a River Embankment

Roberta Ventini[1](✉), Elena Dodaro[2], Marianna Pirone[3], Daniela Giretti[4], Carmine Gerardo Gragnano[2], Vincenzo Fioravante[5], Guido Gottardi[2], and Claudio Mancuso[3]

[1] Ministry of Infrastructures and Transport, Interregional Public Works Department, 50122 Firenze, FI, Italy
roberta.ventini@unina.it
[2] University of Bologna, Viale del Risorgimento 2, 40136 Bologna, BO, Italy
[3] University of Napoli Federico II, Via Claudio 21, 80125 Napoli, NA, Italy
[4] University of Bergamo, Via Pasubio 3, 24044 Dalmine, BG, Italy
[5] University of Ferrara, Via Saragat 1, 44122 Ferrara, FE, Italy

Abstract. Due to the increasingly frequent occurrence of extreme events related to climate change, the in-depth study of exceptional rainfall and flooding on earthen structures is essential. In particular, the associated consequences in terms of economic and human losses have caused the urgency of developing forecasting tools that can ensure a proper evaluation of expected impacts.

In this regard, the paper reports the numerical modelling of a small-scale centrifuge test performed to study the hydro-mechanical response of a river embankment typical of the Alpine and Apennine tributaries of the Po River (Italy). The purpose was to set up a well-validated numerical model that may be used for forecasting the behavior of river embankments under critical scenarios. In particular, the proposed fully coupled numerical modelling considers the partial saturation conditions of a compacted silty sand river embankment subjected to flooding. First, the numerical model parameters were carefully calibrated based on the results of laboratory tests conducted on the soils constituting the physical model. Afterwards, the model was validated by comparing the results of the simulation with the experimental data. It emerged that the numerical modelling proposed enables an accurate prediction of the pore-water pressures under various hydraulic loadings.

Keywords: River embankment · hydraulic loading · transient seepage

1 Introduction

Recent climatic trends make the assessment of the performance of earthen structures during their service life utterly necessary. This requires an in-depth understanding of the partially saturated soil behavior under transient seepage conditions. As stated by Rivera-Hernandez et al. [1], the changes in the statistics of extreme precipitation and

A. Ferrari et al. (Eds.): CNRIG 2023, SSGG, pp. 569–577, 2023.
https://doi.org/10.1007/978-3-031-34761-0_69

floods probability can affect the stability of existing earthen structures and impose a better understanding of transient seepage analysis through river embankments that usually are in unsaturated conditions. The number of river retaining earth structure collapses in Italy has dramatically risen in recent years, affecting small to large catchment areas, measuring from dozens to over 1500 km^2 [2] and becoming a critical concern for the entire country. Therefore, it is urgent to improve the resilience of existing embankments and to define robust and reliable procedures for the design and validation of new earthworks. A valuable tool to accomplish these goals is offered by numerical modelling, however, one of the main issues relies in the determination of appropriate input parameters and boundary conditions. To make sure that numerical simulations provide reliable predictions, they have to be necessarily validated. The results of geotechnical centrifuge tests, which are commonly used to analyse the behavior of earth structures and identify potential failure mechanisms, may represent a realistic benchmark to perform the model validation.

In this paper, the fully coupled numerical modelling of a centrifuge test on a small scaled river embankment is presented. The model parameters were calibrated using results from laboratory tests performed on the soils constituting the physical model. Then, data from the centrifuge test carried out to understand the behavior of an embankment representative of riverbank systems in the Alpine and Apennine tributaries of the Po River during flooding, were used to validate the model. The finite element simulation presented hereafter has been carried out considering physical model dimensions rather than the actual geometry of the full-scale prototype. The stress states experienced during each stage of the test were accurately reproduced by considering different incremental acceleration fields. The capability of the numerical model to predict displacements of the river embankment and changes in pore water pressures within the river embankment and the foundation is proved.

2 Physical Modelling

2.1 Soils and Testing Conditions

The centrifuge test, performed at the *Istituto Sperimentale Modelli Geotecnici* - ISMGEO (Bergamo, Italy), has been carried out on a clayey silty sand river embankment model overlying a homogeneous clayey silt foundation layer, under the enhanced gravity of *50-g*. A compacted mixture of 70% Ticino Sand (TS) and 30% Pontida Clay (PON), was chosen to construct the river embankment to reproduce the main Po river's tributaries, which have recently experienced multiple high-water events. The TS70%-PON30% mixture was compacted in four layers with the Standard Proctor energy at the optimum moisture content of 8.8%, and a dry density of 20.6 kN/m^3. Instead, the foundation was constructed using a solid layer of PON, consolidated under a vertical effective stress of 200 kPa. In terms of degree of saturation, the embankment has a non-zero initial suction distribution due to compaction while the foundation is saturated. The geometry of the physical model is reported in Fig. 1 together with the instrumentation layout for the middle section of the model. The model embankment was 150 mm high (7.5 m at the prototype scale) and 45° and 56° sloped riverside and landside, respectively. The embankment was instrumented with eight miniaturized tensiometers, two linear displacement transducers (L1 and L3) and two roto-translative sensors (LR2 and LR5) while four pressure transducers (named N, P, R, Q) were housed in the foundation layer and a more PPT (M) monitored the riverside water level.

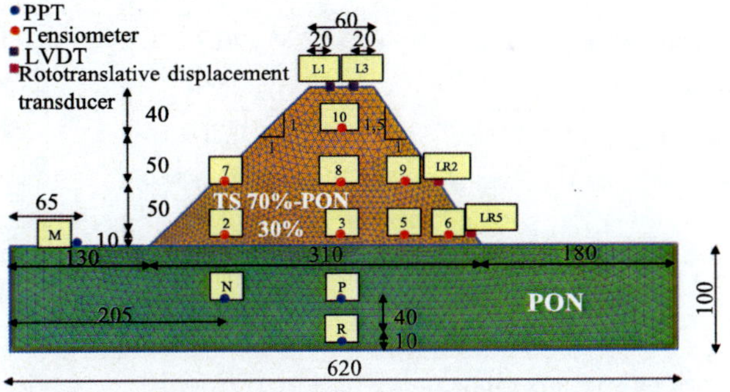

Fig. 1. Geometry of the plane strain section numerically simulated; adopted mesh and indication of the instruments position (length unit in mm).

The experimental campaign for the geotechnical soil characterization consisted of oedometer and triaxial tests, permeameter and evaporation tests, as well as post-compaction suction measurements. The results of these tests allowed determining the strength, deformability and soil water retention curves and hydraulic conductivity curves of the tested materials (Table 1). Further information on the geotechnical properties of the soils can be found in the research done by [3, 4], while more details about the design of the centrifuge tests and the discussion of the experimental results are reported in [5–7].

2.2 Test Procedure

Once reconstituted, the model was accelerated to the target angular velocity (i.e. 50-g) in two steps, as showed in Fig. 2 (black line) and then the acceleration was kept constant until the end of the test. The spin-up of the model caused the generation of an excess of pore water pressures in the foundation, so a self-weight equilibrium stage was performed, to allow the consolidation process. Afterwards, water was supplied to the container from an external tank to simulate river impoundment. Therefore, once a negligible settlement rate was reached, the reservoir water level was increased in two steps to study the hydro-mechanical response of the embankment. The water head was raised to about 90% of the embankment height (H) after an intermediate stage at 0.55 H. The unrealistic hydrometric peak persistence, applied during the flooding stage, was intentionally maintained to achieve a steady state seepage condition in equilibrium with the relevant hydraulic load of 0.9 H. More detailed information on the testing procedure, datasets, and monitoring system can be found in [6].

Table 1. Physical and hydro-mechanical properties of the modelled soils, according to the Hardening soil (HS) and the Modified Cam-Clay (MCC) constitutive models.

Soil		TS70%-PON30%	PON
Parameter		Hardening soil	Modified Cam Clay
Unsaturated unit soil weight, γ_{unsat}	kN/m^3	20.8	17.51
Saturated unit soil weight, γ_{sat}	kN/m^3	22.3	21.01
Initial void index, e_{init}	-	0.30	0.55
Secant stiffness in standard drained triaxial test, E_{50}^{ref}	kN/m^2	$22.52 \cdot 10^3$	-
Tangent stiffness for primary oedometer loading, E_{oed}^{ref}	kN/m^2	$10.00 \cdot 10^3$	-
Unloading / reloading stiffness from drained triaxial test, E_{ur}^{ref}	kN/m^2	$67.56 \cdot 10^3$	-
Power for stress-level dependency of stiffness, m	-	0.5	-
Effective cohesion, c’	kN/m^2	5.00	-
Internal friction angle, ϕ’	°	46.00	-
Coefficient of lateral stress in normal consolidation, K_0^{NC}	-	0.287	0.596
Poisson ratio, υ	-	-	0.20
Cam-Clay compression index, λ	-	-	0.074
Cam-Clay swelling index, κ	-	-	0.055
Tangent of the critical state line, M	-	-	1.33

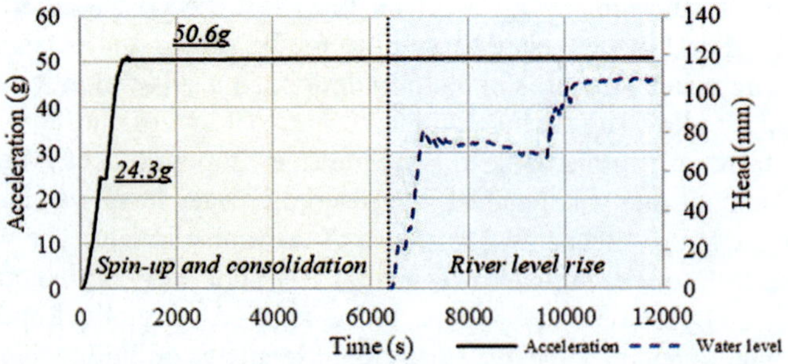

Fig. 2. Time history of the acceleration applied during the stages of the test, hydrograph, and calculation phases.

3 Finite Element Numerical Modelling

3.1 Model Definition: Geometry and Phases

Numerical modeling was carried out by using PLAXIS 2D software [8]. The model scale's geometry shown in Fig. 1 was replicated by iteratively optimizing the mesh density. Specifically, a coarseness factor of 0.7 was chosen for the embankment and 0.4 for the foundation layer.

The boundary conditions were defined using standard constraints, which meant that a full fixity (a rough rigid boundary) was applied along the bottom of the model. Additionally, the lateral boundaries of the foundation were fixed in the horizontal direction, with a smooth rigid boundary. At the end of the centrifuge test, a cortical erosion on the sidewall areas was observed, affecting the contact between the lateral walls of the container and the foundation-embankment model. To model the interaction between the foundation soil and the container walls, experimentally observed, thus, to consider possible flow channels, three interface elements were modeled. These last were characterized by the Young modulus of the steel and a permeability of 10^{-5} m/s. Furthermore, a zero head boundary condition was assigned to the top surface of the foundation on the landside to simulate a water outlet zone. After the consolidation stage, a coupled flow-deformation analysis, a variable hydrometric condition, obtained from PPT M measurements (Fig. 2), was imposed on riverside surfaces that may be affected by water action to reproduce river level fluctuations.

Different phases were simulated to reproduce the test faithfully; the initial stress field (initial phase at 1 g) was calculated with the '*K0* procedure' [8], to consider the initial stresses of foundation body due to its own weight. The preconsolidation stress state of the foundation layer was generated by imposing a pre-overburden pressure (POP) of 200 kPa. The next phase of calculation was the construction of the embankment, which was modeled through a plastic analysis. To reproduce the effective initial condition of the small-scale model, a constant value of matric suction of about 5 kPa in the embankment and the water table at the ground level were assumed. Because the dimensions of the small-scale centrifugal model were considered, the effect of the enhanced gravitational field imposed during the centrifuge test needed to be reproduced. The gradual increase of the angular speed was achieved by setting two multiplier factors in Plaxis ($\Sigma Mweight$ equal to 24.3 and 50.6 g), as illustrated in Fig. 2. A consolidation computation was carried out following each acceleration stage by considering the same time test duration, as shown in Fig. 2. Lastly, a coupled flow-deformation analysis was performed to study the transient groundwater seepage flow caused by the rising of the water level.

3.2 Model Calibration: Soil Tests

The soil test in Plaxis 2D is a powerful tool for simulating the behavior of soil specimens and rock samples under various loading conditions [8]. It provides a wide range of options for modeling soil behavior and a wide range of different constitutive models. Therefore, the laboratory geotechnical tests such as triaxial and oedometer tests available on the soils investigated, were simulated by means of Soil Test to obtain the most suitable numerical parameters to numerically reproduce the mechanical response of the soils. For example, the mechanical parameters of foundation layer (PON) were carried out by simulating the drained triaxial test on a consolidated specimen at a confinement stress of 100 kPa through the Soil Test using the Modified Cam Clay (MCC) model. The comparison between the experimental and numerical results are shown in Fig. 3. The numerical model and the set of parameters adopted are able to capture the stress-strain relationship both in terms of stiffness and peak strength. The parameters adopted for modelling the PON are reported in Table 1. The Hardening Soil Model (HSS) was adopted to model the embankment, the triaxial and oedometer tests were simulated through the Soil Test as it was done for the foundation layer, the parameters best fitting the test are reported in Table 1.

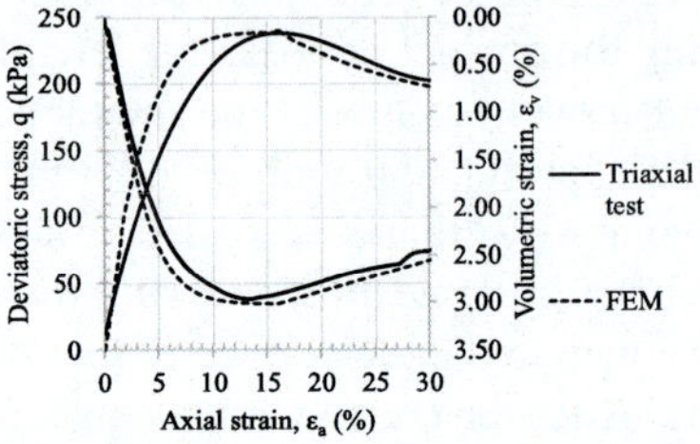

Fig. 3. Testing soil behavior: experimental and numerical results comparison for the PON.

3.3 Model Validation

The model validation was obtained through the comparison between the centrifuge test measurements and the numerical analysis results. From Fig. 4, it is evident the good agreement between the pore-water pressures calculated and measured during the spin-up and the consolidation phases in the foundation layer. However, slight differences in the pre-hydraulic loading pore pressures seem to be a result of the numerical model's ability to dissipate excess pore water pressures more quickly than the physical clayey silt unit. This is probably related to a combination of the value of saturated conductivity adopted for PON and the modelling of the sidewall side as a 'pervious boundary'.

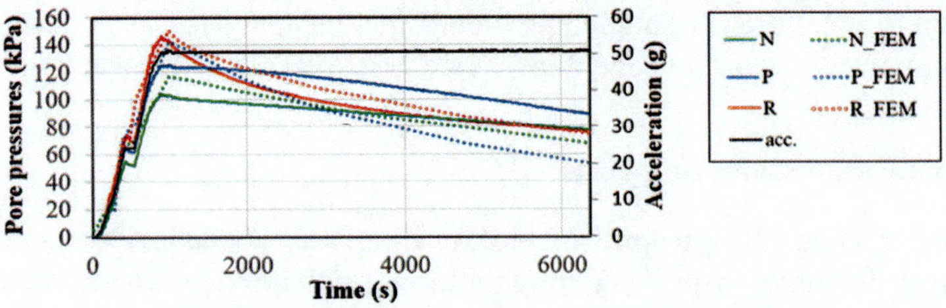

Fig. 4. Pore water pressures recorded in the foundation during the spin-up and the consolidation phases: experimental and numerical results comparison.

The total displacements occurred at embankment crest, as reported in Fig. 5, are well captured by the numerical model. The settlements calculated at L3 in the numerical model are approximately 2 mm higher than in the measurements, while both varied over time in a similar way (Fig. 5). The comparison between experimental data recorded from some tensiometers located in the embankment body of the physical model and the corresponding stress points in the numerical simulations during the flooding stage are reported in Fig. 6. The calculated pore pressure values are very similar to the recorded ones, except data series from tensiometers 10, which suggests a greater tendency of the physical crest to desiccate. This is ascribable to the choice of disregarding the modelling of the evapo-transpiration phenomena, occurring in the surface layers of the embankment.

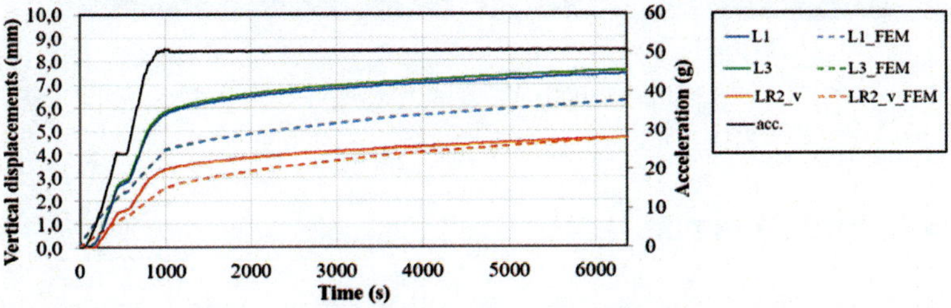

Fig. 5. Spin-up and consolidation phases: vertical displacements recorded.

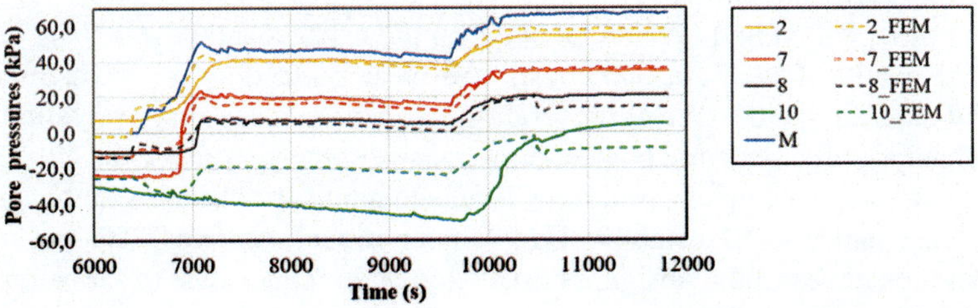

Fig. 6. Hydraulic loading application: pore water pressure recorded in the embankment.

The rise in river level has caused the advancement of the saturation line up to involve the crest of the embankment during the second peak (Fig. 7). The saturation of the embankment has provided an increase in the pore water pressure, which results in a reduction of the soil shear strength. The achievement of the steady state conditions by an unrealistic persistence of hydraulic loading (5440 s at the model scale, 157 days at the prototype scale) has caused the accumulation of plastic deformations on the downstream side without provoking any collapse, as it was confirmed by experimental evidence during the test (Fig. 7).

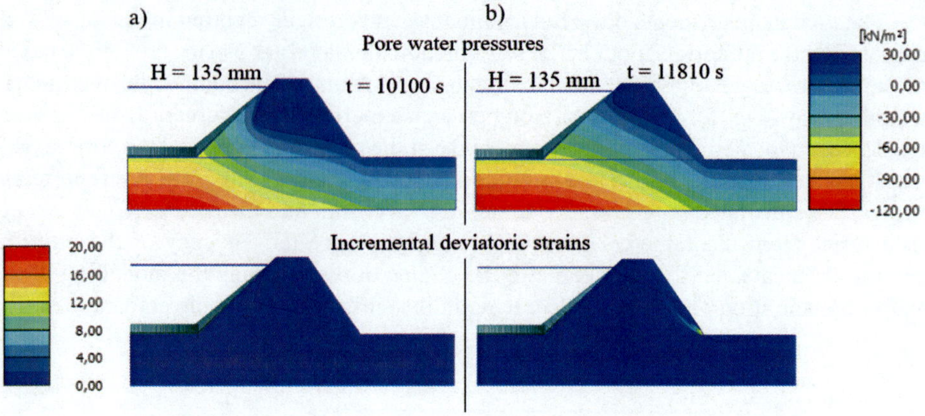

Fig. 7. Pore-water pressures and incremental deviatoric strains shadings: a) at the beginning and b) at the end of the river level rise.

4 Concluding Remarks

Laboratory tests, small-scale physical modelling and numerical analysis of a river embankment subjected to hydraulic loading have been performed for a better understanding of its typical behavior under long persistence of the river water level up to reaching the steady conditions.

It was observed a good agreement between the experimental data of the centrifuge test and the results of the finite element simulation at model dimensions. It emerged that numerical modeling of the centrifuge test at small scale enables a more detailed analysis of the behavior of the structure by shedding light on some features of the test. However, the model can be scaled up to a prototype level and it may then be used for the prediction of failure mechanisms and instability issues under critical scenarios. This can save time, money and resources in the design, construction and maintenance of embankments, as well as of other geotechnical structures.

References

1. Rivera-Hernandez, A.A., Ellithy, G.S., Vahedifard, F.: Integrating field monitoring and numerical modeling to evaluate performance of a levee under climatic and tidal variations. J. Geotech. Geoenviron. Eng. **145**(10), 05019009 (2019)
2. Michelazzo, G., Paris, E., Solari, L.: On the vulnerability of river levees induced by seepage. J. Flood Risk Manag. **11**, 677–686 (2018)
3. Ventini, R., Dodaro, E., Gragnano, C.G., Giretti, D., Pirone, M.: Experimental and numerical investigations of a river embankment model under transient seepage conditions. Geosciences **11**(5), 192 (2021)
4. Ventini, R., Pirone, M., Mancuso, C.: Soil water retention curves of a silty clayey sand compacted at different dry density. In: Proceedings of the 8th International Conference on Unsaturated Soils "Towards Unsaturated Soils Engineering", UNSAT, 2–5 May 2023, Milos, Greece (2023). https://doi.org/10.1051/e3sconf/202338209007

5. Girardi, V., et al.: Numerical study of uplift induced levee failure for the design of a centrifuge test. In: Fifth International Conference on New Developments in Soil Mechanics and Geotechnical Engineering, Near East University, Nicosia, TRNC (2022)
6. Ventini, R., et al.: Analysis of transient seepage through a river embankment by means of centrifuge modelling. In: Proceedings of the 8th International Conference on Unsaturated Soils "Towards Unsaturated Soils Engineering", UNSAT, 2–5 May 2023, Milos, Greece (2023). https://doi.org/10.1051/e3sconf/202338212008
7. Dodaro, E., et al.: On the hydromechanical behaviour of an unsaturated river embankment: centrifuge testing and numerical analysis. In: International Conference on Physical Modelling in Geotechnics; Proc. intern. conf., Daejeon, 19–23 September (2022)
8. PLAXIS 2D Reference Manual. Bentley Systems International Limited, Dublin (2021)

Identification of the Timing of Liquefaction at a Levee Site in Japan Using a Time-Frequency Based Analysis

Paolo Zimmaro[1,2](✉), Maria Giovanna Durante[1], and Ernesto Ausilio[1]

[1] University of Calabria, Arcavacata di Rende, Italy
paolo.zimmaro@unical.it
[2] University of California, Los Angeles, Los Angeles, USA

Abstract. Earthquake-induced liquefaction phenomena are among the most damaging geotechnical effects during earthquake events. Current liquefaction triggering models are based on case histories comprising three pieces of information: (1) ground motion intensity at the site, (2) liquefaction manifestation observation (or lack thereof), and (3) co-located geotechnical field investigation data. In recent years, new case history types, characterized by the presence of liquefaction observations and co-located ground motion recording stations, were identified. The availability of such data enabled new analysis strategies based on time-frequency analyses of ground motions at liquefaction sites through the use of the Stockwell transform. This study presents the implementation of time-frequency analyses to ground motions recorded during the 2011 **M**9.1 Tohoku earthquake in Japan at a well-documented instrumented levee site named Yamazaki. Using this procedure, the timing of liquefaction and possible critical layers were identified, demonstrating the effectiveness of this technique in providing insights into liquefaction case histories.

Keywords: Liquefaction · Stockwell Transform · Instrumented Levee

1 Introduction

Earthquake-induced liquefaction is a geotechnical phenomenon that can produce damage to structures, infrastructure, and geotechnical systems. Earthquake-induced liquefaction occurs in saturated granular materials (typically sands with low relative density characterized by contractive behavior) when shaken hard enough. Current liquefaction prediction models rely upon field observations and/or laboratory test results and are based on so-called "case histories." A case history is defined as the intersection between three pieces of information: (1) ground motion intensity at the site, (2) liquefaction manifestation observation (or lack thereof), and (3) co-located geotechnical field investigation data. Such case histories, sometimes in combination with laboratory test data are used in stress-based frameworks aiming at evaluating liquefaction susceptibility, triggering, and post-liquefaction consequences.

A. Ferrari et al. (Eds.): CNRIG 2023, SSGG, pp. 578–584, 2023.
https://doi.org/10.1007/978-3-031-34761-0_70

A recent research effort conducted globally, focus on development of a high-quality, open source, and community database of liquefaction case histories and related models: the Next-Generation Liquefaction (NGL) project [1, 2]. As part of this research effort, legacy (i.e., used in previous liquefaction models [3–8]) and new (i.e., not used in existing liquefaction models) case histories were identified, collected, compiled, and publish.

Legacy triggering models heavily rely upon post-earthquake observations and inherently link liquefaction surface manifestations to liquefaction triggering. However, when surface manifestations occur, such occurrences may be related to excess pore-pressure build up, but not necessarily to liquefaction triggering. In such occurrences, pore pressure ratios (the ratio between earthquake-induced excess pore pressure and the static effective vertical stress) may reach values lower than one (typically defining liquefaction triggering). On the other hand, when liquefaction surface manifestations do not occur, liquefaction may have been triggered at deeper layers and/or surface evidences impeded by non-liquefiable crusts or due to thinly interbedded layers. A novel category of case histories may help solving such issues and reconcile potential model-data inconsistencies. Such new case histories are characterized by a fourth piece of information: the availability of co-located recording stations at liquefaction sites. Greenfield (2017) collected and compiled such sites. Analyzing this data using advanced time-frequency analysis procedures, based on the Stockwell transform, Kramer et al. [9], Greenfield [10], and Ozener et al. [11], identified the timing of liquefaction at these sites. They also demonstrated that using this approach it is possible to identify "true" liquefaction triggering case histories even in the absence of surface manifestations.

In this paper, we analyze recorded ground motions recorded during the 2011 **M**9.1 Tohoku earthquake at levee site in Japan. This levee was instrumented with a downhole array of recordings, allowing for a thorough data analysis. Post-earthquake observations at this site show that the levee was damaged as a result of liquefaction-related phenomena. The analysis shown in this paper is performed using a time-frequency based analysis with the aim of revealing the timing of liquefaction and identifying the portion of the levee that underwent liquefaction.

2 Performance of River-Protection Levees During the 2011 M9.1 Earthquake in Japan

The 2011 **M**9.1 Tohoku earthquake was one of the most damaging earthquake in human history. It produced damage to a large stock of structures and infrastructure systems, including flood-control levees across the country, and triggered a large tsunami [12]. Building upon available data from the Ministry of Land, Infrastructure, Transport and Tourism (MLIT) and the National Institute for Land and Infrastructure Management (NILIM) in Japan, Kataoka et al. [13] and Zimmaro et al. [14] published an open-source database comprising recorded ground motions from recent earthquakes, geotechnical field investigation data, and geometry and layering information of seven instrumented levee arrays in Japan. These levees are typically equipped with ground motion recording stations at the crest, the bedrock, and at a free-field location on the land-side. Several of these levees were damaged by the 2011 M9.1 Tohoku earthquake and/or by the subsequent tsunami triggered by this event. Some of them are believed to have experienced liquefaction-related damage. Among them, the Yamazaki, Nakashimo, and Kozuka levee sites, were the better-characterized liquefaction case histories [14]. All of them are now part of the NGL database.

In this study, we selected the Yamazaki levee site (Fig. 1), to identify the timing of liquefaction and getting insights into its overall liquefaction performance. This levee is located along the Yoshida river, 15.8 km upstream from the Naruse river mouth [14]. The levee fill comprises sandy soils, which are founded on clayey and sandy layers. It is equipped with three ground motion recording stations: one at the top of the crest and two in the land-side bench, one at the surface, another one within the bedrock at a depth of 22 m). At the time of field investigation (April 2004), the ground water table was approximately 2 m above the base of the levee fill. Such shallow ground water level, would likely produce saturated sandy layers within the fill and in the foundation materials, making it a site susceptible to liquefaction. Following the 2011 **M**9.1 Tohoku earthquake, liquefaction was observed at this site (Fig. 2 and [14] and references therein). The levee experienced, on average, 59 cm of post-event settlements. Diffuse cracking was observed on both sides of the levee. The quantity and quality of data available at this site enabled a complete ground motion data analysis by means of time-frequency based approaches.

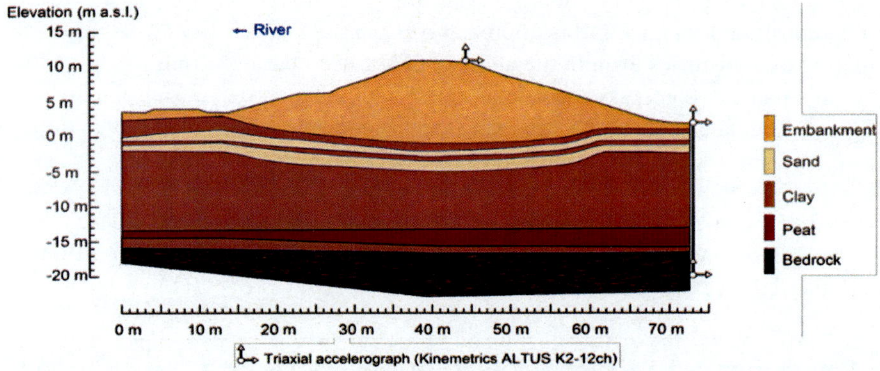

Fig. 1. Cross section and available instrumentation at the Yamazaki levee site (adapted from [14]).

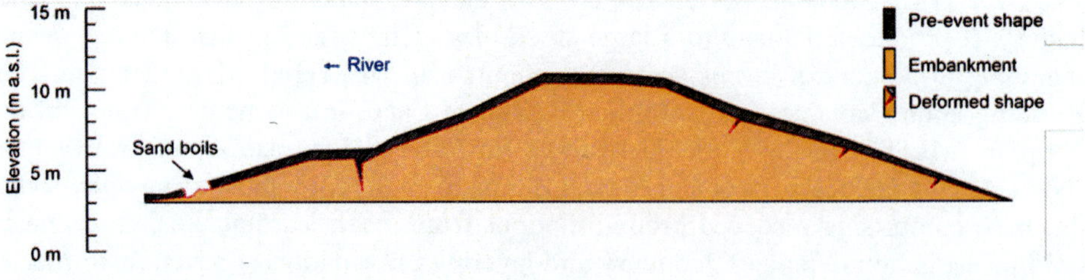

Fig. 2. Sketch of the deformed shape of the Yamazaki levee site following the 2011 **M**9.1 Tohoku earthquake (adapted from [14]).

3 Use of Time-Frequency Analysis Procedures to Identify the Timing of Liquefaction Triggering

When liquefaction triggers, the material involved in this phenomenon experiences substantial softening. As a result of this effect, its ability to transmit high-frequency waves is compromised (e.g., [9]). Such specific feature, makes it attractive trying to identify liquefaction triggering looking at the time-frequency content of motions recorded at liquefaction sites. In recent years, the use of the normalized Stockwell power spectrum (i.e. the spectral amplitude is divided by the peak Stockwell amplitude at each time increment) of the horizontal components of a recorded acceleration time series was shown to be a promising tool to identify the timing of liquefaction [9–11]. This technique was also recently used to identify liquefaction sites where observation did not pervasively allow to discern between liquefaction-related damage and consequences of seismic compression [15]. Such studies show that when liquefaction triggers, the frequency content of the recorded motion in the horizontal direction suddenly drops. This very moment (when there is an abrupt change in the frequency content of the horizontal components) can be identified and deemed to be the timing of liquefaction initiation. When analyzing time-frequency motions at sites potentially experiencing liquefaction, such frequency content shift must be decoupled from the following phenomena (that produce too variations in the frequency content of the motion): (1) S-wave arrival, (2) surface wave arrival, and (3) site effects. Whereas such additional phenomena trigger frequency content variations in both: the horizontal and vertical components, liquefaction triggering only affects the horizontal components. As a result, comparing Stockwell spectra of the horizontal and vertical components can be a useful tool to identify difference contributors of frequency content shifts [11, 15]. Such statement is especially true in the far-field (where, theoretically, the vertical component of the recorded motion mainly comprises P- and surface-waves, [15]).

4 Results

The approach adopted in this study to identify liquefaction initiation (and related relevant features) at the Yamazaki levee site following the 2011 **M**9.1 Tohoku earthquake started by calculating the Stockwell transform for the horizontal and vertical component ground motions at the crest, bench, and bench (−22 m) downhole recordings (Fig. 3). We then looked for any sudden drops in the frequency content of these recordings that were present in the horizontal components, but not in the vertical components. Finally, we compared, side-by-side Stockwell time-frequency plots for recordings at different locations within the levee. By following these steps we identified the timing of liquefaction occurring after 72.5 s from the beginning of the recording. This is evident in Fig. 3 for both crest horizontal recordings. Such frequency content change is not present in the vertical component of the recording at the same location. Even though such sudden drop in the frequency content is evident in both horizontal components, only the NS component is shown here for brevity. The analysis of all time-frequency plots also shows that signs of liquefaction are not present in the recordings at the bench and at the downhole station. This evidence suggests that liquefaction occurred within the saturated portion of the embankment (and perhaps, arguably, in the most surficial layers of the foundation, but only in the central portion of the levee).

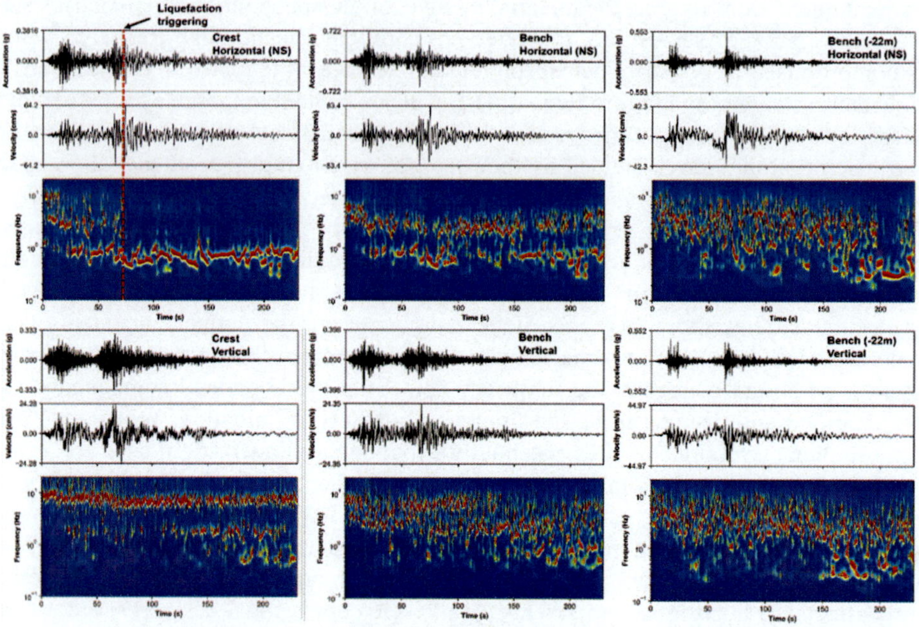

Fig. 3. Acceleration and velocity time series and normalized Stockwell power spectra for the horizontal and vertical components of the 2011 **M**9.1 Tohoku earthquake recorded at the crest, bench, and bench (−22 m) within the Yamazaki levee site in Japan.

To gain additional insights into this specific detail (i.e., whether foundation layers in the central portion of the levee liquefied along with the saturated portion of the embankment), a simple deterministic liquefaction triggering analysis was performed. In this analysis the Standard Penetration Test (SPT)-based method by Boulanger and Idriss (2012) was used, departing from available SPT blow counts at a boring located in the central portion of the levee (Fig. 4). No laboratory information is available at this site. As a result, the only information that can be used to inform this analysis is based on field investigation results.

The factor of safety against liquefaction (FS) was calculated for a total of six data points, all located in saturated sandy layers. The depth to groundwater was assumed equal to 9.2 m below the levee crest. Such value is based on pre-earthquake groundwater elevation data acquired at the time of the boring (April 2004). The results of this analysis are shown in Fig. 4 that reports FS against depth, along with the profile of corrected blow counts $(N_1)_{60,cs}$, the location of the boring, and the adopted groundwater table elevation. Figure 4 clearly shows that FS is always substantially smaller than one for all analyzed saturated sandy layers. These results, along with the outcomes of the time-frequency analysis discussed above, provide convincing evidences that, in addition to the saturated zone of the embankment, the foundation layers in the central portion of the levee liquefied too.

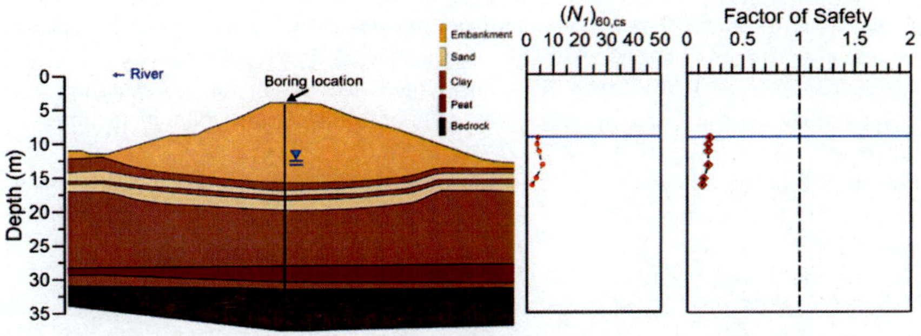

Fig. 4. Cross section of the Yamazaki levee site, location of the analyzed boring, and groundwater table adopted in this study, along with corrected blow counts, and factor of safety against liquefaction calculated with the Boulanger and Idriss (2012) triggering method.

5 Conclusions

Recent pioneering studies [9–11] indicate that the timing of liquefaction can be identified using advanced time-frequency analyses based on the normalized Stockwell transform. In this study, such hypothesis is tested against recordings of the 2011 **M**9.1 Tohoku earthquake at the Yamazaki levee site in Japan. This levee site is present in the NGL database [1, 2] and was compiled as part of a broader international collaboration [13, 14]. Liquefaction surface manifestations were observed at this site following this event. A thorough analysis of time-frequency plots showed that liquefaction occurred 72.5 s after the beginning of the earthquake and that the liquefied layer is located in the saturated portion of the levee fill. The combination of the time-frequency analysis and a simple deterministic evaluation of the factor of safety against liquefaction based on available SPT data from a boring drilled within the levee (in its central part), provides convincing evidences that liquefaction also occurred in the surficial portion of the foundation materials below the crest. This study confirms the effectiveness of such analyses at liquefaction site and suggests that the addition of permanent recordings stations at potentially-liquefiable sites in earthquake-prone zones, including Italy, may help improving current liquefaction models.

References

1. Zimmaro, P., et al.: Next-Generation Liquefaction Database. Next-Generation Liquefaction Consortium (2019). https://doi.org/10.21222/C2J040
2. Brandenberg, S.J., et al.: Next generation liquefaction database. Earthq. Spectra **36**, 939–959 (2020)
3. Boulanger, R.W., Idriss, I.M.: Probabilistic standard penetration test-based liquefaction triggering procedure. J. Geotech. Geoenviron. Eng. **138**, 1185–1195 (2012)
4. Boulanger, R.W., Idriss, I.M.: CPT-based liquefaction triggering procedure. J. Geotech. Geoenviron. Eng. **142**(2), 04015065 (2016)
5. Çetin, K.O., et al.: SPT-based probabilistic and deterministic assessment of seismic soil liquefaction potential. J. Geotech. Geoenviron. Eng. **130**(12), 1314–1340 (2004)

6. Çetin, K.O., et al.: SPT-based probabilistic and deterministic assessment of seismic soil liquefaction triggering hazard. Soil Dyn. Earthq. Eng. **115**, 698–709 (2018)
7. Moss, R.E.S., Seed, R.B., Kayen, R.E., Stewart, J.P., Der Kiureghian, A., Çetin, K.O.: CPT-based probabilistic and deterministic assessment of in situ seismic soil liquefaction potential. J. Geotech. Geoenviron. Eng. **132**, 1032–1051 (2006)
8. Kayen, R.E., et al.: Shear-wave velocity–based probabilistic and deterministic assessment of seismic soil liquefaction potential. J. Geotech. Geoenviron. Eng. **139**, 407–419 (2013)
9. Kramer, S.L., Sideras, S.S., Greenfield, M.W.: The timing of liquefaction and its utility in liquefaction hazard evaluation. Soil Dyn. Earthq. Eng. **91**, 133–146 (2016)
10. Greenfield, M.W.: Effects of long-duration ground motions on liquefaction hazards. Ph.D. Dissertation. University of Washington, Seattle, WA (2017)
11. Ozener, P.T., Greenfield, M.W., Sideras, S.S., Kramer, S.L.: Identification of time of liquefaction triggering. Soil Dyn. Earthq. Eng. **128**, 105895 (2020)
12. National Institute for Land and Infrastructure Management, Japan (NILIM): Quick report on damage to infrastructures by the 2011 off the Pacific coast of Tohoku earthquake. Technical Note of National Institute for Land and Infrastructure Management, No. 646; Technical Note of Public Works Research Institute, No. 4202, p. 496 (2012). http://www.nilim.go.jp/lab/bcg/siryou/tnn/tnn0646pdf/ks0646.pdf. Accessed 22 Jan 2022
13. Kataoka, S., et al.: A database of seismic records at instrumented levee sites in Japan. Earthquake Disaster Management Division, National Institute for Land and Infrastructure Management (2019). https://doi.org/10.21222/C2TC95
14. Zimmaro, P., et al.: Database on seismic response of flood control levees. Earthq. Spectra **36**, 924–938 (2019)
15. Cabas, A., et al.: Geotechnical lessons from the Mw 7.1 2018 Anchorage Alaska earthquake. Earthq. Spectra **37**, 2372–2399 (2021)

Probabilistic Approaches to Data Analysis and Performance Assessment

Reliability-Based Evaluation of the Stability of Underground Cavities in Naples

Filomena de Silva(✉), Massimo Ramondini, and Alessandro Flora

University of Naples Federico II, Naples, Italy
filomena.desilva@unina.it

Abstract. This paper reports stability charts for the roof of anthropic cavities expressing the ratio between the minimum rock uniaxial compression strength ensuring the equilibrium and the acting vertical stress as a function of the cavity span normalized to the roof thickness. Several numerical analyses were carried out with the 2D finite difference Code FLAC, considering the uncertainties due to the spatial variability of the rock uniaxial compression strength and the presence of discontinuities. In the frame of the reliability-based assessment, curves associated with three different probabilities of failure were calculated to delimit four different stability zones. The results showed that the safety margin is slightly influenced by the variability of the rock strength, but largely worsened by the presence of joints. The charts were then applied to several cavities in Naples, whose depth and geometric features were detected during an extensive survey in 2000.

Keywords: underground cavities · spatially variable resistance · rock joint · factor of safety · reliability-based approach

1 Introduction

Local subsoil conditions influence the land transformation, affecting interventions to raise buildings and upgrade infrastructures in densely urbanized area. The stability assessment of the aboveground structures is complicated when networks of shallow underground cavities cross the subsoil, because the structural safety is conditioned to the cavity safety.

Closed form solutions have been derived to evaluate the stability of the roof of cavities in homogeneous rock [1, 2]. The main limit of such solutions is the simplified hypotheses on the mechanical response of the roof.

More recently, stability charts were derived by exploiting the results of more complex numerical models [3–7]. The latter approach allows to consider regional and even city scale "local" features of rock and cavities which control the stability, such as [3] a variable height of rock cover and span of the opening, the presence, the orientation, and the frequency of rock joints, the shear strength of discontinuities and intact rock, the applied load and the ground water conditions.

The aim of this study is to upgrade the charts proposed by Evangelista et al., (2000) [1] for the cavities dug in the tuff in the city of Naples (Italy), by considering the spatial variability of the rock strength and the presence of horizontal joints.

A. Ferrari et al. (Eds.): CNRIG 2023, SSGG, pp. 587–594, 2023.
https://doi.org/10.1007/978-3-031-34761-0_71

This is motivated by the fact that tuff is inhomogeneous and often crossed by horizontal and vertical discontinuities generated by the gas leakage during the formation process [8]. Such aspects are expected to influence the stability of the cavity roof of the whole Campania region characterized by the largest number of sinkholes in Italy, 246 out of 652 [9].

2 Method of Analysis

Figure 1a shows a schematic stability chart of roof cavity based on a power law function [1, 6, 7]. It describes the variation of the minimum value of the uniaxial compressive strength normalized to the acting vertical stress, σ_{cmin}/σ_v, with the ratio between the span and the thickness of the roof, L/t. Figure 1a shows how such curve can be simply used to estimate the factor of safety (FoS $= \sigma_{cc}/\sigma_{cmin}$) of an existing cavity for which the geometry, L_c/t_c, the vertical lithostatic stress in the roof, σ_{vc}, and the uniaxial compression strength σ_{cc} are known.

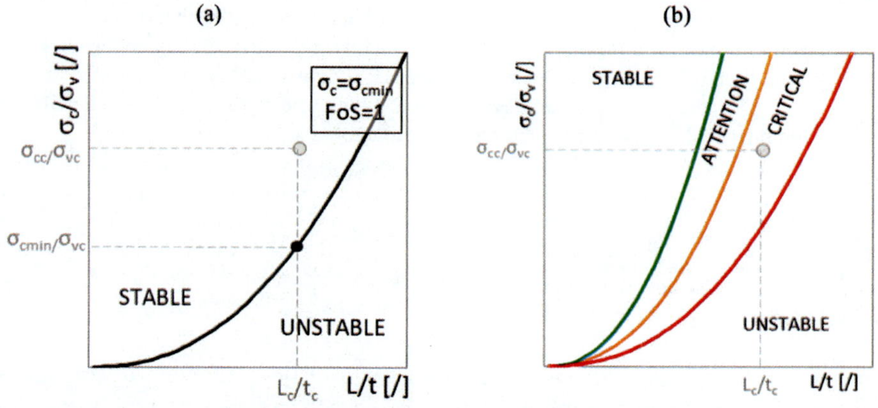

Fig. 1. Scheme of the deterministic (a) and reliability-based (b) stability charts. The geometrical definition of the cavity roof factor of safety (FoS $= \sigma_{cc}/\sigma_{cmin}$) is shown in (a).

When the uncertainties on the factors influencing the stability of the roof cavity are somehow introduced, the inadequacy of a cavity to support the applied loads can be quantified through the probability that σ_{cc}/σ_{vc} is lower than $\sigma_{cmin}/\sigma_{vc}$ conditioned to L_c/t_c. This probability can be expressed as:

$$p_f = p[\sigma_{cc}/\sigma_{vc} < \sigma_{cmin}/\sigma_{vc}|L_c/t_c] = 1 - p\left[\frac{ln\left(\frac{\sigma_{cc}}{\sigma_{vc}}\right) - \mu}{SD}\right] \tag{1}$$

where μ and SD are the mean and the standard deviation of the distribution of σ_{cmin}/σ_v. The mean, μ of the distribution in Eq. (1) varies with L/t according to Eq. (2):

$$\mu = lna + bln\left(\frac{L}{t}\right) \tag{2}$$

in which *a* and *b* are the parameters of a linear law linking σ_{cmin}/σ_v with L/t in a log-log scale.

Equation 1 can be exploited to derive curves characterized by the same p_f to be assumed as thresholds for the stability assessment of the cavity roof. The selected values of the probabilities depend more on the decision makers than on technical judgement. In this study, probabilities of 1%, 10% and 50% were selected just as an example of application. They bound four zones in the stability chart (Fig. 1b) that will be indicated in the following as:

STABLE ZONE: the cavity roof is expected to be safe;
ATTENTION ZONE: the cavity roof needs to be monitored;
CRITICAL ZONE: the stability of the cavity roof needs to be assessed by more refined tools;
UNSTABLE ZONE: the stability of the cavity roof needs to be assessed by more refined tools and interventions may be required to ensure the stability in the meantime.

3 Application to the Cavities Underground Naples

3.1 Description of the Main Features

The shallowest subsoil of Naples is characterized for its largest part by pyroclastic rock and soil, namely tuff in the lower part and ashes on top.

Tuff is a soft rock, characterized by a uniaxial compression strength σ_c that varies with the rock porosity and the included lithic fragments. Since in many areas of the town tuff outcrops or is at a rather shallow depth, it has been excavated since very old times. Nowadays a dense network of more than 900 cavities is known underneath the city, for a total excavated volume of tuff of about 7106 m^3, and new cavities are currently found, indicating an even larger underground world [10–12].

Figure 2a shows a schematic cross section of a typical cave. A layer of cohesionless (mostly pyroclastic) soils is typically on the top of the tuff mass with a thickness p variable between 5 and 25 m. The walls are slightly inclined. The roof is close to be flat and can be crossed by horizontal syngeneic joints, locally called *suoli*.

Figure 2b shows some statistics for 123 cavities inspected during an extensive survey in 2000. The span length L mostly ranges from 2 m to 8 m, but widths as large as 12 m have been measured. The thickness of the tuff on the roof is typically low, between 1 m and 6 m, with a few cases for which even lower values were found. The ratio L/t mostly varies between 1 and 6.

3.2 Performed Analyses

Eight plane strain numerical models associated with the geometric features mostly recurrent in the Neapolitan cavities have been generated through the finite difference code FLAC 2D (Itasca, 2011). The geometric features are shown in Fig. 3 a, while more details on the numerical models are reported by [13].

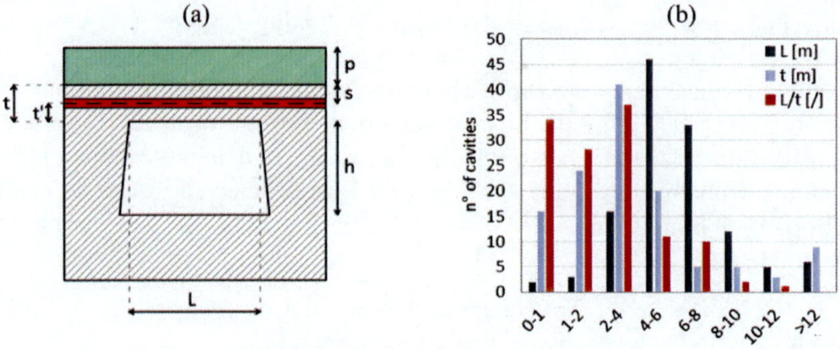

Fig. 2. Typical geometric layout of a Neapolitan cavity (a) and statistics of the available geometric data (b).

Three sets of analyses have been performed to assess the roof safety conditions with:

1) a constant uniaxial compression strength of the tuff;
2) a spatially variable uniaxial compression strength of the tuff;
3) a constant uniaxial compression strength of the tuff and a horizontal joint in the roof.

The mechanical response of the intact rock was simulated through an elastic perfectly plastic behavior with a Mohr-Coulomb failure criterion, defined by a cohesion $c = 866$ kPa and a friction angle $\varphi = 30°$, leading to a $\sigma_c = 3$ MPa.

The spatial variability of the uniaxial compression strength of the tuff (2nd set) was implemented through an *ad-hoc* FLAC routine which assigns randomly to each mesh element a value of the cohesion consistently with a Gaussian distribution characterized by a mean equal to 866 kPa and a standard deviation equal to 240 kPa. Three hundred analyses were performed, i. e. fifty for each geometric scheme, by changing the random assignment of the cohesion to the mesh elements.

For the 3rd set, each cavity roof was assumed to be crossed by a horizontal joint at a distance, s, from the tuff top equal to 0.25 t, 0.50 t and 0.75 t. The Ubiquitous joints model was adopted to simulate the joint behavior. It is a Mohr-Coulomb based model which distributes, within the part of rock mass of interest, planes along which the mechanical properties are reduced. The inclination of such planes was set horizontal, to simulate the sub-horizontal discontinuity of interest. Following the approach proposed by [14], the parameters of the Mohr-Coulomb criterion were calibrated to approximate the [15] empirical equation. The effect of the roughness and weathering were investigated by varying JRC among 5, 10 and 20 and JCS between $\sigma_c = 3$ MPa and $0.50\sigma_c = 1.5$ MPa. The resulting equivalent Mohr- Coulomb criteria are plotted in Fig. 3 b. The third set includes one hundred and forty-four analyses resulting from the eight geometric schemes, the three different locations of the discontinuity (see Fig. 3) and the six hypotheses on the joint resistance.

The value of σ_{cmin} was searched for each analysis through the strength reduction approach implemented in FLAC, which reduces the initial strength parameters until the equilibrium of the cavity under gravity loads is not satisfied.

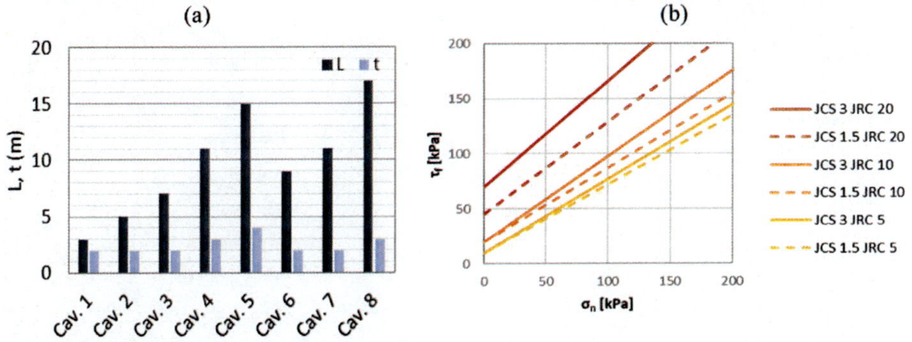

Fig. 3. Geometric features of the eight analyzed cavity schemes (a) and equivalent Mohr-Coulomb model adopted for the joint (b).

3.3 Results

Figure 4 shows the values of σ_{cmin}/σ_V against L/t for the analyzed cases grouped into classes, characterized by uniform rock strength, variable strength or constant strength and joint at 0.5 t. For a given value of L/t, the ratio σ_{cmin}/σ_V increases when the uniaxial compression strength of the rock is assumed variable, and the discontinuity is introduced, meaning that the stability conditions are more critical. Even though the results for joint at 0.25 t and 0.75 t are not reported for sake of brevity, the effect is increasingly more detrimental as the joint approaches the bottom of the roof, due to the reduction of the resistant section.

For joint at 0.5 t, the results are close to be ordered according to the joint resistance (see Fig. 4), because the contribution of the uppermost zone strictly depends on the ability of the joint to transfer the stresses, and such ability is bounded by its resistance.

The values of σ_{cmin}/σ_V obtained for the third set of analyses (see Sect. 3.2) have been plotted against L/t in the log-log plot and then interpolated through a linear regression. The slope and the intercept of the linear regression are therefore ln(a) and b and control the evolution of μ with L/t, as described in Sect. 2. Table 1 lists the resulting values, together with the standard deviation, SD. Note that a instead of ln(a) is provided. The effect of the tuff strength variability appears of minor importance, but it was nevertheless considered by calculating SD as follows:

$$SD = \sqrt{SD_s^2 + SD_j^2} \tag{3}$$

where SDj is the standard deviation of the results of numerical models with the joint, and SDs is the standard deviation of the dataset in which the strength is variable.

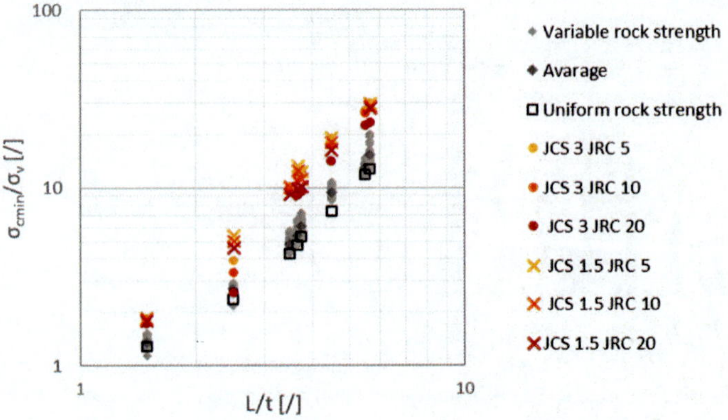

Fig. 4. Results of the analyses considering the variation of the rock resistance and the presence of joint at 0.5 t.

Table 1. Coefficients of the linear regression interpolating all the analyses with the joint.

a	b	SD
0.8797	2.0166	0.43939

Figure 5 a shows the resulting stability curves associated with the three selected probabilities of failure. The curve corresponding to the constant strength is reported in black for comparison. The zone of instability is significantly enlarged when joints and rock variable strength are accounted for.

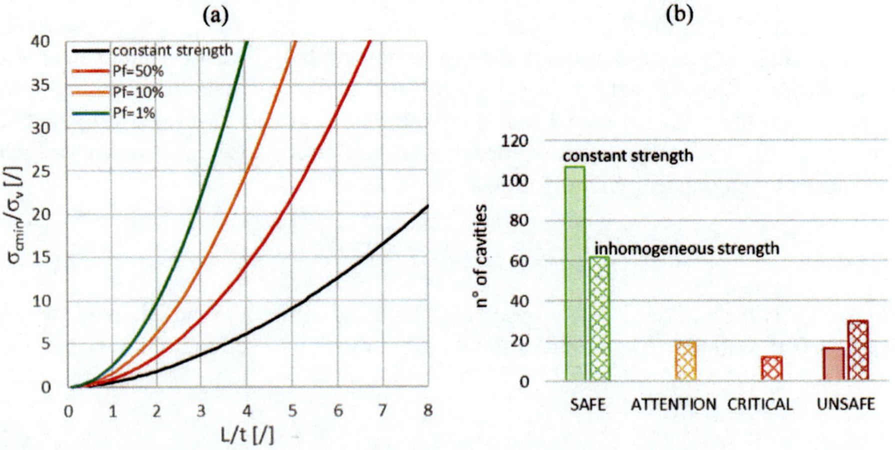

Fig. 5. Stability charts obtained in this study (a) and statistics on the safety of the roof of the Neapolitan cavities (b).

The stability chart was applied to the dataset of cavities in Naples described in Sect. 3.1. Figure 5 b shows the number of cavities falling in the four different safety zones. Almost half of the cavities are stable, while the others are not sufficiently safe with about the 25% fully unstable. Please note that the application of the constant strength curve would provide only 16 out of 123 unsafe cavities (colorful histograms in Fig. 5 b).

4 Conclusions

The paper investigated how the safety of the roof of cavities dug in soft rocks is influenced by the spatial variability of the uniaxial compression strength and by the presence of horizontal discontinuities in the rock mass, usually overlooked or considered in a simplified way in previous studies on the topic. Reliability-based stability curves have been obtained considering different probabilities of failure for the possible random variability of rock strength and horizontal joint location in the roof.

The outcome of this study reveals that the hypothesis of intact rock with a constant resistance can be dangerously unconservative and the possible existence of discontinuities cannot be neglected, even in the preliminary assessment of safety conditions, if the risk of failure has to be kept low. The simple charts herein proposed are well suited to this aim and represent a simple yet sound tool for a first investigation at the urban scale.

Acknowledgements. The authors gratefully acknowledge CUGRI (interuniversity centre between the Universities of Salerno and Napoli Federico II for the prediction and prevention of natural risks) for sharing the dataset of geometric features of the cavities of the city of Napoli. The paper has been written in the framework of the activities carried out to exploit the PRIN project GIANO (Geo-rIsks Assessment and mitigatioN for the prOtection of cultural heritage).

References

1. Evangelista, A., Feola, A., Flora, A., Lirer, S., Maiorano, R.: Numerical anlaysis of roof failure mechanisms of cavities in a soft rock. In: ISRM International Symposium, Melbourne, Australia (2000)
2. Fraldi, M., Guarracino, F.: Limit analysis of collapse mechanisms in cavities and tunnels according to Hoek-Brown failure criterion. Int. J. Rock Mech. Min. Sci. **46**(4), 665–673 (2009)
3. Suchowerska, A.M., Merifield, R.S., Carter, J.P., Clausen, J.: Prediction of underground cavity roof collapse using the Hoek-Brown failure criterion. Comput. Geotech. **44**, 93–103 (2012)
4. Hatzor, Y.H., Wainshtein, I., Mazor, D.B.: Stability of shallow karstic caverns in blocky rock masses. Int. J. Rock Mech. Min. Sci. **47**(8), 1289–1303 (2015)
5. Zimbardo, M., Cannone, C., Ercoli, L., Nocilla, A.: A risk assessment proposal for underground cavities in Hard Soils-Soft Rocks. Int. J. Rock Mech. Min. Sci. **103**, 43–54 (2018)
6. Perrotti, M., Lollino, P., Fazio, N., Pisano, L.: Finite element-based stability charts for underground cavities in soft calcarenites. Int. J. Geomech. **18**(7), 04018071 (2016)
7. de Silva, F., Lusi, T., Ruotolo, M., Flora, A., Ramondini, M., Urciuoli, G.: A simplified approach to assess the stability of tuff cavities accounting for the spatial variability of the shear strength and the presence of joints. In: Proceedings of the Third International Symposium on Geotechnical Engineering for the Preservation of Monuments and Historic Sites, 22–24 June 2022, Napoli, Italy. CRC Press (2022)

8. Evangelista, A., Pellegrino, F.: Caratteristiche geotecniche di alcune rocce tenere italiane. Atti del terzo ciclo di Conferenze di Meccanica e Ingegneria delle Rocce, MIR90, Torino (1990)
9. Parise, M., Vennari, C.: A chronological catalogue of sinkholes in Italy: The first step toward a real evaluation of the sinkhole hazard. In: Proceedings of the 13th Multidisciplinary Conference on Sinkholes and the Engineering and Environmental Impacts of Karst. National Cave and Karst Research Institute, Carlsbad, NM (2013)
10. Evangelista, A., Flora, A., Lirer, S., de Sanctis, F., Lombardi, G.: Studi ed interventi per la tutela di un patrimonio sotterraneo: l'esempio delle cavità di Napoli. In: Proceedings of the XXI Convegno Nazionale di Geotecnica. Patron editore. L'Aquila, 11–14 settembre 2002 (2002a)
11. Evangelista, A., Pellegrino, A., Ramondini, M.: Problematiche connesse alla presenza di cavità nel sottosuolo in un intervento edilizio in ambiente urbano. In: Proceedings of the XXI Convegno Nazionale di Geotecnica su "Opere in ambiente urbano". Patron editore. L'Aquila, 11–14 settembre 2002 (2002b)
12. Scotto di Santolo, A., Forte, G., Di Falco, M., Santo, A.: Sinkhole risk assessment in the metropolitan area of Napoli, Italy. Procedia Eng. **158**, 458–463 (2016)
13. de Silva, F., Lusi, T., Ruotolo, M., Ramondini, M., Flora, A.: Reliability-based roof stability charts for cavities in heterogeneous jointed rock masses. Rivista Italiana di Geotecnica (2023). (under review)
14. Boldini, D., Guido, G.L., Margottini, C., Spizzichino, D.: Stability analysis of a large-volume block in the historical rock-cut city of Vardzia (Georgia). Rock Mech. Rock Eng. **51**(1), 341–349 (2017). https://doi.org/10.1007/s00603-017-1299-7
15. Barton, N.: Review of a shear strength criterion for rock joints. Eng. Geol. **7**(4), 287–332 (1973)

Seismic Hazard Assessment by the Application of a Synthetic Damage-Constrained Parameter

Francesco Castelli[1], Sebastiano D'Amico[2], Salvatore Grasso[3], Valentina Lentini[1], Maria Rossella Massimino[3], and Maria Stella Vanessa Sammito[3](✉)

[1] University "Kore" of Enna, 94100 Enna, Italy
[2] University of Malta, Msida MSD 1306, Malta
[3] University of Catania, 95125 Catania, Italy
mariastella.sammito@phd.unict.it

Abstract. Seismic Microzonation (MS) can be defined as a multidisciplinary process aimed at identifying and mapping the subsoil local response and the instability phenomena in an area. In Italy, the Department of Civil Protection provided national guidelines and standards in order to ensure the homogeneity of the products. According to the Italian guidelines, three levels of detail can be considered (I, II and III) with increasing complexity.

The paper deals with the 3rd level seismic microzonation (MS3) for an area located in the east part of Sicily (Italy). The main aim of MS3 studies is to draw Seismic Microzonation Maps using Amplification Factors (AFj), i.e. synthetic indicators representative of the seismic amplification.

However, AFj are non-dimensional parameters. Therefore, the use of these synthetic indicators provides seismic hazard maps in terms of relative values, not allowing a direct comparison of seismic hazards between different areas. In light of this, a new methodology was applied in order to develop probabilistic seismic hazard maps by means of a synthetic damage-constrained parameter that provides an absolute ranking of seismic hazard.

The produced maps are very important tools that allow the definition of attention zones in order to assist engineers and planners in preventing and limiting seismic risk and prioritizing interventions.

Keywords: Site Characterization · Seismic Microzonation Studies · Amplification Factors · Synthetic damage-constrained parameter

1 Introduction

In this paper MS studies were carried out for the municipality of Riposto located on the eastern flank of Etna Vulcano (see Fig. 1) that is one of the areas in Italy with the highest seismic risk [1, 2].

A deep investigation campaign was performed in order to define the geological and geotechnical models of the soil for each Seismically Homogeneous Microzone (SHM). Another important aspect was the definition of a set of seven input motions to be used

A. Ferrari et al. (Eds.): CNRIG 2023, SSGG, pp. 595–602, 2023.
https://doi.org/10.1007/978-3-031-34761-0_72

in numerical analyses based on the reference response spectra provided by the present Italian seismic code [3].

Numerical analyses were performed using the 1D equivalent-linear site response program STRATA [4]. Amplification factors of spectral acceleration were obtained according to the guidelines and standards for microzonation studies [5]. Finally, in order to define an absolute ranking of seismic hazard, a new methodology was applied by means of a synthetic damage-constrained parameters, H_{SM}, whose classification is linked with the European Macroseismic intensity Scale [6, 7].

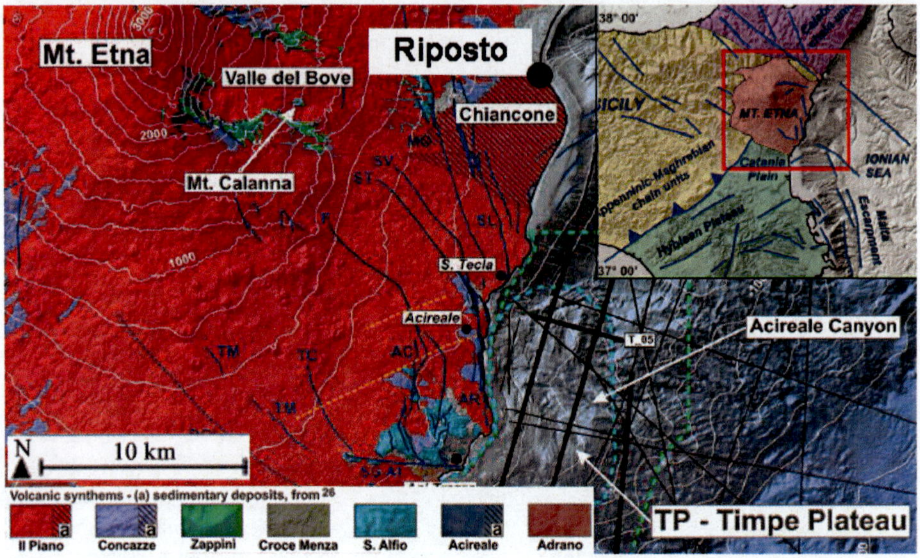

Fig. 1. Etna Vulcano and geophysical dataset with the indication of the study area (From [8], modified).

2 Investigation Campaign and Definition of the Subsoil Models

The definition of SHMs is essential for the construction of the subsoil models. Two SHMs were identified in the municipal area of Riposto (see Fig. 2) based on the subsoil characteristics: the main one covering almost all of the territory (SHM1) and the second one in a small southern zone (SHM2).

To evaluate the geotechnical characteristics of soil, laboratory and in situ tests were performed consisting of n. 19 Boreholes, n. 1 Down Hole Tests (DH), n. 10 Multichannel Array Surface Wave Tests (MASW), n. 10 Electrical Tomography, n. 21 Horizontal to Vertical Spectral Ratio Tests (HVSR) and n. 6 Direct Shear Tests (DST). As presented in Fig. 1, the municipal area of Riposto is mainly composed of the "Chiancone" formation (lithological unit_GW) that is the largest volcaniclastic sequence of the volcano. This formation presents a degraded and altered surface layer of about 1–1.5 m. For the SHM2, a cover lithoid (lithological unit_LC) overlays the "Chiancone" deposits. The values of shear wave velocity, V_S, derived from the Down-Hole test performed in zone of the touristic port (SHM1), are plotted against depth in Fig. 3. From the geophysical tests the equivalent shear wave velocities ($V_{S,eq}$) provide a soil class B ($V_{S,eq}$ = 180–360 m/s) according to the Italian seismic code [3]. To describe the decay of the shear modulus and the increase in damping ratio as a function of the shear strain, dynamic curves from the literature were assumed [9]. Moreover, the subsoil model requires the knowledge of the bedrock that was fixed at the depth of 39 m based on linear interpolation of the shear waves profile.

3 Selection of Input Ground Motions

The application ESM_REXELweb [10] allows to search for suites of waveforms, from the Engineering Strong-Motion Database (ESM) [11], compatible to the reference spectrum in a defined range of periods according to the Italian seismic code [3]. A set of seven accelerograms was selected in a way that their average is in an interval between 10% (lower threshold) and 30% (upper threshold) of the reference spectrum. The range of periods is equal to 0.1 s–1.1 s which is representative of the vibration period of the structures target in seismic microzonation [12].

Figure 4 reports the suites of waveforms compatible to the reference for Italian spectrum (Soil Class A) considering a return period $T_R = 475$ years, a minimum event magnitude $M_{min} = 3.5$, a maximum event magnitude $M_{max} = 6.2$, a minimum epicentral distance $R_{min}[km] = 0$ and a maximum epicentral distance $R_{max}[km] = 20$.

4 Evaluation of Amplification Factors

The average response spectrum obtained by setting a structural damping of 5% from numerical analyses is reported in Fig. 5 for the main microzone SHM1 considering the mean of the values calculated using the set of seven input ground motions. The elastic response spectrum provided by the Italian seismic code [3] is also shown for comparison in Fig. 5.

A main objective of SM3 studies is to draw Seismic Microzonation Maps using Amplification Factors (AFj), i.e. synthetic indicators representative of the seismic amplification [5]. AFj were calculated as the ratio between the average values of the output and input response spectra over j-th period band, using the following formulas:

$$AF_j = \frac{SA_j^{m,o}}{SA_J^{m,i}} \tag{1}$$

$$SA_j^{m,k} = \frac{1}{\Delta T_j} \int_{\Delta T_j} SA^k(T)dt \quad with\ k = i, o \tag{2}$$

where $SA^k(T)$ is the elastic acceleration response spectrum equal to $SA^i(T)$ for the input and $SA^o(T)$ for the output.

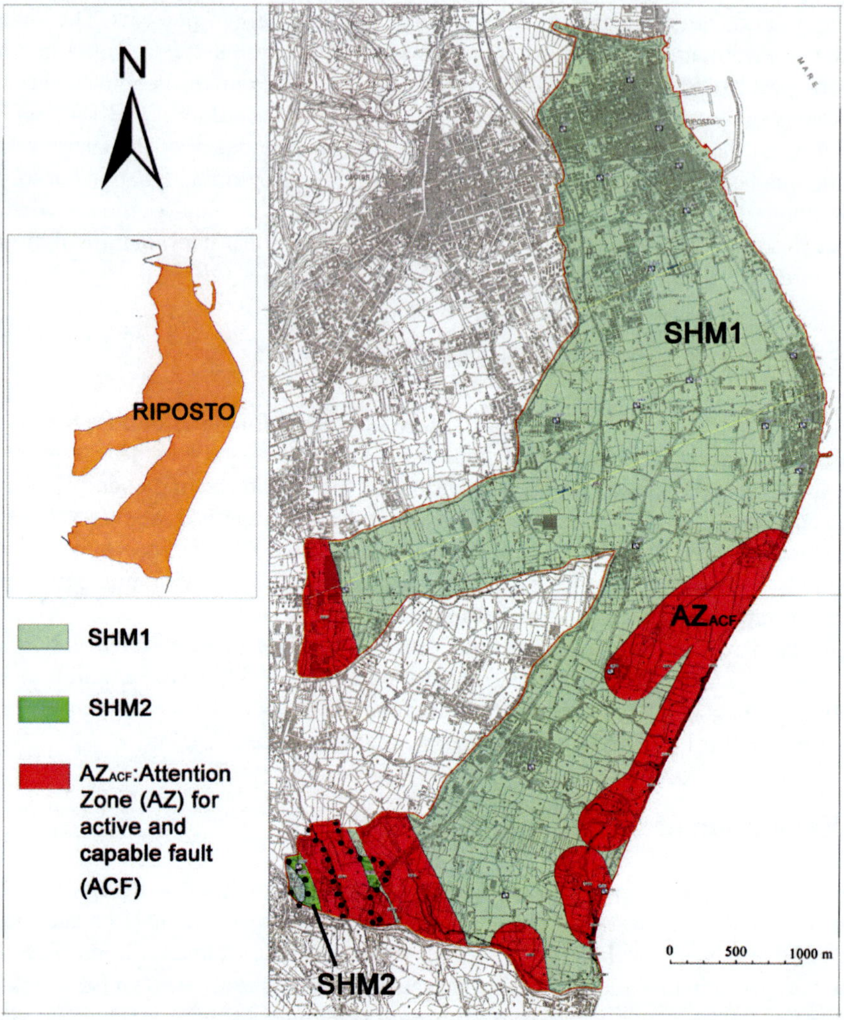

Fig. 2. SHMs Map of the study area: Seismically Homogeneous Microzone 1 (SHM1), Seismically Homogeneous Microzone 2 (SHM2) and Attention Zone for Active and Capable fault (ZA_{ACF}).

The AF_j values have to be computed for the following three period intervals: $\Delta T_1 = [0.1–0.5]$ s, $\Delta T_2 = [0.4–0.8]$ s and $\Delta T_3 = [0.7–1.1]$ s in order to cover the entire period of interest for the buildings. Indeed, these periods are linked to the heights of buildings (about one to three floors for the first interval, three to six floors for the second interval and six to nine floors for the third interval) [7]. The amplification factors, named according to the period intervals, are equal to: FA0105 = 1.9, FA0408 = 0.9 and FA0711 = 1.1 for the area under consideration.

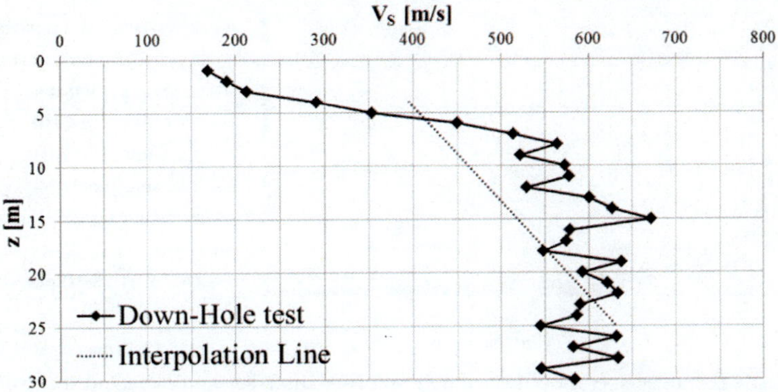

Fig. 3. Vs values from DH test in the municipality of Riposto (SHM1).

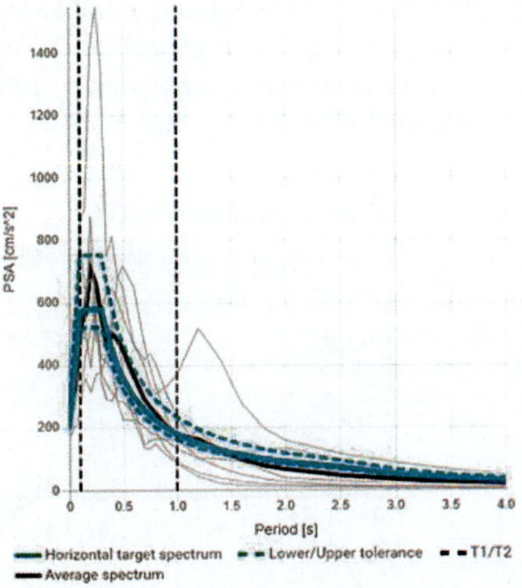

Waveform ID	ML	MW	Epicentral distance [km]
IV.EVRN.00.HN.EMSC-20181226_0000014	4.8	4.9	5.3
IV.T1212.00.HN.EMSC-20161026_0000077	5.4	5.4	15.2
E.SRC0.00.HN.IT-1976-0030	6.0	6.0	15.8
IV.ILLI.00.HN.IT-2010-0032	4.8	4.7	11.4
IT.SVN.00.HG.EMSC-20181226_0000014	4.8	4.9	4.5
IT.ACC.00.HG.EMSC-20161030_0000029	6.1	6.5	18.6
IV.FEMA.00.HN.EMSC-20161026_0000077	5.4	5.4	11.5

Fig. 4. Combination of waveforms obtained from ESM_REXELweb [10].

5 New Methodology for the Seismic Hazard Assessment

FA_j factors are non-dimensional parameters. Therefore, the use of these synthetic indicators provides seismic hazard maps in terms of relative values, not allowing a direct comparison of seismic hazard between different municipalities. Mori et al. [7] introduced a new parameter, named H_{SM}, in order to define an absolute ranking of seismic hazard. Moreover, the H_{SM} parameter allows the classification of the territory according to increasing hazard scale, coupling with a theoretical expected average damage.

H_{SM} parameter values are computed for the three period bands $\Delta T_1 = [0.1–0.5]$ s, $\Delta T_2 = [0.4–0.8]$ s and $\Delta T_3 = [0.7–1.1]$ s, as:

$$H_{SM} = ASI_{UHS}\left(\frac{AF_j}{\Delta T}\right) \tag{3}$$

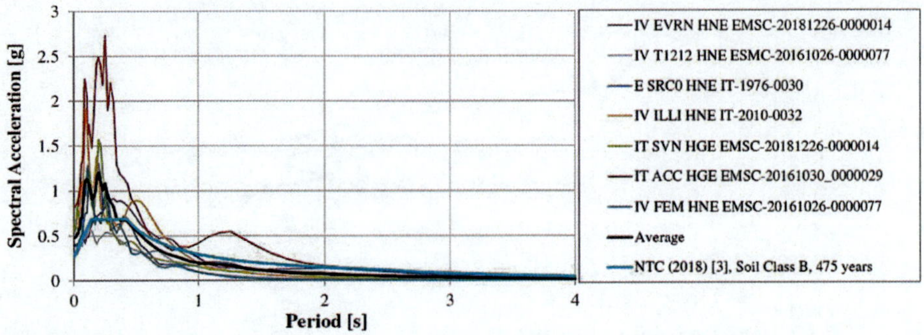

Fig. 5. Elastic response spectra obtained by numerical analyses and expected spectrum provided by the Italian seismic code for soil class B [3].

where: ASI_{UHS} is the Acceleration Spectrum Intensity [6] i.e. the integral value of the Uniform Hazard Spectrum (UHS) in the relative period interval; ΔT is the width of the integration interval (0.4 s); AF_j is the amplification factor calculated by numerical analyses with the input spectrum compatible with the UHS (Fig. 6).

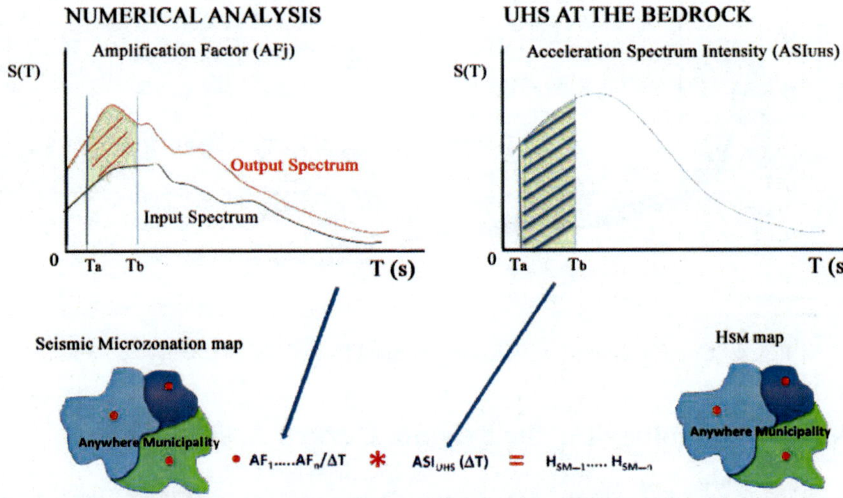

Fig. 6. Calculation of the H_{SM} parameter (From [7]; modified).

The classification of the H_{SM} is based on the correlation with the European Macroseismic intensity Scale (ESM98) [13] using the PGA as linking parameter. The final classification of H_{SM}, reported in Table 1, was obtained linking the threshold of H_{SM} to intermediate I_{EMS98} intensities (6.5 for very-low to low, 8.5 for low to high and 9.5 for high to very high).

This new described methodology was adopted using the results derived from numerical analyses. The following H_{SM} values were calculated for each of the intervals of periods corresponding to different intervals of the fundamental period of the buildings:

Table 1. Final classification of H_{SM} (From [7]).

H_{SM} intervals	H_{SM} [g] classification			
	Very Low	Low	High	Very High
$H_{SM0.1-0.5}$	≤0.21	0.22–0.54	0.55–0.85	≥0.86
$H_{SM0.4-0.8}$	≤0.14	0.15–0.34	0.35–0.55	≥0.56
$H_{SM0.7-1.1}$	≤0.09	0.10–0.22	0.23–0.35	≥0.36

$H_{SM0.1-0.5} = 0.92$, $H_{SM0.4-0.8} = 0.27$, $H_{SM0.7-1.1} = 0.22$, each connected to I_{EMS98} intensities.

According to Table 1 the "low" class was obtained for the intervals of periods $\Delta T_2 = [0.4–0.8]$ s and $\Delta T_3 = [0.7–1.1]$ s. Instead, "Very high" class is associated to the interval of periods $\Delta T_1 = [0.1–0.5]$ s. This interval of periods is also related to the height of the buildings of the studied area (one to three floors) and, consequently, it assumes an extreme relevance in this study.

6 Conclusions

The paper deals with SM studies for the municipality of Riposto (Sicily, Italy). Laboratory and in situ tests in support of SM studies allowed the lithotype characterization of the study area. Numerical analyses were carried out using STRATA code that operates in the frequency domain and takes into account the cyclic soil behaviour applying an equivalent-linear visco-elastic approach. The tool ESM_REXELweb was adopted to select a set of seven accelerograms from the Engineering Strong-Motion Database (ESM, http://esm.mi.ingv.it) compatible on average in the 0.1–1.1 s interval of periods with the Uniform Hazard Spectrum corresponding to 475 years return period at outcropping condition (Soil Class A). Results are reported in terms of 5% damped response spectra.

AF_j amplification factors, as defined by the Italian Guide Lines for microzonation, were calculated for three period intervals: $\Delta T_1 = [0.1–0.5]$ s, $\Delta T_2 = [0.4–0.8]$ s and $\Delta T_3 = [0.7–1.1]$ s, related to the heights of buildings. However, AFj, being non-dimensional parameters, do not allow a direct comparison between different municipalities. Therefore, a new methodology was used in this study by the application of a synthetic damage-constrained parameter, called H_{SM}, that establishes an absolute ranking of seismic hazard on the basis of regional seismic hazard and the amplification factors previously calculated. Three synthetic damage-constrained parameters, one for each of the considered intervals of periods, were calculated, providing a "Very high" class for the area under study characterized by low buildings.

The results obtained from this methodology can help decision makers, engineers and planners to define attention zones in order to prioritize interventions and allocate economic resources for the seismic mitigation.

References

1. Cavallaro, A., Grasso, S., Sammito, M.S.V.: A seismic microzonation study for some areas around the Mt. Etna Volcano on the East Coast of Sicily, Italy. In: Wang, L., Zhang, JM., Wang, R. (eds.) Proceedings of the 4th International Conference on Performance Based Design in Earthquake Geotechnical Engineering. Geotechnical, Geological and Earthquake Engineering, vol. 52, pp. 863–870. Springer, Cham (2022). https://doi.org/10.1007/978-3-031-11898-2_61
2. Grasso, S., Sammito, M.S.V.: Uncertainties in performance based design methodologies for seismic microzonation of ground motion and site effects: state of development and applications for Italy. In: Wang, L., Zhang, JM., Wang, R. (eds.) Proceedings of the 4th International Conference on Performance Based Design in Earthquake Geotechnical Engineering. Geotechnical, Geological and Earthquake Engineering, vol. 52, pp. 412–427. Springer, Cham (2022). https://doi.org/10.1007/978-3-031-11898-2_23
3. NTC D.M.: New Technical Standards for Buildings (2018). https://www.gazzettaufficiale.it/eli/gu/2018/02/20/42/so/8/sg/pdf
4. Kottke, A.R., Rathje, E.M.: Technical Manual for STRATA. PEER Report 2008/10; Univ. of California: Berkeley, CA, USA (2008)
5. Commissione tecnica per la microzonazione sismica: Standard di rappresentazione e archiviazione informatica. Version 4.1_Sicilia. Roma, January (2020). https://www.protezionecivile.gov.it/it/approfondimento/standard-di-rappresentazione-e-archiviazione-informatica-degli-studi-di-ms
6. Von Thun, J.L., Rochim, L.H., Scott, G.A., Wilson, J.A.: Earthquake ground motion for design and analysis of dams. In: Earthquake Engineering and Soil Dynamics II—Recent Advance in Ground-Motion Evaluation, pp. 463–464. Geotechnical Special Publication 20, ASCE, New York (1988)
7. Mori, F., et al.: H_{SM}: a synthetic damage-constrained seismic hazard parameter. Bull. Earthq. Eng. **18**, 5631–5654 (2020)
8. Firetto Carlino, M., Cavallaro, D., Coltelli, M., et al.: Time and space scattered volcanism of Mt. Etna driven by strike-slip tectonics. Sci. Rep. **9**, 12125 (2019)
9. Castelli, F., Cavallaro, A., Ferraro, A., Grasso, S., Lentini, V., Massimino, M.R.: Static and dynamic properties of soils in Catania (Italy). Ann. Geophys. **61**(2), SE221 (2018)
10. Sgobba, S., et al.: REXELweb: a tool for selection of ground-motion records from the Engineering Strong Motion database (ESM). In: 7th International Conference on Earthquake Geotechnical Engineering (ICEGE), Roma, Italy, pp. 4947–4953 (2019)
11. Luzi, L., et al.: ORFEUS Working Group 5: Engineering Strong Motion Database (ESM) (Version 2.0). Istituto Nazionale di Geofisica e Vulcanologia (INGV) (2020). https://doi.org/10.13127/ESM.2
12. Luzi, L., Pacor, F., Lanzano, G., Felicetta, C., Puglia, R., D'Amico, M.: 2016–2017 Central Italy seismic sequence: strong-motion data analysis and design earthquake selection for seismic microzonation purposes. Bull. Earthq. Eng. **18**(12), 5533–5551 (2019). https://doi.org/10.1007/s10518-019-00676-3
13. Grünthal, G.: European Macroseismic Scale 1998 EMS-98, (Cahiers du Centre Eu-ropéen de Géodynamique et de Séismologie; 15). Centre Européen de Géodynamique et de Séismologie, Luxembourg, p. 101 (1998)

Artificial Intelligence-Based Analysis of Numerical Simulations of the Seismic Response of Retaining Walls

Maria Giovanna Durante(✉)

Università della Calabria, Via Pietro Bucci, 87036 Rende, CS, Italy
mgdurante@unical.it

Abstract. Numerical simulations are a great resource to explore the response of retaining systems considering specific soil and structural conditions. The recent availability of significant computational resources enabled the possibility to run extensive parametric analyses to cover a wide range of input parameters. The immediate effect of such extensive parametric studies is the generation of enormous amounts of data that are difficult to analyze using standard approaches. For this reason, artificial intelligence techniques are rapidly becoming common in scientific/engineering fields including geotechnical earthquake engineering. This paper presents some preliminary artificial intelligence techniques applied to an extensive parametric analysis of a retaining wall excited by dynamic inputs. Such analyses are performed with the end goal of defining a novel physics-based methodology for the seismic analysis of retaining structures based on soil-structure interaction principles. The finite element software framework OpenSees is used in the DesignSafe cyberinfrastructure to perform the analyses. The 2D model used in this study considers different soil profiles, while the retaining structure is a cantilever wall with different constraints ranging from a fixed base case to flexible cases. The wall's flexural stiffness is varied relative to the soil stiffness to cover a reasonable range of wall flexibility values. This method represents the theoretical foundation of the EU-funded ReStructure 2.0 Marie Sklodowska-Curie project that is coordinated by the Author.

Keywords: SSI · Retaining walls · geotechnical numerical modelling · machine-learning analysis

1 Introduction

Typical approaches to compute seismic earth pressures on retaining structures are: *(i)* limit equilibrium methods, *(ii)* elastodynamic solutions, that can be subdivided into complete and simplified solutions [1], and *(iii)* numerical simulations. In the limit equilibrium method, the seismic action is simplified as a static horizontal body force acting on an active wedge obtained applying a pseudo-static seismic coefficient (k_h). This approach is usually referred to as Mononobe-Okabe (M-O) method [2, 3]. The original M-O method can result in unrealistically high earth pressures that tend to infinity at high

A. Ferrari et al. (Eds.): CNRIG 2023, SSGG, pp. 603–610, 2023.
https://doi.org/10.1007/978-3-031-34761-0_73

ground acceleration values. Such condition can occur in seismic active regions, like around the Mediterranean. Over the years, several researchers proposed some variations to the original method [4–6] without solving this fundamental flaw. Predictions made using M-O methods rarely agree with *in-situ* observations and experimental measurements [7, 8]. Elastodynamic solutions [9–11] use analytical approaches to implicitly account for factors neglected in the M-O method. Typically, such methods consider a rigid base condition. Furthermore, the earthquake ground motion is applied at the base of the wall. Such approaches usually predict earth pressures even higher than M-O methods. Brandenberg et al. [1] proposed a SSI based solution that removes the rigid base assumption and considers the surface motion as input, and a rigid, massless wall retaining uniform elastic soil. Building upon this approach, subsequent studies [12–14] show a good performance of the method versus experimental data. Furthermore, Durante et al. [15] developed a SSI-based simplified solution to compute the seismic earth pressure increment acting on a retaining wall that matches the frequency-domain results for a cantilevered retaining wall supporting uniform elastic soil. The latter simplified solution has been recently added to the US National Earthquake Hazards Reduction Program (NEHRP) seismic recommended provisions [16]. Numerical simulations are often used to faithfully reproduce site and structure conditions of a specific case study. Different levels of sophistication are available, but they require high-end expertise and a significant amount of time to calibrate the model parameters and to perform the analysis.

This paper presents some preliminary results obtained from a parametric numerical simulation of cantilever walls retaining homogeneous soil excited by dynamic inputs. The simulated data are then used to train a machine learning model. All analyses are performed using the finite element software framework OpenSees [17] running on the DesignSafe cyberinfrastructure [18]. This method represents the theoretical foundation of the EU-funded ReStructure 2.0 Marie Sklodowska-Curie project that is coordinated by the Author [19].

2 Methodology

2.1 Numerical Model

Figure 1 shows the schematic of the problem being investigated. A retaining wall cantilevered from a fixed base supports homogeneous soil overlying an elastic half-space. The two-dimensional mesh is formed by four-node quadrilateral elements using physically stabilized single-point integration (SSPquad element type). The elastic halfspace is modelled allowing all the nodes at the base of the model to move together in the horizontal direction, with the input motion imposed at one node, using Lysmer-Kuhlemeyer dashpots [20], following the Joyner and Chen method [21]. To reproduce the free-field condition at the side of the model, the elements in these locations have increased thickness, and to ensure that they do not affect the wall response, the total extent of the model is set to be equal to six times the height of the wall. The wall is modeled using elastic beam elements with constant flexural stiffness (EI) rigidly attached to the soil continuum immediately behind the wall. The stiffness of the wall relative to the soil is quantified using a dimensionless constant βH, where β represents the Winkler parameters quantifying relative wall/soil stiffness. β is defined by Eq. 1, where k^i_y represents the static

Winkler stiffness intensity for a rigid wall (Kloukinas et al. [22], Eq. 2):

$$\beta = \sqrt[4]{\frac{k_y^i}{4EI}} \tag{1}$$

$$k_y^i = \frac{G}{H}\frac{\pi}{\sqrt{(1-\upsilon)(2-\upsilon)}} \tag{2}$$

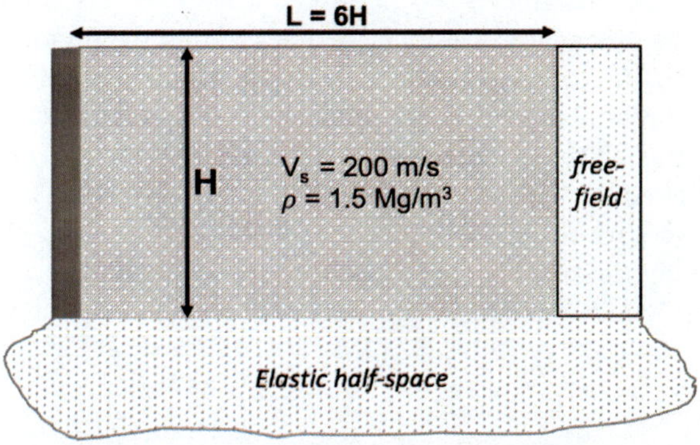

Fig. 1. Schematic of the problem analysed.

Parametric Analysis. An extensive parametric analysis has been conducted using high performance computing power (HPC) available through the DesignSafe cyberinfrastructure, using the parallel interpreter OpenSeesMP. The parametric analysis included two main variables: *(i)* the input motion and *(ii)* the wall flexibility. The input motions used in this paper are recorded ground motions from the NGA West2 database [23]. A suite of 86 ground motions is selected to reproduce the same mean period (T_m) distribution of the entire database, where T_m is defined by Rathje et al. [24]. Several values of wall flexibility configurations are considered in the analysis, obtained considering a βH parameter varying from 0.1 to 4. It should be noted that the flexibility of the wall increases with βh (for a rigid wall, βH = 0). A total of about 3200 analyses are used herein.

2.2 Machine Learning Techniques

Artificial Intelligence (AI) is a relatively new discipline [25] that studies how to train computers to execute specific tasks. Machine Learning (ML) algorithms are a specific type of AI, that requires the data to be organized in a dataset with labelled data and it uses that data to make a prediction. The use of ML algorithms is becoming increasingly popular in various research fields, including geotechnical earthquake engineering [15, 26, 27]. ML can be used in regression problems to predict continuous response value. In AI analyses, each variable considered is called a feature, while the prediction is called a target. In a tree-based method the prediction is made based on the division of the entire space into smaller regions, each able to capture different relationships among the different features. The Random Forest (RF) approach, used in this study, develops multiple decision trees using different parts of the datasets and/or a subset of features. The combination of all the trees generated by the model defines the final prediction model. A proper validation of a ML model requires the subdivision of the dataset into train and test data. The latter is used at the end of the train phase to evaluate the performance of the model with data that was not used to train the algorithm. In this analysis the size of the test data is 20% of the whole dataset. In addition to this division, cross validation is performed during the training of the model. This automated procedure randomly subdivides the dataset into smaller sets (k), and it uses (k-1) sets to train the model and 1 to validate it. This procedure is repeated k times during the training of the model. In this study k is set to 10. For each ML model, a hyperparameter optimization algorithm is used to set the optimal set of model parameters. The variables considered are: *(i)* maximum depth of each tree, *(ii)* number of estimators (trees), *(iii)* maximum number of features considered for node splitting, and *(iv)* the function to measure the quality of a dataset split (criterion). In this study the overall performance of the RF model is evaluated computing the residuals, computed as the difference of natural log seismic earth pressure resultant measured (from numerical analyses) and predicted (using the RF).

3 Results

Figure 2 presents the numerical results by plotting the normalized resultant force in terms of normalized seismic earth pressure resultant $P_E/(k^i{}_y u_{g0} H)$ versus wavelength to height ratio λ/H for various values of βH. This representation is based on the framework developed by Durante et al. [15], where u_{g0} is the surface displacement amplitude, computed as $f_u PGV/\omega_m$, where f_u is an adjustment factor, and λ is computed as $V_s T_m$.

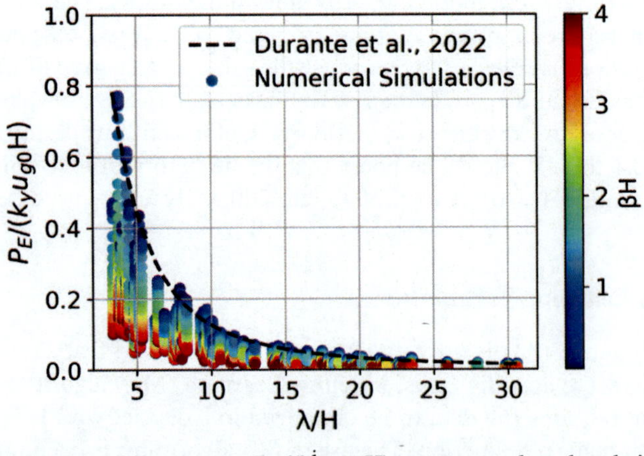

Fig. 2. Dimensionless earth pressure, $P_E/(k^i{}_y u_{g0} H)$ versus wavelength to height ratio λ/H.

The numerical data presented in Fig. 2 shows that P_E decreases with increasing wall flexibility (represented by colors, from blue to red), and with increasing λ/H. The black dashed line in Fig. 2 represents the simplified solution for a rigid wall obtained from the Durante et al. [15] method. As expected, the numerical data for flexible walls do not exceeds the rigid wall condition represented by the dashed line.

The features used in the RF algorithm are: PGV, T_m, u_{g0}, βH and λ/H. Figure 3 shows the results obtained from the RF simulations during the training of the model, using the 80% of the initial dataset. A good agreement is observed not only in the direct comparison of the data (Fig. 3a), but also in the residuals plot (Fig. 3b), where the error is equally distributed among all values considered for βH and λ/H.

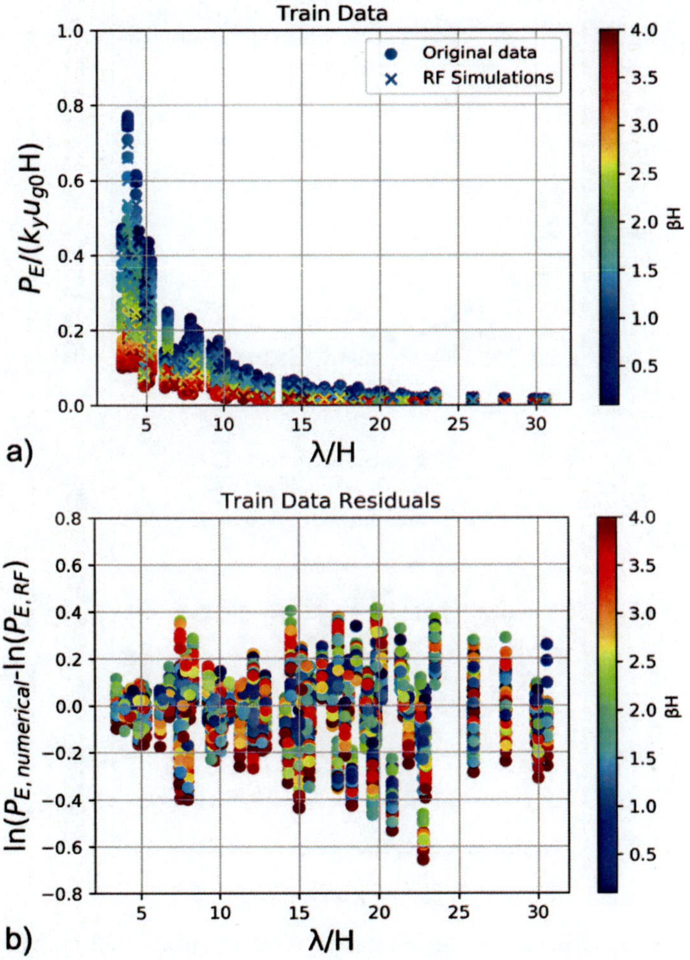

Fig. 3. RF simulation results for the train dataset (80% of initial data): (a) direct comparison between numerical and RF simulated data in terms of dimensionless earth pressure, $P_E/(k^i{}_y u_{g0} H)$ versus wavelength to height ratio λ/H, and (b) model residuals, computed as the difference of natural log P_E from numerical analyses and RF prediction versus wavelength to height ratio λ/H.

Figure 4 shows the results obtained from the RF simulations during the test of the model, using the 20% of the initial dataset, excluded from the training of the algorithm. As observed for the train dataset, a good agreement is observed not only in the direct comparison of the data (Fig. 4a), but also in the residuals plot (Fig. 4b), where the error is equally distributed among all the value considered for βH and λ/H. The similar performance of the algorithm in both train and test data (residual around 20%) demonstrate that the model does not overfit the data.

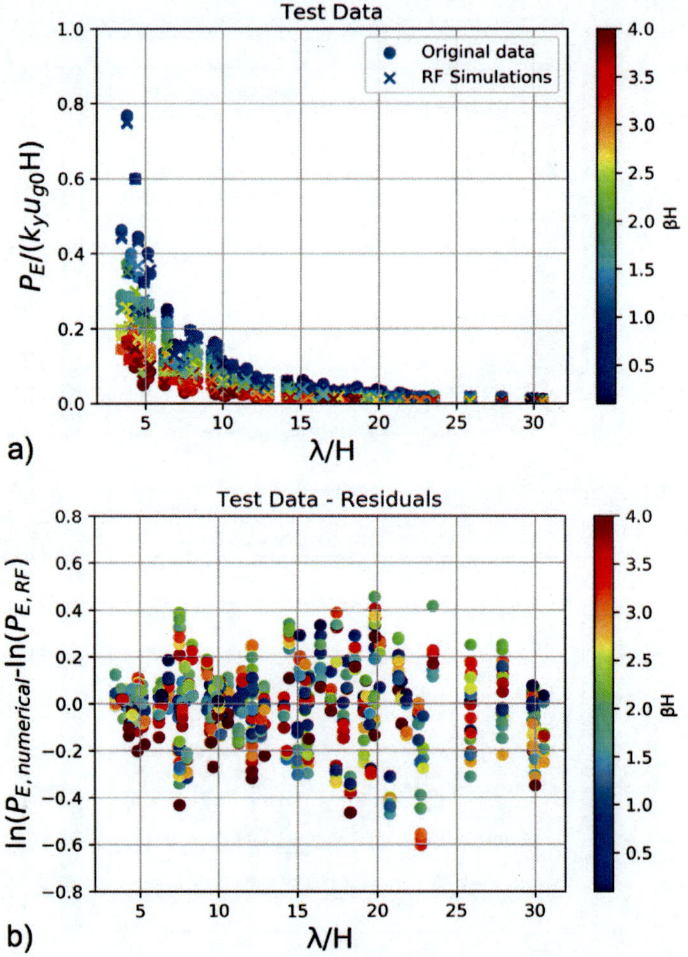

Fig. 4. RF simulation results for the test dataset (20% of initial data): (a) direct comparison between numerical and RF simulated data in terms of dimensionless earth pressure, $P_E/(k^i{}_y u_{g0} H)$ versus wavelength to height ratio λ/H, and (b) model residuals, computed as the difference of natural log P_E from numerical analyses and RF prediction versus wavelength to height ratio λ/H.

4 Conclusions

This paper presented a preliminary study on the application of machine learning techniques to an extensive parametric numerical analysis for a retaining wall excited by several ground motion inputs. The main outcomes can be summarized as follows:

- The numerical simulations confirm that the wall flexibility significantly influences soil-structure interaction and therefore mobilization of seismic earth increments. Furthermore, The P_E reduction observed with λ/H is justified by the fact that when this parameter increases, the deformed shape of the free-field soil profiles becomes vertical, conforming the shape of the rigid wall and producing zero kinematic interaction.
- The RF model is a promising tool, able to reproduce the numerical data. The resulting error (computed as residual values) can be considered acceptable when compared against the significant reduction of the computation power needed.

This paper is part of a more comprehensive on-going Marie Sklodowska-Curie EU funded project. Future studies will include the use of other AI algorithms, more extensive parametric analyses, different soil type, and will release the elastic soil hypothesis, including full non-linear materials.

Acknowledgments. The project leading to this application has received funding from the European Union's Horizon 2020 research and innovation programme under the Marie Sklodowska-Curie Grant Agreement number 101029903 – ReStructure 2.0 – H2020 – MSCA – IF – 2020".

References

1. Brandenberg, S.J., Mylonakis, G., Stewart, J.P.: Kinematic framework for evaluating seismic earth pressures on retaining walls. J. Geotech. Geoenv. Eng. **141**(7), 04015031 (2015)
2. Okabe, S.: General theory on earth pressure and seismic stability of retaining wall and dam. Japanese Society of Civil Engineering **12**(4), 34–41 (1924)
3. Mononobe, N., Matsuo, M.: On the determination of earth pressures during earthquakes. In: Proceedings of World Engineering Congress, October 29 to November 7, Tokyo, Japan (1929)
4. Seed, H.B., Whitman,R.V. : Design of earth retaining structures for dynamic loads. In: ASCE Specialty Conference on Lateral Stresses in the Ground and Design of Earth Retaining Structures, 22–24 June, N.Y., U.S. (1970)
5. Mylonakis, G., Kloukinas, P., Papantonopoulos, C.: An alternative to the Mononobe-Okabe equations for seismic earth pressures. Soil Dyn. Earthq. Eng. **27**, 957–969 (2007)
6. Xu, S.Y., Shamsabadi, A., Taciroglu, E.: Evaluation of active and passive seismic earth pressures considering internal friction and cohesion. Soil Dyn. Earthq. Eng. **70**, 30–47 (2015)
7. Al Atik, L., Sitar, N.: Seismic earth pressures on cantilever retaining structures. J. Geotech. Geoenv. Eng. **136**, 1324–1333 (2010)
8. Hushmand, A., Dashti, S., Davis, C., McCartney, J.S., Hushmand, B.J.S.D.: A centrifuge study of the influence of site response, relative stiffness, and kinematic constraints on the seismic performance of buried reservoir structures. Soil Dyn. Earthq Eng. **88**, 427–438 (2016)
9. Veletsos, A.S., Younan, A.H.: Dynamic modeling and response of soil-wall systems. J. Geotech. Eng. **120**, 2155–2179 (1994)

10. Vrettos, C., Beskos, D.E., Triantafyllidis, T.: Seismic pressures on rigid cantilever walls retaining elastic continuously non-homogeneous soil: an exact solution. Soil Dyn. Earthq. Eng. **82**, 142–153 (2016)
11. Garcia-Suarez, J., Asimaki, D.: Exact seismic response of smooth rigid retaining walls resting on stiff soil. Int. J. Numer. Anal. Methods Geomech. **44**, 1750–1769 (2020)
12. Brandenberg, S.J., Durante, M.G., Mylonakis, G., Stewart, J.P.: Winkler solution for seismic earth pressures exerted on flexible walls by vertically inhomogeneous soil. J. Geotech. Geonviron. Eng. **146**(11), 04020127 (2020)
13. Durante, M.G., Brandenberg, S.J., Dashti, S., Stewart, J.P., Mylonakis, G.: Analysis of seismic earth pressures on flexible underground box structures. In: 7th International Conference on Earthquake Geotechnical Engineering (VII ICEGE), 17–20 June 2019, Rome, Italy (2019)
14. Durante, M.G., Brandenberg, S.J., Stewart, J.P., Mylonakis, G.: Winkler stiffness intensity for flexible walls retaining inhomogeneous soil. In: Geotechnical Earthquake Engineering and Soil Dynamics V: Numerical Modeling and Soil-Structure Interaction, pp. 473–482 (2018)
15. Durante, M.G., Rathje, E.M.: An exploration of the use of machine learning to predict lateral spreading. Earthq. Spectra **37**(4), 2288–2314 (2021)
16. Seismic Lateral Earth Pressures: In NEHRP Recommended Provisions for Seismic Regulations for New Buildings and Other Structures, Part 3: Resource Papers. Building Seismic Safety Council BSSC, Federal Emergency Management Agency, Washington, D.C. (2020)
17. McKenna, F., Fenves, G.L.: The OpenSees command language manual, version 2.5, Pacific Earthquake Engineering Research Center, University of California, Berkeley (2001). http://opensees.berkeley.edu
18. Rathje, E.M., et al.: DesignSafe: a new cyberinfrastructure for natural hazards engineering. Nat. Hazards Rev. **18**(3), 06017001 (2017)
19. ReStructure 2.0: A novel physics-based methodology for the seismic analysis of retaining structures leveraging machine learning techniques (2021). https://cordis.europa.eu/project/id/101029903
20. Lysmer, J., Kuhlemeyer, A.M.: Finite dynamic model for infinite media. J. Eng. Mech. Div. ASCE **95**, 859–877 (1969)
21. Joyner, W.B., Chen, A.T.F.: Calculation of nonlinear ground response in earthquakes. Bull. Seismol. Soc. Am. **65**(5), 1315–1336 (1975)
22. Kloukinas, P., Langoussis, M., Mylonakis, G.: Simple wave solution for seismic earth pressures on non-yielding walls. J. Geotech. Geoenviron. Eng. **138**(12), 1514–1519 (2012)
23. Ancheta, T.D., et al.: NGA-West2 database. Earthq. Spectra **30**, 989–1005 (2014)
24. Rathje, R.M., Faraj, F., Russell, S., Bray, J.D.: Empirical relationships for frequency content parameters of earthquake ground motions. Earthq. Spectra **20**(1), 119–144 (2004)
25. McCarthy, J., Minsky, M.L., Rochester, N., Shannon, C.E.: A proposal for the Dartmouth summer conference on artificial intelligence. In: Conference Announcement (1955)
26. Rathje, E.M., Durante, M.G.: On the use of machine learning techniques to predict lateral spreading displacement in New Zealand. In: 17th World Conference on Earthquake Engineering, 17WCEE, Sendai, Japan - September 13th to 18th 2020 (2020)
27. Cho, Y., Khosravikia, F., Rathje, E.M.: A comparison of artificial neural network and classical regression models for earthquake-induced slope displacements. Soil Dyn. Earthq. Eng. **152**, 107024 (2022)

Evaluation of Seismic Landslide Hazard Based on a New Displacement Semi-empirical Relationship

Fabio Rollo(✉) and Sebastiano Rampello

Dipartimento di Ingegneria Strutturale e Geotecnica, Università di Roma La Sapienza, via Eudossiana 18, 00184 Rome, Italy
fabio.rollo@uniroma1.it

Abstract. This paper presents a new semi-empirical relationship that links the permanent earthquake-induced displacements of slopes to the synthetic ground motion parameter *PGA* or to the couple *PGA* and *PGV*. The displacements are evaluated under the hypothesis of a rigid sliding block performing Newmark's integrations for all the acceleration time histories of the updated Italian seismic database. The relationship reproduces well the displacements for any values of yield seismic coefficient in the whole range of peak ground acceleration. The two parameters expression is more reliable for the study of Italian slopes under seismic loading than that based on the single *PGA* parameter as characterised by a lower standard deviation. The proposed relationship is also combined with a fully probabilistic approach to produce displacement hazard curves and hazard maps for different sites and regions of Italy that represent a useful tool for practicing engineers and national agencies for a preliminary estimate of the seismic performance of a slope.

Keywords: slopes · earthquake-induced displacements · semi-empirical relationships · probabilistic analysis · hazard curves and maps

1 Introduction

A well-established way to evaluate the seismic performance of a slope is to determine the displacements induced at the end of the seismic event. These are often quantified through the method proposed by Newmark (1965) [1], that consists to model the slope with a rigid block sliding on a horizontal plane that experiences permanent displacements only when the critical acceleration, function of the slope resistance, is lower than that of the input motion.

In the last two decades several semi-empirical relationships have been proposed, that link the permanent slope displacements computed through the Newmark's method, using different ground motion databases, to a series of ground motion parameters and the yield seismic coefficient k_y denoting synthetically the seismic slope resistance (e.g. [2–7]). These simplified relationships are often employed to predict the seismic-induced displacements of specific slopes and embankments and are extremely useful when combined

A. Ferrari et al. (Eds.): CNRIG 2023, SSGG, pp. 611–618, 2023.
https://doi.org/10.1007/978-3-031-34761-0_74

with a fully probabilistic-based approach capable to account for the aleatory variability of earthquake ground motion and displacement prediction (e.g. [8]). In light of this, one can develop scalar and vector probabilistic approaches if one or more ground motion parameters are considered. Moreover, the probabilistic analysis can be extended to a regional scale including the ground motion hazard information within the probabilistic approach to produce landslides hazard maps that allows to detect the portions of territory that are prone to earthquake-induced slope instability (e.g. [9, 10]).

In this work the results of the probabilistic approach incorporating a new semi-empirical relationship proposed by Rollo & Rampello (2023) [11] are illustrated. The new semi-empirical relationship is developed with reference to the Italian seismicity, assimilating the slope to a rigid sliding block, and represents an attractive tool for the screening level analysis of slopes at the regional scale. The displacement hazard curves obtained through the vector approach for different Italian sites and yield seismic coefficients are first shown. Finally, a series of hazard maps displaying either the distribution of the return period for different prescribed values of slope displacements and seismic coefficient or the displacement variability are presented, aimed at clarifying the role of the adopted displacement semi-empirical relationship on the evaluation of the seismic hazard at a regional scale.

2 New Semi-empirical Relationship and Probabilistic Framework

The displacement relationships provide the natural log of horizontal displacement d given the natural log of one or more ground motion parameters (GM). In principle, any combination of ground motion parameters can be adopted. However, as discussed by Rollo & Rampello (2021, 2022) [12, 13], the parameters PGA and PGV are more suitable for the development of the probabilistic approach requiring a standard seismic probabilistic hazard analysis (PSHA). In the most general case of two strong motion parameters, the new semi-empirical relationship assumes the form:

$$\ln(d) = a_0 + a_1 \ln\left(1 - \frac{k_y}{PGA}\right) + a_2 \ln\left(\frac{k_y}{PGA}\right) + a_3\left[\ln\left(\frac{k_y}{PGA}\right)\right]^2 + a_4 \ln(PGA) + a_5 \ln(PGV) \quad (1)$$

that depends on the ratio k_y / PGA and on a series of coefficients $a_0 - a_5$ that are calibrated on the considered seismic database. The proposed expression respects the conditions $d \to \infty$ for $k_y/PGA = 0$ and $d = 0$ for $k_y/PGA = 1$ wanted for the case of a rigid block. The coefficients of the Eq. (1) have been calibrated based on the permanent displacements computed with the rigid sliding-block model [1] for the simple scheme of an infinite slope for different values of the yield seismic coefficient $k_y = 0.04, 0.06, 0.08, 0.1, 0.12, 0.15$. The Newmark's computation has been performed for all the recorded acceleration time histories of the Italian strong motion database [14]. The regression coefficients and $\sigma_{\ln}$ values are reported in Table 1. The coefficients as well as the standard deviation modify whether the ground motion parameter PGV is taken or not into account. The two ground motion parameters relationship is characterised by lower standard deviation as it describes better the characteristics of the seismic database and the predicted displacements as compared to the single parameter *PGA*. Figure 1 shows the comparison between the computed displacements and the scalar semi-empirical relationship for different values of yield seismic coefficient.

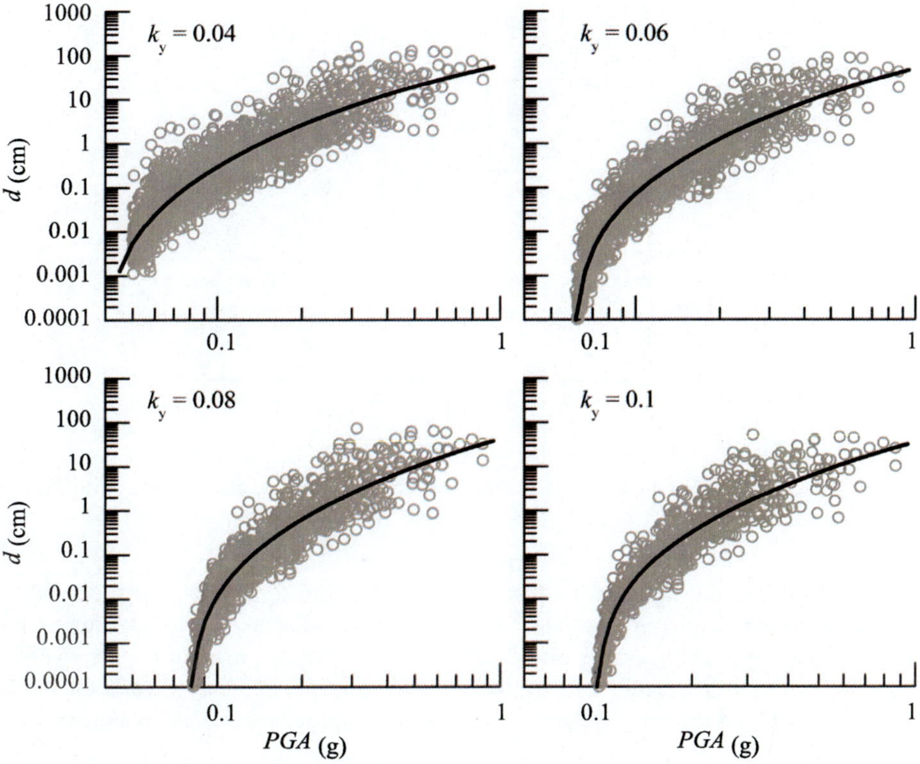

Fig. 1. Comparison between Newmark's displacements and scalar semi-empirical relationship for different values of k_y.

Table 1. Regression coefficients

Approach	a_0	a_1	a_2	a_3	a_4	a_5	σ_{ln}
Scalar	0.698	1.899	−1.987	−0.285	1.101	–	1.001
Vector	−5.124	1.992	−1.736	−0.234	−0.573	1.531	0.547

As shown by Rollo & Rampello (2023) [11], the vector approach provides much lower residuals for all the considered ground motion parameters, with almost constant mean values, demonstrating that the new semi-empirical relationship is reasonable for the study of the Italian seismicity when using *PGA* and *PGV* as the seismic loading parameters. Moreover, it represents a good alternative with respect to a series of existing semi-empirical relationships. Figure 2 shows the Newmark's displacements with the ratio

k_y / PGA while the curve indicates the result of the vector semi-empirical relationship for a value of $PGV = 8$ cm/s, that is the average value for the seismic database.

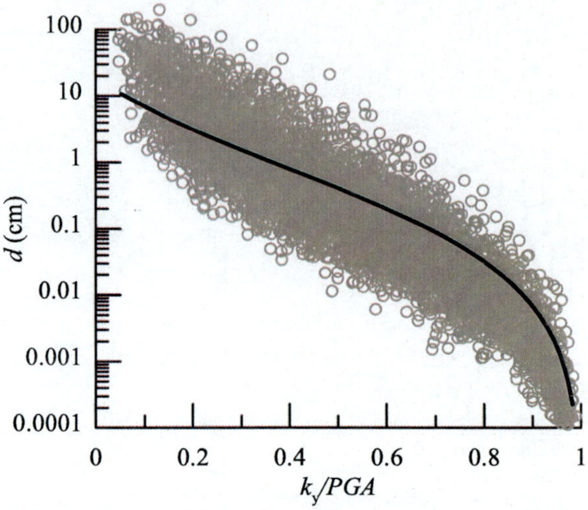

Fig. 2. Displacements versus k_y /*PGA* for the vector semi-empirical relationship (with $PGV =$ 8 cm/s).

The displacement predictive equations are a key tool to develop the probabilistic approach, whose results are synthesised in terms of displacements hazard curves and maps, providing the annual rate of exceedance λ_d for a given level of displacement d. Here, only a brief description of the probabilistic approach is presented and the reader is referred to [11] for further details. For the vector approach, λ_d is calculated as:

$$\lambda_d(x) = \sum_i \sum_j P[d > x | PGA_i, PGV_j] \times P[PGA_i, PGV_j] \quad (2)$$

where the first term is the probability of occurrence of displacements greater than a value x, given the peak ground acceleration PGA_i and the peak ground velocity PGV_j, while the second term is the joint annual probability of PGA_i and PGV_j. The former term requires the disaggregation of the hazard of *PGA* and the correlation coefficient between *PGA* and *PGV*. The correlation coefficient is evaluated through the ground motion prediction equation (GMPE) of Lanzano *et al.* (2019) [15], leading to a value of 0.834. The probabilistic approach has been implemented in the commercial numerical software package MATLAB. Details about the Italian seismic database and the implementation technique can be found in [11, 12].

3 Displacement Hazard Curves and Maps

The displacement hazard curves plot the annual rate of exceedance λ_d against the induced slope displacement and are obtained for different sites in Italy and for different yield seismic coefficients. Figure 3(a) shows the displacement hazard curves for the site of

Amatrice (RI) and three values of k_y: as the seismic resistance of the slope increases the annual rate of exceedance decreases at a given displacement d, as expected. Figure 3(b) shows the results for the sites of Amatrice (RI), Lioni (AV) and Modena in the central, southern and northern Italy, respectively. The displacement hazard curves refer to $k_y = 0.12$ and allow to investigate the site effect: Lioni and Amatrice are located in two more severe seismic regions than Modena and thus the curves are characterised by higher values of λ_d for a given permanent displacement.

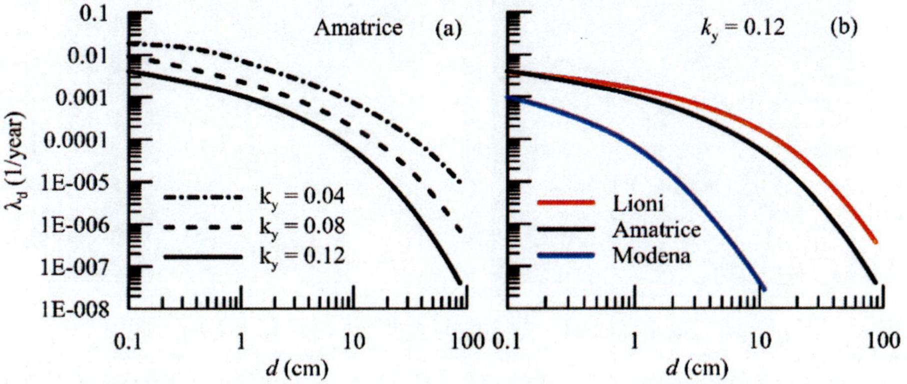

Fig. 3. Displacement hazard curves for (a) the site of Amatrice and different values of yield seismic coefficient and (b) different sites in Italy for a fixed value of k_y.

The results of the probabilistic approach are also described in terms of hazard maps showing the contours of the return periods T_r associated to different levels of earthquake-induced slope displacement and yield seismic coefficients. The maps are obtained considering a grid of points equally spaced of 5 km. Information pertaining to the seismic hazard in terms of *PGA* hazard curves and disaggregation are available in correspondence of these sites on the national territory. For any point of the map, the probabilistic analysis provides the displacement hazard curves for different values of k_y. Therefore, for a given value of k_y and a prescribed displacement, one gets the corresponding value of λ_d (or $T_r = 1/\lambda_d$). The T_r values of the nearby grid points are linearly interpolated to obtain a representation in terms of return period contours, with a logarithmic scale adopted for the sake of convenience. The maps do not account for the real distribution of slope parameters and soils properties. However, they represent a useful tool for a preliminary assessment of the slope seismic hazard. The hazard maps presented here are developed using the vector probabilistic approach for the district of Irpinia, in the Campania region, an area of about 40×40 km^2 located in the South Italy, at 50 km from the city of Naples. This is a mountainous area crossed by the Apennines and characterised by a severe seismic hazard. The coordinates of the map are East and North according to the reference coordinate system WGS84. *PGA* hazard curves and disaggregation corresponding to the grid points of the area under study are extracted from the INGV interactive seismic hazard maps (http://esse1.mi.ingv.it/d2.html).

Figure 4 shows the displacement hazard maps in terms of contours of the return period for a fixed value of the yield seismic coefficient and varying threshold displacements d_y $= 2$ cm (rock-like subsoil) and $d_y = 15$ cm (free-field ductile soil behaviour) according to [16].

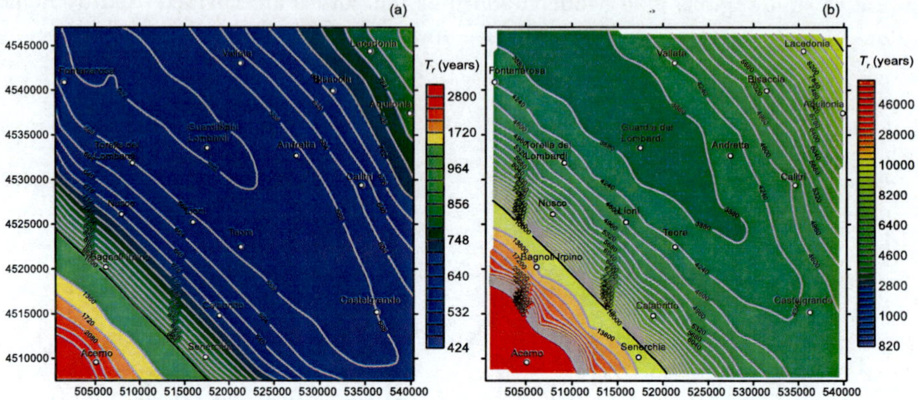

Fig. 4. Displacement hazard maps for the Irpinia district for (a) $d_y = 2$ cm and (b) $d_y = 15$ cm for $k_y = 0.08$.

The distribution of T_r follows the probabilistic seismic hazard and disaggregation information of the Irpinia district: the hazard is more severe in correspondence of the mountainous zone of the Apennines extended from North-Western to South-Eastern corners of the map. In that area, the occurrence of a certain threshold displacement is more frequent and hence characterised by lower values of return period. As expected, the return period increases drastically when higher value of threshold displacement is considered.

The hazard maps can be also plotted in terms of contours of permanent displacements for a fixed value of return period and yield seismic coefficient, as illustrated in Fig. 5. The displacements are determined directly from Eq. (1) stemming from the areal distribution of PGA associated to a return period $T_r = 475$ years obtained from a standard probabilistic seismic hazard analysis (PSHA) and an average value of $PGV = 8$ cm/s. The return period of 475 years corresponds to a 10% probability of exceedance in 50 years, that represents an Ultimate Limit State (ULS) according to the Italian building code (NTC18). As expected, the displacements decrease for increasing values of k_y but in any cases a common pattern of the contours can be recognised.

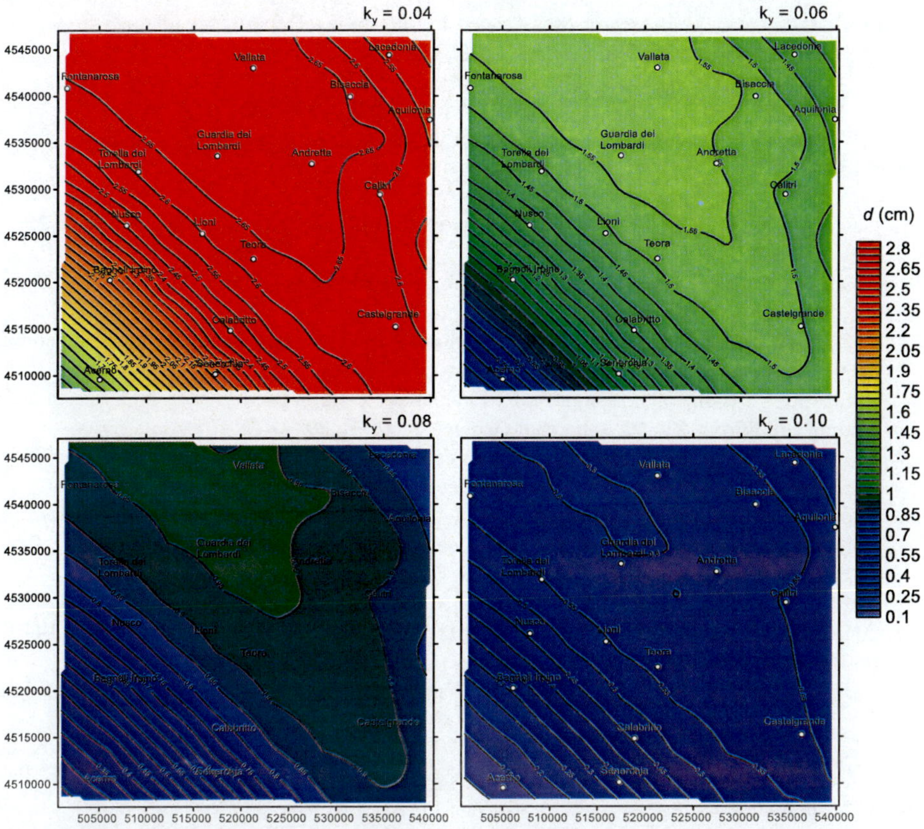

Fig. 5. Displacement hazard maps in terms of displacement contours for the Irpinia district for different values of k_{y} and $T_{\mathrm{r}} = 475$ years.

4 Conclusions

This paper presents a new semi-empirical relationship for the evaluation of the earthquake-induced slope displacements as a function of the ground motion parameters *PGA* and *PGV* and the yield seismic coefficient k_{y}. The new expression was formulated and calibrated with reference to the Italian seismic database, assimilating the slope to a rigid sliding block according to the Newmark's method and well predicts the computed permanent displacements. Moreover, when incorporated within a probabilistic approach, the semi-empirical relationship allows to produce a series of hazard curves and maps providing the annual rate of exceedance to given values of slope displacements and yield seismic coefficient k_{y}. The probabilistic framework can be employed for any locations in Italy and hazard maps expressed either in terms of return periods or in terms of permanent displacements provide with a powerful tool for a more rational, though first estimate, of the regional landslide seismic hazard.

Acknowledgements. The research was funded by the Italian Department of Civil Protection, ReLUIS research project, WP 16: *Geotechnical Engineering*, Task Group 2: *Slope stability.*

References

1. Newmark, N.M.: Effects of earthquakes on dams and embankments. Geotechnique **15**(2), 139–160 (1965)
2. Jibson, R.W.: Regression models for estimating coseismic landslide displacement. Eng. Geol. **91**(2–4), 209–218 (2007)
3. Saygili, G., Rathje, E.M.: Empirical predictive models for earthquake-induced sliding displacements of slopes. J. Geotech. Geoenv. Eng. **134**(6), 790–803 (2008)
4. Rampello, S., Callisto, L., Fargnoli, P.: Evaluation of slope performance under earthquake loading conditions. Rivista Italiana di Geotecnica **44**(4), 29–41 (2010)
5. Biondi, G., Cascone, E., Rampello, S.: Valutazione del comportamento dei pendii in condizioni sismiche. Rivista italiana di geotecnica **45**(1), 9–32 (2011)
6. Song, J., Gao, Y., Rodriguez-Marek, A., Feng, T.: Empirical predictive relationships for rigid sliding displacement based on directionally-dependent ground motion parameters. Eng. Geol. **222**, 124–139 (2017)
7. Du, W., Wang, G., Huang, D.: Evaluation of seismic slope displacements based on fully coupled sliding mass analysis and NGA-West2 database. J. Geotech. Geoenviron. Eng. **144**(8), 06018006 (2018)
8. Lari, S., Frattini, P., Crosta, G.B.: A probabilistic approach for landslide hazard analysis. Eng. Geol. **182**, 3–14 (2014)
9. Sharifi-Mood, M., Olsen, M.J., Gillins, D.T., Mahalingam, R.: Performance-based, seismically-induced landslide hazard mapping of Western Oregon. Soil Dyn. Earthq. Eng. **103**, 38–54 (2017)
10. Li, C., Wang, G., He, J., Wang, Y.: A novel approach to probabilistic seismic landslide hazard mapping using Monte Carlo simulations. Eng. Geology **301**, 106616 (2022)
11. Rollo, F., Rampello, S.: Influence of the displacement predictive relationships on the probabilistic seismic analysis of slopes. J. Geotech. Geoenviron. Eng. **149**(6), 04023033 (2023). https://doi.org/10.1061/JGGEFK/GTENG-11162
12. Rollo, F., Rampello, S.: Probabilistic assessment of seismic-induced slope displacements: an application in Italy. Bull. Earthq. Eng. **19**(11), 4261–4288 (2021). https://doi.org/10.1007/s10518-021-01138-5
13. Rollo, F., Rampello, S.: Probabilistic seismic hazard curves and maps for Italian slopes. In: Wang, L., Zhang, J.-M., Wang, R. (eds.) Proceedings of the 4th International Conference on Performance Based Design in Earthquake Geotechnical Engineering (Beijing 2022), pp. 1348–1355. Springer, Cham (2022). https://doi.org/10.1007/978-3-031-11898-2_116
14. Gaudio, D., Rauseo, R., Masini, L., Rampello, S.: Semi-empirical relationships to assess the seismic performance of slopes from an updated version of the Italian seismic database. Bull. Earthq. Eng. **18**(14), 6245–6281 (2020). https://doi.org/10.1007/s10518-020-00937-6
15. Lanzano, G., et al.: A revised ground-motion prediction model for shallow crustal earthquakes in Italy. Bull. Seismol. Soc. Am. **109**(2), 525–540 (2019)
16. Wilson, R.C., Keefer, D.K.: Predicting areal limits of earthquake-induced landsliding. In: Ziony, E.D. (ed.) Evaluating Earthquake Hazard in the Los Angeles Region. US Geological Survey, Reston, pp. 317–345 (1985)

Maintenance, Reliability and Resilience of Critical Infrastructures

Design of Permeation Grouting Treatments with Eco-Friendly Nanosilica Grouts

Katia Boschi[1,2](✉), Claudio di Prisco[1], and Davide Grassi[2,3]

[1] DICA, Politecnico di Milano, Milan, Italy
katia.boschi@polimi.it
[2] DISAT, Bicocca University, Milan, Italy
[3] MBS Master Builders Solutions Italia S.p.A., Treviso, Italy

Abstract. Nanosilica aqueous suspensions are mechanically flexible and environmentally non-toxic materials. These are injected into soils at low pressures as fast remedials to stop piping induced by excavations, for the sealing of contaminants or to reduce the seismic-induced liquefaction potential. To adopt the technique, it is crucial to firstly predict the temporal evolution of the permeation process highly affected not only by operational parameters, but also by the time-dependent rheological properties of the employed grout. In this paper, a simplified method to design permeation grouting treatments with eco-friendly nanosilica grouts is proposed. Its employment is discussed focusing on how to rationally optimize the injection process in terms of nozzle spacing, injection time, grout compositions and pump pressure. This can be a useful tool in practical applications also to interpret the experimental results obtained from trial in situ tests, always recommended before grouting. The final goal is to guarantee a more efficient design approach for these treatments, currently absent in the scientific literature, taking not only times and costs but also environmental impact into account.

Keywords: Ground Improvement · Permeation Grouting · Colloidal Silica Grouts · Seepage · Time Dependence

1 Introduction

Permeation grouting is a geotechnical stabilisation technique commonly employed to improve ground both mechanical and hydraulic properties. The treatments consist in a fluid-like material injected throughout soil strata under controlled conditions without perturbating the microstructure of the injected soil [1]. Permeation grouting is so considered particularly suitable for remediating foundations of existing buildings or in general when treating urban areas. With respect to the more commonly employed microfine cements, which, due to their particle size distribution [2], cannot permeate both too fine sands and silty soils, chemical grouts, such as mechanically flexible and environmentally non-toxic nanosilica (NS) grouts with a nano-metric characteristic length, begin to be preferred for several applications, e.g. fast remedial to stop piping induced by excavations [3], sealing of contaminants [4] or reduction of the liquefaction potential [5, 6]

A. Ferrari et al. (Eds.): CNRIG 2023, SSGG, pp. 621–628, 2023.
https://doi.org/10.1007/978-3-031-34761-0_75

of sandy soils. In support of the above, it is also worth noticing that NS grouts, which are aqueous suspensions of silica particles (SP) mixed with an activator consisting of sodium chloride (NaCl) water solutions, are characterised by an initial low viscosity, enabling so to cover large distances when injected especially in relatively low permeable soils. Then, these fluid-like materials change their consistency with time, firstly into a gel, then into a solid. The adopted concentration of NaCl and SP, as well as pH and temperature, are important factors influencing the grout rheology time dependence [7], characterised by a Binghamian nature before curing [8, 9], and, consequently, the grout propagation throughout soils. Among them, SP concentration also plays a crucial role in terms of overall costs and achieved mechanical properties of the treated soil [6]. Since NS grout injection causes a considerable reduction of hydraulic conductivity, the mechanical behaviour of the treated soil is reasonable to be assumed as undrained. A quantification of strength improvement is therefore commonly assessed on the basis of the undrained shear strength. In absence of ad-hoc experimental test results, a first estimation of this parameter can be derived by the unconfined compressive strength values already experimentally obtained by [10, 11] and [4].

To design these treatments, it is crucial predicting the temporal evolution of the permeation process and relating the in-situ distance covered by the grout to operational parameters: grout composition, injection pressure, flow rate and spatial distribution of injection sources. Until now, however, the treatment design is currently derived from empirical approaches based on experimental observations and studies, where specific grout and injection techniques are applied to specific soil conditions [12]. Among them, few analytical/numerical models have been conceived and/or employed to simulate permeation grouting applications in porous media [1, 13, 14]. In all these studies, during the injection phase, the grout rheology has been assumed as Newtonian and its intrinsic time dependence (i.e. not related to its interaction with solid matrix, as it is in case of surface adsorption) has been either omitted or not adequately investigated. In contrast, Boschi et al. [8, 9] have proved that the permeation process of NS grouts in soils can be correctly reproduced if and only if the time dependency of the NS grout Binghamian rheology is correctly simulated.

In this work, a design method for permeation grouting treatments, especially referring to NS grouts, is proposed (Sect. 2), by recalling both the grout rheological description and the analytical equations derived by [8] and [9]. These analytical equations predict, when constant flow rate Q is imposed, the temporal evolution of (i) the distance covered in-situ by the grout front and (ii) the evolution of injection pressure p_{inj} for spherical sources of radius r_0. Then, its employment (Sect. 3), in particular for (i) quantifying the finest soils that can be injected with a given NS grout and (ii) optimizing the injection process in terms of grout compositions and operational parameters is discussed.

2 Design Method

2.1 Simplifying Assumptions and Input Data

In common permeation grouting applications, a tubes-à-manchette technique is employed: an oversized hole is drilled, a sleeve-port PVC pipe is inserted inside and a cement-bentonite grout is used to fill the annular space between the tube and the hole wall. Then, from port holes located at regularly spaced intervals within a range of 0.3 to 0.6 m, the injection takes place at one selected depth after the other. Alternatively, in uniform sandy or silty soil strata, the grout can be injected by using tight-fitting steel pipes directly pushed and driven in place to be in tight contact with the surrounding soil. In this case, once reached the deeper zones to be treated, the first injection takes place from holes located at the pipe tip. Then, the pipe is upward lifted, another injection at the pipe tip takes place and so on. For both the adopted techniques, the injected geometry can be approximated as a series of subsequent spheres, deriving from a grout radial seepage starting from each injection source [1, 8]. For the sake of simplicity, in the herein employed analytical equations, each injection is also assumed as independent, developing in an infinite spatial domain and starting from a spherically assumed source of radius r_0. To quantify this last, an equivalent ideal sphere characterised by the same area of the lateral surface of a cylinder better mimicking the in-situ injection source is taken into account [15].

As for the soil to be treated, the necessary input data are (i) porosity n, (ii) intrinsic permeability k ($k = K \cdot \mu/\gamma$, with K the hydraulic conductivity to water, μ water viscosity and γ its weight per unit volume) and (iii) mean pore throat diameter of the soil porous matrix d. In the case of quite poorly-dispersed spherically-shaped granular media, Gueven et al. [16] provided an accurate estimation of d value as a function of n and median particle size diameter D_{50}: $d = (0.1808n + 0.0069) \cdot D_{50}$. Alternatively, Kenney et al. [17] estimated diameter d_p of the largest particles that can permeate throughout the porous matrix of a soil under exam, as a function of its coefficient of uniformity C_u and fine fraction dimensions. Being pore throats the narrowest void areas of the porous matrix coincident with pore restrictions, the d_p value is strictly related to the d one and can be assumed as coincident as a first approximation. The injected soil is also herein assumed as isotropic, homogeneous without any micro-structural change with time; therefore, n, K and d values are kept constant [8].

Table 1. Rheological evolution of different NS grouts; weight fraction of SP and NaCl in %, characteristic time instants, initial and final values of μ_g and τ_0 and fitting parameters.

NSG	SP	NaCl	t_{SR}	$t_{gelling}$	t_{curing}	μ_g (0)	μ_g (t_{curing})	τ_0(0)	a	b	p
	[%]		[min]			[mPa·s]		[mPa]	[-]		
1	3	2	23	115	137	1	1	22	3	9	0
2	5	2	37	60	65	1	9.2	49	0	25	9
3	12.7	1.5	7.5	21.6	24	1.1	6.4	24.1	0	25	3
4	5	3	6	9.5	10	1	44	22.7	0	80	10.5
5	3	3	1	7	8.2	1	1	25	1	9	0

As for injected fluid and its permeation process, in [9, 15] and hereafter, the authors assume: (i) the flow regime to be laminar, (ii) the fluids (injected and interstitial ones) immiscible, (iii) gravity and capillarity effects negligible and (iv) head losses concentrated in the injected grout (with p_0 the constant initial interstitial fluid pressure). A quantification of the complex time-dependent rheological behaviour of NS grouts, considered to be used, is of primary importance to design an injection treatment. The time dependent relation between shear stress τ and shear strain rate $\dot{\gamma}$ for several NS grouts was for the first time accurately characterised by [8] and [9]. These authors performed a series of rheometric tests at constant $\dot{\gamma}$, within the range commonly encountered during in situ treatments ($0 < \dot{\gamma} \leq 100\,\mathrm{s}^{-1}$). All investigated NS grouts (enlisted in Table 1, derived by combining a commercial nanosilica aqueous suspension [18], with a correspondent commercial activator in different proportions) were found to behave as Binghamian fluids before gelling, with viscosity μ and yield stress τ_0 increasing from time t_{SR}. Once gelling time ($t_{gelling}$) is reached, an exponential increase in τ_0 is detected and for $t = t_{curing}$, curing is assumed to take place and any flow within soil pores is inhibited. For $t < t_{curing}$, μ and τ_0 are assumed to evolve with time t as follows:

$$\mu(t) = \mu(0) \cdot \left[1 + \frac{\mu(t_{curing}) - \mu(0)}{\mu(0)} \cdot \left(\frac{t}{t_{curing}}\right)^p\right] \tag{1}$$

$$\tau_0(t) = \tau_0(0) \cdot \left[1 + a \cdot \langle\frac{t - t_{SR}}{t_{curing}}\rangle + \langle e^{b \cdot \left(\frac{t}{t_{gelling}} - 1\right)} - 1\rangle\right], \tag{2}$$

where p, a and b are non-dimensional constitutive parameters changing according to the chemistry of the NS grout as in Table 1.

To practically prevent hydraulic fracturing and grout cavity expansion, injection pressure p_{inj} must be monitored not to overcome a threshold limit value ($p_{inj,MAX}$; [1, 19]). According to Park and Oh [20], such a threshold is equal to 15 kPa/m $\cdot\, d_{inj}$, being d_{inj} the treated depth. In practical applications for quite superficial treatments, for the sake of simplicity, a $p_{inj,MAX} = 10$ bar constant with depth is often employed.

2.2 Design Formulae and Outputs

To obtain closed-form solutions for modelling the single injection from each injection source, Boschi et al. [8, 9] proposed a modified Darcy's law, considering the time dependent Binghamian rheology of grouts. Flow rate Q is imposed to be constant, whereas the temporal evolutions of both radius r_g attained by the injected fluid front and p_{inj} are calculated as follows:

$$\begin{cases} p_{inj}(t) - p_0 = \frac{Q\mu(t)}{4\pi k} \cdot \left(\frac{1}{r_0} - \frac{1}{r_g(t)}\right) + G(t) \cdot \left(r_g(t) - r_0\right) \\ r_g(t) - r_0 \cdot \sqrt[3]{1 + \frac{Q}{V_{injn}}} \cdot t = 0, \end{cases} \tag{3}$$

where $G = 16 \cdot \tau_0(t)/(3d)$ and V_{inj} is the spherical injection source volume. These equations can be employed to design the treatment, mainly in terms of imposed (i) Q, (ii) injection time period Δt_{inj} (coincident with time instant $t_{inj,f}$ when the injection is stopped, being 0 the time instant at which it starts as well as the grout components are mixed) and (iii) NS grout composition to be employed.

System of Eqs. 3 is valid under the following constraints:

$$\begin{cases} p_{inj}(t) \leq p_{inj,MAX} \\ Q \geq \widehat{Q} \cong 2 \cdot \frac{A_{inj} k \gamma_g}{\mu_g(0)} \\ t_{gelling} \leq t_{inj,f} \leq t_{curing}, \end{cases} \tag{4}$$

where γ_g is the grout weight per unit volume, A_{inj} the injection source area and $\widehat{Q}$ the minimum flow rate value to be imposed for the treatment not to lose its effectiveness due to gravity effects. The formula for calculating $\widehat{Q}$ [8, 15], employed in Eq. 4.2, is obtained by assuming $\tau_0 = 0$ and temporal evolution of the grout rheology (grout hydraulic conductivity $K_g = k \cdot \gamma_g / \mu_g(0)$) and provides a conservative estimation of $\widehat{Q}$. Condition 4.3 is introduced to prevent any NS grout dispersion in the groundwater. The fulfillment of Condition 4.3 can be achieved by either modifying the imposed Q value or the NS grout composition.

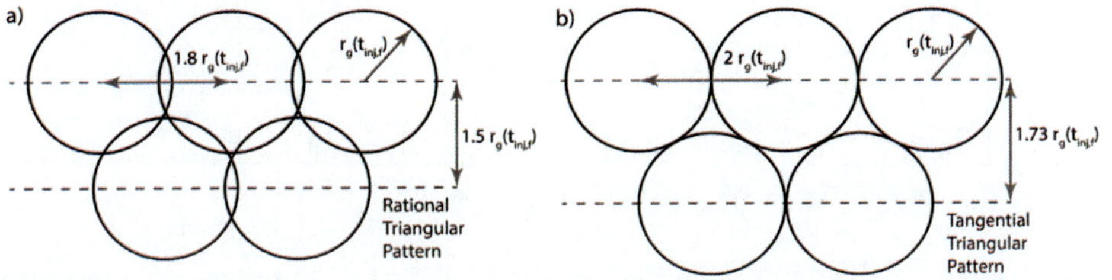

Fig. 1. Layout of two rows of grout holes (both in plan view and in section): a) as proposed by recommendations for tunnelling works typically related to water inflow control [21] and b) less demanding one [1].

Once identified the whole zone to be treated optimizing the treatment beneficial effects (the intervention must be localised in specific positions and its extension must be sustainable [22]), the design of the single injection allows to quantify the overall times and costs of the treatment. Moreover, total costs and times are a function of the employed technique (Sect. 2.1) and the designed injection patterns (common ones reported in Fig. 1), from which spacings along the three directions (S_x, S_y, S_z) are derived.

3 Model Employment

3.1 For a Feasibility Study

System of Eqs. 3 added to Constrains 4 are hereafter employed to identify the soils injectable by NS grout 3 (Table 1). In the reference case hereafter discussed, $r_0 =$ 5.8 cm (a typical value when the tube-à-manchette technique is employed [15]), $p_0 \cong$ 0, $p_{inj,MAX} = 10$ bar and $Q = 5$ l/min (a lower bound in practical applications). The soil is assumed to be mono-dispersed and characterised by a mean value $n = 0.35$. The k and d values are calculated by employing Kozeny-Carman [8] and Gueven et al. [16] formulas, respectively.

This reference case is employed by the authors to discuss the role of D_{50}s in the design of the treatment (Fig. 2). In particular, in Fig. 2a, the temporal evolution of p_{inj} with t is illustrated for different D_{50} values. When D_{50} is sufficiently small, the curves intersect the horizontal line obtained by imposing $p_{inj} = p_{inj,MAX}$, whereas for sufficiently large D_{50} values, the curves intersect the vertical line $t = t_{curing}$. From this Figure, the maximum injection time can be derived and the correspondent maximum r_g calculated (R_g, Eq. 3.2). This last is reported as a function of D_{50} in Fig. 2b.

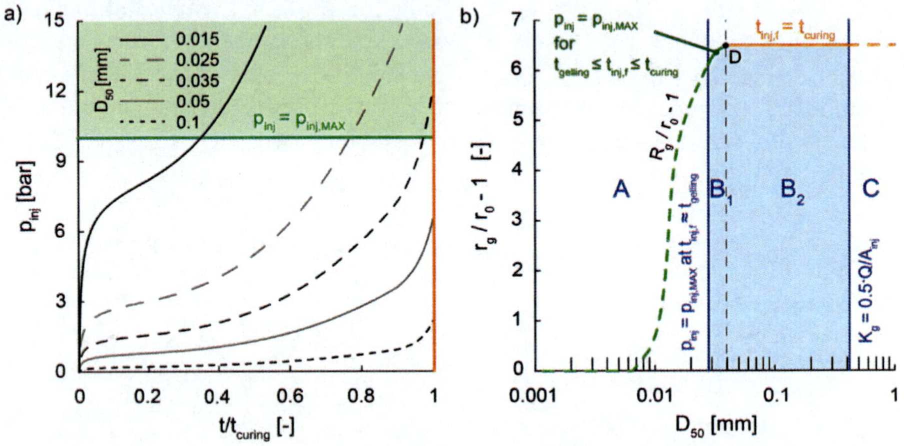

Fig. 2. Predictions of a) p_{inj} temporal evolutions in the case of mono-dispersed soils injected at $Q = 5$ l/min and b) maximum injectable grout front advancement R_g as a function of D_{50}.

In Fig. 2b, four zones are then identified: A, B1, B2 and C, these are defined by three vertical straight lines. Starting from the left, the first line corresponds to the fulfilment of conditions $p_{inj}(t_{inj,f}) = p_{inj,MAX}$ and $t_{inj,f} \cong t_{gelling}$, the second one of $p_{inj}(t_{inj,f}) = p_{inj,MAX}$ and $t_{inj,f} = t_{curing}$ (conditions referred to the most effective treatments), the third one of $Q = \widehat{Q}$. It is worth mentioning that the first fourth conditions listed hereabove depend on the analytical solution expressed by System of Eqs. 3, whereas the last one does not require the solution of 3. In zone A, the soil material is too fine, with respect to the grout rheological properties, to be injectable [23]; in zone C gravity effects become not negligible and the effectiveness of the treatment difficult to be optimized. In zones B1 and B2, the coloured subdomain allows to define the r_gs that can be injected. It is worth mentioning that in case NS grout as well as the injection device are assigned and the soil particle size distribution known, the subdomain of applicability of the treatment can be modified by changing the value of imposed Q.

3.2 Design Charts

The characteristic curves (System of Eqs. 3) shown in Fig. 3 can be employed for a first assessment of $t_{inj,f}$, R_g, Q and maximum reached p_{inj} value P_{inj}, corresponding to the one at $t = t_{inj,f}$. For any soil to be treated, NS grout employed and Q imposed and for a fixed $t_{inj,f}$, R_g and P_{inj} values can be predicted. These abaci can be used by imposing $t_{inj,f} = t_{curing}$ and $P_{inj} = p_{inj,MAX}$ (conditions for the most effective treatment as point D of Fig. 2b), deriving, from Fig. 3a, the Q value and, from Fig. 3b, the correspondent R_g. This approach is correct until the Q value calculated satisfies Constraint 4.2.

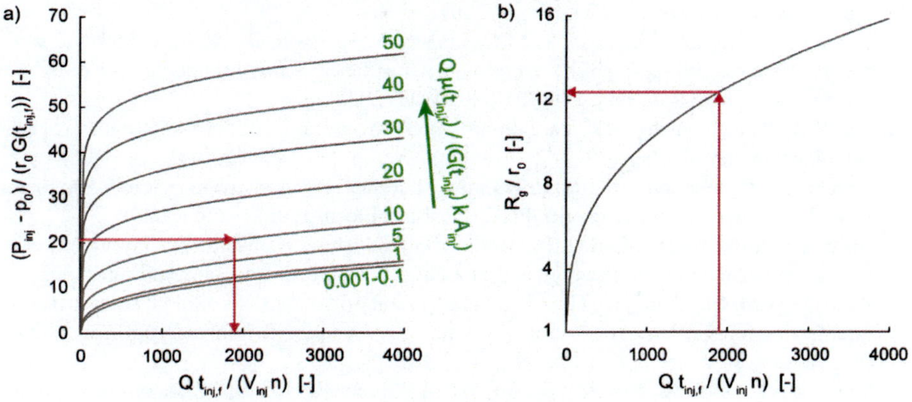

Fig. 3. Characteristic curves for permeation grouting treatments.

4 Conclusions

A simplified design method for permeation grouting treatments referring to nanosilica grouts has been herein proposed, by recalling all the key aspects to be taken into account in terms of imposed flow rate, injection times and pressures, chosen grout composition etc. Derived characteristic curves, predicting the maximum distance covered in-situ by the injected grout front, as well as the maximum injection pressure as a function of operational parameters and employed grout rheological properties, can be a useful tool for a preliminary treatment design. Then, by employing this just-proposed predictive model, permeation grouting with nanosilica grouts turn out to be effectively employed also in fine sands and coarse silty soils, conversely to microfine cements.

Acknowledgements. This research was funded by MBS Master Builders Solutions Italia S.p.A. within a research program aimed at both investigating and modelling the coupled hydro-mechanical processes governing grout injections.

References

1. Han, J.: Principles and Practice of Ground Improvement. Wiley, Hoboken (2015)
2. Boschi, K., Ciantia, M.O., Di Prisco, C.G.: Pressure grouting of microfine cements in soils: micromechanical processes. In: 20th International Conference on Soil Mechanics and Geotechnical Engineering, pp. 621–626. Australian Geomechanics Society (2022)

3. Manassero, V., Di Salvo, G.: Two difficult tunnelling problems solved by using permeation grouting: the excavation of submerged large size tunnels for Roma and Napoli metro projects. In: Grouting and Deep Mixing 2012, pp. 1972–1984 (2012)
4. Persoff, P., Apps, J., Moridis, G., Whang, J.M.: Effect of dilution and contaminants on sand grouted with colloidal silica. J. Geotech. Geoenviron. Eng. **125**(6), 461–469 (1999)
5. Gallagher, P.M., Pamuk, A., Abdoun, T.: Stabilization of liquefiable soils using colloidal silica grout. J. Mater. Civ. Eng. **19**(1), 33–40 (2007)
6. Salvatore, E., Modoni, G., Mascolo, M.C., Grassi, D., Spagnoli, G.: Experimental evidence of the effectiveness and applicability of colloidal nanosilica grouting for liquefaction mitigation. J. Geotech. Geoenviron. Eng. **146**(10), 04020108 (2020)
7. Iler, K.R.: The chemistry of silica. Solubility, polymerization, colloid and surface properties and biochemistry of silica (1979)
8. Boschi, K.: Permeation grouting in granular materials. From micro to macro, from experimental to numerical and viceversa. Ph.D. thesis. Politecnico di Milano (2022)
9. Boschi, K., di Prisco, C., Grassi, D., Modoni, G., Salvatore, E.: Nanosilica grout permeation in sand: experimental investigation and modeling. J. Geotech. Geoenviron. Eng. (2023)
10. Zhao, M., Liu, G., Zhang, C., Guo, W., Luo, Q.: State-of-the-art of colloidal silica-based soil liquefaction mitigation: an emerging technique for ground improvement. Appl. Sci. **10**(1), 15 (2019)
11. Spagnoli, G., Seidl, W., Romero, E., Arroyo, M., Gómez, R., López, J.: Unconfined compressive strength of sand-fines mixtures treated with chemical grouts. In: Geotechnical Aspects of Underground Construction in Soft Ground, pp. 829–835. CRC Press (2021)
12. Bolisetti, T., Reitsma, S., Balachandar, R.: Experimental investigations of colloidal silica grouting in porous media. J. Geotech. Geoenviron. Eng. **135**(5), 697–700 (2009)
13. Bouchelaghem, F., Vulliet, L., Leroy, D., Laloui, L., Descoeudres, F.: Real-scale miscible grout injection experiment and performance of advection–dispersion–filtration model. Int. J. Numer. Anal. Meth. Geomech. **25**(12), 1149–1173 (2001)
14. Coskun, S.B., Tokdemir, T.: Modelling of permeation grouting through soils. J. Appl. Eng. Sci. **10**(1), 11–16 (2020)
15. Boschi, K., Grassi, D., Castellanza, R., di Prisco, C.: Permeation grouting in soils: numerical discussion of a simplified analytical approach. In: Proceedings of the Institution of Civil Engineers – Ground Improvement (2023b)
16. Gueven, I., Frijters, S., Harting, J., Luding, S., Steeb, H.: Hydraulic properties of porous sintered glass bead systems. Granular Matter **19**(2), 1–21 (2017). https://doi.org/10.1007/s10035-017-0705-x
17. Kenney, T.C., Chahal, R., Chiu, E., Ofoegbu, G.I., Omange, G.N., Ume, C.A.: Controlling constriction sizes of granular filters. Can. Geotech. J. **22**(1), 32–43 (1985)
18. MBS Italia. Technical datasheet of MasterRoc MP 325 (2021)
19. Boschi, K., di Prisco, C.G., Ciantia, M.O.: Micromechanical investigation of grouting in soils. Int. J. Solids Struct. **187**, 121–132 (2020)
20. Park, D., Oh, J.: Permeation grouting for remediation of dam cores. Eng. Geol. **233**, 63–75 (2018)
21. AFTES. Recommendations on grouting for underground works. Tunnel. Underground Space Technol. **6**(4), 383–461 (1991)
22. Boschi, K., Arroyo, M., Burbano, D.A., Spagnoli, G.: Permeation grouting of an upstream tailing dam: a feasibility study. In: Proceedings of Tailings and Mine Waste Conference 2022 (2022)
23. Boschi, K., di Prisco, C.G., Grassi, D.: Investigation of microfine cement both rheological properties and permeation in soils. In: Proceedings of World Tunnel Congress 2023 (2023)

Effects of Soil Compaction and Water Retention Properties on the Analysis of Crack Patterns in an Earth Dam

Manuela Cecconi[1](✉), Giovanni Calabresi[2], and Vincenzo Pane[1]

[1] University of Perugia, Via G. Duranti 93, 06125 Perugia, Italy
manuela.cecconi@unipg.it
[2] University of Roma Sapienza, Rome, Italy

Abstract. The paper focuses on the investigation of the main causes responsible for the cracks pattern observed on the crest of an earth dam in Southern Italy. The history of the dam, from its construction to present (about 65 years of life), is quite complex. After the first appearance of a transversal crack on the dam crest, a set of longitudinal cracks gradually developed along a large portion of the crest. Extensive geotechnical investigations and detailed studies begun, leading to important restoration works in the late 90's. Despite of such works, the crack pattern was found to persist, so that the dam has been out of operation for over 15 years. Recently, a limited investigation has been carried out, including a geophysical survey of the whole dam, with the specific aim to detect and interpret the dam anomalies. The results of this campaign highlight the effects of initial soil compaction upon the hydro-mechanical embankment properties and the consequent dam behaviour.

Keywords: Earth dams · monitoring · geophysical tests · soil compaction · retention properties

1 The Abate Alonia Dam: Main Features and Geological Context

The Abate Alonia dam is an earth zoned-type dam located in Southern Italy (Basilicata Region), on the Olivento river [1]. Being among the earliest earth dams built in Italy in the 50's, the dam is a rather peculiar zoned-type dam with a more pervious central core formed by a mixtures of sandy-gravelly-silty alluvial material, placed between two less permeable backfill cores made of clayey-sandy silts; a transversal section is shown in Fig. 1a. The dam is 27.8 m high with a crest length of 1237 m, impounding a reservoir of 22.8 Mm^3 and located in a large sedimentary basin developing along the NW-SE direction. The foundation soils are grey-blue Plio-Pleistocene clays which lie below alluvial deposits of different age mainly consisting of sands and gravels. A cut-off, which constitutes the deepening of the central core in the foundation, intercepts the alluvial deposits by reaching the underlying impervious clayey formation. On the upstream side of the dam, the facing is made of a pervious rockfill.

A. Ferrari et al. (Eds.): CNRIG 2023, SSGG, pp. 629–637, 2023.
https://doi.org/10.1007/978-3-031-34761-0_76

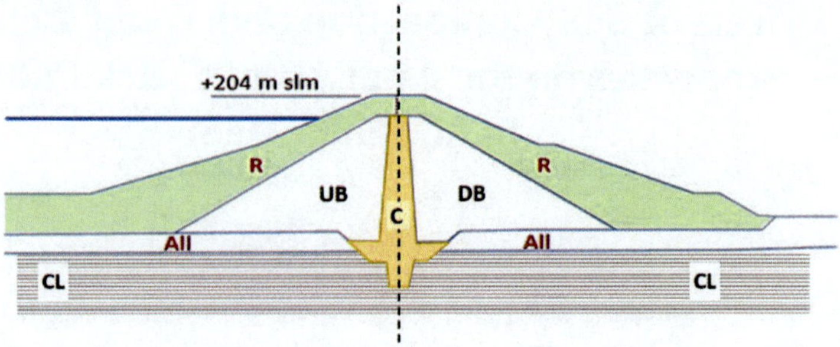

a)

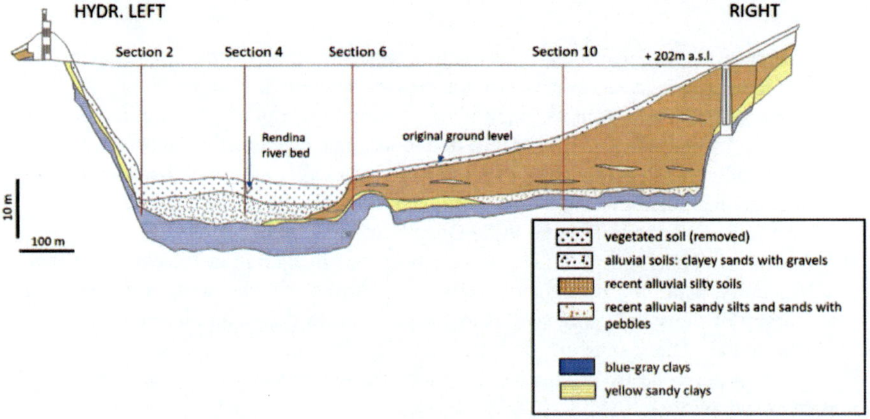

b)

Fig. 1. a) Schematic cross section (Section 4 of the dam) after restoration C: core/cutoff, UB and DB: original upstream/downstream backfills, All: alluvial soils, CL: blue-grey clays, R: enlarged backfills after restoration; b) longitudinal section.

The dam presents a variable height along its longitudinal section, mainly due to the varying elevation of the foundation soils and the geomorphology of the valley. The largest height of the embankment is found on the hydrographic left, corresponding to the bottom of the valley (Fig. 1b), where the blue clay formation outcrops. The recent alluvial deposits below the river bed have a variable thickness of 1 ÷ 7 m above the clayey foundation soils, while the thickness of the more ancient alluvial deposits – brownish in Fig. 1b - progressively increases moving toward the right shoulder.

2 Historical Notes and Observed Damage

The history of the dam, from its construction to the present day - approximately 65 years of life - is rather complex. In 1986, as far as 30 years after the first reservoir filling, there was a first appearance of a transversal crack on the crest, near the left abutment, extending to a depth of about 3 m. This first crack has then been followed – up to recent years - by longitudinal cracks, centimeters in opening, mostly appearing at the contact between the central core and the silty backfills. Extensive geotechnical investigations and studies on the behaviour of the dam thus began, leading to the conviction that the equilibrium conditions of the fine-grained dam's backfills were not satisfactory. However, the available documentation shows that it was not possible at that time to establish with certainty the nature and causes of the cracking phenomena.

Restoration works, which began in 1998, included: *i)* chemical and cement injections inside the core, and between the core and the backfills; *ii)* the reduction of the slope of the backfills and replacement of the external part of the backfills with granular soils; *iii)* a 2 m raising of the crest height, from 202 m a.s.l. to the actual height of 204 m a.s.l. (see Fig. 1a). The restoration works began in 1998 and were completed in 2001. With regard to the injection treatment, they were apparently carried out during the most serious phase of the deformation phenomenon, but their effect was not completely favorable and solving, as shown by inclinometer data obtained from 1999 to 2001.

In 2003, after a flood event that caused the reservoir to reach a height of 199.5 m in less than a month, the reservoir was rapidly emptied. At the end of 2005, for safety reasons, the drawoff tunnel was forced to remain mandatorily open, but a flood event in March 2006 led to a further major reservoir filling and subsequent rapid drawdown. At the beginning of 2006, considering the evolution of the observed cracks pattern, it was decided to held the reservoir below a maximum level of 196 m a.s.l.

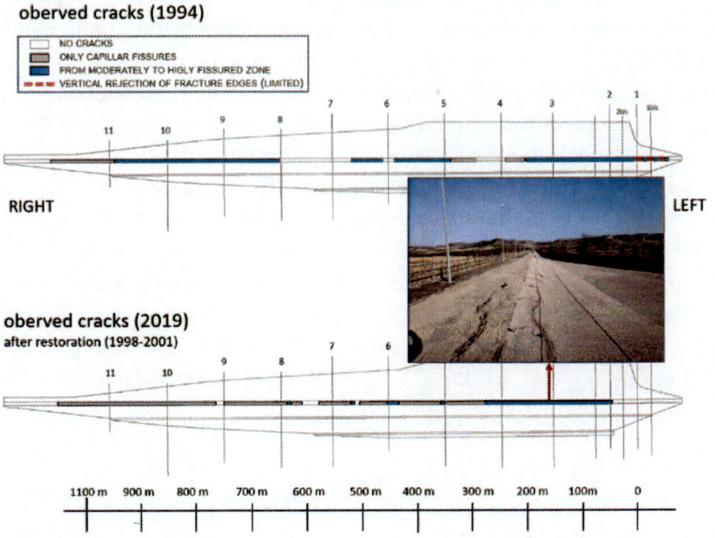

Fig. 2. Development of the observed crack pattern on the longitudinal dam section from topographic surveys in 1994 and 2019.

The dam has been fully instrumented since its construction, and the monitoring covers the last 65 years (1955–2020); it includes piezometers, inclinometers, displacement measurements by means of single/multi-base settlement gauges and topographic surveys. More recently, in 2006 and in 2020, several cross dam sections have been investigated further by means of geophysical tests. Nowadays, the Abate Alonia earth dam has been out of operation for more than 15 years. Comprehensive studies and investigations are currently undertaken in order to assess the possibility to dismiss the dam or – alternatively – carry out a new design for its maintenance and restoration. At the moment, the design activity is ongoing but this topic is beyond the scope of the present paper.

The evolution of the state of damage of the dam is given in Fig. 2 which schematically shows the results of two topographic surveys carried out in 1994 and 2019, respectively; the comparison of the two surveys reveals that the major cracks are located at the left shoulder. Further information regarding the cracks geometry and depth has been inferred from the recent geophysical investigation discussed in the following.

3 Retention Properties of the Embankment Dam

A new experimental campaign has been performed in 2020, consisting of 12 continuous core drilling boreholes carried out in the four investigated cross sections indicated in Fig. 1b. Four in situ-permeability Lefranc tests were performed in the central core, furnishing an average value of the permeability coefficient $k \approx 10^{-4} \div 10^{-5}$ cm/s, i.e., considerably larger than the values reported in the available documentation for the backfills. The dam core is thus more pervious than the backfills; this leads to a first, intrinsic anomaly in the behaviour of the dam, since the dam does not behave as a typical zoned dam with impervious core. During the recent investigation, water contents and index properties have been measured on about a hundred remoulded samples retrieved from the backfills, every 0.5 m in depth (see Fig. 3a). The same determinations were also made on a smaller number of samples from the dam core (every 2.5 of depth). The average values are reported in Table 1, while the plasticity chart for the core and backfills is shown in Fig. 3b. The large number of measurements clearly show that, for all investigated sections, the water content of the backfills (square symbols) is considerably greater than the one measured in the central core (triangles). The same observation holds for the index properties, indicating that the central core data lie in the lower part of the plasticity chart in Fig. 3b. For both the backfills, the measured water content (average values) has been found to be larger than the shrinkage limit, w_S, with values of the ratio w/w_S much larger than 1, and therefore reasonably wet of "optimum", as also depicted in Fig. 4. This implies that, during the embankment construction, the soil compaction has been only partially effective.

Table 1. In situ water content and index properties (average values).

	w_n (%)	w_L (%)	I_P (%)	w_S (%)	w/w_S	γ_d (kN/m^3)	S_r (%)
backfills	$15 \div 25$	$25 \div 52$	$10 \div 30$	13.5	1.48	16.5	73
core	$5 \div 14$	$20 \div 43$	$7 \div 25$	11.6	0.81	19.3	–

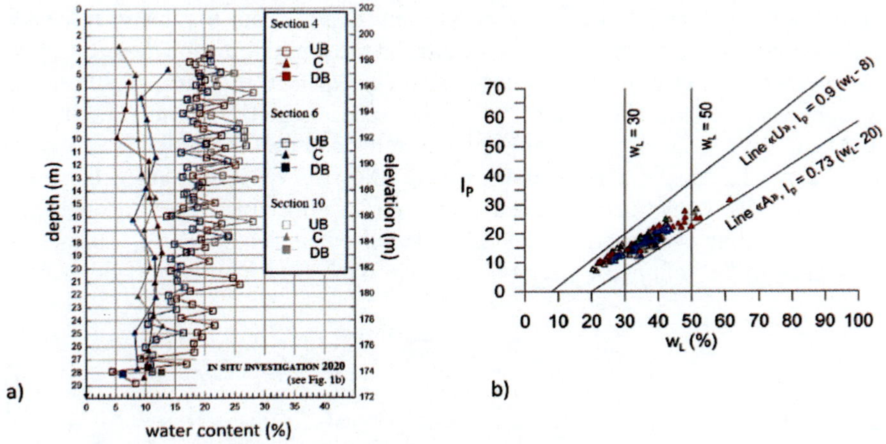

Fig. 3. In situ experimental campaign 2020: a) water content profile in the core and the backfills; b) plasticity chart: C core (open symbols), UB upstream/DB downstream backfills (full symbols).

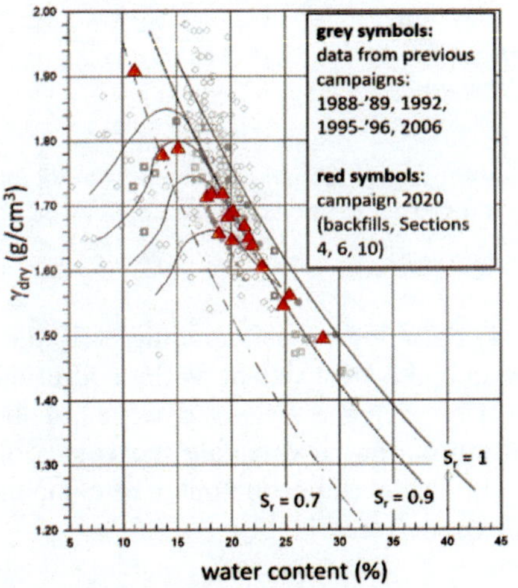

Fig. 4. Dry unit weight (γ_{dry}) *vs.* water content (see [1] for the compaction curves).

This finding is in agreement with the available construction documentation, indicating that soil compaction of both upstream and downstream backfills was carried out by low-energy compaction techniques and presumably without the necessary field control which is mandatory in current design practice.

In addition, considering the fine grained nature of the backfills, saturation and desaturation processes induced by repetitive cycles of reservoir filling/emptying have likely yielded irreversible volume strains, which are likely to be partially responsible for the observed dam behaviour. With the aim to validate this hypothesis and simulate the effects of repeated wetting cycles of the backfills, a limited laboratory testing programme has been carried out; four samples taken at different depths from the upstream backfill - cross Section 2, see Fig. 1b - were reconstituted at the liquid limit and then consolidated in oedometer cells until reaching a water content equal to the shrinkage limit. After fully submerged conditions for 24 h, 2 cycles of shrinkage (by drying) and heave (by wetting) at constant loads were performed. Figure 5a shows the progress of measured axial strains with time; the axial strains produced by the 1st shrinkage cycle are of the order of 3 ÷ 4% while the permanent strains produced by the first shrinkage and the following swelling cycle are relatively high, varying between 0.6 and 1.6%, and likely to be related to the volumetric strains associated to the crack patterns on the dam crest.

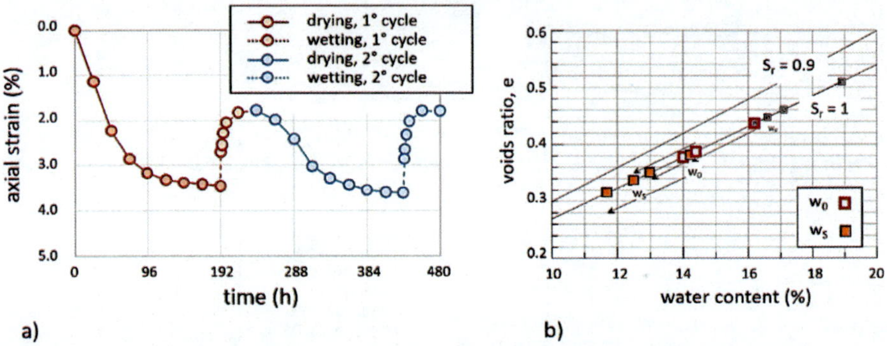

Fig. 5. a) Cycles of drying/wetting on samples retrieved from the upstream backfill (z = 10 m from the dam crest, σ_V = 200 kPa): measured axial strains *vs.* time; b) calculated voids ratio vs. water content.

The available data do not allow to define completely the *shrinkage curve*, useful to estimate the voids ratio and water content as the soil dries as well as the air-entry value of the soil [2, 3]. However, the measured values of the shrinkage limit and of the corresponding void ratio allow to calculate the values of porosity reduction and volumetric strain due to shrinkage, which turn out to be comparable with those measured in the laboratory wetting cycles (see Fig. 5b).

4 Recent Geophysical Investigations for the Detection of Dam Anomalies

In addition to traditional geotechnical monitoring of the dam, geophysical surveys were carried out firstly in 2006 and more recently in 2020 with the specific aim to identify the pattern and extension of the observed crest damage and its possible causes. More precisely, the 2020 survey has been extended to the whole dam, with the purpose of: *i)* locate anomalies due to possible inhomogeneities in the central core and its backfills; *ii)* estimating the depth of the cracks appearing at the crest; *iii)* evaluating the soil stiffness profile and distribution along the longitudinal development of the dam by means of *P*,

S wave velocity measurements. The geophysical investigation consisted of a seismic tomography survey along four transversal sections and along the longitudinal axis of the dam. Along the longitudinal profile of Fig. 6 the V_S contours clearly indicate the top boundary of the foundation blue clays, as well as the presence, in the left portion of the embankment, of a large area of the dam characterized by lower V_S and shear stiffness. A comparison between Section 4 and Section 10 in Fig. 7b also shows the considerable inhomogeneity of the embankment material, as well as the presence of weaker zones (e.g., orange/brownish zone in Fig. 7a where the values of V_S are even lower than 250 m/s, probably related to a lower density in the upstream backfill.

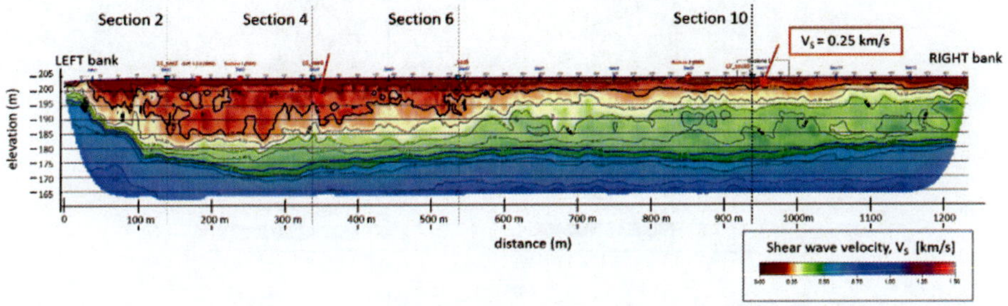

Fig. 6. Shear wave velocity (V_S) contour along the longitudinal axis of the dam (from seismic tomography).

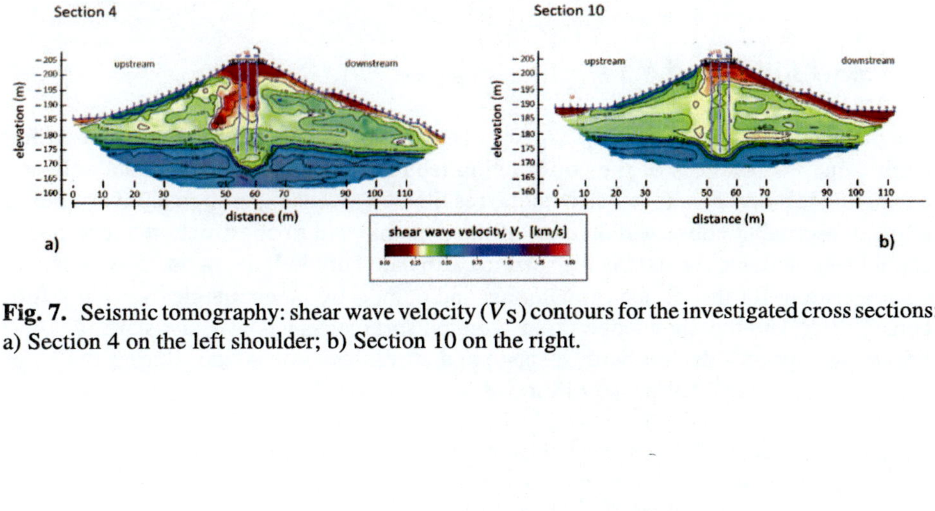

Fig. 7. Seismic tomography: shear wave velocity (V_S) contours for the investigated cross sections: a) Section 4 on the left shoulder; b) Section 10 on the right.

In addition, four Cross-Hole Tomographic (CHT) sections were carried out, each CHT section being carried out across three boreholes spanning along the crest width. One of such sections is given in Fig. 8, indicating – with a lighter pink color – the presence of two symmetrical vertical fissures located between the central core and the backfills and extending down to a depth of ≈4 m, approximately.

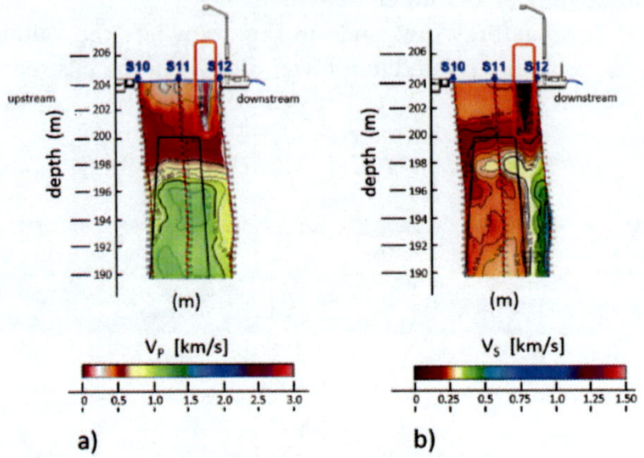

Fig. 8. CHT tomography: wave velocity contours in the dam core (Section 2, left shoulder: a) P-waves; b) S waves

5 Concluding Remarks

The paper summarises only a few aspects of the complex behaviour of the Abate Alonia earth dam. The analysis of the construction techniques and monitoring data collected during the last decades, together with the results of a recent experimental investigation, allow to ascribe the observed damage of the embankment to construction anomalies. It should be noted that the dam is the second earth dam built in Italy, in the '50s, when the design criteria for the choice of materials and compaction technologies were not fully consolidated and the knowledge of unsaturated soil mechanics was just starting. Other important aspects - dealing with seepage and stress-strain behaviour during the entire life of the dam – will be further discussed.

Acknowledgements. This research is part of a Research Agreement (2019–2021) between the Department of Engineering of University of Perugia and the Consorzio di Bonifica della Basilicata. The Authors wish to thank Engs. S. Gravino, M.C. Leone and M. Marchitelli (Consorzio di Bonifica della Basilicata), Dr. M. Furani (Progeo) for helpful discussions and support during the progress of the study. Dr. A. Vecchietti, Geol. V. Cerboni and Eng. D. Bellavita are also greatly acknowledged for their useful collaboration.

References

1. Sollazzo, R.: Diga di terra sul Torrente Rendina. Geotecnica, no. 1 (1958)
2. Fredlund, D.G.: Determination of unsaturated soil property functions for engineering practice. In: Proceedings of the 17th African Regional Conference on Soil Mechanics and Geotechnical Engineering, ISSMGE, Cape Town, 7–9 October (2019)
3. Marinho, F.: Fundamentals of soil shrinkage. In: Proceedings of the PanAm Unsaturated Soils Conference, Dallas, Texas, 12–15 November, pp. 198–222 (2017)

A Novel Method for Assessing Pile Base Resistance in Sand

Raffaele Cesaro(✉), Raffaele Di Laora, and Alessandro Mandolini

Department of Engineering, Università della Campania 'Luigi Vanvitelli', 81031 Aversa, CE, Italy

{raffaele.cesaro,raffaele.dilaora, alessandro.mandolini}@unicampania.it

Abstract. The base resistance of piles in sand is generally assessed through limit equilibrium-based theoretical methods. However, this approach is not capable of accounting for the punching mechanism, resulting in unconservative predictions of the ultimate load, intended as the one corresponding to a certain tolerable settlement. To overcome this issue, in this paper a novel simple analytical method for the prediction of pile tip load-settlement response is proposed. The method is mainly based on the spherical cavity expansion theory, the input soil parameters being: critical state friction angle, relative density, initial mean effective pressure and initial elastic stiffness. The reliability of the proposed method is verified against a database including 50 in-situ pile load tests performed worldwide.

Keywords: Pile bearing capacity · Pile base response · Settlement-based pile design · Spherical cavity expansion theory · Analytical method

1 Introduction

For cast-in-situ piles in sandy soils the pile end-bearing resistance mobilization requires large relative displacement, which may reach values equal to 100% of the pile diameter and beyond. Nevertheless, base resistance of piles in sand is generally assessed by theoretical approaches based on the limit equilibrium method. Several expressions for the well-known bearing capacity factors are available from literature; among these, the ones proposed by Berezantzev et al. (1961, 1965) are perhaps the most commonly used in current practice. The bearing capacity factors estimated by this approach are associated to pile settlements which are always inadmissible from an engineering perspective and therefore their use appears questionable from a practical viewpoint. Furthermore, since the experimental detection of the pile resistance is carried out from the results of load tests run usually until a relative settlement *w/d* at most equal to 10%, there is a mismatch between the theoretical evaluations and the experimental data. To comply with the need for ultimate loads associated to a certain settlement level, designers should assess the entire load-settlement curve, and this operation is generally performed numerically via commercial software. This operation is particularly complex due to the advanced constitutive modelling for soil behaviour, effect of dilatancy and large deformations. To furnish

A. Ferrari et al. (Eds.): CNRIG 2023, SSGG, pp. 638–645, 2023.
https://doi.org/10.1007/978-3-031-34761-0_77

a ready-to-use tool for practitioners, in this paper a novel simple analytical method for the prediction of the load-settlement response at pile base is proposed, allowing for an estimation of the pile base load corresponding to a given displacement. The proposed method is based on the well-known analogy between the spherical cavity expansion and the punching mechanism at pile base.

2 Theoretical Framework

2.1 Cavity Expansion Theory

As stated by Ladanyi and Johnston (1974) and Yu (2000), from observations made in deep punching of metals, Bishop et al. (1945) and Hill (1950) first concluded that the pressure required to produce a deep hole in an elastic-plastic medium is proportional to that necessary to expand a cavity of the same volume and under the same conditions. Moreover, Yasufuku et al. (2001) argue that many authors have observed that the deformation bulb beneath a loaded pile tip strongly resembles a spherical cavity expanded in an infinite medium. Gibson (1950) was the first to correlate cavity expansion limit pressure to tip bearing capacity of a deep foundation, and many authors followed his hint over the years. In order to use cavity expansion theory, to predict the pile base load-settlement behavior, (a) analytical solutions to compute the pressure-expansion behavior of the cavity are needed and (b) semi-analytical correlation are required to relate end bearing resistance to the cavity pressure and pile base settlement to the cavity expansion (Yu 2000). For the step (a) in this work the solution proposed by Yu and Houlsby (1991) is adopted. The latter is based on the hypothesis of elastic-plastic soil with Mohr-Coulomb yield criterion with a non-associated flow rule and no restrictions are imposed on the magnitude of the deformations. Thus, the properties of the soil are defined by the Young's modulus E, Poisson's ratio ν, cohesion c, angle of friction φ and angle of dilation ψ. Since the response prediction of cast-in-situ piles in sandy soils is the aim of this work, the cohesion is always assumed as equal to zero. Due to space limitations, the mathematical formulation of the above solution is not shown herein. Step (b) is analyzed in detail in the next Section.

2.2 Pile Base Response

The correlation between the pile base resistance and the cavity pressure is obtained assuming, as proposed by Gibson (1950), Vesic (1977), Ladanyi and Johnston (1974), Randolph et al. (1994) and Yasufuku et al. (1995, 2001), that during the penetration of the pile a rigid cone of soil is formed at the pile base and that the normal pressure acting on the lateral surface of the cone is equal to the cavity expansion pressure (Fig. 1). Without discussing the validity of such assumption, it is noted that the vertical section of the cone is a triangle with a base angle equal to 45°+φ/2; it follows that the correlation between the pile base normal stress and the cavity pressure is given by:

$$q_b = p_{cavity} \frac{1}{1 - sen\varphi} \tag{1}$$

Furthermore, some simplified hypotheses are adopted in this work:

(a) the initial radius of the cavity is equal to that of the pile
(b) the pile tip settlement is equal to the increase in the cavity radius
(c) the elastic stiffness of the pile base k_b is equal to that of a rigid disk on an elastic half space (Randolph and Wroth 1978):

$$k_b = \frac{2\,d\,G_0}{1-\nu} \tag{2}$$

where G_0 is the shear modulus of the soil and d is the pile diameter.

The hypothesis (c) requires that employing the formulation proposed by Yu and Houlsby (1991) a "modified" stiffness must be adopted. Given the hypothesis (a) and (b) the "modified" stiffness is calculated as:

$$\frac{G_{cavity}}{G_0} = \frac{(1-sen\varphi)}{\pi(1-\nu)} \tag{3}$$

Given the ordinary values of the friction angle and the Poisson's ratio, the above stiffness is typically equal to 0.125 ÷ 0.225 G_0. Note that different assumptions can be made instead of hypothesis (b); for example, one could equate the volume variation in the cavity and beneath pile base. However, a sensitivity study has shown that, since the initial stiffness is fixed and equal to its elastic value, the results are not significantly sensitive to the specific hypothesis adopted.

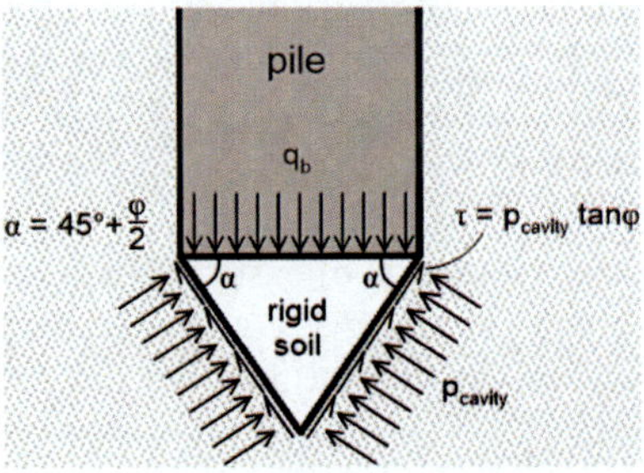

Fig. 1. Correlation between cavity expansion pressure and pile base normal stress.

2.3 Parameters Required

In Yu and Houlsby (1991) solution the friction and dilation angles are assumed constant. Although this is a simplified approximation, since both angles depend on the stress history of the soil in reality, Collins et al. (1992) have shown that adopting average values between the initial state and ultimate state leads to realistic results. Following the suggestions provided by Collins et al. (1992) and the work of Bolton (1986) on the dilatant behavior of sands, Randolph et al. (1994) proposed to estimate the friction and dilatant angles directly to the relative density D_R and the mean effective stress p'_0:

$$\varphi = \varphi_{cv} + 1.5 I_R \tag{4}$$

$$\psi = 1.875 I_R \tag{5}$$

where

$$I_R = 5D_R - 1 \qquad for\ p_0' \le 150\ \text{kPa} \tag{6a}$$

$$I_R = D_R\left[5.4 - \ln\left(p_0'/p_a\right)\right] - 1 \qquad for\ p_0' > 150\ \text{kPa} \tag{6b}$$

and p_a is the atmospheric pressure (=100 kPa).

If follows that the 5 parameters requested as input are:

(a) the mean effective stress p_0'
(b) the critical state friction angle φ_{cv}
(c) the relative density D_R
(d) the soil shear modulus at low strain G_0
(e) the soil Poisson's ratio v

3 Verification from 50 Pile Load Tests

The reliability of the proposed method was checked against 50 well-documented in-situ pile load tests performed worldwide, cast in situ (bored, Continuous Flight Augering and Full Displacement Pile) in sandy soils. For each load test the load-settlement curve at pile base is available and the subsoil condition are well known. The pile load test database involves several piles geometries (diameters from 0.4 m to 2 m and lengths from 6 m to 91 m). In Table 1 for each case history the reference, pile type, pile geometry and soil parameters are reported (the value of the Poisson's ratio is always set equal to 0.2). In Fig. 2 the good performance of the proposed method is shown for two case histories involving very different pile geometries and different installation technologies. Furthermore, in Fig. 3 the predicted bearing capacity factors N_σ, defined as q_b/p_0' (Vesic 1977), are compared with the measured or extrapolated values from the load tests results, for two different values of relative pile base settlement *w/d* equal to 5% and 10%. Also shown in the graph are the predictions by the Berezantzev method (1961). It is noted that the proposed method still produces results which are, in average, slightly unconservative, i.e. about 26% for $w/d = 5\%$ and 30% for $w/d = 10\%$. This may be attributed to that fact that pile base resistance in load tests is detrimentally affected by the disturbance produced by the installation process, not taken into account in the method. In support of this interpretation, most of the database case histories (32/50) involve bored piles for which, as well known, the above-mentioned effect is very crucial. On the other hand, the case histories associated to conservative predictions almost always refer to FDP piles, for which the installation process generally results in an improvement of the soil conditions. However, the error associated to the proposed method is much lower than that obtained from the most widely used methods in the current practice (Cesaro et al. 2022). For example, Berezantzev method (Fig. 3) overestimates, on the average, by about 126% the bearing capacity factor measured, or extrapolated, at a settlement of 10%d.

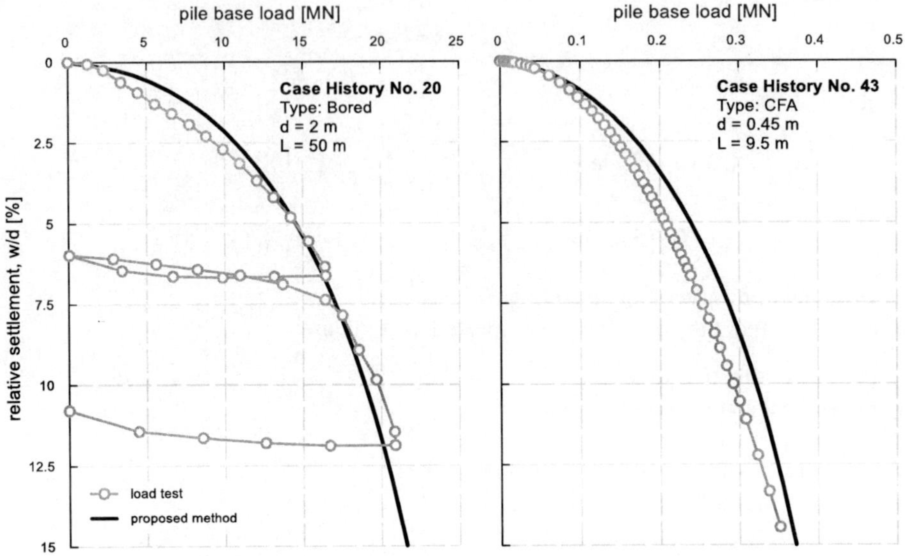

Fig. 2. Load test result against load-settlement curve predicted by the proposed method for two case histories.

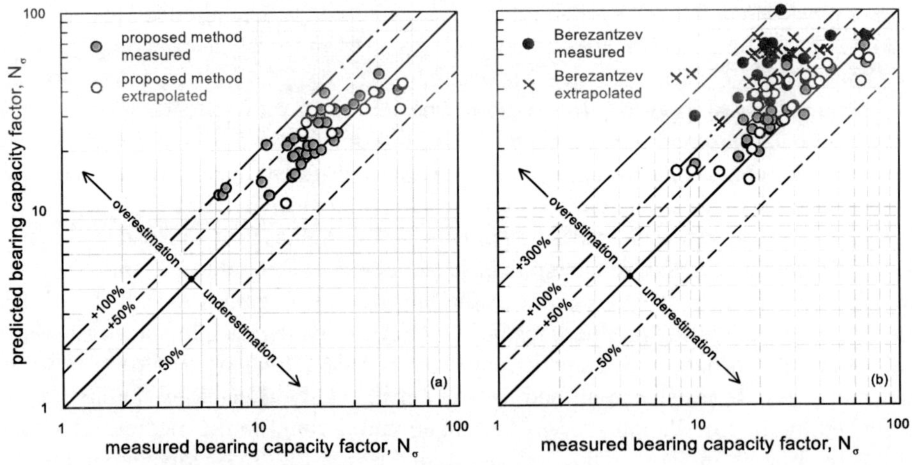

Fig. 3. Measured bearing capacity factors against predicted ones for a relative settlement w/d = 5% (a) and w/d = 10% (b).

Table 1. Load tests database

N.	Case History	Reference	Type	d [m]	L [m]	φ_{cv} [°]	D_R [%]	p'_0 [kPa]	$G_{b,0}$ [MPa]
1	Huston Test site - BB (USA)	Touma & Reese (1972)	Bored	0.76	13.7	33	85	80	70
2	Huston Test site - G1 (USA)	Touma & Reese (1972)	Bored	0.93	18.0	33	50	90	82
3	Huston Test site - G2 (USA)	Touma & Reese (1972)	Bored	0.76	23.0	33	50	117	90
4	Law Court Building (Italy)	Viggiani & Vinale (1983)	Bored	1.50	42.0	32	40	205	72
5	Law Court Building (Italy)	Viggiani & Vinale (1983)	Bored	2.00	42.0	32	40	205	72
6	Bologna-Verona Railway P8 (Italy)	Rocchi et al. (1989)	Bored	1.20	30.0	34	70	180	126
7	Bologna-Verona Railway P9 (Italy)	Rocchi et al. (1989)	Bored	1.20	30.0	34	70	180	126
8	Hanshin Expressway (Japan)	Matsui (1993)	Bored	1.20	36.3	30	40	320	110
9	Georgia Tech Test site (USA)	Mayne & Harris (1993)	Bored	0.76	16.8	30	25	200	95
10	Holiday Inn & Office Tower (Italy)	Mandolini (1994)	CFA	0.60	24.0	34	40	120	85
11	Holiday Inn & Office Tower (Italy)	Mandolini (1994)	CFA	0.60	24.0	34	40	120	85
12	9th JGS (Japan)	Yasufuku et al. (2001)	Bored	0.40	12.0	36	90	74	120
13	27th JGS (Japan)	Yasufuku et al. (2001)	Bored	1.20	38.0	33	70	190	130
14	Mount Pleasant Test Site - MP1 (USA)	Camp et al. (2002)	Bored	2.40	48.2	33	90	184	146
15	Mount Pleasant Test Site - MP2 (USA)	Camp et al. (2002)	Bored	2.40	48.0	33	90	184	146
16	Mount Pleasant Test Site - MP3 (USA)	Camp et al. (2002)	Bored	2.40	33.5	33	90	128	125
17	Mount Pleasant Test Site - MP4 (USA)	Camp et al. (2002)	Bored	1.80	32.6	33	90	125	124
18	Charlestone Test Site - C1 (USA)	Camp et al. (2002)	Bored	2.40	47.9	33	90	183	146
19	Charlestone Test Site - C2 (USA)	Camp et al. (2002)	Bored	2.40	48.2	33	90	184	146
20	Charlestone Test Site - C3 (USA)	Camp et al. (2002)	Bored	2.40	34.3	33	90	131	126
21	Charlestone Test Site - C4 (USA)	Camp et al. (2002)	Bored	1.80	33.1	33	90	126	124
22	Drum Island Test Site - DI1 (USA)	Camp et al. (2002)	Bored	2.40	48.3	33	90	153	135
23	Drum Island Test Site - DI2 (USA)	Camp et al. (2002)	Bored	1.80	35.1	33	90	102	114
24	Baytown Test site (USA)	O'Neil et al. (2002)	CFA	0.46	15.2	33	85	60	85
25	Poggiomarino Treatment Plan (Italy)	Mandolini et al. (2002)	CFA	0.80	24.0	32	50	124	93
26	Poggiomarino Treatment Plan (Italy)	Mandolini et al. (2002)	CFA	0.60	24.0	32	50	124	93
27	Padma River Bridge (Bangladesh)	Castelli et al. (2004)	Bored	1.50	91.0	36	90	547	235
28	ISC 2002 test site - E9 (Portugal)	Fellenius et al. (2007)	Bored	0.60	6.0	32	35	70	65
29	ISC 2002 test site - T1 (Portugal)	Fellenius et al. (2007)	CFA	0.60	6.0	32	35	70	65
30	Piacenza Bridge (Italy)	Becci et al. (2007)	Bored	2.00	50.0	31	50	316	140
31	Piacenza Bridge (Italy)	Becci et al. (2007)	Bored	2.00	55.0	31	50	350	145
32	New education building N.54 (Poland)	Gwidzala et al. (2009)	FDP	0.72	11.0	33	80	90	100
33	New education building N.207 (Poland)	Gwidzala et al. (2009)	FDP	0.53	11.0	33	80	90	100
34	New education building N.236 (Poland)	Gwidzala et al. (2009)	FDP	0.72	11.0	33	100	90	115
35	New education building N.274 (Poland)	Gwidzala et al. (2009)	FDP	0.53	9.5	34	100	78	110
36	Killarney test site (Ireland)	Gavin et al. (2013)	CFA	0.45	15.0	31	40	104	80
37	Killarney test site (Ireland)	Gavin et al. (2013)	CFA	0.80	14.0	31	40	100	80
38	FHWA/LA. 12/495 Report - DS18 (USA)	Abu-Farsakh et al. (2013)	Bored	1.20	29.3	33	35	305	125
39	Ostend oil tank site (Belgium)	Van Impe et al. (2013)	FDP	0.46	21.6	32	60	120	100
40	Portsmouth test site (USA)	Niazi & Mayne (2014)	Bored	1.53	26.0	32	60	135	200
41	Araquarì test site (Brazil)	Nienov (2016)	Bored	1.00	24.0	31	40	92	92
42	Araquarì test site (Brazil)	Nienov (2016)	Bored	1.00	24.0	31	40	92	92
43	B.E.S.T. B2 (Bolivia)	Fellenius (2017)	CFA	0.45	9.5	34	70	40	60
44	B.E.S.T. C2 (Bolivia)	Fellenius (2017)	FDP	0.45	9.5	34	70	47	80
45	Danzica test site (Poland)	Krasinski & Wiszniewski (2021)	CFA	0.40	12.6	32	65	80	97
46	Elbląg test site (Poland)	Krasinski & Wiszniewski (2021)	CFA	0.40	7.5	34	70	50	90
47	John F Fitzgerald Expressway (USA)	FHWA Database	Bored	1.22	45.0	32	20	510	140
48	Agorà (Italy)	Mandolini Private Communication	FDP	0.62	17.0	29	45	100	70
49	Wind Farm WTG01 (Italy)	Mandolini Private Communication	Bored	1.20	28.0	32	35	255	322
50	Wind Farm WTG05 (Italy)	Mandolini Private Communication	Bored	1.20	30.0	30	35	325	365

4 Conclusions

In this paper a novel method for predicting the load settlement curve of piles in sand is proposed. The proposed method is manly based on the cavity expansion theory and the soil parameters needed as input are easy to estimate from ordinary in-situ and in laboratory tests. A databased of 50 pile load tests was developed and the performance of the method was checked against the experimental results. The proposed method allows a performance-based design computing the pile base resistance for a given maximum tolerable settlement, furthermore, offers a much better performance as compared to approaches commonly employed in routine engineering, like the method by Berezantzev et al. (1961). Further research is needed to improve the performance of the method by including the influence of the installation process on the pile base response.

References

Abu-Farsakh, M.Y., Chen, Q., Haque, M.: Calibration of resistance factors for drilled shafts for the new FHWA design method. Report No. FHWA/LA. 12/495, Louisiana Transportation Research Center (2013)

Becci, B., Nova, R., Baù, A., Haykal, R.: Prove di carico su pali di grande diametro mediante l'impiego di celle Osterberg. Rivista Italiana di Geotecnica. **4**, 9–28 (2007)

Berezantzev, V.C., Khristoforov, V., Golubkov, V.: Load bearing capacity and deformation of piled foundations. In: Proceedings of the V ICSMFE, Paris, vol. 2, pp. 11–15 (1961)

Berezantzev, V.C.: Design of deep foundations. In: Proceedings of the VI ICSMFE, Montreal, vol. 2, pp. 234–237 (1965)

Bolton, M.D.: The strength and dilatancy of sands. Geotechnique **36**(1), 65–78 (1986)

Bishop, R.F., Hill, R., Mott, N.F.: The theory of indentation and hardness tests. Proc. Phys. Soc. **57**, 147–159 (1945)

Castelli, R.J., Wilkins, E.: Osterberg load cell test results on base grouted bored piles in Bangladesh. In: Proceedings of the GeoSupport 2004, Orlando, FL, USA, pp. 1–16 (2004)

Camp, W.M., Brown, D.A., Mayne, P.W.: Construction method effects on axial drilled shaft performance. In: Proceedings International Deep Foundations Congress, 193–208 Geotechnical Special Publication No. 116, Orlando, Florida (2002)

Cesaro, R., Di Laora, R., Mandolini, A.: Sulla resistenza alla base di pali trivellati in sabbia. Atti dell'Incontro Annuale dei Ricercatori di Geotecnica – IARG 2022. Edizioni AGI, Roma (2022). ISBN 9788897517108

Collins, L.F., Pender, M.J., Wang, Y.: Cavity expansion in sands under drained loading condition. Int. J. Numer. Anal. Methods Geomech. **16**, 3–23 (1992)

Fellenius, B.H., Santos, J.A., Viana da Fonseca, A.: Analysis of piles in a residual soil—the ISC'2 prediction. Can. Geotech. J. **44**(2), 201–220 (2007)

Fellenius, B.H.: Report on the B.E.S.T. prediction survey of the 3rd CBFP event. In: Proceedings of the 3rd Bolivian International Conference on Deep Foundations, Santa Cruz de la Sierra, Bolivia, 27–29 April, vol. 3, pp. 7–25 (2017)

Gavin, K.G., Cadogan, D., Tolooiyan, A., Casey, P.: The base resistance of non-displacement piles in sand. Part I: Field tests. Proc. Inst. Civ. Eng. **166**(6), 540–548 (2013)

Gibson, R.E.: Correspondence. J. Inst. Civil Eng. **34**, 382–383 (1950)

Gwizdala, K., Krasinski, A., Brzozowski, T.: The assessment of load-settlement curve for Atlas piles correlated with CPT rests. In: Proceedings of the Fifth International Symposium on Deep Foundations on Bored and Auger Piles (BAP V), Ghent, Belgium, pp. 121–126 (2009)

Hill, R.: The Mathematical Theory of Plasticity. Oxford University Press (1950)

Krasiński, A., Wiszniewski, M.: Identification of residual force in static load tests on instrumented screw displacement piles. Studia Geotechnica et Mechanica (2021)

Ladanyi, B., Iohnston, G.H.: Behaviour of circular footings and plate anchors embedded in permafrost. Can. Geotech. J. **11**, 531–553 (1974)

Mandolini, A.: Cedimenti di fondazioni su pali. Ph.D. thesis. Department of Geotechnical Engineering,s University of Napoli Federico II (1994)

Mandolini, A., Ramondini, M., Russo, G., Viggiani, C.: Full scale loading tests on instrumented CFA piles. In: Proceedings of the International Deep Foundations Congress 2002 – International Perspective on Theory, Design, Construction, and Performance, Orlando, USA, pp. 1088–1097. ASCE Geotechnical Special Publication (2002)

Matsui, T.: Case studies on cast-in-place bored piles and some considerations for design. In: Proceedings of the Second International Symposium on Deep Foundations on Bored and Auger Piles (BAP II), Rotterdam, Netherlands (1993)

Mayne, P., Harris, D.: Axial load-displacement behaviour of drilled shaft foundations in piedmont residuum. Research report. Georgie Tech Research Corporation, Geotechnical Engineering Division, Georgia Institute of Technology, Atlanta (1993)

Nienov, F.A.: Desempegnho de estacas escavadas de grande diâmetro em solo arenoso sob carregamento vertical. Ph.D. thesis. Escola De Engenharia. Universidade Federal do Rio Grande do Sul (2016)

Niazi, F.S., Mayne, P.W.: Axial pile response of bidirectional O-cell loading from modified analytical elastic solution and downhole shear wave velocity. Can. Geotech. J. **51**(11), 1284–1302 (2014)

Randolph, M., Dolwin, J., Beck, R.: Design of driven piles in sand. Geotechnique **44**(3), 427–448 (1994)

Randolph, M.F., Wroth, C.P.: Analysis of deformation of vertically loaded piles. J. Geotech. Eng. ASCE **104**(12), 1465–1488 (1978)

Rocchi, F., Albert, L.F., Vacca, O., Nardacci, A., Salvi, M., Montinaro, N.: Prove di carico strumentate a rottura su pali di grande diametro trivellati in sabbia. In: Proceedings XVII Convegno Nazionale di Geotecnica, Taormina, vol. 1, pp. 309–322 (1989)

Touma, F.T., Reese, L.C.: The behaviour of axially loaded drilled shafts in sand. Research report. Center for Highway Research, University of Texas, Austin (1972)

Van Impe, P.O., Van Impe, W.F., Seminck, L.: Discussion of an instrumented screw pile load test and connected pile group load-settlement behaviour. J. Geo-Eng. Sci. **1**(1), 13–36 (2013)

Vesic, A.S.: Design of Pile Foundations: National Cooperative Highway Research Program, Synthesis Highway Practice Report No. 42, Transport Research Board, Washington, DC (1977)

Viggiani, C., Vinale, F.: Comportamento di pali trivellati di grande diametro in terreni piroclastici. Rivista Italiana di Geotecnica **17**(2), 59–84 (1983)

Yasufuku, N., Hyde, A.F.L.: Pile end-bearing capacity in crushable sands. Geotechnique **45**(4), 663–676 (1995)

Yasufuku, N., Ochiai, H., Ohno, S.: Pile end-bearing capacity of sand related to soil compressibility. Soils Found. **41**(4), 59–71 (2001)

Yu, H.S., Houlsby, G.T.: Finite cavity expansion in dilatant soil: loading analysis. Geotechnique **41**, 173–183 (1991)

Yu, H.: Cavity Expansion Methods in Geomechanics. Springer, Heidelberg (2000)

Finite Element Analyses of Piled Foundations: Interaction Domains Under Undrained Conditions

Matteo Corigliano[1](✉), Luca Flessati[2], and Claudio di Prisco[1]

[1] Politecnico di Milano, Milan, Italy
matteo.corigliano@polimi.it
[2] TU Delft, Delft, The Netherlands

Abstract. Most of the bridges in Europe countries are now approaching their design life. Therefore, at present crucial is the choice of the most suitable retrofitting solution taking the current design standards into account. From an economic point of view the costs related to the foundations adaptation are not negligible at all, even because design approaches are in general over-conservative. For instance, in case of piled foundations, the presence of the raft is conventionally disregarded in the calculation of the pile group bearing capacity under general loading. In this work a pile group foundation embedded in a silty-clay soil stratum is studied to emphasise how the use of a non-standard approach may allow to make more sustainable the interventions. An extensive 3D pseudo-static finite element numerical analyses campaign, under general loading, accounting for the non-linear soil mechanical behaviour, was performed. The results were interpreted in terms of interaction domains for the piled foundation system (raft + piles).

Keywords: Bearing capacity · Combined loading · Failure mechanism · Piled foundation · 3D finite element analysis

1 Introduction

Most of the bridges in Europe, are approaching or have already approached their design life. For this reason, they have to be reassessed in terms of safety and serviceability. This is also true for foundations, for which eventual retrofitting measures are often very expensive and technologically challenging.

For most of the tall bridges founded on piles, critical are the aspects related to seismic actions and, in particular, to the moment capacity of the foundation [1] along the bridge transversal direction. According to the current design approaches [2, 3], the moment capacity is calculated by on one side disregarding the presence of the raft and, on the other, assuming the global bearing capacity to be reached when the most loaded pile reaches its bearing capacity (under either tension or compression). The design standards recognize the current practice to be over-conservative and the limit state to occur only when a significant number of piles reaches its bearing capacity [3], but do not suggest any calculation method accounting for (i) the ductile redistribution of vertical forces among

A. Ferrari et al. (Eds.): CNRIG 2023, SSGG, pp. 646–653, 2023.
https://doi.org/10.1007/978-3-031-34761-0_78

piles and (ii) the presence of the raft. Only recently, an approach to estimate the bearing capacity of pile groups under inclined and eccentric loads, accounting the first aspect for, was proposed [4, 5]. This approach, however, disregarding the presence of the foundation raft, is still over-conservative, especially in the typical case of existing bridges founded on a small number of piles. The role of both foundation raft and interaction mechanisms, involving the pile-raft-soil system, is rarely analysed and only very recently, [6], the beneficial role of the raft has been experimentally and numerically shown not to be negligible.

The goal of this paper is to put in evidence that finite element (FE) numerical analyses could provide a significant insight in the pile-soil-raft interaction, allowing a more aware design of both new foundations and retrofitting measures.

To this aim, a case study is numerically discussed by assuming the soil to behave under undrained conditions (Sect. 2). The mechanical response of the system is described by defining the interaction domains and by illustrating the failure mechanisms developing in the soil (Sect. 3). Finally, the foundation system is verified under design seismic actions by employing the interaction domains (Sect. 4).

2 Numerical Model

The viaduct considered in this paper, built at the end of the '70 s, is 13 m high and consists of three spans of 32 m and a 18 m-wide deck. The piers are founded on rectangular piled rafts (Fig. 1). The raft bases are at 4 m depth from the ground surface, whereas the 7 reinforced concrete bored piles, connected to the raft, are characterized by a diameter of 1.2 m and are 12 m long.

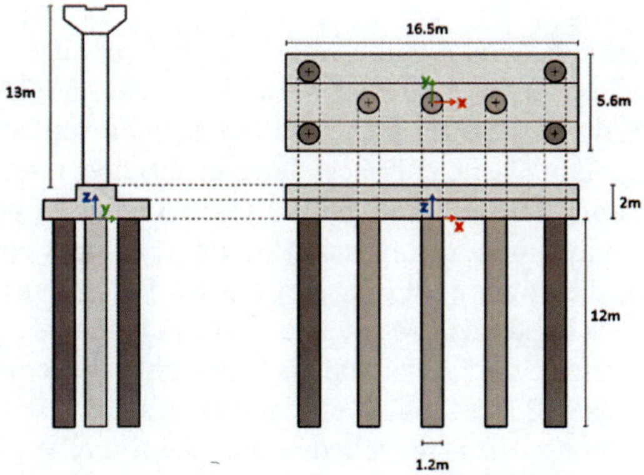

Fig. 1. Geometry of the piled raft foundation.

Recent restoration works made possible a geotechnical characterization of the site. The soil profile consists of (i) 1 m of landfill material, (ii) 5 m of coarse sand and sandy gravel and (iii) an underlying layer composed of fine-grained materials (clayey silt, silty and marly clay). To provide a safe side estimation, the granular material stratum is disregarded and the piled foundation is assumed to be positioned in a saturated homogeneous clay soil stratum of unit weight equal to $\gamma_{sat} = 20\,\text{kN/m}^3$. The results of standard penetration tests [7] highlight a rather constant profile of undrained strength along depth ($S_u = 150$ kPa).

The mechanical response of the system is analysed by performing non-linear pseudo-static 3D FE numerical analyses by employing the commercial code MIDAS/GTS-NX. The geometry and spatial discretization of the FE model (for the sake of symmetry, only one half of the spatial domain has been discretized) are reported in Fig. 2.

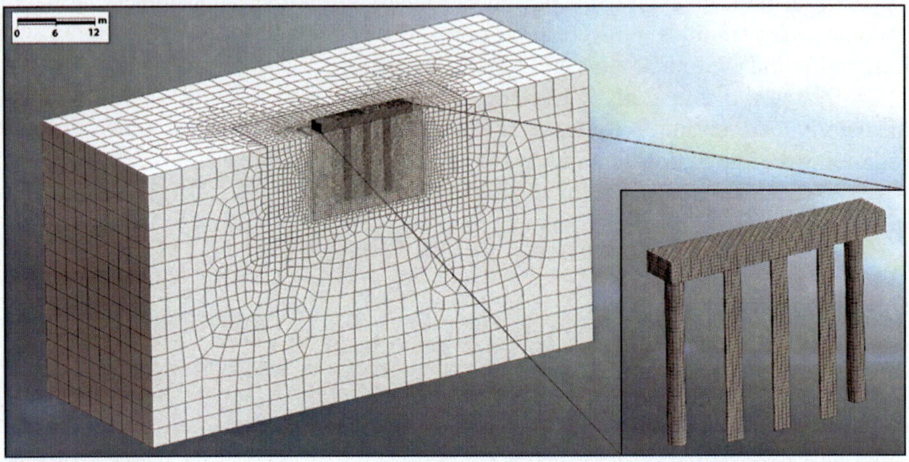

Fig. 2. Finite element discretization for 3D numerical analyses: full model with a zoom on the piled raft only.

Piles and concrete raft (with a unit weight is $25\,\text{kN/m}^3$) are assumed to be elastic with a Young modulus of 30GPa and a Poisson ratio equal to 0.3. The soil is modelled as a 1-phase elastic perfectly plastic material, with a Tresca failure criterion and an associated flow rule. Elastic perfectly plastic interface elements have been set at both soil-pile and soil-raft interfaces. Along the normal direction, these elements are perfectly fragile under tension and "quasi-rigid" under compression (normal stiffness equal to 500 MN/m^3). Along the tangential direction, a nil dilatancy Tresca failure criterion is adopted (the limit tangential stress is imposed to be equal to S_u) and tangential stiffness is assumed to be 50 MN/m^3. Both vertical and horizontal displacements are constrained at the bottom boundary. Along the lateral sides, only vertical displacements are allowed. The lateral soil above the foundation plane is modelled as a uniform surcharge p = 80 kPa.

The initial conditions have been imposed as it follows: (i) the geostatic state of stress condition of the stratum is obtained by linearly increasing gravity, (ii) the pile construction is simulated by progressively changing mechanical properties and unit weight in the piles domain (from soil to piles properties), (iii) the raft construction is simulated by linearly increasing mechanical properties and unit weight in the raft domain. Finally, (iv) the vertical load (V of Fig. 3) is linearly increased up to a prescribed value.

To derive the interaction domain, two different sets of analyses were performed. In the former one, aimed at defining the M_y-V section, 7 prescribed V values are imposed

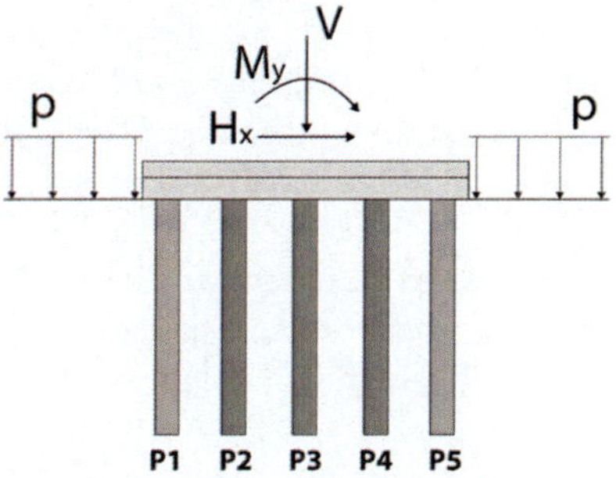

Fig. 3. Geometrical scheme (longitudinal section) with applied loads.

and during the numerical tests $\dot{V} = \dot{H}_x = \dot{H}_y = \dot{M}_y = 0$ whereas modulus of M_y is progressively increased up to failure (Fig. 4a). In the latter one, aimed at defining the M_y-H_x section, $\dot{V} = \dot{H}_y = \dot{M}_y = \dot{M}_x = \dot{e} = 0$ (where $e = M_y/H_x$) but modulus of M_y and H_x are progressively increased up to failure. In this case the prescribed V value coincides with the superstructure weight ($V = V_d = 24$ MN).

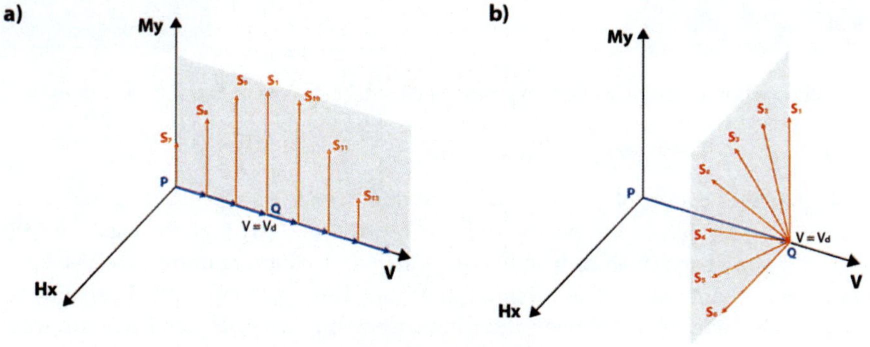

Fig. 4. Load paths in M_y-H_x-V space.

3 Numerical Results

The numerical results are discussed in terms of interaction domains (in Fig. 5 and 6 the straight lines represent the imposed generalized stress paths) and contours of irreversible deviatoric strains (A-D of Fig. 5 and A-H of Fig. 6).

The M_y-V section of the interaction domain, illustrated in Fig. 5, is characterized by a non-negligible upward vertical strength ($V < 0$), due to the pile shaft resistance. This resistance is significantly lower than the compression one due to the contribution of both pile bases and raft. The asymmetry in the interaction domain section is clarified by the plastic mechanisms in the foundation soil plotted in Fig. 5A–D, corresponding to points A-D in M_y-V plane of Fig. 5. In particular, for $V \leq 0$, the raft partially detaches from the soil and plastic strains mainly develop along the piles shaft and in the proximity of the compressed edge of the foundation (Fig. 5A and B). On the contrary, for positive V values, raft detachment does not take place and plastic strains develop both in the proximity of the piles (especially at the base) and at the foundation edges (Fig. 5C and D). It is worth mentioning that the "closed" failure mechanisms of Fig. 5C and D, differing significantly from the one expected in case of shallow foundations, seem to be a sort of punching in which lateral shear of the entire block and bearing capacity of each pile tip dominate the soil system interaction.

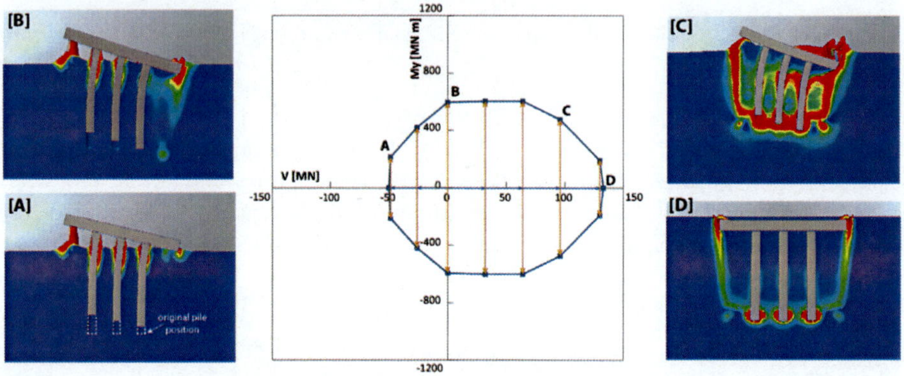

Fig. 5. Interaction diagram sections with load paths in M_y-V plane and plastic strains contours in points A-D.

The M_y-H_x section of the interaction domain (for $V = V_d$) is illustrated in Fig. 6. Analogously to what observed for shallow foundations [8] and for caissons [9], the M_y-H_x section shape is an ellipse whose principal axes are counter-clockwise rotated from the coordinate axes. The orientation of the ellipse principal axes depends on the plastic mechanism developing in the foundation soil: when H_x and M_y are opposite (Fig. 6E–G, corresponding to points E-G), large plastic strains develop in both sides of the foundation, whereas when H_x and M_y are concordant (Fig. 6H), large plastic strains develop only on the compressed side of the foundation.

4 Geotechnical Verification of the Foundation System

The interaction domains illustrated in Sect. 3 are here below employed for the pseudo-static seismic geotechnical verification of the piled foundation. The actions applied to the foundation are calculated, by following NTC2018 [3]: both kinematic interaction and soil inertia are disregarded, whereas the inertial forces transferred by the superstructure are calculated by using an uncoupled approach (the structural dynamic analysis for the viaduct was carried out by assuming the foundations to provide a rigid constraint). Different load combinations were accounted for but, for the sake of brevity, only the most critical one will be discussed.

The verification according to NTC2018 [3] requires the use of a partial coefficient γ_R, defining a reduction of the foundation system strength. The design code, however,

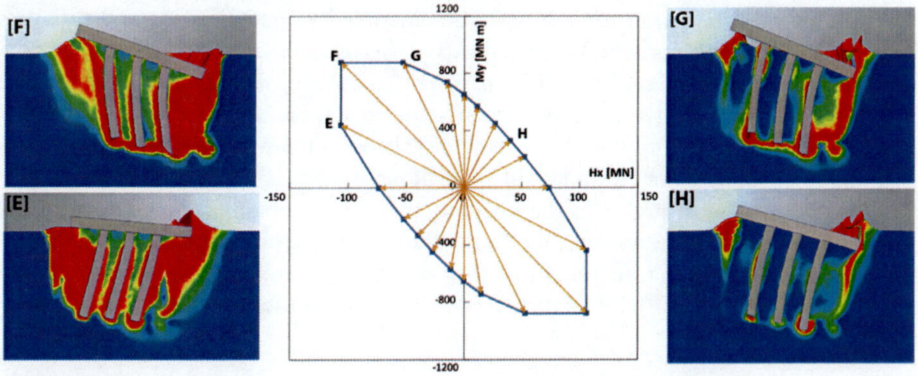

Fig. 6. Interaction diagram sections with load paths in M_y-H_x plane and plastic strains contours in points E-H.

does not provide a γ_R value to be used in case of interaction domain for the entire piled foundation. By following the philosophy of NTC2018 [3], the authors decided to homothetically scale the interaction domain by $\gamma_R = 2.3$ (solid line in Fig. 7). In Fig. 7, the cross, representing the design load, lies inside the scaled interaction domain, implying that, in its current state, the foundation system can be considered to be "safe" and does not require any retrofitting measure, at least in relation to soil limit conditions.

For the sake of completeness, in Fig. 7 are also reported the interaction domain sections obtained by using a conventional approach (dashed lines in Fig. 7), based on the following assumptions: (i) rigid raft not transmitting stresses to the soil, (ii) rigid-perfectly plastic behaviour of the soil, (iii) pile heads rigidly connected to the raft by means of hinges (piles are only axially loaded) and (iv) ultimate loads for each pile along horizontal and vertical directions independent to each other. According to these

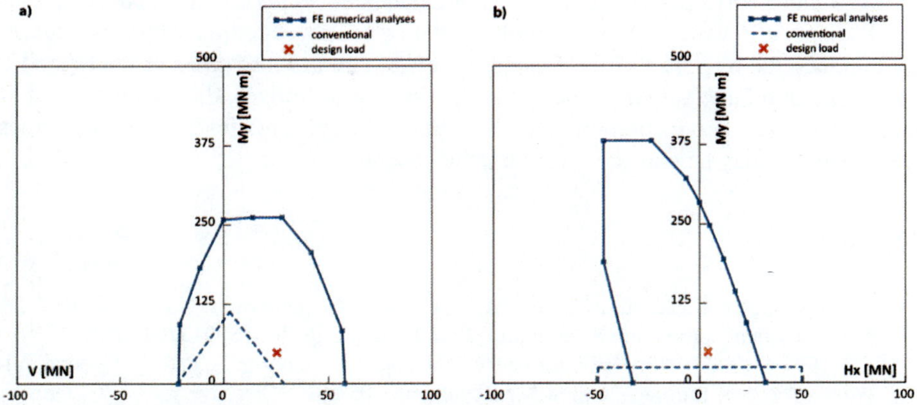

Fig. 7. Interaction diagram sections (partial factors of safety applied) and design load: comparison between numerical analyses results and conventional verification approach in (a) M_y-V plane and in (b) M_y-H_x plane.

assumptions, the interaction domain can be obtained: (i) by imposing a linear distribution for the vertical loads transmitted to each pile (Q_i), (ii) by imposing the vertical balance of momentum and the rotational equilibrium, (iii) by calculating the upward and downward pile bearing capacities [10], and (iv) by using the partial factors of safety of NTC2018 [3]. It is worth mentioning that the rectangular shape in the M_y-H_x plane is due to the assumption of independent ultimate pile load along horizontal and vertical direction.

The comparison in Fig. 7 of the interaction domains, obtained by using the FEM numerical results and the conventional simplified approach, puts clearly in evidence the role played by the raft in affecting the response of the system, in particular in cases like this, where the ratio L/B (where L is the pile length and B the raft width) is sufficiently small.

5 Conclusions

The verification of the safety of existing structures is nowadays a topic of great practical relevance since in Western countries many structures are approaching/have approached their design life. This is particularly critical in case of foundations since their retrofitting is very expensive and requires large investments in terms of time and raw materials.

In this paper, the authors numerically analysed the undrained response of the piled foundation of an existing bridge. The numerical results allowed to define an interaction domain, accounting both piles and raft for, to be used for the geotechnical verification of the foundation system. The definition of the interaction domains has been shown to be useful for the pseudo-static verifications of the foundation under seismic conditions; to this aim, particularly useful is the description of the interaction domain in overturning moment vs horizontal load plane. In this plane the shape of the interaction domain is elliptic, as was shown by other author with reference to shallow foundations or rigid caissons. The description of yielded zones in the soil has allowed to define the geometry of the failure mechanism, not accounting for soil inertia, associated with the different ultimate states corresponding to the different load paths imposed. The analysis of these mechanisms allows to capture the role of the raft-piles coupling, which is expected to dominate the response of the foundation system in all those cases characterized by short piles and large values of the raft width. In case of bridges, this geometry is quite common in the cross-section orthogonal to the bridge axis and, for this reason, the case here analysed may be considered to be quite general.

References

1. Anastasopoulos, I., Gazetas, G., Loli, M., Apostolou, M., Gerolymos, N.: Soil failure can be used for seismic protection of structures. Bull. Earthq. Eng. **8**(2), 309–326 (2010)
2. EN 1997-1 (2004) (English): Eurocode 7: Geotechnical design - Part 1: General rules [Authority: The European Union Per Regulation 305/2011, Directive 98/34/EC, Directive 2004/18/EC]
3. Norme tecniche per le costruzioni (NTC 2018) D. Min. Infrastrutture e Trasporti (2018)
4. Di Laora, R., de Sanctis, L., Aversa, S.: Bearing capacity of pile groups under vertical eccentric load. Acta Geotech. **14**(1), 193–205 (2018). https://doi.org/10.1007/s11440-018-0646-5

5. Di Laora, R., Iodice, C., Mandolini, A.: A closed-form solution for the failure interaction diagrams of pile groups subjected to inclined eccentric load. Acta Geotech. **17**, 3633–3646 (2022). https://doi.org/10.1007/s11440-021-01439-8
6. Sakellariadis, L., Anastasopoulos, I.: On the mechanisms governing the response of pile groups under combined VHM loading. Géotechnique (2022)
7. Corigliano, M., Flessati, L., di Prisco, C.: Numerical push-over analysis of a bridge piled foundation: geotechnical and structural verification. Rivista Italiana di Geotecnica, under review
8. Gottardi, G., Houlsby, G.T., Butterfield, R.: Plastic response of circular footings on sand under general planar loading. Géotechnique **49**(4), 453–469 (1999)
9. Rosati, A., Gaudio, D., di Prisco, C., Rampello, S.: Use of interaction domains for a displacement-based design of caisson foundations. Acta Geotechnica 1–24 (2022)
10. Viggiani, C., Mandolini, A., Russo, G.: Piles and Pile Foundations. CRC Press (2014)

A Simple Parametric Numerical Model to Assist the Design of Repair Works and Maintenance of Tunnels

Simone De Feudis(✉), Alessandra Insana, and Marco Barla

Department of Structural, Geotechnical and Building Engineering (DISEG), Politecnico di Torino, Corso Duca degli Abruzzi 24, 10129 Torino, Italy
simone.defeudis@polito.it

Abstract. Tunnels are resilient infrastructures whose service life is generally much higher than the designed nominal one, thus increasing the number of existing tunnels still in operation with an average service life of more than 50 years. From a more Italian perspective, Italy's complex morphology has made it necessary to excavate a considerable number of tunnels to develop the nowadays railway and motorway traffic lines. However, recent cases of tunnel lining local collapses have highlighted that some existing Italian tunnels have almost reached the end of their service life, revealing their maintenance, refurbishment and/or upgrading works to be an urgent need. Taking advantage of the tunnel refurbishment plan ongoing in Italy, the aim of this study is to support the design of temporary maintenance works by means of simple numerical modelling. In such a way it is possible to consider fundamental geotechnical parameters that can play a key role in the effectiveness and in the applicability of the studied maintenance solutions. The attention is here devoted to interventions with suspended steel ribs to prevent local instabilities of the aged concrete.

Keywords: Tunnel ageing · Existing tunnels · Tunnel refurbishment · Maintenance work · Numerical modelling

1 Status Quo of Existing Tunnels in Italy

Tunnels are resilient infrastructures whose service life is generally much higher than the designed nominal one, thus increasing the number of existing tunnels still in operation with an average service life of more than 50 years [1].

Coping with tunnels ageing consists in the process of ensuring the continuation of service in safe conditions. For this purpose, different strategies can be considered:

- Maintenance works: minor works aimed at guaranteeing the tunnel designed service life, e.g., repair work for preventing local blocks detachment (see Fig. 1a).
- Rehabilitation works: major works aimed at extending the tunnel designed service life, e.g., complete replacement of the tunnel vault after experiencing severe cracks formation and water income (see Fig. 1b).

A. Ferrari et al. (Eds.): CNRIG 2023, SSGG, pp. 654–661, 2023.
https://doi.org/10.1007/978-3-031-34761-0_79

- Upgrading works: major works aimed at changing the tunnel designed use by upgrading it, e.g., tunnel enlargement to host more lines or tracks (see Fig. 1c).
- Disposal: reusing old tunnels changing their designed use by downgrading them, e.g., hosting art exhibitions or bicycle ways (see Fig. 1d).

a b c d

Fig. 1. Examples of strategies throughout coping with tunnel ageing in Italy: a) Provenzale tunnel, Genoa, b) San Fermo tunnel, Como, c) Nazzano tunnel, Rome [2] and d) "Le Gallerie", Trento.

Concerning Italy, due to its complex morphology, its railways and motorways need a great number of tunnels to guarantee safe and fast pathways. It is no coincidence that Italy is one of the first countries in the world for number and length of existing tunnels (see Fig. 2). However, despite the resilience of this kind of underground infrastructures, many Italian existing tunnels have almost reached the end of their service life.

Accordingly, this study is intended to provide a support in the design of maintenance works by means of simple numerical modelling, that allows to further consider fundamental geotechnical parameters which can play a key role in the effectiveness and in the applicability of the maintenance solutions.

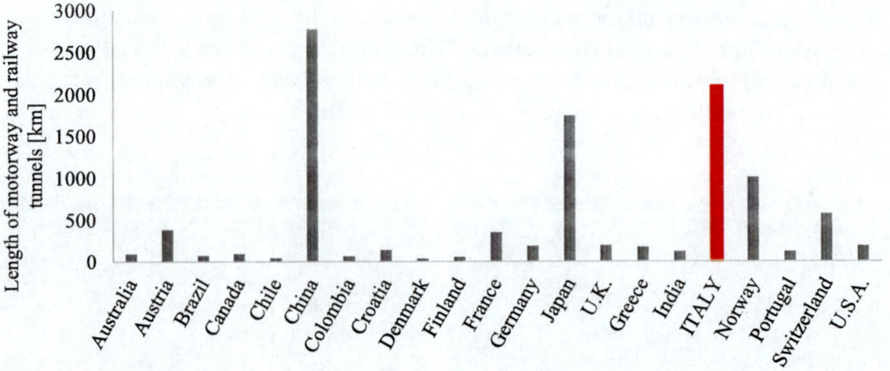

Fig. 2. Length of motorway and railway tunnels among several countries around the world (modified from [1]).

2 Refurbishment Plan for Existing Tunnels in Italy

To assist concessionaires and tunnels owners in the process of inspections, maintenance and renovation of existing tunnels, new methods and procedures have been studied, validated and released in the recent years [3–5]. The tunnel inspection manual envisages that, before applying any repair work, a deep fact-finding survey must be carried out, relying on both as-built documents and in-situ inspections. The aim of the inspections is to check the tunnel conditions by taking note of all the damages and defects. To this aim, supplementary target investigations are often needed (thermography, video-endoscopy, geo-radar, sample coring, laboratory tests, etc.).

Depending on the inspection outcome, the tunnel condition is assessed and an attention class is assigned. The next step is to identify appropriate and/or necessary interventions based on the quantity, the seriousness and the typology of the defects. Temporary interventions, with the aim of securing the tunnel in the short term (from three to ten years), are applied when mobility poses serious constraints to tunnel long closures or to allow getting further insights for defining a suitable rehabilitation intervention, while final interventions are designed where it is possible to make them straightaway.

To support the work of its designers to promptly solve the most common problems, the Italian motorway concessionaire Autostrade per l'Italia SpA who is undergoing an intensive tunnels refurbishment plan to renovate existing tunnels, developed a catalogue of typological prompt interventions. These temporary maintenance solutions change as a function of the harshness of the concrete lining deterioration and several other aspects like:

- the presence or not of the waterproofing system,
- the presence or not of water income,
- whether the tunnel has been excavated in rock or soil masses,
- the presence or not of a significant state of stress in the concrete lining.

With the aim to demonstrate the usefulness of resorting to simple numerical models to support the design of such temporary solutions, the attention in the following will be devoted to a specific temporary solution developed to face the deterioration of the aged concrete lining with the potential formation of unstable blocks with volumes higher than 1 m^3 and presence of water (see Fig. 3). Its realization envisages the implementation of suspended double rectangular profile steel ribs secured to the concrete lining by means of chemical anchors.

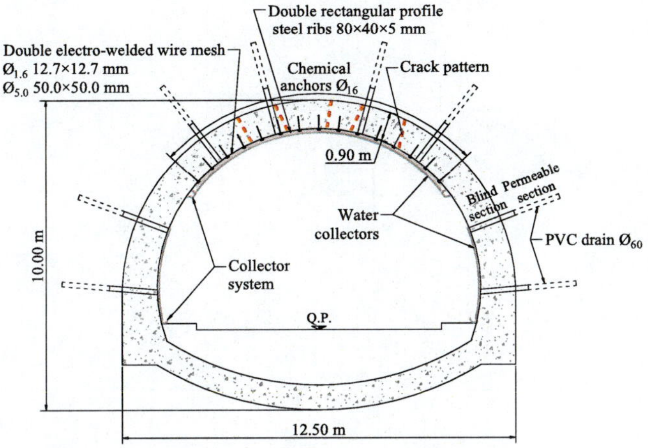

Fig. 3. Sketch of the temporary maintenance solution.

The use of numerical modelling, even kept at a very basic level, is of fundamental support in the design of temporary intervention that are designed for a variety of situations and not for a specific tunnel. Parametric analyses, as the ones that will be discussed in the following, may allow to consider the variability of the ground conditions, of the lining characteristics, of the geometrical aspects to allow for checking the availability of the interventions in a variety of situations. Fundamental aspects like the quality of the rock mass or the initial state of stress can be easily included in the computation and their role unraveled.

3 A Simple Effective Numerical Model

To explore the influence and the implications of some fundamental geotechnical parameters, a plane strain sensitivity analysis has been carried out by means of the software RS^2 [6]. This numerical investigation was meant to highlight if, whether in presence of a specific set of the above-mentioned geotechnical parameters, the chemical anchors slippage condition would have been possible or not. The anchoring resin at the interface between the concrete and the chemical anchors has been modelled through a joint element ruled by the Tresca failure criterion, whose cohesion (c′) represents its strength. The latter has been varied arbitrarily during the analysis, depending on the case considered. The tunnel cross section shown in Fig. 3 was adopted.

The numerical analyses also investigated the role of:

- the earth thrust coefficient (K_0) between 0.5 and 2.0,

- the rock mass quality (GSI, Geological Strength Index [7]) between 20 and 80,
- the overburden of the tunnel(H) equal to 20.0 m and 40.0 m.

With the different parameters above, the sensitivity analysis was intended to investigate typical conditions for rock masses crossed by the Italian motorway network.

To this purpose two numerical models have been built to consider the different overburden. As shown in Fig. 4, for the sake of simplicity, the topographic surface has been considered horizontal.

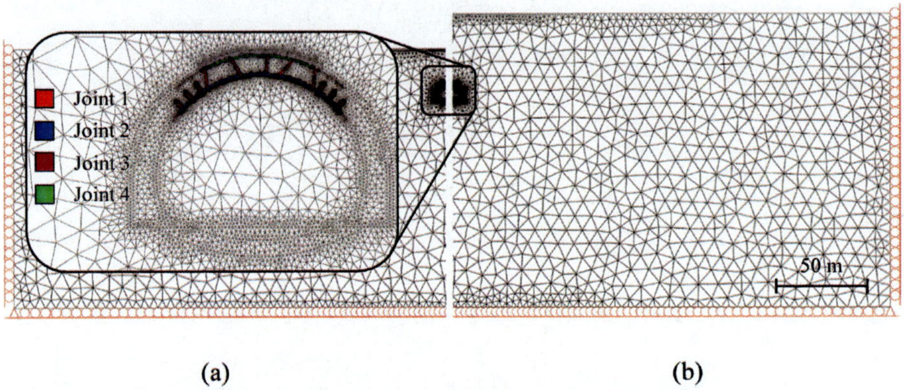

Fig. 4. Parametrical numerical models with two different overburdens: 20 m (a), 40 m (b) and detail of the joint elements adopted in the numerical model.

With respect to the tunnel cross section, a horseshoe shaped concrete lining with a variable thickness has been assumed, with the minimum ones being at the key and the invert (respectively of 0.90 m and 0.80 m) and the maximum one being at the springlines (1.40 m).

Finally, the steel ribs and the chemical anchors have been modelled explicitly through finite elements to assess their deformed setup and investigate the strength of the resin at the concrete-anchor interfaces.

The following stages have been considered to perform the sensitivity analysis:

1. geostatic stress state initialization,
2. head excavation with a relaxation factor of 30%,
3. temporary lining installation and bench excavation with a relaxation factor of 100%,
4. concrete final lining realization,
5. complete decay of the temporary lining,
6. concrete cracking and temporary maintenance work installation.

The last stage is supposed to let concrete wedges detach. To this aim, some joint elements have been adopted in the upper portion of the tunnel vault (see Fig. 4) to well represent the interaction between concrete blocks (joint 1) and between the concrete and the steel ribs (joint 2), the resin (joint 3) and the rock mass (joint 4). A proper parameterization of these elements (see Table 1), as a function of the original mechanical characteristics of each interacting entity, allowed to simulate a likely interaction between the above-mentioned concrete wedges and the maintenance work components. With reference to joint 4, it is noted that a variable GSI value implies different normal stiffnesses.

Table 1. Mechanical parameters of the joint elements adopted in the numerical model (Fig. 4).

	Joint 1	Joint 2	Joint 3	Joint 4
Normal stiffness [MPa/m]	270852	1135430	100000	141711 ÷ 334245
Shear stiffness [MPa/m]	0	0	10000	0
Friction angle [°]	30.0	20.0	0.0	13.4 ÷ 21.1
Cohesion [kPa]	0.0	0.0	100 ÷ 600	0.0
Tensile strength [kPa]	0.0	0.0	0.0	0.0

By way of example, the outcomes of two different analyses are shown in Fig. 5. The two images represent numerical analysis results with GSI = 50, $K_0 = 1.0$, H = 20.0 m and a cohesion, for joint 3, respectively, equal to c′ = 240.0 kPa and c′ = 175.0 kPa. A 30% decrease of c has brought to the slippage of the upper right chemical anchor.

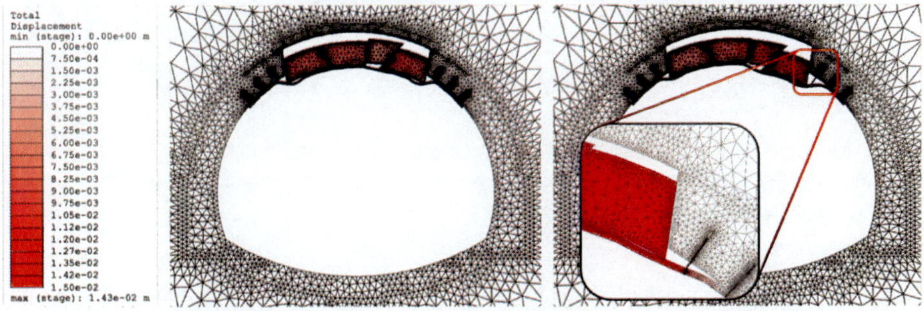

Fig. 5. Outcome of two different analyses with and without plugs slippage.

It is evident that the above-mentioned way of modelling this kind of problem well manages to reproduce the concrete wedges detachment and the structural response of the steel ribs, as well as the chemical anchors one. As anticipated, the aim of the sensitivity analysis was to deepen the knowledge about the interaction between the concrete tunnel lining and the resin of the anchors.

In each examined situations, the attention was devoted to the cohesion at the interface to be enough to avoid the slippage condition for all the chemical anchors simulated in the numerical model.

The results of the sensitivity analysis are shown in Fig. 6.

Based on the results, the effectiveness of the temporary intervention is heavily influenced by the geotechnical parameters. It is evident how the cohesion needed to guarantee the stability of the repair work ($c'_{req.}$) varies substantially in the different situations. Thus,

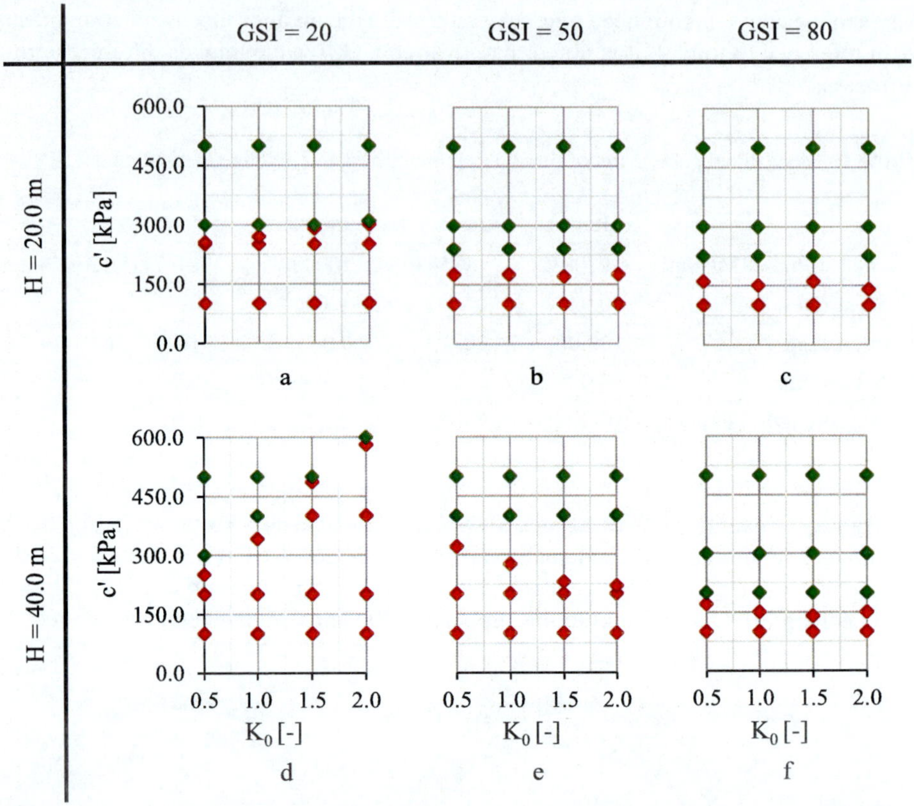

Fig. 6. Slippage (red diamonds) and safe (green diamonds) conditions of the chemical anchors in function of the cohesion (c'), the rock mass quality (GSI), the tunnel overburden (H) and the initial stress state (K_0).

the numerical model may allow to identify the proper conditions for the application of the temporary intervention analyzed here.

For instance, analyzing Fig. 6d, it can be realized how, whether in unfavorable conditions of tunnel overburden (H = 40.0 m) and rock mass quality (GSI = 20), the mandatory anchoring resin strength could increase of even three times depending only on the initial state of stress (K_0).

Finally, for the analyzed cases, the following conclusion can be drawn:

- the lower the quality of the rock mass (GSI), the higher the required cohesion ($c'_{req.}$),
- the higher the tunnel overburden (H), the higher the required cohesion ($c'_{req.}$),
- the influence of the initial state of stress (K_0) is revealed to be more significant with decreasing rock mass quality (GSI) and the increasing overburden (H).

4 Conclusions

This study focused on proving how numerical modelling, even at a very basic level, could assist the design of temporary tunnel interventions. The attention was devoted to a particular temporary repair work designed to cope with ≥ 1 m^3 volume concrete blocks detachment. Its effectiveness has been studied in function of fundamental geotechnical parameters like the quality of the rock mass in which the tunnel was excavated (GSI), the initial stress state (K_0) and the tunnel overburden (H). To this purpose, a numerical model has been built in RS^2 (Rocscience Inc.).

In the case of a 40.0 m deep tunnel excavated in a poor rock mass, characterized by a high stress state, the analysis showed that a higher strength of the resin of the chemical anchors is needed to guarantee the stability of the suspended steel ribs. Therefore, the results illustrated allowed concluding that the above-mentioned geotechnical parameters have a key influence on the temporary maintenance work design, allowing to identify proper conditions for its applications.

Acknowledgements. The work was performed in the framework of a research contract between Autostrade per l'Italia SpA and the Dept. of Structural, Geotechnical and Building Engineering, Politecnico di Torino.

References

1. Beyond a tunnel vision. https://beyondatunnelvision.eu/programme/parallel-sessions/. Tunnelling 4.0: New technologies and future perspectives for maintenance, refurbishment interventions and upgrading. Accessed 15 Dec 2021
2. Lunardi, P., Cangiano, M., Belfiore, A.: Il metodo Nazzano tra passato e futuro Storia e risultati della prima sperimentazione mondiale del sistema di ampliamento delle gallerie in presenza di traffico. Galleria e Grandi Opere Sotterranee **100**, 77–90 (2011)
3. Ministero delle Infrastrutture e dei Trasporti – Direzione generale per la vigilanza sulle concessionarie autostradali. Manuale Ispezione Gallerie, Rev.01 del 25/05/2020 (2020)
4. Ministero delle Infrastrutture e della Mobilità Sostenibili, Consiglio Superiore dei Lavori Pubblici. Linee guida per la classificazione e gestione del rischio, la valutazione della sicurezza ed il monitoraggio delle gallerie esistenti (2021)
5. Barla, M., et al.: A method to define the priority for maintenance and repair works of Italian motorway tunnels. IOP Conf. Ser. Earth Environ. Sci. **883**(1) (2021)
6. Rocscience. 2022. RS^2 11.015. Toronto, Ontario, Canada
7. Hoek, E.: Strength of rock and rock masses. ISRM New J. **2**(2), 4–16 (1994)

Influence of the Seismic Performance of Geotechnical Systems on the Resilience of a Road Network

Chiara Amendola[1](✉), Riccardo Conti[2], Paolo Zimmaro[3,4], Luigi Pariota[5], Mario Marinelli[6], and Filomena de Silva[5]

[1] Aristotle University of Thessaloniki, Thessaloniki, Greece
chiaamen@civil.auth.gr
[2] University of Rome Unicusano, Rome, Italy
[3] University of Calabria, Arcavacata di Rende, Italy
[4] University of California, Los Angeles, Los Angeles, USA
[5] University of Naples Federico II, Naples, Italy
[6] University of Sannio, Benevento, Italy

Abstract. An efficient design of geotechnical systems is crucial to ensure the functionality of a transportation network after the occurrence of an earthquake. This paper illustrates the preliminary results of a novel methodology aimed at identifying geotechnical systems characterized by an unsatisfactory seismic performance, resulting in a compromised functionality of the overall network. To illustrate its potentialities, the presented framework is applied to an example case study. Starting from a scenario ground motion map, the potential damage to each geotechnical system (retaining walls) is quantified through *ad-hoc* generated fragility curves. Existing damage scales are exploited to evaluate the loss of serviceability. Information on partial or full road closures is implemented as part of a typical transportation engineering framework. Such framework is used to estimate variations in the performance of the transportation network by means of specific indicators (*e.g.*, active accessibilities and path generalized costs). The analysis of these indexes is used for the evaluation of the effects of the functionality loss, and of the resulting actions, on the general performance of the transportation network. Thanks to its versatility, the proposed framework can be applied to any road network and geotechnical system.

Keywords: geotechnical systems · fragility curve · seismic resilience · road infrastructure · transportation networks

1 Introduction

Recent earthquakes showed that the vulnerability of road infrastructure systems plays an important role in post-earthquake emergency response and resilience. Existing methods for the assessment of road network resilience are mainly based on indices relevant to post-event accessibility and link importance. Methods based on accessibility, *i.e.* the

A. Ferrari et al. (Eds.): CNRIG 2023, SSGG, pp. 662–670, 2023.
https://doi.org/10.1007/978-3-031-34761-0_80

ease of transport from/to critical facilities, rely upon different factors, but most often on transport costs. Link importance indices classify the link criticality according to its effect on the road network in terms of serviceability, demand and travel time. Most studies disregard the causes producing the loss of functionality to such links. The by-product of such approach is that the connection between geo-hazard and network resilience is lost. This gap can be bridged through fragility models to estimate quickly the expected damage to key components of the network. Few studies examine the resilience of the whole network starting from key components, and are generally focused on bridges and tunnels, due to the lack of fragility models for other geotechnical systems.

This study proposes a novel procedure to quantify the resilience of road networks through fragility curves of key components. It is organized in three steps (Fig. 1):

- *Network definition* in terms of traffic flow on the main roads connecting nodes, as well as classes of key components characterized by homogeneous features;
- *Fragility and hazard analysis*, in which fragility curves are generated for each class and combined with ground motion scenario to estimate damage and functionality loss;
- *Road Network performance analysis* considering the functionality loss of the road (reduced speed, partial or full closure) due to the failed key components and the evaluation of the network resilience, based on the comparison between the network performance pre- and post-earthquake.

In this framework, the reactivation of the functionality is not considered, because the analysis is focused on the network configuration immediately after the earthquake, which is relevant for post-earthquake emergency management procedures.

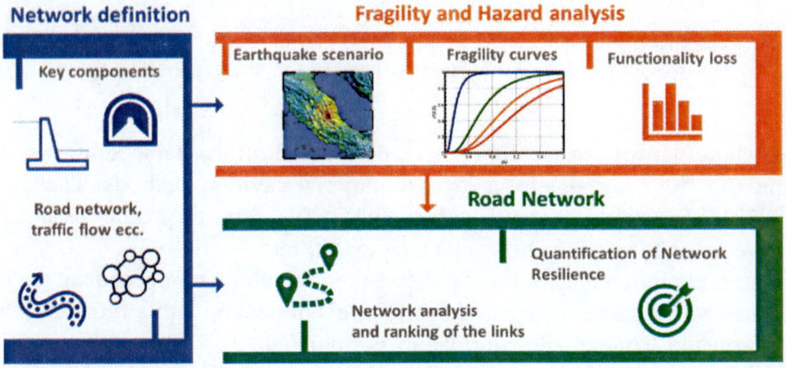

Fig. 1. Flow chart of the procedure to evaluate the seismic resilience of transportation networks.

2 Application to an Ideal Case Study

Figure 2a shows the scheme of the road network analyzed in this study. It is inspired by the existing road network in Basilicata (Italy) and is characterized by three road types (Fig. 2b), as defined by the Italian National Regulation [1]: Motorways (A); Primary

Roads (B); and Secondary Roads (C). As shown in Fig. 2b, the Regulation assigns to each road type specific geometric characteristics and speed limits. These differences were modelled through the BPR link performance function [2]. The model of the pre- and post-event transportation system performances is integrated by an estimation of travel demand levels in the area, based on data collected from the Italian National Institute of Statistics [3]. The estimation is performed during the typical morning peak hour of a working day at the regional scale. The software PTV Visum is used to compute link flows for each simulated scenario, by assigning the estimated demand levels to the transportation network [2].

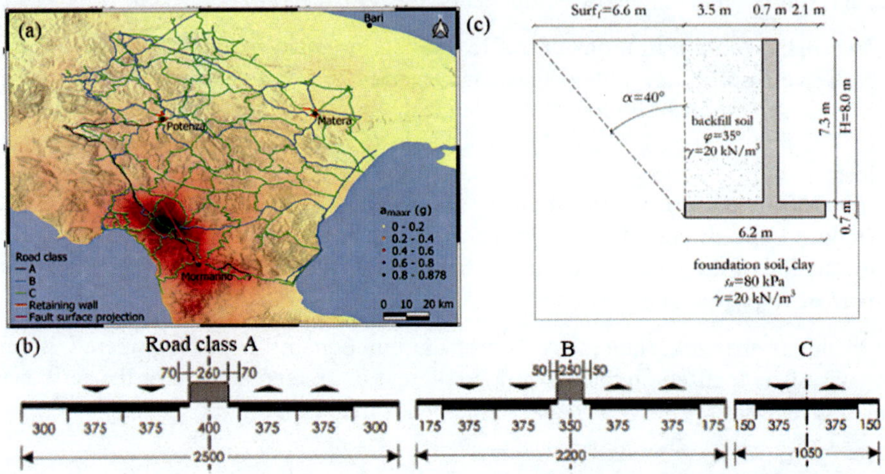

Fig. 2. Analyzed road network on the map of $a_{max,r}$ at bedrock for the simulated earthquake scenario (a); cross sections of the road classes (b) and of the analyzed retaining wall (c).

The whole network is ideally considered located on the same subsoil profile [4]. It comprises a 40 m homogeneous layer of clay overlaying a bedrock. The shear wave velocity, V_S, is assumed to increase with depth [5]. The resulting profile has an equivalent shear wave velocity up to the bedrock equal to 200 m/s.

The ground motion map in Fig. 2a shows the maximum acceleration at the bedrock, $a_{max,r}$, for the analyzed seismic scenario. The event occurs with a moment magnitude **M** = 6.6 on the Rimendiello-Mormanno normal fault [6]. The $a_{max,r}$ was obtained from median predictions of the Italy-adjusted Boore et al. [7] ground motion model and a global correlation model [8] to capture the inherent randomness of ground motion spatial variability.

The post-earthquake performance of the network is controlled by the performance of the retaining wall depicted in Fig. 2c. As mentioned above, this is an ideal representation of the network. Hence, the locations of retaining structures were chosen to be on the main important links of the network (*i.e.*, close to main cities and to the epicentre). Their typology is inferred from recurrent retaining walls along Italian roads, while their geometric features were calculated according to the Italian Building Code [9], given height, H, undrained shear strength of the foundation soil, s_u, and friction angle of the backfill, φ (Fig. 2c). The value assumed for s_u is consistent with the V_S of the shallowest 5 m of soil according to the correlation proposed by Mayne [10]. According to [11], the critical acceleration of the wall, $a_c = 0.15$ g, is associated to a sliding failure mechanism. The corresponding failure surface is shown in Fig. 2c.

3 Seismic Fragility of the Analysed Retaining Walls

3.1 Theoretical Reference Model

Callisto [12] proposed to approximate the capacity of retaining walls through a hyperbolyc function linking the seismic coefficient, k_H, to the displacement at the top of the wall normalized with respect to its height, *u/H*. The capacity curve is controlled by k_c, by the normalized displacement corresponding to the activation of the plastic mechanism, s_F, and by the ratio α between the asymptote of the hyperbolic function and k_c. Given $k_c = 0.15$ g and assuming $s_F = 0.5\%$ and $\alpha = 0.7$ [12], the nonlinear capacity curve for the wall under investigation is shown in Fig. 3a. According to this approach, the curve was exploited to solve the equation of motion of a nonlinear single degree of freedom system, SDoF, equivalent to the wall through an *ad-hoc* generated Matlab routine [11]. The SDoF equation of motion was even solved through the application of the software framework OpenSees based on the constitutive model "*uniaxialMaterial Hysteretic*" [13]. To this aim, the hyperbolyc capacity curve was reduced to the three-linear model in Fig. 3a. It was calibrated to ensure that the 1st branch reproduces the initial stiffness of the hyperbolyc curve, the 3rd branch is flat and corresponds to k_c and the 2nd branch guaranties that the area subtended by the two curves is the same. In all the analyses, a visco-elastic response of the SDoF is assumed along the unloading-reloading branches, with stiffness equal to the initial slope of the capacity curve and damping ratio equal to 10%.

Figure 3a shows the displacements, *u*, accumulated by the wall under 320 input earthquakes, against the maximum free field accelerations, $a_{max,ff}$. The latter were calculated by Brunelli et al. [4] through equivalent linear 1D site response analyses. Details of the analyses are reported in [4]. The results obtained by the nonlinear and three-linear SDoF are compared with the standard Newmark's approach and with empirical correlations available in the literature [11, 14–16]. For $a_{max,ff} < 0.3g$, most of the empirical relationships predict negligible displacements. This is consistent with the results obtained by Newmark's approach, as both rely on the assumption of a rigid-plastic behaviour of the soil-wall system. The equation proposed by [11], calibrated through numerical simulations, provides permanent displacements closer to those predicted by the SDoF.

3.2 Fragility Curves and Functionality Loss of the Network Links

The horizontal dashed lines in Fig. 3a correspond to displacements equal to 2%H, 5%H and 10%H. Following Cosentini and Bozzoni [17], these values were assumed as thresholds for the minor, moderate, and extensive/complete damage states (DS), respectively.

The probability of exceeding the different DS, p_{DSi}, given a value of the maximum acceleration at bedrock, $a_{max,r}$, was calculated as follows:

$$p_{DSi}(u > u_{DSi}|a_{max,r}) = \Phi\left(\frac{log(a_{max,r}/a_{max,rDSi})}{\sigma}\right) \tag{1}$$

where $a_{max,r\,DSi}$ is the value of $a_{max,r}$ causing the attainment of the ith DS, determined through a linear regression in a log-log plot between u and $a_{max,r}$; σ is the lognormal standard deviation, which is modelled by the combination of different uncertainty sources [18]; Φ is the standard cumulative probability function. Equation (1) applied to the results of the nonlinear SDoF leads to the calculation of the fragility curve plotted in Fig. 3b. The same curves were obtained from the three SDoF. Figure 3b shows that the attainement of all DSs is expected from the $a_{max,r}$ scenario in Fig. 2a.

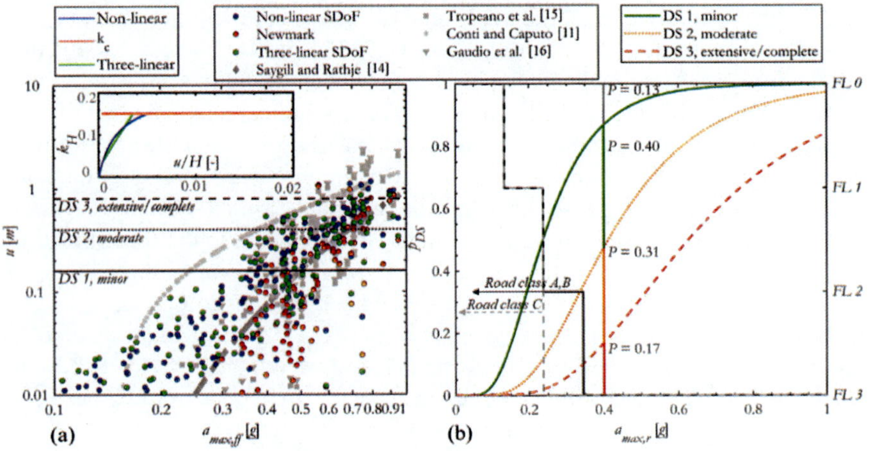

Fig. 3. Comparison of the displacements calculated through different approaches (a) and fragility curves of the wall with the functionality loss function for the analysed Road Classes (b).

The attainment of the different DS produces a functionality loss (FL) of the road above the wall. According to Argyroudis et al. [18], the latter implies: a speed reduction (FL 1); a speed reduction and a partial closure (FL 2); and the full closure (FL 3). In this study, FL 2 is not considered for the road class C because the extension of the failure surface ($Surf_f$ in Fig. 2b) involves both lanes. Following the typical approach applied in the early warning systems or in the risk estimation of buildings, the link is assumed to suffer the FL(i) if the DS(i) overcomes $p_{DS} = 10\%$, with i varying from 1 to 3 or no functionality loss (FL 0) when the threshold is not exceeded. For instance $a_{max,r} = 0.4$g causes a $p_{DS3} = 17\%$, hence it is associated with a FL 3. The functionality loss was calculated for each $a_{max,r}$ and superimposed to Fig. 2b, showing that full closure (FL 3) is estimated starting from moderate acceleration.

4 Road Network

Based on the functionality loss analysis, from the 27 retaining walls considered along the network, only 12 walls were actually affected by the event. The actions carried out as a consequence of the event are summarized in Fig. 4. For the resilience assessment of the road network a joint evaluation of two indices reported in the following has been done. The two indices are defined over a graph $G = (V,L)$, V being the set of nodes, L the set of links, as follows:

– Traffic flow betweenness (*TFB*) of link l defined as [19]:

$$TFB_l = \frac{f_l}{D} \cdot \frac{\sum_{o,d \in V} n_{od}(l)}{N} \tag{2}$$

where f_l is the traffic flow on link l, belonging to the set of links L; D is the total O-D travel demand of the whole road network; $n_{od}(l)$ is 1 if the shortest path between origin o and destination d (belonging to the set of nodes V) goes through link l, 0 otherwise; N is the total number of shortest paths in the network.

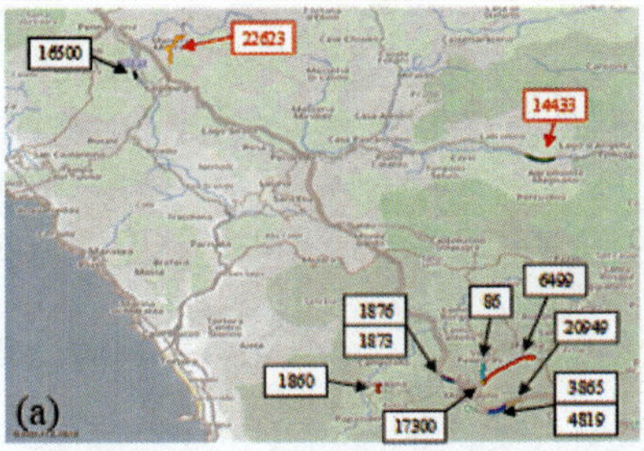

Link ID	LF i	TFB (x 10^{-4})
14433 (+)	1	6.19
14434 (-)	1	6.04
22623 (-)	3	3.24
1873	3	2.69
4819	2	2.69
16500 (+)	3	1.59
22623 (+)	3	1.56
1876	3	1.48
3865	2	1.48
16500 (-)	3	1.17

(b)

Fig. 4. Location of the links affected by the event (a); Actions taken after the event on the links, ordered by descending values of the traffic flow betweenness (TFB) indices (b).

– Accessibility loss (*AL*) of the road network due to the failure of a subset of links $L_f \subset L$, defined as (adapted from [20]):

$$AL = A^{(e)} - A^{(0)} = \sum_{l \in L^{(e)}} f_l^{(e)} \cdot c_l^{(e)} - \sum_{l \in L} f_l \cdot c_l \tag{3}$$

where $L^{(e)} = L/L_f$ is the subset of working links; f_l and $f_l^{(e)}$ are the traffic flows and c_l and $c_l^{(e)}$ are the generalized travel costs on each link l before and after the event, respectively. Thus, $A^{(e)}$ is the accessibility index excluding closed links $l \in L^{(e)}$, expected to be greater than the initial accessibility index $A^{(0)}$ due to an increase in generalized costs of the remaining ones. The traffic flow betweenness is a link-based index useful in evaluating the criticality of the link, generally independent from a specific event, and useful in planning appropriate actions aimed to increase network resilience. Figure 4b shows, for the affected links located as reported in Fig. 4a, the values of TFB indices for non negligible link traffic flow values, clearly identifying the crucial role of the links 14433 and 22623 (+ and – are the different traffic directions on the links). Figure 5 reports the estimated traffic conditions on all the network links before (Fig. 5a) and after (Fig. 5b) the seismic event. From Fig. 5 it is possible to notice that: (i) the event, and the consequent road closures, determine a redistribution of traffic flows on the network that does not necessarily increase traffic congestion; (ii) the redistribution of traffic flows, and the worsening of the traffic conditions (if any) is not necessarily "local", since it could propagate also "far" from the link where the actions are taken; (iii) the traffic load on links subjected to actions (e.g. speed or capacity reduction) does not necessarily decrease (e.g. refer to link 14443).

The reason for that is that a change in a link cost, affects the costs of all the routes traveling through that link, and, in particular, could affect also routes connecting origins and destinations located very far from the link itself. Changes in route cost can increase the likelihood of travelers choosing alternative routes, which may eventually be of interest to other parts of the network.

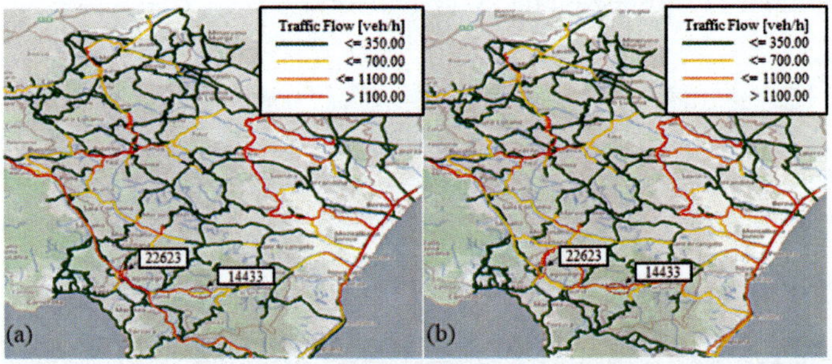

Fig. 5. Traffic flow distribution over the road network before (a) and after the event (b).

On the other hand, in some specific situations (e.g. link 14443) increasing the route costs may not be enough to determine the switch to alternative routes, and eventually, the flow on the link could be increased as a consequence of the same mechanism resulting from actions (e.g. road closure) applied to other links.

This circumstance is exacerbated in a network like the one represented in this case study, characterized by quite a low road density.

As a consequence of the event, it is estimated a reduction of accessibility conditions of 11.4% with reference to the initial value. The values of $A^{(0)}$ and $A^{(e)}$ are $7.9 \cdot 10^7$ and $8.8 \cdot 10^7$, respectively. It is worth noting that this severe reduction results from actions taken on the only 12 links actually affected by the event, which represent a tiny percentage of the link of the network.

5 Conclusions

The objective of the work presented herein is to present, and apply, a framework for evaluating network effects of disruptive events. Specifically the events considered refer to geotechnical systems subjected to seismic actions. The network considered was inspired by the real network of the Basilicata region, Southern Italy, with a realistic estimation of traffic flows, and subjected to a scenario earthquake event. In this study, the whole network was ideally considered located on the same subsoil profile and only one type of retaining wall was studied as geotechnical key component. However, such simplified assumptions can be released. This is the objective of an ongoing research effort performed by the Authors. The analysis of the case study has shown that to effectively forecast the effect of a disruptive earthquake a combination of approaches coming from different engineering fields is needed. Indeed, the detailed knowledge of the characteristics of the geotechnical system are needed to estimate specific effects of the event and consequent actions to be taken, while the deep knowledge of traffic pattern affecting the wider area, as well as of the network structure, are likewise needed to forecast general network functioning after the event.

References

1. D.M. 05/11/2001 n. 6792. Norme funzionali e geometriche per la costruzione delle strade
2. Cascetta, E.: Transportation Systems Engineering: Theory and Methods, vol. 49. Springer, Heidelberg (2013)
3. ISTAT. https://www.istat.it/it/archivio/139381
4. Brunelli, A., de Silva, F., Cattari, S.: On the site-amplification and soil-structure interaction in URM structures: use of fragility curves to assess the simplified code-approach. In: VII COMPDYN 2021, 8th ECCOMAS Thematic Conference, Athens, Greece (2021)
5. d'Onofrio, A., Silvestri, F.: Influence of micro-structure on small-strain stiffness and damping of fine grained soil and effects on local site response. In: International Conference on Recent Advances in Geotechnical Earthquake Engineering and Soil Dynamics, p. 15 (2001)
6. DISS Working Group Database of Individual Seismogenic Sources (DISS), Version 3.3.0: A compilation of potential sources for earthquakes larger than M 5.5 in Italy and surrounding areas. Istituto Nazionale di Geofisica e Vulcanologia (INGV) (2021)
7. Boore, D.M., Stewart, J.P., Seyhan, E., Atkinson, G.M.: NGA-West 2 equations for predicting PGA, PGV, and 5%-damped PSA for shallow crustal earthquakes. Earthq Spectra **30**, 1057–1085 (2014)
8. Baker, J.W., Jayaram, N.: Correlation of spectral acceleration values from NGA ground motion models. Earthq. Spectra **24**(1), 299–317 (2008)
9. D.M. 17/01/2018. Norme Tecniche per le Costruzioni (NTC 2018). Gazzetta Ufficiale, n. 42 del 20/02/2018, Supplemento ordinario n.8
10. Mayne, P.W., Rix, G.J.: Correlations between shear wave velocity and cone tip resistance in natural clays. Soils Found **35**(2), 107–110 (1995)
11. Conti, R., Caputo, G.: A numerical and theoretical study on the seismic behaviour of yielding cantilever walls. Géotechnique **69**(5), 377–390 (2019)
12. Callisto, L.: On the seismic design of displacing earth retaining systems. In: Earth Proceedings of the 7th International Conference on Earthquake Geotechnical Engineering, pp. 239–255 (2019)
13. Mazzoni, S., McKenna, F., Scott, M.H., et al.: Opensees command language manual. Pac. Earthq. Eng. Res. (PEER) Center **264**, 137–158 (2006)
14. Saygili, G., Rathje, E.M.: Empirical predictive models for earthquake-induced sliding displacements of slopes'. J. Geotech. Geoenviron. Eng. **134**(6), 790–803 (2008)
15. Tropeano, G., Silvestri, F., Ausilio, E.: An uncoupled procedure for performance assessment of slopes in seismic conditions. Bull. Earthq. Eng. **15**(9), 3611–3637 (2017). https://doi.org/10.1007/s10518-017-0113-y

16. Gaudio, D., Rauseo, R., Masini, L., Rampello, S.: Semi-empirical relationships to assess the seismic performance of slopes from an updated version of the Italian seismic database. Bull. Earthq. Eng. **18**(14), 6245–6281 (2020)
17. Cosentini, R.M., Bozzoni, F.: Fragility curves for rapid assessment of earthquake-induced damage to earth-retaining walls starting from optimal seismic intensity measures. Soil Dyn. Earthq. Eng. **152**, 107017 (2022)
18. Argyroudis, S., Kaynia, A.M., Pitilakis, K.: Development of fragility functions for geotechnical constructions: application to cantilever retaining walls. Soil Dyn. Earthq. Eng. **50**, 106–116 (2013)
19. Li, F., Jia, H., Luo, Q., Li, Y., Yang, L.: Identification of critical links in a large-scale road network considering the traffic flow betweenness index. PLoS ONE **15**, e0227474 (2020)
20. Maltinti, F., Melis, D., Annunziata, F.: Road network vulnerability: a review of the literature. In: Integrating Sustainability Practices in the Construction Industry, ICSDC 2011, Reston, VA, USA, pp. 677–685. American Society of Civil Engineers (2012)

Centrifuge Experiments Dealing with Monotonic and Cyclic Loads on Pile Foundations in Sand

Maria Iovino[1], Chiara Iodice[2(✉)], Ahmed Alagha[3], and Giulia M. B. Viggiani[3]

[1] University of Napoli "Parthenope", Centro Direzionale Isola C4, 80143 Napoli, Italy
[2] University of Campania "Luigi Vanvitelli", Via Roma 29, 81031 Aversa, Italy
chiara.iodice@unicampania.it
[3] University of Cambridge, JJ Thomson Avenue, Cambridge CB3 0FA, UK

Abstract. A clear understanding of the interaction between superstructures, foundations and the surrounding soil is crucial to enhance the resilience of existing and future Critical Infrastructure by adapting their design and maintenance. Besides the loading components already considered in the current design procedures, the foundation system can be subjected to continuous loads of a cyclic nature (e.g. wind), which are often disregarded in engineering practice even though they may trigger failure mechanisms. This limitation of the current design approach can be attributed to the scarcity of research contributions dealing with cyclic loads on pile foundations. To fill this gap, centrifuge experiments have been carried out at 50 g on piles, isolated and in groups, embedded in Hostun sand and subjected to monotonic and cyclic loads of different amplitudes and frequencies. The details of the model preparation and experimental setup of the innovative tests are presented allowing to identification of key issues in centrifuge experiments dealing with cyclic inclined and eccentric loads on model foundations. These included the manufacturing of new model piles to replicate the behaviour of reinforced concrete piles under horizontal loads which were applied throughout a specific pulley system. Preliminary results are also provided, showing accumulation of displacements during the cyclic loading paths.

Keywords: Cyclic Loadings · Centrifuge Experiments · Pile Foundations · Critical Infrastructure

1 Introduction

The reliability of Critical Infrastructure (CI), such as roads, rail, public buildings and electricity generation facilities, is emerging as one of the paramount issues of this decade. Most often, to address their present and long-term usage, the adaptation and innovation of current design and maintenance approaches are needed. A key point is the assessment of the structural integrity or stability of CI under complex loading conditions, including multi-directional loads and cyclic actions (e.g., wind, extreme weather events, geohazard), which can change during their life span affecting their resilience. This unavoidably depends on the interaction between superstructure, foundation and surrounding soil.

A. Ferrari et al. (Eds.): CNRIG 2023, SSGG, pp. 671–678, 2023.
https://doi.org/10.1007/978-3-031-34761-0_81

With reference to wind industry, for example, in the last 30 years the height of wind turbines has increased to take advantage of larger wind speed and lower turbulence [1]. Employing taller wind turbines requires the use of piled foundations which, according to traditional design approach, are designed by modelling the cyclic wind pressures as equivalent-static actions, thus, neglecting the accumulation of generalized permanent displacements. This lack in design approach can be attributed to very limited research contributions dealing with cyclic loads. It follows that a more reliable approach is necessary to enhance the design of these structures when subjected to non-monotonic loadings. The same arguments can be extended to consider the maintenance of existing CI, often used way beyond their life expectancy, which may be subjected to actions not considered in the original design because of changed loading conditions due to updated requirements by technical codes. The analysis of their response to these changes helps detecting possible failure mechanisms and identifying the required maintenance to enhance reliability and resilience.

Centrifuge testing can certainly provide data to improve our understanding of the complex mechanisms ruling piles deformation and failure [2] and can be used as benchmarks for innovative solutions in terms of design and maintenance of CI. To this end, the behaviour of piled foundation embedded in sand and subjected to different loading types was investigated via experiments carried out at an increased gravity of 50 g in the Turner beam centrifuge [3] at Schofield Centre, University of Cambridge. The details of the model preparation and the experimental setup are presented allowing the identification of key issues in centrifuge tests dealing with cyclic loads on model foundations. Finally, preliminary experimental results are provided.

The experimental program was funded by the European Union H2020 Research and Innovation Programme which supported the execution of experiments proving innovative solutions to enhance the resilience of CI using European centrifuge facilities by the four year project GEOLAB.

2 Model Preparation

2.1 Manufacturing of Miniaturized Reinforced Concrete Piles

Reinforced concrete cylinders 250 mm long and 10 mm in diameter were manufactured to simulate piles. The reinforcement cages were made of galvanized steel mesh (1 mm square opening mesh, 0.4 mm wire diameter), rolled such that the two edges of the mesh formed a cylinder with a diameter of 0.8 mm and a height of 150 mm. Each cage consisted of 14 longitudinal rebars, with the orthogonal wires of the net providing transverse reinforcement (Fig. 1a). The mortar was made of ordinary Portland cement CEM I 52.5 (C) and water (W). The ratio W/C was equal to 0.55 to make the mixture workable enough and easy to flow through the wire mesh. The mould used to cast the piles consisted of three vertical plates with drilled holes of half the pile diameter along their height, placed parallel to one another and bolted to form a total of 24 hollow cylinders for pile casting. Drilled aluminum caps designed to accommodate the load cells were first positioned at the base of these holes with a threaded rod screwed through them to provide an extra rough anchor length (Fig. 1a). Each hole was then filled from the top using a chemistry baker with its spout touching the inner surface of the mould and right after the rebar cages were driven into it (Fig. 1b). This technique allowed easy flow of the mortar and helped avoiding the formation of air-bubbles, eventually vanished while the cages were driven-in. The samples were left to set for 6 days before demoulding, which was facilitated by applying a thin layer of concrete mould release agent to the internal surfaces of the mould before pouring. Finally, a sand film was glued on the piles for simulating the shaft of cast-in-situ reinforced concrete piles (Fig. 1c).

A number of these pile samples were tested in compression, tension and 4-point bending with or without axial compression to derive the envelope of the axial loads and bending moments that produces section failure, which is essential to determine the piles geotechnical horizontal failure load.

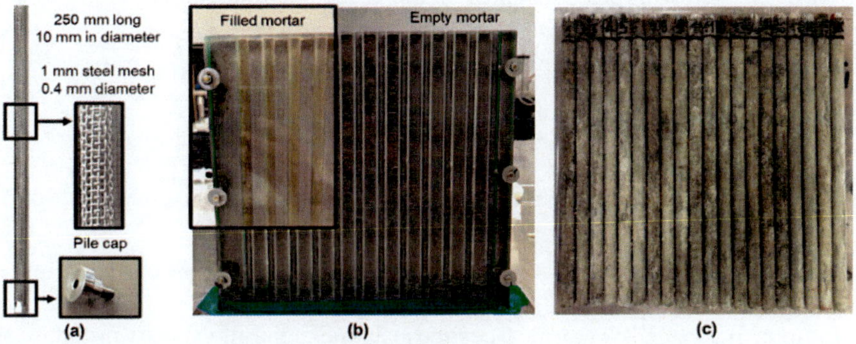

Fig. 1. Manufacturing of miniaturized reinforced concrete piles: (a) steel rebars and pile head connections; (b) pile casting; (c) piles with sand and glue coating.

Pile groups were assembled featuring 8 piles equally spaced along a circle with a diameter of 120 mm and connected by spherical hinges to an aluminum circular rigid raft with a diameter of 138 mm and a variable height (15 mm at pile heads and 5 mm in the central section). The advantages of using such a connection are twofold: the spherical hinges enable a simple interpretation of the load distribution on piles and the rigid raft allows the use of a limited number of instruments to derive the displacement of each pile during the loading process.

2.2 Model Installation and Experimental Setup

The model foundations were embedded in a homogeneous layer of uniformly graded Hostun sand, extensively used in Cambridge for many experimental campaigns [4, 5].

The dry sand was manually poured into a cylindrical steel tub having diameter of 850 mm and height of 400 mm. The sand pouring was stopped to install the model foundations by gently pushing them for a height of 40 mm above piles' tip; after that, the container was filled up to 340 mm in controlled conditions by checking the relative density (about 40%).

The centrifuge models were instrumented to monitor soil and foundation response during the tests. Particularly, each raft was equipped with 3 Linear Variable Differential Transformers (LVDTs) to measure settlements and 4 Micro-Electro-Mechanical-Systems (MEMS) accelerometers for monitoring acceleration and, thus, rotations. Each group of piles was equipped with 5 Load Cells (LCs), positioned immediately below the spherical hinges, to measure axial load on piles while each isolated pile was equipped with 1 LC and 1 LVDT. An additional LVDT was placed onto the soil surface to monitor ground settlements. Moreover, miniature Cone Penetration device (mini-CPT), with a diameter of 6.35 mm and a 60° cone tip, was installed to characterize the soil sample. External loads on single piles were applied using electric actuators by setting the displacement rate at 1 mm/s, while for pile groups pneumatic actuators under load control were used. The actuators were mounted on top of the pile foundations with the support of gantries.

3 Centrifuge Testing

3.1 Test 1

The first series of experiments, schematically depicted in Fig. 2, included tests on:

- single pile in compression up to failure (SPC, Fig. 2a);
- single pile loaded by a centred tension load to estimate its capacity in uplift and then by a compression load up to failure (SPTC, Fig. 2a);
- pile group first loaded by a vertical centred load to estimate its capacity in compression and then by a tension load up to failure (PGCT, Fig. 2b);
- pile group first subjected to a constant dead load and then cyclically loaded by a vertical and eccentric load (PGE, Fig. 2c). This was applied under constant eccentricity (8.51 m at prototype scale) by means of a cantilever beam and oscillated about a constant value. The cyclic component of the external moment consisted of sinusoidal waves with different amplitudes and frequencies. Each wave was made of hundreds of cycles. Finally, PGE was loaded monotonically up to failure.

The arrangement of the foundation models in Test 1 is shown in the plan view and cross sections of Fig. 3, along with the arrangement of instruments.

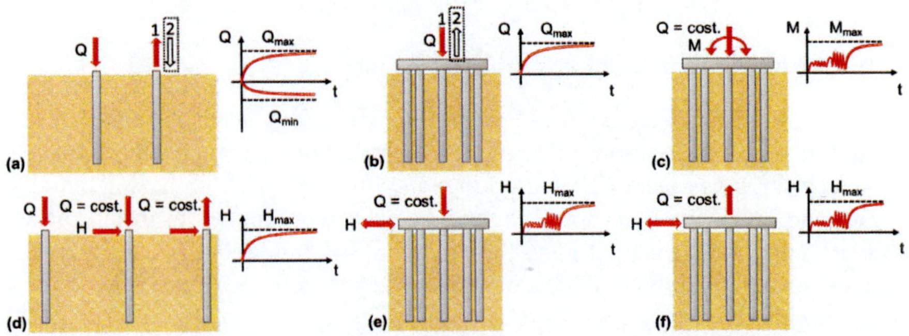

Fig. 2. Schemes of the centrifuge experiments: model foundations and load paths in Test 1 (a, b and c) and Test 2 (d, e and f).

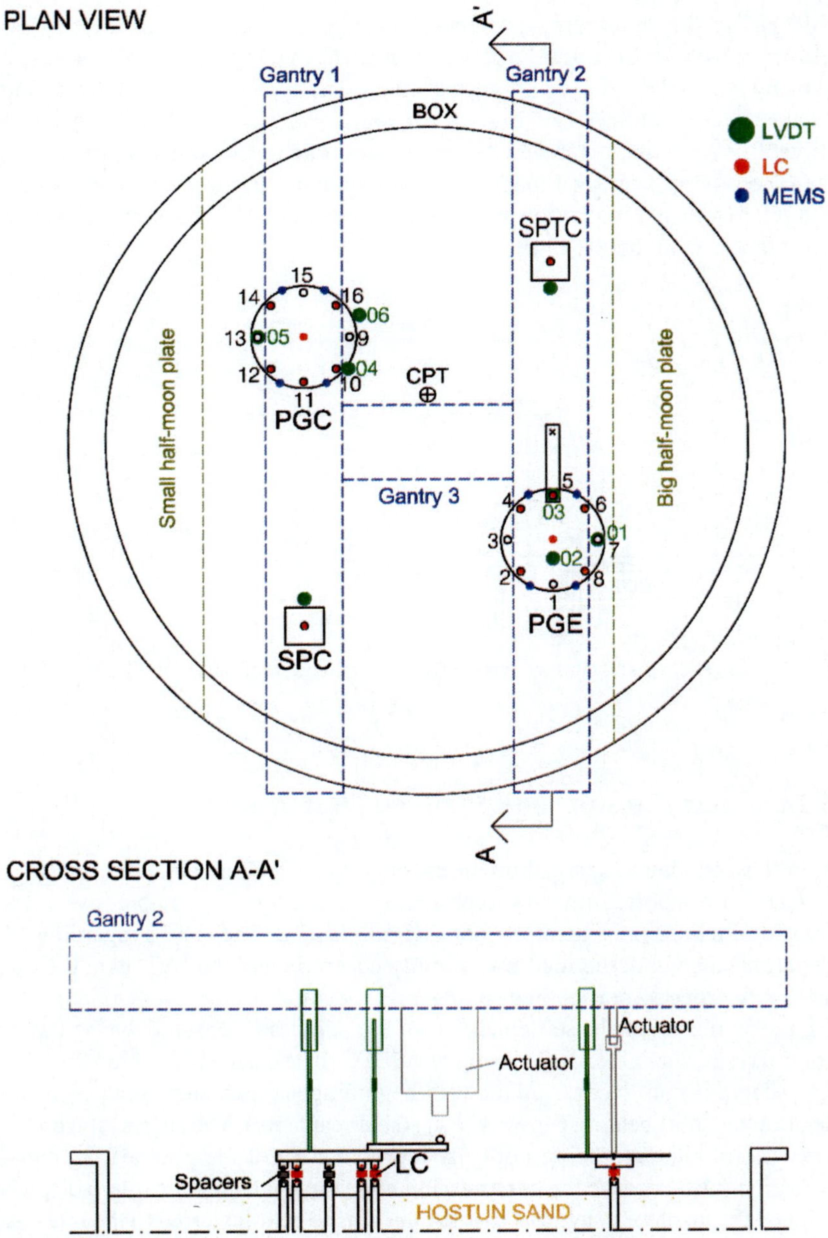

Fig. 3. Test 1: Layout of model foundations and instrument locations. Dimensions are given in mm at model scale.

3.2 Test 2

The second series of experiments included tests on single piles subjected to a constant vertical compressive or tensile dead load plus a horizontal force up to failure (SPCH and SPTH, Fig. 2d) and on pile groups subjected to a constant vertical compressive or tensile dead load plus a cyclical horizontal force (PGCH, Fig. 2e and PGCH, Fig. 2f). An additional SPC test (Fig. 2d) was also performed. The cyclic component of the horizontal load in tests PGCH and PGTH consisted of sinusoidal waves with different amplitudes and frequencies oscillating about constant values. Each wave was made of hundreds of cycles. The groups were then loaded monotonically up to failure.The horizontal loads were applied by using a pulley system specifically designed for this purpose and mounted underneath the actuators (Fig. 4).

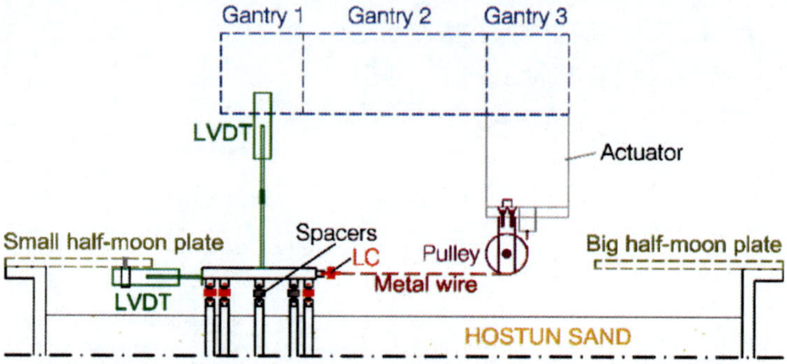

Fig. 4. Test 2: Pulley system for the application of horizontal loads.

4 Preliminary Results and Planned Data Analysis

In the following, due to space limitations, only some preliminary experimental results from Test 1 are reported (at prototype scale). In the figures, compressive forces and downward pile head movement are considered positive. The results regarding SPC are not reported since both the load was slightly eccentric and the LVDT on the pile head did not work properly during the test.

Figure 5a illustrates the settlement, w, of the pile head measured during the application of the vertical axial load, Q, during test SPTC. In the first stage of loading (tension), the Q–w curve is non-linear from the very beginning and exhibits a peak corresponding to the ultimate uplift bearing capacity, $F_{\min}$ (shaft capacity). The second stage of loading (compression), although being not fully representative of the pile bearing capacity in compression (also because the compression was stopped at the maximum displacement allowed by the actuator), suggests that load begins to be transferred to the pile base. The bearing capacity of the pile group can be deduced from the results of test PGCT (Fig. 5b). Note that the compression stage of the test was stopped at a pre-set displacement before the pulling up. At the same displacement, the capacity of the group is approximately 8 times the shaft capacity of the single pile, this means that the base resistance of the pile group has not been mobilized.

Figures 6 and 7 show the force and displacement recorded over the time during tests PGCT and PGE, respectively. Note that only 5 piles out of 8 were instrumented with

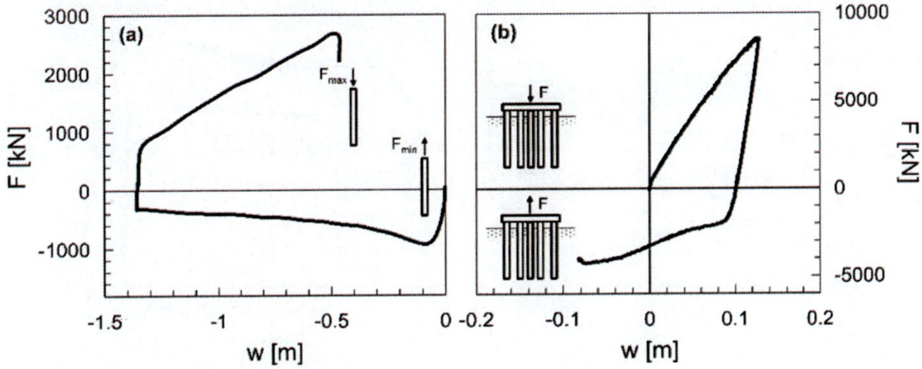

Fig. 5. Load-settlement curves: (a) test SPCT; (b) test PGCT.

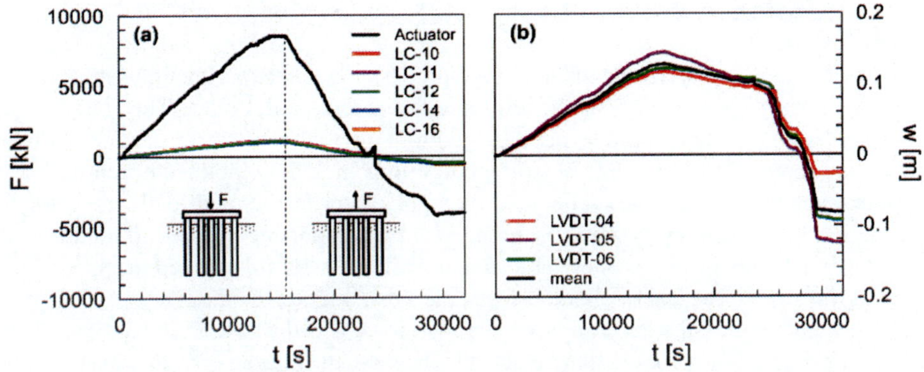

Fig. 6. Test PGCT: (a) force, F, and (b) vertical displacement, w, over the time.

LCs, however, the load at the head of the other piles can be derived by symmetry and by using the data of the applied external load obtained from the LC mounted underneath the pneumatic actuator. The displacements of each pile in the group can be instead determined by writing the equation of the plan defined by the 3 readings of the LVDTs in contact with the rigid cap. This way, information about forces, settlements, accumulation of permanent displacements and stiffnesses can be derived for all the piles belonging to the groups. In test PGCT, the load distribution among the piles is quite uniform (Fig. 6a) while a difference in vertical displacement readings is observed, probably due to a slight rotation of the rigid cap connecting the pile heads (Fig. 6b). Note that the test was stopped at the minimum displacement allowed by the actuator. The application of the cyclic vertical eccentric load on PGE induced significant rotations of the rigid cap, with piles 3–7 undergoing compression and piles 1, 2, 8 tension (Fig. 7a), accumulating displacements (Fig. 7b). Further elaboration of the test results is necessary to better understand the mechanism regulating the behaviour of piled foundation in sand under cyclic loading to properly take into account their effects in design practice.

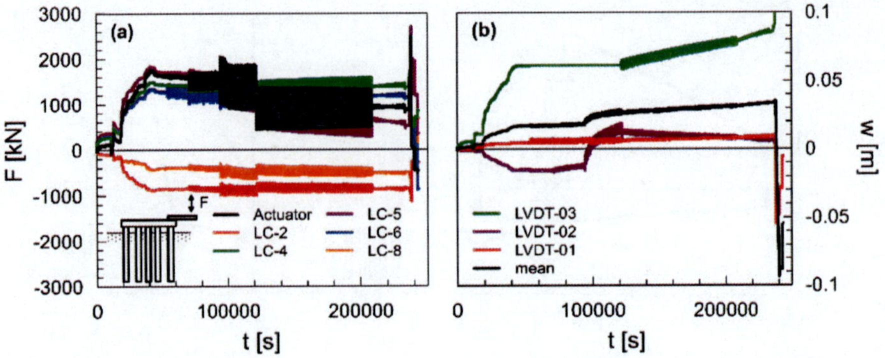

Fig. 7. Test PGE: (a) force, *F*, and (b) vertical displacement, *w*, over the time.

5 Conclusions

This paper presents the details of the model preparation and experimental setup of two series of centrifuge tests at 50 g on reinforced concrete model piles. The foundations consist of single piles and pile groups hinged to a circular raft, not in contact with the soil, and subjected to generalized loading conditions including inclined eccentric loads applied cyclically. The issues related to the application of inclined loads were sorted out by means of a pulley system for the horizontal component and a dead load for the vertical one, while the use of pneumatic actuators in load control allowed to apply cyclic loading histories. The data collected form the centrifuge tests show accumulation of displacements with rates depending on the load level and the number of cycles. These results supply new experimental evidence which add to the limited literature available on the topic, serving as benchmark for the development of innovative approach to strengthen the CI resilience by adapting the traditional design and maintenance to properly consider changing loading conditions.

References

1. Liebreich, M.: State of the Clean Energy Industry. The 'Future of Energy' Summit, Bloomberg New Energy Finance, London (2017)
2. de Sanctis, L., Di Laora, R., Garala, T.K., Madabhushi, S.P.G., Viggiani, G.M.B., Fargnoli, P.: Centrifuge modelling of the behaviour of pile groups under vertical eccentric load. Soils Found. **61**(2), 465–479 (2021)
3. Schofield, A.N.: Cambridge geotechnical centrifuge operations. Geotechnique **30**(3), 227–268 (1980)
4. Mitrani, H.: Liquefaction remediation techniques for existing buildings. Ph.D. dissertation, Cambridge University, Cambridge (2006)
5. Haigh, S.K., Eadington, J., Madabhushi, S.P.G.: Permeability and stiffness of sands at very low effective stresses. Géotechnique **62**(1), 69–75 (2012)

Influence of Vertical Ground Motion on the Seismic Performance of an Earth Dam

Andrea Nardo(✉), Ernesto Cascone, Giovanni Biondi, Giuseppe Di Filippo, and Orazio Casablanca

University of Messina, Contrada di Dio, Messina, Italy
andrea.nardo@studenti.unime.it

Abstract. Earthquake-induced permanent displacements and distortions, as well soil mass movements and cracks, may significantly affect the post-seismic serviceability of earth dams. To minimize these effects the seismic design or retrofitting procedures should aim to improve the resilience in terms of ability to recover and adapt to the severe earthquakes. In this vein, through 2D numerical analyses, the paper investigates the role of the vertical component of the ground motion on the seismic performance of an earth dam. The results of the analyses are described in terms of main features of the seismic-induced plastic mechanisms and magnitude of the corresponding crest settlements showing that in some cases the effect of the vertical acceleration component is not negligible. Finally, an empirical correlation between some seismic parameters and the permanent crest settlements of the dam has been proposed.

Keywords: Earth dams · seismic performance · dam resilience · vertical motion

1 Introduction

Large earth dams are strategic infrastructure as they supply local communities with large water resources, promoting livelihood. In high seismicity areas, the stability conditions during earthquakes and the potential sudden flooding represent a major concern for the safety and sustainability of the areas downstream the dam. Also, seismic-induced displacements and distortions, as well soil mass movements and cracks, may significantly affect the post-seismic serviceability of earth dams and of the overall plant. Thus, for both new and existing dams, design or retrofitting procedures should aim to improve the seismic resilience in terms of ability to recover and adapt to the severe ground motions.

The seismic performance of earth dams is usually examined under only horizontal input motions [e.g. 1–4], however, analyses carried out considering the effect of the vertical component of the ground motion [e.g. 5–7] revealed the need of further investigation and motivated the analyses presented herein.

In this vein the paper focuses on the role of the vertical component of the ground motion on the seismic performance of earth dams. This has been investigated starting from the results of a set of 2D dynamic numerical analyses carried out with reference to the San Pietro Dam and to a large set of earthquake records.

A. Ferrari et al. (Eds.): CNRIG 2023, SSGG, pp. 679–686, 2023.
https://doi.org/10.1007/978-3-031-34761-0_82

The San Pietro dam, along the Osento (Campania, Southern Italy) river, is a zoned earth dam with a 48 m high embankment and a volume of about 2.2 mm^3, retaining 17.7 mm^3 of water with a maximum freeboard of 2 m [1]. It lies on an area of high seismicity according to the most recent probabilistic seismic hazard assessment (*PSHA*) transposed into the latest Italian codes relevant for new and existing dams.

The numerical results presented in this paper focus on the main features of the possible plastic mechanisms that can develop in the dam embankment, possibly compromising the hydraulic tightness of the dam. Finally, an empirical correlation between some seismic parameters representative of the vertical and horizontal input motion and the permanent crest settlements of the dam has been proposed.

2 Input Motions and Numerical Analyses

The horizontal and vertical components of 47 records of large earthquakes occurred worldwide with moment magnitude M_w in the range 4.0–7.3 and recorded on rock outcrop sites with epicentral distances R_{ep} varying between 0.7 and 70 km, have been selected from the *Engineering Strong Motion database*. Figure 1a shows the data coverage of the selected set of records in terms of M_w and R_{ep}: about 21% of the data represent near-field motions. Table 1 lists the values of M_w and R_{ep} of the selected records together with the peak accelerations (PGA^h, PGA^v) and the Arias intensities (I_A^h, I_A^v) of the horizontal and vertical components of the records.

The records were subdivided into four groups (*A-D*), according to the peak horizontal acceleration values (Table 1). The vertical-to-horizontal peak acceleration (PGA^v/PGA^h) and the Arias intensity (I_A^v/I_A^h) ratios vary in wide ranges with values up to 1.42 and 1.20, respectively (Fig. 1b); specifically, PGA^v/PGA^h and I_A^v/I_A^h are less than 1/3 for about 21% and 49% of the data, respectively, while values larger than 2/3 characterize about the 25% and 15% of the selected records.

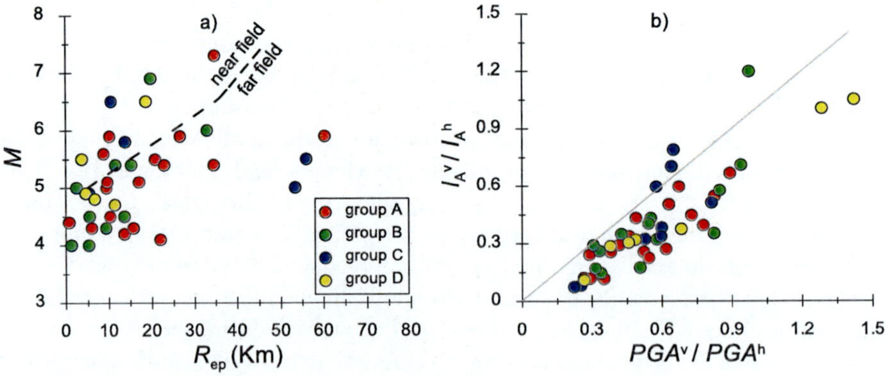

Fig. 1. a) M_W - R_{ep} data coverage of the selected set of records; b) vertical-to-horizontal peak ground acceleration and Arias intensity ratios.

For both the static and seismic loading conditions, finite element (FE) analyses have been performed using Plaxis under plane strain conditions, discretizing the dam and the foundation soils through a FE mesh of 5000 triangular 15-noded elements (Fig. 2). The static analyses, aimed at reproducing the state of stress at the end of the construction stages preceded the non-linear dynamic analyses. These were carried out modelling the reservoir as a hydrostatic load applied on the upstream face of the dam, assuming values of the small-strain shear modulus G_0 evaluated from cross-hole tests and considering an hysteretic soil model, additional Rayleigh damping, excess pore-pressure generation and proper boundary conditions [4]. Using the same set of acceleration records considered herein (Table 1), Casablanca et al. [4] carried out dynamic analyses of the San Pietro dam applying only the horizontal components of the input motion in both the upstream and downstream directions, thus carrying out 94 dynamic analyses. Herein, these 94 analyses have been performed accounting also for the vertical component of each of the selected records.

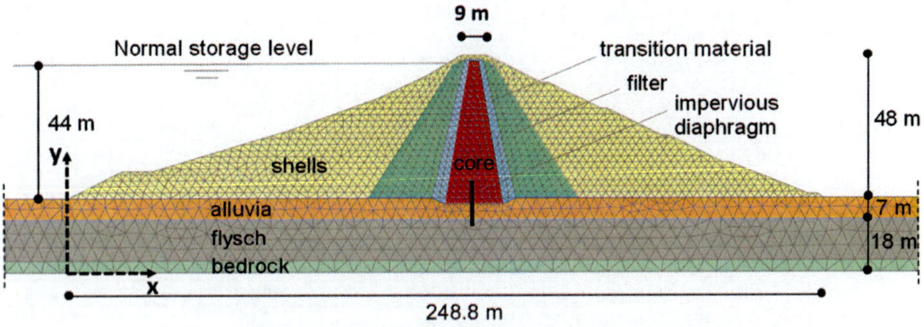

Fig. 2. Finite element model adopted in the numerical analyses.

3 Analysis Results

In order to point out the effect of the vertical component of the ground motion, the analysis results have been examined in terms of predicted plastic mechanisms, associated permanent displacements, vertical crest settlements and their variation with the amplitude and energy content of the vertical and horizontal components of the input motion. Some of the large sets of results are presented herein in terms of main features of the plastic mechanism associated to the dam response and amplitude of the crest settlements induced by the combined vertical and horizontal ground motions.

The results discussed herein refer only to the two near-field records #46 and #42 (Table 1) belonging to the *group D* of the most intense accelerograms. Record #42 (M_w = 6.5, R_{ep} = 5.5 km) is characterized by the largest values of the vertical peak ground acceleration and Arias Intensity ($PGA^v = 0.557g$, $I_A^v \approx 2.02$ m/s) and by very large values of the vertical-to-horizontal peak acceleration and Arias intensity ratios (PGA^v/PGA^h = 1.42, $I_A^v/I_A^h = 1.05$). For record #46 ($M_w = 5.5$, $R_{ep} = 3.6$ km), despite characterized by one of the largest value of the horizontal peak ground acceleration ($PGA^h = 0.666\,g$ with $PGA^v/PGA^h = 0.38$), it is $I_A^v \approx 90$ cm/s with $I_A^v/I_A^h = 0.28$.

For these two records Figs. 3 and 4 show the contours of the horizontal (Fig. 3) and vertical (Fig. 4) permanent displacements at the end of the shaking computed accounting only for the horizontal and for both the horizontal and the vertical components of the motion. Figure 5 shows the contours of the permanent deviatoric deformations.

Table 1. Earthquake records selected as input motions.

Group	n.	Station	Event_ID	Components	M_w (-)	R_{ep} (km)	PGA^h (g)	PGA^v (g)	I_A^h (cm/s)	I_A^v (cm/s)
A ($PGA^h = 0.1$ g ± 5%)	#1	CESM	IT-1997-0137	HNE, HNZ	5.6	8.7	0.096	0.044	5.07	1.72
	#2	T1212	EMSC-20161030_0000135	HNE, HNZ	4.5	10.3	0.096	0.028	1.57	0.38
	#3	T1218	EMSC-20170118_0000034	HNE, HNZ	5.5	20.7	0.097	0.001	5.98	0.001
	#4	PYAD	EMSC-20080716_0000056	HNN, HNZ	4.2	13.6	0.098	0.035	1.50	0.18
	#5	0LAI	IT-2012-0061	HGE, HNZ	5.0	9.4	0.098	0.087	3.21	2.14
	#6	T1213	EMSC-20161030_0000135	HNN, HNZ	4.5	13.6	0.099	0.061	3.69	1.00
	#7	T1218	EMSC-20161026_0000095	HNN, HNZ	5.9	26.5	0.100	0.027	6.97	0.85
	#8	ATN	IT-1984-0004	HNN, HNZ	5.9	10.1	0.100	0.063	6.23	3.14
	#9	C1061	TK-1999-0415	HNN, HNZ	7.3	34.7	0.100	0.052	22.52	5.80
	#10	T0912	IT-2013-0008	HNE, HNZ	4.4	0.7	0.101	0.073	5.39	2.41
	#11	SHAP	EMSC-20150922_0000029	HNE, HNZ	4.3	15.7	0.101	0.079	3.70	1.47
	#12	AC01	EMSC-20141108_0000098	HNN, HNZ	5.1	9.6	0.102	0.068	4.01	2.40
	#13	T1218	EMSC-20161026_0000077	HNN, HNZ	5.4	22.8	0.102	0.030	5.80	0.68
	#14	SHAP	EMSC-20140715_0000041	HNN, HNZ	4.1	21.9	0.103	0.039	2.08	0.53
	#15	ACC	EMSC-20170118_0000027	HGN, HNZ	5.1	17.0	0.104	0.085	5.78	3.17
	#16	MMUR	EMSC-20161026_0000095	HNN, HNZ	5.9	70.0	0.105	0.044	4.29	1.26
	#17	RQT	EMSC-20170118_0000037	HGN, HNZ	5.4	34.5	0.105	0.057	6.41	1.44
	#18	T1213	EMSC-20160825_0000024	HNE, HNZ	4.3	5.9	0.105	0.051	2.70	1.17
B ($PGA^h = 0.2$ g ± 10%)	#19	T1212	EMSC-20161026_0000077	HNN, HNZ	5.4	15.2	0.180	0.060	9.88	2.58
	#20	T1213	EMSC-20161030_0000135	HNE, HNZ	4.5	13.6	0.181	0.061	6.91	1.00
	#21	RQT	EMSC-20160903_0000063	HNE, HNZ	4.3	9.3	0.185	0.058	7.11	1.20
	#22	T1213	EMSC-20161031_0000053	HGN, HNZ	4.0	5.2	0.186	0.107	5.19	1.66
	#23	MRM	IT-2012-0061	HNN, HNZ	5.0	2.4	0.186	0.103	10.69	4.62
	#24	FEMA	EMSC-20160824_0000006	HGE, HNZ	6.0	32.9	0.189	0.081	12.33	4.30
	#25	T1245	EMSC-20161026_0000133	HNN, HNZ	4.5	5.4	0.193	0.105	8.99	3.63
	#26	T1212	EMSC-20161026_0000077	HNN, HNZ	5.4	15.2	0.195	0.060	8.89	2.58
	#27	FEMA	EMSC-20161026_0000077	HNN, HNZ	5.4	11.5	0.198	0.163	23.44	8.27
	#28	SRN	GR-2016-0008	HNE, HNZ	5.0	53.3	0.198	0.186	22.85	16.24
	#29	FEMA	EMSC-20170427_0000119	HNN, HNZ	4.0	1.2	0.201	0.195	5.40	6.47
	#30	T1213	EMSC-20161031_0000053	HNN, HNZ	4.0	5.2	0.212	0.107	9.55	1.66
	#31	ULA	ME-1979-0003	HNE, HNZ	6.9	19.7	0.214	0.181	72.89	42.09
C ($PGA^h = 0.1$ g ± 10%)	#32	T1212	EMSC-20161030_0000029	HNE, HNZ	6.5	10.5	0.278	0.167	70.73	27.03
	#33	T1212	EMSC-20161030_0000029	HGE, HNZ	6.5	10.5	0.280	0.167	80.30	27.03
	#34	SVN	EMSC-20181226_0000014	HNN, HNZ	4.9	4.5	0.283	0.149	29.17	9.41
	#35	SRN	GR-2016-0008	HNN, HNZ	5.0	53.3	0.290	0.186	23.12	16.24
	#36	SRN	GR-2016-0006	HNE, HNZ	5.5	55.9	0.293	0.190	64.36	50.80
	#37	EVRN	EMSC-20181226_0000014	HNE, HNZ	4.9	5.3	0.300	0.076	35.36	2.89
	#38	MI05	IT-2009-0102	HNN, HNZ	5.5	3.6	0.310	0.252	49.83	25.53
	#39	4101	TK-1999-0294	HNN, HNZ	5.8	13.8	0.323	0.072	30.49	2.22
	#40	ILLI	IT-2010-0032	HNE, HNZ	4.7	11.4	0.329	0.190	50.71	30.22
D ($PGA^h = 0.33$-0.67 g)	#41	ILLI	IT-2010-0032	HNE, HNZ	4.7	11.4	0.390	0.190	94.75	30.22
	#42	ACC	EMSC-20161030_0000029	HGE, HNZ	6.5	18.6	0.392	0.557	192.22	201.92
	#43	ACC	EMSC-20161030_0000029	HNN, HNZ	6.5	18.6	0.434	0.557	201.08	201.92
	#44	GMLD	EMSC-20190808_0000045	HNE, HNZ	4.8	6.6	0.450	0.308	123.45	46.24
	#45	SVN	EMSC-20181226_0000014	HGN, HNZ	4.9	4.5	0.559	0.149	86.80	9.41
	#46	MI05	IT-2009-0102	HNE, HNZ	5.5	3.6	0.664	0.252	89.77	25.53
	#47	GMLD	EMSC-20190808_0000045	HNE, HNZ	4.8	6.6	0.673	0.308	152.23	46.24

For both records, the contours depicted Fig. 3 show that, regardless the presence of the vertical input motion, the horizontal displacements of the upstream and downstream shells are of opposite sign, indicating a bulging of the dam.

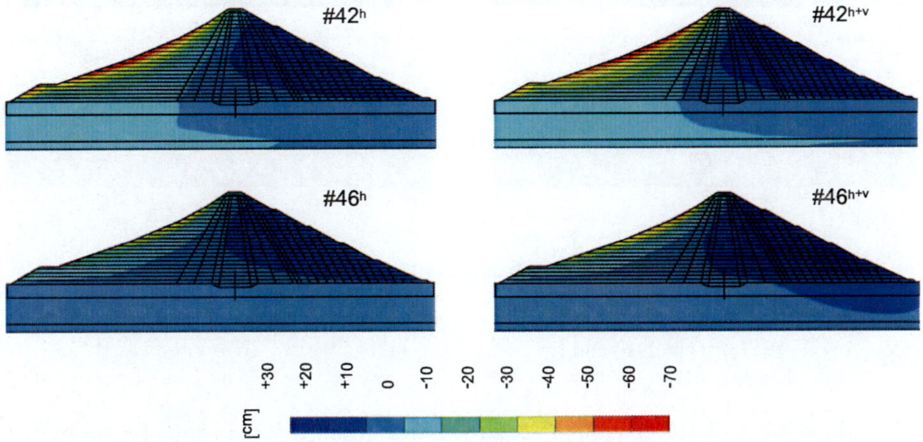

Fig. 3. Records #46 and #42: contours of horizontal permanent displacements computed accounting for the horizontal (h) and for both horizontal and vertical (h + v) input motion.

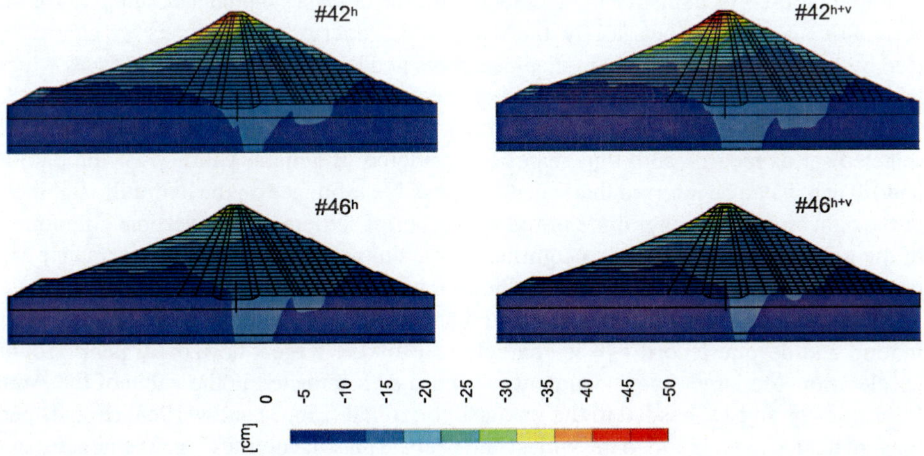

Fig. 4. Records #46 and #42: contours of the vertical permanent displacements computed accounting for the horizontal (h) and for both horizontal and vertical (h + v) input motion.

Also, no relevant effect of the vertical component of the ground motion can be observed in terms of amplitude of permanent horizontal displacements.

It can be observed that the influence of the vertical input motion is more relevant in terms of distribution and amplitude of the vertical component of the permanent displacements (Fig. 4) and size of the plastic mechanism which can be detected starting from the contours of the permanent deviatoric deformations (Fig. 5).

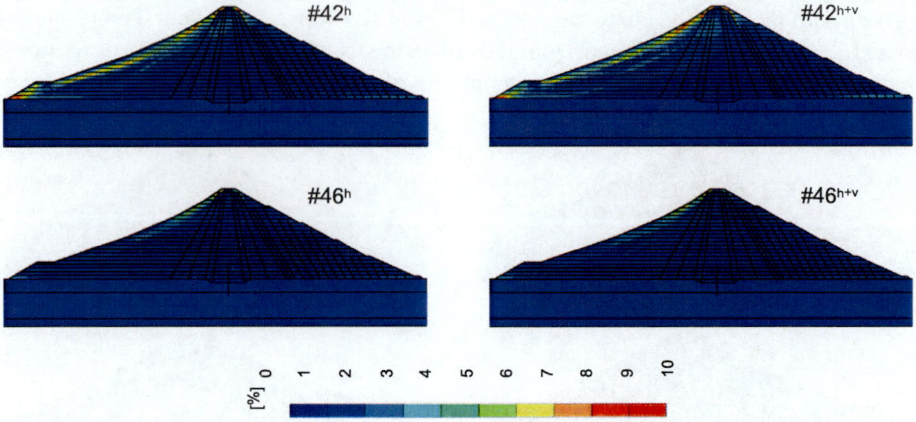

Fig. 5. Records #46 and #42: contours of the permanent deviatoric deformations computed accounting for the horizontal (h) and for both horizontal and vertical (h + v) input motion.

For both the considered records, regardless the presence of the vertical input motion, the potential failure mechanism affects only the upstream shell with the largest deviatoric strains concentrated in the shallow portions of the dam embankment.

Furthermore, an enlargement of the dimension of the plastic volume involved in the potential failure mechanism can be detected for the analyses which account also for the vertical input motion; specifically, the wider and deeper plastic volumes are characterized by larger vertical permanent displacements and by potential sliding surfaces which intersect the core affecting also the uppermost portion of the downstream shell.

Similar results have been obtained for most of the analyses carried out using the selected set of records and, thus, can be considered of general validity for the dam at hand. Then, it can be inferred that, due to its possible influence on the hydraulic tightness of the dam and on the magnitude of the vertical crest settlement, the vertical component of the ground motion can play a significant role on the actual seismic performance and post-seismic serviceability of the San Pietro dam.

The contours shown in Figs. 3, 4 and 5 show also that, regardless the vertical input motion and despite record #46 is characterized by the largest horizontal peak ground acceleration, the larger energy content of record #42, reflected in the value of the Arias Intensities I_A^h and I_A^v, has led to the greatest horizontal (Fig. 3) and vertical (Fig. 4) permanent displacements and to the widest and deepest plastic volume (Fig. 5) characterized by the largest permanent deviatoric strains.

From the contours of Figs. 3, 4 and 5 it is also apparent that the more relevant effects of the vertical input motion can be observed in the case of records #42 characterized by the largest Arias Intensity. The more relevant effect of the energy content of the vertical ground motion in comparison with its peak parameters can be considered as a general outcome and confirm that, also for the vertical ground motion, the Arias intensity has a predominant role in the prediction of the dam overall response and of related permanent crest settlements.

The values of the permanent crest settlements computed accounting only for the horizontal (w_c^h) or both horizontal and vertical (w_c^{h+v}) components of the input motion are shown in Fig. 6a. As in the case of simplified Newmark-type analyses, also for the non-linear dynamic analyses presented herein, the larger values of the ratio w_c^{h+v}/w_c^h occurs in the cases characterized by the smaller w_c^h, implying that more relevant effects of the vertical component of the input motion are expected when the horizontal one, though not crucial for the seismic stability conditions, is relevant for the post-seismic serviceability of the dam. Specifically, for w_c^h in the range 0–15 cm values of the ratio w_c^{h+v}/w_c^h up to 1.5 have been computed for both near- and far-field input motions and values up to about 2.75 can be detected in Fig. 6a for w_c^h less than 5 cm. For the case at hand, the above detected range of values of w_c^h correspond to normalized crest settlements equal to about 0.07% and 0.21%, relevant to check the operational and the damage limit state of the dam [8].

In Fig. 6b the permanent settlement of the dam crest centre w_c^{h+v} is plotted against an intensity measure *IM* defined as a function of the Arias intensity and the mean period of both horizontal (I_A^h, T_m^h) and vertical (I_A^v, T_m^v) components of the seismic records. The plot suggests a good (goodness-of-fit index $R^2 = 0.97$) nonlinear relationship between *IM* and w_c^{h+v}:

$$w_c^{h+v} = 0.6976 \cdot IM - 0.0027 \cdot IM^2 \quad (1)$$

Equation (1) allows predicting the permanent crest settlement w_c^{h+v} considering both horizontal and vertical components of the seismic motion.

It can also be used in the case of only horizontal input motion, *IM* reduces to $IM^h = I_A^h \cdot T_m^h$ alternatively to the Eq. (2) proposed by Casablanca et al. [4]:

$$w_c^{h+v} = 0.6327 \cdot IM - 0.0024 \cdot IM^2 \quad (2)$$

The above equations give the permanent crest settlements in (cm) considering the Arias intensity in (cm/s) and the mean period in (s).

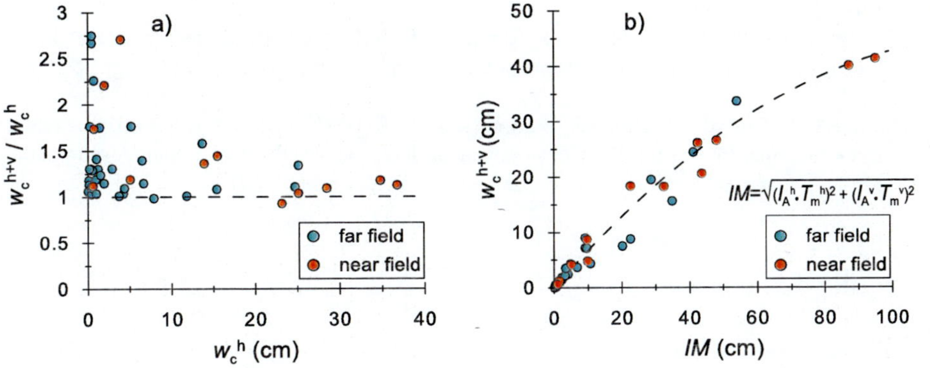

Fig. 6. a) Permanent crest settlements computed accounting only for the horizontal (w_c^h) or for both horizontal and vertical (w_c^{h+v}) components of input motion; b) proposed best-fit equation.

4 Conclusions

The results of the dynamic analyses presented in the paper point out that the influence of the vertical acceleration component is not negligible and assumes a more relevant effect in cases where the horizontal component of the input motion is relatively weak.

The results also allowed formulating a predictive correlation between some seismic parameters of the vertical and horizontal acceleration components and the permanent crest settlements of the dam.

Acknowledgments. This work is part of the research activities carried out in the framework of the research project of major national interest, PRIN n. 2017YPMBWJ, on "*Risk assessment of Earth Dams and River Embankments to Earthquakes and Floods* (REDREEF)" funded by the Italian Ministry of Education University and Research (MIUR).

References

1. Biondi, G., Cascone, E., Aliberti, D., Rampello, S.: Screening-level analyses for the evaluation of the seismic performance of a zoned earth dam. Eng. Geol. **280**, 105954 (2021)
2. Aliberti, D., Cascone, E., Biondi, G.: Seismic performance of the San Pietro dam. Procedia Eng. **158**, 362–367 (2016)
3. Aliberti, D., Biondi, G., Cascone, E., di Prisco, C.: Coupled FE seismic analysis of a zoned earth dam. In: Earthquake Geotechnical Engineering for Protection and Development of Environment and Constructions, pp. 1058–1065. CRC Press (2019)
4. Casablanca, O., Nardo, A., Biondi, G., Di Filippo, G., Cascone, E.: Seismic performance of a zoned earth dam. In: Wang, L., Zhang, J.M., Wang, R. (eds.) Performance Based Design in Earthquake Geotechnical Engineering, vol. 52, pp. 1929–1936. Springer, Cham (2022). https://doi.org/10.1007/978-3-031-11898-2_176
5. Masini, L., Rampello, S., Donatelli, R.: Seismic performance of two classes of earth dams. Earthq. Eng. Struct. Dyn. **50**, 692–711 (2021). https://doi.org/10.1002/eqe.3352
6. Masini, L., Rampello, S.: Influence of input assumptions on the evaluation of the seismic performance of earth dams. J. Earthq. Eng. **26**(9), 4471–4495 (2020). https://doi.org/10.1080/13632469.2020.1835747
7. Cascone, E., Biondi, G., Aliberti, D., Rampello, S.: Effect of vertical input motion and excess pore pressures on the seismic performance of a zoned dam. Soil Dyn. Earthq. Eng. **142**, 106566 (2021)
8. Aliberti, D., Biondi, G., Cascone, E., Rampello, S.: Performance indexes for seismic analyses of earth dams. In: Earthquake Geotechnical Engineering for Protection and Development of Environment and Constructions, pp. 1066–1073. CRC Press (2019)

Effects of Complex Surface Conditions on the Seismic Response of Caisson Foundations

Diana Salciarini(✉), Davide Pauselli, and Giulia Temperoni

Department of Civil and Environmental Engineering, University of Perugia, 06125 Perugia, Italy
diana.salciarini@unipg.it

Abstract. Soil-structure interaction phenomena under seismic loads can be studied by means of several approaches, including full dynamic models that require the use of 2D and 3D FE techniques. Through the latter, it is possible to consider and investigate the effect of the three-dimensionality of the ground on the response of the foundation. In this work, we analyzed the response of caisson foundations for an ideal viaduct, subjected to a seismic load, when the viaduct is located over topographical hollows, simulating configurations that are encountered very frequently in reality (e.g., for the presence of an incise watercourse channel). First, a 3D prototype model was prepared, representative of recurrent conditions mirrored by real-world case studies, adopting simplified conditions of a single caisson and horizontal surface. Then, four different 3D models were considered, reproducing topographical convergences, progressively increasing the lateral slopes' steepness (5°, 10°, 15°, and 20°). The presence of a topographic convergence can produce an amplified seismic acceleration, both at the top of the foundation, both in free-field conditions. The results illustrated that the maximum displacements at the caisson top increase with the increase in the hollow's lateral slope angle. More specifically, going from the horizontal surface case to the case of convergence with 20° of lateral slope angle, the increase is up to 25% in the vertical displacement, while the horizontal displacement increases up to 4 times the one predicted in planar conditions.

Keywords: Soil-Structure Interaction · Finite Element Modelling · Caissons

1 Introduction

Deep foundations like caissons are widely used in relevant engineering works, such as bridges and viaducts. As reported in previous studies (e.g., [1, 2]), a concrete caisson is characterized by a significant stiffness contrast with respect to the surrounding soil, and its massive volume constitutes a major portion of the weight of the entire viaduct.

To ensure increased security levels for infrastructures, numerous recent research efforts have been devoted to developing reliable models for predicting the foundation and superstructure behavior under dynamic loads, considering the soil-structure interaction effect ([3–6] among others). Being the caisson an isolated foundation, its behavior is typically studied considering it as a single element included in a portion of the soil domain

A. Ferrari et al. (Eds.): CNRIG 2023, SSGG, pp. 687–695, 2023.
https://doi.org/10.1007/978-3-031-34761-0_83

of appropriate dimension, under the simplifying assumption of a planar topographical surface [7]. However, the construction of roads and highways on hilly steep terrains or across incise watercourse is undoubtedly very common. This implies the interaction of viaducts and bridges with complex topographical morphology, which can affect the response of the foundation-structure system, especially under dynamic loads. When deep foundations are installed on a steep slope, they are typically subjected to both oblique and vertical loads [8, 9]. In addition, under seismic conditions, the earthquake can be subjected to diffraction phenomena that can modify the wave field and the peak amplification [10].

The objectives of the work are: at first, to conduct analyses of the seismic response of a single caisson foundation, through a complete 3D study, considering a prototype case study with a simple planar surface, as a reference model. Secondly, to compare the results of these first analyses to the case of more complex local topography, to understand the effect on the system response in the three space dimensions.

2 Prototype Case Study

A hypothetical case study, although representative of recurrent conditions mirrored by real-world case studies, has been considered. As for the foundation, typical geometry and dimension have been assigned relying on the analyses of some case studies of concrete caissons located in central Italy: a cylindric shape, a radius of 8.5 m, and a depth of 9 m have been adopted. The pier is an RC column with a rectangular cross-section of 2 × 6 m and a height of 12 m – represented as a beam in the model with appropriate weight and inertia (Fig. 1). As for the soil, a homogeneous substrate constituted of a single layer of fine-grained soil has been assumed. The presence of a 50 m thick, homogeneous, and deformable layer has been preferred for this study to emphasize the seismic response of the caisson and give generality to the study. Finally, the presence of a water table at the base of the caisson has been considered.

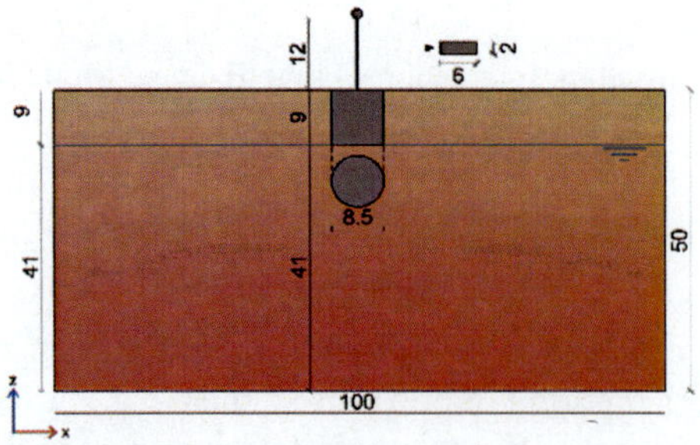

Fig. 1. Sketch of the prototype case study, with the considered dimensions.

2.1 Seismic Input Definition

A seismic input selected among recorded accelerograms has been used for the dynamic analyses performed. This has been defined using the REXEL code [11], which resorts to international databases. Starting from an elastic response spectrum target, the code produces accelerograms satisfying spectrum-compatibility criteria. In this paper, the presentation of the results for the prototype model is limited to a single accelerogram, recorded during the Friuli earthquake (ID 00055), with a moment magnitude of 6.5, an epicentral distance of 23 km, and an expected return time TR = 1980 years. Figure 2 shows the seismic time history, while its main characteristics are summarized in Table 1.

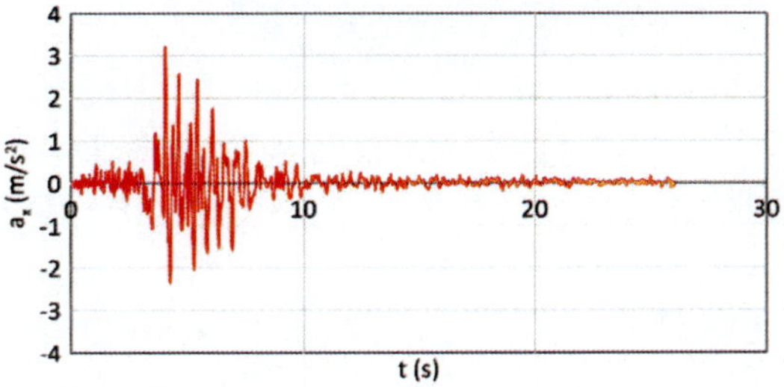

Fig. 2. Seismic input ID 00055 used in the simulation program.

Table 1. Seismic input characteristics.

Record	$D_{(5\text{-}95)\%}$ (s)	F_{max} (Hz)	I_{Arias} (m/s)	PGA (m/s^2)
Friuli Earthquake 000055	4.28	3.846	0.654	3.189

3 The 3D Numerical Modelling

3.1 The Reference Model

A 3D model including the caisson, the pier, and the soil surrounding/underlying the foundation has been built by means of the Plaxis 3D code (Fig. 3), following the geometry already described in Fig. 1. Free field conditions have been assumed at the lateral boundaries, and a compliant base has been assumed at the bottom of the soil domain. The structural elements (pier and caisson) have been modeled as linear visco-elastic elements, with a 5% damping ratio. A lumped mass of 497.4 Mg has been imposed on the top of the pier to simulate the presence of the deck. For the soil layer, the Hardening Soil model with small-strain stiffness (*HS-small*) constitutive law has been adopted [12]. This is able to account for hysteresis in cyclic loading, which leads to damping. The amount of hysteretic damping depends on the applied load amplitude and corresponding strain amplitudes. Table 2 shows the adopted soil properties assumed in the 3D model. Two reference points have been selected at the ground surface of the domain (Fig. 3) to evaluate the model results in terms of accelerations, stresses, strains, and displacements.

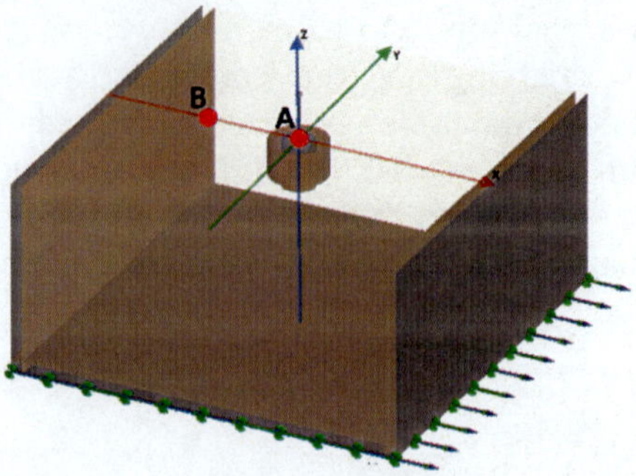

Fig. 3. The 3D model of soil domain, caisson, and pier, loaded at the bottom with the seismic input, used for the simulation program. Red dots indicate the reference points at the surface.

Table 2. Assumed parameters for soil and concrete.

Parameter	Symbol	Clayey soil	Unit
Constitutive model	–	HS-small	–
Dry unit weight	γ_{sat}	20	kN/m^3
Sat unit weight	γ_{unsat}	17	kN/m^3
Mod. sec	E'^{ref}_{50}	20	MPa
Mod. edom	E'^{ref}_{oed}	20	MPa
Mod. unl/rel	E'^{ref}_{ur}	100	MPa
Cohesion	c'	20	kN/m^2
Friction angle	φ'	25	°
Small-strain stiff.	G_0^{ref}	150	MPa
Shear-strain stiff.	$\gamma_{0.7}$	0.1e−3	–
Poisson coeff.	ν'_{ur}	0.25	–
Damp. Rayleigh	ξ	2	%

3.2 Results for the Reference Model

The contours of the residual values of shear strain and vertical displacements are shown in Fig. 4a and Fig. 4b, respectively. From both these figures, ignoring the lateral effects (not affecting the caisson's response), it is apparent that the prominent effects are intensified around the caisson. From Fig. 5a limited amplifications of the seismic motion at the top of the caisson are evident (red line), while more significant ones are obtained in the free-field conditions (blue line). The predicted vertical displacements shown in Fig. 5b are very similar at the two reference points and limited to about 2.5 cm.

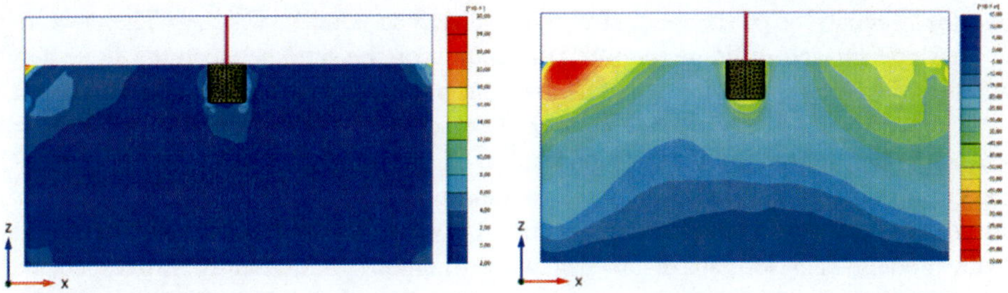

Fig. 4. Contours of the residual values of a) shear strain; and b) vertical displacements.

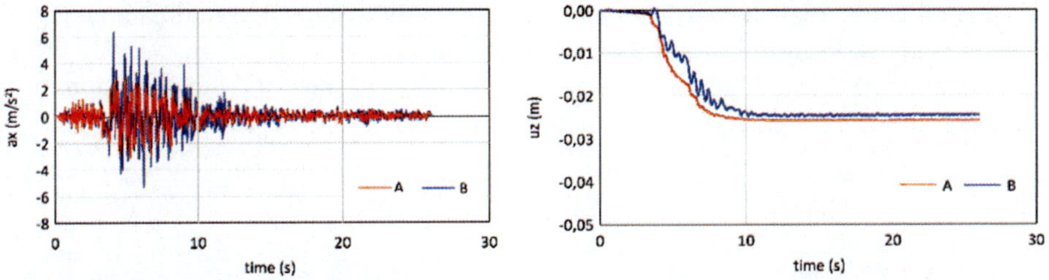

Fig. 5. Model results computed at the two reference-points A and B for a) time-history of surface accelerations; and b) evolution of vertical displacements during the time.

4 Complex Surfaces Effects

4.1 Effect of Topographic Hollows

The presence of topographic convergences is a very common condition for bridges crossing incise water courses, in many Italian areas with complex morphology. Very frequently, the presence of a stream produces erosion that can greatly change the morphology of the territory and the response of the structures. Some examples of these morphological conditions are given in Fig. 6.

Fig. 6. Pictures of a) Viadotto Favazzina, Reggio-Calabria (Calabria, Italy), b) Viadotto Rago, Cosenza (Calabria, Italy), from the web.

In these cases, the seismic input, during the propagation to the surface, may undergo changes (in intensity and frequency) due to the geometric conditions of the slopes [13] and the foundations of the viaduct may be subject to additional stress. The wave field at the slope surface could be modified depending on the ratio between the slope height and the wavelength, which affects the distribution of the peak amplification.

To investigate these phenomena, four 3D models have been prepared, with different geometric configurations for the topographic convergence, considering a progressively increasing slope of the lateral sides (5° pseudo-horizontal, 10°, 15°, and 20°). The analysis was limited to a maximum steepness of 20° to avoid including possible slope failure. In the models, 2 caissons at a distance of 20 m, arranged symmetrically with respect to the axis of the topographical convergence, have been considered. Preliminary studies have shown that this reduced distance produces a limited interaction between the caissons, in terms of stress variation, which can be neglected in the overall results. The soil volume considered in the analysis has been increased laterally to 117 m to maintain negligible the edge effects. Figure 7 shows the patterns of the four models, and especially the 3D discretization in Plaxis for the soil volume, caissons, and pier; and the definition of the control points at the surface for the evaluation of the response of the models.

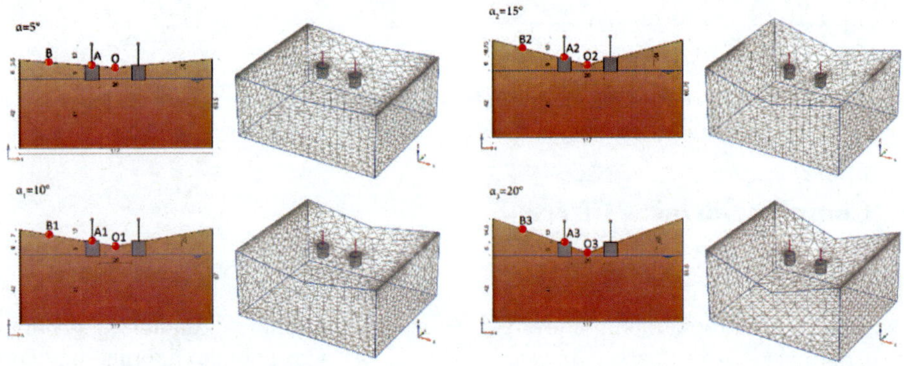

Fig. 7. The four 3D models considered in the parametric analyses, with different steepness of the hollow's lateral slopes: sketches of the four considered configurations with the reference points at the surface and meshed models of soil domain and caissons.

4.2 Results for the Models with the Topographical Convergence

Figure 8 shows the contours of the residual shear strains for the 4 examined cases. In general, the results identify the effect of the interaction between slope topography and seismic waves on acceleration amplification. It is apparent that the shear strains become more relevant with the increase of the topographical convergence, especially in the soil surrounding the caissons. Figure 9 shows the temporal evolution of the horizontal displacements computed at the three reference-points A, B, and O for the four considered models; while Fig. 10 shows the vertical displacements *vs.* time at the same reference points. Since the combined presence of a seismic input and a slope induces horizontal loads to the caissons, the most evident effect is in terms of horizontal displacements at the top of the caissons, which go from 0.027 m of the pseudo-horizontal case (5°) to

0.110 m of the case with 20° slope. The variation of the vertical displacements at the top of the caissons with the increasing steepness is more limited, and it is up to 25%. On the contrary, the topographical effect is more pronounced in the highest portion of the slope, where the amplification of the vertical displacement is of about three times with respect to the pseudo-horizontal condition (from about 0.005 m to 0.150 m).

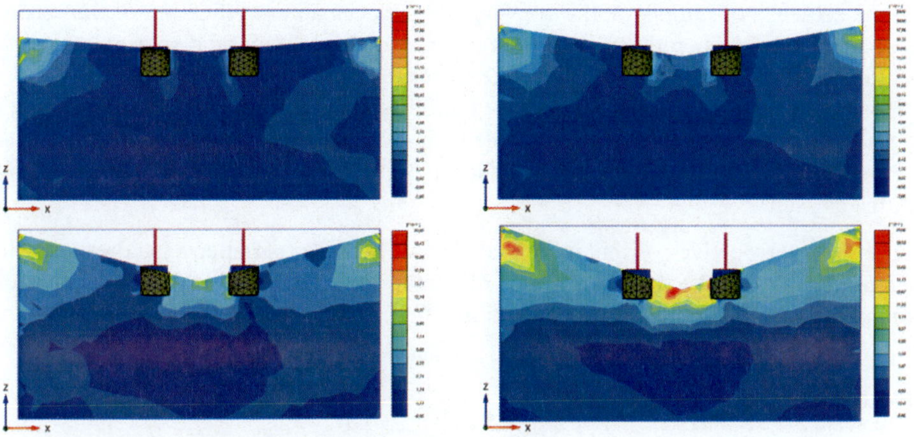

Fig. 8. Contours of the residual values of shear strain for the 4 considered cases.

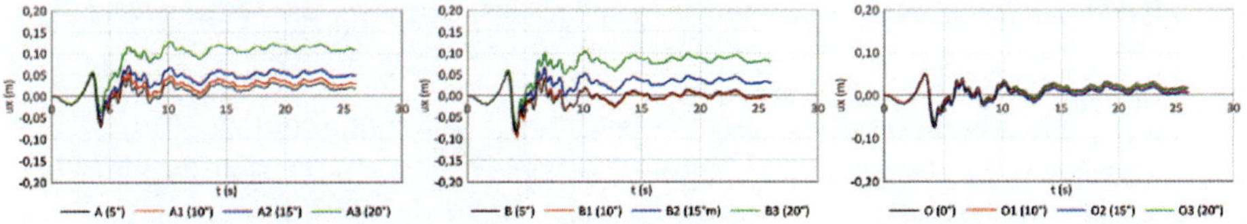

Fig. 9. Temporal evolution of horizontal displacements, computed at the three reference-points A, B, and O for the four considered models

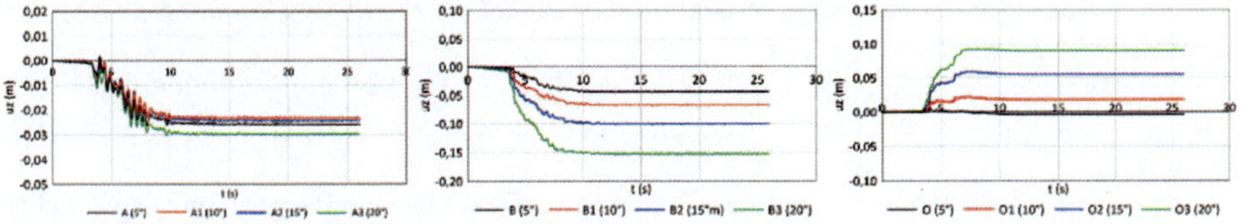

Fig. 10. Temporal evolution vertical displacements, computed at the three reference-points A, B, and O for the four considered models.

5 Conclusions and Perspectives

The effect of a topographic convergence on the seismic response of viaduct caisson foundations has been investigated by means of the construction of different 3D FE models through the Plaxis 3D code. From the study, the following conclusion can be drawn:

- the presence of a topographic convergence induces an amplification of the seismic acceleration; this produces a relevant increase of the horizontal displacements at the top of the caissons and an appreciable increase of the vertical one;
- this effect of the topographic hollow on the seismic amplification, and consequent maximum displacements, is more accentuated in the upper part of the slope, as illustrated by the results at the ground surface in portions of the soil domain located above the caissons.

Future perspectives of this investigation will consider parametric studies with different positions of the foundation with respect to the slope height, as well as different geometric configurations of the foundations.

References

1. De Angelis, A., Mucciacciaro, M., Pecce, M.R., Sica, S.: Influence of SSI on the stiffness of bridge systems founded on caissons. J. Bridge Eng. **22**(8), 04017045 (2017)
2. Gerolymos, N., Gazetas, G.: Static and dynamic response of massive caisson foundations with soil and interface nonlinearities validation and results. Soil Dyn. Earthq. Eng. **26**, 377–394 (2006)
3. Gaudio, D., Rampello, S.: The influence of soil plasticity on the seismic performance of bridge piers on caisson foundations. Soil Dyn. Earthq. Eng. **118**, 120–133 (2019)
4. Mucciacciaro, M., Gerolymos, N., Sica, S.: Seismic response of caisson-supported bridge piers on viscoelastic soil. Soil Dyn. Earthq. Eng. **139**, 106341 (2020)
5. Cattoni, E., Salciarini, D., Tamagnini, C.: A Generalized Newmark Method for the assessment of permanent displacements of flexible retaining structures under seismic loading conditions. Soil Dyn. Earthq. Eng. **117**, 221–233 (2019)
6. Kita, A., Lupattelli, A., Venanzi, I., Salciarini, D., Ubertini, F.: The role of seismic hazard modeling on the simplified structural assessment of an existing concrete gravity dam. Structures **34**, 4560–4573 (2021)
7. Pauselli, D., Salciarini, D., Ubertini, F.: Three-dimensional modeling of soil-structure interaction for a bridge founded on caissons under seismic conditions. Appl. Sci. **12**, 10904 (2022)
8. Chandaluri, V.K., Sawant, V.A.: Effect of slope angle on pile response. Indian J. Sci. Technol. **9**(48) (2017)
9. Zhang, Y.F., Li, J., Li, W., Li, J.M., Liu, H.Y.: Effect of landslides on the displacement of a bridge pile group located on a high and steep slope. Adv. in Civil Eng. 6683967 (2021)
10. Papadimitriou, A.G.: An engineering perspective on topography and valley effects on seismic ground motion. In: Silvestri, F., Moraci, N. (eds.) Earthquake Geotechnical Engineering for Protection and Development of Environment and Constructions: Proceedings of the 7th International Conference on Earthquake Geotechnical Engineering (ICEGE 2019), pp. 426–441. CRC Press, Rome (2019)

11. Iervolino, I., Galasso, C., Cosenza, E.: Rexel: computer aided record selection for code-based seismic structural analysis. Bull. Earthq. Eng. **8**, 339–362 (2009)
12. Brinkgreve, R., Zampich, L., Ragi Manoj, N.: PLAXIS V20: Bentley Systems (2019)
13. Zhang, Z., Fleurisson, J.-A., Pellet, F.: The effects of slope topography on acceleration amplification and interaction between slope topography and seismic input motion. Soil Dyn. Earthq. Eng. **113**, 420–431 (2018)

Seismic Performance of Multi-propped Retaining Structures

Giuseppe Tropeano(✉) and Fabio M. Soccodato

University of Cagliari, Via Marengo 2, 09123 Cagliari, Italy
giuseppe.tropeano@unica.it

Abstract. The performance of retaining walls with multiple levels of props are typically evaluated numerically since it is a typical soil-structure interaction problem which cannot be expressed in a closed form.

Under seismic conditions, the soil-structures interaction problem may require the evaluation of the permanent displacements or deformations developing to dissipate the earthquake energy through the formation of some possible kinematic mechanism. The latter cannot occur for multi-propped retaining structures unless yielding of one or more structural elements is achieved.

In this paper, the dynamic behavior of a multi-propped embedded retaining structures is studied varying the characters of the input seismic motion. For this purpose, the results of several numerical dynamic analyses, carried out under plane-strain conditions and in the time domain, are presented and discussed.

The results of the analyses indicate a very complex response of the system due to the effects of local seismic response and soil-structure interaction phenomena. However, it was possible to identify the main characters of motion that influence the system's response in terms of resulting actions on structural elements. In particular, the analytical relationships presented in this work confirm that the effect of the frequency content of the seismic input compared to the natural frequencies of the soil profile and the geometry of the system are dominant. While peak ground acceleration seems to be appropriate to estimate the maximum (and instantaneous) increase of bending moments, Arias intensity seems to be a more effective parameter in order to evaluate the residual post seismic bending moment.

Keywords: Underground structures · Soil-structure interaction · Numerical analysis · Seismic performance

1 Introduction

The behaviour of embedded, multi-propped retaining structures under seismic actions is quite complex. In fact, the presence of more than one level of props prevents the formation of an instantaneous kinematic mechanisms (rigid motion) related to the full mobilization of soil strength, both behind and in front of the wall [1].

In multi-propped retaining structures, when structural elements do not reach yielding, seismic ground motion causes significant increment of the forces acting on the structure. Simplified relationships that are representative of rigid 'non-displacing' retaining walls,

A. Ferrari et al. (Eds.): CNRIG 2023, SSGG, pp. 696–703, 2023.
https://doi.org/10.1007/978-3-031-34761-0_84

such as [2–4] do not consider the important role played by the embedded part of the wall and by the global dynamic response of the system. Numerical studies carried out by [5–8] and [9] show that forces acting on structures are likely to depend on a number of factors such as the characteristics of ground motion, the problem geometry, the mechanical behaviour of the soil, the soil-structure relative stiffness.

This paper presents an interpretation of some results obtained from a set of dynamic numerical analyses, with specific reference to the effects of the characteristics of seismic input ground motion on forces acting on structures.

2 Numerical Model

2.1 Model and Computational Stages

The parametric analyses have been carried out with reference to the numerical model already considered by [7, 8] and [10]. The geometry and the finite difference computation grid are shown in Fig. 1. The pair of walls ($L = 8$ m) sustains an excavation of height $H = 4$ m and width 16 m.

The layer of dry coarse-grained soil, 30 m in thickness, sits upon a rigid bedrock. An elastic-perfectly plastic model with Mohr-Coulomb strength rule, characterized by mechanical properties corresponding to those of a loose sand, was adopted (Table 1). The shear stiffness at small strains, G_0, is calculated as a function of mean effective stress, p', with the relation:

$$\frac{G_0}{p_{ref}} = K_G \left(\frac{p'}{p_{ref}} \right)^{0.5} \tag{1}$$

where p_{ref} is a reference pressure (100 kPa) and K_G is a stiffness coefficient set equal to 1000. The soil hysteretic behavior was modeled using the shear modulus decay curves for sands given by [11]. The hysteretic damping is, however, computed by applying the generalized Masing criteria implemented in the computer code used in this study.

Table 1. Main soil parameters

	Density ρ [Mg/m^3]	Cohesion c' [kPa]	Frict. Angle φ' [°]	At rest earth pressure coeff. K_0	Poisson ratio ν'
Value	2.04	0.5	30	0.5	0.2

To minimize reflection effects on vertical lateral boundaries of the grid, free field boundary conditions have been used. More details about the numerical model can be found in [7, 8] and [10].

For the walls, linear elastic beam elements with bending stiffness, EI, equal to 2.7×10^5 kN m^2/m were used. The contact between soil and walls was modelled by using elastic-perfectly plastic interface elements, with a friction angle $\delta = 20°$. Props

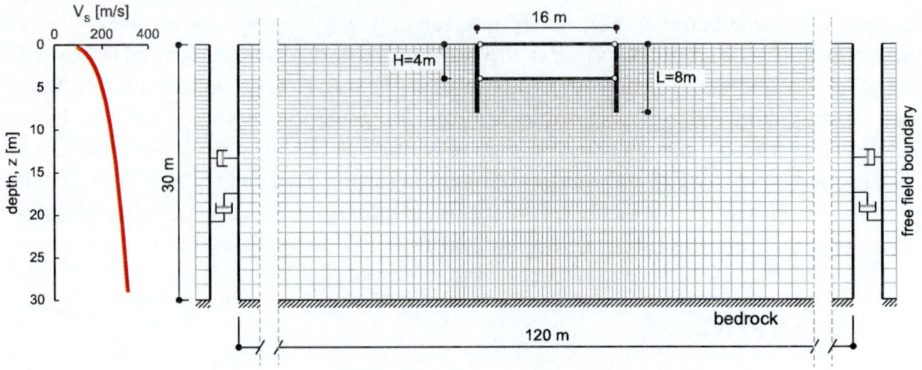

Fig. 1. Model geometry and mesh for 2D analyses.

were also modelled by using linear elastic beam elements, with axial stiffness, *EA*, equal to 1.0×10^6 kN/m; a pin (zero moment) connection was introduced between props and walls.

The permanent levels of props, at the wall top and at the bottom of the excavation, were introduced at the end of the static, cantilevered, excavation stages, before the dynamic stage.

2.2 Seismic Input

The acceleration time histories used (Table 2) are records of strong-motion earthquakes, mainly of Italian events. They are characterized by a different frequency content (Fig. 2) but about the same Arias intensity value ($I_A \approx 0.75$ m/s).

The recordings (S01) have been corrected with a low-pass filter at the cut-off frequency of 15 Hz; additional input signals were obtained by scaling for about 0.7 (S02) and 0.40 (S03) the reference recordings S01.

Table 2. List of acceleration time history S01

ID	Earthquake (date, magnitude M_w)	Station name (*rec. Comp.*)		a_{max} [g]
TMZ E	Friuli (06/05/1976, 6.5)	Tolmezzo	(NS)	0.35
GLR E	Loma Prieta (18/10/1989, 6.9)	Gilroy #1	(NS)	0.440[(1)]
ASS E	Umbria-Marche 2nd (26/09/1997, 6.0)	Assisi	(NS)	0.275
NCR E	Umbria-Marche 3rd (06/10/1997, 5.4)	Nocera Umbra	(WE)	0.330[(2)]

(1) Scaled to 0.9; (2) Scaled to 1.1.

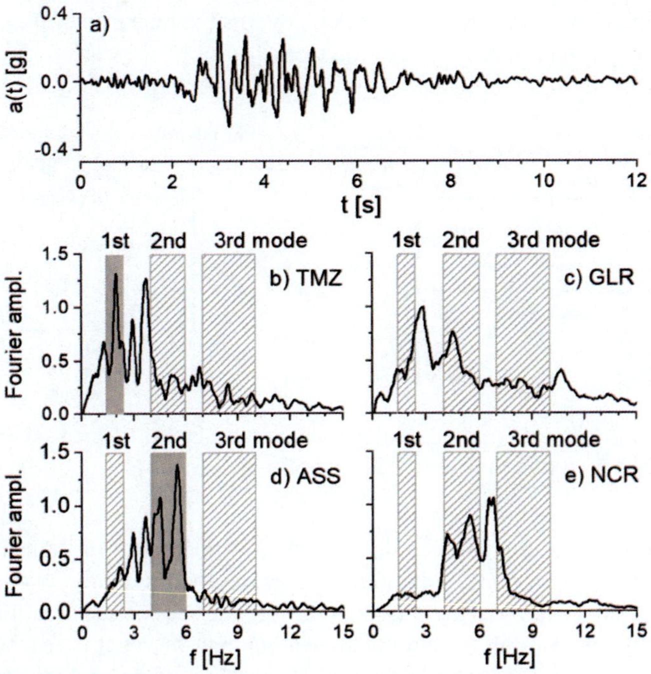

Fig. 2. Seismic signals examples in time domain (a) and in frequency domain (b–e) of waveforms used in this study.

3 Results

3.1 Ground Motion

As reported in [7] and [8] the characteristics of the ground motion obtained from the analyses appear to be strongly influenced by a number of overlapping effects such as: the heterogeneity of the soil stiffness profile, the non-linearity of soil behaviour, the geometry of the system (2D effects) and the soil-structure interaction.

The soil response can be very different depending on the vibration modes excited by the signals. In fact, due to the highly non-linear soil behaviour, the natural periods of the soil column appear to be strongly influenced by soil motion and straining (shaded bandwidths in Fig. 2b–e). Therefore, waveforms that cause soil resonance with the first vibration mode (Fig. 2b) will induce lower amplification than those that excite higher modes and, consequently, a less pronounced increases of forces acting on structures.

3.2 Behaviour of Structures

The response of the soil-structure system to seismic loading appears quite complicated. When structures (walls and props) do not reach yielding conditions (as assumed in this paper), the formation of instantaneous rigid motion mechanisms in the soil is prevented, and earthquake energy cannot be adequately dissipated is plastic soil straining. In this case, the kinematic soil-structure interaction governs the response of the system, giving rise to increasing forces acting on the structures.

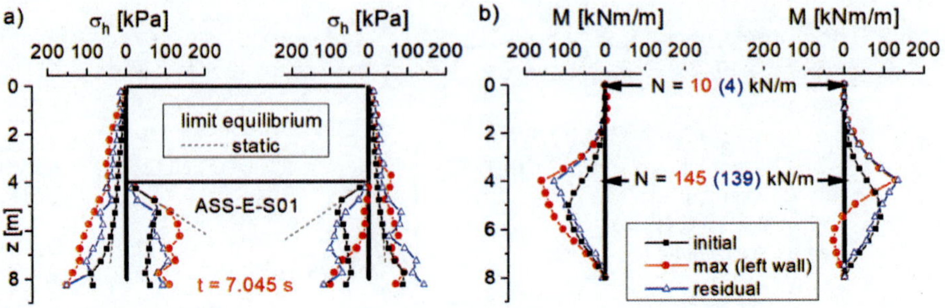

Fig. 3. Horizontal stress, σ_h (a), and resulting actions on structural elements (bending moments, M, and axial load, N) (b) for time history E-ASS-S01.

During seismic loading, forces on structures vary rapidly in both magnitude and sign: for a given time instant, horizontal stresses, σ_h, acting on the two opposite walls are quite different, as in the example shown in the Fig. 3a for acceleration time history E-ASS-S01. In this example, when maximum bending moment is reached in left wall (red lines and symbols) also the horizontal stresses acting behind the left walls reach the maximum value, and they increase significantly from the static initial conditions (black lines and symbols); in front of the wall, stresses also increase, and passive limit state conditions extend to a greater depth. Behind the opposite wall, the increment of horizontal stresses is smaller, and along the excavation side horizontal stresses are very low. The resulting actions on structural elements (bending moments, M, and axial load, N) at this instant of time are shown in Fig. 3b. The upper prop is nearly unloaded, and the maximum bending moment, M_{max}, which is about two times the maximum static moment, is reached in correspondence of the lower prop.

Distributions of soil pressures and bending moments remain asymmetric during the strong motion phase of the shaking, while they tend to be very similar for the two walls in the post-seismic phase (blue lines and symbols in Fig. 3).

Taking the initial static condition as a reference, the maximum increase of bending moment during the earthquake, ΔM_{MAX}, and the maximum bending moment in the post-seismic condition, ΔM_R, are achieved at the level of the lower prop for all the seismic inputs considered.

An example of the time history of bending moment increment, $\Delta M(t)$, obtained from the E-ASS-S01 input, is shown in Fig. 4a. As shown in [7], $\Delta M(t)$ baseline is monotonically increasing and appears to follow a trend similar to the Arias Intensity function (red line) obtained from the free-field response analysis in the free-field condition. This is observed for seismic inputs or waveforms with a significant high frequency content, more specifically in the resonant range of the second mode. For the signals with significant frequency content around the frequency of the first resonant mode, the $\Delta M(t)$ baseline shows a peak during the strong motion phase and a decrease after the maximum bending moment is reached. The maximum (and minimum) instantaneous bending moments increment may be considered as deviations from the baseline trend. In almost all the analyses carried out, the maximum increase, ΔM_{MAX}, occurs at the instant of the peak value of the acceleration at ground surface in free-field conditions, $a_{max,FF}$, (Fig. 4b).

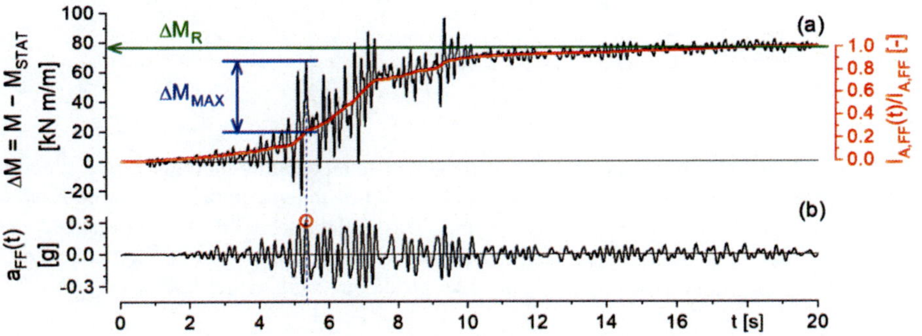

Fig. 4. Time history of bending moments at the section corresponding to the lower prop obtained for E-ASS-S01

In the post-seismic state, Soccodato and Tropeano [7] have shown that the increments of bending moments appear to be linked to the Arias Intensity at ground surface in free field condition, $I_{A,FF}$, to the ratio $T_s^*/T_{m,FF}$ (where T_s^* is the fundamental period of the soil column with height equal to the wall length H and $T_{m,FF}$ is the mean period at ground surface in free field condition) and to the number of equivalent loading cycles, n_{eq} (defined as the ratio of the significant duration, D_{5-95}, to the mean period).

Multivariate analysis of all the results made it possible to determine the functional relationship between the ground motion parameters and the increments of bending moments, summarising the response of the system:

$$m_R \doteq \frac{\Delta M_R H}{EI} \times 1000 = \left(\frac{I_{A,FF} T_m}{H}\right)^a \left(\frac{T_s^*}{T_{m,FF}}\right)^b n_{eq}^c \tag{2}$$

where m_R is the increment of the bending moment in post seismic state normalised by the bending stiffness of the wall, EI/H, and a, b, and c are the regression parameters whose values are listed in Table 3.

Table 3. Parameters, standard deviation and coefficients of determination of Eqs. (2) and (3)

eq.	parameter	value	st.error	eq.	parameter	value	st.error
(2)	a	1.723	0.0267	(3)	d	4.515	1.5628
	b	3.151	0.1877		f	1.5484	0.2879
	c	−0.894	0.0556				
	$\sigma = 0.210$; R^2adj = 0.964				$\sigma = 0.650$; R^2adj = 0.522		

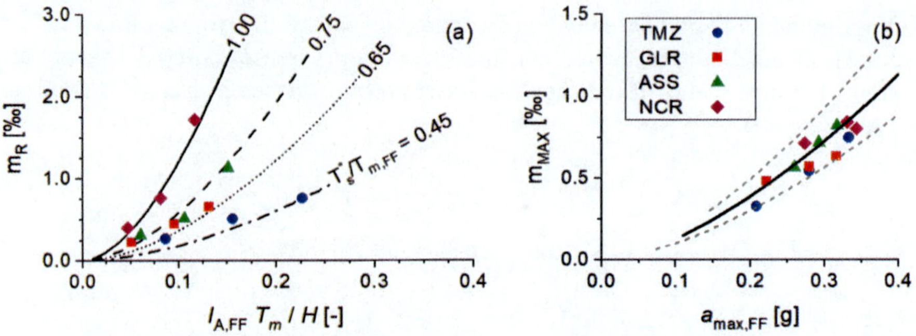

Fig. 5. Relationships between normalized increment residual bending moment, m_R, and normalized Arias intensity in free field condition and period ratio (lines) compared with the results of analyses carried out using natural records (symbols) (a). Relationship between normalized increment of maximum bending moment, m_{MAX}, and peak ground acceleration in free field condition (black line) compared with the results of analyses carried out using natural records (symbols) (b).

Figure 5a shows the comparison of the m_R values obtained from each analysis versus those obtained by Eq. (2). For the maximum increase of bending moment from the baseline trend, ΔM_{MAX}, the correlation with peak ground acceleration at surface in the free-field conditions, $a_{\max,\mathrm{FF}}$, was identified:

$$m_{MAX} \doteq \frac{\Delta M_{MAX} H}{EI} \times 1000 = d \cdot \left(\frac{a_{\max,FF}}{g}\right)^f \tag{3}$$

where m_{MAX} is the value of ΔM_{MAX} normalised to EI/H, and d and f are the two regression parameters whose values are listed in Table 3.

4 Conclusions

This paper presents and discusses the results of a series of time-domain dynamic analyses of a pair of multi-propped retaining walls supporting an excavation in a dry coarse-grained soil excited by seismic loading. The soil model is characterised by stress-dependent stiffness and highly non-linear behaviour. Four strong motion acceleration time histories characterised by similar Arias intensity and significant duration but by different frequency contents were used.

As pointed out in previous works [7] and [8], the results of the analyses indicate a very complex response of the system due to the effects of local seismic response and soil-structure interaction phenomena. However, it was possible to identify the main characters of motion that influence the system's response in terms of resulting actions on structural elements. In this paper, a functional relationship is presented which allows the main ground motion parameters to be related to the maximum and residual increase in the bending moment acting on the walls.

In particular, the analytical relationships presented in this work confirm that the effect of the frequency content of the seismic input compared to the natural frequencies of the soil profile and the geometry of the system are dominant. In addition, the increase of structural actions is a function of the duration of the shaking and of the ground motion at the surface that can be evaluated in free-field conditions. While peak ground acceleration seems to be appropriate to estimate the maximum (and instantaneous) increase of bending moments, Arias intensity seems to be a more effective parameter in order to evaluate the residual post seismic bending moments.

Further developments of this study are aimed to include the effects of soil properties (stiffness and strength) in the proposed relationships.

References

1. Conti, R., Madabhushi, G.S.P., Viggiani, G.M.B.: On the behaviour of flexible retaining walls under seismic actions. Geotechnique **62**, 1081–1094 (2012)
2. Wood, J.H.: Earthquake-induced soil pressures on structures. Ph.D thesis, The California Institute of Technology, Pasadena (1973)
3. Veletsos, A.S., Younan, A.H.: Dynamic soil pressures on rigid vertical walls. Earthq. Eng. Struct. Dyn. **23**, 275–301 (1994)
4. Kloukinas, P., Langousis, M., Mylonakis, G.: Simple wave solution for seismic earth pressures on nonyielding walls. J. Geotech. Geoenviron. Eng. **138**(12), 1514–1519 (2012)
5. Chowdhury, S.S., Deb, K., Sengupta, A.: Behavior of underground strutted retaining structure under seismic condition. Earthq. Struct. **8**(5), 1147–1170 (2015)
6. Bahrami, M., Khodakarami, M.I., Haddad, A.: Seismic behavior and design of strutted diaphragm walls in sand. Comput. Geotech. **108**, 75–87 (2019)
7. Tropeano, G., Soccodato, F.M.: Dynamic analyses of propped retaining structures. In: Proceedings of the 8th European Conference on Numerical Methods in Geotechnical Engineering, NUMGE 2014, vol. 2, pp. 1193–1198 (2014)
8. Soccodato, F.M., Tropeano, G.: The role of ground motion characters on the dynamic performance of propped retaining structures. In: Proceedings of the 6th International Conference on Earthquake Geotechnical Engineering (6ICEGE), 1–4 November 2015, Christchurch, New Zealand (2015)
9. Zucca, M., Crespi, P., Tropeano, G., Simoncelli, M.: On the influence of shallow underground structures in the evaluation of the seismic signals. Ingegneria Sismica **38**(1), 23–35 (2021)
10. Callisto, L., Soccodato, F.M.: Seismic design of cantilevered retaining walls. J. Geotech. Geoenviron. Eng. **136**(2), 344–354 (2010)
11. Seed, H.B., Idriss, I.M.: Soil moduli and damping factors for dynamic analysis. Report No. EERC 70-10, University of California, Berkeley (1970)

Advances in Risk Mitigation Strategies

Predicting the Soil Slip Triggering Through the SLIP Model and ML Approaches Including Vegetation

Salvatore Misiano[1](✉), Michele Placido Antonio Gatto[2], and Lorella Montrasio[2]

[1] University of Parma, Parco Area delle Scienze, 181/A, 43124 Parma, PR, Italy
salvatore.misiano@unipr.it
[2] University of Brescia, Via Branze 38, 25123 Brescia, BS, Italy

Abstract. The paper shows the application of two different methods for the prediction of rainfall-induced shallow landslide, i.e., a physically-based (SLIP) and a machine learning based method. In both cases, the effects of vegetation are considered. A comparison between the prediction capabilities of both methods is shown. Analyses are performed through a MATLAB totally-integrated platform (X-SLIP) on a municipality of the Emilia-Romagna region, Baiso, in northern Italy. Results reveal the overall prediction quality improves when considering vegetation.

Keywords: Rainfall-induced shallow landslides · Machine learning · Slope reinforcement · Climate change · MATLAB

1 Introduction

Shallow landslides (or soil slips) are natural phenomena which involve thin debris layers (topsoil), less than two meters thick, triggered by intense and/or prolonged rainfalls that penetrate the topsoil and cause instability; once triggered, they can evolve into flows of debris or mud, characterised by sustained speeds (up to 9 m/s) and a highly destructive power. Over the past few years, this type of phenomena is rising, even due to climate change [1, 2]. Methods for their prediction and prevention are therefore of great interest.

The prediction of soil slip triggering (which may be considered a prevention technique as well, when used for alert purposes) is performed, among others, through statistical approaches [3–5], physically-based models (PBM) [6–8], or machine learning (ML) techniques [9, 10]. As regards soil slip prevention, one of the existing solutions consists of the reinforcement of slopes through naturalistic techniques, i.e., planting and revegetation, and represents an eco-compatible technique.

SLIP (Shallow Landslides Instability Prediction) is the PBMs that will be adopted in this study. It deals with a model which defines the evolution with time of the safety factor of slopes potentially at risk of soil slip, using the equilibrium limit method, associated with simplified hypotheses on both the rainfall infiltration phenomenon and the

A. Ferrari et al. (Eds.): CNRIG 2023, SSGG, pp. 707–714, 2023.
https://doi.org/10.1007/978-3-031-34761-0_85

mechanical behaviour of the unsaturated soil [7]. The model, applied to different territories, showed excellent predictive skills [11–16]; specifically, its use on a territorial scale was possible thanks to its implementation into the multi-risk platform owned by the Italian Civil Protection. Recently, Gatto and Montrasio [17] developed a proper platform based on MATLAB, named X-SLIP, which was also adopted for risk evaluation [18]. Misiano et al. [19] implemented in X-SLIP several ML algorithms, showing comparable results when applied with only morphology and soil parameters. Due to the necessity to properly model the vegetation effects in the stability analyses, Montrasio et al. [20] have recently introduced G-SLIP (Green-SLIP model), including the vegetation effects in SLIP formulation, and have implemented it in G-X-SLIP platform.

Generally, the introduction of vegetation in stability analyses is a considerable complication, due to the great amount of geometrical and mechanical uncertainties related to plants [20–25]; the aim of this paper is to include it even in ML analyses.

Analyses are conducted with the use of G-XSLIP platform in a municipality of the Emilia-Romagna region, namely, Baiso. This area has been affected by several soil slips, occurred in spring 2005. As a matter of fact, the Emilia-Romagna region is the second most involved in Italy in these events, according to IFFI Italian inventory, which counts almost 700.000 landslides over the years to date.

2 Methods

2.1 Slip

SLIP combines limit equilibrium theory with simplified shear resistance criteria for partially-saturated soils, based on the homogenization of mechanical properties and simplified assumptions about the process of imbibition and drainage of slopes. An illustrative scheme is shown in Fig. 1.

The ratio between stabilizing and destabilizing forces is called Factor of Safety (F_S). F_S is a function of many factors, such as geometrical parameters (inclination of slope β; depth of analysis H), state parameters (specific gravity G_s; porosity n; degree of saturation S_r), mechanical resistance parameters (friction angle φ'; effective cohesion c'), partial saturation parameters (A and λ, which are modeling parameters), a parameter related to the simplified approach (α), drainage coefficient of slope (k_t), unit weight of water (γ_w), and precipitation height (h). The formulation is the following:

$$F_S = \frac{cot\,\beta \cdot tan\,\varphi' \cdot [\Gamma + m \cdot (n_w - 1)]}{\Gamma + m \cdot n_w} + \frac{C' \cdot \Omega}{\Gamma + m \cdot n_w} \tag{1}$$

$$\Gamma = G_s \cdot (1 - n) + n \cdot S_r \tag{2}$$

$$n_w = n \cdot (1 - S_r) \tag{3}$$

$$\Omega = \frac{2}{sin(2\delta) \cdot H \cdot \gamma_w} \tag{4}$$

$$C' = \left[c' + A \cdot S_r \cdot (1 - S_r)^{\lambda} \cdot (1 - m)^{\alpha}\right] \cdot L \tag{5}$$

$$m = \frac{\xi}{n \cdot H \cdot (1 - S_r)} \cdot \sum\nolimits_{i=1}^{\omega} h_i \cdot e^{-k_t \cdot (t_i - t_0)} \tag{6}$$

Theoretically, instability corresponds to $FS \leq 1$, but in practice, due to the simplifications adopted, a more plausible threshold is around 1.5.

The simplicity of the model makes it suitable for large scale applications; in fact, it was used in several case studies [11–16, 18, 19, 26–28] and validated over the years. For a more detailed description, the reader is referred to Montrasio [7].

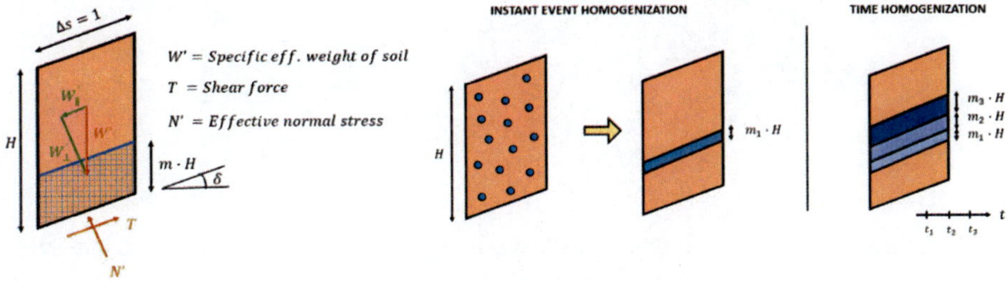

Fig. 1. SLIP scheme

2.2 Machine Learning

Machine Learning makes it possible to create prediction models, by learning from a training dataset. In this study, input data corresponds to susceptibility factors of each rasterized point from a given study area. Output, i.e., prediction result, is basically a binary value, being 1 if landslide occurs or 0 otherwise.

Models adopted are *Random Forest* (*RF*), *Adaptive Boosting* (*AB*), and a *feed-forward artificial neural network* (*ANN*). The first two are some of the classics *ML* approaches, while *ANN* belongs to *Deep Learning* (*DL*), which is a more sophisticated subset of *ML*. A graphic representation of both types of models is reported in Fig. 2a and Fig. 2b. Susceptibility factors considered are: β, θ (aspect angle), c', φ', n, k_t, and m (See Eqs. (1–6)).

Generally, to modify the behaviour of a prediction model, some hyperparameters must be defined. A typical hyperparameter is the *learning rate* (*lr*), i.e., a value governing the learning velocity. Additional hyperparameters could be introduced depending on the ML method. For the approached adopted in this study additional parameters are the number of maximum decision trees and of leaf node for each tree, in the case of ensemble methods, or the activation function to use in each single neuron or the number of hidden layers for neural networks.

2.3 Root Reinforcement

Many authors conducted different types of study related to the contribution of vegetation to soils, from the hydraulic (through the *canopy effect*) and mechanical point of view (through the increase of soil shear strength).

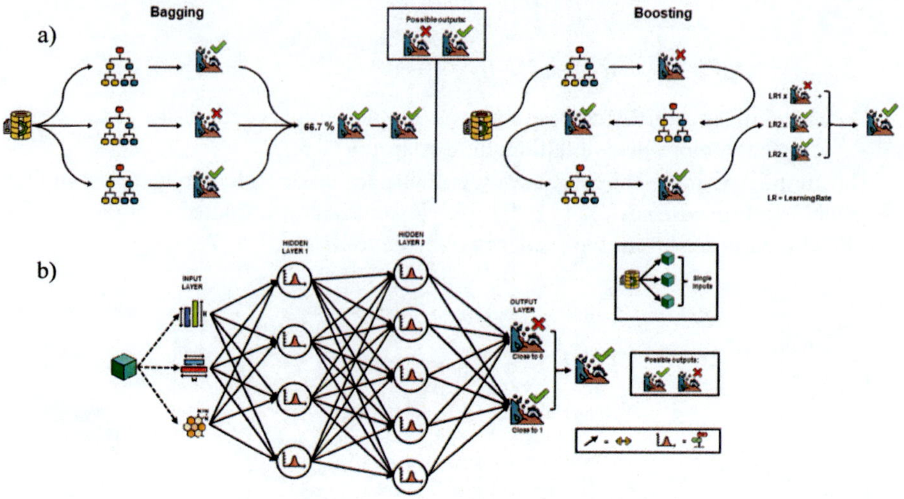

Fig. 2. Scheme of: a) ML, ensemble-methods; b) FF ANN, Feed Forward Artificial Neural Network.

Vegetation was introduced in this specific problem in two main modalities: firstly, adding an extra term c_r to the Mohr-Coulomb criterion and, secondarily, acting on a pre-existing hydraulic property of soil, i.e., by reducing rainfall infiltration through new values of β^*, which determine the percentage of water that infiltrates soil.

The root cohesion c_r can be added in Mohr-Coulomb criterion as follows:

$$\tau = \sigma' \cdot tan\,\varphi' + c' + c_r \tag{7}$$

There are different theories for the evaluation of c_r but among the main ones, the Wu and Waldron model [22, 24] for a system of multiple roots can be mentioned:

$$c_r = k' \cdot RAR \cdot \sum\nolimits_i t_{r,i} \tag{8}$$

where k' is a factor that considers geometry of the root (inclination) and friction angle of soil; *RAR* is the root area ratio, considering density of roots, which varying with depth; $t_{r,i}$ is the tensile resistance of the i-th root.

As regards β^*, value, it is a common subject of agronomic studies and depends on foliar apparatus, size, age, and species. Vegetation parameters derived by different authors are adopted in this study [20, 21, 23, 25].

3 Results

3.1 Study Area

A municipality of Emilia-Romagna region is studied, where many soil slips occurred between 11-04-2005 and 12-04-2005, due to intense rainfalls. The database of occurrences is obtained from ISPRA website, while rain-gauges recordings are provided by the regional Dext3r. Figure 3 shows the results of preliminary processing.

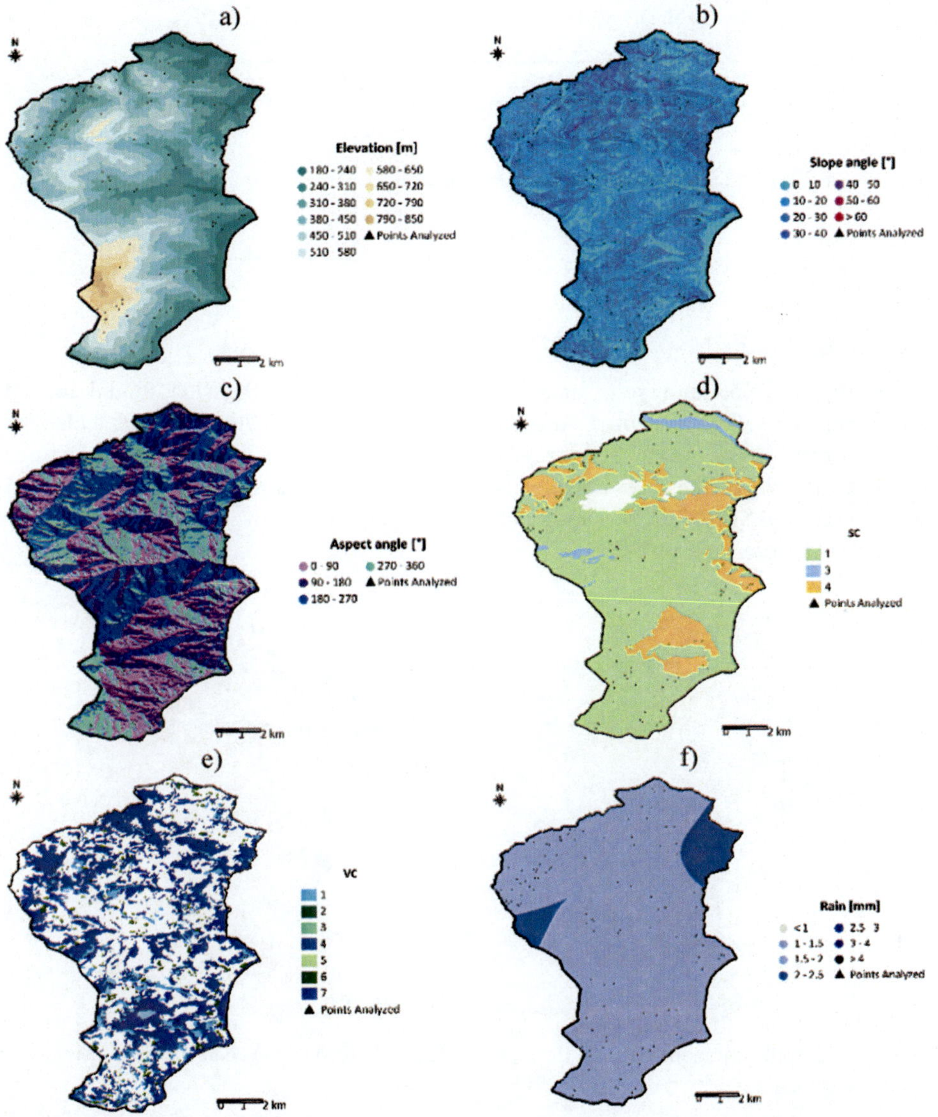

Fig. 3. Map of: a) Elevation b) Slope Angle c) Aspect Angle d) Soil Units e) Vegetation Categories (VC) d) Interpolated rainfall (11-04-2005 22:00)

It can be seen that both soil and vegetation types are grouped into macro classes summarised in Table 1. The description of the grouping procedure recently presented [17, 20] is here omitted.

Table 1. Soil classes and Vegetation classes Tables

SC	c'(kPa)	phi (°)	kt (h^-1)	A (kPa)	n (-)	VC	β (-)	c_r (kPa)
1	5	25	0.018	80	0.46	1	0.55	0
2	10	30	0.00358	100	0.5	2	0.6	12
3	0	35	0.036	40	0.4	3	0.55	21
4	5	20	0.02517	40	0.45	4	0.6	18
						5	0.6	20
						6	0.6	0
						7	0.65	4

3.2 Susceptibility Maps and ROC Curves

To obtain the following results, analyses are performed using the SLIP model and ML algorithms, i.e., *Random Forest*, *Adaptive Boosting* and *ANN*. 70% of the detected soil slips included in the dataset are used to train ML models. For each model, both vegetated and non-vegetated cases are considered. Figure 5 compares the susceptibility maps derived through SLIP (Fig. 4a) and ANN (Fig. 4b); qualitatively, it is evident that ANN overestimates the susceptibility.

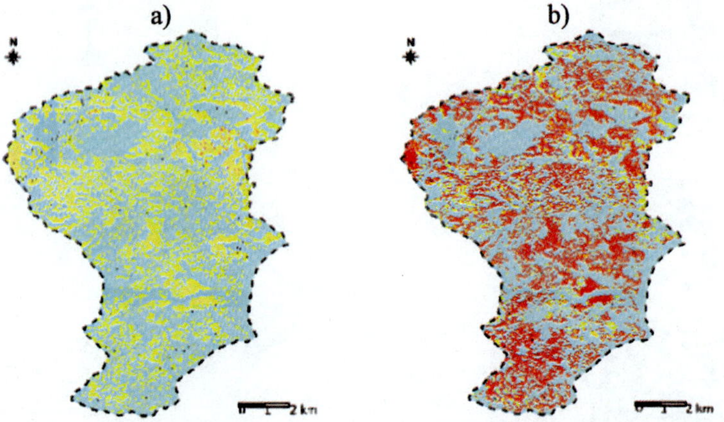

Fig. 4. Susceptibility maps derived through a) SLIP and b) ANN. Vegetation is considered in both analyses.

Figure 5 shows the ROC curves evaluated from results derived through SLIP (Fig. 5a) and ML techniques (Fig. 5b) with and without vegetation. The Area Under Curve (*AUC*) is computed by means of the True Positive Rate *TPR* and the False Positive Rate *FPR* [18–20, 29] and quantify the prediction quality of each model. High values of *AUC* correspond to a model that have both high sensitivity (i.e., the ability of the model to detect true positive points, where a landslide trully occurred and the model correctly predicted it) and specificity (i.e., the ability to detect true negative points, where prediction and reality agree that a landslide not occurred). It can be observed that when vegetation is included in the analyses, *AUC* increases. SLIP and ML predictions are comparable, even though SLIP with vegetation is slightly better.

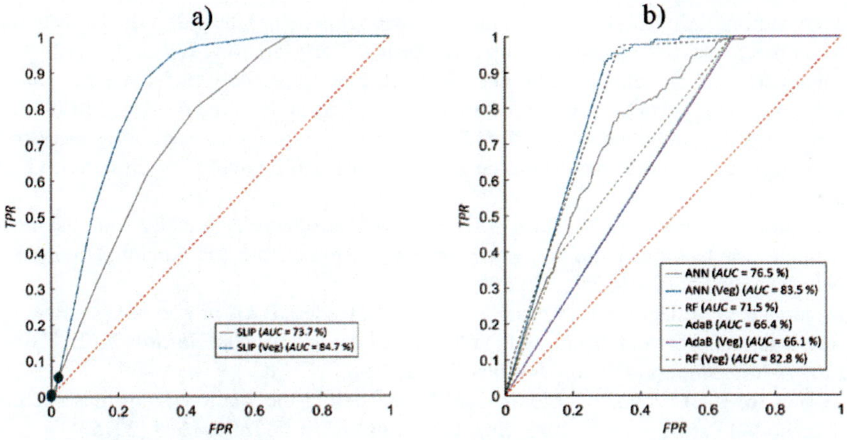

Fig. 5. ROC curves from a) SLIP and b) ML analyses with and without vegetation

4 Conclusions

The implementation of the vegetation in analyses could significantly increase the quality of predictions of both physically-based and ML models. In both cases, the accuracy increased by an average of 7.25%. As an attempt, a hybrid formulation combining the potentialities of both approaches has been recently developed, and it will be shown in future works. This could allow us to further improve the prediction quality achieved with single models. Future developments will be aimed at considering vegetation more accurately and at refining and tuning the abovementioned hybrid formulation.

References

1. Gariano, S.L., Guzzetti, F.: Landslides in a changing climate. Earth-Sc. Rev. **162** (2016)
2. Jakob, M., Owen, T.: Projected effects of climate change on shallow landslides, North Shore Mountains, Vancouver, Canada. Geomorphology **393**, 107921 (2021)
3. Castelli, F., Lentini, V.: Landsliding events triggered by rainfalls in the Enna area (South Italy). Landslide Sc. Pract.: Early Warning Instr. Monit. **2**, 39–47 (2013)
4. Mandaglio, M.C., Gioffrè, D., Pitasi, A., Moraci, N.: Qualitative landslide susceptibility assessment in small areas. Procedia Eng. **158**, 440–445 (2016)
5. Li, L., Lan, H., Guo, C., Zhang, Y., Li, Q., Wu, Y.: A modified frequency ratio method for landslide susceptibility assessment. Landslides **14**(2), 727–741 (2016). https://doi.org/10.1007/s10346-016-0771-x
6. Borga, M., Dalla Fontana, G., Da Ros, D., et al.: Shallow landslide hazard assessment using a physically based model and digital elevation data. Env. Geology **35**, 81–88 (1998)
7. Montrasio, L.: Stability analysis of soil slip. In: Brebbia CA (ed.) Proceedings of International Conference Risk 2000, Wit press, Southampton (2000)
8. Salvatici, T., Tofani, V., Rossi, G., et al.: Application of a physically based model to forecast shallow landslides at a regional scale. Nat. Haz. Earth Syst. Sci. **18**, 1919–1935 (2018)
9. Kadavi, P.R., Lee, C.W., Lee, S.: Application of ensemble-based machine learning models to landslide susceptibility mapping. Remote Sens. **10**(8), 1252 (2018)

10. Merghadi, A., Ali, P.Y., et al.: Machine learning methods for landslide susceptibility studies: a comparative overview of algorithm performance. Earth-Sc. Rev. **207**, 103225 (2020)
11. Montrasio, L., Valentino, R., Losi, G.L.: Rainfall-induced shallow landslides: a model for the triggering mechanism of some case studies in Northern Italy. Land **6**, 241–251 (2009)
12. Montrasio, L., Valentino, R., Losi, G.L.: Towards a real-time susceptibility assessment of rainfall-induced shallow landslides on a regional scale. Nat. Haz. Earth Syst. Sci. **11**, 1927–1947 (2011)
13. Montrasio, L., Valentino, R., Losi, G.L.: Shallow landslides triggered by rainfalls: modeling of some case histories in the Reggiano Apennine (Emilia Romagna Region, Northern Italy). Nat. Hazards **60**, 1231–1254 (2012)
14. Montrasio, L., Valentino, R., Corina, A., Rossi, L., Rudari, R.: A prototype system for space–time assessment of rainfall-induced shallow landslides in Italy. Nat. Hazards **74**(2), 1263–1290 (2014). https://doi.org/10.1007/s11069-014-1239-8
15. Montrasio, L., Terrone, A., Morandi, M.C.: Modeling the shallow landslides occurred in Tizzano Val Parma in April 2013. Eng. Geol. Soc. Territ. **2**, 1605–1609 (2015)
16. Montrasio, L., Schilirò, L., Terrone, A.: Physical and numerical modelling of shallow landslides. Landslides **13**(5), 873–883 (2015). https://doi.org/10.1007/s10346-015-0642-x
17. Gatto, M.P.A., Montrasio, L.: X-SLIP: A SLIP-based multi-approach algorithm to predict the spatial-temporal triggering of rainfall-induced shallow landslides over large areas. Comput. Geotech. **154**, 105175 (2023)
18. Gatto, M.P.A., Lentini, V., Montrasio, L., Castelli, F.: A simplified semi-quantitative procedure based on the SLIP model for landslide risk assessment: the case study of Gioiosa Marea (Sicily, Italy). Landslides (2023). https://doi.org/10.1007/s10346-023-02040-8
19. Misiano, S., Gatto, M.P.A., Montrasio, L.: Qualità predittive dei modelli fisicamente basati e dei modelli di machine learning per la valutazione della suscettibilità alle frane. In: IARG National Proceedings, Caserta, Italia (2022)
20. Montrasio, L., Gatto, M.P.A., Miodini, C.: The role of plants in the prevention of soil-slip: the G-SLIP model and its application on territorial scale through G-X-SLIP platform. Landslides (2023). https://doi.org/10.1007/s10346-023-02031-9
21. Gardner, W.R.: Relation of root distribution to water uptake. Agron. J. **56**, 41–45 (1964)
22. Wu, T.H., McKinnell, P., Swanston, D.N.: Strength of tree roots and landslides on Prince of Wales Island, Alaska. Can. Geotech. J. **16**(1), 19–33 (1979)
23. Aston, A.R.: Rainfall interception by eight small trees. J. Hydrol. **42**, 383–396 (1979)
24. Waldron, L.J., Dakessian, S.: Soil reinforcement by roots: calculation of increased soil shear resistance from root properties. Soil Sci. **132**, 427–435 (1981)
25. Butler, D., Huband, N.: Throughfall and stem-flow in wheat. Agr. For Met. **35**, 329–338 (1985)
26. Montrasio, L., Valentino, R.: Experimental analysis and modelling of shallow landslides. Landslides **4**, 291–296 (2007)
27. Montrasio, L., Valentino, R.: A model for triggering mechanism of shallow landslides. Nat. Hazards Earth Syst. Sci. **8**, 1149–1159 (2008)
28. Montrasio, L., Schilirò, L.: Inferences on modeling rainfall-induced shallow landslides from experimental observations on stratified soils. Italian J. Eng. Geol. Environ. **2**, 77–85 (2018)
29. Fan, J., Upadhye, S., Worster, A.: Understanding receiver operating characteristic (ROC) curves. Can. J. Emerg. Med. **8**(1), 19–20 (2006)

A Semi-quantitative Approach to Assess the Propensity of Rockfall Source Areas to Instability Based on the Susceptibility Index to Failure (SIF): the Case Study of Capo Calavà (Italy)

Maria Lia Napoli[1](✉), Monica Barbero[1], Francesco Castelli[2], Marta Castelli[1], and Valentina Lentini[2]

[1] Department of Structural, Geotechnical and Building Engineering, Politecnico di Torino, C.so Duca degli Abruzzi 24, 10124 Turin, Italy
maria.napoli@polito.it

[2] Università degli Studi di Enna "Kore", Piazza dell'Università, 94100 Enna, Italy

Abstract. This paper develops a practical semi-quantitative methodology to assess the probability of failure of unstable rock blocks for different environments (mountain and coastal-marine) and scales of interest (small and medium-large scales). According to the presence and intensity of several causative factors, a rockfall Susceptibility Index to Failure (SIF) can be defined and assigned to each rockfall source point so as to generate weighted frequency runout maps and, therefore, obtain more reliable rockfall hazard and risk maps.

A novel approach, implemented in a Matlab routine, is also proposed to determine the release activity of different rockfall source areas, with reference to specific elements at risk. Such information makes it possible to identify the most efficient locations for the installation of risk monitoring or mitigation systems.

The proposed methods are applied to a case study in the northern coast of Sicily, Italy, where a susceptibility analysis is carried out in the QGIS environment by means of the QPROTO plugin (QGIS Predictive ROckfall TOol).

Keywords: Rockfalls · Failure Probability · Causative Factors · Susceptibility Index to Failure · Susceptibility Maps

1 Introduction

Rockfalls are widespread phenomena that pose significant threats to people, structure, infrastructures and the environment. They are increasingly occurring both in mountain and marine environments as a consequence of the adverse impacts of climate change and global warming.

Assessing rockfall hazard and risk implies, among the others, the location/geometry of the potentially unstable rock areas to be identified, and their failure probability to

A. Ferrari et al. (Eds.): CNRIG 2023, SSGG, pp. 715–723, 2023.
https://doi.org/10.1007/978-3-031-34761-0_86

be evaluated. To this aim, the major factors affecting instability must be analyzed and quantified [1].

In this paper, a practical semi-quantitative methodology to assess the probability of detachment of unstable rock blocks is developed, resulting in a rockfall Susceptibility Index to Failure (SIF). Coastal-marine and mountain environments are considered specifically. In association with a suitable method, the SIF Index allows weighted frequency runout analyses to be carried out and, therefore, more reliable rockfall hazard and risk maps to be obtained. Moreover, in order to identify the most effective areas for the location of risk stabilization/monitoring systems, a novel approach is also proposed to detect the release activity of the rockfall source areas, with reference to specific elements at risk.

A case study is presented to show the applicability of the approaches proposed.

2 The Rockfall Susceptibility Index to Failure (SIF)

In order to assess the failure probability of potentially unstable rock blocks, the main factors responsible for their instability were selected according to a literature review [1–8]. These factors were subdivided into three tables, as shown in Fig. 1: the first table (A) includes predisposing/triggering factors that can be detected in any territorial context, while the second and third tables (B1 and B2) are mutually exclusive, since they contain predisposing factors specific of mountain and marine-coastal environments, respectively. Each factor "f" was ranked into classes, to which a numerical score "P" was assigned, from the lowest (0) to the highest (3) level of susceptibility to failure. An exception is given by the presence of stabilization/protection works (if considered sufficiently efficient/effective), since they can be assigned a negative score (up to −1) to reduce the probability of detachment of unstable rock blocks.

By evaluating the presence and intensity of such causative factors, a rockfall Susceptibility Index to Failure (SIF) can be defined and assigned to each rockfall source point, according to the following equations:

$$\text{mountain environment: SIF} = \frac{\sum(P_{f\_A} + P_{f\_B1}) - \sum min(P_{f\_A} + P_{f\_B1})]}{\sum max(P_{f\_A} + P_{f\_B1}) - \sum min(P_{f\_A} + P_{f\_B1})]} \quad (1)$$

$$\text{coastal-marine environment: SIF} = \frac{\sum(P_{f\_A} + P_{f\_B2}) - \sum min(P_{f\_A} + P_{f\_B2})]}{\sum max(P_{f\_A} + P_{f\_B2}) - \sum min(P_{f\_A} + P_{f\_B2})]} \quad (2)$$

being:

$P_{f\text{-}A}$: weight assigned to each factor included in Table A;

$P_{f\text{-}Bi}$: weight assigned to each factor included in Table B_i (B1 or B2);

$\sum min(P_A + P_{Bi})$ = sum of the minimum weights that can be assigned to the factors of Tables A and B1 or B2;

$\sum max(P_A + P_{Bi})$ = sum of the maximum weights that can be assigned to the factors of Tables A and B1 or B2.

If one or more factors cannot be evaluated (for example because of a lack of visibility of the slope) the sum of the minimum and maximum weights must not take into account the contribution of these parameters.

PARAMETER \ WEIGHT, P	-1	-0.5	0	0.5	1	2	3
TABLE A - General							
Slope angle			<15°	15°-30°	30°-45°	45°-70°	>70°
Rock mass structural conditions*			Massive rock with no or a few discontinuities (Jn=0.5÷1)	One set of discontinuities (Jn=2÷3)	Two sets of discontinuities (Jn=4÷6)	Three sets of discontinuities, rock mass subdivided into small cubes (Jn=9÷12)	More than three sets of discontinuities, highly fractured rock mass (Jn=15÷20)
Discontinuity aperture *			Closed or slightly opened		1 mm - 1 cm	1 cm - 10 cm	>10 cm
Stability conditions *			Stable		Partially stable	Unstable	
Fracturing degree of the rock mass **			Low		Medium	High	Very high
Expected rockfall events			Few events (1/10 years) - no rockfall scars		Occasional events (3/year)	Many events-visible rockfall scars (6/year)	Numerous and frequent events (9/year)
Precipitation			Low		Moderate	Intense	
Aggravating conditions							
Unstable blocks and/or overhanging sectors			None			Present	
Geological singularities (presence of faults, low resistance interlayers, heterogeneity, etc.)			None		Present		
Seepage/water			No/a few water seeps on slope		Numerous water seeps on slope		
Lateral or foot torrential erosion			None		Present		
Seismicity			Low		Moderate	High	
Stabilization works	Fully efficient/ effective	Partially efficient/ effective	None				
TABLE B1 - Mountain environment							
Lithology			Good quality rock			Soft rock	
Freeze–thaw cycles			None		Present		
TABLE B2 - Marine environment							
Slope orientation			Favorable (roughly shore-normal storm wave fronts)		Adverse (shoreline subparallel to main storm wave fronts)		
Elevation of the source area a.s.l.			High enough not to be affected by the erosive/unstable effects caused by waves, sea spray and tides		Not high enough to exclude erosive/unstable effects caused, even indirectly, by waves, sea spray and tides	Low enough to be affected by the erosive/unstable effects caused by waves, sea spray and tides	
Lithology and sensitivity to the erosive action of the sea			Good quality rocks (metamorphic, volcanic, etc.)		Medium quality rocks (limestones,sandstones conglomerates, etc.)	Rocks of low quality or sensitive to the marine environment	
Tidal effect			Not applicable, altitude of the source area sufficiently high	Low oscillations	Significant oscillations		
Wave energy			Not applicable, altitude of the source area sufficiently high	Moderate	High	Very high	
Cliff foot directly exposed to waves/tides			Not applicable - Protective beaches or engineering structures		No protective beaches or engineering structures		
Coastal retreat rate *			Very limited/limited		Significant		
Karst features			None		Limited	Significant	

** detailed scale only* *** medium-large scale only*

Fig. 1. Parameters controlling rock blocks failure probability: classification and relative scores. The SIF index can be obtained by combining the weights assigned to the parameters of Tables A+B1 in the case of mountain environment, or A+B2 in the case of coastal-marine environment.

3 The Case Study of Capo Calavà (Italy)

Capo Calavà is a promontory located in the northern coast of Sicily, Italy. The site is formed by mica schists and phyllites (in the eastern outcrop) and plutonite (in the western outcrop), and is subjected to diffuse rockfalls. A 300 m long rockfall sheltering tunnel, net fences and draperies, were installed to protect the SS113 road (Fig. 2). Given the importance of the site and the presence of the SS113 road, this area has been chosen as a pilot site within the research project entitled "TEMI MIRATI: Critical infrastructure safety due to fast moving landslide risk", financed with the PNR 2015–2020 (National Research Program).

Detailed survey campaigns were carried out in the last years by expert geologists and engineers, in order to (i) evaluate the rock mass structural conditions, (ii) count and measure the detached rock blocks stopped on (or close to) the tunnel and on the beach below the road, and (iii) analyze the degree of damage/deterioration of the existing mitigation structures.

The case study of Capo Calavà is presented in this paper to show an application of the SIF index and its effect on the rockfall susceptibility assessment. The propagation analysis is carried out in the QGIS environment by means of the QPROTO plugin [9].

Fig. 2. Capo Calavà site, where part of the artificial rockfall sheltering tunnel is also visible.

3.1 Characterization of the Release Areas

The identification of the potential release areas was performed in QGIS from a 2 m × 2 m DTM, on the basis of topographic and morphological features of the site. According to [7], rock outcrops showing inclinations greater than 52° were selected and mapped. Equidistant points, representing rockfall sources, were generated within such areas and assigned the required QPROTO input parameters [9]: elevation and aspect (from the

DTM), energy line angle (36° ÷ 55), lateral spreading angle (10°), visibility distance (800 m), boulder mass (440 kg ÷ 4465 kg), and detachment propensity (i.e. the SIF index, as shown in Fig. 3). To determine the mass of the rock blocks, different design volumes were first associated to the 3623 source points, according to the results of geostructural surveys and in-situ measurements of fallen blocks located on/close to the rockfall sheltering tunnel and on the beach. Then, these results were managed by means of statistical procedures, and the volumes obtained were multiplied by the rock unit weights, equal to 2343 kg/m^3 for the plutonites and to 2215 kg/m^3 for the phyllites. The energy line angle, main parameter of the runout analysis, was assessed for each source point on the basis of the analysis of slope inclination, point elevation and associated volume. Finally, the parameters of the SIF index (listed in Fig. 2) were assigned a score according to both the geostructural survey results and topographic information from the DTM. The SIF index was calculated according to Eq. (2), and the result obtained is shown in Fig. 3.

Fig. 3. Capo Calavà: SIF Index values assigned to each release point.

3.2 Susceptibility Analysis Results

The QPROTO plugin, based on a visibility analysis of the slope [9], was used to carry out two rockfall susceptibility analyses: the former neglecting the detachment propensity of the rock blocks (i.e. assigning a constant SIF index = 1 to all source points), and the second considering it (i.e. assigning the calculated SIF index = 0 ÷ 1 to ach source point). A calibration of the QPROTO input parameters was conducted considering the spatial arrangement of rock blocks detached in the past and accumulated along the slope, on the artificial gallery and on the beach, on the basis of the available orthophoto and on-site data survey results.

Figure 4 and Fig. 5 show the susceptibility maps obtained with and without assigning the SIF Index to the release points.

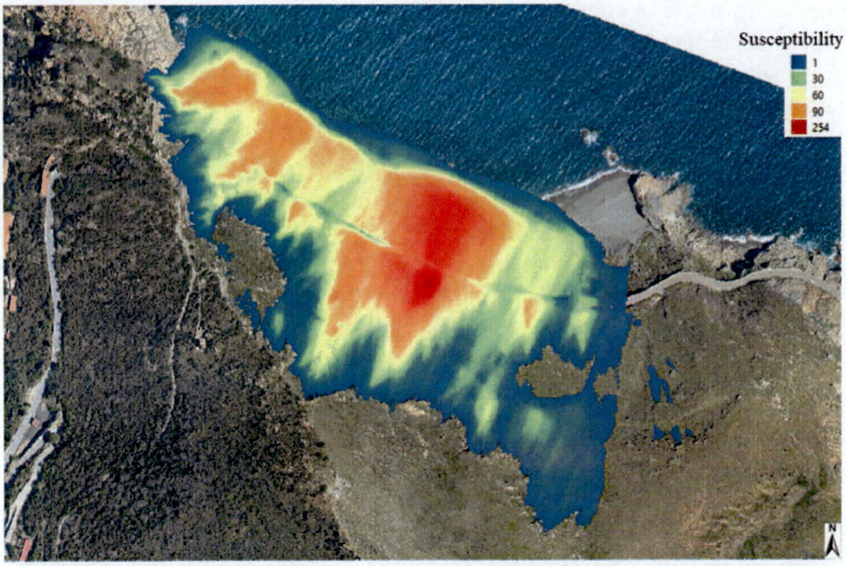

Fig. 4. Susceptibility map obtained without assigning the SIF Index to the source points.

Fig. 5. Susceptibility map obtained by assigning the SIF Index to the source points.

The value of the susceptibility is expressed numerically, and corresponds to the weighted passage frequency of the rock blocks (i.e., the sum of the SIF Indexes of all the source points viewing the considered DTM cell). By comparing the two maps, it is evident how the assignment of a detachment propensity to the different release points affects (reduces in this case) the susceptibility, allowing for more reliable results to be obtained. In fact, through the SIF index, invasion areas with equal passage frequencies (i.e. number of source points viewing these DTM cells) are differentiated on the basis of higher proneness to instability of the corresponding release points, identifying the most critical zones.

3.3 Definition of the Release Activity of the Rockfall Source Areas

A novel approach is proposed that identifies, with reference to specific elements at risk, the rockfall source areas characterized by the greatest release activity (i.e. the areas from which the blocks impacting the elements at risk are most likely to detach). Such information makes it possible to identify the most efficient locations for the installation of monitoring systems on the slope (e.g. geophones, benchmarks) to mitigate the risk.

The method has been implemented in a Matlab routine: once selected the elements at risk, E_i, the code analyzes the correspondences between the rock block passing frequencies on these elements (i.e., output shapefile "Finalpoints" generated by QPROTO, [9]) and the release points. The outcome is the percentage of passing rock blocks on E_i, associated to each source point.

Figure 6 illustrates the release activity map obtained for the Capo Calavà site, where the SS113 road represents the element at risk, E.

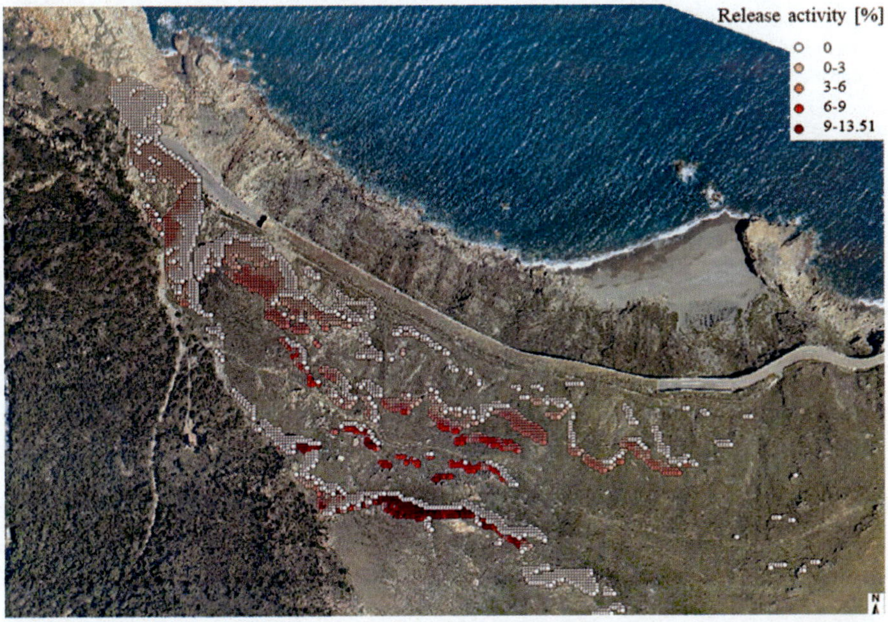

Fig. 6. Release activity of the source areas.

The result obtained indicates that 91% of the source points have a release activity lower than 6% (i.e. can hit up to 6% of the element at risk), and that the most active release area is relatively small (only 170 of the 3623 source points) and with a maximum release activity equal to 13.51%.

4 Conclusions

In this paper, a SIF index is proposed to obtain a weighted rockfall frequency runout map, and a procedure to define the release activity of the source areas with reference to a particular element at risk is set up. Both have been applied to the test site of Capo Calavà, in Sicily. According to the outcomes, geophones and benchmarks, belonging to an integrated monitoring system, were recently installed close to the most active release areas. In the future, the data provided by such monitoring system will be analyzed to further validate the approach proposed in this research.

Acknowledgments. This research was carried out under the framework of the project entitled "TEMI MIRATI: Critical infrastructure safety due to fast moving landslide risk (Project code: ARS01_00158/CUP B76C18001140005)", financed with the PNR 2015–2020 (National Research Program).

References

1. Corominas, J., et al.: Recommendations for the quantitative analysis of landslide risk. Bull. Eng. Geol. Env. **73**(2), 209–263 (2013). https://doi.org/10.1007/s10064-013-0538-8
2. Marques, F.: Regional scale sea cliff hazard assessment at Sintra and Cascais counties, Western Coast of Portugal. Geoscience **8**, 1–22 (2018). https://doi.org/10.3390/geosciences8030080
3. Del Río, L., Gracia, F.J.: erosion risk assessment of active coastal cliffs in temperate environments. Geomorphology **112**, 82–95 (2009). https://doi.org/10.1016/j.geomorph.2009.05.009
4. VV. AA. Prog. Interreg II C "Falaises". Prevenzione dei fenomeni di instabilità delle pareti rocciose. Confronto dei metodi di studio dei crolli nell'arco alpino (2001)
5. Romana, M.R.: A Geomechanical classification for slopes: slope mass rating (1993). ISBN 0080359310
6. Hantz, D., Corominas, J., Crosta, G.B., Jaboyedoff, M.: Definitions and concepts for quantitative rockfall hazard and risk analysis. Geoscience **11** (2021). https://doi.org/10.3390/geosciences11040158
7. Projet n ° 165 PROVIALP (2008). ISBN 9788874790708 (in Italian)

8. MATTM-Regioni Linee Guida per La Difesa Della Costa Dai Fenomeni Di Erosione e Dagli Effetti Dei Cambiamenti Climatici (2018). (in Italian)
9. Castelli, M., Torsello, G., Vallero, G.: Preliminary modeling of rockfall runout : definition of the Input Parameters for the QGIS Plugin QPROTO. Geosciences **11**, 26 (2021). https://doi.org/10.3390/geosciences11020088Academic

Landslide Early Warning Systems as Climate Change Adaptation Measures for Rail Infrastructure

Gaetano Pecoraro[1] (✉), Federico Foria[2], Fabio Villa[3], Andrea Tamburini[3], Serena Pantaneschi[2], Mario Calicchio[2], Gabriele Miceli[2], and Michele Calvello[1]

[1] Department of Civil Engineering, University of Salerno, via Giovanni Paolo II, 132, 84084 Fisciano, SA, Italy
gpecoraro@unisa.it

[2] ETS Srl, via Belice, 9/11, 04100 Latina, LT, Italy

[3] IMAGEO Srl, frazione Combe, 31, 10050 Mattie, TO, Italy

Abstract. Railways represent a significant asset for Italy, with more than 16,000 km of rail lines distributed all over the country. The impacts of extreme weather events can be particularly severe on rail infrastructure because of the highly integrated nature of the rail system. Historical failure of rail infrastructure due to extreme weather events is well documented. In addition, it is to be expected that climate change affects the stability of engineered slopes and have consequences on landslides. Failure of a single asset can result in potential fatalities, large replacement costs, and loss of service. Consequently, Rete Ferroviaria Italiana (RFI), the Italian public limited company for railway transportation, is moving towards risk assessment methods as part of holistic asset management. In this context, ETS Srl has been developing a methodology called MIRETS (Management and Identification of the Risk - ETS), a workflow to connect survey-inspection data for geology, digitalization, diagnostics, and design. Within this framework, a three-phase methodological approach has been developed for the design and the implementation of a landslide early warning system (LEWS) along a strategic railway infrastructure. This study describes the first preliminary results, focusing on the collection and the processing of the input data.

Keywords: Landslide risk · rainfall threshold · rail

1 Introduction

Over the last decades, an increase in the average global temperature is widely recognized. Consequently, the frequency and the intensity of high intensity rainstorms is also increasing, although in places the average yearly cumulated precipitation is not showing significant changes [1]. Therefore, it is expected that rainfall-induced landslides become more severe and occur more frequently, causing inevitable damage and affecting human activities and infrastructure operativity [2]. One of the biggest issues for the safety of the operations along Italian railway lines is the hydrogeological instability: even small volumes of rock or debris on the track can cause train derailment [3].

A. Ferrari et al. (Eds.): CNRIG 2023, SSGG, pp. 724–731, 2023.
https://doi.org/10.1007/978-3-031-34761-0_87

Therefore, the identification of the priorities in the process of landslide risk management is a crucial aspect for the Italian Railway Authority (RFI - Rete Ferroviaria Italiana). In this context, ETS Srl has been developing a methodology called MIRETS (Management and Identification of the Risk - ETS), an integrated workflow to connect survey-inspection data for geology, digitalization, diagnostics and design [4]. ARCHITA, an innovative multi-dimensional mobile mapping system, allows a quick and reliable integrated survey of the existing railway infrastructure [5]. Starting from the data of ARCHITA in the MIRETS framework, a specific module is implemented for the hazard mapping of the slopes along railway corridors [3].

Within this framework, landslide early warning systems (LEWS) can be valuable tools for predicting the occurrence of rainfall-induced landslides at different scales of operation [6, 7]. Indeed, the frequency of landslides along railway corridors is expected to increase with global climate change and population growth, so early warning systems may be useful for landslide risk prevention and mitigation.

A methodological three-phase approach for the design and the implementation of a LEWS on slopes along railway corridors is presented herein. A first application has been developing along a strategic railway line between the provinces of Rome and Cassino in the southern part of Lazio region, Italy. This paper focuses on the description of the proposed methodology and on the first phase of the proposed approach. It can be seen as a first step towards a real-time LEWS for rainfall-induced landslides affecting slopes along railway corridors.

2 Methodological Approach

The algorithm for the development of a real-time EWS for rainfall-induced landslides along railway infrastructures can be schematized into three main blocks: (i) collection and processing of the input data; (ii) calibration and validation of the empirical rainfall thresholds; (iii) LEWS management.

In the first block, the input data (meteorological measurements and landslides occurred along the railway line) are collected. Landslides records of poor quality are removed from the dataset. Then, the railway line is discretized into territorial units, whose extension is adequate to the spatial resolution of the input data.

In the second block, landslides occurred in the period of analysis are correlated to the rainfall events to individuate the most significant rainfall variables. The definition of the rainfall thresholds characterized by diverse non-exceedance probabilities is carried out by means of a statistical analysis of the triggering and non-triggering rainfall events. The set of thresholds is calibrated and validated using statistical indicators derived from a contingency matrix and the best-performing warning model is identified.

In the third block, weather forecasts are compared with the defined rainfall thresholds and, in case of exceedance, a warning is issued and the planned actions are implemented. Moreover, a periodic assessment of the performance of the LEWS is carried out in order to recalibrate the thresholds, if necessary. Finally, a WebGIS platform capable of acquiring and processing real-time monitoring data is also developed. The WebGIS frontend allows users to analyze the data and interact with the LEWS by providing feedbacks on automatic warnings issued (Fig. 1).

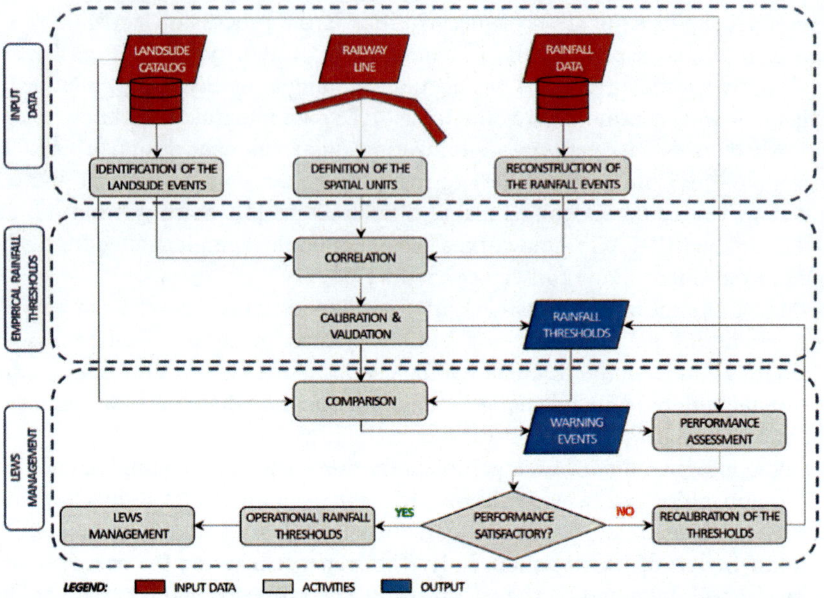

Fig. 1. Workflow of the proposed procedure for the development of a real-time EWS for rainfall-induced landslides along railway infrastructures.

2.1 Collection and Processing of the Input Data

The rainfall measurements considered for the analyses are derived from a gridded rainfall dataset characterized by high spatial (1.5 km) and temporal (1 h) resolution. Information on landslide occurrences is gathered from the "FraneItalia" catalog [8] and from technical reports. Landslide events not classifiable as rainfall-induced or for which the trigger is not known are removed from the dataset. Among the 446 rainfall-induced landslides occurred in the Lazio region between 2010 and 2020, eight landslides affected the railway line of interest (Table 1).

The authors are aware that landslide records are typically affected by spatial uncertainties. For this reason, gridded rainfall data are analyzed using the focal statistics in the GIS environment. This algorithm performs a neighborhood operation, where the value of each cell is the average of all the values encountered in the cells surrounding the processing cell. The sizes of the neighborhood *(n)* tested in this study are limited to 3 and 5, as larger sizes may affect the reliability of the results.

An automated and objective procedure is developed for the reconstruction of the triggering and non-triggering rainfall conditions, following the algorithm proposed by Melillo et al. [9]. Therefore, the individual rainfall events are reconstructed from the records of rainfall measurements following a procedure organized in four successive steps: records of hourly rainfall measurements are associated with documented landslides (Fig. 2a); hourly rainfall measurements lower than a pre-defined threshold, E_R, are considered irrelevant (Fig. 2b); isolated rainfall events are separated considering two different dry periods: D_W, for the warm season; D_C, for the cold season (Fig. 2c); triggering *(T-RE)* and non-triggering rainfall events *(NT-RE)* are identified (Fig. 2d).

Table 1. Landslide events occurred along the railway line of interest between 2010 and 2020.

ID	Date of occurrence	Source of information
2011_01	2011-07-29	FraneItalia
2013_01	2013-11-21	Technical report
2014_01	2014-01-27	Technical report
2014_02	2014-02-02	FraneItalia
2014_03	2014-06-16	FraneItalia
2015_01	2015-02-05	Technical report
2015_02	2015-10-15	FraneItalia
2019_01	2019-02-03	FraneItalia

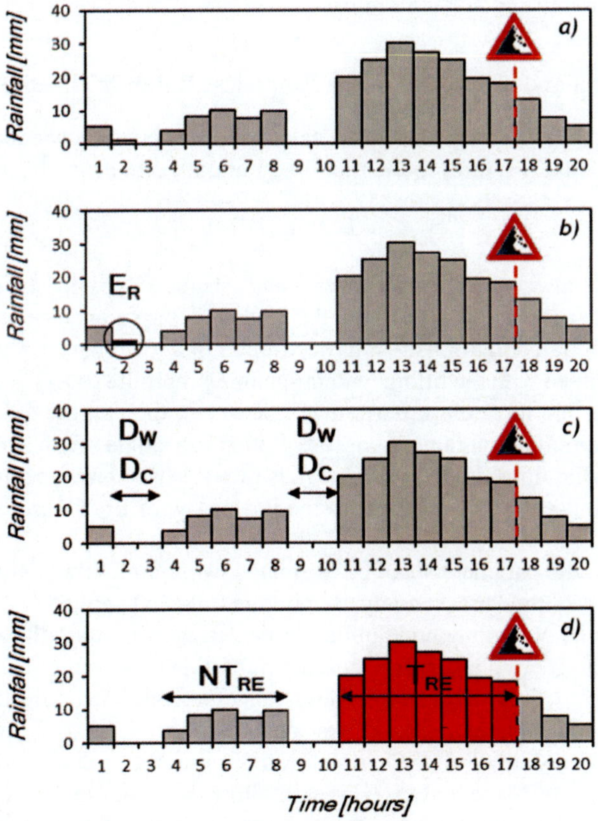

Fig. 2. Example of reconstruction of triggering (T-RE) and non-triggering (NT-RE) rainfall events.

Table 2 lists the parameters used in this study, differentiated considering the "warm" springer-summer period, *CW*, and the "cold" autumn–winter period, *CC*.

Table 2. Parameters used for the application of the algorithm.

Parameter name	Parameter value	Unit of measure
E_R	1	mm
D_W	48	h
D_C	96	h

Considering the meteorological grid cells located along the railway infrastructure, 21.149 rainfall conditions (D, E) in the range of duration $1 \leq D \leq 40$ h and in the range of cumulated rainfall $1 \leq E \leq 180$ mm.

2.2 Calibration and Validation of the Empirical Rainfall Thresholds

The empirical rainfall thresholds will be calculated as a power law equation linking the cumulated rainfall, E (in mm) to the rainfall duration, D (in h):

$$E = \alpha \cdot D^{\gamma} \tag{1}$$

where α is a scaling constant (the intercept) and γ is the slope (the scaling exponent) of the power law curve (Fig. 3). Typically, the method is based on a frequency analysis of the empirical rainfall conditions that have resulted in known landslides. To account for problems associated with the fitting of data spanning multiple orders of magnitude (e.g., the least square minimization criteria may not work), the empirical data are first log-transformed. The empirical rainfall data are plotted in a single graph and the distribution of the rainfall conditions, log(E) vs. log(D), is fitted (least square method) with a linear equation, which is entirely equivalent to the power law of Eq. (1) in linear coordinates [10].

In principle, the parameters can be estimated using three frequentist methods [11]: frequentist P, the method most widely used in literature and applied to triggering rainfall events; frequentist N, the method applied to non-triggering rainfall events; frequentist PN, using both triggering and non-triggering rainfall events. In the scientifici literature, the large majority of the rainfall thresholds have been defined considering triggering events. However, the number of rainfall events that triggered landslide along the railway infrastructure is very small, thus the definition of rainfall thresholds will be carried out using the frequentist N method us. This will allow the definition of rainfall thresholds specific for the railway line of interest and robust from a statistical point of view.

Finally, the calculated thresholds will be calibrated and validated using a set of skill scores derived by a contingency table, identifying: correct alerts (CA), false alerts (FA), missed alerts (MA), and true negatives (TN) [12]. The issuing of a warning concurrently with the occurrence of at least one landslide is assumed as CA. FA and MA are incorrect predictions of the system: the first is related to the issuing of a warning and the simultaneous absence of a landslide; the second refers to the occurrence of a landslide without any warning. TN represent the absence of both landslides and warning occurrences.

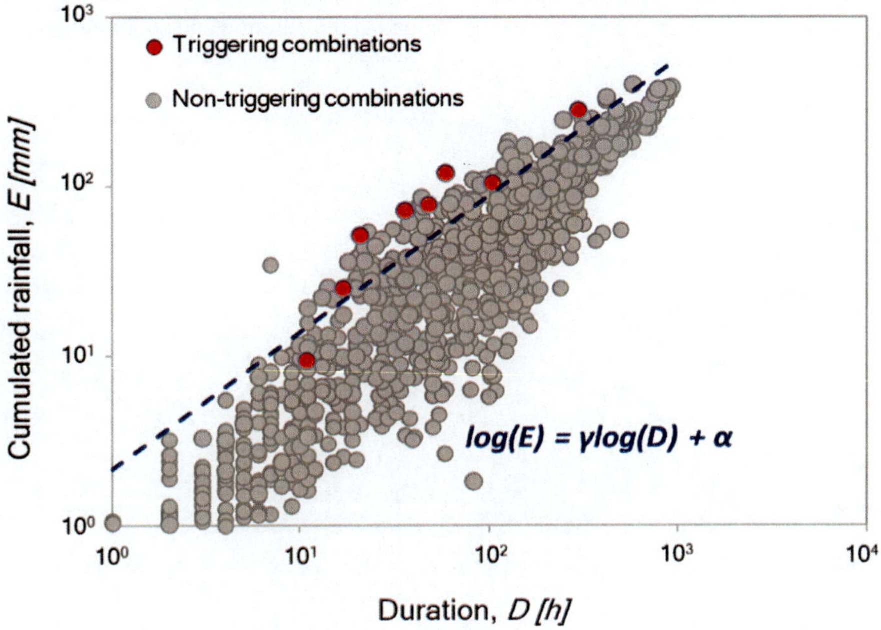

Fig. 3. Example of rainfall threshold determined applying the frequentist method.

2.3 LEWS Management

For each track of the railway line, rainfall forecasts will be compared with the rainfall thresholds on regular intervals to issue the appropriate warning level (Fig. 4). When a threshold value will be exceeded, actions adequate to the level of risk (e.g., in-line inspections; interruptions of the rail traffic) will be undertaken by system managers in charge of landslide risk management.

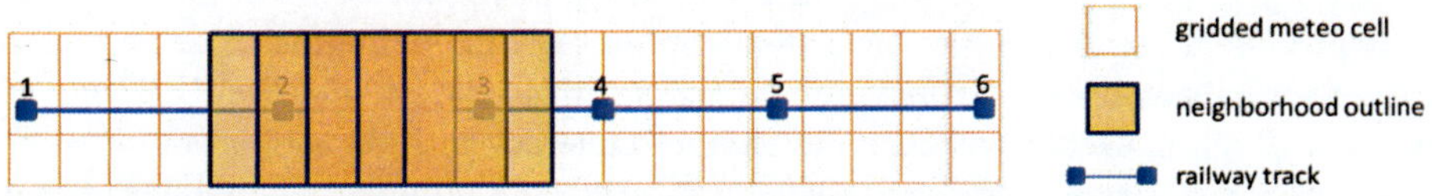

Fig. 4. Example of application of the warning model using the focal statistics with n = 3 along the track 2–3.

Moreover, the system will benefit of a WebGIS structure capable of acquiring and processing real-time monitoring data. The WebGIS frontend will allow users to analyze the data and interact with the system by providing feedbacks on automatic warnings issued by the EWS, supporting the calibration and the optimization of the threshold values and the reference scenarios. When a warning will be issued along the railway line, the user will be able to access the WebGIS platform, visualize the corresponding grid cell (highlighted in purple) and then inspect it in order to visualize (and download, if necessary) the rainfall data series and the values of the duration and cumulated rainfall of the rainfall event (Fig. 5).

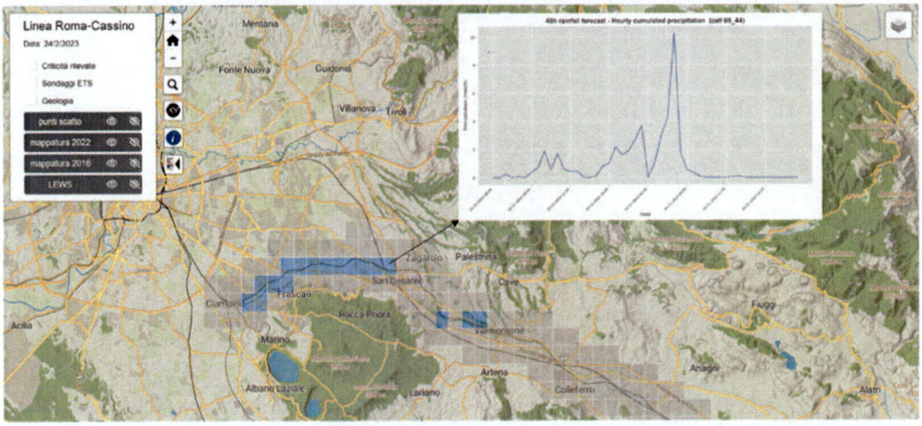

Fig. 5. WebGIS platform developed for real-time data visualization and management.

Moreover, in case of issuing of a warning, an automated email will be sent to the personnel in charge of the management of the system.

Finally, a regular performance assessment of the warning model will be performed in order to assess the predictive capability of the thresholds so that, if necessary, some parameters will be modified, and the thresholds will be recalibrated as well as new data will become available.

3 Concluding Remarks

The methodology presented herein displays the main activities addressed for the design and the implementation of an operational LEWS along a strategic railway line between the provinces of Rome and Cassino in southern Lazio, Italy. This study focuses on the results of the first phase describing the collection and the processing of the input data. This procedure should be seen as a first step towards the definition of an innovative LEWS for railway lines. The operational LEWS will also benefit of a WebGIS frontend, which will allow users to analyze the data and interact with the system by providing feedbacks on automatic warnings issued by the EWS, supporting the calibration and the optimization of the threshold values and the reference scenarios.

Finally, within the general framework of the landslide risk management, an integration with a landslide risk map will also allow to assess the spatio-temporal probability of landslide occurrence in order to rationally plan the funds designate to secure the railway lines.

References

1. Guzzetti, F., Gariano, S.L., Peruccacci, S., Brunetti, M.T.: Geographical landslide early warning systems. Earth Sci. Rev. **200**, 102973 (2020)
2. Martinović, K., Gavin, K., Reale, C., Mangan, C.: Rainfall thresholds as a landslide indicator for engineered slopes on the Irish rail network. Geomorphology **306**, 40–50 (2018)
3. Foria, F., Miceli, G., Tamburini, A., Villa, F., Rech, A., Epifani, F.: Application of Spatial Multi-Criteria Analysis (SMCA) to assess rockfall hazard and plan mitigation strategies along long infrastructures. In:IOP Conference Series: Earth and Environmental Science, vol. 833, p. 012074 (2021)
4. Foria, F., Giordano, R., Avancini, G. Miceli, G.: Mitigation measures for the protection of coastal railways in the Flysch of Western Liguria. In: IOP Conference Series: Earth and Environmental Science, vol. 833, p. 012075 (2021)
5. Foria, F., Avancini, G., Ferraro, R., Miceli, G., Peticchia, E.: ARCHITA: an innovative multidimensional mobile mapping system for tunnels and infrastructures. In: MATEC Web of Conferences, vol. 295, p. 01005 (2019)
6. Piciullo, L., Calvello, M., Cepeda, J.M.: Territorial early warning systems for rainfall-induced landslides. Earth Sci. Rev. **179**, 228–247 (2018)
7. Pecoraro, G., Calvello, M., Piciullo, L.: Monitoring strategies for local landslide early warning systems. Landslides **16**(2), 213–231 (2018). https://doi.org/10.1007/s10346-018-1068-z
8. Calvello, M., Pecoraro, G.: FraneItalia: a catalog of recent Italian landslides. Geoenvironmental Disasters **5**(1), 1–16 (2018). https://doi.org/10.1186/s40677-018-0105-5
9. Melillo, M., Brunetti, M.T., Peruccacci, S., Gariano, S.L., Guzzetti, F.: An algorithm for the objective reconstruction of rainfall events responsible for landslides. Landslides **12**, 311–320 (2015)
10. Brunetti, M.T., Peruccacci, S., Rossi, M., Luciani, S., Valigi, D., Guzzetti, F.: Rainfall thresholds for the possible occurrence of landslides in Italy. Nat. Hazard. **10**, 447–458 (2010)
11. Peres, D.J., Cancelliere, A.: Comparing methods for determining landslide early warning thresholds: potential use of non-triggering rainfall for locations with scarce landslide data availability. Landslides **18**(9), 3135–3147 (2021). https://doi.org/10.1007/s10346-021-01704-7
12. Piciullo, L., Tiranti, D., Pecoraro, G., Cepeda, J.M., Calvello, M.: Standards for the performance assessment of territorial landslide early warning systems. Landslides **17**(11), 2533–2546 (2020). https://doi.org/10.1007/s10346-020-01486-4

Geotechnical Analysis on the Effects of Tiber River Hydraulic Regime in the City Centre of Rome Within the Project Tiber'S

Arianna Pucci[1](✉), Gorizia D'Alessio[1], Ilaria Giannetti[1], Ilaria Moriero[2], Giulia Guida[1], Jose Francisco Guerrero Tello[3], Benedetta Moccia[4], Fabio Russo[4], Maria Marsella[4], and Francesca Casini[1]

[1] Università degli Studi di Roma Tor Vergata, Rome, Italy
arianna.pucci@uniroma2.it
[2] Survey Lab, Spin off di Sapienza Università di Roma, Rome, Italy
[3] INGV, Istituto Nazionale di Geofisica e Vulcanologia, Catania, Sicilia, Italy
[4] DICEA, Dipartimento di Ingegneria Civile Edile e Ambientale, Sapienza Università di Roma, Rome, Italy

Abstract. Seasonal variation in the Tiber hydraulic level results in a fluctuation of the groundwater table (GWT) depth, thus in a change in the effective stress of the superficial layer of made ground (MG). This paper focuses on the evaluation of displacement induced by GWT oscillation in the MG layer by means of hydro-mechanical numerical analyses along a typical geotechnical section of the city centre of Rome. The data predicted are compared with satellite monitored data. This work is conducted within the Tiber'S research project whose aim is to provide open access data on a WebGis platform to promote non-invasive monitoring and predictive risk assessment related to water table oscillations of the Tiber River surrounding areas.

Keywords: Groundwater table oscillations · Unsaturated soils · Satellite monitoring

1 Introduction

Due to climate change, extreme weather events are occurring more frequently, producing river regimes increasingly unstable, with massive flood waves alternating with periods of lean. The level of the river inevitably influences the position of the shallow groundwater level in the areas around. Increased groundwater fluctuations can lead to a build-up of subsidence, which can be particularly detrimental in urbanized areas, where buildings and infrastructures may be subject to damage due to differential settlements [1].

The Tiber'S project aims to construct a webGIS platform able to divulge the construction heritage included in urban areas of the centre of Rome adjacent to the river Tiber, and to monitor and assess the effects and the risks of groundwater fluctuations. The construction of this tool, innovative and scalable, is based on the combination of

A. Ferrari et al. (Eds.): CNRIG 2023, SSGG, pp. 732–740, 2023.
https://doi.org/10.1007/978-3-031-34761-0_88

satellite data, advanced numerical analyses of the soil response, and a punctual historical-technical investigation, starting from the documentation present in the historical archives of the territory. In particular, the project has the mission of enhancing skills and technologies applied to cultural heritage in a synergistic action among the proposers, their own laboratories, the DTC (Lazio with the agencies and the productive fabric of the regional territory.

In the following a brief description of different tasks of the project -- the historical surveys, the hydraulic analysis and the satellite monitoring -- is presented, focusing finally on the geotechnical modelling of the unsaturated layer of MG and the displacement prediction compared to the monitored satellite data.

2 Methodology and Workflow

The methodology of the Tiber'S project is based on a multidisciplinary approach – construction history, geotechnics, hydraulics, and satellite monitoring – via a multi-layered Web GIS platform. The workflow, presented in Fig. 1, resumes the diachronic development of the four main research actions: *i)* historical surveys that aim to collect accurate knowledge of historical data about buildings, monuments and the infrastructures linked to the river Tiber; *ii)* hydraulic analysis that focuses on the elaboration of hydraulic data history concerning the river Tiber level in the urban area; *iii)* geotechnical modelling of a section of interest to study the influence of water table oscillation linked to seasonal changes in the Tiber level, according to the analysed hydraulic trends (phase *ii*); *iv)* monitoring analyses of DinSAR satellite data, concerning the river Tiber urban area, in terms of mean velocity maps and settlement time series.

The outcomes of the main research actions will be, thus, integrated on the WebGis platform (see Sect. 2.2) – supporting cross-disciplinary analysis, and providing interactive maps for risk preventions and dissemination of the results of the project.

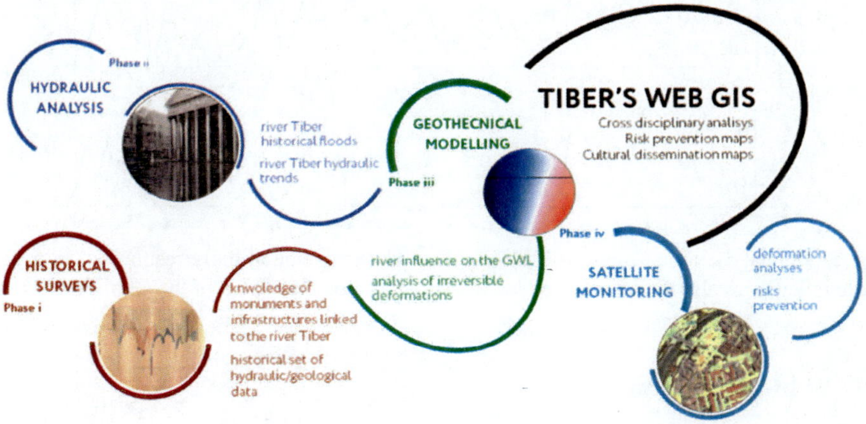

Fig. 1. Diagram of the project workflow

2.1 Historical Surveys

The historical sets of hydraulic and geological data concerning the urban areas next to the river Tiber are collected within the historical survey. The main archival source of the study was the Archive of the former Ufficio Speciale per la Sistemazione del Tevere [Special Office for the River Tiber Management]. In the last decades of the 19th Century, the Special Office was the coordinator of the project and the construction of the modern embankment's structures of the river Tiber, characterized by gravity retaining walls [2, 3]. Thus, the execution project documents here conserved provided crucial data for accurate technical knowledge of the embankment structures of soil characterization and analysis of the hydraulic regime of the river of the time.

2.2 Hydraulic Regimes

The Tiber River regime is daily monitored and recorded by the Civil Protection of Lazio Region. In this work, daily hydraulic levels and discharges recorded at Ripetta hydrometer [4] in the period 1921–2021 were considered. Figure 3 reports the trends with time of the annual maxima (AM), that is the maximum value for each year of observation [5], extracted for both the time series of (a) discharges and (b) hydraulic levels. When the warning level of 13 m for the Ripetta hydrometric station (see Fig. 2b) is exceeded, the Civil Protection actives pre-established plans made to cope the specific flood risks. In the last century, the critical level of 13 m was exceeded nine times before the 1950s. in particular, during the flood event recorded in December 1937, the Ripetta hydrometer measured a hydraulic level of 16.75 m (with a discharge of 2730 m^3/s). In the last 20 years, the most significant flood occurred in December 2008, November 2012 and February 2014, with a stage at Ripetta dealt to 12 m.

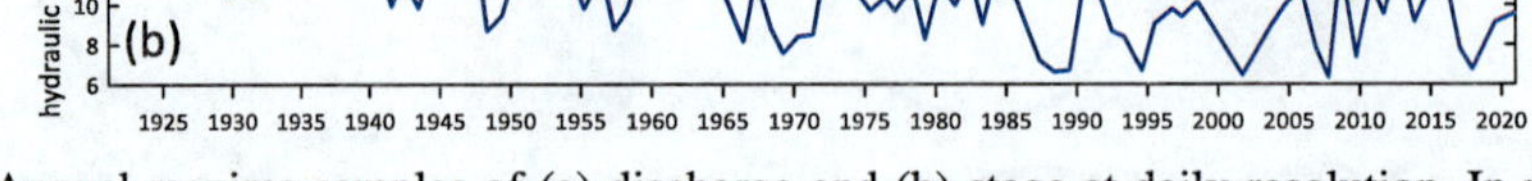

Fig. 2. Annual maxima samples of (a) discharge and (b) stage at daily resolution. In panel (b) warning level are also depicted, provided by the flood protection plan of the Civil Protection.

2.3 WebGIS Platform

In this work an open source WebGIS (Geographic Information System on the Web) is being developed with the aim to provide the spatial analysis and historical data which the users can realize queries and interacting with. The platform is under construction and will be open access once completed and the theme system on the web will allow a simple reading and analysis of the information by the end users, even if not "specialized".

The data acquired for the project have been organized in 4 macro-area:

- Hydrography, containing all the information related to Tiber's floods.
- Monitoring, containing all the information related to satellite monitoring.
- Building, containing all the information about historical buildings.
- Subsoil, containing all the information subsoil characteristics.

3 Case Study

A case study reproducing a typical geotechnical section of the Rione Trastevere (Fig. 3a), in the western bank of the Tiber River, was chosen because of the damages induced by differential settlements experienced by an elongated building in that area [6, 7]. The purpose of this case study is to investigate the effects of seasonal variations in the Tiber water level on the settlements at ground surface, adopting a critical state model extended to unsaturated conditions [8]. The seasonal GWT oscillations lie in the made ground layer (MG in Fig. 3b) consisting in loose and heterogeneous sandy-silty soil, with pieces of brick, tuff and pozzolanic mortar as well.

a)

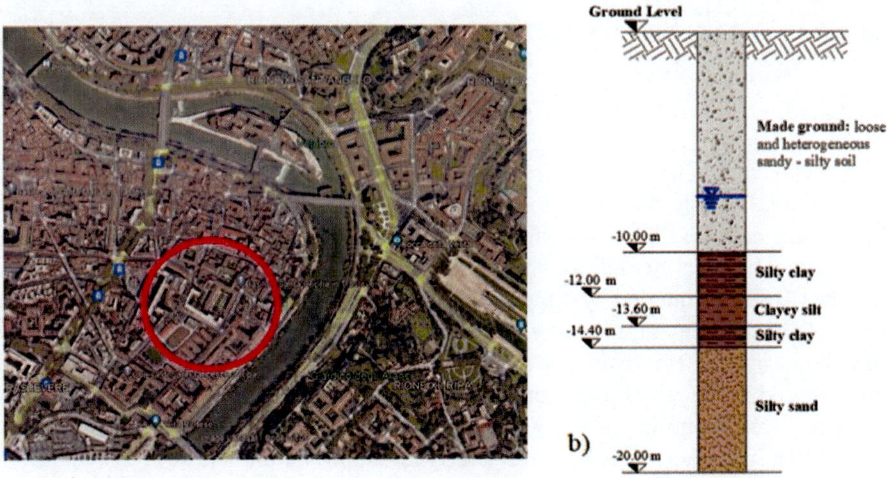

Fig. 3. (a) Rione Trastevere location (b) stratigraphic profile of the investigated area

3.1 Numerical Analyses

Numerical analyses were carried out with a finite element software (COMSOL Multiphysics®) to study the effects of seasonal variations in the Tiber water level according to the period between 2005–2008 on the position of the GWT and on the induced settlements. The GWT level oscillates seasonally within the made ground layer, inducing periodic wetting and drying processes.

The volumetric response of a soil subjected to water content variation needs constitutive models that account for the unsaturated soil theory to be properly predicted. The Barcelona Basic Model (BBM) [8] is adopted to model the soil layers in partially saturated conditions. It is a hardening elastic – plastic constitutive model that allows to reproduce both the volumetric deformation of swelling or shrinkage upon wetting path [9]. The governing equations for the hydro-mechanical problem are equilibrium and the water mass balance equations [7], solved numerically by the finite element method via Comsol Multiphysics® software.

Figure 4 shows the geometry of the two-dimensional 160-m-wide double layers model: for the sake of simplicity, it is assumed that there is only one sand layer below the MG layer and that layers are perfectly horizontal. BBM is adopted for the MG layer, in which the GWT fluctuations lie, and a linear elastic model is adopted for the sandy layer, always fully saturated. Table 1 summarises the model parameters, which were calibrated by oedometric tests in unsaturated conditions for the MG layer [7] and deduced by literature for the sandy one [6].

Table 1. Constitutive model parameters

MG layer (BBM) calibrated by oedometric tests in unsaturated conditions [7, 8]							Sand (Elastic) [6]
λ_0 [-]	λ_S [-]	r [-]	β [kPa^{-1}]	k[-]	k_S[-]	p_C[kPa]	E[MPa]
0.013	0.015	0.300	0.015	0.0019	0.002	25	100

The first step of the analyses is a stationary one, in which gravity is applied to replicate the in situ lithostatic condition and a hydrostatic pore water pressure profile is adopted, in accordance with the initial position of the GWT. Figure 4 shows the hydraulic and mechanical boundary conditions adopted. From a hydraulic point of view, the presence of Tiber River is modelled by imposing a boundary condition on the water pressure at the left edge of the domain, given by the hydrometric height of the river daily recorded by the Ripetta hydrometer in the period 2005–2008 (see Fig. 2). No flux boundary condition are imposed in the remaining boundaries.

The water retention curve (WRC) for the MG layer is defined according to the Van Genuchten formulation [10], calibrated by an empirical method [11] against the grain

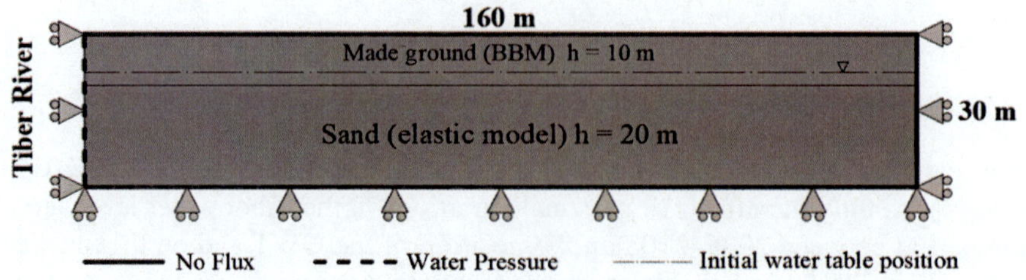

Fig. 4. Geometry of the domain and BCs.

size curve and the initial porosity of the soil:

$$S_r = S_{res} + (1 - S_{res}) \cdot \left[(1 + s/P)^n\right]^{-m} \tag{1}$$

In which $S_{res} = 0.2$ is the residual degree of saturation, s is the current suction and $P = 20$ kPa, $m = 0.6$ and $n = 2.7$ are model parameters depending on the soil type. The hydraulic conductivity k depending on S_r is defined as follows:

$$k(S_r) = K_{sat} \cdot k_{rel}(S_r);\ \text{where}\ k_{rel}(S_r) = Se^{0.5} \cdot \left[1 - \left(1 - Se^{1/m}\right)\right]^2 \tag{2}$$

where, K_{sat} is the hydraulic conductibility at saturation, assumed equal to 5×10^{-5} m/s for the MG layer and $S_e = (S_r - S_{res})/(1 - S_{res})$ is the effective degree of saturation ($K_{sat} = 10^{-4}$ m/s for the sand saturated layer).

3.2 Results and Comparisons with Satellite Data

Figure 5 shows the evolution over time of the total displacements (dashed line) and of the MG layer displacements (solid line) at ground level induced by water table oscillation, for two different distances from the river Tiber, $x = 10$ m and 132 m, as well as the trend of hydrometric heights over time, reproducing the Ripetta hydrometer. In addition, three different time instants were highlighted: t_1, which corresponds to the hydrometric height recorded in February 2005; t_2, which corresponds to the flood occurred in November 2005; and t_3 which corresponds to the one occurred in November 2008.

It can be observed that settlements accumulate when the river level increases, due to the saturation collapse of MG layers. For instance, the flood occurred in November 2005 (t_2, reported zoomed in Fig. 5) caused a rise of about 4 m in hydrometric height, which induced an instantaneous settlement in the MG of about 3 mm at a distance of $x = 10$ m from the river and an almost 2 mm of settlement at a distance of $x = 132$ m from the river with a lag of 10 days. As the distance from the river increases, the displacements become smaller, since the oscillation of the GWT induced by the river level variation gradually attenuates. After the flood peak that occurs at t_2, no significant accumulation of settlements on the MG layer is longer recorded.

Figure 6 shows the profiles of volumetric plastic strain, at (a) $x = 10$ m and (b) $x = 132$ m from the river, along the MG layer depth, for the three different instants of time t_1, t_2 and t_3, introduced in Fig. 5. The range of depths in which plastic deformations mainly develop is between 2 m and 8.50 m, which corresponds to the range of depths where the GWT oscillates. It can also be observed that, as the distance from the river increases, the volumetric plastic strains decrease from a maximum value of almost 0.18% for $x = 10$ m to a second value of 0.08% for $x = 132$ m.

From Fig. 6 it can also be seen that the peak of volumetric plastic strain increases toward ground level as the GWT level increases: for example, at $x = 10$ m from the river (Fig. 6 (a)) and at the time instant t_1, which corresponds to a GWT level of almost 8 m a.s.l. (Fig. 5), the peak of volumetric plastic strain is approximately 6 m from ground level and rises to 4 m when the level of the water table rises to 12 m a.s.l. and 13 m a.s.l. at instants t_2 and t_3 (Fig. 5) respectively.

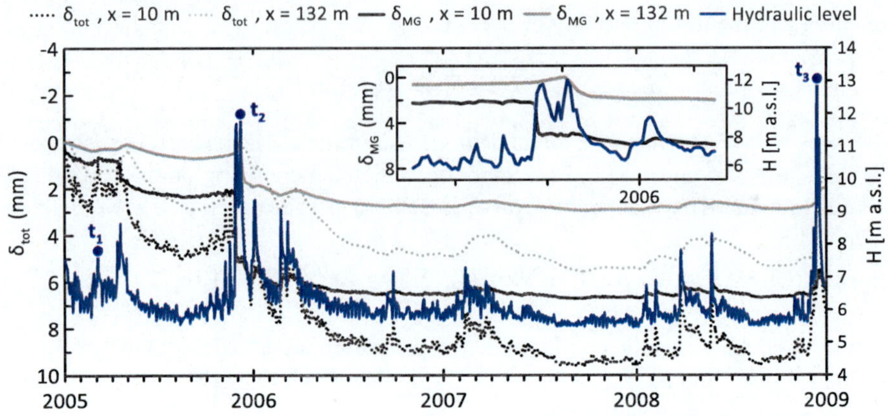

Fig. 5. Displacements (total and MG ones) and hydrometric heights.

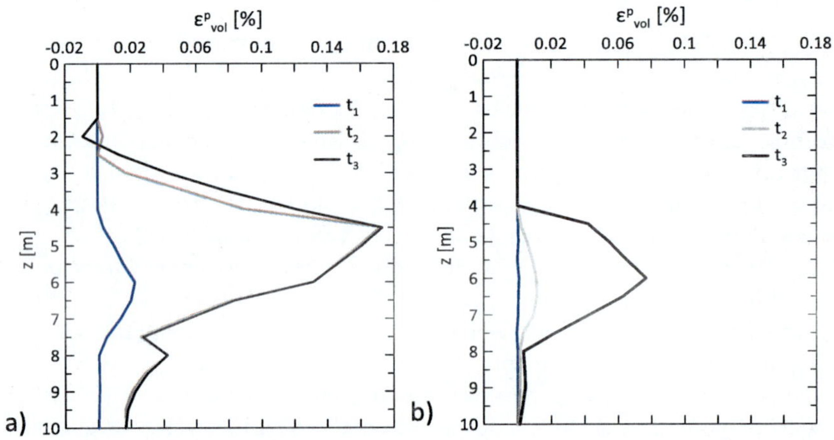

Fig. 6. (a) Volumetric plastic strain at $x = 10$ m and (b) $x = 132$ m from the river.

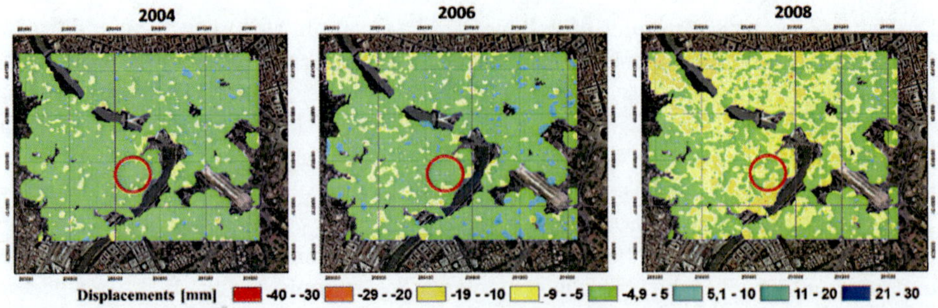

Fig. 7. Development of the cumulative displacements since2002 in three different years.

Figure 7 shows the cumulative displacements at ground level recorded by satellite monitoring since 2002, in three different years with an accuracy of ± 1 mm/year. In the area of interest, highlighted by the circle, a progressive increase in accumulated displacements from 5 mm to 9 mm can be observed especially passing from 2006 to 2008. The results obtained by numerical analysis in terms of total displacements at ground level (Fig. 5) quantitatively agrees with the data recorded by satellite monitoring for the same period.

4 Conclusion

This work describes the first results obtained within the Tiber'S research project on the effect of the Tiber River hydraulic regime on the soil displacements recorded in neighbouring areas. The two – layered model used in the FEM analysis is a simplification of the stratigraphy of the city centre of Rome, accounting for the unsaturated state of the soil above the water table and the oscillation of the latter in relation with the hydraulic level of the Tiber River. The numerical results in terms of accumulated displacements, whilst accounting only for the groundwater oscillations in MG layer, show a promising good agreement with the satellite monitoring data for the investigating years.

References

1. Calabresi, G., Cassini, C., Nisio, P.: Influenza del regime del Tevere sul comportamento di un fabbricato monumentale adiacente. In: Convegno Nazionale di Geotecnica, pp. 25–33. Firenze (1980)
2. Giannetti, I., Casini, F.: The construction and the collapse of the Tiber retaining walls in Rome, Italy (1870–1900). Proc. Inst. Civ. Eng. Eng. Hist. Herit. **175**(2), 48–58 (2021)
3. Casini, F., Pucci, A., Giannetti, I., Guida, G.: Geotechnical and historical aspects on the collapse of the Tiber embankment walls in the centre of Roma (1870–1900). In: Geotechnical Engineering for the Preservation of Monuments and Historic Sites III, pp. 1206–1214. CRC Press, Napoli (2022)
4. Bersani, P., Bencivenga, M.: Le Piene Del Tevere a Roma Dal V Secolo a.C. All'anno 2000. Presidenza del Consiglio dei ministri - Dipartimento per i servizi tecnici nazionali - Servizio idrografico e mareografico nazionale. Roma (2001)
5. Moccia, B., Mineo, C., Ridolfi, E., Russo, F., Napolitano, F.: Probability distributions of daily rainfall extremes in Lazio and Sicily, Italy, and design rainfall inferences. J. Hydrol. Reg. Stud. **33**, 100771 (2021)
6. Ventriglia, U.: La Geologia della Città di Roma. Amministrazione Provinciale di Roma. Roma (1971)
7. D'Alessio, G., Pucci, A., Guida, G., Casini, F.: Numerical study on the effects of ground water table oscillation beneath Palazzaccio courthouse in Rome. In: 8th European Conference on Unsaturated Soils, UNSAT 2023
8. Alonso, E.E., Gens, A., Josa, A.: A constitutive model for partially saturated soils. Géotechnique **40**(3), 405–430 (1990)
9. Rotisciani, G.M., Lalicata, L.M., Desideri, A., Casini, F.: Numerical modelling of the response of an unsaturated silty soil under wetting and gravitational loading processes. In: E3S Web of Conferences, vol. 195. No.02012. EDP Open (2020)

10. Van Genuchten, M.T.: A closed-form equation for predicting the hydraulic conductivity of unsaturated soils. Soil Sci. Soc. Am. J. **44**(5), 892–898 (1980)
11. Arya, L.M., Paris, J.F.: A physicoempirical model to predict the soil moisture characteristic from particle-size distribution and bulk density data. Soil Sci. Soc. Am. J. **45**(6), 1023–1030 (1981)

Assessing the Potential Impact of 1.5 °C Global Warming on the Local Response of a Pyroclastic Cover Susceptible to Shallow Landslides

Marialaura Tartaglia[1](✉), Marianna Pirone[2], Alfredo Reder[3], Guido Rianna[3], and Gianfranco Urciuoli[2]

[1] Formerly DICEA, Department of Civil, Architectural and Environmental Engineering, Università degli Studi di Napoli Federico II, via Claudio 21, 80125 Napoli, Italy
marialaura.tartaglia@unina.it

[2] DICEA, Department of Civil, Architectural and Environmental Engineering, Università degli Studi di Napoli Federico II, via Claudio 21, 80125 Napoli, Italy

[3] Regional Models and Geo-Hydrological Impacts (REMHI) Division, Fondazione Centro, Euro-Mediterraneo sui Cambiamenti Climatici, via T. A. Edison, 81100 Caserta, Italy

Abstract. In the Campania region (southern Italy), shallow landslides frequently involve steep slopes covered by loose pyroclastic soil in unsaturated conditions, causing several fatalities. For these phenomena, the slope response to heavy rainfall depends on the hydraulic state within the soil cover before the rainstorm. A clear indicator of such a state is the water storage (WS, the mean value of the integral of the water content in a unit base soil column). Hence, a threshold in terms of WS may be used in physically based predictive models. This work illustrates a framework to analyse the potential changes, due to global warming, in weather patterns regulating the slope atmosphere interaction (namely, precipitation and temperature) and the associated changes in WS, in pyroclastic slope susceptible to shallow landslides. A pyroclastic slope monitored for two years, placed on Mt. Faito (Lattari Mts. in Southern Italy) close to ancient flow-like landslides has been adopted as the pilot site. The slope hydraulic behaviour under the future temporal horizon and reference period has been modelled by Finite Element (FE) code. The future temporal horizon is identified assuming a fixed increase in global warming: in the specific, 1.5 °C defined in the Paris Agreement (2015) to preserve human and ecosystem livelihood on the Earth. The variations in WS are compared between the 1.5 °C scenario and the reference period 1981–2010. For the investigated case, the results of analyses under an "optimistic" global scenario, show that the number of events potentially leading to slope instability could not be affected significantly.

Keywords: Climate change · shallow landslides · pyroclastic slope

1 Introduction

The Campania region (southern Italy) has been frequently hit by shallow landslides involving steep slopes covered by pyroclastic soil, usually in unsaturated conditions [1–3]. It is well known that the hydraulic state in the subsoil before the critical event

A. Ferrari et al. (Eds.): CNRIG 2023, SSGG, pp. 741–749, 2023.
https://doi.org/10.1007/978-3-031-34761-0_89

represents a predisposing factor of the shallow landslide. In this regard, the water storage (WS) within the loose soil cover (the mean value of the integral of the water content in a unit base soil column) is a clear indicator of the hydraulic condition in the subsoil. Consequently, a threshold in terms of WS may be used in a physically based early warning system [4]. Nowadays, the need to improve the current tool predicting shallow landslide initiation is increased because the expected variations in weather patterns due to climate change could significantly affect the frequency of these phenomena.

In light of these considerations, the present study introduces a predictive numerical model devoted to assessing how the expected variations in weather patterns induced by climate change can affect the hydrological state of a pyroclastic slope susceptible to shallow landslides, assuming as a "proxy" the WS. The predictive model was applied to a pilot site placed on Mt. Faito, in the municipality of Castellammare di Stabia (Na) (Lattari Mts. in Southern Italy) close to ancient flow-like landslides. At the pilot site, wide geotechnical characterization and field data were available by past studies [5, 6].

As the approach aims at comparing the slope hydrological behaviour of the Lattari Mts. study area between a future time horizon and the reference one, two different weather patterns were applied at the upper boundary of the model. According to the World Meteorological Organization, the two-time spans were assumed to be 30 years long to properly account for interannual variability and limit the presence of statistically significant trends in datasets affecting the assumption of processing homogeneous series. Specifically, this study selected 1981–2010 as the reference time span and the 30-year period under which an increase in global warming is ~1.5 °C as the future time horizon. The two sets of weather variables were obtained by exploiting the approach implemented in the IV update of the PESETA Project (Projection of Economic impacts of climate change in Sectors of the European Union based on bottom-up Analysis) carried out by European Joint Research Centre on behalf of the European Commission [7].

2 The Study Area

The Lattari Mts. Study area is located (~15–30 km) near the Somma Vesuvius and Campi Flegrei volcanic complexes. The loose shallow cover consists of pyroclastic products (ash and pumice) from the Plinian eruption of 79 CE of Vesuvius resting on fractured limestone. From the shallowest to the deepest soil, the stratigraphic sequence of the study area consists of the following layers [5]: the topsoil (A1 and A2) classified as silty sand with gravel; layer B as sandy gravel; layer C1 and layer C2 as sandy silt, although C2 is finer than C1. Several undisturbed samples were taken during the 1998 and later 2016 investigation campaigns to investigate the physical and mechanical properties of the different soil layers. Table 1 summarizes all obtained mean physical-mechanical properties.

2.1 The Pilot Site

The pilot site (40°40′32.29″N, 14°28′23.35″E) located on the northern slope of Mt. Faito, in the municipality of Castellammare di Stabia (Naples), at approximately 850 m a.s.l., was chosen as representative of the geological and geotechnical conditions of the Lattari Mts.. It was instrumented to monitor matric suction and volumetric water content in the pyroclastic cover and meteorological conditions. In particular, a total of 40 tensiometers and 42 TDR probes were installed in situ at different depths along with a weather station [6]. Then, constant head tests, evaporation and drying tests on several undisturbed samples were carried out at the laboratory. The average value of the saturated hydraulic conductivity for each soil is reported in Table 1.

Table 1. Physical-hydraulic-mechanical properties. Legend: γ_d: dry soil unit weight; n: soil porosity; φ'_{cvc}: friction angle at the critical state; c′: drained cohesion; k_{sat} saturated hydraulic conductivity.

Soil	γ_d [kN/m^3]	n [-]	φ'_{cvc} [°]	c′ [kPa]	k_{sat} [m/s]
A1	8.80	0.65	38.40	0	1.00E−06
A2	8.10	0.72	38.40	0	6.35E−07
B	4.80	0.80	41.00	0	1.00E−03
C1	7.34	0.66	35.40	0	4.37E−07
C2	14.18	0.46	35.00	5	2.90E−08

3 Numerical Modelling

The hydrologic response of the pyroclastic cover to a climate scenario was simulated via the FEM through the code SEEP/W [8], assuming the soil skeleton to be rigid. It is worth noting that the hydraulic modelling of infiltration phenomena neglecting the soil volume changes correctly takes into account the amount of water stored in the soil sample, as proven by [9].

3.1 Geometry and Soil Properties

The numerical slope model reproduced a longitudinal profile of the pilot site, reconstructed by stratigraphic logs and ERT survey (Fig. 1). A mesh with an element size equal to 5 cm in the first meter of the cover and equal to 10 cm in the second half was adopted. The soil water retention curves of the different soil layers were obtained by coupling site measurements of suction and volumetric water content collected at the same depths at the pilot site [6]. By applying the Van Genuchten model to the experimental points, the soil retention curves for all the lithotypes were obtained (Fig. 2a). For the B soil, the water retention curve was characterized by first and second porosities; therefore, it was modelled through a bimodal function obtained by superposing two Van Genuchten-type functions. Finally, the hydraulic conductivity curves of all the soils were obtained by the modified Mualem–Van Genuchten function by using parameters determined for soil water retention curves (Fig. 2b).

3.2 Model Conditions

Starting from an initial distribution of pore water pressures in agreement with the values measured at the end of autumn, the slope response to both the reference period and future time horizon predicting 1.5 °C global warming was evaluated. It was applied at the upper boundary: the net rainfall, determined by subtracting the rate of evapotranspiration calculated using the Penman-Monteith formula from rainfall. 30-years data for rainfall and temperature adopted to determine evapotranspiration of the two scenarios simulated in the present study come from climate projections. The lateral and lower boundaries were modelled as a waterproof surface with the seepage face review.

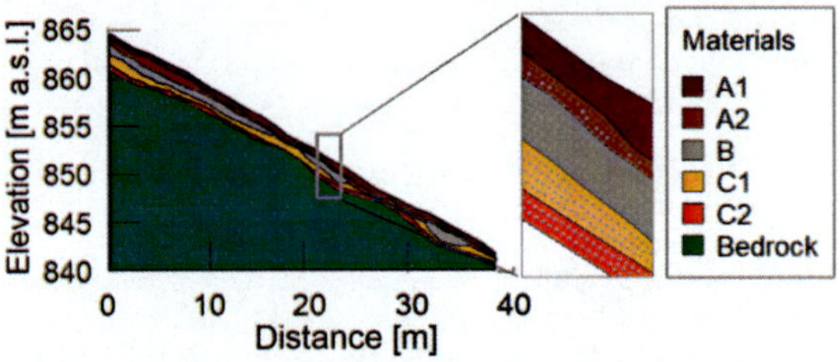

Fig. 1. Slope model with the indication of element size of the mesh and the stratigraphic profile.

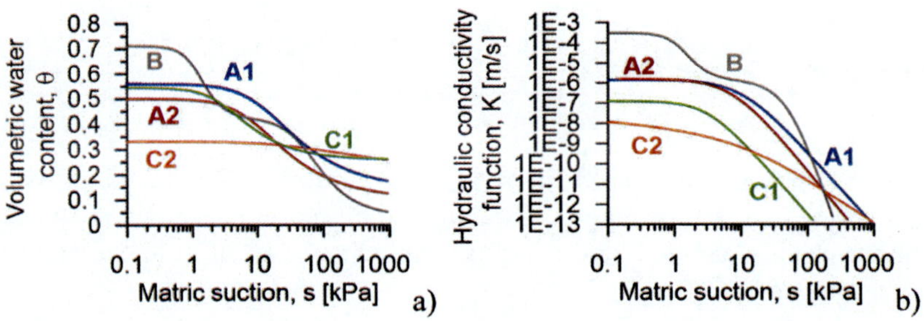

Fig. 2. Hydraulic characteristics of A1, A2, C1, B and C2 soils: a) water-soil retention curves; b) hydraulic conductivity curves.

3.3 Climatic Scenarios

The simulation chain usually adopted to provide climate projections for impact studies includes several elements. First, Integrated Assessment Models (IAMs) estimate future concentrations in climate-altering gases. Global Climate Models (currently, spatial resolutions in the order of 50–80 km), GCM, forced by concentration scenarios, are numerical models able to reproduce atmospheric patterns at the global scale. Then, Regional Climate Models (RCMs) (spatial resolutions in the order of 10–15 km with cutting-edge experiments on limited areas or for limited periods of 1–3 km) allow downscaling simulations at the regional scale. Finally, under the strong assumption that such errors can be assumed as systematic, statistical methods, known as bias correction (BC) approaches, are implemented to reduce the error by matching observed and modelled distribution functions. To account for the different sources of uncertainty, multi-model ensembles of climate projections (at global and/or regional scale) are usually adopted. At the regional level, CORDEX represents the most important initiative to provide localized climate data. The European branch of the CORDEX initiative is acknowledged as EURO-CORDEX, where several climate simulations are made freely available by research institutions up to the spatial resolution of 12 km (0.11°). Among these ones, [10] selects as input for the PESETA IV project, a sub-sample of climate simulations able to return an exhaustive overview of the spread characterizing future assessments. This sub-sample was defined assuming the Equilibrium Climate Sensitivity (ECS) value as a proxy to characterize the response to the increase in climate-altering gases. ECS can be defined as the simulated increase in the temperature calculated in scenario simulations where an increase in the concentration of climate-altering gases is assumed to occur instantly. In the PESETA IV framework, the potential impacts of Climate Change on the Europe economy and population in several sectors have been assessed by considering three potential increases in global warming (i.e., 1.5 °C, 2 °C and 3 °C) assuming, for 1981–2010, an already reached increase of 0.7 °C compared to the preindustrial era. Operatively, for each GCM in the subset, it is identified the year for which further global warming (0.8 °C, 1.3 °C and 2.3 °C for the three scenarios) is attained, and the 30 years around such year are considered for impact analysis.

In the present investigation, the first results by selecting a single climate simulation chain and 1.5 °C global warming (+0.8 °C respect to period 1981–2010) are shown to illustrate the analysis framework. Specifically, the climate simulation chain consists of the GCM CNRM-CERFACS-CM5 (ECS = 3.3 °C, being the nearest to the mean value = 3.2 °C), downscaled with the RCM CLMcom-CLM-CCLM4-8-17. The acronyms return information about the institutions that run the simulation and the name of the adopted model (GCM/RCM). A further advantage of using climate data from the PESETA IV project is that precipitation and temperature data series provided by such a simulation chain are still bias-corrected [10] using the technique developed by [11] against the gridded observational dataset EOBSv10 [12]. For this study, the nine grid points surrounding the slopes of interest have been considered. Of course, such data spatialisation could entail a reduction in the extreme values if compared to the data provided at point scale by the single weather station.

Figure 3a reports the mean monthly temperature values over the reference period and 1.5 °C scenario. The increase recognised at the global level between the two periods (+0.8 °C) is quite consistent with those returned at the local level (+0.7 °C), with a minimum in October (+0.3 °C) and the maximum values in the summer season (June and July with +1.0 °C and + 1.3 °C respectively). Limited increases in crop evapotranspiration, driven by temperature growths, are returned in Fig. 3b. Finally, for what concerns the cumulative monthly precipitation values (Fig. 3c), a small decrease is projected at the annual scale (about 10 mm). At the same time, no clear patterns can be identified on a monthly scale. However, by looking at the mean values of cumulative rainfall, for 1.5 °C scenario significant decreases in March and April (the ratio between the values for 1.5 °C and the reference period are 0.8, 0.6), a sharp increase in August (1.7) can be appreciated. Otherwise, in the other months, the variations range between ±10%. Not substantial variations in spread (summarised by standard deviation) are recognisable for all the variables.

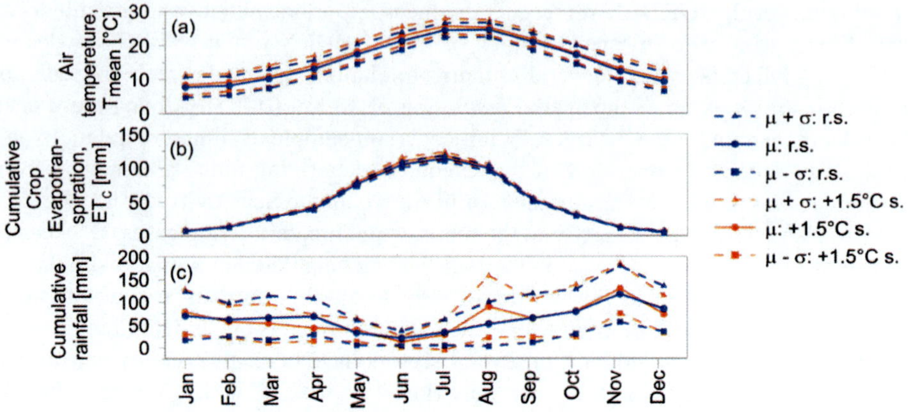

Fig. 3. Mean monthly values averaged on 30 years of temperature (a), cumulative evapotranspiration (b), and cumulative rainfall (c). The values determined by adding and subtracting the standard deviation (σ) from the average (μ) are also overlapped on all the charts (crossed lines). r.s.: reference scenario; +1.5 °C s.: +1.5 °C global warming scenario.

3.4 Results

Figure 4 reports the mean monthly values of maximum daily precipitation and WS determined over 30 years for both the two-time spans by numerical FE model discussed at Sects. 3.1 and 3.2. The variations in monthly values of maximum daily precipitation range between ±10% for most months (Fig. 4a), with a significant increase of 60% in August and a decrease of 30% in April. About the mean monthly values of WS, a mean annual reduction of ~45 mm in the 1.5 °C scenario is assessed due to a reduction of cumulative yearly precipitation and an increase of evaporative demand (Fig. 4b). On the monthly scale, the reductions range between 30 mm in February and November and 60 mm in June.

Summing up, the mean value of WS calculated in the reference period is higher than those corresponding to the +1.5 °C scenario over the entire year, even if the maximum daily rainfall is predicted to increase in August. Anyway, this type of elaboration can provide only some information on the expected trend of WS compared to the reference period. To account for the effective impact of global warming on the frequency of the shallow landslide occurrence, it is necessary to analyse the individual values assumed by WS over 30 years for both periods. To fulfil this purpose, Fig. 5 displays the daily precipitation value (>40 mm/d) and the associated WS over the two time periods. The threshold of 40 mm has been assumed as the minimum past daily rainfall that triggered landslides in the reference geological context amounted to 50 mm. It is clear how an increase in intense daily rainfall over the future time horizon is recognisable (about 27 vs 17) but with a substantial decrease in the mean WS. In particular, the number of intense rainfalls increases in the dry period. It can be explained by considering several insights from previous elaborations. On the one hand, the increase in temperature and then in evapotranspiration entails lower WS (on average); on the other hand, the substantial increase in August precipitation associated to low WS values.

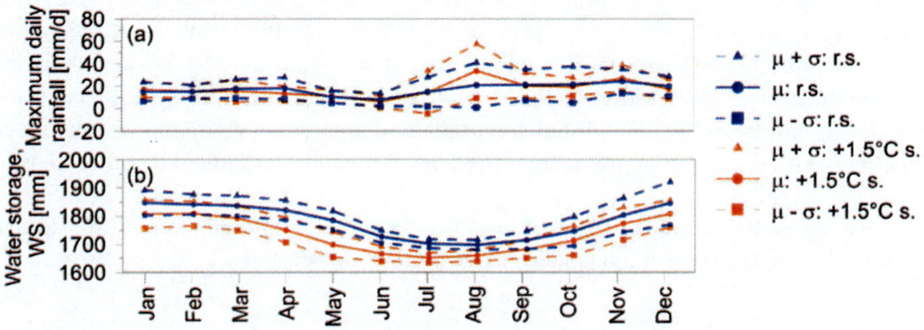

Fig. 4. Mean monthly values averaged on 30 years of maximum daily rainfall (a) and water storage WS determined by numerical FEM analyses (b). The values determined by adding and subtracting the standard deviation (σ) from the average (μ) are also overlapped on all the charts (crossed lines). r.s.: reference scenario; +1.5 °C s.: +1.5 °C global warming scenario.

About the critical $WS_{critical}$ for shallow landslide activation, a value 5–10% higher than the steady value (WWS, wet threshold), corresponding to the state above which water begins to drain outwards from the bottom of the slope, has been determined by back analysing past shallow landslides occurred at the Lattari Mts. by [4]. For the slope analysed, the WWS is 1800 mm; thus, the critical value may likely be equal to 1890 mm. By looking at Fig. 5, both scenarios exceed or overlap the threshold three times by pointing out that 1.5 °C global warming could not make the current situation worse. Hereafter, slope stability analyses must be carried out to associate a value of safety factor to WS and, consequently, to check the critical threshold assumed equal to 1890 mm.

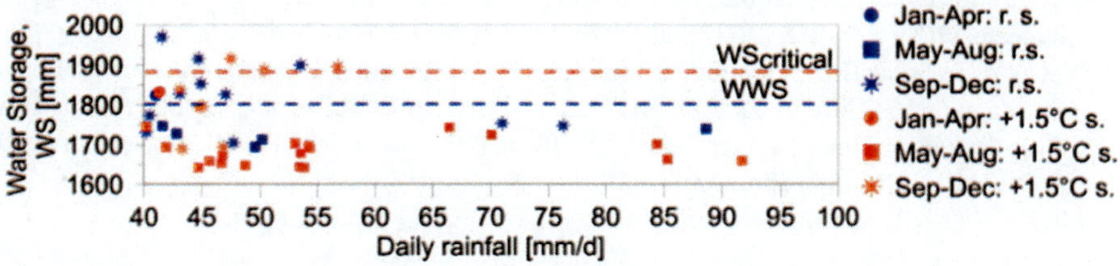

Fig. 5. Daily rainfall higher than 40 mm/d vs WS determined by FE model. The data are plotted by different symbols according to the season during which these occur. r.s.: reference scenario; + 1.5 °C s.: +1.5 °C global warming scenario.

4 Conclusions

The investigation aims at illustrating a framework to assess the potential impacts of climate change on shallow landslide triggering in the geological context of the Lattari Mts. Under the future scenario derived by 1.5 °C global warming, the local variations are quite limited. The main variations could refer to a decrease in the average WS values, an increase in the frequency and in the severity of heavy rainfall events limited to several months (August). Furthermore, the 1.5 °C scenario is considered very optimistic and different sources of uncertainties are disregarded in the investigation. Further research considering more pessimistic global scenarios and a properly estimation of the spread associated with such results are needed to identify the most suitable adaptation options.

Acknowledgements. We thank the EURO-CORDEX initiative and dr. Dosio (JRC) to make available climate data (Sect. 3.3) exploited in the present investigation.

References

1. Pirone, M., Papa, R., Nicotera, M.V., Urciuoli, G.: Analysis of Safety Factor in unsaturated pyroclastic slope. In: Landslides and Engineered Slopes Experience. Theory and Practice, vol. 3, pp. 1647–1654. CRC Press (2016)
2. Cascini, L., Sorbino, G., Cuomo, S., Ferlisi, S.: Seasonal effects of rainfall on the shallow pyroclastic deposits of the Campania region (southern Italy). Landslides **11**(5), 779–792 (2014). https://doi.org/10.1007/s10346-013-0395-3
3. Comegna, L., Damiano, E., Greco, R., Guida, A., Olivares, L., Picarelli, L.: Field hydrological monitoring of a sloping shallow pyroclastic deposit. Can Geotech. J. **53**, 1125–1137 (2016). https://doi.org/10.1139/cgj-2015-0344
4. Tartaglia, M.: Triggering of meteo-induced flow-like landslides in unsaturated soil: Pozzano and Pimonte case histories (Campania region), University of Naples Fedrico II (2022)
5. Forte, G., Pirone, M., Santo, A., Nicotera, M.V., Urciuoli, G.: Triggering and predisposing factors for flow-like landslides in pyroclastic soils: the case study of the Lattari Mts. (southern Italy). Eng. Geol. **257**, 105137 (2019)
6. Dias, A.S., Pirone, M., Nicotera, M.V., Urciuoli, G.: Hydraulic hysteresis of natural pyroclastic soils in partially saturated conditions: experimental investigation and modelling. Acta Geotech. **17**(3), 837–855 (2021). https://doi.org/10.1007/s11440-021-01273-y
7. Feyen, L., Ciscar, J.C., Gosling, S., Ibarreta, D., Soria, A. (eds.): Climate change impacts and adaptation in Europe. JRC PESETA IV final report. EUR 30180EN, Publications Office of the European Union, Luxembourg (2020). ISBN 978-92-76-18123-1, https://doi.org/10.2760/171121, JRC119178
8. SEEP/W (2012) Seepage Modeling with SEEP/W. GEO-SLOPE Int. Ltd.
9. Greco, R., Comegna, L., Damiano, E., Guida, A., Olivares, L., Picarelli, L.: Hydrological modelling of a slope covered with shallow pyroclastic deposits from field monitoring data. Hydrol. Earth Syst. Sci. **17**, 4001–4013 (2013). https://doi.org/10.5194/hess-17-4001-2013
10. Dosio, A.: Mean and extreme climate in Europe under 1.5, 2, and 3 °C global warming, EUR 30194 EN, Publications Office of the European Union, Luxembourg (2020). ISBN 978-92-76-18430-0, https://doi.org/10.2760/826427, JRC120574

11. Piani, C., et al.: Statistical bias correction of global simulated daily precipitation and temperature for the application of hydrological models. J. Hydrol. **395**(3–4), 199–215 (2010)
12. Haylock, M.R., Hofstra, N., Klein Tank, A.M.G., Klok, E.J., Jones, P.D., New, M.: A European daily high-resolution gridded data set of surface temperature and precipitation for 1950–2006. J. Geophys. Res. **113**, D20119 (2008)

Development of Sustainable Approach for Coastal Erosion Risk Mitigation

Mariano Tenuta[1], Stefania Lirer[1](✉), Rosanna De Rosa[2], Paola Donato[2], and Rocco Dominici[2]

[1] Department of Sustainability Engineering, University Guglielmo Marconi, Rome, Italy
s.lirer@unimarconi.it

[2] Department of Biology, Ecology and Earth Sciences, University of Calabria, Rende, Italy

Abstract. Sustainable coastal land management can be achieved by identifying and safeguarding appropriate sediment sources, so-called '*strategic sediment reserves*' that have the capacity to increase coastal resilience safeguarding the sediment balance. This can be achieved through an in-depth large-scale study of sediment production/distribution processes - over the short and long term - together with an innovative monitoring system based on the processing of satellite and aerial images. This approach has been applied at an area of the Puglia region in order to verify its effectiveness as sustainable approach for coastal erosion risk mitigation. A multi-parameter hydrodynamic model (EPM), developed in a GIS environment for the estimation of sediments production, transport and deposition, has been applied at the basin of the Ofanto River which is now considered the main cause of the deficit in sediment balance of the coastal area. The use of the EPM model, properly integrated with the results of a wide experimental campaign, has made it possible to identify the priority areas in the production of the sediment that feeds the coastal area. This kind of approach can be very useful for assessing the impact of anthropic activities (e.g. construction of dams, agricultural cultivations) in the retreat of coastline and for developing planning strategies aimed to the strengthening coastal resilience.

Keywords: coastal erosion · sediment balance · grain size distribution

1 Introduction

1.1 Coastal Erosion

Coastal retreat is a global phenomenon caused by an imbalance between the processes of sediment supply and deposition.

The causes are related to the interaction of natural processes (e.g. sea level variations, natural gradients in longshore sediment transport, storms, the presence of submarine canyons near the coast, sub-sea erosion) and anthropogenic actions (e.g. construction of dams, ports, coastal protection works and sediment extraction processes). These variables, interacting sometimes in very complex ways, modify the whole sediment balance and coastal ecosystems. In addition, the average sea level on a global scale is constantly

A. Ferrari et al. (Eds.): CNRIG 2023, SSGG, pp. 750–757, 2023.
https://doi.org/10.1007/978-3-031-34761-0_90

rising and, with global warming, the annual rate of rise is expected to increase by two to five times the current rate. By 2100, sea levels are expected to be about 50 cm higher than today [1]. Recently the concept of *coastal resilience* has been introduced [2], depending on two key factors: sediments and space for coastal processes. Sustainable coastal land management can be achieved by identifying and safeguarding appropriate sediment sources, so-called *'strategic sediment reserves'* that have the capacity to increase coastal resilience and thus safeguard the sediment balance.

In Italy, Puglia region is the territory most affected by coastal erosion (about 25% of the coastline). Among the various causes (construction of harbor structures and defense works and to the growing urbanization of the coastal area), the drastic reduction of sediments transported by the Ofanto river (it is approximately 170 km long and its drainage basin has an area of approximately 2790 km^2 [3]) to the sea has caused over the years a significant deficit in the sediment balance in the coastal area. Futhermore, more than 20 dams have been built along the river since the 1950s, causing a further reduction in the sediment load reaching the sea [4–6].

In order to identify the *'strategic sediment reserves'* in the Ofanto basin, a multi-parameter hydrodynamic model developed in a GIS environment (for the estimation of sediment production, transport and deposition) has been adopted and integrated with the results of an extensive experimental campaign carried out on the sediment samples retrieved within the Ofanto basin and along the coast. The Erosion Potential Method (EPM), properly calibrated with the experimental data and applied in Geographic Information System (GIS) environment, has made it possible to identify in the Ofanto basin the priority areas in the production of the sediment that feeds the stretch of coast affected by erosion processes.

2 The Erosion Potential Method (EPM)

Among the several semi-quantitative methods developed for the assessment the amount of sediments production, the Erosion Potential Method is the most used in the Mediterranean area. EPM model allows to analyze the erosion severity estimating the total annual sediment yield - for a given basin - through the integration of several factors: climatic condition (average temperature and precipitation), soil erodibility (land use, lithology and basin morphology) and physical features of the basin (surface, average slope, perimeter, average elevation and hydrographic network). The model is generally applied in Geographic Information System (GIS) environment [7–11].

According to this model, the average annual production of sediments (W) (in m^3/y) can be determined using the following equation:

$$W = T \cdot h \cdot \pi \cdot F \cdot Z^{3/2} \tag{1}$$

where h is the mean annual precipitation (mm/y), F is the watershed area (km^2), Z is the erosion coefficient and T is the temperature coefficient calculated from the mean annual air temperature, t (°C). The coefficient Z - of a sub-basin – is quantified by means of the following equation:

$$Z = YX\left(\Upsilon + \sqrt{Im}\right) \tag{2}$$

where Y is the coefficient of the rock/soil resistance, X is the land use coefficient, ϒ is the coefficient of type and extent of erosion and I_m is the average ground slope [12].

The EPM model has been applied to the case study of the Ofanto basin [13], computing the average annual specific production of sediments (W, Eq. 1) with an innovative approach that also integrates the petro-graphic, physical, mineralogical and geochemical properties of the sediment samples [14].

3 Experimental Study

3.1 Soils Sampling and Mineralogical Properties

The experimental study has been carried out with the aim to deeply analyze the mineralogical and physical properties of the soils affecting the whole sediment balance of the Ofanto area (sub-basins and coastline).

Some samples were carried out in 22 sites located along the main course of Ofanto river and its tributaries (Fig. 1), and in 8 sites located along the coastline (Manfredonia and Torre a Mare, Fig. 1).

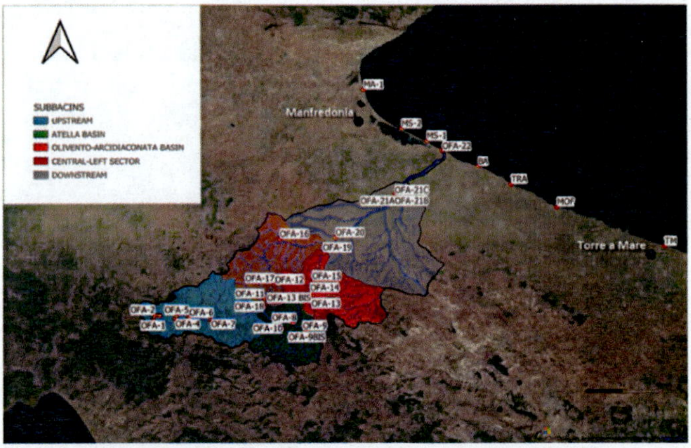

Fig. 1. Ofanto sub-basins and sampling campaign.

The isotopic study [15] revealed that sediments of the Ofanto river come from two volcanic sources (Campania region and Vulture area). The major components of the samples retrived from the Ofanto River and its tributaries (0.25–0.5 mm) are single crystal sand metamorphic, sedimentary and volcanic fine-grained rock fragments. Single crystals are represented by heavy minerals of volcanic origin (mainly pyroxene, and subordinate melanite garnet and amphibole), quartz, feldspar, micas and minor opaque heavy minerals [13, 15]. The results of modal composition indicate that the volcanic component is mainly represented by loose crystals of pyroxene, amphibole and garnet.

3.2 Grain Size Distribution and Physical Properties

Grain size distribution of the sediments is generally affected by many factors such as the distance from the source area, the (mineralogical) nature of the area source, the topography and the transport mechanisms [16]. The hydrodynamic energy of the water in the river makes a natural sorting of the grains according to their weight and size: the flow energy separates the heavy minerals from the light ones, producing heavy-mineral black-sand concentrates named "*placer*" [17]. Placers, easily identifiable by their black color, are located in many sites (from the upstreamsub-basin to the beaches, Fig. 2b) and have been used as "geo-marker" in the sediments balance analysis.

In the upstream and Atella sub-basins (Fig. 2a, Table 1), the retrieved samples can be classified as sandy gravels, while the samples of the Arcidiaconata-Olivento basin and of the central-left sector mainly consist of gravelly sands (Fig. 2a, Table 1). The beaches located at north and south of the river's mouth are made of fine and uniform sands (Fig. 2a, Table 1), characterized by the presence of bars with accumulations of heavy-dark minerals (placers).

Table 1. Mean physical properties of retrieved samples.

Sub-basin (Fig. 1)	Upstream	Atella	Olivento-Arcidiaconata	Central	Coast
D_{50} (mm)	3,43	3,73	0,73	0,49	0.20
C_u (D_{50}/D_{10})	4,3	7,3	2,8	2,9	1,2
G_s	2.65–2.89				

The evolution low of the placers grain size distribution along the Ofanto basin (from the upstream area to the costline), combined with the knowledge of their mineralogical properties, has been applied to all sediments generated within the basin. With this approach, it is assumed, for instance, that the very coarse sand fraction present in the Upstream and Atella basin (and respectively the coarse sand fraction present in the central-left sector and Olivento-Arcidiaconata basin), will be reduced to a medium sand at the mouth of the Ofanto river.

This approach has been considered, as a first attempt, in the further implementation of the EPM [18, 22] in order to identify more accurately the areas of the Ofanto basin that directly affect the grain size properties of the Apulia coast samples. The identification of these areas (*sediments sources*), mainly responsible for the coastal deficit of sediments, is extremely relevant for the planning of mitigation interventions.

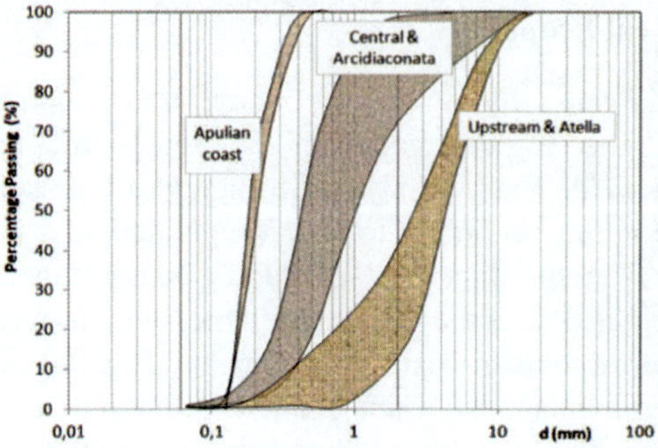

(a)

(b)

Fig. 2. Grain size distribution of the sediments (a) and picture of Manfredonia beaches (b).

4 The Application of EPM for a Sustainable Management of Coastal Areas

The application of the EPM model allows the identification of the strategic areas that have a greater impact in the feeding of a coastal area and, as a consequence, must be safeguarded to avoid an increase of the erosion process. For instance, the construction of a large capacity dam can significantly modify the amount of sediment load transported from the upstream side of the basin (e.g. [19, 20]) to the coastal areas.

Since the 1950s, in the Ofanto basin, twenty dams have been built along the river. The start up of three of them, Conza della Campania, Lampeggiano, and Saetta dams (Fig. 3a) is identified at the beginning of the 1990s. As can be observed in Fig. 3a, the larger drainage area pertains to the Conza dam (upstream sub-basin). The EPM model, applied to the sub-basin of the Conza dam, quantified a loss of volume of sediments of about 250,000 m^3/y linked to the start up of the Conza dam (Fig. 3b). This volume was inevitably subtracted to the global sediment balance also affecting the coastal stretch.

This hypothesis has been verified through the analysis of the coastal retreat velocity (computed in a selected section AA nearby the river's mouth, Fig. 4a) carried out from the processing of the available digital orthophotos of the area [GIS data, 21]. It can be noted that (Fig. 4b) the start-up of the Conza dam (certainly the most relevant in those years) marks a suddenly increase in the coastal erosion process. After an immediate strong increase (between the 1990 and 1994), the coastline slowed its retreat but maintains an almost constant velocity $v \approx 10$ m/y.

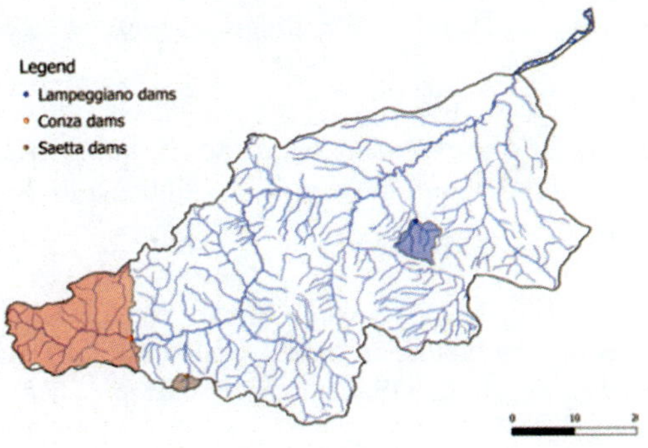

(a)

Dam	Year	Area of drainage basin (km^2)	Sediments production (m^3/y)
Conza	1992	252	251236
Lampeggiano	1993	32	6638
Saetta	1999	9	5180

Land use (eqs. 1,2)	$0.05 < X < 1.0$
Soil resistance (eqs.1,2)	$0.67 < Y < 1.99$
Erosion (eqs. 1,2)	$0.10 < \Upsilon < 1$
Temperature (eq. 1)	$T = 1.23$

(b)

Fig. 3. Location of the three dams built in '90s within the Ofanto basin (a), and volume of sediments computed for their drainage basins (b).

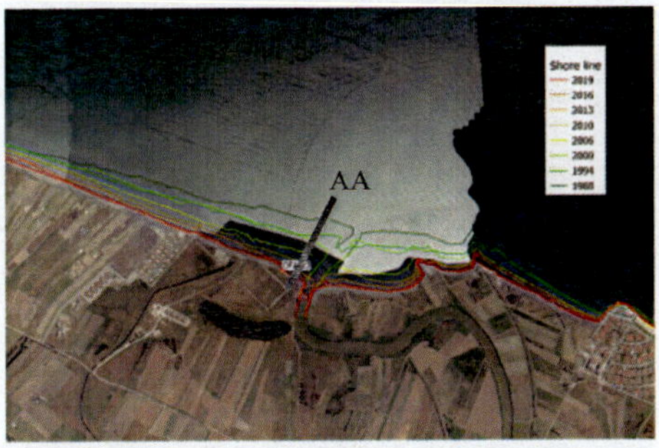

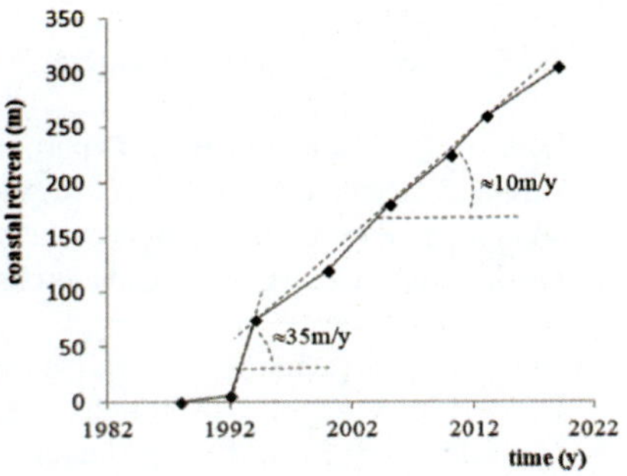

(a) (b)

Fig. 4. (a) Evolution of the coastline of Puglia region (at the river's mouth) obtained from the processing of the digital orthophotos of the area (available at the following link: http://www.pcn.minambiente.it/mattm/servizio-wms/), and (b) velocity of the coastal retreat.

5 Conclusion

In Puglia region the deficit of sediments load transported by the Ofanto River represents the main cause of the erosion processes that affects the coastline. In order to mitigate the effects on coastal areas, it is extremely important to identify which strategic productive areas (*sediments sources*) within the basin should be protected.

The application of the Erosion Potential model (EPM) to the Ofanto basin, calibrated through experimental data collected on the sediment samples, made it possible to quantify the average annual specific production of sediments (W), its spatial distribution and its evolution over time. The PyQGIS YES plug-in allowed the use of EPM model in GIS environment [22]. In such a way, all the available information within the area (administrative limits, hydrographic network, geomorphologic and hydrological assets, lithological maps, land use data, experimental data obtained in specific sampling points, location of coastal defense works, evolution of shore line) can be integrated and managed in a very easily and effective way.

The use of an integrated web-gis platform allows identifying the *sediments sources*, supporting local authorities in making the right choices with regard to the most critical areas (with regard to urbanization works and/or land consumption) and sustainable solutions for coastal erosion risk mitigation.

Acknowledgements. The authors deeply acknowledge Marco Amanti, Mauro Roma, Martina Bussettini and Giovanni Braca of the Higher Institute for Protection and Environmental Research (ISPRA) for having supplied the rasters produced through the BIGBANG project and the shapefiles relating to the 1:100,000 scale geolithological map.

References

1. Cai, F., Su, X., Liu, J., Li, B., Lei, G.: Coastal erosion in China under the condition of global climate change and measures for its prevention. Prog. Nat. Sci. **19**(4), 415–426 (2009)
2. MATTM-Regioni: http://www.erosionecostiera.isprambiente.it/files/linee-guida-nazionali/TNEC_LineeGuidaerosionecostiera_2018.pdf 2018. Accessed 22 Nov 2022
3. AAVV: Studi propedeutici per la predisposizione del Piano Stralcio della Dinamica delle Coste In: Politecnico di Bari, Dipartimento di Ingegneria delle Acque e di Chimica, Bari. Technical Report per l'autorità di bacino della Puglia (2010)
4. Caldara, M.: Aspetti di Geologia ambientale e di morfologia costiera in alcuni tratti del litorale nord barese. Atti Convegno Cave e coste nel territorio del nord-barese. Geologia Suppl. **2**, 39–61 (1996)
5. Caldara, M., Centenaro, E., Mastronuzzi, G., Sansò, P., Sergio, A.: Features and present evolution of Apulian coast (Southern Italy). J. Coast. Res. Suppl. Issue **26**, 55–64 (1998)
6. Mastronuzzi, G., et al.: Palaeotsunami imprints along the coasts of the central Mediterranean Sea. In: Mastronuzzi, G., Pignatelli, C., Sansò, P., Milella, M., Selleri, G. (eds.) Proceedings of the Field Guide-2nd International Tsunami Field Symposium Ostuni: Puglia, Italy, p. 200 (2008)
7. Tangestani, M.H.: Comparison of EPM and PSIAC models in GIS for erosion and sediment yield assessment in a semi-arid environment: Afzar catchment, Fars Province, Iran. J. Asian Earth Sci. **27**, 585–597 (2006)
8. Tosic, R., Dragicevic, S.: Methodology update for determination of the erosion coefficient. Glas. Srp. Geogr. Društva **92**, 11–26 (2012)
9. Barmaki, M., Pazira, E., Hedayat, N.: Investigation of relationships among the environmental factors and water erosion changes using EPM model and GIS. Int. Res. J. Appl. Basic Sci. **3**, 945–949 (2012)
10. Noori, H., Siadatmousavi, M.S., Mojaradi, B.: Assessment of sediment yield using RS and GIS at two sub-basins of Dez Watershed, Iran. Int. Soil Water Conserv. Res. **4**, 199–206 (2016)
11. Aslam, B., Maqsoom, A., Alaloul, W., Musarat, M.A., Jabbar, T., Zafar, A.: Soil erosion susceptibility mapping using a GIS-based multi-criteria decision approach: case of district Chitral, Pakistan. Ain Shams Eng. J. **12**, 1637–1649 (2020)
12. Zemljic, M.: Calcul du debit solide – evaluation de la vegetation comme un des facteurs antierosifs. International Symposium Interpraevent, Villach, Austria (1971)
13. Tenuta, M., Donato, P., Dominici, R., De Rosa, R.: The influence of volcanic supply on the composition of modern river sands: the case study of the Ofanto river, Southern Italy. Geol. Soc. Spec. Publ. **520**(1), SP520-2021 (2021)

14. Dominici, R., Donato, P., Imbrogno, G., Lirer, S., Tenuta, M., De Rosa, R.: Multiparametric calibration of the Erosion Potential Method on petrographic, mineralogical and geochemical parameters. In: Congresso congiunto della Società Geologica Italiana (SGI) e della Società Italiana di Mineralogia e Petrologia (SIMP). Torino, p. 828 (2022)
15. Donato, P., De Rosa, R., Tenuta, M., Iovine, R.S., Totaro, F., D'Antonio, M.: Sr-Nd isotopic composition of pyroxenes as a provenance indicator of a double-volcanic source in sands of the Ofanto river (Southern Italy). Minerals **12**, 232 (2022)
16. Parthasarathy, P., et al.: Sediment dynamics and depositional environment of Coleroon river sediments, Tamil Nadu, southeast coast of India. J. Coast. Sci. **3**(2), 1–7 (2016)
17. Komar, P.D.: The entrainment, transport and sorting of heavy minerals by waves and currents. In: Mange, M.A., Wright, D.T. (eds.) Heavy Minerals in Use. Developments in Sedimentology Series, vol. 58, pp. 3–48. Elsevier, Amsterdam (2007)
18. Tenuta, M., Dominici, R., Donato, P., Imbrogno, G., Lirer, S., De Rosa, R.: Granulometric calibration of the Erosion Potential Method (EPM) model: estimation of sediment production that affects a coastal area. Rend. Online Soc. Geol. It. **59**, 89–96 (2022)
19. Boix-Fayos, C., De Vente, J., Martínez-Mena, M., Barberá, G., Castillo, V.: The impact of land use change and check-dams on catchment sediment yield. Hydrol. Process. **22**, 4922–4935 (2008)
20. De Santis, V., Caldara, M., Marsico, A., Capolongo, D., Pennetta, L.: Evolution of the Ofanto river delta from the 'Little Ice Age' to modern times: implications of large-scale synoptic patterns. Holocene **28**, 1948–1967 (2018)
21. MATTM: http://www.pcn.minambiente.it/mattm/. Accessed 22 Nov 2022
22. Dominici, R., Larosa, R., Viscomi, V., Mao, L., De Rosa, R., Cianflone, G.: Yield erosion sediment (YES): a PyQGIS plug-in for the sediments production calculation based on the erosion potential method. Geosciences **10**, 324 (2020)

Impact of Granular Masses on Sheltering Structures: Definition of the Initial Conditions for the Assessment of Impact Forces

Matteo Zerbi(✉), Pietro Marveggio, and Claudio di Prisco

Politecnico di Milano, Milan, Italy
matteo.zerbi@polimi.it

Abstract. In the engineering practice, in the design of sheltering structures against flow-like landslides, empirical formulas are still commonly employed to estimate the maximum force transmitted by the impacting mass to the structure. In recent years, the development of robust numerical codes, capable of dealing with large displacements, have allowed to simulate the dynamic interaction between flowing granular masses and obstacles and to derive new formulas for a more rational design of protective structures. Nowadays, the propagation phase and the impact are analyzed separately and the initial conditions (initial Froude Number and porosity) for the numerical simulation of the impact are the result of a previous propagation analysis. In this work, a theoretical discussion about such initial conditions is done, taking advantage of what predicted by a simplified rheology for dry granular materials, taking into consideration the role of porosity in affecting the maximum impact force transmitted to the sheltering structure to be designed.

Keywords: flowslides · impact · sheltering structures · rheology

1 Introduction

Flow-like landslides are often catastrophic phenomena, causing every year important social and economic losses. Their associated risk is mitigated by artificial barriers and obstacles aimed at changing the propagation phase, in order to avoid the impact with structures and infrastructures.

In the engineering practice, sheltering structures against flow-like landslides are mainly still designed by using pseudo-static approaches, according to which the maximum impact force is quasi-statically applied to the barrier [2]. Only recently, numerical approaches have been employed to study this problem [4, 12, 13, 20, 21]. In particular, the propagation phase and the impact are analyzed separately and the initial conditions (initial Froude number and porosity) for the numerical simulation of the impact are the result of a previous propagation analysis [15]. Very often, the propagation numerical analyses provide a mean value of velocity at the front and the height profile of the flowing mass. In contrast, the numerical results obtained by Calvetti et al. [3] have clearly shown that the maximum impact force severely depends not only on the Froude Number

A. Ferrari et al. (Eds.): CNRIG 2023, SSGG, pp. 758–765, 2023.
https://doi.org/10.1007/978-3-031-34761-0_91

and front inclination but also on the mass porosity (Sect. 2). This latter governs the nature of the impact phenomenon, since large values of porosity prevent the compression wave propagation and facilitate jet mechanical processes [7].

A proper selection of the initial conditions (in particular in terms of porosity) is mandatory to prevent an overestimation of impact forces and to make more sustainable the design of protection structures.

In this work, the authors have used the $\mu - n - I$ (or $\mu - e - I$) rheology [10, 11]. Indeed, despite not able to take into account of some aspects of granular materials behavior [14, 17], this rheology has become very popular in recent years in the granular flows community due to its simplicity. The $\mu - n - I$ rheology has been used by the authors to define a relation between mass flow height, Froude number and porosity mean values at steady state during landslide propagation, aimed at providing a theoretical tool for the rational use of both numerical codes and simplified formulas for the assessment of impact forces.

2 Role of Porosity in Affecting the Maximum Impact Force

The role of both impact porosity and Froude Number on the maximum dynamic impact force F has been recently numerically studied by Calvetti et al. [3]. The authors performed an extensive number of Discrete Element parametric simulations of granular mass impacting rigid obstacles and provided a design formula neglecting the static contribution, that is:

$$\mathrm{F} = \frac{1}{2}\rho_P g H^2 F^* = \frac{1}{2}\rho_P g H^2 \left(a_1 Fr_m Fr + a_2 Fr^2\right), \tag{1}$$

where Fr is the Froude number [1], ρ_P the particle density, g the gravity acceleration, H the height of the flow mass, a_1 and a_2 non-dimensional coefficients function of front inclination and Fr_m, being Fr_m a variable depending on compaction wave velocity u_m:

$$Fr_m = \frac{u_m}{\sqrt{gH}}. \tag{2}$$

According to their findings, the maximum impact force transmitted by the mass to the obstacle increases with the Froude number but decreases with the porosity, since, as put in evidence by both Redaelli et al. [18] and di Prisco et al. [7], the compression wave speed u_m significantly reduces with the increase in porosity n_0 (Fig. 1).

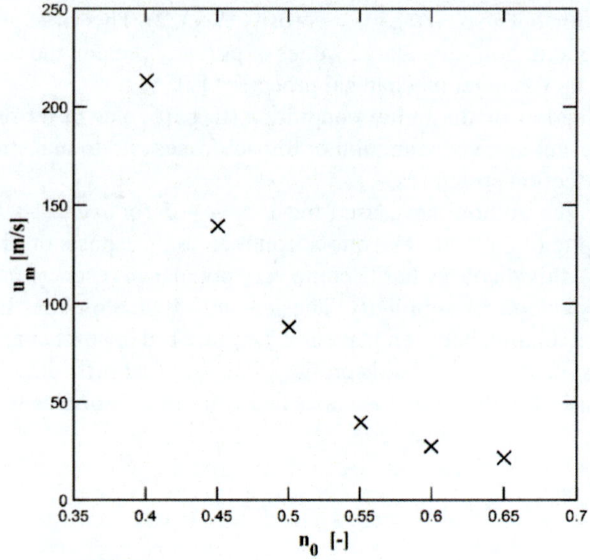

Fig. 1. Compression wave velocity as function of the porosity (from [18])

3 Theoretical Discussion

In this section, the authors use the $\mu - n - I$ rheology to obtain a relationship between mean value of porosity $\overline{n}$ and mean value of Froude Number $\overline{Fr}$. This result is obtained by extrapolating the relationship that can be derived at steady state for a sheet flow (Sects. 3.1 and 3.2). In Sect. 3.3, the influence of material grain size and flow mass height on this relationship is discussed for a given set of constitutive parameters.

3.1 Steady-State $\mu - I$ Solution for a Sheet Flow

The solution of a steady free surface flow down along an inclined plane is provided for a $\mu - I$ rheology by GDR MiDi [10].

According to µ-I rheology [10, 11], under simple shear conditions, the ratio $\mu(\mu = \tau/\sigma$, where τ and σ stand for shear and normal stresses, respectively) at steady state for a dry granular material is not constant but depends on a non-dimensional variable, the inertial number I ($I = d_P\dot{\gamma}\sqrt{\rho_P/\sigma}$, where d_P and ρ_P are the particle diameter and density, respectively and $\dot{\gamma}$ the shear rate):

$$\mu(I) = \mu_0 + \frac{\Delta\mu}{1 + \frac{I_0}{I}} \tag{3}$$

where μ_0 is the tangent of the friction angle at critical state (that is for $I = 0$, [22]) whereas $\Delta\mu$ and I_0 are two additional non-dimensional constitutive parameters, describing the dependence of μ on I (Fig. 2a). As far as the material porosity is concerned, da Cruz et al. [6], Fall et al. [9] and Pouliquen et al. [16] propose a linear dependence of porosity on I as follows:

$$n(I) = n_C + \Delta n I \tag{4}$$

where n_C is the material porosity at critical state (that is for $I = 0$, [22]) and Δn is an additional non-dimensional constitutive model parameter describing the dependence of n on I (Fig. 2b).

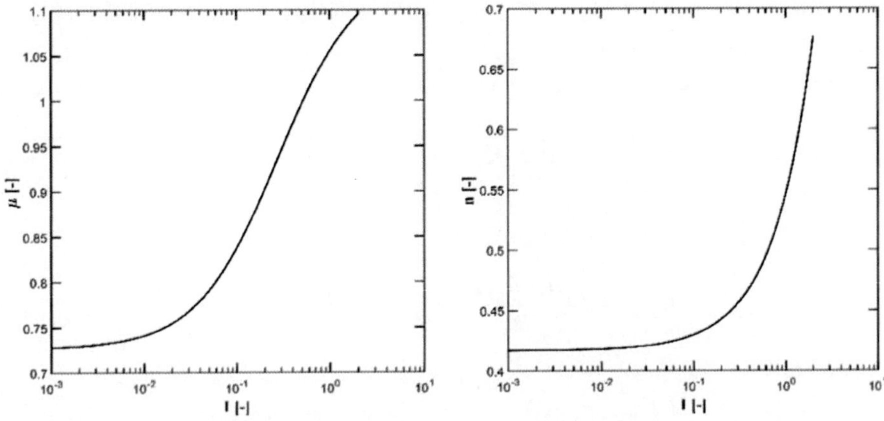

Fig. 2. Dependence of a) friction coefficient and b) porosity on the Inertial number

In a $\mu - I$ sheet flow, the velocity profile depends on the plane inclination θ and flow height H:

$$u(z, \theta, H) = \frac{2}{3}\frac{\mathcal{I}(\theta)}{d}\left(gH^3 \cos\theta\right)^{1/2}\left(1 - (1 - z/H)^{3/2}\right) \tag{5}$$

where

$$\mathcal{I}(\theta) = I_o \frac{\tan\theta - \mu_0}{\mu_0 + \Delta\mu - \tan\theta} \geq 0 \tag{6}$$

and g is the gravity acceleration.

Positive values of $\mathcal{I}(\theta)$ can be obtained only if $\mu_0 < \tan\theta < \mu_0 + \Delta\mu$, out of this range the system will decelerate ($\tan\theta < \mu_0$) or continuously accelerate ($\tan\theta > \mu_0 + \Delta\mu$).

3.2 Sheet Flow Mean Porosity

The relation between the mean porosity $\overline{n}$ and the Froude number mean value $\overline{Fr}$ is obtained by the authors in this section by calculating separately the dependence of $\overline{n}$ and $\overline{Fr}$ on θ and H.

The mean porosity $\overline{n}$ is defined (using Eq. 4) as:

$$\overline{n}(\theta) = \frac{\int_0^H n(z)dz}{H} = \frac{\int_0^H n_C + \Delta n I(z, \theta, H)dz}{H} \tag{7}$$

where $I(z, \theta, H)$ is obtained by solving the following implicit equation:

$$I(z,\theta,H) = d_P \dot{\gamma}(z,\theta,H)\sqrt{\frac{1}{(1-n_C-\Delta n I(z,\theta,H))g\cos\theta H\left(1-\frac{z}{H}\right)}} \tag{8}$$

where, from Eq. 5,

$$\dot{\gamma}(z,\theta,H) = \frac{du(z,\theta,H)}{dz} = \frac{2}{3}\frac{\mathcal{I}(\theta)}{d}\left(gH^3\cos\theta\right)^{1/2}\left(\frac{3}{2H}\left(1-\frac{z}{H}\right)^{1/2}\right) \tag{9}$$

Analogously, $\overline{Fr}$ is defined as

$$\overline{Fr}(\theta, H/d_P) = \frac{\overline{u}(\theta,H)}{\sqrt{gH}} \tag{10}$$

where, from Eq. 5,

$$\overline{u}(\theta,H) = \frac{\int_0^H u(z,\theta,H)dz}{H} = \frac{2}{5}\frac{\mathcal{I}(\theta)}{d}\left(gH^3\cos\theta\right)^{1/2} \tag{11}$$

Therefore

$$\overline{Fr}(\theta, H/d_P) = \frac{1}{5}\mathcal{I}(\theta)(\cos\theta)^{1/2}\frac{H}{d_P} \tag{12}$$

The two dependencies of both $\overline{n}$ and $\overline{Fr}$ on θ obtained for $H/d_P = 15$ and by using the constitutive parameters (from [5, 7, 9, 15]) listed in Table 1 are represented in Fig. 3a.

The numerical interpolation of Eqs. 7 and 12 allows to write the following explicit relation:

$$\overline{n} = n_c e^{\alpha \overline{Fr} d_p/H} \tag{13}$$

where $\alpha = 0.6394$ for the given constitutive parameters of Table 1 (Fig. 3b).

Table 1. Parameters employed for the μ–n–I rheology

$\mu_0[-]$	$\Delta\mu[-]$	$I_0[-]$	$n_C[-]$	$\Delta n[-]$
0.727	0.42	0.279	0.417	0.13

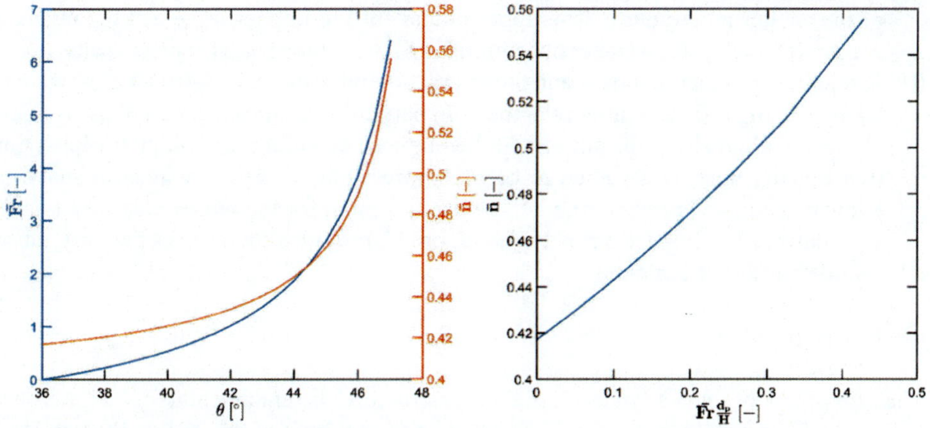

Fig. 3. a) Mean Froude number and porosity as function of the slope angle and b) relationship between mean porosity and Froude number.

3.3 Assessment of Initial Conditions

The theoretical results illustrated in Sect. 3.2 can be useful for performing numerical impact analyses as it follows:

(a) If the propagation phase is not simulated, by fixing the geometrical factors (height H of the landslide and slope angle θ) and average grain size d_P, the initial conditions for numerical impact analyses (Fr_0, n_0) can be set by using Eqs. 12 and 7, assuming that, during the propagation phase along the slope, the steady state conditions are reached. Analytical and numerical solutions [8] indicate that the propagation of an infinite granular mass down to an infinite incline is associated with a progressive acceleration of the mass, reaching asymptotically steady-state conditions. In contrast, if the flowing mass is finite, after an initial acceleration of the front, a subsequent deceleration/arrest takes place. When a granular material is accelerated, a progressive dilation is observed [19]. The steady state porosity is higher than the one associated with unsteady propagation. For these reasons, the initial conditions for the numerical test (Fr_0, n_0) are expected to satisfy the following conditions:

$$0 \leq Fr_0 \leq \overline{Fr}(\theta, H/d_P) \tag{14}$$

$$n_c \leq n_0 \leq \overline{n}(\theta) \tag{15}$$

From a practical point of view, considering that Calvetti et al. [3] (Sect. 2) established that the impact force transmitted by the flowing mass to the protective structure (i) increases with the impact Froude number Fr_0 and (ii) decreases with the impact porosity n_0, the use of the steady state velocity $\overline{Fr}$ will bring to a safe estimation of the impact force, whereas the use of the porosity $\overline{n}$ at steady state will bring to an unsafe estimation of the impact force. The use of an impact Froude number $Fr_0 \geq \overline{Fr}$ or an impact porosity $n_0 \leq n_c$ will bring in every case to an overestimation of the impact force and, as a consequence, to an over dimensioning of the sheltering structure. The upper-bound for the impact force can be obtained combining the steady-state Froude number $\overline{Fr}$ with the critical state porosity.

(b) Often, the propagation phase and the interaction with sheltering structures are simulated by using different numerical tools. In particular, the propagation phase is commonly simulated by using simplified rheologies, neglecting the volumetric behavior (the flowing mass is assumed to be incompressible). These propagation analyses allow to evaluate Froude number to be used as input for the numerical impact analyses but not the impact porosity. Equation 13 provides the impact porosity, once Froude number is assigned.

4 Conclusions

The assessment of impact forces on sheltering structures is fundamental for their rational design. In the last years, some works focused on the analysis of the impact phenomenon by using numerical codes and experimental apparatus. Popular is also the strategy of analysing numerically the propagation phase and using the results of these analyses to numerically simulate the impact phenomenon on rigid/deformable obstacles. According to this approach, crucial is defining the initial conditions for the impact phase, that very often are summarized in mean velocity of the flowing mass, height, front inclination. The absence of data concerning the mass porosity is a weakness point of these approaches, since porosity has demonstrated to severely affect the impact force. In this paper, a theoretical strategy to find a relationship between Froude number and porosity is introduced. This relationship puts in evidence the role of flow height and material grain size in governing the current state of the flowing mass. The theoretical observation is of great interest, in particular when laboratory flume tests are performed to interpret the mechanical response of real landslides.

References

1. Cabrera, M., Kailey, P., Bowman, E.T., Wu, W.: Effects of viscosity in granular flows simulated in a centrifugal acceleration field. In: Physical Modelling in Geotechnics, pp. 1075–1080. CRC Press (2018)
2. Calvetti, F.: Debris flow impact forces on rigid barriers: existing practice versus DEM numerical results. In: Proceedings of ICHONIC 2016, China, pp. 1–14 (2016)
3. Calvetti, F., di Prisco, C.G., Vairaktaris, E.: DEM assessment of impact forces of dry granular masses on rigid barriers. Acta Geotech. **12**(1), 129–144 (2016). https://doi.org/10.1007/s11440-016-0434-z
4. Ceccato, F., Redaelli, I., di Prisco, C., Simonini, P.: Impact forces of granular flow son rigid structures: comparison between discontinuous (DEM) and continuous (MPM) numerical approaches. Comput. Geotech. **103**, 201–217 (2018)
5. Cremonesi, M., Franci, A.: 3D regularized μ(I)-rheology for granular flows simulation. J. Comput. Phys. **378**, 257–277 (2019)
6. da Cruz, F., Emam, S., Prochnow, M., Roux, J., Chevoir, F.: Rheophysics of dense granular materials: discrete simulation of plane shear flows. Phys. Rev. E **72**, 021309 (2015)
7. di Prisco, C., Marveggio, P., Redaelli, I., Zerbi, M.: How simulating phase transition during impacts of dry granular flows against rigid barriers by using a MPM approach (2023). In preparation

8. Faccanoni, G., Mangeney, A.: Exact solutions for granular flows. Int. J. Numer. Anal. Methods Geomech. **37**(10), 1408–1433 (2013)
9. Fall, A., Ovarlez, G., Hautemayou, D., Mèzière, C., Roux, J., Chevoir, F.: Dry granular flows: rheological measurements of the $\mu(I)$-rheology. J. Rheol. **59**, 1065 (2015)
10. MiDi, G.D.R.: On dense granular flows. Eur. Phys. J. **14**, 341–365 (2004)
11. Jop, P., Forterre, Y., Pouliquen, O.: A constitutive law for dense granular flows. Nature **441**, 27–730 (2006)
12. Leonardi, A., Pirulli, M.: Analysis of the load exerted by debris flows on filter barriers: comparison between numerical results and field measurements. Comput. Geotech. **118**, 103311 (2020)
13. Li, X., Xie, Y., Gutierrez, M.: A soft-rigid contact model of MPM for granular flow impact on retaining structures. Comput. Part. Mech. **5**(4), 529–537 (2018)
14. Marveggio, P., Redaelli, I., di Prisco, C.: A new constitutive approach for simulating solid- to fluid-like phase transition in dry and saturated granular media. In: Barla, M., Di Donna, A., Sterpi, D. (eds.) Challenges and Innovations in Geomechanics (IACMAG 2021). LNCE, vol. 125, pp. 491–497. Springer, Cham (2021).https://doi.org/10.1007/978-3-030-64514-4_48
15. Pasqua, A., Leonardi, A., Pirulli, M.: Coupling depth-averaged and 3D numerical models for the simulation of granular flows. Comput. Geotech. **149**(6), 104879 (2022)
16. Pouliquen, O., Cassar, C., Jop, P., Forterre, Y., Nicolas, M.: Flow of dense granular material: towards simple constitutive laws. J. Stat. Mech. Theory Exp. **2006**(07), P07020 (2006)
17. Redaelli, I., di Prisco, C.: Three dimensional steady-state locus for dry monodisperse granular materials: DEM numerical results and theoretical modelling. Int. J. Numer. Anal. Methods Geomech. **43**(16), 2525–2550 (2019)
18. Redaelli, I., di Prisco, C., Calvetti, F.: Temporal evolution of the force exerted by dry granular masses impacting on rigid sheltering structures. In: Calvetti, F., Cotecchia, F., Galli, A., Jommi, C. (eds.) Geotechnical Research for Land Protection and Development (CNRIG 2019). LNCE, vol. 40, pp. 13–22. Springer, Cham (2020). https://doi.org/10.1007/978-3-030-21359-6_2
19. Redaelli, I., di Prisco, C.: DEM numerical tests on dry granular specimens: the role of strain rate under evolving/unsteady conditions. Granul. Matter **23**(2), 1–34 (2021). https://doi.org/10.1007/s10035-021-01091-9
20. Redaelli, I., di Prisco, C., Calvetti, F.: Dry granular masses impacting on rigid obstacles: numerical analysis and theoretical modelling. Acta Geotech. **16**(12), 3923–3946 (2021). https://doi.org/10.1007/s11440-021-01337-z
21. Shen, W., Zhao, T., Zhao, J., Dai, F., Zhou, G.G.: Quantifying the impact of dry debris flow against a rigid barrier by DEM analyses. Eng. Geol. **241**, 86–96 (2018)
22. Vescovi, D., di Prisco, C., Berzi, D.: From solid to granular gases: the steady state for granular materials. Int. J. Numer. Anal. Methods Geomech. **37**(17), 2937–2951 (2013)

Eco-Friendly Solutions for Soil and Rock Stabilization, Environmental Protection and Ecological Transition

Laboratory Tests on Gravel-Rubber Mixtures (GRM): FEM Modelling Versus Experimental Observations

Glenda Abate[1], Gabriele Chiaro[2], and Angela Fiamingo[1](✉)

[1] Department of Civil Engineering and Architecture, University of Catania, via Santa Sofia 62, 95125 Catania, Italy
angela.fiamingo@phd.unict.it

[2] Department of Civil and Natural Resources Engineering, University of Canterbury, Private Bag 8400, Christchurch 8041, New Zealand

Abstract. In the field of earthquake engineering, many studies have been conducted to mitigate dynamic effects on structures, adopting alternative solutions including Geotechnical Seismic Isolation (GSI) systems. The main idea is of improving the mechanical properties of the foundation soil and dissipating the seismic motion before reaching the structure. Among several materials proposed for use in GSI systems, Gravel-Rubber Mixtures (GRMs) represent an effective and eco-friendly solution. Laboratory and in-situ tests performed on GRMs have highlighted their excellent static and dynamic mechanical properties. In addition, the rubber particles adopted for GRMs are recycled from End-of-Life Tires, whose disposal is a critical environmental and socio-economic issue, worldwide; therefore, their use could help to partially solve this problem. This paper presents a non-linear finite element numerical study to simulate the behavior of GRMs undergoing monotonic drained triaxial tests. The non-linear GRMs response is simulated through a hardening elasto-plastic constitutive model, namely the Hardening Soil model with Small-Strain Stiffness. In addition, the elastic-perfectly plastic Mohr-Coulomb model, which is commonly used in many numerical codes, is also used for comparison. The calibration of the model parameters is a key point of the performed analyses; it is carried out comparing the numerical results with available experimental data, performed on GRMs with different volumetric rubber content and confining pressures. The calibrated numerical models represent a useful tool to investigate geotechnical problems involving GRMs (e.g., soil-GRMs-structure interaction), highlighting their adequate constitutive modelling.

Keywords: Gravel-rubber mixtures · Geotechnical Seismic Isolation system · FEM analysis · Non-Linear Modelling · Calibration of Constitutive parameters

1 Introduction

Geotechnical Seismic Isolation (GSI) technique received growing attention in the last decade due to the involvement of synthetic geomaterials and/or natural aggregates as alternative solutions for dissipating seismic energy [1]. Among several GSI techniques,

A. Ferrari et al. (Eds.): CNRIG 2023, SSGG, pp. 769–776, 2023.
https://doi.org/10.1007/978-3-031-34761-0_92

recent laboratory and in situ tests [2–4] have shown that a Gravel-Rubber Mixture (GRM) layer beneath the foundations can reduce and filter the seismic accelerations before reaching the structure [5]. In addition, the use of GRMs makes it possible the recycling of the rubber derived from End-of-Life Tires (ELTs) whose disposal is a critical environmental and socio-economic issue, worldwide [6].

Geotechnical problems are often investigated through numerical analysis [7], but to reproduce the non-linear stress-strain behavior of a material it is necessary to adopt adequate constitutive models. Generally, the elastic-perfectly plastic Mohr-Coulomb (MC) Model is largely used due to the few geotechnical parameters required, but it cannot capture the behavior exhibited by the soil at different stages of static and dynamic loading conditions. In this framework, the elasto-plastic Hardening Soil with Small Strain Stiffness Model (HSsmall) [8] is implemented in many finite element codes. This paper focuses on the calibration of the HSsmall Model parameters to simulate the GRMs mechanical behavior, using data from monotonic drained triaxial tests [9] and bender element tests [10]. The numerical stress-strain/volumetric-strain curves of the GRMs are evaluated and compared with the experimental ones considering the stress paths until the peak deviatoric stress (q_{peak}). The calibration of the soil parameters represents a key step to perform advanced numerical analysis involving GRMs (e.g., soil-GRMs-structure interaction).

2 Modelling of Drained Triaxial Tests on Gravel-Rubber Mixtures

2.1 Gravel-Rubber Mixtures (GRMs) and the Numerical Model

The present paper aims at simulating a series of strain-controlled triaxial drained compression tests (TxCD) performed on two types of GRM, characterized by relative density $D_r = 50\%$, particle size ratio $D_{50,R}/D_{50,G} = 1.2$ and by two different Gravel Fraction (GF) by mass, equal to 55% and 87%, respectively. Figures 1(a–b) show typical gravel and rubber particles. Data concerning the tests are available in literature [9]: specimens of 100 mm in diameter and 200 mm in height were investigated. Confining pressures were varied assuming the values 50 kPa, 100 kPa and 200 kPa.

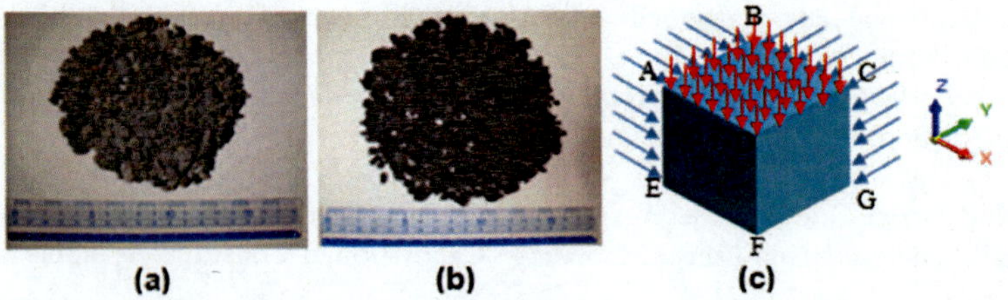

Fig. 1. Testing materials: (a) gravel and (b) rubber chips (after Pasha et al. [9]); (c) developed FEM model for GRMs showing the loading conditions.

The MIDAS FEA NX finite element software was used to model the triaxial tests, simulating a cube of 1 m × 1 m × 1 m unit dimensions (Fig. 1c). As reported in the literature [10–12], the results are not influenced by these large dimensions, as the sample was set as a weightless material.

2.2 Adopted Constitutive Models

The GRM behavior was simulated adopting firstly the elasto-plastic HSsmall constitutive model [8] and performing a very detailed and careful calibration of the model parameters. In a second phase, the elastic-perfectly plastic MC model was considered.

The HSsmall constitutive model is an isotropic hardening elastoplastic hysteretic model. It is based on a combination of the well-known Hardening Soil (HS) model [13] with the small-strain overlay model [14, 15], that requires two additional parameters to describe the hysteretic behavior of the soil at small strain: the shear modulus at small strain $G_0{}^{ref}$ and the threshold shear strain $\gamma_{0.7}$ at which the secant shear modulus G_{sec} is equivalent to 70% of G_0. The HSsmall model has been used recently in many DSSI studies [16, 17]. The relationship between vertical strain ε_a and deviatoric stress q in primary triaxial test is assumed hyperbolic (Eq. 1): when subjected to primary deviator loading, the soil stiffness decreases, and simultaneously irreversible plastic strains increase [13].

$$\varepsilon_a = \frac{q_a}{2E_{50}}\frac{q}{q_a - q} \text{ for } q < q_f \tag{1}$$

The ultimate deviatoric stress, q_f, and the quantity q_a in Eq. 1 are defined as:

$$q_f = \frac{6\sin\phi}{3 - \sin\phi}(\sigma_3 + c\cot\phi) \text{ and } q_a = q_f/R_f \tag{2}$$

where σ_3 is the minor principal effective stress. The above relationship for q_f is derived from the Mohr-Coulomb failure criterion, which involves the strength parameters c and ϕ, cohesion and friction angle respectively. The E_{50} is the secant modulus. As soon as $q = q_f$, the failure criterion is satisfied and perfectly plastic yielding occurs. The ratio between q_f and q_a is given by the failure ratio R_f. The stiffness of the soil is described more accurately than the MC model by means of three different stiffnesses corresponding to the loading condition as: (i) the reference secant modulus from the triaxial test $E_{50}{}^{ref}$; (ii) the reference unloading and reloading modulus $E_{ur}{}^{ref}$; (iii) the reference tangent modulus from oedometer test $E_{oed}{}^{ref}$. These parameters are referred to the reference confining pressure p_{ref} (assumed equal to 100 kPa). The parameter E_{50} and $E_{50}{}^{ref}$ are linked by:

$$E_{50} = E_{50}^{ref}\left(\frac{c\cos\phi + \sigma_3^{'}\sin\phi}{c\cos\phi + p_{ref}\sin\phi}\right)^m \tag{3}$$

where m is the exponent that defines the strain dependence value of the stress state and $\sigma_3^{'}$ is the minor principal effective stress. Similar expressions are assumed for the tangent modulus and the unloading-reloading modulus, respectively (i.e., E_{oed}, and E_{ur}). Details of the yield surface, plastic flow rule and plastic potential functions can be seen in Schanz et al. [13]. Totally, the HS model is characterized by 15 constitutive parameters, summarized in Tables 1 and 2 and discussed in the following paragraph.

As for the MC model, as well-known it is described by the Young's modulus E_0, the Poisson's coefficient ν, the cohesion c, the friction angle ϕ and the dilatancy angle ψ.

3 Calibration of the Model Parameters: Comparison Between Experimental and Numerical Results

The calibration of the 15 HSsmall model parameters was a key point of the performed analyses; Tables 1 and 2 show the values adopted for all the constitutive parameters for the two mixtures with GF equal to 87% and 55%, respectively. The calibration focused essentially on three parameters, chosen by means of a trial-and-error procedure: m, $E_{oed}{}^{ref}$ and R_f. It was carried out comparing the numerical results with the available experimental results in terms of areas under the $q - \varepsilon_a$ and $\varepsilon_v - \varepsilon_a$ curves, considering them until the peak deviatoric stress (q_{peak}). The m parameter can be obtained from the slope of the $log\ E_{50}\text{-}log\ \sigma_3/p_{ref}$ curve; it can assume a value between 0.5 and 1.0, according to the default setting of MIDAS FEA NX. Hence, three different values, 0.5–0.6–0.7 for the GRM with GF = 87% and 0.6–0.7–0.8 for the GRM with GF = 55% were considered for the FEM analyses, choosing the values that allowed to better simulate the tested GRM behavior. The $E_{oed}{}^{ref}$ parameter could be evaluated by the $log\ E_{oed}\text{-}log\ (\sigma_1/p^{ref})$ curve; but oedometer tests were not available for the investigated GRMs and so, due to the uncertainties of this parameter, $E_{oed}{}^{ref}$ was evaluated in relation to $E_{50}{}^{ref}$ assuming it equal to 0.8 $E_{50}{}^{ref}$, $E_{50}{}^{ref}$ and 1.2 $E_{50}{}^{ref}$, choosing the value for which the best matching between numerical and experimental $q - \varepsilon_a$ and $\varepsilon_v - \varepsilon_a$ curves was obtained. Finally, the R_f parameter is generally assumed equal to 0.9 in MIDAS FEA NX, being able to vary between 0.75 and 1, depending on the soil type [18]; three different values equal to 0.9, 0.8 and 0.75 were chosen, evaluating the value for which the better agreement was achieved between the numerical and experimental $q - \varepsilon_a$ and $\varepsilon_v - \varepsilon_a$ curves.

The remaining constitutive parameters were directly evaluated by the triaxial tests results; $\gamma_{0.7}$ was estimated by bender element tests performed on similar GRMs [10].

Totally 81 numerical combinations were analyzed for each simulated GRM, varying the three chosen values for each of the three parameters R_f, m, $E_{oed}{}^{ref}$ and the three adopted confining pressures (50 kPa, 100 kPa and 200 kPa). Figures 2 and 3 report only the results that show a percentage difference between the experimental and numerical area equal to 10% and 20% for the $q - \varepsilon_a$ and $\varepsilon_v - \varepsilon_a$ curves, respectively. It is possible to notice that the number of the combinations that allow a good agreement between experimental and numerical area in terms of $\varepsilon_v - \varepsilon_a$ curves decrease considering the increasing factor between $E_{oed}{}^{ref}$ and $E_{50}{}^{ref}$, and R_f equal to 0.75 ensure a good matching between the experimental and numerical data for every combination and for both the GF. For $GF = 87\%$ (Fig. 2), a good agreement is obtained considering $E_{oed}{}^{ref} = 1.2$ $E_{50}{}^{ref}$, $R_f = 0.75$, $m = 0.5$ for the confining pressure of 50 kPa and 100 kPa; for the confining pressure of 200 kPa, the best agreement is for $E_{oed}{}^{ref} = 0.8$ $E_{50}{}^{ref}$, $R_f = 0.75$, $m = 0.5$. For $GF = 55\%$ (Fig. 2), a good matching is obtained considering $E_{oed}{}^{ref} =$ 0.8 $E_{50}{}^{ref}$, $R_f = 0.75$, $m = 0.6$ for the confining pressure of 50 kPa and 100 kPa; for the confining pressure of 200 kPa, the best agreement is for $E_{oed}{}^{ref} = 0.8$ $E_{50}{}^{ref}$, $R_f =$ 0.75, $m = 0.7$. The best comparisons between experimental and numerical $q - \varepsilon_a$ and $\varepsilon_v - \varepsilon_a$ curves are reported in Fig. 3, that also shows the numerical curves achieved by the MC model. Obviously, the elasto-perfectly plastic MC behavior allow to define the initial tangent stiffness, but it initially overestimates the strength at low-medium strain levels, more markedly at higher confining pressures; then, at large strains it accurately matches the ultimate strength. This allows us to understand the importance of an adequate constitutive modelling.

Table 1. HSsmall model parameters for the mixture with *GF* (gravel fraction) equal to 87%.

HSsmall parameters	$\sigma_3 = 50$ kPa	$\sigma_3 = 100$ kPa	$\sigma_3 = 200$ kPa
G_0^{ref} (kPa) at $p_{ref} = 100$ kPa	5070	5070	5070
m	0.7; 0.5; 0.6	0.7; 0.5; 0.6	0.7; 0.5; 0.6
$\gamma_{0.7}$ (%)	$4.79 \cdot 10^{-5}$	$4.79 \cdot 10^{-5}$	$4.79 \cdot 10^{-5}$
E_{50}^{ref} (kPa) at $p_{ref} = 100$ kPa	10359	10359	10359
E_{oed}^{ref} (kPa)	0.8; 1; 1.2 $\cdot E_{50}^{ref}$	0.8; 1; 1.2 $\cdot E_{50}^{ref}$	0.8; 1; 1.2 $\cdot E_{50}^{ref}$
E_{ur}^{ref} (kPa) at $p_{ref} = 100$ kPa	31077	31077	31077
ν_{ur} (-)	0.3	0.3	0.3
K_0^{NC} (-)	0.31	0.30	0.35
OCR	1	1	1
R_f (-)	0.9; 0.8; 0.75	0.9; 0.8; 0.75	0.9; 0.8; 0.75
c (kPa)	0	0	0
γ (kN/m^3)	25.50	25.50	25.50
ϕ (°)	43.81	44.55	40.48
ψ (°)	7.60	7.61	2.93

Table 2. HSsmall model parameters for the mixture with *GF* (gravel fraction) equal to 55%.

HSsmall parameters	$\sigma_3 = 50$ kPa	$\sigma_3 = 100$ kPa	$\sigma_3 = 200$ kPa
G_0^{ref} (kPa) at $p_{ref} = 100$ kPa	1222	1222	1222
m	0.8; 0.7; 0.6	0.8; 0.7; 0.6	0.8; 0.7; 0.6
$\gamma_{0.7}$ (%)	$8.50 \cdot 10^{-5}$	$8.50 \cdot 10^{-5}$	$8.50 \cdot 10^{-5}$
E_{50}^{ref} (kPa) at $p_{ref} = 100$ kPa	2183	2183	2183
E_{oed}^{ref} (kPa)	0.8; 1; 1.2 $\cdot E_{50}^{ref}$	0.8; 1; 1.2 $\cdot E_{50}^{ref}$	0.8; 1; 1.2 $\cdot E_{50}^{ref}$
E_{ur}^{ref} (kPa) at $p_{ref} = 100$ kPa	6550	6550	6550
ν_{ur} (-)	0.3	0.3	0.3
K_0^{NC} (-)	0.38	0.40	0.43
R_f (-)	0.9; 0.8; 0.75	0.9; 0.8; 0.75	0.9; 0.8; 0.75
OCR	1	1	1
c (kPa)	0	0	0
γ (kN/m^3)	20.30	20.30	20.30
ϕ (°)	37.98	36.87	34.84
ψ (°)	0	0	0

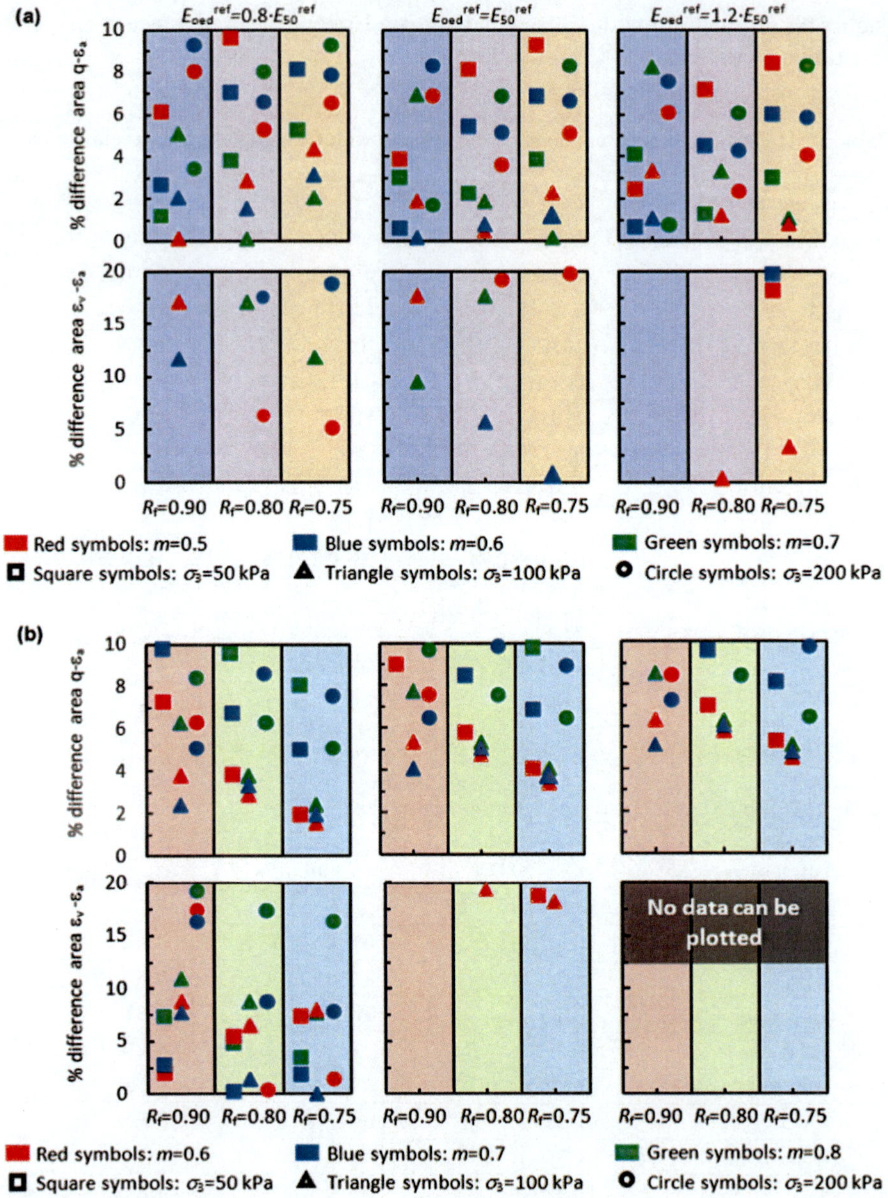

Fig. 2. Influence of the HSsmall parameters (R_f, m, E_{oed}^{ref}) on the percentage difference between the areas under the experimental and numerical curves $q - \varepsilon_a$ and$\varepsilon_v - \varepsilon_a$, for the mixture with *GF* (gravel fraction) equal to 87% (a) and 55% (b), respectively.

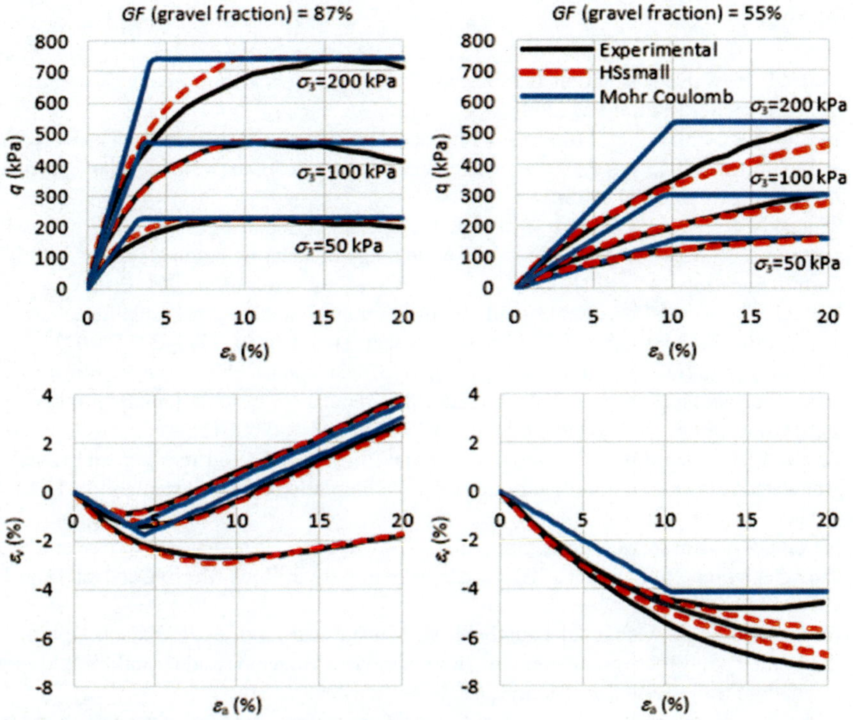

Fig. 3. Comparison between experimental and numerical curves $q-\varepsilon_a$ and $\varepsilon_v-\varepsilon_a$ from monotonic drained triaxial tests on GRMs with *GF* (gravel fraction) equal to 87% and 55%.

4 Conclusions

A non-linear finite element modelling of the behavior of two types of Gravel-Rubber Mixtures (GRMs) subjected to monotonic drained triaxial tests is discusses, focusing the attention on the calibration of the constitutive parameters of the Hardening Soil model with Small-Strain Stiffness, used for modelling the tested GRMs. The calibration is carried out essentially on the three parameters *m*, $E_{oed}{}^{ref}$ and R_f, evaluated by a trial-and-error procedure, choosing the values that allowed the best agreement between numerical and experimental $q-\varepsilon_a$ and $\varepsilon_v-\varepsilon_a$ curves.

For the mixture which shows an initial contracting and subsequently dilatative behavior (*GF* = 83% at the confinement pressure of 50 kPa and 100 kPa), the best results are achieved considering $E_{oed}{}^{ref} = 1.2\ E_{50}{}^{ref}$. While, for the mixture showing a purely contracting behavior (*GF* = 83% at 200 kPa and *GF* = 55% for all the three confinement pressures) the behavior of the mixtures is well defined by setting $E_{oed}{}^{ref} = 0.8\ E_{50}{}^{ref}$.

The so calibrated numerical models represent a useful tool to investigate complex dynamic problems involving GRMs (e.g., soil-GRMs-structure interaction), highlighting the importance of an adequate constitutive modelling.

References

1. Tsang, H.H.: Seismic isolation by rubber–soil mixtures for developing countries. Earthq. Eng. Struct. Dyn. **37**(2), 283–303 (2008)
2. Tasalloti, A., Chiaro, G., Murali, A., Banasiak, L., Palermo, A., Granello, G.: Recycling of end-of-life tires (ELTs) for sustainable geotechnical applications: a New Zealand perspective. Appl. Sci. **11**(17), 7824 (2021)
3. Senetakis, K., Anastasiadis, A., Pitilakis, K.: Dynamic properties of dry sand/rubber (RSM) and gravel/rubber (GRM) mixtures in a wide range of shearing strain amplitudes. Soil Dyn. Earthq. Eng. **33**, 38–53 (2012)
4. Pitilakis, D., et al.: Large-scale field testing of geotechnical seismic isolation of structures using gravel-rubber mixtures. Earthq. Eng. Struct. Dyn. **50**(10), 2712–2731 (2021)
5. Chiaro, G., Palermo, A., Banasiak, L., et al.: Seismic response of low-rise buildings with eco-rubber geotechnical seismic isolation (ERGSI) foundation system: numerical investigation. Bull. Earthq. Eng. (2022). https://doi.org/10.1007/s10518-022-01584-9
6. Tasalloti, A., Chiaro, G., Murali, A., Banasiak, L.: Physical and mechanical properties of granulated rubber mixed with granular soils – a literature review. Sustainability **13**(8), 4309 (2021)
7. Fiamingo, A., Bosco, M., Massimino, M.R.: The role of soil in structure response of a building damaged by the 26 December 2018 earthquake in Italy. J. Rock Mech. Geotech. Eng. **15**(4), 937–953 (2023)
8. Govindasamy, D., Mohamad Ismail, M.A., Mohammad Zaki, M.F., Zainal Abidin, M.H.: Calibration of stiffness parameters for Hardening Soil Model in residual soil from Kenny Hill Formation. Bull. Geol. Soc. Malays. **67**, 119–125 (2019)
9. Pasha, S.M.K., Hazarika, H., Yoshimoto, N.: Physical and mechanical properties of Gravel-Tire Chips Mixture (GTCM). Geosynth. Int. **26**(1), 92–110 (2019)
10. Chiaro, G., Tasalloti, A., Palermo, A., Banasiak, L.: Small-strain shear stiffness and strain-dependent dynamic properties of gravel-rubber mixtures. In: Proc. 17th Symp. Earthquake Eng., India (2022). Forthcoming
11. Sukkarak, R., Likitlersuang, S., Jongpradist, P., Jamsawang, P.: Strength and stiffness parameters for hardening soil model of rockfill materials. Soils Found. **61**(6), 1597–1614 (2021)
12. Golmaei, H.: Application of finite element modeling of isotropic drained monotonic and cyclic triaxial tests, using explicit accumulation model for noncohesive soils under cyclic loading. Int. J. Civ. Struct. Eng. **3**(3), 629 (2013)
13. Schanz, T., Vermeer, P.A., Bonnier, P.G.: The hardening soil model: formulation and verification. In: Beyond 2000 in Computational Geotechnics, pp. 281–296. Routledge (2019)
14. Benz, T.: Small-strain stiffness of soil sand its numerical consequences. Ph.D. thesis, University of Stuttgart, Germany (2006)
15. Benz, T., Vermeer, P.A., Schwab, R.: A small-strain overlay model. Int. J. Numer. Anal. Methods Geomech. **33**, 25–44 (2009)
16. Amorosi, A., Boldini, D., di Lernia, A.: Seismic ground response at Lotung: hysteretic elasto-plastic-based 3D analyses. Soil Dyn. Earthq. Eng. **85**, 44–61 (2016)
17. Shabani, M.J., Shamsi, M., Ghanbari, A.: Seismic response of RC moment frame including topography-soil-structure interaction. Pract. Period. Struct. Des. Constr. **26**(4), 04021046 (2021)
18. Duncan, J.M., Chang, C.Y.: Nonlinear analysis of stress-strain in soils. J. Soil Mech. Found. Div. ASCE **96**(5), 1629–1653 (1970)

Environmentally Sustainable Solutions for Slope Consolidation in the Deruta Historic Center

Erica Cernuto[1](✉), Silvia Settembre[1], Arianna Lupattelli[1], Elisabetta Cattoni[2], Evelina Volpe[3], and Diana Salciarini[1]

[1] Department of Civil and Environmental Engineering, University of Perugia, 06125 Perugia, Italy
erica.cernuto@studenti.unipg.it
[2] eCampus University, 22060 Novedrate, Italy
[3] Research Institute for Geo-Hydrological Protection, National Research Council, 06128 Perugia, Italy

Abstract. A large part of the Italian territory is prone to hydrogeological instability, due to its peculiar geological and geomorphological characteristics. Climate changes and anthropic transformations accentuated these processes, causing situations of ecological and environmental imbalance, and making the necessary interventions for risk mitigation extremely expensive [1]. A possible approach to face this issue, based on an environmentally sustainable criterium, is the use of ground bio-engineering techniques, relying on the use of vegetation. To quantify the improvement in slope stability produced by root reinforcement, the mechanical and hydraulic stabilizing contribution provided by the roots has been investigated separately considering a series of ideal slopes, using the Plaxis 2D code. The slope safety factor has been determined in the case of slopes without vegetation - assumed as reference models - and slopes with vegetation, considering the influence of: i) slope inclination; ii) mechanical characteristics of the soil; iii) contribution of a root system at different depths. From the results of the numerical simulations, it has been possible to observe that: a) in general, the stability of the slope increases in presence of a vegetal cover, with greater benefits for coarse-grained soil; b) the root system improves the behaviour of the slope also in the period of their transient development, and the stability of the slope increases progressively with the deepening of the roots; c) for the soils considered, the mechanical stabilizing effect is greater than the hydraulic one.

Keywords: Slope · landslide · numerical simulations · biotechnical engineering

1 Introduction

Research on the effects that vegetation has on slope stability has developed extensively in recent years. Many Authors ([2–6]) have analysed the role played by vegetation in the mitigation of hydrogeological instability, with reference to the mechanical effect exerted by the plant roots which typically increases the soil shear strength. Other researchers have studied the effects of different slope geometries, soil types, and vegetation mechanical

A. Ferrari et al. (Eds.): CNRIG 2023, SSGG, pp. 777–784, 2023.
https://doi.org/10.1007/978-3-031-34761-0_93

parameters on the slope's factor of safety ([7, 8]). A group of researchers developed a code suitable for rapidly assessing slope stability on large spatial and temporal scales while also considering the effect of plant roots [9], where the reinforcing effect of plant roots is implemented by considering the local plant species and their spatial variations. Other recent studies [10] have tried to derive analytical solutions to investigate the mechanical and hydrological effects of roots with different root architectures and distributions on shallow slope stability. To analyse the contribution of vegetation to slope stability, its hydrological and mechanical roles need to be considered ([2, 11, 12]). Hydrological contribution to slope stability is provided by the roots by reducing the soil water content through transpiration, interception, and evapotranspiration ([6, 13]). The mechanical impact of vegetation on slope stability is due to the mechanical reinforcement of soils through roots (e.g., [2]).

In this paper, we present the results of a set of numerical simulations performed to evaluate the hydro-mechanical contribution provided by roots on slope stability, considering a series of ideal slopes with known characteristics. Numerical analyses carried out through the PLAXIS 2D Finite Element code [14], allowed the determination of the safety factor both in the case of bare slopes, assumed as reference models (without vegetation), and in the case of vegetated slopes. The variables that were considered in the stability analyses are: varying slope inclinations, varying soil mechanical characteristics, absence and presence of vegetation, and varying root system depths. The study made it possible to consider both the complete hydromechanical effect, both to analyse separately the two stabilizing contributions (hydraulic and mechanical). The aim is to assess the relevance of each contribution to stability and to understand which is the prevalent one, under the slope conditions considered in the study.

2 Motivations

A large part of central Italy is characterized by urban areas subjected to high hydrological instability; an example is the City of Deruta located in the province of Perugia. Here, over the years, multiple instabilities have occurred along the slopes around the historic center, often triggered by poor regimentation of surface water, poor soil mechanical characteristics as a result of alteration phenomena, and morphology of the site, characterized by steep slopes. A worsening of the mechanical characteristics of the slopes of the old town is due to the presence of heterometric fired remains from ceramic works, in a silt-sandy matrix, found by in situ investigations. Like many Italian historic centers, the consolidation works must preserve the environmental, naturalistic, and historical characteristics of the area. Hence the need to identify technical solutions for slope consolidation and stabilization that are landscape non-invasive, structurally efficient, cheap, and environmentally friendly. Since the implementation of such interventions often involves the installation of a vegetation cover (Fig. 1), it is necessary to quantitatively assess the stabilizing contribution that they can provide to the slope. The aim of the work, however, is to generalize the understanding on such contribution, and, for this reason, a simulation program based on a parametric study, starting from ideal slopes representative of real-word conditions, has been implemented.

Fig. 1. Technical solutions considered for the slopes around the City of Deruta for their consolidation and stabilization: (a) and (c) current situation; (b) and (d) photo-insertion.

3 Numerical Modelling

A series of numerical analyses were carried out through the Plaxis 2D [14] finite element code. This has allowed considering the hydro-mechanical (HM) effect provided by vegetation roots on the stability of a set of ideal slopes with given characteristics. The program of simulations is constituted of 56 parametric analyses planned to provide technical insights on the variation of the safety factor FS in the stability of the modelled slopes. One of the aims of these simulations is to investigate the relative importance of the two stabilizing contributions exerted by the root system (H and M) by considering them separately. For this purpose, two types of Finite Elements (FE) models were constructed: on the one hand, models that consider the influence of vegetation only on the hydraulic properties of soil (water content, suction, permeability); on the other hand, FE models that consider only the increase in apparent cohesion due to the presence of roots.

3.1 Model Implementation

An ideal reference bare slope model (without vegetation), with uniform soil and variable slope inclination ($\alpha = 10°$, $15°$, $20°$, and $25°$) was initially used to carry out the analyses. The other parameters characterizing the slope were calculated based on simple geometric considerations and they are as follows: $H = 10.23$ m ($\alpha = 10°$), $H = 12.43$ m ($\alpha = 15°$), $H = 14.73$ m ($\alpha = 20°$), $H = 17.19$ m ($\alpha = 25°$); while $L = 54$ m, $h = 6$ m, $l = 15$ m do not vary for the four inclinations considered.

A second model was prepared with a uniform surficial layer simulating soil reinforced by the root system with a constant thickness along the entire slope (Fig. 2). The root thicknesses were considered to be 1 and 2 m. These two different thicknesses are intended to simulate two different stages of growth of the root system, with the aim of simulating the initial stage of growth (thickness equal to 1 m) and the final configuration at the maximum depth of growth (thickness equal to 2 m). Stability analyses have been systematically conducted for all the 56 models prepared. Figure 3 shows the parametric combination used.

The program of simulation is subdivided into two parts. In the right-hand column, a coarse-grained soil (*loamy sand*) has been used, and in the left-hand column, a fine-grained soil has been assigned (*clay*). Each column is subdivided into 4 sectors: the

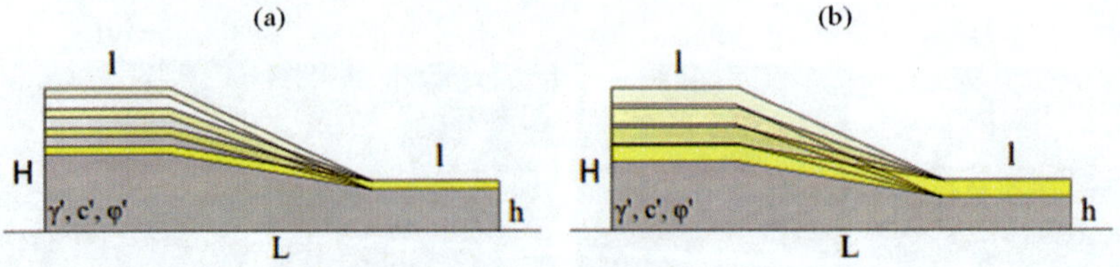

Fig. 2. The 2D model of the slope with uniform vegetation cover: a) root depth of 1 m; b) root depth of 2 m.

Coarse-grained soil				
Hydraulic effect				
Simulation		α [°]	c' [KN/m^2]	ϕ' [°]
Run_01	*Vegetated slope* (s = 2.0 m)	10	1	35
Run_02	*Bare slope*	10	1	35
Run_03	*Vegetated slope* (s = 2.0 m)	15	1	35
Run_04	*Bare slope*	15	1	35
Run_05	*Vegetated slope* (s = 2.0 m)	20	1	35
Run_06	*Bare slope*	20	1	35
Run_07	*Vegetated slope* (s = 2.0 m)	25	1	35
Run_08	*Bare slope*	25	1	35
Mechanical effect				
Simulation		α [°]	c' [KN/m^2]	ϕ' [°]
Run_33	*Vegetated slope* (s = 2.0 m)	10	11	35
Run_02	*Bare slope*	10	1	35
Run_34	*Vegetated slope* (s = 2.0 m)	15	11	35
Run_04	*Bare slope*	15	1	35
Run_35	*Vegetated slope* (s = 2.0 m)	20	11	35
Run_06	*Bare slope*	20	1	35
Run_36	*Vegetated slope* (s = 2.0 m)	25	11	35
Run_08	*Bare slope*	25	1	35
Combined effect				
Simulation		α [°]	c' [KN/m^2]	ϕ' [°]
Run_17	*Vegetated slope* (s = 2.0 m)	10	11	35
Run_18	*Bare slope*	10	1	35
Run_19	*Vegetated slope* (s = 2.0 m)	15	11	35
Run_20	*Bare slope*	15	1	35
Run_21	*Vegetated slope* (s = 2.0 m)	20	11	35
Run_22	*Bare slope*	20	1	35
Run_23	*Vegetated slope* (s = 2.0 m)	25	11	35
Run_24	*Bare slope*	25	1	35
Combined effect				
Simulation		α [°]	c' [KN/m^2]	ϕ' [°]
Run_37	*Vegetated slope* (s = 1.0 m)	10	11	35
Run_38	*Bare slope*	10	1	35
Run_39	*Vegetated slope* (s = 1.0 m)	15	11	35
Run_40	*Bare slope*	15	1	35
Run_41	*Vegetated slope* (s = 1.0 m)	20	11	35
Run_42	*Bare slope*	20	1	35
Run_43	*Vegetated slope* (s = 1.0 m)	25	11	35
Run_44	*Bare slope*	25	1	35

Fine-grained soil				
Hydraulic effect				
Simulation		α [°]	c' [KN/m^2]	ϕ' [°]
Run_09	*Vegetated slope* (s = 2.0 m)	10	5	28
Run_10	*Bare slope*	10	5	28
Run_11	*Vegetated slope* (s = 2.0 m)	15	5	28
Run_12	*Bare slope*	15	5	28
Run_13	*Vegetated slope* (s = 2.0 m)	20	5	28
Run_14	*Bare slope*	20	5	28
Run_15	*Vegetated slope* (s = 2.0 m)	25	5	28
Run_16	*Bare slope*	25	5	28
Mechanical effect				
Simulation		α [°]	c' [KN/m^2]	ϕ' [°]
Run_53	*Vegetated slope* (s = 2.0 m)	10	15	28
Run_10	*Bare slope*	10	5	28
Run_54	*Vegetated slope* (s = 2.0 m)	15	15	28
Run_12	*Bare slope*	15	5	28
Run_55	*Vegetated slope* (s = 2.0 m)	20	15	28
Run_14	*Bare slope*	20	5	28
Run_56	*Vegetated slope* (s = 2.0 m)	25	15	28
Run_16	*Bare slope*	25	5	28
Combined effect				
Simulation		α [°]	c' [KN/m^2]	ϕ' [°]
Run_25	*Vegetated slope* (s = 2.0 m)	10	15	28
Run_26	*Bare slope*	10	5	28
Run_27	*Vegetated slope* (s = 2.0 m)	15	15	28
Run_28	*Bare slope*	15	5	28
Run_29	*Vegetated slope* (s = 2.0 m)	20	15	28
Run_30	*Bare slope*	20	5	28
Run_31	*Vegetated slope* (s = 2.0 m)	25	15	28
Run_32	*Bare slope*	25	5	28
Combined effect				
Simulation		α [°]	c' [KN/m^2]	ϕ' [°]
Run_45	*Vegetated slope* (s = 1.0 m)	10	15	28
Run_46	*Bare slope*	10	5	28
Run_47	*Vegetated slope* (s = 1.0 m)	15	15	28
Run_48	*Bare slope*	15	5	28
Run_49	*Vegetated slope* (s = 1.0 m)	20	15	28
Run_50	*Bare slope*	20	5	28
Run_51	*Vegetated slope* (s = 1.0 m)	25	15	28
Run_52	*Bare slope*	25	5	28

Fig. 3. Program of simulations.

set of runs of the first sector is devoted to assessing the stabilizing contribution due to hydraulic (H) improvement (obtained by the variation of the permeability characteristics of the surficial layer); the set of runs of the second sector refers to the analysis of the stabilizing contribution due to the mechanical (M) reinforcement exerted by the root (simulated by means of a variation of the effective cohesion of the surficial layer); the set of runs of the third sector has analysed the combined hydro-mechanical (HM) beneficial effect exerted by the root considering a root depth of 1 m; and, finally, the set of runs of the fourth sector has examined the same HM effect, but considering a root depth of 2 m. For the sake of comparison, all the simulations have been repeated for the cases of bare *vs.* vegetated slopes. The soil physical-mechanical characteristics have been assigned referring to library materials, available in the Plaxis 2D database.

For the sake of simplicity, a Mohr-Coulomb elastic-perfectly plastic constitutive model was adopted in all the models considered. The main mechanical parameters adopted for the two soil types considered are shown in Table 1 and the hydraulic properties are summarized in Table 2

Table 1. Mechanical parameters of coarse-grained and fine-grained soil.

Material	γ (kN/m^3)	ϕ' (°)	c' (kN/m^2)	c'$_{veg}$ (kN/m^2)
Loamy sand	20	35	1	11
Clay	20	28	5	15

The parameter c'$_{veg}$ is given by the sum of the soil cohesion and the additional cohesion provided by the root system to the soil shear strength, assumed to be 10 kN/m^2 [15].

Table 2. Hydraulic properties for vegetated slope

Material	θ_r	θ_s	k_s (m/s)	γ_a (1/m)	γ_1	γ_n
Loamy sand	0,02000	0,4200	1,450E-6	2,760	-1,060	1,491
Clay	0,01000	0,3600	1,760E-6	2,240	0,000	2,286

4 Results

Numerical simulations carried out on the different slope models examined allowed evaluating: *i)* the stabilizing contribution of vegetation due to the hydraulic effect (H), assuming that the presence of roots produces a variation only in the hydraulic characteristics of the soil in terms of permeability; *ii)* the stabilizing contribution of vegetation due to the mechanical effect (M), assuming that the presence of roots produces a variation only in the shear strength of the soil; *iii)* the stabilizing contribution of vegetation due to the combined hydro-mechanical effect (HM), assuming that the presence of roots produces a variation both in the hydraulic and mechanical characteristics of the soil; and, *iv)* the stabilizing contribution of a root system of vegetation at different depths.

Figure 4 shows the plots of the safety factor *vs.* the slope inclination considering the three different stabilizing effects (H, M, and HM), for coarse-grained soil (Fig. 4a) and fine-grained soil (Fig. 4b). In both plots, the blue solid line represents the results of the reference bare model, while the red lines represent the results of models with vegetation cover of constant thickness equal to 2 m. As expected, it is observed that the vegetated slope has always a higher safety factor than the bare slope, regardless of which stabilizing effect induced by vegetation is considered (H, M, or HM). Comparing the different possible contributions of the vegetation to stability separately, it is observed that the mechanical-type stabilizing effect (M) is greater than the hydraulic-type effect (H), with the line of the combined HM effect that represents an upper bound in the increased stability. Comparing Fig. 4a and Fig. 4b, it is apparent that the expected stabilizing effect in the coarse-grained soil is higher than that of fine-grained soil, as better highlighted in Fig. 5. This is evident both for the stabilizing contribution due to the mechanical reinforcement (Fig. 5a) and for the one due to the hydraulic improvement (Fig. 5b). From the examined set of simulations, the stabilizing effect due to the mechanical improvement is higher than that related to the hydraulic improvement (Fig. 5a *vs.* Figure 5b), especially for gentler steepness (lower than 20°). For steeper slopes, both effects are quite limited.

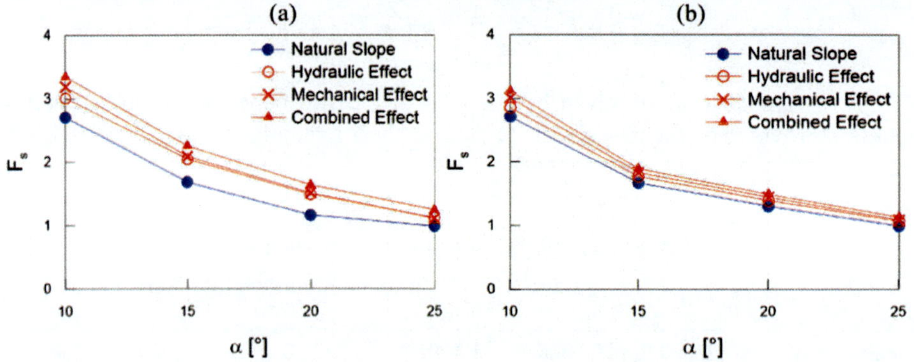

Fig. 4. Safety factor *vs.* slope inclination for coarse-grained (a) and fine-grained (b) soils, comparing the cases of bare vegetated models with the root system at a constant thickness of 2 m.

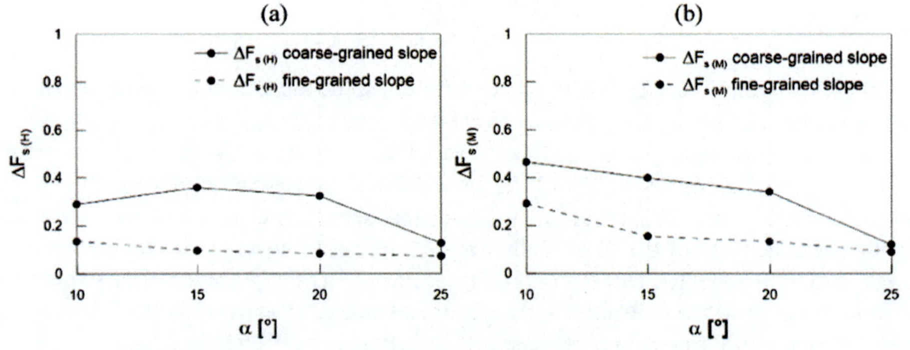

Fig. 5. Increase in the safety factor *vs.* slope inclination for coarse-grained and fine-grained soils, considering separately the: (a) hydraulic effect, and (b) mechanical effect.

For the analysis of the influence of root system depth on slope behaviour, the root depth was varied during modelling, considering the following thicknesses: $s = 1$, and 2 m. The results (Fig. 6) show that the increase ΔFs of the safety coefficient increases for limited slope inclination and is greater of about 50% for root system thickness of

2 m, both in the case of coarse-grained and fine-grained soils. From the plots, it emerges that ΔFs increases linearly with increasing root depth.

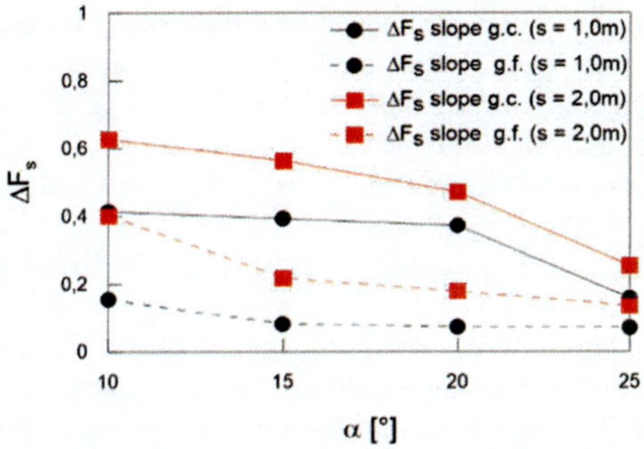

Fig. 6. Plots of ΔFs obtained with the HM model *vs.* slope inclination for coarse-grained and fine-grained soil, as root system depth changes.

5 Conclusions

In this work, the contribution provided by vegetation on slope stability has been evaluated by performing a series of parametric FE analyses considering varying slope geometry, types of soil, and root depth. The analyses conducted allowed the determination of the safety factor variation, in the case of bare slopes assumed as reference models, and slopes with vegetation. The following conclusion can be drawn:

- for the set of parameters considered in the parametric study, the mechanical effect (M) has resulted higher than the hydraulic effect (H);
- regardless of which stabilizing effect is evaluated (H, M, or HM), the presence of vegetation has produced an appreciable increase in slope stability, with greater incidence in coarse-grained soils than in fine-grained soils;
- the stabilizing effect is more relevant for gentler slopes (less than 20°);
- the stabilizing effect increases in a linear-fashion with the thickness of the roots.

References

1. Salciarini, D., et al.: Modeling the effects induced by the expected climatic trends on landslide activity at large scale. Procedia Eng. **158**, 541–545 (2016)
2. Wu, T.H.: Root reinforcement of soil: review of analytical models, test results, and applications to design. Can. Geotech. J. **50**(3), 259–274 (2013)
3. Salciarini, D., Volpe, E., Di Pietro, L., Cattoni, E.: A case-study of sustainable countermeasures against shallow landslides in central Italy. Geosciences **10**(4), 130 (2020)
4. Masi, E.B., Segoni, S., Tofani, V.: Root reinforcement in slope stability models: a review. Geosciences **11**(5), 212 (2021)

5. Volpe, E., Gariano, S.L., Ardizzone, F., Fiorucci, F., Salciarini, D.: A heuristic method to evaluate the effect of soil tillage on slope stability: a pilot case in central Italy. Land **11**(6), 912 (2022)
6. Tagarelli, V., Cotecchia, F.: The effects of slope initialization on the numerical model predictions of the slope-vegetation-atmosphere interaction. Geosciences **10**, 85 (2020)
7. Kokutse, N.K., Temgoua, A.G.T., Kavazović, Z.: Slope stability and vegetation: conceptual and numerical investigation of mechanical effects. Ecol. Eng. **86**, 146–153 (2016)
8. Salciarini, D., Morbidelli, R., Cattoni, E., Volpe, E.: Physical and numerical modelling of the response of slopes under different rainfalls, inclinations and vegetation conditions. Ital. Geotech. J. **2**, 7–21 (2022)
9. Masi, E.B., et al.: Effects of roots cohesion on regional distributed slope stability modelling. CATENA **222**, 106853 (2023)
10. Feng, S., Liu, H.W., Ng, C.W.W.: Analytical analysis of the mechanical and hydrological effects of vegetation on shallow slope stability. Comput. Geotech. **118**, 103335 (2020)
11. Khalilnejad A., Ali F.H, Osman N.: Contribution of the root to slope stability. Geotech. Geol. Eng. 30(2), 277–288 (2012)
12. Ponziani, F. *et al.* (2013). Landwarn: An Operative Early Warning System for Landslides Forecasting Based on Rainfall Thresholds and Soil Moisture. In: Margottini, C., Canuti, P., Sassa, K. (eds) Landslide Science and Practice, vol. 2, pp. 627–634. Springer, Heidelberg (2013). https://doi.org/10.1007/978-3-642-31445-2_82
13. Mulyono A., Subardja A., Ekasari I., Lailati M., Sudirja R., Ningrum W.: The hydromechanics of vegetation for slope stabilization. In: IOP Conference Series: Earth and Environmental Science (2018)
14. Brinkgreve, R., Zampich, L., Ragi Manoj, N.: PLAXIS 2D General Information. CONNECT Edition V20: Bentley Systems (2019)
15. Cecconi, M., Pane, V., Napoli, P., Cattoni, E.: Deep roots planting for surface slope protection. Electron. J. Geotech. Eng. U **17**, 2809–2820 (2012)

Thermal Conductivity of Cement–Based Grouts for Energy Micropiles: Preliminary Experimental Investigation

G. Ciardi[1](✉), K. Oliynyk[1,2], C. Madiai[3], and C. Tamagnini[1]

[1] Department of Civil and Environmental Engineering, University of Perugia, Perugia, Italy
giovanni.ciardi@unipg.it

[2] School of Science and Engineering, University of Dundee, Dundee, Scotland, UK

[3] Department of Civil and Environmental Engineering, University of Florence, Florence, Italy

Abstract. Energy micropiles are a kind of energy geostructure designed to reduce the energy consumption of buildings combining structural and thermal performance. Since a common way to improve the thermal efficiency of an energy pile is to increase the thermal conductivity of the concrete (namely, cement, water and aggregate) used in its realization, the same should hold for the mortar (namely, cement plus water) used in the construction of energy micropiles. In this study, the preliminary results of an experimental investigation campaign are presented. The experimental campaign includes thermal conductivity and workability tests carried out on mortar samples and on liquid grouts, respectively. Both neat mortar and mortar plus additives, selected among commercial products with the purpose of increasing the thermal conductivity of the material, were used in the experiments. Three commercial, chemical additives, originally developed for different purposes (*e.g.*, to produce thixothropic mortars, or for the heating floor industry), were used. Experimental results showed that all tested additives were beneficial in increasing the thermal conductivity of the material over the entire range of investigated water/cement ratios, as well as in lowering the initial viscosity of the mixtures. The preliminary results also indicated that one of the tested additives seemed more promising than the others for increasing the thermal conductivity of the hardened grout.

Keywords: Energy geostructures · Micropile · Experimental investigation · Thermal conductivity

1 Introduction

Energy geostructures have been recently developed to meet the continuously growing demand of renewable energy sources, as well as the need of reducing the greenhouse gas emission. This kind of structures are mainly used for the heating/cooling of spaces by combining thermal and structural performances [1]. The working principle of an energy geostructure is based on the use of the shallow geothermal energy, a green, *unlimited*, free source of energy. The soil, whose temperature is almost constant at the depths reached by

A. Ferrari et al. (Eds.): CNRIG 2023, SSGG, pp. 785–791, 2023.
https://doi.org/10.1007/978-3-031-34761-0_94

common geotechnical engineering structures (excluding a relatively shallow boundary layer at the ground surface, influenced by the seasonal and daily temperature variations), acts as a heat sink/source, exchanging heat with the structure and with the internal spaces of buildings. Examples of energy geostructures are energy piles [2–4], energy tunnels [5, 6] and diaphragm walls [7, 8]. Typically, the heat exchange occurs through a heat–carrier fluid circulating within plastic/aluminum pipes embedded in the geostructures. At present, energy piles represent the most common type of energy geostructure [9].

Micropiles (a drilled– and grouted–in–place foundation structure characterized by small diameter (up to 300 mm), reduced length (up to 15 m), and by a thick steel reinforcement), usually used for retrofitting of existing buildings, could be designed as energy geostructures. Ronchi et al. [10, 11] showed that the specific heat flux obtained from the energy micropile was within the range of typical values of energy piles. Kong et al. [12] showed a case–history of an existing pile–raft foundation retrofitted with energy and standard micropiles. By analyzing both field test data and numerical simulation results, they concluded, among others, that the improved foundation system could provide sufficient heat exchange. Cecinato and Salciarini [13] showed that mortar and soil thermal conductivity, pile length, and pile diameter, are dominant factors in maximizing the micropile energy performance. Ciardi et al. [14] proved that enhancing the thermal conductivity of the mortar can be a suitable strategy to increase the heat transfer occurring at the micropile–soil interface.

This study provides the preliminary results of an experimental laboratory investigation campaign on the thermal performance of energy micropiles. Tests include thermal conductivity tests on mortar samples and workability tests on liquid grout. High thermal conductivity and low initial viscosity are in fact desired when aiming to enhance the thermal performance of an energy micropile and to facilitate its installation. Previous research clearly showed that the thermal properties of cement–based grouts can be modified by using different kinds of additives. For instance, when high–thermal performance materials are required, high–thermal conductivity fillers, such as graphite, can be effectively used [15]. In this study, the experimental tests were performed on samples/grouts made of the mortar commonly used in the construction of micropiles and on samples/grouts made of mortars modified by adding different commercial, chemical additives used to enhance the thermal conductivity of the materials. Three additives – not originally developed for the use in energy geostructures – were tested, to evaluate their suitability.

2 Experimental Tests

2.1 Materials

The experimental campaign, including thermal conductivity and workability tests, was carried out on cement–based mixtures with or without additives. The mixture made of water plus cement without any aggregate – referred to as neat mortar in the following – was taken as reference. Ordinary Portland cement (Colacem CEM II/B–LL 32.5 R) and tap water were used to prepare all the mixtures.

Three chemical additives – all commercially available – were selected and used to increase the thermal conductivity of the neat mortar. These additives are referred to as

A1, A2 and A3 in this study. The additives were added in different proportions to the neat mortar formulation. Additives A1 and A2 are liquid, and both were developed for the heating floor industry. Additive A3 is in powder form and was developed for producing thixothropic mortars with enhanced mechanical characteristics. Additive A3 is produced by the same manufacturer of additive A2.

The different grouts were obtained by mixing cement, water, and additive, when required, at fixed water/cement ratios (w/c), here intended as the ratio between the weight of water and that of the cement. Table 1 summarizes the materials used in this research. In the Table 1, ρ is the material density.

Table 1. Materials used in this research.

Material	ρ (kg/m^3)	Type
Cement	2950	Powder
A1	1030–1040*	Liquid
A2	1075*	Liquid
A3	1050	Powder

* measured at T = 20 °C

2.2 Thermal Conductivity Tests

Thermal conductivity tests were performed using the transient method, a common procedure used for concrete thermal conductivity measurements, allowing the reduction of testing time [16, 17]. An ISOMET 2104 (Applied Precision Ltd.) portable thermal conductivity meter was used for this purpose. This device allows the measurements of thermal conductivity, thermal diffusivity, and volumetric heat capacity of samples by analyzing the temperature response of the sample when subjected to heat flow impulses. In this study, only results in terms of thermal conductivity are discussed.

After a period of 24 h from grout casting in cylindrical plastic molds, the hardened specimens (78 mm in diameter and 42 mm in height) were extracted from the molds and immersed in water to cure. Thermal tests were performed on saturated samples after 7 days from casting: the thermal conductivity is not significantly affected by the age of the specimens, while it is strongly influenced by their moisture content [16], being greater in saturated than in dried state. The tests were carried out in saturated state because this is the state most frequently expected in standard operation conditions of micropiles. To run each test, a circular surface probe with measurement range 0.3–2.0 W/m/°C and precision on thermal conductivity of 10% of reading was placed on the smooth surface of cylindrical samples. The setup of a thermal conductivity tests is shown in Fig. 1. A group of three samples per each of the considered w/c ratios was prepared, to assess the repeatability of the test results. The reported data are the averages for each group.

Fig. 1. Setup of a thermal conductivity test.

2.3 Workability Tests

Workability tests were used to assess the initial viscosity of the mortar. The test was run on liquid grouts, immediately after mixing. It consists of measuring the time required by the grout to fill a fixed volume (1 L) when flowing by gravity, starting from the same height, through a funnel of given geometry. The smaller is the flowing time, the smaller is the initial viscosity of the mixture. Three measurements per each w/c ratio were taken and averaged. Tests were interrupted if the time required for the grout to flow was larger than 5 min.

2.4 Experimental Program

The experimental program is summarized in Table 2. Thermal conductivity tests and workability tests were performed for all mixtures listed in Table 2, except for the mixture belonging to group 3 at $w/c = 0.35$, for which only thermal tests were performed. The content of each additive was suggested by the manufacturers and kept constant throughout the tests. All experiments were carried out at room temperature.

Table 2. Experimental program.

Mixture group#	Additive	Content (mL/kg or %)	w/c (–)
1	No	–	0.30, 0.35, 0.40
2	A1	8	0.30, 0.35, 0.40
3	A2	8	0.35, 0.40
4	A3	5*	0.30, 0.35, 0.40

* by weight of cement powder

3 Results

Test results are shown in Table 3, where k is the thermal conductivity and t is the time required to complete the workability test.

Table 3. Test results.

Mixture#	w/c (–)	k (W/m/°C)	t (s)
1_a	0.30	1.208	> 300
1_b	0.35	1.190	> 300
1_c	0.40	1.188	16.13
2_a	0.30	1.200	43.70
2_b	0.35	1.198	14.26
2_c	0.40	1.217	8.35
3_a	0.35	1.217	–
3_b	0.40	1.247	12.59
4_a	0.30	1.283	40.14
4_b	0.35	1.273	14.60
4_c	0.40	1.250	9.08

The results of thermal conductivity tests are shown in Fig. 2a. As expected, for the neat mortar, the thermal conductivity increases as the w/c ratio decreases (see, *e.g.*, [16, 18]). The most promising additive was A3. Its use increased the thermal conductivity values more significantly than additives A1 and A2 over the whole range of investigated w/c ratios.

Data obtained from additives A2 and A3 are controversial because a decrease of k was obtained as w/c decreases. Moreover, for the additive A1, the conductivity value at $w/c = 0.30$ is less than the corresponding value for the neat mortar. Such controversial data, consistently obtained during measurements, suggest that additives A1 and A2 may not be suitable for the purpose of this research. This is possibly due to the fact that additives A1 and A2 do not work properly in absence of any aggregate.

Figure 2b shows the normalized thermal conductivity increment, $\Delta k = (k - k_r)/k_r$, where k_r is the thermal conductivity of neat mortar at the same w/c ratio. The thermal conductivity increase obtained using additive A3 is generally larger than 5%.

Further investigations are required to explore the effects of using additive A3 in different proportions, or the effectiveness of using different additives (*e.g.*, granular graphite), in terms of i) the possibility of achieving higher increments in thermal conductivity of the mortar, and ii) the increase in the thermal heat flux exchanged between the micropile and the soil, when the thermally enhanced mortars are used.

The results of workability tests are shown in Fig. 3. It is worth noting that the workability tests failed for the neat mortar at $w/c \leq 0.35$. All additives were beneficial in reducing the viscosity of the mixtures, at each w/c ratio considered.

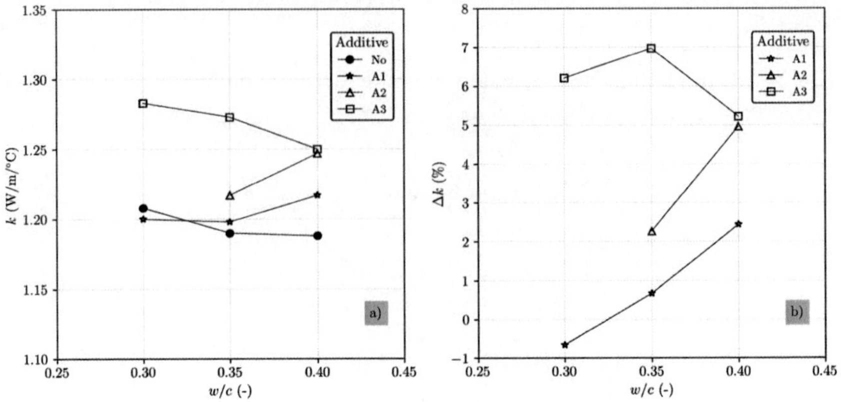

Fig. 2. Measurements of thermal conductivity (a) and thermal conductivity increment (b).

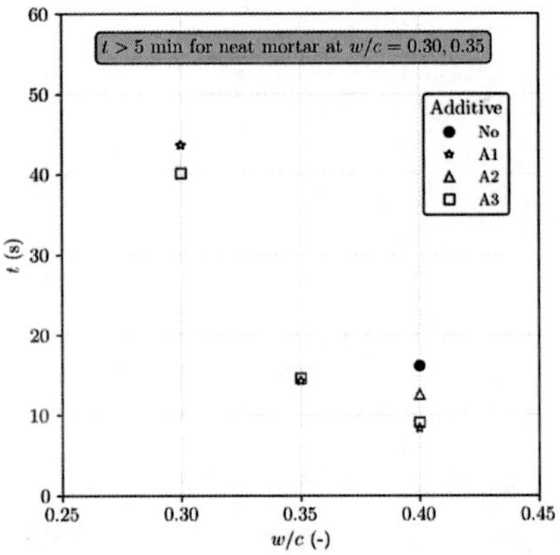

Fig. 3. Results of workability tests.

4 Concluding Remarks and Future Developments

This study shows the preliminary results of an experimental campaign aimed at evaluating the possibility of increasing the thermal conductivity of the mortar used in the construction of energy micropiles by means of additives not specifically designed for this purpose. Increasing the thermal conductivity can in fact be an effective strategy to increase the thermal performance of the energy micropile.

Three additives (A1, A2, A3) were selected among commercial products and thermal conductivity tests and workability tests were performed on specimens of neat and enhanced mortars. The results obtained showed that the best thermal performance was obtained using additive A3, while the results obtained from specimens treated with additives A1 and A2 were somehow controversial. All additives were beneficial in reducing the initial viscosity of the mixtures, over the whole range of investigated water/cement ratios.

Future research should investigate: i) whether the thermal conductivity of the material can be further enhanced; ii) whether the trend exhibited by additives A1 and A2 are confirmed in presence of fine inert in the mortar composition; iii) whether different additives, or additive A3 in different percentages, could be used to enhance the thermal conductivity of the mortar while maintaining good workability and while not compromising the mechanical properties (in terms of stiffness and strength) of the enhanced material; iv) which quantitative benefit in the thermal performance of an energy micropile can be obtained by using thermally–enhanced mortars.

References

1. Brandl, H.: Energy foundations and other thermo–active ground structures. Géotechnique **56**(2), 81–122 (2006)
2. Laloui, L., Nuth, M., Vulliet, L.: Experimental and numerical investigations of the behaviour of a heat exchanger pile. Int. J. Numer. Anal. Meth. Geomech. **30**(8), 763–781 (2006)
3. Amatya, B.L., Soga, K., Bourne-Webb, P.J., Amis, T., Laloui, L.: Thermo–mechanical behaviour of energy piles. Géotechnique **62**(6), 503–519 (2012)
4. Faizal, M., Moradshahi, A., Bouazza, A., McCartney, J.S.: Soil thermal response to temperature cycles and end boundary conditions of energy piles. J. Geotech. Geoenviron. Eng. **148**(5), 04022027 (2022)
5. Barla, M., Di Donna, A.: Energy tunnels: concept and design aspects. Undergr. Space **3**(4), 268–276 (2018)
6. Ma, C., Di Donna, A., Dias, D., Zhang, J.: Numerical investigations of the tunnel environment effect on the performance of energy tunnels. Renew. Energy **172**, 1279–1292 (2021)
7. Sterpi, D., Coletto, A., Mauri, L.: Investigation on the behaviour of a thermo–active diaphragm wall by thermo–mechanical analyses. Geomech. Energy Environ. **9**, 1–20 (2017)
8. Dong, S., et al.: Thermo-mechanical behavior of energy diaphragm wall: physical and numerical modelling. Appl. Therm. Eng. **146**, 243–251 (2019)
9. Laloui, L., Loria, A.R.: Analysis and Design of Energy Geostructures: Theoretical Essentials and Practical Application. Academic Press, Cambridge (2019)
10. Ronchi, F., Salciarini, D., Cavalagli, N., Tamagnini, C.: Numerical model of energy foundation behavior: the prototype of a geothermal micro–pile. Procedia Eng. **158**, 326–331 (2016)
11. Ronchi, F., Salciarini, D., Cavalagli, N., Tamagnini, C.: Thermal response prediction of a prototype energy micro-pile. Geomech. Energy Environ. **16**, 64–82 (2018)
12. Kong, G.Q., Cao, T., Hao, Y.H., Zhou, Y., Ren, L.W.: Thermomechanical properties of an energy micro pile–raft foundation in silty clay. Undergr. Space **6**(1), 76–84 (2021)
13. Cecinato, F., Salciarini, D.: Energy performance assessment of thermo–active micro–piles via numerical modeling and statistical analysis. Geomech. Energy Environ. **29**, 100268 (2022)
14. Ciardi, G., Madiai, C., Tamagnini, C.: Experimental study on the thermal properties of cement–based grouts for energy micropiles. *Submitted*
15. Guo, C., Zhu, J., Zhou, W., Chen, W.: Fabrication and thermal properties of a new heat storage concrete material. J. Wuhan Univ. Technol. Mater. Sci. Ed. **25**(4), 628–630 (2010)
16. Kim, K.H., Jeon, S.E., Kim, J.K., Yang, S.: An experimental study on thermal conductivity of concrete. Cem. Concr. Res. **33**(3), 363–371 (2003)
17. Kim, D., Kim, G., Kim, D., Baek, H.: Experimental and numerical investigation of thermal properties of cement–based grouts used for vertical ground heat exchanger. Renew. Energy **112**, 260–267 (2017)
18. Asadi, I., Shafigh, P., Hassan, Z.F.B.A., Mahyuddin, N.B.: Thermal conductivity of concrete–a review. J. Build. Eng. **20**, 81–93 (2018)

Alkaline Activation of Volcanic Ash as Binder for Soil Improvement

L. T. Costa[1], E. Vitale[1](✉), P. Cappelletti[1], S. F. Graziano[2], C. Rispoli[1], and Giacomo Russo[1]

[1] Department of Earth Science, Environment and Resources, University of Naples Federico II, Naples, Italy
enza.vitale@unina.it

[2] Department of Pharmacy, University of Naples Federico II, Naples, Italy

Abstract. An experimental investigation on chemo-mechanical evolution of alkali-activated binders as alternative and economical solution for soil improvement has been developed. Alkaline activation of volcanic ash from Mount Etna has been analyzed. Reactivity of the volcanic ash has been investigated considering two alkaline activators with different molar concentrations (i.e. 8M and 12M NaOH solutions). Mineralogical evolution of the binders has been monitored at increasing curing times by means of X-ray Diffraction Analysis (XRD) and Scanning Electron Microscopy (SEM). The experimental evidence at microscale has been related to the results of investigations at macroscopic level. Ultrasonic wave velocities and Uniaxial Compressive Strength tests (UCS) have been performed and interpreted considering the mineralogical evolution of the system. Test results showed that the high reactivity of ash, treated with 8M NaOH solution, promotes the formation of minerals (i.e., clays and zeolites) from amorphous substance, determining an increase in compressive strength values over curing time. The use of alkaline solution with higher molar concentration (i.e., 12M NaOH solution) prevents the ongoing of alkaline activation reactions inhibiting the potential mechanical improvement of the binder.

Keywords: Alkali-Activated Binders (AAB) · natural pozzolanic materials · volcanic ash · multi-scale analysis

1 Introduction

The valorization of waste materials, such as natural and artificial pozzolanic materials, for the synthesis of alkali activated binders represents a potential sustainable alternative to the use of traditional binders in the field of soil stabilization. The conventional method for soil improvement is a chemical stabilization in which lime and Portland cement represent the most widely used binders [1–6]. The high production and use of these binders over time competes in a significant carbon footprint along with the consumption of high amounts of natural resources. Therefore, development of new binders based on the activation of natural and artificial pozzolans can be considered as a promising

A. Ferrari et al. (Eds.): CNRIG 2023, SSGG, pp. 792–799, 2023.
https://doi.org/10.1007/978-3-031-34761-0_95

perspective in terms of sustainability since it reduces the carbon footprint and allows reusing by-products. Alkali activated binders are synthesized from the chemical reaction between an amorphous precursor, which is rich in alumina (Al_2O_3) and silica (SiO_2), with a sodium or potassium-based activator [7]. During alkaline activation processes, dissolution of aluminosilicate oxides in the alkali activator promotes a polycondensation reaction which results in the formation of a sodium aluminosilicate gel with cementitious properties [8]. There are several secondary by-products such as fly ash (e.g., ground granulated blast furnace slag, rice husk ash, crushed glass, etc.) which are commonly used as precursor for the synthesis of alkali activated binders for geotechnical purposes [9–14]. Conversely, practical application of natural pozzolanic materials in soil stabilization has been quite limited. Valorization of natural alumino-silica sources (i.e., volcanic pyroclastic deposits) is of practical interest since there is no needs for natural pozzolans to be treated (i.e., calcination) before their use, with advantages in terms of lower cost, lower carbon dioxide emissions and easy access to the available resource. Miraki et al. [15] investigated the potential of using alkali-activated ground granulated blast furnace slag (GGBS) and volcanic ash (VA) as green binders in clayey soil stabilization. The effects of different combinations of VA with GGBS, various liquid/solid ratios, different curing conditions, and different curing periods were investigated. Compressive strength and durability of specimens against wet-dry and freeze-thaw cycles were then studied through mechanical and microstructural testing. Analytical results demonstrated that the coexistence of GGBS and VA in alkaline activation process was more effective due to the synergic formation of N-A-S-H and C-(A)-S-H gels. Gadir et al. 2021 [16] evaluated the feasibility of using alkali activated VA as an alternative soil stabilizer to cement by comparing their shear strength behavior and life cycle assessment. The results showed that the shear strength of treated specimens was improved at higher binder content, longer curing times and lower moisture content, regardless of binder type.

In the present study, an experimental investigation on alkaline activation of volcanic ash from Mount Etna has been reported. Reactivity of the volcanic ash has been investigated considering one alkaline activator with different molar concentrations (i.e., 8M and 12M NaOH solution). Mineralogical evolution of binders has been monitored at increasing curing times by means of X-ray Diffraction (XRD) and Scanning Electron Microscopy (SEM). The experimental evidence at microscale has been related to macroscopic behavior of binders. Ultrasonic wave velocities and Uniaxial Compressive Strength tests (UCS) have been performed and interpreted considering the mineralogical evolution of the binders.

2 Materials and Experimental Procedures

2.1 Materials and Methods

The natural pozzolanic material used in the present work is a volcanic ash from Mount Etna, collected after the recent eruption of February 2021. Etna Volcanic Ash (EVA) was sampled nearby Zafferana Etnea, a town located on the Southeast slope of the Etna volcano, following the cycle of paroxysms that occurred from 22 to 28 February 2021. Two sodium hydroxide (aq) solutions (i.e., NaOH 8M and 12M) were used as alkaline activators and were prepared by dissolving 320 g and 480 g of NaOH pellet (99% assays) into 1 L of distilled water, respectively, and were kept at ambient temperature for at least 24 h for cooling before being used.

Raw material was characterized by the chemical point of view on pressed powder pellets, using an Axios Panalytical X-ray Fluorescence (XrF) spectrometer, equipped with six analyzer crystals, three primary collimators and two detectors (flow counter and scintillator). Analytical uncertainties are 1–2% for major elements and 5–10% for trace elements [17]. The weight Loss on Ignition (LoI) was determined by gravimetric techniques, firing at 1000 ∘C previously dried powders at 110 ∘C overnight.

Mineralogical composition of raw material and alkali activated binders was investigated by means of X-Ray diffraction analysis performed on randomly oriented powder using Malvern Panalytical X'Pert Pro modular diffractometer.

Samples surface was observed through Scanning Electron Microscopy by using Zeiss Merlin VP Compact and JEOL JSM-5310 coupled with Oxford Instrument Microanalysis Unit equipped with an INCA X-Max solid-state detector. Ultrasonic wave velocities were recorded according to UNI EN 14579 [18] using a BOVIAR DSP UTD 1004 Ultrasonic device, with a pair of 55 kHz transducers (diameter = 40 mm) in direct arrangement. To provide an adequate acoustic coupling between samples and transducers, a thin film of hydro soluble gel (GIMA, Italy) was used. Uniaxial Compressive Strength tests (UCS) were performed on cylindrical specimens with a Wykeham Farrance device, at a maximum load of 500 kN and a displacement rate of 1.00 mm/min.

2.2 Sample Preparation

Alkali activated binders were prepared by hand mixing volcanic ash and alkaline solution in fixed proportions. Once sampled, volcanic material was oven dried at 105 °C, milled and sieved to obtain a grain size <75 μm, dimension considered suitable for alkali activation process [19]. Smaller volcanic ash particles with porous microstructure, having higher specific surface area, are more active in an alkaline medium [20]. Alkali Activated Etna Volcanic Ashes (AAEVA), prepared with 8M and 12 M NaOH solutions (i.e. AAEVA_8M and AAEVA_12M), were synthetized by fixing the silica modulus (SiO_2/Na_2O) of the mixture. Silica modulus plays a key role in the reaction kinetics since the dissolution of the amorphous component depends on its SiO_2 content [21]. For the studied volcanic ash, a value of silica modulus approximately equal to 2 was selected ($SiO_2/Na_2O \cong 2$), based on previous studies performed to ensure/guarantee workability and hardening of the mixtures. To obtain a constant silica modulus, alkaline solution/volcanic ash mass ratio was selected equal to 0.37 for 8M samples and equal to 0.5 for 12M samples. In these conditions, for both binders (AAEVA_8M and AAEVA_12M), the deionized water-solid (i.e. volcanic ash) ratio by mass is constant and equal to 0.25. Samples of AAEVA_8M and AAEVA_12M were poured in molds of 38 mm × 76 mm, sealed in plastic bag and then cured at ambient temperature for increasing curing times, namely 7, 14 and 60 days before performing microstructural and mechanical tests.

3 Results

Chemical composition of the volcanic ash mainly consists of silica ($SiO_2 = 47.20\%$ wt.) and alumina ($Al_2O_3 = 17.07\%$ wt.). The raw material is also rich in Fe, Mg and Ca ($Fe_2O_3 = 11.30$ wt.; $MgO = 5.29\%$ wt. And $CaO = 10.98\%$ wt.).

From mineralogical point of view, the investigated material is composed solely by clinopyroxene and plagioclase (both primary phases) along with amorphous substance (volcanic glass shards, due to fragmentation of the erupted magma).

X-ray diffraction patterns of raw and alkali activated volcanic ash prepared with a 8M NaOH solution (AAEVA_8M) are shown in Fig. 1 as function of curing time. Variations in mineralogical composition of volcanic ash is clearly determined by alkaline activation process. While typical crystalline content of Mt. Etna volcanic ash (i.e., clinopyroxene (cpx) and plagioclase (pl) remain present, crystal phases due to secondary mineralization processes, such as clay minerals (smectite-type) and zeolites [21] appear and are consistent with the reorganization of gel formed after alkaline activation process acting on amorphous substance of volcanic ash (i.e., volcanic glass). Zeolites, as confirmed by XRD reflections, were detected in alkali activated samples starting from 14 days of curing, with increasing intensity as curing time increases. A small XRD reflection, corresponding to smectite-type phases was also detected after 60 days of curing. Comparison between X-Ray diffraction patterns of AAEVA_8M and AAEVA_12M after 14 days of curing is shown in Fig. 2. No zeolite were detected in the diffraction pattern of AAEVA_12M. A broad hump in the pattern, corresponding to poorly crystalline clay phase was observed in all the treated samples after 14 days of curing time. Conversely, the presence of sodium carbonate hydrate ($NaCH = (Na_3H(CO_3) \cdot 2H_2O)$) was detected after 14 days of curing only in the sample treated with 12M NaOH solution. This is consistent with the presence of some efflorescence detected on the surface of AAEVA_12M samples, determined by an high alkali concentration which can promote the reaction between free alkalis of the pore solution with humid air containing CO_2 with the resulting formation of white carbonate compounds [23]. SEM observations on untreated and alkali-activated volcanic ash (i.e., AAEVA_8M) after 14 and 60 days of curing are shown in Fig. 3. Untreated sample is composed of particles with different sizes and irregular geometric shapes (Fig. 3a). The progressive transformation of volcanic ash during the reactions was evidenced by SEM observations at increasing time. After 14 days of curing, the sample surface is clearly modified (Fig. 3b). Alteration of amorphous component (glass shards) induced by the alkaline environment is clearly evidenced by a coating of gel-like phases on particles surface and zeolite crystallization. The formation of a structure of zeolite crystals aggregates connected to the gel matrix is more evident at longer curing time (Fig. 3 c), as confirmed by XRD investigation which reveals the presence of these minerals. A comparison of SEM images of 14 days NaOH treated 8M and 12M samples is shown in Fig. 4. The two samples show completely different surface morphology. Compared to 8M treated sample, there is no evidence of zeolite formation for AAEVA_12M, whereas, in agreement with XRD diffraction patterns, crystals of sodium carbonate hydrate (NaCH) have been detected.

The evolution of Unconfined Compression Strength (UCS) over time for AAEVA_8M samples is shown in Fig. 5a. A gradual increase of UCS values is observed as curing time increases. An experimental relationship between UCS and Ultrasonic

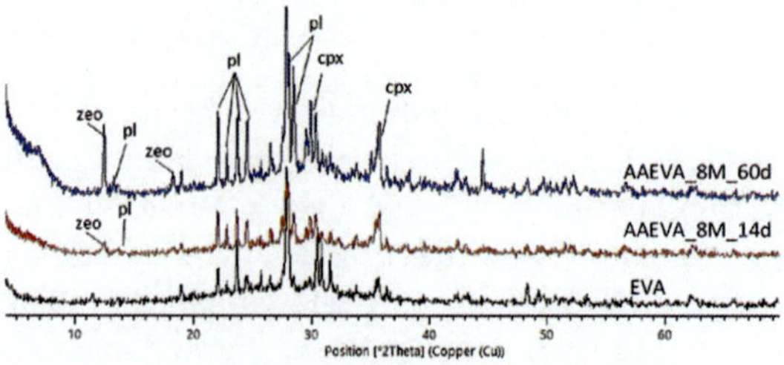

Fig. 1. X-Ray diffraction pattern of raw and alkali activated volcanic ash Etna Volcanic Ash (EVA). Mineral abbreviations from Whitney and Evans [22]

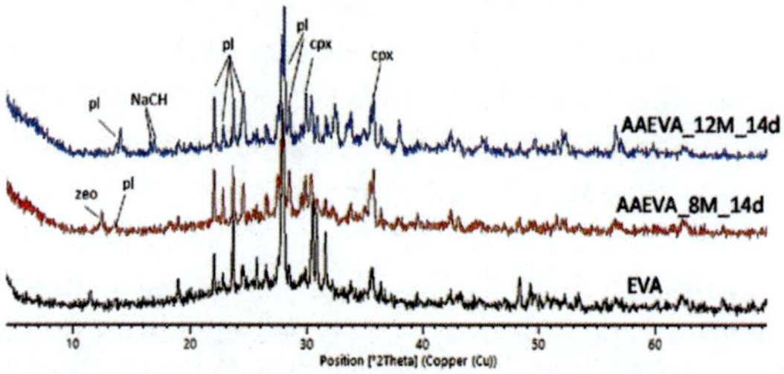

Fig. 2. X-Ray diffraction pattern of raw and alkali activated volcanic ash (AAEVA_8M and AAEVA_12M) after 14 days of curing

Fig. 3. SEM observations: a) raw volcanic ash (EVA); b) AAEVA_8M- 14 days; c) AAEVA_8M-60days

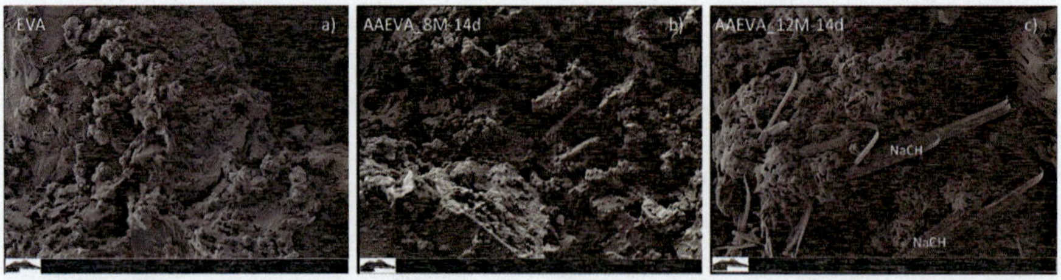

Fig. 4. SEM observations: a) raw and alkali activated volcanic ash Etna Volcanic Ash (EVA) b (8M - 14d) and c (12M – 14d), respectively.

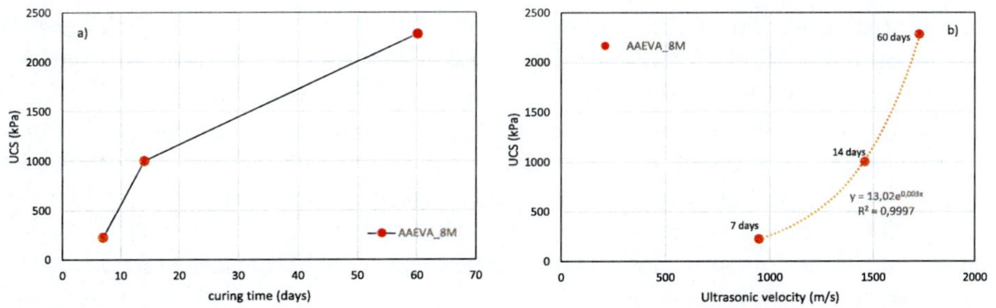

Fig. 5. AAEVA_8M: a) Evolution of UCS as function of time; b) UCS vs. Ultrasonic velocity at increasing curing times

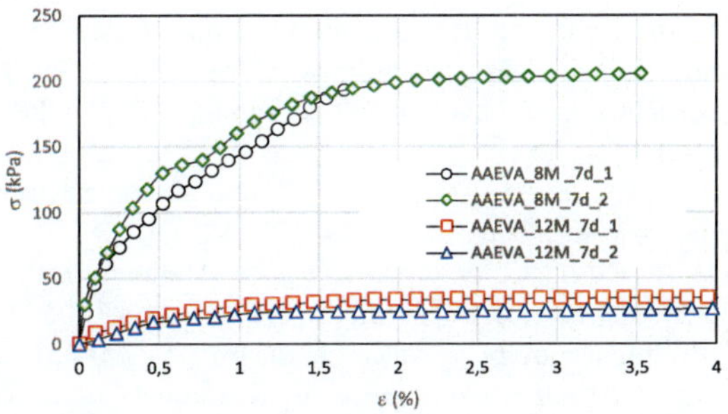

Fig. 6. Stress strain curves of AAEVA_8M and AAEVA_12M after 7 days of curing

Velocity (UV) values is reported in Fig. 5b. The ultrasonic velocity, a physical parameter directly correlated with mechanical performances of the material, tends to increase as UCS increases. A comparison between stress-strain curves of AAEVA_8M and AAEVA_12M after 7 days of curing is reported Fig. 6. At fixed curing time, lower compressive strengths were found for samples treated with higher NaOH molar concentration (i.e., AAEVA_12M). These findings are consistent with efflorescence observed on the surface of AAEVA_12M sample. The use high dosage of alkali concentration in the mixtures results in the formation of efflorescence, accompanied by a detrimental effect on the mechanical performance of the binder [24].

4 Conclusions

An experimental investigation on the alkaline activation of a natural pozzolanic material, as binder alternative to the traditional ones commonly used in the field of soil stabilization, has been developed. Etna Volcanic Ash (EVA) has been activated with two sodium hydroxide solutions prepared at 8M and 12M concentration. Reactivity of binders has been monitored over time by means of microstructural analysis. From a mineralogical point of view, alkali activated volcanic ash treated with 8M NaOH solution (i.e., AAEVA_8M) shows a high reactivity to promote the formation of secondary minerals (i.e., zeolites and smectite) from the volcanic ash, generally formed as result of alkaline activation reactions. From a macroscopic point of view, increase of unconfined compressive strength of AAEVA_8M over curing time is consistent with observed mineralogical evolution of the binder. Conversely, the use of a NaOH solution with higher molarity (i.e., 12M) affects the alkaline activation process, inhibiting the formation of zeolite crystals and thus promoting the precipitation of sodium carbonate hydrate because of the reaction between the excess of free alkalis of the pore solution with humid air containing CO_2. The occurrence of carbonatic compounds has a detrimental effect on the physical and mechanical features of the binder in terms of efflorescence formation and compressive strength reduction.

References

1. Russo, G., Capotosto, A., Croce, P., Modoni, M., Vitale E.: Lime treatment for the reuse of dredged marine sediments. In: Proceedings of the XVI ECSMGE Geotechnical Engineering for Infrastructure and Development, Edinburgh, September 13–17 (2015). ISBN 978–0–7277–6067–8
2. Guidobaldi, G., et al.: Chemo-mineralogical evolution and microstructural modifications of a lime treated pyroclastic soil. Eng. Geol. **245**, 333–343 (2018)
3. Vitale, E., Deneele, D., Russo, G.: Microstructural investigations on plasticity of lime-treated soils. Minerals **10**(5), 386 (2020). https://doi.org/10.3390/min10050386
4. Vitale, E., et al.: Influence of pore water chemistry on hydraulic conductivity of kaolinite suspensions. In: 6th Italian Conference of Researchers in Geotechnical Engineering, CNRIG2016, Procedia Engineering, vol. 158, pp. 81–86 (2016)
5. Tsuchida, T., Egashira, K: The Lightweight Treated Soil Method: New Geomaterials for Soft Ground Engineering in Coastal Areas, Taylor & Francis Routledge (2004)
6. Vitale, E., et al.: Chemo-mechanical behaviour of lightweight cemented soils. Acta Geotech. **15**(4), 933–945 (2019). https://doi.org/10.1007/s11440-019-00797-8
7. Provis, J.L., Bernal, S.A.: Geopolymers and related alkali-activated materials. Annu. Rev. Mater. Res. **44**, 299–327 (2014)
8. Xu, H., Van Deventer, J.S.: Geopolymerisation of multiple minerals. Miner. Eng. **15**(12), 1131–1139 (2002)
9. Cristelo, N., Glendinning, S., Teixeira Pinto, A.: Deep soft soil improvement by alkaline activation. Proc. Inst. Civil Eng.-Ground Improv. **164** (2), 73–82 (2011)
10. Rios, S., Ramos, C., Viana da Fonseca, A., Cruz, N., Rodrigues, C.: Mechanical and durability properties of a soil stabilised with an alkali-activated cement. Eur. J. Environ. Civ. Eng. **23** (2), 245–267 (2019)
11. Vitale, E., Russo, G., Dell'Agli, G., Ferone, C., Bartolomeo, C.: Mechanical behaviour of soil improved by alkali activated binders. Environments **4**, 80 (2017). https://doi.org/10.3390/environments4040080
12. Coudert, E., Deneele, D., Russo, G., Vitale, E., Tarantino, A.: Microstructural evolution and mechanical behaviour of alkali activated fly ash binder treated clay. Constr. Build. Mater. **285**, 122917 (2021). https://doi.org/10.1016/j.conbuildmat.2021.122917, ISSN 0950–0618
13. Vitale, E., Marocco, A., Khatib, M., Russo, G.: Hydro-mechanical behaviour of alkali-activated binder treated soil. In: E3S Web of Conferences, vol. 195 (2020). https://doi.org/10.1051/e3sconf/202019506003
14. Vitale, E., Russo, G., Deneele, D.: Multi-scale analysis on soil improved by alkali activated binders. In: E3S Web of Conferences vol. 92, p. 11003 (2019). https://doi.org/10.1051/e3sconf/20199211003

15. Miraki, H., Shariatmadari, N., Ghadir, P., Jahandari, S., Tao, Z., Siddique, R.: Clayey soil stabilization using alkali-activated volcanic ash and slag. J. Rock Mech. Geotech. Eng. **14**(2), 576–591 (2022)
16. Ghadir, P., Zamanian, M., Mahbubi-Motlagh, N., Saberian, M., Li, J., Ranjbar, N.: Shear strength and life cycle assessment of volcanic ash-based geopolymer and cement stabilized soil: a comparative study. Transp. Geotech. **31**, 100639 (2021). https://doi.org/10.1016/j.trgeo.2021.100639
17. Cucciniello, C., et al.: From olivine nephelinite, basanite and basalt to peralkaline trachyphonolite and comendite in the Ankaratra volcanic complex, Madagascar: 40Ar/39Ar ages, phase compositions and bulk-rock geochemical and isotopic evolution. Lithos, 274–275, 363–382 (2017). https://doi.org/10.1016/j.lithos.2016.12.026
18. UNI EN 14579: Natural Stone Test Methods-Determination of Sound Speed Propagation, CNR-ICR, Rome, (2005)
19. Lancellotti, I., Ponzoni, C., Barbieri, L., Leonelli, C.: Alkali activation processes for incinerator residues management. Waste Manage. **33**, 1740–1749 (2013). https://doi.org/10.1016/j.wasman.2013.04.013
20. Tchakoute, H.K., Elimbi, A., Yanne, E., Djangang, C.N.: Utilization of volcanic ashes for the production of geopolymers cured at ambient temperature. Cem. Concr. Compos. **2013**(38), 75–81 (2013). https://doi.org/10.1016/j.cemconcomp.2013.03.010
21. Firdous, R., Stephan, D.: Effect of silica modulus on the geopolymerization activity of natural pozzolans, Constr. Build. Mater. 219, 31–43 (2019). ISSN 0950–0618, https://doi.org/10.1016/j.conbuildmat.2019.05.161
22. Whitney and Evans - Abbreviations for names of rock-forming minerals - American Mineralogist, 95, 185–187 (2010)
23. Játiva, A., Ruales, E., Etxeberria, M.: Volcanic ash as a sustainable binder material: an extensive review. Materials (Basel). **14**(5), 1302 (2021). https://doi.org/10.3390/ma14051302.PMID:33800514;PMCID:PMC7962944
24. Bondar, D., Lynsdale, C.J., Milestone, N.B., Hassani, N., Ramezanianpour, A.A.: Effect of type, form, and dosage of activators on strength of alkali-activated natural pozzolans. Cem. Concr. Compos. **33**(2), 251–260 (2011)

Hydraulic Conductivity and Compressibility of Soils Treated with Fly Ash

M. Di Sante(✉), I. Bellezza, D. Bernardo, E. Fratalocchi, F. Mazzieri, and F. Pasqualini

Università Politecnica delle Marche, Ancona, Italy
m.disante@univpm.it

Abstract. In the present note, experimental results of geotechnical tests on soil-fly ash mixtures are presented, focusing on hydraulic conductivity and compressibility of the stabilized soils. Two excavated natural soils with different characteristics were treated with 4% of a high pozzolanic fly ash. Samples were obtained by wet mixing of soil and binder and subsequent Standard Proctor compaction. Permeability tests and one-dimensional compression tests were carried out. Results are presented in comparison with those of the same soils treated with lime or with a Portland Cement. SEM observations and Energy Dispersive X-Ray Spectroscopy were performed both on the additive and on the treated soils. Fly ash treatment was found capable to increase the soil hydraulic conductivity and to appreciably reduce the compressibility. Anyway, the fly ash stabilization resulted less effective if compared with that by traditional binders. Microstructure investigation on fly ash treated samples shows the not uniform presence of reaction products suggesting that environmental conditions should be further optimized for a higher reactions development.

Keywords: Soil Stabilization · Fly Ash · Hydraulic conductivity · Compressibility · Microstructure

1 Introduction

Need of sustainability in geotechnical design arose years ago and is constantly growing, fostered by the goals of the European Green Deal.

Soil stabilization surely contributes to this goal allowing the reuse of excavated soil treated with different additives. Several studies (e.g. [1]) indicate that the re-use of excavated soils and rocks saves up to 14 kg CO_2 per ton and 250 MJ/ton of non-renewable energy [1] as transport, landfilling, and collection from quarries are reduced. As far as binders are concerned they are generally classified as traditional binders or by-products from industrial processes, with important differences in terms of sustainability, as discussed in the following.

A. Ferrari et al. (Eds.): CNRIG 2023, SSGG, pp. 800–807, 2023.
https://doi.org/10.1007/978-3-031-34761-0_96

1.1 Soil Stabilization by Traditional Binders

Among the traditional binders used for soil stabilization, lime and cement are the most used. Their production causes significant CO_2 emission (e.g. 1006.5 kg CO_2/ton of lime [2]) and environmental issues linked with the extraction of raw material for their production. On the other hand, their potentiality are well known and the macro-micro modifications to the soil structure have been extensively studied (e.g., [3, 4]). Testing standards and construction procedures have been accordingly finalized.

In particular, lime acts as calcium supplier for clayey soils triggering cation exchange (and consequent aggregation of clayey particles) and reacting with silica and alumina (available from the clay fraction in high pH environment) producing long-term pozzolanic products. Cement can stabilize also sandy soil with very low clay fraction, given its composition that includes calcium silicates and aluminates [5].

Treatment with lime or cement generally improves the mechanical characteristics of a soil (increase in strength and reduction in compressibility), whereas, with reference to the hydraulic conductivity, it can be significantly increased or decreased as a function of the type of binder and soil to be treated and the compaction water content [3, 6–8].

1.2 By-Products for Soil Stabilization

Using by-products as binders for soil stabilization is certainly an environmental-friendly approach. Recent research applied to geomaterials is related to soils treated with rice husk ash, mushroom waste, wet olive pomace, marble powder, ground granulated blast furnace slag or fly ash [9]. All these by-products are able to improve the soil mechanical characteristics but, in some cases, side effects can compromise the improvement (e.g., mushroom waste stabilization is not ensured in the long term; wet olive pomace can cause unfavorable environmental conditions for the pozzolanic reactions). Inclusion of fibers such as sisal, coir or juta is also a new research field to provide tensile strength to soil stabilized compacted systems [10–12].

In the present note, experimental results on soil-fly ash mixtures are presented, focusing on hydraulic conductivity and compressibility. Results are discussed also in comparison with those of the same soils treated with lime or with a Portland cement.

2 Materials and Test Methods

The soils used in the present study (named SGT and BLZ) are two clays (CL accord-ing to USCS-ASTMD2487–93) of low to medium plasticity and with different fine fraction and sand content (Table 1). The former was alternatively treated with 4% of lime and with the same amount of fly ash, while the latter (given its higher sandy content) was alternatively treated with 4% of cement and the same amount of fly ash.

The lime and the cement used are a fine calcic quicklime (QL) classified as CL80-Q dp (UNI EN 459–1) and an ordinary Portland cement (PC), respectively. The Fly Ash (FA) is a high pozzolanic fly ash (supplied by General Admixtures SpA). Its unit weight is 2200 kg/m^3. Microstructural and chemical investigation was performed on the FA by means of SEM observations and Energy Dispersive X-Ray Spectroscopy, EDS

Table 1. Characteristics of the soils

characteristic	SGT soil	BLZ soil
sand (%)	3	37
fine (%)	97	63
clay (<2μm,%)	39	18
Liquid limit (%)	40	29
Plasticity index (%)	20	11

(FESEMSUPRA40–ZEISS), after air dewatering of samples and their gilding (by an Emitech K550 sputter coater). SEM images at several magnification levels are displayed in Fig. 1. EDS highlighted the presence of 45% Silicon, 23% Aluminum,10% Iron, 10% Calcium, 4% Potassium and 7% other components (% by weight, excluding Oxygen).

The X-Ray diffraction spectrum of FA (Philips diffractometer, PW1730 X-ray generator, PW 1050/70 goniometer and CuK radiation) is displayed in Fig. 2: the mineralogical composition include silicon dioxide (quartz), calcium oxide and mullite ($Al_{4.56} Si_{1.44} O_{9.72}$) as main components.

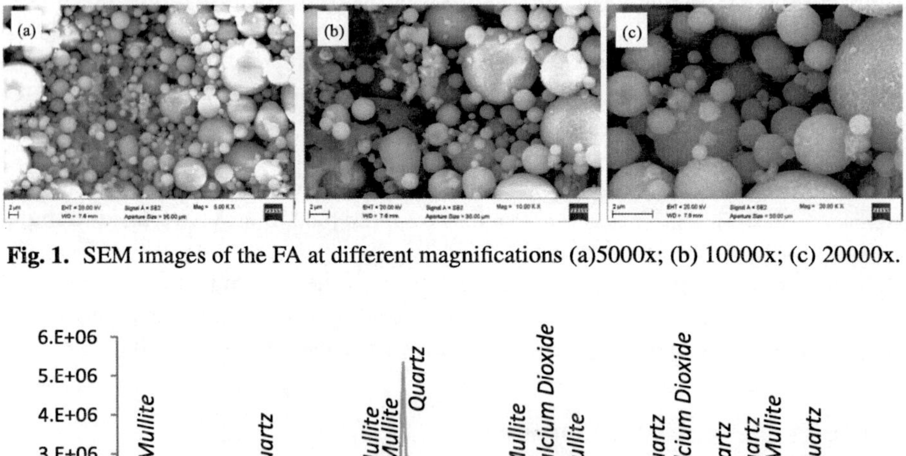

Fig. 1. SEM images of the FA at different magnifications (a)5000x; (b) 10000x; (c) 20000x.

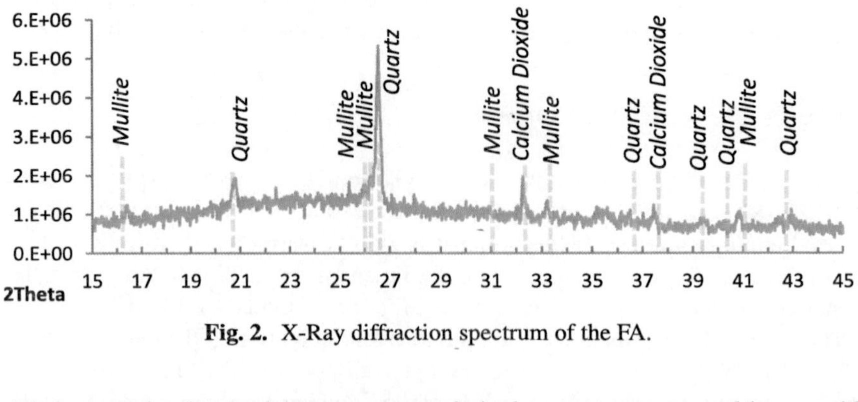

Fig. 2. X-Ray diffraction spectrum of the FA.

All the samples for the laboratory geotechnical tests were prepared by crumbling the air dried soil (to 2 mm in size), adding the amount of water necessary to reach the desired moisture content, then mixing binders and soil at the wet state (as usually done in the field [13]) until a uniform distribution was achieved. The soil mixture was then compacted according to the Standard Proctor procedure (ASTM D698–12) in one layer, obtaining samples of 101 mm in diameter and 40 mm in height. The samples were then cured in sealed plastic bags at a temperature of 20 ± 1 °C.

Permeability tests (ASTM D5084–10) were carried out in flexible walls permeameters at 35 kPa of effective confining stress with tap water as the permeant. The tests on the treated specimens were carried out at least until 28 days of curing.

One-dimensional incremental load consolidation tests (ASTM D24345–11) were performed on specimens after 7 days of curing.

Microstructural investigation of treated and untreated samples was performed by FESEMSUPRA40–ZEISS and TESCAN VEGA3 scanning electron microscopes.

3 Experimental Results

3.1 Compaction

Three samples of the SGT soil and of the BLZ soil added with fly ash were compacted at the water content, w ~ 19% and 21%, respectively. The typical reduction in w after treatment with classical binders occurred also for fly ash treatment: w of SGT soil reduced of 1% and w of BLZ soil reduced of 1.6% (mean values calculated among all the compacted samples). This can be due to the immediate hydration of the CaO in the FA and also suggests the possibility of an immediate start of pozzolanic reactions.

Standard Proctor compaction curves of the untreated soils and of soil +4% classical binders are shown in Fig. 3.

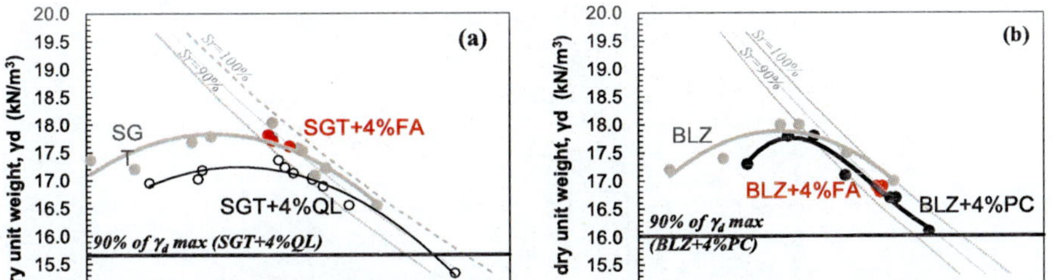

Fig. 3. Proctor compaction curves of treated and untreated soil; (a) SGT soil (b) BLZ soil.

In the same figure the compaction results of soils +4%FA are highlighted in red. For BLZ soil (Fig. 3b), the samples compacted at the water content, w = 21% show the same dry unit weight (16.8 kN/m^3) irrespective of the type and of the presence of the binder while for SGT soil (Fig. 3a) the dry unit weight of the lime treated soil (17 kN/m^3) is slightly lower than that of the other samples (17.8 kN/m^3), w being equal. In Fig. 3 it is possible to verify that compaction characteristics of the soils treated with FA can guarantee a sufficiently high dry unit weight (higher than 90% of the γ_{dmax} of the soil treated with classical binders).

3.2 Hydraulic Conductivity

An increase in hydraulic conductivity, k, due to FA addition resulted for both soils (Fig. 4).

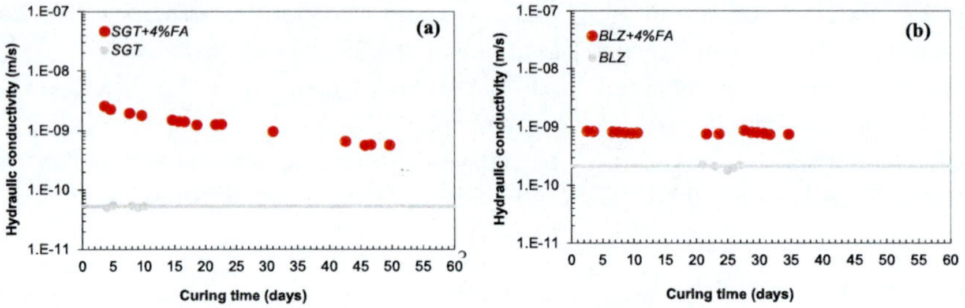

Fig. 4. Hydraulic conductivity of (a) SGT soil and SGT + 4%FA; (b) BLZ soil and BLZ + 4%FA

The major increase, from $5 \cdot 10^{-11}$ to $3 \cdot 10^{-9}$ m/s, is registered at the first days of curing for the SGT soil, which has a higher fine content and a higher clay fraction (Table 1) than the BLZ soil. In addition, the treated SGT soil shows a significant decreasing trend with curing (k values at 40 days attained $6 \cdot 10^{-10}$ m/s); this trend is typical for lime treated soils, due to the growth of pozzolanic products in gel form that partially fill the inter-aggregate porosity resulting from cation exchange [3]. This suggests that CaO in the fly ash added to a soil with a significant clay fraction is capable to trigger a sequence of reaction similar to that of a lime treated soil. In the BLZ soil, which has 37% of sand and whose clay fraction is a half of that of SGT soil (Table 1) no significant reduction with curing is observable.

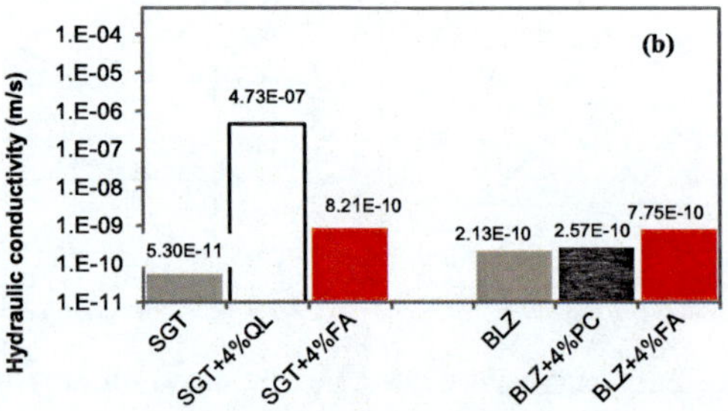

Fig. 5. Long term hydraulic conductivity (mean values after 20 days of curing) of mixtures of soil +4%FA, soil 4% classical binders and untreated soil.

The comparison of the long term k values between FA and classical binders treatments (Fig. 5) shows that:

- FA has a lower capability of increasing the hydraulic conductivity if compared with the same amount of quicklime added to the SGT soil (more clayey than BLZ) at the considered w (increase of 4 order of magnitude vs 1–2 omg);

- with reference to the BLZ soil (sandy), the addition of PC does not affect the hydraulic conductivity of the soil that, on the contrary, is increased by the addition of FA becoming 4 times greater.

3.3 Compressibility

Figure 6a shows the results of the one-dimensional compression tests performed at 7 days of curing on specimens of untreated SGT soil and of FA and QL mixtures; the significant reduction in compressibility caused by FA addition is evident.

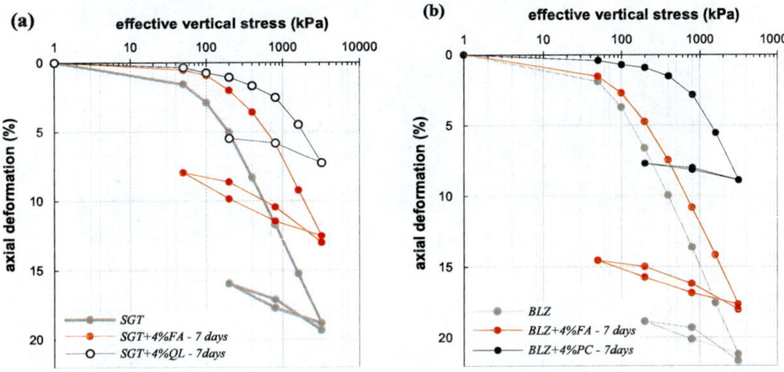

Fig. 6. Compressibility curves of (a) SGT soil: untreated and treated with 4%FA and 4%QL; (b) BLZ soil: treated with 4%FA and 4%PC.

In particular, at the range of 50–100 kPa of effective vertical stress (typical values for designing of earthworks, e.g. road embankments) axial deformation is reduced by both binders to a similar extent. When the vertical confining stress is increased, the untreated soil showed a maximum axial deformation of 19% that is reduced to 13% and 7% after FA and QL addition, respectively. The yield stress follows the same hierarchy increasing from 100 kPa (untreated specimen) to 400 kPa and 800 kPa for the FA and QL treatment, respectively.

Figure 6b shows the comparison between the compressibility curves of the untreated BLZ soil, of BLZ soil + 4%FA and BLZ soil + 4%PC, at 7 days of curing.

In this case, at the range of 50–100 kPa of effective vertical stress, axial deformation of FA treated specimen is similar to the untreated soil and 3–4 times higher than PC treated specimen. When the vertical confining stress is increased, the difference in axial deformation between the two treated specimens increases, reaching 17% at 3200 kPa for soil + 4%FA and 9% for soil + 4%PC. The yield stress for the FA treatment is 100 kPa and for the PC treatment the value approaches 800 kPa.

3.4 Microstructure

The SEM images shown in Fig. 7 were obtained after the end of the k tests (Sect. 3.2).

With reference to the FA treated samples, the presence of fly ash particles and of reaction products are not uniformly distributed across the observed samples. Comparing

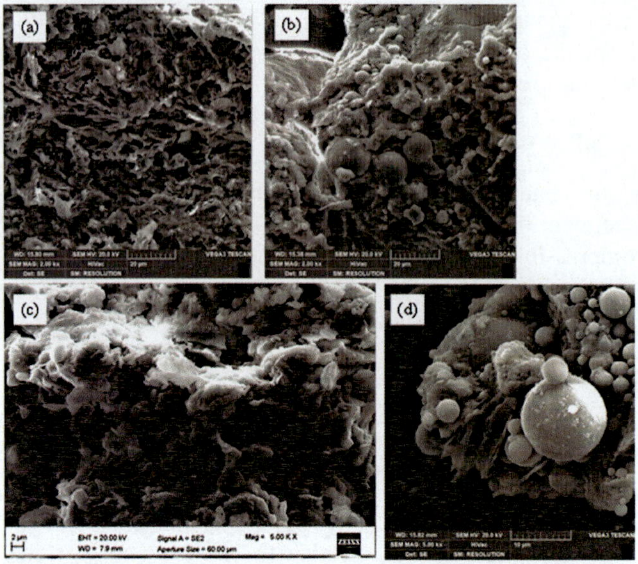

Fig. 7. SEM images of (a) untreated BLZ soil, (b) BLZ soil treated with 4% FA, (c) untreated SGT soil, (b) SGT soil treated with 4% FA, after the permeability tests (Sect. 3.2).

Fig. 7a with Fig. 7b and Fig. 7c with Fig. 7d (couples of untreated and treated samples of BLZ soil and SGT soil, respectively), different appearance is easily observable. In the treated samples (Fig. 7b and 7d) fly ash results in two different forms: unreacted spherical particles with a smooth surface and particles with rough surface. In the first case, the unreacted particles can only act as a filler with the ability to increase the packing effect. In the second case, amorphous pozzolanic products can be observed around particles [13, 14].

Pozzolanic products are not developed to a great extent, reasonably due to two main reasons: the amount of fly ash was too small (4% by weight, selected to be compared with those of classical binders) and the environmental conditions were not optimal for reactions development. In fact, it is known that cementing reactions are fostered by high pH and temperature together with significant amount of dissolved Ca ions in pore water [14]. This conditions could be ensured by the addition of a small amount of quicklime to the soil + FA mix.

4 Concluding Remarks

With reference to the investigated soils, treatment with 4% of FA increased the hydraulic conductivity of about 1 omg and reduced the compressibility. Anyway, the FA stabilization resulted to be less effective if compared with that by the same amount of lime or cement. Micro-structure investigation on FA treated samples showed that reaction products was not uniformly distributed and not developed to a great extent. This demonstrated the need to optimize the stabilization conditions, e.g. addition of small amounts of lime together with FA, to ensure a higher reactions development.

Acknowledgements. The authors thank the former students Andrea Iacoponi e Martina Ferraro for their cooperation in the experimental laboratory activities.

References

1. Blengini, G.: Resources and waste management in Turin (Italy): the role of recycled aggregates in the sustainable supply mix. J. Clean. Prod. **18**, 1021–1030 (2010)
2. Zhang, R., Long, M., Zheng, J.: Comparison of environmental Impacts of two alternative stabilization techniques on expansive soil slopes. Adv. Civ. Eng. **19**, 1–13 (2019)
3. Beetham, P., Dijkstra, T., Dixon, N., Fleming, P., Hutchinson, R., Bateman, J.: Lime stabilisation for earthworks: a UK perspective. Proc. Inst, Civ. Eng. – Ground Improv. **168**(GI2), 81–95 (2014)
4. Di Sante, M., Fratalocchi, E., Mazzieri, F., Pasqualini, E.: Time of reaction in a lime treated clayey soil and influence of curing conditions on its microstructure and behaviour. Appl. Clay Sci. **99**, 100–109 (2014)
5. Winterkorn, H.F., Pamukcu, S.: Soil stabilization and grouting. In: Fang, H.-Y. (ed.) Foundation Engineering Handbook, pp. 317–378. Van Nostrand Reinhold, New York (1991)
6. Bhattacharia, S., Bhatty, J.I., Todres, H.A: Stabilization of clay soils by portland cement or lime - a critical review of literature. In: PCA R&D Serial No. 2066, Portland Cement Association, vol. 60. Skokie Illinois USA (2003)
7. Di Sante, M., Fratalocchi, E., Mazzieri, F.: Effects of variation in compaction water content on geotechnical properties of lime-treated clayey soil. Procedia Eng. **158**, 63–68 (2016)
8. Bellezza, I., Fratalocchi, E.: Effectiveness of cement on hydraulic conductivity of compacted soil–cement mixtures. Ground Improv. **10**(2), 77–90 (2006)
9. Devi, k., Chhachhia, A., Kumar, A.: Soil Improvement using waste materials: a review. J. Build. Mat. Sci. **2**(1), 11–17 (2020)
10. Anggraini, V., Asadi, A., Huat, B.B.K., Nahazanan, H.: Effects of coir fibers on tensile and compressive strength of lime treated soft soil. Measurement **59**, 372–381 (2015)
11. Kumar, S., KumarDutta, R.: Unconfined compressive strength of bentonite-lime-phosphogypsum mixture reinforced with sisal fibers Jord. J. Civ. Eng. **8**(3), 239–250 (2014)
12. Gullu, H., Khudir, A.: Effect of freeze-thaw cycles on unconfined compressivestrength of fine-grained soil treated with jute fiber, steel fiber and lime. Cold Reg. Sci. Tech. **106–107**, 55–65 (2014)
13. Chindaprasirt, P., Jaturapitakkul, C., Sinsiri, T.: Effect of fly ash fineness on microstructure of blended cement paste. Constr. Build. Mat. **21**, 1534–1541 (2007)
14. Fraay, A., Bijen, J.M., de Haan, Y.M.: The reaction of fly ash in concrete: a critical examination. Cem. Concr. Res. **19**, 235–246 (1989)

Numerical Analysis of the Behaviour of Energy Micropiles Used for Heat Storage: A Case Study in Turku (Finland)

Marco Gerola[1(✉)], Arianna Lupattelli[2], Francesco Cecinato[1], Diana Salciarini[2], and Teppo Arola[3]

[1] Department of Earth Sciences "A. Desio", Università degli Studi di Milano, Via Mangiagalli, 34, 20133 Milan, Italy
marco.gerola@unimi.it

[2] Department of Civil and Environmental Engineering, Università degli Studi di Perugia, Via G. Duranti 93, 06125 Perugia, Italy

[3] Geological Survey of Finland (GTK), Vuorimiehentie 5, 02151 Espoo, Finland

Abstract. Energy Micropiles (EMPs) are a promising emerging technology thanks to their dual function of providing both structural support and energy supply to both pre-existing and new buildings. However, further experimental and numerical research is needed to fully understand their thermal and mechanical behaviour. In this work, a geothermal heat storage case study consisting of a group of EMPs installed in Turku (Finland) is presented. Beneath the main historic square of the city an underground car park was built, supported by more than 2000 end-bearing micropiles, including 561 EMPs consisting of a steel tube equipped with one U-shaped heat exchanger. Thermal energy is stored underground using solar collectors placed beneath the square's stone floor. In wintertime, geothermal heat is used to de-ice the square floor and for parking space heating. During site construction, thermal response tests (TRTs) were carried out on EMPs with different lengths and filling materials to assess the mean thermal conductivity of the soil. 3D numerical finite element simulations are presented to back-analyse the TRTs and investigate the effects of the thermal conductivity of the filling material and U-pipe shank spacing in EMP thermal performance. Further analysis is ongoing to assess the thermo-mechanical behaviour of EMPs and the surrounding soil at the Turku site.

Keywords: Micropiles · Energy Geostructures · Heat Storage · Thermal Response Test · Numerical Modelling

1 Introduction

The increasing energy demand, currently dependent mostly on fossil fuels, and climate implications have led to an accelerated growth in the exploitation of renewable energy resources. Storing heat underground from renewable sources for space and water heating is an innovative approach [1, 2].

A. Ferrari et al. (Eds.): CNRIG 2023, SSGG, pp. 808–815, 2023.
https://doi.org/10.1007/978-3-031-34761-0_97

Recently, geotechnical engineering has been involved with thermal storage through the analysis and design of Energy Geostructures (EGs), which can be used for energy applications in addition to providing structural support to any type of built environment. An attractive possibility is represented by Energy Micro-Piles (EMPs) that are micro-piles (MPs) equipped with heat exchangers, and typically used for both existing and new structures [3, 4]. A key aspect in the thermal performance of energy micro-piles is the soil thermal properties [5], that can be evaluated through Thermal Response Tests (TRT). TRT is a well-known in-situ technique performed to assess the soil thermal conductivity and MP-soil system's thermal resistance [6–10].

In this work, a geothermal heat storage case study consisting of a group of EMPs installed in the market square of the city of Turku, in Finland, is presented. Beneath the main market square, an underground car park was built, supported by more than 2000 end-bearing MPs, some of which equipped with heat exchangers. Thermal energy is stored underground using solar collectors, which are placed beneath the square's stone floor, allowing the use of geothermal heat to de-ice the square floor and for underground parking space heating in wintertime. During the site construction, two TRTs were carried out on EMPs with different lengths and filling materials, to assess the mean thermal conductivity of the soil. The main purpose of this work is to back-analyse the TRTs using two different models, and to investigate the effects of the thermal conductivity of the filling material and U-pipe shank spacing in EMP thermal performance.

2 Case Study

2.1 Turku Market Square

The case study is located in Turku (Finland) and includes an underground parking, marketplace, and shopping mall for a total construction area of 30810 m^2.

Heat is collected by means of solar collectors during summertime and stored underground through EMPs. The stored geothermal heat is used to de-ice the square floor and for parking space heating during wintertime.

The foundations are made up of more than 2000 MPs including 561 EMPs. These consist of a tubular steel cage with a thickness of 10 mm, diameters of 140 or 170 mm, and lengths variable from 15 to 55 m. To optimize thermal performance, EMPs have high conductive filling material. The heat exchanger fluid in the pipes is water.

According to local geotechnical studies, clay dominates Turku's subsoil, covering up to 50–60% of the land area. This lies on a Bedrock of plutonic and metamorphic rocks, including granites, gneisses and amphibolites [1].

2.2 In Situ Thermal Response Test

Two TRTs were performed on different EMPs to assess the thermal conductivity of the clayey soil, the initial ground temperature, and the thermal resistance of the EMP.

Both the TRTs lasted 3 days in which the inlet temperature was increased from 10 °C or 15 °C to 34 °C or 38 °C, respectively. The first TRT was made on a water-filled EMP with a length of 23.6 m. Based on the measured temperatures at the inlet and outlet pipes, a soil thermal conductivity of 1.65 W/mK and a thermal resistance of 0.05 mK/W were estimated. The second TRT was made on an EMP characterized by a depth of 36.5 m and wet sand as filling material. The estimated thermal conductivity and thermal resistance values are 1.32 W/mK and 0.073 mK/W, respectively.

3 Numerical Modelling

The two dominant heat exchange phenomena taking place in the pipe-pile-soil system are conduction and convection. The heat exchange between the heat-carrier fluid and pipe walls is mainly convective, while both in the pile filling material and in the soil domain the main heat exchange process is conduction. Radiation was considered negligible in the models.

The transient heat diffusion equation governs the conduction process:

$$\rho c_p \frac{\partial T}{\partial t} = -\nabla \boldsymbol{q} \tag{1}$$

where $q = \lambda \nabla T$ is the heat flux vector described by the Fourier's law, λ is the thermal conductivity of the medium, T is the temperature, ρ is the density of the material, c_p is the specific heat at constant pressure, and t is time.

By also considering the convection process, the transient heat equation becomes:

$$\rho c_p \frac{\partial T}{\partial t} + \rho c_p \boldsymbol{u} \cdot \nabla T = -\nabla \boldsymbol{q} \tag{2}$$

where u is the cross-section average fluid velocity along the tangent of the center line of the pipes. Heat exchange within the pipes can be formulated as a 1D problem by implementing the momentum conservation and the mass conservatnquations into the energy balance equation:

$$A_p \rho_f c_{pf} \frac{\partial T}{\partial t} + A_p \rho_f c_{pf} \boldsymbol{u} \cdot \nabla T = \nabla A_p \lambda_f \nabla T + f_D \frac{\rho_f A_p}{2 d_h} |u|^3 + q'_{wall} \tag{3}$$

where A_p is the pipe cross-section area available for flow, ρ_f is the fluid density, c_{pf}. is the fluid specific heat capacity at constant pressure. The heat dissipated by internal friction within the fluid is represented by the second term on the right-hand side of Eq. (3) and q'_{wall} accounts for the radial heat exchange through pipe walls.

The tested steel EMPs have circular cross sections and were installed with single U-loop heat exchanger pipes. Two 3D numerical finite element (FE) models of EMPs were developed using the *Comsol Multiphysics* code to back-analyse the TRTs following two different modelling approaches, as detailed in Sect. 3.1 and 3.2.

3.1 3D Model with 1D Pipes Representation

A first version of the EMP FE model was set up using the built-in *Comsol Multiphysics* module "heat transfer in pipes" allowing quicker computations by representing heat exchanger pipes in a simplified 1D fashion.

A cylindrical-shaped domain was considered, characterized by radius and height respectively of 1 m and 24.5 m or 37.5 m, depending on which TRT is being simulated. By selecting these values, the lateral and bottom domain boundaries are far enough from the pile (located at the domain center) to avoid any temperature change during the whole TRT simulations.

The fluid flow, the convective process within the pipe and the conduction process through the pipe walls were modelled via 1D elements solving Eq. (3). The surrounding domain, composed of soil, steel MP and pile filling material, was modelled as a 3D continuum.

It can be observed that in this case, contrary to a standard energy pile situation, the pile diameter is of the same order of magnitude as the heat exchanger pipe diameter, leading to a possibly scarce accuracy in reproducing the cross-sectional temperature field when the pipes are reproduced as 1D elements. To overcome this limitation, a numerical artifice was adopted by reducing the EMP cross-section area by subtracting the pipes' cross-section area. This was shown to yield accurate simulation results, as detailed in Sect. 3.2.

As regards boundary conditions, a fixed temperature of 9.1 °C (equal to the undisturbed soil temperature) was assigned to the bottom surface of the domain, while the detected air temperature time history was assigned to the top domain boundary. The air temperature was measured during the TRTs, while the initial soil temperature was measured before the TRTs. The adiabatic boundary condition was assigned to the lateral domain boundary. As regards initial conditions, the temperature measured within the EMP before the beginning of the TRT was set as the initial temperature in the model domain. The TRT was simulated by activating the heat transfer flow within the pipes, assigning the inlet temperature and the fluid velocity measured during the test. The geometry of the FE model is shown in Fig. 1a.

3.2 3D Model with 3D Pipes Representation

A second version of the model was built by modelling the heat exchanger pipes as fully 3D elements.

In this case, the convection process was simulated by assigning the average fluid velocity within the pipe. Only the axial component of the velocity was specified, representing laminar fluid flow. To account for the occurrence of turbulent flow, which was in fact achieved during both TRTs, the thermal conductivity of the heat exchanger fluid was set 100 times higher in the radial directions than the vertical one, as suggested, *e.g.*, by [14]. The geometry of the FE fully 3D model is shown in Fig. 1b.

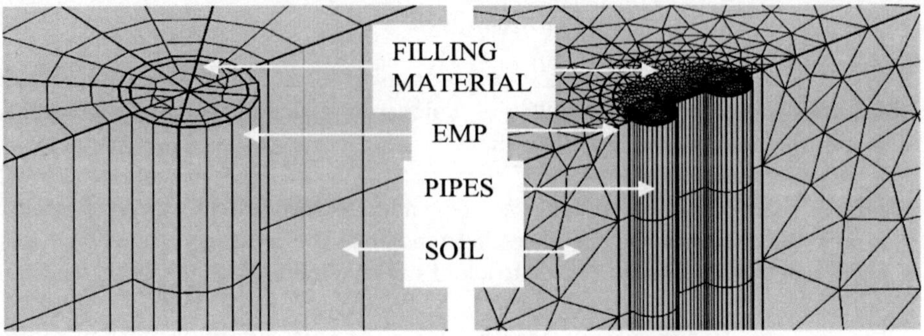

Fig. 1. Details of the FE models geometry and discretization: *a)* 1D pipes representation, and *b)* 3D pipes representation.

3.3 Numerical Analysis of the TRT Results

The thermal properties of the materials adopted in the model are presented in Table 1. The thermal conductivity of the clayey soil was set equal to the one estimated from the results of the TRT. No spacers were installed within the EMP to separate the inlet and outlet pipe. Thus, the shank spacing is likely to vary with depth. In the FE models the inlet-outlet pipe distance was set as constant with depth and was calibrated based on the TRT results (equal to 5.2 cm for the sand filled pile with 3D pipes model).

Both TRTs, executed on EMPs with two different filling materials, were simulated using both the 1D pipe representation and the 3D pipe representation.

Table 1. Thermal properties and density used in both FE models for each material.

Materials	Thermal conductivity	Specific heat capacity	Density
	W/(m K)	J/(kg K)	Kg/m^3
Water	0.6 [12]	4186 [12]	1000 [12]
Sand	2.2 [11]	1190 [11]	2100 [11]
Clay (water-filled)	1.65 [TRT]	1200 [11]	1200 [11]
Clay (sand-filled)	1.32 [TRT]	1200 [11]	1200 [11]
Steel	45 [12]	7850 [3]	7850 [3]
Pipe	0.42 [13]	2000 [13]	960 [13]

Sand-Filled Pile

The results of the two FE models in terms of the difference between the inlet temperature T_{IN} and outlet temperature T_{OUT} are shown in Fig. 2. It can be observed that the numerical simulations effectively reproduce the temperature measurements for the whole TRT duration. The temperature contour lines in the numerical domain cross-section, at a depth of 10 m at the end of the TRT simulation, with both 1D and 3D pipe representation, are shown in Fig. 3. A satisfactory agreement between the two types of temperature distribution can be observed. However, it can be observed that adopting the 3D pipe representation (Fig. 3b) yields more realistic results in terms of temperature homogeneity within the pipe cross sectional area.

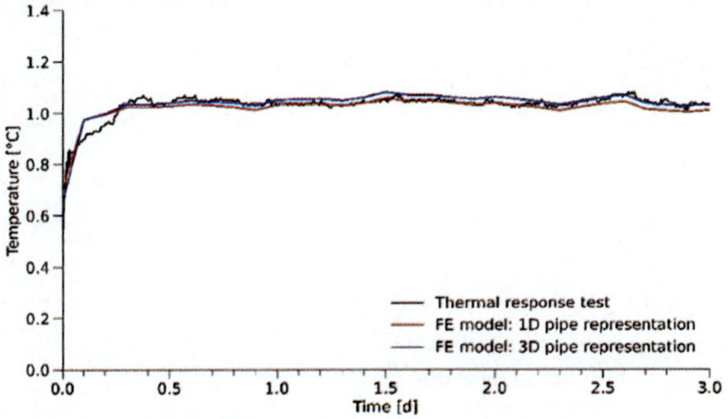

Fig. 2. Comparison between TRT result and FE simulations of the temperature difference between inlet and outlet pipes within the sand-filled EMP.

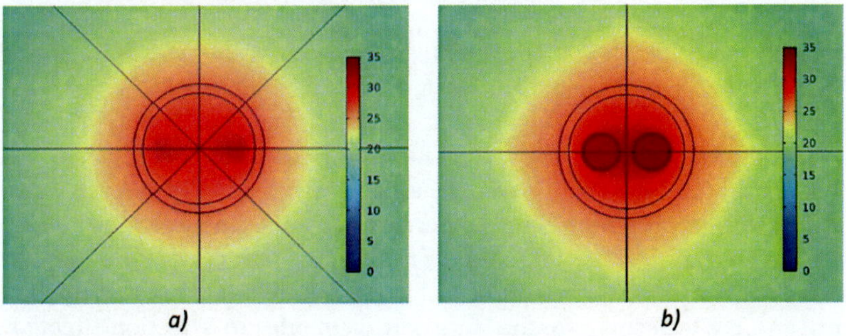

Fig. 3. Temperature contour (°C) at the end of the sand-filled at 10 m of depth TRT with *a)* 1D pipe representation, and *b)* 3D pipe representation.

Water Filled Pile

The results of the two FE models, in terms of $T_{IN} - T_{OUT}$, are shown in Fig. 4. It is worth noting that although the sand-filled EMP is deeper and endowed with a more conductive filling material, the TRTs exhibit a similar $T_{IN} - T_{OUT}$ steady-state value, implying that the sand-filled EMP is less energy efficient. This could be partly due to a locally lower soil thermal conductivity around the sand-filled EMP, as reported in Sect. 2.2, and partly to a non-constant shank spacing, due to the absence of physical spacers. Shank spacing (taken constant in the model, representing an average value) was calibrated based on the TRT results, and it is lower for the sand-filled EMP using both FE models.

These results suggest that the average shank spacing in the sand-filled EMP was smaller, increasing the thermal interaction between the pipes. Also, the sand-filled EMP exhibits a higher thermal resistance since the pipes, due to the smaller shank spacing, are located farther from the surrounding soil.

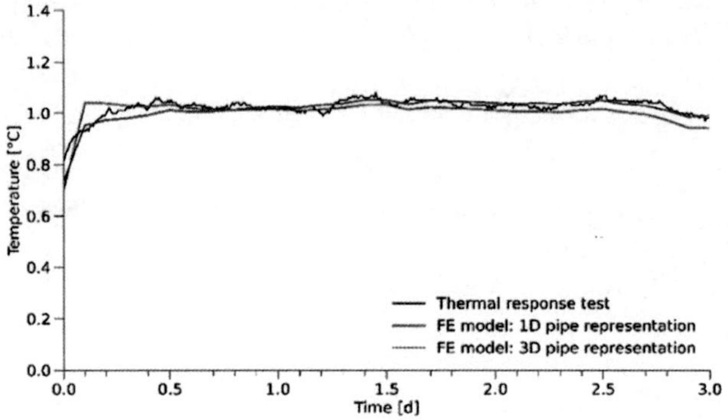

Fig. 4. Comparison between TRT result and FE simulations of the temperature difference between inlet and outlet pipes within the water filled EMP.

4 Conclusions

In this work, a numerical study is presented to back-analyze two TRTs carried out on EMPs with different lengths and filling materials, in the frame of a large geothermal heat storage project involving EMPs supporting an underground car park in Turku (Finland).

Two FE models were developed to back-analyze the TRTs, assuming different modelling hypotheses for the pipes that were simulated by means of 3D or 1D elements. Both FE models effectively reproduce the temperature difference between the inlet and outlet measurements for all the TRTs duration. The 3D pipe representation reproduces more accurately the temperature distribution within the pile and does not require to introduce numerical artifices to account for the large portion of pile cross-section effectively occupied by the pipes. On the other hand, the 1D pipe model is easier to set up and more computationally efficient.

Despite the sand filled EMP being the longest and having the most conductive filling, it seemed to exhibit a lower thermal efficiency than the water filled EMP. This can be ascribed partly to a locally lower thermal conductivity of the surrounding clayey soil and partly to a smaller equivalent shank spacing. This result suggests the need for spacers in EMPs, in order to keep a constant and adequate shank spacing throughout the pile length, avoiding thermal short-circuiting and the consequent overall pile thermal resistance increase.

Acknowledgements. The City of Turku is gratefully acknowledged for providing and permitting the use of the TRT data. M. Gerola and F. Cecinato gratefully acknowledge research funding under the "PON tematica Green – DM 1061/2021" scheme.

References

1. Lautkankare, R., Salomaa, N., Martinkauppi, B., Slobodenyuk, A.: Underground parking lot at Turku market square – Zero energy parking hall and the biggest solar energy storage in the world. In: Proceedings of the 2020|12th Nordic Symposium on Building Physics (NSB 2020), p. 172, Tallinn, Estonia (2020)
2. Baser, T., Mccartney, J.S., Ozdogan-Dolcek, A.: Thermal energy storage in borehole arrays installed in unsaturated soils. Adv. Soil Mech. Geotech. Eng. **7**, 374–390 (2019)
3. Cecinato, F., Salciarini, D.: Energy performance assessment of thermos-active micro-piles via numerical modeling and statistical analysis. Geomech. Energy Environ. **29**, 100268 (2022)
4. Salciarini, D., Cecinato, F.: Numerical modelling of thermo-active micropiles. In: Barla, Marco, Di Donna, Alice, Sterpi, Donatella (eds.) IACMAG 2021. LNCE, vol. 126, pp. 1102–1109. Springer, Cham (2021). https://doi.org/10.1007/978-3-030-64518-2_131
5. Sanner, B., Hellström, G., Spitler, J., Gehlin, S.: Thermal response test – current status and world-wide application. In: Proceedings World Geothermal Congress, Antalya, Turkey (2005)
6. Loveridge, F., Powrie, W.: Pile heat exchangers: thermal behaviour and interactions. Proc. Inst. Civil Eng. – Geotech. Eng. **166**(2), 178–196 (2013)
7. Murphy, K.D., McCartney, J.S., Henry, K.S.: Thermo-mechanical response tests on energy foundations with different heat exchanger configurations. Acta Geotech. **10**(2), 179–195 (2015)
8. Ronchi, F., Salciarini, D., Cavalagli, N., Tamagnini, C.: Numerical model of energy foundations behavior: the prototype of a geothermal micro-pile. Procedia Eng. **158**, 326–331 (2016)
9. Bourne-Webb, P.J., Lupattelli, A., Freitas, T.M.B., Salciarini, D.: The influence of initial shaft resistance mobilisation in the response of seasonally, thermally-activated pile foundations in granular media. Geomech. Energy Environ. **32**, 100299 (2022)
10. Ronchi, F., Salciarini, D., Cavalagli, N., Tamagnini, C.: Thermal response prediction of a prototype energy Micro-Pile. Geomech. Energy . Environ. **16**, 64–82 (2018)
11. Dalla Santa, G., et al.: An updated ground thermal properties database for GSHP applications. Geothermics **85**, 101758 (2020)
12. Laloui, L., Loria, A.F.R.: Analysis and Design of Energy Geostructures. Academic Press, Cambridge (2020)
13. Laloui, L., Di Donna, A.: Energy geostructures: innovation in underground engineering. Wiley, Hoboken (2013)
14. Marcotte, D., Pasquier, P.: On the estimation of thermal resistance in borehole thermal conductivity test. Renew. Energy **33**(11), 2407–2415 (2008)

Design Charts for Induced Partial Saturation: A Promising Mitigation Technique Against Liquefaction

Lucia Mele[1(✉)], Stefania Lirer[2], and Alessandro Flora[1]

[1] University of Napoli Federico II, Naples, Italy
lucia.mele@unina.it
[2] University of Roma Guglielmo Marconi, Rome, Italy

Abstract. Induced Partial Saturation (IPS) is considered one of the most innovative and promising techniques against liquefaction, due to its low-cost, eco sustainability and its applicability in the urbanised areas, where the need to reduce the risk of liquefaction must be addressed taking into account the integrity of the existing buildings. IPS consists of injecting gas/air into pore water in order to increase the compressibility of fluid phase, and thus, increasing soil liquefaction resistance. Although the effectiveness of IPS has been confirmed at small and large scale, its use in practical applications is not yet widespread. This can be basically attributed to a lack of design tools. This paper offers an insight on this topic, providing useful design charts based on the results of in situ (CPT or SPT) tests. The charts allow to choose the degree of saturation to apply in situ to have a desired increase of resistance. A simple application to the case study of Treasure Island (California) has finally been shown.

Keywords: Liquefaction · Induced Partial Saturation (IPS) · Treasure Island

1 Introduction

Earthquake induced soil liquefaction has occurred frequently after seismic events in the world, such as the earthquakes in 1964 Niigata (Japan), 2011 Christchurch (New Zealand), 2012 Emilia Romagna region (Italy) or a most recent case in 2018 Palu (Indonesia). The main cause of liquefaction is the temporary loss of soil shear strength due to the sudden increase in pore water pressure (Δu), which eventually leads to the phenomenon where loose, saturated sandy soil displays fluid-like characteristics [1, 2].

In engineering practice, the evaluation of the potential for the occurrence of liquefaction is usually done by means of simplified approaches in which, at each depth, a safety factor against liquefaction in free field conditions (FS) is computed by comparing the "seismic demand" generated by the earthquake and the "soil capacity" to resist liquefaction [3]. Both of them are usually expressed as cyclic stress ratios (respectively cyclic resistance ratio CRR and cyclic stress ratio CSR). Typically, CRR is linked to q_{c1Ncs} and $(N_1)_{60cs}$, which are respectively the corrected CPT tip resistance and the corrected SPT blow count counting for fine content effects [4]. The soil capacity referred to a

A. Ferrari et al. (Eds.): CNRIG 2023, SSGG, pp. 816–823, 2023.
https://doi.org/10.1007/978-3-031-34761-0_98

magnitude $M_w = 7.5$ and to a vertical effective stress $\sigma'_v = 103$ kPa ($CRR_{M=7.5,\sigma'v=1}$) is given by:

$$CRR_{M=7.5,\sigma'v=1} = exp\left(\frac{N_{1,60cs}}{14.1} + \left(\frac{N_{1,60cs}}{126}\right)^2 - \left(\frac{N_{1,60cs}}{23.6}\right)^3 + \left(\frac{N_{1,60cs}}{25.4}\right)^4 - 2.8\right)SPT \quad (1)$$

$$CRR_{M=7.5,\sigma'v=1} = exp\left(\frac{q_{c1Ncs}}{113} + \left(\frac{q_{c1Ncs}}{1000}\right)^2 - \left(\frac{q_{c1Ncs}}{140}\right)^3 + \left(\frac{q_{c1Ncs}}{137}\right)^4 - 2.8\right)CPT \quad (2)$$

MSF, K_σ and K_α are correcting factors to account respectively the effect of the duration of the seismic event, effective overburden stress and an initial static shear stress on the horizontal plane.

On the other hand, the cyclic stress ratio (CSR) may be found from:

$$CSR = 0.65\frac{a_{max}}{g}\frac{\sigma_v}{\sigma'_v}r_d \quad (3)$$

where σ_v and σ'_v are the vertical total and effective stresses at a depth z, a_{max} is the maximum horizontal acceleration, g is the gravity acceleration and r_d is a reduction factor accounting for soil deformability, whose expression can be found in Boulanger and Idriss [4]. Known CRR and CSR a safety factor (FS) can be quantified for each depth (FS (z) = CRR/CSR).

Soil liquefaction can induce a number of ground failures of varying severity: sand boils, ground settlements, cracks, ground oscillation, lateral spreads and flow failures. The resulting ground deformations cause extensive damages to shallow-founded buildings and other engineering facilities. This is the reason why it is of the outmost relevance to identify mitigation techniques able to satisfy the requisites of soundness and reliability in design, environmental compatibility and cost effectiveness. Existing methods for improving liquefaction resistance include grouting with cement or chemical solutions, foundation densification, gravel columns, foundation replacement and dewatering. While most of these techniques have advanced over the years, their application is problematic because of high construction cost, impact to surrounding infrastructure or effects on the surrounding environment. For example, traditional cement and chemical grouting, though widely used, pollute the surrounding environment and waterways, and risk damaging adjacent buildings when handled improperly. Thus, new techniques and methods for liquefaction mitigation for large-area construction should be developed that are low cost and non-damaging to existing structures while at the same time being environment-friendly, sustainably developed and pollution-free [5]. Induced Partial Saturation seems to meet the aforementioned requirements. IPS consists of introducing gas/air bubbles (environment-friendly) into pore water of saturated granular soils in order to increase the compressibility of fluid phase and thus, decreasing the excess pore pressure build-up. It is sufficient to reduce the degree of saturation by a small amount (generally > 80%) to achieve a considerable increase in soil capacity [6]. In this condition, gas/air bubbles are occluded within a continuous water phase and no mechanical effect of suction ($s = u_a - u_w$, where u_a and u_w are pore pressure of air and water, respectively) on the soil skeleton is expected.

The effectiveness of IPS as mitigation technique against liquefaction has been verified by many authors at laboratory scale [6–8] and in some field tests [9–11]. Although the effectiveness of IPS has been proved, it is still far from being a routine technology in engineering practice because there are still uncertainties about its durability and there is a complete lack of design guidelines. This paper offers a contribution to the last issue, providing useful design charts based on the results of in situ tests. The charts allow to choose the degree of saturation to apply to have a desired increase of soil capacity (ΔCRR). A simple application for the case study of Treasure Island has finally been reported.

2 Design Tool for IPS

2.1 Energetic Approach to Predict Liquefaction Resistance of Non-saturated Sandy Soils

For a long time, several researchers have tried to find the parameter(s) ruling the liquefaction resistance of non-saturated (gassy) soils. Nowadays one of the simplest and most used parameters is the potential volumetric strain (ε_v*) [12], that represents the maximum volumetric strain of the soil when the complete attainment of liquefaction has occurred (i.e. $u_a = u_w = \sigma$, corresponding to σ' = 0). The potential volumetric strain can be obtained by applying Boyle and Mariotte law:

$$\varepsilon_v* = \frac{e_0}{1 + e_0} \cdot (1 - S_{r0}) \cdot \left(1 - \frac{u_{a,0}}{\sigma}\right) \tag{4}$$

where e_0, S_{r0} and $u_{a,0}$ are the initial void ratio, degree of saturation and pore air pressure, respectively, while σ is the total stress.

Recently, Mele et al. [8] proposed to use the specific volumetric energy at liquefaction ($E_{v,liq}$) as the reference parameter. $E_{v,liq}$ is a synthetic and sound state parameter – depending on e, σ'_c and Sr - representing the work spent to reduce soil volume, and can be seen as the sum of three components [8]:

$$E_{v,liq} = E_{v,sk,liq} + E_{w,liq} + E_{air,liq} \tag{5}$$

$E_{v,sk,liq}$, $E_{w,liq}$ and $E_{air,liq}$ represent the specific work done respectively to cause the deformation of the soil skeleton, the flow of water and the flow of air into the pores network. Mele & Flora [13] and Mele et al. [6] showed the potentiality of $E_{v,liq}$ to predict the liquefaction resistance of partially saturated soils, that for sake of brevity has not been discussed in this paper. $E_{v,liq}$ is linked to the increase ΔCRR_{Nliq} of liquefaction resistance from saturated (CRR_s) to non-saturated conditions (CRR_{ns}) via the following equations for triaxial (CTX) and simple shear (CSS) conditions:

$$\Delta CRR_{Nliq} = -105.7 \cdot \left(\frac{E_{v,liq}}{p_a}\right)^2 + 10.2 \cdot \frac{E_{v,liq}}{p_a} \text{ for CTX tests} \tag{6}$$

$$\Delta CRR_{Nliq} = -89.6 \cdot \left(\frac{E_{v,liq}}{p_a}\right)^2 + 7.81 \cdot \frac{E_{v,liq}}{p_a} \text{ for CSS tests} \tag{7}$$

where $\Delta CRR_{Nliq} = CRR_{ns} - CRR_s$ is computed at a given number of equivalent cycles. Further details about the computation of $E_{v,liq}$ may be found in Mele et al., [8].

In Fig. 1, the results of cyclic triaxial tests performed on different sandy soils in different state conditions (Table 1) confirm Eq. (6). Equations (6 and 7) allow to predict, the non-saturated cyclic resistance CRR_{ns} ($CRR_{ns} = CRR_s + \Delta CRR_{Nliq}$) and therefore to plot the non-saturated cyclic resistance curve through a simple upwards translation of the one obtained for saturated soils, in the simplified hypothesis that the non-saturated and saturated cyclic resistance curves are parallel. Even though this latter hypothesis is far from being true for all possible values of N_{cyc}, it is a reasonable simplification in the range of values of engineering interest (i.e., $N_{liq} < 20$).

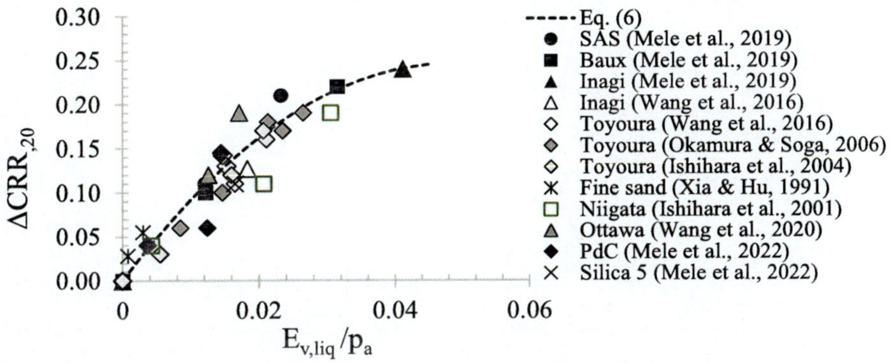

Fig. 1. Relationship between $E_{v,liq}/p_a$ and ΔCRR_{20} (modified from Mele et al., [6]).

Table 1. Range of material properties and state conditions within the energy model has been validated.

D_{50} (mm)	FC (%)	σ'_c (kPa)	Dr (%)	Sr (%)
0.11–0.47	0–20	25–118	30–77	48.0–99.5

The reliability of the energetic approach to predict liquefaction resistance of partially saturated soils, at least in the range of studied state properties (Table 1), has been widely confirmed by Mele & Flora [13] and Mele et al., [6]. Therefore, from a technical point of view, the energy-model can be used as a design tool for IPS technique.

2.2 Design Charts

The energetic model has been adopted to obtain design charts useful for IPS mitigation technique. Following the SPT and CPT based liquefaction triggering approach usually adopted in practice for the liquefaction potential analysis [14] in saturated soils, the increase of soil capacity ΔCRR has been linked to the in situ penetrating tests results (i.e. $(N_1)_{60cs}$ and q_{c1Ncs} for SPT and CPT tests). The liquefaction triggering curves pertaining to saturated soils, (Eqs. 1 and 2), have been translated for each investigated degree of saturation (Sr = 98; 95; 90 and 85%) and void ratio (D_r is linked to $(N_1)_{60cs}$ and q_{c1Ncs}, see for instance [14]), calculating the specific volumetric energy to liquefaction ($E_{v,liq}$), which is in turn correlated to ΔCRR (Eq. 7).

Partially-saturated curves in the planes $(N_1)_{60cs}$ *vs* CRR and q_{c1Ncs} *vs* CRR have been reported in Fig. 2a and 2b, respectively [15]. The upward translation of the saturated curves is possible because for high values of Sr, there is no mechanical effect of suction on soil skeleton (even though air and water pressure are slightly different) so $(N_1)_{60cs}$ and q_{c1Ncs} do not change passing from saturated to gassy conditions. If a soil results liquefiable from a liquefaction susceptibility analysis (CSR > CRR), the charts of Fig. 2 can be used as useful design tool to identify the degree of saturation to apply in situ to have the needed increment of soil capacity. A design example will be shown in the following section (Sect. 3).

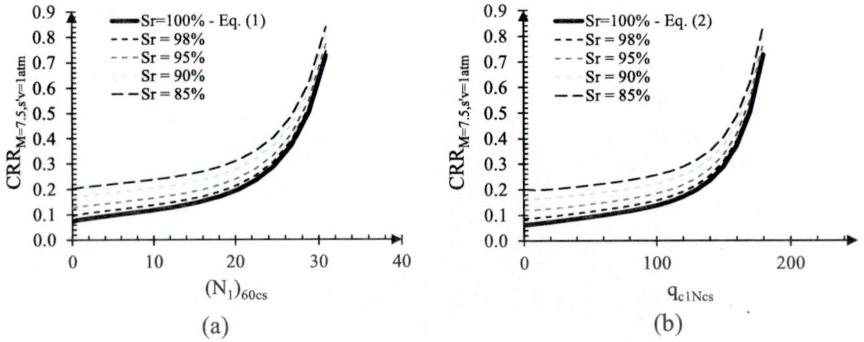

Fig. 2. SPT (a) and CPT (b) based liquefaction triggering curves for partially saturated soils (modified from [15]).

3 Design of IPS for the Case Study of Treasure Island

The case study of Treasure Island (San Francisco, California) has been chosen in order to apply the IPS design procedure described in Sect. 2.2. Treasure Island is a 400 acre man-made island immediately northwest of the Yerba Buena Island, a rock outcrop in San Francisco Bay. During the 1989 Loma Prieta earthquake, the island was affected by soil liquefaction and other liquefaction-related phenomena (sand boils and lateral spreading) [16]. The soils at Treasure Island may be grouped into four broad categories: the fill material (hydraulic fill) until 13 m from ground surface, recent bay sediments (Young Bay Mud) from 13 to 28.8 m, native shoal sands (fine to medium sand) from 28.8 to 41.2 m and older bay sediments (old Bay Clay) from 41.2 to 88 m, at which the bedrock was assumed. The ground water table is at a depth of 4 m from the ground surface (Fig. 3a). CPT tests data are available. For the equivalent corrected CPT tip resistance values, q_{c1ncs} profile has been achieved considering an average FC equal to 15% [17] (Fig. 3b). The evaluation of the potential for the occurrence of liquefaction has been done following the well-known stress-based approach (Sect. 1). The comparison between CRR (Eq. 5) and CSR (Eq. 3, using $a_{max} = 0.234g$ and introducing the magnitude scaling factor, MSF to account for the effect of the considered magnitude equal to 6.1; [18]), highlights (Fig. 3c) that the hydraulic fill layer ($4 < z < 13$ m) are potentially liquefiable ($FS < 1$). Hypothesizing the use of IPS, the charts reported in Fig. 2b can be used to define the needed decrease in degree of saturation to improve the in-situ soil capacity.

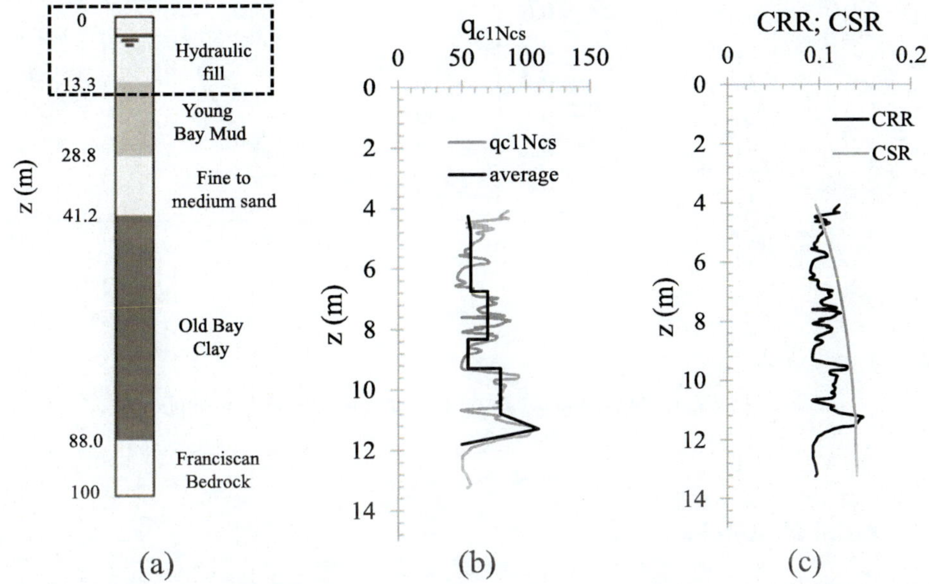

Fig. 3. Stratigraphy profile of Treasure Island (a) q_{c1Ncs} profile (b) and results of susceptibility analysis (c) in the upper hydraulic fill ($z < 13$m).

The partial saturation of the soil below the ground water table can be obtained by injecting pressurized air from sub horizontal well screens deployed in rows at the depth of 9 m from the ground level, where FS is lower than 0.70 (Fig. 4a). In order to guarantee a safety factor FS higher than 1, soil capacity CRR should overcome $CSR_{,max}$. At 9.0 m ($q_{c1Ncs} \approx 55$) CRR should be at least 0.13. The design chart reported in Fig. 4b, shows that Sr should be at least 95–98%.

Assuming an extension of the soil volume to be treated [10] in terms of length ($L_1 = 10$ m), width ($L_2 = 4$ m), and depth ($L_3 = 9$ m), and the mean soil void index (and therefore the void volume V_v), the air volume V_a to be injected can be quantified ($Sr = (V_v - V_a)/V_v$). Considering all air retained by soil, 7.4 m^3 of air should be injected. A slight increase (of about 20%) it is suggested taking into account a certain percentage of injected air lost through the boreholes ($V_{air} \approx 9$ m^3). The air must be pumped into the pipes at a pressure (p) high enough to overcome the water hydrostatic pressure, but not so high to generate soil displacement or erosion (i.e. $p < 90$ kPa for the considered depth). It is worth noting that the degree of saturation can be checked in situ thought the measure of compression wave velocity (V_P) and the soil resistivity (ρ), both sensitive to a change in soil saturation degree.

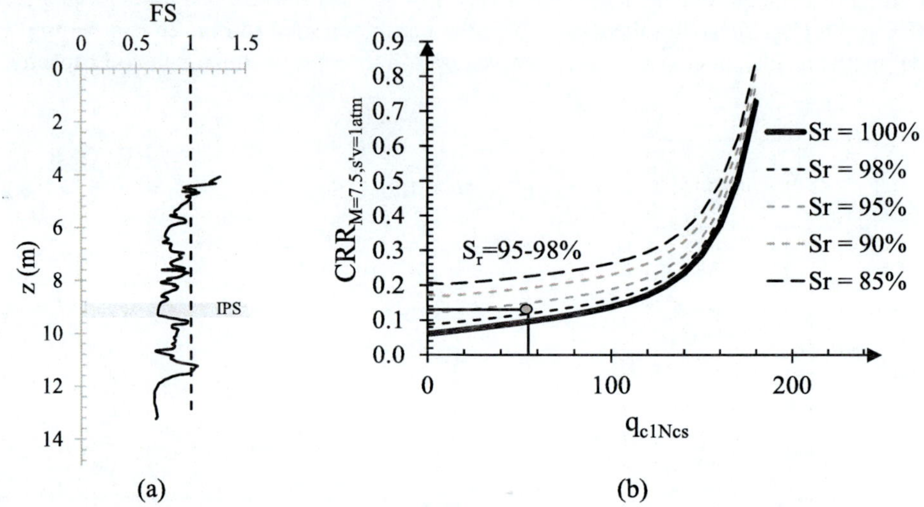

Fig. 4. FS profile (a) and use of design chart to choose S_r to apply (b)

4 Final Remarks

Since IPS technique is low cost, non-damaging to existing structures, environment-friendly and pollution-free is considered one of the most promising mitigation techniques against liquefaction. Although the effectiveness of IPS has been widely demonstrated at small and large scale, it is still far from being a routine technology because of the lack of design tools. This paper aims to offer useful and practical charts, which link soil capacity CRR of gassy soils ($80 \leq Sr\ (\%) \leq 98\%$) to the results of in-situ (CPT or SPT) tests. Once the required increase of liquefaction resistance is known, the charts suggest the degree of saturation to apply in situ. A simple application for the case study of Treasure Island (California) is finally presented. The research is ongoing to improve the knowledge regarding durability and soil-volumes treated.

References

1. Lirer, S., Mele, L.: On the apparent viscosity of granular soils during liquefaction tests. Bull. Earthq. Eng. **17**(11), 5809–5824 (2019). https://doi.org/10.1007/s10518-019-00706-0
2. Mele, L.: An experimental study on the apparent viscosity of sandy soils: from liquefaction triggering to pseudo-plastic behaviour of liquefied sands. Acta Geotech. **17**, 1–19 (2021). https://doi.org/10.1007/s11440-021-01261-2
3. Seed, H.B., Idriss, I.M.: Simplifed procedure for evaluating soil liquefaction potential. J. Soil Mech. Found. Div. ASCE **97**(SM9), 1249–1273 (1971)

4. Boulanger, R.W., Idriss, I.M.: CPT and SPT liquefaction triggering procedures. Report No UCD/GCM-14/01, University of California at Davis, California, USA (2014)
5. Bao, X., Jin, Z., Cui, H., Chen, X., Xie, X.: Soil liquefaction mitigation in geotechnical engineering: an overview of recently developed methods. Soil Dyn. Earthq. Eng. **120**, 273–291 (2019)
6. Mele, L., Lirer, S., Flora, A.: An energetic interpretation of liquefaction laboratory tests on partially saturated soils. J. Geotech. Geoenviron. Eng. **148**(10), 04022082 (2022)
7. Zeybek, A., Madabhushi, S.P.G.: Centrifuge testing to evaluate the liquefaction response of air-injected partially saturated soils beneath shallow foundations. Bull. Earthq. Eng. **15**(1), 339–356 (2016). https://doi.org/10.1007/s10518-016-9968-6
8. Mele, L., Tian, J.T., Lirer, S., Flora, A., Koseki, J.: Liquefaction resistance of unsaturated sands: experimental evidence and theoretical interpretation. Géotechnique **69**(6), 541–553 (2019)
9. Okamura, M., et al.: In-situ desaturation test by air injection and its evaluation through field monitoring and multiphase flow simulation. J. Geotech. Geoenviron. Eng. **137**(7), 643–652 (2011)
10. Flora, A., et al.: Field tests to assess the effectiveness of ground improvement for liquefaction mitigation. In: Earthquake Geotechnical Engineering for Protection and Development of Environment and Constructions, pp. 740–752. CRC Press (2019)
11. Flora, A., Bilotta, E., Chiaradonna, A., Lirer, S., Mele, L., Pingue, L.: A field trial to test the efficiency of induced partial saturation and horizontal drains to mitigate the susceptibility of soils to liquefaction. Bull. Earthq. Eng. **19**(10), 3835–3864 (2020). https://doi.org/10.1007/s10518-020-00914-z
12. Okamura, M., Soga, Y.: Effects of pore fluid compressibility on liquefaction resistance of partially saturated sand. Soils Found. **46**(5), 695–700 (2006)
13. Mele, L., Flora, A.: On the prediction of liquefaction resistance of unsaturated sands. Soil Dyn. Earthq. Eng. **125**, 105689 (2019)
14. Idriss, I.M., Boulanger, R.W.: Soil liquefaction during earthquakes. In: Monograph, Earthquake Engineering Research Institute (EERI), Oakland (2008)
15. Flora, A., Lirer, S., Mele, L.: Mitigation of liquefaction risk by innovative ground improvement techniques. Indian Geotech. J. **52**, 1–11 (2022). https://doi.org/10.1007/s40098-022-00612-4
16. Hanks, T.C., Brady, A.G.: The Loma Prieta earthquake, ground motion, and damage in Oakland, Treasure Island, and San Francisco. Bull Seismol. Soc. Am. **81**(5), 2019–2047 (1991)
17. Mele, L., Lirer, S., Flora, A.: A simple procedure to calibrate a pore pressure energy-based model from in situ tests. Acta Geotech. 1–23 (2022)
18. Chiaradonna, A., d'Onofrio, A., Bilotta, E.: Assessment of post-liquefaction consolidation settlement. Bull. Earthq. Eng. **17**(11), 5825–5848 (2019). https://doi.org/10.1007/s10518-019-00695-0

Performance of Synthetic Lightweight Aggregates for Road Embankment Construction on Improved Soft Soil

Daniela Dominica Porcino(✉), Giuseppe Tomasello, and Marinella Silvana Giunta

University "Mediterranea" of Reggio Calabria, Reggio Calabria, Italy
daniela.porcino@unirc.it

Abstract. The paper presents a study on the performance of innovative synthetic lightweight aggregates (*SLAs*) made from recycled plastic and biomass fly ash when used in road embankment construction on soft soil. It was ascertained that the *SLAs* behave just like a granular soil with compressibility and shear strength properties comparable to those exhibited by other natural lightweight aggregates. Two different applications were considered. The first deals with the use of the *SLAs* as back-fill material of compacted aggregate columns to improve soft soil under the embankment. In this case, the behavior of the embankment in terms of stability, settlements and consolidation time was evaluated using the basic concepts of stone columns method. In the second application, *SLAs* are used as fill material of the embankment constructed in a single stage on a soft soil improved by vertical drains. The performance of this latter solution was compared to that obtained with the adoption of traditional granular materials by a stage construction method. The results obtained allow the conclusion that *SLAs* can be used as an effective alternative to traditional natural aggregates *NAs* in either case. The advantages of using the *SLAs* instead of *NAs* as fill material of the embankment were analyzed not only in terms of geotechnical features but also in terms of improvement in road constructive aspects such as time, costs and environmental burdens.

Keywords: Synthetic lightweight aggregates · recycling · soft soil · road embankment · compacted aggregate columns

1 Introduction

Lightweight aggregates (*LWAs*) manufactured from natural rocks and soils have been used successfully since many decades as light fill in road construction and geotechnical works [1] as a valid alternative to the most traditional natural aggregates (*NAs*) made of sands and gravels that are 1.5 to 2 times heavier than the *LWAs*. Due to the impressive growth of the infrastructure industry all over the world, the overuse of natural resources is becoming a serious issue. Furthermore, reducing and recycling waste materials to avoid exhausting of natural resources whilst reducing the demand for landfills, is a fundamental pillar of the circular economy concept [2]. Previous studies on geotechnical properties of synthetic lightweight aggregates (*SLAs*) manufactured from coal fly ash

A. Ferrari et al. (Eds.): CNRIG 2023, SSGG, pp. 824–832, 2023.
https://doi.org/10.1007/978-3-031-34761-0_99

(*CFA*) and high-density polyethylene (HDPE) were reported by Holmstrom and Swan [3] and Kashi et al. [4]. More recently, Porcino et al. [5] considered the use of biomass fly ash (*BFA*) in lieu of *CFA*. The obtained results show that the *SLAs* composed of *BFA* + HDPE possess most of the properties that are decisive material requirements in geotechnical applications, such as: controlled gradation, low bulk density, high permeability, high strength, moderate deformability. The above-mentioned properties allow to solve a wide range of geotechnical problems related to the construction of roadway and railway embankments on poor soils or hillslopes as well as to reduce the lateral earth pressures applied by the back-fills behind retaining walls and bridge abutments.

This paper focuses on the use of new *SLAs* composed of *BFA* and HDPE as an alternative to traditional *NAs* in geotechnical applications. Taking as benchmark a real case reported in the literature, concerning the construction of a road embankment on a soft soil using traditional *NAs* (see Sect. 3), two possible applications were envisaged, namely: 1) use of *SLAs* as back-fill material for compacted aggregate columns installed under an embankment composed of traditional material to improve soft foundation soils, 2) use of *SLAs* as fill material for a road embankment placed on a soft soil improved by prefabricated vertical drains (*PVD*) to allow construction in a single stage. The results obtained allow the conclusion that *SLAs* can be used as an effective alternative to traditional *NAs* in either case. Further the advantages gained by using the *SLAs* as fill material for the construction of a lightweight embankment (low transportation cost, easiness to handle and compact, time saving, etc.) are presented and discussed.

2 Geotechnical and Environmental Properties of *SLAs*

Two materials, namely biomass fly ash and HDPE, were reused and recovered for *SLAs* production. *SLAs* were manufactured by first mixing biomass fly ash with HDPE and then melt-blending them together in a co-rotating twin-screw extruder. Aggregate fly ash to HDPE ratio, by weight, was 50:50.

Geotechnical characterization on the *SLAs* included: specific gravity measurements, sieve analysis, one-dimensional compression tests, drained triaxial compression tests, constant head permeability tests, and water absorption tests. The specimens were tested in a medium-dense state (i.e. relative density $\approx 65\%$). Physical and mechanical properties obtained by the aforementioned tests are reported in Table 1.

Table 1 shows that the *SLAs* are characterized by low weight, high shear strength, favorable drainage characteristics and satisfactory compressibility features. Furthermore, leaching tests proved the sustainability of the *SLAs* from an environmental point of view according to UNI EN 12457–2 [5, 6].

3 Benchmark Case Study

A case study of a bridge approach embankment constructed on a soft soil [7] was considered as basis for two applications of *SLAs*. The embankment section considered for the analysis was 4 m high with an angle of side slope equal to 34°. Subsoil profile consisted of a highly plastic clay whose thickness was 5.5 m. The lower layer was a dense sand under the soft clay (bearing stratum) (Table 2).

Table 1. Physical and mechanical properties of *SLAs* tested in the present research.

Properties	Value	Properties	Value
$\gamma_{d,min}$ (kN/m^3)	6.05	c'_p (kPa)	8.6
$\gamma_{d,max}$ (kN/m^3)	7.24	φ'_p (°)	46.3
G_S	0.98	c'_{us} (kPa)	2.5
D_{50} (mm)	1.30	φ'_{us} (°)	45.8
C_U	1.55	E'_{50} (MPa)	10.51
Soil classification (USCS)	SP	*RR*	0.016
Soil classification (AASHTO)	A-1-b	*CR*	0.093
k (m/s)	4.35×10^{-4}	*SR*	0.014
w_{24h} (%)	16.3	$C_{\alpha\varepsilon}$ (σ'_v < 200 kPa)	0.0017
γ (kN/m^3)	8.11		

Note: $\gamma_{d,min}$ = minimum dry density; $\gamma_{d,max}$ = maximum dry density; G_S = specific gravity; D_{50} = mean grain size; C_U = uniformity coefficient; SP = poorly graded sands; k = hydraulic conductivity; w_{24h} = absorption coefficient at 24h; γ = bulk unit weight; c'_p = peak cohesion intercept; φ'_p = internal peak friction angle; c'_{us} = ultimate state cohesion intercept; φ'_{us} = internal ultimate state friction angle; E'_{50} = Young's modulus; *RR* = recompression ratio; *CR* = compression ratio; *SR* = swelling ratio; $C_{\alpha\varepsilon}$ = creep coefficient; σ'_v = effective vertical normal stress.

Ground water level was located at a depth equal to 2.5 m below the ground surface. The average measured properties of the subsoil layers deduced from in-situ and laboratory investigation are reported in Table 2.

Stability analyses performed in the present paper by using SLOPE/W code according to Morgenstern and Price method [8] showed that the soft soil could not support the proposed embankment.

In fact, the short-term factor of safety (*FS*) against slope failure was estimated to be less than 1 (Fig. 1). Furthermore, the settlement for the 4 m high embankment was about 487 mm. For this reason, a ground improvement method based on natural gravel stone columns constructions was adopted in the project to improve the soft soil.

3.1 Case 1: Use of *SLAs* as Backfill Material for Compacted Aggregate Columns

Compacted aggregate columns is a ground improvement technique for which two construction processes are generally adopted, namely the traditional vibro-stone columns method and the more recent rammed piers method. In the present paper only the first one has been considered due to the more consolidated theoretical approaches available for this technique to predict the behavior of the columns under the applied surface loading. In a first stage, the real case study presented in Sect. 3 has been taken into account to verify the reliability of the adopted design approach. Then the calculations have been repeated for the hypothetical case where the *SLAs* are used as back-fill material of the

Table 2. Material properties of the embankment, the subsoil and traditional stone columns.

Parameter	Soft soil	Bearing stratum	Embankment	Traditional stone columns
γ (kN/m^3)	18.8	20	22	22
PI (%)	26	-	-	-
c' (kPa)	0	0	14.2	0
φ' (°)	29.25	36	33	42
S_u (kPa)	13.44	-	-	-
e_0	1.674	-	-	-
C_c	0.541	-	-	-
C_r	0.081	-	-	-
OCR	1.42	-	-	-
C_h (m^2/s)	4.46×10^{-7}	-	-	-
M (MPa)	0.18–2.37	-	-	29.23–94.36

Note: γ = bulk unit weight; *PI* = plasticity index; *c'* = cohesion; φ' = friction angle; S_u = undrained shear resistance (inferred by Ladd's et al. [9] relationship); e_0 = initial void ratio; C_c = compression index; C_r = recompression index; *OCR* = overconsolidation ratio; C_h = horizontal consolidation coefficient; *M* = constrained modulus variable with depth (Table 2 adapted from Table 4 in Mohamedzein and Al-Shabani [7]).

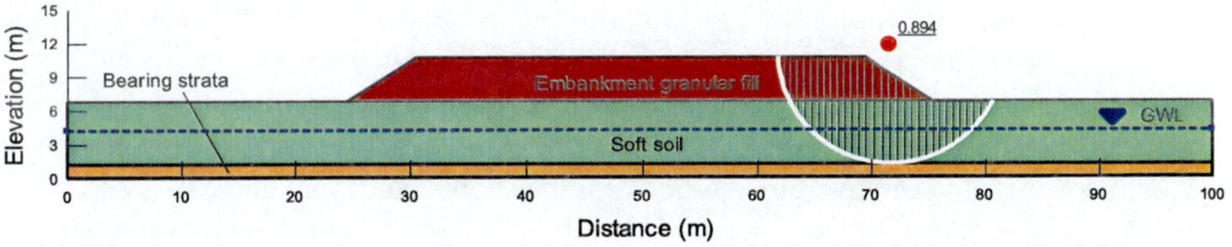

Fig. 1. Stability analysis performed by Slope/W code of the embankment supported on soft soil.

columns and the results obtained have been compared with those gained from the real case.

Stone columns were 0.90 m in diameter and 3.5 m in spacing with an equilateral triangular pattern arrangement. The mechanical properties of natural gravel columns are reported in Table 2. The plane strain model of the columns adopted in the analyses was obtained by transforming the actual stone columns diameter and spacing in 3-D arrangement into equivalent strips having width (*D'*) and spacings (*s'*) evaluated in accordance with the approach suggested by Barksdale and Bachus [10]. The values obtained were the following: *D'* = 0.104 m, *s'* = 1.75 m. The stability analysis of the embankment improved by stone columns provided a short-term safety factor (*FS*) against global failure equal to 1.203 (Fig. 2a). On the other hand, when *SLA* columns were installed (Fig. 2b), a slightly lower, even if still satisfactory, *FS* value was calculated (*FS* = 1.122). The reason of such small reduction can be found in the lower weight of *SLAs* that causes a reduction of the normal stresses acting on the failure plane in the sections where it crosses the columns. Settlement prediction of both traditional and *SLA* columns was made using the analytical method proposed by Priebe [11] based on a combination of elasticity and Rankine earth pressure theories. The concept of a unit cell consisting of a stone column surrounded by soil was used. The final computed settlements and their evolution with time through the radial consolidation theory of Barron [12] are shown in Figs. 3a and 3b for the traditional stone columns and the *SLA* columns, respectively.

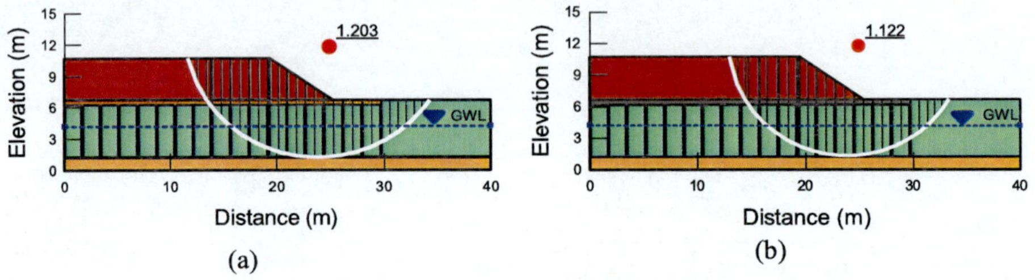

Fig. 2. Stability analyses performed by Slope/W code of the embankment resting on soft soil improved by: (a) natural gravel stone columns and (b) compacted *SLA* columns.

It can be observed that the Priebe method [11] provides a final settlement of about 201 mm for the traditional stone columns, which compares well with the measured value reported in Mohamedzein and Al-Shabani [7] (Fig. 3a). The final settlement slightly increases when soft soil is improved by *SLA* columns ($s_{final} = 230$ mm). The reason of the above difference can be found in the higher compressibility of the *SLAs* compared to the traditional natural aggregate. The settlement improvement factor (i.e. settlement of treated ground over that of untreated one) was estimated equal to 0.47.

It was also successfully verified that the factor of safety against short-term failure mechanisms [13, 14] of the single stone columns (both traditional and *SLA*) was higher than 4.8.

3.2 Case 2: Use of *SLA* as Fill Material for Embankment Constructed on Soft Soil Improved by Prefabricated Vertical Drains

Staged construction embankment in conjunction with prefabricated vertical drains (*PVD*) were adopted and analyzed as an alternative solution for improving soft soils of the real case history presented in Sect. 3. Staged construction consists in the filling of an embankment at controlled steps each of them followed by a specifically designed consolidation time, so as not to cause failure but to permit an increase in shear strength due to consolidation.

In such a way, the obtained strengthening of the foundation soil should be sufficient to support safely the required load. It was assumed that the diameter of *PVD* and their spacing were 80 mm and 2.9 m, respectively. *PVDs* were installed in triangular arrangement (quincunx). To achieve the maximum height of 4 m, three stages of construction were required, namely: a) stage I: $\Delta H_1 = 2.75$ m. The height of the first layer depends on the initial strength of the clay and it is chosen in order to assure that the fill placed is safe against short-term failure; b) stage II: $\Delta H_2 = 0.50$ m. For stages I and II, a waiting

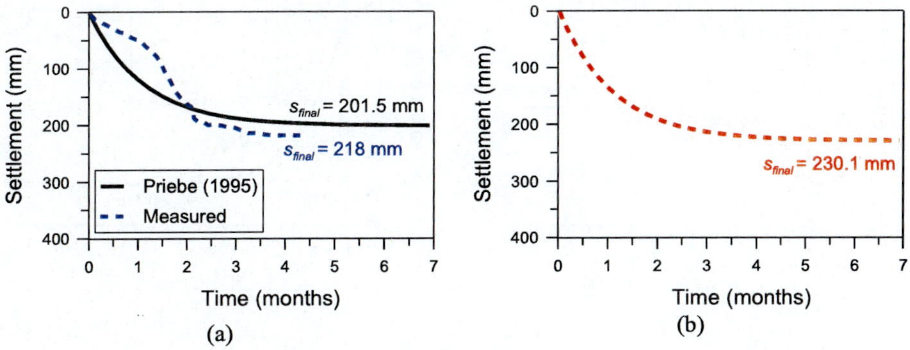

Fig. 3. Settlement-time curves of the analyzed embankment on soft soil improved by (a) natural gravel stone columns, (b) compacted *SLA* columns.

period of about 2 months, required to attain 50% consolidation degree, was allowed; c) stage III: $\Delta H_3 = 0.75$ m, allowing a period of 8.64 months to achieve 95% consolidation. The gain in undrained shear strength caused by the increase in the effective vertical stress $\Delta\sigma'_v$ at the end of each stage was computed by assuming that S_u was approximately constant in recompression whereas in virgin compression the following relationship was suited: $\Delta S_u = \text{cost} \times \Delta\sigma'_v$. Details of all stages and relative factors of safety are reported in Fig. 4a. It is worth noting that *FS* values against slope failure higher than 1.20 were reached at each stage. On the other hand, the advantage of using *SLAs* in place of traditional natural material for the construction of the embankment are clearly shown in Fig. 4 in terms of: a) reduction of the driving forces contributing to instability, so that the full loading of the embankment can be safely placed in a single stage (*FS* = 2.35) (Fig. 4a), b) reduction of the final settlement resulting from the end of primary consolidation of soft foundation soil (Fig. 4b): the final settlement was estimated to be equal to 207 mm which was lower than that obtained with the staged construction (i.e. 487 mm).

The advantages of using the *SLAs* should be considered not only in terms of geotechnical features (stability, settlements, consolidation time) but also in terms of improvements that can be achieved during the road construction in terms of time, costs and environmental burdens. More in details, the use of lightweight aggregates is an effective time-saving solution in situation in which constraints of various kinds make the construction or restoration schedule (i.e. extension of the cross section of existing embankment) unsuited with a staged construction [1].

The solution is also cost-effective in terms of construction cost (cost of material, transportation, laying and compaction) and more in general life cycle cost in a wide range of transportation-related applications.

Environmental benefits are related not only to recycling of wastes (fly ash, plastics) in a circular economy view but also to reduced times of road construction. In fact, road construction has noteworthy effects on air quality especially in areas near to worksites. Particulate matter (PM_{10}) is a significant pollutant that arises from the construction processes and has negative effects on human health and environment [15]. The nature of the proposed *SLAs* contributes to the reduction of PM_{10} emissions during earthworks.

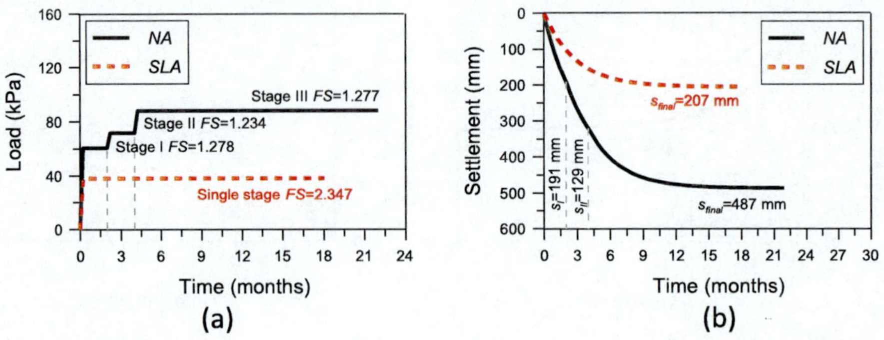

Fig. 4. Stage construction phases (a) and relative settlement-time curves (b) of the embankment on soft soil improved by vertical drains.

However, some technical key aspects should be considered when these lightweight aggregates are used for embankment realization. Particular attention merits the compaction process and the related compacted moist density in laboratory and in place. Experimental evidence of laboratory tests (Proctor tests) suggested that the results should be interpreted differently from those of natural soils because they are affected by some aspects [16]. The first one is that the absorption of lightweight aggregates is greater than natural soils and therefore part of the water added during the test will not affect interparticle physics (bulking, lubrication of the surfaces, etc.). Secondly, lightweight aggregates contain limited fines with respect to the natural aggregates, and this aspect limits the increase in density due to packing of the fines between large particles. In the light of these observations, the objective in compacting lightweight aggregates fill is not to reach the maximum in-place density, but to achieve an optimum density that provides high stability. Optimum field density is commonly achieved by two to four passes of roller compactor, depending on the weight of the compaction machine and the thickness of the layer. Material must be compacted to a minimum 65% relative density as determined by laboratory tests. When the roller compactor acts directly on the lightweight aggregate layer, a good practice could be to insert a geogrid 10 cm below the compacting surface. Excessive particle degradation developed by steel-tracked rolling equipment should be avoided.

4 Conclusions

From the results obtained in the present research, the following conclusive remarks can be made:

– synthetic lightweight aggregates (*SLAs*) made from recycled plastic and biomass fly ash possess most of the properties that are decisive material requirements in geotechnical applications, such as: controlled gradation, low weight, high shear strength, favorable drainage characteristics, satisfactory compressibility, and they are also environment friendly materials;

- the study proved that *SLAs* are suitable sustainable materials to substitute traditional aggregates such as sand, gravel and other recycled materials as filler of compacted columns in ground improvement work. The performance of compacted aggregate columns made of *SLAs* proved to be satisfactory both in terms of stability and settlements, the latter ones predicted by Priebe's method [11];
- due to low weight and high shear strength, the use of *SLAs* as fill material of the embankments on soft soils improved by *PVDs* allows the application of the embankment loading in a single stage and the reduction of consolidation settlements; on the other hand, a stage construction method is necessary when traditional granular materials are adopted for the construction of the embankment;
- the use of *SLAs* results in a time-saving and cost-saving solution bringing also significant environmental benefits due to the recycling of wastes and the reduction of air emissions during earthworks and transportation.

These findings need to be further verified in a full-scale environment to develop design guidelines, specifications and quality assurance testing.

References

1. Marradi, A., Pinori, U., Betti, G.: The use of lightweight materials in road embankment construction. Procedia. Soc. Behav. Sci. **53**, 1001–1010 (2012)
2. Perkins, L., Royal, A.C.D., Jefferson, I., Hills, C.D.: The use of recycled and secondary aggregates to achieve a circular economy within geotechnical engineering. Geotechnics **1**, 416–438 (2021)
3. Holmstrom, O.C., Swan, W.: Geotechnical properties of innovative synthetic lightweight aggregates. In: International Ash Utilization Symposium, pp. 18–20. Centre for Applied Energy and Research, University of Kentuck (1999)
4. Kashi, M.G., Malloy, R.A., Swan, C.W.: Synthetic lightweight aggregates for highway construction. Recycled Mater. Center (2001)
5. Porcino, D.D., Mauriello, F., Bonaccorsi, F., Tomasello, G., Paone, E., Malara, A.: Recovery of biomass Fly Ash and HDPE in innovative synthetic lightweight aggregates for sustainable geotechnical applications. Sustainability **12**, 6552 (2020)
6. Porcino, D.D., Tomasello, G., Mauriello F., Malara, A.: Environmental and geotechnical properties of lightweight aggregates made of reused solid wastes. Environmental Geotechnics (in press). https://doi.org/10.1680/jenge.22.00077
7. Mohamedzein, Y.E.-A., Al-Shabani, I.H.: Performance of an embankment supported on soft soil reinforced by stone columns. Ground Improv. **164**(GI4), 213–224 (2011)
8. Morgenstern, N.R., Price, V.E.: The analysis of the stability of general slip surface. Géotechnique **15**, 79–93 (1965)
9. Ladd, C.C., Foott, R., Ishihara, K., Schlosser, F., Poulos, H.G.: Stress-deformation and strength characteristics: state of the art report. In: Proceedings of the 9th ICSMFE, pp. 421–494, Tokio (1977)
10. Barksdale, R.D., Bachus, R.C.: Design and Construction of Stone Columns (Volume II, Appendices). Federal Highway Administration, Washington, DC, USA, Final Report, no. FHWA7RD-83/027 (1983)
11. Priebe, H.J.: The design of vibroreplacement. Ground Eng. **28**(10), 31–37 (1995)
12. Barron, R.A.: Consolidation of fine-grained soils by drain wells. Trans. ASCE **113**, 718–742 (1948)

13. Hughes, J.M.O., Withers, N.J.: Reinforcing of soft cohesive soils with stone columns. Ground Eng. **3**(7), 42–49 (1974)
14. Brauns, J.: Initial bearing capacity of stone columns and sand piles. In: Internationa Symposium on Soil Reinfoncing and Stabilizing Techniques in Engineering Practice, pp. 497–512, Sidney (1978)
15. Giunta, M.: Assessment of the impact of CO, NO_x and MP_{10} on air quality during road construction and operation phases. Sustainability **12**(24), 10549 (2020)
16. SC-M-203–5: Supplemental Technical Specification for Lightweight Aggregates. SCDOT (2018)

Evaluation of the In-Situ Behaviour of a Lime-Treated Clay in a Real-Scale Experimental Embankment

Marco Rosone, Francesco Moscato, Clara Celauro(✉), and Maurizio Ziccarelli

Department of Engineering, University of Palermo, 90128 Palermo, Italy
clara.celauro@unipa.it

Abstract. The international commitment to achieve carbon neutrality in the next few decades has oriented infrastructure construction towards the preservation of natural and non-renewable resources. In this context, lime treatment of clayey soils involved in earthworks can be considered among the main sustainable options in the field of infrastructure construction. Lime stabilization technique has been extensively tested and validated through laboratory tests, however, the literature is lacking in full-scale tests, where the volumes of soils treated are large and many variables are involved in the execution process. This paper discusses the mechanical response of clayey soil after lime treatment in a trial embankment specifically designed and built for the full-scale evaluation of the stabilisation technique to be used in road construction works for widening an existing roadway to four lanes in Sicily. To this aim, in-situ tests, including plate load tests, in-place density via cone-sand tests, dynamic penetrometer tests and pressure meter tests, were conducted on the different layers of the embankment. The experimental results are useful not only for evaluating the lime stabilisation technique used, but also for understanding how and to what extent the execution processes for embankment construction with large volumes of treated clayey soils can affect the expected mechanical behaviour.

Keywords: Lime treatment · clay · in-situ testing · sustainable infrastructure

1 Introduction

In the last few decades, there has been an increasing interest in sustainable development. Governments, agencies, and academic institutions around the world are involved in developing strategies to achieve the highest standards of sustainable development, aiming to minimise the impact of human activities on the environment. The European Green Deal placed sustainability at the centre of the new strategy, setting as its main goal to make Europe the first climate-neutral continent by 2050. Then, to mitigate climate change and reduce CO_2 emissions, a new action plan was introduced based on the circular economy of materials [1]. Every year, about 50 billion tons of natural and non-renewable resources, such as sand and gravel, are used in the civil engineering industry [2]. In fact, these granular soils are the most preferable materials in road and railways

A. Ferrari et al. (Eds.): CNRIG 2023, SSGG, pp. 833–840, 2023.
https://doi.org/10.1007/978-3-031-34761-0_100

construction due to their high strength and stiffness. Their extraction from borrow pits alone is not the cause of the environmental impact; transport to and from construction sites causes significant CO_2 emissions and premature deterioration of roads. During construction, significant quantities of soils with poor mechanical properties, such as clayey soils, are excavated and usually considered waste materials. With a view to the circular economy, it is mandatory to reuse these materials not only for backfilling but for structural purposes too (e.g. road embankments). The lime treatment represents both an eco-friendly and cost-effective technique, which allows for the stabilization of fine soils located on the construction site [3–5]. Nowadays, the chemical reactions induced by mixing lime and clay were widely investigated and quite well understood [6–8]. First, short-term reactions take place, i.e. lime hydration and cation exchange. Ca^{2+} ions are adsorbed by the clay particles, thus reducing the diffuse water layer surrounding them. Then, the repulsion forces existing between soil particles are reduced and flocculation occurs. This phenomenon immediately changes the geomechanical properties of the clay, reducing plasticity, improving its workability, and changing the microstructure. Moreover, there is an increase in the pH, and the environment becomes highly alkaline due to the high concentration of OH^- that is released during the first phase [9–11]. Pozzolanic reactions occur in a second stage due to the alkaline environment (pH > 12); alumina and silica present in the soil react with Ca^{2+} ions to produce calcium aluminate hydrates (CAH) and calcium silicate hydrates (CSH). These products create cementitious bonds between clay particles; this process is time-dependent and improves the strength and stiffness of the soil [12, 13]. Despite the number of experimental studies aimed at highlighting the peculiarities of the stabilization process conducted at a laboratory scale [7, 14, 15], there are few studies dealing with real-scale tests in the technical literature [16]. In fact, in-situ tests are fundamental to validate results obtained from laboratory tests and allow one to address problems related to, for example, the scale effect of the process on the large volume of soil [17, 18].

The aim of this study is to introduce the results of an extensive experimental programme of in-situ tests carried out on an experimental embankment. The field tests included load plate tests, dynamic penetrometer tests, pressure meter tests, as well as suction and water content measurements. Tests were conducted on the different layers of the embankment treated with different amounts of lime and were also repeated with time.

2 Materials and Methods

During the construction works for widening an existing roadway to four lanes in Sicily, the feasibility of reusing clayey soils from earthworks by lime stabilization was considered. The soils used for the construction of the embankment came from the excavation work of an artificial tunnel. Geotechnical analysis and field studies were conducted to define the homogenous group of soils for the purpose of the treatment. The soils belonged to classes A7–5 and A7–6 according to AASHTO classification. The main geotechnical properties were determined as follows: grain size distribution consisting in gravel fraction $f_{gravel} = 1 \div 6\%$, sandy fraction $f_{sand} = 9 \div 17\%$,silty fraction $f_{silt} = 39 \div 59\%$, clayey fraction $f_{clay} = 24 \div 52\%$, liquid limit $w_l = 51 \div 53\%$, plasticity index

$PI = 24 \div 28\%$, soil specific weight $\gamma_s = 26.3 \div 26.4$ kN/m^3, natural water content $w = 17.6 \div 21.8\%$, dry unit weight $\gamma_d = 15.9 \div 17.4$ kN/m^3 and void ratio $e_0 = 0.52 \div 0.65$. The lime used for stabilisation was a certified quicklime CL 90-Q according to UNI EN 459–1 [19] characterised by high water reactivity and high fineness. Preliminary laboratory tests, including the initial consumption of lime estimation (*ICL* = 1.6%) and standard road geotechnical tests (Proctor Compaction, Immediate Bearing Index and Californian Bearing Ratio tests) on samples treated with different lime contents (2%, 2.8% and 3.5%), were carried out to determine the mix-design suitable for the experimental embankment construction.

Before on-site treatment, the clayey soil was disaggregated using a Motor Grader to achieve a maximum apparent size of the clayey lumps of 31.5 mm. The embankment was built as follows: 50 m in length, 8 m wide (4 strips of 2 m each) at the bottom and 6.2 m at the top, the height was equal to 1.8 m. As depicted in Fig. 1, a total of five layers were realised. The foundation was made in two layers for a total thickness of 0.65 m: the first, obtained by treating the in-situ soils with 2.1% CaO, and the second by adding to the soils to be treated with the 2.3% of lime. The Lower Part of the Embankment (*LPE*) was obtained by two layers of treated soils, the first was 0.43 cm thick and treated with 1.5% CaO, to be representative of possible cases of lime underdosing, while the second was 0.29 m thick and treated with 2.3% CaO; the Upper Part of the Embankment (*UPE*) was obtained by mixing soil with 3.3% CaO for 0.32 m of thickness.

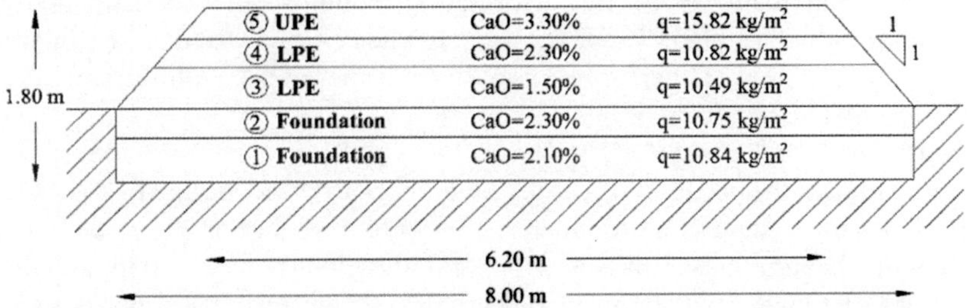

Fig. 1. Cross-section of the experimental embankment with geometric characteristics and lime content (*CaO*). The mean values of lime spreading (*q*) tests are reported for each layer.

Lime spreading tests, thickness tests, grain size analyses, mix homogeneity tests, in-situ density (*CDT*), Immediate Bearing Index (*IBI*) and load plate tests (*Md*) were carried out immediately after the construction of each embankment layer to check the construction process: investigated points are depicted in Fig. 2 and detailed results are reported in Airò Farulla et al. [17]. Moreover, some undisturbed samples were recovered at the end of the construction to assess the time dependent improvements in the hydro-mechanical behaviour of the treated clay [13]. At the end of construction works, the embankment was left uncovered for two years. After the final tests were conducted, the experimental embankment was incorporated into the new road. The experimental programme reported in this study is mainly related to long-term monitoring activities and includes load plate test (*LPT*), penetrometer dynamic penetrometer test (*DPT*), pressuremeter test (*PT*), water content and suction tests. Figure 2 reports the position of the completed experimental programme in the cross section of the embankment. As reported, *LPT* test were conducted during the construction of embankment (t = 0) for each layer. Ten months after the construction *LPT*, *DPT*, *PT*, suction and water content measurements were executed. Twenty-four months after the construction *LPT* tests were finally repeated.

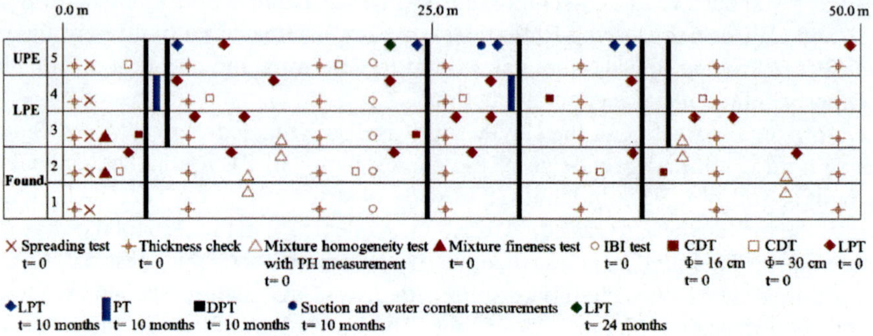

Fig. 2. Layout of field tests in a generic cross section of the embankment.

The Load Plate Test *(LPT)* is a well-known non-destructive in-situ method to determine the load-displacement curve and the stiffness of natural and compacted soils. For the tested embankment layers, the deformation modulus *Md* was estimated according to Italian Standard [19]. These data are frequently associated to the in-situ Cone Density tests (*CDT*), were performed two kinds of these tests with different diameters respectively of 16 and 30 cm.

The Dynamic Probe Test (*DPT*) is generally used [8] because it provides a continuum characterization of the in situ-soil. The tests were conducted according to EN ISO 22476–3[20], by driving a steel cone (diameter $d = 35.68$ mm and angle $\alpha = 60°$) into the ground with a beating mass (mass $m = 30$ kg, falling height $h = 0.20$ m) and recording the number of blows for 0.10 m of penetration. Seven tests were performed on the experimental embankment to analyze the difference in the lime treatment among the layers and their upper, intermediate, and bottom layer parts.

Pressure meter Tests (*PT*) were performed by means of a standard Mènard pressure meter. Three tests were carried out at depth of about 0.5 m at the places reported in Fig. 2. The probe was positioned in a pilot hole, manually perforated with the dimension compatible with the diameter of the probe. During the test, the probe was pressurized with water and it deformed the artificial vertical cavity. The pressure inside the measuring cell was kept constant for about 60 s and the increase in volume required to maintain a constant pressure was recorded. Analysis of data allowed the estimation of the limit pressure p_L, defined as the pressure required for doubling the initial volume of the cavity, and the pressure meter modulus E_M calculated in the elastic trend of the pressure-volume curve. Water content and total suction measurements were immediately conducted on small samples recovered from the pilot hole made for *PT* tests. Total suction measurements were conducted on-site by means of a dew-point water potentiometer (WP4-T, Decagon Devices).

3 Analysis of Results and Discussion

During the construction, several *LPT*s were made for all layers of the body of the embankment immediately after the compaction stage [17]. Other *LPTs* were performed after ten and twenty-four months to evaluate the variation with time of the performance of the upper layer of the embankment (*UPE*). The *PLTs* results are represented in Fig. 3 in terms of deformation modulus *Md* vs depth from the top surface of the embankment; data are collected for layers and time. The results show the very high stiffness of the geomaterial due to the lime stabilization process already in the short term (t = 0) and a dependency on lime content and dry unit weight of soil can be also pointed out. After 10 months and two years from the treatment the deformation modulus *Md* increased, as expected from lime- and water-consuming pozzolanic reactions with time. Rosone et al. [11] reported the total suction and water contents of samples recovered from the embankment (Fig. 2). The water content in the first 0.45 m was reduced 21% to 16% and, consequently, the total suction varied from 1.35 MPa to 4.4 MPa. So, data pointed out on these significant variations in water content and total suction should be partially attributed to the environmental actions, too.

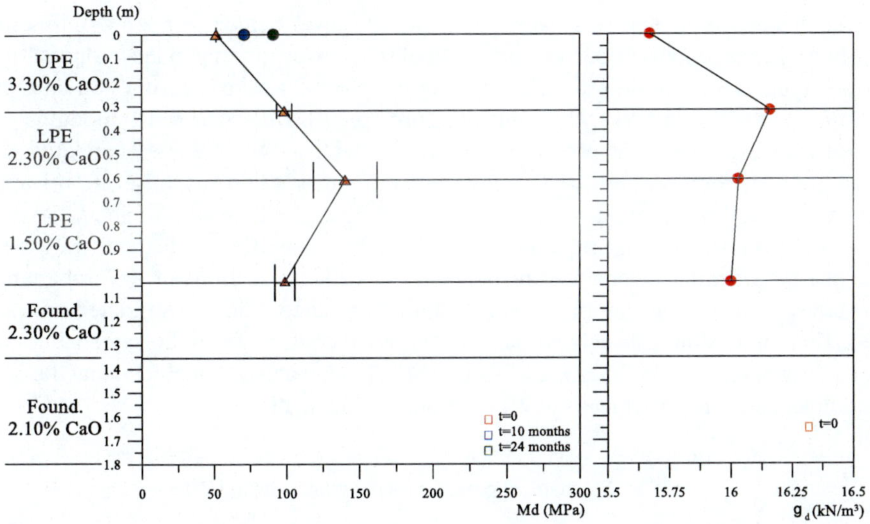

Fig. 3. Deformation modules *Md* and in-situ density tests result at different layers and times.

The dynamic resistance to penetration (*Rpd*) was evaluated by *DPTs* using the so-called "Dutch formula" [18]. The dynamic resistance to penetration profiles were obtained along the embankment layers and are represented for two tests in is reported in Fig. 4. As expected, the results proved that the highest strength occurs in the upper part of the embankment treated with a higher lime content (3.3% CaO). Moreover, from Fig. 4 it is quite evident that the upper part of the layers is always more resistant than the intermediate and the lower part. The thicker is the layer, more pronounced is this tendency, thus proving that both the compaction action and the in-situ mixing with lime is slightly less effective at the bottom of each layer. In fact, a new increase is systematically observed as the penetrometer tip encounters a new layer.

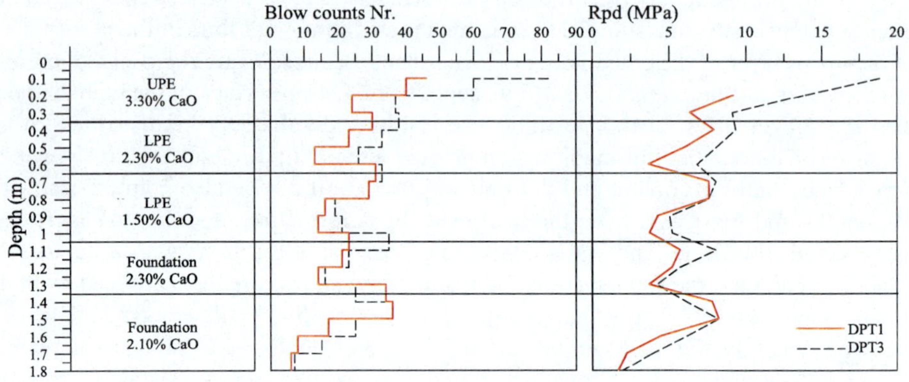

Fig. 4. Blows and resistance penetration profile for DPT1 and DPT3.

All the pressuremeter tests investigated the LPE layer treated with 2.3% of lime. This was because the probe was pushed to a depth of 0.5, corresponding to the midpoint of this layer. Figure 5 shows the pressuremeter curves plotted as the pressure and corrected-volume values. From these, the limit pressure (p_L) and pressuremeter modulus (E_M) values were evaluated. The pressuremeter parameters, shown in Table 1, are quite close to each other, representing the stiffness and the stress-strain response of soil treated with the same lime content. In particular, for PT1 and PT3 tests E_M and p_L are very nearby, these tested points are in the same strip C but in a different distance progressive; this proved a similar mechanical response in two distant sections of the embankment. Moreover, considering that the elastic modulus obtained from consolidated undrained triaxial compression tests carried out on samples recovered from the same layer of the embankment resulted in the range 25 ÷ 100 MPa [11], it can be concluded that the elastic modulus determined by means of *PTs* are quite consistent.

Table 1. Mechanical parameters calculated from PT tests.

PT	Distance(m)	Strip	p_L(MPa)	E_M(MPa)
PT1	7.0	C	1.5	46
PT2	27.5	B	2.0	110
PT3	27.5	C	1.4	77

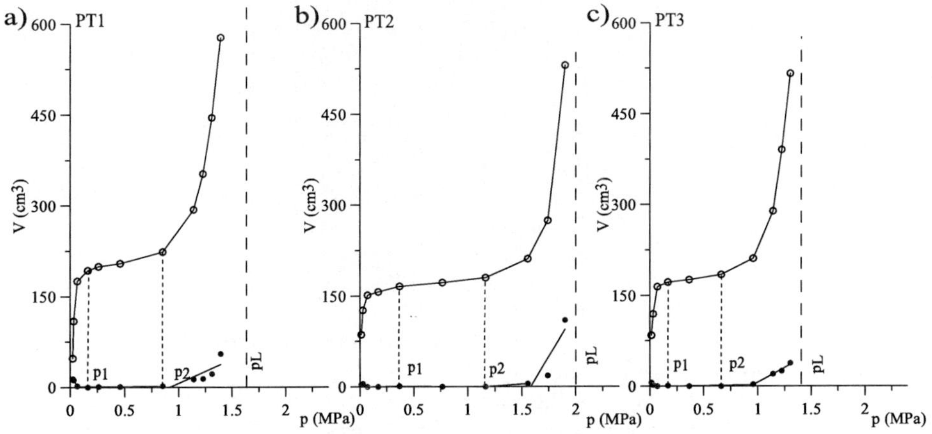

Fig. 5. Results of pressuremeter tests in terms of corrected volume V vs pressure p: (a) for PT1 (b) for PT2 (c) for PT3.

4 Concluding Remarks

An in-situ tests programme was conducted on an experimental embankment specially built to evaluate the lime stabilization technique for road construction works in Sicily. The embankment was made by overlaying layers of in-situ soil treated with different amounts of lime. During the construction, several tests were carried out on each layer immediately after the compaction. At the end of construction, the embankment was left uncovered for two years and some tests were repeated with time. The results obtained from the in-situ tests support the experimental studies conducted at a laboratory scale. Data collected pointed out the variability of the mechanical properties within each layer, despite good repeatability obtained along the embankment (i.e. at different sections). The data dispersion with the layer thickness also proves that attention should be paid to the execution by accurate control of the working parameters of earthworks equipment. Therefore, field tests and long-term monitoring of full-scale trials are essential for the geotechnical characterization of earthworks, and also for validating the mix design carried out at a laboratory scale.

References

1. Wolf, S., Teitge, J., Mielke, J., Schütze, F., Jaeger, C.: The European green deal more than climate neutrality. Intereconomics **56**, 99–107 (2021)
2. Bonoli, A., Zanni, S., Serrano-Bernardo, F.: Sustainability in building and construction within the framework of circular cities and European new green deal. Contrib. Concr. Recycl. Sustain. **13**, 1–16 (2021)
3. Celauro, C., Corriere, F., Guerrieri, M., Lo Casto, B.: Environmentally appraising different pavement and construction scenarios: a comparative analysis for a typical local road. Transp. Res. D Transp. Environ. **34**, 41–51 (2015)
4. Amhadi, T.S., Assaf, G.J.: Overview of soil stabilization methods in road construction. Sustain. Civil Infrastruct. 21–33 (2019)

5. Das, G., Razakamanantsoa, A., Herrier, G., Saussaye, L., Lesueur, D., Deneele, D.: Evaluation of the long-term effect of lime treatment on a silty soil embankment after seven years of atmospheric exposure: mechanical, physicochemical, and microstructural studies. Eng. Geol. 281, (2021)
6. Locat, J., Bérubé, M.-A., Choquette, M.: Laboratory investigations on the lime stabilization of sensitive clays: shear strength development1. Can. Geotec. J. **27**, 294–304 (1990)
7. Bell, F.G.: Lime stabilization of clay minerals and soils. Eng. Geol. **42**, 223–237 (1996)
8. Rosone, M., Airò Farulla, C., Celauro, C., Ferrari, A.: Volumetric behaviour of lime treated high plasticity clay subjected to suction controlled drying and wetting cycles. In: Ferrari, A., Laloui, L. (eds.) ATMSS 2017. SSGG, pp. 165–172. Springer, Cham (2017). https://doi.org/10.1007/978-3-319-52773-4_18
9. Al-Mukhtar, M., Khattab, S., Alcover, J.F.: Microstructure and geotechnical properties of lime-treated expansive clayey soil. Eng Geol. **139–140**, 17–27 (2012)
10. Jawad, I.T., Taha, M.R., Majeed, Z.H., Khan, T.A.: Soil stabilization using lime: advantages, disadvantages and proposing a potential alternative. Res. J. Appl. Sci. Eng. Technol. **8**, 510–520 (2014)
11. Rosone, M., Ferrari, A., Celauro, C.: On the hydro-mechanical behaviour of a lime-treated embankment during wetting and drying cycles. Geomech. Energy Environment. **14**, 48–60 (2018)
12. Rosone, M., Megna, B., Celauro, C.: Analysis of the chemical and microstructural modifications effects on the hydro-mechanical behaviour of a lime-treated clay. Int. J. Geotech. Eng. **15**, 447–460 (2021)
13. Liu, M.D., Indraratna, B., Horpibulsuk, S., Suebsuk, J.: Variations in strength of lime-treated soft clays. In: Proceedings of the Institution of Civil Engineers: Ground Improvement, pp. 217–223 (2012)
14. Russo, G., Modoni, G.: Fabric changes induced by lime addition on a compacted alluvial soil. Geotech. Lett. **3**, 93–97 (2013)
15. Rosone, M., Celauro, C., Ferrari, A.: Microstructure and shear strength evolution of a lime-treated clay for use in road construction. Int. J. Pavement Eng. **21**, 1147–1158 (2020)
16. Herrier, G., Chevalier, C., Froumentin, M., Cuisinier, O., Bonelli, S.: Lime treated soil as an erosion-resistant material for hydraulic earthen structures. In: 6th International Conference on Scourand Erosion, Paris (2012)
17. Airò Farulla, C., Celauro, B., Celauro, C., Rosone, M.: Field test of lime treatment of clayey soils for railways and road works. Ingegneria Ferroviaria **69**, 729–752 (2014)
18. Sanglerat, G.: The Penetrometer and Soil Exploration, Amsterdam (2012)
19. CNR: Determinazione dei moduli di deformazione Md e Md1 mediante prova di carico a doppio ciclo con piastra circolare. Bollettino Ufficiale 146/1992. (1992)
20. EN ISO 22476-3: Geotechnical investigation and testing - Field testing. Part 2: Dynamic probing test. (2005)

On the Efficiency of GFRP Anchors in Soft Rocks

L. Sandrini[1(✉)], Matteo Oryem Ciantia[1], R. Castellanza[2], and I. Bridi[3]

[1] School of Science and Engineering, University of Dundee, Small's Lane - Fulton Building, Dundee DD1 4HR, UK
libero.sandrini@unimib.it

[2] Dipartimento di Scienze dell'Ambiente e della Terra, Università degli Studi di Milano-Bicocca, Piazza della Scienza 1/U4, 20126 Milan, Italy

[3] Freelance Geologist, Via delle regole, 73, 38123 Trento, Italy

Abstract. The failure of geo-structures such as underground caves and vertical cliffs in soft rocks are frequent hazards that may cause damage to infrastructure and people when developing in inhabited centres. Canosa di Puglia, is a city located in southern Italy, characterised by hundreds of artificial cavities with high cultural value excavated in a soft Calcarenite. Because of chemical weathering, water infiltration and the increase of the in-service loads, the risk of sinkhole formation is high. The rapid evolution of the failure mechanism is in fact detrimental as it does not give any warning signs that may be used to mobilise countermeasures. The most common mitigation technique used in this context is cavity filling. On top of being expensive, such approach is highly environmentally unfriendly as the volumes of cement required and consequent CO_2 released are large. Since most of these cavities are of high cultural value alternative mitigation measures preserving their originality is required. This paper proposes a new reinforcement method developed to answer to this need. Through an experimental campaign it is shown that by using Glass Fiber Reinforced Polymers (GFRP) bars combined with thixotropic materials a greater efficiency can be guaranteed with respect to the usual filling methods or to the use of steel bars. The paper discusses in detail the development and choice of the most efficient and performing materials by means of in situ pull tests on anchors installed in a cavity located near the city centre of Canosa di Puglia.

Keywords: Hazard · soft-rocks · mitigation · cultural heritage · fibre glass

1 Introduction

The geology of the Apulian region is characterized by a large presence of Calcarenites and Limestones. The Apulian calcarenites are peculiar Plio-Pleistocenic formations [1–3] highly susceptible to hydro-chemo-mechanical weathering [4]. Because of their homogeneous nature, they are widely used as construction material in historic buildings and monuments. The physical-mechanical characterization of calcarenite bricks is for

A. Ferrari et al. (Eds.): CNRIG 2023, SSGG, pp. 841–847, 2023.
https://doi.org/10.1007/978-3-031-34761-0_101

this reason usually performed to verify the weathering level of building and ornamental materials [5, 6]. The extraction activities undertaken centuries ago for the recovery of building material have contributed to the creation of a dense network of cavities in many areas of the Apulian region (Fig. 1). After their abandonment and because of anthropization, this complex network of tunnels, to date, is often found below towns. In recent years, several instability events, including catastrophic ones, have been documented in [7]. This has led municipalities and local authorities to commission a rigorous stability assessment of the underground environment.

To date, because of its simplicity, the most common remedial measure used in the Apulian region is cavity filling with Portland cement mixes. Considering that a typical cavity of small-medium dimensions, using this approach uses tons of non-environmentally friendly material. In addition, some cavities are UNESCO protected sites (eg. The *Sassi di Matera* in Matera or *Grotta Palazzese* in Polignano a Mare) and cavity filling, regardless of CO_2 emissions, cannot be employed.

Recently research performed jointly between the University of Milano Bicocca and the University of Dundee led to a more environmental friendly approach able to preserve the aesthetic of the site [8, 9]. The principle is based on the approach proposed by [9] utilising a combination of Glass Fiber Reinforced Polymers (GFRP) bar reinforcement and transparent permeable consolidants.

Fig. 1. Cavitiy n°63 in Canosa di Puglia

In this work a case study located in Canosa di Puglia where the novel reinforcement approach using GFRP reinforcement is presented. To this end a set of laboratory pull-out tests of GFRP anchors in calcarenite will be used. Before detailing the case study, the field and experimental study based approach, detailing how the consolidation approach alternative to cement grout mixes and steel bars reinforcements is designed, is presented.

2 Aesthetic Rock-Mass Improvement Concept

From a conceptual point of view, as detailed by [10], the safety factor (F_s) of an ideal artificial cave can be considered as evolving with time. In Fig. 2, F_s of an underground cave at the time of excavation, i.e. initial conditions, is represented by F_0 and corresponds to the stability of the geostructure with initial (unweathered) mechanical properties of the rock material. As the process of alteration develops over time, the F_s of the cave, $F_s(t)$, tends to decrease due to rock mechanical weakening and the corresponding variation law could potentially defined by performing numerical analyses implementing different sets of rock mechanical properties corresponding to different stages of the degradation process. [11–13] show how $F_s(t)$ can be treated at the serviceability limit state or the ultimate limit state through numerical analyses.

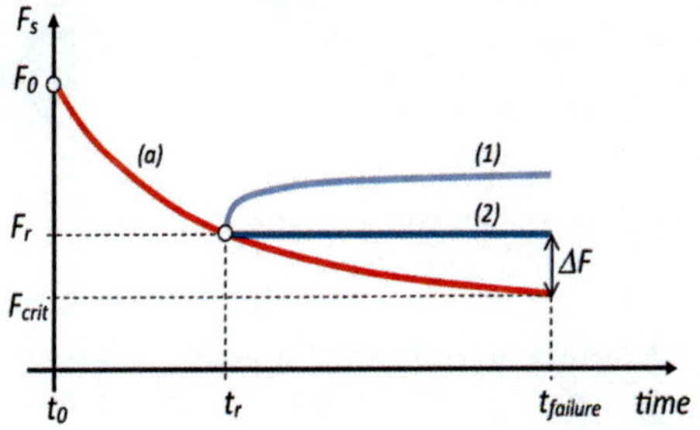

Fig. 2. Increase of safety factor ΔF with proposed intervention [10]

If at time t_r, the safety factor (F_r) is determined to be critical and a remedy is required, two possible approaches can be followed: (1) structural interventions, aimed at increasing rock mass strength, or (2) conservative interventions, aimed at preventing any further mechanical-weakening weathering process. The first can generally lead to an increment of the stability factor (curve (1)), whereas the latter is intended to maintain the cave stability constant over time (curve (2)). Often, the second option involves preserving air ventilation, reducing water infiltration, preventing chemical dissolution processes, the creation of waterproof rock surfaces with specific chemical consolidant materials. These latter are recommended where environmental preservation is required, such as for cultural heritage sites. Conservative interventions should be accompanied by monitoring activities aimed at maintaining the environmental conditions allowing for the increase of *Fs*. This should be done by performing in-situ inspections, and testing (in situ or in the laboratory on the reference material) at regular time intervals demonstrated that the water infiltration and long-term dissolution effects are particularly detrimental for the mechanical performance of the calcarenite. The use of cement based consolidant is impractical due to its high viscosity and its negative aesthetic impact. To preserve the natural beauty of the calcarenite, [9] proposed a novel more sustainable approach that uses chemical consolidants that stabilizes the existing rock microstructure. In the presence of faults and cracks the this can be combined with the use Glass Fiber Reinforced Polymer (GFRP) bars that are more durable and, thanks to their light colour and corrosion resistant properties, have a less visual impact with respect to steel bars (Fig. 3). The same methodological approach proposed by [10] can hence be used employing this new reinforcement technique in order to stabilise cultural heritage protected sites.

In this work the above describe environmentally friendly stabilization approach is employed in a cavity in Canosa di Puglia. As shown in Fig. 4, the cavity is situated beneath an urbanised area of the city. It corresponds to cavity no. 63 according to municipal

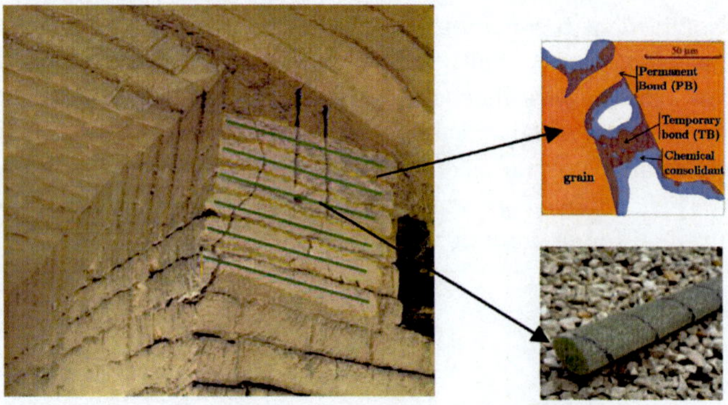

Fig. 3. Schematic showing the stabilization method composed by permeation of consolidant material and the use of GFRP reinforcement bars with grout material [9]

census. By means of 3D laser scanning using Light Detection and Ranging (LIDAR) remote sensing a full detailed reconstruction of the cavity space was performed. Point clouds and image processing were then used to identify cavity surface to construct a 3D numerical model based on the Finite Element Method (FEM) (Fig. 5).

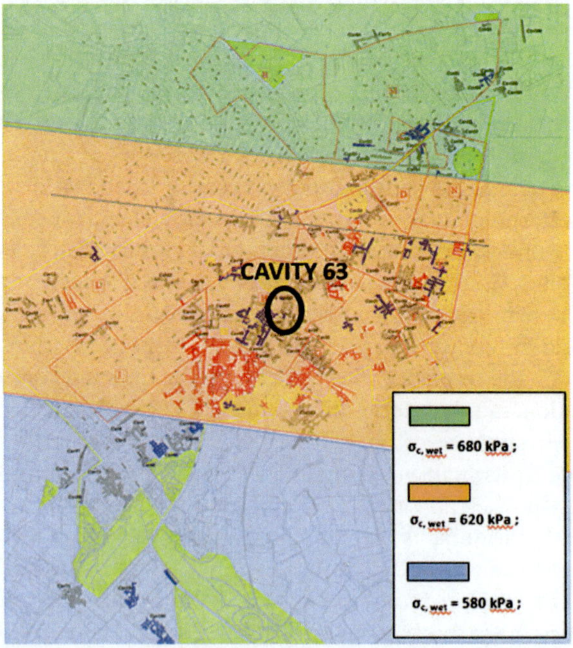

Fig. 4. Map of cavities of Canosa di Puglia superimposed with the average saturated compression strength. Cavity no. 63 is the one considered in the study.

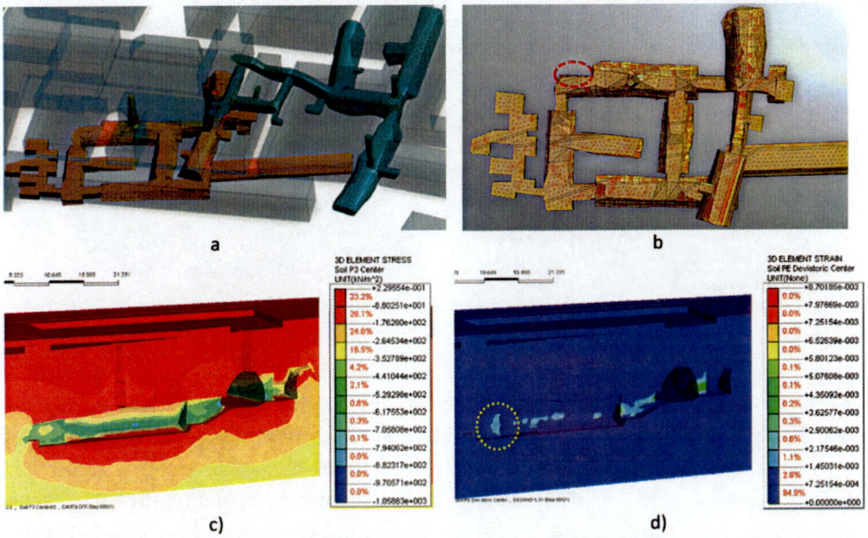

Fig. 5. a) 3D geometry solids b) FEM mesh of cavity no. 63 highlighting problematic zone; c) principal stress in compression d) Plastic strain at a global safety factor equal to 4 by adopting a Strength Reduction Method.

As shown in Fig. 5 the LIDAR system allowed a detailed reconstruction and identification of potential failure zones in the rock mass then confirmed by visual inspection (Fig. 5c). 38 mm diameter samples from the cavity wall were cored and numerous laboratory tests were carried out. In particular Unconfined Compression Test (UCS), indirect tensile tests, drained triaxial tests were used for a mechanical characterization. The resulting mean UCS of 1.1 MPa in dry condition and 0.62 in wet condition (Fig. 4), a Young Modulus of 0.4 GPa and critical state friction angle (ϕ)of 28° were used to calibrate the yield surface of the Mohr Coulomb constitutive model. A c-ϕ (c is the cohesion) reduction analysis with the wet strength of the calcarenite rock for the entire cavity produced a global $F_s = 4$. Therefore, the global stability of the cave was confirmed. Nevertheless, the FEM numerical simulations highlighted some zones characterised with plastic regions. The combination of these numerical result joint to a concurrent visual inspection was used as a criterion to identify local critical zones. In particular, a relatively large plastic region in the FEM model was found in correspondence of a partition wall (Fig. 5d). The local critical zone was then confirmed by the detailed visual inspection which revealed two potential unstable wedges (Fig. 6). The top wedge had a volume of ~1450 m^3 while the second one of ~1300 m^3. Rock anchors were deemed the most suitable solution and following [9], laboratory pull tests were performed to characterise the GFRP rock interface properties and identify the most appropriate grouting material. Several consolidants were tested and it was found that MasterInject 222 grout gave the best results. This consolidant was hence used as it was the best compromise in terms of adherence to the bar, resistance to compression, colour, workability, and environmental impact.

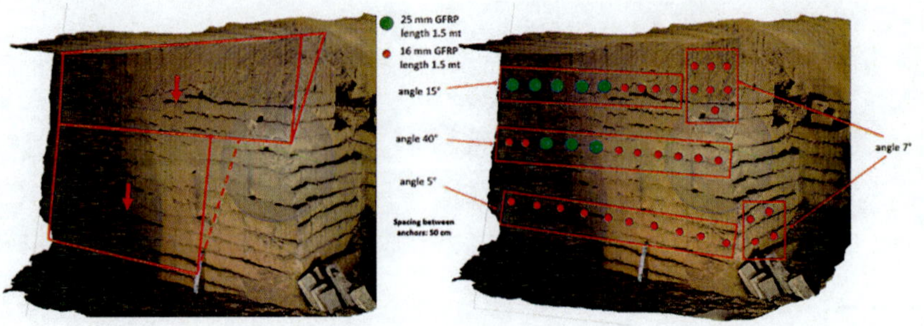

Fig. 6. a) unstable wedges in the wall; b) anchors position

Pull tests data joint to the intact rock characterization were used in the reinforcement design. Each wedge (Fig. 6a) was considered separately and typical limit equilibrium design of passive type of anchors was performed in order to attain a local $F_s > 1.5$. The final design consisted in 42 GFRP bars with two different diameters disposed as shown in Fig. 6b. The final reinforcement design, shown in Fig. 6b and Fig. 7, was used.

Fig. 7. Picture of reinforced partition wall

3 Conclusions

In this work a novel reinforcement method to stabilise cavities in the urban environment protected by cultural heritage is presented. The procedure is presented by means of a real case study located in Canosa di Puglia (Italy). By means of laser scanning of the cavity, laboratory characterization of the calcarenite and in situ anchor pull tests, a 3D FEM model is developed and used to assess the F_s. The model results show potential instability in a localised region of the cavity. Visual inspection was used as a criterion to confirm potential local instability. The laboratory characterization joint to the pull tests are hence used to design the localised GFRP rock anchor reinforcement measure. It is shown that by using GFRP bars combined with calcium materials the usual costly and environmental unfriendly cavity filling methods can be avoided.

References

1. Andriani, G.F., Walsh, N.: Physical properties and textural parameters of calcarenitic rocks: qualitative and quantitative evaluations. Eng. Geol. **67**(1–2), 5–15 (2002). https://doi.org/10.1016/S0013-7952(02)00106-0
2. Ciaranfi, N., Pieri, P., Ricchetti, G.: Note alla carta geologica delle Murge e del Salento (Puglia centromeridionale. Mem. Soc. Geol. It. **41**, 450–458 (1988)
3. Ricchetti, G.: Contributo alla conoscenza strutturale della fossa bradanica. Geol. Appl. Idrogeol. **14**, 489–492 (1979)
4. Ciantia, M.O., Castellanza, R.: Modelling weathering effects on the mechanical behaviour of rocks. Eur. J. Environ. Civ. Eng. **20**(9), 1054–1082 (2016). https://doi.org/10.1080/19648189.2015.1030086
5. Anania, L., et al.: The stones in monumental masonry buildings of the "val di Noto" area: new data on the relationships between petrographic characters and physical-mechanical properties. Constr. Build. Mater. **33**, 122–132 (2012)
6. Andriani, G.F., Walsh, N.: Rocky coast geomorphology and erosional processes: a case study along the Murgia coastline South of Bari, Apulia — SE Italy. Geomorphology **87**(3), 224–238 (2007). https://doi.org/10.1016/J.GEOMORPH.2006.03.033
7. Delle Rose, M., Parise, M., G. A.: Evaluating the Impact of Quarrying on Karst Aquifers of Salento (Southern Italy). Geological Society, London, Special Publications (2007)
8. Sandrini, L., et al.: GFRP anchoring systems for soft-rock geostructures with high cultural and environmental value, 1173–1183 (2022). https://doi.org/10.1201/9781003308867-94
9. Sandrini, L.: Transparent consolidants and GFRP anchors as aesthetic solution for soft- rock stabilization, 1–8 (2022)
10. Castellanza, R., Lollino, P., Ciantia, M.: A methodological approach to assess the hazard of underground cavities subjected to environmental weathering. Tunn. Undergr. Space Technol. **82**, 278–292 (2018). https://doi.org/10.1016/J.TUST.2018.08.041
11. Ciantia, M.O., Castellanza, R., Fernandez-Merodo, J.A.: A 3D numerical approach to assess the temporal evolution of settlement damage to buildings on cavities subject to weathering. Rock Mech. Rock Eng. **51**(9), 2839–2862 (2018). https://doi.org/10.1007/s00603-018-1468-3
12. Ciantia, M.O.: Numerical assessment of sinkhole-induced damage to buildings. In: Proceeding of 29th March 2023, Cardiff (2023)
13. Mánica, M.A., Ciantia, M.O., Gens, A.: On the Stability of underground caves in calcareous rocks due to long-term weathering. Rock Mech. Rock Eng. **53**(9), 3885–3901 (2020). https://doi.org/10.1007/s00603-020-02142-y

Author Index

A. Ferrari et al. (Eds.): CNRIG 2023, SSGG, pp. 849–852, 2023.
https://doi.org/10.1007/978-3-031-34761-0